Die Grundlehren der mathematischen Wissenschaften

in Einzeldarstellungen
mit besonderer Berücksichtigung
der Anwendungsgebiete

Band 154

Herausgegeben von

J. L. Doob · A. Grothendieck · E. Heinz · F. Hirzebruch
E. Hopf · H. Hopf · W. Maak · S. MacLane · W. Magnus
M. M. Postnikov · F. K. Schmidt · D. S. Scott · K. Stein

Geschäftsführende Herausgeber

B. Eckmann und B. L. van der Waerden

Ivan Singer

Bases in Banach Spaces I

Springer-Verlag New York · Heidelberg · Berlin 1970

Prof. Dr. Ivan Singer

Institute of Mathematics, Academy of the Socialist Republic of Romania, Bucharest

Printed in Germany. Title No. 5137

Preface

This monograph attempts to present the results known today on bases in Banach spaces and some unsolved problems concerning them. Although this important part of the theory of Banach spaces has been studied for more than forty years by numerous mathematicians, the existing books on functional analysis (e. g. M. M. Day [43], A. Wilansky [263], R. E. Edwards [54]) contain only a few results on bases.

A survey of the theory of bases in Banach spaces, up to 1963, has been presented in the expository papers [241], [242] and [243], which contain no proofs; although in the meantime the theory has rapidly developed, much of the present monograph is based on those expository papers. Independently, a useful bibliography of papers on bases, up to 1963, was compiled by B. L. Sanders [219].

Due to the vastness of the field, the monograph is divided into two volumes, of which this is the first (see the table of contents). Some results and problems related to those treated herein have been deliberately planned to be included in Volume II, where they will appear in their natural framework (see [242], [243]).

We hope that the present monograph will be useful both for specialists in the field and for those who want to apply basis theory to other problems. In order to make the book accessible to a larger circle of readers, we have indicated, for the results of functional analysis which we used, a reference to a treatise containing the proof of the respective result; when we applied results which are not contained in such treatises, but only in journals, we have usually mentioned them as lemmas, giving also their proofs.

The bibliography given at the end does not aim at being complete, but wants merely to give useful orientation to the reader. Since some of the results have deep roots in classical analysis and some have been discovered and rediscovered, independently, by several authors, we did not intend to trace down completely the history of all results. Some of our unpublished results and remarks have also been included in the present monograph, without any special mention. The references given here concern the material of Volume I; the bibliography for Volume II will be given separately in that volume.

It is a great pleasure to acknowledge here the generous help of our friend, Professor Aleksander Pelczynski, with whom we had numerous stimulating conversations and correspondence during the preparation of the present monograph. Also, we have profited from valuable remarks in discussions and letters made by our colleagues and friends Professors Czeslaw Bessaga, William J. Davis, David W. Dean, Ciprian Foiaş, Gliceria Godini, Vladimir I. Gurariĭ, Mihail I. Kadec, Bor-Luh Lin, Joram Lindenstrauss, Charles W. McArthur, James R. Retherford and William Ruckle (in alphabetical order). We are indebted to Dr. Clifford Kottman for reading the entire manuscript and making valuable suggestions for its improvement.

Part of this monograph was completed in the Institute of Mathematics of the Academy of the Socialist Republic of Romania (over a period of several years) and various parts of it were written while the author was Visiting Professor at the University College of Swansea (4 months), Florida State University (3 months), Pennsylvania State University (2 months), University of Iowa (6 months) and Ohio State University (1 month). We wish to express here our gratitude to Professor Miron Nicolescu, President of the Academy of the Socialist Republic of Romania and Director of the Institute of Mathematics and to the Chairmen of the Departments of Mathematics of the above Universities, Professors Jeffrey D. Weston (Swansea), Orville G. Harrold (F.S.U.), Raymond G. Ayoub (P.S.U.), Robert H. Oehmke (U.I.) and Arnold E. Ross (O.S.U.) for ensuring excellent working conditions. We extend our thanks to all colleagues in these universities who attended our seminars on selected topics of the theory of bases in Banach spaces, for their stimulating interest and comments.

Finally, our thanks are due to Dr. Klaus Peters of Springer Verlag for solving promptly and efficiently the various problems which appeared during the preparation of this monograph.

June, 1969 IVAN SINGER

After this book has been typeset, we felt it necessary to make the following additional remarks:

1. The reader is warned to distinguish carefully between ϕ (small phi $=\varphi$) and Φ (capital phi), which are used for different purposes. Also, throughout the book $\emptyset$ denotes the empty set and $A\backslash B$ the set-theoretic difference $\{x \in A \mid x \notin B\}$.

2. We should note also the appearance, while the present book was in press, of the introductory text book [276] on bases.

February, 1970 IVAN SINGER

Contents

Chapter I. The Basis Problem. Some Properties of Bases in Banach Spaces

Chapter II. Special Classes of Bases in Banach Spaces

I. Classes of Bases not Involving Unconditional Convergence

Chapter I

The Basis Problem. Some Properties of Bases in Banach Spaces

§ 1. Definition of a basis in a Banach space. The basis problem. Relations between bases in complex and real Banach spaces

The scalar field K for all (general or concrete) linear spaces considered in the sequel will be either the field of complex numbers or the field of real numbers.

Definition 1.1. A sequence $\{x_n\}$ in an infinite dimensional Banach space E is called a *basis* of E if for every $x \in E$ there exists a *unique* sequence of scalars $\{\alpha_n\} \subset K$ such that

$$x = \sum_{i=1}^{\infty} \alpha_i x_i \tag{1.1}$$

(i.e. such that $\lim_{n\to\infty} \left\| x - \sum_{i=1}^{n} \alpha_i x_i \right\| = 0$).

A system of n elements $\{x_j\}_{j=1}^{n}$ in a Banach space E of dimension $n < \infty$ is called a *basis* of E if it is a basis (in the usual algebraic sense) of the underlying linear space, i.e., if for every $x \in E$ there exists a unique system of n scalars $\{\alpha_j\}_{j=1}^{n} \subset K$ such that $x = \sum_{i=1}^{n} \alpha_i x_i$.

For the sake of brevity, throughout the sequel we shall make the following convention: we shall unify the infinite and finite dimensional cases, by writing always $\sum_{i=1}^{\infty} \alpha_i x_i$, instead of $\sum_{i=1}^{\infty} \alpha_i x_i$ when $\dim E = \infty$ and $\sum_{i=1}^{n} \alpha_i x_i$ when $\dim E < \infty$. In other words, by $\sum_{i=1}^{\infty} \alpha_i x_i$ we shall actually mean $\sum_{i=1}^{\dim E} \alpha_i x_i$.

If a Banach space E has a basis $\{x_n\}$, then it is separable, since the set of all finite linear combinations $\sum_{i=1}^{n} r_i x_i$, where the r_i are (complex or real) rational numbers and $n=1,2,\ldots$, is a countable dense set in E. It is not known whether the converse is true or not, i.e.:

Problem 1.1 (called "the basis problem"). *Does every separable Banach space possess a basis?*

By the Banach-Mazur theorem on the universality of the space $C([0,1])$ for separable Banach spaces[1], the basis problem is equivalent to the following:

Problem 1.2. Does every subspace[2] of $C([0,1])$ possess a basis?

In connection with this problem it is natural to ask

Problem 1.3. Does there exist an infinite dimensional separable Banach space F, not isomorphic[3] to $L^2([0,1])$, such that every subspace of F has a basis?

Since by a theorem of Banach and Mazur every separable Banach space is equivalent[4] to a quotient space of l^1, the basis problem is also equivalent to the following:

Problem 1.4. Does every quotient space of l^1 possess a basis?

In connection with this problem, it is natural to ask

Problem 1.5. Does there exist an infinite dimensional separable Banach space F, not isomorphic to $L^2([0,1])$, such that every quotient space of F has a basis?

As we have already mentioned, the scalar field K for all (general or concrete) linear spaces which we consider throughout this book, can be either the field of complex numbers or the field of real numbers. Since there are some well known relations between complex and real Banach spaces, it is natural to ask how the bases of complex Banach spaces are related to the bases of the corresponding real Banach spaces and conversely. We shall now examine some aspects of this question.

a) *Restriction of the field of scalars.* Let E be a complex Banach space. Then E is also a Banach space over the subfield R of real numbers. This Banach space is denoted by $E_{(r)}$ and it is called *the real Banach space obtained from E by the restriction of the field of scalars*, or

---

[1] See e.g. [10], p. 185, theorem 9. In [10] only real Banach spaces are considered. Whenever we shall refer the reader to [10], we shall understand, without any special mention, that the respective result of [10] can be extended to complex Banach spaces with standard methods.

[2] Unless otherwise stated, by "subspace" we shall mean: closed linear subspace.

[3] We shall use the term "isomorphic" in the sense of Banach, i.e.: linearly homeomorphic.

[4] See e.g. [12], theorem e) or [133], p. 283, theorem (1). We shall use the term "equivalent" in the sense of Banach, i.e.: isometrically isomorphic.

shortly, *the real Banach space associated to E*. A useful relation between the bases of E and $E_{(r)}$ is the following:

Proposition 1.1. *A sequence $\{x_n\}\subset E$ is a basis of the complex Banach space E if and only if the sequence $\{z_n\}$ defined by*[1]

$$z_{2n-1}=x_n, \qquad z_{2n}=i x_n \qquad (n=1,2,\ldots) \tag{1.2}$$

is a basis of $E_{(r)}$ (but $E_{(r)}$ can have also other bases which are not of the form (1.2)).

Proof. Assume that $\{x_n\}$ is a basis of the complex Banach space E. Then every $x\in E$ has an expansion $x=\sum_{j=1}^{\infty}(\beta_j+i\gamma_j)x_j=\sum_{j=1}^{\infty}(\beta_j z_{2j-1}+\gamma_j z_{2j})$ with respect to $\{z_n\}$. Furthermore, this expansion is unique, since the relation $\sum_{j=1}^{\infty}(\beta_j z_{2j-1}+\gamma_j z_{2j})=0$ implies $\sum_{j=1}^{\infty}(\beta_j+i\gamma_j)x_j=0$, whence, since $\{x_n\}$ is a basis of E, $\beta_j+i\gamma_j=0$ $(j=1,2,\ldots)$ and thus $\beta_j=\gamma_j=0$ $(j=1,2,\ldots)$. Consequently, $\{z_n\}$ is a basis of $E_{(r)}$.

Conversely, if the sequence $\{z_n\}\subset E$ defined by (1.2) is a basis of $E_{(r)}$, then a similar argument shows that every $x\in E$ has a unique expansion $x=\sum_{j=1}^{\infty}(\beta_j+i\gamma_j)x_j$, i.e. that $\{x_n\}$ is a basis of E.

On the other hand, let E be a one-dimensional complex Banach space. Then every basis of E is of the form $\{x_1\}$, where $x_1\neq 0$. However, for any $x\in E$, $x\neq 0$, the couple $z_1=x$, $z_2=(1+i)x$ is a basis of $E_{(r)}$, which is clearly not of the form $\{x_1\}\cup\{ix_1\}$. This completes the proof of proposition 1.1.

In view of the relations between the coefficients β_j and γ_j above, or, equivalently, between the coefficient functionals associated to the bases $\{x_n\}$ and $\{z_n\}$ above (see § 3, definition 3.1), let us mention the relationship between the elements of the conjugate spaces E^* and $(E_{(r)})^*$. If $f\in E^*$, then for the functional g defined by[2]

$$g(x)=\operatorname{Re} f(x) \qquad (x\in E_{(r)})$$

we have $g\in(E_{(r)})^*$ and

$$\operatorname{Im} f(x)=\operatorname{Re}[-if(x)]=-\operatorname{Re} f(ix)=-g(ix),$$

[1] Here $i=\sqrt{-1}$, but in general we shall use the letter i to denote positive integers. Whenever we shall use i as $\sqrt{-1}$, we shall make a special mention.

[2] For any complex number $\xi=\eta+i\zeta$ (η, ζ real) we use the notations $\operatorname{Re}\xi=\eta$, $\operatorname{Im}\xi=\zeta$, $\bar{\xi}=\eta-i\zeta$.

and thus $f(x)=g(x)-ig(ix)$ is uniquely determined by g. Conversely, if $g\in(E_{(r)})^*$, then for the functional f defined by

$$f(x)=g(x)-ig(ix) \quad (x\in E)$$

we have $f\in E^*$, since

$$f(ix)=g(ix)-ig(-x)=g(ix)+ig(x)=if(x) \quad (x\in E).$$

Furthermore, since $E_{(r)}$ is a real Banach space and $g\in(E_{(r)})^*$, $g(x)$ is real for all $x\in E$, whence

$$\operatorname{Re} f(x)=g(x), \quad \operatorname{Im} f(x)=-g(ix) \quad (x\in E).$$

Thus there exists a one to one real-linear[1] mapping $f\to g$ (where $g(x)=\operatorname{Re} f(x)$) of E^* onto $(E_{(r)})^*$; in other words, this is a one to one linear mapping of $(E^*)_{(r)}$ onto $(E_{(r)})^*$. This mapping is also an isometry. Indeed, if $f(x)=re^{i\theta}$ ($r>0$ and θ real), then, since $|f(x)|$ is real, we have

$$|f(x)|=f(e^{-i\theta}x)=g(e^{-i\theta}x)\leqslant\|g\|\,\|e^{-i\theta}x\|=\|g\|\,\|x\|,$$

whence, since this holds for all $x\in E$, we obtain $\|f\|\leqslant\|g\|$. On the other hand,

$$|g(x)|=|\operatorname{Re} f(x)|\leqslant|f(x)|\leqslant\|f\|\,\|x\| \quad (x\in E),$$

whence $\|g\|\leqslant\|f\|$. Consequently, $\|f\|=\|g\|$, which proves our assertion.

In the above, to every complex Banach space E there has been associated a real Banach space $E_{(r)}$. Conversely, it is natural to ask, which real Banach spaces F have the property that there exists a complex Banach space E with $E_{(r)}=F$. It is convenient to consider the following more general question: for which real Banach spaces F does there exist a complex Banach space E such that $E_{(r)}$ is isomorphic to F? In other words: for which real Banach spaces F does there exist a real-linear isomorphism of F onto a complex Banach space E? If a real Banach space F has this property, then we say that F *admits a complex structure*[2]. A characterization of such spaces is the following: A real Banach space F admits a complex structure *if and only if there exists an automorphism*[3] *u of F such that* $u^2(x)=-x$ $(x\in F)$. Indeed, if v is an isomorphism of F onto $E_{(r)}$, then $u=v^{-1}iv$ is an automorphism of F satisfying $u^2(x)=-x$ $(x\in F)$; conversely, if u is an automorphism of F satisfying $u^2(x)=-x$ $(x\in F)$, then, putting $(\alpha+i\beta)x=\alpha x+\beta u(x)$, $|||x|||=\sup_{0\leqslant\theta\leqslant 2\pi}\|e^{i\theta}x\|$ $(x\in F)$, and $E=F$ endowed with this multiplication

---

[1] I.e. additive and homogeneous with respect to real scalars.

[2] For instance, we have seen in the above that for every complex Banach space E the real Banach space $F_1=(E_{(r)})^*$ admits a complex structure.

[3] I.e. an isomorphism (linear homeomorphism) of F onto F.

and norm, the mapping $v: x \to x$ will be an isomorphism of F onto $E_{(r)}$ (since $\|x\| \leqslant |||x||| \leqslant (1+\|u\|)\|x\|$ for all $x \in F$).

There are real Banach spaces F which do not have this property, e.g. if $\dim F = n < \infty$, then the necessary and sufficient condition for F to admit a complex structure is that n be even; in this case $\dim E = \frac{n}{2}$. There are also infinite dimensional real Banach spaces which do not admit a complex structure, e.g. the space J constructed in Ch. II, § 4, example 4.1[1].

b) *Extension of the field of scalars.* Let G be a real Banach space. Then G can be embedded (by a real-linear isometry) into a complex Banach space E, by the following procedure. Let F be the cartesian square $G \times G$, endowed with the norm $\|\{y,z\}\| = (\|y\|^2 + \|z\|^2)^{\frac{1}{2}}$ $(y \in G, z \in G)$. Then the mapping

$$u: \{y,z\} \to \{-z,y\} \qquad (y \in G, z \in G)$$

is an automorphism of F satisfying $u^2(\{y,z\}) = -\{y,z\}$, whence, by a) above, F admits a complex structure[2]. Let E be the complex Banach space obtained from F as in a), i.e. the space F endowed with the multiplication by complex scalars

$$(\alpha + i\beta)\{y,z\} = \alpha\{y,z\} + \beta u(\{y,z\}) = \{\alpha y - \beta z, \alpha z + \beta y\}$$

and with the norm

$$|||\{y,z\}||| = \sup_{0 \leqslant \theta \leqslant 2\pi} \|e^{i\theta}\{y,z\}\|.$$

Then the mapping $y \to \{y,0\}$ is a real-linear isometry of G into E, since by

$$\begin{aligned}\|(\alpha+i\beta)\{y,0\}\|_{G\times G} &= \|\{\alpha y, \beta y\}\|_{G\times G} = (\|\alpha y\|^2 + \|\beta y\|^2)^{\frac{1}{2}} \\ &= (\alpha^2+\beta^2)^{\frac{1}{2}}\|y\| = |\alpha + i\beta|\,\|y\| \qquad (y \in G;\ \alpha, \beta \text{ real}),\end{aligned}$$

we have

$$|||\{y,0\}||| = \sup_{|\alpha+i\beta|=1} \|(\alpha+i\beta)\{y,0\}\| = \|y\| \quad \text{for all} \quad y \in G.$$

The space E is called *the complex Banach space obtained from G by the extension of the field of scalars*, or shortly, *the complexification of E*.

Since $i\{z,0\} = \{0,z\}$ $(z \in G)$, we have

$$\{y,z\} = \{y,0\} + \{0,z\} = \{y,0\} + i\{z,0\} \qquad (y \in G, z \in G),$$

---

[1] For a proof see [45].

[2] In particular, it follows that a sufficient condition for a real Banach space F to admit a complex structure, is that F be isomorphic to the cartesian square of a real Banach space G.

and this decomposition is obviously unique. Thus, *identifying the space* G *with its isometrical image* $\{G,0\}$ *in* E, *every* $x\in E$ *can be uniquely written in the form* $x=y+iz$, *where* $y\in G$, $z\in G$. Furthermore, the mappings $x\to y$ and $x\to z$ are projections of norm 1 onto G, since

$$|||\{y,0\}|||=\|y\|\leqslant(\|y\|^2+\|z\|^2)^{\frac{1}{2}}=\|\{y,z\}\|_{G\times G}\leqslant|||\{y,z\}||| \qquad (y\in G,\, z\in G).$$

A useful relation between the bases of G and E is the following:

Proposition 1.2. *Let* G *a real Banach space and let* E *be the complexification of* G. *A sequence* $\{y_n\}\subset G$ *is a basis of* G *if and only if it is a basis of* E *(but* E *can have also other bases which* $\not\subset G$*).*

Proof. Assume that $\{y_n\}$ is a basis of G and let $x=y+iz\in E$ be arbitrary. Then there exist expansions with real scalars

$$y=\sum_{j=1}^{\infty}\alpha_j y_j, \qquad z=\sum_{j=1}^{\infty}\beta_j y_j,$$

whence $y+iz=\sum_{j=1}^{\infty}(\alpha_j+i\beta_j)y_j$. Furthermore, let us show that this expansion is unique. Firstly, we observe that if $\sum_{j=1}^{\infty}(\alpha_j+i\beta_j)y_j$ converges, say to y_0+iz_0, then $\sum_{j=1}^{\infty}\alpha_j y_j$ and $\sum_{j=1}^{\infty}\beta_j y_j$ converge and we have $y_0=\sum_{j=1}^{\infty}\alpha_j y_j$, $z_0=\sum_{j=1}^{\infty}\beta_j y_j$. Indeed, we have

$$\sum_{j=1}^{n}(\alpha_j+i\beta_j)y_j=\sum_{j=1}^{n}\alpha_j y_j+i\sum_{j=1}^{n}\beta_j y_j \qquad (n=1,2,\ldots),$$

whence, since the mapping $y+iz\to y$ of E onto G is of norm 1,

$$\left\|y_0-\sum_{j=1}^{n}\alpha_j y_j\right\|\leqslant\left\|\left(y_0-\sum_{j=1}^{n}\alpha_j y_j\right)+i\left(z_0-\sum_{j=1}^{n}\beta_j y_j\right)\right\|$$
$$=\left\|(y_0+iz_0)-\sum_{j=1}^{n}(\alpha_j+i\beta_j)y_j\right\|\to 0 \quad\text{as}\quad n\to\infty,$$

and similarly $z_0=\sum_{j=1}^{\infty}\beta_j y_j$. Now, if $\sum_{j=1}^{\infty}(\alpha_j+i\beta_j)y_j=0$, then, by the preceding, both $\sum_{j=1}^{\infty}\alpha_j y_j$ and $\sum_{j=1}^{\infty}\beta_j y_j$ converge to 0, whence, since $\{y_n\}$ is a basis of G, $\alpha_j=\beta_j=0$ $(j=1,2,\ldots)$ and thus $\alpha_j+i\beta_j=0$ $(j=1,2,\ldots)$, which proves the uniqueness of the expansions $\sum_{j=1}^{\infty}(\alpha_j+i\beta_j)y_j$. Consequently, $\{y_n\}$ is a basis of E.

Conversely, if $\{y_n\} \subset G$ is a basis of E, then every $y \in G$ has an expansion $y = \sum_{j=1}^{\infty} (\alpha_j + i\beta_j) y_j$ (α_j, β_j real), whence, by the above, $y = \sum_{j=1}^{\infty} \alpha_j y_j + i \sum_{j=1}^{\infty} \beta_j y_j$. Since every $x_0 \in E$ can be uniquely written in the form $x_0 = y_0 + i z_0$, where $y_0, z_0 \in G$, it follows that $\sum_{j=1}^{\infty} \beta_j y_j = 0$, $y = \sum_{j=1}^{\infty} \alpha_j y_j$. Since $\{y_n\}$ is a basis of E, this expansion is unique, and thus $\{y_n\}$ is a basis of G.

On the other hand, let G be a one-dimensional real Banach space. Then the complexification E of G is a one-dimensional[1] complex Banach space and any element $x \in E$, $x \neq 0$, is a basis of E. Hence there exist bases of E which are not in G. This completes the proof of proposition 1.2.

In view of the relations between the coefficients α_j and $\alpha_j + i\beta_j$ above, or, equivalently, between the coefficient functionals associated to the bases $\{y_n\}$ of G and $\{y_n\}$ of E (see § 3, definition 3.1), let us mention the relationship between the elements of the conjugate spaces G^* and E^*. If $f \in E^*$, then for the functionals h_1, h_2 defined by

$$h_1(y) = \operatorname{Re} f(y), \qquad h_2(y) = \operatorname{Im} f(y) \qquad (y \in G)$$

we have $h_1 \in G^*$, $h_2 \in G^*$ and

$$\begin{aligned} f(x) &= f(y) + i f(z) = h_1(y) + i h_2(y) + i[h_1(z) + i h_2(z)] \\ &= [h_1(y) - h_2(z)] + i[h_2(y) + h_1(z)] \qquad (x = y + iz \in E), \end{aligned}$$

and thus f is uniquely determined by the couple $\{h_1, h_2\} \in G^* \times G^*$. Conversely, if $\{h_1, h_2\} \in G^* \times G^*$, then for the functional f defined by

$$f(x) = [h_1(y) - h_2(z)] + i[h_2(y) + h_1(z)] \qquad (x = y + iz \in E)$$

we have $f \in E^*$. Furthermore, since for $x = y + iz \in G$ we have $y = x$, $z = 0$, we obtain

$$f(y) = h_1(y) + i h_2(y) \qquad (y \in G),$$

whence $\operatorname{Re} f(y) = h_1(y)$, $\operatorname{Im} f(y) = h_2(y)$ for all $y \in G$ (because $h_1(y), h_2(y)$ are real for all $y \in G$).

Thus there exists a one to one real-linear mapping $f \to \{h_1, h_2\}$ (where $h_1(y) = \operatorname{Re} f(y)$, $h_2(y) = \operatorname{Im} f(y)$, $y \in G$) of E^* onto $G^* \times G^*$; in other words, this is a one to one linear mapping of $(E^*)_{(r)}$ onto $G^* \times G^*$.

---

[1] Whenever we write $\dim G$, $\dim E$, we understand the dimension with respect to the field of scalars of G or E respectively.

This mapping is also an isomorphism[1], when $G^* \times G^*$ is endowed with the norm $\|\{h_1,h_2\}\|=(\|h_1\|^2+\|h_2\|^2)^{\frac{1}{2}}$. Indeed, we have

$$[h_1(y)-h_2(z)]\leqslant\|h_1\|\,\|y\|+\|h_2\|\,\|z\|\leqslant(\|h_1\|^2+\|h_2\|^2)^{\frac{1}{2}}(\|y\|^2+\|z\|^2)^{\frac{1}{2}}$$
$$\leqslant(\|h_1\|^2+\|h_2\|^2)^{\frac{1}{2}}\,|||x||| \qquad (x=y+iz\in E),$$

and, similarly,

$$[h_2(y)+h_1(z)]\leqslant(\|h_1\|^2+\|h_2\|^2)^{\frac{1}{2}}\,|||x||| \qquad (x=y+iz\in E),$$

whence

$$|f(x)|=([h_1(y)-h_2(z)]^2+[h_2(y)+h_1(z)]^2)^{\frac{1}{2}}$$
$$\leqslant\sqrt{2}(\|h_1\|^2+\|h_2\|^2)^{\frac{1}{2}}\,|||x||| \qquad (x\in E),$$

and thus $\|f\|\leqslant\sqrt{2}\|\{h_1,h_2\}\|$. On the other hand, since $h_1(y), h_2(y)$ are real for all $y\in G$, we have

$$|h_1(y)|\leqslant([h_1(y)]^2+[h_2(y)]^2)^{\frac{1}{2}}=|f(y)|\leqslant\|f\|\,\|y\| \qquad (y\in G),$$

whence $\|h_1\|\leqslant\|f\|$. Similarly, $\|h_2\|\leqslant\|f\|$, whence

$$\|\{h_1,h_2\}\|=(\|h_1\|^2+\|h_2\|^2)^{\frac{1}{2}}\leqslant\sqrt{2}\|f\|.$$

Consequently,

$$\frac{1}{\sqrt{2}}\|f\|\leqslant\|\{h_1,h_2\}\|\leqslant\sqrt{2}\|f\| \qquad (f\in E^*),$$

which proves our assertion.

In the above, to every real Banach space G there has been associated a complex Banach space E. Conversely, it is natural to ask, which complex Banach spaces E have the property that there exists a real Banach space G whose complexification is E. It is convenient to consider the following more general question: for which complex Banach spaces E does there exist a real Banach space G such that the complexification of G is isomorphic[2] to E? A characterization of such spaces is

---

[1] Hence, by part a), this mapping induces an isomorphism $g\to\{h_1,h_2\}$ of $(E_{(r)})^*$ onto $G^*\times G^*$, defined by

$$g(x)=h_1(y)-h_2(z) \qquad (x=y+iz\in E_{(r)}).$$

[2] The answer cannot be obtained by the restriction of the field of scalars to the reals. Indeed, as shown by the example of finite dimensional complex Banach spaces E, the complexification of $E_{(r)}$ need not be isomorphic to E. On the other hand, let us also mention that, as shown by the example of finite dimensional real Banach spaces G, the real Banach space $E_{(r)}$ associated to the complexification E of a real Banach space G need not be isomorphic to G.

the following: A complex Banach space E is isomorphic to the complexification of a real Banach space *if and only if there exists an involution on E, i.e. an antilinear*[1] *automorphism w of E such that $w^2(x)=x$ $(x\in E)$.* Indeed, if v is an isomorphism of E onto the complexification E_1 of a real Banach space G, then $\tau\colon y+iz\to y-iz$ $(y\in G, z\in G)$ is an involution on E_1, whence $w=v^{-1}\tau v$ is an involution on E; conversely, if w is an involution on E, then E is isomorphic to the complexification of the real Banach space $G=\{x\in E|w(x)=x\}$ (since every $x\in E$ can be uniquely written in the form $x=y+iz$, where $y=\dfrac{x+w(x)}{2}\in G$, $z=\dfrac{x-w(x)}{2i}\in G$ and since $\|x\|\leqslant\sqrt{2}(\|y\|^2+\|z\|^2)^{\frac{1}{2}}\leqslant(1+\|w\|)\|x\|$ for all $x\in E$).

If $\{x_n\}$ is a basis of a complex Banach space E admitting an involution w, then $\{w(x_n)\}$ is also a basis of E; this remark is useful in the study of bases of $E\times E$.

If E is a complex Banach space with $\dim E=n<\infty$, then there exists a natural involution $\{\xi_i\}_1^n\to\{\bar{\xi}_i\}_1^n$ on E, whence E is isomorphic to the complexification of the n-dimensional real Banach space $\{\{\xi_i\}_1^n\in E|\xi_1,\dots,\xi_n=\text{real}\}$. More generally, *if a complex Banach space E has a basis $\{x_n\}$ such that the mapping $\sum_{j=1}^{\infty}\alpha_j x_j\to\sum_{j=1}^{\infty}(\operatorname{Re}\alpha_j)x_j$ is continuous, then E is isomorphic to the complexification of the real Banach space* $\left\{\sum_{j=1}^{\infty}\alpha_j x_j\in E|\alpha_n=\text{real }(n=1,2,\dots)\right\}$, since the mapping w: $\sum_{j=1}^{\infty}(\alpha_j+i\beta_j)x_j\to\sum_{j=1}^{\infty}(\alpha_j-i\beta_j)x_j$ is an involution on E. Furthermore, if E is one of the usual complex Banach spaces, e.g. $C([0,1])$, $L^p([0,1])$ $(1\leqslant p<\infty)$, c_0 or l^p $(1\leqslant p\leqslant<\infty)$, then the natural involution $x(t)=y(t)+iz(t)\to\overline{x(t)}=y(t)-iz(t)$ shows that $C([0,1])$ is isomorphic to the complexification of the real Banach space $C([0,1])$, and similarly, each of the other spaces is isomorphic to the complexification of the corresponding real space $L^p([0,1])$, c_0 or l^p respectively. This remark, together with proposition 1.2 above, shows that *every basis of the real Banach space $C([0,1])$ is also a basis of the complex Banach space $C([0,1])$*, and similarly for $L^p([0,1])$, c_0 or l^p respectively.

Let us also mention that some other specifically complex Banach spaces are also isomorphic to complexifications of suitable real Banach spaces. For instance, in the Banach space A of all complex functions $x(\xi)$ of a complex variable ξ, which are analytic in $\{\xi||\xi|<1\}$ and continuous in $\{\xi||\xi|\leqslant 1\}$, endowed with the norm $\|x\|=\max_{|\xi|\leqslant 1}|x(\xi)|$, the mapping w: $x(\xi)\to\overline{x(\bar{\xi})}$ is an involution, whence A is isomorphic

[1] I.e. such that $w(\lambda x)=\bar{\lambda}w(x)$ for all $x\in E$ and all (complex) scalars λ.

to the complexification of the real Banach space $G=\{x\in A|x|_{[-1,1]}$ is real$\}$. In the Banach space $L(E,E)$ of all continuous linear mappings of E into E, endowed with the norm $\|u\|=\sup_{\substack{x\in E\\ \|x\|\leqslant 1}}\|u(x)\|$, where E is a complex Banach space admitting an involution w, the mapping $w_1: u\to wuw$ is an involution[1], whence $L(E,E)$ is isomorphic to the complexification of the real Banach space $\{u\in L(E,E)|u(G)\subseteq G\}$, where $G=\{x\in E|w(x)=x\}$.

§2. Some examples of bases in concrete Banach spaces. Some separable Banach spaces in which no basis is known

The first idea which arises quite naturally in connection with the basis problem is to examine whether or not the various concrete separable Banach spaces occurring in practice possess a basis. In this section we shall give some examples of such bases and some examples of separable Banach spaces in which no basis is known. We shall use, for all concrete Banach spaces occurring in the sequel, the standard notations. Whenever we shall write l^p or L^p, we shall understand that $p<\infty$. Unless otherwise stated, the scalar field K for all concrete Banach spaces considered in the sequel, can be either the field of complex numbers, or the field of real numbers. Actually, there will be constructed only bases consisting of real functions or real sequences respectively, but these will be bases of the corresponding spaces considered either with complex scalars, or with real scalars. This fact is quite natural, in the light of the considerations of § 1.

Example 2.1. In the spaces c_0 and l^p $(p\geqslant 1)$ the sequence[2]

$$x_n=\{\delta_{nj}\}_{j=1}^{\infty}\qquad (n=1,2,\ldots)\tag{2.1}$$

constitutes a basis.

Indeed, for $x=\{\xi_n\}\in c_0$ we have $\left\|x-\sum_{i=1}^{n}\xi_i x_i\right\|=\sup_{n+1\leqslant i<\infty}|\xi_i|\to 0$ when $n\to\infty$, while for $x=\{\xi_n\}\in l^p$ we have $\left\|x-\sum_{i=1}^{n}\xi_i x_i\right\|=\left[\sum_{i=n+1}^{\infty}|\xi_i|^p\right]^{\frac{1}{p}}$

---

[1] Unfortunately, this is not an involution for the natural structure of Banach algebra of $L(E,E)$, since $w_1(uv)=wuvw=wuwwvw=w_1(u)w_1(v)$, which, in general, $\neq w_1(v)w_1(u)$.

[2] Throughout this book, δ_{ij} will be the Kronecker delta, i.e.

$$\delta_{ij}=\begin{cases}1 & \text{for } i=j,\\ 0 & \text{for } i\neq j\end{cases}\qquad (i,j=1,2,\ldots).$$

$\to 0$ when $n\to\infty$. On the other hand, $\sum_{i=1}^{\infty}\xi_i x_i=\sum_{i=1}^{\infty}\eta_i x_i$ implies, in c_0, $\lim_{n\to\infty}\max_{1\leqslant i\leqslant n}|\xi_i-\eta_i|=0$, while in l^p it implies $\lim_{n\to\infty}\left[\sum_{i=1}^{n}|\xi_i-\eta_i|^p\right]^{\frac{1}{p}}=0$, whence, in both cases, $\xi_i=\eta_i$ $(i=1,2,\ldots)$.

The basis (2.1) is called *the natural basis* or *the unit vector basis* of the spaces c_0 and $l^p\,(p\geqslant 1)$.

From the above it follows that in the space c the sequence

$$x_0=\{1,1,\ldots\},\qquad x_n=\{\delta_{nj}\}_{j=1}^{\infty}\qquad (n=2,3,\ldots) \tag{2.2}$$

constitutes a basis, namely, every $x=\{\xi_n\}\in c$ has a unique expansion of the form $x=\left(\lim_{n\to\infty}\xi_n\right)x_0+\sum_{i=1}^{\infty}\left(\xi_i-\lim_{n\to\infty}\xi_n\right)x_i$.

Example 2.2. In the Banach space $C([0,1])$ the sequence

$$x_0(t)\equiv 1,\qquad x_1(t)=t,$$

$$x_{2^k+l}(t)=\begin{cases}0 & \text{for}\quad t\notin\left(\dfrac{2l-2}{2^{k+1}},\dfrac{2l}{2^{k+1}}\right),\\[2ex] 1 & \text{for}\quad t=\dfrac{2l-1}{2^{k+1}},\\[2ex] \text{linear in} & \left[\dfrac{2l-2}{2^{k+1}},\dfrac{2l-1}{2^{k+1}}\right]\quad\text{and}\quad\left[\dfrac{2l-1}{2^{k+1}},\dfrac{2l}{2^{k+1}}\right]\end{cases} \tag{2.3}$$

$$(l=1,2,\ldots,2^k;\qquad k=0,1,2,\ldots)$$

constitutes a basis. More generally, let $\{a_n\}\subset[0,1]$ be an arbitrary sequence dense in $[0,1]$ and for each $n\geqslant 2$ let ϑ_n denote that one of the n subsegments of $[0,1]$ determined by $0,\ a_1,a_2,\ldots,a_{n-1},1$ (rearranged in increasing order) which contains a_n. Then the sequence

$$x_n(t)=\begin{cases}0 & \text{for}\quad t\notin\vartheta_{n-1},\\ 1 & \text{for}\quad t=a_{n-1},\\ \text{linear for the other } t\end{cases}\qquad (n=2,3,\ldots) \tag{2.4}$$

constitutes a basis of the space $C([0,1])$.

Indeed, for $x\in C([0,1])$ satisfying $x(0)=x(1)=0$ let

$$\alpha_2=x(a_1),\qquad \alpha_n=x(a_{n-1})-\sum_{i=2}^{n-1}\alpha_i x_i(a_{n-1})\qquad (n=3,4,\ldots), \tag{2.5}$$

$$s_n(x)=\sum_{i=2}^{n}\alpha_i x_i\qquad (n=2,3,\ldots). \tag{2.6}$$

Then, since $x_2, \ldots, x_n$ are linear on each subsegment of $[0,1]$ determined by $\{a_i\}_{i=1}^{n-1}$ (rearranged in increasing order), so is s_n. Furthermore, by (2.4) we have $x_{j+1}(a_j)=1$ and $x_i(a_j)=0$ for $i=j+2, j+3, \ldots, n$ $(j=1,\ldots,n-1)$, whence, taking also into account (2.5), we obtain

$$[s_n(x)](a_j) = \sum_{i=2}^{j} \alpha_i x_i(a_j) + \alpha_{j+1} x_{j+1}(a_j) + \sum_{i=j+2}^{n} \alpha_i x_i(a_j) \tag{2.7}$$

$$= \sum_{i=2}^{j} \alpha_i x_i(a_j) + \left[x(a_j) - \sum_{i=2}^{j} \alpha_i x_i(a_j) \right] = x(a_j) \quad (j=1,\ldots,n-1).$$

Consequently, $s_n(x)$ is the polygonal (piecewise linear) function interpolating the values of x in the points $a_{-1}=0, a_1, a_2, \ldots, a_{n-1}, 1=a_1$, i.e., if a_{i_1}, a_{i_2} are any two consecutive points of $\{a_i\}_{i=-1}^{n-1}$ (in the sense of the natural order of $[0,1]$), then

$$s_n[\lambda a_{i_1} + (1-\lambda) a_{i_2}] = \lambda x(a_{i_1}) + (1-\lambda) x(a_{i_2}) \quad (0 \leqslant \lambda \leqslant 1). \tag{2.8}$$

Let $\varepsilon > 0$ be arbitrary. Since x is uniformly continuous on $[0,1]$, there exists an $\eta = \eta(\varepsilon) > 0$ such that $|x(t') - x(t'')| < \varepsilon$ whenever $t', t'' \in [0,1]$, $|t' - t''| < \eta$. Since the sequence $\{a_j\}$ is dense in $[0,1]$, there exists a positive integer $N = N[\eta(\varepsilon)]$ such that for $n > N$ we have $\max |a_{i_1} - a_{i_2}| < \eta$, where the max is taken over all couples of consecutive points of $\{a_i\}_{i=-1}^{n-1}$. Now, let $t \in [0,1]$ be arbitrary. Then there exists a λ with $0 \leqslant \lambda \leqslant 1$, such that $t = \lambda a_{i_1} + (1-\lambda) a_{i_2}$, where a_{i_1}, a_{i_2} are consecutive points of $\{a_i\}_{i=-1}^{n-1}$ $(n > N[\eta(\varepsilon)])$ satisfying $[a_{i_1}, a_{i_2}] \ni t$. Then, by (2.8),

$$\begin{aligned} |x(t) - s_n(t)| &= |x(t) - \lambda x(a_{i_1}) - (1-\lambda) x(a_{i_2})| \\ &= |\lambda [x(t) - x(a_{i_1})] + (1-\lambda)[x(t) - x(a_{i_2})]| \\ &\leqslant \max_{t', t'' \in [a_{i_1}, a_{i_2}]} |x(t') - x(t'')| < \varepsilon \qquad (n > N[\eta(\varepsilon)]), \end{aligned}$$

and thus, since $N[\eta(\varepsilon)]$ is independent of $t \in [0,1]$,

$$\|x - s_n\| < \varepsilon \quad (n > N[\eta(\varepsilon)]).$$

Consequently, we have

$$x = \lim_{n \to \infty} s_n = \sum_{i=2}^{\infty} \alpha_i x_i.$$

Now we drop the restriction $x(0) = x(1) = 0$. Let $x \in C([0,1])$ be arbitrary. Then for the $\bar{x} \in C([0,1])$ defined by

$$\bar{x}(t) = x(t) - x(0) - [x(1) - x(0)]t \quad (t \in [0,1])$$

we have $\bar{x}(0) = \bar{x}(1) = 0$, whence, by the above, $\bar{x}$ has an expansion of the form $\bar{x} = \sum_{i=2}^{\infty} \alpha_i x_i$, whence, putting $\alpha_0 = x(0)$, $\alpha_1 = x(1) - x(0)$, we obtain

$$x = \sum_{i=0}^{\infty} \alpha_i x_i .$$

On the other hand, assume now that we have $\sum_{i=0}^{\infty} \alpha_i x_i = 0$. Then

$$\sum_{i=0}^{\infty} \alpha_i x_i(t) = 0 \qquad (t \in [0,1]),$$

whence, putting $t = 0, 1, a_1, a_2, \ldots$ and taking into account (2.4), we obtain, successively, $\alpha_n = 0$ $(n = 0, 1, 2, \ldots)$, which proves that (2.4) is a basis of $C([0,1])$.

In particular, the basis (2.3) is called the *Schauder basis* of $C([0,1])$.

The above construction can be generalized to the case of the space $C(Q)$, where Q is an arbitrary metric compact space (see Vol. II, Chapter V).

Example 2.3. The sequence of equivalence classes[1] $\{\tilde{y}_n\}$, where y_n are the Haar functions, i.e. the functions defined on $[0,1]$ by

$$y_1(t) \equiv 1,$$

$$y_{2^k + l}(t) = \begin{cases} \sqrt{2^k} & \text{for} \quad t \in \left[\dfrac{2l-2}{2^{k+1}}, \dfrac{2l-1}{2^{k+1}}\right), \\ -\sqrt{2^k} & \text{for} \quad t \in \left[\dfrac{2l-1}{2^{k+1}}, \dfrac{2l}{2^{k+1}}\right), \\ 0 & \text{for the other } t \end{cases} \tag{2.9}$$

$(l = 1, 2, \ldots, 2^k;\ k = 0, 1, 2, \ldots)$.

constitutes a basis of the space $L^p([0,1])\ (p \geqslant 1)$.

Indeed, let $a_1 < \cdots < a_{n-1}$ be the points of discontinuity of the functions $y_1, \ldots, y_n$ and let $a_0 = 0$, $a_n = 1$. Then from (2.9) it follows, by induction on n, that for $a_k < t < a_{k+1}$ we have

$$\sum_{i=1}^{n} y_i(t)\, y_i(\tau) = \begin{cases} \dfrac{1}{a_{k+1} - a_k} & \text{for} \quad a_k < \tau < a_{k+1}, \\ 0 & \text{for} \quad \tau < a_k \quad \text{and} \quad a_{k+1} < \tau . \end{cases} \tag{2.10}$$

Indeed, for $n = 1$ formula (2.10) is obvious. Assume now that (2.10) is true for some $n \geqslant 1$ and let $a_1' < \cdots < a_n'$ be the points of discontinuity

[1] We denote by $\tilde{x}$ the equivalence class of the function x.

of $y_1,\dots,y_{n+1}$ and $a'_0=0$, $a'_{n+1}=1$. Then there exists a j with $1\leqslant j\leqslant n$ such that

$$a'_1=a_1,\dots,a'_{j-1}=a_{j-1},a'_{j+1}=a_j,a'_{j+2}=a_{j+1},\dots,a'_{n+1}=a_n=1, \tag{2.11}$$

$$y_{n+1}(t)=\begin{cases} \sqrt{2^m} & \text{for} \quad t\in[a'_{j-1},a'_j)=\left[\frac{2l-2}{2^{m+1}},\frac{2l-1}{2^{m+1}}\right), \\ -\sqrt{2^m} & \text{for} \quad t\in[a'_j,a'_{j+1})=\left[\frac{2l-1}{2^{m+1}},\frac{2l}{2^{m+1}}\right), \\ 0 & \text{for the other } t, \end{cases} \tag{2.12}$$

where $n+1=2^m+l$, with $1\leqslant l\leqslant 2^m$.

Now let $0\leqslant k\leqslant n$ and let $a'_k<t<a'_{k+1}$. If $k\neq j-1,j$, then by (2.12) we have $y_{n+1}(t)=0$, whence, by the induction hypothesis and by (2.11) we obtain

$$\sum_{i=1}^{n+1} y_i(t)\,y_i(\tau)=\sum_{i=1}^{n} y_i(t)\,y_i(\tau)=\begin{cases}\frac{1}{a'_{k+1}-a'_k} & \text{for} \quad a'_k<\tau<a'_{k+1}, \\ 0 & \text{for} \quad \tau<a'_k \quad \text{and} \quad a'_{k+1}<\tau,\end{cases}$$

i.e., (2.10) with $n+1$ instead of n. If $k=j-1$, then by (2.12) we have $y_{n+1}(t)=\sqrt{2^m}$ and

$$y_{n+1}(t)\,y_{n+1}(\tau)=\begin{cases} 2^m & \text{for} \quad a'_k\leqslant\tau<a'_{k+1}, \\ -2^m & \text{for} \quad a'_{k+1}\leqslant\tau<a'_{k+2}, \\ 0 & \text{for} \quad \tau<a'_k \quad \text{and} \quad a'_{k+2}<\tau,\end{cases}$$

whence, taking into account that, by the induction hypothesis,

$$\sum_{i=1}^{n} y_i(t)\,y_i(\tau)=\begin{cases}\frac{1}{a'_{k+2}-a'_k}=2^m & \text{for} \quad a'_k<\tau<a'_{k+2} \\ 0 & \text{for} \quad \tau<a'_k \quad \text{and} \quad a'_{k+2}<\tau,\end{cases}$$

we obtain

$$\sum_{i=1}^{n+1} y_i(t)\,y_i(\tau)=\begin{cases} 2^m+2^m=2^{m+1}=\frac{1}{a'_{k+1}-a'_k} & \text{for} \quad a'_k<\tau<a'_{k+1}, \\ 2^m-2^m=0 & \text{for} \quad a'_{k+1}<\tau<a'_{k+2}, \\ 0 & \text{for} \quad \tau<a'_k \quad \text{and} \quad a'_{k+2}<\tau,\end{cases}$$

i.e., (2.10) with $n+1$ instead of n. Finally, if $k=j$, then by (2.12) we have $y_{n+1}(t)=-\sqrt{2^m}$ and similarly to the preceding case we obtain (2.10) for $n+1$ instead of n, which completes the induction. This proves (2.10).

Now, for $\tilde{x} \in L^p([0,1])$ let[1]

$$\alpha_i = \int_0^1 x(\tau) y_i(\tau) d\tau \qquad (i=1,2,\ldots), \tag{2.13}$$

$$p_n(x) = \sum_{i=1}^{n} \alpha_i y_i \qquad (n=1,2,\ldots). \tag{2.14}$$

Then, by (2.10) we have, for $a_k < t < a_{k+1}$,

$$[p_n(x)](t) = \sum_{i=1}^{n} y_i(t) \int_0^1 x(\tau) y_i(\tau) d\tau = \frac{1}{a_{k+1} - a_k} \int_{a_k}^{a_{k+1}} x(\tau) d\tau, \tag{2.15}$$

whence

$$\int_0^1 |[p_n(x)](t)|^p dt = \sum_{k=0}^{n-1} \int_{a_k}^{a_{k+1}} |[p_n(x)](t)|^p dt = \sum_{k=0}^{n-1} |a_{k+1} - a_k| \left| \frac{\int_{a_k}^{a_{k+1}} x(\tau) d\tau}{a_{k+1} - a_k} \right|^p.$$

However, by the Hölder inequality for integrals, we have

$$\left| \int_{a_k}^{a_{k+1}} x(\tau) d\tau \right|^p \leqslant |a_{k+1} - a_k|^{p-1} \int_{a_k}^{a_{k+1}} |x(\tau)|^p d\tau \qquad (k=0,1,\ldots,n-1).$$

Consequently,

$$\int_0^1 |[p_n(x)](t)|^p dt \leqslant \int_0^1 |x(t)|^p dt.$$

Since $\tilde{x}$ has been an arbitrary element of $L^p([0,1])$, it follows that for the linear operators s_n defined on $L^p([0,1])$ by

$$s_n(\tilde{x}) = \widetilde{p_n(x)} = \sum_{i=1}^{n} \alpha_i \tilde{y}_i \qquad (\tilde{x} \in L^p([0,1])), \tag{2.16}$$

we have

$$\|s_n\| \leqslant 1 \qquad (n=1,2,\ldots). \tag{2.17}$$

On the other hand, for every $x \in C([0,1])$ it follows from (2.15) that $[p_n(x)](t)$ converges uniformly to $x(t)$ on $[0,1] \setminus A$, where A denotes the countable set of all discontinuity points of the functions (2.9). Hence we have, in the norm of $L^p([0,1])$,

$$\lim_{n \to \infty} \|s_n(\tilde{x}) - \tilde{x}\| = 0 \qquad (x \in C([0,1])). \tag{2.18}$$

Since the set $\{\tilde{x} | x \in C([0,1])\}$ is dense in $L^p([0,1])$, from (2.17) and (2.18) we obtain

$$\lim_{n \to \infty} \|s_n(\tilde{x}) - \tilde{x}\| = 0 \qquad (\tilde{x} \in L^p([0,1])),$$

i.e. every $\tilde{x} \in L^p([0,1])$ has an expansion of the form $\tilde{x} = \sum_{i=1}^{\infty} \alpha_i \tilde{y}_i$.

---

[1] We denote by x an arbitrary function from the equivalence class $\tilde{x}$.

On the other hand, by (2.9) we have the orthogonality relations

$$\int_0^1 y_i(t)\,y_j(t)\,dt=0 \qquad (i\neq j;\ i,j=1,2,\ldots). \tag{2.19}$$

Indeed, it is obvious that $\int_0^1 y_1(t)\,y_{2^k+l}(t)\,dt=0$ $(l=1,2,\ldots,2^k;$ $k=0,1,2,\ldots)$ and that for $1\leqslant l_1,\ l_2\leqslant 2^k,\ l_1\neq l_2$, we have

$$\int_2^1 y_{2^k+l_1}(t)\,y_{2^k+l_2}(t)\,dt=0,$$

since already $y_{2^k+l_1}y_{2^k+l_2}=0$. On the other hand, if $k_1>k_2$, then for any l_1,l_2 with $1\leqslant l_i\leqslant 2^{k_i}$ $(i=1,2)$, the set $\{t\in[0,1]\,|\,y_{2^k+l_1}(t)\neq 0\}$ is contained in an interval in which the function $y_{2^{k_2}+l_2}(t)$ has a constant value $\lambda\,(=\sqrt{2^{k_2}},\ -\sqrt{2^{k_2}}$ or $0)$, whence

$$\int_0^1 y_{2^{k_1}+l_1}(t)\,y_{2^{k_2}+l_2}(t)\,dt=\lambda\int_0^1 y_{2^{k_1}+l_1}(t)\,dt=\lambda\left[\frac{\sqrt{2^{k_1}}}{2^{k_1+1}}-\frac{\sqrt{2^{k_1}}}{2^{k_1+1}}\right]=0,$$

which proves (2.19). Now, from (2.19) it follows that the coefficients α_i in the above expansions are uniquely determined, which proves that $\{\tilde{y}_n\}$ is a basis of $L^p([0,1])$. This basis is called *the Haar basis* of $L^p([0,1])$.

Similarly to the extension (2.4) of the Schauder basis (2.3), one can also define Haar functions $\{y_n\}$ with respect to a dense sequence $\{a_n\}\subset[0,1]$ and prove that the sequence $\{\tilde{y}_n\}$ constitutes a basis of $L^p([0,1])$.

The above arguments and constructions can be extended to prove the existence of bases in separable Orlicz spaces and in the more general separable L^λ spaces, where λ is an arbitrary levelling length function. These spaces are defined as follows.

Let T be a set, $\mathscr{F}$ a relatively complemented, countably additive collection of subsets of T and $v=v(e)$ a non-negative countably additive measure defined for all $e\in\mathscr{F}$ (σ-finiteness is not assumed). Let $\lambda(y)\leqslant\infty$ be a non-negative function defined for all non-negative measurable y on T and let L^λ consist of those measurable x for which $\lambda(|x|)<\infty$ (where $|x|(t)=|x(t)|$ $(t\in T)$).

The function λ is called a *length function* if the following conditions are satisfied: a) $\lambda(y)=0$ whenever $y(t)=0$ for almost all $t\in T$; b) $\lambda(y_1)\leqslant\lambda(y_2)$ whenever $y_1(t)\leqslant y_2(t)$ for almost all $t\in T$; c) $\lambda(y+y')\leqslant\lambda(y)+\lambda(y')$; d) $\lambda(\alpha y)=\alpha\lambda(y)$ $(\alpha>0)$, and e) the relations $y_1(t)\leqslant y_2(t)\leqslant\ldots$ $(t\in T)$ imply $\lambda(\sup y_n)=\sup\lambda(y_n)$. In this case L^λ is a Banach space. A length function λ is called a *levelling length function* if it satisfies the following conditions: f) $\lambda(y)=\sup_{\substack{e\in\mathscr{F}\\ v(e)<\infty}}\lambda(y\chi_e)$, where χ_e denotes the characteristic function of the set e, and g) $\lambda(y) \geqslant \lambda(\bar{y})$ whenever $\bar{y}$ coincides with y except on some e with $0 < \nu(e) < \infty$, and $\bar{y}(t) = \frac{1}{\nu(e)} \int_e y(\tau)\, d\nu(\tau)$ $(t \in e)$. As we have mentioned above, every separable L^λ space, where λ is a levelling length function, has a basis.

However, there are some concrete separable Banach spaces in which no basis is known, e.g. the following:

Problem 2.1. Does a separable L^λ space, where λ is *not* a levelling length function, posses a basis?

Problem 2.2. Does the Banach space H^1 possess a basis?

We recall that H^1 is the subspace of the complex Banach space $L^1(\Omega)$, where Ω is the unit circumference $\{\zeta \,\|\, |\zeta| = 1\}$, constituted by all $x \in L^1(\Omega)$ such that $\int_{-\pi}^{\pi} e^{in\vartheta} x(\vartheta)\, d\vartheta = 0$ $(n = 1, 2, \ldots)$.

Let A be the Banach space of complex functions considered at the end of § 1.

Problem 2.3. Does the Banach space A possess a basis?

In order to be able to recognize whether or not a given sequence $\{x_n\}$ in a Banach space E constitutes a basis of the space E, it is useful to know the properties possessed by bases in Banach spaces. In the next sections we shall present some properties of bases and various characterizations of bases in Banach spaces.

§ 3. The coefficient functionals associated to a basis. Bounded bases. Normalized bases

Definition 3.1. Let $\{x_n\}$ be a basis of a Banach space E. The sequence of linear functionals $\{f_n\}$ defined by

$$f_j(x) = \alpha_j \qquad \left(x = \sum_{i=1}^{\infty} \alpha_i x_i \in E,\ j = 1, 2, \ldots\right) \tag{3.1}$$

is called *the sequence of coefficient functionals associated to the basis* $\{x_n\}$, or, shortly, *the associated sequence of coefficient functionals* (we shall write: a.s.c.f.).

Thus, if $\{x_n\}$ is a basis of the space E and $\{f_n\}$ the a.s.c.f., then every $x \in E$ has a unique expansion of the form

$$x = \sum_{i=1}^{\infty} f_i(x)\, x_i. \tag{3.2}$$

Our next aim is to prove the fundamental fact that the coefficient functionals associated to a basis of a Banach space E are continuous on the space E. Let us first prove

Proposition 3.1. *Let $\{x_n\}$ be a sequence in a Banach space E, such that $x_n \neq 0$ $(n=1,2,\ldots)$ and let A_1 be the linear space of sequences of scalars*

$$A_1=\left\{\{\alpha_n\}\subset K \;\middle|\; \sum_{i=1}^{\infty} \alpha_i x_i \text{ converges}\right\} \tag{3.3}$$

endowed with the norm

$$\|\{\alpha_n\}\| = \sup_{1\leqslant n<\infty} \left\| \sum_{i=1}^{n} \alpha_i x_i \right\|. \tag{3.4}$$

Then A_1 is a Banach space.

Proof. The number (3.4) is finite, since the sequence $\left\{\left\|\sum_{i=1}^{n}\alpha_i x_i\right\|\right\}$ is convergent. Since all $x_n \neq 0$ $(n=1,2,\ldots)$, (3.4) is a norm on the linear space (3.3).

Let $\{\alpha_n^{(k)}\}$ $(k=1,2,\ldots)$ be a Cauchy sequence in A_1. Then for every $\varepsilon>0$ there exists a positive integer $N(\varepsilon)$ such that

$$\|\{\alpha_n^{(k)}\}-\{\alpha_n^{(m)}\}\| = \sup_{1\leqslant n<\infty} \left\| \sum_{i=1}^{n} (\alpha_i^{(k)}-\alpha_i^{(m)}) x_i \right\| < \varepsilon \qquad (k,m>N(\varepsilon)).$$

Hence

$$\|(\alpha_n^{(k)}-\alpha_n^{(m)})x_n\| \leqslant \left\| \sum_{i=1}^{n} (\alpha_i^{(k)}-\alpha_i^{(m)}) x_i \right\| + \left\| \sum_{i=1}^{n-1} (\alpha_i^{(k)}-\alpha_i^{(m)}) x_i \right\| < 2\varepsilon$$
$$(k,m>N(\varepsilon);\ n=1,2,\ldots),$$

whence, since all $x_n \neq 0$ $(n=1,2,\ldots)$, it follows that

$$|\alpha_n^{(k)}-\alpha_n^{(m)}| < \frac{2\varepsilon}{\|x_n\|} \qquad (k,m>N(\varepsilon);\ n=1,2,\ldots).$$

Consequently, for each $n\geqslant 1$ the sequence of scalars $\alpha_n^{(k)}$ $(k=1,2,\ldots)$ is convergent to a scalar α_n. Hence, from the inequalities

$$\left\| \sum_{i=1}^{n} (\alpha_i^{(k)}-\alpha_i^{(m)}) x_i \right\| < \varepsilon \quad (k,m>N(\varepsilon);\ n=1,2,\ldots)$$

we obtain, for $m\to\infty$,

$$\left\| \sum_{i=1}^{n} (\alpha_i^{(k)}-\alpha_i) x_i \right\| \leqslant \varepsilon \qquad (k>N(\varepsilon),\ n=1,2,\ldots).$$

Then

$$\left\| \sum_{i=n+1}^{n+l} \alpha_i x_i \right\| \leqslant 2\varepsilon + \left\| \sum_{i=n+1}^{n+l} \alpha_i^{(k)} x_i \right\| \qquad (k > N(\varepsilon);\ n, l = 1, 2, \ldots),$$

whence, since each series $\sum_{i=1}^{\infty} \alpha_i^{(k)} x_i$ is convergent and since E is complete, it follows that $\sum_{i=1}^{\infty} \alpha_i x_i$ converges, i.e. $\{\alpha_n\} \in A_1$. Moreover, by the above we have

$$\|\{\alpha_n^{(k)}\} - \{\alpha_n\}\| = \sup_{1 \leqslant n < \infty} \left\| \sum_{i=1}^{n} (\alpha_i^{(k)} - \alpha_i) x_i \right\| \leqslant \varepsilon \qquad (k > N(\varepsilon)),$$

which completes the proof of proposition 3.1.

Proposition 3.2. *Let E be a Banach space with a basis $\{x_n\}$ and let $\{f_n\}$ be the a.s.c.f. Then*

a) *The Banach space A_1 introduced in proposition 3.1 is isomorphic to E, by the mapping*

$$\{\alpha_n\} \to \sum_{i=1}^{\infty} \alpha_i x_i. \tag{3.5}$$

b) *The numbers*

$$|||x||| = \sup_{1 \leqslant n < \infty} \left\| \sum_{i=1}^{n} f_i(x) x_i \right\| \qquad (x \in E) \tag{3.6}$$

define a norm on the space E, equivalent to the initial norm of E.

Proof. a) Since $\{x_n\}$ is a basis, we have $x_n \neq 0$ $(n = 1, 2, \ldots)$ (by virtue of the uniqueness of the expansions $\sum_{i=1}^{\infty} \alpha_i x_i$), whence, by proposition 3.1, A_1 is a Banach space. The mapping (3.5) of A_1 into E is obviously linear and of norm 1. Since $\{x_n\}$ is a basis of E, (3.5) is one to one and maps A_1 onto E (because of the uniqueness and existence of the expansions $\sum_{i=1}^{\infty} \alpha_i x_i$). Hence, by the inversion theorem of Banach[1], (3.5) is an isomorphism of A_1 onto E.

b) By part a), there exists a constant $C \geqslant 1$ such that

$$\left\| \sum_{i=1}^{\infty} \alpha_i x_i \right\| \leqslant \sup_{1 \leqslant n < \infty} \left\| \sum_{i=1}^{n} \alpha_i x_i \right\| \leqslant C \left\| \sum_{i=1}^{\infty} \alpha_i x_i \right\| \qquad \left(\sum_{i=1}^{\infty} \alpha_i x_i \in E \right),$$

[1] See e.g. [10], p. 41, theorem 5.

and it remains to observe that $\alpha_n = f_n(x)$ $(n=1,2,\ldots)$ for all $x = \sum_{i=1}^{\infty} \alpha_i x_i \in E$.

Theorem 3.1. *Let $\{x_n\}$ be a basis of a Banach space E. Then the coefficient functionals f_n associated to the basis $\{x_n\}$ are continuous on E, i.e. we have $f_n \in E^*$ $(n=1,2,\ldots)$. Moreover, there exists a constant M such that*

$$1 \leqslant \|x_n\| \, \|f_n\| \leqslant M \qquad (n=1,2,\ldots). \tag{3.7}$$

Proof. Let $|||x|||$ be the norm on E defined by (3.6). Then there exists, by proposition 3.2 b), a constant $C \geqslant 1$ such that

$$|||x||| \leqslant C\|x\| \qquad (x \in E).$$

Since all $x_n \neq 0$ $(n=1,2,\ldots)$, it follows that

$$\begin{aligned} |f_n(x)| = \frac{\|f_n(x)x_n\|}{\|x_n\|} &\leqslant \frac{\left\|\sum_{i=1}^{n} f_i(x)x_i\right\| + \left\|\sum_{i=1}^{n-1} f_i(x)x_i\right\|}{\|x_n\|} \\ &\leqslant \frac{2|||x|||}{\|x_n\|} \leqslant \frac{2C}{\|x_n\|}\|x\| \qquad (x \in E, n=1,2,\ldots), \end{aligned} \tag{3.8}$$

whence $f_n \in E^*$ and $\|f_n\| \leqslant \dfrac{2C}{\|x_n\|}$ $(n=1,2,\ldots)$. On the other hand, by the definition of f_n we have $1 = f_n(x_n) \leqslant \|f_n\| \, \|x_n\|$ $(n=1,2,\ldots)$, which completes the proof.

Let us point out that in the above proof of proposition 3.2, and hence also in that of theorem 3.1, the inversion theorem of Banach (which amounts to the open mapping theorem or the closed graph theorem) has played a decisive role.

From (3.7) we infer

Corollary 3.1. *Let E be a Banach space with a basis $\{x_n\}$ and let $\{f_n\} \subset E^*$ be the a. s. c. f. Then*

a) *We have* $\inf_{1 \leqslant n < \infty} \|x_n\| > 0$ *if and only if* $\sup_{1 \leqslant n < \infty} \|f_n\| < \infty$.

b) *We have* $\sup_{1 \leqslant n < \infty} \|x_n\| < \infty$ *if and only if* $\inf_{1 \leqslant n < \infty} \|f_n\| > 0$.

In connection with this corollary, let us also mention some equivalent conditions for a sequence $\{x_n\}$ to be bounded from below or from above:

Lemma 3.1. *Let $\{x_n\}$ be a sequence in a Banach space E. Then*

a) *The following conditions are equivalent:*

1°. $\inf_{1 \leqslant n < \infty} \|x_n\| > 0$.

2°. *The convergence of* $\sum_{i=1}^{\infty} \alpha_i x_i$ *implies* $\lim_{n \to \infty} \alpha_n = 0$.

3°. *The convergence of* $\sum_{i=1}^{\infty} \alpha_i x_i$ *implies* $\sup_{1 \leqslant n < \infty} |\alpha_n| < \infty$.

b) *The following conditions are equivalent:*

1°. $\sup_{1 \leqslant n < \infty} \|x_n\| < \infty$.

2°. $\sum_{i=1}^{\infty} |\alpha_i| < \infty$ *implies that* $\sum_{i=1}^{\infty} \alpha_i x_i$ *converges.*

Proof. a) If we have 1° and $\sum_{i=1}^{\infty} \alpha_i x_i$ converges, then $|\alpha_n| \leqslant \frac{1}{\inf_{1 \leqslant j < \infty} \|x_j\|} \|\alpha_n x_n\| \to 0$ as $n \to \infty$. Thus, 1° ⇒ 2°.

The implication 2° ⇒ 3° is obvious.

Finally, if $\inf_{1 \leqslant n < \infty} \|x_n\| = 0$, then, taking indices n_k with $\|x_{n_k}\| \leqslant \frac{1}{2^{2k}}$ $(k=1,2,\ldots)$ and taking $\alpha_{n_k} = 2^k$ $(k=1,2,\ldots)$ and $\alpha_n = 0$ $(n \neq n_1, n_2, \ldots)$, we obtain that $\sum_{i=1}^{\infty} \alpha_i x_i$ converges but $\sup_{1 \leqslant n < \infty} |\alpha_n| = \infty$. Thus, 3° ⇒ 1°.

b) The implication 1° ⇒ 2° is obvious, since $\left\| \sum_{i=n+1}^{n+p} \alpha_i x_i \right\| \leqslant \sup_{1 \leqslant j < \infty} \|x_j\| \sum_{i=n+1}^{n+p} |\alpha_i|$.

Conversely, if $\sup_{1 \leqslant n < \infty} \|x_n\| = \infty$, then, taking indices n_k with $\|x_{n_k}\| > 2^k$ $(k=1,2,\ldots)$ and taking $\alpha_{n_k} = \frac{1}{2^k}$ $(k=1,2,\ldots)$ and $\alpha_n = 0$ $(n \neq n_1, n_2, \ldots)$, we obtain that $\sum_{i=1}^{\infty} |\alpha_i| < \infty$, but $\|\alpha_{n_k} x_{n_k}\| > 1$ $(k=1,2,\ldots)$, whence $\sum_{i=1}^{\infty} \alpha_i x_i$ does not converge. Thus, 2° ⇒ 1°, which completes the proof.

In the terminology of § 8, condition a) 2° means that $\{x_n\}$ dominates the unit vector basis of c_0, while condition b) 2° means that the unit vector basis of l^1 dominates $\{x_n\}$. Therefore, using the results of § 8, one obtains other equivalent conditions.

Definition 3.2. A basis $\{x_n\}$ of a Banach space E is called a *bounded basis* if $0 < \inf_{1 \leqslant n < \infty} \|x_n\| \leqslant \sup_{1 \leqslant n < \infty} \|x_n\| < \infty$. The basis $\{x_n\}$ is said to be a *normalized basis* if $\|x_n\| = 1$ $(n=1,2,\ldots)$.

From the definition 1.1 of a basis it follows that if $\{x_n\}$ is a basis of a Banach space E and $\{\lambda_n\}$ an arbitrary sequence of scalars such that $\lambda_n \neq 0$ $(n=1,2,\ldots)$, then $\{\lambda_n x_n\}$ is also a basis of the space E; hence, in particular, $\left\{ \frac{x_n}{\|x_n\|} \right\}$ is a normalized basis of E.

Theorem 3.2. *Let $\{x_n\}$ be a bounded basis of a Banach space E. Then there exists an equivalent norm $\|x\|'$ on E, in which the basis $\{x_n\}$ is normalized.*

Proof. Let

$$E^{(n)}=[x_1,\dots,x_{n-1},x_{n+1},\dots] \qquad (n=1,2,\dots) \tag{3.9}$$

(the closed linear subspace of E spanned by $x_1,\dots,x_{n-1},x_{n+1},\dots$) and let $\{f_n\}\subset E^*$ be the a. s. c. f. to the basis $\{x_n\}$. Then for any $\sum_{i\neq n}\alpha_i x_i\in E^{(n)}$ we have, by (3.7),

$$\left\|x_n-\sum_{i\neq n}\alpha_i x_i\right\| \geqslant \frac{1}{\|f_n\|}\left|f_n\left(x_n-\sum_{i\neq n}\alpha_i x_i\right)\right| = \frac{1}{\|f_n\|}$$

$$\geqslant \frac{1}{M}\|x_n\| \geqslant \frac{1}{M}\inf_{1\leqslant k<\infty}\|x_k\| \qquad (n=1,2,\dots),$$

where $M\geqslant 1$ is a constant. Consequently, taking an α such that $0<\alpha<\frac{1}{M}\inf_{1\leqslant k<\infty}\|x_k\|$, we shall have

$$\operatorname{dist}(x_n,E^{(n)})>\alpha \qquad (n=1,2,\dots). \tag{3.10}$$

Put

$$S_E'=\overline{\text{co}}\left[\{\beta x_k \mid |\beta|=1, k=1,2,\dots\}\cup\alpha S_E\right]=\overline{\text{co}}\,A, \tag{3.11}$$

where $S_E=\{x\in E \mid \|x\|\leqslant 1\}$ is the unit ball of E and where $\overline{\text{co}}\,A$ denotes the closed convex hull of the set

$$A=\{\beta x_k \mid |\beta|=1, k=1,2,\dots\}\cup\alpha S_E. \tag{3.12}$$

We claim that the Minkowski functional $\|x\|'=\inf\limits_{\substack{\lambda>0\\ \frac{1}{\lambda}x\in S_E'}}\lambda$ of the circled[1] closed convex set S_E' is a norm on E, having the required properties.

Indeed, $\|x\|'$ is a norm on E, equivalent to the initial norm on E, since by (3.11) and $\alpha<\frac{1}{M}\inf_{1\leqslant k<\infty}\|x_k\|\leqslant\inf_{1\leqslant k<\infty}\|x_k\|\leqslant\sup_{1\leqslant k<\infty}\|x_k\|$ we have

$$\alpha S_E\subset S_E'\subset\left(\sup_{1\leqslant k<\infty}\|x_k\|\right)S_E$$

---

[1] We recall that a set $D\subset E$ is said to be *circled*, if $\beta D\subset D$ for each scalar β with $|\beta|=1$.

and thus it remains only to prove that

$$\|x_n\|' = 1 \qquad (n=1,2,\ldots). \tag{3.13}$$

Fix an arbitrary n and put

$$\Omega_n = \overline{\mathrm{co}}\,[\{\beta x_k \,|\, |\beta| = 1,\, k = 1,\ldots,n-1,n+1,\ldots\} \cup \alpha S_E]. \tag{3.14}$$

Then obviously

$$\Omega_n \subset \{x \in E \,|\, \operatorname{dist}(x, E^{(n)}) \leqslant \alpha\},$$

whence, by (3.10),

$$x_n \notin \Omega_n. \tag{3.15}$$

Furthermore, we have

$$A \subset \{\beta x_n \,|\, |\beta| = 1\} \cup \Omega_n \subset \overline{\mathrm{co}}\, A = S'_E,$$

whence

$$S'_E = \overline{\mathrm{co}}\,[\{\beta x_n \,|\, |\beta| = 1\} \cup \Omega_n]. \tag{3.16}$$

Now, by (3.15) and since Ω_n is a circled closed convex set, there exists[1] an $f \in E^*$ such that

$$f(x_n) \geqslant \sup_{x \in \Omega_n} |f(x)|.$$

Consequently, by (3.16) and $x_n \in S'_E$,

$$f(x_n) = \sup_{x \in S'_E} |f(x)| = \|f\|',$$

whence

$$1 \geqslant \|x_n\|' \geqslant \frac{1}{\|f\|'} f(x_n) = 1,$$

and thus we have (3.13), which completes the proof.

§ 4. Biorthogonal systems. The partial sum operators. Some characterizations of regular biorthogonal systems. Applications

Definition 4.1. Let E be a Banach space. A pair of sequences (x_n, f_n), where $\{x_n\} \subset E$, $\{f_n\} \subset E^*$, is called a *biorthogonal system* if

$$f_i(x_j) = \delta_{ij} \qquad (i,j = 1,2,\ldots). \tag{4.1}$$

---

[1] See e.g. [270], p. 109, theorem 4.

The biorthogonal system (x_n, f_n) is said to be *E-complete* if the sequence $\{x_n\}$ is complete[1] in E.

From theorem 3.1 it follows that *if $\{x_n\}$ is a basis of a Banach space E and $\{f_n\}$ the a.s.c.f., then (x_n, f_n) is an E-complete biorthogonal system.* However, the converse of this statement is not true, as shown by

Example 4.1. Let $E=C_{2\pi}$, the space of all continuous functions on the real axis $(-\infty,\infty)$ and having period 2π (i.e. $x(t+2\pi)=x(t)$ for all $t\in(-\infty,\infty)$), with the norm $\|x\|=\max_{t\in(-\infty,\infty)}|x(t)|$ and let

$$x_0(t)\equiv\tfrac{1}{2},\ x_{2n-1}(t)=\sin nt,\ x_{2n}(t)=\cos nt \quad (t\in(-\infty,\infty),\ n=1,2,\ldots),$$

$$f_{2n}(x)=\frac{1}{\pi}\int_{-\pi}^{\pi}x(\tau)\cos n\tau\,d\tau,\quad f_{2n+1}(x)=\frac{1}{\pi}\int_{-\pi}^{\pi}x(\tau)\sin(n+1)\tau\,d\tau \tag{4.2}$$

$$(x\in E,\ n=0,1,2,\ldots).$$

Then (x_n,f_n) is an E-complete biorthogonal system, since by the theorem of Weierstrass the sequence $\{x_n\}$ is complete in E. However, from the existence of continuous functions whose Fourier series are not uniformly convergent it follows that $\{x_n\}$ is not a basis of E.

Remark 4.1. Actually, in every infinite dimensional separable Banach space E there exist E-complete biorthogonal systems (x_n,f_n) such that $\{x_n\}$ is not a basis of E. Indeed, for infinite dimensional Banach spaces with bases this follows immediately from proposition 4.3 below and its proof. For separable Banach spaces which have no basis (if such spaces exist at all), the assertion follows from the fact that in every separable Banach space E there exist E-complete biorthogonal systems (x_n,f_n) (embed E isometrically into $C([0,1])$ and orthogonalize the image of a finitely linearly independent complete sequence $\{y_n\}\subset E$ in $C([0,1])$, by the classical procedure of E. Schmidt). Let us also mention that in Vol. II, Ch. III, we shall see many other examples of E-complete biorthogonal systems (x_n,f_n) such that $\{x_n\}$ is not a basis of E (T-bases which are not bases, etc.).

From the remark preceding example 4.1 it follows that we have to seek for the bases of a Banach space E only among the sequences $\{x_n\}\subset E$

---

[1] We use this term in the following sense: the set of all finite linear combinations $\sum_{i=1}^{n}\alpha_i x_i$ ($\alpha_i\in K$, $i=1,\ldots,n$; $n=1,2,\ldots$) is dense in E. Banach [10] has used the term "fundamental" for such sequences and we also have done so in the expository papers [241]—[243]. However, since Cauchy sequences are also called sometimes fundamental sequences, we prefer to use here the term "complete sequence". This term is frequently used, at least for sequences in Hilbert spaces.

belonging to biorthogonal systems (x_n, f_n). Therefore it is convenient to give

Definition 4.2. Let E be a Banach space. A biorthogonal system (x_n, f_n) $(\{x_n\} \subset E, \{f_n\} \subset E^*)$ is said to be *regular*, if the sequence $\{x_n\}$ is a basis of the space E; otherwise (x_n, f_n) is said to be *irregular*.

Definition 4.3. Let E be a Banach space and $(x_n, f_n)(\{x_n\} \subset E, \{f_n\} \subset E^*)$ a biorthogonal system. For every $x \in E$, the (convergent or divergent) series $\sum_{i=1}^{\infty} f_i(x) x_i$ is called *the formal expansion of x corresponding to the biorthogonal system* (x_n, f_n), and the following notation is used:

$$x \sim \sum_{i=1}^{\infty} f_i(x) x_i. \tag{4.3}$$

Definition 4.4. Let E be a Banach space and (x_n, f_n) $(\{x_n\} \subset E, \{f_n\} \subset E^*)$ a biorthogonal system. The sequence of continuous linear operators $\{s_n\}$, where

$$s_n(x) = \sum_{i=1}^{n} f_i(x) x_i \qquad (x \in E, n = 1, 2, \ldots), \tag{4.4}$$

is called *the sequence of partial sum operators associated to the biorthogonal system* (x_n, f_n). If $\{x_n\}$ is a basis of E and $\{f_n\}$ the a.s.c.f., then $\{s_n\}$ is called *the sequence of partial sum operators associated to the basis* $\{x_n\}$.

Theorem 4.1. *Let E be a Banach space and (x_n, f_n) $(\{x_n\} \subset E, \{f_n\} \subset E^*)$ a biorthogonal system. Then the following statements are equivalent:*

1°. *(x_n, f_n) is a regular biorthogonal system.*

2°. *For every $x \in E$ the formal expansion (4.3) is convergent and its sum is x, i.e. we have $\lim_{n \to \infty} s_n(x) = x$ for all $x \in E$.*

3°. *(x_n, f_n) is E-complete and $\sup_{1 \leqslant n < \infty} \|s_n(x)\| < \infty$ for all $x \in E$.*

4°. *(x_n, f_n) is E-complete and there exists a constant $M \geqslant 1$ such that*

$$\|s_n\| \leqslant M \qquad (n = 1, 2, \ldots). \tag{4.5}$$

Proof. If we have 1°, then $\{x_n\}$ is a basis of E and $\{f_n\}$ is the a.s.c.f. (since if we denote by $\{g_n\}$ the a.s.c.f., then $g_i(x_j) = f_i(x_j)$ for all $i, j = 1, 2, \ldots$, whence $g_i(x) = f_i(x)$ for all $x \in E$, $i = 1, 2, \ldots$). Hence we have 2°. The implication $2^\circ \Rightarrow 3^\circ$ is obvious. The implication $3^\circ \Rightarrow 4^\circ$ is a consequence of the principle of uniform boundedness[1]. Finally, if (x_n, f_n) is a biorthogonal system, then for every finite linear combination $p = \sum_{j=1}^{m} \alpha_j x_j$ and all $n \geqslant m$ we have

[1] See e.g. [10], p. 80, theorem 5.

$$s_n(p) = \sum_{i=1}^{n} f_i\left(\sum_{j=1}^{m} \alpha_j x_j\right) x_i = \sum_{i=1}^{n} \sum_{j=1}^{m} \alpha_j \delta_{ij} x_i = \sum_{j=1}^{m} \alpha_j x_j = p,$$

whence it follows[1] that 4° implies 1°. This completes the proof.

The implication 1° ⇒ 4° is also a consequence of § 3, proposition 3.2. On the other hand, let us point out that in the proof of the implication 3° ⇒ 1° the principle of uniform boundedness plays a decisive role.

From the implication 4° ⇒ 1° of theorem 4.1 there follows

Corollary 4.1. *Let E be a Banach space and (x_n, f_n) ($\{x_n\} \subset E$, $\{f_n\} \subset E^*$) an irregular E-complete biorthogonal system. Then for every number $a \geqslant 1$ there exists a finite linear combination $p = \sum_{j=1}^{m} \alpha_j x_j$ such that $\|p\| = 1$ and $\sup_{1 \leqslant n \leqslant m} \|s_n(p)\| = a$.*

Proof. There exists a finite linear combination $q = \sum_{j=1}^{m} \beta_j x_j$ such that $\|q\| = 1$ and $\sup_{1 \leqslant n < \infty} \|s_n(q)\| > a$, since otherwise the uniform boundedness of $\sup_{1 \leqslant n < \infty} \|s_n(x)\|$ on a dense subset of the unit sphere $\{x \in E \,|\, \|x\| = 1\}$ would imply (4.5), in contradiction with the irregularity of (x_n, f_n). By biorthogonality we have $(1-\lambda)x_1 + \lambda q \neq 0$ $(0 \leqslant \lambda \leqslant 1)$. Let

$$\phi(\lambda) = \sup_{1 \leqslant n \leqslant m} \left\| s_n\left(\frac{(1-\lambda)x_1 + \lambda q}{\|(1-\lambda)x_1 + \lambda q\|}\right)\right\| \qquad (0 \leqslant \lambda \leqslant 1).$$

Then $\phi(\lambda)$ is continuous on $[0,1]$ (as the supremum of a finite family of continuous functions) and we have, by biorthogonality,

$$\phi(0) = \sup_{1 \leqslant n \leqslant m} \left\| s_n\left(\frac{x_1}{\|x_1\|}\right)\right\| = \left\|\frac{x_1}{\|x_1\|}\right\| = 1,$$

$$\phi(1) = \sup_{1 \leqslant n \leqslant m} \|s_n(q)\| = \sup_{1 \leqslant n < \infty} \|s_n(q)\| > a.$$

Hence there exists a λ_0 with $0 \leqslant \lambda_0 < 1$, such that $\phi(\lambda_0) = a$. Then for $p = \dfrac{(1-\lambda_0)x_1 + \lambda_0 q}{\|(1-\lambda_0)x_1 + \lambda_0 q\|}$ we have $\|p\| = 1$ and $\sup_{1 \leqslant n \leqslant m} \|s_n(p)\| = a$.

Let us give now some applications of theorem 4.1.

Proposition 4.1. *Let E be a Banach space with a basis $\{x_n\}$. Let $\{i_n\}$ be an increasing sequence of positive integers, $\{l_n\}$ the increasing sequence*

---

[1] We use the following simple fact: if $\{u_n\}$ is a sequence of continuous linear operators on E such that $\|u_n\| \leqslant M$ $(n = 1, 2, \ldots)$ and that $\lim_{n \to \infty} u_n(p) = p$ for all p in a dense subset of E, then $\lim_{n \to \infty} u_n(x) = x$ for all $x \in E$.

of positive integers complementary to $\{i_n\}$ *and* ω *the canonical mapping of E onto*[1] $E/[x_{i_n}]$. *Then*

a) $\{x_{i_n}\}$ *is a basis of the space* $[x_{i_n}]$.

b) $\{\omega(x_{l_n})\}$ *is a basis of the quotient space* $E/[x_{i_n}]$.

Proof. Let $\{f_n\}\subset E^*$ be the sequence of coefficient functionals associated to the basis $\{x_n\}$.

a) Since $f_{l_m}(x_{i_n})=0\ (m,n=1,2,\ldots)$, we have $f_{l_m}(x)=0$ for all $x\in[x_{i_n}]$, $m=1,2,\ldots$. Hence $x=\sum_{n=1}^{\infty} f_{i_n}(x)x_{i_n}$ for every $x\in[x_{i_n}]$ and we can apply the implication $2^\circ\Rightarrow 1^\circ$ of theorem 4.1 to the biorthogonal system[2] $(x_{i_n}, f_{i_n}|_{[x_{i_n}]})$.

b) Define

$$\phi_m[\omega(x)]=f_{l_m}(x) \qquad (x\in E,\ m=1,2,\ldots). \tag{4.6}$$

If $\omega(x)=\omega(x')$, then $x-x'\in[x_{i_n}]$, whence $f_{l_m}(x-x')=0\ (m=1,2,\ldots)$, which proves that ϕ_m is well defined on $E/[x_{i_n}]$. Obviously, ϕ_m is linear and $\phi_m[\omega(x_{l_n})]=\delta_{mn}\ (m,n=1,2,\ldots)$. Since for every $x\in E$ and $\varepsilon>0$ there exists an $x'\in\omega(x)$ with $\|x'\|\leqslant\|\omega(x)\|+\varepsilon$, it follows from (4.6) that ϕ_m is continuous on $E/[x_{i_n}]$ (we have even $\|\phi_m\|=\|f_{l_m}\|\ (m=1,2,\ldots)$). Now, since $\omega(x_{i_n})=0\ (n=1,2,\ldots)$, we have $\omega(x)=\sum_{n=1}^{\infty}\phi_n[\omega(x)]\,\omega(x_{l_n})$ for every $x=\sum_{i=1}^{\infty} f_i(x)x_i\in E$, and thus we can apply the implication $2^\circ\Rightarrow 1^\circ$ of theorem 4.1 to the biorthogonal system $(\omega(x_{l_n}),\phi_n)$.

Definition 4.5. A sequence $\{x_n\}$ in a Banach space E is said to be a *basic sequence* if $\{x_n\}$ is a basis of the closed linear subspace $[x_n]$ of E.

According to proposition 4.1 a), every subsequence of a basis is a basic sequence. It is natural to ask whether a certain converse is also true, namely:

Problem 4.1. Let $\{y_n\}$ be an infinite basic sequence in a separable Banach space E. Does there exist a basis $\{x_n\}$ of E with the property that for each n there is an index i_n such that $x_{i_n}=y_n$? What if E has a basis?

An affirmative answer to the first of these questions would imply an affirmative answer to the basis problem, since every Banach space contains an infinite basic sequence (see § 14, remark made after table 14.1). Moreover, the answer to the second question is also believed to be negative, namely, it is conjectured that the basic sequence

[1] We denote by $[x_{i_n}]$ the (closed linear) subspace of E spanned by sequence $\{x_{i_n}\}$.

[2] We denote by $f_{i_n}|_{[x_{i_n}]}$ the restriction of f_{i_n} to $[x_{i_n}]$.

$$y_n(t) = \sin 2^n \pi t \qquad (t \in [0,1],\ n = 1,2,\ldots) \tag{4.7}$$

in $C([0,1])$ cannot be extended to a basis of $C([0,1])$.

We shall show now that if $[y_n]$ admits a complementary subspace F with a basis, then the answer to problem 4.1 is affirmative. Indeed, let us first prove

Proposition 4.2. *Let G, F be two Banach spaces with bases $\{y_n\}$ and $\{z_n\}$, respectively. Then the sequence $\{x_n\} \subset G \times F$ defined by*

$$x_{2n-1} = \{y_n, 0\}, \quad x_{2n} = \{0, z_n\} \qquad (n = 1,2,\ldots) \tag{4.8}$$

is a basis of $G \times F$.

Proof. Let $\{y,z\} \in G \times F$ be arbitrary. Then, since $\{y_n\}$, $\{z_n\}$ are bases of G and F, respectively, we can write $y = \sum_{j=1}^{\infty} \alpha_j y_j$, $z = \sum_{j=1}^{\infty} \beta_j z_j$. Since by the definition of the norm in $G \times F$ we have

$$\left\| \{y,0\} - \sum_{j=1}^{n} \alpha_j \{y_j, 0\} \right\| = \left\| \left\{ y - \sum_{j=1}^{n} \alpha_j y_j, 0 \right\} \right\| = \left\| y - \sum_{j=1}^{n} \alpha_j y_j \right\|, \tag{4.9}$$

it follows that $\sum_{j=1}^{\infty} \alpha_j \{y_j, 0\}$ converges to $\{y,0\}$ and, similarly, $\sum_{j=1}^{\infty} \beta_j \{0, z_j\}$ converges to $\{0,z\}$. Consequently,

$$\begin{aligned} \{y,z\} = \{y,0\} + \{0,z\} &= \sum_{j=1}^{\infty} \alpha_j \{y_j, 0\} + \sum_{j=1}^{\infty} \beta_j \{0, z_j\} \\ &= \sum_{j=1}^{\infty} \alpha_j x_{2j-1} + \sum_{j=1}^{\infty} \beta_j x_{2j}. \end{aligned}$$

To show the uniqueness of these expansions, observe first that if $\sum_{j=1}^{\infty} \gamma_j x_j$ converges, say to $\{y,z\}$, then, by

$$\begin{aligned} \left\| y - \sum_{j=1}^{n} \gamma_{2j-1} y_j \right\| &= \left\| \left\{ y - \sum_{j=1}^{n} \gamma_{2j-1} y_j, 0 \right\} \right\| \\ &\leqslant \left\| \left\{ y - \sum_{j=1}^{n} \gamma_{2j-1} y_j,\ z - \sum_{j=1}^{n} \gamma_{2j} z_j \right\} \right\| = \left\| \{y,z\} - \sum_{j=1}^{2n} \gamma_j x_j \right\|, \end{aligned}$$

we have $\sum_{j=1}^{\infty} \gamma_{2j-1} y_j = y$ and, similarly, $\sum_{j=1}^{\infty} \gamma_{2j} z_j = z$. Consequently, if $\sum_{j=1}^{\infty} \gamma_j x_j = \{0,0\}$, then $\sum_{j=1}^{\infty} \gamma_{2j-1} y_j = 0$, $\sum_{j=1}^{\infty} \gamma_{2j} z_j = 0$, whence, since $\{y_n\}$,

$\{z_n\}$ are bases of G and F, respectively, $\gamma_j=0\,(j=1,2,\ldots)$, which completes the proof of proposition 4.2.

Definition 4.6. Let G, F be two Banach spaces with bases $\{y_n\}$ and $\{z_n\}$, respectively. The basis (4.8) of $G\times F$ is called the *cartesian product* of the bases $\{y_n\}$ and $\{z_n\}$; sometimes, for brevity, we shall denote this sequence by $\{\{y_n,0\}\}\cup\{\{0,z_n\}\}$ or by $\{y_n\}\times\{z_n\}$.

The desired affirmative answer to problem 4.1 in the particular case when $[y_n]$ admits a complementary subspace F with a basis, follows now from proposition 4.2, taking into account the following well known

Lemma 4.1. *Let G, F be two complementary subspaces of a Banach space E, i.e., $E=G\oplus F$ (i.e., $E=G+F$, $G\cap F=\{0\}$). Then E is isomorphic to $G\times F$.*

Proof. The mapping $\{y,z\}\to y+z$ of $G\times F$ into $G\oplus F$ is obviously linear, one to one and onto. By virtue of

$$\|y+z\|\leqslant\|y\|+\|z\|\leqslant\sqrt{2}\sqrt{\|y\|^2+\|z\|^2}=\sqrt{2}\,\|\{y,z\}\|\qquad(y\in G,\ z\in F)$$

this mapping is also continuous, whence, by the inversion theorem of Banach, an isomorphism, which completes the proof.

We shall see another important instance when the answer to problem 4.1 is affirmative in § 7, theorem 7.2.

Proposition 4.3. *Let E be a Banach space with a bounded basis $\{x_n\}$ and let $\{\alpha_n\}$ be a sequence of scalars such that $\alpha_n\neq 0$ $(n=1,2,\ldots)$. The sequence $\{y_n\}\subset E$ defined by*

$$y_n=\sum_{i=1}^{n}\alpha_i x_i\qquad(n=1,2,\ldots)\tag{4.10}$$

is a basis of the space E if and only if the sequence $\left\{\frac{\|y_n\|}{|\alpha_{n+1}|}\right\}$ is bounded.

Proof. Let $\{f_n\}\subset E^*$ be the sequence of coefficient functionals associated to the basis $\{x_n\}$. Then for the sequence of functionals $\{g_n\}\subset E^*$ defined by

$$g_n=\frac{1}{\alpha_n}f_n-\frac{1}{\alpha_{n+1}}f_{n+1}\qquad(n=1,2,\ldots)\tag{4.11}$$

we have $g_i(y_j)=\delta_{ij}\,(i,j=1,2,\ldots)$ and

$$\sum_{i=1}^{n}g_i(x)y_i=\sum_{i=1}^{n}f_i(x)x_i-\frac{f_{n+1}(x)}{\alpha_{n+1}}y_n\qquad(x\in E,\ n=1,2,\ldots).\tag{4.12}$$

Assume now that $\left\{\frac{\|y_n\|}{|\alpha_{n+1}|}\right\}$ is not bounded. Then there exists a subsequence $\beta_{i_n} = \frac{\|y_{i_n}\|}{|\alpha_{i_n+1}|}$ $(n=1,2,\ldots)$ such that $\beta_{i_n} \to \infty$ and that $\sum_{n=1}^{\infty} \frac{1}{\sqrt{\beta_{i_n}}} x_{i_n}$ converges, say to $z \in E$ (we have used the fact that $\sup_{1 \leqslant n < \infty} \|x_n\| < \infty$). Then $f_i(z) = 0$ for $i \neq i_n$ and $f_{i_n}(z) = \frac{1}{\sqrt{\beta_{i_n}}}$ $(n=1,2,\ldots)$. Therefore $\left\|\frac{f_{i_n}(z)}{\alpha_{i_n+1}} y_{i_n}\right\| = \sqrt{\beta_{i_n}} \to \infty$, whence, by (4.12) and by $\sum_{i=1}^{n} f_i(z) x_i \to z$, $\sum_{i=1}^{\infty} g_i(z) y_i$ fails to converge, and thus $\{y_n\}$ is not a basis of E.

Conversely, assume now that $\left\{\frac{\|y_n\|}{|\alpha_{n+1}|}\right\}$ is bounded. Since $\{x_n\}$ is a basis of E with $\inf_{1 \leqslant n < \infty} \|x_n\| > 0$, whe have $f_n(x) \to 0$ $(x \in E)$, whence, by (4.12) and $\sum_{i=1}^{n} f_i(x) x_i \to x$, we obtain $\sum_{i=1}^{n} g_i(x) y_i \to x$ $(x \in E)$. Thus we can apply the implication $2° \Rightarrow 1°$ of theorem 4.1 to the biorthogonal system (y_n, g_n).

In view of the next application of theorem 4.1 it is convenient to give

Definition 4.7. Let $\{x_n\}$ be a basis of a Banach space E, with $\inf_{1 \leqslant n < \infty} \|x_n\| > 0$. We shall call *block-perturbation* of $\{x_n\}$ any sequence $\{z_k\} \subset E$ of the form

$$z_k = \begin{cases} x_k & \text{for } k \neq p_n \\ x_{p_n} + y_n & \text{for } k = p_n \end{cases} \qquad (n=1,2,\ldots), \tag{4.13}$$

where

$$y_n = \sum_{i=m_{n-1}+1}^{p_n - 1} \alpha_i x_i + \sum_{i=p_n+1}^{m_n} \alpha_i x_i, \quad \|y_n\| \leqslant M < \infty \qquad (n=1,2,\ldots), \tag{4.14}$$

and where $\{m_n\}$, $\{p_n\}$ are increasing sequences of positive integeres such that $m_0 = 0$, $m_{n-1} + 1 \leqslant p_n \leqslant m_n$ $(n=1,2,\ldots)$.

Proposition 4.4. *Let $\{x_n\}$ be a basis of a Banach space E, with $\inf_{1 \leqslant n < \infty} \|x_n\| > 0$. Then every block perturbation $\{z_k\}$ of $\{x_n\}$ is a basis of E.*

Proof. Let $\{z_k\}$ be of the form (4.13) with $\{y_n\}$ satisfying (4.14). Then $\{z_k\}$ admits a biorthogonal sequence $\{h_n\} \subset E^*$ given by

$$h_k = \begin{cases} f_k - \alpha_k f_{p_n} & \text{for } k \neq p_n,\ m_{n-1} + 1 \leqslant k \leqslant m_n \\ f_{p_n} & \text{for } k = p_n \end{cases} \qquad (n=1,2,\ldots),$$

where $\{f_n\}$ is the a. s. c. f. to the basis $\{x_n\}$. Hence, for all $x \in E$,

$$\sum_{k=1}^{l} h_k(x) z_k = \begin{cases} \sum_{j=1}^{l} f_j(x) x_j - f_{p_n}(x) \sum_{i=m_{n-1}+1}^{l} \alpha_i x_i & \text{for } m_{n-1}+1 \leqslant l \leqslant p_n - 1, \\ \sum_{j=1}^{l} f_j(x) x_j + f_{p_n}(x) \sum_{i=l+1}^{m_n} \alpha_i x_i & \text{for } p_n \leqslant l \leqslant m_n \end{cases}$$
$$(n=1,2,\ldots).$$

Since $\{x_n\}$ is a basis of E, there exists, by the implication $1^\circ \Rightarrow 4^\circ$ of theorem 4.1, a constant $C \geqslant 1$ such that

$$\|s_n(x)\| \leqslant C\|x\| \qquad (x \in E, n=1,2,\ldots),$$

whence

$$\left\| \sum_{i=m_{n-1}+1}^{l} \alpha_i x_i \right\| \leqslant C\|y_n\| \leqslant CM \qquad (m_{n-1}+1 \leqslant l \leqslant p_n - 1; n=1,2,\ldots),$$

$$\left\| \sum_{i=l+1}^{m_n} \alpha_i x_i \right\| \leqslant \|y_n\| + C\|y_n\| \leqslant (1+C)M \qquad (p_n \leqslant l \leqslant m_n; n=1,2,\ldots).$$

Since the basis $\{x_n\}$ satisfies $\inf_{1 \leqslant n < \infty} \|x_n\| > 0$, by § 3, lemma 3.1 a) we also have $\lim_{n \to \infty} f_{p_n}(x) = 0$ for all $x \in E$. Consequently, for every $\varepsilon > 0$ and $x \in E$ there exists an integer $N(\varepsilon, x) > 0$ such that

$$\left\| \sum_{k=1}^{l} h_k(x) z_k - \sum_{j=1}^{l} f_j(x) x_j \right\| < \varepsilon \qquad (l > N(\varepsilon, x)).$$

Hence $x = \sum_{k=1}^{\infty} h_k(x) z_k$ for all $x \in E$, and thus, by virtue of the implication $2^\circ \Rightarrow 1^\circ$ of theorem 4.1, $\{z_k\}$ is a basis of E, which completes the proof.

§ 5. Some characterizations of regular E-complete biorthogonal systems. Multipliers

Let E be a Banach spece and (x_n, f_n) $(\{x_n\} \subset E, \{f_n\} \subset E^*)$ an E-complete biorthogonal system. We shall use the notation

$$|||x||| = \sup_{1 \leqslant n < \infty} \left\| \sum_{i=1}^{n} f_i(x) x_i \right\| \qquad (x \in E), \tag{5.1}$$

where $|||x||| = \infty$ is also possible. Let

$$\mathscr{E}_0 = \{x \in E \mid \lim_{n \to \infty} s_n(x) = x\}, \tag{5.2}$$

$$\mathscr{E}_1 = \{x \in E \mid \lim_{n \to \infty} s_n(x) \text{ exists}\}, \tag{5.3}$$

$$\mathscr{E}_2 = \{x \in E \mid \sup_{1 \leqslant n < \infty} \|s_n(x)\| < \infty\} = \{x \in E \mid |||x||| < \infty\}, \tag{5.4}$$

$$\begin{aligned} \mathscr{E}_3 = \{x \in E \mid \lim_{n \to \infty} \|s_n(x)\| \text{ exists and } < \infty\} \\ \cup \{x \in E \mid \lim_{n \to \infty} \|s_n(x)\| \text{ does not exist}\}, \end{aligned} \tag{5.5}$$

where s_n are the associated partial sum operators (see § 4, definition 4.4). Then we have the inclusions

$$\mathscr{E}_0 \subset \mathscr{E}_1 \subset \mathscr{E}_2 \subset \mathscr{E}_3. \tag{5.6}$$

If $\{x_n\}$ is complete in E, the sets $\mathscr{E}_0, \mathscr{E}_1, \mathscr{E}_2, \mathscr{E}_3$ are dense in E, and if $\{x_n\}$ is a basis of E, we have $\mathscr{E}_0 = \mathscr{E}_1 = \mathscr{E}_2 = \mathscr{E}_3 = E$.

Proposition 5.1. *Let E be a Banach space and (x_n, f_n) ($\{x_n\} \subset E$, $\{f_n\} \subset E^*$) a biorthogonal system. We have $\mathscr{E}_0 = \mathscr{E}_1$ if and only if the sequence $\{f_n\} \subset E^*$ is total[1] on E.*

Proof. Assume that $\mathscr{E}_0 = \mathscr{E}_1$ and let $x \in E$, $f_n(x) = 0$ $(n = 1, 2, \ldots)$. Then $s_n(x) = 0$ $(n = 1, 2, \ldots)$, whence, by $\mathscr{E}_0 = \mathscr{E}_1$, it follows that $x = 0$.

Conversely, assume that $\{f_n\}$ is total on E, and let $x \in \mathscr{E}_1$. Then, by biorthogonality and the continuity of the f_n, we have

$$f_j\left[x - \lim_{n \to \infty} \sum_{i=1}^{n} f_i(x) x_i\right] = f_j(x) - f_j(x) = 0 \qquad (j = 1, 2, \ldots),$$

whence, since $\{f_n\}$ is total on E, $x - \lim_{n \to \infty} \sum_{i=1}^{n} f_i(x) x_i = 0$, i.e. $x \in \mathscr{E}_0$.

Proposition 5.2. *Let E be a Banach space and (x_n, f_n) ($\{x_n\} \subset E$, $\{f_n\} \subset E^*$) an irregular E-complete biorthogonal system. Then*

a) *The set $\mathscr{E}_2$ is of the first category.*

b) *The set $E \setminus \mathscr{E}_3 = \{x \in E \mid \lim_{n \to \infty} \|s_n(x)\| = \infty\}$ is of the first category.*

c) *If A is a subset of E, such that every $x \in A$ is the limit of a sequence $\{y_n\} \subset E$ satisfying $\sup_{1 \leqslant n < \infty} |||y_n||| < \infty$, then A is of the first category.*

Proof. a) Since (x_n, f_n) is irregular, for every positive integer m the set $\left\{x \in E \,\middle|\, \sup_{1 \leqslant n < \infty} \|s_n(x)\| \leqslant m\right\}$ is nowhere dense (by the implication $4° \Rightarrow 1°$ of theorem 4.1) and closed, whence the set

$$\mathscr{E}_2 = \bigcup_{m=1}^{\infty} \left\{x \in E \,\middle|\, \sup_{1 \leqslant n < \infty} \|s_n(x)\| \leqslant m\right\}$$

is of the first category.

---

[1] In the sense of Banach [10], i.e. $\{x \in E \mid f_n(x) = 0 \ (n = 1, 2, \ldots)\} = \{0\}$.

b) Define a sequence of functions $\{\beta_n\}$ on $\mathscr{E}_0 \cup (E \setminus \mathscr{E}_3)$ by

$$\beta_n(x) = \frac{\|s_n(x)\|}{1+\|s_n(x)\|} \qquad (x \in \mathscr{E}_0 \cup (E \setminus \mathscr{E}_3), n=1,2,\ldots).$$

Then each β_n is continuous on $\mathscr{E}_0 \cup (E \setminus \mathscr{E}_3)$, $0 \leqslant \beta_n(x) < 1$ $(x \in \mathscr{E}_0 \cup (E \setminus \mathscr{E}_3))$, and we have $\lim_{n\to\infty} \beta_n(x) = \beta(x)$ $(x \in \mathscr{E}_0 \cup (E \setminus \mathscr{E}_3))$, where

$$\beta(x) = \begin{cases} \dfrac{\|x\|}{1+\|x\|} & \text{for } x \in \mathscr{E}_0, \\ 1 & \text{for } x \in E \setminus \mathscr{E}_3. \end{cases} \tag{5.7}$$

Assume now that the set $E \setminus \mathscr{E}_3$ is of the second category. Then, by a well known extension of a theorem of R. Baire[1], the function β must have at least one point of continuity in $\mathscr{E}_0 \cup (E \setminus \mathscr{E}_3)$, say x_0.

Since (x_n, f_n) is E-complete, $\mathscr{E}_0$ is dense in E. Let $\{y_n\}$ be a sequence in $\mathscr{E}_0$ such that $\lim_{n\to\infty} y_n = x_0$. Then, taking into account (5.7),

$$\beta(x_0) = \lim_{n\to\infty} \beta(y_n) = \lim_{n\to\infty} \frac{\|y_n\|}{1+\|y_n\|} = \frac{\|x_0\|}{1+\|x_0\|}. \tag{5.8}$$

On the other hand, $E \setminus \mathscr{E}_3$ is dense in E, since for $x \in E \setminus \mathscr{E}_3$ and $y \in \mathscr{E}_0$ we have $x+y \in E \setminus \mathscr{E}_3$ and since $\mathscr{E}_0$ is dense in E. Let $\{z_n\}$ be a sequence in $E \setminus \mathscr{E}_3$ such that $\lim_{n\to\infty} z_n = x_0$. Then, taking into account (5.7),

$$\beta(x_0) = \lim_{n\to\infty} \beta(z_n) = 1,$$

which contradicts (5.8).

c) Define a sequence of functions $\{\gamma_n\}$ on E by

$$\gamma_n(x) = \frac{|||s_n(x)|||}{1+|||s_n(x)|||} \qquad (x \in E, n=1,2,\ldots).$$

Then each γ_n is continuous on E, $0 \leqslant \gamma_1(x) \leqslant \gamma_2(x) \leqslant \cdots < 1$ $(x \in E)$, and we have $\lim_{n\to\infty} \gamma_n(x) = \gamma(x)$, where

$$\gamma(x) = \begin{cases} \dfrac{|||x|||}{1+|||x|||} & \text{for } x \in \mathscr{E}_2, \\ 1 & \text{for } x \in E \setminus \mathscr{E}_2. \end{cases} \tag{5.9}$$

Since by a) above the set $E \setminus \mathscr{E}_2$ is dense in E, it follows that we have $\gamma(x)=1$ for every point of continuity x of γ. Consequently, if x

---

[1] We recall this result: Let $\{\beta_n\}$ be a sequence of continuous real functions on a metric space $\mathscr{E}$, such that $|\beta_n(x)| \leqslant M$ $(x \in \mathscr{E}, n=1,2,\ldots)$ and that $\lim_{n\to\infty} \beta_n(x) = \beta(x)$ $(x \in \mathscr{E})$; then the set of all points of discontinuity of β is of the first category.

is a point of continuity of γ and $\{z_n\}$ a sequence in E such that $\lim_{n\to\infty} z_n = x$, then

$$\lim_{n\to\infty} \gamma(z_n) = \gamma(x) = 1,$$

whence, by (5.9) and (5.4), $\sup_{1\leqslant n<\infty} |||z_n||| = \infty$. Hence A is contained in the set of points of discontinuity of γ, which is of the first category by virtue of the extension of the theorem of Baire, used in the proof of b) above. This completes the proof.

Proposition 5.2a), b) says, in other words, that if (x_n, f_n) is an irregular biorthogonal system, then almost everywhere on E (in the sense of Baire category) $\sup_{1\leqslant n<\infty} \|s_n(x)\| = \infty$ and almost everywhere $\lim_{n\to\infty} \|s_n(x)\|$ does not exist.

Remark 5.1. In the proof of b) we did not use the hypothesis that (x_n, f_n) is irregular. Actually, for regular (x_n, f_n) b) is trivially valid, since then $E\backslash\mathscr{E}_3$ is void. However, in a) and c) it is essential to assume that (x_n, f_n) is irregular. Let us also mention the following short proof of b); For every positive integer n the set

$$\mathscr{F}_n = \{x \in E \mid \|s_{n+k}(x)\| \geqslant \|x\| + 1 \qquad (k=0,1,2,\ldots)\}$$

is closed and nowhere dense (since $\mathscr{E}_0$ is dense in E), whence, by

$$E\backslash\mathscr{E}_3 \subset \bigcup_{n=1}^{\infty} \mathscr{F}_n,$$

$E\backslash\mathscr{E}_3$ is of the first category, which completes the proof.

Theorem 5.1. *Let E be a Banach space and (x_n, f_n) ($\{x_n\} \subset E$, $\{f_n\} \subset E^*$) an E-complete biorthogonal system. Then the following statements are equivalent:*

1°. (x_n, f_n) *is regular.*

2°. $\mathscr{E}_2 = E$.

3°. $\mathscr{E}_1 = \mathscr{E}_2$.

4°. $\mathscr{E}_2 = \mathscr{E}_3$.

For every $x \in E$ let $\{p_n(x)\}$ be a sequence of finite linear combinations

$$p_n(x) = \sum_{i=1}^{m_n} \alpha_i^{(n)}(x)\, x_i \qquad (n=1,2,\ldots) \tag{5.10}$$

such that[1] $\lim_{n\to\infty} p_n(x) = x$. *Then the above statements are equivalent to the following:*

5°. $\sup_{1\leqslant n<\infty} |||p_n(x)||| < \infty \qquad (x\in E)$.

---

[1] E.g. we can choose $p_n(x)$ to be an element of $P_{(n)} = [x_j]_{j=1}^{n}$ for which $\inf_{p\in P_{(n)}} \|x-p\|$ is attained. It is well known that there exists at least one such element (see e.g. [246], Ch. I, § 2, corollary 2.2).

Proof. The equivalence $1^\circ \Leftrightarrow 2^\circ$ is nothing else but the equivalence $1^\circ \Leftrightarrow 3^\circ$ of theorem 4.1. The implications $1^\circ \Rightarrow 3^\circ$ and $1^\circ \Rightarrow 4^\circ$ being immediate (since 1° implies $\mathscr{E}_1 = \mathscr{E}_2 = \mathscr{E}_3 = E$), it remains to prove that $3^\circ \Rightarrow 1^\circ$, $4^\circ \Rightarrow 1^\circ$ and $1^\circ \Leftrightarrow 5^\circ$.

$3^\circ \Rightarrow 1^\circ$. Assume that (x_n, f_n) is irregular. Then, by the implication $4^\circ \Rightarrow 1^\circ$ of § 4, theorem 4.1, we can successively construct an increasing sequence of positive integers $\{m_n\}$ and a sequence $\{y_n\} \subset E$ with the following properties;

$$y_n = \sum_{i=m_{n-1}+1}^{m_n} f_i(y_n) x_i \qquad (n=1,2,\dots; m_0=0), \tag{5.11}$$

$$\|y_n\| \leqslant \frac{1}{2^n} \qquad (n=1,2,\dots), \tag{5.12}$$

$$\max_{m_{n-1}+1 \leqslant k \leqslant m_n} \left\| \sum_{i=m_{n-1}+1}^{k} f_i(y_n) x_i \right\| = M_n \geqslant 1 \qquad (n=1,2,\dots). \tag{5.13}$$

We claim that the element

$$x = \sum_{j=1}^{\infty} \frac{1}{M_j} y_i \in E \tag{5.14}$$

satisfies $x \in \mathscr{E}_2 \setminus \mathscr{E}_1$, whence $\mathscr{E}_1 \neq \mathscr{E}_2$, which proves that $3^\circ \Rightarrow 1^\circ$. In fact, by (5.14), (5.11) and biorthogonality we have, for $m_{n-1}+1 \leqslant k \leqslant m_n$,

$$s_k(x) = \sum_{j=1}^{n-1} \frac{1}{M_j} y_j + \frac{1}{M_n} \sum_{i=m_{n-1}+1}^{k} f_i(y_n) x_i,$$

whence, by (5.13) and (5.12),

$$\sup_{1 \leqslant k < \infty} \|s_k(x)\| \leqslant \sup_{1 \leqslant n < \infty} \sum_{j=1}^{n-1} \frac{1}{M_j} \|y_j\| + 1 \leqslant \sup_{1 \leqslant n < \infty} \sum_{j=1}^{n-1} \frac{1}{2^j M_j} + 1 \leqslant 2,$$

i.e. $x \in \mathscr{E}_2$. On the other hand, for suitable integers k_n with $m_{n-1}+1 \leqslant k_n \leqslant m_n$ $(n=1,2,\dots; m_0=0)$ we have

$$\left\| \sum_{i=m_{n-1}+1}^{k_n} f_i(y_n) x_i \right\| = M_n \qquad (n=1,2,\dots),$$

whence, by (5.14), (5.11) and biorthogonality,

$$\left\| s_{k_n}(x) - s_{m_{n-1}}(x) \right\| = \frac{1}{M_n} \left\| \sum_{i=m_{n-1}+1}^{k_n} f_i(y_n) x_i \right\| = 1 \qquad (n=1,2,\dots),$$

which proves that $x \notin \mathscr{E}_1$. Thus $3^\circ \Rightarrow 1^\circ$.

$4° \Rightarrow 1°$. Assume that (x_n, f_n) is irregular. Then, by proposition 5.2a) and b), the set $\mathscr{E}_2$ is of the first category, while $\mathscr{E}_3$ is of the second category. Hence $\mathscr{E}_2 \neq \mathscr{E}_3$, which proves that $4° \Rightarrow 1°$.

$1° \Rightarrow 5°$. If (x_n, f_n) is regular, then there exists, by § 3, proposition 3.2, a constant $C \geqslant 1$ such that

$$|||p_n(x)||| \leqslant C\|p_n(x)\| \leqslant C(\|x\| + \|x - p_n(x)\|) \quad (x \in E, n = 1, 2, \ldots)$$

which, together with $\lim_{n \to \infty} p_n(x) = x$ $(x \in E)$, implies $5°$.

$5° \Rightarrow 1°$. If (x_n, f_n) is irregular, then, by virtue of proposition 5.2c), we cannot have $5°$. This completes the proof of theorem 5.1.

Remark 5.2. One can also give a direct proof of the implication $4° \Rightarrow 1°$ of theorem 5.1, similar ot the above proof of the implication $3° \Rightarrow 1°$. In fact, if (x_n, f_n) is irregular, we can construct an increasing sequence of positive integers $\{m_n\}$ and a sequence $\{y_n\} \subset E$ satisfying (5.11), (5.12) and

$$\max_{m_{n-1}+1 \leqslant k \leqslant m_n} \left\| \sum_{i=m_{n-1}+1}^{k} f_i(y_n) x_i \right\| = M_n \geqslant n \quad (n = 1, 2, \ldots),$$

whence for the element $x = \sum_{j=1}^{\infty} y_j \in E$ we shall have $x \in \mathscr{E}_3 \setminus \mathscr{E}_2$ (because $\|s_{m_n}(x)\| = \left\| \sum_{j=1}^{n} y_j \right\| < 1$ $(n = 1, 2, \ldots)$, whence $x \in \mathscr{E}_3$, but for k_n such that $\left\| \sum_{i=m_{n-1}+1}^{k_n} f_i(y_n) x_i \right\| = M_n \geqslant n$ we have $\|s_{k_n}(x)\| = \left\| \sum_{j=1}^{n-1} y_j + \sum_{i=m_{n-1}+1}^{k_n} f_i(y_n) x_i \right\| \geqslant n - 1$ $(n = 1, 2, \ldots)$, whence $x \notin \mathscr{E}_2$).

Remark 5.3. From theorem 5.1 and the inclusions (5.6) one can derive new conditions equivalent to $1°$ (e.g. $\mathscr{E}_0 = \mathscr{E}_2$, etc.).

In the preceding we have considered $\mathscr{E}_1, \mathscr{E}_2$ only as subsets of E. However, they are also linear subspaces of E. Let us introduce on these linear spaces the norm $|||x|||$ defined by (5.1), and denote by E_1, E_2 the normed linear spaces obtained in this way.

Proposition 5.3. *Let E be a Banach space, (x_n, f_n) $(\{x_n\} \subset E, \{f_n\} \subset E^*)$ a biorthogonal system, and E_1, E_2 the normed linear spaces defined above. Then*

a) *E_1 is a Banach space. Consequently, E_1 is a closed linear subspace of E_2 and if, $E_1 \neq E_2$, the sequence $\{x_n\}$ is not complete in E_2.*

b) *$\{x_n\}$ is a basis of E_1.*

c) *If $\{f_n\}$ is complete in E^*, then E_2 is a Banach space.*

d) *If $\{x_n\}$ is complete in E and $\{f_n\}$ is complete in E^*, and if the sequence $\{\psi_n\} \subset E_2^*$ defined by*

$$\psi_n(x)=f_n(x) \qquad (x\in E_2,\ n=1,2,\ldots) \tag{5.15}$$

is complete in E_2^*, *then* $\{x_n\}$ *is a basis of* E.

e) *If* $\{x_n\}$ *is complete in* E *and* $\{f_n\}$ *is total on* E, *and if* E_1 *is reflexive, then* $\{x_n\}$ *is a basis of* E.

Proof. a) Since (x_n, f_n) is biorthogonal, all $x_n \neq 0$ $(n=1,2,\ldots)$, whence, by § 3, proposition 3.1, the space A_1 introduced in proposition 3.1 is a Banach space. Since the mapping

$$\{\alpha_n\}\rightarrow \sum_{i=1}^{\infty} \alpha_i x_i$$

is a linear isometry of A_1 onto E_1, it follows that E_1 is complete.

b) Let

$$\phi_n(x)=f_n(x) \qquad (x\in E_1,\ n=1,2,\ldots). \tag{5.16}$$

Since by (5.1)

$$|\phi_n(x)|=|f_n(x)| = \frac{1}{\|x_n\|}\,\|f_n(x)x_n\| \leqslant \frac{2}{\|x_n\|}\,|||x||| \qquad (x\in E_1),$$

we have $\phi_n \in E_1^*$ $(n=1,2,\ldots)$. Since (x_n, f_n) is a biorthogonal system, it follows that (x_n, ϕ_n) is a biorthogonal system. Taking into account (5.1), (5.16) and (5.3), we have

$$\begin{aligned}\left|\left|\left| x-\sum_{i=1}^{n}\phi_i(x)x_i\right|\right|\right| &= \sup_{1\leqslant k<\infty}\left\|\sum_{j=1}^{k} f_j\left[x-\sum_{i=1}^{n} f_i(x)x_i\right]x_j\right\| \\ &= \sup_{n+1\leqslant k<\infty}\left\|\sum_{j=n+1}^{k} f_j(x)x_j\right\| \rightarrow 0 \qquad (x\in E_1),\end{aligned}$$

whence, by the implication $2^\circ \Rightarrow 1^\circ$ of § 4, theorem 4.1, $\{x_n\}$ is a basis of E_1.

c) Let us first prove that if $\{f_n\}$ is complete in E^*, then

$$\|x\| \leqslant |||x||| \qquad (x\in E_2). \tag{5.17}$$

Let $x\in E_2$. Since $\sup_{1\leqslant n<\infty}\left\|\sum_{i=1}^{n} f_i(x)x_i\right\| < \infty$ and since by biorthogonality $\lim_{n\to\infty} f_j\left[\sum_{i=1}^{n} f_i(x)x_i\right] = f_j(x)$ $(j=1,2,\ldots)$, from the completeness of

$\{f_n\}$ in E^* it follows that the sequence $\left\{\sum_{i=1}^{n} f_i(x)x_i\right\}$ is convergent to x for the weak topology $\sigma(E,E^*)$. Consequently, we have (5.17).

Now let $\{y_k\}$ be a Cauchy sequence in E_2. Then by (5.17) and the completeness of E, there exists an $x\in E$ such that $\lim_{k\to\infty} y_k = x$. Since

$$\left\|\sum_{i=1}^{n} f_i(x)x_i\right\| = \lim_{k\to\infty}\left\|\sum_{i=1}^{n} f_i(y_k)x_i\right\| \leqslant \sup_{1\leqslant k<\infty} |||y_k||| \qquad (n=1,2,\ldots)$$

(which is $<\infty$ because $\{y_k\}$ is a Cauchy sequence in E_2), we have $x\in E_2$. From the inequalities

$$\left\|\sum_{i=1}^{n} [f_i(y_k)-f_i(y_m)]x_i\right\| < \varepsilon \qquad (k,m>N(\varepsilon);\ n=1,2,\ldots)$$

we obtain, for $m\to\infty$,

$$\left\|\sum_{i=1}^{n} [f_i(y_k)-f_i(x)]x_i\right\| \leqslant \varepsilon \qquad (k>N(\varepsilon);\ n=1,2,\ldots),$$

whence $\lim_{k\to\infty} |||y_k - x||| = 0$. Thus E_2 is complete.

d) Since $\{f_n\}$ is complete in E^*, we have (5.17), whence $\psi_n\in E_2^*$ $(n=1,2,\ldots)$. Let $x\in E_2$. Since $\sup_{1\leqslant n<\infty}\left\|\sum_{i=1}^{n} f_i(x)x_i\right\| < \infty$ and since by biorthogonality $\lim_{n\to\infty}\psi_j\left[\sum_{i=1}^{n} f_i(x)x_i\right] = \lim_{n\to\infty} f_j\left[\sum_{i=1}^{n} f_i(x)x_i\right] = f_j(x) = \psi_j(x)$ $(j=1,2,\ldots)$, from the completeness of $\{\psi_n\}$ in E_2^* it follows that the sequence $\left\{\sum_{i=1}^{n} f_i(x)x_i\right\}$ is convergent to x for the weak topology $\sigma(E_2,E_2^*)$. Thus x belongs[1] to the (strongly) closed linear subspace of E_2 spanned by the sequence $\left\{\sum_{i=1}^{n} f_i(x)x_i\right\}$, whence, since $\sum_{i=1}^{n} f_i(x)x_i\in E_1$ $(n=1,2,\ldots)$, it follows that $x\in E_1$. Consequently, $E_2=E_1$, whence, by the completeness of $\{x_n\}$ in E and by the implication $3^\circ \Rightarrow 1^\circ$ of theorem 5.1, $\{x_n\}$ is a basis of E.

e) Since E_1 is reflexive and since by b) above $\{x_n\}$ is a basis of E_1, the sequence $\{\phi_n\}\subset E_1^*$ defined by (5.16) is complete in E_1^* (in fact, otherwise there would exist[2] an $y=\sum_{i=1}^{\infty}\phi_i(y)x_i\in E_1\setminus\{0\}$ such that

---

[1] See e.g. [10], p. 134, theorem 2.

[2] See e.g. [10], p. 58, theorem 7.

$\phi_n(y)=0$ $(n=1,2,\ldots)$, which is impossible). Let $x\in E_2$. Then, since $\sup_{1\leqslant n<\infty}\left\|\sum_{i=1}^{n} f_i(x)x_i\right\|<\infty$ and since by biorthogonality $\lim_{n\to\infty}\phi_j\left[\sum_{i=1}^{n} f_i(x)x_i\right]$ $=\lim_{n\to\infty} f_j\left[\sum_{i=1}^{n} f_i(x)x_i\right]=f_j(x)$ $(j=1,2,\ldots)$, from the completeness of $\{\phi_n\}$ in E_1^* it follows that the limit $\lim_{n\to\infty}\phi\left[\sum_{i=1}^{n} f_i(x)x_i\right]$ exists for every $\phi\in E_1^*$. Hence, since E_1 is reflexive, there exists an element $x'\in E_1$ such that $\lim_{n\to\infty}\phi\left[\sum_{i=1}^{n} f_i(x)x_i\right]=\phi(x')$ $(\phi\in E_1^*)$. Thus, in particular, $f_j(x)$ $=\lim_{n\to\infty}\phi_j\left[\sum_{i=1}^{n} f_i(x)x_i\right]=\phi_j(x')$ $(j=1,2,\ldots)$, whence, by (5.16), $f_j(x)=f_j(x')$ $(j=1,2,\ldots)$. Since $\{f_n\}$ is total on E, it follows that $x=x'$, whence $x\in E_1$. Consequently, $E_2=E_1$, whence, by the completeness of $\{x_n\}$ in E and by the implication $3^\circ\Rightarrow 1^\circ$ of theorem 5.1, $\{x_n\}$ is a basis of E. This completes the proof.

Remark 5.4. One has the following result, the proof of which is similar to that of § 3, proposition 3.1:

Let $\{x_n\}$ be a sequence in a Banach space E, such that $x_n\neq 0$ $(n=1,2,\ldots)$, and let A_2 be the linear space of sequences of scalars

$$A_2=\left\{\{\alpha_n\}\subset K \;\middle|\; \sup_{1\leqslant n<\infty}\left\|\sum_{i=1}^{n}\alpha_i x_i\right\|<\infty\right\}, \tag{5.18}$$

endowed with the norm $\|\{\alpha_n\}\|=\sup_{1\leqslant n<\infty}\left\|\sum_{i=1}^{n}\alpha_i x_i\right\|$. *Then A_2 is a Banach space.* (Moreover, we shall see in § 12, theorem 12.5c), that A_2 is isomorphic to a certain conjugate Banach space if $\{x_n\}$ is a basis[1] of E). However, one cannot use this result to give a proof of c) similar to the above proof of a), since if (x_n, f_n) is a biorthogonal system such that $[f_n]=E^*$, then, in general, the mapping $x\to\{f_n(x)\}$ is an isometry of E_2

[1] Actually, this result remains valid for any sequence $\{x_n\}\subset E$ with $x_n\neq 0$ $(n=1,2,\ldots)$. Indeed, by § 8, proposition 8.1a), the unit vectors $e_n=\{\delta_{nj}\}_{j=1}^{\infty}$ $(n=1,2,\ldots)$ are a basis of $A_1=\left\{\{\alpha_n\}\subset K \;\middle|\; \sum_{i=1}^{\infty}\alpha_i x_i \text{ converges}\right\}$ and by $\left\|\sum_{i=1}^{n}\alpha_i e_i\right\|$ $=\sup_{1\leqslant k\leqslant n}\left\|\sum_{i=1}^{k}\alpha_i x_i\right\|$ we have $A_2=\left\{\{\alpha_n\}\subset K \;\middle|\; \sup_{1\leqslant n<\infty}\left\|\sum_{i=1}^{n}\alpha_i x_i\right\|<\infty\right\}=\left\{\{\alpha_n\}\right.$ $\left.\subset K \;\middle|\; \sup_{1\leqslant n<\infty}\left\|\sum_{i=1}^{n}\alpha_i e_i\right\|<\infty\right\}$ and $\|\{\alpha_n\}\|=\sup_{1\leqslant n<\infty}\left\|\sum_{i=1}^{n}\alpha_i x_i\right\|=\sup_{1\leqslant n<\infty}\left\|\sum_{i=1}^{n}\alpha_i e_i\right\|_{A_1}$ $(\{\alpha_n\}\in A_2)$.

onto a proper closed linear subspace of A_2, i.e. for an $\{\alpha_n\}\in A_2$ there does not exist, in general, an $x\in E$ such that $f_n(x)=\alpha_n$ $(n=1,2,\ldots)$. In fact, e.g. for $E=c_0$, $\{x_n\}$ = the natural basis of c_0 and $\{f_n\}$ = the a.s.c.f., we have $E_2=E=c_0$ and $A_2=m$.

We shall now characterize regular biorthogonal systems among E-complete total biorthogonal systems in terms of properties of the set of multipliers.

Definition 5.1. Let E be a Banach space and (x_n,f_n) $(\{x_n\}\subset E, \{f_n\}\subset E^*)$ a biorthogonal system such that $\{f_n\}$ is total on E. A sequence of scalars $\{\gamma_n\}$ is called a *multiplier of an element* $x\in E$ if there exists an element $x_{\{\gamma_n\}}\in E$ such that $x_{\{\gamma_n\}}\sim\sum_{k=1}^{\infty}\gamma_k f_k(x)x_k$, i.e., such that[1]

$$f_k(x_{\{\gamma_n\}})=\gamma_k f_k(x) \qquad (k=1,2,\ldots); \tag{5.19}$$

since $\{f_n\}$ is total on E, this $x_{\{\gamma_n\}}$ is uniquely determined by x. The set of all multipliers $\{\gamma_n\}$ of x is denoted by $M(x,(x_n,f_n))$. A sequence of scalars $\{\gamma_n\}$ is called a *multiplier of E* if it is a multiplier of each $x\in E$. The set of all multipliers $\{\gamma_n\}$ of E is denoted by $M(E,(x_n,f_n))$. Thus,

$$M(E,(x_n,f_n))=\bigcap_{x\in E}M(x,(x_n,f_n)). \tag{5.20}$$

Theorem 5.2. *Let E be a Banach space and (x_n,f_n) $(\{x_n\}\subset E, \{f_n\}\subset E^*)$ an E-complete biorthogonal system such that $\{f_n\}$ is total on E. The following statements are equivalent:*

1°. (x_n,f_n) *is regular.*

2°. $M(E,(x_n,f_n))\supset bv$, *the set of all sequences of scalars $\{\gamma_n\}$ such that $\sum_{i=1}^{\infty}|\gamma_i-\gamma_{i+1}|<\infty$.*

3°. $M(E,(x_n,f_n))$ *contains every non-increasing sequence $\{\gamma_n\}$ (i.e., such that $\gamma_1\geqslant\gamma_2\geqslant\cdots$) tending to zero.*

Proof. 1°⇒2°. Assume that (x_n,f_n) is regular and let $\{\gamma_n\}\in bv$ be arbitrary. Then, for any $x\in E$ and $n=1,2,\ldots$

$$\begin{aligned}\sum_{i=1}^{n}\gamma_i f_i(x)x_i&=\sum_{i=1}^{n}\gamma_i\left(\sum_{j=1}^{i}f_j(x)x_j-\sum_{j=1}^{i-1}f_j(x)x_j\right)\\&=\sum_{i=1}^{n}(\gamma_i-\gamma_{i+1})\sum_{j=1}^{i}f_j(x)x_j+\gamma_{n+1}\sum_{j=1}^{n}f_j(x)x_j.\end{aligned}$$

---

[1] Observe that if (x_n,f_n) is regular, then, obviously, $x_{\{\gamma_n\}}=\sum_{k=1}^{\infty}\gamma_k f_k(x)x_k$.

Since we have, by § 4, theorem 4.1,

$$\left\| \sum_{i=1}^{n} (\gamma_i - \gamma_{i+1}) \sum_{j=1}^{i} f_j(x) x_j \right\| \leqslant \sum_{i=1}^{n} |\gamma_i - \gamma_{i+1}| \left\| \sum_{j=1}^{i} f_j(x) x_j \right\|$$
$$\leqslant C\|x\| \sum_{i=1}^{n} |\gamma_i - \gamma_{i+1}|,$$

$$\left\| \gamma_{n+1} \sum_{j=1}^{n} f_j(x) x_j \right\| \leqslant C\|x\| \, |\gamma_{n+1}| \leqslant C\|x\| \left(\sum_{i=n+1}^{\infty} |\gamma_i - \gamma_{i+1}| + \lim_{n\to\infty} |\gamma_n| \right),$$

it follows that

$$\left\| \sum_{i=1}^{n} \gamma_i f_i(x) x_i \right\| \leqslant C\|x\| \left(\sum_{i=1}^{\infty} |\gamma_i - \gamma_{i+1}| + \lim_{n\to\infty} |\gamma_n| \right) \qquad (x\in E,\ n=1,2,\ldots). \tag{5.21}$$

Since $\lim_{n\to\infty} \sum_{i=1}^{n} \gamma_i f_i(x_k) x_i = \gamma_k x_k$ $(k=1,2,\ldots)$ and since $[x_n]=E$, by (5.21) we infer that $\sum_{i=1}^{\infty} \gamma_i f_i(x) x_i$ converges for all $x\in E$. Furthermore, for each $x\in E$ the element

$$x_{\{\gamma_n\}} = \sum_{i=1}^{\infty} \gamma_i f_i(x) x_i \tag{5.22}$$

obiously satisfies (5.19). Thus, $\{\gamma_n\}\in M(E, (x_n, f_n))$.

$2^\circ \Rightarrow 1^\circ$. Assume that we have 2°. Then for each $x\in E$ we can define a mapping $u_x\colon bv\to E$, by

$$u_x(\{\gamma_n\}) = x_{\{\gamma_n\}} \sim \sum_{k=1}^{\infty} \gamma_k f_k(x) x_k \qquad (\{\gamma_n\}\in bv). \tag{5.23}$$

Since for $x\in E$, $\{\gamma_n^{(1)}\}$, $\{\gamma_n^{(2)}\}$, $\{\gamma_n\}\in M(E, (x_n, f_n))$ and $\alpha=$scalar, we have

$$f_k(x_{\{\gamma_n^{(1)}\}+\{\gamma_n^{(2)}\}}) = (\gamma_k^{(1)} + \gamma_k^{(2)}) f_k(x) = f_k(x_{\{\gamma_n^{(1)}\}} + x_{\{\gamma_n^{(2)}\}}) \qquad (k=1,2,\ldots),$$
$$f_k(x_{\alpha\{\gamma_n\}}) = \alpha\gamma_k f_k(x) = f_k(\alpha x_{\{\gamma_n\}}) \qquad (k=1,2,\ldots)$$

and since $\{f_n\}$ is total on E, it follows that

$$x_{\{\gamma_n^{(1)}\}+\{\gamma_n^{(2)}\}} = x_{\{\gamma_n^{(1)}\}} + x_{\{\gamma_n^{(2)}\}} \qquad \left(x\in E,\ \{\gamma_n^{(1)}\}, \{\gamma_n^{(2)}\}\in M(E,(x_n,f_n))\right), \tag{5.24}$$

$$x_{\alpha\{\gamma_n\}} = \alpha x_{\{\gamma_n\}} \qquad \left(x\in E,\ \{\gamma_n\}\in M(E,(x_n,f_n)),\ \alpha=\text{scalar}\right), \tag{5.25}$$

whence each mapping u_x is linear. We shall now prove that each u_x is closed (we recall that with the usual vector operations and with the norm

$$\|\{\gamma_n\}\|_{bv} = \sum_{i=1}^{\infty} |\gamma_i - \gamma_{i+1}| + \lim_{n\to\infty} |\gamma_n| \qquad (\{\gamma_n\}\in bv), \tag{5.26}$$

bv is a Banach space). Let $\{\gamma_n^{(m)}\}_{n=1}^{\infty}$ $(m=1,2,\ldots)$ be a sequence in bv, converging to an element $\{\gamma_n\}\in bv$ and such that $\lim_{m\to\infty} u_x(\{\gamma_n^{(m)}\}_{n=1}^{\infty}) = z\in E$. Then, since $f_k\in E^*$, we have

$$\gamma_k^{(m)} f_k(x) = f_k(x_{\{\gamma_n^{(m)}\}}) = f_k[u_x\{\gamma_n^{(m)}\}] \to f_k(z) \quad \text{as} \quad m\to\infty \qquad (k=1,2,\ldots). \tag{5.27}$$

On the other hand, since

$$|\gamma_k^{(m)} - \gamma_k| \leqslant \|\{\gamma_n^{(m)}\} - \{\gamma_n\}\|_{bv} \to 0 \quad \text{as} \quad m\to\infty \qquad (k=1,2,\ldots),$$

we have $\lim_{m\to\infty} \gamma_k^{(m)} = \gamma_k$ $(k=1,2,\ldots)$. Consequently,

$$\gamma_k^{(m)} f_k(x) \to \gamma_k f_k(x) \quad \text{as} \quad m\to\infty \qquad (k=1,2,\ldots),$$

whence, by (5.27), we get

$$f_k(z) = \gamma_k f_k(x) \qquad (k=1,2,\ldots),$$

and therefore $z = x_{\{\gamma_n\}} = u_x(\{\gamma_n\})$, which proves that u_x is closed. Hence, by the closed graph theorem, each u_x is continuous on bv.

Now, if e_i denotes the i-th unit vector $\{\delta_{ni}\}_{n=1}^{\infty} = \{\underbrace{0,\ldots,0}_{i-1},1,0,\ldots\}$ in bv, we have

$$f_k[u_x(e_i)] = f_k(x_{e_i}) = \delta_{ki} f_k(x) = f_k(f_i(x)x_i) \qquad (x\in E;\, i,k=1,2,\ldots),$$

whence, since $\{f_n\}$ is total on E,

$$u_x(e_i) = f_i(x)x_i \qquad (x\in E, i=1,2,\ldots). \tag{5.28}$$

Consequently, since $\|e_1\|_{bv}=1$, $\left\|\sum_{i=1}^{m} e_i\right\|_{bv} = 2$ $(m=2,3,\ldots)$, we get

$$\left\|\sum_{i=1}^{m} f_i(x)x_i\right\| = \left\|u_x\left(\sum_{i=1}^{m} e_i\right)\right\| \leqslant \|u_x\| \left\|\sum_{i=1}^{m} e_i\right\|_{bv} \leqslant 2\|u_x\| \qquad (x\in E, m=1,2,\ldots),$$

whence, by § 4, theorem 4.1 (implication $3^\circ \Rightarrow 1^\circ$), (x_n, f_n) is regular.

The implication $2^\circ \Rightarrow 3^\circ$ is obvious, since every non-increasing sequence tending to 0 belongs to bv.

$3^\circ \Rightarrow 2^\circ$. Every $\{\gamma_n\}\in bv$ can be written as the sum $\left\{\gamma_n - \lim_{k\to\infty}\gamma_k\right\} + \left\{\lim_{k\to\infty}\gamma_k\right\}$, where the first sequence is in bv and tends to 0 and the second sequence is constant. Furthermore, every real sequence $\{\gamma_n\}\in bv$ tending to 0 can be written as the difference $\left\{\sum_{i=n}^{\infty}(\gamma_i-\gamma_{i+1})^+\right\} - \left\{\sum_{i=n}^{\infty}(\gamma_i-\gamma_{i+1})^-\right\}$, (where $\alpha_i^+=\alpha_i$, $\alpha_i^-=0$ if $\alpha_i\geqslant 0$ and $\alpha_i^+=0$, $\alpha_i^-=-\alpha_i$ if $\alpha_i\leqslant 0$) of two non-increasing sequences tending to 0. On the other hand, by (5.24) and (5.25), $M(E,(x_n,f_n))$ is a linear space. Hence, by 2° and since the constant sequences are in $M(E,(x_n,f_n))$, it follows that every real $\{\gamma_n\}\in bv$, whence also every $\{\gamma_n\}\in bv$, is in $M(E,(x_n,f_n))$, which completes the proof of theorem 5.2.

In the above proof we have seen that for any biorthogonal system (x_n,f_n) such that $\{f_n\}$ is total on E, the set of all multipliers $M(E,(x_n,f_n))$, with the natural vector operations, is a linear space. Let us now give some more properties of $M(E,(x_n,f_n))$.

Proposition 5.4. *Let E be a Banach space and (x_n,f_n) ($\{x_n\}\subset E$, $\{f_n\}\subset E^*$) a biorthogonal system such that $\{f_n\}$ is total on E. Then*

a) *With the natural vector operations, coordinatewise multiplication and the norm*

$$\|\{\gamma_n\}\| = \sup_{\substack{x\in E\\ \|x\|\leqslant 1}} \|x_{\{\gamma_n\}}\| \qquad (\{\gamma_n\}\in M(E,(x_n,f_n))), \tag{5.29}$$

$M(E,(x_n,f_n))$ is a commutative Banach algebra, containing the identity $\{1,1,1,\ldots\}$.

b) *The mapping $\{\gamma_n\}\to v_{\{\gamma_n\}}$, where*

$$v_{\{\gamma_n\}}(x) = x_{\{\gamma_n\}} \sim \sum_{k=1}^{\infty} \gamma_k f_k(x) x_k \qquad (x\in E), \tag{5.30}$$

is an isometrical algebraic isomorphism of $M(E,(x_n,f_n))$ into $L(E,E)$ (the Banach algebra of all continuous linear mappings of E into E, with $\|u\| = \sup_{\substack{x\in E\\ \|x\|\leqslant 1}} \|u(x)\|$), satisfying

$$v_{e_n}(x) = f_n(x) x_n \qquad (x\in E,\, n=1,2,\ldots), \tag{5.31}$$

where e_n denotes the n-th unit vector $\{\delta_{ni}\}_{i=1}^{\infty}$ in $M(E,(x_n,f_n))$. Furthermore,

$$v_{\{\gamma_n\}}(x_j) = \gamma_j x_j \qquad (j=1,2,\ldots), \tag{5.32}$$

i.e., each x_j is a proper vector of $v_{\{\gamma_n\}}$, corresponding to the proper value γ_j.

c) *We have $M(E,(x_n,f_n))\subset m(=l^\infty)$ and the inclusion mapping is continuous, namely*

$$\sup_{1\leqslant j<\infty} |\gamma_j| \leqslant \|\{\gamma_n\}\|_{M(E,(x_n,f_n))} \qquad (\{\gamma_n\}\in M(E,(x_n,f_n))). \tag{5.33}$$

Proof. a) Let $\{\gamma_n^{(1)}\}$, $\{\gamma_n^{(2)}\}\in M(E,(x_n,f_n))$. Then for every $x\in E$ there exists an element $x_{\{\gamma_n^{(2)}\}}\in E$ such that

$$f_k(x_{\{\gamma_n^{(2)}\}}) = \gamma_k^{(2)} f_k(x) \qquad (k=1,2,\ldots)$$

and for this $x_{\{\gamma_n^{(2)}\}}$ there exists an element $(x_{\{\gamma_n^{(2)}\}})_{\{\gamma_n^{(1)}\}}\in E$ such that

$$f_k((x_{\{\gamma_n^{(2)}\}})_{\{\gamma_n^{(1)}\}}) = \gamma_k^{(1)} f_k(x_{\{\gamma_n^{(2)}\}}) = \gamma_k^{(1)}\gamma_k^{(2)} f_k(x) \qquad (k=1,2,\ldots)$$

whence $\{\gamma_n^{(1)}\gamma_n^{(2)}\}\in M(E,(x_n,f_n))$. Thus, with coordinatewise multiplication, $M(E,(x_n,f_n))$ is a linear algebra, which obviously is commutative and contains the identity $\{1,1,1,\ldots\}$. Note that, since $\{f_n\}$ is total on E, we also have the relations

$$x_{\{\gamma_n^{(1)}\gamma_n^{(2)}\}} = (x_{\{\gamma_n^{(2)}\}})_{\{\gamma_n^{(1)}\}} \qquad (x \in E; \{\gamma_n^{(1)}\}, \{\gamma_n^{(2)}\} \in M(E,(x_n,f_n)). \tag{5.34}$$

Now, if $\|\{\gamma_n\}\| = 0$, then $x_{\{\gamma_n\}} = 0$ $(x \in E, \|x\| \leqslant 1)$, whence

$$\gamma_k f_k(x) = f_k(x_{\{\gamma_n\}}) = 0 \qquad (x \in E, \|x\| \leqslant 1, k = 1,2,\ldots)$$

and hence, putting $x = \dfrac{x_k}{\|x_k\|}$ $(k=1,2,\ldots)$, we get $\gamma_k = 0$ $(k=1,2,\ldots)$. This, together with (5.24) and (5.25), proves that (5.29) is a norm on $M(E,(x_n,f_n))$.

Now let $\{\gamma_n^{(m)}\}_{n=1}^{\infty}$ $(m=1,2,\ldots)$ be a Cauchy sequence in $M(E,(x_n,f_n))$, i.e., such that

$$\left\|\{\gamma_n^{(m)}\} - \{\gamma_n^{(l)}\}\right\| = \sup_{\substack{x \in E \\ \|x\| \leqslant 1}} \left\|x_{\{\gamma_n^{(m)}\}} - x_{\{\gamma_n^{(l)}\}}\right\| < \varepsilon \qquad (m,l > N(\varepsilon)). \tag{5.35}$$

Then for every fixed k and $x \in E$ with $f_k(x) \neq 0$, $\|x\| = 1$,

$$\begin{aligned} |f_k(x)|\,|\gamma_k^{(m)} - \gamma_k^{(l)}| &= |\gamma_k^{(m)} f_k(x) - \gamma_k^{(l)} f_k(x)| = \left|f_k\left(x_{\{\gamma_n^{(m)}\}} - x_{\{\gamma_n^{(l)}\}}\right)\right| \\ &\leqslant \|f_k\|\,\left\|x_{\{\gamma_n^{(m)}\}} - x_{\{\gamma_n^{(l)}\}}\right\| < \varepsilon \|f_k\| \qquad (m,l > N(\varepsilon)), \end{aligned}$$

whence the limits $\lim\limits_{m \to \infty} \gamma_k^{(m)} = \gamma_k$ $(k=1,2,\ldots)$ exist and for every $x \in E$ we have

$$\gamma_k^{(m)} f_k(x) \to \gamma_k f_k(x) \quad \text{as} \quad m \to \infty \qquad (k=1,2,\ldots). \tag{5.36}$$

On the other hand, taking any $x \in E$ and putting

$$y_m = x_{\{\gamma_n^{(m)}\}} \qquad (m=1,2,\ldots),$$

we see from (5.35) that $\{y_m\}$ is a Cauchy sequence in E, whence $\lim\limits_{m \to \infty} y_m = y \in E$ exists and

$$\gamma_k^{(m)} f_k(x) = f_k\left(x_{\{\gamma_n^{(m)}\}}\right) = f_k(y_m) \to f_k(y) \quad \text{as} \quad m \to \infty \qquad (k=1,2,\ldots).$$

Therefore, taking into account (5.36), we obtain

$$f_k(y) = \gamma_k f_k(x) \qquad (k=1,2,\ldots),$$

whence $\{\gamma_n\} \in M(E,(x_n,f_n))$ and $y = x_{\{\gamma_n\}}$. Furthermore, since $\lim\limits_{m \to \infty} x_{\{\gamma_n^{(m)}\}} = \lim\limits_{m \to \infty} y_m = y = x_{\{\gamma_n\}}$, from (5.35) for $l \to \infty$ we get

$$\left\|x_{\{\gamma_n^{(m)}\}} - x_{\{\gamma_n\}}\right\| \leqslant \varepsilon \qquad (x \in E, \|x\| \leqslant 1; m,l > N(\varepsilon)),$$

whence $\lim\limits_{m \to \infty} \{\gamma_n^{(m)}\} = \{\gamma_n\}$, which proves that $M(E,(x_n,f_n))$ is a Banach space for the norm (5.29).

Finally, by (5.29) and formula (5.38) below we have

$$\frac{1}{\|x\|}\|x_{\{\gamma_n\}}\| = \left\|\left(\frac{x}{\|x\|}\right)_{\{\gamma_n\}}\right\| \leqslant \|\{\gamma_n\}\| \qquad (x \in E, x \neq 0, \{\gamma_n\} \in M(E,(x_n,f_n))),$$

whence, by (5.34), we get

$$\|x_{\{\gamma_n^{(1)}\gamma_n^{(2)}\}}\| = \|(x_{\{\gamma_n^{(2)}\}})_{\{\gamma_n^{(1)}\}}\| \leqslant \|\{\gamma_n^{(1)}\}\| \, \|x_{\{\gamma_n^{(2)}\}}\|$$
$$\leqslant \|\{\gamma_n^{(1)}\}\| \, \|\{\gamma_n^{(2)}\}\| \, \|x\| \qquad (x \in E, \{\gamma_n^{(1)}\}, \{\gamma_n^{(2)}\} \in M(E,(x_n,f_n)))$$

and consequently, by (5.29),

$$\|\{\gamma_n^{(1)}\}\{\gamma_n^{(2)}\}\| \leqslant \|\gamma_n^{(1)}\| \, \|\gamma_n^{(2)}\| \qquad (\{\gamma_n^{(1)}\}, \{\gamma_n^{(2)}\} \in M(E, x_n, f_n))),$$

which proves that $M(E,(x_n,f_n))$ is a Banach algebra.

b) For each $\{\gamma_n\} \in M(E,(x_n,f_n))$ the mapping $v_{\{\gamma_n\}}$ defined by (5.30) is in $L(E,E)$. Indeed, by

$$f_k((x+y)_{\{\gamma_n\}}) = \gamma_k f_k(x+y) = f_k(x_{\{\gamma_n\}} + y_{\{\gamma_n\}}) \qquad (x, y \in E; k=1,2,\ldots)$$

and since $\{f_n\}$ is total on E, we have

$$(x+y)_{\{\gamma_n\}} = x_{\{\gamma_n\}} + y_{\{\gamma_n\}} \qquad (x, y \in E) \tag{5.37}$$

and similarly

$$(\alpha x)_{\{\gamma_n\}} = \alpha x_{\{\gamma_n\}} \qquad (x \in E, \alpha = \text{scalar}), \tag{5.38}$$

whence $v_{\{\gamma_n\}}$ is linear. Furthermore, each $v_{\{\gamma_n\}}$ is closed, whence continuous, since if $\lim_{j\to\infty} z_j = z \in E$, $\lim_{j\to\infty} v_{\{\gamma_n\}}(z_j) = y \in E$, then, by $f_k \in E^*$ $(k=1,2,\ldots)$,

$$\gamma_k f_k(z_j) = f_k((z_j)_{\{\gamma_n\}}) = f_k[v_{\{\gamma_n\}}(z_j)] \to f_k(y) \quad \text{as} \quad j \to \infty \qquad (k=1,2,\ldots),$$
$$\gamma_k f_k(z_j) \to \gamma_k f_k(z) \quad \text{as} \quad j \to \infty \qquad (k=1,2,\ldots),$$

whence $f_k(y) = \gamma_k f_k(z)$ $(k=1,2,\ldots)$, i.e., $y = z_{\{\gamma_n\}} = v_{\{\gamma_n\}}(z)$, which proves our assertion.

Now, by (5.24), (5.25), (5.34), (5.29), (5.28), and (5.23), the mapping $\{\gamma_n\} \to v_{\{\gamma_n\}}$ is an isometrical algebraic isomorphism of $M(E,(x_n,f_n))$ into $L(E,E)$, satisfying (5.31). Finally, from the relations

$$f_k[v_{\{\gamma_n\}}(x_j)] = f_k((x_j)_{\{\gamma_n\}}) = \gamma_k f_k(x_j) = \gamma_k \delta_{kj} = f_k(\gamma_j x_j) \qquad (k=1,2,\ldots)$$

we infer (5.32) (since $\{f_n\}$ is total on E).

c) By (5.30) and (5.32) we have

$$\left(\frac{x_j}{\|x_j\|}\right)_{\{\gamma_n\}} = v_{\{\gamma_n\}}\left(\frac{x_j}{\|x_j\|}\right) = \gamma_j \frac{x_j}{\|x_j\|} \qquad (j=1,2,\ldots),$$

whence, by (5.29), we get

$$\|\{\gamma_n\}\| \geqslant \left\|\gamma_j \frac{x_j}{\|x_j\|}\right\| = |\gamma_j| \qquad (j=1,2,\ldots),$$

i.e., (5.33), which completes the proof of proposition 5.4.

Let us also observe that by the isometry $\{\gamma_n\}\to v_{\{\gamma_n\}}$ and by (5.31) we have

$$\|e_n\| = \|v_{e_n}\| = \sup_{\substack{x\in E\\ \|x\|\leqslant 1}} |f_n(x)|\,\|x_n\| = \|f_n\|\,\|x_n\| \geqslant 1 \qquad (n=1,2,\dots); \tag{5.39}$$

if (x_n, f_n) is regular, then, by (5.39) and § 3, theorem 3.1, we also have $\sup_{1\leqslant n<\infty} \|e_n\| < \infty$.

Remark 5.5. Similarly to (5.23), one can define, for each $x\in E$, a mapping $u_x: M(E,(x_n,f_n))\to E$, by

$$u_x(\{\gamma_n\}) = v_{\{\gamma_n\}}(x) = x_{\{\gamma_n\}} \qquad (\{\gamma_n\}\in M(E,(x_n,f_n))). \tag{5.40}$$

Then, by (5.24) and (5.25), each u_x is linear and by

$$|u_x(\{\gamma_n\})| \leqslant \|v_{\{\gamma_n\}}\|\,\|x\| = \|\{\gamma_n\}\|\,\|x\| \qquad (\{\gamma_n\}\in M(E,(x_n,f_n)))$$

each u_x is continuous.

This remark and the above proof of theorem 5.2 suggest the following corollary of proposition 5.4:

Corollary 5.1. *Let E be a Banach space and* (x_n, f_n) $(\{x_n\}\subset E, \{f_n\}\subset E^*)$ *an E-complete biorthogonal system such that* $\{f_n\}$ *is total on E. The following statements are equivalent:*

1°. (x_n, f_n) *is regular.*

2°. *There exists a constant* $C\geqslant 1$ *such that*

$$\left\|\sum_{i=1}^{n} e_i\right\| \leqslant C \qquad (n=1,2,\dots), \tag{5.41}$$

where e_n *is the n*-th *unit vector* $\{\delta_{nj}\}_{j=1}^{\infty}$ *in* $M(E,(x_n,f_n))$.

3°. *There exists a constant* $C\geqslant 1$ *such that*

$$\left\|\sum_{i=1}^{n} \gamma_i e_i\right\| \leqslant C\|\{\gamma_n\}\| \qquad (\{\gamma_n\}\in M(E,(x_n,f_n)),\ n=1,2,\dots). \tag{5.42}$$

Proof. 1°$\Leftrightarrow$2°. Since by proposition 5.4 the mapping $\{\gamma_n\}\to v_{\{\gamma_n\}}$ is a linear isometry we have, taking also into account (5.31),

$$\left\|\sum_{i=1}^{n} e_i\right\| = \left\|v_{\sum_{i=1}^{n} e_i}\right\| = \left\|\sum_{i=1}^{n} v_{e_i}\right\| = \sup_{\substack{x\in E\\ \|x\|\leqslant 1}} \left\|\sum_{i=1}^{n} v_{e_i}(x)\right\|$$

$$= \sup_{\substack{x\in E\\ \|x\|\leqslant 1}} \left\|\sum_{i=1}^{n} f_i(x)\,x_i\right\| \qquad (n=1,2,\dots), \tag{5.43}$$

whence, by § 4, theorem 4.1, the equivalence 1°$\Leftrightarrow$2° follows.

$1^\circ \Rightarrow 3^\circ$. If we have 1°, then, by § 4, theorem 4.1, there exists a constant $C \geqslant 1$ such that for any $\{\gamma_n\} \in M(E,(x_n, f_n))$ and $n=1,2,\ldots$ we have

$$\left\|\sum_{i=1}^{n} \gamma_i e_i\right\| = \left\|v_{\sum_{i=1}^{n}\gamma_i e_i}\right\| = \sup_{\substack{x \in E\\ \|x\| \leqslant 1}} \left\|\sum_{i=1}^{n} \gamma_i v_{e_i}(x)\right\| = \sup_{\substack{x \in E\\ \|x\| \leqslant 1}} \left\|\sum_{i=1}^{n} \gamma_i f_i(x) x_i\right\|$$

$$= \sup_{\substack{x \in E\\ \|x\| \leqslant 1}} \left\|\sum_{i=1}^{n} f_i(x_{\{\gamma_n\}}) x_i\right\| \leqslant C \sup_{\substack{x \in E\\ \|x\| \leqslant 1}} \|x_{\{\gamma_n\}}\| = C\|\{\gamma_n\}\|,$$

i.e., (5.42).

Finally, the implication $3^\circ \Rightarrow 2^\circ$ is obvious (taking $\gamma_1 = \gamma_2 = \cdots = 1$). This completes the proof.

Corollary 5.2. *Let $\{x_n\}$ be a basis of a Banach space E, with the a.s.c.f. $\{f_n\} \subset E^*$. Then*

a) *The unit vectors $e_m = \{\delta_{mi}\}_{i=1}^{\infty}$ $(m=1,2,\ldots)$ form a basic sequence in $M(E, (x_n, f_n))$.*

b) *We have*

$$M(E, (x_n, f_n)) = \left\{\{\gamma_n\} \subset K \;\middle|\; \sup_{1 \leqslant m < \infty} \left\|\sum_{i=1}^{m} \gamma_i e_i\right\|_{M(E,(x_n,f_n))} < \infty\right\}. \quad (5.44)$$

Proof. a) Let h_i denote the coordinate functionals on $M(E, (x_n, f_n))$, i.e.,

$$h_i(\{\gamma_n\}) = \gamma_i \qquad (\{\gamma_n\} \in M(E, (x_n, f_n)),\ i=1,2,\ldots).$$

Then each h_i is linear and by (5.42) we have

$$|h_i(\{\gamma_n\})| = |\gamma_i| = \frac{\|\gamma_i e_i\|}{\|e_i\|} \leqslant \frac{2}{\|e_i\|}\left(\left\|\sum_{j=1}^{i} \gamma_j e_j\right\| + \left\|\sum_{j=1}^{i-1} \gamma_j e_j\right\|\right)$$

$$\leqslant \frac{2C}{\|e_i\|} \|\{\gamma_n\}\| \qquad (\{\gamma_n\} \in M(E, (x_n, f_n), j=1,2,\ldots),$$

and thus each h_i is continuous on $M(E, (x_n, f_n))$, whence also on $[e_n]$. Since obviously $h_i(e_j) = \delta_{ij}$ $(i,j=1,2,\ldots)$, by (5.42) and § 4, theorem 4.1 it follows that $\{e_n\}$ is a basic sequence in $M(E, (x_n, f_n))$.

b) The inclusion $\subset$ in (5.44) is an immediate consequence of (5.42).

Conversely, let $\{\gamma_n\} \subset K$ be such that $\sup_{1 \leqslant m < \infty} \left\|\sum_{i=1}^{m} \gamma_i e_i\right\| = C' < \infty$. Then

$$\sup_{\substack{x \in E\\ \|x\| \leqslant 1}} \left\|\sum_{i=1}^{n} \gamma_i f_i(x) x_i\right\| = \sup_{\substack{x \in E\\ \|x\| \leqslant 1}} \left\|\sum_{i=1}^{n} \gamma_i v_{e_i}(x)\right\| = \sup_{\substack{x \in E\\ \|x\| \leqslant 1}} \left\|v_{\sum_{i=1}^{n}\gamma_i e_i}(x)\right\|$$

$$= \left\|\sum_{i=1}^{n} \gamma_i e_i\right\| \leqslant C' < \infty \qquad (n=1,2,\ldots).$$

Hence, since $\lim_{n\to\infty} \sum_{i=1}^{n} \gamma_i f_i(x_k) x_i = \gamma_k x_k$ $(k=1,2,\ldots)$ and since $[x_n]=E$, it follows that $\lim_{n\to\infty} \sum_{i=1}^{n} \gamma_i f_i(x) x_i = \sum_{i=1}^{\infty} \gamma_i f_i(x) x_i = x_{\{\gamma_n\}}$ exists for all $x\in E$. Since this $x_{\{\gamma_n\}}$ obviously satisfies (5.19), we have $\{\gamma_n\}\in M(E,(x_n,f_n))$, which completes the proof of corollary 5.2.

In general, $\{e_n\}$ need not be a basis of $M(E,(x_n,f_n))$, as shown by

Example 5.1. Let $E=c_0$ and let $\{x_n\}$ be the unit vector basis of E. Then $M(E,(x_n,f_n))=m$ with the norm $\|\{\gamma_n\}\| = \sup_{\substack{|\xi_1|,|\xi_2|,\ldots\leqslant 1\\ \xi_j\to 0}} \sup_{1\leqslant n<\infty} |\gamma_n \xi_n| = \sup_{1\leqslant n<\infty} |\gamma_n|$ (i.e., with its usual norm), which is a non-separable space and hence it has no basis.

Now we shall drop the hypothesis that $\{f_n\}$ is total on E and give another characterization of regular E-complete biorthogonal systems, suggested by the above results.

Theorem 5.3. *Let E be a Banach space and (x_n,f_n) $(\{x_n\}\subset E, \{f_n\}\subset E^*)$ an E-complete biorthogonal system. Then (x_n,f_n) is regular if and only if there exists a continuous linear mapping $\{\gamma_n\}\to v_{\{\gamma_n\}}$ of bv into $L(E,E)$, such that*

$$v_{e_n}(x)=f_n(x)x_n \qquad (x\in E,\ n=1,2,\ldots), \tag{5.45}$$

where e_n is the n-th unit vector $\{\delta_{ni}\}_{i=1}^{\infty}$ in bv. Moreover, in this case for every $\{\gamma_n\}\in bv$ such that $\lim_{n\to\infty}\gamma_n=0$ the mapping $v_{\{\gamma_n\}}$ is compact.

Proof. Assume that (x_n,f_n) is regular. Then $\{f_n\}$ is total on E and hence, by proposition 5.4, there exists a continuous linear mapping $\{\gamma_n\}\to v_{\{\gamma_n\}}$ of $M(E,(x_n,f_n))$ into $L(E,E)$ (with $v_{\{\gamma_n\}}$ defined by (5.30)), satisfying (5.45). Furthermore, by theorem 5.2, we have $bv\subset M(E,(x_n,f_n))$ and by (5.21) the inclusion mapping $bv\to M(E,(x_n,f_n))$ is continuous, whence the restriction of the mapping $\{\gamma_n\}\to v_{\{\gamma_n\}}$ to bv is continuous.

Conversely, if the condition is satisfied, then

$$v_{\sum_{i=1}^{n} e_i}(x) = \sum_{i=1}^{n} v_{e_i}(x) = \sum_{i=1}^{n} f_i(x)x_i = s_n(x) \qquad (x\in E,\ n=1,2,\ldots),$$

whence, since $\left\|\sum_{i=1}^{n} e_i\right\|_{bv} \leqslant 2$ $(n=1,2,\ldots)$ and since the mapping $\{\gamma_n\}\to v_{\{\gamma_n\}}$ of bv into $L(E,E)$ is continuous, say of norm C, we get

$$\|s_n\| = \|v_{\sum_{i=1}^{n} e_i}\| \leqslant C \left\|\sum_{i=1}^{n} e_i\right\|_{bv} \leqslant 2C \qquad (n=1,2,\ldots),$$

and therefore, by § 4, theorem 4.1, (x_n,f_n) is regular.

Finally, assume again that the condition is satisfied and let $\{\gamma_n\} \in bv$ such that $\lim_{n \to \infty} \gamma_n = 0$. Then, with C as above,

$$\left\| v_{\{\gamma_n\}} - \sum_{i=1}^{m} \gamma_i v_{e_i} \right\| = \left\| v_{\{\gamma_n\}} - v_{\sum_{i=1}^{m} \gamma_i e_i} \right\| = \left\| v_{\{\gamma_n\} - \sum_{i=1}^{m} \gamma_i e_i} \right\| \leqslant C \left\| \{\gamma_n\} - \sum_{i=1}^{m} \gamma_i e_i \right\|$$

$$= C \sum_{i=m+1}^{\infty} |\gamma_i - \gamma_{i+1}| \to 0 \quad \text{as} \quad m \to \infty$$

whence, since by (5.45) $\sum_{i=1}^{m} \gamma_i v_{e_i}$ is of finite rank, we infer that $v_{\{\gamma_n\}}$ is compact, which completes the proof.

From proposition 5.4 and theorem 5.3 we obtain

Corollary 5.3. *A Banach space E has a basis if and only if there exists a continuous multiplicative linear mapping* $\{\gamma_n\} \to v_{\{\gamma_n\}}$ *of* bv *into* $L(E,E)$, *such that*

$$\dim v_{e_n}(E) = 1 \qquad (n = 1, 2, \ldots), \tag{5.46}$$

$$[v_{e_n}(E)]_{n=1}^{\infty} = E, \tag{5.47}$$

where e_n *is the n-th unit vector* $\{\delta_{ni}\}_{i=1}^{\infty}$ *in* bv.

Proof. The necessity part is an immediate consequence of proposition 5.4 and theorem 5.3.

Conversely, assume that the condition of corollary 5.2 is satisfied and by (5.46) let $x_n \in E$, $x_n \neq 0$ be such that

$$v_{e_n}(E) = \{\alpha x_n \mid \alpha = \text{scalar}\} \qquad (n = 1, 2, \ldots). \tag{5.48}$$

Then, by (5.47), $[x_n] = E$. Furthermore, by $v_{e_n} \in L(E,E)$ and (5.48) there exist continuous linear functionals $f_n \in E^*$ such that

$$v_{e_n}(x) = f_n(x) x_n \qquad (x \in E,\ n = 1, 2, \ldots). \tag{5.49}$$

Since $e_i e_j = \delta_{ij} e_i$ $(i, j = 1, 2, \ldots)$ in bv and since the mapping $\{\gamma_n\} \to v_{\{\gamma_n\}}$ is multiplicative and linear, we have

$$v_{e_i} v_{e_j} = v_{e_i e_j} = v_{\delta_{ij} e_i} = \delta_{ij} v_{e_i} \qquad (i, j = 1, 2, \ldots). \tag{5.50}$$

Putting $j = i$, we get

$$f_i(x) v_{e_i}(x_i) = v_{e_i}(f_i(x) x_i) = v_{e_i}(v_{e_i}(x)) = v_{e_i}(x) = f_i(x) x_i \qquad (x \in E,\ i = 1, 2, \ldots),$$

whence $v_{e_i}(x_i) = x_i$ $(i = 1, 2, \ldots)$, whence, by (5.49), $f_i(x_i) x_i = x_i$ $(i = 1, 2, \ldots)$, and hence $f_i(x_i) = 1$ $(i = 1, 2, \ldots)$. Therefore we also have, for $j \neq i$, $f_i(x_j) x_i = v_{e_i}(x_j) = v_{e_i}(v_{e_j}(x_j)) = \delta_{ij} v_{e_i}(x_j) = 0$, whence $f_i(x_j) = 0$. Thus (x_n, f_n) is an E-complete biorthogonal system satisfying the condition of theorem 5.3, and hence by this theorem, $\{x_n\}$ is a basis of E. This completes the proof of corollary 5.2.

§ 6. Some types of linear independence of sequences

In the present section we shall consider three fundamental types of linear independence of sequences, which will be frequently used throughout the sequel. Later we shall also consider some other types of linear independence (see Ch. II, § 11 and § 16 and Vol. II, Ch. III).

Definition 6.1. A sequence $\{x_n\}$ in a Banach space E is said to be

a) *finitely linearly independent*, if every finite subsequence of $\{x_n\}$ is linearly independent;

b) *ω-linearly independent*, if

$$\{\alpha_n\} \subset K, \quad \sum_{i=1}^{\infty} \alpha_i x_i = 0, \quad \text{imply} \quad \alpha_i = 0 \quad (i=1,2,\ldots); \tag{6.1}$$

c) *minimal*, if[1]

$$x_n \notin [x_1, \ldots, x_{n-1}, x_{n+1}, \ldots] \qquad (n=1,2,\ldots; x_0=0). \tag{6.2}$$

Obviously, every minimal sequence is ω-linearly independent and every ω-linearly independent sequence is finitely linearly independent. For finite sequences the converse statements are also valid, but for infinite sequences this is no longer true, as shown by the following examples:

Example 6.1. a) Let $E = C([0,1])$ and let

$$x_n(t) = t^{n-1} \qquad (t \in [0,1]; n=1,2,\ldots). \tag{6.3}$$

Then $\{x_n\}$ is complete in E and ω-linearly independent, but not[2] minimal.

b) More generally, one can obtain examples of complete ω-linearly independent but not minimal sequences as follows. Let $\{x_n\}$ be an ω-linearly independent complete sequence in a Banach space E, which is not a basis of E (see e. g. § 4, example 4.1) and let x be an element of E which does not admit an expansion of the form $x = \sum_{i=1}^{\infty} \alpha_i x_i$. Then the sequence $\{y_n\} \subset E$ defined by

$$y_1 = x, \qquad y_n = x_{n-1} \qquad (n=2,3,\ldots) \tag{6.4}$$

is complete in E and ω-linearly independent, but not minimal.

---

[1] We recall that by $[x_1, \ldots, x_{n-1}, x_{n+1}, \ldots]$ we denote the (closed linear) subspace of E spanned by the elements $x_1, \ldots, x_{n-1}, x_{n+1}, \ldots$

[2] E.g. by virtue of the theorem of Müntz (see [174], Ch. III, § 3, theorem 2).

Example 6.2. Let $\{x_n\}$ be a basis of a Banach space E, with the a. s. c. f. $\{f_n\}$ and let x be an element of E such that $f_n(x) \neq 0$ $(n=1,2,\ldots)$, e. g. one can take

$$x = \sum_{i=1}^{\infty} \frac{1}{2^i \|x_i\|} x_i. \tag{6.5}$$

Then the sequence $\{y_n\} \subset E$ defined by (6.4) is complete in E and finitely linearly independent, but not ω-linearly independent.

Obviously, every basis of a Banach space E is a minimal sequence. The converse is not true, as shown by § 4, example 4.1. Furthermore, a sequence $\{x_n\} \subset E$ is a basis of E if and only if $\{x_n\}$ is ω-linearly independent but no supersequence of $\{x_n\}$ is ω-linearly independent.

Let us recall now a lemma from the general theory of normed linear spaces, which will be useful in the sequel. For an arbitrary Banach space B we shall use the notations

$$S_B = \{x \in B \,|\, \|x\| \leqslant 1\}, \tag{6.6}$$

$$\sigma_B = \{x' \in B \,|\, \|x'\| = 1\}. \tag{6.7}$$

Lemma 6.1. *Let F, G be two closed linear subspaces of a normed linear space E. Then*

a) *We have*

$$\operatorname{dist}(\sigma_F, \sigma_G) \geqslant \operatorname{dist}(\sigma_F, G) \geqslant \tfrac{1}{2} \operatorname{dist}(\sigma_F, \sigma_G). \tag{6.8}$$

b) *If G_1 is a closed linear subspace of G and $x \neq 0$ an element of G such that*[1] $G = [x] + G_1$, *then*

$$\operatorname{dist}(\sigma_F, \sigma_{G_1}) \geqslant \operatorname{dist}(\sigma_F, \sigma_G) \geqslant \frac{1}{3} \operatorname{dist}(\sigma_F, \sigma_{G_1}) \operatorname{dist}\left(\frac{x}{\|x\|}, F + G_1\right). \tag{6.9}$$

Proof. a) Since $\sigma_G \subset G$, the first inequality of (6.8) is obvious. In order to prove the second one, we shall first prove that

$$\operatorname{dist}(S_F, \sigma_G) \geqslant \tfrac{1}{2} \operatorname{dist}(\sigma_F, \sigma_G). \tag{6.10}$$

For an arbitrary $\varepsilon > 0$ choose $y \in S_F$ and $z' \in \sigma_G$ such that

$$\|y + z'\| < \operatorname{dist}(S_F, \sigma_G) + \varepsilon. \tag{6.11}$$

If $1 - \frac{1}{2} \operatorname{dist}(\sigma_F, \sigma_G) \geqslant \|y\|$, then we have

$$\|y + z'\| \geqslant 1 - \|y\| \geqslant \tfrac{1}{2} \operatorname{dist}(\sigma_F, \sigma_G).$$

If $1 - \frac{1}{2} \operatorname{dist}(\sigma_F, \sigma_G) < \|y\|$, then from the inequalities

$$\|y + z'\| \geqslant \left| \left\| y - \frac{y}{\|y\|} \right\| - \left\| \frac{y}{\|y\|} + z' \right\| \right|,$$

---

[1] For $A, B \subset E$ we use the notation

$$A + B = \{a + b \,|\, a \in A, b \in B\}.$$

$$\left\| y - \frac{y}{\|y\|} \right\| = 1 - \|y\| < \frac{1}{2} \operatorname{dist}(\sigma_F, \sigma_G),$$

$$\left\| \frac{y}{\|y\|} + z' \right\| \geqslant \operatorname{dist}(\sigma_F, \sigma_G),$$

it follows that

$$\|y + z'\| > \operatorname{dist}(\sigma_F, \sigma_G) - \tfrac{1}{2} \operatorname{dist}(\sigma_F, \sigma_G) = \tfrac{1}{2} \operatorname{dist}(\sigma_F, \sigma_G).$$

Thus in all cases we have, taking into account (6.11),

$$\operatorname{dist}(S_F, \sigma_G) > \tfrac{1}{2} \operatorname{dist}(\sigma_F, \sigma_G) - \varepsilon,$$

which, since $\varepsilon > 0$ is arbitrary, proves (6.10). By changing the roles of F and G in (6.10), we also have

$$\operatorname{dist}(\sigma_F, S_G) \geqslant \tfrac{1}{2} \operatorname{dist}(\sigma_F, \sigma_G). \tag{6.12}$$

Now, take arbitrary $y' \in \sigma_F$ and $z \in G$. If $\|z\| \geqslant 1$, we have, taking into account (6.10),

$$\|y' + z\| = \|z\| \left\| \frac{y'}{\|z\|} + \frac{z}{\|z\|} \right\| \geqslant \operatorname{dist}(S_F, \sigma_G) \geqslant \frac{1}{2} \operatorname{dist}(\sigma_F, \sigma_G).$$

If $\|z\| < 1$, we have, by (6.12),

$$\|y' + z\| \geqslant \operatorname{dist}(\sigma_F, S_G) \geqslant \tfrac{1}{2} \operatorname{dist}(\sigma_F, \sigma_G).$$

Thus, in all cases, we have

$$\|y' + z\| \geqslant \tfrac{1}{2} \operatorname{dist}(\sigma_F, \sigma_G),$$

which proves the second inequality of (6.8).

b) Since $\sigma_{G_1} \subset \sigma_G$, the first inequality of (6.9) is obvious. In order to prove the second one, take arbitrary $y' \in \sigma_F, z' \in \sigma_G$. Since $G = [x] + G_1$ and $x \neq 0$, we have $z' = \lambda \dfrac{x}{\|x\|} + z_1$ for a suitable scalar λ and a suitable $z_1 \in G_1$.

If $|\lambda| > \frac{1}{3} \operatorname{dist}(\sigma_F, \sigma_{G_1})$, we have

$$\|y' + z'\| = |\lambda| \left\| \frac{x}{\|x\|} + \frac{y' + z_1}{|\lambda|} \right\| > \frac{1}{3} \operatorname{dist}(\sigma_F, \sigma_{G_1}) \operatorname{dist}\left(\frac{x}{\|x\|}, F + G_1 \right).$$

If $|\lambda| \leqslant \frac{1}{3} \operatorname{dist}(\sigma_F, \sigma_{G_1})$, we have

$$\|y' + z'\| \geqslant \left| \|y' + z_1\| - \left\| \lambda \frac{x}{\|x\|} \right\| \right| = \left| \|y' + z_1\| - |\lambda| \right|. \tag{6.13}$$

Now, by the inequalities

$$\|y' + z_1\| \geqslant \left| \left\| z_1 - \frac{z_1}{\|z_1\|} \right\| - \left\| \frac{z_1}{\|z_1\|} + y' \right\| \right| = \left| |1 - \|z_1\|| - \left\| \frac{z_1}{\|z_1\|} + y' \right\| \right|,$$

$$\big|1-\|z_1\|\big| = \big|\,\|z'\|-\|z_1\|\,\big| \leqslant \|z'-z_1\| = |\lambda| \leqslant \tfrac{1}{3}\operatorname{dist}(\sigma_F,\sigma_{G_1}),$$

$$\left\|\frac{z_1}{\|z_1\|}+y'\right\| \geqslant \operatorname{dist}(\sigma_F,\sigma_{G_1}),$$

we have

$$\|y'+z_1\| \geqslant \operatorname{dist}(\sigma_F,\sigma_{G_1}) - \tfrac{1}{3}\operatorname{dist}(\sigma_F,\sigma_{G_1}) = \tfrac{2}{3}\operatorname{dist}(\sigma_F,\sigma_{G_1}).$$

Hence, taking into account (6.13) and $1=\left\|\frac{x}{\|x\|}\right\| \geqslant \operatorname{dist}\left(\frac{x}{\|x\|}, F+G_1\right)$, we obtain

$$\|y'+z'\| \geqslant \left|\frac{2}{3}\operatorname{dist}(\sigma_F,\sigma_{G_1})-|\lambda|\right| \geqslant \frac{2}{3}\operatorname{dist}(\sigma_F,\sigma_{G_1}) - \frac{1}{3}\operatorname{dist}(\sigma_F,\sigma_{G_1})$$

$$=\frac{1}{3}\operatorname{dist}(\sigma_F,\sigma_{G_1}) \geqslant \frac{1}{3}\operatorname{dist}(\sigma_F,\sigma_{G_1})\operatorname{dist}\left(\frac{x}{\|x\|}, F+G_1\right).$$

Thus, in all cases we have

$$\operatorname{dist}(\sigma_F,\sigma_G) \geqslant \frac{1}{3}\operatorname{dist}(\sigma_F,\sigma_{G_1})\operatorname{dist}\left(\frac{x}{\|x\|}, F+G_1\right),$$

which completes the proof of lemma 6.1.

Let us give now some characterizations of minimal sequences. We shall use, for a sequence $\{x_n\}\subset E$, the notations

$$P_{(n)} = [x_1,\dots,x_n] \qquad (n=1,2,\dots), \tag{6.14}$$

$$P^{(n)} = [x_{n+1},x_{n+2},\dots] \qquad (n=1,2,\dots), \tag{6.15}$$

$$\sigma_{(n)} = \{x\in P_{(n)} \,|\, \|x\|=1\} = \sigma_{P_{(n)}} \qquad (n=1,2,\dots), \tag{6.16}$$

$$\sigma^{(n)} = \{x\in P^{(n)} \,|\, \|x\|=1\} = \sigma_{P^{(n)}} \qquad (n=1,2,\dots), \tag{6.17}$$

$$P=\left\{\sum_{i=1}^{n}\alpha_i x_i \,\middle|\, \alpha_1,\dots,\alpha_n\in K;\, n=1,2,\dots\right\} = \bigcup_{n=1}^{\infty} P_{(n)}, \tag{6.18}$$

$$S_k\left(\sum_{i=1}^{n}\alpha_i x_i\right) = \begin{cases} 0 & \text{for } k=0 \\ \sum_{i=1}^{k}\alpha_i x_i & \text{for } 1\leqslant k\leqslant n \\ \sum_{i=1}^{n}\alpha_i x_i & \text{for } k\geqslant n \end{cases} \qquad \left(\sum_{i=1}^{n}\alpha_i x_i\in P\right), \tag{6.19}$$

$$R_k(p) = p - S_k(p) \qquad (p\in P,\, k=0,1,2,\dots). \tag{6.20}$$

Theorem 6.1. *For a sequence $\{x_n\}$ in a Banach space E the following statements are equivalent:*

1°. *$\{x_n\}$ is minimal.*

2°. *There exists a sequence $\{f_n\}\subset E^*$ such that (x_n, f_n) is a biorthogonal system.*

3°. *There exists a sequence of constants $\delta_n>0$ $(n=1,2,\ldots)$ such that we have*

$$\sum_{i=1}^{n} |\alpha_i|\,\delta_i \leqslant \left\| \sum_{i=1}^{n} \alpha_i x_i \right\| \tag{6.21}$$

for all finite sequences of scalars $\alpha_1, \alpha_2, \ldots, \alpha_n$.

4°. *The relation $\lim_{n\to\infty} \sum_{i=1}^{m_n} \alpha_i^{(n)} x_i = 0$, where $\alpha_i^{(n)} \in K$ $(i=1,2,\ldots,m_n;$ $n=1,2,\ldots)$, implies $\lim_{n\to\infty} \alpha_i^{(n)}=0$ $(i=1,2,\ldots)$.*

5°. *If $\lim_{n\to\infty} \sum_{i=1}^{m_n} \alpha_i^{(n)} x_i$ exists, then so do the limits $\lim_{n\to\infty} \alpha_i^{(n)}$ $(i=1,2,\ldots)$.*

If all $x_n \neq 0$ $(n=1,2,\ldots)$[1]*, then the above statements are equivalent to the following statements:*

6°. *We have*[2]

$$\bar{P} = [x_1, x_2, \ldots] = P_{(n)} \oplus P^{(n)} \qquad (n=1,2,\ldots). \tag{6.22}$$

7°. *There exists a sequence of endomorphisms*[3] *$\{u_n\} \subset L(E,E)$ such that*

$$u_n(x) = x \qquad (x \in P_{(n)};\ n=1,2,\ldots), \tag{6.23}$$

$$u_n(x) = 0 \qquad (x \in P^{(n)};\ n=1,2,\ldots). \tag{6.24}$$

8°. *For each positive integer n there exists a constant C_n, $1 \leqslant C_n < \infty$, such that we have*

$$\left\| \sum_{i=1}^{n} \alpha_i x_i \right\| \leqslant C_n \left\| \sum_{i=1}^{n+m} \alpha_i x_i \right\| \tag{6.25}$$

for all positive integers m and all $\alpha_1, \alpha_2, \ldots, \alpha_{n+m} \in K$ $(m=1,2,\ldots)$.

9°. *We have*

$$\operatorname{dist}(\sigma_{(n)}, P^{(n)}) > 0 \qquad (n=1,2,\ldots). \tag{6.26}$$

10°. *We have*

$$\operatorname{dist}(\sigma_{(n)}, \sigma^{(n)}) > 0 \qquad (n=1,2,\ldots), \tag{6.27}$$

---

[1] Of course, this condition is satisfied whenever we have 1°.

[2] I.e. $\bar{P} = P_{(n)} + P^{(n)}$ and $P_{(n)} \cap P^{(n)} = \{0\}$ $(n=1,2,\ldots)$.

[3] I.e. continuous linear mappings of E into E. For Banach spaces E, F we denote by $L(E,F)$ the Banach space of all continuous linear mappings of E into F endowed with the usual norm $\|u\| = \sup_{\substack{x\in E\\ \|x\|\leqslant 1}} \|u(x)\|$.

11°. *For each positive integer n there exists a constant* C'_n, $1 \leqslant C'_n < \infty$, *with the following property: for every* $p_0 \in P_{(n)}$ *there exists an* $f \in E^*$ *such that*

$$f(p_0) = \|p_0\|, \tag{6.28}$$

$$f(y) = 0 \qquad (y \in P^{(n)}), \tag{6.29}$$

$$1 \leqslant \|f\| \leqslant C'_n. \tag{6.30}$$

12°. *We have*

$$\sup_{\substack{p \in P \\ \|p\| \leqslant 1}} \|S_n(p)\| < \infty \qquad (n = 1, 2, \ldots). \tag{6.31}$$

13°. *We have*

$$\sup_{\substack{p \in P \\ \|p\| \leqslant 1}} \|R_n(p)\| < \infty \qquad (n = 1, 2, \ldots). \tag{6.32}$$

Proof. The equivalence $1^\circ \Leftrightarrow 2^\circ$ is an immediate consequence of a well known corollary of the Hahn-Banach theorem[1].

$2^\circ \Rightarrow 3^\circ$. Let

$$\delta_n = \frac{1}{2^{n+1}\|f_n\|} \qquad (n = 1, 2, \ldots). \tag{6.33}$$

Then, for all finite sequences of scalars $\alpha_1, \alpha_2, \ldots, \alpha_n$ we have

$$|\alpha_j|\,\delta_j = \frac{\left|f_j\left(\sum_{i=1}^{n} \alpha_i x_i\right)\right|}{2^{j+1}\|f_j\|} \leqslant \frac{1}{2^{j+1}}\left\|\sum_{i=1}^{n} \alpha_i x_i\right\| \qquad (j = 1, 2, \ldots, n),$$

whence

$$\sum_{i=1}^{n} |\alpha_i|\,\delta_i \leqslant \left(\sum_{i=1}^{n} \frac{1}{2^{i+1}}\right)\left\|\sum_{i=1}^{n} \alpha_i x_i\right\| < \frac{1}{2}\left\|\sum_{i=1}^{n} \alpha_i x_i\right\|.$$

The implication $3^\circ \Rightarrow 4^\circ$ follows from the fact that for any fixed i_0 we have, by 3°,

$$|\alpha_{i_0}^{(n)}|\,\delta_{i_0} \leqslant \sum_{i=1}^{m_n} |\alpha_i^{(n)}|\,\delta_i \leqslant \left\|\sum_{i=1}^{m_n} \alpha_i^{(n)} x_i\right\| \qquad (i_0 \leqslant m_n < \infty).$$

$4^\circ \Rightarrow 5^\circ$. Assume that we have 4° and that $\lim_{n\to\infty} \sum_{i=1}^{m_n} \alpha_i^{(n)} x_i$ exists. Then $\lim_{n,k\to\infty}\left(\sum_{i=1}^{m_{n+k}} \alpha_i^{(n+k)} x_i - \sum_{i=1}^{m_n} \alpha_i^{(n)} x_i\right) = 0$ whence, by 4°, $\lim_{n,k\to\infty} |\alpha_i^{(n+k)} - \alpha_i^{(n)}| = 0$ $(i = 1, 2, \ldots)$, whence the limits $\lim_{n\to\infty} \alpha_i^{(n)}$ $(i = 1, 2, \ldots)$ exist.

$5^\circ \Rightarrow 2^\circ$. Let

$$\phi_j\left(\sum_{i=1}^{n} \alpha_i x_i\right) = \alpha_j \qquad \left(\sum_{i=1}^{n} \alpha_i x_i \in P;\, j = 1, 2, \ldots, n\right).$$

---

[1] See e.g. [10], p. 58, theorem 6.

If $x\in\bar{P}$, $x=\lim_{n\to\infty} p_n$, $p_n=\sum_{i=1}^{m_n}\alpha_i^{(n)}x_i$ $(n=1,2,\ldots)$, then, by 5°, we can put

$$\phi_j(x)=\lim_{n\to\infty}\alpha_j^{(n)} \qquad (j=1,2,\ldots).$$

Then ϕ_j is a continuous linear functional on $\bar{P}$, satisfying $\phi_j(x_i)=\delta_{ij}$ $(i,j=1,2,\ldots)$. Extending each ϕ_j to an $f_j\in E^*$ $(j=1,2,\ldots)$, we obtain 2°. Thus $1°\Leftrightarrow 2°\Leftrightarrow 3°\Leftrightarrow 4°\Leftrightarrow 5°$.

$2°\Rightarrow 6°$. If we have 2°, let $\{s_n\}$ be the sequence of partial sum operators associated to the biorthogonal system (x_n,f_n). Then $w_n=s_n|_{\bar{P}}$ is a continuous linear projection of $\bar{P}$ onto $P_{(n)}$, along $P^{(n)}$ $(n=1,2,\ldots)$, whence we have (6.22).

$6°\Rightarrow 7°$. If we have 6°, let w_n be a continuous linear projection of $\bar{P}$ onto $P_{(n)}$, along $P^{(n)}$ $(n=1,2,\ldots)$. Then, extending each w_n to an $u_n\in L(E,E)$ (this is possible, since w_n is of finite-dimensional range), we have (6.23) and (6.24).

$7°\Rightarrow 8°$. By (6.23) and (6.24) we have

$$\left\|\sum_{i=1}^{n}\alpha_i x_i\right\|=\left\|u_n\left(\sum_{i=1}^{n+m}\alpha_i x_i\right)\right\|\leqslant\|u_n\|\left\|\sum_{i=1}^{n+m}\alpha_i x_i\right\|,$$

i.e. (6.25) with $C_n=\|u_n\|$ $(n=1,2,\ldots)$.

$8°\Rightarrow 9°$. By (6.25) we have, for every $\sum_{i=1}^{n}\alpha_i x_i\in\sigma_{(n)}$ and $\alpha_{n+1},\alpha_{n+2},\ldots,\alpha_{n+m}\in K$,

$$\left\|\sum_{i=1}^{n}\alpha_i x_i-\sum_{i=n+1}^{n+m}\alpha_i x_i\right\|\geqslant\frac{1}{C_n}\left\|\sum_{i=1}^{n}\alpha_i x_i\right\|=\frac{1}{C_n},$$

whence (6.26).

The equivalence $9°\Leftrightarrow 10°$ is an immediate consequence of lemma 6.1 a) applied to $F=P_{(n)}$, $G=P^{(n)}$.

$9°\Rightarrow 11°$. Let $p_0\in P_{(n)}$. If $p_0=0$, there exists an $f\in E^*$ satisfying (6.28), (6.29) and $\|f\|=$arbitrary. If $p_0\neq 0$, we have $\frac{p_0}{\|p_0\|}\in\sigma_{(n)}$, whence, by 9°, $\operatorname{dist}\left(\frac{p_0}{\|p_0\|},P^{(n)}\right)\geqslant\operatorname{dist}(\sigma_{(n)},P^{(n)})=d_n>0$. Then there exists[1] an $f\in E^*$ satisfying (6.28), (6.29) and $\|f\|=\frac{1}{d_n}$.

$11°\Rightarrow 12°$. By (6.28), (6.29) and (6.30), for every $p=\sum_{i=1}^{m}\alpha_i x_i\in P$ and $n\leqslant m$ we have, putting $p_0=\sum_{i=1}^{n}\alpha_i x_i=S_n(p)$,

$$\|S_n(p)\|=\|p_0\|=|f(p_0)|=|f(p)|\leqslant C_n'\|p\|.$$

[1] See e.g. [10], p. 57, lemma.

The equivalence $12^\circ \Leftrightarrow 13^\circ$ is obvious by (6.20).

$12^\circ \Rightarrow 3^\circ$. Assume now that all $x_n \neq 0\ (n=1,2,\ldots)$ and that we have 12°, and put

$$\delta_n = \frac{\|x_n\|}{2^{n+1}(\lambda_n + \lambda_{n-1})} \qquad (n=1,2,\ldots),$$

where $\lambda_n = \sup_{\substack{p\in P\\ \|p\|\leqslant 1}} \|S_n(p)\|\ (n=1,2,\ldots)$. Then $\delta_n > 0$ and for every finite sequence of scalars $\alpha_1, \alpha_2, \ldots, \alpha_n$ we have

$$|\alpha_j|\,\delta_j = \frac{\|\alpha_j x_j\|}{2^{j+1}(\lambda_j+\lambda_{j-1})} \leqslant \frac{\left\|S_j\left(\sum_{i=1}^n \alpha_i x_i\right)\right\| + \left\|S_{j-1}\left(\sum_{i=1}^n \alpha_i x_i\right)\right\|}{2^{j+1}(\lambda_j+\lambda_{j-1})}$$
$$\leqslant \frac{1}{2^{j+1}}\left\|\sum_{i=1}^n \alpha_i x_i\right\| \qquad (j=1,2,\ldots,n),$$

whence

$$\sum_{i=1}^n |\alpha_i|\,\delta_i \leqslant \left(\sum_{i=1}^n \frac{1}{2^{i+1}}\right)\left\|\sum_{i=1}^n \alpha_i x_i\right\| < \frac{1}{2}\left\|\sum_{i=1}^n \alpha_i x_i\right\|,$$

which completes the proof of theorem 6.1.

Remark 6.1. If $\{x_n\}$ is complete in E, the sequence $\{u_n\} \subset L(E,E)$ in 7° is uniquely determined and coincides with the sequence $\{s_n\}$ of partial sum operators associated to the biorthogonal system (x_n, f_n). Indeed, this follows e.g. from 6°.

Remark 6.2. Condition 4° is of similar type as those occurring in the definition of finitely linear independence and ω-linear independence, but more restrictive. Therefore the sequences $\{x_n\}$ satisfying 4° have been called, by A. I. Markushevich [156], *strongly linearly independent* sequences. Note that if all $x_n \neq 0\ (n=1,2,\ldots)$, condition 4° is equivalent to the continuity of the linear mappings $S_n\ (n=1,2,\ldots)$ on P, while 12° is nothing else but the boundedness of the $S_n\ (n=1,2,\ldots)$ on P.

§ 7. Intrinsic characterizations of bases. The norm and the index of a sequence. The index of a Banach space. Extension of block basic sequences

In § 4 we have seen some characterizations of those minimal sequences $\{x_n\}$ in a Banach space E, which are bases of E, while in § 5 we have seen some characterizations of those complete minimal sequences $\{x_n\} \subset E$ which are bases of E. All these characterizations have explicitly used the sequence $\{f_n\}\subset E^*$ of the biorthogonal system (x_n, f_n). We shall now give necessary and sufficient conditions for a complete sequence $\{x_n\}\subset E$, $x_n\neq 0$ $(n=1,2,\ldots)$ to be a basis of E, in terms only of the properties of the sequence $\{x_n\}$. Thus these will be intrinsic characterizations of bases.

We shall use, for a sequence $\{x_n\}\subset E$, the notations $P_{(n)}, P^{(n)}$, $\sigma_{(n)}$, $\sigma^{(n)}$, P, S_k and R_k of the preceding section (see (6.14)–(6.20)) and the notations

$$E^{(n)}=[x_1,\ldots,x_{n-1},x_{n+1},\ldots]=P_{(n-1)}+P^{(n)} \qquad (n=1,2,\ldots), \tag{7.1}$$

$$|||p|||=\sup_{1\leqslant k<\infty}\|S_k(p)\|=\sup_{1\leqslant k\leqslant n}\left\|\sum_{i=1}^{k}\alpha_i x_i\right\| \qquad \left(p=\sum_{i=1}^{n}\alpha_i x_i\in P\right). \tag{7.2}$$

Theorem 7.1. *Let E be a Banach space and $\{x_n\}$ a complete sequence in E such that $x_n\neq 0$ $(n=1,2,\ldots)$.*[1] *Then the following statements are equivalent:*

1°. *$\{x_n\}$ is a basis of the space E.*

2°. *For every sequence $\{p_n\}\subset P$ converging to 0 we have*

$$\sup_{1\leqslant n<\infty}|||p_n|||<\infty. \tag{7.3}$$

3°. *There exists a sequence of endomorphisms $\{u_n\}\subset L(E,E)$ satisfying* (6.23), (6.24) *and*

$$1\leqslant C_1=\sup_{1\leqslant n<\infty}\|u_n\|<\infty. \tag{7.4}$$

In this case, the sequence $\{u_n\}$ is uniquely determined and coincides with the sequence $\{s_n\}$ of partial sum operators associated to the basis $\{x_n\}$.

4°. *There exists a constant C_2 with $1\leqslant C_2<\infty$, such that we have*

$$\left\|\sum_{i=1}^{n}\alpha_i x_i\right\|\leqslant C_2\left\|\sum_{i=1}^{n+m}\alpha_i x_i\right\| \tag{7.5}$$

for all positive integers n,m and all $\alpha_1,\alpha_2,\ldots,\alpha_{n+m}\in K$.

5°. *We have*

$$C_3=\inf_{1\leqslant n<\infty}\operatorname{dist}(\sigma_{(n)},P^{(n)})>0. \tag{7.6}$$

6°. *We have*

$$\inf_{1\leqslant n<\infty}\operatorname{dist}(\sigma_{(n)},\sigma^{(n)})>0. \tag{7.7}$$

---

[1] Of course, these conditions are satisfied whenever $\{x_n\}$ is a basis of E.

7°. *We have*

$$\inf_{1 \leqslant n < \infty} \operatorname{dist}\left(\frac{x_n}{\|x_n\|}, E^{(n)}\right) > 0, \tag{7.8}$$

$$\inf_{1 \leqslant n, k < \infty} \operatorname{dist}(\sigma_{(n)}, \sigma^{(n+k)}) > 0. \tag{7.9}$$

8°. *We have* (7.8) *and there exists a positive integer* k_0 *such that*

$$\inf_{1 \leqslant n < \infty} \operatorname{dist}(\sigma_{(n)}, \sigma^{(n+k_0)}) > 0. \tag{7.10}$$

9°. *There exists a constant* $C_4, 1 \leqslant C_4 < \infty$, *with the following property: for every* n *and every* $p_0 \in P_{(n)}$ *there exists an* $f \in E^*$ *satisfying* (6.28), (6.29) *and*

$$1 \leqslant \|f\| \leqslant C_4. \tag{7.11}$$

10°. *We have*

$$C_5 = \sup_{1 \leqslant n < \infty} \sup_{\substack{p \in P \\ \|p\| \leqslant 1}} \|S_n(p)\| = \sup_{\substack{p \in P \\ \|p\| \leqslant 1}} |||p||| < \infty. \tag{7.12}$$

11°. *We have*

$$\sup_{1 \leqslant n < \infty} \sup_{\substack{p \in P \\ \|p\| \leqslant 1}} \|R_n(p)\| < \infty. \tag{7.13}$$

Furthermore, for the above constants we have

$$1 \leqslant C_1 = \inf C_2 = \frac{1}{C_3} = \inf C_4 = C_5 \leqslant \infty, \tag{7.14}$$

where $< \infty$ *holds if and only if* $\{x_n\}$ *is a basis of the space* E.

Proof. $1° \Rightarrow 3°$. If $\{x_n\}$ is a basis of E, then, by the implication $1° \Rightarrow 4°$ of § 4, theorem 4.1, the sequence $\{s_n\}$ of partial sum operators associated to the basis $\{x_n\}$ satisfies (6.23), (6.24) and (7.4) and, by virtue of § 6, remark 6.1, it is the only sequence having these properties.

$3° \Rightarrow 4°$. By (6.23), (6.24) and (7.4) we have

$$\left\|\sum_{i=1}^{n} \alpha_i x_i\right\| = \left\|u_n\left(\sum_{i=1}^{n+m} \alpha_i x_i\right)\right\| \leqslant C_1 \left\|\sum_{i=1}^{n+m} \alpha_i x_i\right\|.$$

$4° \Rightarrow 5°$. By (7.5) we have, for every $\sum_{i=1}^{n} \alpha_i x_i \in \sigma_{(n)}$ and $\alpha_{n+1}, \alpha_{n+2}, \ldots, \alpha_{n+m} \in K$,

$$\left\|\sum_{i=1}^{n} \alpha_i x_i - \sum_{i=n+1}^{n+m} \alpha_i x_i\right\| \geqslant \frac{1}{C_2} \left\|\sum_{i=1}^{n} \alpha_i x_i\right\| = \frac{1}{C_2}.$$

The equivalence $5^\circ \Leftrightarrow 6^\circ$ is an immediate consequence of § 6, lemma 6.1 a) applied to $F=P_{(n)}, G=P^{(n)}$.

$6^\circ \Rightarrow 7^\circ$. Assume that we have 6° and take an arbitrary $x\in E^{(n)}=P_{(n-1)}+P^{(n)}$. Then $x=y_{(n-1)}+z^{(n)}$ for suitable $y_{(n-1)}\in P_{(n-1)}, z^{(n)}\in P^{(n)}$, whence

$$\left\|\frac{x_n}{\|x_n\|}-x\right\|$$

$$=\left\|\frac{x_n}{\|x_n\|}-y_{(n-1)}\right\|\left\|\frac{\frac{x_n}{\|x_n\|}-y_{(n-1)}}{\left\|\frac{x_n}{\|x_n\|}-y_{(n-1)}\right\|}-\frac{z^{(n)}}{\left\|\frac{x_n}{\|x_n\|}-y_{(n-1)}\right\|}\right\|.$$

By § 6, lemma 6.1 a), applied to $F=P^{(n-1)}$ and $G=P_{(n-1)}$, we have

$$\left\|\frac{x_n}{\|x_n\|}-y_{(n-1)}\right\|\geqslant \operatorname{dist}(\sigma^{(n-1)},P_{(n-1)})\geqslant \frac{1}{2}\operatorname{dist}(\sigma^{(n-1)},\sigma_{(n-1)})\geqslant \frac{C}{2},$$

where $C=\inf_{1\leqslant n<\infty}\operatorname{dist}(\sigma_{(n)},\sigma^{(n)})>0$ by 6°. On the other hand, by § 6, lemma 6.1 a), applied to $F=P_{(n)}, G=P^{(n)}$, we have

$$\left\|\frac{\frac{x_n}{\|x_n\|}-y_{(n-1)}}{\left\|\frac{x_n}{\|x_n\|}-y_{(n-1)}\right\|}-\frac{z^{(n)}}{\left\|\frac{x_n}{\|x_n\|}-y_{(n-1)}\right\|}\right\|\geqslant \operatorname{dist}(\sigma_{(n)},P^{(n)})\geqslant\frac{C}{2}.$$

Consequently, we have

$$\left\|\frac{x_n}{\|x_n\|}-x\right\|\geqslant\frac{C^2}{4}\qquad (x\in E^{(n)},\ n=1,2,\ldots),$$

which proves (7.8).

Since $\sigma^{(n+k)}\subset\sigma^{(n)}\ (n,k=1,2,\ldots)$, we also have

$$\operatorname{dist}(\sigma_{(n)},\sigma^{(n+k)})\geqslant\operatorname{dist}(\sigma_{(n)},\sigma^{(n)})\geqslant C\qquad (n,\ k=1,2,\ldots),$$

whence (7.9). Thus $6^\circ\Rightarrow 7^\circ$.

The implication $7^\circ\Rightarrow 8^\circ$ is obvious.

$8^\circ\Rightarrow 6^\circ$. For a fixed positive integer n, applying § 6, lemma 6.1 b) successively to $F=P_{(n)}, G=P^{(n+k-1)}, G_1=P^{(n+k)}, x=x_{n+k}\ (k=1,2,\ldots,k_0)$ and taking into account that $\operatorname{dist}\left(\frac{x_{n+k}}{\|x_{n+k}\|},P_{(n)}+P^{(n+k)}\right)\geqslant\operatorname{dist}\left(\frac{x_{n+k}}{\|x_{n+k}\|},E^{(n+k)}\right)$, we obtain

$$\operatorname{dist}(\sigma_{(n)},\sigma^{(n+k-1)}) \geqslant \frac{1}{3}\operatorname{dist}(\sigma_{(n)},\sigma^{(n+k)})\operatorname{dist}\left(\frac{x_{n+k}}{\|x_{n+k}\|},E^{(n+k)}\right)$$
$$(k=1,2,\ldots,k_0),$$

whence, by multiplication,

$$\operatorname{dist}(\sigma_{(n)},\sigma^{(n)}) \geqslant \frac{1}{3^{k_0}}\operatorname{dist}(\sigma_{(n)},\sigma^{(n+k_0)}) \prod_{k=1}^{k_0} \operatorname{dist}\left(\frac{x_{n+k}}{\|x_{n+k}\|},E^{(n+k)}\right). \tag{7.15}$$

Assume now that we have 8° and put $C' = \inf_{1\leqslant n<\infty} \operatorname{dist}\left(\frac{x_n}{\|x_n\|}, E^{(n)}\right)$, which is >0 by (7.8). Then, by (7.15), we have

$$\inf_{1\leqslant n<\infty} \operatorname{dist}(\sigma_{(n)},\sigma^{(n)}) \geqslant \left(\frac{C'}{3}\right)^k \inf_{1\leqslant n<\infty} \operatorname{dist}(\sigma_{(n)},\sigma^{(n+k_0)}),$$

whence, by (7.10), we obtain (7.7). Thus $8^\circ \Rightarrow 6^\circ$.

$5^\circ \Rightarrow 9^\circ$. Let $p_0 \in P_{(n)}$. If $p_0=0$, there exists an $f\in E^*$ satisfying (6.28), (6.29) and $\|f\|=$arbitrary. If $p_0\neq 0$, we have $\frac{p_0}{\|p_0\|}\in\sigma_{(n)}$, whence, by 5°, $\operatorname{dist}\left(\frac{p_0}{\|p_0\|}, P^{(n)}\right)\geqslant C_3$. Then there exists[1] an $f\in E^*$ satisfying (6.28), (6.29) and

$$\|f\| = \frac{1}{\operatorname{dist}\left(\frac{p_0}{\|p_0\|}, P^{(n)}\right)} \leqslant \frac{1}{C_3}.$$

$9^\circ \Rightarrow 10^\circ$. By (6.28), (6.29) and (7.11), for every $p=\sum_{i=1}^m \alpha_i x_i \in P$ and $n\leqslant m$ we have, putting $p_0=\sum_{i=1}^n \alpha_i x_i = S_n(p)$,

$$\|S_n(p)\| = \|p_0\| = |f(p_0)| = |f(p)| \leqslant C_4 \|p\|.$$

The equivalence $10^\circ \Leftrightarrow 11^\circ$ is obvious by (6.20).

$10^\circ \Rightarrow 2^\circ$. If $\{p_n\}\subset P$ is a sequence converging to 0, we have $\sup_{1\leqslant n<\infty} \|p_n\|<\infty$, whence, by 10°,

$$\sup_{1\leqslant n<\infty} |||p_n||| \leqslant C_5 \sup_{1\leqslant n<\infty} \|p_n\| < \infty.$$

$2^\circ \Rightarrow 10^\circ$. If 10° is not satisfied, there exists a sequence $\{p_n\}\subset P$ with $\|p_n\|\leqslant 1$ $(n=1,2,\ldots)$ and $\lim_{n\to\infty} |||p_n||| = \infty$. Then for $q_n = \frac{p_n}{\sqrt{|||p_n|||}}$ we have $\|q_n\|\to 0$ and $|||q_n|||\to\infty$, which contradicts 2°.

---

[1] See e.g. [10], p. 57, lemma.

$10° \Rightarrow 1°$. If we have $10°$, then, since $x_n \neq 0$ $(n=1,2,\ldots)$, from the implication $12° \Rightarrow 2°$ of § 6, theorem 6.1 it follows that there exists a sequence $\{f_n\} \subset E^*$ such that (x_n, f_n) is a biorthogonal system. Then for the partial sum operators s_n associated to (x_n, f_n) we have, by biorthogonality,

$$s_n(p) = S_n(p) \qquad (p \in P, n=1,2,\ldots), \tag{7.16}$$

whence, by $10°$ and since $\{x_n\}$ is complete in E, $\sup_{1 \leqslant n < \infty} \|s_n\| < \infty$. Consequently, by the implication $4° \Rightarrow 1°$ of § 4, theorem 4.1, $\{x_n\}$ is a basis of E.

Furthermore, from the above proofs of the implications $3° \Rightarrow 4° \Rightarrow 5° \Rightarrow 9° \Rightarrow 10°$ it follows that

$$C_1 \geqslant \inf C_2 \geqslant \frac{1}{C_3} \geqslant \inf C_4 \geqslant C_5 \geqslant 1.$$

Let us prove that $C_5 = C_1$. If $\{x_n\}$ is not a basis of E, then, by the above, we have $C_5 = C_1 = \infty$. On the other hand, if $\{x_n\}$ is a basis of E, then, by the proof of $1° \Rightarrow 3°$ given above, we have $C_1 = \sup_{1 \leqslant n < \infty} \|s_n\| < \infty$, where $\{s_n\}$ is the sequence of partial sum operators associated to the basis $\{x_n\}$. Hence, by (7.12) and (7.16), $C_5 = C_1 < \infty$, which completes the proof of theorem 1.5.

Remark 7.1. Let us also mention the following alternative proof of the implication $10° \Rightarrow 1°$. If we have $10°$, then, since $x_n \neq 0$ $(n=1,2,\ldots)$, from the implication $12° \Rightarrow 2°$ of § 6, theorem 6.1 it follows that there exists a sequence $\{f_n\} \subset E^*$ such that (x_n, f_n) is a biorthogonal system. Then, since $\{x_n\}$ is complete in E, from $10°$ and the implication $5° \Rightarrow 1°$ of § 5, theorem 5.1 it follows that $\{x_n\}$ is a basis of E.

Remark 7.2. A comparing of the conditions $7° - 13°$ of §6, theorem 6.1 with the conditions $3° - 6°$ and $9° - 11°$ of theorem 7.1 shows that the bases of a Banach space E could be called "uniformely minimal" sequences. Furthermore, with the condition $4°$ of § 6, theorem 6.1 (condition of "strong linear independence", see remark 6.2), it is natural to compare condition $2°$ of theorem 7.1. However, note that in $2°$ of theorem 7.1 it is also assumed separately that $x_n \neq 0$ $(n=1,2,\ldots)$, while in $4°$ of § 6, theorem 6.1 this is not assumed, but follows as a consequence. Finally, let us mention that condition (7.8) is equivalent to the existence of a sequence $\{f_n\} \subset E^*$ such that (x_n, f_n) is a biorthogonal system satisfying $\sup_{1 \leqslant n < \infty} \|x_n\| \, \|f_n\| < \infty$ (this follows from Ch. II, § 2, corollary 2.1, applied to $\left\{\frac{x_n}{\|x_n\|}\right\}$). Of course, every basis of E satisfies this condition (see § 3, theorem 3.1, formula (3.7)).

Definition 7.1. Let $\{x_n\}$ be a complete sequence in a Banach space E, such that $x_n \neq 0$ $(n=1,2,\ldots)$. Then the number (7.14) is called *the norm of the sequence* $\{x_n\}$; we shall denote this number by $v = v_{\{x_n\}}$. The number $\gamma = \gamma_{\{x_n\}} = \frac{1}{v}$ is called *the index of the sequence* $\{x_n\}$.

These numbers admit a geometrical interpretation. Indeed, we recall that if F, G are subspaces of a Banach space E and $\sigma_F = \{x \in F \mid \|x\| = 1\}$, the number

$$\widehat{(F;G)} = \operatorname{dist}(\sigma_F, G) \tag{7.17}$$

is called *the inclination* of F to G. With the aid of this notion and of the notations (6.14), (6.15), we can write

$$\gamma_{\{x_n\}} = \inf_{1 \leqslant n < \infty} \widehat{(P_{(n)}; P^{(n)})}. \tag{7.18}$$

By theorem 7.1, we have

$$1 \leqslant v_{\{x_n\}} \leqslant \infty, \qquad 0 \leqslant \gamma_{\{x_n\}} \leqslant 1; \tag{7.19}$$

the sequence $\{x_n\}$ is a basis of E if and only if $v_{\{x_n\}} < \infty$, or, equivalently, $\gamma_{\{x_n\}} > 0$.

Definition 7.2. Let E be a Banach space. The number

$$\Gamma(E) = \sup \gamma_{\{x_n\}}, \tag{7.20}$$

where the sup is taken over all complete sequences $\{x_n\} \subset E$ with $x_n \neq 0$ $(n=1,2,\ldots)$, is called *the index of the space E.*

By (7.19) we have, for all Banach spaces E,

$$0 \leqslant \Gamma(E) \leqslant 1, \tag{7.21}$$

and the space E has a basis if and only if $\Gamma(E) > 0$. It is natural to ask whether or not there exist Banach spaces with bases such that $\Gamma(E) < 1$, i. e. Banach spaces E such that $0 < \Gamma(E) < 1$. We shall see in Chapter II, § 1, that the answer is affirmative for finite dimensional Banach spaces, but unknown for infinite dimensional Banach spaces.

Let us give now some corollaries of theorem 7.1.

Corollary 7.1. *Let E be a Banach space and $\{x_n\}$ a complete sequence in E such that*

$$\prod_{n=1}^{\infty} \widehat{([x_1, \ldots, x_n]; [x_{n+1}])} > 0, \tag{7.22}$$

where $[x_{n+1}]$ is the one-dimensional subspace of E spanned by x_{n+1}. Then $\{x_n\}$ is a basis of E, of index $\gamma_{\{x_n\}} \geqslant \beta$.

Proof. Put $\beta_n = (\widehat{[x_1,\ldots,x_n];[x_{n+1}]})$ $(n=1,2,\ldots)$, $\beta = \prod_{n=1}^{\infty} \beta_n$. Then $0<\beta\leqslant 1$ and for any finite sequence of scalars $\alpha_1,\ldots,\alpha_{n+m}$ we have, by (7.22),

$$\left\|\sum_{i=1}^{n+m} \alpha_i x_i\right\| = \left\|\sum_{i=1}^{n+m-1} \alpha_i x_i + \alpha_{n+m} x_{n+m}\right\| \geqslant \beta_{n+m-1} \left\|\sum_{i=1}^{n+m-1} \alpha_i x_i\right\|$$

$$\geqslant \beta_{n+m-1}\beta_{n+m-2}\left\|\sum_{i=1}^{n+m-2} \alpha_i x_i\right\| \geqslant \cdots$$

$$\geqslant \beta_{n+m-1}\beta_{n+m-2}\cdots\beta_n\left\|\sum_{i=1}^{n} \alpha_i x_i\right\| \geqslant \beta\left\|\sum_{i=1}^{n} \alpha_i x_i\right\|,$$

whence, by the implication $4° \Rightarrow 1°$ of theorem 7.1, $\{x_n\}$ is a basis of E, of index $\gamma_{\{x_n\}} \geqslant \beta$, which completes the proof.

The following is a useful tool for proving that certain normed linear spaces are complete:

Corollary 7.2. *Let E be a normed linear space and $\{x_n\}$ a complete sequence in E, with $x_n \neq 0$ $(n=1,2,\ldots)$, satisfying any one of the equivalent conditions of theorem 7.1. Then E is complete if (and, obviously, only if) every Cauchy series of the form $\sum_{i=1}^{\infty} \alpha_i x_i$ is convergent to an element of E.*

Proof. By virtue of theorem 7.1, $\{x_n\}$ is a basis of the completion $E^\wedge$ of E. Let $y \in E^\wedge$ be arbitrary. Then y has an expansion of the form $y = \sum_{i=1}^{\infty} \alpha_i x_i$ and therefore $\sum_{i=1}^{\infty} \alpha_i x_i$ is a Cauchy series. Hence, by our hypothesis, $\sum_{i=1}^{\infty} \alpha_i x_i$ is convergent to an element of E, and thus $y \in E$. Since $y \in E^\wedge$ has been arbitrary, it follows that E is complete, which concludes the proof.

The next corollary gives a method of "piecewise" construction of new bases in Banach spaces with bases.

Corollary 7.3. *Let $\{x_n\}$ be a basis of a Banach space E, $\{m_n\}$ an increasing sequence of positive integers, $m_0=0$, and $\{z_n\}$ a sequence in E such that $\{z_i\}_{i=m_{n-1}+1}^{m_n}$ is a basis of $[x_{m_{n-1}+1},\ldots,x_{m_n}]$ $(n=1,2,\ldots)$, of norm*

$$\nu_{\{z_i\}_{i=m_{n-1}+1}^{m_n}} \leqslant M < \infty \qquad (n=1,2,\ldots). \tag{7.23}$$

Then $\{z_n\}$ is a basis of E.

Proof. It is obvious that $[z_n]=E$. Let l, k be arbitrary positive integers and let n, q be positive integers with $n \leqslant q$, such that

$m_{n-1}+1 \leqslant l \leqslant m_n$, $m_{q-1}+1 \leqslant l+k \leqslant m_q$. Furthermore, let $\alpha_1, \ldots, \alpha_{l+k}$ be arbitrary scalars, let $\sum_{i=m_{j-1}+1}^{m_j} \beta_i x_i$ be the representation of $\sum_{i=m_{j-1}+1}^{m_j} \alpha_i z_i$ with respect to the basis $\{x_n\}_{n=1}^{\infty}$ $(j=1,\ldots,q-1)$ and let $\sum_{i=m_{q-1}+1}^{m_q} \beta_i x_i$ be the representation of $\sum_{i=m_{q-1}+1}^{l+k} \alpha_i z_i$. Then

$$\begin{aligned}
\left\| \sum_{i=1}^{l} \alpha_i z_i \right\| &\leqslant \left\| \sum_{i=1}^{m_{n-1}} \alpha_i z_i \right\| + \left\| \sum_{i=m_{n-1}+1}^{l} \alpha_i z_i \right\| \\
&\leqslant \left\| \sum_{i=1}^{m_{n-1}} \beta_i x_i \right\| + M \left\| \sum_{i=m_{n-1}+1}^{m_n} \alpha_i z_i \right\| \\
&= \left\| \sum_{i=1}^{m_{n-1}} \beta_i x_i \right\| + M \left\| \sum_{i=m_{n-1}+1}^{m_n} \beta_i x_i \right\| \\
&\leqslant v_{\{x_n\}} \left\| \sum_{i=1}^{m_q} \beta_i x_i \right\| + M(v_{\{x_n\}}+1) \left\| \sum_{i=1}^{m_n} \beta_i x_i \right\| \\
&\leqslant 2 v_{\{x_n\}} (v_{\{x_n\}}+1) M \left\| \sum_{i=1}^{m_q} \beta_i x_i \right\| = 2 v_{\{x_n\}} (v_{\{x_n\}}+1) M \left\| \sum_{i=1}^{l+k} \alpha_i z_i \right\|,
\end{aligned}$$

whence, by theorem 7.1 (implication $4^\circ \Rightarrow 1^\circ$), $\{z_n\}$ is a basis of E, which completes the proof.

Let us observe that if we omit in theorem 7.1 the hypothesis $[x_n]=E$, then we obtain various characterizations of basic sequences[1]. We shall use now this remark to prove

Corollary 7.4. *Let $\{x_n\}$ be a basis of a Banach space E, let $\{m_n\}$ be an increasing sequence of positive integers, $m_0=0$, and let*

$$y_n = \sum_{i=m_{n-1}+1}^{m_n} \alpha_i x_i, \qquad y_n \neq 0 \qquad (n=1,2,\ldots). \tag{7.24}$$

Then $\{y_n\}$ is a basic sequence, with $v_{\{y_n\}} \leqslant v_{\{x_n\}}$.

Proof. By the implication $1^\circ \Rightarrow 4^\circ$ of theorem 7.1 and by (7.24) we have

$$\begin{aligned}
\left\| \sum_{j=1}^{n} \gamma_j y_j \right\| &= \left\| \sum_{j=1}^{n} \gamma_j \sum_{i=m_{j-1}+1}^{m_j} \alpha_i x_i \right\| = \left\| \sum_{j=1}^{n} \sum_{i=m_{j-1}+1}^{m_j} \gamma_j a_i x_i \right\| \\
&\leqslant v_{\{x_n\}} \left\| \sum_{j=1}^{n+q} \sum_{i=m_{j-1}+1}^{m_j} \gamma_j \alpha_i x_i \right\| = v_{\{x_n\}} \left\| \sum_{j=1}^{n+q} \gamma_j y_j \right\|
\end{aligned}$$

[1] See § 4, definition 4.5.

for all positive integers n, q and all $\gamma_1, \dots, \gamma_{n+q} \in K$. Consequently, by the implication $4^\circ \Rightarrow 1^\circ$ of theorem 7.1, $\{y_n\}$ is a basic sequence, with $v_{\{y_n\}} \leqslant v_{\{x_n\}}$, which completes the proof.

Definition 7.3. Let $\{x_n\}$ be a basis of a Banach space E. A sequence $\{y_n\} \subset E$ is said to be a *block basic sequence* (*with respect to the basis* $\{x_n\}$), if it is of the form (7.24), where $\{m_n\}$ is an increasing sequence of positive integers and $m_0 = 0$.

By corollary 7.4, every block basic sequence is necessarily a basic sequence. The block basic sequences will have many applications in the sequel. In the present section we shall show only that for this important class of basic sequences the problem of their extension to a basis of the whole space E (§ 4, problem 4.1) has an affirmative answer. For this purpose, we need

Lemma 7.1. *Let F and G be two $(n-1)$-dimensional subspaces of an n-dimensional Banach space E, where $n < \infty$. Then there exists an isomorphism T from F onto G such that*

$$\tfrac{1}{3}\|x\| \leqslant \|T(x)\| \leqslant 3\|x\| \qquad (x \in F). \tag{7.25}$$

Proof. There exists[1] a projection u of F onto its $(n-2)$-dimensional subspace $F \cap G$, such that $\|I_F - u\| = 1$ (whence $\|u\| \leqslant \|I_F - u\| + \|I_F\| = 2$) and similarly, a projection v of G onto $F \cap G$, such that $\|I_G - v\| = 1$ (whence $\|v\| \leqslant 2$). Take $y_0 \in F$, $\|y_0\| = 1$, such that $u(y_0) = 0$, and $z_0 \in G$, $\|z_0\| = 1$, such that $v(z_0) = 0$. Then the mapping $T: F \to G$ defined by

$$T(x + \alpha y_0) = x + \alpha z_0 \qquad (x \in F \cap G, \alpha \in K) \tag{7.26}$$

is obviously linear and we have, for any $x + \alpha y_0 \in F$,

$$\begin{aligned}\|T(x+\alpha y_0)\| &= \|x + \alpha z_0\| \leqslant \|x\| + |\alpha| = \|x\| + \|\alpha y_0\| \\ &= \|u(x+\alpha y_0)\| + \|(I_F - u)(x + \alpha y_0)\| \leqslant 3\|x + \alpha y_0\|.\end{aligned}$$

Similarly, for any $x + \alpha y_0 \in F$ we have

$$\begin{aligned}\|x+\alpha y_0\| &\leqslant \|x\| + |\alpha| = \|x\| + \|\alpha z_0\| = \|v(x+\alpha z_0)\| + \|(I_G - v)(x+\alpha z_0)\| \\ &\leqslant 3\|x + \alpha z_0\| = 3\|T(x+\alpha y_0)\|,\end{aligned}$$

whence (7.25), which completes the proof of lemma 7.1.

---

[1] Indeed, in any reflexive Banach space B there exists a projection u onto any hyperplane $H = \{x \in B \mid f(x) = 0\}$, such that $\|I_B - u\| = 1$, where I_B denotes the identical mapping of B onto itself (take $x_1 \in B$ with $\|x_1\| = 1$, $f(x_1) = \|f\|$ and put $u(x) = x - \dfrac{f(x)}{\|f\|} x_1$ for all $x \in B$).

Now we can prove

Theorem 7.2. *Let $\{x_n\}$ be a basis of a Banach space E and let*

$$y_n = \sum_{i=m_{n-1}+1}^{m_n} \alpha_i x_i \neq 0 \qquad (n=1,2,\dots;\ m_0=0) \tag{7.27}$$

be a block basic sequence with respect to $\{x_n\}$. Then there exists a basis $\{z_n\}$ of E such that

$$z_{m_n} = y_n \qquad (n=1,2,\dots), \tag{7.28}$$

$$[z_i]_{i=m_{n-1}+1}^{m_n} = [x_i]_{i=m_{n-1}+1}^{m_n} \qquad (n=1,2,\dots). \tag{7.29}$$

Consequently, for the sequences of coefficient functionals $\{f_n\}$, $\{h_n\} \subset E^*$ associated to the bases $\{x_n\}$ and $\{z_n\}$ respectively, we have

$$[h_n] = [f_n]. \tag{7.30}$$

Proof. For each n there exists[1] an $(m_n - m_{n-1} - 1)$-dimensional subspace G_n of $E_n = [x_i]_{i=m_{n-1}+1}^{m_n}$ which admits a projection $U_n\colon E_n \to G_n$ such that $U_n(y_n) = 0$, $\|U_n\| \leqq 2$. Furthermore, by lemma 7.1 above, for each n there exists an isomorphism T_n from $F_n = [x_i]_{i=m_{n-1}+1}^{m_n-1}$ onto G_n such that

$$\tfrac{1}{3}\|x\| \leqq \|T_n(x)\| \leqq 3\|x\| \qquad (x \in F_n). \tag{7.31}$$

Put

$$z_i = \begin{cases} T_n(x_i) & \text{for } m_{n-1}+1 \leqq i \leqq m_n - 1 \\ y_n & \text{for } i = m_n \end{cases} \qquad (n=1,2,\dots). \tag{7.32}$$

Then for any scalars $\beta_{m_{n-1}+1}, \dots, \beta_{m_n}$ we have $U_n\left(\sum_{i=m_{n-1}+1}^{m_n} \beta_i z_i\right) = \sum_{i=m_{n-1}+1}^{m_n-1} \beta_i z_i$ (since $U_n(z_{m_n}) = U_n(y_n) = 0$, and $z_i = T_n(x_i) \in G_n$ for $m_{n-1}+1 \leqq i \leqq m_n - 1$), whence taking into account (7.31), it follows that for any such scalars and any l with $m_{n-1}+1 \leqq l \leqq m_n - 1$ we have

$$\begin{aligned}
\left\|\sum_{i=m_{n-1}+1}^{l} \beta_i z_i\right\| &= \left\|\sum_{i=m_{n-1}+1}^{l} \beta_i T_n(x_i)\right\| \leqq 3\left\|\sum_{i=m_{n-1}+1}^{l} \beta_i x_i\right\| \\
&\leqq 3 v_{\{x_j\}} \left\|\sum_{i=m_{n-1}+1}^{m_n-1} \beta_i x_i\right\| \leqq 9 v_{\{x_j\}} \left\|\sum_{i=m_{n-1}+1}^{m_n-1} \beta_i T_n(x_i)\right\| \\
&= 9 v_{\{x_j\}} \left\|U_n\left(\sum_{i=m_{n-1}+1}^{m_n} \beta_i z_i\right)\right\| \leqq 18 v_{\{x_j\}} \left\|\sum_{i=m_{n-1}+1}^{m_n} \beta_i z_i\right\|,
\end{aligned}$$

---

[1] Indeed, in any Banach space B, for any $x_0 \in B$ there exists a hyperplane $G \subset B$ which admits a projection $U\colon B \to G$ such that $U(x_0) = 0$, $\|U\| \leqq 2$ (take $f \in B^*$ with $\|f\| = \frac{1}{\|x_0\|}$, $f(x_0) = 1$, and put $G = \{x \in B \mid f(x) = 0\}$, $U(x) = x - f(x)x_0$ for all $x \in B$).

and therefore $\{z_i\}_{i=m_{n-1}+1}^{m_n}$ is a basis of $[x_i]_{i=m_{n-1}+1}^{m_n}$, of norm

$$v_{\{z_i\}_{i=m_{n-1}+1}^{m_n}} \leqslant 18\, v_{\{x_j\}} \qquad (n=1,2,\ldots).$$

Consequently, by corollary 7.3, $\{z_n\}$ is a basis of E, which, by (7.32), satisfies (7.28) and the inclusion $\subset$ in (7.29), whence also the equality in (7.29).

Now, for $\{h_n\}\subset E^*$ with $h_i(z_j)=\delta_{ij}$ we have, by (7.29),

$$h_i(x_j)=0 \qquad (i=m_{n-1}+1,\ldots,m_n;\ j\neq m_{n-1}+1,\ldots,m_n;\ n=1,2,\ldots),$$

that is,

$$[h_i]_{i=m_{n-1}+1}^{m_n} \subset ([x_j]_{j\neq m_{n-1}+1,\ldots,m_n})^{\perp} \qquad (n=1,2,\ldots). \tag{7.33}$$

On the other hand, for $\{f_n\}\subset E^*$ with $f_i(x_j)=\delta_{ij}$ we obtain, as in Ch. II, § 4, formula (4.7),

$$[f_i]_{i=m_{n-1}+1}^{m_n} = ([x_j]_{j\neq m_{n-1}+1,\ldots,m_n})^{\perp} \qquad (n=1,2,\ldots). \tag{7.34}$$

From (7.33) and (7.34) we infer

$$[h_i]_{i=m_{n-1}+1}^{m_n} = [f_i]_{i=m_{n-1}+1}^{m_n} \qquad (n=1,2,\ldots), \tag{7.35}$$

whence (7.30), which completes the proof of theorem 7.2.

Remark 7.3. Actually, proposition 4.4 of § 4, on block perturbations of bases, may be also regarded as a theorem on extension of block basic sequences of a particular form.

§ 8. Domination and equivalence of sequences. Equivalent, affinely equivalent and permutatively equivalent bases

Definition 8.1. Let E, F be two Banach spaces. A sequence $\{x_n\}\subset E$ is said to *dominate* a sequence $\{y_n\}\subset F$ provided that for all sequences $\{\alpha_n\}$ of scalars

$$\sum_{i=1}^{\infty} \alpha_i x_i \text{ converges} \Rightarrow \sum_{i=1}^{\infty} \alpha_i y_i \text{ converges}. \tag{8.1}$$

In this case we shall use the notation $\{x_n\}\succ\{y_n\}$.

We shall say that $\{x_n\}\subset E$ *dominates strictly* the sequence $\{y_n\}\subset F$ and we shall write $\{x_n\}\gg\{y_n\}$, provided that there exists a continuous linear mapping $u\in L([x_n],[y_n])$ such that

$$u(x_n)=y_n \qquad (n=1,2,\ldots). \tag{8.2}$$

The sequences $\{x_n\}\subset E$ and $\{y_n\}\subset F$ are said to be *equivalent* if we have simultaneously $\{x_n\}\succ\{y_n\}\succ\{x_n\}$, and *strictly equivalent* if we have $\{x_n\} \gg \{y_n\} \gg \{x_n\}$. In these cases we shall use the notations $\{x_n\} \sim \{y_n\}$ and $\{x_n\} \approx \{y_n\}$ respectively.

Finally, we shall say that the sequences $\{x_n\} \subset E$, $\{y_n\} \subset F$ are *fully equivalent*, and we shall write $\{x_n\} \approxeq \{y_n\}$, provided that there exists an isomorphism[1] u of E *onto* F satisfying (8.2).

It is obvious that strict domination $\Rightarrow$ domination and that full equivalence $\Rightarrow$ strict equivalence $\Rightarrow$ equivalence. As we shall see below, the converse implications are not valid. It is also immediate that if $\{x_n\} \approx \{y_n\}$, then the mappings $u \in L([x_n], [y_n])$ and $v \in L([y_n], [x_n])$ in the definition of strict equivalence are such that u is an isomorphism of $[x_n]$ onto $[y_n]$ and v is the inverse isomorphism, carrying $[y_n]$ onto $[x_n]$. Hence, *if both* $\{x_n\}$ *and* $\{y_n\}$ *are complete in* E *and* F *respectively, we have* $\{x_n\} \approx \{y_n\}$ *if and only if* $\{x_n\} \approxeq \{y_n\}$.

Theorem 8.1. *Let* E, F *be two Banach spaces and let* $\{x_n\} \subset E$, $\{y_n\} \subset F$. *Then*

a) *The following statements are equivalent:*

1°. $\{x_n\} \succ \{y_n\}$.

2°. *There exist a positive integer* n_0 *and a constant* $C>0$ *such that we have*

$$\left\| \sum_{i=n_0}^{n_0+m} \alpha_i y_i \right\| \leqslant C \sup_{n_0 \leqslant k \leqslant n_0+m} \left\| \sum_{i=n_0}^{k} \alpha_i x_i \right\| \tag{8.3}$$

for all finite sequences of scalars $\alpha_{n_0}, \alpha_{n_0+1}, \ldots, \alpha_{n_0+m}$ $(m=1,2,\ldots)$.

b) *The following statements are equivalent:*

1°. $\{x_n\} \gg \{y_n\}$.

2°. *There exists a constant* $C'>0$ *such that we have*

$$\left\| \sum_{i=1}^{n} \alpha_i y_i \right\| \leqslant C' \left\| \sum_{i=1}^{n} \alpha_i x_i \right\| \tag{8.4}$$

for all finite sequences of scalars $\alpha_1, \alpha_2, \ldots, \alpha_n$.

3°. *For every* $\psi \in [y_n]^*$ *the system of equations*

$$\phi(x_n) = \psi(y_n) \qquad (n=1,2,\ldots) \tag{8.5}$$

has a (unique[2]*) solution* $\phi \in [x_n]^*$.

If $\{x_n\}$ *is minimal, these statements are equivalent to the following:*

4°. *There exists a positive integer* n_0 *such that* $\{x_n\}_{n_0}^{\infty} \gg \{y_n\}_{n_0}^{\infty}$.

If (x_n, ϕ_n) $(\{\phi_n\} \subset [x_n]^*)$ *and* (y_n, ψ_n) $(\{\psi_n\} \subset [y_n]^*)$ *are biorthogonal systems such that* $\{\psi_n\}$ *is total on* $[y_n]$, *these statements are equivalent to the following:*

---

[1] We recall that by "isomorphism" we mean: linear homeomorphism.

[2] Since $\{x_n\}$ is complete in $[x_n]$, the solution of (8.5) is unique.

5°. *For every* $x \in [x_n]$ *the system of equations*

$$\phi_n(x) = \psi_n(y) \qquad (n=1,2,\ldots) \tag{8.6}$$

has a (unique)[1] *solution* $y \in [y_n]$.

If $\{x_n\}$ *is a basis of* $[x_n]$ *and* $\{y_n\} \subset F$ *is arbitrary, these statements are equivalent to the following:*

6°. $\{x_n\} \succ \{y_n\}$.

c) *The following statements are equivalent:*

1°. $\{x_n\} \sim \{y_n\}$.

2°. *There exist a positive integer* n_0 *and a constant* $C>0$ *such that we have* (8.3) *and*

$$\left\| \sum_{i=n_0}^{n_0+m} \alpha_i x_i \right\| \leqslant C \sup_{n_0 \leqslant k \leqslant n_0+m} \left\| \sum_{i=n_0}^{k} \alpha_i y_i \right\| \tag{8.7}$$

for all finite sequences of scalars $\alpha_{n_0}, \alpha_{n_0+1}, \ldots, \alpha_{n_0+m}$ $(m=1,2,\ldots)$.

d) *The following statements are equivalent:*

1°. $\{x_n\} \approx \{y_n\}$.

2°. *There exists a constant* $C'>0$ *such that we have* (8.4) *and*

$$\left\| \sum_{i=1}^{n} \alpha_i x_i \right\| \leqslant C' \left\| \sum_{i=1}^{n} \alpha_i y_i \right\| \tag{8.8}$$

for all finite sequences of scalars $\alpha_1, \alpha_2, \ldots, \alpha_n$.

3°. *The system of equations* (8.5) *has a (unique) solution* $\phi \in [x_n]^*$ *for each* $\psi \in [y_n]^*$ *and a (unique) solution* $\psi \in [y_n]^*$ *for each* $\phi \in [x_n]^*$.

If $\{x_n\}, \{y_n\}$ *are minimal, these statements are equivalent to the following:*

4°. *There exists a positive integer* n_0 *such that* $\{x_n\}_{n_0}^{\infty} \approx \{y_n\}_{n_0}^{\infty}$.

If (x_n, ϕ_n) $(\{\phi_n\} \subset [x_n]^*)$ *and* (y_n, ψ_n) $(\{\psi_n\} \subset [y_n]^*)$ *are biorthogonal systems such that* $\{\phi_n\}$, $\{\psi_n\}$ *are total on* $[x_n]$ *and* $[y_n]$ *respectively, these statements are equivalent to the following:*

5°. *The system of equations* (8.6) *has a (unique) solution* $y \in [y_n]$ *for each* $x \in [x_n]$ *and a (unique) solution* $x \in [x_n]$ *for each* $y \in [y_n]$.

If $\{x_n\}, \{y_n\}$ *are bases of* $[x_n]$ *and* $[y_n]$ *respectively, these statements are equivalent to the following:*

6°. $\{x_n\} \sim \{y_n\}$.

Proof. a) The implication $2° \Rightarrow 1°$ is obvious.

$1° \Rightarrow 2°$. Assume that 2° is not satisfied. Then we can successively find an increasing sequence of positive integers $\{m_n\}$ and a sequence of scalars $\{\alpha_n\}$ such that

[1] Since $\{\psi_n\}$ is total on $\{y_n\}$, the solution of (8.6) is unique.

$$1 = \left\| \sum_{i=m_{n-1}+1}^{m_n} \alpha_i y_i \right\| \geqslant 2^n \max_{m_{n-1}+1 \leqslant k \leqslant m_n} \left\| \sum_{i=m_{n-1}+1}^{k} \alpha_i x_i \right\|$$

$$(n=1,2,\ldots;\ m_0=0).$$

Then $\sum_{i=1}^{\infty} \alpha_i x_i$ is convergent and $\sum_{i=1}^{\infty} \alpha_i y_i$ divergent, i.e. 1° is not satisfied.

b) 1°⇒2°. If $u \in L([x_n],[y_n])$ satisfies (8.2), then for any $\alpha_1, \alpha_2, \ldots, \alpha_n \in K$ we have $\left\| \sum_{i=1}^{n} \alpha_i y_i \right\| = \left\| u\left(\sum_{i=1}^{n} \alpha_i x_i \right) \right\| \leqslant \|u\| \left\| \sum_{i=1}^{n} \alpha_i x_i \right\|$, i.e. (8.4) with $C' = \|u\|$.

2° ⇒ 1°. Assume that we have 2° and for any $p = \sum_{i=1}^{n} \alpha_i x_i \in P$ put $u_0(p) = \sum_{i=1}^{n} \alpha_i y_i$. Then u_0 is a linear mapping of the dense subspace P of $[x_n]$ into $[y_n]$ (u_0 is well defined on P, since $\sum_{i=1}^{n} \alpha_i x_i = 0$ implies, by (8.4), that $\sum_{i=1}^{n} \alpha_i y_i = 0$), satisfying $u_0(x_n) = y_n$ $(n=1,2,\ldots)$. Since by (8.4) this mapping u_0 is also continuous on P, it can be extended to an $u \in L([x_n], [y_n])$ satisfying (8.2).

1° ⇒ 3°. If $u \in L([x_n], [y_n])$ satisfies (8.2), then for every $\psi \in [y_n]^*$ the system of equations (8.5) has the (unique) solution $\phi = u^*(\psi)$, where $u^* \in L([y_n]^*, [x_n]^*)$ is the adjoint of u.

3° ⇒ 1°. Assume now that we have 3° and for any $p = \sum_{i=1}^{n} \alpha_i x_i \in P$ put $u_0(p) = \sum_{i=1}^{n} \alpha_i y_i$. Then $u_0(p)$ is well defined. Indeed, assume the contrary, i.e. that there exists a finite sequence of scalars $\alpha_1, \alpha_2, \ldots, \alpha_n$ such that $\sum_{i=1}^{n} \alpha_i x_i = 0$, $\sum_{i=1}^{n} \alpha_i y_i \neq 0$. Then there exists a $\psi \in [y_n]^*$ such that $\psi\left(\sum_{i=1}^{n} \alpha_i y_i \right) \neq 0$. By 3° to this $\psi \in [y_n]^*$ corresponds a $\phi \in [x_n]^*$ such that we have (8.5), whence $\phi(0) = \phi\left(\sum_{i=1}^{n} \alpha_i x_i \right) = \sum_{i=1}^{n} \alpha_i \phi(x_i) = \sum_{i=1}^{n} \alpha_i \psi(y_i)$ $= \psi\left(\sum_{i=1}^{n} \alpha_i y_i \right) \neq 0$, which is impossible. Thus u_0 is a well defined linear mapping of P into $[y_n]$ satisfying $u_0(x_n) = y_n$ $(n=1,2,\ldots)$. Now, for every $\psi \in [y_n]^*$ let us denote by $T(\psi)$ the functional $\phi \in [x_n]^*$ corresponding to ψ by 3°. Then $T\colon \psi \to T(\psi)$ is a linear mapping of $[y_n]^*$ into $[x_n]^*$. Indeed, for arbitrary $\psi, \psi' \in [y_n]^*$ and scalars $\alpha, \alpha' \in K$ we have

$$\begin{aligned}[T(\alpha\psi + \alpha'\psi')](x_i) &= (\alpha\psi + \alpha'\psi')(y_i) = \alpha\psi(y_i) + \alpha'\psi'(y_i) \\ &= \alpha[T(\psi)](x_i) + \alpha'[T(\psi')](x_i) \\ &= [\alpha T(\psi) + \alpha' T(\psi')](x_i) \quad (i=1,2,\ldots),\end{aligned}$$

whence, since $\{x_n\}$ is complete in $[x_n]$,

$$T(\alpha\psi+\alpha'\psi')=\alpha T(\psi)+\alpha' T(\psi').$$

We claim that T is also continuous on $[y_n]^*$ for the norm topology. Indeed, let $\{\psi_n\}\subset[y_n]^*$, $\psi\in[y_n]^*$ and $\phi\in[x_n]^*$ be such that

$$\lim_{n\to\infty}\psi_n=\psi, \qquad \lim_{n\to\infty}T(\psi_n)=\phi.$$

Then

$$\psi(y_i)=\lim_{n\to\infty}\psi_n(y_i)=\lim_{n\to\infty}[T(\psi_n)](x_i)=\phi(x_i) \qquad (i=1,2,\ldots),$$

whence $\phi=T(\psi)$. Consequently, by the closed graph theorem[1], T is continuous on $[y_n]^*$.

Now, since we have

$$|\psi[u_0(p)]|=|[T(\psi)](p)|\leqslant\|T\|\,\|\psi\|\,\|p\| \qquad (p\in P,\ \psi\in[y_n]^*),$$

where $\|T\|<\infty$ by the above, it follows that u_0 is also continuous on P. Hence u_0 can be extended to a mapping $u\in L([x_n],[y_n])$ satisfying (8.2). Thus we have 1°.

The implication 1° ⇒ 4° is obvious.

4° ⇒ 1°. Assume that $\{x_n\}$ is minimal and that we have 4°. For any $p=\sum_{i=1}^{n}\alpha_i x_i\in P$ put

$$u_1(p)=\begin{cases}\sum_{i=1}^{n_0-1}\alpha_i y_i & \text{if } n\geqslant n_0-1,\\ \sum_{i=1}^{n}\alpha_i y_i & \text{if } n\leqslant n_0-1,\end{cases}$$

$$u_2(p)=\begin{cases}\sum_{i=n_0}^{n}\alpha_i y_i & \text{if } n\geqslant n_0,\\ 0 & \text{if } n\leqslant n_0-1,\end{cases}$$

where n_0 is as in 4°. Then u_1, u_2 are linear mappings of P into F (they are well defined on P, since $\{x_n\}$ is minimal) and we have $u_1=v_1 w_1$, $u_2=v_2 w_2$, where w_1 denotes the linear projection of P onto $P_{(n_0-1)}$ along $P^{(n_0-1)}$, v_1 the mapping $\sum_{i=1}^{n_0-1}\alpha_i x_i\to\sum_{i=1}^{n_0-1}\alpha_i y_i$ of $P_{(n_0-1)}$ into $[y_n]$,

---

[1] See e.g. [10], p. 41, theorem 7.

w_2 the linear projection of P onto $P^{(n_0-1)}$ along $P_{(n_0-1)}$ and v_2 the mapping $\sum_{i=n_0}^{n} \alpha_i x_i \to \sum_{i=n_0}^{n} \alpha_i y_i$. Here w_1, w_2 are continuous by the implication $1^\circ \Rightarrow 6^\circ$ of § 6, theorem 6.1, v_1 is continuous by the linear independence of $x_1, x_2, \ldots, x_{n_0-1}$ and v_2 is continuous by 4°. Consequently, u_1 and u_2, whence also $u_0 = u_1 + u_2$, are continuous on P. Extending this u_0 to an $u \in L([x_n], [y_n])$, we obtain a continuous linear mapping of $[x_n]$ into $[y_n]$ satisfying (8.2).

$1^\circ \Rightarrow 5^\circ$. Assume that (x_n, ϕ_n) $(\{\phi_n\} \subset [x_n]^*)$ and (y_n, ψ_n) $(\{\psi_n\} \subset [y_n]^*)$ are biorthogonal systems such that $\{\psi_n\}$ is total on $[y_n]$ and that we have 1°. Then, if $u \in L([x_n], [y_n])$ satisfies (8.2), we have

$$\phi_i(x_j) = \delta_{ij} = \psi_i(y_j) = \psi_i[u(x_j)] \qquad (i, j = 1, 2, \ldots),$$

whence, since $\{x_n\}$ is complete in $[x_n]$,

$$\phi_i(x) = \psi_i[u(x)] \qquad (x \in [x_n], i = 1, 2, \ldots).$$

Consequently, for every $x \in [x_j]$, the element $u(x) \in [y_n]$ is a solution of (8.6).

$5^\circ \Rightarrow 1^\circ$. Assume that (x_n, ϕ_n) $(\{\phi_n\} \subset [x_n]^*)$ and (y_n, ψ_n) $(\{\psi_n\} \subset [y_n]^*)$ are biorthogonal systems such that $\{\psi_n\}$ is total on $[y_n]$ and that we have 5°. For every $x \in [x_n]$, let us denote by $u(x)$ the element $y \in [y_n]$ corresponding to x by 5°. Then $u: x \to u(x)$ is a linear mapping of $[x_n]$ into $[y_n]$. Indeed, for arbitrary $z_1, z_2 \in [x_n]$ and scalars $\alpha_1, \alpha_2 \in K$ we have, by the linearity of ϕ_i and ψ_i $(i = 1, 2, \ldots)$,

$$\begin{aligned}\psi_i[u(\alpha_1 z_1 + \alpha_2 z_2)] &= \phi_i(\alpha_1 z_1 + \alpha_2 z_2) = \alpha_1 \phi_i(z_1) + \alpha_2 \phi_i(z_2) \\ &= \alpha_1 \psi_i[u(z_1)] + \alpha_2 \psi_i[u(z_2)] \\ &= \psi_i[\alpha_1 u(z_1) + \alpha_2 u(z_2)] \qquad (i = 1, 2, \ldots),\end{aligned}$$

whence, since $\{\psi_n\}$ is total on $[y_n]$,

$$u(\alpha_1 z_1 + \alpha_2 z_2) = \alpha_1 u(z_1) + \alpha_2 u(z_2).$$

We claim that u is continuous on $[x_n]$, i.e. that $u \in L([x_n], [y_n])$. Indeed, let $\{z_n\} \subset [x_n]$, $z \in [x_n]$ and $z' \in [y_n]$ be such that

$$\lim_{n \to \infty} z_n = z, \qquad \lim_{n \to \infty} u(z_n) = z'.$$

Then, by the continuity of ϕ_i and ψ_i $(i = 1, 2, \ldots)$,

$$\phi_i(z) = \lim_{n \to \infty} \phi_i(z_n) = \lim_{n \to \infty} \psi_i[u(z_n)] = \psi_i(z') \qquad (i = 1, 2, \ldots),$$

whence $z' = u(z)$. Consequently, by the closed graph theorem, $u \in L([x_n], [y_n])$.

Finally, from the relations

$$\psi_i[u(x_j)] = \phi_i(x_j) = \delta_{ij} = \psi_i(y_j) \qquad (i,j=1,2,\ldots)$$

it follows, since $\{\psi_n\}$ is total on $[y_n]$, that u satisfies (8.2). Thus we have 1°.

The implication 1°⇒6° is obvious.

6°⇒4°. Assume now that $\{x_n\}$ is a basis of $[x_n]$ and that we have 6°. Then, by the implication 1°⇒2° of part a), proved above, there exist a positive integer n_0 and a constant $C>0$ such that we have (8.3). Hence, by the implication 1°⇒4° of § 7, theorem 7.1, applied to the basis $\{x_n\}$ of $[x_n]$ and to $\alpha_1=\cdots=\alpha_{n_0-1}=0$, there exists a constant C' such that we have

$$\left\|\sum_{i=n_0}^{n_0+m} \alpha_i y_i\right\| \leqslant C' \left\|\sum_{i=n_0}^{n_0+m} \alpha_i x_i\right\|$$

for all finite sequences of scalars $\alpha_{n_0}, \alpha_{n_0+1}, \ldots, \alpha_{n_0+m}$. Consequently, by the implication 2°⇒1° of part b), proved above, we have $\{x_n\}_{n_0}^{\infty} \gg \{y_n\}_{n_0}^{\infty}$.

Finally, c) and d) are immediate consequences of a) and b) respectively. This completes the proof of theorem 8.1.

Remark 8.1. Let us also mention the following alternative proof of the implication 6°⇒1° of part b). If $\{x_n\}$ is a basis of $[x_n]$, condition 6° ensures the convergence of the series

$$u(x) = \sum_{i=1}^{\infty} \alpha_i y_i$$

for all $x=\sum_{i=1}^{\infty} \alpha_i x_i \in [x_n]$. The mapping $u\colon [x_n]\to[y_n]$ defined in this way is linear and satisfies (8.2). Furthermore, since we have $u(x)=\lim_{n\to\infty} u_n(x)$ $(x\in E)$, where $u_n(x)=\sum_{i=1}^{n} \alpha_i y_i \left(x=\sum_{i=1}^{\infty} \alpha_i x_i \in [x_n],\, n=1,2,\ldots\right)$, and since by § 3, theorem 3.1, each u_n is continuous, from the Banach-Steinhaus theorem it follows that u is continuous. This completes the proof.

In the implication 6°⇒1° of b) it is essential to assume that $\{x_n\}$ is a basis of $[x_n]$, and in the implication 6°⇒1° of d) it is essential to assume that both $\{x_n\}$ and $\{y_n\}$ are bases of $[x_n]$ and $[y_n]$ respectively. Indeed, this follows e. g. from the following general result (see also theorem 8.2 below):

Proposition 8.1. *Let $\{x_n\}$ be a sequence in a Banach space E, such that $x_n\neq 0$ $(n=1,2,\ldots)$, and let A_1 be the Banach space of sequences of scalars introduced in § 3, proposition 3.1. Then*

a) *The unit vectors $e_n=\{\delta_{nj}\}_{j=1}^{\infty}$ $(n=1,2,\ldots)$ constitute a basis of A_1.*

b) *We have $\{x_n\}\sim\{e_n\}$ and $\{e_n\}\gg\{x_n\}$.*

Proof. a) If $\{\alpha_n\}\in A_1$, i. e. if $\sum_{i=1}^{\infty}\alpha_i x_i$ converges, then

$$\left\|\{\alpha_n\}-\sum_{i=1}^{m}\alpha_i e_i\right\|=\sup_{m+1\leqslant k<\infty}\left\|\sum_{i=m+1}^{k}\alpha_i x_i\right\|\to 0 \quad\text{for}\quad m\to\infty,$$

whence $\sum_{i=1}^{\infty}\alpha_i e_i$ converges to $\{\alpha_n\}$. On the other hand, if $\sum_{i=1}^{\infty}\alpha_i e_i=0$, then, by the above,

$$\|\{\alpha_n\}\|=\sup_{1\leqslant n<\infty}\left\|\sum_{i=1}^{n}\alpha_i x_i\right\|\leqslant\lim_{n\to\infty}\sup_{1\leqslant k\leqslant n}\left\|\sum_{i=1}^{k}\alpha_i x_i\right\|$$
$$=\lim_{n\to\infty}\left\|\sum_{i=1}^{n}\alpha_i e_i\right\|=\left\|\sum_{i=1}^{\infty}\alpha_i e_i\right\|=0,$$

whence $\alpha_n=0$ $(n=1,2,\ldots)$. Thus, every $\{\alpha_n\}\in A_1$ has a unique expansion $\sum_{i=1}^{\infty}\alpha_i e_i$, i. e. $\{e_n\}$ is a basis of A_1.

b) We have seen in the above proof of a) that the convergence of $\sum_{i=1}^{\infty}\alpha_i x_i$ implies that of $\sum_{i=1}^{\infty}\alpha_i e_i$. The converse implication follows from the inequality $\left\|\sum_{i=n+1}^{n+m}\alpha_i x_i\right\|\leqslant\sup_{n+1\leqslant k\leqslant n+m}\left\|\sum_{i=n+1}^{k}\alpha_i x_i\right\|=\left\|\sum_{i=n+1}^{n+m}\alpha_i e_i\right\|$ and from the completeness of E. Thus $\{x_n\}\sim\{e_n\}$. Finally, by $\left\|\sum_{i=1}^{n}\alpha_i x_i\right\|$ $\leqslant\sup_{1\leqslant k\leqslant n}\left\|\sum_{i=1}^{k}\alpha_i x_i\right\|=\left\|\sum_{i=1}^{n}\alpha_i e_i\right\|$ and the implication b) $2^\circ\Rightarrow$ b) 1° of theorem 8.1, we also have $\{e_n\}\succ\!\!\succ\{x_n\}$. This completes the proof.

Proposition 8.1 shows that if a sequence $\{x_n\}\subset E$, with $x_n\neq 0$ $(n=1,2,\ldots)$, is not a basis of $[x_n]$, then $\{x_n\}\sim\{e_n\}$ (and hence $\{x_n\}\succ\{e_n\}$) but $\{x_n\}\not\approx\{e_n\}$[1] (actually, $\{x_n\}\not\succ\!\!\succ\{e_n\}$). Thus domination and equivalence do not imply strict domination and strict equivalence respectively.

We shall now show that the situation when domination implies strict domination, i. e., the implication b) $6^\circ\Rightarrow$ b) 1° of theorem 8.1, characterizes, in a certain sense, bases.

Theorem 8.2. *Let E be a Banach space and $\{x_n\}$ a complete sequence in E, such that $x_n\neq 0$ $(n=1,2,\ldots)$. The sequence $\{x_n\}$ is a basis of E if and only if*

$$\{\beta_n\}\subset K,\quad \{x_n\}\succ\{\beta_n\}\Rightarrow\{x_n\}\succ\!\!\succ\{\beta_n\}, \tag{8.9}$$

[1] Indeed, if $\{x_n\}\approx\{e_n\}$, then there exists an isomorphism u of A_1 onto $[x_n]$, such that $u(e_n)=x_n$ $(n=1,2,\ldots)$, whence $\{x_n\}$ is a basis of $[x_n]$.

i. e., if and only if for every sequence of scalars $\{\beta_n\}$ *with the property that the series* $\sum_{i=1}^{\infty} \alpha_i \beta_i$ *converges for all* $\{\alpha_n\}$ *such that* $\sum_{i=1}^{\infty} \alpha_i x_i$ *converges, the system of equations*

$$f(x_n) = \beta_n \qquad (n=1,2,\ldots) \tag{8.10}$$

has a solution $f \in E^*$.

Proof. The necessity is nothing else but the particular case $F=K$ of the implication b) $6^\circ \Rightarrow$ b) 1° of theorem 8.1.

Conversely, assume that condition (8.9) is satisfied and consider the space A_1 of proposition 8.1. Since $\{e_n\}$ is a basis of A_1, for any $h \in A_1^*$ we have

$$h(\{\alpha_n\}) = h\left(\sum_{i=1}^{\infty} \alpha_i e_i\right) = \sum_{i=1}^{\infty} \alpha_i h(e_i) \qquad (\{\alpha_n\} \in A_1),$$

whence $\{x_n\} \succ \{h(e_n)\}$. Consequently, by condition (8.9), $\{x_n\} \gg \{h(e_n)\}$, i. e., there exists an $f \in E^*$ such that

$$f(x_n) = h(e_n) \qquad (n=1,2,\ldots). \tag{8.11}$$

Thus, for any $h \in A_1^* = [e_n]^*$ the system of equations (8.10) has a solution $f \in E^* = [x_n]^*$, whence, by the implication b) $3^\circ \Rightarrow$ b) 1° of theorem 8.1, we infer $\{x_n\} \gg \{e_n\}$. Since by proposition 8.1 $\{e_n\} \gg \{x_n\}$, it follows that $\{x_n\} \approx \{e_n\}$, whence $\{x_n\}$ is a basis of E, which completes the proof.

Proposition 8.1 above also shows that the usual equivalence of sequences does not conserve any one of the following properties: 1. completeness, 2. linear independence of any type, 3. being a basis. This is also true for sequences of the same space E, as shown by

Example 8.1. Let $\{x_n\}$ be a basis of a Banach space E and let $\{y_n\} \subset E$ be the sequence defined by

$$y_1 = x_1, \quad y_2 = 2x_1, \quad y_n = x_n \qquad (n=3,4,\ldots). \tag{8.12}$$

Then $\{y_n\}$ is not complete in E (and hence not a basis of E), not finitely linearly independent (and hence not ω-linearly independent and not minimal) and we have $\{x_n\} \sim \{y_n\}$, $\{x_n\} \not\approx \{y_n\}$, $\{y_n\} \not\gg \{x_n\}$ (since $v \in L([y_n], E)$, $v(y_1) = x_1$, $v(y_2) = x_2$ would imply $x_2 = v(y_2) = v(2x_1) = 2v(x_1) = 2v(y_1) = 2x_1$, which contradicts the assumption that $\{x_n\}$ is a basis of E).

Similarly, even a stronger condition than the strict equivalence of sequences, in which we require the existence of an $u \in L(E,F)$ satisfying (8.2) and of a $v \in L(F,E)$ satisfying $v(y_n) = x_n$ $(n=1,2,\ldots)$, does not imply their full equivalence and does not conserve completeness (and hence also the property of being a basis), as shown by

Example 8.2. Let $E=c_0$, $\{x_n\}=$ the natural basis of c_0, and let $\{y_n\}\subset E$ be the sequence defined by

$$y_n = x_{2n-1} \qquad (n=1,2,\ldots). \tag{8.13}$$

Then $\{y_n\}$ is not complete in E and we have $\{x_n\}\approx\{y_n\}$ in the above stronger sense, but $\{x_n\}\not\approx\{y_n\}$. Indeed, the correspondence $x_n\to x_{2n-1}$ generates a continuous linear mapping u of c_0 into c_0 satisfying (8.2), while the correspondence $x_{2n-1}\to x_n$, $x_{2n}\to 0$ generates a continuous linear mapping v of c_0 onto c_0 satisfying $v(y_n)=x_n$ $(n=1,2,\ldots)$, but the (unique) continuous linear mapping $u\in L(c_0,c_0)$ satisfying (8.2) maps c_0 onto the proper subspace $[x_{2n-1}]$ of c_0.

However, the strict equivalence conserves all types of linear independence of sequences considered in § 6 (and also those in Ch. II, § 11 and § 16) since they are, in a certain sense, "properties of $\{x_n\}$ with respect to $[x_n]$" (and not "with respect to E"). More generally, the strict domination also conserves these properties, as shown by

Proposition 8.2. *Let E,F be two Banach spaces and let $\{x_n\}\subset E$, $\{y_n\}\subset F$, $\{y_n\}\gg\{x_n\}$. Then*

a) *If all $x_n\neq 0$ $(n=1,2,\ldots)$, then all $y_n\neq 0$ $(n=1,2,\ldots)$.*

b) *If $\{x_n\}$ is finitely linearly independent, so is $\{y_n\}$.*

c) *If $\{x_n\}$ is ω-linearly independent, so is $\{y_n\}$.*

d) *If $\{x_n\}$ is minimal, so is $\{y_n\}$.*

Proof. Let $v\in L([y_n],[x_n])$ be such that

$$v(y_n)=x_n \qquad (n=1,2,\ldots). \tag{8.14}$$

a) If $y_n=0$, then by (8.14), $x_n=v(0)=0$.

b) If $\alpha_1,\alpha_2,\ldots,\alpha_n\in K$ are such that $\sum_{i=1}^{n}\alpha_i y_i=0$, then, by (8.14), we have $\sum_{i=1}^{n}\alpha_i x_i=v\left(\sum_{i=1}^{n}\alpha_i y_i\right)=v(0)=0$, whence, if $\{x_n\}$ is finitely linearly independent, $\alpha_1=\alpha_2=\cdots=\alpha_n=0$.

c) The proof is similar to that of part b).

d) If $\{x_n\}$ is minimal, there exists, by the implication $1°\Rightarrow 2°$ of § 6, theorem 6.1, a sequence $\{f_n\}\subset E^*$ such that $f_i(x_j)=\delta_{ij}$ $(i,j=1,2,\ldots)$. Then for the sequence $\{g_n\}\subset F^*$ defined by $g_n=v^*(f_n)$ $(n=1,2,\ldots)$ we have, by (8.14),

$$g_i(y_j)=[v^*(f_i)](y_j)=f_i[v(y_j)]=f_i(x_j)=\delta_{ij} \qquad (i,j=1,2,\ldots),$$

whence, by the implication $2°\Rightarrow 1°$ of § 6, theorem 6.1, $\{y_n\}$ is minimal, which completes the proof. Let us mention that d) can be also easily derived e.g. from the equivalence $1°\Leftrightarrow 3°$ of § 6, theorem 6.1 and the implication b) $1°\Rightarrow$ b) $2°$ of theorem 8.1.

Finally, it is obvious that the full equivalence conserves all properties of sequences considered above, since they are invariant under isomorphic mappings of E onto another space F. One can also add to these properties the more restrictive types of completeness of sequences, known under the names of $\{a_n\}$-completeness and completeness of order p. We recall that a sequence $\{x_n\}$ in a Banach space E is said to be *$\{a_n\}$-complete in E*, where $\{a_n\}$ is a given sequence of non negative numbers, provided that for every $x \in E$ and every $\varepsilon > 0$ there exists a finite sequence of scalars $\alpha_1, \alpha_2, \ldots, \alpha_n$ such that

$$\left\| x - \sum_{i=1}^{n} \alpha_i x_i \right\| < \varepsilon, \qquad \sum_{i=1}^{n} |\alpha_i| a_i < \varepsilon. \tag{8.15}$$

We also recall that the sequence $\{x_n\} \subset E$ is said to be *complete of order p in E*, where $1 \leqslant p \leqslant \infty$, provided that for every $x \in E$ and every $\varepsilon > 0$ there exists a finite sequence of scalars $\alpha_1, \alpha_2, \ldots, \alpha_n$ such that

$$\left\| x - \sum_{i=1}^{n} \alpha_i x_i \right\| < \varepsilon, \qquad \left(\sum_{i=1}^{n} |\alpha_i| \right)^{\frac{1}{q}} < \varepsilon \qquad \left(\frac{1}{p} + \frac{1}{q} = 1 \right), \tag{8.16}$$

where in the case $p = \infty$ the second inequality is to be replaced by $\max\limits_{1 \leqslant i \leqslant n} |\alpha_i| < \varepsilon$. It is obvious from these definitions that both $\{a_n\}$-completeness and completeness of order p imply completeness in the usual sense. It is also immediate that if $\{x_n\} \subset E$ and $\{y_n\} \subset F$ are fully equivalent and if $\{x_n\}$ is complete in any one of the above senses, then so is $\{y_n\}$. Indeed, e.g. if $\{x_n\}$ is $\{a_n\}$-complete, u an isomorphism of E onto F satisfying (8.2), y an arbitrary element of F and $\varepsilon > 0$, then for the $\alpha_1, \alpha_2, \ldots, \alpha_n \in K$ satisfying $\left\| u^{-1}(y) - \sum_{i=1}^{n} \alpha_i x_i \right\| < \frac{\varepsilon}{\|u\|}$, $\sum_{i=1}^{n} |\alpha_i| a_i < \varepsilon$, we have

$$\left\| y - \sum_{i=1}^{n} \alpha_i y_i \right\| = \left\| u\left[u^{-1}(y) - \sum_{i=1}^{n} \alpha_i x_i \right] \right\| < \|u\| \frac{\varepsilon}{\|u\|} = \varepsilon,$$

which shows that $\{y_n\}$ is $\{a_n\}$-complete in F.

A similar remark is also valid for most of the properties of sequences which we shall consider in the sequel, since they are invariant under isomorphic mappings of E onto another space F. However, an exception is e.g. the property of being a normalized sequence (see § 3, definition 3.2).

In finite dimensional Banach spaces all bounded bases $\{x_n\}$ (see § 3, definition 3.2) are equivalent. It is natural to ask whether the converse is also true[1], i.e. whether this property characterizes finite dimensional

[1] It is obvious that a boundedness condition is necessary, since in infinite dimensional Banach spaces a bounded basis $\{x_n\}$ is not equivalent to the basis $\{y_n\}$ defined by $y_n = n x_n$ $(n = 1, 2, \ldots)$.

Banach spaces among Banach spaces with a basis. We shall see in Ch. II, § 23, that the answer is affirmative and, moreover, that in every infinite dimensional Banach space with a basis there exist a continuum of mutually non-equivalent normalized bases.

Even for bases the condition of being equivalent (in the sense of definition 8.1) is sometimes too strong, since a transformation of the form $y_n=\lambda_n x_n$ $(n=1,2,\ldots)$, where

$$0<\inf_{1\leqslant n<\infty}|\lambda_n|\leqslant\sup_{1\leqslant n<\infty}|\lambda_n|<\infty, \tag{8.17}$$

leads to a basis $\{y_n\}$, which in general is not equivalent to the basis $\{x_n\}$. Therefore the following less restrictive condition of affine equivalence seems to be useful:

Definition 8.2. A basis $\{x_n\}$ of a Banach space E is said to *affinely dominate* (respectively, to be *affinely equivalent* to) a basis $\{y_n\}$ of a Banach space F, if there exists a sequence of scalars $\{\lambda_n\}$ satisfying (8.17), such that the basis $\{\lambda_n x_n\}$ of E dominates (respectively, is equivalent to) the basis $\{y_n\}$ of F in the usual sense, i.e. such that

$$\sum_{i=1}^{\infty}\lambda_i\alpha_i x_i \text{ converges} \Rightarrow \sum_{i=1}^{\infty}\alpha_i y_i \text{ converges} \tag{8.18}$$

(respectively, $\sum_{i=1}^{\infty}\lambda_i\alpha_i x_i$ converges $\Leftrightarrow \sum_{i=1}^{\infty}\alpha_i y_i$ converges).

Problem 8.1. Do there exist in every infinite-dimensional Banach space with a basis two normalized bases which are not affinely equivalent?

Of course, one can also consider, similarly to definition 8.1, affine domination and affine equivalence of arbitrary sequences (instead of bases) as well as the corresponding "strict" notions[1].

The following notion of equivalence is also useful (see Ch. II, § 13 and § 18):

Definition 8.3. We shall say that a basis $\{x_n\}$ of a Banach space E is *permutatively equivalent*[2] to a basis $\{y_n\}$ of a Banach space F, and we shall write $\{x_n\}\overset{\pi}{\sim}\{y_n\}$, if there exists a permutation σ of the set $\mathcal{N}=\{1,2,3,\ldots\}$ such that the sequence $\{x_{\sigma(n)}\}$ is a basis of E and that the bases $\{x_{\sigma(n)}\}$ and $\{y_n\}$ are equivalent.

---

[1] It is easy to verify that $\{x_n\}$ is affinely equivalent to $\{y_n\}$ in the sense of definition 8.2 only if $\{x_n\}$ and $\{y_n\}$ affinely dominate each other.

[2] Naturally, one can also define permutative domination, but we shall not use it in the sequel.

Remark 8.2. The relation of permutative equivalence of bases is reflexive, symmetric and transitive. Indeed, reflexivity is obvious. Assume now that $\{x_n\} \overset{\pi}{\sim} \{y_n\}$ and let σ be a permutation as in definition 8.3. Then, by theorem 8.1, implication d) $6^\circ \Rightarrow$ d) 1°, $\{x_{\sigma(n)}\} \approx \{y_n\}$, i.e., there exists an isomorphism u of E onto F such that $u(x_{\sigma(n)}) = y_n$ $(n=1,2,\dots)$. Consequently, $u(x_n) = y_{\sigma^{-1}(n)}$ $(n=1,2,\dots)$, whence $\{y_{\sigma^{-1}(n)}\}$ is a basis of F equivalent to the basis $\{x_n\}$ of E, i.e., $\{y_n\} \overset{\pi}{\sim} \{x_n\}$. Finally, if $\{x_n\} \overset{\pi}{\sim} \{y_n\} \overset{\pi}{\sim} \{z_n\}$, say $\{x_{\sigma_1(n)}\} \sim \{y_n\}$, $\{y_{\sigma_2(n)}\} \sim \{z_n\}$, then $u(x_{\sigma_1(n)}) = y_n$, $v(y_{\sigma_2(n)}) = z_n$ $(n=1,2,\dots)$ for suitable isomorphisms u, v, whence $vu(x_{\sigma_1\sigma_2(n)}) = v(y_{\sigma_2(n)}) = z_n$ $(n=1,2,\dots)$ and thus $\{x_n\} \overset{\pi}{\sim} \{z_n\}$. We shall use script letters $\mathscr{X}, \mathscr{Y}, \dots$ to denote the equivalence classes with respect to the relation of permutative equivalence.

We shall need to generalize the notion of cartesian product of two bases (§ 4, definition 4.6) to an infinity of factors. We recall that for a sequence $\{E_n\}$ of Banach spaces we denote by $(E_1 \times E_2 \times \dots)_{l^2}$ the Banach space of all sequences $y = \{y_n\}$ such that $y_n \in E_n$ $(n=1,2,\dots)$ and $\|y\| = \left(\sum_{n=1}^{\infty} \|y_n\|^2\right)^{\frac{1}{2}} < \infty$.

Proposition 8.3. *Let $\{E_n\}$ be a sequence of Banach spaces and for each n let E_n have a normalized basis $\{x_k^{(n)}\}$, such that the norms of these bases satisfy $\sup_{1 \leq n < \infty} v_{\{x_k^{(n)}\}} < \infty$. Then the sequence $\{e_k\}$ defined by*

$$e_{N^2+j} = \begin{cases} \{\underbrace{0,\dots,0}_{N}, x_j^{(N+1)}, 0, 0, \dots\} & \text{for } j = 1, \dots, N+1;\ N = 0, 1, 2, \dots, \\ \{\underbrace{0,\dots,0}_{j-N-2}, x_{N+1}^{(j-N-1)}, 0, 0, \dots\} & \text{for } j = N+2, \dots, 2N+1;\ N = 0, 1, 2, \dots \end{cases} \tag{8.19}$$

is a normalized basis of the space $E = (E_1 \times E_2 \times \cdots)_{l^2}$, of norm

$$v_{\{e_k\}} \leq \sup_{1 \leq n < \infty} v_{\{x_k^{(n)}\}}. \tag{8.20}$$

Proof. Let $y = \{y_n\} \in (E_1 \times E_2 \times \dots)_{l^2}$ be arbitrary. Then, since $y_n \in E_n$ and since $\{x_k^{(n)}\}$ is a basis of E_n $(n=1,2,\dots)$, there exist scalars $\alpha_k^{(n)}$ $(k, n = 1, 2, \dots)$ such that $y_n = \sum_{k=1}^{\infty} \alpha_k^{(n)} x_k^{(n)}$ $(n=1,2,\dots)$. Put

$$\beta_i = \alpha_k^{(n)}, \tag{8.21}$$

where k, n are defined by the natural one to one correspondence $e_i \leftrightarrow x_k^{(n)}$ induced by (8.19) (i.e., by $e_i = \{0, \dots, 0, x_k^{(n)}, 0, \dots\}$) and let $\varepsilon > 0$ be arbitrary. Then there exist an $N = N(\varepsilon)$ such that $\sum_{i=N+1}^{\infty} \|y_i\|^2 < \frac{\varepsilon^2}{4}$ and an

$m=m[N(\varepsilon)]$ such that $\left\| y_n - \sum_{k=1}^{m} \alpha_k^{(n)} x_k^{(n)} \right\| < \frac{\varepsilon^2}{4N}$ $(n=1,\dots,N)$. Since by (8.21) $\left\{ \sum_{k=1}^{m} \alpha_k^{(1)} x_k^{(1)}, \dots, \sum_{k=1}^{m} \alpha_k^{(N)} x_k^{(N)}, 0, 0, \dots \right\}$ is of the form $\sum_{i=1}^{M} \gamma_i e_i$, where $M<\infty$ and where some $\gamma_i=\beta_i$ and the other $\gamma_i=0$, we obtain

$$\begin{aligned}\left\| y - \sum_{i=1}^{M} \gamma_i e_i \right\| &\leq \| y - \{y_1, \dots, y_N, 0, 0, \dots\} \| \\ &+ \left\| \{y_1, \dots y_N, 0, 0, \dots\} - \left\{ \sum_{k=1}^{m} \alpha_k^{(1)} x_k^{(1)}, \dots, \sum_{k=1}^{m} \alpha_k^{(N)} x_k^{(N)}, 0, 0, \dots \right\} \right\| \\ &= \left(\sum_{i=N+1}^{\infty} \|y_i\|^2 \right)^{\frac{1}{2}} + \left(\sum_{n=1}^{N} \left\| y_n - \sum_{k=1}^{m} \alpha_k^{(n)} x_k^{(n)} \right\|^2 \right)^{\frac{1}{2}} < \varepsilon,\end{aligned}$$

which proves that $[e_k]=(E_1 \times E_2 \times \dots)_{l^2}$.

Now let β_i be arbitrary scalars and define $\alpha_k^{(n)}$ by (8.21). Then for any integers N and j with $1 \leq j \leq 2N+1$ we have

$$\sum_{i=1}^{N^2+j} \beta_i e_i = \begin{cases} \left\{ \sum_{k=1}^{N} \alpha_k^{(1)} x_k^{(1)}, \dots, \sum_{k=1}^{N} \alpha_k^{(N)} x_k^{(N)}, \sum_{k=1}^{j} \alpha_k^{(N+1)} x_k^{(N+1)}, 0, 0, \dots \right\} \\ \qquad \text{if } 1 \leq j \leq N+1, \\ \left\{ \sum_{k=1}^{N+1} \alpha_k^{(1)} x_k^{(1)}, \dots, \sum_{k=1}^{N+1} \alpha_k^{(j-N-1)} x_k^{(j-N-1)}, \sum_{k=1}^{N} \alpha_k^{(j-N)} x_k^{(j-N)}, \right. \\ \left. \dots, \sum_{k=1}^{N} \alpha_k^{(N)} x_k^{(N)}, \sum_{k=1}^{N+1} \alpha_k^{(N+1)} x_k^{(N+1)}, 0, 0, \dots \right\} \\ \qquad \text{if } N+2 \leq j \leq 2N+1 \end{cases}$$

(where $\sum_{k=1}^{N+1} \alpha_k^{(N+1)} x_k^{(N+1)}$ is the $(N+1)$-th coordinate, i.e., for $j=2N+1$ the term $\sum_{k=1}^{N} \alpha_k^{(j-N)} x_k^{(j-N)} = \sum_{k=1}^{N} \alpha_k^{(N+1)} x_k^{(N+1)}$ does not occur), whence

$$\left\| \sum_{i=1}^{N^2+j} \beta_i e_i \right\| = \begin{cases} \left(\sum_{n=1}^{N} \left\| \sum_{k=1}^{N} \alpha_k^{(n)} x_k^{(n)} \right\|^2 + \left\| \sum_{k=1}^{j} \alpha_k^{(N+1)} x_k^{(N+1)} \right\|^2 \right)^{\frac{1}{2}} \\ \qquad \text{if } 1 \leq j \leq N+1, \\ \left(\sum_{n=1}^{j-N-1} \left\| \sum_{k=1}^{N+1} \alpha_k^{(n)} x_k^{(n)} \right\|^2 + \sum_{n=j-N}^{N} \left\| \sum_{k=1}^{N} \alpha_k^{(n)} x_k^{(n)} \right\|^2 \right. \\ \qquad \left. + \left\| \sum_{k=1}^{N+1} \alpha_k^{(N+1)} x_k^{(N+1)} \right\|^2 \right)^{\frac{1}{2}} \\ \qquad \text{if } N+2 \leq j \leq 2N+1, \end{cases}$$

whence it follows that for any integers N_1 and j_1 with $1 \leqslant j_1 \leqslant 2N_1+1$, such that $N_1^2+j_1 \geqslant N^2+j$, we have

$$\left\| \sum_{i=1}^{N^2+j} \beta_i e_i \right\| \leqslant \sup_{1 \leqslant n < \infty} v_{\{x_k^{(n)}\}} \left\| \sum_{i=1}^{N_1^2+j_1} \beta_i e_i \right\|.$$

Consequently, by § 7, theorem 7.1, $\{e_k\}$ is a basis of $(E_1 \times E_2 \times \cdots)_{l^2}$, of norm satisfying (8.20), which (since obviously $\|e_k\|=1$, $k=1,2,\ldots$) completes the proof of proposition 8.3.

Definition 8.4. Let $\{E_n\}$ be a sequence of Banach spaces and for each n let E_n have a normalized basis $\{x_k^{(n)}\}$, such that the norms of these bases satisfy $\sup_{1 \leqslant n < \infty} v_{\{x_k^{(n)}\}} < \infty$. The basis (8.19) of $(E_1 \times E_2 \times \cdots)_{l^2}$ is called the *cartesian product* of the bases $\{x_k^{(n)}\}$ $(n=1,2,\ldots)$ and we shall denote it by $\{x_k^{(1)}\} \times \{x_k^{(2)}\} \times \cdots$.

Let us define now the operations of cartesian product and infinite power for equivalence classes of bases (with respect to the relation of permutative equivalence).

Definition 8.5. Let E, F be two Banach spaces with bases $\{x_n\}$ and $\{y_n\}$, respectively, and let $\mathscr{X}$ and $\mathscr{Y}$ denote the equivalence class of all bases which are permutatively equivalent to the bases $\{x_n\}$ and $\{y_n\}$, respectively. We shall call *cartesian product* of the classes $\mathscr{X}$ and $\mathscr{Y}$ and we shall denote by $\mathscr{X} \times \mathscr{Y}$, the equivalence class (with respect to the relation of permutative equivalence) of the basis $\{x_n\} \times \{y_n\}$ of $E \times F$ (see § 4, definition 4.6). We shall call *infinite power* of the class $\mathscr{X}$ and we shall denote by $\mathscr{X}^\infty$ the equivalence class of the basis $\{e_k\}$ of $(E_1 \times E_2 \times \cdots)_{l^2}$ defined by (8.19), where $E_n = E$ and $x_k^{(n)} = x_k$ $(n, k=1,2,\ldots)$.

Remark 8.3. The definition of the classes $\mathscr{X} \times \mathscr{Y}$ and $\mathscr{X}^\infty$ does not depend on the particular choice of the representatives from the classes $\mathscr{X}$ and $\mathscr{Y}$. Indeed, let $\{x_n'\} \overset{\pi}{\sim} \{x_n\}$ and $\{y_n'\} \overset{\pi}{\sim} \{y_n\}$ be bases of E', E and F', F, respectively. Then there exist permutations σ_1, σ_2 of $\mathscr{N} = \{1,2,3,\ldots\}$ such that $\{x'_{\sigma_1(n)}\}$ and $\{y'_{\sigma_2(n)}\}$ are bases of E' and F', respectively, and $\{x'_{\sigma_1(n)}\} \sim \{x_n\}$, $\{y'_{\sigma_2(n)}\} \sim \{y_n\}$, whence also $\{x'_{\sigma_1(n)}\} \approx \{x_n\}$, $\{y'_{\sigma_2(n)}\} \approx \{y_n\}$. Let $\{z_k\} = \{x_n\} \times \{y_n\}$, $\{z_k'\} = \{x_n'\} \times \{y_n'\}$, and define a permutation σ of $\mathscr{N}$ by

$$\sigma(2n-1) = 2\sigma_1(n) - 1, \quad \sigma(2n) = 2\sigma_2(n) \quad (n=1,2,\ldots). \tag{8.22}$$

Then $\{z'_{\sigma(k)}\} = \{x'_{\sigma_1(n)}\} \times \{y'_{\sigma_2(n)}\}$, whence $\{z'_{\sigma(k)}\}$ is a basis of $E' \times F'$. Furthermore, if u and v are isomorphisms of E' and F' onto E and F, respectively, with $u(x'_{\sigma_1(n)}) = x_n$, $v(y'_{\sigma_2(n)}) = y_n$ $(n=1,2,\ldots)$, then the mapping

$$(u \times v)(\{x, y\}) = \{u(x), v(y)\} \quad (x \in E, y \in F) \tag{8.23}$$

is an isomorphism of $E' \times F'$ onto $E \times F$, satisfying $(u \times v)(z'_{\sigma(k)}) = z_k$ $(k=1,2,\ldots)$, whence $\{z'_{\sigma(k)}\} \approx \{z_k\}$, whence $\{z'_k\} \overset{\pi}{\sim} \{z_k\}$. A similar argument shows that $\{x'_n\} \overset{\pi}{\sim} \{x_n\}$ implies $\{x'_n\} \times \{x'_n\} \times \cdots \overset{\pi}{\sim} \{x_n\} \times \{x_n\} \times \cdots$.

Proposition 8.4. *Let E, F, G be three Banach spaces with bases $\{x_n\}$, $\{y_n\}$ and $\{z_n\}$, respectively, and let $\mathscr{X}, \mathscr{Y}$ and $\mathscr{Z}$ be the equivalence class of all bases which are permutatively equivalent to the bases $\{x_n\}$, $\{y_n\}$ and $\{z_n\}$, respectively. Then*

a) $\mathscr{X} \times \mathscr{Y} = \mathscr{Y} \times \mathscr{X}$;

b) $\mathscr{X} \times (\mathscr{Y} \times \mathscr{Z}) = (\mathscr{X} \times \mathscr{Y}) \times \mathscr{Z}$;

c) $\mathscr{X} \times \mathscr{X}^\infty = \mathscr{X}^\infty$;

d) $(\mathscr{X} \times \mathscr{Y})^\infty = \mathscr{X}^\infty \times \mathscr{Y}^\infty$.

Proof. a) Let $\{z_k\} = \{x_n\} \times \{y_n\}$, i.e., $z_{2n-1} = \{x_n, 0\}$, $z_{2n} = \{0, y_n\}$ $(n=1,2,\ldots)$, and define a permutation σ of $\mathscr{N} = \{1,2,3,\ldots\}$ by

$$\sigma(2n-1) = 2n, \qquad \sigma(2n) = 2n-1 \qquad (n=1,2,\ldots). \tag{8.24}$$

Then $z_{\sigma(2n-1)} = \{0, y_n\}$, $z_{\sigma(2n)} = \{x_n, 0\}$ $(n=1,2,\ldots)$, whence $\{z_{\sigma(k)}\} \approx \{y_n\} \times \{x_n\} \subset F \times E$, by the mapping $\{x, y\} \to \{y, x\}$ $(x \in E, y \in F)$, whence $\{x_n\} \times \{y_n\} \overset{\pi}{\sim} \{y_n\} \times \{x_n\}$.

b) Since $\{x_n\} \times (\{y_n\} \times \{z_n\})$ is the basis

$$\{x_1, \{0,0\}\}, \quad \{0, \{y_1, 0\}\}, \quad \{x_2, \{0,0\}\}, \quad \{0, \{0, z_1\}\}, \ldots$$

and $(\{x_n\} \times \{y_n\}) \times \{z_n\}$ is the basis

$$\{\{x_1, 0\}, 0\}, \quad \{\{0,0\}, z_1\}, \quad \{\{0, y_1\}, 0\}, \quad \{\{0,0\}, z_2\}, \ldots$$

we have $\{x_n\} \times (\{y_n\} \times \{z_n\}) \overset{\pi}{\sim} (\{x_n\} \times \{y_n\}) \times \{z_n\}$, via a suitable permutation σ of $\mathscr{N}$ and the mapping $\{x, \{y, z\}\} \to \{\{x, y\}, z\}$ $(x \in E, y \in F, z \in G)$.

c) Since $\{x_n\} \times (\{x_n\} \times \{x_n\} \times \cdots)$ is the basis

$$\{x_1, \{0, 0, \ldots\}\}, \quad \{0, \{x_1, 0, 0, \ldots\}\}, \quad \{x_2, \{0, 0, \ldots\}\}, \quad \{0, \{0, x_1, 0, \ldots\}\}, \ldots$$

and $(\{x_n\} \times \{x_n\} \times \cdots)$ is the basis

$$\{x_1, 0, 0, \ldots\}, \quad \{0, x_1, 0, \ldots\}, \quad \{0, x_2, 0, \ldots\}, \quad \{x_2, 0, 0, \ldots\}, \ldots$$

we have $\{x_n\} \times (\{x_n\} \times \{x_n\} \times \cdots) \overset{\pi}{\sim} (\{x_n\} \times \{x_n\} \times \cdots)$, via a suitable permutation σ of $\mathscr{N}$ and the mapping $\{x, \{y, z, \ldots\}\} \to \{x, y, z, \ldots\}$ $(x, y, z, \ldots \in E)$; this mapping is an isomorphism by Ch. II, § 18, lemma 18.5.

d) Since $(\{x_n\} \times \{y_n\}) \times (\{x_n\} \times \{y_n\}) \times \cdots$ is the basis

$$\begin{aligned} &\{\{x_1, 0\}, \{0, y_1\}, \{x_2, 0\}, \{0, y_2\}, \ldots\} \times \{\{x_1, 0\}, \{0, y_1\}, \{x_2, 0\}, \{0, y_2\}, \ldots\} \times \cdots \\ &\quad = \{\{x_1, 0\}, \{0,0\}, \ldots\}, \quad \{\{0,0\}, \{x_1, 0\}, \{0,0\}, \ldots\}, \\ &\quad\quad \{\{0,0\}, \{0, y_1\}, \{0,0\}, \ldots\}, \quad \{\{0, y_1\}, \{0,0\}, \ldots\}, \ldots \end{aligned}$$

and $(\{x_n\}\times\{x_n\}\times\cdots)\times(\{y_n\}\times\{y_n\}\times\cdots)$ is the basis

$$\{\{x_1,0,0,\ldots\},\{0,0,\ldots\}\},\quad \{\{0,0,\ldots\},\{y_1,0,0,\ldots\}\},$$
$$\{\{0,x_1,0,0,\ldots\},\{0,0,\ldots\}\},\quad \{\{0,0,\ldots\},\{0,y_1,0,0,\ldots\}\},\ldots$$

we have $(\{x_n\}\times\{y_n\})\times(\{x_n\}\times\{y_n\})\times\cdots\overset{\pi}{\sim}(\{x_n\}\times\{x_n\}\times\cdots)\times(\{y_n\}\times\{y_n\}\times\cdots)$, via a suitable permutation σ of $\mathcal{N}$ and the mapping $\{\{t_1,s_1\},\{t_2,s_2\},\ldots\}\to\{\{t_1,t_2,\ldots\},\{s_1,s_2,\ldots\}\}$ $(t_1,t_2,\ldots\in E,\ s_1,s_2,\ldots\in F)$; by Ch. II, § 18, lemma 18.5, this mapping is an isomorphism, which completes the proof of proposition 8.4.

§ 9. Stability theorems of Paley-Wiener type

In this section as well as in § 10 we shall see that if a sequence $\{y_n\}$ in a Banach space E is "sufficiently near" to a given sequence $\{x_n\}$ in E, then $\{x_n\}\approx\{y_n\}$ and, if $[x_n]=E$, then $\{x_n\}\approxeq\{y_n\}$. Hence it will follow that various properties of sequences $\{x_n\}$ in a Banach space E are "stable" in the sense that they are conserved by every sequence $\{y_n\}$ "sufficiently near" to the sequence $\{x_n\}$.

In the literature, the emphasis in the formulations of stability theorems is put on the conservation of certain properties of sequences "sufficiently near" to a given sequence $\{x_n\}$. However, in our opinion, the main assertions of the stability theorems are those of the strict equivalence, or full equivalence if $[x_n]=E$, of "sufficiently near" sequences $\{x_n\}$ and $\{y_n\}$, while all assertions of "stability" are consequences of these assertions. In the sequel we shall carry through this point of view, in the formulations of stability theorems.

The implications b) γ) and b) δ) of the following theorem are called the *Paley-Wiener theorem* (see the Notes and remarks for more details):

Theorem 9.1. *Let* $\{x_n\}$ *and* $\{y_n\}$ *be sequences in a Banach space E. Assume that there exists a constant* λ, $0\leqslant\lambda<1$, *such that we have*

$$\left\|\sum_{i=1}^{n}\alpha_i(x_i-y_i)\right\|\leqslant\lambda\left\|\sum_{i=1}^{n}\alpha_i x_i\right\| \tag{9.1}$$

for all finite sequences of scalars $\alpha_1,\alpha_2,\ldots,\alpha_n$. *Then*

a) *We have* $\{x_n\}\approx\{y_n\}$. *Consequently, we have the following equivalences:*

α) $x_n\neq 0$ $(n=1,2,\ldots)$ *if and only if* $y_n\neq 0$ $(n=1,2,\ldots)$.

β) $\{x_n\}$ *is finitely linearly independent if and only if* $\{y_n\}$ *is finitely linearly independent.*

γ) $\{x_n\}$ *is* ω-*linearly independent if and only if* $\{y_n\}$ *is* ω-*linearly independent.*

δ) $\{x_n\}$ is minimal if and only if $\{y_n\}$ is minimal.

ε) $\{x_n\}$ is a basic sequence if and only if $\{y_n\}$ is a basic sequence.

b) *If $\{x_n\}$ is complete in E, we have $\{x_n\} \approxeq \{y_n\}$. Consequently, we have the following implications*[1]*:*

α) If $\{x_n\}$ is $\{a_n\}$-complete in E ($a_n \geqslant 0, n=1,2,\ldots$), so is $\{y_n\}$.

β) If $\{x_n\}$ is complete of order p in E ($1 \leqslant p \leqslant \infty$), so is $\{y_n\}$.

γ) If $\{x_n\}$ is complete in E, so is $\{y_n\}$.

δ) If $\{x_n\}$ is a basis of E, so is $\{y_n\}$.

Proof. a) From (9.1) it follows that we have

$$\left| \left\| \sum_{i=1}^{n} \alpha_i x_i \right\| - \left\| \sum_{i=1}^{n} \alpha_i y_i \right\| \right| \leqslant \lambda \left\| \sum_{i=1}^{n} \alpha_i x_i \right\|,$$

whence

$$(1-\lambda) \left\| \sum_{i=1}^{n} \alpha_i x_i \right\| \leqslant \left\| \sum_{i=1}^{n} \alpha_i y_i \right\| \leqslant (1+\lambda) \left\| \sum_{i=1}^{n} \alpha_i x_i \right\| \tag{9.2}$$

for all finite sequences of scalars $\alpha_1, \alpha_2, \ldots, \alpha_n$. Hence, by the implication d) 2° ⇒ d) 1° of § 8, theorem 8.1, we have $\{x_n\} \approx \{y_n\}$. The equivalences α) – ε) are now a consequence of § 8, definition 8.1.

b) Assume that $[x_n] = E$. Then, by part a) proved above, there exists an (unique) isomorphism u of E into E (onto $[y_n]$), satisfying (8.2). For this isomorphism we have, by (9.1), $\|I_E - u\| \leqslant \lambda$ (where I_E denotes the identical mapping of E onto E). Since $\lambda < 1$, it follows that the inverse mapping $u^{-1} = \sum_{k=0}^{\infty} (I_E - u)^k$ exists and thus u is an isomorphism of E onto E. Hence $\{x_n\} \approxeq \{y_n\}$. The implications α) – δ) are now consequences of the properties of full equivalence (see § 8). This completes the proof of theorem 9.1.

Remark 9.1. If we have (9.1) with $0 \leqslant \lambda < \frac{1}{2}$, then, since by (9.2) and (9.1) we have

$$\left\| \sum_{i=1}^{n} \alpha_i (y_i - x_i) \right\| \leqslant \lambda \left\| \sum_{i=1}^{n} \alpha_i x_i \right\| \leqslant \frac{\lambda}{1-\lambda} \left\| \sum_{i=1}^{n} \alpha_i y_i \right\|$$

for all finite sequences of scalars $\alpha_1, \alpha_2, \ldots, \alpha_n$ and since $0 \leqslant \frac{\lambda}{1-\lambda} < 1$, from part b) above it follows that we have $\{x_n\} \approxeq \{y_n\}$ whenever $\{y_n\}$

---

[1] Actually, from theorem 9.2 below it follows (see remark 9.4) that *if we have* (9.1) *and if $\{y_n\}$ is complete in E, then still $\{x_n\} \approxeq \{y_n\}$. Consequently, the converse implications to* α), β), γ), δ) *of* b) *are also true.*

is complete in E. We shall see below that here we may drop the restriction $\lambda<\frac{1}{2}$, i.e. the same conclusion holds if we have (9.1) with $0\leqslant\lambda<1$.

Remark 9.2. Actually, b) *is equivalent to* γ) *of* b). Indeed, we have seen above that b) implies γ) of b). Conversely, if $\{x_n\}$ is complete in E, then, since by a) we have $\{x_n\}\approx\{y_n\}$, from γ) of b) it follows that $\{x_n\}\approxeq\{y_n\}$. In view of this remark let us also give the following alternative proof of the implication γ) of b):

Assume that $\{x_n\}$ is complete in E but $\{y_n\}$ is not complete in E. Then $G=[y_n]\neq E$, whence, by a well known lemma of F. Riesz[1], there exists an $x\in E\setminus G$ such that $\|x\|=1, \operatorname{dist}(x,G)>\lambda$, whence

$$\|x\|<\frac{1}{\lambda}\operatorname{dist}(x,G). \tag{9.3}$$

For this x, by a corollary of the Hahn-Banach theorem[2], there exists an $f\in E^*$ such that

$$f(y_i)=0 \quad (i=1,2,\ldots),$$
$$f(x)=1,$$
$$\|f\|=\frac{1}{\operatorname{dist}(x,G)}.$$

Since the sequence $\{x_n\}$ is complete in E, there exists a sequence $\left\{\sum_{i=1}^{m_n}\alpha_i^{(n)}x_i\right\}$ such that

$$\lim_{n\to\infty}\sum_{i=1}^{m_n}\alpha_i^{(n)}x_i=x.$$

Then, taking into account (9.1), we obtain

$$1=|f(x)|=\lim_{n\to\infty}\left|f\left(\sum_{i=1}^{m_n}\alpha_i^{(n)}x_i\right)\right|=\lim_{n\to\infty}\left|f\left[\sum_{i=1}^{m_n}\alpha_i^{(n)}(x_i-y_i)\right]\right|$$
$$\leqslant\overline{\lim_{n\to\infty}}\|f\|\left\|\sum_{i=1}^{m_n}\alpha_i^{(n)}(x_i-y_i)\right\|\leqslant\frac{\lambda}{\operatorname{dist}(x,G)}\overline{\lim_{n\to\infty}}\left\|\sum_{i=1}^{m_n}\alpha_i x_i\right\|=\frac{\lambda\|x\|}{\operatorname{dist}(x,G)},$$

which contradicts (9.3) and completes the proof.

A comparing of the two proofs above of b) γ) shows the power of the operator technique used in the first proof. This operator technique is useful in stability problems, as we shall see in the sequel.

---

[1] See e.g. [10], p. 83, lemma.
[2] See e.g. [10], p. 57, lemma.

In the limit case $\lambda=1$ theorem 9.1 is no longer valid, as shown by

Example 9.1. Let $\{x_n\}$ be a basis of a Banach space E and let $y_n=0$ $(n=1,2,\ldots)$. Then $\{x_n\}$, $\{y_n\}$ satisfy (9.1) with $\lambda=1$, but $\{y_n\}$ is not even complete in E.

However, one can prove e.g. the following positive result:

Proposition 9.1. *Let $\{x_n\}$ and $\{y_n\}$ be sequences in a reflexive Banach space E, such that $\{x_n\}\succ\{y_n\}$ and that*

$$\left\|\sum_{i=1}^{\infty}\alpha_i(x_i-y_i)\right\|<\left\|\sum_{i=1}^{\infty}\alpha_i x_i\right\| \tag{9.4}$$

for all sequences of scalars $\{\alpha_n\}$ such that $\sum_{i=1}^{\infty}\alpha_i x_i\in E$. Then, if $\{x_n\}$ is a basis of E, $\{y_n\}$ is complete in E.

Proof. Assume that $\{y_n\}$ is not complete in E. Then $G=[y_n]\neq E$, whence, since E is reflexive, there exists an $x\in E\setminus G$ such that

$$\|x\|=1=\operatorname{dist}(x,G).$$

Then, by a corollary of the Hahn-Banach theorem, there exists an $f\in E^*$ satisfying $f(y_i)=0$ $(i=1,2,\ldots)$, $f(x)=1$ and $\|f\|=1$. On the other hand, since $\{x_n\}$ is a basis of E, we have an expansion $x=\sum_{i=1}^{\infty}\alpha_i x_i$, whence

$$1=|f(x)|=\left|f\left(\sum_{i=1}^{\infty}\alpha_i x_i\right)\right|=\left|f\left[\sum_{i=1}^{\infty}\alpha_i(x_i-y_i)\right]\right|$$

$$\leqslant\|f\|\left\|\sum_{i=1}^{\infty}\alpha_i(x_i-y_i)\right\|<\left\|\sum_{i=1}^{\infty}\alpha_i x_i\right\|=\|x\|=1,$$

which is impossible. This completes the proof.

The next theorem and its corollary show that condition (9.1) of theorem 9.1 can be replaced by a weaker condition which is symmetric with respect to the sequences $\{x_n\}$, $\{y_n\}$ and which implies that $\{x_n\}\approx\{y_n\}$, respectively, that $\{x_n\}\approxeq\{y_n\}$ if either $[x_n]=E$ or $[y_n]=E$.

Theorem 9.2. *Let $\{x_n\}$ and $\{y_n\}$ be sequences in a Banach space E. Assume that there exists a constant λ, $0\leqslant\lambda<1$, such that we have*

$$\left\|\sum_{i=1}^{n}\alpha_i(x_i-y_i)\right\|\leqslant\lambda\left(\left\|\sum_{i=1}^{n}\alpha_i x_i\right\|+\left\|\sum_{i=1}^{n}\alpha_i y_i\right\|\right) \tag{9.5}$$

for all finite sequences of scalars $\alpha_1,\alpha_2,\ldots,\alpha_n$. Then

a) *We have $\{x_n\}\approx\{y_n\}$. Consequently, we have the equivalences* a) $\alpha)-\varepsilon)$ *of theorem 9.1.*

b) *If either* $\{x_n\}$ *or* $\{y_n\}$ *is complete in E, then* $\{x_n\} \approxeq \{y_n\}$. *Consequently, we have the implications* b) $\alpha)-\delta)$ *of theorem 9.1 as well as the converse implications.*

Proof. a) From (9.5) it follows that we have

$$\frac{1-\lambda}{1+\lambda}\left\|\sum_{i=1}^{n}\alpha_i x_i\right\| \leqslant \left\|\sum_{i=1}^{n}\alpha_i y_i\right\| \leqslant \frac{1+\lambda}{1-\lambda}\left\|\sum_{i=1}^{n}\alpha_i x_i\right\| \tag{9.6}$$

for all finite sequences of scalars $\alpha_1, \alpha_2, \ldots, \alpha_n$. Hence, by the implication d) $2^\circ \Rightarrow$ d) 1° of § 8, theorem 8.1, we have $\{x_n\} \approx \{y_n\}$.

b) Assume that $[x_n]=E$. Then, by part a) proved above, there exists an (unique) isomorphism u of E into E (onto $[y_n]$), satisfying (8.2). For this isomorphism we have, by (9.5),

$$\|x-u(x)\| \leqslant \lambda(\|x\|+\|u(x)\|) \quad (x \in E). \tag{9.7}$$

We shall now show that u maps E *onto* E, which will complete the proof. To this end, it will be sufficient to show that u^* is one to one; indeed, then $\overline{u(E)}=E$ (since otherwise there would exist an $f \in E^*$, $f \neq 0$, with $[u^*(f)](x)=f[u(x)]=0$ for all $x \in E$, whence u^* would not be one to one), whence, $u(E)$ being complete (because E is complete and u is an isomorphism), $u(E)=\overline{u(E)}=E$.

Put

$$\varepsilon = \frac{1-\lambda}{4}. \tag{9.8}$$

We claim that

$$\|u(x)-\alpha x\| \geqslant \varepsilon\|x\| \quad (x \in E, -\infty < \alpha < \varepsilon). \tag{9.9}$$

Indeed, assume, a contrario, that there exist an $x \in E$ and an $\alpha \in (-\infty, \varepsilon)$ such that

$$\|u(x)-\alpha x\| < \varepsilon\|x\|. \tag{9.10}$$

Put $\beta=\varepsilon-\alpha$ and $y=u(x)-\alpha x$. Then $\beta>0$ and $u(x)=y+\alpha x$ $=y+(\varepsilon-\beta)x$, whence, by (9.7), (9.8), and (9.10),

$$\begin{aligned}\|u(x)-x\| &\leqslant (1-4\varepsilon)(\|x\|+\|u(x)\|) \leqslant (1-4\varepsilon)(\|x\|+\|y\|+(\varepsilon-\beta)\|x\|)\\ &< (1-4\varepsilon)(1+2\varepsilon+\beta)\|x\| = [(1-2\varepsilon+\beta)-(8\varepsilon^2+4\varepsilon\beta)]\|x\|\end{aligned}$$

and, on the other hand, again by (9.10),

$$\begin{aligned}\|u(x)-x\| &\geqslant \|x\|-\|u(x)\| \geqslant \|x\|-[\|y\|+(\varepsilon-\beta)\|x\|]\\ &> \|x\|-(\varepsilon+\varepsilon-\beta)\|x\| = (1-2\varepsilon+\beta)\|x\|,\end{aligned}$$

a contradiction which proves the claim (9.9).

Now let

$$D=\left\{\rho \text{ real} \,\middle|\, \|u^*(f)-\alpha f\| > \frac{\varepsilon}{3}\|f\| \quad \text{for all} \quad f \in E^*, \alpha<\rho\right\}. \tag{9.11}$$

Then $D \neq \emptyset$, since for $\rho_0 = -\left(\|u\| + \frac{\varepsilon}{3}\right)$ and any $\alpha < \rho_0$ we have $|\alpha| > \|u\| + \frac{\varepsilon}{3}$, whence

$$\|u^*(f) - \alpha f\| \geqslant (|\alpha| - \|u\|)\|f\| > \frac{\varepsilon}{3}\|f\| \qquad (f \in E^*),$$

and thus $\rho_0 \in D$.

Furthermore, let

$$e = \sup_{\rho \in D} \rho. \tag{9.12}$$

We shall prove that $e \geqslant \varepsilon$, whence, by (9.11) with $\alpha = 0$ (and with any $\rho \in D$ such that $0 < \rho < e$),

$$\|u^*(f)\| \geqslant \frac{\varepsilon}{3}\|f\| \qquad (f \in E^*), \tag{9.13}$$

which will complete the proof.

Assume, a contrario, that $e < \varepsilon$ and let $\alpha < e$ be arbitrary. Then, by (9.9), $u - \alpha I_E$ is an isomorphism of E into E. On the other hand, since there exists a $\rho \in D$ with $\alpha < \rho < e$, we have

$$\|u^*(f) - \alpha f\| > \frac{\varepsilon}{3}\|f\| \qquad (f \in E^*), \tag{9.14}$$

whence $(u - \alpha I_E)^* = u^* - \alpha I_{E^*}$ is one to one and consequently the isomorphism $u - \alpha I_E$ maps E *onto* E (by the same argument as that made above for u). Furthermore, by (9.9),

$$\|(u^* - \alpha I_{E^*})^{-1}\| = \|[(u - \alpha I_E)^*]^{-1}\| = \|[(u - \alpha I_E)^{-1}]^*\| = \|(u - \alpha I_E)^{-1}\| \leqslant \frac{1}{\varepsilon},$$

whence

$$\|u^*(f) - \alpha f\| \geqslant \varepsilon\|f\| \qquad (f \in E^*), \tag{9.15}$$

a crucial sharpening of (9.14). In particular, for $\alpha = e - \frac{\varepsilon}{3} < e$ we obtain

$$\begin{aligned}\|u^*(f) - ef\| &= \left\|u^*(f) - \alpha f - \frac{\varepsilon}{3} f\right\| \geqslant \|u^*(f) - \alpha f\| - \frac{\varepsilon}{3}\|f\| \\ &\geqslant \varepsilon\|f\| - \frac{\varepsilon}{3}\|f\| = \frac{2\varepsilon}{3}\|f\| \qquad (f \in E^*),\end{aligned}$$

and hence for any β with $e \leqslant \beta < e + \frac{\varepsilon}{3}$ we have

$$\begin{aligned}\|u^*(f) - \beta f\| &= \|u^*(f) - ef - (\beta - e)f\| \geqslant \|u^*(f) - ef\| - (\beta - e)\|f\| \\ &> \frac{2\varepsilon}{3}\|f\| - \frac{\varepsilon}{3}\|f\| = \frac{\varepsilon}{3}\|f\| \qquad (f \in E^*),\end{aligned}$$

which, together with (9.14), shows that $e+\frac{\varepsilon}{3}\in D$. However, this contradicts the definition (9.12) of e and thus the proof of theorem 9.2 is complete.

Remark 9.3. If we have (9.5) with $0\leqslant\lambda<\frac{1}{3}$, one can give the following simple proof of b): For the isomorphism u of E onto $[y_n]$, satisfying (8.2), we have (9.7) and, by virtue of (9.6),

$$\|u\|\leqslant\frac{1+\lambda}{1-\lambda},$$

whence, taking into account $0\leqslant\lambda<\frac{1}{3}$,

$$\|I_E-u\|\leqslant\lambda(1+\|u\|)\leqslant\lambda\left(1+\frac{1+\lambda}{1-\lambda}\right)<1.$$

Consequently, there exists the inverse mapping $u^{-1}=\sum_{k=0}^{\infty}(I_E-u)^k$, and thus u is an isomorphism of E onto E and $\{x_n\}\approx\{y_n\}$. This completes the proof.

Remark 9.4. Condition (9.1) implies (9.5), and thus theorem 9.1 is also a consequence of theorem 9.2; moreover, it follows that the implications $\alpha)-\delta)$ in theorem 9.1 b) can be replaced by equivalences.

Corollary 9.1. *Let $\{x_n\}$ and $\{y_n\}$ be sequences in a Banach space E, satisfying one of the following two conditions:*

1°. *There exist three constants k,λ_1,λ_2 with $k>0, 0\leqslant\lambda_1,\lambda_2<\min(1,2^{1-\frac{1}{k}})$, such that we have*

$$\left\|\sum_{i=1}^{n}\alpha_i(x_i-y_i)\right\|\leqslant\left(\lambda_1^k\left\|\sum_{i=1}^{n}\alpha_i x_i\right\|^k+\lambda_2^k\left\|\sum_{i=1}^{n}\alpha_i y_i\right\|^k\right)^{\frac{1}{k}} \tag{9.16}$$

for all finite sequences of scalars $\alpha_1,\alpha_2,\ldots,\alpha_n$.

2°. *There exist three constants λ',μ,ν with $0\leqslant\lambda',\mu,\nu<1$, such that we have*

$$\left\|\sum_{i=1}^{n}\alpha_i(x_i-y_i)\right\|^2\leqslant\lambda'\left\|\sum_{i=1}^{n}\alpha_i x_i\right\|^2+2\mu\left\|\sum_{i=1}^{n}\alpha_i x_i\right\|\left\|\sum_{i=1}^{n}\alpha_i y_i\right\|+\nu\left\|\sum_{i=1}^{n}\alpha_i y_i\right\|^2 \tag{9.17}$$

for all finite sequences of scalars $\alpha_1,\alpha_2,\ldots,\alpha_n$.

Then we have a) *and* b) *of theorem 9.2.*

Proof. If we have (9.16), then for $\lambda=\max(\lambda_1,\lambda_2)\max\left(1,2^{\frac{1}{k}-1}\right)$ we have (9.5). If we have (9.17), then for $\lambda=[\max(\lambda',\mu,\nu)]^{\frac{1}{2}}$ we have (9.5). Hence in both cases we can apply theorem 9.2.

Remark 9.5. Actually, the conditions occurring in theorem 9.2 and corollary 9.1 are equivalent. In fact, we have seen that $1^\circ \Rightarrow (9.5)$ and $2^\circ \Rightarrow (9.5)$. On the other hand, $(9.5) \Rightarrow 1^\circ$ with $k=1, \lambda_1=\lambda_2=\lambda$, and $(9.5) \Rightarrow 2^\circ$ with $\lambda'=\mu=\nu=\lambda^2$. Consequently, *theorem* 9.2$\Leftrightarrow$*corollary* 9.1, 1° $\Leftrightarrow$*corollary* 9.1, 2°.

In the limit case $\lambda=1$ theorem 9.2 is no longer valid, as shown again by example 9.1. Corollary 9.1, 1° with $k \geqslant 1$ is no longer valid in the limit cases $\lambda_1=1$ or $\lambda_2=1$ and corollary 9.1, 2° is not valid if $\lambda'=1$ or $\nu=1$ (take $y_n=0$ $(n=1,2,\ldots)$, respectively $E=l^2$, $x_n=e_{n+1}-e_n$, $y_n=e_{n+1}$ $(n=1,2,\ldots)$, where $\{e_n\}$ is the unit vector basis of l^2). Corollary 9.1, 1° with $0<k\leqslant 1$ is no longer valid if $\lambda_1=\lambda_2=2^{1-\frac{1}{k}}$ (take $[x_n]=E$ and $y_n=u(x_n)$ $(n=1,2,\ldots)$, where $u:E\to E$ is a linear isometry such that $u(E)\neq E$).

In the above conditions (9.1), (9.5), (9.16) and (9.17) the coefficients of x_i and y_i have been the same. However, one can replace (9.1) (and similarly, (9.5), (9.16) and (9.17) respectively) by a condition in which this is no longer required. In fact, we have

Theorem 9.3. *Let* $\{x_n\}$ *and* $\{y_n\}$ *be sequences in a Banach space* E. *Assume that there exists a constant* λ, $0\leqslant\lambda<1$, *such that for every finite sequence of scalars* $\alpha_1,\alpha_2,\ldots,\alpha_n$ *there exist scalars* $\beta_1,\beta_2,\ldots,\beta_n$ *satisfying*

$$\left\|\sum_{i=1}^{h}(\alpha_i y_i-\beta_i x_i)\right\|\leqslant\lambda\left(\left\|\sum_{i=1}^{h}\beta_i x_i\right\|+\left\|\sum_{i=1}^{h}\alpha_i y_i\right\|\right)\qquad(h=1,2,\ldots,n).\tag{9.18}$$

Then, if $\{x_n\}$ *is a basis of* $[x_n]$, *the sequence* $\{y_n\}$ *is a basis of* $[y_n]$.

Proof. By (9.18) we have

$$\begin{aligned}\left\|\sum_{i=1}^{h}\alpha_i y_i\right\|&\leqslant\left\|\sum_{i=1}^{h}(\alpha_i y_i-\beta_i x_i)\right\|+\left\|\sum_{i=1}^{h}\beta_i x_i\right\|\\&\leqslant(1+\lambda)\left\|\sum_{i=1}^{h}\beta_i x_i\right\|+\lambda\left\|\sum_{i=1}^{h}\alpha_i y_i\right\|,\end{aligned}$$

whence

$$\left\|\sum_{i=1}^{h}\alpha_i y_i\right\|\leqslant\frac{1+\lambda}{1-\lambda}\left\|\sum_{i=1}^{h}\beta_i x_i\right\|\qquad(h=1,2,\ldots,n),$$

and similarly

$$\left\|\sum_{i=1}^{h}\beta_i x_i\right\|\leqslant\frac{1+\lambda}{1-\lambda}\left\|\sum_{i=1}^{h}\alpha_i y_i\right\|\qquad(h=1,2,\ldots,n).$$

On the other hand, since $\{x_n\}$ is a basis of $[x_n]$, there exists, by § 7, theorem 7.1, a constant $C \geqslant 1$ such that we have

$$\left\| \sum_{i=1}^{h} \beta_i x_i \right\| \leqslant C \left\| \sum_{i=1}^{n} \beta_i x_i \right\| \qquad (h=1,2,\ldots,n).$$

Consequently, we have

$$\begin{aligned} \left\| \sum_{i=1}^{h} \alpha_i y_i \right\| &\leqslant \frac{1+\lambda}{1-\lambda} \left\| \sum_{i=1}^{h} \beta_i x_i \right\| \leqslant \frac{1+\lambda}{1-\lambda} C \left\| \sum_{i=1}^{n} \beta_i x_i \right\| \\ &\leqslant \left(\frac{1+\lambda}{1-\lambda} \right)^2 C \left\| \sum_{i=1}^{n} \alpha_i y_i \right\| \qquad (h=1,2,\ldots,n) \end{aligned}$$

for every finite sequence of scalars $\alpha_1, \alpha_2, \ldots, \alpha_n$. Hence, by § 7, theorem 7.1, $\{y_n\}$ is a basis of $[y_n]$, which completes the proof.

Condition (9.18) is a weakening of (9.1) (and even of (9.5)), but we lose some of the conclusions of theorem 9.1, namely, (9.18) does not imply that $[x_n]$ and $[y_n]$ are isomorphic (hence (9.18) does not imply that $\{x_n\} \approx \{y_n\}$), as shown by

Example 9.2. Let $E = c_0 \times l^1$, endowed with the norm $\|\{x,y\}\| = \max(\|x\|_{c_0}, \|y\|_{l^1})$. If $\{e_n^{(i)}\}$ $(i=0,1)$ denote the unit vector bases of c_0 and l^1, respectively, let

$$x_n = \{e_n^{(0)}, 0\}, \qquad y_n = \{0, e_n^{(1)}\} \qquad (n=1,2,\ldots). \tag{9.19}$$

Then $\{x_n\}$ and $\{y_n\}$ are basic sequences in E, satisfying (9.18), but $[x_n]$ is not isomorphic to $[y_n]$ (and hence $\{x_n\} \not\approx \{y_n\}$).

Indeed, for any finite sequence of scalars $\alpha_1, \ldots, \alpha_n$ let

$$\beta_i = \sum_{j=1}^{i} |\alpha_j| \qquad (i=1,\ldots,n). \tag{9.20}$$

Then we have

$$\begin{aligned} \left\| \sum_{i=1}^{h} (\alpha_i y_i - \beta_i x_i) \right\| &= \left\| \left\{ -\sum_{i=1}^{h} \beta_i e_i^{(0)}, \sum_{i=1}^{h} \alpha_i e_i^{(1)} \right\} \right\| \\ &= \max\left(\sup_{1 \leqslant i \leqslant h} |\beta_i|, \sum_{i=1}^{h} |\alpha_i| \right) = \sum_{i=1}^{h} |\alpha_i| \\ &= \frac{1}{2} \left(\left\| \sum_{i=1}^{h} \beta_i x_i \right\| + \left\| \sum_{i=1}^{h} \alpha_i y_i \right\| \right) \qquad (h=1,\ldots,n), \end{aligned}$$

i.e., (9.18) with $\lambda = \frac{1}{2}$. However, $[x_n] \equiv c_0$ and $[y_n] \equiv l^1$ and thus $[x_n]$ is not isomorphic to $[y_n]$.

§ 10. Other stability theorems

The following theorem is a consequence of § 9, theorem 9.1, but we state it separately because of its importance for applications.

Theorem 10.1. *Let E be a Banach space,* (x_n, f_n) $(\{x_n\} \subset E, \{f_n\} \subset E^*)$ *a biorthogonal system, and* $\{y_n\}$ *a sequence in E. If*

$$\sum_{i=1}^{\infty} \|x_i - y_i\| \, \|f_i\| = \lambda < 1, \tag{10.1}$$

then we have (9.1). *Consequently,* $\{x_n\} \approx \{y_n\}$ *(and hence* $\{y_n\}$ *is minimal) and we also have* b) *of theorem* 9.2.

Proof. If we have (10.1), then, for any finite sequence of scalars $\alpha_1, \alpha_2, \ldots, \alpha_n$ we have, putting $x = \sum_{i=1}^{n} \alpha_i x_i$,

$$\left\| \sum_{i=1}^{n} \alpha_i (x_i - y_i) \right\| = \left\| \sum_{i=1}^{n} f_i(x)(x_i - y_i) \right\| \leqslant \lambda \|x\| = \lambda \left\| \sum_{i=1}^{n} \alpha_i x_i \right\|,$$

and thus we can apply § 9, theorem 9.2.

Remark 10.1. In particular, condition (10.1) is satisfied if $\sup_{1 \leqslant n < \infty} \|f_n\| < \infty$ and

$$\sum_{i=1}^{\infty} \|x_i - y_i\| < \frac{1}{\sup_{1 \leqslant n < \infty} \|f_n\|} \tag{10.2}$$

(since $\sum_{i=1}^{\infty} \|x_i - y_i\| \, \|f_i\| \leqslant \sup_{1 \leqslant n < \infty} \|f_n\| \sum_{i=1}^{\infty} \|x_i - y_i\|$) and conversely, from this particular case of theorem 10.1 it follows already theorem 10.1 in the general case (considering the sequences $\{x_n \|f_n\|\}$, $\{y_n \|f_n\|\}$).

Remark 10.2. The conclusion of theorem 10.1 also holds under weaker assumptions, namely, condition (10.1) can be replaced by any one of the weaker conditions

$$\sup_{1 \leqslant n < \infty} \sup_{\substack{x \in E \\ \|x\| \leqslant 1}} \left\| \sum_{i=1}^{n} f_i(x)(x_i - y_i) \right\| \leqslant \lambda < 1, \tag{10.3}$$

$$\sup_{1 \leqslant n < \infty} \sup_{\substack{x \in E \\ \|x\| \leqslant 1}} \sum_{i=1}^{n} \|x_i - y_i\| \, |f_i(x)| \leqslant \lambda < 1, \tag{10.4}$$

or if $\sup_{1 \leqslant n < \infty} \|f_n\| < \infty$, by

$$\sup_{\substack{f \in E^* \\ \|f\| \leqslant 1}} \sum_{i=1}^{\infty} |f(x_i - y_i)| = \mu < \frac{1}{\sup_{1 \leqslant n < \infty} \|f_n\|}. \tag{10.5}$$

Indeed, it is obvious that (10.4)⇒(10.3) and the above proof of theorem 10.1 shows that (10.3)⇒(9.1). Finally, if $\sup_{1\leqslant n<\infty}\|f_n\|<\infty$, let $\alpha_1,\ldots,\alpha_n$ be arbitrary scalars and take a $g_n\in E^*$ with $\|g_n\|=1$ such that $g_n\left(\sum_{i=1}^n \alpha_i(x_i-y_i)\right)=\left\|\sum_{i=1}^n \alpha_i(x_i-y_i)\right\|$. Then, by $\alpha_j=f_j\left(\sum_{i=1}^n \alpha_i x_i\right)$ $(j=1,\ldots,n)$ and by (10.5),

$$\left\|\sum_{i=1}^n \alpha_i(x_i-y_i)\right\|=\sum_{i=1}^n \alpha_i g_n(x_i-y_i)\leqslant \max_{1\leqslant j\leqslant n}|\alpha_j|\sum_{i=1}^n |g_n(x_i-y_i)|$$

$$\leqslant \sup_{1\leqslant j<\infty}\|f_j\|\left\|\sum_{i=1}^n \alpha_i x_i\right\|\sum_{i=1}^n |g_n(x_i-y_i)|\leqslant \mu \sup_{1\leqslant j<\infty}\|f_j\|\left\|\sum_{i=1}^n \alpha_i x_i\right\|,$$

which, since $\mu\sup_{1\leqslant j<\infty}\|f_j\|<1$ (by (10.5)), completes the proof.

If we weaken (10.1) by replacing $\lambda<1$ with $\lambda<\infty$, we obtain

Corollary 10.1. *Let E be a Banach space,* (x_n, f_n) $(\{x_n\}\subset E, \{f_n\}\subset E^*)$ *a biorthogonal system and* $\{y_n\}$ *a sequence in E. If*

$$\sum_{i=1}^{\infty}\|x_i-y_i\|\,\|f_i\|=\lambda<\infty, \tag{10.6}$$

then there exists a positive integer n_0 *such that*

a) $\{x_n\}_{n_0}^{\infty}\approx\{y_n\}_{n_0}^{\infty}$ *(and hence* $\{y_n\}_{n_0}^{\infty}$ *is minimal).*

b) *If* $\{x_n\}_{n_0}^{\infty}$ *is a basis of* $[x_n]_{n_0}^{\infty}$ *(in particular, if* $\{x_n\}$ *is a basis of E), then* $\{y_n\}_{n_0}^{\infty}$ *is a basis of* $[y_n]_{n_0}^{\infty}$.

c) *If* $\{x_n\}$ *is a basis of E, then the sequence* $\{x_n\}_1^{n_0-1}\cup\{y_n\}_{n_0}^{\infty}$ *is a basis of E, fully equivalent to* $\{x_n\}$*. Hence, in this case*[1]

$$\operatorname{codim}_E[y_n]_{n_0}^{\infty}=n_0-1,\qquad \operatorname{codim}_E[y_n]=k\leqslant n_0-1. \tag{10.7}$$

Proof. If we have (10.6), then there exists a positive integer n_0 such that $\sum_{i=n_0}^{\infty}\|x_i-y_i\|\,\|f_i\|<1$. Hence, by theorem 10.1 applied to the sequences $\{x_n\}_{n_0}^{\infty}$, $\{y_n\}_{n_0}^{\infty}$, we have a) and b). Furthermore, from theorem 10.1 applied to the sequences $\{x_n\}$, $\{x_n\}_1^{n_0-1}\cup\{y_n\}_{n_0}^{\infty}$ we have c), which completes the proof.

In the situation of c) above, we also have the following results:

Theorem 10.2. *Let E be a Banach space with a basis* $\{x_n\}$ *and let* $\{y_n\}$ *be a sequence in E, satisfying* (10.6), *where* $\{f_n\}$ *is the sequence of coefficient functionals associated to the basis* $\{x_n\}$*. Then*

---

[1] We recall that if E is a Banach space and G a closed linear subspace of E, then, by definition, $\operatorname{codim}_E G=\dim E/G$.

a) *The following statements are equivalent:*

1°. $\{y_n\}$ *is complete in* E.

2°. $\{y_n\}$ *is* ω*-linearly independent.*

3°. $\{y_n\}$ *is minimal.*

4°. $\{y_n\}$ *is a basis of* E.

5°. $\{x_n\} \approx \{y_n\}$.

b) *The sequence* $\{y_n\}$ *can be transformed to become a basis of* E, *by changing suitable* k *elements of it, where* $k = \operatorname{codim}_E[y_n] < \infty$.

c) *Among the possible relations of* ω*-linear dependence*

$$\sum_{i=1}^{\infty} \alpha_i y_i = 0 \tag{10.8}$$

there exist k *relations*

$$\sum_{i=1}^{\infty} \alpha_{ji} y_i = 0 \qquad (j = 1, 2, \ldots, k), \tag{10.9}$$

where $k = \operatorname{codim}_E[y_n]$, *such that any other relation* (10.8) *is a linear combination of the relations* (10.9).

Proof. a) 2° ⇒ 5°. By (10.6) the mapping

$$s(x) = \sum_{i=1}^{\infty} f_i(x)(x_i - y_i) \qquad (x \in E) \tag{10.10}$$

is well defined on E and it is an endomorphism of norm $\leqslant \lambda$. Consequently, the mapping $u = I_E - s$ (where I_E is the identical mapping of E onto E) is continuous. Since we have

$$u(x) = x - s(x) = \sum_{i=1}^{\infty} f_i(x) y_i \qquad (x \in E), \tag{10.11}$$

and since $\{y_n\}$ is ω-linearly independent, u is *one to one*. On the other hand, the mapping s is compact, since by (10.6) we have $\lim_{n\to\infty} \|s - v_n\| = 0$, where $v_n(x) = \sum_{i=1}^{n} f_i(x)(x_i - y_i)$ $(x \in E, n = 1, 2, \ldots)$. Consequently[1], $u = I_E - s$ maps E *onto* E. Hence, by the inversion theorem of Banach, u is an isomorphism of E onto E, which satisfies, by (10.11), $u(x_n) = y_n$ $(n = 1, 2, \ldots)$. Thus $\{x_n\} \approx \{y_n\}$.

---

[1] We apply here the following well known result (see e.g. [10], p. 154, theorem 14): If for a compact s the equation $x - s(x) = 0$ admits the unique solution $x = 0$, then for every $y \in E$ the equation $y = x - s(x)$ has a solution x.

The implications $5^\circ \Rightarrow 4^\circ \Rightarrow 3^\circ \Rightarrow 2^\circ$ and $5^\circ \Rightarrow 1^\circ$ are obvious.

$1^\circ \Rightarrow 2^\circ$. Assume that 2° is not satisfied, i.e. that we have (10.8) with $\sup_{1 \leqslant n < \infty} |\alpha_n| \neq 0$. Then there exists a positive integer j with $1 \leqslant j \leqslant n_0 - 1$ such that $\alpha_j \neq 0$, where n_0 is as in corollary 10.1 (since otherwise $\sup_{1 \leqslant n < \infty} |\alpha_n| = 0$, because $\{y_n\}_{n_0}^{\infty}$ is minimal). Hence $y_j = \sum_{i \neq j} \left(-\frac{\alpha_i}{\alpha_j}\right) y_i \in [y_n]_{n \neq j}$ and consequently $[y_n] = [y_n]_{n \neq j}$. Since by corollary 10.1 c) we have $\operatorname{codim}_E [y_n]_{n_0}^{\infty} = n_0 - 1$ and since $1 \leqslant j \leqslant n_0 - 1$, it follows that $\operatorname{codim}_E [y_n] \geqslant 1$, i.e. that 1° is not satisfied.

b) If $k = \operatorname{codim}_E [y_n] = 0$, then, by the implication $1^\circ \Rightarrow 4^\circ$ of part a), proved above, $\{y_n\}$ is a basis of E. If $1 \leqslant k \leqslant n_0 - 1$, where n_0 is as in corollary 10.1, then among the elements $y_1, y_2, \ldots, y_{n_0 - 1}$ there exist $n_0 - k - 1$ elements, say $y'_1, y'_2, \ldots, y'_{n_0 - k - 1}$, such that $[y_n] = [\{y'_n\}_1^{n_0 - k - 1} \cup \{y_n\}_{n_0}^{\infty}]$ and that $\{y'_n\}_1^{n_0 - k - 1} \cup \{y_n\}_{n_0}^{\infty}$ is a basis of $[y_n]$. Thus, replacing the k elements, say $y'_{n_0 - k}, y'_{n_0 - k + 1}, \ldots, y'_{n_0 - 1}$ of the set $\{y_1, y_2, \ldots, y_{n_0 - 1}\} \setminus \{y'_1, y'_2, \ldots, y'_{n_0 - k - 1}\}$, by a basis $\{z_n\}_1^k$ of an arbitrary subspace G of E such that $G \oplus [y_n] = E$, we obtain a basis $\{z_n\}_1^k \cup \{y'_n\}_1^{n_0 - k - 1} \cup \{y_n\}_{n_0}^{\infty}$ of E.

c) If $k = \operatorname{codim}_E [y_n] = 0$, then, as we have seen above, there exists no relation (10.8) with $\sup_{1 \leqslant n < \infty} |\alpha_n| \neq 0$. If $1 \leqslant k \leqslant n_0 - 1$, where n_0 is as in corollary 10.1, rearrange the finite sequence $\{y_n\}_1^{n_0 - 1}$ into $\{y'_n\}_1^{n_0 - k - 1} \cup \{y'_n\}_{n_0 - k}^{n_0 - 1}$ as in part b) above. Then, since $y'_{n_0 - k}, \ldots, y'_{n_0 - 1} \in [\{y'_n\}_1^{n_0 - k - 1} \cup \{y_n\}_{n_0}^{\infty}]$ and since $\{y'_n\}_1^{n_0 - k - 1} \cup \{y_n\}_{n_0}^{\infty}$ is a basis of $[\{y'_n\}_1^{n_0 - k - 1} \cup \{y_n\}_{n_0}^{\infty}]$, there exist k relations of the form

$$y'_{n_0 - j} = \sum_{i=1}^{n_0 - k - 1} \beta_{n_0 - j, i} y'_i + \sum_{i = n_0}^{\infty} \beta_{n_0 - j, i} y_i \qquad (j = 1, \ldots, k), \tag{10.12}$$

which, obviously, can be rewritten in the form (10.9). Consider now an arbitrary relation of the form (10.8). Then

$$-\sum_{j=1}^{k} \alpha'_{n_0 - j} y'_{n_0 - j} = \sum_{i=1}^{n_0 - k - 1} \alpha'_i y'_i + \sum_{i = n_0}^{\infty} \alpha_i y_i, \tag{10.13}$$

where $\{\alpha'_n\}_1^{n_0 - 1}$ is the rearrangement of $\{\alpha_n\}_1^{n_0 - 1}$ corresponding to $\{y'_n\}_1^{n_0 - 1}$, and since $\{y'_n\}_1^{n_0 - k - 1} \cup \{y_n\}_{n_0}^{\infty}$ is a basis of $[\{y'_n\}_1^{n_0 - k - 1} \cup \{y_n\}_{n_0}^{\infty}]$, the coefficients α'_i, α_i in the second member of (10.13) are uniquely determined. Hence, taking into account (10.12), we obtain

$$\alpha'_i = -\sum_{j=1}^{k} \alpha'_{n_0 - j} \beta_{n_0 - j, i} \qquad (i = 1, \ldots, n_0 - k - 1),$$

$$\alpha_i = -\sum_{j=1}^{k} \alpha'_{n_0 - j} \beta_{n_0 - j, i} \qquad (i = n_0, n_0 + 1, \ldots).$$

Consequently,

$$0=\sum_{i=1}^{\infty}\alpha_i y_i=\sum_{j=1}^{k}\alpha'_{n_0-j}y'_{n_0-j}-\sum_{i=1}^{n_0-k-1}\sum_{j=1}^{k}\alpha'_{n_0-j}\beta_{n_0-j,i}y'_i-\sum_{i=n_0}^{\infty}\sum_{j=1}^{k}\alpha'_{n_0-j}\beta_{n_0-j,i}y_i$$

$$=\sum_{j=1}^{k}\alpha'_{n_0-j}\left[y'_{n_0-j}-\sum_{i=1}^{n_0-k-1}\beta_{n_0-j,i}y'_i-\sum_{i=n_0}^{\infty}\beta_{n_0-j,i}y_i\right],$$

which completes the proof.

Remark 10.3. In particular, condition (10.6) of corollary 10.1 and theorem 10.2 is satisfied if $\sup_{1\leqslant n<\infty}\|f_n\|<\infty$ and

$$\sum_{i=1}^{\infty}\|x_i-y_i\|<\infty \tag{10.14}$$

(and in the particular case when $0<\inf_{1\leqslant n<\infty}\|f_n\|\leqslant\sup_{1\leqslant n<\infty}\|f_n\|<\infty$, conditions (10.6) and (10.14) are equivalent, since in this case we also have $\sum_{i=1}^{\infty}\|x_i-y_i\|\leqslant\frac{1}{\inf_{1\leqslant n<\infty}\|f_n\|}\sum_{i=1}^{\infty}\|x_i-y_i\|\,\|f_i\|$) and conversely, this particular case of corollary 10.1 or theorem 10.2 implies already corollary 10.1 or theorem 10.2 in the general case (considering the sequences $\{x_n\|f_n\|\}$, $\{y_n\|f_n\|\}$).

Remark 10.4. The conclusions of corollary 10.1 and theorem 10.2 also hold under weaker assumptions, namely, condition (10.6) can be replaced by any one of the weaker conditions

$$\lim_{n\to\infty}\sup_{\substack{x\in E\\\|x\|\leqslant 1}}\left\|\sum_{i=n+1}^{\infty}f_i(x)(x_i-y_i)\right\|=0, \tag{10.15}$$

$$\sum_{i=1}^{\infty}\|x_i-y_i\|f_i \quad \text{is unconditionally convergent,} \tag{10.16}$$

or, if $\sup_{1\leqslant n<\infty}\|f_n\|<\infty$, by

$$\sum_{i=1}^{\infty}(x_i-y_i) \qquad \text{is unconditionally convergent.} \tag{10.17}$$

Indeed, (10.16) implies (by Ch. II, § 16, lemma 16.1)

$$\lim_{n\to\infty}\sup_{\substack{x\in E\\\|x\|\leqslant 1}}\sum_{i=n+1}^{\infty}\|x_i-y_i\|\,|f_i(x)|=0,$$

which obviously implies (10.15). Furthermore, if we have (10.15), then there exists a positive integer n_0 such that

$$\sup_{n_0 \leqslant n < \infty} \sup_{\substack{x \in E \\ \|x\| \leqslant 1}} \left\| \sum_{i=n_0}^{n} f_i(x)(x_i - y_i) \right\| < 1,$$

i.e., we have (10.3) for the pairs of sequences $\{x_n\}_{n_0}^{\infty}$, $\{y_n\}_{n_0}^{\infty}$ and $\{x_n\}$, $\{x_n\}_1^{n_0-1} \cup \{y_n\}_{n_0}^{\infty}$, whence, by remark 10.2, the desired conclusion follows. Finally, (10.17) implies (by Ch. II, § 16, lemma 16.1)

$$\lim_{n \to \infty} \sup_{\substack{f \in E^* \\ \|f\| \leqslant 1}} \sum_{i=n+1}^{\infty} |f(x_i - y_i)| = 0,$$

whence, if $\sup_{1 \leqslant n < \infty} \|f_n\| < \infty$, there exists a positive integer n_0 such that

$$\sup_{\substack{f \in E^* \\ \|f\| \leqslant 1}} \sum_{i=n_0}^{\infty} |f(x_i - y_i)| = \mu < \frac{1}{\sup_{1 \leqslant n < \infty} \|f_n\|},$$

i.e., we have (10.5) for the pairs $\{x_n\}_{n_0}^{\infty}$, $\{y_n\}_{n_0}^{\infty}$ and $\{x_n\}$, $\{x_n\}_1^{n_0-1} \cup \{y_n\}_{n_0}^{\infty}$.

Let us also observe that in the case when $\{x_n\}$ is a basis of E, condition $\sup_{1 \leqslant n < \infty} \|f_n\| < \infty$ in remarks 10.1–10.4 can be replaced by the equivalent condition $\inf_{1 \leqslant n < \infty} \|x_n\| > 0$ (by § 3, corollary 3.1 a)).

The following theorem and its corollary 10.2 below will be called *the Krein-Milmann-Rutman theorem* (see the Notes and remarks for motivation):

Theorem 10.3. *Let E be a Banach space and let $\{x_n\}$ be a minimal sequence in E. Then there exists a sequence of constants $\gamma_n > 0$ $(n = 1, 2, \ldots)$ with the following properties: if a sequence $\{y_n\} \subset E$ satisfies*

$$\|x_n - y_n\| \leqslant \gamma_n \qquad (n = 1, 2, \ldots), \tag{10.18}$$

then we have (9.1) *with $\lambda < 1$. Consequently, $\{x_n\} \approx \{y_n\}$ (and hence $\{y_n\}$ is minimal) and we also have* b) *of* § 9, *theorem* 9.2.

Proof. Since $\{x_n\}$ is minimal, there exists, by the implication $1° \Rightarrow 3°$ of § 6, theorem 6.1, a sequence of constants $\delta_n > 0$ $(n = 1, 2, \ldots)$ such that we have $\sum_{i=1}^{n} |\alpha_i| \delta_i \leqslant \left\| \sum_{i=1}^{n} \alpha_i x_i \right\|$ for all finite sequences of scalars $\alpha_1, \alpha_2, \ldots, \alpha_n$. Then, putting $\gamma_n = \lambda \delta_n$ $(n = 1, 2, \ldots)$, where $\lambda < 1$, we have, by (10.18),

$$\left\| \sum_{i=1}^{n} \alpha_i (x_i - y_i) \right\| \leqslant \sum_{i=1}^{n} |\alpha_i| \gamma_i \leqslant \lambda \left\| \sum_{i=1}^{n} \alpha_i x_i \right\|,$$

and thus we can apply § 9, theorem 9.2.

Remark 10.5. Let us also mention the following alternative proof of theorem 10.3. Since $\{x_n\}$ is minimal, there exists, by the implication $1^\circ \Rightarrow 2^\circ$ of § 6, theorem 6.1, a sequence $\{f_n\} \subset E^*$ such that (x_n, f_n) is a biorthogonal system. Putting in (10.18) $\gamma_n = \dfrac{1}{2^{n+1}\|f_n\|}$ $(n=1,2,\ldots)$, we have then (10.1) with $\lambda \leqslant \frac{1}{2}$, and thus we can apply theorem 10.1.

An important immediate consequence of theorem 10.3 is the following:

Corollary 10.2. *If E is a Banach space with a basis $\{x_n\}$, then in every dense subset D of E there exists a sequence $\{y_n\} \subset D$ which is a basis of the space E. In particular, there exists a basis $\{y_n\}$ of E of the form*

$$y_n = \sum_{i=1}^{m_n} \alpha_i^{(n)} x_i \qquad (n=1,2,\ldots). \tag{10.19}$$

We shall consider now the problem, which sequences $\gamma_n > 0$ $(n=1,2,\ldots)$ have, for a given minimal sequence $\{x_n\} \subset E$, the property occurring in theorem 10.3.

Theorem 10.4. *Let E be a Banach space and (x_n, f_n) $(\{x_n\} \subset E, \{f_n\} \subset E^*)$ a biorthogonal system. In order that a sequence $\gamma_n > 0$ $(n=1,2,\ldots)$ have the property that every sequence $\{y_n\} \subset E$ satisfying* (10.18) *also satisfies* (9.1) *with $\lambda < 1$ (and hence is strictly equivalent to $\{x_n\}$) it is sufficient that*

$$b = \sup_{|\varepsilon_i|=1} \sup_{1 \leqslant n < \infty} \Big\| \sum_{i=1}^{n} \varepsilon_i \gamma_i f_i \Big|_{[x_j]} \Big\| < 1. \tag{10.20}$$

In order that every sequence $\{y_n\} \subset E$ satisfying (10.18) *be finitely linearly independent (in particular, in order that every $\{y_n\} \subset E$ satisfying* (10.18) *be minimal, or strictly equivalent to $\{x_n\}$, or satisfy* (9.1)), *it is necessary that*

$$b \leqslant 1. \tag{10.21}$$

Proof. Sufficiency. Assume that (10.20) holds. Let $\{y_n\} \subset E$ satisfy (10.18) and let $\alpha_1, \ldots, \alpha_n$ be arbitrary scalars. Then, putting $x = \sum_{i=1}^{n} \alpha_i x_i$ and $\varepsilon_i = \operatorname{sign} f_i(x)$ for $f_i(x) \neq 0$, $\varepsilon_i = 1$ for $f_i(x) = 0$, we have $|\varepsilon_i| = 1$ and

$$\Big\| \sum_{i=1}^{n} \alpha_i (x_i - y_i) \Big\| = \Big\| \sum_{i=1}^{n} f_i(x)(x_i - y_i) \Big\| \leqslant \sum_{i=1}^{n} |f_i(x)| \, \|x_i - y_i\| \leqslant \sum_{i=1}^{n} |f_i(x)| \gamma_i$$

$$= \sum_{i=1}^{n} \varepsilon_i f_i(x) \gamma_i \leqslant \Big\| \sum_{i=1}^{n} \varepsilon_i \gamma_i f_i \Big|_{[x_j]} \Big\| \, \|x\| \leqslant b \Big\| \sum_{i=1}^{n} \alpha_i x_i \Big\|,$$

i.e., (9.1) with $\lambda = b < 1$.

Necessity. Assume that $b>1$. Then there exists a functional $f_0=\sum_{i=1}^{n}\varepsilon_i^{(0)}\gamma_i f_i$ with $|\varepsilon_i^{(0)}|=1$ $(i=1,\ldots,n)$ such that $\|f_0|_{[x_j]}\| = \sup_{\substack{p\in P\\ \|p\|\leqslant 1}} |f_0(p)|>1$, whence also a polynomial $p=\sum_{i=1}^{m}\alpha_i x_i\in P$ such that $|f_0(p)|\geqslant 1$, $\|p\|\leqslant 1$. Put

$$z_0=\frac{1}{f_0(p)}p=\sum_{i=1}^{m}\frac{\alpha_i}{f_0(p)}x_i=\sum_{i=1}^{m}\beta_i x_i. \tag{10.22}$$

Then $f_0(z_0)=1$ and $\|z_0\|=\dfrac{1}{|f_0(p)|}\|p\|\leqslant\dfrac{1}{|f_0(p)|}\leqslant 1$. Hence, putting

$$y_i=\begin{cases} x_i-\varepsilon_i^{(0)}\gamma_i z_0 & \text{for } i=1,\ldots,n,\\ x_i & \text{for } i=n+1,n+2,\ldots\end{cases} \tag{10.23}$$

we obtain

$$\|x_i-y_i\|\leqslant\|\varepsilon_i^{(0)}\gamma_i z_0\|=\|\gamma_i z_0\|\leqslant\gamma_i \qquad (i=1,2,\ldots)$$

and

$$\sum_{i=1}^{m}\beta_i y_i=\begin{cases}\sum_{i=1}^{n}\beta_i(x_i-\varepsilon_i^{(0)}\gamma_i z_0)+\sum_{i=n+1}^{m}\beta_i x_i=\sum_{i=1}^{m}\beta_i x_i-z_0\sum_{i=1}^{n}\beta_i\varepsilon_i^{(0)}\gamma_i\\ \qquad =z_0-z_0 f_0(z_0)=0 \ \text{ if } m>n,\\[2ex] \sum_{i=1}^{m}\beta_i(x_i-\varepsilon_i^{(0)}\gamma_i z_0)=\sum_{i=1}^{m}\beta_i x_i-z_0\sum_{i=1}^{m}\beta_i\varepsilon_i^{(0)}\gamma_i\\ \qquad =z_0-z_0 f_0(z_0)=0 \ \text{ if } m\leqslant n.\end{cases}$$

Thus, $\{y_n\}$ satisfies (10.18), but $\{y_n\}$ is not finitely linearly independent. This completes the proof of theorem 10.4.

Corollary 10.3. *Let E be a Banach space, (x_n,f_n) ($\{x_n\}\subset E$, $\{f_n\}\subset E^*$) an E-complete biorthogonal system such that $[f_n]$ contains no subspace isomorphic to c_0 and let $\gamma_n>0$ $(n=1,2,\ldots)$. The following statements are equivalent:*

1°. *There exists a positive integer n_0 such that every sequence $\{y_n\}\subset E$ satisfying* (10.18) *also satisfies* (9.1) *"from n_0 on", i.e.,*

$$\left\|\sum_{i=n_0}^{n}\alpha_i(x_i-y_i)\right\|\leqslant\lambda\left\|\sum_{i=n_0}^{n}\alpha_i x_i\right\| \tag{10.24}$$

for all scalars $\alpha_{n_0},\alpha_{n_0+1},\ldots,\alpha_n$ and some constant $\lambda<1$.

2°. *There exists a positive integer* n_0 *such that for every sequence* $\{y_n\} \subset E$ *satisfying* (10.18) *we have* $\{x_n\}_{n_0}^{\infty} \approx \{y_n\}_{n_0}^{\infty}$.

3°. *There exists a positive integer* n_0 *such that for every sequence* $\{y_n\} \subset E$ *satisfying* (10.18) *the sequence* $\{y_n\}_{n_0}^{\infty}$ *is minimal.*

4°. $\sum_{i=1}^{\infty} \gamma_i f_i$ *is unconditionally convergent.*

Proof. The implication $1° \Rightarrow 2°$ is a consequence of § 9, theorem 9.1. Since $\{x_n\}_{n_0}^{\infty}$ is minimal, the implication $2° \Rightarrow 3°$ is obvious.

$3° \Rightarrow 4°$. If we have 3°, then by $[x_j] = E$ and by the necessity part of theorem 10.4 we have

$$\sup_{|\varepsilon_i|=1} \sup_{n_0 \leqslant n < \infty} \left\| \sum_{i=n_0}^{n} \varepsilon_i \gamma_i f_i \right\| = \sup_{|\varepsilon_i|=1} \sup_{n_0 \leqslant n < \infty} \left\| \sum_{i=n_0}^{n} \varepsilon_i \gamma_i f_i \,\Big|_{[x_j]_{n_0}^{\infty}} \right\| \leqslant 1$$

and therefore

$$\sup_{|\varepsilon_i|=1} \sup_{1 \leqslant n < \infty} \left\| \sum_{i=1}^{n} \varepsilon_i \gamma_i f_i \right\| \leqslant \sum_{i=1}^{n_0-1} |\gamma_i| \, \|f_i\| + 1 < \infty. \tag{10.25}$$

Hence, by Ch. II, § 15, lemma 15.1 and lemma 15.8, $\sum_{i=1}^{\infty} \gamma_i f_i$ is unconditionally convergent.

$4° \Rightarrow 1°$. If we have 4°, then, by Ch. II, § 16, lemma 16.1, there exists a positive integer n_0 such that

$$\sup_{\substack{\Phi \in E^{**} \\ \|\Phi\| \leqslant 1}} \sum_{i=n_0}^{\infty} |\Phi(\gamma_i f_i)| \leqslant \tfrac{1}{2}.$$

Let $|\varepsilon_i| = 1$ $(i = 1, 2, \ldots)$ and n be arbitrary. Take a $\Psi \in E^{**}$ with $\|\Psi\| = 1$ such that $\Psi\left(\sum_{i=n_0}^{n} \varepsilon_i \gamma_i f_i\right) = \left\| \sum_{i=n_0}^{n} \varepsilon_i \gamma_i f_i \right\|$. Then

$$\left\| \sum_{i=n_0}^{n} \varepsilon_i \gamma_i f_i \right\| = \sum_{i=n_0}^{n} \varepsilon_i \gamma_i \Psi(f_i) \leqslant \sum_{i=n_0}^{n} |\Psi(\gamma_i f_i)| \leqslant \sup_{\substack{\Phi \in E^{**} \\ \|\Phi\| \leqslant 1}} \sum_{i=n_0}^{\infty} |\Phi(\gamma_i f_i)| \leqslant \tfrac{1}{2},$$

whence $b_{n_0} = \sup_{|\varepsilon_i|=1} \sup_{n_0 \leqslant n < \infty} \left\| \sum_{i=n_0}^{n} \varepsilon_i \gamma_i f_i \right\| \leqslant \frac{1}{2} < 1$. Consequently, by the sufficiency part of theorem 10.4 and by $[x_n] = E$, we have (10.24), which completes the proof of corollary 10.3.

In the necessity part of theorem 10.4 the condition $b \leqslant 1$ cannot be replaced by $b < 1$, as shown by

Example 10.1. Let $E = c_0$, $\{x_n\}$ = the unit vector basis of c_0, $\{f_n\}$ = the a.s.c.f. to $\{x_n\}$ and

$$\gamma_n = \frac{1}{2^n} \quad (n=1,2,\ldots). \tag{10.26}$$

Then $b = \sup_{|\varepsilon_i|=1} \sup_{1 \leqslant n < \infty} \left\| \sum_{i=1}^{n} \frac{\varepsilon_i}{2^i} f_i \right\| = \sum_{i=1}^{\infty} \frac{1}{2^i} = 1$, but for every sequence $\{y_n\} \subset E$ satisfying (10.18) we still have $\{x_n\} \approx \{y_n\}$.

Indeed, assume that $\{y_n\} \subset E$ satisfies (10.18) with $\gamma_n = \frac{1}{2^n}$. Observe first that for every $\sum_{i=1}^{\infty} \alpha_i x_i \in c_0 \backslash \{0\}$ there exists an index i_0 such that $|\alpha_{i_0}| < \max_{1 \leqslant i < \infty} |\alpha_i|$ and hence

$$\left\| \sum_{i=1}^{\infty} \alpha_i (x_i - y_i) \right\| \leqslant \sum_{i=1}^{\infty} |\alpha_i| \, \|x_i - y_i\| \leqslant \sum_{i=1}^{\infty} \frac{|\alpha_i|}{2^i} < \max_{1 \leqslant i < \infty} |\alpha_i| = \left\| \sum_{i=1}^{\infty} \alpha_i x_i \right\|. \tag{10.27}$$

Furthermore, for any scalars $\alpha_2, \ldots, \alpha_n$ we have

$$\left\| \sum_{i=2}^{n} \alpha_i (x_i - y_i) \right\| \leqslant \sum_{i=2}^{n} |\alpha_i| \, \|x_i - y_i\| \leqslant \sum_{i=2}^{n} \frac{|\alpha_i|}{2^i} \leqslant \sum_{i=2}^{n} \frac{1}{2^i} \max_{2 \leqslant j \leqslant n} |\alpha_j| \leqslant \frac{1}{2} \left\| \sum_{i=2}^{n} \alpha_i x_i \right\|$$

and hence, by § 9, theorem 9.1, $\{x_n\}_2^\infty \approx \{y_n\}_2^\infty$, i.e., there exists an isomorphism u of $[x_n]_2^\infty$ onto $[y_n]_2^\infty$ such that

$$u(x_n) = y_n \quad (n=2,3,\ldots). \tag{10.28}$$

Therefore, since $\{x_n\}$ is a basis of E, in order to prove that $\{x_n\} \approx \{y_n\}$ it is sufficient to prove that $y_1 \notin [y_n]_2^\infty$. Assume, a contrario, that $y_1 \in [y_n]_2^\infty$. Then, since by (10.28) $\{y_n\}_2^\infty$ is a basis of $[y_n]_2^\infty$, we have an expansion $y_1 = \sum_{i=2}^{\infty} \beta_i y_i$, whence, again by (10.28), $\sum_{i=2}^{\infty} \beta_i x_i = u^{-1}(y_1) \in [x_n]_2^\infty$. Consequently, taking into account (10.27), we obtain

$$\left\| x_1 - \sum_{i=2}^{\infty} \beta_i x_i \right\| = \left\| x_1 - y_1 - \sum_{i=2}^{\infty} \beta_i (x_i - y_i) \right\| < \left\| x_1 - \sum_{i=2}^{\infty} \beta_i x_i \right\|,$$

a contradiction. This proves that $\{x_n\} \approx \{y_n\}$.

Finally, assume that $[y_n] \neq E = c_0$. Then there exists a continuous linear functional

$$f(x) = \sum_{i=1}^{\infty} \eta_i \xi_i \quad (x = \{\xi_n\} \in c_0) \tag{10.29}$$

such that $\|f\| = 1$, $f(y_n) = 0$ $(n = 1,2,\ldots)$. Hence, by (10.18) for $\gamma_n = \frac{1}{2^n}$,

$$|\eta_n| = |f(x_n)| = |f(x_n - y_n)| \leqslant \|x_n - y_n\| \leqslant \frac{1}{2^n} \quad (n=1,2,\ldots),$$

which, together with $\sum_{n=1}^{\infty}|\eta_n|=\|f\|=1$, implies

$$|\eta_n|=\|x_n-y_n\| = \frac{1}{2^n} \qquad (n=1,2,\ldots). \tag{10.30}$$

However, if $y_1=\sum_{i=1}^{\infty}\beta_i^{(1)}x_i$, then

$$\left|1-|\beta_1^{(1)}|\right|\leqslant|1-\beta_1^{(1)}|=|f_1(x_1-y_1)|\leqslant\|x_1-y_1\|=\tfrac{1}{2},$$

whence $|\beta_1^{(1)}|\geqslant\frac{1}{2}$ and hence, by (10.30), $|\eta_1\beta_1^{(1)}|\geqslant\frac{1}{4}$. Furthermore, since $\lim_{n\to\infty}f_n(x_1-y_1)=0$ and $\|x_1-y_1\|=\frac{1}{2}\neq 0$, we have

$$\left|\sum_{i=2}^{\infty}\eta_i\beta_i^{(1)}\right|\leqslant\sum_{i=2}^{\infty}|\eta_i\beta_i^{(1)}|=\sum_{i=2}^{\infty}\frac{1}{2^i}|\beta_i^{(1)}|=\sum_{i=2}^{\infty}\frac{1}{2^i}|f_i(y_1)|$$

$$=\sum_{i=2}^{\infty}\frac{1}{2^i}|f_i(x_1-y_1)|<\sum_{i=2}^{\infty}\frac{1}{2^i}\|x_1-y_1\|=\frac{1}{2}\sum_{i=2}^{\infty}\frac{1}{2^i}=\frac{1}{4}.$$

Consequently,

$$0=|f(y_1)|=\left|\sum_{i=1}^{\infty}\eta_i\beta_i^{(1)}\right|\geqslant|\eta_1\beta_1^{(1)}|-\left|\sum_{i=2}^{\infty}\eta_i\beta_i^{(1)}\right|>0,$$

a contradiction, which completes the proof of the assertions of example 10.1.

In theorem 10.5 below we shall see that such an example is no longer possible if E is reflexive and that in the sufficiency part of theorem 10.4 the condition $b<1$ cannot be replaced by $b\leqslant 1$.

Let us first prove

Lemma 10.1. *Let $\sum_{i=1}^{\infty}x_i$ be an unconditionally convergent series in a Banach space E. Then the set*

$$M=\left\{\sum_{i=1}^{n}\varepsilon_i x_i\,\middle|\,|\varepsilon_i|=1 \quad (i=1,\ldots,n), \quad n=1,2,\ldots\right\} \tag{10.31}$$

is conditionally compact.

Proof. Let $\varepsilon>0$ be arbitrary. We shall find a finite ε-net for M, which will complete the proof.

By Ch. II, § 16, lemma 16.1 (implication $1^\circ\Rightarrow 6^\circ$) there exists a positive integer $N=N(\varepsilon)$ such that

$$\sup_{\substack{f\in E^*\\ \|f\|\leqslant 1}}\ \sum_{i=N+1}^{\infty}|f(x_i)|<\frac{\varepsilon}{2}. \tag{10.32}$$

Let $\{\varepsilon_i\}$ with $|\varepsilon_i|=1$ $(i=1,2,\ldots)$ be arbitrary and let p be an arbitrary positive integer. Take a $g\in E^*$ with $\|g\|=1$ such that $g\left(\sum_{i=N+1}^{N+p}\varepsilon_i x_i\right)=\left\|\sum_{i=N+1}^{N+p}\varepsilon_i x_i\right\|$. Then, by (10.32),

$$\left\|\sum_{i=N+1}^{N+p}\varepsilon_i x_i\right\|=\sum_{i=N+1}^{N+p}\varepsilon_i g(x_i)\leqslant\sum_{i=N+1}^{N+p}|g(x_i)|\leqslant\sum_{i=N+1}^{\infty}|g(x_i)|<\frac{\varepsilon}{2}. \tag{10.33}$$

On the other hand, the subspace $E_N=[x_1,\ldots,x_N]$ of E is finite-dimensional and the set

$$M_1=\left\{\sum_{i=1}^{n}\varepsilon_i x_i\ \middle|\ |\varepsilon_i|=1\quad (i=1,\ldots,n),\quad n=1,\ldots,N\right\} \tag{10.34}$$

in this subspace is bounded (e.g. by the number $\sum_{i=1}^{N}\|x_i\|$) and hence conditionally compact. Let $y_1,\ldots,y_l$ be a finite $\frac{\varepsilon}{2}$-net for M_1 and let $\sum_{i=1}^{n}\varepsilon_i x_i\in M$ be arbitrary. If $n\leqslant N$, then $\sum_{i=1}^{n}\varepsilon_i x_i\in M_1$ and hence there exists a y_j such that $\left\|\sum_{i=1}^{n}\varepsilon_i x_i-y_j\right\|<\frac{\varepsilon}{2}<\varepsilon$. If $n>N$, $n=N+p$, then $\sum_{i=1}^{N}\varepsilon_i x_i\in M_1$ and hence there exists a y_j such that $\left\|\sum_{i=1}^{N}\varepsilon_i x_i-y_j\right\|<\frac{\varepsilon}{2}$, which, together with (10.33), implies

$$\left\|\sum_{i=1}^{N+p}\varepsilon_i x_i-y_j\right\|\leqslant\left\|\sum_{i=1}^{N}\varepsilon_i x_i-y_j\right\|+\left\|\sum_{i=N+1}^{N+p}\varepsilon_i x_i\right\|<\varepsilon.$$

Thus, $y_1,\ldots,y_l$ is a finite ε-net for M, which completes the proof of lemma 10.1.

Theorem 10.5. *Let E be a reflexive Banach space, (x_n,f_n) $(\{x_n\}\subset E$, $\{f_n\}\subset E^*)$ a biorthogonal system and $\gamma_n>0$ $(n=1,2,\ldots)$. The following statements are equivalent:*

1°. *Every sequence $\{y_n\}\subset E$ satisfying* (10.18) *also satisfies* (9.1) *with $\lambda<1$.*

2° *Every sequence $\{y_n\}\subset E$ satisfying* (10.18) *is strictly equivalent to $\{x_n\}$.*

3°. *Every sequence $\{y_n\}\subset E$ satisfying* (10.18) *is minimal.*

4°. *For the number b defined by* (10.20) *we have*

$$b<1. \tag{10.35}$$

Proof. The implication $1^\circ \Rightarrow 2^\circ$ is a consequence of § 9, theorem 9.1. Since $\{x_n\}$ is minimal, the implication $2^\circ \Rightarrow 3^\circ$ is obvious.

$3^\circ \Rightarrow 4^\circ$. Assume now that we have 3°, but not 4°, i.e., $b \geqslant 1$. Then, by 3° and theorem 10.4, $b \leqslant 1$ and hence $b=1$. Therefore the set

$$\left\{ \sum_{i=1}^{n} \varepsilon_i \gamma_i f_i \Big|_{[x_j]} \;\Big|\; |\varepsilon_i| = 1 \quad (i=1,\dots,n), \quad n=1,2,\dots \right\} \tag{10.36}$$

is bounded, whence, by Ch. II, § 15, lemma 15.1 (implication $4^\circ \Rightarrow 6^\circ$) and Ch. II, § 15, lemma 15.8, the series $\sum_{i=1}^{\infty} \gamma_i f_i \Big|_{[x_j]}$ is unconditionally convergent. Consequently, by lemma 10.1, the set (10.36) is conditionally compact and hence there exists a functional $f_0 \in E^*$ with $f_0|_{[x_j]} = \lim_{n\to\infty} \sum_{i=1}^{m_n} \varepsilon_i^{(n)} \gamma_i f_i \Big|_{[x_j]}$, where $|\varepsilon_i^{(n)}| = 1$, such that

$$\| f_0|_{[x_j]} \| = \sup_{|\varepsilon_i|=1} \sup_{1 \leqslant k < \infty} \left\| \sum_{i=1}^{k} \varepsilon_i \gamma_i f_i \Big|_{[x_j]} \right\| = b = 1. \tag{10.37}$$

Since $\varepsilon_j^{(n)} \gamma_j = \sum_{i=1}^{m_n} \varepsilon_i^{(n)} \gamma_i f_i(x_j) \to f_0(x_j)$ as $n \to \infty$, the limits $\lim_{n\to\infty} \varepsilon_j^{(n)} = \varepsilon_j^{(0)} = \frac{1}{\gamma_j} f_0(x_j)$ $(j=1,2,\dots)$ exist and $|\varepsilon_j^{(0)}| = 1$ $(j=1,2,\dots)$. Hence, by (10.37) and since by our assumption $[x_j]$ is reflexive, there exists an element $z_0 = \lim_{n\to\infty} \sum_{j=1}^{l_n} \alpha_j^{(n)} x_j \in [x_j]$ such that

$$\|z_0\| = 1 = f_0(z_0) = \lim_{n\to\infty} \sum_{j=1}^{l_n} \alpha_j^{(n)} f_0(x_j) = \lim_{n\to\infty} \sum_{j=1}^{l_n} \alpha_j^{(n)} \varepsilon_j^{(0)} \gamma_j. \tag{10.38}$$

Put

$$y_i = x_i - \varepsilon_i^{(0)} \gamma_i z_0 \qquad (i=1,2,\dots). \tag{10.39}$$

Then, by (10.39) and (10.38),

$$\|x_i - y_i\| = \|\varepsilon_i^{(0)} \gamma_i z_0\| = \gamma_i \qquad (i=1,2,\dots),$$

$$\lim_{n\to\infty} \sum_{i=1}^{l_n} \alpha_i^{(n)} y_i = \lim_{n\to\infty} \sum_{i=1}^{l_n} \alpha_i^{(n)} x_i - z_0 \lim_{n\to\infty} \sum_{i=1}^{l_n} \alpha_i^{(n)} \varepsilon_i^{(0)} \gamma_i = z_0 - z_0 = 0$$

and, since $\|z_0\| = 1$, there exists at least one index i such that

$$\lim_{n\to\infty} \alpha_i^{(n)} = f_i(z_0) \neq 0.$$

Thus, $\{y_n\}$ satisfies (10.18) but $\{y_n\}$ is not minimal (by § 6, theorem 6.1, implication $1^\circ \Rightarrow 4^\circ$), a contradiction with the assumption that we have 3°.

Finally, by theorem 10.4, the implication $4^\circ \Rightarrow 1^\circ$ is valid for any Banach space E. This completes the proof of theorem 10.5.

For stability properties another terminology is also used. Namely, it is convenient to give the following definitions.

Definition 10.1. Let $\{x_n\}$ and $\{y_n\}$ be sequences in a Banach space E and let

$$\rho_n = \|x_n - y_n\| \qquad (n=1,2,\ldots). \tag{10.40}$$

The sequence $\{y_n\}$ is said to be

a) *PW-near to* $\{x_n\}$, if there exists a constant λ, $0 \leqslant \lambda < 1$, such that we have (9.1) for all finite sequences of scalars $\alpha_1, \alpha_2, \ldots, \alpha_n$;

b) *PH-near to* $\{x_n\}$, if there exist three constants k, λ_1, λ_2 with $k>0$, $0 \leqslant \lambda_1, \lambda_2 < \min(1, 2^{1-\frac{1}{k}})$ such that we have (9.16) (as we have observed in § 9, this is equivalent to the existence of a constant λ, $0 \leqslant \lambda < 1$, such that we have (9.5)) for all finite sequences of scalars $\alpha_1, \alpha_2, \ldots, \alpha_n$;

c) *N-near to* $\{x_n\}$, if there exist three constants λ', μ, ν with $0 \leqslant \lambda', \mu, \nu < 1$ such that we have (9.17) for all finite sequences of scalars $\alpha_1, \alpha_2, \ldots, \alpha_n$;

d) *in the* $\{\gamma_n\}$*-neighbourhood of* $\{x_n\}$, where $\{\gamma_n\}$ is a sequence of positive numbers, if we have

$$\rho_n \leqslant \gamma_n \qquad (n=1,2,\ldots); \tag{10.41}$$

e) *near to* $\{x_n\}$, if we have

$$\lim_{n\to\infty} \rho_n = 0; \tag{10.42}$$

f) *strictly near to* $\{x_n\}$, if we have

$$\sum_{i=1}^{\infty} \rho_i < \infty; \tag{10.43}$$

g) *weakly near to* $\{x_n\}$, if we have

$$\lim_{n\to\infty} \sup_{\substack{f\in E^* \\ \|f\| \leqslant 1}} \sum_{i=n+1}^{\infty} |f(x_i - y_i)| = 0, \tag{10.44}$$

or, equivalently (by Ch. II, § 16, lemma 16.1), if $\sum_{i=1}^{\infty} (x_i - y_i)$ is unconditionally convergent.

If (x_n, f_n) $(\{x_n\} \subset E, \{f_n\} \subset E^*)$ is a biorthogonal system, a sequence $\{y_n\} \subset E$ is said to be

h) *KL-near to* $\{x_n\}$, if

$$\sum_{i=1}^{\infty} \rho_i \|f_i\| < \infty; \tag{10.45}$$

i) *strongly KL-near to* $\{x_n\}$, if

$$\sum_{i=1}^{\infty} \rho_i \|f_i\| = \lambda < 1. \tag{10.46}$$

Definition 10.2. A property $\mathscr{P}$ of a sequence $\{x_n\}$ in a Banach space E is said to be

a) *PW-stable*, if every sequence $\{y_n\} \subset E$, which is PW-near to $\{x_n\}$, has property $\mathscr{P}$;

b) *PH-stable*, if every sequence $\{y_n\} \subset E$, which is PH-near to $\{x_n\}$, has property $\mathscr{P}$;

c) *N-stable*, if every sequences $\{y_n\} \subset E$, which is N-near to $\{x_n\}$, has property $\mathscr{P}$:

d) *stable*, if there exists a $\{\gamma_n\}$-neighbourhood V of $\{x_n\}$ such that every sequence $\{y_n\} \in V$ has property $\mathscr{P}$;

e) *strongly stable*, if every sequence $\{y_n\} \subset E$, which is near to $\{x_n\}$, has property $\mathscr{P}$;

f) *strictly stable*, if every sequence $\{y_n\} \subset E$, which is strictly near to $\{x_n\}$, has property $\mathscr{P}$;

g) *weakly stable*, if every sequence $\{y_n\} \subset E$, which is weakly near to $\{x_n\}$, has property $\mathscr{P}$.

If (x_n, f_n) $(\{x_n\} \subset E, \{f_n\} \subset E^*)$ is a biorthogonal system, a property $\mathscr{P}$ of $\{x_n\}$ is said to be

h) *strictly KL-stable*, if every sequence $\{y_n\} \subset E$ which is KL-near to $\{x_n\}$, has property $\mathscr{P}$;

i) *KL-stable*, if every sequence $\{y_n\} \subset E$, which is strongly KL-near to $\{x_n\}$, has property $\mathscr{P}$.

Then § 9, theorem 9.1 shows that the properties of a sequence $\{x_n\} \subset E$ of being α) constituted of non-zero elements, β) finitely linearly independent, γ) ω-linearly independent, δ) minimal, ε) a basic sequence, ζ) complete (of any type) in E, and η) a basis of E are PW-stable. Theorem 9.2 and corollary 9.1 show that these properties of $\{x_n\}$ are also PH-stable and N-stable. Theorem 10.1 shows that if (x_n, f_n) is a biorthogonal system, the properties $\delta)-\eta)$ of $\{x_n\}$ are KL-stable. Theorem 10.2 and remark 10.4 show that if $\{x_n\}$ is a basis of E with the a.s.c.f. $\{f_n\}$, then the property of $\{x_n\}$ of being a basis is strictly KL-stable and, if $\{x_n\}$ is a bounded basis, weakly stable, in various classes of sequences. Consequently, the property of being a bounded basis is strictly stable in the same classes of sequences. Finally, theorem 10.3 shows that the properties $\delta)-\eta)$ are stable.

Every sequence $\{y_n\} \subset E$ is in the $\{\gamma_n\}$-neighbourhood of any sequence $\{x_n\} \subset E$, for $\gamma_n = \|x_n - y_n\|$ $(n = 1, 2, \ldots)$. We also have, obviously, the implications: PW-near $\Rightarrow$ N-near; strictly near $\Rightarrow$ weakly near $\Rightarrow$ near and, if $\{x_n\}$ is a minimal sequence, strongly KL-near

$\Rightarrow KL$-near. By § 9, remark 9.5 we have N-near $\Leftrightarrow PH$-near. From theorem 10.1 it follows that strongly KL-near $\Rightarrow PW$-near, while from the proof of theorem 10.3, given in remark 10.5, it follows that if V denotes the $\left\{\frac{1}{2^{n+1}\|f_n\|}\right\}$-neighbourhood of $\{x_n\}$, then $\{y_n\}\in V \Rightarrow \{y_n\}$ is strongly KL-near to $\{x_n\}$, whence also PW-near; on the other hand, obviously, PW-near $\Rightarrow \{y_n\}$ belongs to the $\{\|x_n\|\}$-neighbourhood of $\{x_n\}$.

Concerning any property $\mathscr{P}$ of a sequence $\{x_n\}$, we have, obviously, the implications: strongly stable $\Rightarrow$ weakly stable $\Rightarrow$ strictly stable $\Rightarrow$ stable (the last of them follows taking $\{\gamma_n\}$ to be a sequence of positive numbers satisfying $\sum_{i=1}^{\infty} \gamma_i < \infty$) and strictly KL-stable $\Rightarrow KL$-stable. From the remarks made above it also follows that PH-stable $\Leftrightarrow N$-stable $\Rightarrow PW$-stable $\Rightarrow KL$-stable $\Rightarrow$ stable and that, if $\{x_n\}$ is a bounded basis, strictly KL-stable $\Leftrightarrow$ strictly stable.

Finally, let us mention that certain stability properties characterize bases among complete minimal sequences $\{x_n\}$ with $\inf_{1\leqslant n<\infty} \|x_n\| > 0$. For this purpose, let us give

Definition 10.3. A complete minimal sequence $\{x_n\}$ in a Banach space E is called

a) *strictly stable*, if every ω-linearly independent sequence $\{y_n\}\subset E$, which is strictly near to $\{x_n\}$, is complete in E;

b) *weakly stable*, if every ω-linearly independent sequence $\{y_n\}\subset E$, which is weakly near to $\{x_n\}$, is complete in E.

Theorem 10.6. *Let $\{x_n\}$ be a complete minimal sequence in a Banach space E, with $\inf_{1\leqslant n<\infty} \|x_n\| > 0$. The following statements are equivalent:*

1°. *$\{x_n\}$ is a basis of E.*
2°. *$\{x_n\}$ is strictly stable.*
3°. *$\{x_n\}$ is weakly stable.*

Moreover, if one of these statements holds, then every ω-linearly independent sequence $\{y_n\}\subset E$, which is weakly near (or, in particular, strictly near) to $\{x_n\}$, is a basis of E, equivalent to $\{x_n\}$.

Proof. The implication $1° \Rightarrow 3°$ and also the last statement of the theorem, are contained in theorem 10.2a) (implication $2° \Rightarrow 5°$), taking also into account remark 10.4.

The implication $3° \Rightarrow 2°$ is obvious.

$2° \Rightarrow 1°$. Assume that $\{x_n\}$ is weakly stable, but not a basis of E. Then there exists an $x_0 \in E$ such that $\sum_{i=1}^{\infty} f_i(x_0)x_i$ does not converge, where $\{f_n\} \subset E^*$, $f_i(x_j) = \delta_{ij}$ $(i, j = 1, 2, \ldots)$. Hence there exists an index n such that $f_n(x_0) \neq 0$. Put

$$y_i = x_i - \frac{f_n(x_i)}{f_n(x_0)} x_0 = \begin{cases} x_i & \text{for} \quad i = 1, \ldots, n-1, n+1, n+2, \ldots, \\ x_n - \dfrac{1}{f_n(x_0)} x_0 & \text{for} \quad i = n. \end{cases} \tag{10.47}$$

Then $\sum_{i=1}^{\infty} \|x_i - y_i\| = \frac{\|x_0\|}{|f_n(x_0)|} \sum_{i=1}^{\infty} |f_n(x_i)| = \frac{\|x_0\|}{|f_n(x_0)|} < \infty$, i.e., $\{y_n\}$ is strictly near to $\{x_n\}$. Furthermore, we show that $\{y_n\}$ is ω-linearly independent. Indeed, assume that $\sum_{i=1}^{\infty} \alpha_i y_i = 0$. Then, applying f_i to the relations

$$\sum_{j=1}^{\infty} \alpha_j x_j = \sum_{j=1}^{\infty} \alpha_j (x_j - y_j) = \frac{x_0}{f_n(x_0)} \sum_{j=1}^{\infty} \alpha_j f_n(x_j) = \frac{x_0}{f_n(x_0)} \alpha_n,$$

we obtain

$$\alpha_i = \frac{f_i(x_0)}{f_n(x_0)} \alpha_n \qquad (i = 1, 2, \ldots). \tag{10.48}$$

Since $\sum_{i=1}^{\infty} \alpha_i x_i$ converges and $\sum_{i=1}^{\infty} f_i(x_0) x_i$ does not converge, the equalities (10.48) are possible only if $\alpha_n = 0$, whence $\alpha_i = 0$ $(i = 1, 2, \ldots)$, which proves that $\{y_n\}$ is ω-linearly independent.

Since by our hypothesis $\{x_n\}$ is strictly stable, it follows that $\{y_n\}$ must be complete in E. However, this contradicts the relations

$$f_n(y_i) = f_n(x_i) - \frac{f_n(x_i)}{f_n(x_0)} f_n(x_0) = 0 \qquad (i = 1, 2, \ldots).$$

Thus, $2^\circ \Rightarrow 1^\circ$, which completes the proof of theorem 10.6.

Some other types of nearness of sequences and some other stability theorems will be given in Ch. II, § 11 and § 16.

§ 11. An application to the basis problem

The basis problem (see § 1, problem 1.1) seems to have a negative answer, i.e. it is probable that there exists a separable Banach space which has no basis. In this section we shall give a sufficient condition for the existence of a separable Banach space having no basis, which suggests a possible method of constructing such a space.

For a Banach space E and a closed linear subspace G of E we shall denote by $\mathscr{P}(E,G)$ the set of all continuous linear projections of E onto G (i.e. the set $\{u \in L(E,E) \mid u^2 = u,\ u(E) = G\}$). We shall denote the statement "G is a closed linear subspace of the Banach space E" by

$$G \prec E. \tag{11.1}$$

For an arbitrary Banach space E let us consider the functions

$$\phi_n(E) = \inf_{\substack{G \prec E \\ \dim G = n}} \inf_{u \in \mathscr{P}(E,G)} \|u\| \qquad (n = 1, 2, \ldots). \tag{11.2}$$

We have then

$$\phi_1(E) = 1, \qquad \phi_n(E) \geqslant 1 \qquad (n = 2, 3, \ldots). \tag{11.3}$$

If E is a Hilbert space, we have

$$\phi_n(E) = 1 \qquad (n = 1, 2, \ldots). \tag{11.4}$$

Furthermore, from § 4, theorem 4.1 it follows that *for every Banach space E with a basis $\{x_n\}$, we have*

$$\sup_{1 \leqslant n < \infty} \phi_n(E) < \infty, \tag{11.5}$$

since the partial sum operator s_n is a projection of E onto the n-dimensional subspace $P_{(n)} = [x_1, \ldots, x_n]$ of E $(n = 1, 2, \ldots)$.[1] Consequently, *if there exists a separable Banach space E with* $\sup_{1 \leqslant n < \infty} \phi_n(E) = \infty$, *then this space E has no basis*, and thus it yields a negative solution of the basis problem. However, this sufficient condition for the existence of a separable Banach space having no basis does not yield a method for constructing such a space. Making use of § 10, corollary 10.2 one can give the following more constructive result:

Theorem 11.1. *Let $\{\alpha_k\}$ be an increasing sequence of positive numbers such that* $\lim_{k \to \infty} \alpha_k = \infty$. *If there exists an increasing sequence of Banach spaces*

$$E_3 \prec E_4 \prec E_5 \prec \cdots, \tag{11.6}$$

$$\dim E_n = n \qquad (n = 3, 4, \ldots), \tag{11.7}$$

such that for every pair of positive integers k, n with $n = n(k) \geqslant k+1$ there exists a positive integer $N = N(k,n)$ with the properties

---

[1] Thus, in particular, for every Banach space E having a basis $\{x_n\}$ of norm $\nu = \sup_{1 \leqslant n < \infty} \|s_n\| = 1$, we have (11.4). Furthermore, since $\|s_n^*\| = \|s_n\|$, for the conjugate space E^* of a Banach space E having a basis we have $\sup_{1 \leqslant n < \infty} \phi_n(E^*) < \infty$ and, if $\{x_n\}$ is of norm $\nu = 1$, then $\phi_n(E^*) = 1$ $(n = 1, 2, \ldots)$.

$$N \geqslant n, \tag{11.8}$$

$$\phi_{k+1}(E_N) > \alpha_k, \tag{11.9}$$

then there exists a separable Banach space having no basis.

Proof. By (11.6) there exists a natural norm on the linear space $\bigcup_{n=3}^{\infty} E_n$. We claim that the completion $E = \left(\bigcup_{n=3}^{\infty} E_n\right)^{\wedge}$ of the normed linear space $\bigcup_{n=3}^{\infty} E_n$ is a separable Banach space having no basis.

Indeed, E is separable by (11.7). Assume now that E has a basis. Then, by § 10, corollary 10.2, $\bigcup_{n=3}^{\infty} E_n$ must contain a basis $\{x_n\}$ of E. Since $0 < \alpha_1 < \alpha_2 < \cdots$, $\lim_{k \to \infty} \alpha_k = \infty$, for the sequence of partial sum operators $\{s_n\}$ associated to the basis $\{x_n\}$ there exists, by § 4, theorem 4.1, an α_k such that

$$\|s_n(x)\| \leqslant \alpha_k \|x\| \qquad (x \in E,\ n = 1, 2, \ldots). \tag{11.10}$$

Furthermore, since $\{x_n\} \subset \bigcup_{n=3}^{\infty} E_n$, there exists a positive integer $n = n(k) \geqslant k+1$ such that $P_{(k+1)} = [x_1, \ldots, x_{k+1}] \prec E_n$. Let $N = N(k, n)$ be a positive integer with the properties (11.8) and (11.9). Then, by (11.8) and (11.6) we have $E_n \prec E_N$, whence $P_{(k+1)} \prec E_N$. Since $\dim P_{(k+1)} = k+1$, from (11.9) it follows that we have $\|u\| > \alpha_k$ for all $u \in \mathscr{P}(E_N, P_{(k+1)})$. However, for $u = s_{k+1}|_{E_N}$ this contradicts (11.10), completing the proof of theorem 11.1.

Remark 11.1. For the numbers k, N and α_k satisfying (11.9) we have

$$\frac{2^N}{N} > \frac{2^{k+1}}{k+1} \alpha_k . \tag{11.11}$$

Indeed, according to a result of F. Bohnenblust [28] we have

$$\phi_n(E_{n+1}) \leqslant 2 \frac{n}{n+1} \qquad (n = 1, 2, \ldots), \tag{11.12}$$

whence, by induction,

$$\phi_n(E_N) \leqslant 2^{N-n} \frac{n}{N} \qquad (n = 1, 2, \ldots;\ N > n), \tag{11.13}$$

which for $n = k+1$ gives, taking into account (11.9), the inequality (11.11).

Remark 11.2. From Ch. II, § 2, theorem 2.2 and the Hahn-Banach theorem it follows that for every Banach space E we have

$$\phi_n(E) \leqslant n \qquad (n=1,2,\ldots). \tag{11.14}$$

Consequently, condition (11.9) in theorem 11.1 is non-void only if we assume that

$$\alpha_k < k+1 \qquad (k=1,2,\ldots). \tag{11.15}$$

Concerning the possibility of practical application of this sufficient (but not necessary) condition for the existence of a separable Banach space having no basis we remark, however, that the effective calculation and even a suitable lower evalution of $\phi_{k+1}(E_N)$ seems to be difficult. We shall see in Chapter II, § 1, that there exists an E_3 with $\phi_2(E_3) > 1$, but for the calculation of $\phi_2(E_4)$ for an $E_4 \succ E_3$ one has to consider new two-dimensional subspaces of E_4.

§ 12. Properties of strong duality. Application: bases and sequence spaces

Let E be a Banach space with a basis $\{x_n\}$ and let $\{f_n\} \subset E^*$ be the a.s.c.f. Then it is natural to ask, what can we say about the sequence $\{f_n\}$. We shall call *properties of duality* any relations between bases $\{x_n\}$ of Banach spaces E and their a.s.c.f. $\{f_n\} \subset E^*$. When both E and E^* are endowed with their norm-topologies, we shall call these relations *properties of strong duality*; when E and E^* are endowed with the weak topology $\sigma(E, E^*)$ and the weak* topology $\sigma(E^*, E)$ respectively, we shall use the term: *properties of weak duality*. In the present section we shall turn our attention to properties of strong duality.

If $\{x_n\}$ is a basis of a Banach space E, then, in general, the a.s.c.f. $\{f_n\}$ need not be a basis of E^*, since it may even happen that E^* is non-separable (and hence E^* has no basis at all), as shown by

Example 12.1. Let $E = l^1$. Then E has a basis, but the conjugate space[1] $E^* \equiv m$ is non-separable.

However, one can also give the following positive result:

Theorem 12.1. *Let $\{x_n\}$ be a basis of a Banach space E and let $\{f_n\} \subset E^*$ be the a.s.c.f. Then $\{f_n\}$ is a basic sequence (in E*) and we have*

$$f = \sum_{i=1}^{\infty} f(x_i) f_i \qquad (f \in [f_n]). \tag{12.1}$$

---

[1] For two Banach spaces F and G we denote the assertion "F is isometrically isomorphic to G" by $F \equiv G$.

Proof. For the adjoint s_n^* of the n-th partial sum operator s_n associated to the basis $\{x_n\}$ we have

$$[s_n^*(g)](x)=g\left[\sum_{i=1}^{n} f_i(x)x_i\right]=\left[\sum_{i=1}^{n} g(x_i)f_i\right](x) \qquad (x\in E,\ g\in E^*,\ n=1,2,\ldots),$$

whence

$$s_n^*(g)=\sum_{i=1}^{n} g(x_i)f_i \qquad (g\in E^*,\ n=1,2,\ldots). \tag{12.2}$$

Hence, for every finite linear combination $g=\sum_{j=1}^{m}\beta_j f_j$ we have, by biorthogonality,

$$s_n^*(g)=\sum_{i=1}^{n} g(x_i)f_i=\sum_{i=1}^{m}\beta_i f_i=g \qquad (n=m,m+1,\ldots). \tag{12.3}$$

Now, let $f\in[f_n]$ and $\varepsilon>0$. Then there exists a finite linear combination $g=\sum_{j=1}^{m_\varepsilon}\beta_j^{(\varepsilon)}f_j$ such that

$$\|f-g\|<\frac{\varepsilon}{\nu+1},$$

where $\nu=\sup_{1\leqslant n<\infty}\|s_n\|<\infty$. Hence, taking into account $\|s_n^*\|=\|s_n\|$ $(n=1,2,\ldots)$ and (12.3), we obtain

$$\begin{aligned}\|s_n^*(f)-f\|&\leqslant\|s_n^*(f)-s_n^*(g)\|+\|s_n^*(g)-g\|+\|g-f\|\\&<\nu\frac{\varepsilon}{\nu+1}+\frac{\varepsilon}{\nu+1}=\varepsilon \qquad (n=m_\varepsilon,m_\varepsilon+1,\ldots),\end{aligned}$$

which, by (12.2), proves (12.1). On the other hand, if $\sum_{i=1}^{\infty}\alpha_i f_i=0$, then, by biorthogonality, we have

$$\alpha_j=\sum_{i=1}^{\infty}\alpha_i f_i(x_j)=0 \qquad (j=1,2,\ldots).$$

Thus, $\{f_n\}$ is a basis of $[f_n]$, which completes the proof of theorem 12.1.

In the sequel we shall denote by π the canonical mapping of E into E^{**}, i.e.

$$[\pi(x)](f)=f(x) \qquad (x\in E,\ f\in E^*). \tag{12.4}$$

Corollary 12.1. *Let E be a Banach space and let (x_n,f_n) $(\{x_n\}\subset E$, $\{f_n\}\subset E^*)$ be a biorthogonal system. If $\{f_n\}$ is a basis of the space E^*, then $\{x_n\}$ is a basis of the space E.*

Proof. By biorthogonality, the a.s.c.f. to the basis $\{f_n\}$ of E^* is nothing else but $\{\pi(x_n)\}$. Hence, by theorem 12.1 applied to the basis

$\{f_n\}$ of E^*, $\{\pi(x_n)\}$ is a basis of $\pi(E)$. Since π is a linear isometry, it follows that $\{x_n\}$ is a basis of E.

Corollary 12.2. *Let $\{x_n\}$ be a basis of a reflexive Banach space E and let $\{f_n\}\subset E^*$ be the a.s.c.f. Then $\{f_n\}$ is a basis of the space E^*.*

Proof. Since E is reflexive, we have $\pi(E)=E^{**}$, whence, since π is a linear isometry, $\{\pi(x_n)\}$ is a basis of E^{**}. Hence from corollary 12.1 applied to the biorthogonal system $(f_n, \pi(x_n))$ it follows that $\{f_n\}$ is a basis of E^*.

In Ch. II, § 4, we shall give necessary and sufficient conditions for a basis $\{x_n\}$ to have the property that the a.s.c.f. $\{f_n\}$ is a basis of E^*.

We have seen in example 12.1 that if a Banach space E has a basis, its conjugate space E^* need not have a basis. However the converse problem is unsolved:

Problem 12.1. If the conjugate space E^* (of a Banach space E) has a basis, does the space E possess a basis?

A partial result is given in corollary 12.1 above. The following related questions are also unsolved:

Problem 12.2. a) If the conjugate space E^* (of a Banach space E) is separable, does E^* possess a basis? b) If E^* is separable, does E possess a basis?

Naturally, problems 12.1 and 12.2 can be also considered as particular cases of § 1, problem 1.1. Problem 12.2b) is intermediate between problems 12.1 and 1.1 (since E^* has a basis $\Rightarrow E^*$ is separable $\Rightarrow E$ is separable).

For a linear subspace V of the conjugate space E^* (of a Banach space E) we shall denote by ϕ the canonical mapping of E into V^*, i.e. the continuous linear mapping of E into V^* defined by

$$[\phi(x)](f)=f(x) \qquad (x\in E,\ f\in V). \tag{12.5}$$

In other words, we have

$$\phi(x)=\pi(x)|_V \qquad (x\in E), \tag{12.6}$$

where, as before, π denotes the canonical mapping (12.4) of E into E^{**}.

We shall also use the notation

$$V^{\perp}=\{\Phi\in E^{**} \mid \Phi(f)=0 \text{ for all } f\in V\}. \tag{12.7}$$

Then for any *total* subspace V of E^* we have, obviously,

$$\pi(E)\cap V^{\perp}=\{0\}. \tag{12.8}$$

If E is a Banach space with a basis $\{x_n\}$, the a.s.c.f. $\{f_n\}\subset E^*$ is total on E, whence we have (12.8), ϕ is one to one, and the norm-closed linear subspace $V=[f_n]$ of E^* is dense in E^* for the weak* topology $\sigma(E^*,E)$. However, in the next theorem we shall give considerably stronger results.

Let us recall that the *characteristic* of a linear subspace V of a conjugate Banach space E^* is the greatest number $r=r(V)$ such that the unit cell $\{f\in V \mid \|f\|\leqslant 1\}$ of V is $\sigma(E^*,E)$-dense in the r-cell $\{f\in E^* \mid \|f\|\leqslant r\}$ of E^*. Obviously, $0\leqslant r(V)\leqslant 1$, but it may also happen that $r(V)=0$ for a $\sigma(E^*,E)$-dense linear subspace V of E^*. However, in the next theorem we shall see that if $V=[f_n]$, where $\{f_n\}$ is the a.s.c.f. to a basis $\{x_n\}$, then this pathological case cannot occur.

Theorem 12.2. *Let E be a Banach space with a basis $\{x_n\}$, let $V=[f_n]$ be the norm-closed linear subspace of E^* spanned by the a.s.c.f. $\{f_n\}\subset E^*$ to $\{x_n\}$ and let $v=v_{\{x_n\}}$ be the norm of the basis $\{x_n\}$, i.e., $v=\sup_{1\leqslant n<\infty}\|s_n\|$, where $\{s_n\}$ is the associated sequence of partial sum operators. Then*

a) *We have*

$$r(V)\geqslant\frac{1}{v}>0, \tag{12.9}$$

i.e., the unit cell $\{f\in V \mid \|f\|\leqslant 1\}$ of V is $\sigma(E^*,E)$-dense at least in the $\frac{1}{v}$-cell $\left\{f\in E^* \,\middle|\, \|f\|\leqslant\frac{1}{v}\right\}$ of E^*.

b) *The closure Σ_E of the unit cell $S_E=\{x\in E \mid \|x\|\leqslant 1\}$ of E for the weak topology $\sigma(E,V)$ is bounded, namely*

$$\Sigma_E\subset vS_E. \tag{12.10}$$

c) *We have*

$$\inf_{\substack{x\in E\\ x\neq 0}}\ \sup_{\substack{f\in V\\ \|f\|\leqslant 1}}\left|f\left(\frac{x}{\|x\|}\right)\right|\geqslant\frac{1}{v}>0. \tag{12.11}$$

d) *For the canonical mapping π of E into E^{**} we have*

$$\inf_{\substack{x\in E,\, x\neq 0\\ \Phi\in V^\perp}}\frac{\|\pi(x)+\Phi\|}{\|\pi(x)\|}\geqslant\frac{1}{v}>0. \tag{12.12}$$

e) *The canonical mapping ϕ of E into V^* is an isomorphism, satisfying*

$$\|x\|_V\leqslant\|x\|\leqslant\frac{1}{r(V)}\|x\|_V \qquad (x\in E), \tag{12.13}$$

where we use the notation

$$\|x\|_V=\|\phi(x)\|=\sup_{\substack{f\in V\\ \|f\|\leqslant 1}}|f(x)| \qquad (x\in E). \tag{12.14}$$

If $\{x_n\}$ is a monotone basis[1], *the above inequalities and inclusions become equalities. If $r(V)=1$ (in particular, if $\{x_n\}$ is a monotone basis), then ϕ is an isometry.*

Proof. By well known results[2] on the characteristic $r(V)$, we have

$$r(V) = \frac{1}{\sup\limits_{x\in\Sigma_E}\|x\|} = \inf_{\substack{x\in E\\ x\neq 0}} \sup_{\substack{f\in V\\ \|f\|\leq 1}} \left|f\left(\frac{x}{\|x\|}\right)\right| = \inf_{\substack{x\in E,\, x\neq 0\\ \Phi\in V^{\perp}}} \frac{\|\pi(x)+\Phi\|}{\|x\|}, \tag{12.15}$$

and thus the statements a), b), c), d) are equivalent, while e) follows from c), via $r(V)\leqslant\left\|\frac{x}{\|x\|}\right\|_V$ $(x\in E)$. Therefore it will be sufficient to prove any one of the assertions a), b), c), d). Let us prove b).

Let $\{y_d\}_{d\in\Delta}$ be a net in S_E which is $\sigma(E,V)$-convergent to an $x\in\Sigma_E$. Then $\lim\limits_{d\in\Delta} f_j(y_d)=f_j(x)$ $(j=1,2,\ldots)$, whence, in the norm-topology,

$$\lim_{d\in\Delta} s_n(y_d)=s_n(x) \qquad (n=1,2,\ldots).$$

Let $\varepsilon>0$ be arbitrary. Then there exists a $d_0=d_0(\varepsilon,n)\in\Delta$ such that

$$\|s_n(x)\|-\|s_n(y_d)\|\leqslant\|s_n(x)-s_n(y_d)\|<\varepsilon \qquad (d\geqslant d_0),$$

whence, by $\|y_d\|\leqslant 1$ $(d\in\Delta)$ and $\sup\limits_{1\leqslant n<\infty}\|s_n\|=v<\infty$,

$$\|s_n(x)\|\leqslant\|s_n(y_d)\|+\varepsilon\leqslant v+\varepsilon \qquad (n=1,2,\ldots).$$

Consequently, since $\|x\|\leqslant\sup\limits_{1\leqslant n<\infty}\|s_n(x)\|$ $(x\in E)$ and since $\varepsilon>0$ and $x\in\Sigma_E$ were arbitrary, we get

$$\|x\|\leqslant v \qquad (x\in\Sigma_E),$$

i.e., (12.10), which completes the proof of theorem 12.2.

Remark 12.1. The weaker assertion $r(V)>0$ can be also derived from § 14, theorem 14.1, formula (14.8) and a theorem of S. Banach[3].

Remark 12.2. In general, $r(V)\neq\frac{1}{v}$. Indeed, if E is reflexive, then for every basis $\{x_n\}$ of E we have, by corollary 12.2, $V=E^*$ (where $V=[f_n]$, $f_i(x_j)=\delta_{ij}$), whence $r(V)=1$, while it is easy to construct bases of E for which $\frac{1}{v}$ is arbitrarily small.

Let us also mention that in a certain sense the constant $\frac{1}{v}$ in theorem 12.2 is the best possible lower bound for $r(V)$. Indeed, *if $\{x_n\}$ is a monotone basis (i.e. with the norm $v=1$), then $r(V)=1=\frac{1}{v}$.*

---

[1] I.e., $v=1$. Such bases will be studied in Ch. II, § 1.

[2] See [47] or [33], p. 121, exercise 14.

[3] See [10], p. 213, theorem 2.

Now we shall show that $r(V)$ *can have any value* λ *such that* $0<\lambda\leqslant 1$.

Example 12.2. Let $0<\lambda\leqslant 1$ and let $\{x_n\}$ be the following basis of the space $E=l^1$:

$$x_1=\frac{1}{\lambda}e_1,\qquad x_n=-\frac{1}{\lambda}e_1+e_n\qquad(n=2,3,\ldots),\tag{12.16}$$

where $\{e_n\}$ denotes the natural basis of the space $E=l^1$. Then $r(V)=\lambda$.

Indeed, let $\{f_n\}$ and $\{h_n\}$ be the a.s.c.f. to the bases $\{x_n\}$ and $\{e_n\}$ respectively. Then

$$f_1(x)=\lambda h_1(x)+\sum_{i=2}^{\infty}h_i(x)\quad(x\in l^1),\qquad f_n=h_n\quad(n=2,3,\ldots).\tag{12.17}$$

We claim that for every $f=\sum_{i=1}^{\infty}\alpha_i f_i\in V$ (see theorem 12.1) with $\|f\|\leqslant 1$ we have

$$|\alpha_1|\leqslant 1.\tag{12.18}$$

In fact, by (12.17),

$$\begin{aligned}f(x)&=\alpha_1\left[\lambda h_1(x)+\sum_{i=2}^{\infty}h_i(x)\right]+\sum_{i=2}^{\infty}\alpha_i h_i(x)\\&=\lambda\alpha_1 h_1(x)+\sum_{i=2}^{\infty}(\alpha_1+\alpha_i)h_i(x)\quad(x\in l^1),\end{aligned}$$

whence

$$|\alpha_1|-|\alpha_n|\leqslant|\alpha_1+\alpha_n|\leqslant\|f\|\leqslant 1\qquad(n=2,3,\ldots).\tag{12.19}$$

On the other hand, by $f=\sum_{i=1}^{\infty}\alpha_i f_i$ and $\|f_n\|=1$ $(n=1,2,\ldots)$, we have $\lim_{n\to\infty}\alpha_n=0$, which, together with (12.19), implies (12.18).

Consequently, for every $x=\sum_{i=1}^{\infty}\xi_i e_i\in l^1$ and $f=\sum_{i=1}^{\infty}\alpha_i f_i\in V$ with $\|f\|\leqslant 1$, we have

$$|f(x)|=\left|\alpha_1\lambda\xi_1+\sum_{i=2}^{\infty}(\alpha_1+\alpha_i)\xi_i\right|\leqslant\lambda|\xi_1|+\sum_{i=2}^{\infty}|\xi_i|.$$

Considering, for any fixed $x=\sum_{i=1}^{\infty}\xi_i e_i\in l^1$, the sequence of functionals[1]

[1] We recall that

$$\operatorname{sign}\alpha=\begin{cases}\dfrac{\bar{\alpha}}{|\alpha|} & \text{for}\quad\alpha\neq 0,\\[4pt] 0 & \text{for}\quad\alpha=0.\end{cases}$$

$$g_n(z)=(\operatorname{sign}\xi_1)f_1(z)+\sum_{i=2}^{n}(-\operatorname{sign}\xi_1+\operatorname{sign}\xi_i)f_i(z)$$

$$=\lambda(\operatorname{sign}\xi_1)h_1(z)+\sum_{i=2}^{n}(\operatorname{sign}\xi_i)h_i(z)+(\operatorname{sign}\xi_1)\sum_{i=n+1}^{\infty}h_i(z)\in V$$

$$(z\in l^1,\ n=1,2,\ldots),$$

it follows that

$$\sup_{\substack{f\in V\\ \|f\|\leqslant 1}}|f(x)|=\lambda|\xi_1|+\sum_{i=2}^{\infty}|\xi_i| \qquad \left(x=\sum_{i=1}^{\infty}\xi_i e_i\in l^1\right).$$

Hence, for every $x=\sum_{i=1}^{\infty}\xi_i e_i\in l^1$ with $\|x\|=\sum_{i=1}^{\infty}|\xi_i|=1$,

$$\sup_{\substack{f\in V\\ \|f\|\leqslant 1}}|f(x)|=1-(1-\lambda)|\xi_1|\geqslant 1-(1-\lambda)=\lambda,$$

and the value λ is attained for $x=e_1\in l^1$. Since by the theorem of Dixmier mentioned above, we have $r(V)=\inf_{\substack{x\in E\\ x\neq 0}}\sup_{\substack{f\in V\\ \|f\|\leqslant 1}}\left|f\left(\frac{x}{\|x\|}\right)\right|$, it follows that we have $r(V)=\lambda$, which proves our assertion.

Let us give an application of theorem 12.2e).

Proposition 12.1. *Let E,F be Banach spaces with bases $\{x_n\}$ and $\{y_n\}$, respectively, and let $\{f_n\}\subset E^*$, $\{h_n\}\subset F^*$ be the a.s.c.f. to $\{x_n\}$ and $\{y_n\}$, respectively. Then*

a) $\{x_n\}\succ\{y_n\}$ *if and only if* $\{h_n\}\succ\{f_n\}$.

b) $\{x_n\}\sim\{y_n\}$ *if and only if* $\{f_n\}\sim\{h_n\}$.

Proof. a) If $\{x_n\}\succ\{y_n\}$, then, by § 8, theorem 8.1 b) (implication $6^\circ\Rightarrow 1^\circ$), there exists a continuous linear mapping $u\colon E\to F$ such that

$$u(x_n)=y_n \qquad (n=1,2,\ldots).$$

Hence, for the adjoint mapping $u^*\colon F^*\to E^*$ we have

$$[u^*(h_i)](x_j)=h_i[u(x_j)]=h_i(y_j)=\delta_{ij}=f_i(x_j) \qquad (i,j=1,2,\ldots),$$

whence, since $[x_n]=E$,

$$u^*(h_i)=f_i \qquad (i=1,2,\ldots).$$

Thus $\{h_n\}\succ\!\!\succ\{f_n\}$ and hence $\{h_n\}\succ\{f_n\}$.

Conversely, assume now that $\{h_n\}\succ\{f_n\}$. Since the a.s.c.f. to the basis $\{h_n\}$ of $[h_n]$ and to the basis $\{f_n\}$ of $[f_n]$ are nothing else then $\{\phi_F(y_n)\}$ and $\{\phi_E(x_n)\}$, respectively, where ϕ_F,ϕ_E denote the canonical mappings of F into $[h_n]^*$ and E into $[f_n]^*$, respectively, we have, by the "only if" part of a) proved above, $\{\phi_F(x_n)\} \succ \{\phi_E(y_n)\}$. Since by theorem 12.2e) both ϕ_E and ϕ_F are isomorphisms, it follows that $\{x_n\} \succ \{y_n\}$.

Finally, b) is an immediate consequence of a), which completes the proof of proposition 12.1.

Theorem 12.1 suggests naturally the question, what are the relations between the norms of the bases $\{x_n\}$ of E and $\{f_n\}$ of $[f_n]$. The following theorem gives upper and lower evaluations of $v_{\{f_n\}}$ by means of $v_{\{x_n\}}$.

Theorem 12.3. *Let $\{x_n\}$ be a basis of a Banach space E, let $\{f_n\} \subset E^*$ be the a.s.c.f. to $\{x_n\}$ and let $V = [f_n]$. Then*

$$1 \leqslant r(V)\, v_{\{x_n\}} \leqslant v_{\{f_n\}} \leqslant v_{\{x_n\}}. \tag{12.20}$$

Hence, in particular, if $r(V) = 1$,[1] *then*

$$v_{\{f_n\}} = v_{\{x_n\}}. \tag{12.21}$$

Proof. From formula (12.2) for the adjoint s_n^* of the partial sum operator s_n, it follows that

$$s_n^* |_V(f) = \sum_{i=1}^{n} f(x_i) f_i = \sum_{i=1}^{n} [\phi(x_i)](f) f_i \qquad (f \in V,\ n = 1, 2, \ldots), \tag{12.22}$$

where ϕ denotes, as before, the canonical mapping of E into V^*. Hence, since $\{\phi(x_n)\} \subset V^*$ is the a.s.c.f. to the basis $\{f_n\}$ of $V = [f_n]$,

$$v_{\{f_n\}} = \sup_{1 \leqslant n < \infty} \|s_n^* |_V\| \leqslant \sup_{1 \leqslant n < \infty} \|s_n\| = v_{\{x_n\}}, \tag{12.23}$$

and taking also into account (12.13),

$$\begin{aligned} v_{\{f_n\}} &= \sup_{1 \leqslant n < \infty} \|s_n^* |_V\| = \sup_{1 \leqslant n < \infty} \sup_{\substack{f \in V \\ \|f\| \leqslant 1}} \sup_{\substack{x \in E \\ \|x\| \leqslant 1}} \left|[s_n^*(f)](x)\right| \\ &= \sup_{1 \leqslant n < \infty} \sup_{\substack{f \in V \\ \|f\| \leqslant 1}} \sup_{\substack{x \in E \\ \|x\| \leqslant 1}} \left|f[s_n(x)]\right| = \sup_{1 \leqslant n < \infty} \sup_{\|x\| \leqslant 1} \|s_n(x)\|_V \\ &\geqslant r(V) \sup_{1 \leqslant n < \infty} \sup_{\|x\| \leqslant 1} \|s_n(x)\| = r(V)\, v_{\{x_n\}}. \end{aligned} \tag{12.24}$$

Consequently, by (12.9), (12.24) and (12.23) we have (12.20), which completes the proof of theorem 12.3.

Let us also mention separately the following part of formula (12.24), which may be useful for other applications:

$$v_{\{f_n\}} = \sup_{1 \leqslant n < \infty} \sup_{\|x\| \leqslant 1} \|s_n(x)\|_V. \tag{12.25}$$

---

[1] This happens e.g. when the basis $\{x_n\}$ is monotone (by theorem 12.2) or when E is reflexive.

The inequalities (12.10) are, in a certain sense, the best possible, since for monotone bases they all become equalities. Let us show that they are, in general, strict inequalities. For this purpose we shall give an example in which simultaneously all of them are strict.

Example 12.3. Let $\{x_n\}$ be the following basis of the space $E=l^1$:

$$x_1=\frac{1}{\lambda}e_1,\quad x_2=e_2-e_3,\quad x_n=-\frac{1}{\lambda}e_1+e_n\quad (n=3,4,\ldots),\tag{12.26}$$

where $\{e_n\}$ denotes the natural basis of the space l^1 and λ is an arbitrary number such that $0<\lambda\leqslant 1$ (this basis differs from (12.16) only in the term x_2), let $\{f_n\}\subset E^*$ be the a.s.c.f. to $\{x_n\}$ and let $V=[f_n]$. Then

$$r(V)=\lambda,\qquad v_{\{x_n\}}=2+\frac{1}{\lambda},\qquad v_{\{f_n\}}=3,\tag{12.27}$$

and thus, for $0<\lambda<1$,

$$1<2\lambda+1=r(V)\,v_{\{x_n\}}<3=v_{\{f_n\}}<2+\frac{1}{\lambda}=v_{\{x_n\}},$$

i.e., all inequalities in (12.20) are strict.

Indeed, if $\{h_n\}$ is the a.s.c.f. to the natural basis $\{e_n\}$ of l^1, we have

$$\begin{gathered}f_1(x)=\lambda h_1(x)+\sum_{i=2}^{\infty}h_i(x)\qquad (x\in l^1),\\ f_2=h_2,\, f_3=h_2+h_3,\, f_n=h_n\qquad (n=4,5,\ldots).\end{gathered}\tag{12.28}$$

Thus $V=[f_n]$ is the same as for the basis (12.16), whence, by example 12.2, $r(V)=\lambda$.

Furthermore, by (12.26), for any positive integer n and any scalars $\alpha_1,\ldots,\alpha_n$ we have

$$\begin{aligned}\left\|\sum_{i=1}^{n}\alpha_i x_i\right\|&=\left\|\left(\frac{\alpha_1}{\lambda}-\sum_{i=3}^{n}\frac{\alpha_i}{\lambda}\right)e_1+\alpha_2 e_2+(\alpha_3-\alpha_2)\,e_3+\sum_{i=4}^{n}\alpha_i e_i\right\|\\&=\left|\frac{\alpha_1}{\lambda}-\sum_{i=3}^{n}\frac{\alpha_i}{\lambda}\right|+|\alpha_2|+|\alpha_3-\alpha_2|+\sum_{i=4}^{n}|\alpha_i|.\end{aligned}$$

Hence, since for $\sum_{i=3}^{n}|\alpha_i|\neq 0$ $(n\geqslant 3)$, $m=1,2,\ldots$ we have

$$\begin{aligned}\left|\frac{\alpha_1}{\lambda}-\sum_{i=3}^{n}\frac{\alpha_i}{\lambda}\right|-\frac{1}{\lambda}\sum_{i=n+1}^{n+m}|\alpha_i|\leqslant\left|\frac{\alpha_1}{\lambda}-\sum_{i=3}^{n+m}\frac{\alpha_i}{\lambda}\right|&\leqslant\frac{1}{\lambda}\left|\frac{\alpha_1}{\lambda}-\sum_{i=3}^{n+m}\frac{\alpha_i}{\lambda}\right|\\&+\left(\frac{1}{\lambda}-1\right)\left(|\alpha_2|+|\alpha_3-\alpha_2|+\sum_{i=4}^{n}|\alpha_i|\right),\end{aligned}$$

we obtain

$$\begin{aligned}\left\|\sum_{i=3}^{n}\alpha_i x_i\right\| &= \left|\frac{\alpha_1}{\lambda}-\sum_{i=3}^{n}\frac{\alpha_i}{\lambda}\right|+|\alpha_2|+|\alpha_3-\alpha_2|+\sum_{i=4}^{n}|\alpha_i| \\ &\leqslant \frac{1}{\lambda}\left(\left|\frac{\alpha_1}{\lambda}-\sum_{i=3}^{n+m}\frac{\alpha_i}{\lambda}\right|+|\alpha_2|+|\alpha_3-\alpha_2|+\sum_{i=4}^{n+m}|\alpha_i|\right)=\frac{1}{\lambda}\left\|\sum_{i=1}^{n+m}\alpha_i x_i\right\|\end{aligned} \tag{12.29}$$

whenever $\sum_{i=3}^{n}|\alpha_i|\neq 0$ $(n\geqslant 3)$, $m=1,2,\ldots$. Since obviously

$$\|\alpha_1 x_1\| = \frac{|\alpha_1|}{\lambda}\leqslant\frac{|\alpha_1|}{\lambda}+2|\alpha_2|=\left\|\sum_{i=1}^{2}\alpha_i x_i\right\|, \tag{12.30}$$

$$\|\alpha_1 x_1\| = \frac{|\alpha_1|}{\lambda}\leqslant\frac{1}{\lambda}\left(\left|\frac{\alpha_1}{\lambda}-\sum_{i=3}^{k}\frac{\alpha_i}{\lambda}\right|+|\alpha_2|+|\alpha_3-\alpha_2|+\sum_{i=4}^{k}|\alpha_i|\right)=\frac{1}{\lambda}\left\|\sum_{i=1}^{k}\alpha_i x_i\right\| \tag{12.31}$$

$(k\geqslant 3)$, it remains to consider $\left\|\sum_{i=1}^{2}\alpha_i x_i\right\|$. Taking in the inequalities

$$\left\|\sum_{i=1}^{2}\alpha_i x_i\right\|\leqslant C\left\|\sum_{i=1}^{k}\alpha_i x_i\right\| \tag{12.32}$$

$(k\geqslant 2)$ the scalars $\alpha_1=\alpha_3=1$, $\alpha_2=\dfrac{n-1}{n}$, $\alpha_4=\cdots=\alpha_k=0$, we get

$$\begin{aligned}\frac{1}{\lambda}+\frac{2(n-1)}{n} &= \left\|\sum_{i=1}^{2}\alpha_i x_i\right\|\leqslant C\left\|\sum_{i=1}^{k}\alpha_i x_i\right\| \\ &= C\left(\left|\frac{1}{\lambda}-\frac{1}{\lambda}\right|+\frac{n-1}{n}+\frac{1}{n}\right)=C,\end{aligned}$$

whence, for $n\to\infty$, we obtain $\dfrac{1}{\lambda}+2\leqslant C$, and thus, taking into account (12.29), (12.30) and (12.31),

$$v_{\{x_n\}}=\min C\geqslant\frac{1}{\lambda}+2.$$

However, we have here the equality sign, since from

$$\begin{aligned}\frac{|\alpha_1|}{\lambda}-\frac{|\alpha_2|}{\lambda} &\leqslant \frac{|\alpha_1-\alpha_2|}{\lambda}\leqslant\left|\frac{\alpha_1}{\lambda}-\sum_{i=3}^{k}\frac{\alpha_i}{\lambda}\right|+\frac{|\alpha_3-\alpha_2|}{\lambda}+\sum_{i=4}^{k}\frac{|\alpha_i|}{\lambda} \\ &\leqslant\left(\frac{1}{\lambda}+2\right)\left(\left|\frac{\alpha_1}{\lambda}-\sum_{i=3}^{k}\frac{\alpha_i}{\lambda}\right|+|\alpha_3-\alpha_2|+\sum_{i=4}^{k}|\alpha_i|\right)\end{aligned}$$

we obtain

$$\left\|\sum_{i=1}^{2}\alpha_i x_i\right\| = \frac{|\alpha_1|}{\lambda} + 2|\alpha_2| \leqslant \left(\frac{1}{\lambda}+2\right)\left(\left|\frac{\alpha_1}{\lambda} - \sum_{i=3}^{k}\frac{\alpha_i}{\lambda}\right| + |\alpha_2| + |\alpha_3-\alpha_2| + \sum_{i=4}^{k}|\alpha_i|\right) = \left(\frac{1}{\lambda}+2\right)\left\|\sum_{i=1}^{k}\alpha_i x_i\right\|$$

$(k \geqslant 2)$. Therefore $v_{\{x_n\}} = \dfrac{1}{\lambda} + 2.$

Finally, let us compute $v_{\{f_n\}}$. By (12.28), for any positive integer n and any scalars $\alpha_1, \ldots, \alpha_n$ we have

$$\left\|\sum_{i=1}^{n}\alpha_i f_i\right\| = \sup_{\substack{x\in E=l^1\\ \|x\|\leqslant 1}}\left|\alpha_1\lambda h_1(x) + (\alpha_1+\alpha_2+\alpha_3)h_2(x) + \sum_{i=3}^{n}(\alpha_1+\alpha_i)h_i(x) + \alpha_1\sum_{i=n+1}^{\infty}h_i(x)\right| = \max_{3\leqslant i\leqslant n}(|\alpha_1|, |\alpha_1+\alpha_2+\alpha_3|, |\alpha_1+\alpha_i|). \tag{12.33}$$

Hence for $\sum_{i=3}^{n}|\alpha_i| \neq 0$ $(n \geqslant 3)$, $m=1,2,\ldots$ we have

$$\left\|\sum_{i=1}^{n}\alpha_i f_i\right\| = \max_{3\leqslant i\leqslant n}(|\alpha_1|, |\alpha_1+\alpha_2+\alpha_3|, |\alpha_1+\alpha_i|) \leqslant \max_{3\leqslant i\leqslant n+m}(|\alpha_1|, |\alpha_1+\alpha_2+\alpha_3|, |\alpha_1+\alpha_i|) = \left\|\sum_{i=1}^{n+m}\alpha_i f_i\right\|. \tag{12.34}$$

Since obviously

$$\|\alpha_1 f_1\| = |\alpha_1| \leqslant \left\|\sum_{i=1}^{k}\alpha_i f_i\right\| \tag{12.35}$$

$(k \geqslant 2)$, it remains to consider $\left\|\sum_{i=1}^{2}\alpha_i f_i\right\|$. Taking in the inequalities

$$\left\|\sum_{i=1}^{2}\alpha_i f_i\right\| \leqslant C\left\|\sum_{i=1}^{k}\alpha_i f_i\right\|$$

$(k \geqslant 2)$ the scalars $\alpha_1=1, \alpha_2=2, \alpha_3=-2, \alpha_4=\cdots=\alpha_k=0$, we get

$$\max(1,3) = \left\|\sum_{i=1}^{2}\alpha_i f_i\right\| \leqslant C\left\|\sum_{i=1}^{k}\alpha_i f_i\right\| = C\max(1,1,1),$$

whence $3 \leqslant C$ and thus, taking also into account (12.34) and (12.35),

$$v_{\{f_n\}} = \min C \geqslant 3.$$

However, we have here the equality sign, since

$$\left\|\sum_{i=1}^{2}\alpha_i f_i\right\| = \max(|\alpha_1|, |\alpha_1+\alpha_2|) \leqslant |\alpha_1|+|\alpha_2| \leqslant 3\max\left(|\alpha_1|, \frac{|\alpha_2|}{2}\right)$$

$$\leqslant 3 \max_{3\leqslant i\leqslant k}(|\alpha_1|, |\alpha_1+\alpha_2+\alpha_3|, |\alpha_1+\alpha_i|) = 3\left\|\sum_{i=1}^{k}\alpha_i f_i\right\|$$

$(k \geqslant 2)$; indeed, if $\frac{|\alpha_2|}{2} \leqslant |\alpha_1|$, then $|\alpha_1|+|\alpha_2| \leqslant |\alpha_1|+2|\alpha_1| = 3|\alpha_1|$, while if $\frac{|\alpha_2|}{2} \geqslant |\alpha_1|$, then $|\alpha_1|+|\alpha_2| \leqslant \frac{|\alpha_2|}{2}+|\alpha_2| = \frac{3|\alpha_2|}{2}$ and, on the other hand, $\frac{|\alpha_2|}{2} = \frac{|(\alpha_1+\alpha_2+\alpha_3)-(\alpha_1+\alpha_3)|}{2} \leqslant \max(|\alpha_1+\alpha_2+\alpha_3|,$ $|\alpha_1+\alpha_3|)$. Therefore $v_{\{f_n\}} = 3$, which completes the proof of the assertions of example 12.3.

This example shows that there exist bases $\{x_n\}$ with $v_{\{f_n\}} = 3$ (where $\{f_n\} \subset E^*$ is the a.s.c.f. to $\{x_n\}$) and with $v_{\{x_n\}}$ taking any arbitrarily large preassigned value. It is natural to ask whether the limit case $v_{\{x_n\}} = \infty$ is also possible, i.e., whether there exists a biorthogonal system (x_n, f_n) with $[x_n] = E$ and $\{f_n\}$ total on E, such that $\{f_n\}$ is a basic sequence, but $\{x_n\}$ is *not* a basis of E. We shall show that the answer is affirmative (this also shows that a certain converse of the first statement of theorem 12.1 is not valid).

Example 12.4. Let[1] $E = (E_1 \times E_2 \times \cdots)_{l^1} \equiv l^1$, where $E_j = l^1$ $(j = 1, 2, \ldots)$ and for each j let $\{x_n^{(j)}\}$ be the basis (12.26) of $E_j = l^1$, with $\lambda = \frac{1}{j}$. Since the set $\bigcup_{j=1}^{\infty}\bigcup_{n=1}^{\infty}\{\underbrace{0, \ldots, 0}_{j-1}, x_n^{(j)}, 0, \ldots\}$ in E is countable, let $\{x_n\}$ be an arbitrary numbering of it, and let $\{f_n\}$ be the corresponding numbering of the functionals $\{\underbrace{0, \ldots, 0}_{j-1}, f_n^{(j)}, 0, \ldots\} \in (E_1^* \times E_2^* \times \cdots)_m \equiv E^* \equiv m$, where $\{f_n^{(j)}\}$ is the basic sequence (12.28) in $E_j^* \equiv m$, with $\lambda = \frac{1}{j}$.

---

[1] We recall that $(E_1 \times E_2 \times \cdots)_{l^1}$ (respectively, $E_1 \times E_2 \times \cdots)_m$) is the space of all sequences $\{x_n\}$ with $x_n \in E_n$ $(n = 1, 2, \ldots)$, for which $\{\|x_n\|\} \in l^1$ (respectively, $\{\|x_n\|\} \in m$), endowed with the norm $\|\{x_n\}\| = \sum_{n=1}^{\infty}\|x_n\|$ (respectively, $\|\{x_n\}\| = \sup_{1\leqslant n<\infty}\|x_n\|$). There exists a natural linear isometry $(l^1 \times l^1 \times \cdots)_{l^1} \equiv l^1$ (see Ch. II, § 18, lemma 18.5c)).

Then (x_n, f_n) is a biorthogonal system, $[x_n]=E$, $\{f_n\}$ is total on E and it is a basic sequence in E^*, but $\{x_n\}$ is not a basis of E.

Indeed, it is obvious that $[x_n]=E$ and that $\{f_n\}$ is total on E. Since by (12.27) we have $v_{\{x_n^{(j)}\}}=2+j$ $(j=1,2,\ldots)$, it follows, by § 7, theorem 7.1, that $\{x_n\}$ is not a basis of E. On the other hand, since $\{x_n^{(j)}\}$ is an unconditional basis[1] of E_j, $\{f_n^{(j)}\}$ is an unconditional basic sequence in E_j^* (by Ch. II, § 17, theorem 17.7) and by (12.33) its unconditional norm $v_{\{f_n^{(j)}\}}^{(u)}$ does not depend on j. Since for $\{g_1, g_2, \ldots\} \in (E_1^* \times E_2^* \times \cdots)_m$ we have $\|\{g_1, g_2, \ldots\}\| = \sup_{1 \leqslant j < \infty} \|g_j\|$, it follows that $\{f_n\}$ is a basic sequence in E^*, which completes the proof of the assertions of example 12.4. Let us also mention that $r(V)=0$ (where $V=[f_n]$).

We don't know whether the converse of the last statement of theorem 12.3 is true, i.e., whether for a basis $\{x_n\}$ of a Banach space E, with the a.s.c.f. $\{f_n\}$, the equality $v_{\{f_n\}}=v_{\{x_n\}}$ implies $r(V)=1$ (where $V=[f_n]$).

If $\{x_n\}$ is a basis of a Banach space E, with the a.s.c.f. $\{f_n\}$, then $\{f_n\}$ is a basis of $V=[f_n]$ (by theorem 12.1), with the a.s.c.f. $\{\phi(x_n)\} \subset V^*$, where ϕ denotes the canonical mapping of E into V^*, whence, by virtue of theorem 12.3, we have

$$1 \leqslant r(W)\, v_{\{f_n\}} \leqslant v_{\{\phi(x_n)\}} \leqslant v_{\{f_n\}}, \tag{12.36}$$

where $W=[\phi(x_n)]=\phi(E) \subset V^*$. It is natural to ask whether one can say more about the constant $v_{\{\phi(x_n)\}}$. We shall now show that this is indeed the case, namely, that we always have $r(W)=1$, whence the last two inequalities in (12.36) are actually equalities.

Theorem 12.4. *Let V be a linear subspace[2] of a conjugate Banach space E^*, let ϕ be the canonical mapping of E into V^* and let $W=\phi(E) \subset V^*$. Then*

$$r(W)=1. \tag{12.37}$$

Hence, in particular, if $\{x_n\}$ is a basis of a Banach space E, $\{f_n\}$ the a.s.c.f. to $\{x_n\}$, $V=[f_n] \subset E^*$ and $W=[\phi(x_n)]=\phi(E) \subset V^*$, then

$$r(W)=1, \qquad v_{\{\phi(x_n)\}}=v_{\{f_n\}}. \tag{12.38}$$

Proof. By formula (12.15) applied to $W=\phi(E) \subset V^*$ and the relations $\|\phi(x)\| = \|x\|_V \leqslant \|x\|$ $(x \in E)$, we have

---

[1] For the notions of unconditional basis, unconditional basic sequence, unconditional norm of a basic sequence, see Ch. II, § 14, § 17.

[2] We need not assume that $r(V)>0$. Let us mention that the semi-norm $\|x\|_V$ is a norm on E if and only if V is total on E (or, equivalently, ϕ is one-to-one).

$$r(W)=\inf_{\substack{f\in V\\ f\neq 0}}\ \sup_{\substack{\Psi\in W\\ \|\Psi\|\leqslant 1}}\left|\Psi\left(\frac{f}{\|f\|}\right)\right|=\inf_{\substack{f\in V\\ f\neq 0}}\ \sup_{\substack{x\in E\\ \|x\|_V\leqslant 1}}\left|\frac{f}{\|f\|}(x)\right|$$

$$\geqslant\inf_{\substack{f\in V\\ f\neq 0}}\ \sup_{\substack{x\in E\\ \|x\|\leqslant 1}}\left|\frac{f}{\|f\|}(x)\right|=\inf_{\substack{f\in V\\ f\neq 0}}\left\|\frac{f}{\|f\|}\right\|=1,$$

whence, since the characteristic of any subspace is always $\leqslant 1$, we infer (12.37).

The second assertion of theorem 12.4 follows from the first one, taking into account that $[\phi(x_n)]=\phi(E)$ whenever $\{x_n\}$ is a basis of E and ϕ the canonical mapping of E into $V^*=[f_n]^*$, and applying the last statement of theorem 12.3 to the pair $(\{f_n\}, \{\phi(x_n)\})$. This completes the proof of theorem 12.4.

Remark 12.3. Formula (12.37) is equivalent to the statement that the canonical mapping u of V into $W^*=\phi(E)^*$, defined by

$$u(f)[\phi(x)]=[\phi(x)](f)=f(x) \qquad (f\in V,\ \phi(x)\in W)$$

is an isometry, i.e.,

$$\sup_{\substack{x\in E\\ \|x\|_V\leqslant 1}}|f(x)|=\|u(f)\|=\|f\|=\sup_{\substack{x\in E\\ \|x\|\leqslant 1}}|f(x)| \qquad (f\in V), \tag{12.39}$$

and this latter formula follows also directly from the relations $\|x\|_V\leqslant\|x\|$ $(x\in E)$ and $\|u\|\leqslant 1$.

Combining theorems 12.3 and 12.4, we obtain the following relations between the norms of $\{x_n\}$ and $\{\phi(x_n)\}$:

Corollary 12.3. *Let $\{x_n\}$ be a basis of a Banach space E, $\{f_n\}$ the a.s.c.f. to $\{x_n\}$, $V=[f_n]\subset E^*$, and ϕ the canonical mapping of E into V^*. Then*

$$1\leqslant r(V)\,v_{\{x_n\}}\leqslant v_{\{\phi(x_n)\}}\leqslant v_{\{x_n\}}. \tag{12.40}$$

Remark 12.4. By (12.25) and (12.22) we have the following formula for the computation of $v_{\{\phi(x_n)\}}$:

$$v_{\{\phi(x_n)\}}=\sup_{1\leqslant n<\infty}\ \sup_{\substack{f\in V\\ \|f\|\leqslant 1}}\|s_n^*(f)\|_W=\sup_{1\leqslant n<\infty}\ \sup_{\substack{x\in E\\ \|x\|_V\leqslant 1}}\|s_n(x)\|_V. \tag{12.41}$$

With the aid of this, it is easy to obtain again, directly, the second equality in formula (12.38). Indeed, by (12.41), $\|x\|_V\leqslant\|x\|$ $(x\in E)$ and (12.25) we have

$$v_{\{\phi(x_n)\}}\geqslant\sup_{1\leqslant n<\infty}\ \sup_{\substack{x\in E\\ \|x\|\leqslant 1}}\|s_n(x)\|_V=v_{\{f_n\}},$$

whence, since by theorem 12.3 we have $\nu_{\{\phi(x_n)\}} \leqslant \nu_{\{f_n\}}$, we infer $\nu_{\{\phi(x_n)\}} = \nu_{\{f_n\}}$.

Proposition 12.2. *Let E be a Banach space with a basis $\{x_n\}$, $V=[f_n]$ the norm-closed linear subspace of E^* spanned by the a.s.c.f. $\{f_n\}$ to $\{x_n\}$, and ν the norm of the basis $\{x_n\}$, i.e., $\nu = \sup_{1\leqslant n<\infty} \|s_n\| < \infty$, where $\{s_n\}$ is the associated sequence of partial sum operators. Then*

a) *We have*

$$\|f\| \leqslant \sup_{1\leqslant n<\infty} \left\| \sum_{i=1}^{n} f(x_i) f_i \right\| \leqslant \nu \|f\| \qquad (f \in E^*). \tag{12.42}$$

Conversely, if for a sequence of scalars $\{\alpha_n\} \subset K$ we have $\sup_{1\leqslant n<\infty} \left\| \sum_{i=1}^{n} \alpha_i f_i \right\| < \infty$, then there exists an $f \in E^*$ such that

$$f(x_n) = \alpha_n \qquad (n=1,2,\ldots). \tag{12.43}$$

b) *We have*

$$\|\Psi\| \leqslant \sup_{1\leqslant n<\infty} \left\| \sum_{i=1}^{n} \Psi(f_i) x_i \right\|_V \leqslant \sup_{1\leqslant n<\infty} \left\| \sum_{i=1}^{n} \Psi(f_i) x_i \right\| \leqslant \nu \|\Psi\| \qquad (\Psi \in V^*). \tag{12.44}$$

Conversely, if for a sequence of scalars $\{\alpha_n\} \subset K$ we have $\sup_{1\leqslant n<\infty} \left\| \sum_{i=1}^{n} \alpha_i x_i \right\| < \infty$, then there exists a $\Psi \in V^*$ such that

$$\Psi(f_n) = \alpha_n \qquad (n=1,2,\ldots). \tag{12.45}$$

c) *We have*

$$\sup_{1\leqslant n<\infty} \left\| \sum_{i=1}^{n} \Phi(f_i) x_i \right\| \leqslant \nu \|\Phi\| \qquad (\Phi \in E^{**}). \tag{12.46}$$

Conversely, if for a sequence $\{\alpha_n\} \subset K$ we have $\sup_{1\leqslant n<\infty} \left\| \sum_{i=1}^{n} \alpha_i x_i \right\| < \infty$, then there exists a $\Phi \in E^{**}$ such that

$$\Phi(f_n) = \alpha_n \qquad (n=1,2,\ldots). \tag{12.47}$$

Proof. a) Let $f \in E^*$ and $\varepsilon > 0$ be arbitrary and let $x = x_{\varepsilon,f} \in E$ be such that $\|x\| = 1$, $|f(x)| \geqslant \|f\| - \varepsilon$. Then

$$\|f\| \leqslant |f(x)| + \varepsilon = \left| f\left(\sum_{i=1}^{\infty} f_i(x) x_i \right) \right| + \varepsilon = \lim_{n\to\infty} \left| \left(\sum_{i=1}^{n} f(x_i) f_i \right)(x) \right| + \varepsilon$$

$$\leqslant \sup_{1\leqslant n<\infty} \left| \left(\sum_{i=1}^{n} f(x_i) f_i \right)(x) \right| + \varepsilon \leqslant \sup_{1\leqslant n<\infty} \left\| \sum_{i=1}^{n} f(x_i) f_i \right\| + \varepsilon,$$

whence, since $\varepsilon>0$ and $f\in E^*$ were arbitrary,

$$\|f\|\leqslant \sup_{1\leqslant n<\infty}\left\|\sum_{i=1}^{n} f(x_i)f_i\right\| \qquad (f\in E^*).$$

On the other hand, by (12.2),

$$\sup_{1\leqslant n<\infty}\left\|\sum_{i=1}^{n} f(x_i)f_i\right\| = \sup_{1\leqslant n<\infty}\|s_n^*(f)\| \leqslant \sup_{1\leqslant n<\infty}\|s_n^*\|\,\|f\|$$
$$= \sup_{1\leqslant n<\infty}\|s_n\|\,\|f\| = v\|f\| \qquad (f\in E^*),$$

and thus we have (12.42).

Conversely, assume now that $\sup_{1\leqslant n<\infty}\left\|\sum_{i=1}^{n}\alpha_i f_i\right\|=M<\infty$. Then for any finite sequence of scalars $\beta_1,\ldots,\beta_n$ we have

$$\left|\sum_{i=1}^{n}\beta_i\alpha_i\right| = \left|\sum_{i=1}^{n}\alpha_i f_i\left(\sum_{j=1}^{n}\beta_j x_j\right)\right| \leqslant \left\|\sum_{i=1}^{n}\alpha_i f_i\right\|\left\|\sum_{j=1}^{n}\beta_j x_j\right\| \leqslant M\left\|\sum_{j=1}^{n}\beta_j x_j\right\|,$$

whence, by a well known theorem of E. Helly[1], there exists an $f\in E^*$ satisfying (12.43).

b) and c). The first inequality in (12.44) follows from part a) applied to the basis $\{f_n\}$ of $V=[f_n]$, since it has the a.s.c.f. $\{\phi(x_n)\}$ (whence $\left\|\sum_{i=1}^{n}\Psi(f_i)\phi(x_i)\right\|=\left\|\sum_{i=1}^{n}\Psi(f_i)x_i\right\|_V$) and the second inequality in (12.44) is obvious.

Furthermore, by (12.2) we have

$$(s_n^{**}(\Phi))(f)=\Phi\left(\sum_{i=1}^{n} f(x_i)f_i\right)=f\left(\sum_{i=1}^{n}\Phi(f_i)x_i\right) \quad (\Phi\in E^{**}, f\in E^*, n=1,2,\ldots),$$

whence, by (12.4),

$$s_n^{**}(\Phi)=\pi\left(\sum_{i=1}^{n}\Phi(f_i)x_i\right) \qquad (\Phi\in E^{**},\ n=1,2,\ldots). \tag{12.48}$$

where π is the canonical mapping of E into E^{**}. Since π is an isometry, it follows that we have

$$\left\|\sum_{i=1}^{n}\Phi(f_i)x_i\right\| = \|s_n^{**}(\Phi)\| \leqslant \|s_n^{**}\|\,\|\Phi\| = \|s_n\|\,\|\Phi\| \leqslant v\|\Phi\|$$
$$(\Phi\in E^{**},\ n=1,2,\ldots),$$

i.e., (12.46). In particular, if $\Psi\in V^*$, then, taking an arbitrary extension $\Phi\in E^{**}$ of Ψ, with norm $\|\Phi\|=\|\Psi\|$, we obtain the last inequality in (12.44).

---

[1] See e.g. [10], p. 55, theorem 4.

Conversely, assume now that $\sup_{1\leqslant n<\infty} \left\|\sum_{i=1}^{n} \alpha_i x_i\right\| < \infty$. Then $\sup_{1\leqslant n<\infty} \left\|\sum_{i=1}^{n} \alpha_i \phi(x_i)\right\| \leqslant \sup_{1\leqslant n<\infty} \left\|\sum_{i=1}^{n} \alpha_i x_i\right\| < \infty$, whence, by part a) applied to the basis $\{f_n\}$ of $V=[f_n]$, there exists a $\Psi \in V^*$ satisfying (12.45). Taking an arbitrary extension $\Phi \in E^{**}$ of Ψ, we shall also have (12.47), which completes the proof of proposition 12.2.

Let us also mention the following alternative proof of the converse part in a): If $\sup_{1\leqslant n<\infty} \left\|\sum_{i=1}^{n} \alpha_i f_i\right\| < \infty$, then, since E is separable, there exist[1] an $f \in E^*$ and a sequence $\{n_j\}$ of positive integers such that

$$\lim_{j\to\infty} \sum_{i=1}^{n_j} \alpha_i f_i(x) = f(x) \qquad (x \in E),$$

whence, taking $x = x_n$ $(n=1,2,\ldots)$, we obtain (12.43), which completes the proof.

Theorem 12.5. *Let E be a Banach space with a basis $\{x_n\}$, $V=[f_n]$ the norm-closed linear subspace of E^* spanned by the a.s.c.f. $\{f_n\}$ to $\{x_n\}$, ϕ the canonical mapping of E into V^* and π the canonical mapping of E into E^{**}. Then*

a) *The space E^* is isomorphic, by the mapping*

$$\eta: f \to \{f(x_n)\} \qquad (f \in E^*), \tag{12.49}$$

to the Banach space of sequences of scalars

$$A_4 = \left\{ \{\alpha_n\} \subset K \;\middle|\; \sup_{1\leqslant n<\infty} \left\|\sum_{i=1}^{n} \alpha_i f_i\right\| < \infty \right\}, \tag{12.50}$$

in which the norm is defined by

$$\|\{\alpha_n\}\| = \sup_{1\leqslant n<\infty} \left\|\sum_{i=1}^{n} \alpha_i f_i\right\|. \tag{12.51}$$

b) *The restriction $\eta|_V$ of the isomorphism* (12.49) *to V maps V onto the Banach space*

$$A_3 = \left\{ \{\alpha_n\} \subset K \;\middle|\; \sum_{i=1}^{\infty} \alpha_i f_i \ \textit{converges (strongly)} \right\}, \tag{12.52}$$

in which the norm is defined by (12.51).

[1] See e.g. [10], p. 123, theorem 3.

c) *The space* $V^*(\equiv E^{**}/V^\perp)$ *is isomorphic, by the mapping*

$$\tau\colon \Psi \to \{\Psi(f_n)\} \qquad (\Psi \in V^*), \tag{12.53}$$

to the Banach space of sequences of scalars

$$A_2 = \left\{ \{\alpha_n\} \subset K \;\middle|\; \sup_{1 \leqslant n < \infty} \left\| \sum_{i=1}^{n} \alpha_i x_i \right\| < \infty \right\}, \tag{12.54}$$

defined in § 5, *remark* 5.4.

d) *The restriction* $\tau|_{\phi(E)}$ *of the isomorphism* (12.53) *to* $\phi(E)$ *maps* $\phi(E)$ *onto the Banach space*

$$A_1 = \left\{ \{\alpha_n\} \subset K \;\middle|\; \sum_{i=1}^{\infty} \alpha_i x_i \quad \text{converges} \right\}, \tag{12.55}$$

introduced in § 3, *proposition* 3.1.

If $\{x_n\}$ *is a monotone basis, the isomorphisms in* a)–d) *above are isometries.*

e) *We have*

$$\phi(E) = \left\{ \Psi \in V^* \;\middle|\; \sum_{i=1}^{\infty} \Psi(f_i) x_i \quad \text{converges} \right\}, \tag{12.56}$$

and $\phi(E)$ *is norm-closed in* V^*.

f) *We have*

$$\pi(E) \oplus V^\perp = \left\{ \Phi \in E^{**} \;\middle|\; \sum_{i=1}^{\infty} \Phi(f_i) x_i \quad \text{converges} \right\}, \tag{12.57}$$

and $\pi(E) \oplus V^\perp$ *is norm-closed in* E^{**}.

Proof. a) is an immediate consequence of proposition 12.2a).

b) For every $f \in V$ we have $\{f(x_n)\} \in A_3$ (by (12.1)). Conversely, if $\{\alpha_n\} \in A_3$, i.e., $\sum_{i=1}^{\infty} \alpha_i f_i$ converges to an $f \in V$, then $\alpha_n = \sum_{i=1}^{\infty} \alpha_i f_i(x_n) = f(x_n)$ $(n = 1, 2, \ldots)$, whence $\{\alpha_n\} = \eta(f)$.

c) is an immediate consequence of proposition 12.2b). It also follows from observing that by proposition 12.2c), $\Phi \to \{\Phi(f_n)\}$ $(\Phi \in E^{**})$ is a continuous linear mapping of E^{**} onto A_2, having the kernel $V^\perp$, whence it induces, by the open mapping theorem[1], an isomorphism of $E^{**}/V^\perp$ onto A_2, and taking into account that V^* is canonically isomorphic to $E^{**}/V^\perp$, by the mapping[2] $\Psi \to \{\Phi \in E^{**} | \Phi|_V = \Psi\}$ $(\Psi \in V^*)$.

[1] See e.g. [10], p. 40, theorem 4.

[2] See e.g. [43], Ch. II, § 1, lemma 2.

Moreover, c) also follows from a) above applied to the basis $\{f_n\}$ of $V=[f_n]$, taking into account theorem 12.2e). Conversely, let us observe that a) can be derived from c) above, since by c) applied to the basis $\{f_n\}$ of $[f_n]$, the space $[\phi(x_n)]^*=\phi(E)^*$ is isomorphic, by the mapping $h\to\{h(\phi(x_n))\}=\{(\phi^*(h))(x_n)\}$ $(h\in\phi(E)^*)$ to $\left\{\{\alpha_n\}\subset K \,\middle|\, \sup_{1\leqslant n<\infty}\left\|\sum_{i=1}^{n}\alpha_i f_i\right\|<\infty\right\}=A_4$ and since (by theorem 12.2e)) E^* is isomorphic to $\phi(E)^*$, by the mapping $(\phi^*)^{-1}\colon \phi^*(h)\to h$ $(h\in\phi(E)^*)$.

e) Since $\{x_n\}$ is a basis with the a.s.c.f. $\{f_n\}$, for every $\Psi=\phi(x)\in\phi(E)$ the series $\sum_{i=1}^{\infty}\Psi(f_i)x_i=\sum_{i=1}^{\infty}f_i(x)x_i$ converges. Conversely, if for a $\Psi\in V^*$ the series $\sum_{i=1}^{\infty}\Psi(f_i)x_i$ converges, say to $x\in E$, then, since $\{x_n\}$ is a basis with the a.s.c.f. $\{f_n\}$, $\sum_{i=1}^{\infty}\Psi(f_i)x_i=x=\sum_{i=1}^{\infty}f_i(x)x_i$, whence $\Psi(f_n)=f_n(x)$ $(n=1,2,\dots)$ and consequently $\Psi(f)=f(x)$ for all $f\in V$, i.e., $\Psi=\phi(x)\in\phi(E)$, which proves (12.56).

Furthermore, $\phi(E)$ is complete, whence norm-closed in V^*, since by theorem 12.2e) it is isomorphic to E. However, one can also derive the norm-closedness of $\phi(E)$ directly from (12.56). Indeed, if $\{\Psi_n\}\subset\phi(E)$, $\Psi\in V^*$, $\|\Psi_n-\Psi\|\to 0$ as $n\to\infty$, and $\varepsilon>0$, then we have, by proposition 12.2b),

$$\left\|\sum_{i=n+1}^{n+l}\Psi(f_i)x_i\right\|\leqslant\left\|\sum_{i=n+1}^{n+l}\Psi_k(f_i)x_i\right\|+\left\|\sum_{i=1}^{n+l}(\Psi-\Psi_k)(f_i)(x_i)\right\|$$
$$+\left\|\sum_{i=1}^{n}(\Psi-\Psi_k)(f_i)(x_i)\right\|\leqslant\left\|\sum_{i=n+1}^{n+l}\Psi_k(f_i)x_i\right\|+2v\|\Psi-\Psi_k\|,$$

which is $<\varepsilon$ for $n>N=N(\varepsilon)$, if we take $k_0=k_0(\varepsilon)$ such that $\|\Psi-\Psi_{k_0}\|<\frac{\varepsilon}{4v}$ and $N=N(\varepsilon)$ such that $\left\|\sum_{i=n+1}^{n+l}\Psi_{k_0}(f_i)x_i\right\|<\frac{\varepsilon}{2}$ for $n>N$, $l=1,2,\dots$. Hence, by (12.56), $\Psi\in\phi(E)$, which proves that $\phi(E)$ is norm-closed in V^*.

d) For every $\Psi\in\phi(E)$ we have, by e), $\{\Psi(f_n)\}\in A_1$. Conversely, for $\{\alpha_n\}\in A_1\subset A_2$ let $\Psi=\tau^{-1}(\{\alpha_n\})$. Then $\{\Psi(f_n)\}=\tau(\Psi)=\{\alpha_n\}$ and thus the series $\sum_{i=1}^{\infty}\Psi(f_i)x_i=\sum_{i=1}^{\infty}\alpha_i x_i$ converges, whence, by e), $\Psi\in\phi(E)$.

Let us also observe that from d) it follows again that $\phi(E)$ is norm-closed in E^*, since by d) $\phi(E)$ is complete.

f) For any $\Phi\in\pi(E)\oplus V^{\perp}$ we have $\Phi|_V=\Psi\in\phi(E)$, whence, by e), the series $\sum_{i=1}^{\infty}\Phi(f_i)x_i=\sum_{i=1}^{\infty}\Psi(f_i)x_i$ converges. Conversely, if $\sum_{i=1}^{\infty}\Phi(f_i)x_i$

converges, then $\sum_{i=1}^{\infty} \Psi(f_i) x_i$ converges, where $\Psi = \Phi\big|_V \in V^*$, whence, by e), $\Phi\big|_V = \Psi = \phi(x) \in \phi(E)$ for a suitable $x \in E$. Then, by (12.6), $\Phi\big|_V = \pi(x)\big|_V$, whence $\Phi - \pi(x) \in V^\perp$ and thus $\Phi \in \pi(E) \oplus V^\perp$. This proves (12.57).

Finally, the norm-closedness of $\pi(E) \oplus V^\perp$ is well known[1] to be a consequence of $r(V) > 0$ (which holds by theorem 12.2a)), but it can be derived also directly from (12.57), with an argument similar to that used in the above proof of e), making use of proposition 12.2c). This completes the proof of theorem 12.5.

As another application of proposition 12.2, we shall give now a relation between bases and sequence spaces.

A *sequence space* (or a *coordinate space*) is a set S of sequences of scalars which is closed under coordinatewise addition and scalar multiplication, i.e., a linear subspace of the space of all sequences.

Definition 12.1. We shall say that a sequence space S is *associated to a basis of a Banach space*, if there exists a Banach space E with a basis $\{x_n\}$ such that S coincides with the set $A_1 = \left\{\{\alpha_n\} \subset K \,\middle|\, \sum_{i=1}^{\infty} \alpha_i x_i \text{ converges}\right\}$.

We shall consider the problem of characterization of sequence spaces associated to a basis. For this purpose we recall some notions of the theory of sequence spaces.

The *γ-dual* of a sequence space S is the sequence space S^γ defined by

$$S^\gamma = \left\{\{\beta_n\} \subset K \,\middle|\, \sup_{1 \leq n < \infty} \left|\sum_{i=1}^{n} \beta_i \alpha_i\right| < \infty \quad \text{for all} \quad \{\alpha_n\} \in S\right\}. \tag{12.58}$$

For every sequence space S we have, obviously, $S \subset S^{\gamma\gamma}$. A sequence space S is said to be *γ-perfect*, if $S^{\gamma\gamma} = S$.

A sequence space S is called a *BK-space*, if it is a Banach space and the coordinate functionals are continuous on S, i.e., the relations $x_n = \{\alpha_j^{(n)}\}$, $x = \{\alpha_j\} \in S$, $\lim_{n\to\infty} x_n = x$ imply $\lim_{n\to\infty} \alpha_j^{(n)} = \alpha_j$ $(j = 1, 2, \ldots)$.

In the sequel we shall need

Lemma 12.1. *If T is a BK-space containing all unit vectors e_n, then its γ-dual T^γ can be normed so as to become a BK-space, by*

$$\|\{\beta_j\}\| = \sup_{\substack{\{\alpha_j\} \in T \\ \|\{\alpha_j\}\| \leq 1}} \sup_{1 \leq n < \infty} \left|\sum_{i=1}^{n} \beta_i \alpha_i\right| \qquad (\{\beta_j\} \in T^\gamma). \tag{12.59}$$

[1] See e.g. [47].

Proof. Let us first show that $\|\{\beta_j\}\| < \infty$ for all $\{\beta_j\} \in T^\gamma$. For any fixed $\{\beta_j\} \in T^\gamma$ consider on T the linear functionals

$$F_n = \sum_{i=1}^{n} \beta_i f_i \qquad (n=1,2,\ldots), \tag{12.60}$$

where f_i is the i-th coordinate functional on T, i.e.

$$f_i(\{\alpha_j\}) = \alpha_i \qquad (i=1,2,\ldots). \tag{12.61}$$

Since T is a BK-space, the functionals f_i, whence also F_n, are continuous on T and by the definition of T^γ we have

$$\sup_{1 \leqslant n < \infty} |F_n(\{\alpha_j\})| = \sup_{1 \leqslant n < \infty} \left| \sum_{i=1}^{n} \beta_i \alpha_i \right| < \infty \qquad (\{\alpha_j\} \in T),$$

whence, by the principle of uniform boundedness,

$$\|\{\beta_j\}\| = \sup_{\substack{\{\alpha_j\} \in T \\ \|\{\alpha_j\}\| \leqslant 1}} \sup_{1 \leqslant n < \infty} |F_n(\{\alpha_j\})| = \sup_{1 \leqslant n < \infty} \|F_n\| < \infty. \tag{12.62}$$

Obviously, $\|\{\beta_j\}\|$ is a norm on T^γ. To show that T^γ with this norm is a Banach space, let $\{\beta_j^{(n)}\}$ be a Cauchy sequence in T^γ, i.e.,

$$\sup_{\substack{\{\alpha_j\} \in T \\ \|\{\alpha_j\}\| \leqslant 1}} \sup_{1 \leqslant n < \infty} \left| \sum_{i=1}^{n} (\beta_i^{(k)} - \beta_i^{(k+p)}) \alpha_i \right| < \varepsilon \qquad (k > N(\varepsilon),\ p=1,2,\ldots). \tag{12.63}$$

Taking in these inequalities $\{\alpha_j\} = \dfrac{e_n}{\|e_n\|}$ we obtain

$$|\beta_n^{(k)} - \beta_n^{(k+p)}| < \varepsilon \|e_n\| \qquad (k > N(\varepsilon),\ p=1,2,\ldots;\ n=1,2,\ldots),$$

whence the limits $\lim_{k \to \infty} \beta_n^{(k)} = \beta_n$ $(n=1,2,\ldots)$ exist. Taking $p \to \infty$ in (12.63), we obtain

$$\sup_{\substack{\{\alpha_j\} \in T \\ \|\{\alpha_j\}\| \leqslant 1}} \sup_{1 \leqslant n < \infty} \left| \sum_{i=1}^{n} (\beta_i^{(k)} - \beta_i) \alpha_i \right| \leqslant \varepsilon \qquad (k > N(\varepsilon)), \tag{12.64}$$

whence we infer that $\{\beta_j^{(k)}\} - \{\beta_j\} \in T^\gamma$ $(k > N(\varepsilon))$. Consequently, $\{\beta_j\} \in T^\gamma$ and $\lim_{k \to \infty} \|\{\beta_j^{(k)}\} - \{\beta_j\}\| = 0$, which proves that T^γ is complete.

Finally, to show that T^γ is a BK-space, let $\{\beta_j^{(k)}\}$, $\{\beta_j\} \in T^\gamma$ satisfy (12.64). Then for $\{\alpha_j\} = \dfrac{e_n}{\|e_n\|}$ it follows that

$$|\beta_n^{(k)} - \beta_n| < \varepsilon \|e_n\| \qquad (k > N(\varepsilon),\ n=1,2,\ldots),$$

whence $\lim_{k \to \infty} \beta_n^{(k)} = \beta_n$ $(n=1,2,\ldots)$, which completes the proof.

This being said, we can give

Theorem 12.6. *A sequence space S is associated to a basis of a Banach space if and only if S contains all unit vectors e_n and there exists a γ-perfect BK-space T such that $[e_n]_T = S$. In this case, $T = S^{\gamma\gamma}$.*

Proof. Assume that S is associated to a basis $\{x_n\}$ of a Banach space E, i.e., $S = A_1 = \left\{\{\alpha_n\} \subset K \,\middle|\, \sum_{i=1}^{\infty} \alpha_i x_i \text{ converges}\right\}$. We shall show that the Banach sequence space $A_2 = \left\{\{\alpha_n\} \subset K \,\middle|\, \sup_{1 \leqslant n < \infty} \left\| \sum_{i=1}^{n} \alpha_i x_i \right\| < \infty\right\}$, with $\|\{\alpha_n\}\| = \sup_{1 \leqslant n < \infty} \left\| \sum_{i=1}^{n} \alpha_i x_i \right\|$, introduced in § 5, remark 5.4, is a γ-perfect BK-space; since $S = A_1 = [e_n]_{A_2}$, this will prove the necessity of the condition of theorem 12.6.

Let $\{\alpha_j^{(k)}\}, \{\alpha_j\} \in A_2$ be such that $\lim_{k\to\infty} \{\alpha_j^{(k)}\} = \{\alpha_j\}$, i.e.,

$$\lim_{k\to\infty} \sup_{1 \leqslant n < \infty} \left\| \sum_{j=1}^{n} (\alpha_j^{(k)} - \alpha_j) x_j \right\| = 0.$$

Then for each $i = 1, 2, \ldots$ we have

$$|\alpha_i^{(k)} - \alpha_i| = \frac{\|(\alpha_i^{(k)} - \alpha_i) x_i\|}{\|x_i\|} = \frac{1}{\|x_i\|} \left\| \sum_{j=1}^{i} (\alpha_j^{(k)} - \alpha_j) x_j - \sum_{j=1}^{i-1} (\alpha_j^{(k)} - \alpha_j) x_j \right\|$$

$$\leqslant \frac{2}{\|x_i\|} \sup_{1 \leqslant n < \infty} \left\| \sum_{j=1}^{n} (\alpha_j^{(k)} - \alpha_j) x_j \right\| \to 0 \quad \text{as} \quad k \to \infty,$$

and thus A_2 is a BK-space.

To show that A_2 is γ-perfect, observe first that

$$A_2^{\gamma} = \{\{f(x_n)\} \mid f \in E^*\}. \tag{12.65}$$

Indeed, if $\{\beta_n\} \in A_2^{\gamma}$, then, since $\{f_n(x)\} \in A_2$ for all $x \in E$ (because $\{x_n\}$ is a basis of E), we have, by the definition of A_2^{γ}, $\sup_{1 \leqslant n < \infty} \left| \sum_{i=1}^{n} \beta_i f_i(x) \right| < \infty$ for all $x \in E$, whence, by the principle of uniform boundedness, $\sup_{1 \leqslant n < \infty} \left\| \sum_{i=1}^{n} \beta_i f_i \right\| < \infty$ and hence, by proposition 12.2a), there exists an $f \in E^*$ such that $f(x_n) = \beta_n$ $(n = 1, 2, \ldots)$; since $[x_n] = E$, this f is obviously unique. Conversely, if $f \in E^*$, then $\sup_{1 \leqslant n < \infty} \left| \sum_{i=1}^{n} f(x_i) \alpha_i \right| \leqslant \|f\| \sup_{1 \leqslant n < \infty} \left\| \sum_{i=1}^{n} \alpha_i x_i \right\| < \infty$ for all $\{\alpha_n\} \in A_2$, i.e., $\{f(x_n)\} \in A_2^{\gamma}$, which proves (12.65).

Now, if $\{\alpha_n\} \in A_2^{\gamma\gamma}$, then, by (12.65) and the definition of $A_2^{\gamma\gamma}$, we have $\sup_{1 \leqslant n < \infty} \left| f\left(\sum_{i=1}^{n} \alpha_i x_i\right) \right| = \sup_{1 \leqslant n < \infty} \left| \sum_{i=1}^{n} \alpha_i f(x_i) \right| < \infty$ for all $f \in E^*$, whence, by the principle of uniform boundedness, $\sup_{1\leqslant n<\infty}\left\|\sum_{i=1}^{n}\alpha_i x_i\right\|<\infty$, i.e., $\{\alpha_n\}\in A_2$. Thus $A_2^{\gamma\gamma}\subset A_2$, whence A_2 is γ-perfect. This proves that the condition is necessary.

Let us prove now that we also have $S^\gamma=A_2^\gamma$, whence $S^{\gamma\gamma}=A_2^{\gamma\gamma}=A_2$. By virtue of $S=A_1\subset A_2$ we have $S^\gamma\supset A_2^\gamma$. Assume, conversely, that $\{\beta_n\}\in S^\gamma$. Then, since $\{f_n(x)\}\in A_1=S$ for all $x\in E$ (because $\{x_n\}$ is a basis of E), we have, by the definition of S^γ, $\sup_{1\leqslant n<\infty}\left|\sum_{i=1}^{n}\beta_i f_i(x)\right|<\infty$ for all $x\in E$, whence, as in the above proof of (12.65), we infer that there exists an $f\in E^*$ with $f(x_n)=\beta_n$ $(n=1,2,\ldots)$ and hence, by (12.65), $\{\beta_n\}\in A_2^\gamma$, which proves $S^\gamma=A_2^\gamma$ and $S^{\gamma\gamma}=A_2$.

To prove the sufficiency of the condition of theorem 12.6, assume that it is satisfied, i.e., that S contains all unit vectors e_n and there exists a γ-perfect BK-space T such that $[e_n]_T=S$. We shall show that $\{e_n\}$ is a basic sequence in T, whence S is associated to the basis $\{x_n\}=\{e_n\}$ of the Banach space $E=S$.

By lemma 12.1 applied to the γ-dual space T^γ endowed with the norm (12.59), the space $T=T^{\gamma\gamma}$ is a BK-space also for the norm

$$|||\{\alpha_j\}|||=\sup_{\substack{\{\beta_j\}\in T^\gamma\\ \|\{\beta_j\}\|\leqslant 1}}\ \sup_{1\leqslant n<\infty}\left|\sum_{i=1}^{n}\beta_i\alpha_i\right|\qquad(\{\alpha_n\}\in T).\tag{12.66}$$

We claim that this norm is equivalent to the initial norm on T. Indeed, let us denote by T_1 the space T endowed with the norm (12.66) and let $u\colon x\to x$ be the identical mapping of T onto T_1. In order to show that u is continuous, let $x_n=\{\alpha_j^{(n)}\}$, $x=\{\alpha_j\}\in T$ be such that $\lim_{n\to\infty}x_n=x$, $\lim_{n\to\infty}u(x_n)=y=\{\beta_j\}\in T_1$. Then, since both T and T_1 are BK-spaces, we must have $\lim_{n\to\infty}\alpha_j^{(n)}=\alpha_j$, $\lim_{n\to\infty}\alpha_j^{(n)}=\beta_j$ $(j=1,2,\ldots)$, whence $\beta_j=\alpha_j$ $(j=1,2,\ldots)$, i.e., $y=u(x)$ and thus, by the closed graph theorem, u is continuous. A similar argument shows that u^{-1} is also continuous, which proves that (12.66) is equivalent to the initial norm on T.

Now, in the norm (12.66) we have, obviously,

$$\begin{aligned}\left|\left|\left|\sum_{i=1}^{n}\alpha_i e_i\right|\right|\right| &=\sup_{\substack{\{\beta_j\}\in T^\gamma\\ \|\{\beta_j\}\|\leqslant 1}}\ \sup_{1\leqslant k\leqslant n}\left|\sum_{i=1}^{k}\beta_i\alpha_i\right|\leqslant\sup_{\substack{\{\beta_j\}\in T^\gamma\\ \|\{\beta_j\}\|\leqslant 1}}\ \sup_{1\leqslant k\leqslant n+m}\left|\sum_{i=1}^{k}\beta_i\alpha_i\right|\\ &=\left|\left|\left|\sum_{i=1}^{n+m}\alpha_i e_i\right|\right|\right|\end{aligned}$$

for all finite sequences of scalars $\alpha_1,\ldots,\alpha_{n+m}$. Consequently, by § 7, theorem 7.1, $\{e_n\}$ is a basic sequence in T endowed with the norm (12.66), whence also in T with its initial norm. This proves the sufficiency part of theorem 12.6.

Finally, let us show that if the above condition is satisfied, we have necessarily $T=S^{\gamma\gamma}$. By $S\subset T$ we have $S^\gamma\supset T^\gamma$, whence $S^{\gamma\gamma}\subset T^{\gamma\gamma}=T$. Conversely, let $\{\alpha_j\}\in T$. Then, since by the above the norm (12.66) is equivalent to the initial norm on T, there exists a constant $C>0$ such that

$$\left\|\sum_{i=1}^{n}\alpha_i e_i\right\| \leqslant C\left|\left|\left|\sum_{i=1}^{n}\alpha_i e_i\right|\right|\right| = C\sup_{\substack{\{\beta_j\}\in T^\gamma\\ \|\{\beta_j\}\|\leqslant 1}}\ \sup_{1\leqslant k\leqslant n}\left|\sum_{i=1}^{k}\beta_i\alpha_i\right| \leqslant C|||\{\alpha_j\}||| \qquad (n=1,2,\ldots),$$

whence $\sup_{1\leqslant n<\infty}\left\|\sum_{i=1}^{n}\alpha_i e_i\right\|<\infty$, i.e., $\{\alpha_n\}\in A_2$ (for the basis $\{e_n\}$ of S). Since we have proved in the above that $S^{\gamma\gamma}=A_2$, it follows that $T\subset A_2 = S^{\gamma\gamma}$, whence $T=S^{\gamma\gamma}$, which completes the proof of theorem 12.6.

As an application of sequence spaces associated to bases, we shall now consider the problem, which are the infinite matrices $a=(a_{ij})$ which "preserve bases". Let us first recall one more notion of the theory of sequence spaces.

The *β-dual* of a sequence space S is the sequence space S^β defined by

$$S^\beta = \left\{\{\beta_n\}\subset K \;\middle|\; \sum_{i=1}^{\infty}\beta_i\alpha_i \text{ converges for all } \{\alpha_n\}\in S\right\}. \tag{12.67}$$

For every sequence space S we have, obviously, $S^\beta\subset S^\gamma$. Hence, by lemma 12.1 and the argument of the second part of its proof, we obtain

Lemma 12.2. *If T is a BK-space containing all unit vectors e_n, then its β-dual T^β can be normed so as to become a BK-space, by* (12.59).

In the sequel we shall assume that T^β is endowed with this norm.

Proposition 12.3. a) *Let $\{y_n\}$ be a sequence in a Banach space E, such that $y_n\neq 0$ $(n=1,2,\ldots)$ and let $A_1=A_1(\{y_n\})$ be the Banach space of sequences of scalars introduced in* § 3, *proposition* 3.1. *Then $A_1(\{y_n\})^\beta \cong A_1(\{y_n\})^*$, by the mapping $\{\beta_n\}\to h$, where*

$$h(\{\alpha_n\})=\sum_{i=1}^{\infty}\beta_i\alpha_i \qquad (\{\alpha_n\}\in A_1(\{y_n\})). \tag{12.68}$$

The inverse mapping is given by

$$h\to\{h(e_n)\}, \tag{12.69}$$

where e_n are the unit vectors in $A_1(\{y_n\})$.

b) *If* $\{x_n\}$ *is a basis of* E, *then* $A_1(\{x_n\})^\beta \cong E^*$, *by the mapping* $\{\beta_n\} \to f$, *where*

$$f\left(\sum_{i=1}^{\infty} \alpha_i x_i\right) = \sum_{i=1}^{\infty} \beta_i \alpha_i \quad \left(\sum_{i=1}^{\infty} \alpha_i x_i \in E\right). \tag{12.70}$$

The inverse mapping is given by

$$f \to \{f(x_n)\}. \tag{12.71}$$

Proof. a) The mapping $\{\beta_n\} \to h$ is obviously linear. Furthermore, if for h defined by (12.68) we have $h=0$, then $\beta_n = h(e_n) = 0$ $(n=1,2,\ldots)$. Thus, the mapping $\{\beta_n\} \to h$ is one to one.

Now let $h \in A_1(\{y_n\})^*$ be arbitrary. Then for $\beta_n = h(e_n)$ $(n=1,2,\ldots)$ we have, taking into account that $\{e_n\}$ is a basis of $A_1(\{y_n\})$ (by § 8, proposition 8.1),

$$h(\{\alpha_n\}) = h\left(\sum_{i=1}^{\infty} \alpha_i e_i\right) = \sum_{i=1}^{\infty} \alpha_i h(e_i) = \sum_{i=1}^{\infty} \alpha_i \beta_i \quad (\{\alpha_n\} \in A_1(\{y_n\})). \tag{12.72}$$

Thus, $\{\beta_n\} \to h$ maps $A_1(\{y_n\})^\beta$ *onto* $A_1(\{y_n\})^*$ and the inverse mapping is given by (12.69).

Finally, we have

$$\|h\| = \sup_{\substack{\{\alpha_n\} \in A_1(\{y_n\}) \\ \|\{\alpha_n\}\| \leqslant 1}} \left|\sum_{i=1}^{\infty} \beta_i \alpha_i\right| \leqslant \sup_{\substack{\{\alpha_n\} \in A_1(\{y_n\}) \\ \|\{\alpha_n\}\| \leqslant 1}} \sup_{1 \leqslant n < \infty} \left|\sum_{i=1}^{n} \beta_i \alpha_i\right| = \|\{\beta_n\}\|,$$

whence, by the inversion theorem of Banach, the mapping $\{\beta_n\} \to h$ is an isomorphism of $A_1(\{y_n\})^\beta$ onto $A_1(\{y_n\})^*$.

b) If $\{x_n\}$ is a basis of E, then, by § 3, proposition 3.2, $A_1(\{x_n\}) \cong E$, by the mapping

$$u_{\{x_n\}}: \{\alpha_n\} \to \sum_{i=1}^{\infty} \alpha_i x_i, \tag{12.73}$$

and hence $E^* \cong A_1(\{x_n\})^*$, by the mapping $u_{\{x_n\}}^*: f \to h$, where

$$h(\{\alpha_n\}) = [u_{\{x_n\}}^*(f)](\{\alpha_n\}) = f[u_{\{x_n\}}(\{\alpha_n\})] = \sum_{i=1}^{\infty} \alpha_i f(x_i)$$
$$(\{\alpha_n\} \in A_1(\{x_n\})). \tag{12.74}$$

Consequently, taking into account part a) above, it follows that $E^* \cong A_1(\{x_n\})^\beta$, by the mapping $f \to \{h(e_n)\} = \{f(x_n)\}$, i.e., by (12.71), which, since the inverse of (12.71) is obviously the mapping $\{\beta_n\} \to f$ defined by (12.70), concludes the proof of proposition 12.3.

Remark 12.5. For the sake of completeness we observe that the isomorphism $(u^*_{\{x_n\}})^{-1}: A_1(\{x_n\})^* \to E^*$ is nothing else than the mapping $h \to f$, where

$$f\left(\sum_{i=1}^{\infty} \alpha_i x_i\right) = \sum_{i=1}^{\infty} \alpha_i f(x_i) = \sum_{i=1}^{\infty} \alpha_i h(e_i) \qquad \left(\sum_{i=1}^{\infty} \alpha_i x_i \in E\right), \tag{12.75}$$

which, together with part a) above, implies again that the mapping $\{\beta_n\} \to f$ defined by (12.70) is an isomorphism of $A_1(\{x_n\})^\beta$ onto E^*, inverse to (12.71).

Let us now formulate the problem of characterizing the infinite matrices which "preserve bases". Let $\{x_n\}$ be a basis of a Banach space E and let $\{y_n\} \subset E$, $y_n \neq 0$. Then there exists an (unique) infinite matrix $a = (a_{ij})$ such that

$$y_j = \sum_{i=1}^{\infty} a_{ij} x_i \qquad (j = 1, 2, \ldots), \tag{12.76}$$

or, with the matrix notations $\mathscr{X} = (x_n)$, $\mathscr{Y} = (y_n)$, such that

$$\mathscr{Y} = \mathscr{X} a. \tag{12.77}$$

We shall consider the problem, what conditions on $a = (a_{ij})$ are necessary and sufficient in order that $\{y_n\}$ be a basis of E. The following proposition will be useful:

Proposition 12.4. *Let $\{x_n\}$ be a basis of a Banach space E and let $\{y_n\} \subset E$, $y_n \neq 0$ $(n = 1, 2, \ldots)$ be defined by (12.76). Then*

a) *The matrix $a = (a_{ij})$ induces a continuous linear mapping $v_a: A_1(\{y_n\}) \to A_1(\{x_n\})$ defined by*

$$v_a(\{\alpha_n\}) = \left\{\sum_{j=1}^{\infty} a_{ij} \alpha_j\right\}_{i=1}^{\infty} \qquad (\{\alpha_n\} \in A_1(\{y_n\})), \tag{12.78}$$

or, with the matrix notation $\alpha = \begin{pmatrix} \alpha_1 \\ \alpha_2 \\ \vdots \end{pmatrix}$, *by*[1]

$$v_a(\alpha) = a\alpha. \tag{12.79}$$

b) *The transposed matrix $a^t = (a_{ji})$ induces a linear mapping $v_{a^t}: A_1(\{x_n\})^\beta \to A_1(\{y_n\})^\beta$ defined by*

$$v_{a^t}(\{\beta_n\}) = \left\{\sum_{i=1}^{\infty} a_{ij} \beta_i\right\}_{j=1}^{\infty} \qquad (\{\beta_n\} \in A_1(\{x_n\})^\beta), \tag{12.80}$$

---

[1] In (12.79) and (12.81) we consider $v_a(\alpha)$ and $v_{a^t}(\beta)$ as one-column matrices.

or with the matrix notation $\beta=\begin{pmatrix}\beta_1\\ \beta_2\\ \vdots\end{pmatrix}$, *by*

$$v_{a^t}(\beta)=a^t\beta, \tag{12.81}$$

and for the adjoint $v_a^*\colon A_1(\{x_n\})^*\to A_1(\{y_n\})^*$ *of* v_a *we have*

$$\{[v_a^*(h)](e_j^y)\}_{j=1}^{\infty}=v_{a^t}(\{h(e_n^x)\})\quad (h\in A_1(\{x_n\})^*), \tag{12.82}$$

where in the left side e_n^y *denote the unit vectors in* $A_1(\{y_n\})$ *and in the right side* e_n^x *denote the unit vectors in* $A_1(\{x_n\})$.

c) *Consequently, we have*

$$v_{a^t}=w_{\{y_n\}}^{-1}v_a^*\,w_{\{x_n\}}, \tag{12.83}$$

where $w_{\{x_n\}}, w_{\{y_n\}}$ *denote respectively the isomorphisms* $A_1(\{x_n\})^\beta\cong A_1(\{x_n\})^*$ *and* $A_1(\{y_n\})^\beta\cong A_1(\{y_n\})^*$ *of proposition* 12.3 a).

Proof. a) If $\{\alpha_n\}\in A_1(\{y_n\})$, then, by (12.76),

$$f_i\Big(\sum_{j=1}^{\infty}\alpha_j y_j\Big)=\sum_{j=1}^{\infty}\alpha_j f_i(y_j)=\sum_{j=1}^{\infty}\alpha_j a_{ij}\qquad (i=1,2,\ldots)$$

where $\{f_n\}\subset E^*$ is the a.s.c.f. to $\{x_n\}$. Hence each series $\sum_{j=1}^{\infty}a_{ij}\alpha_j$ converges and $\Big\{\sum_{j=1}^{\infty}a_{ij}\alpha_j\Big\}_{i=1}^{\infty}\in A_1(\{x_n\})$, namely,

$$\sum_{i=1}^{\infty}\Big(\sum_{j=1}^{\infty}a_{ij}\alpha_j\Big)x_i=\sum_{j=1}^{\infty}\alpha_j y_j. \tag{12.84}$$

Furthermore, the mapping v_a defined by (12.78) is obviously linear and for every $\{\alpha_n\}\in A_1(\{y_n\})$ we have

$$\begin{aligned}\|v_a(\{\alpha_n\})\|&=\sup_{1\leqslant n<\infty}\Big\|\sum_{i=1}^{n}\Big(\sum_{j=1}^{\infty}a_{ij}\alpha_j\Big)x_i\Big\|\leqslant v_{\{x_n\}}\Big\|\sum_{i=1}^{\infty}\Big(\sum_{j=1}^{\infty}a_{ij}\alpha_j\Big)x_i\Big\|\\ &=v_{\{x_n\}}\Big\|\sum_{j=1}^{\infty}\alpha_j y_j\Big\|\leqslant v_{\{x_n\}}\sup_{1\leqslant n<\infty}\Big\|\sum_{j=1}^{n}\alpha_j y_j\Big\|=v_{\{x_n\}}\|\{\alpha_n\}\|,\end{aligned}$$

whence $\|v_a\|\leqslant\|v_{\{x_n\}}\|$.

b) Let $h\in A_1(\{x_n\})^*$ be arbitrary. Then, since by (12.76) $\{a_{ij}\}_{i=1}^{\infty}\in A_1(\{x_n\})$ $(j=1,2,\ldots)$ and since $\{e_n^x\}$ is a basis of $A_1(\{x_n\})$, we have, taking also into account (12.78),

$$\begin{aligned}\{[v_a^*(h)](e_j^y)\}_{j=1}^{\infty}&=\{h[v_a(e_j^y)]\}_{j=1}^{\infty}=\{h(\{a_{ij}\}_{i=1}^{\infty})\}_{j=1}^{\infty}\\ &=\Big\{h\Big(\sum_{i=1}^{\infty}a_{ij}e_i^x\Big)\Big\}_{j=1}^{\infty}=\Big\{\sum_{i=1}^{\infty}a_{ij}h(e_i^x)\Big\}_{j=1}^{\infty}.\end{aligned} \tag{12.85}$$

Hence the series $\sum_{i=1}^{\infty} a_{ij} h(e_i^x)$ $(j=1,2,\dots)$ converge for all $h \in A_1(\{x_n\})^*$ and, since $v_a^*(h) \in A_1(\{y_n\})^*$, from proposition 12.3a) we infer that $\left\{\sum_{i=1}^{\infty} a_{ij} h(e_i^x)\right\}_{j=1}^{\infty} \in A_1(\{y_n\})^\beta$ for all $h \in A_1(\{x_n\})^*$. Consequently, again by proposition 12.3a), the series $\sum_{i=1}^{\infty} a_{ij} \beta_i$ $(j=1,2,\dots)$ converge and $\left\{\sum_{i=1}^{\infty} a_{ij} \beta_i\right\}_{j=1}^{\infty} \in A_1(\{y_n\})^\beta$ for all $\{\beta_n\} \in A_1(\{x_n\})^\beta$. Thus, $v_{a^t}: A_1(\{x_n\})^\beta \to A_1(\{y_n\})^\beta$ is well defined by (12.80). Furthermore, by (12.80) we have

$$v_{a^t}(\{h(e_n^x)\}) = \left\{\sum_{i=1}^{\infty} a_{ij} h(e_i^x)\right\}_{j=1}^{\infty} \qquad (h \in A_1(\{x_n\})^*),$$

which, together with (12.85), gives (12.82).

Finally, the linearity and continuity of v_{a^t} can be seen directly, but they also follow from formula (12.83), which will be proved below.

c) By the definition (12.69) of the isomorphisms $w_{\{y_n\}}^{-1}$, $w_{\{x_n\}}^{-1}$ and by (12.82) we have

$$v_{a^t} w_{\{x_n\}}^{-1}(h) = v_{a^t}(\{h(e_n^x)\}) = \{[v_a^*(h)](e_j^y)\}_{j=1}^{\infty} = w_{\{y_n\}}^{-1} v_a^*(h) \qquad (h \in A_1(\{x_n\})^*),$$

whence, putting $h = w_{\{x_n\}}(\{\beta_n\})$ $(\{\beta_n\} \in A_1(\{x_n\})^\beta)$, we infer (12.83), which completes the proof of proposition 12.4.

Now we can prove

Theorem 12.7. *Let $\{x_n\}$ be a basis of a Banach space E, let $\{y_n\} \subset E$, $y_n \neq 0$ $(n=1,2,\dots)$ and let $a=(a_{ij})$ be the (unique) infinite matrix satisfying (12.76). Furthermore, let $v_a: A_1(\{y_n\}) \to A_1(\{x_n\})$ and $v_{a^t}: A_1(\{x_n\})^\beta \to A_1(\{y_n\})^\beta$ be the mappings defined in proposition 12.4. The following statements are equivalent:*

1°. $\{y_n\}$ is a basis of the space E.

2°. $[y_n] = E$ and there exists a constant $C \geqslant 1$ such that we have, for any finite sequence of scalars $\alpha_1, \dots, \alpha_{n+m}$

$$\left\| \sum_{i=1}^{\infty} \left(\sum_{j=1}^{n} a_{ij} \alpha_j \right) x_i \right\| \leqslant C \left\| \sum_{i=1}^{\infty} \left(\sum_{j=1}^{n+m} a_{ij} \alpha_j \right) x_i \right\|. \tag{12.86}$$

3°. v_a maps $A_1(\{y_n\})$ one to one onto $A_1(\{x_n\})$.

4°. $a=(a_{ij})$ has an inverse $b=(b_{ij})$ such that

$$\{b_{ij}\}_{j=1}^{\infty} \in A_1(\{x_n\})^\beta \qquad (i=1,2,\dots), \tag{12.87}$$

$$\left\{\sum_{j=1}^{\infty} b_{ij} \gamma_j\right\}_{i=1}^{\infty} \in A_1(\{y_n\}) \qquad (\{\gamma_n\} \in A_1(\{x_n\})). \tag{12.88}$$

5°. *v_{a^t} maps $A_1(\{x_n\})^\beta$ one to one onto $A_1(\{y_n\})^\beta$.*

6°. *$v_a(A_1(\{y_n\}))=A_1(\{x_n\})$ and $v_{a^t}(A_1(\{x_n\})^\beta)=A_1(\{y_n\})^\beta$.*

7°. (i) *v_a is one to one on $A_1(\{y_n\})$*; (ii) *$v_a(A_1(\{y_n\}))$ is closed in $A_1(\{x_n\})$*; (iii) *v_{a^t} is one to one on $A_1(\{x_n\})^\beta$.*

8°. (i) *v_a is one to one on $A_1(\{y_n\})$*; (ii) *$v_{a^t}(A_1(\{x_n\})^\beta)$ is closed in $A_1(\{y_n\})^\beta$*; (iii) *v_{a^t} is one to one on $A_1(\{x_n\})^\beta$.*

9°. (i) $A_1(\{y_n\})^\beta=\{\{f(y_n)\}\mid f\in E^*\}$; (ii) *$v_{a^t}$ is one to one on $A_1(\{x_n\})^\beta$.*

Proof. By (12.84) we have

$$\sum_{i=1}^{\infty}\left(\sum_{j=1}^{n} a_{ij}\alpha_j\right)x_i=\sum_{j=1}^{n}\alpha_j y_j.$$

and hence the equivalence 1°⇔2° is a consequence of § 7, theorem 7.1.

1°⇔3°. Let us first observe that in any case we have

$$u_{\{y_n\}}=u_{\{x_n\}}v_a, \tag{12.89}$$

where $u_{\{x_n\}}$ denotes the isomorphism (12.73) of $A_1(\{x_n\})$ onto E and $u_{\{y_n\}}$ denotes the continuous linear mapping $\{\alpha_n\}\to\sum_{j=1}^{\infty}\alpha_j y_j$ of $A_1(\{y_n\})$ into E. Indeed, by (12.78) and (12.84),

$$\begin{aligned} u_{\{x_n\}}v_a(\{\alpha_n\})&=u_{\{x_n\}}\left(\left\{\sum_{j=1}^{\infty}a_{ij}\alpha_j\right\}_{i=1}^{\infty}\right)=\sum_{i=1}^{\infty}\left(\sum_{j=1}^{\infty}a_{ij}\alpha_j\right)x_i\\ &=\sum_{j=1}^{\infty}\alpha_j y_j=u_{\{y_n\}}(\{\alpha_n\})\qquad(\{\alpha_n\}\in A_1(\{y_n\})).\end{aligned}$$

Consequently[1], v_a is onto if and only if for each $x\in E$ there exist scalars α_j such that $x=\sum_{j=1}^{\infty}\alpha_j y_j$ and v_a is one to one if and only if $\{y_n\}$ is ω-linearly independent.

1°⇒4°. If $\{y_n\}$ is a basis of E, there exists a (unique) matrix $b=(b_{ij})$ such that

$$x_j=\sum_{i=1}^{\infty}b_{ij}y_i\qquad(j=1,2,\ldots). \tag{12.90}$$

Hence, by proposition 12.4a) applied for $\{x_n\}$ interchanged with $\{y_n\}$, for every $\{\gamma_n\}\in A_1(\{x_n\})$ each series $\sum_{j=1}^{\infty}b_{ij}\gamma_j$ converges and

---

[1] Let us also mention the following alternative argument: by (12.89) we have 3°, or equivalently (by the inversion theorem of Banach), v_a is an isomorphism of $A_1(\{y_n\})$ onto $A_1(\{x_n\})$, if and only if $u_{\{y_n\}}$ is an isomorphism of $A_1(\{y_n\})$ onto E, which holds (since by § 8, proposition 8.1, $\{e_n\}$ is a basis of $A_1(\{y_n\})$ and since $u_{\{y_n\}}(e_j)=y_j$ for $j=1,2,\ldots$) if and only if $\{y_n\}$ is a basis of E.

$\left\{\sum_{j=1}^{\infty} b_{ij}\gamma_j\right\}_{i=1}^{\infty} \in A_1(\{y_n\})$, i.e., we have (12.87) and (12.88). Furthermore, by (12.84) for $\{x_n\}$ interchanged with $\{y_n\}$,

$$\sum_{i=1}^{\infty}\left(\sum_{j=1}^{\infty} b_{ij}\gamma_j\right) y_i = \sum_{j=1}^{\infty} \gamma_j x_j \qquad (\{\gamma_j\} \in A_1(\{x_n\})). \tag{12.91}$$

Since by (12.76) we have $\{\gamma_j\} = \{a_{jk}\}_{j=1}^{\infty} \in A_1(\{x_n\})$, applying (12.91) to this particular sequence we obtain

$$\sum_{i=1}^{\infty}\left(\sum_{j=1}^{\infty} b_{ij}a_{jk}\right) y_i = \sum_{j=1}^{\infty} a_{jk} x_j = y_k \qquad (k=1,2,\ldots),$$

whence, since $\{y_n\}$ is a basis of E,

$$\sum_{j=1}^{\infty} b_{ij}a_{jk} = \delta_{ik} \qquad (i, k=1,2,\ldots),$$

i.e., ba = the identity matrix. Similarly, from (12.84) for $\{\alpha_j\} = \{b_{jk}\}_{j=1}^{\infty} \in A_1(\{y_n\})$ (by (12.90)), taking also into account that $\{x_n\}$ is a basis of E, we obtain that ab = the identity matrix, and thus b is the inverse of a.

$4^\circ \Rightarrow 3^\circ$. Assume that we have 4°. If for a sequence $\{\alpha_n\} \in A_1(\{y_n\})$ we have $v_a(\{\alpha_n\})=0$, i.e., $a\alpha = \left(\sum_{j=1}^{\infty} a_{ij}\alpha_j\right)_{i=1}^{\infty} = 0$, then, by 4°, $\alpha = ba\alpha = 0$, i.e., $\alpha_n = 0$ $(n=1,2,\ldots)$. Thus, $v_a\colon A_1(\{y_n\}) \to A_1(\{x_n\})$ is one to one.

Now let $\gamma = \{\gamma_n\} \in A_1(\{x_n\})$ be arbitrary. Then, by 4°, $b\gamma \in A_1(\{y_n\})$ and $\gamma = ab\gamma = v_a(b\gamma)$. Thus, v_a maps $A_1(\{y_n\})$ onto $A_1(\{x_n\})$.

$3^\circ \Leftrightarrow 5^\circ$. We have 3° if and only if v_a^* maps $A_1(\{x_n\})^*$ one to one onto $A_1(\{y_n\})^*$, which, by proposition 12.4, happens if and only if we have 5°.

$3^\circ \Rightarrow 6^\circ$ by $3^\circ \Rightarrow 5^\circ$.

The proof of the implication $6^\circ \Rightarrow 3^\circ$ is similar to the above proof of the implication $5^\circ \Rightarrow 3^\circ$.

$3^\circ \Leftrightarrow 7^\circ$. $v_a(A_1(\{y_n\}))$ is dense in $A_1(\{x_n\})$ if and only if[1] v_a^* is one to one on $A_1(\{x_n\})^*$, which, by proposition 12.4, happens if and only if we have (iii) of 7°. Consequently, v_a maps $A_1(\{y_n\})$ onto $A_1(\{x_n\})$ if and only if we have (ii) and (iii) of 7°.

$3^\circ \Rightarrow 8^\circ$. If we have 3°, then we obviously have (i) if 8° and, by the implication $3^\circ \Rightarrow 5^\circ$ proved above, we also have (ii) and (iii) of 8°.

$8^\circ \Rightarrow 7^\circ$. If we have 8°, then by (ii) and by proposition 12.4, $v_a^*(A_1(\{x_n\})^*)$ is closed in $A_1(\{y_n\})^*$, whence there follows[2] (ii) of 7°.

---

[1] See e.g. [50], p. 479, lemma 8.

[2] See e.g. [50], p. 488, theorem 4.

$1° \Leftrightarrow 9°$. By § 8, theorem 8.2, $\{x_n\}$ is a basis of E if and only if we have (i) of 9° and $[y_n]=E$. Now, if v_{a^t} is not one to one on $A_1(\{x_n\})^\beta$, then there exists a sequence $\{\beta_n\}\in A_1(\{x_n\})^\beta$, $\{\beta_n\}\neq 0$, such that $v_{a^t}(\{\beta_n\}) = \left\{\sum_{i=1}^{\infty} a_{ij}\beta_i\right\}_{j=1}^{\infty} = 0$. Hence, for the functional $f\in E^*$ defined by (12.70) we have $f\neq 0$ and

$$f(y_j)=f\left(\sum_{i=1}^{\infty} a_{ij}x_i\right) = \sum_{i=1}^{\infty} a_{ij}\beta_i=0 \qquad (j=1,2,\ldots)$$

and therefore $[y_n]\neq E$. On the other hand, if v_{a^t} is one to one on $A_1(\{x_n\})^\beta$ and $f\in E^*$, $f(y_j)=0$ $(j=1,2,\ldots)$, then $\{f(x_n)\}\in A_1(\{x_n\})^\beta$ and

$$v_{a^t}(\{f(x_n)\}) = \left\{\sum_{i=1}^{\infty} a_{ij}f(x_i)\right\}_{j=1}^{\infty} = \left\{f\left(\sum_{i=1}^{\infty} a_{ij}x_i\right)\right\}_{j=1}^{\infty} = \{f(y_j)\}_{j=1}^{\infty}=0,$$

whence $f=0$ and therefore $[y_n]=E$. This completes the proof of theorem 12.7.

Remark 12.6. In the above proof we have also shown that the following statements are equivalent:

1°. $[y_n]=E$.

2°. $v_a(A_1(\{y_n\}))$ is dense in $A_1(\{x_n\})$.

3°. v_{a^t} is one to one on $A_1(\{x_n\})^\beta$.

This observation, together with theorem 12.7, yields necessary and sufficient conditions in order that $\{y_n\}$ be a basic sequence in E.

Definition 12.2. We shall say that a sequence space S is *a multiplier algebra for a basis of a Banach space*, if there exists a Banach space E with a basis $\{x_n\}$ such that $S=M(E,(x_n,f_n))$, where $\{f_n\}\subset E^*$ is the a.s.c.f. to $\{x_n\}$.

We shall consider the problem of characterization of such sequence spaces.

We recall that a sequence space S is called a *BK-algebra* if S is a BK-space and S is closed under coordinatewise multiplication, i.e., S is a subalgebra of the algebra of all sequences. In a BK-algebra S the functions $x\to yx$ and $x\to xy$ are continuous in x for fixed y by the continuity of the coordinate functionals and the closed graph theorem. Consequently[1], every BK-algebra is a Banach algebra (in general, without identity).

Theorem 12.8. *A sequence space T is the multiplier algebra for a basis of a Banach space if and only if T is a γ-perfect BK-algebra containing all unit vectors e_n and the identity $e=\{1,1,1,\ldots\}$.*

---

[1] See [51], p. 860—861.

Proof. Assume that $T=M(E,(x_n,f_n))$, where $\{x_n\}$ is a basis of a Banach space E, with the a.s.c.f. $\{f_n\}$. Then $e_n\in T$ $(n=1,2,\ldots)$, $e\in T$ and by § 5, proposition 5.4a), T is a Banach algebra for the norm (5.29). Furthermore, by § 5, proof of corollary 5.2a), the coordinate functionals are continuous on T. Finally, by § 5, corollary 5.2, T is the space $A_2=A_2(\{e_n\})$ for the basis $\{e_n\}$ of $S=[e_n]_T$, whence, by the above proof of the necessity part of theorem 12.6, T is γ-perfect (moreover, $T=S^{\gamma\gamma}$).

Conversely, assume now that T is a γ-perfect BK-algebra containing all unit vectors e_n and the identity $e=\{1,1,1,\ldots\}$. Then, by the above proof of the sufficiency part of theorem 12.6, $\{e_n\}$ is a basic sequence in T (moreover, we also have $T=[e_n]_T^{\gamma\gamma}$ $=\left\{\{\alpha_n\}\subset K \;\middle|\; \sup_{1\leqq n<\infty}\left\|\sum_{i=1}^{n}\alpha_i e_i\right\|<\infty\right\}$). We shall prove that

$$T=M(S,(e_n,h_n)), \tag{12.92}$$

where $S=[e_n]_T$ and where $\{h_n\}$ is the a.s.c.f. to the basis $\{e_n\}$ of S, which will complete the proof.

Let $\{\gamma_n\}\in T$ be arbitrary. Then, since T is a BK-algebra, the mapping $v_{\{\gamma_n\}}$ defined by

$$v_{\{\gamma_n\}}(\{\alpha_n\})=\{\gamma_n\alpha_n\}=\{\gamma_n\}\{\alpha_n\} \qquad (\{\alpha_n\}\in T) \tag{12.93}$$

is a continuous linear mapping of T into T, as we have observed above. Since for any finite linear combination $\sum_{i=1}^{m}\alpha_i e_i\in S$ we have $v_{\{\gamma_n\}}\left(\sum_{i=1}^{m}\alpha_i e_i\right)$ $=\sum_{i=1}^{m}\gamma_i\alpha_i e_i$, it follows that $v_{\{\gamma_n\}}(S)\subset S=[e_n]_T$ and hence, since $\{e_n\}$ is a basis of S, $\sum_{i=1}^{\infty}\gamma_i\alpha_i e_i$ converges for every $\sum_{i=1}^{\infty}\alpha_i e_i\in S$. Consequently, $\{\gamma_n\}\in M(S,(e_n,h_n))$ and thus $T\subset M(S,(e_n,h_n))$.

Now let $\{\gamma_n\}\in M(S,(e_n,h_n))$ and $\{\beta_n\}\in T^{\gamma}$ be arbitrary. Then, by the above proof of theorem 12.6, formula (12.65), there exists a functional $h\in T^*$ such that

$$h(e_n)=\beta_n \qquad (n=1,2,\ldots). \tag{12.94}$$

Furthermore, by § 5, proposition 5.4b), for the mapping $v_{\{\gamma_n\}}$ defined by (12.93) we have $v_{\{\gamma_n\}}|_S\in L(S,S)$. Since $e\in T$, we have $bv\subset[e]^{\gamma\gamma}$ $\subset T^{\gamma\gamma}=T$ (because for any $\{\alpha_n\}\in bv$ and $\{\beta'_n\}\in[e]^{\gamma}$ we have $\sum_{i=1}^{\infty}|\alpha_i-\alpha_{i+1}|<\infty$, $\sup_{1\leqq i<\infty}\left|\sum_{j=1}^{i}\beta'_j\right|<\infty$, whence, by $\sum_{i=1}^{n}\alpha_i\beta'_i$

$=\sum_{i=1}^{n}(\alpha_i-\alpha_{i+1})\sum_{j=1}^{i}\beta'_j+\alpha_{n+1}\sum_{j=1}^{n}\beta'_j$, we get $\sup_{1\leqslant n<\infty}\left|\sum_{i=1}^{n}\alpha_i\beta'_i\right|<\infty$, i.e., $\{\alpha_n\}\in[e]^{\gamma\gamma}$). Hence, since the sequence $\left\{\sum_{i=1}^{m}e_i\right\}$ is bounded in bv and since the inclusion map of BK-spaces is continuous (by the same argument as that used in the above proof of formula (12.66)), it follows that $\left\{\sum_{i=1}^{m}e_i\right\}\subset S\subset T$ is bounded in S. Consequently,

$$\left|\sum_{i=1}^{n}\gamma_i\beta_i\right|=\left|\sum_{i=1}^{n}\gamma_i h(e_i)\right|=\left|h\left(\sum_{i=1}^{n}\gamma_i e_i\right)\right|=\left|h\left[v_{\{\gamma_n\}}\left(\sum_{i=1}^{n}e_i\right)\right]\right|$$
$$\leqslant\|h\|\,\|v_{\{\gamma_n\}}|_S\|\sup_{1\leqslant m<\infty}\left\|\sum_{i=1}^{m}e_i\right\|\qquad(n=1,2,\ldots),$$

whence $\{\gamma_n\}\in T^{\gamma\gamma}=T$. Thus $M(S,(e_n,h_n))\subset T$, which, together with the opposite inclusion proved above, gives (12.92). This completes the proof of theorem 12.8.

§ 13. Bases in topological linear spaces. Weak bases and bounded weak bases in Banach spaces. Weak* bases and bounded weak* bases in conjugate Banach spaces

The notion of a basis has the following natural extension to general topological linear spaces:

Definition 13.1. A sequence $\{x_n\}$ is a topological linear space U is called a *basis* of U if for every $x\in U$ there exists a *unique* sequence of scalars $\{\alpha_n\}\subset K$ such that

$$x=\sum_{i=1}^{\infty}\alpha_i x_i, \tag{13.1}$$

convergence of the series being that of the topology on U.

A topological linear space U which possesses a basis $\{x_n\}$ is clearly separable[1], since the set of all finite linear combinations $\sum r_i x_i$, where the r_i are rational scalars, is countable and dense in U. It is natural to ask whether or not the converse is also true, i.e. whether or not every separable topological linear space U possesses a basis. The answer is negative, as shown by

---

[1] I.e. (see e.g. [50], p. 21, definition 11) there exists a countable subset of U, which is dense in U; see also the Notes and remarks to this §.

Example 13.1. Let U be a separable complete metric linear space, such that the only continuous linear functional on U is $f \equiv 0$ (e.g. one can take $U = S([0,1])$ = the space of all equivalence classes of measurable functions on $[0,1]$, endowed with the metric $\rho(\tilde{x},\tilde{y}) = \int_0^1 \frac{|x(t)-y(t)|}{1+|x(t)-y(t)|}\,dt$, or $U = L^p([0,1])$ with $0<p<1$, where a fundamental system of neighbourhoods of 0 is given by the sets $\left\{x \in U \,\middle|\, \int_0^1 |x(t)|^p\,dt < \varepsilon\right\}$, $\varepsilon > 0$). Then U has no basis. In fact, let us sketch the proof of this statement. With a method similar to that used in the proof of § 3, theorem 3.1, one can prove[1] that the coefficient functionals f_n associated to any basis $\{x_n\}$ of U (see § 14, definition 14.1) are continuous on U. Hence, by our hypothesis, $f_n \equiv 0$ $(n=1,2,\ldots)$. However, this is impossible.

We shall not present here the theory of bases in topological linear spaces, since it would take us beyond the scope of this book. We shall consider only the particular cases when U is a Banach space endowed with one of the usual associated locally convex topologies; it will turn out that the bases of these spaces are closely related to the bases of Banach spaces endowed with their norm topologies.

Definition 13.2. A sequence $\{x_n\}$ is a Banach space E is called a *weak basis (bounded weak basis)* of E, if $\{x_n\}$ is a basis of the topological linear space U obtained by endowing E with the weak topology $\sigma(E,E^*)$ (respectively, with the bounded weak topology[2]). A sequence $\{h_n\}$ in the conjugate space E^* (of a Banach space E) is called a *weak* basis (bounded weak* basis)* of E^*, if $\{h_n\}$ is a basis of the topological linear space U obtained by endowing E^* with the weak* topology $\sigma(E^*,E)$ (respectively, with the bounded weak* topology). For the sake of brevity, we shall use for such bases the terms w-basis, bw-basis, w^*-basis and bw^*-basis respectively.

Lemma 13.1. a) *Let $\{y_n\}$ be a sequence in a Banach space E and let $y \in E$. Then, $w\text{-}\lim_{n\to\infty} y_n = y$ if and only if $bw\text{-}\lim_{n\to\infty} y_n = y$.*

b) *Let $\{g_n\}$ be a sequence in the conjugate space E^* (of a Banach space E) and let $g \in E^*$. Then, $w^*\text{-}\lim_{n\to\infty} g_n = g$ if and only if $bw^*\text{-}\lim_{n\to\infty} g_n = g$.*

---

[1] See the Notes and remarks to § 3.

[2] We recall that the bounded weak topology (bw-topology) for E is the strongest topology which coincides with the weak topology $\sigma(E,E^*)$ (w-topology) on each t-cell $\{x \in E \mid \|x\| \leq t\}$ of E. The bounded weak* topology (bw^*-topology) for E^* is the strongest topology which coincides with the weak* topology $\sigma(E^*,E)$ (w^*-topology) on each t-cell of E^*.

Proof. a) If $bw\text{-}\lim_{n\to\infty} y_n = y$, then, since the bw-topology is stronger than the w-topology, we have $w\text{-}\lim_{n\to\infty} y_n = y$.

Conversely, if $w\text{-}\lim_{n\to\infty} y_n = y$, then there exists a $t>0$ such that $\|y_n\|<t$ $(n=1,2,\dots)$ and $\|y\|<t$. Let $\mathscr{U}$ be an arbitrary open bw-neighbourhood of y and put $tS_E = \{x\in E \mid \|x\| \leqslant t\}$. Then $\mathscr{U}\cap tS_E$ is relatively w-open in tS_E. Hence there is a w-neighbourhood $\mathscr{V}$ of y such that $\mathscr{V}\cap tS_E \subset \mathscr{U}\cap tS_E$. Since $w\text{-}\lim_{n\to\infty} y_n = y$, there exists a positive integer N such that $y_n\in\mathscr{V}\cap tS_E$ for all $n\geqslant N$, whence $y_n\in\mathscr{U}$ for all $n\geqslant N$ and so $bw\text{-}\lim_{n\to\infty} y_n = y$.

The proof of part b) is similar.

Theorem 13.1. *Let be a Banach space, $\{x_n\}$ a sequence in E, and $\{h_n\}$ a sequence in E^*. Then*

a) *The following statements are equivalent:*

1°. *$\{x_n\}$ is a w-basis of E.*

2°. *$\{x_n\}$ is a bw-basis of E.*

3°. *$\{x_n\}$ is a basis of E.*

b) *The following statements are equivalent:*

1°. *$\{h_n\}$ is a w*-basis of E*.*

2°. *$\{h_n\}$ is a bw*-basis of E*.*

Each of these latter statements implies the following:

3°. *$\{h_n\}$ is a basic sequence.*

4°. *$r([h_n])>0$ (where $r([h_n])$ = the characteristic of the subspace $[h_n]$ of E^*).*

Proof. The equivalences 1°⇔2° of a) and b) are immediate consequences of lemma 13.1.

a) 3°⇒1°. If $\{x_n\}$ is a basis of E, every $x\in E$ has an expansion $x=\sum_{i=1}^{\infty}\alpha_i x_i$, whence also a w-expansion $f\left(\sum_{i=1}^{n}\alpha_i x_i\right)\to f(x)$ $(f\in E^*)$. This w-expansion is unique, since from the relations $f\left(\sum_{i=1}^{n}\alpha_i x_i\right)\to 0$ $(f\in E^*)$ we obtain, taking $f=f_j$ (where $\{f_n\}$ is the sequence of coefficient functionals associated to the basis $\{x_n\}$), that $\alpha_j=0$ $(j=1,2,\dots)$. Thus $\{x_n\}$ is a w-basis of E.

If we have a) 1°, then the w-closed linear span of $\{x_n\}$ coincides with E, whence[1] also $[x_n]=E$. Therefore, in order to prove a) 1°⇒3°, it is sufficient to prove that a) 1°⇒$\{x_n\}$ is a basic sequence. However, this latter implication, as well as the implication b) 1°⇒3°, are consequences of the following more general result:

[1] See e.g. [10], p. 134, theorem 2.

Proposition 13.1. *Let E be a Banach space, and let M be a subset of the conjugate space E^*, having the following two properties:*

($\mathscr{P}_1$) *If $\{z_n\} \subset E$ and $f(z_n) \to 0$ for all $f \in M$, then $\sup_{1 \leqslant n < \infty} \|z_n\| < \infty$.*

($\mathscr{P}_2$) *For every $\varepsilon > 0$ there exists a $\delta > 0$ such that the relations $\|x_n\| \leqslant \delta$ $(n = 1, 2, \ldots)$, $f(x_n) \to f(x_0)$ $(f \in M)$ imply $\|x_0\| \leqslant \varepsilon$.*

Then every "basis with respect to M" of the space E, i.e. every sequence $\{x_n\} \subset E$ such that for each $x \in E$ there exists a unique sequence of scalars $\{\alpha_n\} \subset K$ satisfying $f\left(\sum_{i=1}^{n} \alpha_i x_i\right) \to f(x)$ $(f \in M)$, is a basic sequence.

Proof. For the sake of brevity, we shall say that a sequence $\{z_n\} \subset E$ is (M)-convergent to $x \in E$, and we shall write $z_n \xrightarrow{(M)} x$, or $(M)\text{-}\lim_{n \to \infty} z_n = x$, if we have $f(z_n) \to f(x)$ $(f \in M)$. From ($\mathscr{P}_2$) it follows that the (M)-limit is unique, i.e. that the relations $z_n \xrightarrow{(M)} x$, $z_n \xrightarrow{(M)} x'$ imply $x = x'$ (since $z_n - z_n \xrightarrow{(M)} x - x'$ and $\|z_n - z_n\| \leqslant \delta$ for all $\delta > 0$). If $\sum_{i=1}^{n} y_i \xrightarrow{(M)} x$, then we shall write $(M) \sum_{i=1}^{\infty} y_i = x$.

Assume now that $\{x_n\}$ is a basis with respect to M of the space E, and let $A_1^{(M)}$ be the linear space of sequences of scalars

$$A_1^{(M)} = \left\{ \{\alpha_n\} \subset K \;\middle|\; \sum_{i=1}^{\infty} \alpha_i x_i \quad (M)\text{-converges to an element of } E \right\}, \tag{13.2}$$

endowed with the norm

$$\|\{\alpha_n\}\| = \sup_{1 \leqslant n < \infty} \left\| \sum_{i=1}^{n} \alpha_i x_i \right\|. \tag{13.3}$$

By ($\mathscr{P}_1$) the number (13.3) is finite for every $\{\alpha_n\} \in A_1^{(M)}$. Since $\{x_n\}$ is a basis with respect to M of the space E, we have $x_n \neq 0$ $(n = 1, 2, \ldots)$, whence (13.3) is indeed a norm on $A_1^{(M)}$.

We claim that $A_1^{(M)}$ is a Banach space. In fact, let $\{\alpha_n^{(k)}\}$ $(k = 1, 2, \ldots)$ be a Cauchy sequence in $A_1^{(M)}$. Then for every $\varepsilon > 0$ there exists a positive integer $N(\varepsilon)$ such that for $k, m > N(\varepsilon)$ we have

$$\left\| \sum_{i=1}^{n} \alpha_i^{(k)} x_i - \sum_{i=1}^{n} \alpha_i^{(m)} x_i \right\| < \min(\varepsilon, \delta) \qquad (n = 1, 2, \ldots), \tag{13.4}$$

where $\delta > 0$ is chosen as in ($\mathscr{P}_2$). Hence there exists, by the arguments used in the proof of § 3, proposition 3.1, a sequence of scalars $\{\alpha_n\}$ such that

$$\lim_{m \to \infty} \alpha_n^{(m)} = \alpha_n \qquad (n = 1, 2, \ldots). \tag{13.5}$$

Let us prove that $\{\alpha_n\} \in A_1^{(M)}$. Put

$$y_k = (M) \sum_{i=1}^{\infty} \alpha_i^{(k)} x_i \qquad (k = 1, 2, \ldots).$$

Then, by (13.4) and ($\mathscr{P}_2$) we have

$$\|y_k - y_m\| \leqslant \varepsilon \qquad (k, m > N(\varepsilon)),$$

whence, by the completeness of the space E, there exists an $x \in E$ such that

$$\lim_{k \to \infty} y_k = x. \tag{13.6}$$

We shall show that $x = (M) \sum_{i=1}^{\infty} \alpha_i x_i$, which will prove that $\{\alpha_n\} \in A_1^{(M)}$. We have the inequalities

$$\begin{aligned} \left| f\left(x - \sum_{i=1}^{n} \alpha_i x_i \right) \right| \leqslant |f(x - y_k)| + \left| f\left(y_k - \sum_{i=1}^{n} \alpha_i^{(k)} x_i \right) \right| \\ + \left| f\left(\sum_{i=1}^{n} \alpha_i^{(k)} x_i - \sum_{i=1}^{n} \alpha_i x_i \right) \right| \qquad (f \in M;\ n, k = 1, 2, \ldots). \end{aligned} \tag{13.7}$$

Let $\eta > 0$ and $f \in M$ be arbitrary, and put

$$\varepsilon = \frac{\eta}{4\|f\|}. \tag{13.8}$$

Then by (13.6) there exists a $k > N(\varepsilon)$ (where $N(\varepsilon)$ is as in the above), such that $\|x - y_k\| < \varepsilon$, whence, by (13.8),

$$|f(x - y_k)| \leqslant \|f\| \, \varepsilon = \frac{\eta}{4}. \tag{13.9}$$

On the other hand, since $k > N(\varepsilon)$, from (13.4) for $m \to \infty$ we obtain, by (13.5),

$$\left\| \sum_{i=1}^{n} \alpha_i^{(k)} x_i - \sum_{i=1}^{n} \alpha_i x_i \right\| \leqslant \min(\varepsilon, \delta) \qquad (n = 1, 2, \ldots), \tag{13.10}$$

whence, by (13.8),

$$\left| f\left(\sum_{i=1}^{n} \alpha_i^{(k)} x_i - \sum_{i=1}^{n} \alpha_i x_i \right) \right| \leqslant \frac{\eta}{4} \qquad (n = 1, 2, \ldots). \tag{13.11}$$

Furthermore, since $y_k = (M) \sum_{i=1}^{\infty} \alpha_i^{(k)} x_i$, there exists, by the definition of (M)-convergence, a positive integer n_0 such that we have

$$\left|f\left(y_k - \sum_{i=1}^{n} \alpha_i^{(k)} x_i\right)\right| \leqslant \frac{\eta}{4} \qquad (n>n_0). \tag{13.12}$$

Thus, by (13.7), (13.9), (13.11) and (13.12) we have

$$\left|f\left(x - \sum_{i=1}^{n} \alpha_i x_i\right)\right| \leqslant \frac{3}{4}\eta<\eta \qquad (n>n_0),$$

which proves that $x=(M) \sum_{i=1}^{\infty} \alpha_i x_i$ and that $\{\alpha_n\} \in A_1^{(M)}$.

Now, by (13.10) (which is actually true for all $k>N(\varepsilon)$) and (13.3) we have

$$\|\{\alpha_n^{(k)}\}-\{\alpha_n\}\| \leqslant \min(\varepsilon,\delta) \qquad (k>N(\varepsilon)),$$

whence $\lim_{k\to\infty}\{\alpha_n^{(k)}\}=\{\alpha_n\}$ in $A_1^{(M)}$. This proves that $A_1^{(M)}$ is a Banach space.

Hence, by arguments similar to those used in the proof of § 3, proposition 3.2a), the mapping

$$\{\alpha_n\} \to (M) \sum_{i=1}^{\infty} \alpha_i x_i \tag{13.13}$$

is an isomorphism of $A_1^{(M)}$ onto E. Consequently, as in the proof of § 3, theorem 3.1, the coefficient functionals $f_n(x)=\alpha_n$ $(x\in E,\ n=1,2,\ldots)$ are continuous on E, i.e. $f_n \in E^*$ $(n=1,2,\ldots)$. Thus, by the implication $3^\circ \Rightarrow 1^\circ$ of § 4, theorem 4.1, $\{x_n\}$ is a basic sequence, which completes the proof of proposition 13.1.

Now, applying proposition 13.1 for $M=E^*$, we obtain the implication a) $1^\circ \Rightarrow 3^\circ$ of theorem 13.1 (taking into account that for a w-basis $\{x_n\}$ of E we have $[x_n]=E$, as remarked above). Finally, applying proposition 13.1 for E^* instead of E, $\{h_n\}$ instead $\{x_n\}$, and for M = the canonical image of E in E^{**}, we obtain the implication b) $1^\circ \Rightarrow 3^0$.

b) $1^\circ \Rightarrow 4^\circ$. Let us denote by V the subspace $[h_n]$ of E^*, by V^1 the set of all w^*-limits of all w^*-convergent subsequences of V, and by $V^{(1)}$ the union of the w^*-closures of the t-cells $\{f\in V \mid \|f\| \leqslant t\}$ of V, for all $t<\infty$. If we have 1°, then every $f\in E^*$ has a w^*-expansion $f(x)=\lim_{n\to\infty} \sum_{i=1}^{n} \alpha_i h_i(x)$ $(x\in E)$, whence $V^1=E^*$. Since[1] $V^1 \subset V^{(1)}$, it follows that $V^{(1)}=E^*$, whence, by a theorem of J. Dixmier[2], we have $r(V)>0$, i.e. 4°. This completes the proof of theorem 13.1.

---

[1] See e.g. [47], theorem 1(a).

[2] See e.g. [47], theorem 6.

One cannot replace b) 3° by the assertion: $\{h_n\}$ is a basis of E^*. In fact, it may happen that E^* has a w^*-basis, but it is not even separable, as shown by

Example 13.2. Let $E=l^1$. Then the unit vectors in $E^*=m$ constitute a w^*-basis of E^* (see § 14, theorem 14.1), but E^* is not separable (and hence it has no basis at all).

Conversely, it is natural to ask whether or not a basis $\{h_n\}$ of E^* is also a w^*-basis of E^*. The answer is again negative, as shown by

Example 13.3. Let $E=c_0$ and define $\{h_n\}\subset E^*=l^1$ by

$$h_1=f_1, \quad h_n=f_{n-1}-f_n \quad (n=2,3,\ldots), \tag{13.14}$$

where $\{f_n\}$ is the natural basis of l^1. Then $\{h_n\}$ is a basis of E^*, but not a w^*-basis of E^*.

Indeed, observe first that by § 4, proposition 4.3, the sequence

$$y_n=\{\underbrace{1,\ldots,1}_{n},0,0,\ldots\} \quad (n=1,2,\ldots) \tag{13.15}$$

is a basis of $E=c_0$. Identifying canonically l^1 with $E^*=c_0^*$, the sequence $\{h_{n+1}\}$ is obviously the sequence of coefficient functionals associated to the basis $\{y_n\}$ of $E=c_0$, whence, by § 12, theorem 12.1, $\{h_{n+1}\}$ is a basic sequence. Now, $h_1\notin[h_{n+1}]$, since for the functional $\Phi_0(f)=\sum_{i=1}^{\infty}\zeta_i$ $(f=\{\zeta_n\}\in E^*\equiv l^1)$ we have $\Phi_0(h_1)=1$, $\Phi_0(h_{n+1})=0\,(n=1,2,\ldots)$ and $\{h_n\}$ is complete in E^*, since the relations $\Phi\in E^{**}$, $\Phi(h_n)=0$ $(n=1,2,\ldots)$ imply $\Phi=0$. Consequently, $\{h_n\}$ is a basis of E^*. Furthermore, although every $f\in E^*$ has a w^*-expansion $f(x)=\sum_{i=1}^{\infty}\alpha_i h_i(x)$ $(x\in E)$ (since $\{h_n\}$ is a basis of E^*), this w^*-expansion is not unique, since by $h_1(x)-\sum_{i=2}^{n}h_i(x)$ $=f_n(x)$ $(x\in E,\ n=2,3,\ldots)$ we have $h_1(x)-\sum_{i=2}^{\infty}h_i(x)=0$ $(x\in E)$. Thus, $\{h_n\}$ is not a w^*-basis of E^*.

Since there exists no locally convex space U satisfying the hypotheses of example 13.1, it is natural to ask whether or not every separable locally convex space (in particular[1], every w^*-separable conjugate Banach space) possesses a basis (respectively, a w^*-basis). The answer is negative, as shown by

Example 13.4. The conjugate space $E^*=m^*$ of the Banach space $E=m$ is w^*-separable, since the sequence $\{f_n\}\subset E^*$ defined by

---

[1] Note that every conjugate Banach space E^* endowed with the w^*-topology $\sigma(E^*,E)$ is a sequentially complete locally convex space.

$$f_n(x)=\xi_n \qquad (x=\{\xi_k\}\in E,\ n=1,2,\ldots) \tag{13.16}$$

is total on E (whence its w^*-closed linear hull is E^*). However, the space $E^*=m^*$ has no w^*-basis. In fact, assume the contrary, i.e. that m^* has a w^*-basis $\{f_n\}$. Then, by definition, for every $f\in m^*$ there exists a unique sequence of scalars $\{\alpha_n\}\subset K$ such that

$$\lim_{n\to\infty}\left[f(x)-\sum_{i=1}^{n}\alpha_i f_i(x)\right]=0 \qquad (x\in m).$$

However, by a result of Grothendieck[1], from these relations it follows that

$$\lim_{n\to\infty}\Phi\left(f-\sum_{i=1}^{n}\alpha_i f_i\right)=0 \qquad (\Phi\in m^{**}).$$

Consequently, m^* is w-separable, whence[2] also separable for the norm-topology. However, this is impossible (since m itself is not separable). Thus m^* has no w^*-basis.

Problem 13.1. a) Does the conjugate space E^* of a separable Banach space E possess a w^*-basis? b) What about a separable conjugate Banach space E^*? c) What about a conjugate Banach space E^* having a basis?

Concerning a), we shall see in § 14 that if E has a basis, then E^* has a w^*-basis. However, the converse problem is unsolved:

Problem 13.2. a) If the conjugate space E^* (of a Banach space E) has a w^*-basis, does the space E possess a basis? b) Under the same hypothesis, is the space E separable?

§ 14. Schauder bases in topological linear spaces. Properties of weak duality for bases in Banach spaces

Definition 14.1. Let $\{x_n\}$ be a basis of a topological linear space U. The sequence of linear functionals $\{f_n\}$ defined by

$$f_j(x)=\alpha_j \qquad \left(x=\sum_{i=1}^{\infty}\alpha_i x_i\in U,\ j=1,2,\ldots\right) \tag{14.1}$$

is called *the sequence of coefficient functionals associated to the basis* $\{x_n\}$ (we shall write: a.s.c.f.).

---

[1] Let us recall this result of Grothendieck (see e.g. [43], Ch. VI, § 4, proposition (4)(a), or [88], p. 168): If $\{g_n\}\subset m^*$ and $\lim_{n\to\infty} g_n(x)=0$ for all $x\in m$, then $\lim_{n\to\infty}\Phi(f_n)=0$ for all $\Phi\in m^{**}$.

[2] See e.g. [10], p. 134, theorem 2.

Thus, if $\{x_n\}$ is a basis of U and $\{f_n\}$ the a.s.c.f., then every $x \in U$ has a unique expansion of the form

$$x = \sum_{i=1}^{\infty} f_i(x) x_i. \tag{14.2}$$

Definition 14.2. A basis $\{x_n\}$ of a topological linear space U is said to be a *Schauder basis* of U, if all coefficient functionals f_n $(n=1,2,\ldots)$ are continuous on U (i.e. if $f_n \in U^*$ for $n=1,2,\ldots$).

By § 3, theorem 3.1, every basis of a Banach space E is a Schauder basis of this space. However, we shall see that in other locally convex spaces[1] this is, in general, no longer true. Among the properties of bases which are not Schauder bases, let us mention the following:

Proposition 14.1. *Let $\{x_n\}$ be a basis of a topological linear space U, which is not a Schauder basis of U, and let $\{f_n\}$ be the a.s.c.f. If $f_1 \notin U^*$, then*

a) $x_1 \in [x_n]_2^{\infty} = U$.

b) $\{x_n\}_2^{\infty}$ *is not a basis of* $[x_n]_2^{\infty} = U$.

Proof. a) If $x_1 \notin [x_n]_2^{\infty}$, then there exists a linear[2] functional f on U such that

$$f(x_1)=1, \quad f(y)=0 \quad (y \in [x_n]_2^{\infty}).$$

Since $\{x_n\}$ is a basis of U, every $x \in U$ has a unique expansion $x = \alpha_1 x_1 + \sum_{i=2}^{\infty} \alpha_i x_i$ and hence the null-subspace of f is $[x_n]_2^{\infty}$. Since this subspace is closed, f is continuous on U, i.e. $f \in U^*$. On the other hand, for every $x = \sum_{i=1}^{\infty} f_i(x) x_i \in U$ we have

$$f(x) = f\left(\sum_{i=1}^{\infty} f_i(x) x_i\right) = f_1(x) f(x_1) + f\left(\sum_{i=2}^{\infty} f_i(x) x_i\right) = f_1(x)$$

and hence $f_1 = f \in U^*$, which contradicts the assumption $f_1 \notin U^*$. Thus $x_1 \in [x_n]_2^{\infty}$, whence $U = [x_n] = [x_1, [x_n]_2^{\infty}] = [x_n]_2^{\infty}$.

[1] When considering more general topological linear spaces, it is obvious that those spaces on which the only continuous linear functional if $f \equiv 0$, do not admit any Schauder basis.

[2] A priori we don't know that f is continuous, since U is an arbitrary topological linear space.

b) We claim that x_1 has no expansion with respect to $\{x_n\}_2^\infty$, whence $\{x_n\}_2^\infty$ is not a basis of $[x_n]_2^\infty = U$. Indeed, if $x_1 = \sum_{i=2}^{\infty} \beta_i x_i$, then $x_1 - \sum_{i=2}^{\infty} \beta_i x_i = 0$, and thus 0 has two distinct expansions $\sum_{i=1}^{\infty} 0 . x_i$ and $x_1 - \sum_{i=2}^{\infty} \beta_i x_i$, which contradicts the assumption that $\{x_n\}$ is a basis of U. This completes the proof of proposition 14.1.

In the sequel we shall be concerned again with the particular cases when U is a Banach space E endowed with one of the locally convex topologies considered in § 13. Since a linear functional on E is continuous if and only if it is w-continuous, which happens if and only if it is bw-continuous, it follows that in a Banach space E the notions of Schauder basis, w-Schauder basis and bw-Schauder basis are equivalent. Moreover, by theorems 3.1 and 13.1 these notions are also equivalent to the notions of basis, w-basis and bw-basis. In conjugate Banach spaces endowed with norm-, w^*- and bw^*-topologies, the situation is different. Since a linear functional Φ on E^* is w^*-continuous if and only if it is bw^*-continuous on E^*, it follows that in E^* the notions of w^*-Schauder basis and bw^*-Schauder basis are equivalent. However, a w^*-Schauder basis of E^* need not be a (Schauder) basis of E^*, as shown e.g. by the sequence considered in § 13, example 13.2. Conversely, a (Schauder) basis of E^* need not be a w^*-Schauder basis of E^*, as shown by § 13, example 13.3. Furthermore, a w^*-basis of E^* need not be a w^*-Schauder basis of E^*, as shown by

Example 14.1. Let $E = c_0$ and let $\{f_n\}$ be the sequence in $E^* \equiv l^1$ defined by

$$f_1 = e_1, \quad f_n = (-1)^{n+1} e_1 + e_n \quad (n = 2, 3, \ldots), \tag{14.3}$$

where $\{e_n\}$ denotes the natural basis of l^1. Then $\{f_n\}$ is a basis of E^* in the norm-topology and a w^*-basis of E^*, but not a w^*-Schauder basis of this space.

Indeed, let $\{\Phi_n\} \subset E^{**}$ be the a.s.c.f. to the basis $\{e_n\}$ of E^* and put

$$\chi_1(f) = \Phi_1(f) + \sum_{j=2}^{\infty} (-1)^j \Phi_j(f) \quad (f \in E^*), \quad \chi_n = \Phi_n \quad (n = 2, 3, \ldots). \tag{14.4}$$

Then (f_n, χ_n) is a biorthogonal system and we have

$$\sum_{i=1}^{n} \chi_i(f) f_i = \left[\Phi_1(f) + \sum_{j=2}^{\infty} (-1)^j \Phi_j(f) \right] e_1 + \sum_{j=2}^{n} \Phi_j(f) [(-1)^{j+1} e_1 + e_j]$$

$$= \sum_{j=1}^{n} \Phi_j(f) e_j + \left[\sum_{j=n+1}^{\infty} (-1)^j \Phi_j(f) \right] e_1 \quad (f \in E^*, n = 1, 2, \ldots),$$

whence, since $\sum_{j=1}^{\infty} |\Phi_j(f)| < \infty \ (f \in E^* \equiv l^1)$, for every $\varepsilon > 0$ and $f \in E^*$ there exists an integer $N(\varepsilon, f) > 0$ such that

$$\left\| \sum_{j=1}^{n} \chi_j(f) f_j - \sum_{j=1}^{n} \Phi_j(f) e_j \right\| < \varepsilon \qquad (n > N(\varepsilon, f)).$$

Hence $f = \sum_{j=1}^{\infty} \chi_j(f) f_j$ for all $f \in E^*$, and thus, by virtue of the implication $2^\circ \Rightarrow 1^\circ$ of § 4, theorem 4.1, $\{f_n\}$ is a basis of E^* in the norm-topology.

Furthermore, from the above it follows that every $f \in E^*$ also has a w^*-expansion

$$f(x) = \sum_{j=1}^{\infty} \chi_j(f) f_j(x) \qquad (x \in E),$$

and we shall now show that this expansion is unique. Assume that for a sequence of scalars $\{\alpha_n\}$ we have

$$\sum_{j=1}^{\infty} \alpha_j f_j(x) = 0 \qquad (x \in E). \tag{14.5}$$

Since for $x = b_n \ (n = 1, 2, \ldots)$, where $\{b_n\}$ is the natural basis of $E = c_0$, we have, by (14.3),

$$f_j(b_1) = (-1)^{j+1}, \qquad f_j(b_n) = \delta_{jn} \qquad (n = 2, 3, \ldots;\ j = 1, 2, \ldots),$$

from (14.5) for $x = b_n \ (n = 1, 2, \ldots)$ it follows that

$$\sum_{j=1}^{\infty} (-1)^{j+1} \alpha_j = 0, \qquad \alpha_2 = \alpha_3 = \cdots = \alpha_n = \cdots = 0,$$

whence $\alpha_n = 0 \ (n = 1, 2, \ldots)$. Thus $\{f_n\}$ is a w^*-basis of E^*.

On the other hand, by (14.4) we have $\chi_1 \notin \pi(E)$, whence χ_1 is not w^*-continuous, and thus $\{f_n\}$ is not a w^*-Schauder basis of E^*. This completes the proof of our assertion.

We can now summarize the preceding results concerning a sequence $\{h_n\} \subset E^*$ in the following diagram of implications and non-implications:

$$\begin{array}{ccc}
\left[\begin{array}{c}\text{Schauder basis}\\ (\Leftrightarrow \text{basis})\end{array}\right] & \begin{array}{c}\not\Longrightarrow\\ \not\Longleftarrow\end{array} & \left[\begin{array}{c}w^*\text{-Schauder basis}\\ (\Leftrightarrow b\,w^*\text{-Schauder basis})\end{array}\right]\\
\Downarrow \ \not\Uparrow & \begin{array}{c}\not\nwarrow\\ \not\searrow\end{array} & \Downarrow \ \not\Uparrow\\
\left[\begin{array}{c}\text{Basic sequence}\\ \text{with } r([h_n]) > 0\end{array}\right] & \begin{array}{c}\Longleftarrow\\ \not\Longrightarrow\end{array} & \left[\begin{array}{c}w^*\text{-basis}\\ (\Leftrightarrow b\,w^*\text{-basis})\end{array}\right]
\end{array} \tag{14.6}$$

We have the following theorem of weak duality:

Theorem 14.1. *A sequence $\{f_n\}$ in a conjugate Banach space E^* is a w^*-Schauder basis of E^* if and only if E has a basis $\{x_n\}$ whose a.s.c.f. is $\{f_n\}$.*

Proof. Assume that $\{f_n\}$ is a w^*-Schauder basis of E^*. Then the associated coefficient functionals

$$\alpha_n = \phi_n(f) \quad (n=1,2,\ldots)$$

are all w^*-continuous on E^*, whence there exists a sequence $\{x_n\} \subset E$ such that

$$\phi_n(f) = f(x_n) \quad (f \in E^*,\ n=1,2,\ldots).$$

Then, from $\phi_i(f_j) = \delta_{ij}$ we infer that

$$f_j(x_i) = \delta_{ij} \quad (i,j=1,2,\ldots),$$

i.e. that (x_n, f_n) is a biorthogonal system. Since $\{f_n\}$ is a w^*-basis of E^*, with the a.s.c.f. $\{\phi_n\}$, we have

$$\lim_{n\to\infty} \sum_{i=1}^{n} f(x_i) f_i(x) = \lim_{n\to\infty} \sum_{i=1}^{n} \phi_i(f) f_i(x) = f(x) \quad (x \in E,\ f \in E^*). \tag{14.7}$$

Hence, for the sequence of partial sum operators $s_n(x) = \sum_{i=1}^{n} f_i(x) x_i$ $(x \in E,\ n=1,2,\ldots)$ we have $\sup_{1\leqslant n<\infty} \|s_n(x)\| < \infty$ $(x \in E)$. Thus, by the implication $3^\circ \Rightarrow 1^\circ$ of § 4, theorem 4.1, $\{x_n\}$ is a basic sequence. We claim that $[x_n] = E$. In fact, otherwise there exists an $x_0 \in E$, $x_0 \notin [x_n]$, whence also an $f \in E^*$ with $f_0(x_0) = 1$, $f_0(x_n) = 0$ $(n=1,2,\ldots)$. Thus, by (14.7), $0 = \sum_{i=1}^{n} f_0(x_i) f_i(x_0) \to f_0(x_0) = 1$, which is impossible. This proves that $[x_n] = E$, and thus $\{x_n\}$ is a basis of E, with the a.s.c.f. $\{f_n\}$.

Conversely, assume now that $\{x_n\}$ is a basis of E with the a.s.c.f. $\{f_n\}$. Then

$$f(x) = \sum_{i=1}^{\infty} f(x_i) f_i(x) \quad (x \in E,\ f \in E^*), \tag{14.8}$$

i.e. every $f \in E^*$ has a w^*-representation by $\{f_n\}$. In order to prove the uniqueness of this w^*-representation, assume that for a sequence of scalars $\{\alpha_n\} \subset K$ we have

$$\lim_{n\to\infty} \sum_{i=1}^{n} \alpha_i f_i(x) = 0 \quad (x \in E).$$

Then for $x = x_j$ $(j=1,2,\ldots)$, taking into account the biorthogonality relations $f_i(x_j) = \delta_{ij}$, we obtain

$$\alpha_j = 0 \quad (j = 1, 2, \ldots),$$

and thus $\{f_n\}$ is a w^*-basis of E^*. Finally, the coefficient functionals

$$\phi_n(f) = f(x_n) \quad (f \in E^*,\ n = 1, 2, \ldots)$$

of the w^*-representation (14.8) are obviously w^*-continuous on E^*, and thus $\{f_n\}$ is a w^*-Schauder basis of E^*, which completes the proof of theorem 14.1.

An immediate consequence of theorem 14.1 is the following:

Corollary 14.1. *The conjugate space E^* of a Banach space E has a w^*-Schauder basis if and only if the space E has a basis. Consequently, the problem whether the conjugate space E^* of an arbitrary separable Banach space E possesses a weak* Schauder basis is equivalent to the basis problem (problem 1.1).*

However, the general w^*-Schauder basis problem (i.e.: does every w^*-separable conjugate Banach space E^* possess a w^*-Schauder basis?) and hence also the Schauder basis problem in locally convex spaces (i. e.: does every separable locally convex space U possess a Schauder basis?) has a negative answer, as shown by § 13, example 13.4.

Consider now the following assertions concerning a Banach space E:

1°. E is separable.
2°. E has a basis ($\Leftrightarrow E^*$ has a w^*-Schauder basis).
3°. E^* is w^*-separable.
4°. E^* is separable.
5°. E^* has a basis.
6°. E^* has a w^*-basis.

Then the preceding results concerning these assertions can be summarized in the following table of implications:

Table 14.1.

	1°	2°	3°	4°	5°	6°
1°	+	?	+	$-l^1$	$-l^1$	?
2°	+	+	+	$-l^1$	$-l^1$	+
3°	$-m$	$-m$	+	$-l^1, m$	$-l^1, m$	$-m$
4°	+	?	+	+	?	?
5°	+	?	+	+	+	?
6°	?	?	+	$-l^1$	$-l^1$	+

Here the symbol + (respectively −) in the n-th row and m-th column means that the implication $n^\circ \Rightarrow m^\circ$ is (is not) valid, while the symbol ? means that the problem of the validity of the implication is unsolved. The only implications which have not been proved in the preceding, are $1^\circ \Rightarrow 3^\circ$ and $4^\circ \Rightarrow 1^\circ$, but they are well known[1]. On the right side of each symbol − we have also written the concrete spaces E which yield the desired counter-examples (naturally, there exist also other counter-examples). The symbols ? are nothing else but the problems 1.1, 12.1, 12.2, 13.1 and 13.2; we have not formulated separately problems concerning w^*-Schauder bases in E^*, since they are equivalent to the corresponding problems concerning bases in E (by corollary 14.1). We have not considered the property of existence of a basic sequence in E^*, since *such sequences exist in every Banach space* (indeed, this follows from Ch. II, § 15, proposition 15.1, since every separable Banach space E can be embedded isomorphically in a space with a basis, namely, in $C([0,1])$ and since obviously there exists a sequence $\{y_n\} \subset E$ such that $\|y_n\|=1$, $f_i(y_n)=0$ for $i=1,\dots,n$; $n=1,2,\dots$). One can also combine some of the above hypotheses, e.g., the question whether 2° and 4° together imply 5° is nothing else than the first question in Ch. II, § 5, problem 5.1 a).

It is natural to raise the problem of the extension of the known results for bases in Banach spaces to bases in locally convex spaces and, in particular, to w^*-Schauder bases in conjugate Banach spaces. We shall not treat this problem in detail, since this would take us beyond the scope of the present book, but we shall give here an example which shows that this extension is not always possible. For this purpose, let us consider the Krein-Milman-Rutman stability theorem (see corollary 10.2). It is natural to give

Definition 14.3. We shall say that a topological linear space U having a basis (Schauder basis) has *the KMR property for bases (Schauder bases)* if every dense linear subspace of U contains a basis (Schauder basis) of the space U.

By § 10, corollary 10.2 and § 3, theorem 3.1, every Banach space with a basis has the KMR property for (Schauder) bases. However, for locally convex spaces this is, in general, no longer true, as shown by

Example 14.2. Let $E=c_0$ and let G be a $\sigma(E^*,E)$-dense norm-closed linear subspace of $E^* \equiv l^1$, of characteristic $r(G)=0$[2]. Then E^* has a w^*-Schauder basis (by corollary 14.1), but G contains no w^*-basis of the space E^*, and thus E^* endowed with the w^*-topology does not have the KMR-property for w^*-bases or w^*-Schauder bases. In fact, assum-

---

[1] See e.g. [10], p. 124, theorem 4 and p. 189, theorem 12.

[2] Such a subspace exists, see e.g. [159], [47]; see also [33], p.121, exercise 14f).

ing the contrary, let $\{g_n\} \subset G$ be a w^*-basis of the space E^*. Then, by the implication $1^\circ \Rightarrow 4^\circ$ of § 13, theorem 13.1 b), $r([g_n]) \neq 0$. On the other hand, from $[g_n] \subset G$ and $r(G)=0$ it follows that $r([g_n]) \leqslant r(G)=0$, i.e. $r([g_n])=0$. This contradiction proves our assertion.

Remark 14.1. It is natural to ask whether or not with the method of example 14.2 one can also obtain an example of a separable Banach space having no basis, i.e. a negative solution of the basis problem. However, the answer is negative. In fact, all known subspaces $G \subset E^*$ with $r(G)=0$ possess bases and thus the statement 4° of § 13, theorem 13.1 b) does not remain valid if we replace in it the w^*-basis $\{h_n\}$ of E^* by a basic sequence $\{h_n\}$ in E^*.

§ 15. (*e*)-Schauder bases and (*b*)-Schauder bases in topological linear spaces

Definition 15.1. Let $\{x_n\}$ be a Schauder basis of a topological linear space U, with the a.s.c.f. $\{f_n\}$. The sequence of continuous linear operators $\{s_n\}$, where

$$s_n(x) = \sum_{i=1}^{n} f_i(x) x_i \qquad (x \in U,\ n=1,2,\ldots), \tag{15.1}$$

is called *the sequence of partial sum operators associated to the basis* $\{x_n\}$.

In the case when U is a Banach space, we have, by the implication $1^\circ \Rightarrow 4^\circ$ of § 4, theorem 4.1, $\sup_{1 \leqslant n < \infty} \|s_n\| < \infty$. A corresponding property can be defined for Schauder bases in general topological linear spaces in several ways, of which the following two seem to be useful:

Definition 15.2. Let $\{x_n\}$ be a Schauder basis of a topological linear space U, and let $\{s_n\}$ be the associated sequence of partial sum operators. $\{x_n\}$ is said to be

a) an (*e*)-Schauder basis of U, if 0 is a point of equicontinuity of the sequence $\{s_n\}$ (or, equivalently[1], if $\{s_n\}$ is an equicontinuous subset of the linear space $L(U,U)$ of all continuous linear mappings of U into itself).

b) a (*b*)-Schauder basis of U, if $\{s_n\}$ is a bounded subset of the space $L(U,U)$ endowed with the topology of uniform convergence on bounded subsets of U.

Since every equicontinuous subset of $L(U,U)$ is bounded[2], every (*e*)-Schauder basis of U is a (*b*)-Schauder basis of U. In the case when U is a Banach space, every basis of U is an (*e*)-Schauder basis and a

---

[1] See e.g. [32], p. 9, proposition 7.

[2] See e.g. [33], p. 26, proposition 7.

(b)-Schauder basis, as we have observed above. However, for general topological linear spaces this is no longer valid. To illustrate this, we shall consider here the case of Banach spaces endowed with w-topologies and conjugate Banach spaces with w^*-topologies.

Proposition 15.1. *Let E be an infinite-dimensional Banach space. Then*

a) *E has no w-(e)-Schauder basis.*

b) *E^* has no w^*-(e)-Schauder basis.*

c) *If E has a basis, then every basis of E is a w-(b)-Schauder basis of E.*

d) *If E^* has a basis which is also a w^*-Schauder basis of E^*, then every such basis of E^* is a w^*-(b)-Schauder basis of E^*.*

Proof. b) Assume that $\{f_n\}$ is a w^*-Schauder basis of E^*. Then, by § 14, theorem 14.1, E has a basis $\{x_n\}$ whose a.s.c.f. is $\{f_n\}$. Let

$$x_0=\sum_{i=1}^{\infty}\frac{1}{2^i\|x_i\|}x_i. \tag{15.2}$$

Then

$$f_n(x_0)=\frac{1}{2^n\|x_n\|}\qquad(n=1,2,\ldots). \tag{15.3}$$

Consider now the w^*-neighbourhood

$$W=W_{x_0;1}(0)=\{f\in E^*\mid |f(x_0)|<1\},$$

and let $V=V_{y_1,\ldots,y_m;\varepsilon}(0)$ be an arbitrary w^*-neighbourhood of 0. Since $[y_1,\ldots,y_m]\neq E$, there exists an $f\in E^*$, $f\neq 0$, such that $f(x)=0$ for all $x\in[y_1,\ldots,y_m]$, whence $\alpha f\in V$ for all scalars $\alpha\in K$. On the other hand, since $f\neq 0$ and since $f=w^*\text{-}\lim\limits_{n\to\infty}\sum_{i=1}^{n}f(x_i)f_i$, there exists a positive integer n such that $f(x_n)\neq 0$. Let n_0 be the least positive integer with this property. Then, denoting by $\mathscr{S}_n$ the sequence of partial sum operators associated to the w^*-Schauder basis $\{f_n\}$ of E^* (i.e. $\mathscr{S}_n(f)=\sum_{i=1}^{n}f(x_i)f_i$, $f\in E^*$, $n=1,2,\ldots$), and putting

$$\alpha=\frac{2^{n_0+1}\|x_{n_0}\|}{|f(x_{n_0})|},$$

we have, taking into account (15.3),

$$\begin{aligned}|[\mathscr{S}_{n_0}(\alpha f)](x_0)|&=\left|\sum_{i=1}^{n_0}\alpha f(x_i)f_i(x_0)\right|=|\alpha|\,|f(x_{n_0})|\,|f_{n_0}(x_0)|\\&=\frac{2^{n_0+1}\|x_{n_0}\|\,|f(x_{n_0})|}{|f(x_{n_0})|\,2^{n_0}\|x_0\|}=2,\end{aligned}$$

and thus $\mathscr{S}_{n_0}(\alpha f) \notin W$, which proves that $\{f_n\}$ is not a w-(e)-Schauder basis of E^*.

a) Since the intersection of the null spaces of a finite number of non-zero continuous linear functionals on E contains a non-zero element, an argument similar to that used in b) above proves a).

d) Let $\{f_n\}$ be a basis of E^* which is also a w^*-Schauder basis of E^* and let $\{\mathscr{S}_n\}$ be the associated sequence of partial sum operators. Then, by § 4, theorem 4.1, we have $\sup_{1 \leqslant n < \infty} \|\mathscr{S}_n\| < \infty$. Let A be an arbitrary w^*-bounded subset of E^* and $V = V_{y_1, \ldots, y_m; \varepsilon}(0)$ an arbitrary w^*-neighbourhood of 0 in E^*. We have to prove that there exists a $\lambda > 0$ such that $\bigcup_{n=1}^{\infty} \mathscr{S}_n(A) \subset \lambda V$, i.e. such that

$$|[\mathscr{S}_n(f)](y_i)| < \lambda \varepsilon \quad (f \in A,\ n = 1, 2, \ldots;\ i = 1, \ldots, m). \tag{15.4}$$

Since A is a w^*-bounded subset of E^*, it is also strongly bounded, i.e. $\sup_{f \in A} \|f\| < \infty$. Consequently, taking $\lambda > \frac{1}{\varepsilon} \sup_{1 \leqslant n < \infty} \|\mathscr{S}_n\| \sup_{f \in A} \|f\| \max_{1 \leqslant i \leqslant n} \|y_i\|$, we have (15.4), which proves that $\{f_n\}$ is a w^*-(b)-Schauder basis of E^*.

c) Since every w-bounded subset of a Banach space E is strongly bounded, an argument similar to that used in d) above proves c). This completes the proof of proposition 15.1.

§ 16. Some remarks on bases in normed linear spaces

In the preceding three sections we have defined bases, Schauder bases, (e)-Schauder bases and (b)-Schauder bases in general topological linear spaces U, and we have considered the particular cases when U is a Banach space endowed with one of the usual locally convex topologies: w, bw, w^*, bw^* (and of course, the norm topology). Now we shall consider another important particular case, when U is a normed linear space N (with its norm topology). We shall see that the results for bases in Banach spaces are no longer valid for bases in general normed linear spaces.

Firstly, let us observe that in normed linear spaces there exist bases which are not Schauder bases, as shown by

Example 16.1. Let $\{x_n\}$ be an ω-linearly independent but non-minimal sequence in a Banach space E, such that $[x_n] = E$ (see § 6, example 6.1) and let N be the linear space

$$N = \left\{ \sum_{i=1}^{\infty} \alpha_i x_i \in E \;\middle|\; \{\alpha_n\} \subset K,\ \sum_{i=1}^{\infty} \alpha_i x_i \text{ converges} \right\}, \tag{16.1}$$

endowed with the norm induced by E. Then $\{x_n\}$ is a basis of N, but not a Schauder basis of N. Indeed, if there exists a sequence $\{\phi_n\} \subset N^*$ such that $\phi_i(x_j) = \delta_{ij}$ $(i,j=1,2,\ldots)$, then, since $\bar{N}=E$, one can extend each ϕ_n, by continuity, to an $f_n \in E^*$ and we have, obviously, $f_i(x_j)=\delta_{ij}$ $(i,j=1,2,\ldots)$, in contradiction with the assumption that $\{x_n\}$ is not minimal.

The same example also shows that a basis $\{x_n\}$ of a normed linear space N need not be a basis of the completion $E=N^\wedge$ of N.

Furthermore, in normed linear spaces there exist Schauder bases which are neither (e)-Schauder bases, nor (b)-Schauder bases, as shown by

Example 16.2. Let E be a separable Banach space, let (x_n, f_n) $(\{x_n\} \subset E, \{f_n\} \subset E^*)$ be an E-complete biorthogonal system such that $\{x_n\}$ is not a basis of E (see e.g. § 4, example 4.1; we have seen in § 4, remark 4.1 that every separable Banach space E admits such a biorthogonal system) and let N be the linear space

$$N = \{x \in E \mid \lim_{n \to \infty} s_n(x) = x\} \tag{16.2}$$

(where $s_n(x) = \sum_{i=1}^{n} f_i(x) x_i$, $x \in E$, $n=1,2,\ldots$), endowed with the norm induced by E. Then $\{x_n\}$ is a Schauder basis of N, but not a (b)-Schauder basis of N. In fact, if $\sup_{1 \leq n < \infty} \|s_n|_N\| < \infty$, then, since $\bar{N}=E$, we also have $\sup_{1 \leq n < \infty} \|s_n\| < \infty$, whence, by § 4, theorem 4.1, $\{x_n\}$ is a basis of E, in contradiction with our hypothesis. Since for every normed linear space N the notions of (b)-Schauder basis and (e)-Schauder basis are obviously equivalent, it follows that $\{x_n\}$ is not an (e)-Schauder basis of N.

The same example also shows that a Schauder basis $\{x_n\}$ of a normed linear space N need not be a basis of the completion $E=N^\wedge$ of N. Furthermore, the arguments used above also show that a basis $\{x_n\}$ of a normed linear space N is a basis of the completion $E=N^\wedge$ of N if and only if $\{x_n\}$ is a (b)-Schauder basis (or, equivalently, an (e)-Schauder basis) of N. From this latter remark it follows that the following existence problems are equivalent:

1°. The basis problem (problem 1.1).

2°. Does every separable normed linear space N posses a basis?

3°. Does every separable normed linear space N possess a Schauder basis?

4°. Does every separable normed linear space N possess a (b)-Schauder basis (an (e)-Schauder basis)?

In fact, if the answer to 1° is affirmative, then so is the answer to 4° (by the preceding remark), whence so is the answer to 3°, whence so is the answer to 2°. On the other hand, it is obvious that if the answer to 2° is affirmative, then so is the answer to 1°.

Finally, let us observe that problems 2° and 3° above can be decomposed into the sum of problem 1° and respectively one of the following unsolved problems:

Problem 16.1. a) Does every *incomplete* separable normed linear space N possess a basis? b) What about a Schauder basis?

§ 17. Continuous linear operators in Banach spaces with bases

Theorem 17.1. *Let E be a Banach space with a basis $\{x_n\}$, let $\{f_n\}$ be the a.s.c.f. to $\{x_n\}$, and let F be a Banach space. Then the Banach space $L(E,F)$ of all continuous linear mappings of E into F is isomorphic, by the mapping $u\to\{u(x_n)\}$, to the space of sequences of elements*

$$B=\left\{\{y_n\}\subset F \,\middle|\, \sup_{1\leqslant n<\infty}\ \sup_{\substack{x\in E\\ \|x\|\leqslant 1}} \left\|\sum_{i=1}^{n} f_i(x)\,y_i\right\|_F<\infty\right\}, \tag{17.1}$$

endowed with the norm

$$\|\{y_n\}\|=\sup_{1\leqslant n<\infty}\ \sup_{\substack{x\in E\\ \|x\|\leqslant 1}} \left\|\sum_{i=1}^{n} f_i(x)\,y_i\right\|_F. \tag{17.2}$$

Namely, we have

$$\|u\|\leqslant\|\{u(x_n)\}\|\leqslant v_{\{x_n\}}\|u\| \qquad (u\in L(E,F)), \tag{17.3}$$

where $v_{\{x_n\}}$ is the norm of the basis $\{x_n\}$.

The inverse mapping $\{y_n\}\to u$ is given by the formula

$$u(x)=\sum_{i=1}^{\infty} f_i(x)\,y_i \qquad (x\in E). \tag{17.4}$$

If $\{x_n\}$ is a monotone basis[1] *of E, this correspondence is an isometry and we have*

$$\|u\|=\|\{y_n\}\|=\lim_{n\to\infty}\ \sup_{\substack{x\in E\\ \|x\|\leqslant 1}} \left\|\sum_{i=1}^{n} f_i(x)\,y_i\right\| \qquad (u\in L(E,F)). \tag{17.5}$$

Proof. Let us first observe that (17.2) is indeed a norm on B, since if $\|\{y_n\}\|=0$, then from $\sup_{\substack{x\in E\\ \|x\|\leqslant 1}}\|f_1(x)\,y_1\|=0$ we infer $y_1=0$, whence by $\sup_{\substack{x\in E\\ \|x\|\leqslant 1}}\left\|\sum_{i=1}^{2} f_i(x)\,y_i\right\|=\sup_{\substack{x\in E\\ \|x\|\leqslant 1}}\|f_2(x)\,y_2\|=0$, we get $y_2=0$, whence, continuing in this way, we obtain $y_n=0$ $(n=1,2,\ldots)$.

[1] I.e. $v_{\{x_n\}}=1$. Such bases will be studied in more detail in Ch. II, § 1.

Now let $u \in L(E,F)$ be arbitrary. We have then, putting $y_n = u(x_n)$ $(n=1,2,\ldots)$ and $u_n(x) = \sum_{i=1}^{n} f_i(x) y_i$ $(x \in E,\ n=1,2,\ldots)$,

$$u(x) = u\left(\sum_{i=1}^{\infty} f_i(x) x_i\right) = \sum_{i=1}^{\infty} f_i(x) u(x_i) = \sum_{i=1}^{\infty} f_i(x) y_i = \lim_{n\to\infty} u_n(x) \qquad (x \in E),$$

whence, by the principle of uniform boundedness[1],

$$\sup_{1\leqslant n<\infty} \sup_{\substack{x\in E\\ \|x\|\leqslant 1}} \left\| \sum_{i=1}^{n} f_i(x) y_i \right\| = \sup_{1\leqslant n<\infty} \|u_n\| < \infty,$$

and thus $\{y_n\} = \{u(x_n)\} \in B$ and

$$\|u\| \leqslant \sup_{1\leqslant n<\infty} \|u_n\| = \sup_{1\leqslant n<\infty} \sup_{\substack{x\in E\\ \|x\|\leqslant 1}} \left\| \sum_{i=1}^{n} f_i(x) y_i \right\| = \|\{u(x_n)\}\|.$$

On the other hand, for every $x \in E$ and $n=1,2,\ldots$ we have

$$\left\| \sum_{i=1}^{n} f_i(x) y_i \right\| = \left\| u\left(\sum_{i=1}^{n} f_i(x) x_i\right) \right\| \leqslant \|u\| \left\| \sum_{i=1}^{n} f_i(x) x_i \right\| \leqslant v_{\{x_n\}} \|u\|\, \|x\|$$

whence

$$\|u(\{x_n\})\| \leqslant v_{\{x_n\}} \|u\|,$$

and thus we also have (17.3). This proves that the mapping $u \to \{u(x_n)\}$ is an isomorphism of $L(E,F)$ into B.

In the particular case when the basis $\{x_n\}$ is monotone, we have

$$\|u\| \leqslant \varliminf_{n\to\infty} \|u_n\| \leqslant \varlimsup_{n\to\infty} \|u_n\| \leqslant \sup_{1\leqslant n<\infty} \|u_n\| = \|\{u(x_n)\}\| \leqslant v_{\{x_n\}} \|u\| = \|u\|,$$

whence

$$\|u\| = \lim_{n\to\infty} \sup_{\substack{x\in E\\ \|x\|\leqslant 1}} \left\| \sum_{i=1}^{n} f_i(x) y_i \right\| = \sup_{1\leqslant n<\infty} \sup_{\substack{x\in E\\ \|x\|\leqslant 1}} \left\| \sum_{i=1}^{n} f_i(x) y_i \right\| = \|\{y_n\}\|,$$

and thus the mapping $u \to \{u(x_n)\}$ is now an isometry of $L(E,F)$ into B, and we have (17.5).

Now let $\{y_n\} \in B$ be arbitrary. We shall show that formula (17.4) defines a mapping $u \in L(E,F)$ (satisfying, obviously, $u(x_n) = y_n$ for all $n=1,2,\ldots$), which will complete the proof. Put

$$u_n(x) = \sum_{i=1}^{n} f_i(x) y_i \qquad (x \in E,\ n=1,2,\ldots). \tag{17.6}$$

[1] See e.g. [10], p. 80, theorem 5.

Then, since $\{y_n\} \in B$, we have $\sup_{1 \leqslant j < \infty} \|u_j\| = \|\{y_n\}\| < \infty$. Furthermore, by $f_i(x_j) = \delta_{ij}$ we have $u_{n+m}\left(\sum_{i=1}^{n} \alpha_i x_i\right) = u_n\left(\sum_{i=1}^{n} \alpha_i x_i\right)$ for all $n, m = 1, 2, \ldots$ and all scalars $\alpha_1, \ldots, \alpha_n$, whence $\lim_{j \to \infty} u_j\left(\sum_{i=1}^{n} \alpha_i x_i\right)$ exists. Since the set of all $\sum_{i=1}^{n} \alpha_i x_i$ is dense in E and since F is complete, it follows that the operator $u(x) = \lim_{x \to \infty} u_n(x)$ $(x \in E)$, i.e. (17.4), is well defined and in $L(E, F)$, which completes the proof of theorem 17.1.

Remark 17.1. In the particular case when $F = K$ (the field of scalars), from theorem 17.1 above we obtain again theorem 12.5a) of § 12; indeed, in this case for any $\{\alpha_n\} \subset K$ we have $\|\{\alpha_n\}\| = \sup_{1 \leqslant n < \infty} \sup_{\substack{x \in E \\ \|x\| \leqslant 1}} \left|\sum_{i=1}^{n} f_i(x) \alpha_i\right| = \sup_{1 \leqslant n < \infty} \left\|\sum_{i=1}^{n} \alpha_i f_i\right\|$.

Theorem 17.2. *Let E be a Banach space with a basis $\{x_n\}$, let $\{f_n\}$ be the a.s.c.f. to $\{x_n\}$, and let F be a Banach space. Then the Banach space $L(F, E)$ of all continuous linear mappings of F into E is isomorphic, by the mapping $v \to \{v^*(f_n)\}$, to the space of sequences of functionals*

$$D = \left\{\{g_n\} \subset F^* \;\middle|\; \sum_{i=1}^{\infty} g_i(y) x_i \text{ converges in } E \text{ for every } y \in F\right\}, \tag{17.7}$$

endowed with the norm

$$\|\{g_n\}\| = \sup_{1 \leqslant n < \infty} \sup_{\substack{y \in F \\ \|y\| \leqslant 1}} \left\|\sum_{i=1}^{n} g_i(y) x_i\right\|_E. \tag{17.8}$$

Namely, we have

$$\|v\| \leqslant \|\{v^*(f_n)\}\| \leqslant v_{\{x_n\}} \|v\| \qquad (v \in L(F, E)), \tag{17.9}$$

where $v_{\{x_n\}}$ is the norm of the basis $\{x_n\}$.

The inverse mapping $\{g_n\} \to v$ is given by the formula

$$v(y) = \sum_{i=1}^{\infty} g_i(y) x_i \qquad (y \in F). \tag{17.10}$$

If $\{x_n\}$ is a monotone basis of E, this correspondence is an isometry and we have

$$\|v\| = \|\{g_n\}\| = \lim_{n \to \infty} \sup_{\substack{y \in F \\ \|y\| \leqslant 1}} \left\|\sum_{i=1}^{n} g_i(y) x_i\right\| \qquad (v \in L(F, E)). \tag{17.11}$$

Proof. Let us first observe that (17.8) is indeed a norm on D, since by the principle of uniform boundedness it is finite for each $\{g_n\} \in D$ and since if $\|\{g_n\}\| = 0$ then from $\sum_{i=1}^{n} g_i(y) x_i = 0$ and the linear independence of $x_1, \ldots, x_n$ we infer $g_i(y) = 0$ $(y \in F,\ \|y\| \leqslant 1,\ i = 1, \ldots, n;\ n = 1, 2, \ldots)$, whence $g_n = 0$ $(n = 1, 2, \ldots)$.

Now let $v \in L(F, E)$ be arbitrary. We have then, putting $g_n = v^*(f_n)$ $(n = 1, 2, \ldots)$,

$$v(y) = \sum_{i=1}^{\infty} f_i[v(y)] x_i = \sum_{i=1}^{\infty} g_i(y) x_i \qquad (y \in F),$$

and thus $\{g_n\} = \{v^*(f_n)\} \in D$ and

$$\|v\| = \sup_{\substack{y \in F \\ \|y\| \leqslant 1}} \left\| \sum_{i=1}^{\infty} g_i(y) x_i \right\| \leqslant \sup_{1 \leqslant n < \infty} \sup_{\substack{y \in F \\ \|y\| \leqslant 1}} \left\| \sum_{i=1}^{n} g_i(y) x_i \right\| = \|\{v^*(f_n)\}\|.$$

On the other hand, since $\left\| \sum_{i=1}^{n} f_i(x) x_i \right\| \leqslant v_{\{x_n\}} \|x\|$ $(x \in E, n = 1, 2, \ldots)$, we have

$$\|\{v^*(f_n)\}\| = \sup_{1 \leqslant n < \infty} \sup_{\substack{y \in F \\ \|y\| \leqslant 1}} \left\| \sum_{i=1}^{n} f_i[v(y)] x_i \right\| \leqslant v_{\{x_n\}} \sup_{\substack{y \in F \\ \|y\| \leqslant 1}} \|v(y)\| = v_{\{x_n\}} \|v\|,$$

and thus we also have (17.9). This proves that the mapping $v \to \{v^*(f_n)\}$ is an isomorphism of $L(F, E)$ into D.

In the particular case when the basis $\{x_n\}$ is monotone, we have $\sup_{\substack{y \in F \\ \|y\| \leqslant 1}} \left\| \sum_{i=1}^{n} g_i(y) x_i \right\| \leqslant \sup_{\substack{y \in F \\ \|y\| \leqslant 1}} \left\| \sum_{i=1}^{n+1} g_i(y) x_i \right\|$ $(n = 1, 2, \ldots)$ and $v_{\{x_n\}} = 1$, whence

$$\|v\| \leqslant \|\{v^*(f_n)\}\| = \|\{g_n\}\| = \lim_{n \to \infty} \sup_{\substack{y \in F \\ \|y\| \leqslant 1}} \left\| \sum_{i=1}^{n} g_i(y) x_i \right\| \leqslant v_{\{x_n\}} \|v\| = \|v\|,$$

and thus the mapping $v \to \{v^*(f_n)\}$ is an isometry of $L(F, E)$ into D and we have (17.11).

Now let $\{g_n\} \in D$ be arbitrary. Then, by the Banach-Steinhaus theorem applied to the sequence of operators $v_n(y) = \sum_{i=1}^{n} g_i(y) x_i$ $(y \in F, n = 1, 2, \ldots)$, the operator $v(y) = \lim_{n \to \infty} v_n(y)$ $(y \in F)$, i.e. (17.10), is in $L(F, E)$, which completes the proof of theorem 17.2.

Remark 17.2. One may also consider the spaces B and D above endowed with the norms $\|\{y_n\}\| = \sup_{\substack{x \in E \\ \|x\| \leqslant 1}} \left\| \sum_{i=1}^{\infty} f_i(x) y_i \right\|$ and $\|\{g_n\}\|$

$= \sup_{\substack{y \in F \\ \|y\| \leqslant 1}} \left\| \sum_{i=1}^{\infty} g_i(y) x_i \right\|$ respectively, and in this case, obviously, the mappings $u \to \{u(x_n)\}$ and $v \to \{v^*(f_n)\}$ become isometries. However, the norms (17.2) and (17.8) are more suitable for applications.

Remark 17.3. One cannot replace in theorem 17.2 the space D by the space $\left\{ \{g_n\} \subset F^* \,\middle|\, \sup_{1 \leqslant n < \infty} \sup_{\substack{y \in F \\ \|y\| \leqslant 1}} \left\| \sum_{i=1}^{n} g_i(y) x_i \right\| < \infty \right\}$, since for the elements $\{g_n\}$ of this latter space some series $\sum_{i=1}^{\infty} g_i(y) x_i$ may be divergent, as shown by the example $F = E = c_0$, $\{x_n\}$ = the unit vector basis of E, $g_i(y) = \eta_1 + \eta_i$ ($y = \{\eta_n\} \in c_0$, $i = 1, 2, \ldots$) and $y_0 = \{\eta_n^0\} \in F$ with $\eta_1^0 \neq 0$; indeed, in this example, for the operators $v_n(y) = \sum_{i=1}^{n} g_i(y) x_i = \sum_{i=1}^{n} (\eta_1 + \eta_i) x_i$ we have $\|v_n\| = 2$ ($n = 1, 2, \ldots$) but $\lim_{n \to \infty} v_n(y_0) \notin c_0$.

Let us give now some applications of theorems 17.1 and 17.2 to concrete spaces.

Corollary 17.1. *Let F be an arbitrary Banach space. Then $L(l^1, F)$ is equivalent*[1]*, by the mapping $u \to \{u(x_n)\}$, where $\{x_n\}$ is the unit vector basis of l^1, to the space of sequences of elements*

$$B = \left\{ \{y_n\} \subset F \,\middle|\, \sup_{1 \leqslant n < \infty} \|y_n\| < \infty \right\} = l_F^{\infty} \tag{17.12}$$

endowed with the norm

$$\|\{y_n\}\| = \sup_{1 \leqslant n < \infty} \|y_n\|. \tag{17.13}$$

The inverse mapping $\{y_n\} \to u$ is given by the formula

$$u(x) = \sum_{i=1}^{\infty} \xi_i y_i \qquad (x = \{\xi_n\} \in l^1). \tag{17.14}$$

Proof. Take in theorem 17.1 $E = l^1$ and $\{x_n\}$ = the unit vector basis of E. This basis in monotone (whence $L(E, F) \equiv B$) and its a. s. c. f. $\{f_n\}$ is nothing else but the sequence of coordinate functionals

$$f_i(x) = \xi_i \qquad (x = \{\xi_n\} \in l^1). \tag{17.15}$$

Formula (17.2) becomes

$$\|\{y_n\}\| = \sup_{1 \leqslant n < \infty} \sup_{\sum_{j=1}^{\infty} |\xi_j| \leqslant 1} \left\| \sum_{i=1}^{n} \xi_i y_i \right\| = \sup_{1 \leqslant n < \infty} \|y_n\|,$$

i. e. (17.13), which completes the proof.

[1] I. e. linearly isometric.

Corollary 17.2. *Let F be an arbitrary Banach space. Then $L(F,l^1)$ is equivalent, by the mapping $v \to \{v^*(f_n)\}$, where f_n are the coordinate functionals on l^1, to the space of sequences of functionals*

$$D = \left\{ \{g_n\} \subset F^* \,\middle|\, \sum_{i=1}^{\infty} g_i \text{ is weakly* unconditionally Cauchy} \right\} \tag{17.16}$$

endowed with the norm

$$\|\{g_n\}\| = \sup_{1 \leqslant n < \infty} \sup_{\substack{y \in F \\ \|y\| \leqslant 1}} \sum_{i=1}^{n} |g_i(y)| = \sup_{1 \leqslant n < \infty} \sup_{|\beta_1|, |\beta_2|, \ldots \leqslant 1} \left\| \sum_{i=1}^{n} \beta_i g_i \right\|. \tag{17.17}$$

The inverse mapping $\{g_n\} \to v$ is given by the formula

$$v(y) = \{g_n(y)\} \qquad (y \in F). \tag{17.18}$$

Proof. Take in theorem 17.2 $E = l^1$ with its (monotone) unit vector basis $\{x_n\}$. Then by (17.7) we have

$$D = \left\{ \{g_n\} \subset F^* \,\middle|\, \sum_{i=1}^{\infty} |g_i(y)| < \infty \quad (y \in F) \right\},$$

i. e. (17.16), and by (17.8) we have

$$\|\{g_n\}\| = \sup_{1 \leqslant n < \infty} \sup_{\substack{y \in F \\ \|y\| \leqslant 1}} \sum_{i=1}^{n} |g_i(y)|,$$

i. e. the first equality of (17.17). Now,

$$\sup_{|\beta_1|, |\beta_2|, \ldots \leqslant 1} \left\| \sum_{i=1}^{n} \beta_i g_i \right\| = \sup_{|\beta_1|, |\beta_2|, \ldots \leqslant 1} \sup_{\substack{y \in F \\ \|y\| \leqslant 1}} \left| \sum_{i=1}^{n} \beta_i g_i(y) \right|$$

$$\leqslant \sup_{\substack{y \in F \\ \|y\| \leqslant 1}} \sum_{i=1}^{n} |g_i(y)|, \tag{17.19}$$

and, on the other hand

$$\sum_{i=1}^{n} |g_i(y)| = \left| \sum_{i=1}^{n} [\operatorname{sign} g_i(y)] g_i(y) \right| \leqslant \sup_{|\beta_1|, |\beta_2|, \ldots \leqslant 1} \left| \sum_{i=1}^{n} \beta_i g_i(y) \right| \qquad (y \in F),$$

whence

$$\sup_{\substack{y \in F \\ \|y\| \leqslant 1}} \sum_{i=1}^{n} |g_i(y)| \leqslant \sup_{|\beta_1|, |\beta_2|, \ldots \leqslant 1} \left\| \sum_{i=1}^{n} \beta_i g_i \right\|. \tag{17.20}$$

From (17.19) and (17.20) we get the second equality of (17.17), which completes the proof of corollary 17.2.

Corollary 17.3. *Let $1 < p < \infty$ and let F be an arbitrary Banach space. Then $L(l^p, F)$ is equivalent, by the mapping $u \to \{u(x_n)\}$, where $\{x_n\}$ is the unit vector basis of l^p, to the space of sequences of elements*

$$B=\left\{\{y_n\}\subset F \,\middle|\, \lim_{n\to\infty} \sup_{\sum_{j=1}^{n}|\xi_j|^p\leqslant 1} \left\|\sum_{i=1}^{n}\xi_i y_i\right\| \text{ exists and } <\infty\right\} \tag{17.21}$$

endowed with the norm

$$\|\{y_n\}\|=\lim_{n\to\infty} \sup_{\sum_{j=1}^{n}|\xi_j|^p\leqslant 1} \left\|\sum_{i=1}^{n}\xi_i y_i\right\| = \lim_{n\to\infty} \sup_{\substack{g\in F^*\\ \|g\|\leqslant 1}} \left(\sum_{i=1}^{n}|g(y_i)|^q\right)^{\frac{1}{q}}, \tag{17.22}$$

where $\frac{1}{p}+\frac{1}{q}=1$.

The inverse mapping $\{y_n\}\to u$ *is given by the formula*

$$u(x)=\sum_{i=1}^{\infty}\xi_i y_i \qquad (x=\{\xi_n\}\in l^p). \tag{17.23}$$

Proof. Taking in theorem 17.1 $E=l^p$ and $\{x_n\}=$ the unit vector basis of E, we obtain (17.21) and the first equality of (17.22). Furthermore, from

$$\sup_{\sum_{j=1}^{n}|\xi_j|^p\leqslant 1} \left\|\sum_{i=1}^{n}\xi_i y_i\right\| = \sup_{\sum_{j=1}^{n}|\xi_j|^p\leqslant 1} \sup_{\substack{g\in F^*\\ \|g\|\leqslant 1}} \left|\sum_{i=1}^{n}\xi_i g(y_i)\right| = \sup_{\substack{g\in F^*\\ \|g\|\leqslant 1}} \left(\sum_{i=1}^{n}|g(y_i)|^q\right)^{\frac{1}{q}}$$

we obtain the second equality of (17.22), which completes the proof.

Corollary 17.4. *Let* $1<p<\infty$ *and let F be an arbitrary Banach space. Then* $L(F,l^p)$ *is equivalent, by the mapping* $v\to\{v^*(f_n)\}$, *where* f_n *are the coordinate functionals on* l^p, *to the space of sequences of functionals*

$$D=\left\{\{g_n\}\subset F^* \,\middle|\, \left(\sum_{i=1}^{\infty}|g_i(y)|^p\right)^{\frac{1}{p}}<\infty \quad \text{for all } y\in F\right\} \tag{17.24}$$

endowed with the norm

$$\|\{g_n\}\|=\lim_{n\to\infty} \sup_{\substack{y\in F\\ \|y\|\leqslant 1}} \left(\sum_{i=1}^{n}|g_i(y)|^p\right)^{\frac{1}{p}} = \lim_{n\to\infty} \sup_{\sum_{j=1}^{n}|\beta_j|^q\leqslant 1} \left\|\sum_{i=1}^{n}\beta_i g_i\right\|, \tag{17.25}$$

where $\frac{1}{p}+\frac{1}{q}=1$.

The inverse mapping $\{g_n\}\to v$ *is given by the formula*

$$v(y)=\{g_n(y)\} \qquad (y\in F). \tag{17.26}$$

Proof. Taking in theorem 17.2 $E=l^p$ and $\{x_n\}=$ the unit vector basis of E, we obtain (17.24) and the first equality of (17.25). The second equality of (17.25) follows as in the above proof of corollary 17.3.

Corollary 17.5. *Let F be an arbitrary Banach space. Then* $L(c_0,F)$ *is equivalent, by the mapping* $u\to\{u(x_n)\}$, *where* $\{x_n\}$ *is the unit vector basis of* c_0, *to the space of sequences of elements*

$$B=\left\{\{y_n\}\subset F \,\middle|\, \sum_{i=1}^{\infty} y_i \quad \textit{is weakly unconditionally Cauchy}\right\} \tag{17.27}$$

endowed with the norm

$$\|\{y_n\}\| = \sup_{1\leqslant n<\infty} \sup_{|\xi_1|,|\xi_2|,\ldots\leqslant 1} \left\|\sum_{i=1}^{n} \xi_i y_i\right\| = \sup_{1\leqslant n<\infty} \sup_{\substack{g\in F^* \\ \|g\|\leqslant 1}} \sum_{i=1}^{n} |g(y_i)|\,. \tag{17.28}$$

The inverse mapping $\{y_n\}\to u$ *is given by the formula*

$$u(x)=\sum_{i=1}^{\infty} \xi_i y_i \qquad (x=\{\xi_n\}\in c_0). \tag{17.29}$$

Proof. Take in theorem 17.1 $E=c_0$ with its (monotone) unit vector basis $\{x_n\}$. Then by (17.1) we have

$$B=\left\{\{y_n\}\subset F \,\middle|\, \sup_{1\leqslant n<\infty} \sup_{|\xi_1|,|\xi_2|,\ldots\leqslant 1} \left\|\sum_{i=1}^{n} \xi_i y_i\right\| <\infty\right\}, \tag{17.30}$$

and by (17.2) we have the first equality of (17.28). The second equality of (17.28) follows as in the above proof of corollary 17.2. Finally, from (17.30) and the second equality of (17.28) we infer (17.27), which completes the proof of corollary 17.5.

Remark 17.4. Since in conjugate Banach spaces the weakly* and weakly unconditionally Cauchy series coincide[1], from corollaries 17.2–17.5 it follows that $L(F,l^1)\equiv L(c_0,F^*)$ and $L(F,l^p)\equiv L(l^q,F^*)$, where $1<p<\infty$, $\frac{1}{p}+\frac{1}{q}=1$. However, this is well known and obvious via $L(F,E^*)\equiv(F\otimes_\gamma E)^*\equiv(E\otimes_\gamma F)^*\equiv L(E,F^*)$ (for any Banach spaces F,E).

Corollary 17.6. *Let F be an arbitrary Banach space. Then $L(F,c_0)$ is equivalent, by the mapping $v\to\{v^*(f_n)\}$, where f_n are the coordinate functionals on c_0, to the space of sequences of functionals*

$$D=\{\{g_n\}\subset F^* \mid \{g_n\} \ \textit{weakly* converges to } 0\} \tag{17.31}$$

endowed with the norm

$$\|\{g_n\}\| = \sup_{1\leqslant n<\infty} \|g_n\|\,. \tag{17.32}$$

The inverse mapping $\{g_n\}\to v$ *is given by the formula*

$$v(y)=\{g_n(y)\} \qquad (y\in F). \tag{17.33}$$

Proof. Take in theorem 17.2 $E=c_0$ and $\{x_n\}=$ the unit vector basis of E.

[1] See e. g. Ch. II, § 15, lemma 15.1, equivalence 1°⇔6°.

Theorem 17.3. *Let E be a Banach space with a basis $\{x_n\}$. Then E has the approximation property in the sense of Grothendieck, i.e. the identity operator $I_E: E \to E$ can be approximated, uniformly on every compact subset of E, by linear operators of finite rank*[1]. *Namely, the sequence of partial sum operators $\{s_n\}$ associated to the basis $\{x_n\}$ approximates I_E uniformly on each compact set $M \subset E$.*

Proof. Let $\varepsilon > 0$ be arbitrary, let $\{y_i\}_{i=1}^{l} \subset E$ be a finite $\frac{\varepsilon}{2(1+v_{\{x_n\}})}$-net for the compact set $M \subset E$, where $v_{\{x_n\}} = \sup\limits_{1 \leqslant n < \infty} \|s_n\|$, and let $x \in M$ be arbitrary. Then there exist an y_j such that $\|x - y_j\| < \frac{\varepsilon}{2(1+v_{\{x_n\}})}$ and a positive integer $N = N(\varepsilon)$ such that $\|y_i - s_n(y_i)\| < \frac{\varepsilon}{2}$ for all $n > N(\varepsilon)$ and $i = 1, \ldots, l$, whence

$$\begin{aligned} \|x - s_n(x)\| &\leqslant \|x - y_j\| + \|y_j - s_n(y_j)\| + \|s_n(y_j) - s_n(x)\| \\ &\leqslant \|x - y_j\|(1 + v_{\{x_n\}}) + \|y_j - s_n(y_j)\| < \frac{\varepsilon}{2} + \frac{\varepsilon}{2} = \varepsilon \qquad (n > N(\varepsilon)), \end{aligned}$$

and thus, since $x \in M$ and $\varepsilon > 0$ have been arbitrary,

$$\lim_{n \to \infty} \sup_{x \in M} \|x - s_n(x)\| = 0,$$

which completes the proof of theorem 17.3.

Remark 17.5. Since the basis problem is unsolved, no example is known of a separable Banach space which does not have the approximation property, but neither has it been proved that every separable space has this property. This is one of the famous classical problems of functional analysis, of great importance for applications, as shown also by corollary 17.7 below.

Remark 17.6. It is also not known whether the converse of theorem 17.3 above is valid, i.e. whether a separable Banach space having the approximation property must have a basis or not. However, for sequentially complete separable locally convex spaces the answer to this latter question is negative, as shown by example 13.4 of § 13. Indeed, as we have seen there, the space m^* endowed with the weak* topology $\sigma(m^*, m)$ is locally convex, separable and sequentially complete, but has no basis. On the other hand, let us show that every conjugate Banach space E^* endowed with the weak* topology $\sigma(E^*, E)$ has the approximation

[1] I. e. of finite-dimensional range.

property[1]. Let $M \subset E^*$ be $\sigma(E^*, E)$-compact and let $V = V_{x_1,\ldots,x_n;\varepsilon}(0)$ be a $\sigma(E^*, E)$-neighborhood of 0 in E^*. Then there exists a continuous linear mapping $v: E \to E$ of finite rank, satisfying

$$v(x_i) = x_i \qquad (i = 1, \ldots, n),$$

e.g. one can take a norm continuous linear projection of E onto $[x_i]_{i=1}^n$. Hence

$$|[v^*(f) - f](x_i)| = |f[v(x_i)] - f(x_i)| = 0 < \varepsilon \qquad (f \in M,\ i = 1, \ldots, n),$$

and thus $v^*(f) - f \in V$ $(f \in M)$. Since $v^*: E^* \to E^*$ is continuous for $\sigma(E^*, E)$, $\sigma(E^*, E)$ and of finite rank, this proves our assertion.

From theorem 17.3 we infer

Corollary 17.7. *Let E be a Banach space with a basis $\{x_n\}$ and let F be an arbitrary Banach space. Then every compact linear operator $u \in L(F, E)$ can be uniformly approximated by linear operators of finite rank.*

Proof. Let $u \in L(F, E)$ be an arbitrary compact linear mapping of F into E. Then the set $M = u(S_F) \subset E$ is compact[2], whence, by theorem 17.3,

$$\|u - s_n u\| = \sup_{\substack{y \in F \\ \|y\| \leqslant 1}} \|u(y) - s_n[u(y)]\| = \sup_{x \in M} \|x - s_n(x)\| \to 0 \quad \text{as} \quad n \to \infty,$$

which, since $s_n u \in L(F, E)$ is of finite rank (by $\dim s_n[u(F)] \leqslant n < \infty$), completes the proof of corollary 17.7.

Remark 17.7. The importance of corollary 17.7 (and hence also of the spaces with bases) for applications is obvious. The problem, whether every separable Banach space E has the property ocurring in corollary 17.7 is known[3] to be equivalent to the problem whether every subspace of c_0 has this property and also to[4] the problem mentioned in remark 17.5.

§ 18. Bases of tensor products

In this section we shall consider the problem, whether the tensor product $E \otimes_\alpha F$ (i.e. the completion of the algebraic tensor product $E \otimes F$ with respect to the cross norm α) of two Banach spaces E, F

---

[1] We recall that a locally convex space U is said to have the approximation property (in the sense of A. Grothendieck [89], p. 165), if for every compact subset M of U and every neighborhood V of 0 in U there exists a continuous linear mapping $u: U \to U$ of finite rank, such that $u(x) - x \in V$ for all $x \in M$.

[2] We use, as before, the notation $S_F = \{y \in F \mid \|y\| \leqslant 1\}$.

[3] See [89], p. 170, proposition 37.

[4] See [89], p. 164, proposition 35.

with bases, has a basis. We shall use essentially the terminology and notations of R. Schatten [223], assuming that the reader is familiar with some elementary notions and results of the theory of normed tensor products.

Theorem 18.1. *Let E, F be two Banach spaces with bases $\{x_n\}$, $\{y_n\}$ respectively, let $\{f_n\}\subset E^*$, $\{g_n\}\subset F^*$ be the respective a.s.c.f. and let α be a uniform crossnorm on $E\otimes F$. Then the system of all products $x_i\otimes y_j$, arranged into a single sequence $\{z_k\}$ by the numbering*[1]

$$z_1=x_1\otimes y_1,\ z_k=\begin{cases} x_i\otimes y_{n+1} & \text{for}\quad k=n^2+i,\ i=1,\dots,n+1,\\ x_{n+1}\otimes y_{n+1-i} & \text{for}\quad k=n^2+n+1+i,\ i=1,\dots,n\end{cases} \quad (n=1,2,\dots), \tag{18.1}$$

is a basis of $E\otimes_\alpha F$, having as a.s.c.f. the system $\{f_i\otimes g_j\}$ arranged in the same way into a single sequence $\{h_k\}$. Hence, in particular, $\{z_k\}$ is a basis of $E\otimes_\gamma F$ and of $E\otimes_\lambda F$.

Proof. Let us first show that $[z_k]=E\otimes_\alpha F$. For any $x\in E, y\in F$ we have

$$\begin{aligned} R_n&=x\otimes y-\sum_{i,j=1}^{n} f_i(x)g_j(y)x_i\otimes y_j=x\otimes y-s_n^1(x)\otimes s_n^2(y)\\ &=x\otimes[y-s_n^2(y)]+[x-s_n^1(x)]\otimes s_n^2(y),\end{aligned}$$

where s_n^1, s_n^2 denote the partial sum operators associated to the bases $\{x_n\}$ and $\{y_n\}$ respectively. Consequently, since α is a crossnorm on $E\otimes F$ and $\{x_n\}$, $\{y_n\}$ are bases,

$$\alpha(R_n)\leqslant \|x\|\,\|y-s_n^2(y)\|+\|x-s_n^1(x)\|\,\|s_n^2(y)\|\to 0 \quad \text{as}\quad n\to\infty.$$

This argument extends by linearity to finite sums of elements of the form $x\otimes y$, and hence the linear subspace spanned by $\{z_k\}=\{x_i\otimes y_j\}$ is dense in $E\otimes F$, whence also in $E\otimes_\alpha F$. Thus, $[z_k]=E\otimes_\alpha F$.

Furthermore, for the partial sum operators s_n associated to the biorthogonal system (z_k,h_k) ($\{z_k\}\subset E\otimes_\alpha F$, $\{h_k\}\subset(E\otimes_\alpha F)^*$) we have, by (18.1),

$$s_{n^2}=s_n^1\otimes s_n^2, \tag{18.2}$$

$$s_{n^2+l}=s_n^1\otimes s_n^2+s_l^1\otimes(s_{n+1}^2-s_n^2) \qquad (l=1,\dots,n+1), \tag{18.3}$$

$$s_{n^2+n+1+l}=s_{n+1}^1\otimes s_{n+1}^2-(s_{n+1}^1-s_n^1)\otimes s_{n-l}^2 \qquad (l=1,\dots,n;\ s_0^2=0). \tag{18.4}$$

Indeed, it is sufficient to prove that these equalities hold for all elements of the form $x\otimes y$, where $x\in E, y\in F$. Now, we have

[1] Note that this numbering is different from the classical diogonal numbering (used e.g. in the proof that the set of all rationals is countable).

$$s_{n^2}(x\otimes y)=\sum_{k=1}^{n^2} h_k(x\otimes y)z_k=\sum_{i,j=1}^{n}(f_i\otimes g_j)(x\otimes y)x_i\otimes y_j$$
$$=\sum_{i=1}^{n} f_i(x)x_i\otimes\sum_{j=1}^{n} g_j(y)y_j=s_n^1(x)\otimes s_n^2(y)=(s_n^1\otimes s_n^2)(x\otimes y),$$

whence we infer (18.2). The proof of (18.3) and (18.4) is similar, taking also into account (18.1).

Finally, since α is a uniform crossnorm on $E\otimes F$, for any positive integers n,m we have $\|s_n^1\otimes s_m^2\|\leqslant\|s_n^1\|\,\|s_m^2\|$, whence, since $\{x_n\}$, $\{y_n\}$ are bases of E and F respectively, it follows that $\{s_n^1\otimes s_m^2\}$ is uniformly bounded. Hence, by (18.2), (18.3) and (18.4), $\{s_n\}$ is uniformly bounded, which, by § 4, theorem 4.1, proves the first assertion of theorem 18.1. The second assertion follows now from the equalities $f_i\otimes g_j(x_k\otimes y_l)=f_i(x_k)g_j(y_l)=\delta_{ik}\delta_{jl}$ $(i,j,k,l=1,2,\dots)$, which completes the proof of theorem 18.1.

The sequence $\{z_k\}=\{x_i\otimes y_j\}$ above may be called the *tensor product* of the bases $\{x_n\}$ and $\{y_n\}$.

Let us give now some applications of the particular cases $\alpha=\gamma$ and $\alpha=\lambda$ of theorem 18.1.

Corollary 18.1. *Let T be a locally compact space, μ a positive Radon measure on T such that the space $L^1(T,\mu)$ has a basis $\{x_n\}$ and F a Banach space with a basis $\{y_n\}$. Then the Banach space $L_F^1(T,\mu)$ (of all equivalence classes of F-valued ν-integrable functions on T) has a basis consisting of the system of all products $z_k(t)=x_i(t)y_j$ $(t\in T)$, numbered as in* (18.1).

Proof. This follows from theorem 18.1 and from the well known[1] canonical isometry $L_F^1(T,\mu)\equiv L^1(T,\mu)\otimes_\gamma F$.

Corollary 18.2. *Let T, S be locally compact spaces and μ,ν positive Radon measures on T and S respectively, such that the spaces $L^1(T,\mu)$, $L^1(S,\nu)$ have bases $\{x_n\}$ and $\{y_n\}$ respectively. Then the Banach space $L^1(T\times S,\ \mu\otimes\nu)$ has a basis consisting of all equivalence classes of functions of the form $z_k(t,s)=x_i(t)y_j(s)$ $(t\in T,\ s\in S)$, numbered as in* (18.1).

Proof. This follows from theorem 18.1 and from the canonical isometry[2] $L^1(T\times S,\ \mu\otimes\nu)\equiv L^1(T,\mu)\otimes_\gamma L^1(S,\nu)$.

Corollary 18.3. *Let Q be a compact metric space, $\{x_n\}$ a basis of the space $C(Q)$ and F a Banach space with a basis $\{y_n\}$. Then the Banach space $C_F(Q)$ (of all F-valued continuous functions on Q) has a basis consisting of the system of all functions of the form $z_k(q)=x_i(q)y_j$ $(q\in Q)$, numbered as in* (18.1).

---

[1] See [89], p. 59, theorem 2.
[2] See [89], p. 61, corollary 4.

Proof. This follows from theorem 18.1 and from the canonical isometry[1] $C_F(Q) \equiv C(Q) \otimes_\lambda F$.

A similar result is also valid for the space $(c_0)_F$ of all sequences of elements in F converging to 0.

Corollary 18.4. *Let Q, P be compact metric spaces and let $\{x_n\}$, $\{y_n\}$ be bases of the spaces $C(Q)$ and $C(P)$ respectively. Then the Banach space $C(Q \times P)$ has a basis consisting of all functions of the form $z_k(q,p) = x_i(q) y_j(p)$ $(q \in Q, p \in P)$, numbered as in* (18.1).

Proof. This follows from theorem 18.1 and from the canonical isometry[2] $C(Q \times P) \equiv C(Q) \otimes_\lambda C(P)$.

Corollary 18.5. *Let E, F be Banach spaces with bases $\{x_n\}$ and $\{y_n\}$ respectively. Then the Banach space $\mathscr{C}(E^*, F)$ of all compact linear mappings of E^* into F has a basis consisting of all linear operators of rank* 1 *of the form $u_k = x_i \otimes y_j$ (i.e. $u_k(f) = f(x_i) y_j$ for all $f \in E^*$), numbered as in* (18.1).

Proof. This follows from theorem 18.1 and from the canonical isometry[3] $\mathscr{C}(E^*, F) \equiv E \otimes_\lambda F$.

A similar result is also valid for the space $\mathscr{C}(E, F)$, if E^* has a basis $\{h_n\}$. These results may be considered as sharpenings of corollary 17.7 of § 17 on the approximability in the norm of $L(E, F)$ of the operators $u \in \mathscr{C}(E, F)$ by linear operators of finite rank when F has a basis.

It is natural to ask whether theorem 18.1 remains valid if α is not necessarily uniform, or, more generally:

Problem 18.1. Let E, F be two Banach spaces with bases and let α be a crossnorm on $E \otimes F$. Does $E \otimes_\alpha F$ have a basis?

It is clear that for the sequence $\{z_k\}$ defined by (18.1) we have $[z_k] = E \otimes_\alpha F$. The difficulty lies in establishing that $\{s_n\}$ is uniformly bounded with respect to α.

§ 19. Best approximation in Banach spaces with bases

We recall that if G is a set in a Banach space E and $x \in E$, then any element $y_0 \in G$ with the property

$$\|x - y_0\| = \inf_{g \in G} \|x - y\| \tag{19.1}$$

[1] See [89], p. 90.

[2] See [89], p. 90.

[3] See e.g. [89], p. 165, condition (A_4) or [223], Ch. III, theorem 3.5.

is called an *element of best approximation* of x (by means of the elements of G). In the particular case when G is an n-dimensional linear subspace of E, where $n<\infty$, i.e.

$$G=[x_1,\ldots,x_n] \quad (x_1,\ldots,x_n \text{ linearly independent}), \tag{19.2}$$

the elements of G are called *polynomials* and it is well known[1] that every $x\in E$ has at least one polynomial of best approximation. We shall also consider the particular case when G is a closed linear subspace of finite codimension n, i.e.

$$G\oplus[x_1,\ldots,x_n]=E \quad (x_1,\ldots,x_n \text{ linearly independent}); \tag{19.3}$$

in this case the elements of G are called *polynomial complements*.

Even in the concrete cases in which the existence and uniqueness of the elements of best approximation is ensured (e.g. in the case of best approximation of continuous functions on a segment $[a,b]$ by algebraic polynomials of degree $\leqslant n$), the effective computation of the elements of best approximation presents difficulties, since the mapping $\pi_G: x\to y_0$ (where $y_0=\pi_G(x)$ is the element of best approximation of x) is, in general, non-linear. In the present section we shall see that in every Banach space E with a basis $\{x_n\}$ one can introduce a new norm, equivalent to the initial norm of E, such that in this new norm the mapping π_G of best approximation by means of the elements of the subspaces (19.2) and (19.3) corresponding to the basis $\{x_n\}$ becomes linear and the computation of the elements of best approximation becomes easy. From this it will follow for the original norm of E that in spaces with bases one can "replace" the respective mappings π_G by continuous linear mappings of E onto G which give the same order of approximation.

Definition 19.1. The norm in a Banach space E with a basis $\{x_n\}$ is called a *T-norm* (with respect to the basis $\{x_n\}$), if

a) for every $x\in E$ and $n=1,2,\ldots$ there exists a unique polynomial $y_0=\pi_{P_{(n)}}(x)\in P_{(n)}=[x_1,\ldots,x_n]$ of best approximation of x;

b) this polynomial coincides with the n-th partial sum of the expansion of the element x with respect to the basis $\{x_n\}$, i.e.

$$\pi_{P_{(n)}}(x)=s_n(x) \quad (x\in E, n=1,2,\ldots). \tag{19.4}$$

We shall denote the T-norms by ((((.

Definition 19.2. The norm in a Banach space E with a basis $\{x_n\}$ is called a *K-norm* (with respect to the basis $\{x_n\}$), if

a) for every $x\in E$ and $n=1,2,\ldots$ there exists a unique polynomial complement $y_0=\pi_{P^{(n)}}(x)\in P^{(n)}=[x_{n+1},x_{n+2},\ldots]$ of best approximation of x;

[1] See e.g. [246], Ch. I, § 2, corollary 2.2.

b) this polynomial complement coincides with the n-th rest of the expansion of the element x with respect to the basis $\{x_n\}$, i. e.

$$\pi_{P^{(n)}}(x) = r_n(x) = x - s_n(x) \quad (x \in E, n = 1,2,\ldots). \tag{19.5}$$

We shall denote the K-norms by)))).

Definition 19.3. The norm in a Banach space E with a basis $\{x_n\}$ is called a *TK-norm* (with respect to the basis $\{x_n\}$) if it is simultaneously a T-norm and a K-norm with respect to this basis.

We shall denote the TK-norms by (( )).

Let us first give a useful characterization for each of these norms.

Proposition 19.1. *Let E be a Banach space with a basis $\{x_n\}$. Then*

a) *The norm in E is a T-norm if and only if*

$$\left\|\sum_{i=l}^{\infty} \alpha_i x_i\right\| < \left\|\sum_{i=l-1}^{\infty} \alpha_i x_i\right\| \tag{19.6}$$

for every sequence of scalars $\{\alpha_n\}_{l-1}^{\infty} \subset K$ with $\alpha_{l-1} \neq 0$, for which the series in (19.6) *are convergent.*

b) *The norm in E is a K-norm if and only if*

$$\left\|\sum_{i=1}^{n} \alpha_i x_i\right\| < \left\|\sum_{i=1}^{n+1} \alpha_i x_i\right\| \tag{19.7}$$

for any scalars $\alpha_1, \ldots, \alpha_{n+1} \in K$ with $\alpha_{n+1} \neq 0$[1].

c) *If the norm in E is a TK-norm, we have*

$$\left\|\sum_{i=l}^{n} \alpha_i x_i\right\| < \left\|\sum_{i=l-1}^{n+1} \alpha_i x_i\right\| \tag{19.8}$$

for any scalars $\alpha_{l-1}, \alpha_l, \ldots, \alpha_n, \alpha_{n+1} \in K$ with $|\alpha_{l-1}| + |\alpha_{n+1}| \neq 0$.

Proof. a) Assume that the norm in E is a T-norm and let $\sum_{i=l}^{\infty} \alpha_i x_i$ be convergent. Then $\sum_{i=l-1}^{\infty} \alpha_i x_i$ has a unique element of best approximation in $P_{(l-1)} = [x_1, \ldots, x_{l-1}]$, namely

$$\pi_{P_{(l-1)}}\left(\sum_{i=l-1}^{\infty} \alpha_i x_i\right) = s_{l-1}\left(\sum_{i=l-1}^{\infty} \alpha_i x_i\right) = \alpha_{l-1} x_{l-1},$$

whence, since $0 \in P_{(l-1)}$,

[1] The bases $\{x_n\}$ satisfying (19.7) (i. e. such that the norm in E is a K-norm with respect to $\{x_n\}$) are called strictly monotone (see Ch. II, § 1 for other results on such bases).

$$\left\|\sum_{i=l}^{\infty}\alpha_i x_i\right\| = \left\|\sum_{i=l-1}^{\infty}\alpha_i x_i - \pi_{P_{(l-1)}}\left(\sum_{i=l-1}^{\infty}\alpha_i x_i\right)\right\| < \left\|\sum_{i=l-1}^{\infty}\alpha_i x_i\right\|,$$

i. e. we have (19.6).

Conversely, assume now that we have (19.6). Then for every $x=\sum_{i=1}^{\infty}\alpha_i x_i \in E$ and $p=\sum_{i=1}^{n}\beta_i x_i \in P_{(n)}$ with $p \neq s_n(x)$ (where $1 \leqslant n < \infty$), we have

$$\|x-s_n(x)\| = \left\|\sum_{i=n+1}^{\infty}\alpha_i x_i\right\| < \left\|\sum_{i=n+1}^{\infty}\alpha_i x_i - \sum_{i=1}^{n}(\beta_i-\alpha_i)x_i\right\| = \|x-p\|,$$

and thus the norm in E is a T-norm.

b) Assume that the norm in E is a K-norm and let $\alpha_1,\dots,\alpha_{n+1}$ be scalars such that $\alpha_{n+1}\neq 0$. Then $\sum_{i=1}^{n+1}\alpha_i x_i$ has a unique element of best approximation in $P^{(n)}=[x_{n+1},x_{n+2},\dots]$, namely

$$\pi_{P^{(n)}}\left(\sum_{i=1}^{n+1}\alpha_i x_i\right) = \sum_{i=1}^{n+1}\alpha_i x_i - s_n\left(\sum_{i=1}^{n+1}\alpha_i x_i\right) = \alpha_{n+1}x_{n+1},$$

whence, since $0\in P^{(n)}$,

$$\left\|\sum_{i=1}^{n}\alpha_i x_i\right\| = \left\|\sum_{i=1}^{n+1}\alpha_i x_i - \pi_{P^{(n)}}\left(\sum_{i=1}^{n+1}\alpha_i x_i\right)\right\| < \left\|\sum_{i=1}^{n+1}\alpha_i x_i\right\|,$$

i. e. we have (19.7).

Conversely, assume now that we have (19.7) and let $x=\sum_{i=1}^{\infty}\alpha_i x_i \in E$ and $p=\sum_{i=n+1}^{\infty}\beta_i x_i \in P^{(n)}$ be arbitrary, with $p\neq r_n(x)=\sum_{i=n+1}^{\infty}\alpha_i x_i$. Then there exists a smallest index, say $n+m$, such that $\beta_{n+m}\neq\alpha_{n+m}$. Hence, applying (19.7) successively, we obtain

$$\begin{aligned}\|x-r_n(x)\| &= \left\|\sum_{i=1}^{n}\alpha_i x_i\right\| = \left\|\sum_{i=1}^{n}\alpha_i x_i - \sum_{i=n+1}^{n+m-1}(\beta_i-\alpha_i)x_i\right\| \\ &< \left\|\sum_{i=1}^{n}\alpha_i x_i - \sum_{i=n+1}^{n+m}(\beta_i-\alpha_i)x_i\right\| \leqslant \left\|\sum_{i=1}^{n}\alpha_i x_i - \sum_{i=n+1}^{n+m+1}(\beta_i-\alpha_i)x_i\right\| \leqslant \cdots \\ &\leqslant \left\|\sum_{i=1}^{n}\alpha_i x_i - \sum_{i=n+1}^{\infty}(\beta_i-\alpha_i)x_i\right\| = \|x-p\|,\end{aligned}$$

and thus the norm in E is a K-norm.

c) is an immediate consequence of the necessity parts of a) and b), which completes the proof of proposition 19.1.

There exist T-norms which are not K-norms, as shown by

Example 19.1. The numbers

$$((x((= \max_{1 \leqslant n < \infty} \left(\frac{1}{n} \sum_{i=1}^{n} |\xi_i| + \sup_{n+1 \leqslant j < \infty} |\xi_j| \right) \qquad (x = \{\xi_n\} \in c_0) \qquad (19.9)$$

define a norm on c_0, equivalent to the initial norm of c_0. This norm $((x(($ is a T-norm, but not a K-norm, with respect to the unit vector basis $\{x_n\}$ of c_0.

Indeed, $((x(($ is a T-norm equivalent to the initial norm of c_0, by theorem 19.1 below. On the other hand, we have

$$((x_1 + x_2((= \max(1+1, \tfrac{1}{2}(1+1), \tfrac{1}{3}(1+1), \ldots) = 2,$$

$$((x_1 + x_2 + x_3((= \max(1+1, \tfrac{1}{2}(1+1)+1, \tfrac{1}{3}(1+1+1), \tfrac{1}{4}(1+1+1), \ldots) = 2,$$

whence, by proposition 19.1 b), $((x(($ is not a K-norm.

There exist K-norms, even satisfying (19.8), which are not T-norms, as shown by

Example 19.2. For every integer $n \geqslant 2$ let $\Pi_{1,n}$ denote the collection of all permutations of the set $\{2, 3, \ldots, n-1, n+1, n+2, \ldots\}$. Then the numbers

$$))x)) = \sup_{2 \leqslant n < \infty} \sup_{\sigma \in \Pi_{1,n}} \left(\frac{|\xi_1|}{n 2^n} + \sum_{i=2}^{\infty} \frac{|\xi_{\sigma(i)}|}{2^i} \right) \qquad (x = \{\xi_n\} \in c_0) \qquad (19.10)$$

define a norm on c_0, equivalent to the initial norm of c_0. This norm $))x))$ is a K-norm, satisfying (19.8), but not a T-norm, with respect to the unit vector basis $\{x_n\}$ of c_0.

Indeed, it is obvious that $))x))$ is a norm on c_0. Let us show that this norm is equivalent to the initial norm $\|x\| = \sup_{1 \leqslant j < \infty} |\xi_j|$ of c_0. Since $\frac{|\xi_1|}{n 2^n} \leqslant \frac{1}{8} |\xi_1|$ and $\sum_{\substack{i=2 \\ i \neq n}}^{\infty} \frac{|\xi_{\sigma(i)}|}{2^i} \leqslant \frac{1}{2} \sup_{2 \leqslant j < \infty} |\xi_j|$ for all $n \geqslant 2$ and $\sigma \in \Pi_{1,n}$ we obtain

$$))x)) \leqslant \tfrac{5}{8} \|x\| \qquad (x \in c_0). \qquad (19.11)$$

On the other hand, we have

$$|\xi_1| \leqslant 8 \left(\frac{|\xi_1|}{2 \cdot 2^2} + \sum_{i=3}^{\infty} \frac{|\xi_i|}{2^i} \right) \leqslant 8))x)) \qquad (x = \{\xi_n\} \in c_0),$$

$$|\xi_2| \leqslant 4 \left(\frac{|\xi_1|}{3 \cdot 2^3} + \frac{|\xi_2|}{2^2} + \sum_{i=4}^{\infty} \frac{|\xi_i|}{2^i} \right) \leqslant 4))x)) \qquad (x = \{\xi_n\} \in c_0),$$

and choosing, for each $j > 2$, a positive integer $n > 2$ with $n \neq j$ and a permutation $\sigma_j \in \Pi_{1,n}$ with $\sigma_j(2) = j$, we obtain

$$|\xi_j| \leqslant 4 \left(\frac{|\xi_1|}{n 2^n} + \sum_{\substack{i=2 \\ i \neq n}}^{\infty} \frac{|\xi_{\sigma_j(i)}|}{2^i} \right) \leqslant 4))x)) \qquad (x = \{\xi_n\} \in c_0),$$

whence

$$\tfrac{1}{8}\|x\| \leqslant |||x||| \qquad (x \in c_0), \tag{19.12}$$

which, together with (19.11), proves that the norms $|||x|||$ and $\|x\|$ are equivalent.

Let us show now that for every pair d_1, d_2 of finite sets of indices with $d_1 \subset d_2$ and every finite sequence of scalars $\{\alpha_i\}_{i\in d_2}$ with $\sum_{i\in d_2\setminus d_1} |\alpha_i| \neq 0$ we have

$$\Big|\Big|\Big|\sum_{i\in d_1} \alpha_i x_i\Big|\Big|\Big| < \Big|\Big|\Big|\sum_{i\in d_2} \alpha_i x_i\Big|\Big|\Big| \tag{19.13}$$

whence, in particular, $|||x|||$ is a K-norm, satisfying (19.8). Since d_1 is finite, the norm $\Big|\Big|\Big|\sum_{i\in d_1} \alpha_i x_i\Big|\Big|\Big|$ is attained for some $n_0 \geqslant 2$ and $\sigma_0 \in \Pi_{1,n}$. Since $\sum_{i\in d_2\setminus d_1} |\alpha_i| \neq 0$, a larger sum must be obtained when this same n_0 and σ_0 are used for $\sum_{i\in d_2} \alpha_i x_i$, whence we obtain (19.13).

Finally, let us show that

$$\Big|\Big|\Big|\sum_{m=1}^{\infty} \frac{1}{m} x_m\Big|\Big|\Big| = \Big|\Big|\Big|\sum_{m=2}^{\infty} \frac{1}{m} x_m\Big|\Big|\Big| = \sum_{m=2}^{\infty} \frac{1}{m 2^m}, \tag{19.14}$$

whence, by proposition 19.1 a), $|||x|||$ is not a T-norm. We have, obviously,

$$\Big|\Big|\Big|\sum_{m=1}^{\infty} \frac{1}{m} x_m\Big|\Big|\Big| \geqslant \Big|\Big|\Big|\sum_{m=2}^{\infty} \frac{1}{m} x_m\Big|\Big|\Big|. \tag{19.15}$$

Furthermore, let $n \geqslant 2$ be fixed. If for a pair $i, i+j \in \{2, 3, \ldots, n-1, n+1, \ldots\}$ and a $\sigma \in \Pi_{1,n}$ we have $\frac{1}{\sigma(i)} < \frac{1}{\sigma(i+j)}$, then for the permutation $\sigma' \in \Pi_{1,n}$ defined by

$$\sigma'(i) = \sigma(i+j), \sigma'(i+j) = \sigma(i), \quad \sigma'(k) = \sigma(k) \quad (k \neq i, i+j)$$

we have, taking into account that $\frac{\alpha}{2^i} + \frac{\beta}{2^{i+j}} > \frac{\beta}{2^i} + \frac{\alpha}{2^{i+j}}$ for $\alpha > \beta \geqslant 0$,

$$\sum_{\substack{m=2\\ m\neq n}}^{\infty} \frac{1}{\sigma'(m) 2^m} > \sum_{\substack{m=2\\ m\neq n}}^{\infty} \frac{1}{\sigma(m) 2^m}.$$

Consequently, for every $n \geqslant 2$ and $\sigma \in \Pi_{1,n}$ we have

$$\frac{1}{n 2^n} + \sum_{\substack{m=2\\ m\neq n}}^{\infty} \frac{1}{\sigma(m) 2^m} \leqslant \frac{1}{n 2^n} + \sum_{\substack{m=2\\ m\neq n}}^{\infty} \frac{1}{m 2^m} = \sum_{m=2}^{\infty} \frac{1}{m 2^m}$$

$$= \sup_{2\leqslant n<\infty} \sum_{\substack{m=2\\ m\neq n}}^{\infty} \frac{1}{m 2^m} \leqslant \sup_{2\leqslant n<\infty} \sup_{\tau\in\Pi_{1,n}} \sum_{\substack{m=2\\ m\neq n}}^{\infty} \frac{1}{\tau(m) 2^m} = \Big|\Big|\Big|\sum_{m=2}^{\infty} \frac{1}{m} x_m\Big|\Big|\Big|,$$

which, together with (19.15), implies (19.14), completing the proof of the assertions of example 19.2.

Theorem 19.1. *Let E be a Banach space with a basis $\{x_n\}$ and let $\{f_n\} \subset E^*$ be the a.s.c.f. Then*

a) *One can introduce on E a T-norm equivalent to the initial norm on E, by the formula*

$$((x((= \max_{1 \leqslant n < \infty} \left(\frac{1}{n} \sum_{i=1}^{n} \| f_i(x) x_i \| + \left\| \sum_{i=n+1}^{\infty} f_i(x) x_i \right\| \right) \tag{19.16}$$

and also another equivalent T-norm on E, by the formula

$$((x((= \sum_{i=1}^{\infty} \frac{1}{2^i} \| f_i(x) x_i \| + \max_{1 \leqslant n < \infty} \left\| \sum_{i=n}^{\infty} f_i(x) x_i \right\| . \tag{19.17}$$

b) *One can introduce on E a K-norm equivalent to the initial norm on E, by the formula*

$$))x)) = \sum_{i=1}^{\infty} \frac{1}{2^i} \| f_i(x) x_i \| + \sup_{1 \leqslant n < \infty} \left\| \sum_{i=1}^{n} f_i(x) x_i \right\| . \tag{19.18}$$

c) *One can introduce on E a TK-norm equivalent to the initial norm on E, by the formula*

$$((x)) = \sum_{i=1}^{\infty} \frac{1}{2^i} \| f_i(x) x_i \| + \sup_{1 \leqslant n, m < \infty} \left\| \sum_{i=n}^{m} f_i(x) x_i \right\| . \tag{19.19}$$

Proof. a_1) Let us first observe that the max in (19.16) is indeed attained for some n, since $\{x_n\}$ being a basis, both summands in (19.16) tend to 0 as $n \to \infty$.

Let us show now that $((x(($ defined by (19.16) is a norm on E. Obviously $((x((\geqslant 0$ and $((0((= 0$. If we have $((x((= 0$, then by

$$((x((\geqslant \| f_1(x) x_1 \| + \left\| \sum_{i=2}^{\infty} f_i(x) x_1 \right\| \geqslant \left\| \sum_{i=1}^{\infty} f_i(x) x_i \right\| = \|x\| \quad (x \in E) \tag{19.20}$$

we have $x = 0$. Finally, the inequality $((x+y((\leqslant ((x((+ ((y(($ is obvious from $((x+y((= \frac{1}{n_0} \sum_{i=1}^{n_0} \| f_i(x+y) x_i \| + \left\| \sum_{i=n_0+1}^{\infty} f_i(x+y) x_i \right\|$ (for a suitable n_0), and $((\alpha x((= |\alpha| ((x(($ is obvious directly from (19.16).

Furthermore, $((x(($ is equivalent to the initial norm on E by (19.20) and by the relations

$$((x((= \frac{1}{n_0} \sum_{i=1}^{n_0} \|f_i(x)x_i\| + \left\|\sum_{i=n_0+1}^{\infty} f_i(x)x_i\right\|$$

$$\leq \max_{1 \leq i \leq n_0} \|f_i(x)x_i\| + \left\|x - \sum_{i=1}^{n_0} f_i(x)x_i\right\|$$

$$\leq \max_{1 \leq i \leq n_0} \left(\left\|\sum_{j=1}^{i} f_j(x)x_j\right\| + \left\|\sum_{j=1}^{i-1} f_j(x)x_j\right\|\right) + \|x\| + \left\|\sum_{i=1}^{n_0} f_i(x)x_i\right\|$$

$$\leq 3 \sup_{1 \leq n < \infty} \left\|\sum_{i=1}^{n} f_i(x)x_i\right\| + \|x\| \leq (3v_{\{x_n\}} + 1)\|x\| \quad (x \in E).$$

Finally, let us prove that $((x((\text{ is a } T\text{-norm}$. Let $\{\alpha_i\}_{i=l-1}^{\infty}$ be a sequence of scalars with $\alpha_{l-1} \neq 0$, such that $\sum_{i=l}^{\infty} \alpha_i x_i$ converges. We have then, for a suitable n_0 with $l \leq n_0 < \infty$,

$$\left(\left(\sum_{i=l}^{\infty} \alpha_i x_i\right(\left(= \frac{1}{n_0} \sum_{i=l}^{n_0} \|\alpha_i x_i\| + \left\|\sum_{i=n_0+1}^{\infty} \alpha_i x_i\right\| < \frac{1}{n_0} \sum_{i=l-1}^{n_0} \|\alpha_i x_i\| + \left\|\sum_{i=n_0+1}^{\infty} \alpha_i x_i\right\|$$

$$\leq \max_{l-1 \leq n < \infty} \left(\frac{1}{n} \sum_{i=l-1}^{n} \|\alpha_i x_i\| + \left\|\sum_{i=n+1}^{\infty} \alpha_i x_i\right\|\right) = \left(\left(\sum_{i=l-1}^{\infty} \alpha_i x_i\right((,\right.$$

whence, by proposition 19.1 a), $((x(($ is a T-norm.

a_2) Consider now the $((x(($ defined by (19.17). Since $\{x_n\}$ is a basis, the max in (19.17) is indeed attained for some n and it is also obvious that $((x(($ is a norm on E. This norm is equivalent to the initial norm on E, since for every $x \in E$ we have

$$\|x\| = \left\|\sum_{i=1}^{\infty} f_i(x)x_i\right\| \leq ((x((\leq \max_{1 \leq i < \infty} \|f_i(x)x_i\| + \max_{1 \leq n < \infty} \left\|\sum_{i=n}^{\infty} f_i(x)x_i\right\|$$

$$\leq \max_{1 \leq i < \infty} \left(\left\|\sum_{j=i}^{\infty} f_j(x)x_j\right\| + \left\|\sum_{j=i+1}^{\infty} f_j(x)x_j\right\|\right) + \max_{1 \leq n < \infty} \left\|\sum_{i=n}^{\infty} f_i(x)x_i\right\|$$

$$\leq 3 \max_{1 \leq n < \infty} \left\|\sum_{i=n}^{\infty} f_i(x)x_i\right\| \leq 3\left(\|x\| + \sup_{1 \leq n < \infty} \left\|\sum_{i=1}^{n} f_i(x)x_i\right\|\right)$$

$$\leq 3(1 + v_{\{x_n\}})\|x\|.$$

Let us prove, finally, that $((x(($ is a T-norm. Let $\{\alpha_i\}_{i=l-1}^{\infty}$ be a sequence of scalars with $\alpha_{l-1} \neq 0$, such that $\sum_{i=l}^{\infty} \alpha_i x_i$ converges. Then

$$\left(\left(\sum_{i=l}^{\infty} \alpha_i x_i\right(\left(= \sum_{i=l}^{\infty} \frac{1}{2^i} \|\alpha_i x_i\| + \max_{1 \leq n < \infty} \left\|\sum_{i=n}^{\infty} \alpha_i x_i\right\|\right.$$

$$< \sum_{i=l-1}^{\infty} \frac{1}{2^i} \|\alpha_i x_i\| + \max_{l-1 \leq n < \infty} \left\|\sum_{i=n}^{\infty} \alpha_i x_i\right\| = \left(\left(\sum_{i=l-1}^{\infty} \alpha_i x_i\right((,\right.$$

whence, by proposition 19.1 a), $((x(($ is a T-norm.

b) It is immediate that $))x))$ defined by (19.18) is a norm on E. This norm is equivalent to the initial norm on E, since for every $x \in E$ we have

$$\|x\| \leqslant))x)) \leqslant \max_{1 \leqslant i < \infty} \|f_i(x) x_i\| + \sup_{1 \leqslant n < \infty} \left\| \sum_{i=1}^{n} f_i(x) x_i \right\|$$

$$\leqslant \max_{1 \leqslant i < \infty} \left(\left\| \sum_{j=1}^{i} f_j(x) x_j \right\| + \left\| \sum_{j=1}^{i-1} f_j(x) x_j \right\| \right) + \sup_{1 \leqslant n < \infty} \left\| \sum_{i=1}^{n} f_i(x) x_i \right\|$$

$$\leqslant 3 v_{\{x_n\}} \|x\| .$$

Finally, for any scalars $\alpha_1, \ldots, \alpha_{n+1}$ with $\alpha_{n+1} \neq 0$ we have

$$\Big)\Big)\sum_{i=1}^{n} \alpha_i x_i \Big)\Big) = \sum_{i=1}^{n} \frac{1}{2^i} \|\alpha_i x_i\| + \max_{1 \leqslant k \leqslant n} \left\| \sum_{i=1}^{k} \alpha_i x_i \right\|$$

$$< \sum_{i=1}^{n+1} \frac{1}{2^i} \|\alpha_i x_i\| + \max_{1 \leqslant k \leqslant n+1} \left\| \sum_{i=1}^{k} \alpha_i x_i \right\| = \Big)\Big)\sum_{i=1}^{n+1} \alpha_i x_i \Big)\Big),$$

whence, by proposition 19.1 b), $))x))$ is a K-norm.

c) can be proved similarly to a_2) and b) above, and thus the proof of theorem 19.1 is complete.

We have given each of the above equivalent norms (instead of giving only (19.19)) since they are useful for applications; in particular, the norm (19.16) has been already used in example 19.1. Let us mention that one can also prove that in the conditions of theorem 19.1 it is possible to introduce another equivalent TK-norm on E, by the formula

$$((x)) = \sup_{1 \leqslant n < \infty} \left\{ \left(\left(\sum_{i=1}^{n} f_i(x) x_i \left(\left(+ \right) \right) \sum_{i=n+1}^{\infty} f_i(x) x_i \right) \right) \right\}, \tag{19.21}$$

where $((x(($ is the T-norm (19.16) and $))x))$ is the K-norm (19.18).

Remark 19.1. Theorem 19.1 admits, in a certain sense, a converse. Namely, the above definition of T-norms also has sense when (x_n, f_n) is a biorthogonal system with $[x_n] = E$, and if for such a system one can introduce on E a T-norm with respect to $\{x_n\}$, equivalent to the initial norm on E, then $\{x_n\}$ must be a basis of E. Indeed, even a more general result is also true. Namely, for an arbitrary sequence $\{x_n\} \subset E$ with $x_n \neq 0$ $(n = 1, 2, \ldots)$ and $[x_n] = E$, if one can introduce on E a *weak T-norm* with respect to $\{x_n\}$, equivalent to the initial norm on E, i.e. an equivalent norm $(x($ such that for every polynomial $p = \sum_{i=1}^{k} \alpha_i x_i$ and every positive integer $n \leqslant k$ the polynomial $\sum_{i=1}^{n} \alpha_i x_i \in P_{(n)} = [x_1, \ldots, x_n]$ is in this new norm a (not necessarily unique) polynomial of best approximation of p by means of the elements of $P_{(n)}$, then $\{x_n\}$ is a basis of E. For if $\alpha_1,\dots,\alpha_n, \alpha_{n+1},\dots,\alpha_{n+m}=\alpha_k$ are arbitrary scalars, then, since $(x($ is a weak T-norm,

$$\Big(\sum_{i=1}^{n}\alpha_i x_i\Big(\leqslant \Big(\sum_{i=1}^{n+m}\alpha_i x_i\Big(+ \Big(\sum_{i=1}^{n+m}\alpha_i x_i - \sum_{i=1}^{n}\alpha_i x_i\Big(\leqslant 2\Big(\sum_{i=1}^{n+m}\alpha_i x_i\Big(,$$

whence, by § 7, theorem 7.1, $\{x_n\}$ is a basis of E endowed with the norm $(x($, and therefore also of E with its initial norm. Defining similarly *weak K-norms*, *weak TK-norms*, the same argument shows that if one can introduce on E such a norm with respect to $\{x_n\}$ (where $[x_n]=E$, $x_n\neq 0$ for $n=1,2,\dots$), equivalent to the initial norm on E, then $\{x_n\}$ is a basis of E.

Let us now turn our attention to E endowed with its initial norm. We shall denote by $e_n(x)$ the best approximation of $x\in E$ by means of the elements of $P_{(n)}=[x_1,\dots,x_n]$, i. e.

$$e_n(x) = \inf_{p\in P_{(n)}} \|x-p\| \qquad (x\in E,\, n=1,2,\dots). \tag{19.22}$$

From theorem 19.1 it follows

Corollary 19.1. *Let E be a Banach space with a basis $\{x_n\}$. Then there exists a constant c, $0<c\leqslant 1$, depending only on the basis $\{x_n\}$, such that*

$$c\,\|x-s_n(x)\| \leqslant e_n(x) \leqslant \|x-s_n(x)\| \qquad (x\in E,\, n=1,2,\dots), \tag{19.23}$$

where $s_n(x)$ are the partial sum operators associated to the basis $\{x_n\}$.

Proof. Let $x\in E$ and n be arbitrary and let $p_0\in P_{(n)}$ be an element of best approximation of x; such an element exists, since $\dim P_{(n)}=n<\infty$. Then $e_n(x)=\|x-p_0\|$. By virtue of theorem 19.1, let $((x(($ be a T-norm on E, equivalent to the initial norm of E, and let $C_1, C_2>0$ be constants such that $C_1\|x\|\leqslant((x((\leqslant C_2\|x\|$ for all $x\in E$. Then

$$e_n(x) = \|x-p_0\| \geqslant \frac{1}{C_2}((x-p_0((\geqslant \frac{1}{C_2}((x-s_n(x)((\geqslant \frac{C_1}{C_2}\|x-s_n(x)\|$$

and thus we have the first inequality of (19.23) with $c=\dfrac{C_1}{C_2}$. On the other hand, the second inequality of (19.23) is obvious by (19.22), which completes the proof.

Corollary 19.1 shows that if $\{x_n\}$ is a basis of E and $P_{(n)}=[x_1,\dots,x_n]$, then $e_n(x)$ and $\|x-s_n(x)\|$ are "of the same order". Consequently, if we want to classify the elements $x\in E$ by the rapidity of the convergence to 0 of the sequences $\{e_n(x)\}$, then we can use the sequences $\{\|x-s_n(x)\|\}$, which can be computed more easily.

Remark 19.2. Corollary 19.1 also admits, in a certain sense, a converse. Namely, if for a biorthogonal system (x_n, f_n) with $[x_n]=E$ we have the first inequality of (19.23), then

$$\|s_n(x)\| \leqslant \|x\| + \|x-s_n(x)\| \leqslant \|x\| + \frac{1}{c}\|e_n(x)\| \leqslant \left(1+\frac{1}{c}\right)\|x\| \qquad (x\in E, n=1,2,\ldots),$$

whence, by § 4, theorem 4.1, $\{x_n\}$ is a basis of E. Moreover, introducing for a sequence $\{x_n\}\subset E$ with $x_n\neq 0$ $(n=1,2,\ldots)$ and $[x_n]=E$ the operators

$$S_n\left(\sum_{i=1}^{k}\alpha_i x_i\right)=\begin{cases}\sum_{i=1}^{n}\alpha_i x_i & \text{for } n=1,\ldots,k\\ \sum_{i=1}^{k}\alpha_i x_i & \text{for } n=k+1,k+2,\ldots\end{cases}$$

the same argument with S_n instead of s_n and with § 7, theorem 7.1, shows that if for a sequence $\{x_n\}\subset E$ with $[x_n]=E$ and $x_n\neq 0$ $(n=1,2,\ldots)$ there exists a constant c, $0<c\leqslant 1$, such that

$$\|p-S_n(p)\| \leqslant \frac{1}{c}e_n(p) \qquad \left(p\in\bigcup_{k=1}^{\infty}P_{(k)}, n=1,2,\ldots\right), \tag{19.24}$$

then $\{x_n\}$ is a basis of E.

§ 20. Polynomial bases. Strict polynomial bases. Γ systems and Λ systems

Definition 20.1. Let $\{x_n\}$ be a finitely linearly independent sequence[1] in a Banach space E, with $[x_n]=E$. We call *polynomial basis* of the space E (with respect to $\{x_n\}$) any basis of E of the form

$$y_n=\sum_{i=1}^{m_n}\alpha_i^{(n)}x_i \qquad (n=1,2,\ldots), \tag{20.1}$$

and *strict polynomial basis* of E (with respect to $\{x_n\}$) any basis of E of the form (20.1), where

$$m_n=n, \qquad \alpha_n^{(n)}\neq 0 \qquad (n=1,2,\ldots). \tag{20.2}$$

Obviously, every strict polynomial basis is a polynomial basis (with respect to the same sequence) and every basis is a strict polynomial basis with respect to itself.

[1] I. e. such that every finite subsequence of $\{x_n\}$ is linearly independent (see § 6).

Definition 20.2. Let $\{x_n\}$ be a finitely linearly independent sequence in a Banach space E and let $p=\sum_{i=1}^{n}\alpha_i x_i$ be a polynomial with respect to $\{x_n\}$. The *order* of the polynomial p is the number

$$\operatorname{ord} p=\operatorname{ord}\sum_{i=1}^{n}\alpha_i x_i=\max_{\alpha_j\neq 0} j. \tag{20.3}$$

By virtue of condition (20.2), a strict polynomial basis $\{y_n\}$ with respect to $\{x_n\}$ contains polynomials of all orders, namely, $\operatorname{ord} y_n=n$ $(n=1,2,\ldots)$, and we have

$$[y_1,\ldots,y_n]=[x_1,\ldots,x_n] \quad (n=1,2,\ldots). \tag{20.4}$$

We have seen in § 10, corollary 10.2 of the Krein-Milman-Rutman theorem, that if the space E has a basis, the problem of existence of polynomial bases with respect to any finitely linearly independent sequence $\{x_n\}\subset E$ with $[x_n]=E$ has an affirmative answer. On the contrary, we shall see below that the problem of existence of *strict* polynomial bases with respect to a finitely linearly independent complete sequence $\{x_n\}$ in a space with a basis has, in general, a negative answer.

Let us first give some necessary and sufficient conditions for the existence of strict polynomial bases with respect to a finitely linearly independent complete sequence $\{x_n\}$.

Theorem 20.1. *Let $\{x_n\}$ be a finitely linearly independent sequence in a Banach space E, with $[x_n]=E$. The following statements are equivalent:*

1°–10°. *There exists a sequence of endomorphisms* $\{v_n\}\subset L(E,E)$ *such that*

a) $v_n(x)\in P_{(n)}=[x_1,\ldots,x_n]$ $\quad(x\in E,\ n=1,2,\ldots)$,

b) $v_n(p)=p$ $\quad(p\in P_{(n)},\ n=1,2,\ldots)$,

c) $\lim_{n\to\infty} v_n(x)=x$ $\quad(x\in E)$,

and satisfying any one of the following conditions:

d$_1$) $\operatorname{ord} v_n(x)\leqslant \operatorname{ord} v_{n+1}(x)$ $\quad(x\in E,\ n=1,2,\ldots)$,

where we make the convention $\operatorname{ord} 0=0$;

d$_2$) $v_i[v_n(x)]=v_i(x)$ $\quad(x\in E,\ i=1,\ldots,n;\ n=1,2,\ldots)$;

d$_3$) $\lim_{1\leqslant k_1<\cdots<k_m\to\infty} r_{k_1}r_{k_2}\ldots r_{k_m}(x)=0$ $\quad(x\in E)$,

where $r_n(x)=x-v_n(x)$ $(x\in E,\ n=1,2,\ldots)$;

d$_4$) $\sup_{1\leqslant k_1<\cdots<k_m<\infty}\|r_{k_1}r_{k_2}\ldots r_{k_m}\|<\infty$;

d$_5$) *there exists an equivalent norm* $|||x|||$ *on E such that*

$$|||v_n|||=\sup_{|||x|||\leqslant 1}|||v_n(x)|||\leqslant 1 \quad (n=1,2,\ldots); \tag{20.5}$$

d_6) *there exists an equivalent norm* $|||x|||$ *on* E *such that*

$$|||v_n(x)||| \leqslant |||v_{n+1}(x)||| \qquad (x \in E,\ n = 1, 2, \ldots); \tag{20.6}$$

d_7) *there exists an equivalent norm* $|||x|||$ *on* E *such that*

$$\left|\left|\left| v_n\left(\sum_{i=1}^{n+1} \alpha_i x_i\right)\right|\right|\right| < \left|\left|\left| v_{n+1}\left(\sum_{i=1}^{n+1} \alpha_i x_i\right)\right|\right|\right| = \left|\left|\left| \sum_{i=1}^{n+1} \alpha_i x_i \right|\right|\right| \tag{20.7}$$

for any scalars $\alpha_1, \ldots, \alpha_{n+1}$ *with* $\alpha_{n+1} \neq 0$;

d_8) *there exists an equivalent norm* $|x|$ *on* E *such that*

$$|I_E - v_n| = \sup_{|x| \leqslant 1} |x - v_n(x)| \leqslant 1 \qquad (n = 1, 2, \ldots); \tag{20.8}$$

d_9) *there exists an equivalent norm* $|x|$ *on* E *such that for every* $x \in E$, $v_n(x) \in P_{(n)}$ *be a polynomial of best approximation of* x *in this new norm*;

d_{10}) *there exists an equivalent norm* $|x|$ *on* E *such that for every* $x \in E$, $v_n(x) \in P_{(n)}$ *be the unique polynomial of best approximation of* x *in this new norm.*

By statement n° *we mean here that there exists a sequence* $\{v_j\} \subset L(E, E)$ *satisfying* a), b), c) *and* d_n) $(n = 1, \ldots, 10)$.

11°. *E has a strict polynomial basis with respect to* $\{x_n\}$.

Proof. We shall prove the implications

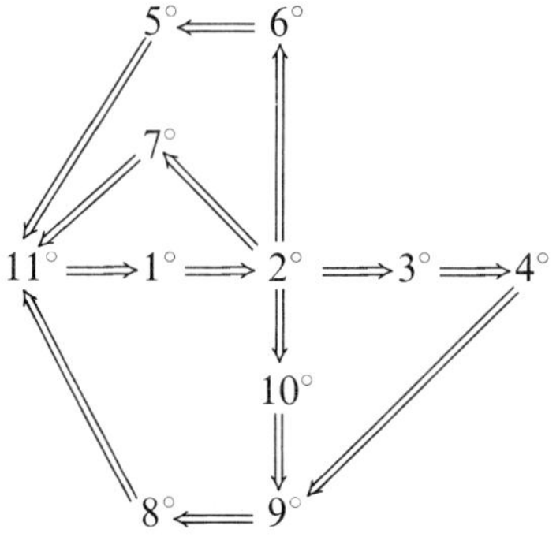

The implication $11^\circ \Rightarrow 1^\circ$ is obvious, since if $\{y_n\}$ is a strict polynomial basis with respect to $\{x_n\}$, one can take $v_j =$ the partial sum operator s_j associated to the basis $\{y_n\}$ $(j = 1, 2, \ldots)$.

Assume now that we have 1°. Then by a), b) we have

$$v_n[x - v_n(x)] = v_n(x) - v_n[v_n(x)] = 0 \qquad (x \in E,\ n = 1, 2, \ldots),$$

whence, by d_1) (with the convention ord $0 = 0$), we obtain

$$0 \leqslant \operatorname{ord} v_i[x - v_n(x)] \leqslant \operatorname{ord} v_n[x - v_n(x)] = \operatorname{ord} 0 = 0$$
$$(x \in E,\ i = 1, \ldots, n;\ n = 1, 2, \ldots),$$

whence

$$v_i[x - v_n(x)] = 0 \qquad (x \in E,\ i = 1, \ldots, n;\ n = 1, 2, \ldots),$$

i.e. d_2). Thus, $1^\circ \Rightarrow 2^\circ$.

Assume now that we have 2°. Put

$$|||x||| = \sum_{i=1}^{\infty} \frac{1}{2^i} \|v_{i+1}(x) - v_i(x)\| + \sup_{1 \leqslant j < \infty} \|v_j(x)\| \qquad (x \in E). \tag{20.9}$$

By c) we have $|||x||| < \infty$ and $|||x|||$ is a norm on E. This norm is equivalent to the initial norm of E, since by c) and the principle of uniform boundedness we have $\sup_{1 \leqslant j < \infty} \|v_j\| < \infty$ and

$$\|x\| = \lim_{n \to \infty} \|v_n(x)\| \leqslant |||x||| \leqslant 3 \sup_{1 \leqslant j < \infty} \|v_j(x)\| \leqslant \Big(3 \sup_{1 \leqslant j < \infty} \|v_j\|\Big) \|x\| \qquad (x \in E).$$

Finally, for any scalars $\alpha_1, \ldots, \alpha_{n+1}$ with $\alpha_{n+1} \neq 0$ we have, by d_2) and a), b),

$$\left|\left|\left| v_n\left(\sum_{j=1}^{n+1} \alpha_j x_j\right) \right|\right|\right| = \sum_{i=1}^{n-1} \frac{1}{2^i} \left\| v_{i+1}\left(\sum_{j=1}^{n+1} \alpha_j x_j\right) - v_i\left(\sum_{j=1}^{n+1} \alpha_j x_j\right) \right\| + \sup_{1 \leqslant k \leqslant n} \left\| v_k\left(\sum_{j=1}^{n+1} \alpha_j x_j\right) \right\|$$

$$\left|\left|\left| v_{n+1}\left(\sum_{j=1}^{n+1} \alpha_j x_j\right) \right|\right|\right| = \sum_{i=1}^{n} \frac{1}{2^i} \left\| v_{i+1}\left(\sum_{j=1}^{n+1} \alpha_j x_j\right) - v_i\left(\sum_{j=1}^{n+1} \alpha_j x_j\right) \right\| + \sup_{1 \leqslant k \leqslant n+1} \left\| v_k\left(\sum_{j=1}^{n+1} \alpha_j x_j\right) \right\|,$$

whence, since $v_{n+1}\left(\sum_{j=1}^{n+1} \alpha_j x_j\right) - v_n\left(\sum_{j=1}^{n+1} \alpha_j x_j\right) = \sum_{j=1}^{n+1} \alpha_j x_j - \sum_{j=1}^{n} \alpha_j x_j - v_n(\alpha_{n+1} x_{n+1}) = \alpha_{n+1}[x_{n+1} - v_n(x_{n+1})] \neq 0$ (by $\alpha_{n+1} \neq 0$ and $v_n(x_{n+1}) \in P_{(n)}$, $x_{n+1} \notin P_{(n)}$), we infer (20.7). Thus, $2^\circ \Rightarrow 7^\circ$.

Assume now again that we have 2°. Put

$$|||x||| = \sup_{1 \leqslant j < \infty} \|v_j(x)\| \qquad (x \in E). \tag{20.10}$$

Then, by d_2), for any $x \in E$ and $n = 1, 2, \ldots$ we have

$$|||v_n(x)||| = \sup_{1 \leqslant j < \infty} \|v_j v_n(x)\| = \sup_{1 \leqslant j \leqslant n} \|v_j(x)\| \leqslant \sup_{1 \leqslant j \leqslant n+1} \|v_j(x)\| = |||v_{n+1}(x)|||,$$

i.e. (20.6). Thus, $2^\circ \Rightarrow 6^\circ$.

Assume now that we have 6°. Then, by c), we have

$$|||v_n(x)||| \leqslant |||v_{n+1}(x)||| \leqslant \cdots \leqslant \lim_{n \to \infty} |||v_n(x)||| = |||x||| \qquad (x \in E,\ n = 1, 2, \ldots),$$

i.e. (20.5). Thus, $6^\circ \Rightarrow 5^\circ$.

Now assume again that we have 2°. Then for any $x \in E$ and any positive integers n, m with $n \leqslant m$ we have, by d_2),

$$r_n r_m(x) = r_m(x) - v_n[r_m(x)] = x - v_m(x) - v_n(x) + v_n v_m(x) = x - v_m(x) = r_m(x),$$

whence, by c), we get

$$\lim_{1 \leqslant k_1 < \cdots < k_m \to \infty} r_{k_1} r_{k_2} \dots r_{k_m}(x) = \lim_{k_m \to \infty} r_{k_m}(x) = 0,$$

i.e. d_3). Thus, $2° \Rightarrow 3°$.

The implication $3° \Rightarrow 4°$ is obvious (the set $\{r_{k_1} r_{k_2} \dots r_{k_m}(x)\}_{1 \leqslant k_1 < \cdots < k_m}$ is countable, and the limit in d_3) is the limit of this sequence in the usual sense), by the principle of uniform boundedness.

Assume now that we have 4°. Put $v_0(x) = 0$, $r_0(x) = x$ $(x \in E)$ and

$$|x| = \sup_{0 \leqslant k_1 < \cdots < k_m} \|r_{k_1} r_{k_2} \dots r_{k_m}(x)\| \quad (x \in E).[1] \tag{20.11}$$

Then, by d_4), $|x| < \infty$. Furthermore, $|x|$ is a norm on E, which is equivalent to the initial norm of E, since by d_4) we have

$$\|x\| \leqslant |x| \leqslant \Big(\sup_{0 \leqslant k_1 < \cdots < k_m} \|r_{k_1} r_{k_2} \dots r_{k_m}\|\Big) \|x\| \quad (x \in E).$$

Since for any $x \in E$ and any positive integers n, m with $n \leqslant m$ we have, by a) and b),

$$r_m r_n(x) = r_n(x) - v_m[r_n(x)] = x - v_n(x) - v_m(x) + v_m v_n(x) = x - v_m(x) = r_m(x), \tag{20.12}$$

it follows that for any $x \in E$ and any positive integer n,

$$\begin{aligned} |x - v_n(x)| = |r_n(x)| &= \sup_{0 \leqslant k_1 < \cdots < k_m} \|r_{k_1} r_{k_2} \dots r_{k_m} r_n(x)\| \\ &\leqslant \sup_{0 \leqslant k_1 < \cdots < k_m} \|r_{k_1} r_{k_2} \dots r_{k_m}(x)\| = |x|, \end{aligned}$$

whence, again by b), we get

$$|x - v_n(x)| = |x - p - v_n(x) + p| = |x - p - v_n(x - p)| \leqslant |x - p| \quad (p \in P_{(n)}),$$

i.e. d_9). Thus, $4° \Rightarrow 9°$.

Assume now again that we have 2°. Put

$$|x| = \|v_1(x)\| + \sum_{i=1}^{\infty} \frac{1}{2^i} \|v_{i+1}(x) - v_i(x)\| + \sup_{1 \leqslant i < \infty} \|x - v_i(x)\| \quad (x \in E). \tag{20.13}$$

---

[1] In particular, if we also have d_2), this norm reduces to

$$|x| = \sup_{0 \leqslant j < \infty} \|r_j(x)\| = \sup_{1 \leqslant j < \infty} (\|x\|, \|r_j(x)\|) \quad (x \in E).$$

Then by c), $|x|<\infty$. Furthermore, $|x|$ is a norm on E, which is equivalent to the initial norm of E, since by c) and the principle of uniform boundedness we have $\sup_{1\leqslant n<\infty}\|v_n\|<\infty$ and

$$\|x\|\leqslant\|v_1(x)\|+\|x-v_1(x)\|\leqslant|x|\leqslant\sup_{1\leqslant n<\infty}\|v_n(x)\|+2\sum_{i=1}^{\infty}\frac{1}{2^i}\sup_{1\leqslant n<\infty}\|v_n(x)\|$$
$$+\|x\|+\sup_{1\leqslant n<\infty}\|v_n(x)\|\leqslant\left(4\sup_{1\leqslant n<\infty}\|v_n\|+1\right)\|x\| \qquad (x\in E).$$

Finally, let $x\in E$ and $p\in P_{(n)}$ be arbitrary. We have then, by d_2),

$$|x-v_n(x)|=\left\|v_1[x-v_n(x)]\right\|+\sum_{i=1}^{\infty}\frac{1}{2^i}\left\|v_{i+1}[x-v_n(x)]-v_i[x-v_n(x)]\right\|$$
$$+\sup_{1\leqslant i<\infty}\left\|x-v_n(x)-v_i[x-v_n(x)]\right\|$$
$$=\sum_{i=n}^{\infty}\frac{1}{2^i}\|v_{i+1}(x)-v_i(x)\|+\sup_{n\leqslant i<\infty}\|x-v_i(x)\|,$$

$$|x-p|=\|v_1(x-p)\|+\sum_{i=1}^{\infty}\frac{1}{2^i}\|v_{i+1}(x-p)-v_i(x-p)\|$$
$$+\sup_{1\leqslant i<\infty}\|x-p-v_i(x-p)\|$$

whence, since by $p\in P_{(n)}$ we have $v_{i+1}(p)=v_i(p)=p$ $(i=n,n+1,\ldots)$, it follows that

$$|x-v_n(x)|\leqslant|x-p|,$$

where the equality holds only if

$$v_1(x-p)=0,\quad v_{i+1}(x-p)-v_i(x-p)=0 \qquad (i=1,\ldots,n-1),$$

which gives $0=v_1(x-p)=v_2(x-p)=\cdots=v_n(x-p)=v_n(x)-v_n(p)=v_n(x)-p$, i.e. $p=v_n(x)$. Thus, $2^\circ\Rightarrow10^\circ$.

The implication $10^\circ\Rightarrow9^\circ$ is obvious[1]. The implication $9^\circ\Rightarrow8^\circ$ is also immediate, since if we have d_9), then, by $0\in P_{(n)}$,

$$|x-v_n(x)|\leqslant|x| \qquad (x\in E,\ n=1,2,\ldots).$$

Thus, it remains to prove the implications $5^\circ\Rightarrow11^\circ$, $7^\circ\Rightarrow11^\circ$ and $8^\circ\Rightarrow11^\circ$. Let us first prove that if we have any one of 5° or 7°, then the sequence[2]

---

[1] Let us observe that the implication $10^\circ\Rightarrow1^\circ$ is also immediate, since if $\operatorname{ord}v_n(x)>\operatorname{ord}v_{n+1}(x)$, then, by d_{10}), $e_n(x)=|x-v_n(x)|<|x-v_{n+1}(x)|=e_{n+1}(x)\leqslant e_n(x)$, which is impossible.

[2] Obviously, the sequence $\{x_n\}$ itself is not necessarily a basis of E (take e.g. $E=l^2$, $x_n=e_1+\cdots+e_n$, where $\{e_n\}$ is the unit vector basis of l^2, and $v_n=$ the partial sum operator s_n with respect to $\{e_n\}$).

$$y_1 = x_1, \quad y_n = x_n - v_{n-1}(x_n) \quad (n=2,3,\ldots) \tag{20.14}$$

is a strict polynomial basis of E. By a) it is sufficient to prove that $\{y_n\}$ is a basis of E.

Assume first 5°. Then, since $v_n(y_{n+1}) = v_n(x_{n+1}) - v_n(x_{n+1}) = 0$, we have

$$\left|\left|\left|\sum_{i=1}^{n} \alpha_i y_i\right|\right|\right| = \left|\left|\left|v_n\left(\sum_{i=1}^{n+1} \alpha_i y_i\right)\right|\right|\right| \leqslant \left|\left|\left|\sum_{i=1}^{n+1} \alpha_i y_i\right|\right|\right|$$

for any scalars $\alpha_1, \ldots, \alpha_{n+1}$ and hence, by § 7, theorem 7.1, $\{y_n\}$ is a basis of E endowed with the norm $|||x|||$, whence also of E endowed with its initial norm.

Assume now 7°. Then, again since $v_n(y_{n+1}) = 0$, we have

$$\begin{aligned}\left|\left|\left|\sum_{i=1}^{n} \alpha_i y_i\right|\right|\right| &= \left|\left|\left|v_n\left(\sum_{i=1}^{n+1} \alpha_i y_i\right)\right|\right|\right| = \left|\left|\left|v_n\left(\sum_{i=1}^{n+1} \beta_i x_i\right)\right|\right|\right| \\ &< \left|\left|\left|v_{n+1}\left(\sum_{i=1}^{n+1} \beta_i x_i\right)\right|\right|\right| = \left|\left|\left|\sum_{i=1}^{n+1} \beta_i x_i\right|\right|\right| = \left|\left|\left|\sum_{i=1}^{n+1} \alpha_i y_i\right|\right|\right|\end{aligned}$$

for any scalars $\alpha_1, \ldots, \alpha_{n+1}$ with $\alpha_{n+1} \neq 0$ (and suitable $\beta_1, \ldots, \beta_{n+1}$), whence, by § 7, theorem 7.1, $\{y_n\}$ is a basis of E endowed with the norm $|||x|||$, whence also of E with its initial norm.

Assume, finally, that we have 8°. We shall prove that in this case the sequence

$$y_1 = x_1, \quad y_n = r_1 r_2 \ldots r_{n-1}(x_n) \quad (n=2,3,\ldots) \tag{20.15}$$

is a strict polynomial basis of E, which will complete the proof. By a) it is again sufficient to prove that $\{y_n\}$ is a basis of E.

By virtue of d_8) we have

$$|r_1 \ldots r_n| \leqslant |r_1| \ldots |r_n| \leqslant 1 \quad (n=1,2,\ldots). \tag{20.16}$$

On the other hand, by (20.12) we have, for any $n=2,3,\ldots$

$$\begin{aligned} r_1 \ldots r_n(y_i) &= r_1 \ldots r_n r_1 \ldots r_{i-1}(x_i) \\ &= \begin{cases} r_1 \ldots r_n(x_i) = r_1 \ldots r_{n-1}[x_i - v_n(x_i)] = 0 & (1 \leqslant i \leqslant n) \\ r_1 \ldots r_{i-1}(x_i) = y_i & (n+1 \leqslant i < \infty) \end{cases}\end{aligned}$$

and obviously $r_1(y_1) = y_1 - v_1(y_1) = x_1 - v_1(x_1) = 0$. Consequently,

$$\left|\sum_{i=1}^{n} \alpha_i y_i\right| = \left|(I_E - r_1 \ldots r_n)\left(\sum_{i=1}^{n+m} \alpha_i y_i\right)\right| \leqslant 2\left|\sum_{i=1}^{n+m} \alpha_i y_i\right|$$

for any scalars $\alpha_1, \ldots, \alpha_{n+m}$, whence, by § 7, theorem 7.1, $\{y_n\}$ is a basis of E endowed with the norm $|x|$ and hence also of E endowed with its initial norm, which completes the proof of theorem 20.1.

By the duality between v_n and $I_E - v_n$ and between polynomials and polynomial complements, one can also give other equivalent conditions, corresponding to 6°, 7° and 9°, 10°.

Problem 20.1. Is it possible to omit in theorem 20.1 the conditions $\mathrm{d}_1)-\mathrm{d}_{10})$? In other words: if there exists a sequence of endomorphisms $\{v_n\} \subset L(E,E)$ satisfying a), b) and c), does E have a strict polynomial basis with respect to $\{x_n\}$?

For any closed linear subspace G of the Banach space E, we shall put

$$\lambda(G) = \lambda_E(G) = \inf_{u \in \mathscr{P}(E,G)} \|u\| \tag{20.17}$$

where $\mathscr{P}(E,G)$ denotes the set of all continuous linear projections of E onto G (i. e. the set $\{u \in L(E,E) \mid u^2 = u,\ u(E) = G\}$).

Proposition 20.1. *Let $\{x_n\}$ be a finitely linearly independent sequence in a Banach space E, with $[x_n] = E$ and let $P_{(n)} = [x_1, \ldots, x_n]$ $(n = 1, 2, \ldots)$. The following statements are equivalent:*

1°. *There exists a sequence of endomorphisms $\{v_n\} \subset L(E,E)$ satisfying* a), b) *and* c) *of theorem* 20.1.

2°. *There exist a sequence of endomorphisms $\{v_n\} \subset L(E,E)$ satisfying* a) *and a constant $C \geqslant 1$ such that*

$$\|x - v_n(x)\| \leqslant C\, e_n(x) \qquad (x \in E,\ n = 1, 2, \ldots), \tag{20.18}$$

where, as before, $e_n(x) = \inf_{p \in P_{(n)}} \|x - p\|$.

3°. *There exists a sequence of endomorphisms $\{v_n\} \subset L(E,E)$ satisfying* a), b) *and*

c′) $\sup_{1 \leqslant n < \infty} \|v_n\| < \infty$.

4°. *We have*

$$\sup_{1 \leqslant n < \infty} \lambda(P_{(n)}) < \infty. \tag{20.19}$$

Proof. The implication 1° ⇒ 3° is a consequence of the principle of uniform boundedness.

Assume now that we have 3° and let $x \in E$ and n be arbitrary. Then for any polynomial of best approximation $p_n(x) \in P_{(n)}$ of x we have

$$\begin{aligned} \|x - v_n(x)\| &= \|x - p_n(x) - v_n(x - p_n(x))\| \leqslant \|x - p_n(x)\| + \|v_n\|\, \|x - p_n(x)\| \\ &= (1 + \|v_n\|)\, e_n(x) \leqslant \Big(1 + \sup_{1 \leqslant j < \infty} \|v_j\|\Big)\, e_n(x), \end{aligned}$$

i. e. (20.18) with $C = 1 + \sup_{1 \leqslant j < \infty} \|v_j\|$. Thus, 3° ⇒ 2°.

Assume now that we have 2°. Then by $e_n(p) = 0$ $(p \in P_{(n)})$ we have b) and by $[x_n] = E$ we have $\lim_{n \to \infty} e_n(x) = 0$ $(x \in E)$, whence c). Thus, 2° ⇒ 1°, which, together with the above, shows that 1° ⇔ 2° ⇔ 3° with the same sequence $\{v_n\}$.

The implication $3^\circ \Rightarrow 4^\circ$ is obvious, since if we have 3°, then $\lambda(P_{(n)}) \leqslant \|v_n\|$ $(n=1,2,\ldots)$.

Assume, finally, that we have 4°. Then, since by (20.17) there exist $v_n \in \mathscr{P}(E, P_{(n)})$ such that[1] $\|v_n\| \leqslant \lambda(P_{(n)})+1$ $(n=1,2,\ldots)$, we have 3°. Thus, $4^\circ \Rightarrow 3^\circ$, which completes the proof of proposition 20.1.

A natural way to obtain an affirmative answer to problem 20.1 above would be to show that for any given sequence of endomorphisms $\{v_n\} \subset L(E,E)$ satisfying a), b), c), we also must have d_n) for some n with $1 \leqslant n \leqslant 10$. In particular, for d_9) this conjecture seems also to be supported by proposition 20.1 (implication $1^\circ \Rightarrow 2^\circ$ proved above with the same $\{v_n\}$). However, the following example disproves this conjecture:

Example 20.1. Let $\{x_n\}$ be a normalized basis of a Banach space E, with the a. s. c. f. $\{f_n\}$. Put

$$v_n(x) = s_n(x) - 2f_{n+1}(x)x_n = \sum_{i=1}^{n-1} f_i(x)x_i + [f_n(x) - 2f_{n+1}(x)]x_n \qquad (20.20)$$

$$(x \in E,\ n=1,2,\ldots).$$

Then $\{v_n\} \subset L(E,E)$ satisfies a), b), c), but none of the conditions $\mathrm{d}_1) - \mathrm{d}_{10}$).

Indeed, obviously $\{v_n\}$ satisfies the conditions a), b). Furthermore, since $\|x_n\|=1$ and $|f_n(x)| = \|f_n(x)x_n\| \to 0$ $(x \in E)$, we have

$$\|x - v_n(x)\| \leqslant \|x - s_n(x)\| + 2|f_{n+1}(x)| \to 0 \qquad (x \in E),$$

i. e. $\{v_n\}$ also satisfies c).

However $\{v_n\}$ does not satisfy d_9). Indeed, for $x_0 = \sum_{i=1}^{\infty} \frac{1}{2^i} x_i \in E$ we have

$$v_n(x_0) = \sum_{i=1}^{n} \frac{1}{2^i} x_i - \frac{2}{2^{n+1}} x_n = \sum_{i=1}^{n-1} \frac{1}{2^i} x_i = s_{n-1}(x_0) \in P_{(n-1)} \qquad (n=1,2,\ldots)$$

(where $s_0(x_0)=0$), and therefore from d_9) it would follow that in the norm $|x|$ of d_9) we have $e_n(x_0) = e_{n-1}(x_0)$ $(n=1,2,\ldots)$, whence $\|x_0\| = e_0(x_0) = \lim_{n\to\infty} e_n(x_0) = 0$, which contradicts $f_1(x_0) = \frac{1}{2} \neq 0$.

It is also easy to see that $\{v_n\}$ does not satisfy d_8), since d_8) implies d_9) with the same sequence $\{v_n\}$ (and same norm $|x|$). Indeed, if we have d_8), then, by b),

[1] Since $\dim P_{(n)} = n < \infty$, it is also easy to show, by a compactness argument, that actually there exist $v_n \in \mathscr{P}(E, P_{(n)})$ such that $\|v_n\| = \lambda(P_{(n)})$ $(n=1,2,\ldots)$ (see Ch. II, § 1).

$$|x-v_n(x)| = |x-p-v_n(x)+p| = |x-p-v_n(x-p)|$$
$$= |(I_E - v_n)(x-p)| \leqslant |x-p|$$

for all $p \in P_{(n)}$, which proves our assertion.

Moreover, $\{v_n\}$ does not satisfy any one of the conditions d_n) for $n=1,2,3,4,10$, since in the above proof of theorem 20.1 the implications $1^\circ \Rightarrow 2^\circ \Rightarrow 3^\circ \Rightarrow 4^\circ \Rightarrow 9^\circ$ and $2^\circ \Rightarrow 10^\circ \Rightarrow 9^\circ$ have been actually proved for *the same* $\{v_n\}$ (and same norms).

Finally, $\{v_n\}$ does not satisfy any one of d_5), d_6), d_7). Indeed, in the above proof of theorem 20.1 we have seen that if any one of d_5), d_6), or d_7) holds, then the sequence $\{y_n\}$ defined by (20.14) must be a basis of E. However, since now $v_{n-1}(x_n) = s_{n-1}(x_n) - 2f_n(x_n)x_{n-1} = -2x_{n-1}$, we have

$$y_1 = x_1, \quad y_n = x_n - 2x_{n-1} \qquad (n=2,3,\ldots),$$

but this sequence is not even minimal. For if a functional[1]

$$h_1(x) = \sum_{i=1}^{\infty} \beta_i f_i(x) \qquad (x \in E) \tag{20.21}$$

satisfied $h_1(y_1)=1$, $h_1(y_n)=0$ $(n=2,3,\ldots)$, then by $h_1(y_1)=\beta_1$, $h_1(y_n) = \beta_n - 2\beta_{n-1}$ $(n=2,3,\ldots)$ we would obtain

$$\beta_n = 2^{n-1} \qquad (n=1,2,\ldots),$$

whence, for the element $x_0 = \sum_{i=1}^{\infty} \frac{1}{2^i} x_i \in E$, $h_1(x_0) = \sum_{i=1}^{\infty} 2^{i-1} \frac{1}{2^i} = \sum_{i=1}^{\infty} \frac{1}{2} = \infty$, an absurdity. This completes the proof of the assertions of example 20.1.

Let us observe, however, that this example does not give a negative answer to problem 20.1 above, since there exists *another* sequence of endomorphisms of E, namely $\{s_n\} \subset L(E,E)$, having all properties a), b), c) and d_n) $(n=1,\ldots,10)$ and $\{x_n\}$ is a strict polynomial basis with respect to itself.

We shall turn now our attention to finitely linearly independent complete sequences $\{x_n\} \subset E$, with respect to which there exists no strict polynomial basis.

Definition 20.3. A finitely linearly independent sequence $\{x_n\}$ in a Banach space E is called a

a) *sub-Γ system*, if $\sup\limits_{1 \leqslant n < \infty} \lambda(P_{(n)}) = \infty$, where $P_{(n)} = [x_1, \ldots, x_n]$ $(n=1,2,\ldots)$ and where $\lambda(P_{(n)})$ are the numbers defined by (20.17);

b) *sub-Λ system*, if $\lim\limits_{n \to \infty} \lambda(P_{(n)}) = \infty$;

[1] Since $\{x_n\}$ is a basis of E, every functional $h_1 \in E^*$ can be written in the form (20.21), with $\beta_i = h_1(x_i)$ $(i=1,2,\ldots)$.

c) *Γ system*, if it is a sub-Γ system and $[x_n]=E$;

d) *Λ system* (or *Lozinsky-Haršiladze system*) if it is a sub-Λ system and $[x_n]=E$.

By virtue of theorem 20.1 and proposition 20.1, in order that there exist a strict polynomial basis with respect to $\{x_n\}$ it is necessary (and, if the answer to problem 20.1 would be affirmative, then also sufficient) that $\{x_n\}$ be a non-Γ system with $[x_n]=E$.

By definition 20.3 we have obviously the implications $\Lambda \Rightarrow \Gamma \Rightarrow$ sub-Γ and $\Lambda \Rightarrow$ sub-$\Lambda \Rightarrow$ sub-Γ. It is easy to see that none of the converse implications is true. Indeed, any Γ-system (Λ-system) $\{x_n\}$ of a hyperplane F in a Banach space E is a sub-Γ but non-Γ (respectively, sub-Λ but non-Λ) system of the whole space E, since $\inf_{v \in \mathscr{P}(E,P_{(n)})} \|v\| \geqslant \inf_{v \in \mathscr{P}(E,P_{(n)})} \|v|_F\| \geqslant \inf_{w \in \mathscr{P}(F,P_{(n)})} \|w\| \to \infty$ as $n\to\infty$ and since $[x_n]=F\neq E$. On the other hand, it is well known[1] that in each l_n^1 one can choose a subspace $G_{k_n} \subset l_n^1$ in such a way that $\lambda(G_{k_n}) \to \infty$ as $n\to\infty$. Hence, choosing a basis $x_1^n, \ldots, x_n^n$ of l_n^1 such that $x_1^n, \ldots, x_{k_n}^n$ be a basis of G_{k_n}, the sequence

$$\bar{x}_1^1, \bar{x}_1^2, \bar{x}_2^2, \ldots, \bar{x}_1^n, \ldots, \bar{x}_n^n, \ldots \tag{20.22}$$

where $\bar{x}_j^n = \{\underbrace{0,\ldots,0}_{n-1}, x_j^n, 0, 0, \ldots\}$ $(j=1,\ldots,n;\, n=1,2,\ldots)$, will be a Γ (whence also sub-Γ) system in $E=(l_1^1 \times l_2^1 \times \cdots \times l_n^1 \times \cdots)_{l^1} \equiv l^1$, which is, however, a non-sub-Λ (whence also non-Λ) system in E. In fact, if there existed projections v_n of E onto $[\bar{x}_1^1, \bar{x}_1^2, \bar{x}_2^2, \ldots, \bar{x}_1^{n-1}, \ldots, \bar{x}_{n-1}^{n-1}, \bar{x}_1^n, \ldots, \bar{x}_{k_n}^n] = (l_1^1 \times l_2^1 \times \cdots \times l_{n-1}^1 \times G_{k_n} \times \{0\} \times \cdots)_{l^1}$ of norms $\|v_n\| \leqslant M < \infty$ $(n=1,2,\ldots)$, then the restrictions $v_n|_{[\bar{x}_1^n,\ldots,\bar{x}_n^n]} = v_n|_{(\underbrace{\{0\}\times\cdots\times\{0\}}_{n-1} \times l_n^1 \times \{0\} \times \cdots)_{l^1}}$ would induce projections w_n of l_n^1 onto G_{k_n}, of norms $\|w_n\| = \|v_n|_{[\bar{x}_1^n,\ldots,\bar{x}_n^n]}\| \leqslant \|v_n\| \leqslant M$ $(n=1,2,\ldots)$, in contradictions with our choice of the G_{k_n}. Thus, $\lambda([x_1^1, \ldots, x_{n-1}^{n-1}, x_1^n, \ldots, x_{k_n}^n]) \to \infty$ i.e. (20.22) is a Γ system. However, the natural projections $E \to [\bar{x}_1^1, \ldots, \bar{x}_{n-1}^{n-1}, \bar{x}_1^n, \ldots, \bar{x}_n^n] = (l_1^1 \times \cdots \times l_n^1 \times \{0\} \times \cdots)_{l^1}$ defined by $\{\bar{x}_1, \ldots, \bar{x}_n, \ldots\} \to \{\bar{x}_1, \ldots, \bar{x}_n, 0, 0, \ldots\}$, are of norm 1, whence $\lambda([\bar{x}_1^1, \ldots, \bar{x}_{n-1}^{n-1}, \bar{x}_1^n, \ldots, \bar{x}_n^n]) = 1$ $(n=1,2,\ldots)$, and thus (20.22) is not a Λ system.

---

[1] See e.g. [170], [252]. We recall that l_n^1 is the space of all n-tuples $x=\{\xi_1,\ldots,\xi_n\}$ with the norm $\|x\| = \sum_{i=1}^{n} |\xi_i|$ and $(l_1^1 \times l_2^1 \times \cdots)_{l^1}$ is the space of all sequences $\{x_n\}$ with $x_n \in l_n^1$ $(n=1,2,\ldots)$ for which $\{\|x_n\|\} \in l^1$, endowed with the norm $\|\{x_n\}\| = \sum_{n=1}^{\infty} \|x_n\|_{l^1}$. There exists a natural linear isometry $(l_1^1 \times l_2^1 \times \cdots)_{l^1} \equiv l^1$, given by the formula

$$\{\{\xi_1\}, \{\xi_2, \xi_3\}, \{\xi_4, \xi_5, \xi_6\}, \ldots\} \to \{\xi_n\}.$$

Now it is natural to ask whether Λ systems exist at all. The answer is affirmative, namely, it is a classical result of S. M. Lozinsky and F. I. Haršiladze[1] that for the sequence

$$x_n(t) = t^{n-1} \qquad (t \in [0,1], n = 1,2,\ldots) \tag{20.23}$$

we have, in the space $E = C([0,1])$,

$$\lambda(P_{(n)}) \geqslant \frac{\ln n}{8\sqrt{\pi}} \qquad (n = 1,2,\ldots), \tag{20.24}$$

whence, since (by the theorem of Weierstrass) $[x_n] = E$, the sequence $\{x_n\}$ is a Λ system in E (consequently, in $E = C([0,1])$ there exists no basis consisting of algebraic polynomials of all degrees). On the other hand, in $E = l^2$ there exist no Λ systems, since for any sequence $\{x_n\} \subset E$ we have $\lambda(P_{(n)}) = 1$ $(n = 1,2,\ldots)$ (because there exists a projection of norm 1 onto any subspace G of l^2, namely, the orthogonal projection). Therefore it is natural to ask, which separable Banach spaces possess Λ systems. We shall give below some characterizations of such spaces.

Lemma 20.1. *Let* $E = G \oplus Y$, *where* $\dim G < \infty$ *and let*[2] $\sup\limits_{\substack{F \prec Y \\ \dim F < \infty}} \lambda_E(F) < \infty$. *Then* $\sup\limits_{\substack{F \prec E \\ \dim F < \infty}} \lambda_E(F) < \infty$.

Proof. Since $E = G \oplus Y$, there exists a projection u of E onto G, such that $u(Y) = 0$.

Now let F_0 be an arbitrary subspace of E, with $\dim F_0 < \infty$. Then $(I_E - u)(F_0)$ is a finite-dimensional subspace of Y, whence there exists a projection v_1 of Y onto $(I_E - u)(F_0)$, such that

$$\|v_1\| \leqslant \lambda_Y[(I_E - u)(F_0)] + 1. \tag{20.25}$$

We claim that $v = u + v_1(I_E - u)$ is then a projection of E onto its finite-dimensional subspace

$$D = G \oplus (I_E - u)(F_0).$$

Indeed, for any $x \in E$ we have $u(x) \in G$ and $v_1(I_E - u)(x) \in (I_E - u)(F_0)$, whence $v(x) \in D$ and for any $x \in D$ we have $v_1(I_E - u)(x) = (I_E - u)(x)$, whence $v(x) = u(x) + (I_E - u)(x) = x$, which proves the claim.

Observe now that $F_0 \subset D$, since for any $x \in F_0$ we have $x = u(x) + (I_E - u)(x) \in G \oplus (I_E - u)(F_0) = D$ and that, by $\dim D < \infty$,

---

[1] See e.g. [174], Appendix 3.

[2] We recall (see § 11) that the notation $F \prec X$ stands for "F is a (closed linear) subspace of the Banach space X".

$$\operatorname{codim}_D F_0 = \dim D - \dim F_0 \leqslant \dim D - \dim[(I_E - u)(F_0)] = \dim G,$$

whence there exists a projection w of D onto F_0, of norm[1]

$$\|w\| \leqslant 2^{\operatorname{codim}_D F_0} \leqslant 2^{\dim G}. \tag{20.26}$$

Consequently, wv is a projection of E onto F_0, of norm[2]

$$\begin{aligned} \|wv\| &\leqslant \|w\| \, \|v\| \leqslant 2^{\dim G}(\|u\| + \|I_E - u\| \, \|v_1\|) \\ &\leqslant 2^{\dim G}\{\|u\| + \|I_E - u\| (\lambda_Y[(I_E - u)(F_0)] + 1)\} \\ &\leqslant 2^{\dim G}\Big[\|u\| + (1 + \|u\|)\Big(\sup_{\substack{F \prec Y \\ \dim F < \infty}} \lambda_E(F) + 1\Big)\Big] = M < \infty, \end{aligned}$$

which, since F_0 has been an arbitrary subspace of E with $\dim F_0 < \infty$ and since M is independent of F_0, completes the proof of lemma 20.1.

Theorem 20.2. *Let E be a Banach space. The following statements are equivalent*:

1°. *There exists a sequence of subspaces* $\{G_n\}$ *of E with* $\dim G_n < \infty$, *such that* $\lim_{n \to \infty} \lambda(G_n) = \infty$.

2°. *E has a sub-Γ system.*

3°. *E has a sub-Λ system.*

If E is separable, these statements are equivalent to each of the following statements:

4°. *E has a Γ system.*

5°. *E has a Λ system.*

Proof. Assume that we have 1°. Let E_1 be a finite dimensional subspace of E. Choose $\phi_1, \ldots, \phi_{n_1} \in E_1^*$ with $\|\phi_i\| = 1$ $(i = 1, \ldots, n_1)$ so that the relations $x \in E_1$, $|\phi_i(x)| \leqslant 1$ $(i = 1, \ldots, n_1)$ imply $\|x\| \leqslant 2$. This can be done e.g. by taking a finite ε-net $\{z_1, \ldots, z_{n_1}\}$ of $S = \{x \in E_1 \mid \|x\| = 1\}$, where $0 < \varepsilon \leqslant \frac{1}{2}$, and functionals $\phi_i \in E_1^*$ such that $\phi_i(z_i) = 1$, $\|\phi_i\| = 1$ $(i = 1, \ldots, n_1)$, since then for any $x \in E_1$ with $|\phi_i(x)| \leqslant 1$ $(i = 1, \ldots, n_1)$, $\|x\| > 1$, and any index j with $1 \leqslant j \leqslant n_1$ such that $\left\| \frac{x}{\|x\|} - z_j \right\| < \varepsilon$, we have

[1] Indeed, this follows from the fact in any reflexive (in particular, in any finite dimensional) Banach space B there exists a projection of norm $\leqslant 2$ onto any hyperplane $H = \{x \mid f(x) = 0\}$ (take $x_1 \in B$ with $\|x_1\| = 1$, $f(x_1) = \|f\|$ and put $w_1(x) = x - \frac{f(x)}{\|f\|} x_1$ for all $x \in B$).

[2] We use here the fact that for any $F \prec Y$ we have $\lambda_Y(F) = \inf_{w_1 \in \mathscr{P}(Y,F)} \|w_1\| \leqslant \inf_{w_2 \in \mathscr{P}(E,F)} \|w_2|_F\| \leqslant \inf_{w_2 \in \mathscr{P}(E,F)} \|w_2\| = \lambda_E(F)$.

$$0<1-\frac{1}{\|x\|}\leqslant 1-\left|\phi_j\left(\frac{x}{\|x\|}\right)\right|\leqslant\left|1-\phi_j\left(\frac{x}{\|x\|}\right)\right|$$

$$=\left|\phi_j\left(z_j-\frac{x}{\|x\|}\right)\right|\leqslant\left\|z_j-\frac{x}{\|x\|}\right\|<\varepsilon,$$

whence $\|x\|<\dfrac{1}{1-\varepsilon}\leqslant 2$. Let $f_1,\dots,f_{n_1}\in E^*$ be Hahn-Banach extensions of the ϕ_i's and let $Y_1=\{x\in E\,|\,f_i(x)=0\ (i=1,\dots,n_1)\}$. Then $\operatorname{codim} Y_1\leqslant n_1<\infty$, $E_1\cap Y_1=\{0\}$, and the natural projection $u_1: x+y\to x$ of $E_1\oplus Y_1$ onto E_1 has norm $\|u_1\|\leqslant 2$, since for any $x\in E_1$, $y\in Y_1$ with $\|x+y\|\leqslant 1$ we have $|\phi_i(x)|=|f_i(x)|=|f_i(x+y)|\leqslant\|f_i\|\,\|x+y\|\leqslant 1\ (i=1,\dots,n_1)$, whence $\|x\|\leqslant 2$.

By 1° and lemma 20.1, choose a finite dimensional subspace E_2 of Y_1 with $\lambda(E_2)\geqslant 2$. Furthermore, as above, choose $f_{n_1+1},\dots,f_{n_2}\in E^*$ with $\|f_i\|=1$ so that the relations $x\in E_1\oplus E_2$, $|f_i(x)|\leqslant 1$ $(i=1,\dots,n_2)$ imply $\|x\|\leqslant 2$. Then for $Y_2=\{x\in E\,|\,f_i(x)=0\ (i=1,\dots,n_2)\}$ we have $Y_2\subset Y_1$, $\operatorname{codim}_E Y_2\leqslant n_2<\infty$, and the natural projection u_2 of $E_1\oplus E_2\oplus Y_2$ onto $E_1\oplus E_2$ has norm $\|u_2\|\leqslant 2$.

Continuing in this way, we obtain two sequences of subspaces $\{E_n\}$, $\{Y_n\}$ of E such that $\dim E_n<\infty$, $\operatorname{codim}_E Y_n<\infty$, $\lambda(E_n)\geqslant n$, E_{n+1}, $Y_{n+1}\subset Y_n$ and that the natural projection u_n of $E_1\oplus\cdots\oplus E_n\oplus Y_n$ onto $E_1\oplus\cdots\oplus E_n$ has norm $\|u_n\|\leqslant 2$ $(n=1,2,\dots)$.

Observe now that for any n, any subspace $B\subset E_{n+1}$ and any projection u of E onto $E_1\oplus\cdots\oplus E_n\oplus B$, we have $\|u\|\geqslant\dfrac{n}{6}$. Indeed, then $(I_E-u_{n-1})u_n u$ is a projection of E onto E_n, whence $n\leqslant\lambda(E_n)\leqslant\|(I_E-u_{n-1})u_n u\|\leqslant 3.2\,\|u\|$, whence the assertion follows.

Consequently, if $x_1^n,\dots,x_{k_n}^n$ (where $k_n=\dim E_n$) is any basis of E_n, the sequence

$$x_1^1,\dots,x_{k_1}^1,x_1^2,\dots,x_{k_2}^2,\dots,x_1^n,\dots,x_{k_n}^n,\dots \tag{20.27}$$

is a sub-Λ system in E. Thus, $1°\Rightarrow 3°$.

The implications $3°\Rightarrow 2°\Rightarrow 1°$ are obvious.

Assume now that E is separable and that we have 3°, i.e. that there exists in E a sub-Λ system $\{x_n\}$ and let us prove that in this case we have 5°. If $[x_n]=E$, then $\{x_n\}$ is already a Λ system and thus we have 5°. Assume that $[x_n]\neq E$ and let $\{y_n\}$ be a sequence in $E\setminus[x_n]$, such that $[x_i,y_j]=E$ and that $\{x_i,y_j\}$ is finitely linearly independent[1]. For each

[1] It is easy to see that such a sequence exists. Indeed, since E is separable, let $\{z_n\}$ be a dense sequence in E. Let i_1 be the smallest index for which $z_{i_1}\notin[x_n]$, i_2 the smallest index for which $z_{i_2}\notin[x_n,z_{i_1}]=[x_n]\oplus[z_{i_1}]$ and continue in this way. Then take $y_j=z_{i_j}$ $(j=1,2,\dots)$.

k choose a projection u_k of E onto $[y_1,\ldots,y_k]$ such that $u_k(x_j)=0$ $(j=1,2,\ldots)$[1] and put $r_k=I_E-u_k$. Then for any projection $v_{n,k}$ of E onto $[x_1,\ldots,x_n,y_1,\ldots,y_k]$ the operator $r_k v_{n,k}$ is a projection of E onto $[x_1,\ldots,x_n]$ with $\lambda([x_1,\ldots,x_n])\leqslant\|r_k v_{n,k}\|\leqslant\|r_k\|\,\|v_{n,k}\|$, whence

$$\|v_{n,k}\| \geqslant \frac{\lambda([x_1,\ldots,x_n])}{\|r_k\|}. \tag{20.28}$$

Now, since $\{x_n\}$ is a sub-Λ system, for each k there exists a positive integer n_k such that $n\geqslant n_k$ implies $\lambda([x_1,\ldots,x_n])\geqslant\|r_k\|\,k$. We claim that the sequence $\{z_n\}\subset E$ defined by

$$z_n=\begin{cases} x_{n-k+1} & \text{for}\quad n_{k-1}+k\leqslant n\leqslant n_k+k-1 \qquad (k=1,2,\ldots)\\ y_k & \text{for}\quad n=n_k+k \qquad (k=1,2,\ldots)\end{cases} \tag{20.29}$$

(where $n_0=0$), i.e. the sequence

$$x_1,\ldots,x_{n_1},y_1,x_{n_1+1},\ldots,x_{n_2},y_2,\ldots \tag{20.30}$$

is a Λ system in E. Indeed, $\{z_n\}=\{x_i,y_j\}$ is finitely linearly independent and $[z_n]=[x_i,y_j]=E$. Furthermore, if $n_{k-1}+k\leqslant n\leqslant n_k+k-1$ and v is an arbitrary projection of E onto $[z_1,\ldots,z_n]=[x_1,\ldots,x_{n-k+1},y_1,\ldots,y_{k-1}]$, then v is a $v_{n-k+1,k-1}$, whence by (20.28), $n-k+1\geqslant n_{k-1}+1$ and the choice of n_{k-1}, we infer

$$\|v\|=\|v_{n-k+1,k-1}\|\geqslant\frac{\lambda([x_1,\ldots,x_{n-k+1}])}{\|r_{k-1}\|}\geqslant\frac{\|r_{k-1}\|\,(k-1)}{\|r_{k-1}\|} = k-1\,.$$

Similarly, if $n=n_k+k$ and v is an arbitrary projection of E onto $[z_1,\ldots,z_n]=[x_1,\ldots,x_{n-k},y_1,\ldots,y_k]$, then v is a $v_{n-k,k}$, whence by (20.28), $n-k=n_k$ and the choice of n_k, we infer

$$\|v\|=\|v_{n-k,k}\|\geqslant\frac{\lambda([x_1,\ldots,x_{n-k}])}{\|r_k\|}\geqslant\frac{\|r_k\|\,k}{\|r_k\|}=k\,,$$

which proves that $\{z_n\}$ is a Λ system in E. Thus, $3^\circ\Rightarrow5^\circ$.

Finally, the implications $5^\circ\Rightarrow4^\circ\Rightarrow1^\circ$ are obvious, which completes the proof of theorem 20.2.

Corollary 20.1. *If a separable Banach space E has a subspace F possessing a Λ system*[2]*, then the space E itself has a Λ system, namely, every Λ system of F can be extended to a Λ system of E.*

---

[1] Such a projection exists. Indeed, extend the coefficient functionals h_i $(i=1,\ldots,k)$ associated to the basis $\{y_1,\ldots,y_k\}$ of $[y_1,\ldots,y_k]$ to $[x_j]\oplus[y_1,\ldots,y_k]$ by 0 on $[x_j]$ and linearity and then to the whole space E by the Hahn-Banach theorem. Denoting these extensions again by h_i, put $u_k(x)=\sum_{i=1}^{k}h_i(x)y_i$ $(x\in E)$.

[2] In analogy with the term "basic sequence", one may call "Λ sequence" (in E) a sequence $\{x_n\}\subset E$ such that $\{x_n\}$ is a Λ system in $[x_n]=F\subset E$.

Proof. Observe that every Λ system $\{x_n\}$ in $F \subset E$ is a sub-Λ system in E, since for any projection v of E onto $P_{(n)} = [x_1, \ldots, x_n]$ the restriction $v|_F$ is a projection of F onto $P_{(n)}$, of norm $\leqslant \|v\|$. Hence, by theorem 20.2 (implication $3° \Rightarrow 5°$) and its proof given above, $\{x_n\}$ can be extended to a Λ system (20.29) of E, which completes the proof.

Since the spaces l^p $(1 \leqslant p < \infty)$ and c_0 satisfy[1] condition 1° of theorem 20.2, we obtain

Corollary 20.2. *The spaces* $L^p(T, \nu)$, where $1 \leqslant p < \infty, p \neq 2$ *and where* (T, ν) *is a positive measure space such that* $L^p(T, \nu)$ *is separable and the spaces* $C(Q)$, *where* Q *is a compact metric space, have* Λ *systems.*

The converse of the second statement of corollary 20.1 is not true, i.e. a subsequence of a Λ system need not be a Λ system in its closed linear span, it may be even a basic sequence. Moreover, we have

Proposition 20.2. *Every finitely linearly independent complete sequence* $\{x_n\}$ *in a separable Banach space* F *(hence, in particular, every basis* $\{x_n\}$ *of* F*) can be extended to a* Λ *system of a suitable separable superspace* E.

Proof. Let G be a separable Banach space having a Λ system $\{y_n\}$ and let $E = F \times G$, endowed with the norm $\|\{x, y\}\| = \|x\| + \|y\|$. Then the sequence $\{z_n\} \subset E$ defined by

$$z_{2n-1} = \{x_n, 0\}, \qquad z_{2n} = \{0, y_n\} \qquad (n = 1, 2, \ldots) \tag{20.31}$$

is a Λ system in E. In fact, $\{z_n\}$ is obviously finitely linearly independent and complete in E. Furthermore, if v_{2n} is an arbitrary projection of E onto $[z_1, \ldots, z_{2n}]$, and u is the natural projection of $E = F \times G$ onto $\{0\} \times G$, then the restriction $u v_{2n}|_{\{0\} \times G}$ induces a projection w of G onto $[y_1, \ldots, y_n]$, with

$$\lambda_G([y_1, \ldots, y_n]) \leqslant \|w\| = \|u v_{2n}|_{\{0\} \times G}\| \leqslant \|u\| \, \|v_{2n}\| = \|v_{2n}\| .$$

Since a similar relation also holds for any v_{2n+1} and since $\{y_n\}$ is a Λ system in G, it follows that $\{z_n\}$ is a Λ system in E, which completes the proof.

For an arbitrary Banach space E, introduce now the functions[2]

$$\psi_n(E) = \sup_{\substack{G \prec E \\ \dim G = n}} \lambda(G) = \sup_{\substack{G \prec E \\ \dim G = n}} \inf_{u \in \mathscr{P}(E, G)} \|u\| \qquad (n = 1, 2, \ldots). \tag{20.32}$$

---

[1] See e.g. [170], [252].

[2] For comparison, let us recall that in § 11 we have considered the functions $\phi_n(E) = \inf_{\substack{G \prec E \\ \dim G = n}} \lambda(G) = \inf_{\substack{G \prec E \\ \dim G = n}} \inf_{u \in \mathscr{P}(E, G)} \|u\|$

We deduce from theorem 20.2 the following property of this sequence of functions:

Corollary 20.3. *If for a Banach space E we have* $\sup_{1 \leqslant n < \infty} \psi_n(E) = \infty$, *then* $\lim_{n \to \infty} \psi_n(E) = \infty$.

Proof. Take $G_n \prec E$ with $\dim G_n = n$, such that $\lambda(G_n) \geqslant \psi_n(E) - 1$ $(n = 1, 2, \ldots)$. Then, by our hypothesis, $\sup_{1 \leqslant n < \infty} \lambda(G_n) = \infty$, whence there exists a subsequence $\{G_{n_k}\}$ such that $\lim_{k \to \infty} \lambda(G_{n_k}) = \infty$. Consequently, by theorem 20.2, E has a sub-Λ system $\{x_n\}$. Putting $P_{(n)} = [x_1, \ldots, x_n]$ $(n = 1, 2, \ldots)$, we have then $\psi_n(E) \geqslant \lambda(P_{(n)}) \to \infty$ as $n \to \infty$, whence $\lim_{n \to \infty} \psi_n(E) = \infty$, which completes the proof.

Notes and remarks

§ 1. The notion of a basis of a Banach space (see definition 1.1) was introduced by J. Schauder [224]. The basis problem (see problem 1.1) was raised explicitly in the book of S. Banach ([10], p. 111–112 and p. 245) and it remains one of the important unsolved problems of functional analysis (see e.g. I. Kaplansky [130] and G. Köthe [134]). The fact that this problem is equivalent to problem 1.2 was observed by S. Banach ([10], p. 238). Problem 1.3 was also raised by S. Banach ([10], p. 238 and p. 245).

For the terms "restriction of the field of scalars" and "extension of the field of scalars", see N. Bourbaki [30], § 5 and [31], § 2. In [30] it is also shown that the complexification E of a real space G is "smallest" in a certain sense, and that this "smallest" E is uniquely determined, up to an isomorphism. The notation $E_{(r)}$ for the real Banach space associated to a complex Banach space E is used in the monograph of M. M. Day [43], while in Bourbaki [30] the notation E_R occurs. For the characterization, given in § 1, of real Banach spaces which admit a complex structure, see J. Dieudonné [45]. For the term "complexification" and for the characterization, given in § 1, of complex spaces which are isomorphic to the complexification of a suitable real Banach space, see e.g. A. Lichnerowicz [147], Ch. V[1]. For the involution on the space A,

[1] Actually, both in Bourbaki [30], [31] and Lichnerowicz [147] these problems are considered only for linear spaces without topology, and the corresponding mappings are required only to be linear. However, the additional assumption of continuity of these mappings can be made without alteration of the corresponding results and proofs. On the other hand, it is easy to see that on every complex linear space there exists an involution (without assumption of continuity) whence every such space is the complexification of a suitable real linear space (without topology).

mentioned at the end of § 1, see e.g. M. A. Naĭmark [171]. Ch. III, § 14, p. 196.

§ 2. Example 2.1 was given by J. Schauder [224] (see also S. Banach [10], p. 112). Example 2.2 was constructed by J. Schauder [224]. As we have mentioned in § 2, the basis (2.3) is called the Schauder basis of $C([0,1])$. Various other bases of $C([0,1])$ and bases of the spaces $C(Q)$, where Q is a metric compact space, will be studied in Vol. II.

Example 2.3 was also given by J. Schauder [225]. The fact that the Haar functions constitute a basis in separable Orlicz spaces, was proved by W. Orlicz ([185], p. 122–126); see also the book of M. A. Krasnoselskiĭ and Ya. B. Rutickiĭ [138], p. 122–126).

The spaces L^{λ} with respect to a levelling length function were introduced by H. W. Ellis and I. Halperin [55], who also constructed generalized Haar bases for separable spaces of this type. Problem 2.1 was raised by I. Halperin ([103], p. 247). For a study of Haar bases in modulared function spaces $L_{M(\xi,t)}$ see J. Ishii and T. Shimogaki [111].

Problem 2.3 was raised by S. Banach ([10], p. 238). Problems 2.2 and 2.3 have been studied by F. S. Vaher, who has claimed (verbal communication, International Congress of Mathematicians, Moscow, 1966) to have proved that the space A has no basis, thus solving the basis problem in the negative. However, the proof was not complete and to our knowledge it has not been given to print so far. On the other hand, at the same time, E. Akutowicz [1] published the opposite result, that the space A, and also the space H^1, has a basis, but it has since been discovered that the proof is not correct. Thus, the problem whether these Banach spaces of analytic functions have a basis or not, still remains open.

S. Banach ([10], p. 238) also raised the following problem: Let B be the space of all real-valued continuous functions on the unit square $0 \leqslant t \leqslant 1$, $0 \leqslant s \leqslant 1$, admitting continuous partial derivatives of order 1, endowed with the norm $\|x\| = \max\limits_{\substack{0 \leqslant t \leqslant 1 \\ 0 \leqslant s \leqslant 1}} |x(t,s)| + \max\limits_{\substack{0 \leqslant t \leqslant 1 \\ 0 \leqslant s \leqslant 1}} |x'_t(t,s)| + \max\limits_{\substack{0 \leqslant t \leqslant 1 \\ 0 \leqslant s \leqslant 1}} |x'_s(t,s)|$; does B possess a basis? This problem has recently been solved in the affirmative by S. Schonefeld [226] and Z. Ciesielski [275].

§ 3. In his definition of a basis in a Banach space, J. Schauder [224] separately required the continuity of the coefficient functionals f_n. S. Banach ([10], p. 111) proved that this condition is satisfied by all bases in a Banach space, i.e. that we have theorem 3.1. The idea of the proof of theorem 3.1 (through propositions 3.1 and 3.2), due to S. Banach [10], has become a useful tool in various extensions of this theorem. E.g. this theorem has been extended with a similar method to bases in locally convex complete metric linear spaces (i.e. Fréchet spaces) by F. Newns [176], in complete metric linear spaces (without condition of local convexity) by V. N. Nikolskiĭ ([180], p. 125 and p. 135), C. Bessaga and A. Pelczyński ([24], theorem 1), M. G. Arsove ([4], theorem 2; however, all these extensions were already known by S. Banach, as shown by a remark of [10], p. 239), in certain inductive limits of Fréchet spaces by M. G. Arsove and R. E. Edwards ([5], theorem 12); more generally, B. S. Mityagin observed ([169], p. 92) that whenever $\{x_n\}$ is a basis of a topological linear space for which the open mapping theorem or the closed graph theorem is valid, the corresponding coefficient functionals are continuous. For other extensions of theorem 3.1 (e.g. to T-bases), see Vol. II, Chapter III and the corresponding Notes and remarks.

Theorem 3.2 was communicated to us by C. Bessaga.

§ 4. Example 4.1 of an E-complete biorthogonal system (x_n, f_n) such that $\{x_n\}$ is not a basis of E, is well known, e.g. it is given as such in a paper of A. I. Markushevich [156]. Theorem 4.1, which gives useful characterizations of bases, is essentially due to S. Banach ([10], p. 107–108, theorems 2 and 4). Corollary 4.1 was proved by S. R. Foguel [60]. Propositions 4.1 and 4.3 are due to B. R. Gelbaum ([71], § 2, theorem 1 and § 4, theorem 1(d)).

Basic sequences (see definition 4.5) were already considered by S. Banach [10]. The term "basic sequence" was introduced by C. Bessaga and A. Pelczyński [22]. A detailed study of basic sequences will be made in Vol. II, Ch. III.

Problem 4.1 and the conjecture concerning the sequence (4.7) are due to A. Pelczyński [194]. Proposition 4.2 is well known (see e.g., A. Pelczyński [195], lemma 11).

The notion of block-perturbation of a basis (see definition 4.7) and proposition 4.4 were given in [196] (see [196], § 2, definition 4 and lemma 2); let us mention that V. G. Vinokurov [261] considered perturbations of the form $z_{2j-1}=x_{2j-1}$, $z_{2j}=\alpha_{2j-1}x_{2j-1}+\alpha_{2j}x_{2j}$ $(j=1,2,\ldots)$, where $\sup_{1\leqslant j<\infty}|\alpha_{2j-1}|<\infty$, $\inf_{1\leqslant j<\infty}|\alpha_{2j}|>0$, and established that they constitute a basis of E ([261], theorem 4).

§ 5. The sufficiency part of proposition 5.1 was observed by S. Banach ([10], p. 106). Proposition 5.2 is due to S. R. Foguel [60] (for proposition 5.2b) and 5.2c) see [60], theorems 1 and 2).

The equivalence $1^\circ\Leftrightarrow 3^\circ$ of theorem 5.1 was proved by M. M. Grinblium ([85], theorem 3) and the equivalences $1^\circ\Leftrightarrow 4^\circ$, $1^\circ\Leftrightarrow 5^\circ$ of this theorem were proved by S. R. Foguel ([60], theorems 3 and 4); the equivalence $1^\circ\Leftrightarrow 2^\circ$ of theorem 5.1 is nothing else but the equivalence $1^\circ\Leftrightarrow 3^\circ$ of § 4, theorem 4.1. The direct proof of the implication $4^\circ\Rightarrow 1^\circ$ of theorem 5.1, mentioned in remark 5.2, was given in [229].

Proposition 5.3 was proved by M. M. Grinblium ([85], theorems 1, 2, 4, 5).

Multipliers were used a long time ago in analysis (see e.g., W. Orlicz [183], S. Banach [10]). The equivalence $1° \Leftrightarrow 3°$ of theorem 5.2 was proved by M. I. Kadec ([126], theorem 1) and the equivalence $1° \Leftrightarrow 2°$ was proved independently, by S. Yamazaki ([267], § 5, theorem 1). In the case when $\{x_n\}$ is a basis with the a.s.c.f. $\{f_n\}$, proposition 5.4 is due to S. Yamazaki ([267], § 3, theorem 1 and § 2, lemma 2) and in the general case, essentially, to R. J. McGivney and W. Ruckle ([166], corollary 3.3).

Theorem 5.3 and corollary 5.3 were given by S. Yamazaki ([267], § 5, theorems 2, 3 and § 6, theorem 1).

§ 6. The three types of linear independence of sequences considered in § 6 were studied long-ago in infinite-dimensional Hilbert spaces (see e.g. the book of S. Kaczmarz and H. Steinhaus [117]). Example 6.1 a) was given by V. N. Nikolskiĭ ([180], p. 127).

Lemma 6.1 is essentially due to S. Yamazaki (see [265], the proofs of theorems 1 and 2). The equivalences $1° \Leftrightarrow 2° \Leftrightarrow 4° \Leftrightarrow 5°$ of theorem 6.1 were proved by A. I. Markushevich [156]. The equivalence $1° \Leftrightarrow 2°$ can be also found in the book of S. Kaczmarz and H. Steinhaus ([117], Ch. VIII, § 1). The equivalence $1° \Leftrightarrow 3°$ was proved in the paper [61] (see [61], theorem 1). The equivalence $4° \Leftrightarrow 12°$ is due to V. N. Nikolskiĭ ([180], § 8, theorem 1.2)).

§ 7. The equivalences $1° \Leftrightarrow 5° \Leftrightarrow 9°$ of theorem 7.1 were given by M. M. Grinblium ([82], theorems A and A′) under the additional hypothesis that $\{x_n\}$ is a minimal sequence; however, this hypothesis is superfluous. The equivalences $1° \Leftrightarrow 6° \Leftrightarrow 7° \Leftrightarrow 8°$ are essentially due to S. Yamazaki ([265], theorems 1 and 2′). The equivalences $1° \Leftrightarrow 10° \Leftrightarrow 11°$ were proved by V. N. Nikolskiĭ ([180], § 8, theorems 1.1) and 2.1)). The equivalences $1° \Leftrightarrow 3°$ and $1° \Leftrightarrow 4°$ are also known (see e. g. M. Z. Solomiak [253] and R. C. James [113], respectively).

The notion of index of a sequence (see definition 7.1) was introduced by M. M. Grinblium [82], while the norm of a sequence (see definition 7.1) was considered by C. Bessaga [20]. The index $\Gamma(E)$ of a Banach space E (see definition 7.2) was introduced by M. M. Grinblium [82], who also raised [82] the problem of the existence of Banach spaces such that $0 < \Gamma(E) < 1$, mentioned after definition 7.2.

Corollary 7.1 is due to V. I. Gurariĭ (verbal communication). Corollary 7.2 was communicated to us by C. Bessaga. Corollary 7.3 was given by V. I. Gurariĭ ([93], theorem 2) and, independently, by M. Zippin ([273], lemma 2).

Block basic sequences have been considered e. g., by R. C. James [113]. For such sequences, C. Bessaga and A. Pelczyński [22] have used the term "block basis", while the term "block basic sequence" was suggested in [237].

The fact that all (closed) hyperplanes of a Banach space are isomorphic to each other was observed by C. Bessaga and A. Pelczyński [25], by considering the intersection of any two hyperplanes; lemma 7.1 was proved by M. Zippin [273], by using this idea of C. Bessaga and A. Pelczynski. The first part of theorem 7.2 was proved by M. Zippin [273]; however, as we observed in remark 7.3, a theorem on extension of block basic sequences of a particular form had been given already in [196]. The second part of theorem 7.2 was observed in [44] and, independently, by J. R. Holub [108].

§ 8. Pairs of sequences $\{x_n\}$, $\{y_n\}$ satisfying (8.1) were considered by S. Banach ([9], p. 1638; see also [10], p. 112). Fully equivalent sequences were studied by L. A. Gurevich [99], who used the term "equivalent sequences" for them. Equivalent bases were studied by M. G. Arsove [3], who used the term "similar bases"; some authors still use this latter term. The terms "domination" and "strict domination" were introduced in [251].

The equivalences 1°a)$\Leftrightarrow 2^\circ$a) and 1°c)$\Leftrightarrow 2^\circ$c) of theorem 8.1 were proved by M. G. Arsove ([3], lemma 1 and theorem 3). For the particular case of complete sequences $\{x_n\}$, $\{y_n\}$ in $E=F$, the equivalence 1°d)$\Leftrightarrow 2^\circ$d) was proved by L. A. Gurevich ([99], theorem 2). Under the additional hypotheses that $[x_n]=[y_n]=E=F$, that E is sequentially weakly complete, and that the mapping $T: \psi \rightarrow \phi$ of $[y_n]^*$ into $[x_n]^*$ defined by (8.5) is an isomorphism of $[y_n]^*$ onto $[x_n]^*$, the equivalence 1°d)$\Leftrightarrow 3^\circ$d) was proved by L. A. Gurevich ([99], theorem 3); however, all these additional hypotheses are superfluous. In the particular case when $\{x_n\}$, respectively both $\{x_n\}$ and $\{y_n\}$, are bases of E and F respectively, the equivalences 1°b)$\Leftrightarrow 4^\circ$b) and 1°d)$\Leftrightarrow 4^\circ$d) are essentially due to M. G. Arsove ([3], theorems 5 and 4). The equivalence 1°b)$\Leftrightarrow 5^\circ$b) is essentially due to S. Banach ([10], p. 112, theorem 7; for a particular case see also S. Banach [9], theorem II); actually, S. Banach assumed that for every $x\in[x_n]$ the system of equations (8.6) has exactly one solution $y\in[y_n]$, but this is obviously equivalent to the assumption that for every $x\in[x_n]$ the system (8.6) has a solution $y\in[y_n]$ and that $\{\psi_n\}$ is total on $[y_n]$ (indeed, assuming the uniqueness of the solution of (8.6), take $x=0$ in (8.6)). The equivalence 1°b)$\Leftrightarrow 6^\circ$b) is essentially due to M. G. Arsove (see [3], the proof of theorem 5). Finally, for the equivalence 1°d)$\Leftrightarrow 6^\circ$d), see C. Bessaga and A. Pelczyński ([22], proposition 1.2) and M. G. Arsove ([3], theorem 1). Various extensions of parts of theorem 8.1 have been given, to more general spaces (see e. g. M. G. Arsove [3]) and for other types of sequences in these spaces (see e. g. M. G. Arsove and R. E. Edwards [5]).

Let us also observe that $\{x_n\} \sim \{y_n\}$ (where $\{x_n\} \subset E$, $\{y_n\} \subset F$) if and only if

$$\left\{\{\alpha_n\} \subset K \,\middle|\, \sup_{1 \leqslant n < \infty} \left\| \sum_{i=1}^{n} \alpha_i x_i \right\| < \infty \right\} = \left\{\{\alpha_n\} \subset K \,\middle|\, \sup_{1 \leqslant n < \infty} \left\| \sum_{i=1}^{n} \alpha_i y_i \right\| < \infty \right\}$$

(indeed, this is an immediate consequence of theorem 8.1 d), equivalence $1° \Leftrightarrow 2°$); under the (extraneous) assumption that $\{x_n\}$, $\{y_n\}$ are basic sequences, this remark was made by V. D. Milman ([167], lemma 2).

Proposition 8.1 a) was given by C. Bessaga and A. Pelczyński ([22], lemma 1,2°).

Theorem 8.2 on characterization of bases is due to M. M. Grinblium [86].

The fact that the full equivalence conserves the completeness of sequences, their minimality and the property of being a basis, was observed by L. A. Gurevich ([99], theorem 1). The notions of $\{a_n\}$-completeness and completeness of order p have been introduced by P. Davis and Ky Fan [36 a].

The notion of affine equivalence (see definition 8.2) was introduced in the paper [196] (see [196], definition 5[1]). Problem 8.1 was raised in the same paper (see [196], problem 1).

Permutatively equivalent bases have been considered by A. Pelczyński [190], who called them "c-equivalent bases". The term "permutatively equivalent" was suggested in [241] and recently used also by A. Pelczyński [195].

Remark 8.2, proposition 8.3, definition 8.5, remark 8.3 and proposition 8.4 were given by A. Pelczyński [195].

§ 9. In the particular case when E is a Hilbert space, the implications b) γ) and b) δ) of theorem 9.1 were given by R. E. A. C. Paley and N. Wiener [188]. R. P. Boas [27] observed that the proof given by R. E. A. C. Paley and N. Wiener remains valid in an arbitrary Banach space E. The implication [$\{x_n\}$ is minimal $\Rightarrow \{y_n\}$ is minimal] (i. e. half of the equivalence a) δ)) has been given by A. I. Markushevich [156]; a short proof of this implication was given in [61]. The implications b) γ) and b) δ) were reproved, using the technique of inversion of operators by means of geometric series, by K. I. Babenko [7]. Theorem 9.1 a) and its equivalence a) δ), were proved (and b) γ), b) δ) reproved) by F. W. Schäfke [222], who also made remark 9.1, for the converse of b) γ) and b) δ) if $0 \leqslant \lambda < \frac{1}{2}$. The implications b) α), b) β) of theorem 9.1 are due to P. Davis and Ky Fan ([36], theorem 4). Various extensions

---

[1] Actually, the notion introduced in [196] is slightly different, since in [196] it is assumed $\lambda_n \neq 0$ $(n = 1, 2, \ldots)$ instead of (8.17).

of the Paley-Wiener theorem have been given: a) to more general spaces, e. g. to complete metric linear spaces (see M. G. Arsove [4], where other references are also given), and b) to generalizations of bases, e. g. to T-bases (see Vol. II, Ch. III). On the other hand, theorems of Paley-Wiener type are also known for special classes of bases in Hilbert spaces (see Vol. II, Ch. V).

Example 9.1 was given by R. J. Duffin and J. J. Eachus [48]. For proposition 9.1 see R. J. Duffin and J. J. Eachus [48], the footnote to theorem B, where this result is given with the remark that it can be proved by making use of an ergodic theorem.

In the particular case when E is a Hilbert space, theorem 9.2b) γ) was given by H. Pollard ([198], theorem 1.1), corollary 9.1, part 2° b) α), γ) by B. Sz. Nagy [173] and part 1° b) α) by S. H. Hilding [107]. Theorem 9.2 for Banach spaces is due to J. R. Retherford [203], [209], who also observed the equivalence of conditions (9.16), (9.17) of corollary 9.1 and (9.5) of theorem 9.2. Theorem 9.3 was also given in the paper [203] of J. R. Retherford. Example 9.2 is due to W. J. Davis (see J. R. Retherford [209]). For some related results see also W. J. Davis [38].

§ 10. The fact that under the assumptions of theorem 10.1 the conclusions of theorem 9.1 hold, was proved by M. G. Krein and L. A. Liusternik [142] and reproved by C. Bessaga and A. Pelczyński [22]. Recently, C. W. McArthur and J. R. Retherford ([162], p. 120) have observed that, by the inequality

$$\Big\|\sum_{i\in I}\alpha_i x_i\Big\| \leqslant 4\sup_{i\in I}|\alpha_i| \sup_{I'\subset I}\Big\|\sum_{i\in I'} x_i\Big\|,$$

where I is any finite subset of $\mathcal{N}=\{1,2,3,\dots\}$, $\{x_i\}_{i\in I}\subset E$ and $\{\alpha_i\}_{i\in I}\subset K$ (in the case of real scalars the constant 4 may be replaced by 2), the conclusion of theorem 10.1 also holds under the weaker assumption

$$\sup_{\substack{I\subset\mathcal{N}\\ I\,\text{finite}}}\Big\|\sum_{i\in I}(x_i-y_i)\Big\| \leqslant \frac{\lambda}{4\sup\limits_{1\leqslant n<\infty}\|f_n\|},$$

where $0\leqslant\lambda<1$.

The implication a) 2° ⇒ a) 4° of theorem 10.2 was given by M. G. Krein and L. A. Liusternik [142]; see also M. Š. Altman [2] for a) 3° ⇒ a) 4°. The implication a) 1° ⇒ a) 4° is nothing else but the particular case $k=0$ of part b), while parts b) and c) are due to M. G. Krein ([140], the footnote on p. 333).

The last part of remark 10.4, for the case of the implication a) 2° ⇒ 4° of theorem 10.2, is due to B. E. Veic [259].

The fact that under the hypotheses of theorem 10.3 there exist $\gamma_n>0$ $(n=1,2,\dots)$ such that for $\{y_n\}\subset E$ satisfying (10.18) the conclusions of theorem 9.1 hold, as well as corollary 10.2, were proved by M. G. Krein,

D. P. Milman and M. A. Rutman ([143], theorems B and A). Corollary 10.2 solves in the affirmative a problem raised by S. Banach, S. Mazur and S. Ulam (cf. "Scottish book", problem 108). Both the above mentioned part of theorem 10.3 and corollary 10.2 are called "the Krein-Milman-Rutman theorem". This theorem was also reproved by C. Bessaga and A. Pelczyński [22]; for a weaker form of this theorem see also M. M. Grinblium [83], [84]. The fact that the Krein-Milman-Rutman theorem is actually a consequence of the older Paley-Wiener theorem was proved in the paper [61]. C. Bessaga and A. Pelczyński [21], [24] proved that the Krein-Milman-Rutman theorem cannot be extended to B_0-spaces (in the sense of S. Mazur and W. Orlicz [157]). Furthermore, this theorem cannot be carried over to w^*-bases in conjugate Banach spaces (see § 14 and the corresponding Notes and remarks), nor to monotone bases (Ch. II, § 1, remark 1.1).

Theorem 10.3 suggests the problem of introducing suitable topologies in the set of all bases of a Banach space. In connection with this problem let us mention the following definition of the distance between two basic sequences $\{x_n\}, \{y_n\} \subset E$, due to S. Banach [11]:

$$\operatorname{dist}(\{x_n\}, \{y_n\}) = \sum_{i=1}^{\infty} \frac{1}{2^i} \frac{\|x_i - y_i\|}{1 + \|x_i - y_i\|}.$$

S. Banach proved ([11], theorem 1) that if E is a Hilbert space and $\mathscr{F}$ the set of all orthogonal basic sequences of E endowed with the above metric, then the set $\mathscr{B}$ of all orthogonal bases of E is a dense G_δ set of the second category in $\mathscr{F}$.

V. Ya. Kozlov [136] introduced the following distance between two classes of related bases of a Banach space E:

$$\operatorname{dist}(\mathscr{X}, \mathscr{Y}) = \sup_{1 \leq n < \infty} \sup_{x \in E} \left\| \sum_{i=1}^{n} f_i(x) x_i - \sum_{i=1}^{n} g_i(x) y_i \right\| = \sup_{1 \leq n < \infty} \|s_n^{(1)} - s_n^{(2)}\|,$$

where $\{x_n\} \in \mathscr{X}$, $\{y_n\} \in \mathscr{Y}$ and $\{f_n\}$, $\{s_n^{(1)}\}$ and $\{g_n\}$, $\{s_n^{(2)}\}$ are the sequences of coefficient functionals and partial sum operators associated to the bases $\{x_n\}$ and $\{y_n\}$ respectively; two bases $\{x_n\}$, $\{x'_n\}$ of E are called[1] *related*, if $f_n(x) x_n = f'_n(x) x'_n$ $(n = 1, 2, \ldots)$, where $\{f_n\}, \{f'_n\} \subset E^*$ are the a.s.c.f. to the bases $\{x_n\}$ and $\{x'_n\}$ respectively. V. Ya. Kozlov proved ([136], theorem 2) that with the above metric, the set $\mathscr{A}$ of all classes of related bases of E becomes a complete metric space. Furthermore, V. Ya. Kozlov [136] also observed that if E is a Hilbert space, the space $\mathscr{A}$ is unbounded and all orthonormal bases of E are situated on a sphere of radius 1.

[1] V. Ya. Kozlov [136] used the term "equivalent bases" ,which has in the present monograph a different meaning. The term "related bases", in a slightly more general sense, has been used by L. W. Baric [17].

Let us also mention the following problem, communicated to us by G. Neubauer: what are the properties of the set of all subspaces with a basis of a Banach space E, in the metric space consisting of the set of all subspaces of E, endowed with one of the usual metrics? (for such metrics, see e.g., E. Berkson [19]).

Theorem 10.4, in a slightly different form, is due to V. D. Milman ([167], theorem 1) and so are part of example 10.1 ([167], p. 402, example 3) and of corollary 10.3 ([167], theorem 5).

Lemma 10.1, with a different proof, was given by I. M. Gelfand ([73], p. 244, theorem 5). Theorem 10.5 is due, essentially, to V. D. Milman ([167], theorem 3).

For a part of the terminology introduced in definitions 10.1 and 10.2 see N. Bari [13], [14] and J. R. Retherford [203]. In these definitions PW, PH, N and KL stand for Paley-Wiener, Pollard-Hilding, Nagy, and Krein-Liusternik, respectively.

Theorem 10.6 is a particular case of [251], theorem 7.

§ 11. The functions $\phi_n(E)$ (see formula (11.2)) were introduced by H. F. Bohnenblust [29]. Theorem 11.1, suggesting a possible method for constructing a separable Banach space having no basis, was given in [243] (see [243], p. 722, theorem 1.11).

§ 12. Theorem 12.1 is essentially due to S. Banach ([10], p. 107, theorem 3); it was also reproved by L. A. Gurevich [100]. Corollary 12.1 was given by S. Karlin ([131], theorem 3), who also raised problem 12.1 in [131]. Corollary 12.2 was proved by M. M. Grinblium and L. A. Gurevich ([87], theorem II). Problem 12.2b) was raised in the paper [245] (see [245], problem 4).

Theorem 12.2 was given in [240], theorem 1. Remark 12.1 was made by V. F. Gaposhkin and M. I. Kadec [67]. Remark 12.2 was made in [240], remarks 2 and 3. Example 12.2 was constructed in [240].

Theorems 12.3, 12.4, examples 12.3, 12.4, remarks 12.3, 12.4 and corollary 12.3 were given in the paper [249].

Proposition 12.2a), c) in a slightly less "quantitative" form (namely, with

$$\sup_{1 \leqslant n < \infty} \left\| \sum_{i=1}^{n} f(x_i) f_i \right\| < \infty, \quad \sup_{1 \leqslant n < \infty} \left\| \sum_{i=1}^{n} \Phi(f_i) x_i \right\| < \infty$$

instead of (12.42) and 12.46)), is due to A. Wilansky ([262], lemmas 2 and 1).

Theorem 12.5 was given in [239], theorem 4 and corollary 4 and [240], theorem 1 and its corollary; actually, part a) was proved in [239], corollary 4 under the hypothesis that $\{x_n\}$ is a shrinking basis of E and the remark that this latter assumption is superfluous, was made by J. R. Retherford [205].

Definition 12.1 and theorem 12.6, with a somewhat different proof, are due to W. Ruckle ([214], theorem 3.1).

Theorem 12.7 has been given by W. Ruckle [213] and L. W. Baric and W. Ruckle [18].

Theorem 12.8 is due to R. J. McGivney and W. Ruckle ([166], theorem 6.1); the proof of the sufficiency part, presented here, was communicated to us by W. Ruckle.

§ 13. The fact that the notion of a basis admits an obvious extension to complete metric linear spaces, was mentioned in the book of S. Banach ([10], p. 239); the passage to general topological linear spaces (see definition 13.1) was made by M. G. Arsove and R. E. Edwards [5]. S. Banach also remarked ([10], p. 239) that the space $S([0,1])$ has no basis; the more general example 13.1 (including the spaces $L^p([0,1])$ with $0<p<1$) was given in the paper [245] (see [245], theorem 1).

Weak bases in Banach spaces (see definition 13.2) were considered in the book of S. Banach ([10], p. 238). The term "w^*-basis" occurs in the book of M. M. Day ([43], Ch. IV, § 3, theorem 1) without a precise definition; actually, M. M. Day calls w^*-bases those sequences in conjugate Banach spaces, which we have called w^*-Schauder bases (see § 14). A study of w^*-bases in conjugate Banach spaces was made in the papers [234], [240], [245]. The bw^*-bases in conjugate Banach spaces were considered by J. R. Retherford [204], who also gave lemma 13.1 b), with the mention that it is due to R. D. McWilliams (see [204], lemme 5.2).

The implication a) $1^\circ \Rightarrow$ a) 3° of theorem 13.1 is called ([5], p. 97) "the weak basis theorem". The weak basis theorem was stated without proof in the book of S. Banach ([10], p. 238). A proof was sketched by S. Karlin [131]. However, in the book of M. M. Day ([43], Ch. IV, § 3, theorem 2), this theorem is stated only under the additional hypothesis that either E is weakly complete, or the coefficient functionals f_n associated to the w-basis $\{x_n\}$ are continuous on E for the norm topology. A complete proof of the weak basis theorem in its full generality was given by C. Bessaga and A. Pelczyński [24], where it is also mentioned that this theorem is actually due to S. Mazur; see also B. R. Gelbaum ([69], p. 194–195) and A. Wilansky ([263], p. 212). There exist various extensions of the weak basis theorem to more general spaces (see e. g. C. Bessaga and A. Pelczyński [24] and M. G. Arsove and R. E. Edwards [5]). The implication b) $1^\circ \Rightarrow$ b) 3° of theorem 13.1 and the more general proposition 13.1 were proved in the same paper of C. Bessaga and A. Pelczyński ([24], theorem 3).

Example 13.3 of a basis in a conjugate Banach space, which is not a w^*-basis, was given in the paper [245]. In the same paper (see [245], theorem 2) example 13.4 of a w^*-separable conjugate Banach space having no w^*-basis (and thus of a separable locally convex space having no basis) was given and problems 13.1 and 13.2 were raised (see [245], problems 2 and 3).

In the present book we have considered only w^*-separability of E^* in the usual sense of separability of topological spaces (i. e. the admitting of a countable dense subset; see e. g. [50], p. 21, definition 11). However, A. Pelczyński called our attention to the interest of considering a stronger condition of "sequential" w^*-separability, since ([10], p. 124, theorem 4) the conjugate space E^* of any separable Banach space E is sequentially w^*-separable (i. e. there exists a countable sequence $\{f_n\}$ in E^* such that every $f \in E^*$ is the w^*-limit of a suitable subsequence $\{f_{n_k}\}$ of $\{f_n\}$), while the space m^* is w^*-separable but not sequentially w^*-separable. Obviously, every topological linear space with a basis is sequentially separable. In the paper [234] there was given an example of a sequentially w^*-separable conjugate Banach space E^* having no w^*-Schauder basis (see § 14, definition 14.2), but we don't know of any example of a sequentially w^*-separable conjugate space E^* having no w^*-basis (and more generally, of any example of a sequentially separable locally convex space U having no basis).

§ 14. The notion of Schauder basis for a general topological linear space (see definition 14.2) was considered by M. G. Arsove and R. E. Edwards [5]. The problem: in which spaces does theorem 3.1 remain valid (i. e.: in which spaces is every basis a Schauder basis?) has been studied earlier by several authors (see the Notes and remarks to § 3).

The idea of the proof of proposition 14.1 is similar to one of C. Goffman and D. Waterman concerning topological linear spaces on which the only continuous linear functional is $f \equiv 0$ ([76], the proofs of lemma 2 and corollary 1).

Example 14.1 of a w^*-basis which is not a w^*-Schauder basis, was given in the paper [234] (see [234], § 1). The basis of l^1 occuring in example 14.1 was considered previously by B. R. Gelbaum ([69], p. 188, example 3) for a slightly different purpose. The equivalence of the notions of w^*-Schauder basis and bw^*-Schauder basis was established by J. R. Retherford ([204], theorem 5.2). Theorem 14.1 and corollary 14.1 were proved in the paper [234] (see [234], theorem 3 and corollary 3).

The Krein-Milman-Rutman property (see definition 14.3) for B_0-spaces was defined by C. Bessaga and A. Pelczyński (see the Notes and remarks to § 10) and for general topological linear spaces it was defined in the paper [240], where example 14.2 of a conjugate space with a w^*-basis, which does not possess the Krein-Milman-Rutman property in the w^*-topology was also given (see [240], § 2, section 3).

§ 15. The notion of (e)-Schauder basis of a topological linear space (see definition 15.2) and proposition 15.1 a), b) were given by Ch. W. McArthur and J. R. Retherford [161].

§ 16. Although the study of bases in general normed linear spaces might present some interest, we don't know of any bibliographical reference concerning this subject.

§ 17. Part of theorems 17.1 and 17.2 (namely, the general forms (17.4), (17.10) of continuous linear mappings $u: E \to F$ and $v: F \to E$, where E has a basis $\{x_n\}$ and the equalities (17.5), (17.11) for $\|u\|$, $\|v\|$ in the case when $\{x_n\}$ is monotone) were given by I. A. Ezrohi [56]. For the particular case when $F=K$ (the field of scalars), mentioned in remark 17.1, see the Notes and remarks to § 12, theorem 12.5a). See also W. Ruckle [216].

Corollaries 17.1 – 17.6 are well known (see e. g., I. M. Gelfand [73] for corollaries 17.1 and 17.2) and so are theorem 17.3 and corollary 17.7 (see e. g., L. A. Liusternik and V. I. Sobolev [152]).

Remark 17.6 was made in [245], p. 455.

§ 18. In the particular cases when $\alpha=\gamma$ and $\alpha=\lambda$, theorem 18.1 has been proved by B. R. Gelbaum and J. Gil de Lamadrid ([72], theorems 1 and 2') and in the general case when α is an arbitrary uniform crossnorm on $E \otimes F$, by J. Gil de Lamadrid [75]. For results of the type of corollaries 18.1 – 18.4 see L. A. Gurevich [101] and Z. Semadeni [227]. Corollary 18.5 is due to B. R. Gelbaum and J. Gil de Lamadrid [72].

Problem 18.1 was raised by B. R. Gelbaum and J. Gil de Lamadrid [72].

§ 19. T-norms, K-norms and TK-norms with respect to a basis were introduced by V. N. Nikolskiĭ [179]; here T stands for "Čebyšev" and K for the Russian word for "canonical".

Proposition 19.1c) and the necessity parts of proposition 19.1a) and b) were proved by V. N. Nikolskiĭ ([180], § 4 and § 5); the sufficiency parts are due to J. R. Retherford and R. C. James ([210], p. 111).

Example 19.1 is due to J. R. Retherford and R. C. James ([210], example (3.2)).

Example 19.2 is a slightly changed version of an example due to J. R. Retherford and R. C. James ([210], example (2.5)), namely, in their example the second sup of formula (19.10) was taken over the set $\Pi(\mathcal{N} \setminus \{1\})$ of all permutations of $\mathcal{N} \setminus \{1\} = \{2, 3, \dots\}$; however, their proof was based on the inequalities

$$\frac{1}{n 2^n} + \sum_{\substack{m=2 \\ m \neq n}}^{\infty} \frac{1}{\sigma(m) 2^m} \leqslant \frac{1}{n 2^n} + \sum_{\substack{m=2 \\ m \neq n}}^{\infty} \frac{1}{m 2^m} \qquad (\sigma \in \Pi(\mathcal{N} \setminus \{1\}))$$

which are not correct (take e. g., $n=2$ and take the permutation $\sigma \in \Pi(\mathcal{N} \setminus \{1\})$ defined by $\sigma(2)=100$, $\sigma(n)=n-1$ for $3 \leqslant n \leqslant 100$ and $\sigma(n)=n$ for $n \geqslant 101$).

Theorem 19.1, with the norm (19.16) in part a), was proved by V. N. Nikolskiĭ ([180], § 4 and § 5). The norm (19.17) was given in [247]; in the particular case when $E = c_0$ and $\{x_n\}$ = the unit vector basis of E, this T-norm was also found, independently, by R. O. Davies [36]. For a related T-norm, which is also strictly convex, see V. Istratescu [112]. The proof of theorem 19.1 presented here, based on the sufficiency parts of proposition 19.1 a) b), is somewhat simpler than the original proof of V. N. Nikolskiĭ; a similar proof has been also found by J. R. Retherford.

Remark 19.1 and corollary 19.1 are due to V. N. Nikolskiĭ ([180], § 6 and § 7). Remark 19.2 was given in [243], theorem 5.4.

§ 20. Polynomial bases occur e.g., in the Krein-Milman-Rutman theorem (§ 10, corollary 10.2) and strict polynomial bases already inthe classical Gram-Schmidt "orthogonalization" procedure of finitely linearly independent complete sequences in separable Hilbert spaces. The term "polynomial basis" was used in [61] and the term "strict polynomial basis" was introduced in [243], definition 5.8.

The equivalence $1° \Leftrightarrow 11°$ of theorem 20.1 was proved by V. N. Nikolskiĭ ([180], § 7) and that paper also contains, implicitly, the equivalence $1° \Leftrightarrow 2°$. The equivalences $3° \Leftrightarrow 11°$ and $9° \Leftrightarrow 10° \Leftrightarrow 11°$ are also due, essentially, to V. N. Nikolskiĭ ([181], theorems 2, 3 and 4). The implications $5° \Rightarrow 11°$ and $6° \Rightarrow 11°$ were observed in [243], proposition 5.4.

Problem 20.1, for the implication $1° \Rightarrow 11°$ of theorem 20.1, was raised by V. N. Nikolskiĭ ([180], § 7), with the comment that the answer is probably negative.

The equivalence $1° \Leftrightarrow 2°$ of proposition 20.1 was observed by V. N. Nikolskiĭ [182]. The part of example 20.1 concerning property d_1), is a simplified version of an example due to V. N. Nikolskiĭ [182].

As we mentioned in § 20, S. M. Lozinsky and F. I. Haršiladze proved (see [174], Appendix 3) that the sequence (20.23) in $E = C([0,1])$ satisfies (20.24) and therefore it is a Λ system. The term "Λ system" (or "Lozinsky-Haršiladze system") was introduced by M. I. Kadec [125], who also proved in [125] that in the spaces $L^p([0,1])$, where $1 < p \neq 2$, there exist Λ systems. The problem, whether the non-existence of Λ systems characterizes spaces isomorphic to l^2 among separable Banach spaces, was raised in [247] and was studied in [40].

Sub-Γ systems, sub-Λ systems and Γ systems were introduced in [40]. Lemma 20.1, theorem 20.2, corollaries 20.1, 20.2 and proposition 20.2 were proved in [40].

Corollary 20.3 was communicated to us, as a remark to the results of [40], by S. B. Stečkin.

Chapter II

Special Classes of Bases in Banach Spaces

In the present chapter we shall study various particular classes[1] of bases in Banach spaces.

Along with every new class of bases in Banach spaces which we introduce, it is natural to consider also the set of all bases which do not belong to this class; for the sake of brevity, we shall denominate the bases of this set by the prefix "non-" followed by the name of the original class, e. g. the bases which are not monotone, will be called non-monotone bases (we shall make only one exception, namely, the non-unconditional bases will be called conditional bases).

One of the main problems which we shall consider for each special class of bases, is *the existence problem*, i. e., the following problem: *does there exist in every separable Banach space a basis belonging to the respective class?* In finite dimensional Banach spaces the solutions of these existence problems are known and, with a few exceptions (e. g. monotone bases or normal bases) they are obvious; therefore, in most cases, we shall not mention them separately. In infinite dimensional Banach spaces the answer to the corresponding existence problems is either negative or unknown (an affirmative answer would also imply an affirmative answer to the basis problem). In the first case, we shall give an example of a separable Banach space which has no basis belonging to the respective class. In the second case, there arises the more restricted problem of the existence of bases of that class in infinite dimensional Banach spaces *with bases*. If this latter restricted problem has an affirmative answer, then the original existence problem is obviously equivalent to the basis problem and we shall make no mention of it; moreover, if the answer to the restricted existence problem is trivial, we shall not mention it either (however, in some cases the solution of the restricted existence problem may be difficult, see e. g. § 23, theo-

[1] We use here the word "class" in the sense: family, collection (not in the sense: equivalence class).

rem 23.2). Finally[1], if the restricted existence problem is also unsolved, we shall state it as a separate problem.

We shall also study other problems concerning the special classes of bases, such as characterizations of bases belonging to these classes, duality properties, etc.

Due to the importance of unconditional bases, we have divided the present Chapter into the following two parts: I) Classes of bases not involving unconditional convergence. II) Unconditional bases and some classes of unconditional bases.

Finally, let us mention that one can obtain other classes of bases by taking intersections of the classes considered in Ch. II, in the case when these intersections are non-void (e. g., normal monotone bases, shrinking unconditional bases, etc.). With a few exceptions, we leave to the reader the study of the corresponding problems (e. g., existence problems) for these classes of bases.

I. Classes of Bases not Involving Unconditional Convergence

§ 1. Monotone and strictly monotone bases

Definition 1.1. A basis $\{x_n\}$ of a Banach space E is said to be *monotone*, if we have

$$\left\|\sum_{i=1}^{n} \alpha_i x_i\right\| \leqslant \left\|\sum_{i=1}^{n+m} \alpha_i x_i\right\| \tag{1.1}$$

for all finite sequences of scalars $\alpha_1,\dots,\alpha_{n+m}\in K$. The basis $\{x_n\}$ is said to be *strictly monotone*, if we have

$$\left\|\sum_{i=1}^{n} \alpha_i x_i\right\| < \left\|\sum_{i=1}^{n+m} \alpha_i x_i\right\| \tag{1.2}$$

for all finite sequences $\alpha_1,\dots,\alpha_{n+m}\in K$ with $\sum_{i=n+1}^{n+m} |\alpha_i| \neq 0$.

For instance, the natural basis of l^p $(p\geqslant 1)$ is strictly monotone, while the natural basis of c_0 is monotone, but not strictly monotone.

---

[1] Obviously, under the above hypotheses the restricted existence problem cannot have a negative answer, since otherwise the original existence problem would also have a negative answer and thus the answer to the original problem would be known.

From Ch. I, § 7, theorem 7.1[1], it follows that *a basis* $\{x_n\}$ *of a Banach space* E *is monotone if and only if it is of norm* $\nu_{\{x_n\}} = \sup_{1 \leqslant n < \infty} \|s_n\| = 1$. Hence: a) The Schauder basis of $C([0,1])$ is monotone (but not strictly), because $s_n(x) = \sum_{i=0}^{n} f_i(x) x_i$ is a polygonal function which interpolates $x \in C([0,1])$ in the points $0, a_1, .., a_{n-1}, 1$ of $[0,1]$ (see Ch. I, § 2); b) The Haar basis of $L^p([0,1])$ $(p \geqslant 1)$ is monotone, by Ch. I, § 2, formula (2.17) (actually, it is strictly monotone for $p > 1$, but not so for $p = 1$). In Vol. II we shall also see that every space $C(Q)$ (Q compact metric) has a monotone basis.

The notions of monotone basis and strictly monotone basis admit a geometrical interpretation. In fact, let us recall that an $x \in E$ is said to be *orthogonal (strictly orthogonal)* to a $y \in E$, and we write $x \perp y$ $(x \perp \perp y)$, if we have $\|x + \alpha y\| \geqslant \|x\|$ for all scalars α, or, equivalently, $\operatorname{dist}(x, [y]) = \operatorname{dist}(x, 0)$ (respectively, $\|x + \alpha y\| > \|x\|$ for all $\alpha \neq 0$). A subspace F of E is said to be *orthogonal (strictly orthogonal)* to a subspace G of E, and we write $F \perp G (F \perp \perp G)$, if we have $x \perp y$ for all $x \in F$, $y \in G$ (respectively, $x \perp \perp y$ for all $x \in F$, $y \in G$, $y \neq 0$). Hence, *a basis* $\{x_n\}$ *of* E *is monotone (strictly monotone) if and only if we have* $[x_1, \ldots, x_n] \perp [x_{n+1}, \ldots, x_{n+m}]$ *for all positive integers* n, m *(respectively,* $[x_1, \ldots, x_n] \perp \perp [x_{n+1}, \ldots, x_{n+m}]$ *for all positive integers* n, m*)*.

Let us now consider the particular case when E is a Hilbert space H. We recall

Lemma 1.1. *Let* H *be a Hilbert space and let* $x, y \in H$, $y \neq 0$. *The following statements are equivalent:*

1°. $x \perp y$.

2°. $x \perp \perp y$.

3°. $(x, y) = 0$[2].

Proof. 1° ⇒ 3°. If $(x, y) \neq 0$, then for $\alpha = -\dfrac{(x,y)}{(y,y)}$ we have

$$\|x + \alpha y\|^2 = \left(x - \frac{(x,y)}{(y,y)} y,\, x - \frac{(x,y)}{(y,y)} y\right) = (x,x) - 2\frac{|(x,y)|^2}{(y,y)} + \frac{|(x,y)|^2}{(y,y)^2}(y,y)$$

$$= (x,x) - \frac{|(x,y)|^2}{(y,y)} < (x,x) = \|x\|^2,$$

whence x non $\perp y$.

---

[1] By means of Ch. I, § 7, theorem 7.1, we have at once several characterizations of monotone bases.

[2] We denote by $(,)$ the scalar product in the Hilbert space H.

$3^\circ \Rightarrow 2^\circ$. If $(x,y)=0$, then for every scalar $\alpha \neq 0$ we have

$$\|x+\alpha y\|^2 = (x+\alpha y, x+\alpha y) = \|x\|^2 + |\alpha|^2 \|y\|^2 > \|x\|^2,$$

whence $x \perp \perp y$.

Finally, $2^\circ \Rightarrow 1^\circ$ is obvious, which completes the proof.

An immediate consequence of lemma 1.1 and of the preceding remark is the following:

Proposition 1.1. *Let H be a Hilbert space with a basis $\{x_n\}$. The following statements are equivalent:*

1°. *The basis $\{x_n\}$ is monotone.*

2°. *The basis $\{x_n\}$ is strictly monotone.*

3°. *The basis $\{x_n\}$ is orthogonal in the usual Hilbert space sense, i.e. $(x_i,x_j)=0$ for all $i \neq j$ $(i,j=1,2,\ldots)$.*

Let us turn now our attention to existence problems. We shall first show that the answer to the problem of existence of monotone and strictly monotone bases in finite dimensional Banach spaces is negative, by constructing a 3-dimensional Banach space which has no monotone (and hence no strictly monotone) basis. Let us recall

Lemma 1.2. *Let E be a Banach space and G a closed linear subspace of E. If there exists a projection u of E onto G with $\|u\|=1$, then for every $x \in G$ there exists a "maximal functional" $f_x \in E^*$ (i.e. an $f_x \in E^*$ with $\|f_x\|=1$, $f_x(x)=\|x\|$) contained in the subspace*

$$(\operatorname{Ker} u)^\perp = \{f \in E^* \mid f(y)=0 \quad \text{for all} \quad y \in \operatorname{Ker} u\} \tag{1.3}$$

of E, where $\operatorname{Ker} u = \{y \in E \mid u(y)=0\}$.

Proof. Let $x \in G$. Then for all $y \in \operatorname{Ker} u$ we have

$$\|x\| = \|u(x+y)\| \leqslant \|x+y\|,$$

whence $\operatorname{dist}(x, \operatorname{Ker} u) = \|x\|$. Consequently, there exists an $f_x \in E^*$ such that $\|f_x\|=1$, $f_x(x)=\|x\|$ and $f(y)=0$ for all $y \in \operatorname{Ker} u$, which completes the proof.

Actually, we shall apply this lemma only in the particular case when $E=$ a real Banach space, $\dim E=3$, $\dim G=2$. In this case, for any projection u of E onto G we have $\dim \operatorname{Ker} u=1$, whence $\dim(\operatorname{Ker} u)^\perp=2$, and a maximal functional f_x (for $\|x\|=1$) is nothing else but the normal vector to the support plane $\{y \in E \mid f_x(y)=1\}$ at x of the unit cell $S_E=\{y \in E \mid \|y\| \leqslant 1\}$. Thus, if there exists a projection u of E onto G with $\|u\|=1$, then by lemma 1.2, the following condition must be satisfied:

Condition (A): *Through each point x with $\|x\|=1$ of the central section $G \cap S_E$ of the centrally symmetric convex body S_E one can construct a support plane H_x of S_E in such a way that the normals to all these planes H_x be coplanar.*

Now we can prove

Theorem 1.1. *Let D be a regular dodecahedron. Then one can cut the vertices of D by planes, so as to obtain a centrally symmetric convex body S with the following property: the 3-dimensional real Banach space E in which the unit cell $\{x \in E \mid \|x\| \leqslant 1\}$ is the convex body S, has no monotone (and hence no strictly monotone) basis.*

Proof. The regular dodecahedron D has six pairs of faces parallel two by two and the normals to any three of these pairs are non coplanar. Now, let us consider the central sections $G \cap D$ of D, where G is any plane through the center of symmetry. They have, clearly, the following properties:

a) If a central section $G \cap D$ of D passes through a pair of parallel edges of D, then this section satisfies condition (A) (see figure 1.1).

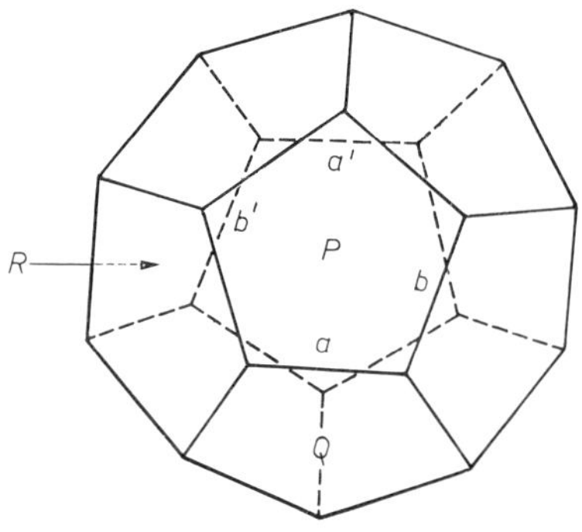

Fig. 1.1

b) If a central section does not pass through any pair of parallel edges of D, then this section intersects the interiors of at least six faces of D, whence it does not satisfy condition (A).

c) If a central section intersects the interiors of two adjacent edges of D, then it intersects the interiors of at least eight faces of D. Indeed, assume e.g. that the section intersects the interiors of the edges a, b of the face P; then it also intersects the interiors of the edges a', b' symmetric to a, b. Now, going along this section from a towards b', we must intersect the interior of the faces Q and R (see figure 1.1), because a is an edge of Q and b' is an edge of R. Since Q and R have no common edge, we must intersect the interior of at least one more face, and thus the section intersects the interior of at least eight faces of D.

Let us cut the vertices of D by planes H so that the following conditions be satisfied:

(i) The polyhedron S which we obtain remains centrally symmetric.

(ii) Each plane H cuts only one vertex, but in such a way that all new faces which we obtain (i.e. all the intersections of these cutting planes H with D) are disjoint two by two.

(iii) The normals to any three pairs of symmetric faces of S are non-coplanar.

(iv) No plane G passing through the center of symmetry can contain a pair of non-symmetric edges of S.

We shall now show that *no central section* $G \cap S$ *of* S *(*G *being any plane through the center of symmetry) satisfies condition* (A). We have to consider three cases:

α) If $G \cap S$ does not pass through any pair of parallel edges of S, then, by (ii), $G \cap D$ does not pass through any pair of parallel edges of D, whence, as we have observed in b), $G \cap D$ intersects the interiors of at least six faces of D and thus the polygon which bounds $G \cap D$ has at least six sides. However, from (ii) it follows that for any plane G' through the center of symmetry, the number of the sides of the polygon which bounds $G' \cap S$ is $\geqslant$ than the number of the sides of the polygon which bounds $G' \cap D$. Thus the polygon which bounds $G \cap S$ has at least six sides and, by our hypothesis on G, none of these sides coincides with an edge of S. Hence $G \cap S$ intersects the interiors of at least six faces of S and thus, by (i) and (iii)[1], it does not satisfy condition (A).

β) If $G \cap S$ passes through a pair of parallel edges of S which are contained in edges of D, then, by (ii) $G \cap S$ intersects, besides the interiors of two pairs of faces of D, the interiors of two pairs of new faces of S. Hence, by (i) and (iii), the section $G \cap S$ does not satisfy condition (A).

γ) If $G \cap S$ passes through a pair of parallel edges of S which are not contained in edges of D, then, since any such edge of S intersects the interiors of two adjacent edges of D, from c) it follows that the polygon which bounds $G \cap D$ has at least eight sides. Hence, as we have observed in α) above, the polygon which bounds $G \cap S$ has at least eight sides and, by our hypothesis on G, one pair of these sides coincides with a pair of parallel edges of S. However, by (iv), no other side of the polygon which bounds $G \cap S$ can coincide with an edge of S and thus this polygon

[1] We use the following fact: if a plane H' supports S at an interior point of a face Q of S, then H' contains the whole face Q and thus the normal to H' coincides with the normal to Q.

has at least six sides which are not edges of S. Hence $G \cap S$ intersects the interiors of at least six faces of S and thus, by (i) and (iii), it does not satisfy condition (A).

Thus we have proved that no central section $G \cap S$ of S (where $\dim G = 2$) satisfies condition (A). Consequently, by lemma 1.2, we have $\|u\| > 1$ for all projections u of E onto any 2-dimensional subspace G of E. However, if $\{x_1, x_2, x_3\}$ would be a monotone basis of E, then, by (1.1), the partial sum operator $s_2 \colon \sum_{i=1}^{3} \alpha_i x_i \to \sum_{i=1}^{2} \alpha_i x_i$ would be a projection of E onto $G = [x_1, x_2]$, with $\|s_2\| = 1$. This shows that E has no monotone (and hence no strictly monotone) basis and thus the proof of theorem 1.1 is complete.

Now we are also able to show that the answer to the problem of the existence of Banach spaces E with index $\Gamma(E)$ satisfying $0 < \Gamma(E) < 1$ is affirmative, as announced in Ch. I, § 7.

Example 1.1. For the 3-dimensional Banach space E defined in theorem 1.1 we have

$$0 < \Gamma(E) < 1. \tag{1.4}$$

In fact, since E has a basis, we have $\Gamma(E) > 0$. On the other hand, assume that $\Gamma(E) = \sup_{\{x_j\}} \gamma_{\{x_j\}} = \dfrac{1}{\inf\limits_{\{x_j\}} \sup\limits_{1 \leqslant n \leqslant 3} \|s_n\|} = 1$, the first sup, respectively the inf being taken over all bases $\{x_j\}_{j=1}^{3}$ of E. Then for every $\varepsilon > 0$ there exists a basis $\{x_j^{(\varepsilon)}\}_{j=1}^{3}$ of E such that $\sup\limits_{1 \leqslant n \leqslant 3} \|s_n^{(\varepsilon)}\| < 1 + \varepsilon$. We may assume, without loss of generality, that $\|x_j^{(\varepsilon)}\| = 1$ $(j = 1, 2, 3)$. Then, since the unit sphere $\sigma_E = \{x \in E \mid \|x\| = 1\}$ is compact, each of the sets $\{x_1^{(\varepsilon)}\}_{\varepsilon > 0}$, $\{x_2^{(\varepsilon)}\}_{\varepsilon > 0}$, $\{x_3^{(\varepsilon)}\}_{\varepsilon > 0}$ admits at least one limit point, say x_1, x_2 and x_3 respectively. The corresponding partial sum operators $s_n \colon \sum_{i=1}^{3} \alpha_i x_i \to \sum_{i=1}^{n} \alpha_i x_i$ $(n = 1, 2, 3)$ are then of norm $\|s_n\| = 1$, whence $\{x_j\}_{j=1}^{3}$ is a monotone basis of E, which contradicts theorem 1.1. Consequently, we have $\Gamma(E) < 1$, which completes the proof of (1.4). Obviously, this argument remains valid for any finite dimensional Banach space E having no monotone basis.

Since for any Banach space E of dimension 3 we have[1] $\phi_2(E) = \dfrac{1}{\Gamma(E)}$,

---

[1] Indeed, for any $G \prec E$ with $\dim G = 2$ and $u \in \mathscr{P}(E, G)$ take a monotone basis $\{x_1, x_2\}$ of G and an $x_3 \in \operatorname{Ker} u \setminus \{0\}$; then $\{x_j\}_{j=1}^{3}$ is a basis of E with $\sup\limits_{1 \leqslant n \leqslant 3} \|s_n\| \leqslant \|u\|$. Conversely, taking, for any basis $\{x_j\}_{j=1}^{3}$ of $E, G = [x_1, x_2]$ and $u = s_2$, we have $u \in \mathscr{P}(E, G)$, $\|u\| \leqslant \sup\limits_{1 \leqslant n \leqslant 3} \|s_n\|$. Thus, $\inf\limits_{\{x_j\}} \sup\limits_{1 \leqslant n \leqslant 3} \|s_n\| = \inf\limits_{\substack{G \prec E \\ \dim G = 2}} \inf\limits_{u \in \mathscr{P}(E, G)} \|u\|$, i.e. $\dfrac{1}{\Gamma(E)} = \phi_2(E)$.

where $\phi_2(E) = \inf_{\substack{G \prec E \\ \dim G = 2}} \inf_{u \in \mathscr{P}(E,G)} \|u\|$ is the function introduced in Ch. I, § 11, it follows that for the space E constructed above we have

$$\phi_2(E) > 1. \tag{1.5}$$

One can also show that for every integer $n \geqslant 3$ there exists an n-dimensional Banach space E_n such that

$$\phi_1(E_n) = 1, \quad \phi_j(E_n) > 1 \quad (j = 2, 3, \ldots, n-1); \tag{1.6}$$

for instance, one can choose E_n to be a suitable subspace of a space l_k^p with $k > 2(2n-3)$ and $p \neq$ integer[1]. However, we shall prefer to give another construction of such spaces E_n, in theorem 1.2 below.

Let us also remark that the iteration of the method of cutting the vertices of a centrally symmetric polyhedron by suitable hyperplanes might perhaps present some interest for the construction of an increasing sequence of finite-dimensional Banach spaces having the properties required in Ch. I, § 11, theorem 11.1, i.e. for the construction of a separable Banach space having no basis.

In order to prove theorem 1.2 below we shall need several lemmas. In the sequel we shall denote by E^n the real n-dimensional euclidean space l_n^2.

Lemma 1.3. *Let $z_1, \ldots, z_n \in E^n$ with $\|z_i\| = 1$ $(i = 1, \ldots, n)$ be arbitrary. Then for every $\varepsilon > 0$ there exist n elements $y_i \in S(z_i, \varepsilon)$ with $\|y_i\| = 1$ $(i = 1, \ldots, n)$ which are linearly independent.*

Proof. We shall first show that for every $\varepsilon > 0$ there exist elements $z_i' \in S\left(z_i, \frac{\varepsilon}{2}\right)$ $(i = 1, \ldots, n)$ which are linearly independent. Let $x_1, \ldots, x_n$ be an arbitrary basis of E^n and let

$$z_i'(t) = t x_i + (1-t) z_i \quad (0 \leqslant t \leqslant 1,\ i = 1, \ldots, n). \tag{1.7}$$

Since $x_1, \ldots, x_n$ is a basis, we can write

$$z_i = \sum_{j=1}^{n} a_{ij} x_j, \quad z_i'(t) = \sum_{j=1}^{n} b_{ij}(t) x_j \quad (0 \leqslant t \leqslant 1,\ i = 1, \ldots, n), \tag{1.8}$$

whence, by (1.7),

$$\sum_{j=1}^{n} b_{ij}(t) x_j = z_i'(t) = t x_i + (1-t) \sum_{j=1}^{n} a_{ij} x_j = \sum_{j=1}^{n} [t \delta_{ij} + (1-t) a_{ij}] x_j,$$

---

[1] See H. F. Bohnenblust [29]. We recall that l_k^p $(p \geqslant 1)$ is the space of all k-tuples $x = \{\xi_i\}_{i=1}^k$ endowed with the norm $\|x\| = \left(\sum_{i=1}^{k} |\xi_i|^p\right)^{\frac{1}{p}}$.

and thus, since $x_1, \ldots, x_n$ is a basis,

$$b_{ij}(t) = t\,\delta_{ij} + (1-t)\,a_{ij} \qquad (0 \leqslant t \leqslant 1;\ i,j=1,\ldots,n). \tag{1.9}$$

Put

$$D(t) = \det(b_{ij}(t)) \qquad (0 \leqslant t \leqslant 1). \tag{1.10}$$

Then $D(1) = \det(\delta_{ij}) = 1 \neq 0$. Since $D(t)$ is an algebraic polynomial in t (and thus it has a finite number of zeros), it follows that for every $\varepsilon > 0$ there exists a t_ε with $0 < t_\varepsilon < \dfrac{\varepsilon}{2\sqrt{n} \max\limits_{1 \leqslant i,j \leqslant n} |\delta_{ij} - a_{ij}|}$ such that

$$\det(b_{ij}(t_\varepsilon)) = D(t_\varepsilon) \neq 0.$$

Then, by (1.8) $z_1'(t_\varepsilon), \ldots, z_n'(t_\varepsilon)$ are linearly independent and by (1.9) we have $|a_{ij} - b_{ij}(t_\varepsilon)| = |t_\varepsilon \delta_{ij} - t_\varepsilon a_{ij}| = |t_\varepsilon(\delta_{ij} - a_{ij})| < \dfrac{\varepsilon}{2\sqrt{n}}$, whence

$$\|z_i - z_i'(t_\varepsilon)\| = \sqrt{\sum_{j=1}^{n} (a_{ij} - b_{ij}(t_\varepsilon))^2} \leqslant \sqrt{\sum_{j=1}^{n} \frac{\varepsilon^2}{4n}} = \frac{\varepsilon}{2} \qquad (i=1,\ldots,n),$$

and thus $z_i' = z_i'(t_\varepsilon)$ $(i=1,\ldots,n)$ satisfy our assertions.

Now, put

$$y_i = \frac{z_i'}{\|z_i'\|} \qquad (i=1,\ldots,n). \tag{1.11}$$

Then $y_1, \ldots, y_n$ are of norm 1 and linearly independent. Furthermore, by $\|z_i\| = 1$, $\|z_i' - z_i\| \leqslant \dfrac{\varepsilon}{2}$ $(i=1,\ldots,n)$ we have

$$\|z_i' - y_i\| = \left\| z_i' - \frac{z_i'}{\|z_i'\|} \right\| = \left| 1 - \frac{1}{\|z_i'\|} \right| \|z_i'\| = \big|\, \|z_i'\| - 1 \,\big|$$
$$= \big|\, \|z_i'\| - \|z_i\| \,\big| \leqslant \|z_i' - z_i\| \leqslant \frac{\varepsilon}{2} \qquad (i=1,\ldots,n),$$

whence

$$\|z_i - y_i\| \leqslant \|z_i - z_i'\| + \|z_i' - y_i\| \leqslant \frac{\varepsilon}{2} + \frac{\varepsilon}{2} = \varepsilon \qquad (i=1,\ldots,n),$$

which completes the proof of lemma 1.3.

Definition 1.2. A system of N elements $z_1, \ldots, z_N$ in a linear space is said to be *n-independent*, where $n \leqslant N$, if every subsystem consisting of n elements (shortly: every n-subsystem) is linearly independent.

Lemma 1.4. *Let* $z_1, \dots, z_N \in E^n$ $(n \leqslant N)$ *be n-independent. Then there exists an* $\varepsilon = \varepsilon(z_1, \dots, z_N) > 0$ *such that every system* $z'_1, \dots, z'_N$ *with* $z'_i \in S(z_i, \varepsilon)$ $(i = 1, \dots, N)$ *is also N-independent.*

Proof. Let $x_1, \dots, x_n$ be an arbitrary basis of E^n. Then we can write

$$z_i = \sum_{j=1}^{n} a_{ij} x_j \qquad (i = 1, \dots, N). \tag{1.12}$$

Let $z_{k_1}, \dots, z_{k_n}$ be an arbitrary subsystem consisting of n elements. Then by our hypothesis $z_{k_1}, \dots, z_{k_n}$ are linearly independent and therefore

$$D = \begin{vmatrix} a_{k_1 1} & \dots & a_{k_1 n} \\ \dots & \dots & \dots \\ a_{k_n 1} & \dots & a_{k_n n} \end{vmatrix} \neq 0.$$

Since D is a continuous function, on E^{n^2}, of the coordonates $a_{k_1 1}, a_{k_1 2}, \dots, a_{k_n n}$, there exists an $\varepsilon'_1 > 0$ such that whenever $\sqrt{\sum_{i,j=1}^{n} (a_{k_i j} - a'_{k_i j})^2} < \varepsilon'_1$, we have

$$D' = \begin{vmatrix} a'_{k_1 1} & \dots & a'_{k_1 n} \\ \dots & \dots & \dots \\ a'_{k_n 1} & \dots & a'_{k_n n} \end{vmatrix} \neq 0. \tag{1.13}$$

Since all norms on E^{n^2} are equivalent, it follows that there exists an $\varepsilon' > 0$ such that whenever

$$\max_{1 \leqslant i \leqslant n} \|z_{k_i} - z'_{k_i}\| = \max_{1 \leqslant i \leqslant n} \sqrt{\sum_{j=1}^{n} (a_{k_i j} - a'_{k_i j})^2} < \varepsilon',$$

we have (1.13), i.e. the system $z'_{k_1}, \dots, z'_{k_n}$ is linearly independent[1]. Thus for every n-subsystem $S' = \{z_{k_1}, \dots, z_{k_n}\}$ of $S = \{z_1, \dots, z_N\}$ we find an $\varepsilon' > 0$. Since the number of all subsystems S' is finite, the number $\varepsilon = \min_{S'} \varepsilon' > 0$ will satisfy our requirements, which completes the proof of lemma 1.4.

Lemma 1.5. *Let* $z_1, \dots, z_N \in E^n$ $(n \leqslant N)$ *with* $\|z_i\| = 1$ $(i = 1, \dots, N)$ *be arbitrary. Then for every* $\varepsilon > 0$ *there exist N elements* $y_i \in S(z_i, \varepsilon)$ *with* $\|y_i\| = 1$ $(i = 1, \dots, N)$, *which are n-independent.*

Proof. Let us number all n-subsystems $\{z_{k_1}, \dots, z_{k_n}\}$, denoting them by $S_1 = \{z_1, \dots, z_n\}, \dots, S_r$. By lemma 1.3, for every $\varepsilon > 0$ there exist n elements $y_i \in S(z_i, \varepsilon)$ with $\|y_i\| = 1$ $(i = 1, \dots, n)$ which are linearly independent. Assume now that by an arbitrarily small displacement of

[1] Actually, this also follows from Ch. I, § 10, theorem 10.3.

the whole system $S=\{z_1,\dots,z_N\}$ we have obtained that all n-subsystems $S_1,\dots,S_p$, where $p<r$, are of norm 1 and linearly independent. By lemma 1.3, the n-subsystem S_{p+1} can be also made of norm 1 and linearly independent, by an arbitrarily small displacement. By the above proof of lemma 1.4, this displacement can be chosen in such a way that the linear independence of the systems $S_1,\dots,S_p$ be conserved. This completes the proof of lemma 1.5.

Lemma 1.6. *Let* $z_1,\dots,z_N\in E^n$ $(n\leqslant N)$ *with* $\|z_i\|=1$ $(i=1,\dots,N)$ *and* $\varepsilon>0$ *be arbitrary. Then there exist N elements* $y_1,\dots,y_N\in E^n$ *with* $\|y_i\|=1$ $(i=1,\dots,N)$ *and an* $\varepsilon'>0$ *such that*

$$S(y_i,\varepsilon')\subset S(z_i,\varepsilon)\qquad (i=1,\dots,N) \tag{1.14}$$

and that every system $y'_1,\dots,y'_N$ *with* $y'_i\in S(y_i,\varepsilon')$ $(i=1,\dots,N)$ *is n-independent.*

Proof. By lemma 1.5, from the system $z_1,\dots,z_N$ one can obtain, by an $\frac{\varepsilon}{2}$-displacement, a system $y_1,\dots,y_N$ with $\|y_i\|=1$ $(i=1,\dots,N)$, which is n-independent. For this system there exists, by lemma 1.4, an ε' with $0<\varepsilon'<\frac{\varepsilon}{2}$, such that every system $y'_1,\dots,y'_N$ with $y'_i\in S(y_i,\varepsilon')\,(i=1,\dots,N)$ is n-independent. Furthermore, for any $y'_i\in S(y_i,\varepsilon')$ $(i=1,\dots,N)$ we have

$$\|y'_i-z_i\|\leqslant\|y'_i-y_i\|+\|y_i-z_i\|\leqslant\varepsilon'+\frac{\varepsilon}{2}<\varepsilon\qquad (i=1,\dots,N),$$

and thus we have (1.14), which completes the proof of lemma 1.6.

Definition 1.3. A system $R_1,\dots,R_N$ of m-dimensional linear manifolds[1] in $E^n(m<n)$ is said to be *total*, if for every linear manifold $Q\subset E^n$ of dimension $n-m$ there exists an R_i such that $R_i\cap Q$ consists of at most one element.

In the particular case when $R_1,\dots,R_N$ are one-dimensional linear subspaces of E, this reduces, obviously, to the totality of any system $x_i\in R_i\setminus\{0\}$ $(i=1,\dots,N)$ in E^n in the usual sense (Ch. I, § 5), or, equivalently, to $[R_i]_{i=1}^N=E^n$.

Definition 1.4. If $G_1\neq\{0\}$, $G_2\neq\{0\}$ are closed linear subspaces of a normed linear space E, the number

$$\theta(G_1,G_2)=\max\Big(\sup_{\substack{x\in G_1\\ \|x\|=1}}\operatorname{dist}(x,G_2),\ \sup_{\substack{y\in G_2\\ \|y\|=1}}\operatorname{dist}(y,G_1)\Big) \tag{1.15}$$

[1] I.e. translated linear subspaces.

is called the opening of the subspaces G_1, G_2. This is not a metric in the set $\mathscr{S}(E)$ of all closed linear subspaces $G \neq \{0\}$ of E (since in general it does not satisfy the triangle inequality), but it is well known that there also exists a metric $\tilde{\theta}(G_1, G_2)$ on $\mathscr{S}(E)$ such that

$$\theta(G_1, G_2) \leqslant \tilde{\theta}(G_1, G_2) \leqslant 2\theta(G_1, G_2) \quad (G_1, G_2 \in \mathscr{S}(E)), \tag{1.16}$$

e.g. one can take

$$\tilde{\theta}(G_1, G_2) = \max\big(\sup_{x \in \sigma_{G_1}} \operatorname{dist}(x, \sigma_{G_2}), \sup_{y \in \sigma_{G_2}} \operatorname{dist}(y, \sigma_{G_1})\big), \tag{1.17}$$

where $\sigma_G = \{x \in G \mid \|x\| = 1\}$. We leave to the reader the easy verification that this $\tilde{\theta}$ satisfies the triangle inequality and (1.16).

We shall use in the sequel the opening $\theta(G_1, G_2)$. We shall denote for a linear subspace $G_0 \subset E$ and an $\varepsilon > 0$,

$$V(G_0, \varepsilon) = \{G \text{ linear subspace of } E \mid \theta(G, G_0) < \varepsilon\}. \tag{1.18}$$

Lemma 1.7. *Let* $G_1, \ldots, G_{\frac{n(n-1)}{2}} \subset E^n$ *be an arbitrary* $\frac{n(n-1)}{2}$*-system of* $(n-2)$*-dimensional linear subspaces of* E^n*. Then for every* $\varepsilon > 0$ *there exist* $(n-2)$*-dimensional linear subspaces* $G'_1, \ldots, G'_{\frac{n(n-1)}{2}}$ *of* E^n *with* $G'_i \in V(G_i, \varepsilon)\left(i = 1, \ldots, \frac{n(n-1)}{2}\right)$ *which form a total system.*

Proof. Choose in each subspace G_i a basis $y_1^i, \ldots, y_{n-2}^i$ $\left(i = 1, \ldots, \frac{n(n-1)}{2}\right)$. Let $x_1, \ldots, x_n$ be a basis of E^n and let

$$y_k^i = \sum_{j=1}^{n} a_{kj}^{(i)} x_j \quad \left(k = 1, \ldots, n-2;\ i = 1, \ldots, \frac{n(n-1)}{2}\right). \tag{1.19}$$

Consider the system of $\frac{n(n-1)}{2}$ homogeneous linear equations

$$\begin{vmatrix} a_{11}^{(i)} & \ldots a_{1n}^{(i)} \\ \ldots & \ldots \\ a_{n-2,1}^{(i)} & \ldots a_{n-2,n}^{(i)} \\ b_{11} & \ldots b_{1n} \\ b_{21} & \ldots b_{2n} \end{vmatrix} = 0 \quad \left(i = 1, \ldots, \frac{n(n-1)}{2}\right) \tag{1.20}$$

with respect to $\frac{n(n-1)}{2}$ unknowns $\begin{vmatrix} b_{1p} & b_{1q} \\ b_{2p} & b_{2q} \end{vmatrix}$ $(1 \leqslant p < q \leqslant n)$. Here the coefficients are all subdeterminants[1] of order $n-2$ of the $n \times (n-2)$-matrices

$$\begin{pmatrix} a_{11}^{(i)} \dots a_{1n}^{(i)} \\ \dots\dots\dots\dots \\ a_{n-2,1}^{(i)} \dots a_{n-2,n}^{(i)} \end{pmatrix} \qquad \left(i = 1, \dots, \frac{n(n-1)}{2}\right), \tag{1.21}$$

taken with their corresponding signs from the developments of the $n \times n$ determinants (1.20). We claim that for every $\varepsilon' > 0$ there exist numbers $a_{kj}^{\prime(i)}$ with $|a_{kj}^{\prime(i)} - a_{kj}^{(i)}| < \varepsilon'$ $\left(k = 1, \dots, n-2;\ j = 1, \dots, n;\ i = 1, \dots, \frac{n(n-1)}{2}\right)$ such that the determinant of the system of equations obtained from (1.20) by replacing the $a_{kj}^{(i)}$ by the $a_{kj}^{\prime(i)}$, i.e. of the system of equations

$$\begin{vmatrix} a_{11}^{\prime(i)} \dots a_{1n}^{\prime(i)} \\ \dots\dots\dots\dots \\ a_{n-2,1}^{\prime(i)} \dots a_{n-2,n}^{\prime(i)} \\ b_{11} \dots b_{1n} \\ b_{21} \dots b_{2n} \end{vmatrix} = 0 \qquad \left(i = 1, \dots, \frac{n(n-1)}{2}\right) \tag{1.22}$$

be $\neq 0$. In fact, for $i = 1$ consider instead of (1.21) the matrix

$$(a_{kj}^{\prime(1)}(t))_{\substack{k=1,\dots,n-2 \\ j=1,\dots,n}} = (t\,\delta_{kj} + (1-t)\,a_{kj}^{(1)})_{\substack{k=1,\dots,n-2 \\ j=1,\dots,n}}. \tag{1.23}$$

For $t = 1$ this matrix becomes $(\delta_{kj})_{\substack{k=1,\dots,n-2 \\ j=1,\dots,n}}$, i.e. all minors of order $n-2$ are $=0$, except the first one, which is $=1$. Similarly, for each i with $1 < i \leqslant \frac{n(n-1)}{2}$ one can replace the matrix (1.21) by a matrix of the form

$$(a_{kj}^{\prime(i)}(t))_{\substack{k=1,\dots,n-2 \\ j=1,\dots,n}} = (t\,\beta_{kj}^{(i)} + (1-t)\,a_{kj}^{(i)})_{\substack{k=1,\dots,n-2 \\ j=1,\dots,n}}, \tag{1.24}$$

where $(\beta_{kj}^{(i)})_{\substack{k=1,\dots,n-2 \\ j=1,\dots,n}}$ is obtained from $(\delta_{kj})_{\substack{k=1,\dots,n-2 \\ j=1,\dots,n}}$ by a suitable permutation of the columns, in such a way that for $t = 1$ all coefficients of the unknowns $\begin{vmatrix} b_{1p} & b_{1q} \\ b_{2p} & b_{2q} \end{vmatrix}$ in the corresponding i-th equation (obtained from the i-th equation of (1.20) by replacing the $a_{kj}^{(i)}$ by the

[1] For each fixed i, these subdeterminants are nothing else but the Plücker-Grassman coordinates of the subspace G_i.

$\beta_{kj}^{(i)} = a_{kj}'^{(i)}(1)$) become 0, except the $\left(\frac{n(n-1)}{2} - i\right)$-th one, which becomes ± 1.

Now, let $D(t)$ be the determinant of the system of equations obtained from (1.20) by replacing the $a_{kj}^{(i)}$ by the numbers $a_{kj}'^{(i)}(t)$. Then $D(t)$ is an algebraic polynomial in t and $D(1) = \pm 1 \neq 0$, whence there exist arbitrarily small numbers $t > 0$ such that $D(t) \neq 0$. Since $t > 0$ can be arbitrarily small, and since

$$|a_{kj}'^{(i)}(t) - a_{kj}^{(i)}| = t\,|\beta_{kj}^{(i)} - a_{kj}^{(i)}|,$$

it follows that the numbers $a_{kj}'^{(i)}(t)$ can be taken ε'-near to $a_{kj}^{(i)}$ and so that $D(t) \neq 0$. This proves our assertion, taking $a_{kj}'^{(i)}$ to be these $a_{kj}'^{(i)}(t)$.

Now, since the determinant of the system of equations (1.22) is $\neq 0$, it follows that this system has only the trivial solution

$$\begin{vmatrix} b_{1p} & b_{1q} \\ b_{2p} & b_{2q} \end{vmatrix} = 0 \qquad (1 \leqslant p < q \leqslant n). \tag{1.25}$$

However, this implies that the system of subspaces $G_i' = [y_1'^{(i)}, \ldots, y_{n-2}'^{(i)}]$ $\left(i = 1, \ldots, \frac{n(n-1)}{2}\right)$, where

$$y_k'^i = \sum_{j=1}^{n} a_{kj}'^{(i)} x_j \qquad \left(k = 1, \ldots, n-2;\ i = 1, \ldots, \frac{n(n-1)}{2}\right), \tag{1.26}$$

is total. Indeed, assume the contrary, i.e. that there exists a linear manifold $Q \subset E^n$ of dimension 2 such that $Q \cap G_i'$ consists of more than one element for each i. Translating this Q into the origin we obtain a linear subspace $F \subset E^n$ of dimension 2, such that

$$F \cap G_i' \neq \{0\} \qquad \left(i = 1, \ldots, \frac{n(n-1)}{2}\right). \tag{1.27}$$

Now, let

$$z_1 = \sum_{j=1}^{n} b_{1j} x_j, \qquad z_2 = \sum_{j=1}^{n} b_{2j} x_j \tag{1.28}$$

be an arbitrary basis of F. Then, since by (1.27) there exist elements $x_i' \in F \cap G_i' \setminus \{0\}$, we can write $x_i' = \sum_{k=1}^{n-2} c_k^{(i)} y_k'^i = \sum_{l=1}^{2} d_l z_l$, with $\sum_{k=1}^{n-2} |c_k^{(i)}| \neq 0$, $\sum_{l=1}^{2} |d_l| \neq 0$, and therefore the elements $y_1'^i, \ldots, y_{n-2}'^i, z_1, z_2$ are linearly dependent $\left(i = 1, \ldots, \frac{n(n-1)}{2}\right)$ whence we have (1.22). Since by the above the system (1.22) has only the trivial solution (1.25), it follows that z_1, z_2

are linearly dependent, in contradiction with the assumption that they are a basis of F. This proves our assertion that the system $G'_1, \ldots, G'_{\frac{n(n-1)}{2}}$ is total.

Now, by the definition (1.15) of the opening and by the compactness of spheres in finite dimensional spaces we have, for suitable elements $x_0^i = \sum_{k=1}^{n-2} \alpha_k^{(0)} y_k^i \in G_i$, $y_0^i = \sum_{k=1}^{n-2} \beta_k^{(0)} y_k'^i \in G'_i$ with $\|x_0^i\| = \|y_0^i\| = 1$ $\left(i=1, \ldots, \frac{n(n-1)}{2}\right)$,

$$\begin{aligned}\theta(G_i, G'_i) &= \max\left(\sup_{\substack{x\in G_i\\ \|x\|=1}} \operatorname{dist}(x, G'_i),\ \sup_{\substack{y\in G'_i\\ \|y\|=1}} \operatorname{dist}(y, G_i)\right)\\ &= \max\left(\operatorname{dist}(x_0^i, G'_i),\ \operatorname{dist}(y_0^i, G_i)\right)\\ &\leqslant \max\left(\left\|\sum_{k=1}^{n-2} \alpha_k^{(0)} y_k^i - \sum_{k=1}^{n-2} \alpha_k^{(0)} y_k'^i\right\|, \left\|\sum_{k=1}^{n-2} \beta_k^{(0)} y_k'^i - \sum_{k=1}^{n-2} \beta_k^{(0)} y_k^i\right\|\right)\\ &\leqslant \max\left(\sum_{k=1}^{n-2} |\alpha_k^{(0)}|\, \|y_k^i - y_k'^i\|,\ \sum_{k=1}^{n-2} |\beta_k^{(0)}|\, \|y_k^i - y_k'^i\|\right). \end{aligned} \tag{1.29}$$

Since by the above $a_{kj}'^{(i)}$ can be taken arbitrarily near to $a_{kj}^{(i)}$, from (1.19) it follows that $y_k'^i$ can be taken arbitrarily near to y_k^i. Furthermore, since $\|x_0^i\| = \|y_0^i\| = 1$ and since all norms on E^{n-2} are equivalent, there exists a constant M_{n-2} such that $\sum_{k=1}^{n-2} |\alpha_k^{(0)}|,\ \sum_{k=1}^{n-2} |\beta_k^{(0)}| \leqslant M_{n-2}$. These facts, together with (1.29), imply that G'_i can be taken arbitrarily θ-near to G_i for each $i=1, \ldots, \frac{n(n-1)}{2}$, which completes the proof of lemma 1.7.

Lemma 1.8. *Let* $G_1, \ldots, G_{\frac{n(n-1)}{2}}$ *be a total* $\frac{n(n-1)}{2}$*-system of* $(n-2)$*-dimensional linear subspaces of* E^n. *Then there exists an* $\varepsilon = \varepsilon\left(G_1, \ldots, G_{\frac{n(n-1)}{2}}\right) > 0$ *such that every system of* $(n-2)$*-dimensional linear subspaces* $G'_1, \ldots, G'_{\frac{n(n-1)}{2}}$ *with* $G'_i \in V(G_i, \varepsilon)$ $\left(i=1, \ldots, \frac{n(n-1)}{2}\right)$ *is also total.*

Proof. Choose $y_1^i, \ldots, y_{n-2}^i$ $\left(i=1, \ldots, \frac{n(n-1)}{2}\right)$ and $x_1, \ldots, x_n$, $a_{kj}^{(i)}$ $\left(k=1, \ldots, n-2;\ j=1, \ldots, n;\ i=1, \ldots, \frac{n(n-1)}{2}\right)$ as in the proof of the preceding lemma. We claim that since $G_1, \ldots, G_{\frac{n(n-1)}{2}}$ is total, the determinant D of the system of equations (1.20) is $\neq 0$. In fact, assume the contrary, i.e. that $D=0$. Then the system (1.20) has a non-trivial solution $\begin{vmatrix} b_{1p} & b_{1q} \\ b_{2p} & b_{2q} \end{vmatrix}$ $(1 \leqslant p < q \leqslant n)$, whence the elements $z_1, z_2 \in E^n$ defined by (1.28) are linearly independent and the elements $y_1^i, \ldots, y_{n-2}^i$, z_1, z_2 are linearly dependent $\left(i=1, \ldots, \frac{n(n-1)}{2}\right)$. Consequently, for the 2-dimensional linear subspace $F=[z_1, z_2]$ of E^n we have then $F \cap G_i \neq \{0\}$ $\left(i=1, \ldots, \frac{n(n-1)}{2}\right)$, which contradicts the assumption that $G_1, \ldots, G_{\frac{n(n-1)}{2}}$ is total. Therefore $D \neq 0$.

Observe now that if the subspaces G_i, G_i' are arbitrarily θ-near, then, assuming also $\|y_k^i\|=1$ $(k=1, \ldots, n-2)$, one can choose in G_i' a basis $y_k'^i = \sum_{j=1}^n a_{kj}' x_j$ $(k=1, \ldots, n-2)$ such that the numbers $a_{kj}'^{(i)}$ are arbitrarily near to the $a_{kj}^{(i)}$. Indeed, if $\theta(G_i, G_i') < \varepsilon$, then, by (1.15), one can find in G_i' elements $z_1'^i, \ldots, z_{n-2}'^i$ such that

$$\|y_k^i - z_k'^i\| = \operatorname{dist}(y_k^i, G_i') \leqslant \sup_{\substack{y \in G_i \\ \|y\|=1}} \operatorname{dist}(y, G_i') \leqslant \theta(G_i, G_i') < \varepsilon \qquad (k=1, \ldots, n-2),$$

and then, by lemma 1.3, one can find a basis $y_1'^i, \ldots, y_{n-2}'^i$ of G_i' such that $y_k'^i = \sum_{j=1}^n a_{kj}'^{(i)} x_j \in S(z_k'^i, \varepsilon)$ $(k=1, \ldots, n-2)$, whence

$$\|y_k^i - y_k'^i\| \leqslant \|y_k^i - z_k'^i\| + \|z_k'^i - y_k'^i\| < 2\varepsilon \qquad (k=1, \ldots, n-2),$$

and therefore (since all norms on E^n are equivalent) the numbers $a_{kj}'^{(i)}$ can be assumed arbitrarily near to the $a_{kj}^{(i)}$.

Since the determinant D of the system of equations (1.20) is $\neq 0$ and since D is a continuous function of the $a_{kj}^{(i)}$, it follows that for a sufficiently small $\varepsilon > 0$ and for $\theta(G_i, G_i') < \varepsilon$ $\left(i=1, \ldots, \frac{n(n-1)}{2}\right)$, one can choose a basis $y_k'^i = \sum_{j=1}^n a_{kj}'^{(i)} x_j$ $(k=1, \ldots, n-2)$ of G_i $\left(i=1, \ldots, \frac{n(n-1)}{2}\right)$ such that the determinant D' of the corresponding system of equations (1.22) be still $\neq 0$. Then, as we have seen in the proof of the preceding lemma, the system $G_1', \ldots, G_{\frac{n(n-1)}{2}}'$ is total, which completes the proof of lemma 1.8.

Lemma 1.9. *Let $G_1, \ldots, G_{\frac{n(n-1)}{2}} \subset E^n$ be an arbitrary $\frac{n(n-1)}{2}$-system of $(n-2)$-dimensional linear subspaces of E^n and let $\varepsilon > 0$ be arbitrary.*

Then there exist $\frac{n(n-1)}{2}$ *linear subspaces* $G'_1, \ldots, G'_{\frac{n(n-1)}{2}}$ *of* E^n *and an* $\varepsilon' > 0$ *such that*

$$V(G'_i, \varepsilon') \subset V(G_i, \varepsilon) \quad \left(i = 1, \ldots, \frac{n(n-1)}{2}\right) \tag{1.30}$$

and that every system of linear subspaces $\tilde{G}'_1, \ldots, \tilde{G}'_{\frac{n(n-1)}{2}}$ *with* $\tilde{G}'_i \in V(G'_i, \varepsilon')$ $\left(i = 1, \ldots, \frac{n(n-1)}{2}\right)$ *is total.*

Proof. By lemma 1.7, from the system $G_1, \ldots, G_{\frac{n(n-1)}{2}}$ one can obtain, by an $\frac{\varepsilon}{4}$-displacement (in the sense of θ), a system $G'_1, \ldots, G'_{\frac{n(n-1)}{2}}$ of $(n-2)$-dimensional subspaces, which is total. For this system there exists, by lemma 1.8, an ε' with $0 < \varepsilon' < \frac{\varepsilon}{4}$, such that every system $\tilde{G}'_1, \ldots, \tilde{G}'_{\frac{n(n-1)}{2}}$ of $(n-2)$-dimensional subspaces with $\tilde{G}'_i \in V(G'_i, \varepsilon')$ $\left(i = 1, \ldots, \frac{n(n-1)}{2}\right)$ is total. Furthermore, for any $\tilde{G}'_i \in V(G'_i, \varepsilon')$ we have, by (1.16) and since $\tilde{\theta}$ satisfies the triangle inequality,

$$\begin{aligned}\theta(\tilde{G}'_i, G_i) &\leqslant \tilde{\theta}(\tilde{G}'_i, G_i) \leqslant \tilde{\theta}(\tilde{G}'_i, G'_i) + \tilde{\theta}(G'_i, G_i)\\ &\leqslant 2\theta(\tilde{G}'_i, G'_i) + 2\theta(G'_i, G_i) < 2\left(\varepsilon' + \frac{\varepsilon}{4}\right) < \varepsilon,\end{aligned}$$

and thus we have (1.30), which completes the proof of lemma 1.9.

Lemma 1.10. *Let* G_1, G_2 *be two linear subspaces of* E^n *and let* Q_1, Q_2 *be their orthogonal complements respectively. Then*

$$\theta(Q_1, Q_2) = \theta(G_1, G_2). \tag{1.31}$$

Proof. By a well known corollary of the Hahn-Banach theorem, we have

$$\operatorname{dist}(x, G_i) = \max_{\substack{f \in Q_i \\ \|f\| = 1}} |f(x)| \qquad (x \in E^n, \ i = 1, 2),$$

and thus, since in this formula the roles of G_i and Q_i can be also interchanged,

$$\theta(G_1, G_2) = \max_{\substack{f \in Q_1, h \in Q_2, x \in G_1, y \in G_2 \\ \|f\| = \|h\| = \|x\| = \|y\| = 1}} (|f(x)|, |h(y)|) = \theta(Q_1, Q_2),$$

which completes the proof.

For a system of elements $y_1, \ldots, y_k \in E^n$ we shall denote by $\mathscr{D}(y_1, \ldots, y_k)$ the orthogonal complement in E^n of the subspace $[y_1, \ldots, y_k]$.

Lemma 1.11. *Let $z_1, \ldots, z_N \in E^n$ $(N \geqslant n(n-1))$ with $\|z_i\| = 1$ $(i=1, \ldots, N)$ and $\varepsilon > 0$ be arbitrary. Then there exist N elements $y_i \in S(z_i, \varepsilon)$ with $\|y_i\| = 1$ $(i=1, \ldots, N)$, such that*

1°. The system $y_1, \ldots, y_N$ is n-independent.

2°. Every system of $\frac{n(n-1)}{2}$ subspaces (of dimension $n-2$)

$$\mathscr{D}(y_{p_1}, y_{q_1}), \ldots, \mathscr{D}\left(y_{p_{\frac{n(n-1)}{2}}}, y_{q_{\frac{n(n-1)}{2}}}\right), \tag{1.32}$$

where $p_1, \ldots, p_{\frac{n(n-1)}{2}}, q_1, \ldots, q_{\frac{n(n-1)}{2}}$ are distinct indices taken from $\{1, \ldots, N\}$, is total.

Proof. By lemma 1.6 we may assume, without restriction of the generality, that the system $z_1, \ldots, z_N$ is n-independent and that for the given $\varepsilon > 0$ every system $y_1', \ldots, y_N'$ with $y_i' \in S(z_i, \varepsilon)$ $(i=1, \ldots, N)$ is n-independent.

Let $p_1^{(1)}, \ldots, p_{\frac{n(n-1)}{2}}^{(1)}, q_1^{(1)}, \ldots, q_{\frac{n(n-1)}{2}}^{(1)}$ be the first choice of $n(n-1)$ distinct indices taken from $\{1, \ldots, N\}$. Then for the $\frac{n(n-1)}{2}$-system of $(n-2)$-dimensional subspaces

$$\mathscr{D}\left(z_{p_1^{(1)}}, z_{q_1^{(1)}}\right), \ldots, \mathscr{D}\left(z_{p_{\frac{n(n-1)}{2}}^{(1)}}, z_{q_{\frac{n(n-1)}{2}}^{(1)}}\right) \tag{1.33}$$

(which need not be total) there exist, by lemma 1.9, a total system of $(n-2)$-dimensional subspaces $\mathscr{D}_1, \ldots, \mathscr{D}_{\frac{n(n-1)}{2}}$ and an $\varepsilon' > 0$, such that

$$V(\mathscr{D}_i, \varepsilon') \subset V\left(\mathscr{D}\left(z_{p_i^{(1)}}, z_{q_i^{(1)}}\right), \frac{\varepsilon}{2}\right) \quad \left(i=1, \ldots, \frac{n(n-1)}{2}\right) \tag{1.34}$$

and that every system of subspaces choosen from $V(\mathscr{D}_i, \varepsilon')$ $\left(i=1, \ldots, \frac{n(n-1)}{2}\right)$ is also total.

Now let Q_i be the (2-dimensional) orthogonal complement in E^n of $\mathscr{D}_i$ $\left(i=1, \ldots, \frac{n(n-1)}{2}\right)$. Then, by lemma 1.10 and (1.34),

$$\sup_{\substack{y \in [z_{p_i^{(1)}}, z_{q_i^{(1)}}] \\ \|y\| = 1}} \operatorname{dist}(y, Q_i) \leqslant \theta\left(Q_i, \left[z_{p_i^{(1)}}, z_{q_i^{(1)}}\right]\right) = \theta\left(\mathscr{D}_i, \mathscr{D}\left(z_{p_i^{(1)}}, z_{q_i^{(1)}}\right)\right) < \frac{\varepsilon}{2},$$

whence there exist elements $z_{p_i^{(1)}}^1 \in Q_i$, $z_{q_i^{(1)}}^1 \in Q_i$ such that

$$z^1_{p_i^{(1)}} \in S\left(z_{p_i^{(1)}}, \frac{\varepsilon}{2}\right), \quad z^1_{q_i^{(1)}} \in S\left(z_{q_i^{(1)}}, \frac{\varepsilon}{2}\right) \quad \left(i=1, \ldots, \frac{n(n-1)}{2}\right). \tag{1.35}$$

Then the system $z^1_{p_1^{(1)}}, \ldots, z^1_{p^{(1)}_{\frac{n(n-1)}{2}}}, z^1_{q_1^{(1)}}, \ldots, z^1_{q^{(1)}_{\frac{n(n-1)}{2}}}$ is n-independent, whence, in particular, each couple $z^1_{p_i^{(1)}}, z^1_{p_i^{(1)}}$ is linearly independent. Thus $Q_i = [z^1_{p_i^{(1)}}, z^1_{q_i^{(1)}}]$ $\left(i=1, \ldots, \frac{n(n-1)}{2}\right)$, whence

$$\mathscr{D}_i = \mathscr{D}(z^1_{p_i^{(1)}}, z^1_{q_i^{(1)}}) \quad \left(i=1, \ldots, \frac{n(n-1)}{2}\right) \tag{1.36}$$

and therefore the system of subspaces

$$\mathscr{D}(z^1_{p_1^{(1)}}, z^1_{q_1^{(1)}}), \ldots, \mathscr{D}\left(z^1_{p^{(1)}_{\frac{n(n-1)}{2}}}, z^1_{q^{(1)}_{\frac{n(n-1)}{2}}}\right) \tag{1.37}$$

is total and every system of subspaces chosen from $V(\mathscr{D}(z^1_{p_i^{(1)}}, z^1_{q_i^{(1)}}), \varepsilon')$ $\left(i=1, \ldots, \frac{n(n-1)}{2}\right)$ is also total; replacing, if necessary, $z^1_{p_i^{(1)}}, z^1_{q_i^{(1)}}$ by $\frac{1}{\|z^1_{p_i^{(1)}}\|} z^1_{p_i^{(1)}}$, $\frac{1}{\|z^1_{q_i^{(1)}}\|} z^1_{q_i^{(1)}}$ and $\frac{\varepsilon}{2}$ by $\frac{\varepsilon}{4}$ in (1.34), one may assume (see the end of the proof of lemma 1.3) that $\|z^1_{p_i^{(1)}}\| = \|z^1_{q_i^{(1)}}\| = 1$ $\left(i=1, \ldots, \frac{n(n-1)}{2}\right)$. Since by lemma 1.10 we have

$$\theta(\mathscr{D}(\tilde{z}_{p_i^{(1)}}, \tilde{z}_{q_i^{(1)}}), \mathscr{D}(z^1_{p_i^{(1)}}, z^1_{q_i^{(1)}})) = \theta([\tilde{z}_{p_i^{(1)}}, \tilde{z}_{q_i^{(1)}}], [z^1_{p_i^{(1)}}, z^1_{q_i^{(1)}}]),$$

it follows (see the proof of lemma 1.7, formula (1.29)) that there exists an $\varepsilon'_1 > 0$ such that every system of subspaces $\mathscr{D}(\tilde{z}_{p_1^{(1)}}, \tilde{z}_{q_1^{(1)}}), \ldots,$ $\mathscr{D}\left(\tilde{z}_{p^{(1)}_{\frac{n(n-1)}{2}}}, \tilde{z}_{q^{(1)}_{\frac{n(n-1)}{2}}}\right)$ with $\|\tilde{z}_{p_i^{(1)}} - z^1_{p_i^{(1)}}\| < \varepsilon'_1$, $\|\tilde{z}_{q_i^{(1)}} - z^1_{q_i^{(1)}}\| < \varepsilon'_1$ $\left(i=1, \ldots, \frac{n(n-1)}{2}\right)$ is total; one may assume, without loss of generality, that $\varepsilon'_1 < \frac{\varepsilon}{4}$.

Putting $z^1_i = z_i$ for all $i \neq p_1^{(1)}, \ldots, p^{(1)}_{\frac{n(n-1)}{2}}, q_1^{(1)}, \ldots, q^{(1)}_{\frac{n(n-1)}{2}}$, we have thus obtained a new system of elements

$$z^1_1, \ldots, z^1_N \tag{1.38}$$

with $\|z^1_i\| = 1$, $z^1_i \in S\left(z_i, \frac{\varepsilon}{2}\right)$ $(i=1, \ldots, N)$ and an ε'_1 with $0 < \varepsilon'_1 < \frac{\varepsilon}{4}$, having the property that the system of $\frac{n(n-1)}{2}$ subspaces (1.37), corresponding to our first choice of $\frac{n(n-1)}{2}$ distinct indices $p_1^{(1)}, \ldots, p^{(1)}_{\frac{n(n-1)}{2}}$,

$q_1^{(1)}, \ldots, q_{\frac{n(n-1)}{2}}^{(1)}$, is total and that every system of subspaces $\mathscr{D}(\tilde{z}_{p_1^{(1)}}, \tilde{z}_{q_1^{(1)}}), \ldots,$ $\mathscr{D}\left(\tilde{z}^1_{p^{(1)}_{\frac{n(n-1)}{2}}}, \tilde{z}^1_{q^{(1)}_{\frac{n(n-1)}{2}}}\right)$ with $\|\tilde{z}_{p_i^{(1)}} - z^1_{p_i^{(1)}}\| < \varepsilon_1'$, $\|\tilde{z}_{q_i^{(1)}} - z^1_{q_i^{(1)}}\| < \varepsilon_1'$ $\left(i = 1, \ldots, \frac{n(n-1)}{2}\right)$ is also total.

Let now $p_1^{(2)}, \ldots, p^{(2)}_{\frac{n(n-1)}{2}}, q_1^{(2)}, \ldots, q^{(2)}_{\frac{n(n-1)}{2}}$ be a second choice of $n(n-1)$ distinct indices taken from $\{1, \ldots, N\}$. Then, starting from the new system (1.38) and from ε_1', one can find, by the above arguments, a system of elements

$$z_1^2, \ldots, z_N^2 \tag{1.39}$$

with $\|z_i^2\| = 1$, $z_i^2 \in S(z_i^1, \varepsilon_1')$ $(i = 1, \ldots, N)$ and an ε_2'' with $0 < \varepsilon_2'' < \frac{\varepsilon_1'}{4}$, such that the system of subspaces

$$\mathscr{D}(z^2_{p_1^{(2)}}, z^2_{q_1^{(2)}}), \ldots, \mathscr{D}\left(z^2_{p^{(2)}_{\frac{n(n-1)}{2}}}, z^2_{q^{(2)}_{\frac{n(n-1)}{2}}}\right) \tag{1.40}$$

is total and that every system of subspaces $\mathscr{D}(\tilde{z}_{p_1^{(2)}}, \tilde{z}_{q_1^{(2)}}), \ldots, \mathscr{D}\left(\tilde{z}_{p^{(2)}_{\frac{n(n-1)}{2}}}, \tilde{z}_{q^{(2)}_{\frac{n(n-1)}{2}}}\right)$ with $\|\tilde{z}_{p_i^{(2)}} - z^2_{p_i^{(2)}}\| < \varepsilon_2''$, $\|\tilde{z}_{q_i^{(2)}} - z^2_{q_i^{(2)}}\| < \varepsilon_2''$ $\left(i = 1, \ldots, \frac{n(n-1)}{2}\right)$ is also total. By virtue of $z_i^2 \in S(z_i^1, \varepsilon_1')$, the system of subspaces corresponding to the first choice of indices

$$\mathscr{D}(z^2_{p_1^{(1)}}, z^2_{q_1^{(1)}}), \ldots, \mathscr{D}\left(z^2_{p^{(1)}_{\frac{n(n-1)}{2}}}, z^2_{q^{(1)}_{\frac{n(n-1)}{2}}}\right) \tag{1.41}$$

remains total. Furthermore, we have, by $\varepsilon_1' < \frac{\varepsilon}{4}$,

$$\|z_i^2 - z_i\| \leqslant \|z_i^2 - z_i^1\| + \|z_i^1 - z_i\| \leqslant \varepsilon_1' + \frac{\varepsilon}{2} < \frac{3}{4}\varepsilon \quad (i = 1, \ldots, N), \tag{1.42}$$

whence $z_i^2 \in S(z_i, \varepsilon)$ $(i = 1, \ldots, N)$.

Continue this procedure, starting from the third choice of $n(n-1)$ distinct indices taken from $\{1, \ldots, N\}$, the new system (1.39) and ε_2''. Since the number of all possible choices of $n(n-1)$ distinct indices taken from $\{1, \ldots, N\}$ is finite, say M, repeating the above arguments M times, we obtain a system of elements

$$y_1 = z_1^M, \ldots, y_N = z_N^M$$

with $\|y_i\| = 1$ $(i = 1, \ldots, N)$, such that for every choice of $n(n-1)$ distinct indices $p_1, \ldots, p_{\frac{n(n-1)}{2}}, q_1, \ldots, q_{\frac{n(n-1)}{2}}$ taken from $\{1, \ldots, N\}$ the system of $\frac{n(n-1)}{2}$ subspaces (1.32) is total and that $y_i \in S(z_i, \varepsilon)$ $(i=1, \ldots, N)$, whence $y_1, \ldots, y_N$ is n-independent. This completes the proof of lemma 1.11.

Now we can prove

Theorem 1.2. *For every integer* $n \geqslant 3$ *there exists a Banach space* E *with* $\dim E = n$ *such that for every couple of subspaces* G, F *of* E *with* $\dim G \geqslant 2$ *we have*

$$\widehat{(G; F)} < 1. \tag{1.43}$$

Consequently, no subspace E_0 *of* E *with* $\dim E_0 \geqslant 3$ *has a monotone basis and the space* $E_n = E$ *satisfies* (1.6).

Proof. For $\delta = \frac{1}{8n^3}$ consider on $Fr\, S_{E^n} = \{x \in E^n \mid \|x\| = 1\}$ a δ-net $z_1, \ldots, z_N$ with $N \geqslant n(n-1)$. Then, by lemma 1.11, there exist N elements $y_i \in S(z_i, \delta)$ with $\|y_i\| = 1$ $(i=1, \ldots, N)$ such that we have 1° and 2° of lemma 1.11.

For each $y \in E^n$ we shall denote by $\mathscr{D}(y)$ the orthogonal complement of y in E^n, i.e. $\mathscr{D}(y) = \{x \in E^n \mid (x, y) = 0\}$, and by T^y the strip

$$T^y = \{x + t y \mid x \in \mathscr{D}(y),\ -1 \leqslant t \leqslant 1\}. \tag{1.44}$$

Put

$$S = \bigcap_{i=1}^{N} T^{y_i}. \tag{1.45}$$

Then S is a centrally symmetric convex body in E^n and therefore one may consider the n-dimensional Banach space E whose unit cell S_E is S. We shall show that this space E satisfies the requirements of theorem 1.2. The idea of the proof is the following: Using the assumptions on δ, $y_1, \ldots, y_N$ and the totality property 2° of lemma 1.11, we shall show that for any subspace $G \subset E$ with $\dim G \geqslant 2$ there exist at least n elements $x_{i_1}, \ldots, x_{i_n} \in G \cap Fr\, S = Fr\, S_G$ such that each x_{i_k} belongs only to one of the sets $Fr\, T^{y_{j_k}}$ $(k=1, \ldots, n)$, with $j_k \neq j_l$ for $k \neq l$, i.e. only to one of the sets

$$P^{(\pm) y_{j_k}} = \pm y_{j_k} + \mathscr{D}(y_{j_k}) \qquad (k=1, \ldots, n), \tag{1.46}$$

with $j_k \neq j_l$ for $k \neq l$, where the notation $(\pm)$ means that one has to take only one of these two signs. From this we shall easily deduce that the existence of a subspace $F \subset E$ with $\widehat{(G; F)} = 1$ would contradict the n-independence of the system $y_1, \ldots, y_N$ ensured by 1° of lemma 1.11, which will complete the proof.

We shall proceed in several steps. Let us prove, firstly, that we have

$$S_{E^n} \subset S \subset \frac{1}{\sqrt{1-4\delta^2}} S_{E^n}. \tag{1.47}$$

We shall continue to denote by $\|x\|$ the euclidean norm. Since each strip T^{y_i} contains S_{E^n} (by $\|y_i\|=1$), we have $S \supset S_{E^n}$. To prove the other inclusion in (1.47), take an arbitrary $x \in S$. Then, since $z_1, \ldots, z_N$ is a δ-net for $Fr\, S_{E^n}$, there exists a z_i such that $\left\| \frac{x}{\|x\|} - z_i \right\| < \delta$, whence, by $y_i \in S(z_i, \delta)$,

$$\left\| \frac{x}{\|x\|} - y_i \right\| \leqslant \left\| \frac{x}{\|x\|} - z_i \right\| + \|z_i - y_i\| < 2\delta. \tag{1.48}$$

Since $E=[y_i] \oplus \mathscr{D}(y_i)$, there is a unique decomposition $x=\beta y_i + x_0$, where $x_0 \in \mathscr{D}(y_i)$. Since $x \in S \subset T^{y_i}$, we have $|\beta| \leqslant 1$. Furthermore[1], since $x-\beta y_i = x_0 \perp [y_i]$, we have, taking into account (1.48),

$$\|x_0\| = \|x-\beta y_i\| = \min_{-\infty<t<\infty} \|x - t y_i\| \leqslant \|x - \|x\| y_i\| < 2\delta \|x\|,$$

whence, by $x - x_0 = \beta y_i \perp \mathscr{D}(y_i) \ni x_0$,

$$\begin{aligned} 1 \geqslant |\beta| = \|\beta y_i\| = \|x - x_0\| &= \sqrt{\|x\|^2 - \|x_0\|^2} \geqslant \sqrt{\|x\|^2 - 4\delta^2 \|x\|^2} \\ &= \|x\| \sqrt{1-4\delta^2}, \end{aligned}$$

and thus $\|x\| \leqslant \frac{1}{\sqrt{1-4\delta^2}}$, which, since $x \in S$ has been arbitrary, proves the second inclusion in (1.47).

By (1.45), every $x \in Fr\, S$ belongs to at least one of the sets $P^{(\pm) y_i} = \pm y_i + \mathscr{D}(y_i)$, where the notation $(\pm)$ means that one has to take only one of these two signs. With the aid of (1.47) we shall now show that if $x_1, x_2 \in P^{y_l} \cap Fr\, S$ or if $x_1, x_2 \in P^{-y_l} \cap Fr\, S$, then

$$\|x_1 - x_2\| < 6\delta. \tag{1.49}$$

In fact, since $F=[y_l] \oplus \mathscr{D}(y_l)$, we have unique decompositions $x_1 = \beta_1 y_l + x_0^{(1)}$, $x_2 = \beta_2 y_l + x_0^{(2)}$, where $x_0^{(1)}, x_0^{(2)} \in \mathscr{D}(y_l)$. Since either $x_1, x_2 \in P^{y_l}$ or $x_1, x_2 \in P^{-y_l}$, we have either $\beta_1=\beta_2=1$, or $\beta_1=\beta_2=-1$. Taking into account $x_i - \beta_i y_l = x_0^{(i)} \perp [y_l]$, $x_i \in S$, $|\beta_i|=1$ $(i=1,2)$ and (1.47), we obtain

[1] Throughout this proof, we use the symbol $\perp$ in the sense of the euclidean norm.

$$\|x_0^{(i)}\| = \|x_i - \beta_i y_l\| = \sqrt{\|x_i\|^2 - \|\beta_i y_l\|^2} = \sqrt{\|x_i\|^2 - 1} \leqslant \sqrt{\frac{1}{1-4\delta^2} - 1}$$

$$= \frac{2\delta}{\sqrt{1-4\delta^2}} < 3\delta \qquad (i=1,2)$$

(because $\sqrt{1-4\delta^2} = \sqrt{1 - \frac{1}{16n^6}} > \sqrt{1 - \frac{5}{9}} = \frac{2}{3}$). Consequently, since $\beta_1 = \beta_2$,

$$\|x_1 - x_2\| = \|x_0^{(1)} - x_0^{(2)}\| \leqslant \|x_0^{(1)}\| + \|x_0^{(2)}\| < 6\delta,$$

which proves (1.49).

Now let G be an arbitrary subspace of E with $\dim G = 2$. Then its unit cell $S_G = G \cap S$ is a centrally symmetric convex polygon which contains, by (1.47), a euclidean circumference. Divide an arbitrary quarter of this circumference into n^3 equal parts and denote the intersection points of $Fr\, S$ with the rays starting from the origin and passing through the corresponding division points, except the last one, by $s_1, \dots, s_{n^3}$. Each of these intersection points may belong to several faces $P^{(\pm)y_i}$ of $Fr\, S$, but we shall show that in sufficiently small neighbourhoods of these points one can find at least n other points $x_{i_1}, \dots, x_{i_n}$ of $Fr\, S$ such that each x_{i_k} belongs only to one face (1.46), with $j_k \neq j_l$ for $k \neq l$.

In fact, we have $\|s_i\| \geqslant 1$ and $\sphericalangle s_i 0 s_j = \frac{|i-j|}{n^3} \frac{\pi}{2}$, whence

$$\|s_i - s_j\| \geqslant \|s_i\| \sin \sphericalangle s_i 0 s_j \geqslant \sin \frac{|i-j|}{n^3} \frac{\pi}{2} > \sin \frac{\pi}{2n^3} > \frac{\pi}{4n^3} = 2\pi\delta$$

$$(i, j = 1, \dots, n;\ i \neq j).$$

Consequently, there exists an $\varepsilon > 0$ such that for every $\tilde{s}_i \in S(s_i, \varepsilon)$ and $\tilde{s}_j \in S(s_j, \varepsilon)$, $i \neq j$, we have

$$\|\tilde{s}_i - \tilde{s}_j\| > 6\delta, \tag{1.50}$$

$$\sphericalangle \tilde{s}_i 0 \tilde{s}_j < \frac{\pi}{2}. \tag{1.51}$$

Consider now the first $\frac{n(n-1)}{2}$ elements $s_1, \dots, s_{\frac{n(n-1)}{2}}$. We claim that there exists an element $x_{i_1} \in Fr\, S_G$, $x_{i_1} \in S(s_{i_1}, \varepsilon)$, such that $x_{i_1} \in P^{(\pm)y_j}$ only for one value $j = j_1$ of the index j. Indeed, assume the contrary, i.e. that each $x_i \in S(s_i, \varepsilon) \cap Fr\, S_G$ $\left(i = 1, \dots, \frac{n(n-1)}{2}\right)$ belongs to at least two distinct faces $P^{(\pm)y_j}$, i.e. to an intersection

$$P^{(\pm)y_p} \cap P^{(\pm)y_q}. \tag{1.52}$$

Then, since for each i the set $S(s_i, \varepsilon) \cap Fr\, S_G$ is infinite and the number of all possible intersections (1.52) is finite, it follows that for each $i = 1, \ldots, \frac{n(n-1)}{2}$ there exist a pair of indices $p_i \neq q_i$ and a pair of signs such that the $(n-2)$-dimensional linear manifold

$$P^{(\pm)y_{p_i}} \cap P^{(\pm)y_{q_i}} \tag{1.53}$$

contains at least two distinct elements $s_i^1, s_i^2 \in S(s_i, \varepsilon) \cap Fr\, S_G$. Let us show that here all indices $p_1, \ldots, p_{\frac{n(n-1)}{2}}, q_1, \ldots, q_{\frac{n(n-1)}{2}}$ are distinct. If $q_i = q_j$ for $i \neq j$, we have to consider two cases:

a) $s_i^1 \in P^{(\pm)y_{p_i}} \cap P^{y_{q_i}}$, $s_j^1 \in P^{(\pm)y_{p_i}} \cap P^{y_{q_i}}$, whence $s_i^1, s_j^1 \in P^{y_{q_i}}$. In this case, by (1.49) we must have $\|s_i^1 - s_j^1\| < 6\delta$, and by (1.50) we must have $\|s_i^1 - s_j^1\| > 6\delta$, which is impossible.

b) $s_i^1 \in P^{(\pm)y_{p_i}} \cap P^{y_{q_i}}$, $s_j^1 \in P^{(\pm)y_{p_i}} \cap P^{-y_{q_i}}$, whence $s_i^1 \in P^{y_{q_i}}$, $s_j^1 \in P^{-y_{q_i}}$. In this case divide the angle $\sphericalangle s_i^1 0 s_j^1$ in the plane G with the aid of a ray $0s$ from the origin parallel to $P^{y_{q_i}} \cap G$ and to $P^{-y_{q_i}} \cap G$, into two angles $\alpha_1 = \sphericalangle s_i^1 0 s$, $\alpha_2 = \sphericalangle s 0 s_j^1$. We have then, by (1.47),

$$\sin \alpha_1 = \frac{1}{\|s_i^1\|} \geqslant \sqrt{1 - 4\delta^2} = \sqrt{1 - \frac{1}{16n^6}} > \frac{\sqrt{2}}{2},$$

and similarly, $\sin \alpha_2 = \frac{1}{\|s_j^1\|} > \frac{\sqrt{2}}{2}$, whence $\alpha_1, \alpha_2 > \frac{\pi}{4}$. Hence

$$\sphericalangle s_i^1 0 s_j^1 = \alpha_1 + \alpha_2 > \frac{\pi}{2},$$

which contradicts (1.51).

We have thus proved that $q_i \neq q_j$ for $i \neq j$ in (1.53). Similarly, we have $p_i \neq p_j$ for $i \neq j$. Finally, from the same arguments it follows that we also have $p_i \neq q_j$ for $i \neq j$, and thus all indices $p_1, \ldots, p_{\frac{n(n-1)}{2}}, q_1, \ldots, q_{\frac{n(n-1)}{2}}$ in (1.53) are distinct.

Thus, the system of $\frac{n(n-1)}{2}$ linear manifolds (1.53) is not total (since the intersection of G with each of these linear manifolds contains at least two distinct elements s_i^1, s_i^2), whence the system of $\frac{n(n-1)}{2}$ linear subspaces

$$\mathscr{D}(y_{p_1}, y_{q_1}), \ldots, \mathscr{D}\left(y_{p_{\frac{n(n-1)}{2}}}, y_{q_{\frac{n(n-1)}{2}}}\right)$$

obtained by translating these linear manifolds into the origin, is also not total, which contradicts property 2° of lemma 1.11. This proves our assertion that for the first $\frac{n(n-1)}{2}$ elements $s_1, \ldots, s_{\frac{n(n-1)}{2}}$ above there exists an element $x_{i_1} \in S(s_{i_1}, \varepsilon) \cap Fr\, S_G$ such that $x_{i_1} \in P^{(\pm) y_j}$ only for one value $j = j_1$ of the index j.

By a similar argument, for the second group of $\frac{n(n-1)}{2}$ elements $s_{\frac{n(n-1)}{2}+1}, \ldots, s_{n(n-1)}$ there exists an element $x_{i_2} \in S(s_{i_2}, \varepsilon) \cap Fr\, S_G$ which belongs only to one $P^{(\pm) y_{j_2}}$. Since $n^3 > n\frac{n(n-1)}{2}$, continuing in this manner we obtain at least n elements $x_{i_1}, \ldots, x_{i_n} \in Fr\, S_G$, $x_{i_k} \in S(s_{i_k}, \varepsilon)$ $(k = 1, \ldots, n)$, such that each x_{i_k} belongs only to one $P^{(\pm) y_{j_k}}$ $(k = 1, \ldots, n)$. Here all indices $j_1, \ldots, j_n$ are distinct. Indeed, if $j_k = j_l$ for $k \neq l$, then $x_{i_k}, x_{i_l} \in P^{(\pm) y_{j_k}}$, which, as we have seen above, is impossible by (1.47) $-$(1.51).

This being established, the conclusion of the theorem follows easily. Indeed, assume a contrario that there exists a subspace $F \subset E$ with $\widehat{(G; F)} = 1$ and let $x \in F \setminus \{0\}$ be arbitrary. Then, denoting by $|||z|||$ the norm in E, we have

$$|||y + tx||| \geqslant 1 \qquad (y \in G, \ |||y||| = 1, \ -\infty < t < \infty),$$

whence, in particular, for $y = x_{i_k}$ we obtain

$$|||x_{i_k} + tx||| \geqslant 1 \qquad (-\infty < t < \infty), \tag{1.54}$$

and thus each line $L_k = x_{i_k} + \{tx \mid -\infty < t < \infty\}$ $(k = 1, \ldots, n)$ supports the cell S. Since by the above each x_{i_k} belongs only to one face $P^{(\pm) y_{j_k}}$ of S, it follows that there is also a segment $x_{i_k} + \{tx \mid -t_k \leqslant t \leqslant t_k\}$ of L which belongs to that face $P^{(\pm) y_{j_k}}$. Hence, taking into account that by the definition of $P^{(\pm) y_{j_k}}$ we have $(y_{j_k}, y) = 0$ for all $y \in P^{(\pm) y_{j_k}}$, and considering any $t \neq 0$ with $-t_k \leqslant t \leqslant t_k$, we get

$$(y_{j_k}, x) = \frac{1}{t}\left[(y_{j_k}, x_{i_k} + tx) - (y_{j_k}, x_{i_k})\right] = 0 \qquad (k = 1, \ldots, n), \tag{1.55}$$

which, since all j_k are distinct, contradicts the n-independence of the system $y_1, \ldots, y_N$, ensured by 1° of lemma 1.11. This proves that for every couple of subspaces $G, F \subset E$ with $\dim G = 2$, whence also for any $G, F \subset E$ with $\dim G \geqslant 2$, we have (1.43).

Assume now that a subspace E_0 of E with $\dim E_0 = m \geqslant 3$ has a monotone basis $x_1, \ldots, x_m$. Then for $G = [x_1, x_2]$, $F = [x_3]$ we have $(\widehat{G; F}) = 1$, which contradicts (1.43). Finally, the fact that the space $E_n = E$ satisfies (1.6) follows from $\dim E < \infty$ with a similar argument to that used in example 1.1. This completes the proof of theorem 1.2.

We observe that the above proof of theorem 1.2 is, essentially, an extension of the previously given proof of theorem 1.1. In fact, for $n=3$, in that proof of theorem 1.1 we have shown that if G is any plane through the origin, $G \cap S$ intersects the interiors of at least 6 faces of S, or, equivalently, the interiors of at least $3 = n$ pairwise non-symmetric faces of S (existence of $x_{i_1}, x_{i_2}, x_{i_3}$!) whence, by condition (iii) (which corresponds to the 3-independence of $y_1, \ldots, y_N$, since these y_i are the normals to the faces $P^{(\pm) y_i}$ of S), it has followed the non-existence of a projection u of E onto G with $\|u\| = 1$ (which, since $\dim E = 3$ and $\dim G = 2$, is equivalent to the non-existence of a subspace $F \subset E$ with $\dim F = 1$, such that $(\widehat{G; F}) = 1$). There has only been a difference in the construction of S in the two proofs (condition (iv) versus the totality of $\mathscr{D}(y_{p_1}, y_{q_1})$, $\mathscr{D}(y_{p_2}, y_{q_2})$, $\mathscr{D}(y_{p_3}, y_{q_3})$).

Since for any n-dimensional Banach space E and any $k = 1, \ldots, n-1$ we have, obviously,

$$\frac{1}{\Gamma(E)} = \inf_{\{x_j\}} \sup_{1 \leqslant i \leqslant n} \|s_i\| \geqslant \inf_{\substack{G \prec E \\ \dim G = k}} \inf_{u \in \mathscr{P}(E, G)} \|u\| = \phi_k(E), \tag{1.56}$$

it follows that for n-dimensional space E constructed in theorem 1 above we have

$$0 < \Gamma(E) < 1. \tag{1.57}$$

This shows that for the numbers $\Gamma(n)$ defined by

$$\Gamma(n) = \inf_{B_n} \Gamma(B_n), \tag{1.58}$$

where the inf is taken over all Banach spaces B_n of dimension n, we have $\Gamma(n) < 1$ $(n = 3, 4, \ldots)$. It is natural to ask how large is the number $\sup_{3 \leqslant n < \infty} \Gamma(n) = \sup_{3 \leqslant n < \infty} \inf_{B_n} \sup_{[x_j] = B_n} \gamma_{\{x_j\}}$. If this number would be 1, then for each $\varepsilon > 0$ every Banach space B of a sufficiently large finite dimension $n = n(\varepsilon)$ would have a basis $\{x_j\}$ such that $\gamma_{\{x_j\}} > 1 - \varepsilon$. However, we shall show that this is not the case; on the contrary, we have $\sup_{3 \leqslant n < \infty} \Gamma(n) < 1$, i.e. there is a constant $C < 1$, independent of n (namely, $C = \sup_{3 \leqslant n < \infty} \Gamma(n)$), such that for every positive integer n there exists a Banach space B_n of dimension n, with the property that for every basis $\{x_j\}_{j=1}^n$ of B_n we have $\gamma_{\{x_j\}} \leqslant C$. This fact might perhaps also be of some interest for the problem of constructing a separable Banach space having no basis.

For an arbitrary n-dimensional Banach space E put

$$U(E) = U_{n-1}(E) = \sup_{\{x_j\}_{j=1}^n} (\widehat{[x_1, \dots, x_{n-1}]; [x_n]}) \tag{1.59}$$

where the sup is taken over all bases $\{x_j\}_{j=1}^n$ of E. We have then, for every n-dimensional Banach space E,

$$\Gamma(E) = \sup_{\{x_j\}} \gamma_{\{x_j\}} = \sup_{\{x_j\}} \inf_{1 \leqslant i \leqslant n} (\widehat{[x_1, \dots, x_i]; [x_{i+1}, \dots, x_n]}) \leqslant U(E). \tag{1.60}$$

On the other hand, since for any complementary subspaces G, F of E we have $(\widehat{G; F}) = \frac{1}{\|u\|}$, where $u \in \mathscr{P}(E, G)$, $u(F) = 0$,[1] we can write

$$U(E) = \sup_{\substack{G \prec E \\ \dim G = n-1}} \sup_{x \in E} (\widehat{G; [x]}) = \frac{1}{\inf\limits_{\substack{G \prec E \\ \dim G = n-1}} \inf\limits_{u \in \mathscr{P}(E,G)} \|u\|} = \frac{1}{\phi_{n-1}(E)}, \tag{1.61}$$

and consequently, for the n-dimensional Banach space $E_n = E$ constructed in theorem 1.2, we have

$$U(E_n) < 1. \tag{1.62}$$

Define now the number $U(n)$ by

$$U(n) = \inf_{B_n} U(B_n), \tag{1.63}$$

where the inf is taken over all Banach spaces B of dimension n. Then, by (1.60) and (1.62), we have

$$\Gamma(n) \leqslant U(n) < 1 \qquad (n = 3, 4, \dots) \tag{1.64}$$

and we shall prove below that $\sup\limits_{3 \leqslant n < \infty} U(n) = C < 1$, which, of course, will also prove the desired result on $\sup\limits_{3 \leqslant n < \infty} \Gamma(n)$.

Lemma 1.12. *Let B_{n_1}, B_{n_2} be two finite dimensional Banach spaces and let $B_{n_1} \times B_{n_2}$ be their cartesian product endowed with the norm $\|\{z_1, z_2\}\| = \sqrt{\|z_1\|^2 + \|z_2\|^2}$. Then we have*

$$U(B_{n_1} \times B_{n_2}) \leqslant \max(U(B_{n_1}), U(B_{n_2})). \tag{1.65}$$

---

[1] Indeed, $(\widehat{G;F}) = \inf\limits_{\substack{y \in G \\ z \in F}} \frac{\|y+z\|}{\|y\|} = \frac{1}{\sup\limits_{\substack{y \in G \\ z \in F}} \frac{\|y\|}{\|y+z\|}} = \frac{1}{\|u\|}$.

Proof. Let G be an arbitrary hyperplane in $B_{n_1} \times B_{n_2}$ and let $z = \{z_1, z_2\} \in B_{n_1} \times B_{n_2}$. Put

$$G_1 = \{x \in B_{n_1} \mid \{x, 0\} \in G\}, \qquad G_2 = \{y \in B_{n_2} \mid \{0, y\} \in G\}. \tag{1.66}$$

Then, since the image of G_i by the canonical isometrical embedding $B_{n_i} \to B_{n_1} \times B_{n_2}$ is $G \cap (B_{n_1} \times \{0\})$, respectively $G \cap (\{0\} \times B_{n_2})$, the co-dimension of G_i in B_{n_i} is[1] at most 1, whence, by the definition of $U(B_{n_i})$,

$$(\widehat{G_i; [z_i]}) \leqslant U(B_{n_i}) \qquad (i = 1, 2). \tag{1.67}$$

Now, since $\dim G_i$, $\dim [z_i] < \infty$, the value $\operatorname{dist}(\sigma_{G_i}, [z_i]) = (\widehat{G_i; [z_i]})$ is attained[2] for a suitable pair $y_i \in \sigma_{G_i}$, $\alpha_i z_i \in [z_i]$, and thus

$$\|y_i - \alpha_i z_i\| \leqslant U(B_{n_i}) \qquad (i = 1, 2),$$

whence for $x_i = -\frac{1}{\alpha_i} y_i \in G_i$ $(i = 1, 2)$ we obtain

$$\|x_i + z_i\| \leqslant U(B_{n_i}) \|x_i\| \qquad (i = 1, 2). \tag{1.68}$$

Consequently, for $x = \{x_1, x_2\} \in B_{n_1} \times B_{n_2}$ we have

$$\begin{aligned} \|x + z\| &= \|\{x_1 + z_1, x_2 + z_2\}\| = (\|x_1 + z_1\|^2 + \|x_2 + z_2\|^2)^{\frac{1}{2}} \\ &\leqslant (U(B_{n_1})^2 \|x_1\|^2 + U(B_{n_2})^2 \|x_2\|^2)^{\frac{1}{2}} \\ &\leqslant \max(U(B_{n_1}), U(B_{n_2}))(\|x_1\|^2 + \|x_2\|^2)^{\frac{1}{2}} = \max(U(B_{n_1}), U(B_{n_2})) \|x\|, \end{aligned}$$

whence

$$(\widehat{G; [z]}) = \operatorname{dist}(\sigma_G, [z]) \leqslant \left\| \frac{x}{\|x\|} + \frac{1}{\|x\|} z \right\| \leqslant \max(U(B_{n_1}), U(B_{n_2})),$$

whence, since the hyperplane G in $B_{n_1} \times B_{n_2}$ and the element $z \in B_{n_1} \times B_{n_2}$ were arbitrary, we infer

$$U(B_{n_1} \times B_{n_2}) = \sup_{\substack{G \prec B_{n1} \times B_{n2} \\ \operatorname{codim} G = 1}} \ \sup_{z \in B_{n1} \times B_{n2}} (\widehat{G; [z]}) \leqslant \max(U(B_{n_1}), U(B_{n_2})),$$

which completes the proof of lemma 1.12.

---

[1] We use here the following obvious remark: if $H = \{x \in E \mid f(x) = 0\}$ is a hyperplane in a normed linear space E, then for any subspace F of E the intersection $F \cap H = \{x \in F \mid f|_F(x) = 0\}$ either coincides with F (if $f|_F = 0$), or is a hyperplane in F (if $f|_F \neq 0$).

[2] See e.g. [246], Appendix I, theorem 2.3.

Theorem 1.3. *We have*

$$U(n_1+n_2)\leqslant\max(U(n_1),U(n_2))\qquad(n_1,n_2=3,4,\ldots).\tag{1.69}$$

Consequently,

$$\sup_{3\leqslant n<\infty}\Gamma(n)\leqslant\sup_{3\leqslant n<\infty}U(n)\leqslant\max(U(3),U(4),U(5))<1.\tag{1.70}$$

Proof. Let $n_1,n_2\geqslant 3$ and $\varepsilon>0$ be arbitrary. Then, by the definition (1.63) of $U(n_i)$, there exist Banach spaces $B_{n_1}^{\varepsilon}, B_{n_2}^{\varepsilon}$ of dimension n_1 and n_2 respectively, such that

$$U(B_{n_i}^{\varepsilon})\leqslant U(n_i)+\varepsilon\qquad(i=1,2).\tag{1.71}$$

Consequently, by lemma 1.12,

$$\begin{aligned}U(n_1+n_2)&=\inf_{B_{n_1+n_2}}U(B_{n_1+n_2})\leqslant U(B_{n_1}^{\varepsilon}\times B_{n_2}^{\varepsilon})\\&\leqslant\max(U(B_{n_1}^{\varepsilon}),U(B_{n_2}^{\varepsilon}))\leqslant\max(U(n_1),U(n_2))+\varepsilon,\end{aligned}$$

whence, since $\varepsilon>0$ and $n_1,n_2\geqslant 3$ have been arbitrary, we obtain (1.69).

The second inequality in (1.70) follows now by induction from (1.69) and from the obvious remark that every integer $n\geqslant 6$ can be written in one of the forms $n=3k+3$, $n=3k+4$ or $n=3k+5$, where k is a positive integer. On the other hand, the first and third inequalities in (1.70) are established by (1.64), which completes the proof of theorem 1.3.

Let us pass now to the infinite dimensional case. We shall show that the problem of existence of monotone bases in every infinite dimensional Banach space with a basis, has also a negative answer.

For a function $x\in C([0,1])$ we shall use the notation

$$M_x=\{t\in[0,1]\mid |x(t)|=\|x\|\}.\tag{1.72}$$

In the sequel we shall consider only real-valued functions.

Lemma 1.13. *For every $x\in C([0,1])$ and $\varepsilon>0$ there exists a $\delta=\delta(\varepsilon,x)>0$ such that*

$$M_{x+y}\subset\bigcup_{t\in M_x}(t-\varepsilon,t+\varepsilon)\cap[0,1]\qquad(y\in C([0,1]),\ \|y\|<\delta).\tag{1.73}$$

Proof. Let us denote the set in the right side of (1.73) by $M_x(\varepsilon)$. We claim that

$$\eta=\|x\|-\max_{t\in\complement M_x(\varepsilon)}|x(t)|>0,\tag{1.74}$$

where $\complement M_x(\varepsilon)=[0,1]\setminus M_x(\varepsilon)$. Indeed, since $\complement M_x(\varepsilon)$ is closed and bounded, whence compact, and since $x|_{\complement M_x(\varepsilon)}$ is continuous, $\max_{t\in\complement M_x(\varepsilon)}|x(t)|$ is attained in at least one point $t_0\in\complement M_x(\varepsilon)$. Now, if $\eta=0$, then $|x(t_0)| = \max_{t\in\complement M_x(\varepsilon)}|x(t)|=\|x\|$, whence $t_0\in M_x$, a contradiction which proves that $\eta>0$, i.e. (1.74).

Now take $\delta=\frac{\eta}{2}$. Then for every $y\in C([0,1])$ with $\|y\|<\delta=\frac{\eta}{2}$ we have, by (1.74),

$$\begin{aligned}\max_{t\in \complement M_x(\varepsilon)}|x(t)+y(t)| &\leqslant \max_{t\in \complement M_x(\varepsilon)}|x(t)|+\max_{t\in[0,1]}|y(t)|\\ &<\|x\|-\eta+\frac{\eta}{2}=\|x\|-\frac{\eta}{2}=\|x\|-\delta\\ &<\|x\|-\|y\|\leqslant\|x+y\|,\end{aligned}$$

whence we get $M_{x+y}\subset M_x(\varepsilon)$, i.e. (1.73), which completes the proof of lemma 1.13.

Lemma 1.14. *Let $x\in C([0,1])$ be differentiable on $[0,1]$ and let*

$$0<t_1=\min_{t\in M_x}t\leqslant t_2=\max_{t\in M_x}t<1. \tag{1.75}$$

Assume, furthermore, that there exists a differentiable function $y\in C([0,1])$ such that on some interval $(a,b)\supset M_x$ we have $y(t)>0$ and $y'(t)$ has constant sign. Then there exists an $\alpha_0>0$ such that for $0<\alpha<\alpha_0$ we have

$$M_{x+\alpha y}\subset(t_2,b) \quad \textit{if} \quad y'(t)>0 \quad \textit{on} \quad (a,b) \quad \textit{and} \quad x(t_2)>0, \tag{1.76}$$

$$M_{x+\alpha y}\subset(a,t_1) \quad \textit{if} \quad y'(t)<0 \quad \textit{on} \quad (a,b) \quad \textit{and} \quad x(t_1)>0. \tag{1.77}$$

Proof. Let us consider the case when $y'(t)>0$ on (a,b) and $x(t_2)>0$. Since $M_x\subset(a,b)$, by lemma 1.13 there exists an $\alpha_0>0$ such that for $0<\alpha<\alpha_0$ we have $M_{x+\alpha y}\subset(a,b)$. Now let $t_0\in(a,t_2)$ be arbitrary. Then, by $t_2\in M_x$ we have

$$x(t_2)\geqslant x(t_0)$$

and by $t_2>t_0, y'(t)>0$ $(t\in(a,b))$ we have

$$y(t_2)>y(t_0),$$

whence for $0<\alpha(<\alpha_0)$,

$$x(t_2)+\alpha y(t_2)>x(t_0)+\alpha y(t_0).$$

Since $t_0\in(a,t_2)$ has been arbitrary, it follows that

$$M_{x+\alpha y}\subset[t_2,b).$$

However, $t_2\notin M_{x+\alpha y}$, since by $t_2\in M_x$ and $\alpha>0, y'(t_2)>0$ we have

$$x'(t_2)+\alpha y'(t_2)=\alpha y'(t_2)\neq 0.$$

Thus $M_{x+\alpha y}\subset(t_2,b)$, which proves (1.76). The case when $y'(t)<0$ on (a,b) and $x(t_1)>0$ can be treated in a similar way, which completes the proof of lemma 1.14.

Lemma 1.15. *Let* $x, y \in C([0,1])$. *We have* $(\widehat{[x];[y]}) < 1$ *if and only if either*

$$\operatorname{sign} x(t) = \operatorname{sign} y(t) \qquad (t \in M_x) \tag{1.78}$$

or

$$\operatorname{sign} x(t) = -\operatorname{sign} y(t) \qquad (t \in M_x). \tag{1.79}$$

Proof. Necessity. Assume that we have neither (1.78) nor (1.79). Then two cases are possible:

a) There exists a $t_0 \in M_x$ such that $y(t_0) = 0$. In this case for every scalar α we have

$$\|x + \alpha y\| \geqslant |x(t_0) + \alpha y(t_0)| = |x(t_0)| = \|x\|.$$

b) There exist $t_1, t_2 \in M_x$ such that $\operatorname{sign} x(t_1) = \operatorname{sign} y(t_1)$, $\operatorname{sign} x(t_2) = -\operatorname{sign} y(t_2)$. In this case for every $\alpha \geqslant 0$ we have

$$\|x + \alpha y\| \geqslant |x(t_1) + \alpha y(t_1)| \geqslant |x(t_1)| = \|x\|,$$

and for every $\alpha \leqslant 0$ we have

$$\|x + \alpha y\| \geqslant |x(t_2) + \alpha y(t_2)| \geqslant |x(t_2)| = \|x\|.$$

Thus, in all cases, for every scalar α we have $\|x + \alpha y\| \geqslant \|x\|$, whence $(\widehat{[x];[y]}) = 1$.

Sufficiency. We may assume, without loss of generality, that $\|x\| = 1$. Assume that we have (1.78). Then, since M_x is compact and $|y|\,|_{M_x}$ is continuous,

$$\min_{t \in M_x} |y(t)| = \mu > 0. \tag{1.80}$$

For $\delta > 0$ denote, as in the proof of lemma 1.13,

$$M_x(\delta) = \bigcup_{t \in M_x} (t - \delta, t + \delta) \cap [0,1]. \tag{1.81}$$

Since M_x is closed and x, y are continuous, there exists a $\delta > 0$ such that

$$\inf_{t \in M_x(\delta)} |y(t)| > \frac{\mu}{2}, \tag{1.82}$$

$$\operatorname{sign} x(t) = \operatorname{sign} y(t) \qquad (t \in M_x(\delta)). \tag{1.83}$$

Let

$$\lambda = \max_{t \in \complement M_x(\delta)} |x(t)|, \tag{1.84}$$

$$\alpha = \frac{1 - \lambda}{2\|y\|}. \tag{1.85}$$

By (1.74) we have $0 \leqslant \lambda < \|x\| = 1$, whence $\alpha > 0$. Let us estimate $|x(t) - \alpha y(t)|$.

If $t \in M_x(\delta)$, then by $\alpha > 0$, (1.83), (1.82) and (1.85) we have

$$|x(t)-\alpha y(t)| = | \, |x(t)| - \alpha |y(t)| \, | \leqslant \max\{|x(t)|-\alpha|y(t)|, -|x(t)|+\alpha|y(t)|\}$$
$$\leqslant \max\left\{1-\alpha\frac{\mu}{2}, \alpha\|y\|\right\} = \max\left\{1-\frac{\alpha\mu}{2}, \frac{1-\lambda}{2}\right\},$$

while if $t \in \complement M_x(\delta)$, then by $\alpha > 0$, (1.84) and (1.85) we have

$$|x(t)-\alpha y(t)| \leqslant |x(t)| + \alpha |y(t)| \leqslant \lambda + \frac{1-\lambda}{2\|y\|}\|y\| = \frac{1+\lambda}{2}.$$

Consequently, since $\alpha > 0$, $\mu > 0$ and $0 \leqslant \lambda < 1$, we get

$$\|x-\alpha y\| = \max_{t\in[0,1]} |x(t)-\alpha y(t)| \leqslant \max\left\{1-\frac{\alpha\mu}{2}, \frac{1-\lambda}{2}, \frac{1+\lambda}{2}\right\} < 1 = \|x\|,$$

whence $(\widehat{[x];[y]}) < 1$, which proves the sufficiency of condition (1.78).

Assume now that we have (1.79). Then for the function $-y \in C([0,1])$ we have (1.78), whence, by the above,

$$(\widehat{[x];[y]}) = (\widehat{[x];[-y]}) < 1,$$

which completes the proof of lemma 1.15.

Theorem 1.4. *Let $\{x_n\} \subset C([0,1])$ be an infinite sequence of diferentiable real-valued functions on $[0,1]$, such that*

1°. The set of all stationary points[1] *of any linear combination $\sum_{i=1}^{n} \alpha_i x_i(t)$ has no limit points in $(0,1)$ and*

2°. There exists a function $y \in [x_n]$ such that both $y(t)$ and $y'(t)$ have constant sign on $(0,1)$.

Then we have

$$\gamma_{\{x_n\}} < 1. \tag{1.86}$$

Consequently, if a closed linear subspace E_0 of $C([0,1])$, with $\dim E_0 = \infty$, consists of non-constant analytic functions on $(0,1)$ and contains a function y such that both $y(t)$ and $y'(t)$ have constant sign on $(0,1)$, then E_0 has no monotone basis.

Proof. Since the system of two homogeneous linear equations

$$\alpha_1 x_1(0) + \alpha_2 x_2(0) + \alpha_3 x_3(0) = 0,$$
$$\alpha_1 x_1(1) + \alpha_2 x_2(1) + \alpha_3 x_3(1) = 0,$$

[1] We recall that a point t is said to be a stationary point of a differentiable function z if $z'(t) = 0$.

always has a non-trivial solution $\alpha_1, \alpha_2, \alpha_3$, there exists a function $z_1 = \sum_{i=1}^{3} \alpha_i x_i \in [x_1, x_2, x_3] \setminus \{0\}$ such that $z_1(0) = z_1(1) = 0$, whence

$$0 < t_0 = \max_{t \in M_{z_1}} t < 1.$$

Assume now that $z_1(t_0) > 0$ (since otherwise one can take the function $-z_1$ instead of z_1) and that for the function $y \in [x_n]$ of condition 2° we have $y(t) > 0$, $y'(t) > 0$ on $(0, 1)$ (the other cases can be treated similarly).

We claim that there exists a τ with $t_0 < \tau < 1$, such that

$$z_1'(t) < 0 \qquad (t \in (t_0, \tau)). \tag{1.87}$$

Indeed, since $z_1(t_0) = \max_{t \in [0,1]} |z_1(t)|$, in every interval (t_0, t_1) there must exist a $t_2^{(0)} \in (t_0, t_1)$ with $z_1'(t_2^{(0)}) < 0$. Assume now that there exists no τ such that we have (1.87). Then in every interval $(t_0, t_2^{(0)})$ there exists a $t_1^{(0)} \in (t_0, t_2^{(0)})$ such that $z_1'(t_1^{(0)}) \geqslant 0$. Hence, by a well known theorem of Darboux[1], in every $(t_1^{(0)}, t_2^{(0)})$, where $t_1^{(0)}, t_2^{(0)}$ are as above, there exists a $t \in (t_1^{(0)}, t_2^{(0)})$ such that $z_1'(t) = 0$, which contradicts the assumption 1°. Thus we have (1.87).

Since z_1 is continuous and $z_1(t_0) > 0$, we may also assume that

$$z_1(t) > 0 \qquad (t \in (t_0, \tau)). \tag{1.88}$$

Since $y(t), y'(t) > 0$ on $(0, 1)$ and $z_1(t_0) > 0$, by lemma 1.14 (with $t_2 = t_0$, $b = \tau$) we have $M_{z_1 + \alpha y} \subset (t_0, \tau)$ for sufficiently small $\alpha > 0$. Furthermore, since $z_1, y \in [x_k]$, for every $\delta > 0$ there exist an integer $n = n(\delta)$ and a function $z_2 = z_2^\delta \in [x_1, \ldots, x_n]$ such that $\|z_2 - (z_1 + \alpha y)\| < \delta$; obviously, we may assume that $n \geqslant 3$. Since $z_2 = (z_1 + \alpha y) + [z_2 - (z_1 + \alpha y)]$, by lemma 1.13 one can take $\varepsilon > 0$ and $\delta = \delta(\varepsilon, z_1 + \alpha y) > 0$ so small that

$$M_{z_2} \subset \bigcup_{t \in M_{z_1 + \alpha y}} (t - \varepsilon, t + \varepsilon) \subset (t_0, \tau). \tag{1.89}$$

Put

$$t_0' = \min_{t \in M_{z_2}} t. \tag{1.90}$$

Then, by (1.89) we have $t_0 < t_0' < \tau$, whence, by $\alpha > 0$, (1.88) and $y(t) > 0$ $(t \in (0, 1))$, we get $z_1(t_0') + \alpha y(t_0') > 0$, and therefore, taking δ sufficiently small, we may assume that

$$z_2(t_0') > 0. \tag{1.91}$$

---

[1] See e. g. G. Fichtengolz [58], p. 283.

Now let $x \in [x_{n+1}, \ldots, x_m]$ $(n<m)$ be arbitrary. Then, by condition 1° and the theorem of Rolle, we have $\operatorname{sign} x(t) = \text{const} \neq 0$ on some interval (τ_0, t_0'), where $\tau_0 \in (t_0, t_0')$ (indeed, otherwise $x(t)$ would have an increasing infinite sequence of zeros in (t_0, t_0') and between any two of them there would exist a zero of $x'(t)$, contradicting condition 1°). By (1.90), (1.89), (1.91), (1.88), (1.87) and lemma 1.14, the function

$$z_3 = z_2 + \beta z_1 \in [x_1, \ldots, x_n] \tag{1.92}$$

satisfies $M_{z_3} \subset (\tau_0, t_0')$ for sufficiently small $\beta > 0$. Furthermore, by the continuity of z_2 and by (1.91) we may assume that

$$z_2(t) > 0 \qquad (t \in (\tau_0, t_0')),$$

which, together with $\beta > 0, (\tau_0, t_0') \subset (t_0, \tau)$ and (1.88), implies

$$z_3(t) > 0 \qquad (t \in M_{z_3}). \tag{1.93}$$

Hence, since $\operatorname{sign} x(t) = \text{const} \neq 0$ on $M_{z_3} \subset (\tau_0, t_0')$, by virtue of lemma 1.15 we obtain

$$\widehat{([z_3]; [x])} < 1. \tag{1.94}$$

Consequently, by (1.92) and $x \in [x_{n+1}, \ldots, x_m]$,

$$\gamma_{\{x_n\}} \leqslant \widehat{([x_1, \ldots, x_n]; [x_{n+1}, \ldots, x_m])} \leqslant \widehat{([z_3]; [x])} < 1,$$

i. e. we have (1.86).

Assume now that E_0 is a closed linear subspace of $C([0,1])$, with $\dim E_0 = \infty$, consisting of non-constant analytic functions on $(0,1)$ and containing a function y such that both $y(t)$ and $y'(t)$ have constant sign on $(0,1)$, and let $\{x_n\}$ be an arbitrary sequence in E_0, such that $[x_n] = E_0$. Then, since every linear combination $\sum_{i=1}^{n} \alpha_i x_i(t)$ is an analytic function, condition 1° is satisfied; in fact, otherwise $\sum_{i=1}^{n} \alpha_i x_i'(t)$ would be 0 on a convergent sequence in $[0,1]$, whence, by the uniqueness theorem for analytic functions, it would be identically 0 on $[0,1]$, contradicting the hypothesis that $\sum_{i=1}^{n} \alpha_i x_i(t)$ is non-constant on $[0,1]$. Furthermore, since $[x_n] = E_0$, condition 2° is also satisfied. Hence, by (1.86), $\{x_n\}$ is not a monotone basis of E_0, which completes the proof of theorem 1.4.

We shall show now that there exist subspaces E_0 of $C([0,1])$, even with a basis, which satisfy the conditions of theorem 1.4, and thus there exist infinite dimensional Banach spaces E_0 with a basis, having no monotone basis.

Lemma 1.16. *Let* $x, y \in C([0,1])$ *with* $\|x\| = \|y\| = 1$ *and* $\varepsilon > 0$. *If there exists a* $t_0 \in [0,1]$ *such that*

$$|x(t_0)| \geqslant 1-\varepsilon, \qquad |y(t_0)| \leqslant \varepsilon, \tag{1.95}$$

then for all scalars α *we have*

$$\|x+\alpha y\| \geqslant 1-3\varepsilon = (1-3\varepsilon)\|x\|. \tag{1.96}$$

Proof. If $|\alpha| \leqslant 2$, we have

$$\begin{aligned}\|x+\alpha y\| &\geqslant |x(t_0)+\alpha y(t_0)| \geqslant \big|\,|x(t_0)| - |y(t_0)|\,\big| \geqslant 1-\varepsilon-|\alpha|\varepsilon \\ &= 1-\varepsilon(1+|\alpha|) \geqslant 1-3\varepsilon,\end{aligned}$$

while if $|\alpha| \geqslant 2$, we have

$$\|x+\alpha y\| \geqslant \big|\,\|x\| - |\alpha|\,\|y\|\,\big| = \big|1-|\alpha|\,\big| \geqslant 1 > 1-3\varepsilon.$$

With the aid of this lemma, we prove the following proposition on selection of basic sequences in $C([0,1])$:

Proposition 1.2. *Let* $\{x_n\} \subset C([0,1])$, $\|x_n\| = 1$ $(n=1,2,\ldots)$. *If for every couple* $\varepsilon > 0$, $\delta > 0$ *there exists a positive integer* $N = N(\varepsilon, \delta)$ *such that for each* $n > N$ *the set* $A(x_n, \varepsilon) = \{t \in [0,1] \mid |x_n(t)| < \varepsilon\}$ *be a* δ*-net for* $[0,1]$, *then there exists an infinite subsequence* $\{x_{n_k}\}$ *of* $\{x_n\}$, *which is a basic sequence.*

Proof. Let $\{\beta_n\}$ be an arbitrary sequence of numbers such that $0 < \beta_n < 1$ $(n=1,2,\ldots)$, $\prod_{i=1}^{\infty} \beta_i > 0$. Put

$$\varepsilon_n = \frac{1-\beta_n}{3} \qquad (n=1,2,\ldots). \tag{1.97}$$

Take $x_{n_1} = x_1$. Since this function is uniformly continuous on $[0,1]$, there exists a $\delta_1 > 0$ such that $|x_{n_1}(t') - x_{n_1}(t'')| < \varepsilon_1$ whenever $|t'-t''| < \delta_1$. Take an arbitrary $n_2 > N(\varepsilon_1, \delta_1)$ and consider x_{n_2}. We claim that

$$\widehat{([x_{n_1}]; [x_{n_2}])} \geqslant \beta_1. \tag{1.98}$$

Indeed, take an arbitrary $t_1 \in M_{x_{n_1}}$. Since $n_2 > N(\varepsilon_1, \delta_1)$, there exists by the definition of $N(\varepsilon_1, \delta_1)$, a $t_2 \in [0,1]$ such that

$$|t_1 - t_2| < \delta_1, \qquad |x_{n_2}(t_2)| < \varepsilon_1.$$

Then $|x_{n_1}(t_2)| \geqslant |x_{n_1}(t_1)| - \varepsilon_1 = 1-\varepsilon_1$ and thus, by lemma 1.16 applied to $x = x_{n_1}$, $y = x_{n_2}$, $t_0 = t_2$ and $\varepsilon = \varepsilon_1$, we get

$$\|x_{n_1} + \alpha x_{n_2}\| \geqslant (1-3\varepsilon_1)\|x_{n_1}\| = \beta_1 \|x_{n_1}\|$$

for all scalars α, whence we infer (1.98).

Now, the unit sphere $\sigma_{[x_{n_1},x_{n_2}]}=\{x\in[x_{n_1},x_{n_2}]\mid \|x\|=1\}$ is compact, whence, by the theorem of Arzelà, equicontinuous, and thus there exists a $\delta_2>0$ such that $|x(t')-x(t'')|<\varepsilon_2$ whenever $x\in\sigma_{[x_{n_1},x_{n_2}]}$, $|t'-t''|<\delta_2$. Take an arbitrary $n_3>N(\varepsilon_2,\delta_2)$ and consider x_{n_3}. We claim that

$$(\widehat{[x_{n_1},x_{n_2}];[x_{n_3}]})\geqslant\beta_2. \tag{1.99}$$

Indeed, take an arbitrary $x\in\sigma_{[x_{n_1},x_{n_2}]}$. Then, repeating the above argument for x, x_{n_3}, ε_2 and δ_2 instead of x_{n_1}, x_{n_2}, ε_1 and δ_1 respectively, it follows that

$$\|x+\alpha x_{n_3}\|\geqslant\beta_2\|x\|$$

for all scalars α, whence, since $x\in\sigma_{[x_{n_1},x_{n_2}]}$ has been arbitrary, we infer (1.99).

Continuing this process indefinitely, we obtain a subsequence $\{x_{n_k}\}$ of $\{x_n\}$ such that

$$(\widehat{[x_{n_1},x_{n_2},\ldots,x_{n_k}];[x_{n_{k+1}}]})\geqslant\beta_k \qquad (k=1,2,\ldots). \tag{1.100}$$

Then, by Ch. I, § 7, corollary 7.1, $\{x_n\}$ must be a basic sequence, which completes the proof of proposition 1.2.

Corollary 1.1. *Let*

$$x_n(t)=t^n \qquad (t\in[0,1],\, n=1,2,\ldots). \tag{1.101}$$

Then one can select an infinite subsequence $\{x_{n_k}\}$ of $\{x_n\}$, which is a basic sequence in $E=C([0,1])$. Consequently, the subspace $E_0=[x_{n_k}]$ satisfies the conditions of theorem 1.4 *and thus it has no monotone basis.*

Proof. The sequence (1.101) satisfies the conditions of proposition 1.2; indeed, $\|x_n\|=1$ and, since $A(x_n,\varepsilon)=[0,\varepsilon^{\frac{1}{n}})$ $(n=1,2,\ldots)$, one can take e.g. $N(\varepsilon,\delta)=\left[\dfrac{\log(1-\delta)}{\log\varepsilon}\right]+1$.

Now, since $x_{n_k}(t)=t^{n_k}$ is a basic sequence in $C([0,1])$, every element x of the subspace $E_0=[x_{n_k}]$ admits an expansion $x(t)=\sum_{k=1}^{\infty}a_k t^{n_k}$ converging uniformly on $[0,1]$ and thus it is an analytic function on $(0,1)$. Furthermore, E_0 consists of non-constant functions on $(0,1)$, since by (1.101) we have $x_{n_k}(0)=0$ for all $k=1,2,\ldots$ Since any function $x_{n_k}(t)=t^{n_k}$ can be taken as the y of theorem 1.4, the conditions of theorem 1.4 are fulfilled. This completes the proof of corollary 1.1.

Thus, the problem of existence of monotone bases in every infinite dimensional Banach space with a basis, is solved in the negative. However, the following problem remains open:

Problem 1.1. Does there exist an infinite dimensional Banach space with a basis, such that $0<\Gamma(E)<1$?

Remark 1.1. The same argument as that used at the end of the proof of theorem 1.4 also shows that the *hyperplane*

$$E_1=[t,t^2,t^3,\ldots]=\{x\in C([0,1])\mid x(0)=0\} \tag{1.102}$$

of $C([0,1])$ *has no monotone basis consisting of analytic functions on* $(0,1)$, whence, in particular, it has no monotone basis consisting of algebraic polynomials. Since the sequence obtained from the usual Schauder basis of $C([0,1])$ by deleting its first term is a monotone basis[1] of E_1 and since the set of all algebraic polynomials satisfying $x(0)=0$ is dense in E_1 (by the theorem of Weierstrass), this shows that *a stability theorem of Krein-Milman-Rutman type* (Ch. I, § 10) *is no longer valid for monotone bases.*

In the footnote of Ch. I, § 11, a necessary condition was given for a Banach space E to have a monotone basis. The following proposition gives a necessary and sufficient condition:

Proposition 1.3. *A Banach space E has a monotone basis if and only if there exists a sequence* $\{G_n\}$ *of closed linear subspaces of E with the following properties:*

a) $\dim G_n=n$ $(n=1,2,\ldots)$;
b) $G_n\subset G_{n+1}$ $(n=1,2,\ldots)$;
c) $\bigcup_{n=1}^{\infty} G_n$ *is dense in* E;
d) *there exists a projection u of E onto* G_n, *with* $\|u_n\|=1$ $(n=1,2,\ldots)$.

Proof. If E has a monotone basis $\{x_n\}$, then for $G_n=[x_1,x_2,\ldots,x_n]$ $(n=1,2,\ldots)$ we have a), b), c) and d) with $u=s_n$ $(n=1,2,\ldots)$.

Conversely, assume that E has a sequence $\{G_n\}$ of subspaces satisfying a), b), c), d). Take an arbitrary element x_1 of G_1, different from 0, and for each $n=1,2,\ldots$ take an element x_{n+1} of G_{n+1}, different from 0 and such that $u_n(x_{n+1})=0$. Then the sequence $\{x_n\}$ is complete in E and we have

---

[1] Indeed, we have already observed that the Schauder basis of $C([0,1])$ is monotone and therefore, if we delete its first term, it remains a monotone basic sequence. Furthermore, all these functions vanish in $t=0$ (as second term of the Schauder basis, i.e. as first term of our sequence, we take the function $x(t)=t$ instead of $x(t)=1-t$), i.e. they belong to E_1, whence they span the whole hyperplane E_1.

$$\left\| \sum_{i=1}^{n} \alpha_i x_i \right\| = \left\| u_n \left(\sum_{i=1}^{n+1} \alpha_i x_i \right) \right\| \leqslant \left\| \sum_{i=1}^{n+1} \alpha_i x_i \right\|$$

for all finite sequences of scalars $\alpha_1, \ldots, \alpha_{n+1} \in K$, whence also (1.1). Consequently, by Ch. I, § 7, theorem 7.1, $\{x_n\}$ is a monotone basis of E, which completes the proof.

Although there exist, as we have seen in the preceding, Banach spaces which have no monotone basis, every basis of a Banach space E can be "monotonized" by replacing the norm of E with a suitable equivalent norm. Indeed, this follows from Ch. I, § 3, proposition 3.2 b), since for the equivalent norm $|||x||| = \sup_{1 \leqslant n < \infty} \left\| \sum_{i=1}^{n} f_i(x) x_i \right\|$ and for any finite sequence of scalars $\alpha_1, \ldots, \alpha_{n+m}$ we have

$$\left|\left|\left| \sum_{i=1}^{n} \alpha_i x_i \right|\right|\right| = \sup_{1 \leqslant k \leqslant n} \left\| \sum_{i=1}^{k} \alpha_i x_i \right\| \leqslant \sup_{1 \leqslant k \leqslant n+m} \left\| \sum_{i=1}^{k} \alpha_i x_i \right\| = \left|\left|\left| \sum_{i=1}^{n+m} \alpha_i x_i \right|\right|\right| .$$

Moreover, since $|||x_n||| = \|x_n\|$ $(n=1,2,\ldots)$, from this remark and Ch. I, § 1, theorem 1.2 it follows that *for every bounded basis $\{x_n\}$ of a Banach space E there exists an equivalent norm $\|x\|'$ on E in which the basis $\{x_n\}$ is normalized and monotone.*

As shown by the example of the unit vector basis in c_0, the basis $\{x_n\}$ need not be strictly monotone in the above norm $|||x|||$. However, by Ch. I, § 19, theorem 19.1 b) and proposition 19.1 b), *every basis of a Banach space E can be also "strictly monotonized"*.

Remark 1.2. Not only every basis $\{x_n\}$, but also every sequence $\{x_n\}$ such that $x_n \neq 0$ $(n=1,2,\ldots)$ is equivalent[1] to a strictly monotone basis. Indeed, if $A_1 = \left\{ \{\alpha_n\} \subset K \;\middle|\; \sum_{i=1}^{\infty} \alpha_i x_i \text{ converges} \right\}$ is the Banach space of sequences of scalars introduced in Ch. I, § 3, proposition 3.1, then

$$\left.\left.\left) \right) \sum_{i=1}^{\infty} \alpha_i x_i \right) \right) = \sup_{1 \leqslant n < \infty} \left\| \sum_{i=1}^{n} \alpha_i x_i \right\| + \sum_{i=1}^{\infty} \frac{1}{2^{i+1}} \|\alpha_i x_i\| \tag{1.103}$$

is an equivalent norm on A_1, in which the unit vector basis $e_n = \{\delta_{nj}\}_{j=1}^{\infty}$ $(n=1,2,\ldots)$ of A_1 is strictly monotone. On the other hand, by Ch. I, § 8, proposition 8.1 b), we have $\{x_n\} \sim \{e_n\}$ and obviously this remains valid for A_1 endowed with the equivalent norm $\left.\left.\right)\right)\sum_{i=1}^{\infty} \alpha_i x_i \left.\left.\right)\right)$. Thus $\{x_n\}$ is equivalent to a strictly monotone basis.

[1] For the equivalence of sequences see Ch. I, § 8.

From the remark that every basis can be "monotonized" and from proposition 1.3 we infer

Corollary 1.2. *A Banach space E has a basis if and only if there exist a sequence $\{G_n\}$ of subspaces of E and a norm on E, equivalent to the initial norm of E, such that we have* a), b), c) *and* d) *of proposition* 1.3 *in this new norm.*

Let us observe that there is an obvious connection between proposition 1.3, corollary 1.2 and the results of Ch. I, § 20. In particular, problem 20.1 of Ch. I, § 20 admits (by Ch. I, § 20, proposition 20.1, equivalence $1^\circ \Leftrightarrow 3^\circ$) the following equivalent formulation: If a Banach space E has a sequence of subspaces $\{G_n\}$ with the properties a), b), c) of proposition 1.3 and

d') there exists a projection v_n of E onto G_n $(n=1,2,\dots)$ such that $\sup_{1\leqslant n<\infty} \|v_n\| < \infty$, then does E have a basis?

The answer to the problem of existence of non-monotone (and hence also of non strictly monotone) bases in Banach spaces with bases is affirmative. In fact, let us first remark that *for every normalized monotone basis* $\{x_n\}$ *we have*

$$\|x_n - x_m\| \geqslant \|x_{\min(n,m)}\| = 1 \qquad (n,m=1,2,\dots;\, n\neq m). \tag{1.104}$$

Now, if $\{x_n\}$ is a monotone basis of a (finite or infinite dimensional) Banach space E, then $\left\{\frac{x_n}{\|x_n\|}\right\}$ is a normalized monotone basis of E, whence, by the above remark, for any $y\in\sigma_E\cap[x_1,x_2] = \{x\in[x_1,x_2] \mid \|x\|=1\}$ such that

$$0<\left\|\frac{x_1}{\|x_1\|} - y\right\| < 1,$$

the sequence

$$y_1 = \frac{x_1}{\|x_1\|}, \qquad y_2 = y, \qquad y_n = \frac{x_n}{\|x_n\|} \qquad (n=3,4,\dots) \tag{1.105}$$

is a normalized non-monotone basis of E. This proves our assertion.

We conclude this paragraph with the following proposition on duality properties of monotone bases:

Proposition 1.4. *Let $\{x_n\}$ be a monotone basis of a Banach space E and let $\{f_n\}\subset E^*$ be the a.s.c.f. to $\{x_n\}$. Then*

a) *$\{f_n\}$ is a monotone basis of $[f_n]$.*

b) *We have*

$$\|x_1\|\,\|f_1\| = 1, \qquad 1\leqslant \|x_n\|\,\|f_n\| \leqslant 2 \qquad (n=2,3,\dots). \tag{1.106}$$

Proof. a) is a consequence of Ch. I, § 12, theorem 12.3 and of the remark made at the beginning of the present section, according to which a basis is monotone if and only if it is of norm $v=1$.

b) Assume that $\{x_n\}$ is a monotone basis of E. Then

$$\left|f_1\left(\sum_{i=1}^{\infty}\alpha_i x_i\right)\right| = |\alpha_1| = \frac{1}{\|x_1\|}\|\alpha_1 x_1\| \leqslant \frac{1}{\|x_1\|}\left\|\sum_{i=1}^{\infty}\alpha_i x_i\right\| \quad \left(\sum_{i=1}^{\infty}\alpha_i x_i \in E\right),$$

$$\left|f_n\left(\sum_{i=1}^{\infty}\alpha_i x_i\right)\right| = |\alpha_n| = \frac{1}{\|x_n\|}\|\alpha_n x_n\| \leqslant \frac{1}{\|x_n\|}\left(\left\|\sum_{i=1}^{n}\alpha_i x_i\right\| + \left\|\sum_{i=1}^{n-1}\alpha_i x_i\right\|\right)$$

$$\leqslant \frac{2}{\|x_n\|}\left\|\sum_{i=1}^{\infty}\alpha_i x_i\right\| \quad \left(\sum_{i=1}^{\infty}\alpha_i x_i \in E;\ n=2,3,\ldots\right),$$

whence $\|f_1\| \leqslant \frac{1}{\|x_1\|}$, $\|f_n\| \leqslant \frac{2}{\|x_n\|}$ $(n=2,3,\ldots)$. On the other hand, we have $\|f_n\|\,\|x_n\| \geqslant |f_n(x_n)| = 1$ $(n=1,2,\ldots)$, which completes the proof of proposition 1.4.

The converse of proposition 1.4 a) is not true, since e.g. the a.s.c.f. $\{f_n\}$ to the basis $\{x_n\}$ of l^1, given in Ch. I, § 12, example 12.2, is monotone, but $\{x_n\}$ is not monotone; a similar statement is also valid for the basis of c_0 given in Ch. I, § 13, formula (13.15).

For strictly monotone bases a result similar to proposition 1.4 a) is no longer valid, since e.g. the natural basis $\{x_n\}$ of l^1 is strictly monotone, while the a.s.c.f. $\{f_n\}$ to $\{x_n\}$ is monotone but not strictly monotone. Also, a result of converse type is again not valid, since e.g. the natural basis $\{x_n\}$ of c_0 is monotone but not strictly, while the a.s.c.f. $\{f_n\}$ to $\{x_n\}$ is strictly monotone.

§ 2. Normal bases

Definition 2.1. A basis $\{x_n\}$ of a Banach space E is said to be *normal*, if both $\{x_n\}$ and the a.s.c.f. $\{f_n\} \subset E^*$ are normalized, i.e. if

$$\|x_n\| = \|f_n\| = 1 \qquad (n=1,2,\ldots). \tag{2.1}$$

For instance, the natural bases of c_0 and $l^p (p \geqslant 1)$ are normal bases. *The Schauder basis* $\{x_n\}_0^\infty$ *in* $C([0,1])$ *is normalized but not normal*, since the a.s.c.f. is

$$f_0(x)=x(0), \qquad f_1(x)=x(1)-x(0),$$

$$f_{2^k+l}(x)=x\left(\frac{2l-1}{2^{k+1}}\right)-\frac{1}{2}x\left(\frac{2l-2}{2^{k+1}}\right)-\frac{1}{2}x\left(\frac{2l}{2^{k+1}}\right) \tag{2.2}$$

$$(x\in C([0,1]);\ l=1,\ldots,2^k;\ k=0,1,2,\ldots),$$

whence $\|f_0\|=1$, $\|f_n\|=2$ $(n=1,2,\ldots)$. Indeed, f_0 and f_1 have been computed in Ch. I, § 2, example 2.2, while for the other equalities it is sufficient to verify that $f_i(x_j)=\delta_{ij}$. Now, for any fixed pair (k,l) with $1\leqslant l\leqslant 2^k$, $0\leqslant k<\infty$, it is obvious that

$$f_{2^k+l}(x_0)=f_{2^k+l}(x_1)=0,$$

$$f_{2^k+l}(x_{2^h+m})=0 \qquad (m=1,\ldots,2^h;\ h=k+1,\ k+2,\ldots),$$

$$f_{2^k+l}(x_{2^k+m})=\delta_{2^k+l,\,2^k+m} \qquad (m=1,\ldots,2^k).$$

On the other hand, if $h<k$, then $\left[\frac{2l-2}{2^{k+1}},\frac{2l}{2^{k+1}}\right]$ is entirely contained in one of the 2^{h+1} segments $\left[\frac{2j-2}{2^{h+1}},\frac{2j-1}{2^{h+1}}\right]$, $\left[\frac{2j-1}{2^{h+1}},\frac{2j}{2^{h+1}}\right]$ $(j=1,\ldots,2^h)$. Since on each of these segments every x_{2^h+m} $(m=0,1,\ldots,2^h)$ is linear, it follows that every x_{2^h+m} $(m=0,1,\ldots,2^h)$ is linear on $\left[\frac{2l-2}{2^{k+1}},\frac{2l}{2^{k+1}}\right]$, whence, by (2.2),

$$f_{2^k+l}(x_{2^h+m})=0 \qquad (m=1,\ldots,2^h;\ h=1,\ldots,k-1),$$

which completes the proof of $f_i(x_j)=\delta_{ij}$.

Furthermore, *the normalized Haar basis* $\{\tilde{z}_n^{(p)}\}=\left\{\frac{\tilde{y}_n}{\|\tilde{y}_n\|}\right\}$ in $L^p([0,1])$, i.e.[1] the sequence $\{\tilde{z}_n^{(p)}\}$, where

$$z_1^{(p)}(t)\equiv 1,$$

$$z_{2^k+l}^{(p)}(t)=\begin{cases}\sqrt[p]{2^k} & \text{for } t\in\left[\frac{2l-2}{2^{k+1}},\frac{2l-1}{2^{k+1}}\right),\\ -\sqrt[p]{2^k} & \text{for } t\in\left[\frac{2l-1}{2^{k+1}},\frac{2l}{2^{k+1}}\right),\\ 0 & \text{for the other } t\end{cases} \tag{2.3}$$

$$(l=1,2,\ldots,2^k;\ k=0,1,2,\ldots),$$

[1] See Ch. I, § 2, example 2.3.

is a normal basis of $L^p([0,1])$ $(p \geqslant 1)$, since the a.s.c.f. $\{h_n\}$ to $\{\tilde{z}_n^{(p)}\}$ is

$$h_i(\tilde{x}) = \int_0^1 x(\tau) z_i^{(q)}(\tau)\, d\tau \qquad (\tilde{x} \in L^p([0,1]),\ i=1,2,\ldots), \tag{2.4}$$

where $\frac{1}{p} + \frac{1}{q} = 1$, whence $\|h_i\| = \|\tilde{z}_i^{(q)}\| = 1$ $(i=1,2,\ldots)$. Indeed, it is sufficient to verify that $h_i(\tilde{z}_j^{(p)}) = \delta_{ij}$. Since obviously $h_i(\tilde{z}_i^{(p)}) = 1$ $(i=1,2,\ldots)$, it remains to prove that $h_i(\tilde{z}_j^{(p)}) = 0$ for $i \neq j$, and for this purpose it is sufficient to observe that *the Haar system* $\{y_n\}$ *is orthogonal*, i.e.

$$\int^1 y_i(t) y_j(t)\, dt = 0 \qquad (i,j=1,2,\ldots;\ i \neq j), \tag{2.5}$$

by virtue of Ch. I, § 2, formula (2.19).

The notion of normal basis admits several geometrical interpretations. For this purpose, let us first recall

Lemma 2.1. *Let E be a Banach space, $f \in E^*$, $H = \{y \in E \mid f(y) = 0\}$ and $x \in E$. Then*

$$\operatorname{dist}(x,H) = \frac{|f(x)|}{\|f\|}. \tag{2.6}$$

Proof. For every $y \in H$ we have

$$\|x-y\| \geqslant \frac{1}{\|f\|} |f(x-y)| = \frac{|f(x)|}{\|f\|},$$

whence $\operatorname{dist}(x,H) \geqslant \frac{|f(x)|}{\|f\|}$. On the other hand, for $0 < \varepsilon < \|f\|$ there exists a $z \in E$ such that $|f(z)| > (\|f\| - \varepsilon)\|z\|$. Multiplying with $\left|\frac{f(x)}{f(z)}\right|$ and defining $y \in H$ by

$$y = x - \frac{f(x)}{f(z)} z, \tag{2.7}$$

we obtain $|f(x)| > (\|f\| - \varepsilon)\|x-y\|$, i.e. $\operatorname{dist}(x,H) \leqslant \|x-y\| < \frac{|f(x)|}{\|f\| - \varepsilon}$, whence, since $0 < \varepsilon < \|f\|$ has been arbitrary, $\operatorname{dist}(x,H) \leqslant \frac{|f(x)|}{\|f\|}$. This completes the proof.

Let us mention that lemma 2.1 can be also derived from a corollary of the Hahn-Banach theorem which we have already used in the preceding.

Corollary 2.1. *Let E be a Banach space and* (x_n, f_n) $(\{x_n\} \subset E, \{f_n\} \subset E^*)$ *an E-complete biorthogonal system. Then*

$$\operatorname{dist}(x_n, E^{(n)}) = \frac{1}{\|f_n\|} \qquad (n=1,2,\ldots), \tag{2.8}$$

where, as in Ch. I, § 3, formula (3.9) *and Ch. I, § 7, formula* (7.1), $E^{(n)}$ *is the closed hyperplane of E spanned by the sequence* $\{x_1, \ldots, x_{n-1}, x_{n+1}, \ldots\}$, *i.e.*

$$E^{(n)} = [x_1, \ldots, x_{n-1}, x_{n+1}, \ldots] \qquad (n=1,2,\ldots). \tag{2.9}$$

Proof. Apply lemma 2.1 to $H = E^{(n)} = \{y \in E \mid f_n(y) = 0\}$ $(n=1,2,\ldots)$.

Now we can give some geometric characterizations of normal bases.

Theorem 2.1. *For a basis* $\{x_n\}$ *of a Banach space E the following statements are equivalent:*

1°. $\{x_n\}$ *is normal.*

2°. *We have*

$$\|x_n\| = 1 \quad \text{and} \quad x_n \perp E^{(n)} \qquad (n=1,2,\ldots), \tag{2.10}$$

where $E^{(n)}$ *is defined by* (2.9).

3°. *The unit cell* $S_E = \{x \in E \mid \|x\| \leq 1\}$ *contains* x_n *on its boundary* $\sigma_E = \{x \in E \mid \|x\| = 1\}$ *and has at* x_n *a support hyperplane* H_n *parallel to the hyperplane* $E^{(n)}$ $(n=1,2,\ldots)$.

In this case, each hyperplane H_n *is uniquely determined.*

4°. *The unit cell* S_E *lies between the "hyperoctahedron"*

$$M_1 = \left\{ \sum_{i=1}^{\infty} \alpha_i x_i \in E \,\middle|\, \sum_{i=1}^{\infty} |\alpha_i| \leq 1 \right\} \tag{2.11}$$

and the "hyperparallelepiped"

$$M_2 = \left\{ \sum_{i=1}^{\infty} \alpha_i x_i \in E \,\middle|\, \max_{1 \leq i < \infty} |\alpha_i| \leq 1 \right\} \tag{2.12}$$

(as shown by (2.11) *and* (2.12), *both terms are understood with respect to the "system of coordinates"* $\{x_n\}$).

5°. *We have* $\|x_n\| = 1$ $(n=1,2,\ldots)$ *and*

$$\|\alpha_n x_n\| \leq \left\| \sum_{i=1}^{\infty} \alpha_i x_i \right\| \qquad \left(\sum_{i=1}^{\infty} \alpha_i x_i \in E,\ n=1,2,\ldots \right). \tag{2.13}$$

Proof. 1°⇔2°. By corollary 2.1, we have $\|f_n\| = 1 = \|x_n\|$ if and only if $\operatorname{dist}(x_n, E^{(n)}) = 1 = \operatorname{dist}(x_n, 0)$, which is equivalent, by the definition of orthogonality, to $x_n \perp E^{(n)}$, $\|x_n\| = 1$ $(n=1,2,\ldots)$.

$1^\circ \Rightarrow 3^\circ$. If $\|x_n\| = \|f_n\| = 1$, the hyperplane H_n defined by

$$H_n = \{y \in E \mid f_n(y) = 1\} \tag{2.14}$$

supports S_E at x_n and it is parallel to $E^{(n)}$ $(n = 1, 2, \ldots)$.

$3^\circ \Rightarrow 1^\circ$ and the uniqueness of H_n. Assume that for a positive integer n and a $g \in E^*$ with $\|g\| = 1$ the hyperplane

$$H = \{y \in E \mid g(y) = 1\} \tag{2.15}$$

supports[1] S_E at x_n and is parallel to $E^{(n)}$. Then, since $x_n \in H$, we have $g(x_n) = 1$ and, since H is parallel to $E^{(n)}$, we have $g(x_i) = 0$ $(i = 1, \ldots, n-1, n+1, \ldots)$. Consequently, since $[x_n] = E$, we have $g = \lambda f_n$, where $\lambda \neq 0$, whence, since $\lambda = \lambda f_n(x_n) = g(x_n) = 1$, we get $g = f_n$, i.e. H coincides with the hyperplane H_n defined by (2.14), which proves the uniqueness of H_n. Since $\|g\| = 1$, it also follows that $\|f_n\| = 1$. Furthermore, since $x_n \in S_E$, we have $\|x_n\| \leqslant 1$. On the other hand, since $x_n \in H$, we have $1 = g(x_n) \leqslant \|g\| \, \|x_n\| = \|x_n\|$, whence, finally, $\|x_n\| = 1$. Thus we have 1°.

$1^\circ \Rightarrow 4^\circ$. If $\|x_n\| = \|f_n\| = 1$ $(n = 1, 2, \ldots)$, for every $x = \sum_{i=1}^{\infty} \alpha_i x_i \in E$ we have

$$\begin{aligned} \max_{1 \leqslant i < \infty} |\alpha_i| &= \max_{1 \leqslant i < \infty} \left| f_i\left(\sum_{j=1}^{\infty} \alpha_j x_j \right) \right| \leqslant \max_{1 \leqslant i < \infty} \|f_i\| \left\| \sum_{j=1}^{\infty} \alpha_j x_j \right\| \\ &= \left\| \sum_{i=1}^{\infty} \alpha_i x_i \right\| \leqslant \sum_{i=1}^{\infty} |\alpha_i| \, \|x_i\| = \sum_{i=1}^{\infty} |\alpha_i|, \end{aligned} \tag{2.16}$$

whence

$$M_1 \subset S_E \subset M_2 . \tag{2.17}$$

$4^\circ \Rightarrow 5^\circ$. Assume that we have (2.17). Then, by $x_n \in M_1 \subset S_E$ we have $\|x_n\| \leqslant 1$. Furthermore, let $\sum_{i=1}^{\infty} \alpha_i x_i \in E$ be arbitrary. Then by

$$\sum_{j=1}^{\infty} \frac{\alpha_j}{\left\| \sum_{i=1}^{\infty} \alpha_i x_i \right\|} x_j \in S_E \subset M_2$$

[1] We recall that every support hyperplane H of S_E can be written in the form (2.15) with a suitable $g \in E^*$ of norm $\|g\| = 1$ (see e.g. [246], Ch. I, § 1, lemma 1.4).

we have $\max_{1 \leqslant j < \infty} \frac{|\alpha_j|}{\left\| \sum_{i=1}^{\infty} \alpha_i x_i \right\|} \leqslant 1$, whence, since $\|x_n\| \leqslant 1$, we obtain

$$\|\alpha_n x_n\| \leqslant |\alpha_n| \leqslant \left\| \sum_{i=1}^{\infty} \alpha_i x_i \right\| \qquad (n=1,2,\ldots).$$

In particular, for $\alpha_1 = \cdots = \alpha_{n-1} = \alpha_{n+1} = \cdots = 0, \alpha_n = 1$, we obtain $\|x_n\| = 1$ $(n=1,2,\ldots)$.

$5^\circ \Rightarrow 1^\circ$. Assume that we have 5°. Then

$$1 = \|x_n\| \leqslant \inf_{\{\alpha_j\}} \left\| x_n - \sum_{i \neq n} \alpha_i x_i \right\| = \operatorname{dist}(x_n, E^{(n)}) \leqslant \|x_n\| = 1 \qquad (n=1,2,\ldots),$$

whence, by corollary 2.1, $\|f_n\| = 1$ $(n=1,2,\ldots)$, which completes the proof of theorem 2.1.

Remark 2.1. The hyperoctahedron M_1 occurring in 4° above is nothing else but the closed circled convex hull of the sequence $\{x_n\}$, while the hyperparallelepiped M_2 of 4° is nothing else but the intersection of all closed half-spaces determined by H_n and S_E, where H_n are the support hyperplanes of S_E occurring in 3°.

Remark 2.2. From definition 2.1 it is obvious that *a basis $\{x_n\}$ of a Banach space E is normal if and only if every finite section $\{x_j\}_{j=1}^m$ of $\{x_n\}$ is a normal basis of the finite-dimensional Banach space $[x_j]_{j=1}^m$* $(m=1,2,\ldots)$. Thus, applying theorem 2.5 to each finite section $\{x_j\}_{j=1}^m$, we obtain several other geometric characterizations of normal bases $\{x_n\}$.

In the particular case when E is a Hilbert space H, we have

Proposition 2.1. *A basis $\{x_n\}$ of a Hilbert space H is normal if and only if it is a normalized orthogonal basis of H.*

Proof. By the equivalence $1^\circ \Leftrightarrow 2^\circ$ of theorem 2.1, $\{x_n\}$ is a normal basis of H if and only if $\|x_n\| = 1$ $(n=1,2,\ldots)$ and $x_n \perp x_j$ for all $j \neq n$ $(j, n = 1,2,\ldots)$. However, by virtue of § 1, lemma 1.1, this happens if and only if $\{x_n\}$ is a normalized orthogonal basis of H.

Now we shall prove that the answer to the problem of existence of normal bases in finite-dimensional Banach spaces is affirmative.

Theorem 2.2. *Let E be an n-dimensional Banach space. Then E has a normal basis.*

Proof. Introduce a Cartesian coordinate system into E and let

$$D(y_1, \ldots, y_n) = \det(\eta_i^{(k)})_{i,k=1,\ldots,n}, \tag{2.18}$$

where

$$y_k = \{\eta_1^{(k)}, \ldots, \eta_n^{(k)}\} \in E \qquad (k=1,\ldots,n).$$

Since $\dim E = n$, the unit sphere $\sigma_E = \{x \in E \,|\, \|x\| = 1\}$, whence also the product $(\sigma_E)^n = \sigma_E \times \cdots \times \sigma_E$ (n times), is compact. Consequently, since $D(y_1, \ldots, y_n)$ is continuous on $(\sigma_E)^n$, we can choose $x_1, \ldots, x_n \in E$ with $\|x_j\| = 1$ $(j = 1, \ldots, n)$, which maximize $|D(y_1, \ldots, y_n)|$ on $(\sigma_E)^n$. This being done, put

$$f_i(x) = \frac{D(x_1, \ldots, x_{i-1}, x, x_{i+1}, \ldots, x_n)}{D(x_1, \ldots, x_n)} \qquad (x \in E;\, i = 1, \ldots, n). \quad (2.19)$$

Then

$$f_i(x_j) = \delta_{ij} \qquad (i, j = 1, \ldots, n),$$

and by our choice of $x_1, \ldots, x_n$ we have

$$\|f_i\| = \sup_{\|x\|=1} |f_i(x)| = \sup_{\|x\|=1} \left| \frac{D(x_1, \ldots, x_{i-1}, x, x_{i+1}, \ldots, x_n)}{D(x_1, \ldots, x_n)} \right| = 1 \quad (i = 1, \ldots, n),$$

which shows that $\{x_j\}_{j=1}^n$ is a normal basis of E. This completes the proof.

As shown by the above proof, the condition of maximizing $|D(y_1, \ldots, y_n)|$ on $(\sigma_E)^n$ is sufficient for $\{x_j\}_{j=1}^n$ to be a normal basis of E. However, this condition is not necessary, as shown by

Example 2.1. The basis

$$x_1 = \{\tfrac{1}{2}, \tfrac{1}{2}\}, \qquad x_2 = \{-\tfrac{1}{2}, \tfrac{1}{2}\} \quad (2.20)$$

of the space $E = l_2^1$ (of all pairs of scalars $x = \{\xi_1, \xi_2\}$, endowed with the usual vector operations and the norm $\|x\| = |\xi_1| + |\xi_2|$) is normal, since the associated coefficient functionals are $f_1 = \{1, 1\}$, $f_2 = \{-1, 1\}$. However, for $z_1 = \{1, 0\}$, $z_2 = \{0, 1\} \in \sigma_E$ we have

$$D(x_1, x_2) = \begin{vmatrix} \frac{1}{2} & \frac{1}{2} \\ -\frac{1}{2} & \frac{1}{2} \end{vmatrix} = \tfrac{1}{2} < \begin{vmatrix} 1 & 0 \\ 0 & 1 \end{vmatrix} = D(z_1, z_2),$$

and thus x_1, x_2 do not maximize $D(y_1, y_2)$ on $\sigma_E \times \sigma_E$.

Remark 2.3. In the particular case of real scalars, the above condition of maximization (and also the above proof) admits a geometrical interpretation. Indeed, in this case $2^n |D(y_1, \ldots, y_n)|$ is the volume of the parallelepiped $M_2^{(n)} = \left\{ \sum_{j=1}^n \alpha_j y_j \in E \,\middle|\, \max_{1 \leqslant i \leqslant n} |\alpha_i| \leqslant 1 \right\}$ with axes $y_1, \ldots, y_n$ and thus the above condition requires to maximize the volume of $M_2^{(n)}$ for $\|y_j\| = 1$ $(j = 1, \ldots, n)$.

Problem 2.1. Does every infinite dimensional Banach space with a basis possess a normal basis? What about the space $C([0,1])$? What about a reflexive infinite dimensional Banach space with a basis?

Although these questions are unsolved, it is known that every bounded basis of a Banach space E "can be made normal" by replacing the norm of E with a suitable equivalent norm. In other words, we have

Theorem 2.3. *Let $\{x_n\}$ be a bounded basis of a Banach space E. Then there exists an equivalent norm on E, in which the basis $\{x_n\}$ is normal.*

Proof. By Ch. I, § 3, theorem 3.2, we may assume, without loss of generality, that $\|x_n\|=1$ $(n=1,2,\ldots)$. Put

$$|||x|||=\sup_{1\leqslant n,m<\infty}\left\|\sum_{i=n}^{m} f_i(x)x_i\right\| \qquad (x\in E), \tag{2.21}$$

where $\{f_n\}\subset E^*$ is the a.s.c.f. to $\{x_n\}$. Then $|||x|||$ is a norm on E and, by Ch. I, § 3, proposition 3.2 b) and $\left\|\sum_{i=n}^{m} f_i(x)x_i\right\|\leqslant\left\|\sum_{i=1}^{m} f_i(x)x_i\right\| + \left\|\sum_{i=1}^{n-1} f_i(x)x_i\right\|$, it is equivalent to the initial norm of E. Furthermore,

$$|||x_j|||=\sup_{1\leqslant n,m<\infty}\left\|\sum_{i=n}^{m} f_i(x_j)x_i\right\|=\|x_j\|=1 \qquad (j=1,2,\ldots),$$

and hence also

$$|||\alpha_j x_j|||=\|\alpha_j x_j\|\leqslant\sup_{1\leqslant n,m<\infty}\left\|\sum_{i=n}^{m}\alpha_i x_i\right\|=\left|\left|\left|\sum_{i=1}^{\infty}\alpha_i x_i\right|\right|\right| \quad \left(\sum_{i=1}^{\infty}\alpha_i x_i\in E, j=1,2,\ldots\right).$$

Therefore, by theorem 2.1 (implication $5^\circ\Rightarrow 1^\circ$), $\{x_n\}$ is normal in the norm (2.21), which completes the proof of theorem 2.3.

On the other hand, obviously, the answer to the problem of existence of non-normal bases in Banach spaces with bases is affirmative, since if $\{x_n\}$ is a normal basis of a Banach space E, then $\{2x_n\}$ is a non-normal basis of E. Moreover, the answer to the problem of existence of normalized non-normal bases in Banach spaces with bases is also affirmative. In fact, let us first remark *for every normal basis $\{x_n\}$ we have*

$$\|x_n-x_m\|\geqslant|f_n(x_n-x_m)|=1 \qquad (n,m=1,2,\ldots;\, n\neq m) \tag{2.22}$$

where $\{f_n\}$ is the a.s.c.f. to $\{x_n\}$. Now, if $\{x_n\}$ is a normal basis of a (finite- or infinite-dimensional) Banach space E, let y be any element of $\sigma_E\cap[x_1,x_2]=\{x\in[x_1,x_2]\,|\,\|x\|=1\}$ such that

$$0<\|x_1-y\|<1.$$

Then, by the above remark, the sequence

$$y_1=x_1, \quad y_2=y, \quad y_n=x_n \quad (n=3,4,\ldots) \tag{2.23}$$

is a normalized non-normal basis of E. This proves our assertion.

We conclude this section with some remarks on duality properties of normal bases.

Proposition 2.2. *If $\{x_n\}$ is a normal basis of a Banach space E, the a.s.c.f. $\{f_n\}$ is a normal basis of the space $[f_n]$.*

Proof. By definition 2.1, we have $\|f_n\|=1$ $(n=1,2,\ldots)$. On the other hand, the a.s.c.f. (in $[f_n]^*$) to the basis $\{f_n\}$ of $[f_n]$ is $\{\phi(x_n)\}$, where ϕ denotes the canonical mapping of E into $[f_n]^*$ (see Ch. I, § 12). Hence, by definition 2.1,

$$\|\phi(x_n)\|=\sup_{\substack{f\in[f_j]\\ \|f\|\leqslant 1}}|f(x_n)|\leqslant\|x_n\|=1 \qquad (n=1,2,\ldots).$$

On the other hand, $f_n\in[f_j]$, $\|f_n\|=1$, whence

$$\|\phi(x_n)\|\geqslant|\phi(x_n)(f_n)|=|f_n(x_n)|=1 \qquad (n=1,2,\ldots),$$

and thus $\|\phi(x_n)\|=1$ $(n=1,2,\ldots)$, which shows that $\{f_n\}$ is a normal basis of $[f_n]$. This completes the proof.

The converse of proposition 2.2 is not valid, as shown by

Example 2.2. Let $E=c_0$ and let $\{x_n\}$ be the basis[1] of the space E defined by

$$x_n=\{\underbrace{2,\ldots,2}_{n},0,0,\ldots\} \qquad (n=1,2,\ldots). \tag{2.24}$$

Then $\|x_n\|=2$ $(n=1,2,\ldots)$, whence $\{x_n\}$ is not a normal basis of E. However, the a.s.c.f. $\{f_n\}$ is a normal basis of $[f_n]$. Indeed, we have

$$f_n=\{\underbrace{0,\ldots,0}_{n-1},\tfrac{1}{2},-\tfrac{1}{2},0,0,\ldots\} \qquad (n=1,2,\ldots), \tag{2.25}$$

whence $\|f_n\|=1$ $(n=1,2,\ldots)$. Furthermore, the a.s.c.f. (in $[f_n]^*$) to the basis $\{f_n\}$ of $[f_n]$ is $\{\phi(x_n)\}$, where ϕ is the canonical mapping of E into $[f_n]^*$. By (2.25) we have

$$[f_n]=\left\{f=\{\eta_n\}\in l^1 \;\middle|\; \sum_{i=1}^{\infty}\eta_i=0\right\}, \tag{2.26}$$

whence, for any $f=\{\eta_n\}\in[f_n]$ with $\|f\|\leqslant 1$,

$$|f(x_n)|=\left|\sum_{i=1}^{n}2\eta_i\right|=\left|\sum_{i=1}^{n}\eta_i-\sum_{i=n+1}^{\infty}\eta_i\right|\leqslant\sum_{i=1}^{\infty}|\eta_i|=\|f\|\leqslant 1 \qquad (n=1,2,\ldots),$$

---

[1] We have $x_n=\sum_{i=1}^{n}2e_i$ $(n=1,2,\ldots)$, where $\{e_n\}$ is the natural basis of $E=c_0$, whence, by Ch. I, § 4, proposition 4.3, $\{x_n\}$ is a basis of E.

and therefore

$$\|\phi(x_n)\| = \sup_{\substack{f\in[f_n]\\ \|f\|\leqslant 1}} |f(x_n)| \leqslant 1 \qquad (n=1,2,\ldots).$$

On the other hand, since $f_n \in [f_j]$, $\|f_n\|=1$, we have

$$\|\phi(x_n)\| \geqslant |\phi(x_n)(f_n)| = |f_n(x_n)| = 1 \qquad (n=1,2,\ldots),$$

and thus $\|\phi(x_n)\|=1$ $(n=1,2,\ldots)$, which shows that $\{f_n\}$ is a normal basis of $[f_n]$. This completes the proof of our assertion.

§ 3. Positive bases

Definition 3.1. A basis $\{x_n\}$ of a Banach space E is said to be *positive*, if for every linear isometry $T: E\to E$, with $T(x_j)=\sum_{i=1}^{\infty} a_{ij}x_i$ $(j=1,2,\ldots)$, there exists a second linear isometry $T_+: E\to E$ such that

$$T_+(x_j) = \sum_{i=1}^{\infty} |a_{ij}|\, x_i \qquad (j=1,2,\ldots). \tag{3.1}$$

Let us give some examples of positive bases.

Lemma 3.1. *A continuous linear mapping* $T: c_0\to c_0$, *with* $T(x_j)=\sum_{i=1}^{\infty} a_{ij}x_i$ $(j=1,2,\ldots; \{x_n\}=$ *the natural basis of* $c_0)$ *is an isometry if and only if*

$$\sup_{1\leqslant i<\infty} |a_{ij}| = 1 \qquad (j=1,2,\ldots), \tag{3.2}$$

$$\sum_{j=1}^{\infty} |a_{ij}| \leqslant 1 \qquad (i=1,2,\ldots). \tag{3.3}$$

Proof. Let $T: c_0\to c_0$ be a linear isometry, with $T(x_j)=\sum_{i=1}^{\infty} a_{ij}x_i$ $(j=1,2,\ldots)$, where $\{x_n\}$ is the natural basis of c_0. Then we have

$$\sup_{1\leqslant i<\infty} |a_{ij}| = \|T(x_j)\| = \|x_j\| = 1 \qquad (j=1,2,\ldots),$$

i.e. (3.2). For a fixed i put $\varepsilon_j = \operatorname{sign} a_{ij}$ $(j=1,2,\ldots)$. Then

$$\sum_{j=1}^{n} |a_{ij}| = \left|\sum_{j=1}^{n} a_{ij}\varepsilon_j\right| \leqslant \sup_{1\leqslant i<\infty} \left|\sum_{j=1}^{n} a_{ij}\varepsilon_j\right| = \left\|\sum_{i=1}^{\infty}\sum_{j=1}^{n} a_{ij}\varepsilon_j x_i\right\|$$

$$= \left\|\sum_{j=1}^{n} \varepsilon_j \sum_{i=1}^{\infty} a_{ij}x_i\right\| = \left\|\sum_{j=1}^{n} \varepsilon_j T(x_j)\right\| = \left\|T\left(\sum_{j=1}^{n}\varepsilon_j x_j\right)\right\| = \left\|\sum_{j=1}^{n}\varepsilon_j x_j\right\| = 1$$

$(n=1,2,\ldots)$, whence we infer (3.3) and thus the conditions are necessary.

Conversely, let $T: c_0 \to c_0$ be a continuous linear mapping, with $T(x_j) = \sum_{i=1}^{\infty} a_{ij} x_i$ $(j=1,2,\ldots)$, satisfying (3.2), (3.3) and let $\alpha_1, \ldots, \alpha_n$ be arbitrary scalars. Then there is an index $\leqslant n$, say j_0, such that $|\alpha_{j_0}| = \sup_{1\leqslant j\leqslant n} |\alpha_j| = \left\| \sum_{j=1}^{n} \alpha_j x_j \right\|$. By (3.2) we have $\sup_{1\leqslant i<\infty} |a_{ij_0}|\,|\alpha_{j_0}| = |\alpha_{j_0}|$ and hence, for every $\varepsilon>0$ there exists an index $i_0 = i_0(\varepsilon)$ such that $|a_{i_0 j_0}\alpha_{j_0}| > |\alpha_{j_0}| - \varepsilon$. Then, by virtue of (3.3), $\sum_{j=1}^{n} |a_{i_0 j}| \leqslant 1$, and therefore

$$|\alpha_{j_0}| - \varepsilon + \sum_{\substack{j=1\\ j\neq j_0}}^{n} |a_{i_0 j}|\,|\alpha_j| < \sum_{j=1}^{n} |a_{i_0 j}|\,|\alpha_j| \leqslant |\alpha_{j_0}| \sum_{j=1}^{n} |a_{i_0 j}| \leqslant |\alpha_{j_0}|,$$

whence

$$\left| \sum_{\substack{j=1\\ j\neq j_0}}^{n} a_{i_0 j}\alpha_j \right| \leqslant \sum_{\substack{j=1\\ j\neq j_0}}^{n} |a_{i_0 j}|\,|\alpha_j| < \varepsilon.$$

Consequently,

$$\left| \sum_{j=1}^{n} a_{i_0 j}\alpha_j \right| \geqslant |a_{i_0 j_0}\alpha_{j_0}| - \left| \sum_{\substack{j=1\\ j\neq j_0}}^{n} a_{i_0 j}\alpha_j \right| > |\alpha_{j_0}| - 2\varepsilon = \left\| \sum_{j=1}^{n} \alpha_j x_j \right\| - 2\varepsilon,$$

whence

$$\left\| T\left(\sum_{j=1}^{n} \alpha_j x_j \right) \right\| = \left\| \sum_{j=1}^{n} \alpha_j \sum_{i=1}^{\infty} a_{ij} x_i \right\| = \left\| \sum_{i=1}^{\infty} \sum_{j=1}^{n} a_{ij}\alpha_j x_i \right\|$$
$$= \sup_{1\leqslant i<\infty} \left| \sum_{j=1}^{n} a_{ij}\alpha_j \right| \geqslant \left| \sum_{j=1}^{n} a_{i_0 j}\alpha_j \right| > \left\| \sum_{j=1}^{n} \alpha_j x_j \right\| - 2\varepsilon,$$

whence, since $\varepsilon>0$ was arbitrary,

$$\left\| T\left(\sum_{j=1}^{n} \alpha_j x_j \right) \right\| \geqslant \left\| \sum_{j=1}^{n} \alpha_j x_j \right\|. \tag{3.4}$$

On the other hand, by (3.3) we have

$$\left\| T\left(\sum_{j=1}^{n} \alpha_j x_j \right) \right\| = \left\| \sum_{j=1}^{n} \alpha_j \sum_{i=1}^{\infty} a_{ij} x_i \right\| = \left\| \sum_{i=1}^{\infty} \sum_{j=1}^{n} a_{ij}\alpha_j x_i \right\|$$
$$= \sup_{1\leqslant i<\infty} \left| \sum_{j=1}^{n} a_{ij}\alpha_j \right| \leqslant \sup_{1\leqslant j\leqslant n} |\alpha_j| = \left\| \sum_{j=1}^{n} \alpha_j x_j \right\|,$$

which, together with (3.4), gives

$$\left\| T\left(\sum_{j=1}^{n} \alpha_j x_j\right)\right\| = \left\| \sum_{j=1}^{n} \alpha_j x_j \right\|.$$

Since $\alpha_1, \ldots, \alpha_n$ were arbitrary scalars and $[x_n]=E$, it follows that T is an isometry, which completes the proof of lemma 3.1.

Proposition 3.1. *The natural basis of c_0 is positive.*

Proof. Since the conditions (3.2), (3.3) remain invariant when passing from $\{a_{ij}\}$ to $\{|a_{ij}|\}$, the assertion follows from lemma 3.1.

Remark 3.1. A similar argument (with obvious simplifications) shows that the natural basis of the finite dimensional space l_n^∞ is positive.

Lemma 3.2. *Let $1 \leqslant p \neq 2$ and let $y=\{\eta_n\} \in l^p$, $z=\{\zeta_n\} \in l^p$ be such that*

$$\|y+z\|^p + \|y-z\|^p = 2(\|y\|^p + \|z\|^p). \tag{3.5}$$

Then

$$\eta_i \zeta_i = 0 \qquad (i=1,2,\ldots). \tag{3.6}$$

Proof. Let $p>2$. We shall first show that for any scalars η, ζ we have

$$|\eta+\zeta|^p + |\eta-\zeta|^p \geqslant 2(|\eta|^p + |\zeta|^p), \tag{3.7}$$

with the equality sign holding if and only if $\eta\zeta=0$. Let $|\eta|$ and $|\zeta|$ be fixed and let us compute the minimum of

$$d = |\eta+\zeta|^p + |\eta-\zeta|^p. \tag{3.8}$$

If $\eta=0$, this minimum is $2|\zeta|^p$ and if $\zeta=0$ this minimum is $2|\eta|^p$. Assume now that $\eta\zeta \neq 0$ and let $|\eta| = \alpha > 0$, $\zeta = \frac{\eta}{\alpha} \beta e^{i\theta}$, $\beta>0$. Then

$$\begin{aligned} d=d(\theta) &= \left|\eta + \frac{\eta}{\alpha}\beta e^{i\theta}\right|^p + \left|\eta - \frac{\eta}{\alpha}\beta e^{i\theta}\right|^p = |\alpha+\beta e^{i\theta}|^p + |\alpha-\beta e^{i\theta}|^p \\ &= [(\alpha+\beta\cos\theta)^2 + \beta^2 \sin^2\theta]^{\frac{p}{2}} + [(\alpha-\beta\cos\theta)^2+\beta^2\sin^2\theta]^{\frac{p}{2}} \\ &= (\alpha^2+\beta^2+2\alpha\beta\cos\theta)^{\frac{p}{2}} + (\alpha^2+\beta^2-2\alpha\beta\cos\theta)^{\frac{p}{2}}, \end{aligned}$$

whence the minimum of $d(\theta)$ is $2(\alpha^2+\beta^2)^{\frac{p}{2}} = 2(|\eta|^2+|\zeta|^2)^{\frac{p}{2}}$, being attained for $\theta = \pm\frac{\pi}{2}$. But, since $|\eta|>0$, $|\zeta|>0$, $\frac{p}{2}>1$, we have

$$(|\eta|^2+|\zeta|^2)^{\frac{p}{2}} > (|\eta|^2)^{\frac{p}{2}} + (|\zeta|^2)^{\frac{p}{2}} = |\eta|^p + |\zeta|^p,$$

whence we infer (3.7), with the equality sign holding if and only if $\eta\zeta=0$.

Now let $y=\{\eta_n\}\in l^p$, $z=\{\zeta_n\}\in l^p$ be arbitrary. Then, by the above we have

$$\begin{aligned}\|y+z\|^p+\|y-z\|^p &= \sum_{i=1}^{\infty}|\eta_i+\zeta_i|^p+\sum_{i=1}^{\infty}|\eta_i-\zeta_i|^p \\ &\geqslant 2\left(\sum_{i=1}^{\infty}|\eta_i|^p+\sum_{i=1}^{\infty}|\zeta_i|^p\right)=2(\|y\|^p+\|z\|^p),\end{aligned} \tag{3.9}$$

with the equality sign holding if and only if we have (3.6).

In the case when $1\leqslant p<2$, the proof is similar, with the difference that for these values of p we have the opposite inequalities in (3.7) and (3.9) ($d(\theta)$ attains its maximum for $\theta=0,\pi$). This completes the proof of lemma 3.2.

Lemma 3.3. *Let* $1\leqslant p\neq 2$. *If* $T: l^p\to l^p$ *is a linear isometry, with* $T(x_j)=\sum_{i=1}^{\infty}a_{ij}x_i$ $(j=1,2,\dots;\{x_n\}=$ *the natural basis of* l^p), *then*

$$a_{ij}a_{im}=0 \qquad (i,j,m=1,2,\dots;j\neq m). \tag{3.10}$$

Proof. Each pair $x_j, x_m (j\neq m)$ satisfies (3.5), whence each pair $T(x_j)$, $T(x_m)$ also satisfies (3.5), whence, by lemma 3.2, we have (3.10), which completes the proof.

Remark 3.2. If we define the "support" of $T(x_j)$ by

$$\operatorname{Supp} T(x_j)=\{i\mid a_{ij}\neq 0\} \qquad (j=1,2,\dots), \tag{3.11}$$

then (3.7) can be expressed in the equivalent form

$$\operatorname{Supp} T(x_j)\cap\operatorname{Supp} T(x_m)=\emptyset \qquad (j,m=1,2,\dots;j\neq m), \tag{3.12}$$

which may be called *the disjoint support condition* (for the pair $(E,\{x_n\})$).

Proposition 3.2. *Let* $1\leqslant p\neq 2$. *Then the natural basis of* l^p *is positive.*

Proof. Let $T: l^p\to l^p$ be a linear isometry, with $T(x_j)=\sum_{i=1}^{\infty}a_{ij}x_i$ $(j=1,2,\dots)$, where $\{x_n\}$ is the natural basis of l^p and let $\alpha_1,\dots,\alpha_n$ be arbitrary scalars. Then, by lemma 3.3, for each i there is, among the n numbers $a_{i1},\dots,a_{in}$, at most one $\neq 0$, whence $\left|\sum_{j=1}^{n}|a_{ij}|\alpha_j\right|=\left|\sum_{j=1}^{n}a_{ij}\alpha_j\right|$, whence

$$\begin{aligned}\left\|\sum_{i=1}^{\infty}\left(\sum_{j=1}^{n}|a_{ij}|\alpha_j\right)x_i\right\| &= \left(\sum_{i=1}^{\infty}\left|\sum_{j=1}^{n}|a_{ij}|\alpha_j\right|^p\right)^{\frac{1}{p}}=\left(\sum_{i=1}^{\infty}\left|\sum_{j=1}^{n}a_{ij}\alpha_j\right|^p\right)^{\frac{1}{p}} \\ &= \left\|\sum_{i=1}^{\infty}\left(\sum_{j=1}^{n}a_{ij}\alpha_j\right)x_i\right\|.\end{aligned}$$

Consequently, by proposition 3.4 below, $\{x_n\}$ is a positive basis of E, which completes the proof.

Let us consider now the case when $p=2$.

Lemma 3.4. *If $\{x_n\}$ is an orthonormal basis of a (finite or infinite dimensional) Hilbert space $E=H$, then $\{x_n\}$ is not positive.*

Proof. Put

$$y_1 = \frac{x_1+x_2}{\|x_1+x_2\|}, \qquad y_2 = \frac{x_1-x_2}{\|x_1-x_2\|}, \qquad y_n=x_n \quad (n=3,4,\ldots). \tag{3.13}$$

Then $\{y_n\}$ is an orthonormal basis of E, whence there is a linear isometry $T\colon E\to E$ such that

$$T(x_n)=y_n \quad (n=1,2,\ldots). \tag{3.14}$$

Then

$$\begin{aligned}\|T_+(x_1+x_2)\| &= \left\|\frac{x_1+x_2}{\|x_1+x_2\|}+\frac{x_1+x_2}{\|x_1-x_2\|}\right\| \\ &= \|x_1+x_2\|\left\|\frac{1}{\|x_1+x_2\|}+\frac{1}{\|x_1-x_2\|}\right\| \\ &= \sqrt{2}\left(\frac{1}{\sqrt{2}}+\frac{1}{\sqrt{2}}\right)=2>\sqrt{2}=\|x_1+x_2\|,\end{aligned}$$

i.e. T_+ is not an isometry, which completes the proof.

Now we can show that the problem of existence of positive bases in (finite or infinite dimensional) Banach spaces with bases has a negative answer, even in "very good" spaces.

Proposition 3.3. *A (finite or infinite dimensional) Hilbert space $E=H$ has no positive basis.*

Proof. Assume that $E=H$ has a positive basis $\{x_n\}$; we may assume, without loss of generality, that $\{x_n\}$ is normalized (replacing, if necessary, $\{x_n\}$ by $\left\{\frac{x_n}{\|x_n\|}\right\}$, the basis still remains positive). Let i,j be arbitrary indices and put

$$x = \frac{x_i-x_j}{\|x_i-x_j\|}. \tag{3.15}$$

Then x_1 and x may be considered as first elements of different orthonormal bases, whence there exists a linear isometry $T\colon E\to E$ such that $T(x_1)=x$. Then, since $\{x_n\}$ is a positive basis,

$$\left\|\frac{x_i+x_j}{\|x_i-x_j\|}\right\| = \|T_+(x_1)\| = \|T(x_1)\| = \left\|\frac{x_i-x_j}{\|x_i-x_j\|}\right\|,$$

whence $(x_i, x_j)=0$. Since i,j have been arbitrary, it follows that $\{x_n\}$ is orthonormal, contradicting lemma 3.4. This completes the proof of proposition 3.3.

Remark 3.1 and proposition 3.3 show that, similarly to monotone bases, the notion of positive basis divides even the finite-dimensional Banach spaces into two classes (according to the existence or non-existence of a positive basis of the space).

We have the following characterization of positive bases:

Proposition 3.4. *A basis $\{x_n\}$ of a Banach space E is positive if and only if for every linear isometry $T: E \to E$ with $T(x_j) = \sum_{i=1}^{\infty} a_{ij} x_i$ $(j=1,2,\ldots)$ and any scalars $\alpha_1, \ldots, \alpha_n$ the series $\sum_{i=1}^{\infty} \left(\sum_{j=1}^{n} |a_{ij}| \alpha_j \right) x_i$ converges and*

$$\left\| \sum_{i=1}^{\infty} \left(\sum_{j=1}^{n} |a_{ij}| \alpha_j \right) x_i \right\| = \left\| \sum_{i=1}^{\infty} \left(\sum_{j=1}^{n} a_{ij} \alpha_j \right) x_i \right\|. \tag{3.16}$$

Proof. If $\{x_n\}$ is positive, we have

$$\left\| \sum_{i=1}^{\infty} \left(\sum_{j=1}^{n} |a_{ij}| \alpha_j \right) x_i \right\| = \left\| T_+ \left(\sum_{j=1}^{n} \alpha_j x_j \right) \right\| = \left\| T \left(\sum_{j=1}^{n} \alpha_j x_j \right) \right\| = \left\| \sum_{i=1}^{\infty} \left(\sum_{j=1}^{n} a_{ij} \alpha_j \right) x_i \right\|,$$

i.e. (3.16).

Conversely, assume that the condition is satisfied and let $T: E \to E$ be a linear isometry with $T(x_j) = \sum_{i=1}^{\infty} a_{ij} x_i$ $(j=1,2,\ldots)$. Then, taking $\alpha_1 = \cdots = \alpha_{k-1} = \alpha_{k+1} = \cdots = \alpha_n = 0$, $\alpha_k = 1$, it follows that the series $\sum_{i=1}^{\infty} |a_{ik}| x_i$ converges $(k=1,2,\ldots)$. Put

$$T_+(x_j) = \sum_{i=1}^{\infty} |a_{ij}| x_i \qquad (j=1,2,\ldots), \tag{3.17}$$

and extend T_+ by linearity to the (dense) linear subspace of E spanned by $\{x_n\}$. Then for any scalars $\alpha_1, \ldots, \alpha_n$ we have, by (3.16) and since T is a linear isometry,

$$\left\| T_+ \left(\sum_{j=1}^{n} \alpha_j x_j \right) \right\| = \left\| \sum_{i=1}^{\infty} \left(\sum_{j=1}^{n} |a_{ij}| \alpha_j \right) x_i \right\| = \left\| \sum_{i=1}^{\infty} \left(\sum_{j=1}^{n} a_{ij} \alpha_j \right) x_i \right\|$$
$$= \left\| T \left(\sum_{j=1}^{n} \alpha_j x_j \right) \right\| = \left\| \sum_{j=1}^{n} \alpha_j x_j \right\|,$$

and thus T_+ is a linear isometry on a dense subspace of E, whence it can be extended to a linear isometry $T_+: E \to E$. This completes the proof of proposition 3.4.

As shown by definition 3.1, the "positivity" of bases of a space E depends very much on the form of the linear isometries $T: E \to E$. In particular, if in a Banach space E with a basis the only linear isometries $T: E \to E$ are $T = I_E$ (the identical mapping of E onto E) and $T = -I_E$, then obviously *every basis of E is positive*. Since for each $n \geqslant 2$ there exist n-dimensional Banach spaces with this property (e.g. for $n \geqslant 3$ one can take $\sigma_E = \{x \in E \mid \|x\| = 1\}$ to contain $(k+2)$-gons A_k $(k = 1, \ldots, n)$ such that $A_k \cap A_{k+1}$ consists of one single point x_k $(k = 1, \ldots, n-1)$, with $x_1, \ldots, x_{n-1}$ linearly independent, and that except for the non-extremal points of these polygons every other point of σ_E is an exposed point of S_E), it follows that the answer to the problem of existence of non-positive bases in finite-dimensional Banach spaces is negative.

Problem 3.1. Does there exist in every infinite dimensional Banach space with a basis a non-positive basis?

The difficulty consists in the fact that (by definition 3.1) we have to consider also the linear isometries T of E *into* E.

Problem 3.1 is related to the following:

Problem 3.2. Is every positive basis necessarily unconditional[1]?

If the answer to problem 3.2 would be affirmative, then, since in every infinite dimensional Banach space with a basis there exists a conditional basis (by § 23, theorem 23.2), the answer to problem 3.1 would be also affirmative.

The classes of bases considered in § 1–§ 3 depend on the metric properties of the Banach space E. All other classes of bases which we shall consider in this Chapter (except orthogonal, hyperorthogonal, strictly orthogonal and strictly hyperorthogonal bases in § 20) depend only on the topological linear structure of E, i.e. they are invariant under an isomorphism of E onto another Banach space E_1.

§ 4. k-shrinking bases

We have seen in Ch. I, § 12, that for a basis $\{x_n\}$ of a Banach space E, the a.s.c.f. $\{f_n\} \subset E^*$ is a basic sequence in E^*, but need not be a basis of E^*. Therefore it is natural to give

[1] For the definition of unconditional and conditional bases see § 14.

Definition 4.1. Let k be a non-negative integer. A basis $\{x_n\}$ of a Banach space E is called *k-shrinking*, if for the a.s.c.f. $\{f_n\}\subset E^*$ we have[1]

$$\operatorname{codim}_{E^*}[f_n]=k. \tag{4.1}$$

The 0-shrinking bases, i.e. the bases $\{x_n\}$ such that $[f_n]=E^*$, are also called *shrinking bases.*

For instance, the natural bases of c_0 and l^p $(p>1)$ and the Haar basis of $L^p([0,1])$ $(p>1)$ are shrinking bases. We shall also give, for arbitrary $k>0$, an example of a k-shrinking basis (see example 4.1 below).

Lemma 4.1. *Let X be a Banach space and k a non-negative integer. Then*

a) *A closed linear subspace G of X is of codimension $\leqslant k$ if and only if*

α) *for every $(k+1)$-dimensional linear subspace F_{k+1} of X, the intersection $F_{k+1}\cap G$ contains a non-zero element x.*

b) *We have* $\operatorname{codim}_X G\geqslant k$ *if and only if*

β) *there exists a $(k+1)$-dimensional linear subspace F'_{k+1} of X such that the intersection $F'_{k+1}\cap G$ contains a non-zero element x', which is unique up to a homothety.*

c) *Consequently,* $\operatorname{codim}_X G=k$ *if and only if we have simultaneously* α) *and* β).

Proof. a) Assume that $\operatorname{codim}_X G\leqslant k$. Then there exist h linearly independent functionals $\phi_1,\dots,\phi_h$, where $h=\operatorname{codim}_X G\leqslant k$, such that

$$G=\{x\in X\mid \phi_j(x)=0 \quad (j=1,\dots,h)\}. \tag{4.2}$$

Let F_{k+1} be an arbitrary (k+1)-dimensional linear subspace of X and let $y_1,\dots,y_{k+1}$ be a basis of F_{k+1}. Then $x=\sum_{i=1}^{k+1}\alpha_i y_i\in G$ is equivalent to

$$\sum_{i=1}^{k+1}\alpha_i\phi_j(y_i)=0 \qquad (j=1,\dots,h),$$

i.e. to a homogeneous system of $h\leqslant k$ linear equation with $k+1$ unknowns. Since this system always has a non-zero solution $\{\alpha_1,\dots,\alpha_{k+1}\}$, we have α).

The converse implication $\alpha)\Rightarrow\operatorname{codim}_X G\leqslant k$ is obvious.

b) We have $\operatorname{codim}_X G\geqslant k$ if and only if $\operatorname{codim}_X G\nleqslant k-1$. By part a) proved above, this happens if and only if there exists a k-dimensional linear subspace F'_k of X such that $F'_k\cap G=\{0\}$.

[1] For the notation $\operatorname{codim}_X G$ see Ch. I, § 10, the footnote to formula (10.7).

Now, if $F'_k \cap G = \{0\}$, then for $0 \neq x' \in G$ and $F'_{k+1} = F'_k \oplus [x']$, the intersection $F'_{k+1} \cap G$ contains a non zero element x', which is unique up to a homothety, i.e. we have β).

Conversely, if we have β), then for any k-dimensional linear subspace F'_k of F'_{k+1}, such that $x' \notin F'_k$, we have $F'_k \cap G = \{0\}$. This completes the proof of lemma 4.1.

Let E be a Banach space. For a biorthogonal system (x_n, f_n) ($\{x_n\} \subset E$, $\{f_n\} \subset E^*$) we shall use the notation

$$\|f\|_n = \|f\,|_{[x_{n+1}, x_{n+2}, \ldots]}\| \qquad (f \in E^*,\ n = 1, 2, \ldots). \tag{4.3}$$

If E is n-dimensional ($1 \leqslant n < \infty$), we shall put

$$\|f\|_n = \|f\|_{n+1} = \cdots = 0 \qquad (f \in E^*). \tag{4.4}$$

Proposition 4.1. *Let E be a Banach space and* (x_n, f_n) ($\{x_n\} \subset E$, $\{f_n\} \subset E^*$) *an E-complete biorthogonal system. Then*

$$\|f\|_n = \operatorname{dist}(f, [f_1, \ldots, f_n]) \qquad (f \in E^*,\ n = 1, 2, \ldots). \tag{4.5}$$

Proof. By the well known[1] formula

$$\operatorname{dist}(f, [f_1, \ldots, f_n]) = \sup_{\substack{x \in [f_1, \ldots, f_n]_\perp \\ \|x\| \leqslant 1}} |f(x)| \qquad (f \in E^*,\ n = 1, 2, \ldots), \tag{4.6}$$

where $[f_1, \ldots, f_n]_\perp = \{y \in E \mid f_i(y) = 0 \ (i = 1, \ldots, n)\}$, and by (4.3), it is sufficient to prove that

$$[f_1, \ldots, f_n]_\perp = [x_{n+1}, x_{n+2}, \ldots]. \tag{4.7}$$

The inclusion

$$[f_1, \ldots, f_n]_\perp \supset [x_{n+1}, x_{n+2}, \ldots]$$

is obvious by biorthogonality. Conversely, assume that $x \in [f_1, \ldots, f_n]_\perp$, i.e. that

$$f_1(x) = \cdots = f_n(x) = 0.$$

Then, since $E = [x_j]_1^n \oplus [x_{n+1}, x_{n+2}, \ldots]$ (by Ch. I, § 6, the implication $2° \Rightarrow 6°$ of theorem 6.1), we have $x = \sum_{i=1}^{n} f_i(x) x_i + y = y$, with a suitable $y \in [x_{n+1}, x_{n+2}, \ldots]$. Thus we also have the converse inclusion

$$[f_1, \ldots, f_n]_\perp \subset [x_{n+1}, x_{n+2}, \ldots],$$

whence the equality (4.7). This completes the proof.

[1] See e.g. [246], p. 20. This formula also follows e.g. from the canonical isometry $E^*/[f_1, \ldots, f_n] = E^*/([f_1, \ldots, f_n]_\perp)^\perp \equiv ([f_1, \ldots, f_n]_\perp)^*$.

Proposition 4.2. *Let E be a Banach space,* (x_n, f_n) $(\{x_n\} \subset E, \{f_n\} \subset E^*)$ *a biorthogonal system, and*

$$V' = \left\{f \in E^* \mid \lim_{n\to\infty} \|f_n\|_n = 0\right\}, \tag{4.8}$$

$$V'' = \Big\{f \in E^* \mid \lim_{n\to\infty} f(y_n) = 0 \ \textit{for all}\ \{y_n\} \subset E \ \textit{such that}$$

$$\sup_{1\leqslant n<\infty} \|y_n\| < \infty,\ \lim_{n\to\infty} f_i(y_n) = 0\ (i = 1, 2, \ldots)\Big\}, \tag{4.9}$$

$$V''' = \Big\{f \in E^* \mid \lim_{n\to\infty} f(y_n) = 0 \ \textit{for all}\ \{y_n\} \subset E \ \textit{such that}$$

$$\sup_{1\leqslant n<\infty} \|y_n\| < \infty,\ f_i(y_n) = 0\ (n > i)\Big\}, \tag{4.10}$$

$$V^{(4)} = \Big\{f \in E^* \mid \lim_{n\to\infty} f(y_n) = 0 \ \textit{for all}\ \{y_n\} \subset E \ \textit{such that}$$

$$\sup_{1\leqslant n<\infty} \|y_n\| < \infty, \quad y_n = \sum_{k=m_{n-1}+1}^{m_n} \alpha_k x_k \neq 0 \quad (n = 1, 2, \ldots),$$

$$\textit{where} \quad 0 = m_0 < m_1 < m_2 < \cdots\Big\}. \tag{4.11}$$

Then

$$[f_j] = V' = V'' = V''' = V^{(4)}. \tag{4.12}$$

Proof. By proposition 4.1 we have $[f_j] = V'$. The inclusions $[f_j] \subset V'' \subset V''' \subset V^{(4)}$ being obvious, it will be sufficient to prove that

$$V^{(4)} \subset V'. \tag{4.13}$$

Let $f \notin V'$, i.e. $\lim_{n\to\infty} \|f\|_n \neq 0$. Then there exist a number $\varepsilon_0 > 0$, an increasing sequence of positive integers $\{m_n\}$, $m_0 = 0$, a sequence of scalars $\{\alpha_n\}$ and a sequence of elements $\{y_n\} \subset E$ such that

$$y_n = \sum_{k=m_{n-1}+1}^{m_n} \alpha_k x_k \quad (n = 1, 2, \ldots), \tag{4.14}$$

$$\|y_n\| \leqslant 1 \qquad (n = 1, 2, \ldots), \tag{4.15}$$

$$f(y_n) > \varepsilon_0 \qquad (n = 1, 2, \ldots). \tag{4.16}$$

Indeed, since $\|f|_{[x_j]}\| = \|f\|_0 \geqslant \|f\|_1 \geqslant \|f\|_2 \geqslant \cdots$, there exists an $\varepsilon > 0$ such that $\|f\|_n \geqslant \varepsilon\,(n = 0, 1, 2, \ldots)$. Put $m_0 = 0$, choose $z_1 \in [x_j]$ with $\|z_1\| \leqslant \frac{1}{2}$, $f(z_1) > \frac{\varepsilon}{2}$ and then choose m_1, y_1 so that $y_1 = \sum_{i=1}^{m_1} \beta_i^{(1)} x_i$,

$\|y_1 - z_1\| \leqslant \frac{1}{2}$, $f(y_1) > \frac{\varepsilon}{2}$. Continuing by induction, choose $z_n \in [x_j]_{j=m_{n-1}+1}^{\infty}$ with $\|z_n\| \leqslant \frac{1}{2}$, $f(z_n) > \frac{\varepsilon}{2}$ and then m_n, y_n so that $y_n = \sum_{i=m_{n-1}+1}^{m_n} \beta_i^{(n)} x_i$, $\|y_n - z_n\| \leqslant \frac{1}{2}$, $f(y_n) > \frac{\varepsilon}{2}$. Then, putting $\alpha_i = \beta_i^{(n)}$ $(i = m_{n-1}+1, \ldots, m_n;\ n = 1, 2, \ldots)$ and $\varepsilon_0 = \frac{\varepsilon}{2}$, we have (4.14), (4.15) and (4.16).

Hence $f \notin V^{(4)}$ and thus we have (4.13), which completes the proof.

From the above we obtain the following characterizations of k-shrinking bases:

Theorem 4.1. *Let $\{x_n\}$ be a basis of a Banach space E, with the a.s.c.f. $\{f_n\}$, and let k be a non-negative integer. The following statements are equivalent:*

1°. *$\{x_n\}$ is k-shrinking.*

2°. *In every $(k+1)$-dimensional linear subspace B_{k+1} of E^* there exists a non-zero element $f \in B_{k+1}$ such that $\lim_{n\to\infty} \|f\|_n = 0$ and there exists a $(k+1)$-dimensional linear subspace B'_{k+1} of E^* for which this element f is unique up to a homothety.*

3°. *In every $(k+1)$-dimensional linear subspace B_{k+1} of E^* there exists a non-zero element $f \in B_{k+1}$ such that $\lim_{n\to\infty} f(y_n) = 0$ for all bounded sequences $\{y_n\} \subset E$ satisfying $\lim_{n\to\infty} f_i(y_n) = 0$ $(i = 1, 2, \ldots)$ and there exists a $(k+1)$-dimensional linear subspace B'_{k+1} of E^* for which this element f is unique up to a homothety.*

4°. *In every $(k+1)$-dimensional linear subspace B_{k+1} of E^* there exists a non-zero element $f \in B_{k+1}$ such that $\lim_{n\to\infty} f(y_n) = 0$ for all bounded sequences $\{y_n\} \subset E$ satisfying $f_i(y_n) = 0$ $(n > i)$, and there exists a $(k+1)$-dimensional linear subspace B'_{k+1} of E^* for which this element f is unique up to a homothety.*

5°. *In every $(k+1)$-dimensional linear subspace B_{k+1} of E^* there exists a non-zero element $f \in B_{k+1}$ such that $\lim_{n\to\infty} f(y_n) = 0$ for all block basic sequences $y_n = \sum_{k=m_{n-1}+1}^{m_n} \alpha_k x_k \neq 0$ $(n = 1, 2, \ldots;\ m_0 = 0)$ with $\sup_{1 \leqslant n < \infty} \|y_n\| < \infty$, and there exists a $(k+1)$-dimensional linear subspace B'_{k+1} of E^* for which this element f is unique up to a homothety.*

Proof. This follows immediately from lemma 4.1 and proposition 4.2.

In view of the special importance of the particular case $k = 0$, let us give separately

Theorem 4.2. *Let $\{x_n\}$ be a basis of a Banach space E, with the a.s.c.f. $\{f_n\} \subset E^*$. The following statements are equivalent:*

1°. *$\{x_n\}$ is shrinking (i.e. $[f_n]=E^*$).*

2°. $\lim_{n\to\infty} \|f\|_n = 0$ $(f \in E^*)$.

3°. *Every sequence $\{y_n\} \subset E$ with $\sup_{1\leqslant n<\infty} \|y_n\| < \infty$, $\lim_{n\to\infty} f_i(y_n)=0$ $(i=1,2,\ldots)$, converges weakly to zero.*

4°. *Every sequence $\{y_n\} \subset E$ with $\sup_{1\leqslant n<\infty} \|y_n\| < \infty$, $f_i(y_n)=0$ $(n>i)$, converges weakly to zero.*

5°. *Every block basic sequence $y_n = \sum_{k=m_{n-1}+1}^{m_n} \alpha_k x_k$ $(n=1,2,\ldots;\, m_0=0)$ with $\sup_{1\leqslant n<\infty} \|y_n\| < \infty$, converges weakly to zero.*

6°. *$\{f_n\}$ is a basis of E^*.*

7°. *The second conjugate space E^{**} is isomorphic, by the mapping*

$$\tau: \Phi \to \{\Phi(f_n)\} \qquad (\Phi \in E^{**}) \tag{4.17}$$

to the Banach space of sequences of scalars

$$A_2 = \left\{ \{\alpha_n\} \subset K \;\middle|\; \sup_{1\leqslant n<\infty} \left\| \sum_{i=1}^{n} \alpha_i x_i \right\| < \infty \right\}, \tag{4.18}$$

defined in Ch. I, § 5, remark 5.4.

8°. *We have*

$$\pi(E) = \left\{ \Phi \in E^{**} \;\middle|\; \sum_{i=1}^{\infty} \Phi(f_i) x_i \text{ converges} \right\}, \tag{4.19}$$

(where π denotes the canonical mapping of E into E^{**}), and the image of $\pi(E)$ in A_2 by the isomorphism (4.17) *is the space*

$$A_1 = \left\{ \{\alpha_n\} \subset K \;\middle|\; \sum_{i=1}^{\infty} \alpha_i x_i \text{ converges} \right\} \tag{4.20}$$

defined in Ch. I, § 3, proposition 3.1.

If $\{x_n\}$ is a monotone basis, the isomorphism (4.17) *is an isometry.*

Proof. The equivalences $1° \Leftrightarrow \cdots \Leftrightarrow 5°$ are nothing else but the particular case $k=0$ of theorem 4.1.

The implication $6° \Rightarrow 1°$ is obvious. The implication $1° \Rightarrow 6°$ is a consequence of Ch. I, § 12, theorem 12.1.

Assume now that we have 1°. Then, by Ch. I, § 12, theorem 12.5 c), e) (or f)) and d), we have 7° and 8°.

Conversely, assume that we do not have 1°. Then, by the Hahn-Banach theorem, there exists a $\Phi \in E^{**}$, $\Phi \neq 0$, such that $\Phi(f_n)=0$ $(n=1,2,\ldots)$, whence τ is not one-to-one. Thus $7° \Rightarrow 1°$.

Finally, if we have 8°, then, by Ch. I, § 12, theorem 12.5f), we obtain $[f_n]^{\perp}=\{0\}$, whence, by the Hahn-Banach theorem, $[f_n]=E^*$, i. e. 1°, which completes the proof.

Other characterizations of shrinking bases will be given in § 5, § 6, § 12.

Now we can give the example announced at the beginning of this section.

Example 4.1. Let J be the *real* linear space of all sequences of real numbers $\{\xi_n\}$ with $\lim_{n\to\infty}\xi_n=0$, satisfying

$$\|\{\xi_n\}\|=\sup\left[\sum_{i=1}^{n}(\xi_{k_{2i-1}}-\xi_{k_{2i}})^2+(\xi_{k_{2n+1}})^2\right]^{\frac{1}{2}}<\infty, \tag{4.21}$$

where the sup is taken over all positive integers n and finite increasing sequences of positive integers $k_1, k_2, \ldots, k_{2n+1}$. Then J endowed with the norm (4.21) is a Banach space and the sequence $\{x_n\}\subset J$ defined by

$$x_n=\{\delta_{nj}\}_{j=1}^{\infty}\qquad (n=1,2,\ldots) \tag{4.22}$$

is a shrinking monotone basis of J. The a. s. c. f. $\{f_n\}\subset J^*$ is a 1-shrinking basis of the space J^*. The corresponding basis $\{(f_n,0)\}\cup\{(0,f_n)\}$ of[1] $J^*\times J^*$ is 2-shrinking, etc. Furthermore, the sequence $\{y_n\}\subset J$ defined by

$$y_n=\sum_{i=1}^{n}x_i \tag{4.23}$$

is a 1-shrinking basis of J, the sequence $\{(y_n,0)\}\cup\{(0,y_n)\}$ is a 2-shrinking basis of J, etc.

We shall verify these assertions in several steps.

a) (4.21) is a norm on J. In fact, this is obvious, except for the triangle inequality of the norm. Let $x=\{\xi_n\}\in J, y=\{\eta_n\}\in J$. Then for any $\varepsilon>0$ there exists an increasing sequence of positive integers $k_1,k_2,\ldots,k_{n+1}$ such that

$$\begin{aligned}\|x+y\|&<\varepsilon+\left[\sum_{i=1}^{n}(\xi_{k_{2i-1}}+\eta_{k_{2i-1}}-\xi_{k_{2i}}-\eta_{k_{2i}})^2+(\xi_{k_{2n+1}}+\eta_{k_{2n+1}})^2\right]^{\frac{1}{2}}\\&\leqslant\varepsilon+\left[\sum_{i=1}^{n}(\xi_{k_{2i-1}}-\xi_{k_{2i}})^2+\xi_{k_{2n+1}}^2\right]^{\frac{1}{2}}+\left[\sum_{i=1}^{n}(\eta_{k_{2i-1}}-\eta_{k_{2i}})^2+\eta_{k_{2n+1}}^2\right]^{\frac{1}{2}}\\&\leqslant\varepsilon+\|x\|+\|y\|,\end{aligned}$$

whence $\|x+y\|\leqslant\|x\|+\|y\|$.

[1] For the definition of $E\times F$ see Ch. I, § 1. We recall that by $\{(f_n,0)\}\cup\{(0,f_n)\}$ we understand the sequence $g_{2n-1}=\{(f_n,0)\}$, $g_{2n}=\{(0,f_n)\}$ $(n=1,2,\ldots)$.

b) J is complete. In fact, this is an immediate consequence of (4.21), and also follows from the fact to be proved in step e) below that J is isometric with a closed linear subspace of J^{**}.

c) $\{x_n\}$ is a monotone basis of J. Indeed, for any $x=\{\xi_n\}\in J$ we have

$$\left\|x-\sum_{j=1}^{n}\xi_j x_j\right\| = \|\{\underbrace{0,0,\ldots,0}_{n},\ \xi_{n+1},\xi_{n+2},\ldots\}\|$$

$$=\sup\left[\sum_{i=1}^{m}(\xi_{k_{2i-1}}-\xi_{k_{2i}})^2+\xi_{k_{2m+1}}^2\right]^{\frac{1}{2}},$$

where the sup is taken over all integers $m\geqslant 1$ and finite increasing sequences of integers $k_1,k_2,\ldots,k_{2m+1}$ with $k_1\geqslant n+1$. Assume that this sup does nit converge to 0 as $n\to\infty$. Then there exist an increasing sequence of positive integers $\{n_j\}$, $n_0=0$, integers $n_{j-1}+1\leqslant k_1^{(j)}<k_2^{(j)}<\cdots<k_{2m_j+1}^{(j)}\leqslant n_j$ and an $\varepsilon>0$ such that

$$\left[\sum_{i=1}^{m_j}(\xi_{k_{2i-1}^{(j)}}-\xi_{k_{2i}^{(j)}})^2+\xi_{k_{2m_j+1}^{(j)}}^2\right]^{\frac{1}{2}}\geqslant\varepsilon \qquad (j=1,2,\ldots)$$

whence, by $\lim_{n\to\infty}\xi_n=0$ and (4.21), $\|\{\xi_n\}\|=\infty$. This contradiction proves that $x=\sum_{i=1}^{\infty}\xi_i x_i$. This expansion is unique, since $\sum_{i=1}^{\infty}\xi_i x_i=0$ obviously implies $\xi_n=0$ $(n=1,2,\ldots)$. Thus $\{x_n\}$ is a basis of J. Furthermore,

$$\left\|\sum_{i=1}^{n+m}\alpha_i x_i\right\|=\|\{\alpha_1,\ldots,\alpha_{n+m},0,0,\ldots\}\|\geqslant\|\{\alpha_1,\ldots,\alpha_n,0,0,\ldots\}\|=\left\|\sum_{i=1}^{n}\alpha_i x_i\right\|$$

for all finite sequences of scalars $\alpha_1,\ldots,\alpha_{n+m}$ and thus, by § 1, definition 1.1, $\{x_n\}$ is a monotone basis of J.

d) $\{x_n\}$ is shrinking. In fact, by the implication $2^\circ\Rightarrow 1^\circ$ of theorem 4.1, it is sufficient to prove that

$$\lim_{n\to\infty}\|f\|_n=0 \qquad (f\in J^*), \tag{4.24}$$

where $\|f\|_n$ is defined by (4.3). Assume a contrario that there exists an $f\in J^*$ for which (4.24) is not satisfied. Then, as we have seen in the proof of proposition 4.2, there exist an $\varepsilon>0$ and a sequence $\{z_n\}\subset E$ with $z_n=\sum_{i=m_{n-1}+1}^{m_n}\beta_i x_i$ $(m_0=0)$, $\|z_n\|\leqslant 1$ $(n=1,2,\ldots)$ such that

$$f(z_n)>\varepsilon \qquad (n=1,2,\ldots). \tag{4.25}$$

Now put

$$z=\sum_{n=1}^{\infty}\frac{1}{n}z_n=\sum_{n=1}^{\infty}\frac{1}{n}\sum_{i=m_{n-1}+1}^{m_n}\beta_i x_i=\sum_{i=1}^{\infty}\alpha_i x_i, \tag{4.26}$$

where

$$\alpha_i = \frac{1}{n}\beta_i \qquad (i=m_{n-1}+1,\dots,m_n;\ n=1,2,\dots). \tag{4.27}$$

Then $z \in J$, since, for any sum of type $\sum_{i=1}^{l} (\alpha_{k_{2i-1}} - \alpha_{k_{2i}})^2 + \alpha_{k_{2l+1}}^2$ it is true that for each term $(\alpha_{k_{2i-1}} - \alpha_{k_{2i}})^2$ either: a) $m_{n-1}+1 \leqslant k_{2i-1} < k_{2i} \leqslant m_n$ for a certain n or b) $m_{n-1}+1 \leqslant k_{2i-1} \leqslant m_n$, $m_{n'-1}+1 \leqslant k_{2i} \leqslant m_{n'}$, where $n' \neq n$. The sum of the terms $(\alpha_{k_{2i-1}} - \alpha_{k_{2i}})^2$ falling in case b) is less then or equal to

$$\sum_{n=1}^{\infty} \left(\frac{1}{n} + \frac{1}{n+1}\right)^2 < 4 \sum_{n=1}^{\infty} \frac{1}{n^2},$$

since for $m_{n-1}+1 \leqslant i \leqslant m_n$ $(n=1,2,\dots)$ we have, by (4.27) and (4.21),

$$|\alpha_i| = \frac{1}{n}|\beta_i| \leqslant \frac{1}{n}\left\| \sum_{j=m_{n-1}+1}^{m_n} \beta_j x_j \right\| = \frac{1}{n}\|z_n\| \leqslant \frac{1}{n},$$

and since $k_1 < k_2 \cdots < k_{2l+1}$.

On the other hand, the sum of the remaining terms $(\alpha_{k_{2i-1}} - \alpha_{k_{2i}})^2$ is less then or equal to

$$\sum_{n=1}^{\infty} \frac{\|z_n\|^2}{n^2},$$

since for $m_{n-1}+1 \leqslant k_{2p-1} < k_{2p} < \cdots < k_{2q-1} < k_{2q} \leqslant m_n$ we have, by (4.27) and (4.21),

$$\sum_{i=p}^{q} (\alpha_{k_{2i-1}} - \alpha_{k_{2i}})^2 = \frac{1}{n^2} \sum_{i=p}^{q} (\beta_{k_{2i-1}} - \beta_{k_{2i}})^2 \leqslant \frac{\|z_n\|^2}{n^2}.$$

Consequently,

$$\sum_{i=1}^{l} (\alpha_{k_{2i-1}} - \alpha_{k_{2i}})^2 + \alpha_{k_{2l+1}}^2 \leqslant \sum_{n=1}^{\infty} \frac{\|z_n\|^2}{n^2} + 4 \sum_{n=1}^{\infty} \frac{1}{n^2} + 1 \leqslant 5 \sum_{n=1}^{\infty} \frac{1}{n^2} + 1,$$

whence $\|z\| \leqslant \left(5 \sum_{n=1}^{\infty} \frac{1}{n^2} + 1\right)^{\frac{1}{2}}$. However, by (4.25) and (4.26) we have

$$f\left(\sum_{j=1}^{n} \frac{1}{j} z_j\right) > \varepsilon \sum_{j=1}^{n} \frac{1}{j} \qquad (n=1,2,\dots),$$

whence $f(z) = \infty$. This contradiction proves (4.24).

e) We have

$$\operatorname{codim}_{J^{**}} \pi(J) = 1, \tag{4.28}$$

where π denotes the canonical mapping of J into J^{**}. Indeed, let $\{f_n\} \subset J^*$ be the a.s.c.f. to the basis $\{x_n\}$. We shall prove that

$$\Phi \in J^{**} \Rightarrow \lim_{n\to\infty} \Phi(f_n) \quad \text{exists}, \tag{4.29}$$

$$\pi(J) = \{\Phi \in J^{**} \mid \lim_{n\to\infty} \Phi(f_n) = 0\}, \tag{4.30}$$

and that there exists a $\Phi_0 \in J^{**}$ satisfying

$$\Phi_0(f_n) = 1 \qquad (n = 1, 2, \ldots), \tag{4.31}$$

whence $J^{**} = \pi(J) \oplus [\Phi_0]$, which will complete the proof of (4.28).

Let Φ be an arbitrary element of J^{**}. Since by d) $[f_n] = J^*$, by Ch. I, § 12, theorem 12.1 we have

$$\Phi(f) = \sum_{i=1}^{\infty} \Phi(f_i) f(x_i) \qquad (f \in J^*).$$

However, for $n = 1, 2, \ldots$ we have

$$\begin{aligned} \left|\sum_{i=1}^{n} \Phi(f_i) f(x_i)\right| &= \left|f\left(\sum_{i=1}^{n} \Phi(f_i) x_i\right)\right| \leqslant \|f\| \left\|\sum_{i=1}^{n} \Phi(f_i) x_i\right\| \\ &= \|f\| \sup\left[\sum_{i=1}^{m} (\Phi(f_{k_{2i-1}}) - \Phi(f_{k_{2i}}))^2 + (\Phi(f_{k_{2m+1}}))^2\right]^{\frac{1}{2}}, \end{aligned}$$

where the sup is taken over all positive integers m and finite increasing sequences of positive integers $k_1, k_2, \ldots, k_{2m+1}$ with $\Phi(f_{k_j})$ replaced by 0 if $k_j > n$. Hence, for $n \to \infty$,

$$|\Phi(f)| \leqslant \|f\| \sup\left[\sum_{i=1}^{m} (\Phi(f_{k_{2i-1}}) - \Phi(f_{k_{2i}}))^2 + (\Phi(f_{k_{2m+1}}))^2\right]^{\frac{1}{2}},$$

and thus

$$\|\Phi\| \leqslant \sup\left[\sum_{i=1}^{m} (\Phi(f_{k_{2i-1}}) - \Phi(f_{k_{2i}}))^2 + (\Phi(f_{k_{2m+1}}))^2\right]^{\frac{1}{2}}, \tag{4.32}$$

where the sup is taken over all positive integers m and finite increasing sequences of positive integers $k_1, k_2, \ldots, k_{2m+1}$.

On the other hand, since by c) $\{x_n\}$ is a monotone basis of J, by Ch. I, § 12, proposition 12.2c) we have

$$\left\|\sum_{i=1}^{n} \Phi(f_i) x_i\right\| \leqslant \|\Phi\| \qquad (n = 1, 2, \ldots),$$

whence

$$\|\Phi\| \geqslant \left[\sum_{i=1}^{m} (\Phi(f_{k_{2i-1}}) - \Phi(f_{k_{2i}}))^2 + (\Phi(f_{k_{2m+1}}))^2\right]^{\frac{1}{2}} \tag{4.33}$$

for any positive integer m and finite increasing sequence of integers $k_1, k_2, \ldots, k_{2m+1}$. Combining (4.32) and (4.33) gives

$$\|\Phi\| = \sup\left[\sum_{i=1}^{m} (\Phi(f_{k_{2i-1}}) - \Phi(f_{k_{2i}}))^2 + (\Phi(f_{k_{2m+1}}))^2\right]^{\frac{1}{2}}, \tag{4.34}$$

where the sup is taken over all positive integers m and finite increasing sequences of positive integers $k_1, k_2, \ldots, k_{2m+1}$.

Now we can prove (4.30). If $\Phi \in \pi(J)$, then $\Phi = \pi(x)$ for some $x \in J$, whence, by the definition of J, $\Phi(f_n) = f_n(x) \to 0$ as $n \to \infty$. Thus we have the inclusion $\subset$ in (4.30). Conversely, let $\Phi \in J^{**}$ be such that $\lim_{n\to\infty} \Phi(f_n) = 0$. Then, by (4.34) and the definition of J, the sequence $x = \{\Phi(f_n)\}$ belongs to J. Hence, by c), $x = \sum_{i=1}^{\infty} \Phi(f_i) x_i = \sum_{i=1}^{\infty} f_i(x) x_i$ and thus $\Phi(f_i) = f_i(x)$ $(i = 1, 2, \ldots)$. This, together with d), proves that $\Phi = \pi(x)$, whence the inclusion $\supset$ in (4.30), whence the equality (4.30).

Furthermore, we have (4.29). Indeed, otherwise for some fixed $\varepsilon > 0$ and for n arbitrarily large it would be possible to have $|\Phi(f_n) - \Phi(f_m)| > \varepsilon$ for $m > n$, which contradicts (4.34).

Finally, since by (4.21) we have

$$\left\| \sum_{i=1}^{n} x_i \right\| = 1 \qquad (n = 1, 2, \ldots), \tag{4.35}$$

from Ch. I, § 12, proposition 12.2c) it follows that there exists a $\Phi_0 \in J^{**}$ satisfying (4.31), which completes the proof of (4.28).

f) $\{f_n\}$ is a 1-shrinking basis of J^*. In fact, by d) and Ch. I, § 12, theorem 12.1, $\{f_n\}$ is a basis of J^* with the a.s.c.f. $\{\pi(x_n)\}$. Since[1] by c) $[\pi(x_n)] = \pi(J)$, from e) it follows that $\{f_n\}$ is 1-shrinking.

g) The basis $\{(f_n, 0)\} \cup \{(0, f_n)\}$ of $J^* \times J^*$ is 2-shrinking etc. Indeed, by e) there exists a $\Phi_0 \in J^{**}$ such that every $\Phi \in J^{**}$ can be uniquely written in the form $\Phi = \pi(x) + \lambda \Phi_0$, with $x \in J$ and $-\infty < \lambda < \infty$.

[1] Indeed, since $\pi(x_n) \in \pi(J)$ $(n = 1, 2, \ldots)$ and since $\pi(J)$ is a closed linear subspace of J^{**}, we have $[\pi(x_n)] \subset \pi(J)$. On the other hand, if $\Phi \in \pi(J)$, then for some $x \in J$ we have $\Phi = \pi(x) = \pi\left(\sum_{i=1}^{\infty} f_i(x) x_i\right) = \sum_{i=1}^{\infty} f_i(x) \pi(x_i) \in [\pi(x_n)]$ and thus $\pi(J) \subset [\pi(x_n)]$. Hence, finally, $\pi([x_n]) = \pi(J)$.

Hence every $(\Phi,\Psi)\in J^{**}\times J^{**}$ can be uniquely written in the form $(\pi(x),\pi(y))+(\lambda\Phi_0,\mu\Phi_0)$, with $x,y\in J$ and $-\infty<\lambda,\mu<\infty$, and thus

$$\operatorname{codim}_{J^{**}\times J^{**}}\pi(J)\times\pi(J)=2. \tag{4.36}$$

However, for any pair of Banach spaces E, F we have the linear isometry[1] $E^*\times F^*\equiv(E\times F)^*$, by the mapping $(f,g)\rightarrow\phi$, where

$$\phi(x,y)=f(x)+g(y)\qquad(x\in E,\ y\in F).$$

Now, since $\{f_n\}$ is a basis of J^* with the a.s.c.f. $\{\pi(x_n)\}$ (see f)), $\{(f_n,0)\}\cup\{(0,f_n)\}$ is a basis of $J^*\times J^*$ with the a.s.c.f. $\{(\pi(x_n),0)\}\cup\{(0,\pi(x_n))\}$ spanning the subspace $\pi(J)\times\pi(J)$ (we identify $(J^*\times J^*)^*$ with $J^{**}\times J^{**}$ by the above isometry), whence, by (4.36), $\{(f_n,0)\}\cup\{(0,f_n)\}$ is 2-shrinking, etc.

h) By (4.35) and Ch. I, § 4, proposition 4.3, the sequence $\{y_n\}\subset J$ defined by (4.23) is a basis of J, with the a.s.c.f.

$$g_n=f_n-f_{n+1}\qquad(n=1,2,\ldots). \tag{4.37}$$

Since $f_n\in[g_j]$ for $n=2,3,\ldots$ and since $f_1\notin[g_j]$ (because $\Phi_0(f_1)=1$, $\Phi_0(g_n)=\Phi_0(f_n)-\Phi_0(f_{n+1})=0$ for $n=1,2,\ldots$ by (4.31)), we have, taking into account d),

$$\operatorname{codim}_{J^*}[g_n]=\operatorname{codim}_{[f_n]}[g_n]=1,$$

which shows that $\{y_n\}$ is 1-shrinking. With an argument similar to that used in g) it follows that $\{(y_n,0)\}\cup\{(0,y_n)\}$ is 2-shrinking, etc. This completes the proof of the assertions stated in example 4.1.

The answer to the problem of existence of k-shrinking bases is negative, as shown by

Example 4.2. Let E be a Banach space with a basis, whose conjugate space E^* is non-separable (e.g. $E=l^1$, or $E=L^1([0,1])$, or $E=C([0,1])$). Then for any non-negative integer k, the space E has no k-shrinking basis.

In the case when $k=0$, the answer to the problem of the existence of non-k-shrinking bases is also negative, as shown by

Example 4.3. Let E be a reflexive Banach space with a basis. Then all bases of E are shrinking.

In fact, this is an immediate consequence of Ch. I, § 12, corollary 12.2, but can be seen also directly, as follows. If $\{x_n\}$ is a non-shrinking basis of E, then for the a.s.c.f. $\{f_n\}\subset E^*$ we have $[f_n]\neq E^*$, whence there exists a $\Phi\in E^{**}$, $\Phi\neq 0$, such that $\Phi(f_n)=0$ $(n=1,2,\ldots)$. Since $\{f_n\}$ is

[1] See e.g. [10], p. 192, theorem 14 and the remark to this theorem.

total on E, it follows that $\Phi \notin \pi(E)$, and thus E cannot be reflexive. This proves our assertion.

In Vol. II, Ch. IV we shall see that if k is a positive integer, then every infinite dimensional Banach space with a basis possesses a non-k-shrinking basis.

§ 5. Retro-bases in conjugate Banach spaces

Definition 5.1. Let E^* be the conjugate space of a Banach space E. A basis $\{f_n\}$ of E^* is called *retro-basis*, if for the a.s.c.f. $\{\Phi_n\} \subset E^{**}$ we have $\{\Phi_n\} \subset \pi(E)$ (where π denotes, as before, the canonical mapping of E into E^{**}).

For instance, the natural basis of $E^* = l^p\ (p \geqslant 1)$ is a retro-basis. On the other hand, the basis $\{f_n\}$ of $E^* = l^1$ defined in Ch. I, § 14, example 14.1, formula (14.3), is a non-retro-basis, since for the a.s.c.f. $\{\chi_n\} \subset E^{**}$ we have $\chi_1 \notin \pi(E)$ (by Ch. I, § 14, formula (14.4)). Furthermore the natural basis $\Phi_0 \cup \{\pi(x_n)\}$ of J^{**} (§ 4, example 4.1) is a non-retro-basis of J^{**}. In these two examples only the first coefficient functional $\notin \pi(E)$, and all other coefficient functionals $\in \pi(E)$. However, there also exist non-retro-bases such that all associated coefficient functionals $\Phi_n \notin \pi(E)$ $(n = 1, 2, \ldots)$, e.g. the basis $\{h_n\}$ of $E^* = l^1$ defined in Ch. I, § 13, example 13.3, formula (13.14).

Some characterizations of retro-bases among bases of conjugate Banach spaces are collected in

Proposition 5.1. *For a basis $\{f_n\}$ of a conjugate Banach space E^* the following statement are equivalent:*

1°. *$\{f_n\}$ is a retro-basis of E^*.*

2°. *There exists a (shrinking) basis $\{x_n\}$ of the space E such that $\{f_n\}$ is the a.s.c.f. to $\{x_n\}$.*

3°. *$\{f_n\}$ is a w^*-Schauder basis of E^*.*

4°. *All subspaces*

$$M_n = [f_1, \ldots, f_{n-1}, f_{n+1}, f_{n+2}, \ldots] \tag{5.1}$$

of E are closed for the w^*-topology $\sigma(E^*, E)$.

Proof. 1° ⇒ 2°. If $\{f_n\}$ is a retro-basis of E^*, for the a.s.c.f. $\{\Phi_n\} \subset E^{**}$ we have $\Phi_n = \pi(x_n)$ with some $x_n \in E$ $(n = 1, 2, \ldots)$, whence $f_i(x_j) = \delta_{ij}$ $(i, j = 1, 2, \ldots)$. Then, by Ch. I, § 12, corollary 12.1, $\{x_n\}$ is a basis of E, which is obviously shrinking and has the a.s.c.f. $\{f_n\}$.

The equivalence 2° ⇔ 3° is an immediate consequence of Ch. I, § 14, theorem 14.1.

$2^\circ \Rightarrow 4^\circ$. If we have 2°, then $f_i(x_j)=\delta_{ij}$ $(i,j=1,2,\ldots)$ and by Ch. I, § 12, theorem 12.1, we also have

$$f=\sum_{i=1}^{\infty} f(x_i) f_i \qquad (f \in E^*).$$

Consequently, for each n the subspace M_n of E^* defined by (5.1) is nothing else but the $\sigma(E^*, E)$-closed hyperplane $\{f \in E^* \mid f(x_n)=0\}$.

$4^\circ \Rightarrow 1^\circ$. If we have 4°, then $f_n \notin M_n = \bar{M}_n$ = the $\sigma(E^*, E)$-closure of M_n $(n=1,2,\ldots)$, whence, by a well known corollary of the Hahn-Banach theorem (on separation of closed subspaces from outside points), applied in E^* endowed with the topology $\sigma(E^*, E)$, there exists an $x_n \in E$ such that $f_n(x_j)=\delta_{nj}$ $(n,j=1,2,\ldots)$. Consequently, for the a.s.c.f. $\{\Phi_n\} \subset E^{**}$ to $\{f_n\}$ we have $\Phi_n=\pi(x_n) \in \pi(E)$ $(n=1,2,\ldots)$, which completes the proof.

The equivalence $1^\circ \Leftrightarrow 2^\circ$ of proposition 5.1 above shows that the study of retro-bases reduces to that of shrinking bases. Therefore in the sequel we shall not study retro-bases of E^*, but only shrinking bases of E.

The equivalence $1^\circ \Leftrightarrow 3^\circ$ of proposition 5.1 reduces the study of retro-bases $\{f_n\}$ in conjugate Banach spaces E^* to that of bases $\{f_n\} \subset E^*$ which are also w^*-Schauder bases of E^*. The implication $1^\circ \Rightarrow 3^\circ$ shows that a retro-basis $\{f_n\}$ of E^* is necessarily a w^*-Schauder basis of E^*, but the converse of this statement is not true (if we do not assume that $[f_n]=E^*$), as shown e.g. by the sequence of coordinate functionals $\{f_n\}$ in $E^*=(l^1)^* \equiv m$ (see Ch. I, § 14, theorem 14.1).

Problem 5.1. a) Does every infinite dimensional conjugate Banach space E^* with a basis $\{f_n\}$ possess a retro-basis? b) If E has a basis and E^* is separable, does E^* possess a basis? What about a retro-basis?

An affirmative answer to a) would imply an affirmative answer to Ch. I, § 12, problem 12.1.

The answer to the problem of existence of non-retro-bases is negative, as shown by

Example 5.1. Let E be a reflexive Banach space with a basis. Then E^* has a basis and all bases in E^* are retro-bases.

Indeed, by Ch. I, § 12, corollary 12.2, E^* has a basis. On the other hand, from $E^{**}=\pi(E)$ it follows that every basis $\{f_n\}$ of E^* is a retro-basis.

One may ask whether a normal basis of a conjugate Banach E^* is necessarily a retro-basis of E^*. We shall now show that the answer is negative.

Let us recall that a norm $|||f|||$ on a conjugate Banach space E^* is said to be *the dual norm* of a norm $|||x|||$ on E if

$$|||f||| = \sup_{\substack{x \in E \\ |||x||| \leqslant 1}} |f(x)| \qquad (f \in E^*). \tag{5.2}$$

Lemma 5.1. *Let E be a Banach space and let* $|||f|||$ *be a norm on the conjugate space* E^*, *equivalent to the initial norm on* E^*. *If the set*

$$A = \{f \in E^* \mid |||f||| \leqslant 1\} \tag{5.3}$$

is closed for the weak topology* $\sigma(E^*, E)$, *then there exists on E a norm* $|||x|||$ *equivalent to the initial norm on E, such that* $|||f|||$ *is the dual norm of* $|||x|||$.

Proof. Since $|||f|||$ is equivalent to the initial norm $\|f\|$ on E^*, there exist two constants $C_1, C_2 > 0$ such that

$$C_1 \|f\| \leqslant |||f||| \leqslant C_2 \|f\| \qquad (f \in E^*). \tag{5.4}$$

Put

$$|||x||| = \sup_{\substack{f \in E^* \\ |||f||| \leqslant 1}} |f(x)| \qquad (x \in E). \tag{5.5}$$

Then $|||x|||$ is a norm on E, equivalent to the initial norm $\|x\|$ on E, since by (5.5) and (5.4) we have

$$\frac{1}{C_2}\|x\| = \frac{1}{C_2} \sup_{\substack{f \in E^* \\ \|f\| \leqslant 1}} |f(x)| \leqslant |||x||| \leqslant \frac{1}{C_1} \sup_{\substack{f \in E^* \\ \|f\| \leqslant 1}} |f(x)| = \frac{1}{C_1}\|x\| \qquad (x \in E).$$

Furthermore, by (5.5) we have $|f(x)| \leqslant |||f||| \, |||x|||$ for all $x \in E$, $f \in E^*$, whence

$$\sup_{\substack{x \in E \\ |||x||| \leqslant 1}} |f(x)| \leqslant |||f||| \qquad (f \in E^*). \tag{5.6}$$

On the other hand, since by our hypothesis A is a $\sigma(E^*, E)$-closed circled convex set in E^*, for every $f_0 \in E^*$ with $|||f_0||| > 1$ there exists, by a theorem of S. Mazur[1], an $x_0 \in E$ such that $f_0(x_0) > 1$, $|||x_0||| = \sup_{\substack{f \in E^* \\ |||f||| \leqslant 1}} |f(x_0)| \leqslant 1$. Consequently, in (5.6) we have the equality sign, i.e. $|||f|||$ is the dual norm of $|||x|||$ (since if there existed an $f \in E^*$ with $\sup_{\substack{x \in E \\ |||x||| \leqslant 1}} |f(x)| < |||f|||$, then for $f_0 = \left(\sup_{\substack{x \in E \\ |||x||| \leqslant 1}} |f(x)|\right)^{-1} f$ one would have $|||f_0||| > 1$, $\sup_{\substack{x \in E \\ |||x||| \leqslant 1}} |f_0(x)| = 1$, contradicting the above), which completes the proof of lemma 5.1.

[1] See e.g. [270], p. 108, theorem 3.

Proposition 5.2. *Let $\{x_n\}$ be a shrinking basis of a Banach space E, with the a.s.c.f. $\{f_n\} \subset E^*$. Then*

$$|||f||| = \sup_{1 \leqslant n < m < \infty} \left\| \sum_{i=n}^{m} f(x_i) f_i \right\| \quad (f \in E^*) \tag{5.7}$$

is an equivalent norm on E, which is the dual norm of a suitable equivalent norm on E.*

Proof. Since $\{x_n\}$ is shrinking, $\{f_n\}$ is a basis of E^*, whence

$$\|f\| = \left\| \sum_{i=1}^{\infty} f(x_i) f_i \right\| \leqslant |||f||| \leqslant 2 \sup_{1 \leqslant m < \infty} \left\| \sum_{i=1}^{m} f(x_i) f_i \right\| \leqslant 2 v_{\{f_n\}} \|f\| \; (f \in E^*),$$

and thus $|||f|||$ is a norm on E^*, equivalent to the initial norm on E^*. Therefore, by lemma 5.1, it will be sufficient to prove that the set $A = \{f \in E^* \mid |||f||| \leqslant 1\}$ is closed for the weak* topology $\sigma(E^*, E)$.

Let $\{g_d\}_{d \in \Delta}$ be a net in A, converging in the weak* topology $\sigma(E^*, E)$ to an element $g \in E^*$. Let $\varepsilon > 0$ be arbitrary and fix n, m with $1 \leqslant n \leqslant m < \infty$. Then for any $d \in \Delta$ we have, by $|||g_d||| \leqslant 1$,

$$\left\| \sum_{i=n}^{m} g(x_i) f_i \right\| \leqslant \left\| \sum_{i=n}^{m} g_d(x_i) f_i \right\| + \left\| \sum_{i=n}^{m} (g - g_d)(x_i) f_i \right\|$$
$$\leqslant |||g_d||| + \left\| \sum_{i=n}^{m} (g - g_d)(x_i) f_i \right\| \leqslant 1 + \left\| \sum_{i=n}^{m} (g - g_d)(x_i) f_i \right\|.$$

Furthermore, since $g_d(x_i) \to g(x_i)$ $(i = 1, 2, \ldots)$, we may choose a $d_0 = d_0(m, n) \in \Delta$ such that

$$\left\| \sum_{i=n}^{m} (g - g_d)(x_i) f_i \right\| < \varepsilon \quad (d \geqslant d_0).$$

Consequently, since $\varepsilon > 0$ and n, m have been arbitrary, we get

$$\left\| \sum_{i=n}^{m} g(x_i) f_i \right\| \leqslant 1 \quad (n \leqslant m; \; n = 1, 2, \ldots),$$

whence $|||g||| \leqslant 1$, i.e. $g \in A$, which completes the proof of proposition 5.2.

Now we can give the example annouced above.

Example 5.2. Let J be the space of § 4, example 4.1, let $\{x_n\}$ be the unit vector basis (4.22) of J (which is shrinking, as shown there), let $\{f_n\}$ be the a.s.c.f. to $\{x_n\}$, let $|||f|||$ be the equivalent norm on J^* defined by (5.7) and let E be the space J endowed with the equivalent norm $|||x|||$

of proposition 5.2. We claim that the sequence $\{g_n\} \subset E^*$ defined by

$$g_1 = f_1, \quad g_n = f_{n-1} - f_n \quad (n = 2, 3, \ldots) \tag{5.8}$$

is a normal basis of E^*, but not a retro-basis of E^*.

Indeed, from step h) of the proof of the assertions of § 4, example 4.1, it follows that $\{g_n\}$ is a basis of J^*, whence also of J^* endowed with the equivalent norm $|||f|||$, i.e. of E^*. Furthermore, obviously the a.s.c.f. $\{\Psi_n\} \subset E^{**}$ to the basis $\{g_n\}$ of E^* is

$$\Psi_1 = \Phi_0, \quad \Psi_n = -\Phi_0 + \sum_{i=1}^{n-1} \pi(x_i) \quad (n = 2, 3, \ldots), \tag{5.9}$$

where $\Phi_0 \in E^{**}$ is the functional satisfying (4.31) and where π denotes the canonical mapping of E into E^{**}. Thus $\Psi_n \notin \pi(E)$ $(n = 1, 2, \ldots)$, i.e. the basis $\{g_n\}$ of E^* is "very non-retro".

Finally let us prove that the basis $\{g_n\}$ is normal, i.e. that $|||g_n||| = |||\Psi_n||| = 1$ $(n = 1, 2, \ldots)$, where $|||\Psi_n||| = \sup_{\substack{f \in J^* \\ |||f||| \leqslant 1}} |\Psi_n(f)|$. For any $x = \{\xi_n\} \in J$ we have $|f_n(x)| = |\xi_n| \leqslant \|x\|$, which, together with $|f_n(x_n)| = 1 = \|x_n\|$, gives $\|f_n\| = 1$ $(n = 1, 2, \ldots)$. Furthermore, for any $x = \{\xi_n\} \in J$ we have $|(f_{n-1} - f_n)(x)| = |\xi_{n-1} - \xi_n| \leqslant \|x\|$ and therefore $\|f_{n-1} - f_n\| \leqslant 1$ $(n = 2, 3, \ldots)$. Hence

$$\begin{aligned} |||g_1||| &= |||f_1||| = \|f_1\| = 1, \\ |||g_n||| &= |||f_{n-1} - f_n||| = \sup_{1 \leqslant k \leqslant m < \infty} \left\| \sum_{i=k}^{m} (f_{n-1} - f_n)(x_i) f_i \right\| \\ &= \max(\|f_{n-1}\|, \|f_{n-1} - f_n\|, \|f_n\|) = 1 \qquad (n = 2, 3, \ldots), \end{aligned}$$

and thus also $|||\Psi_n||| = |||\Psi_n|||\ |||g_n||| \geqslant |\Psi_n(g_n)| = 1$ $(n = 1, 2, \ldots)$. Now let $f \in J^*$ with $|||f||| \leqslant 1$ and k, m be arbitrary. Then, by (4.35),

$$\left| \sum_{i=k}^{m} f(x_i) \right| = \left| \sum_{i=k}^{m} f(x_i) f_i \left(\sum_{j=1}^{m} x_j \right) \right| \leqslant \left\| \sum_{i=k}^{m} f(x_i) f_i \right\| \leqslant |||f||| \leqslant 1,$$

whence, taking into account (4.31),

$$\begin{aligned} |\Psi_n(f)| &= \left| -\Phi_0(f) + \sum_{i=1}^{n-1} [\pi(x_i)](f) \right| = \left| -\Phi_0 \left(\sum_{i=1}^{\infty} f(x_i) f_i \right) + \sum_{i=1}^{n-1} f(x_i) \right| \\ &= \left| \sum_{i=n}^{\infty} f(x_i) \right| \leqslant 1, \end{aligned}$$

and thus $|||\Psi_n||| \leqslant 1$ $(n=1,2,\ldots)$, which, together with the opposite inequality observed above, gives $|||\Psi_n|||=1$ $(n=1,2,\ldots)$. This completes the proof of the assertions of example 5.2.

Let us also mention, briefly, another interesting example of a normal non-retro-basis of a conjugate Banach space.

Example 5.3. Let $E=C(\omega^\omega)$ be the space of all continuous functions on the (compact metrizable) space ω^ω of all ordinal numbers $\leqslant \omega^\omega$ endowed with the order topology (with "open intervals" as neighbourhoods). Then it is well known[1] that $E^* \equiv l^1$. Let $\{f_n\} \subset E^*$ be the image in E^* of the natural basis of l^1 by this equivalence. Then obviously $\{f_n\}$ is a normal basis of E^*. Furthermore, the a.s.c.f. $\{\Phi_n\} \subset E^{**}$ to $\{f_n\}$ is equivalent to the natural basis of c_0, but $E=C(\omega^\omega)$ is not[2] isomorphic to c_0, and thus $\{f_n\}$ is not a retro-basis of E^*.

§ 6. k-boundedly complete bases

Definition 6.1. Let k be a non-negative integer. A basis $\{x_n\}$ of a Banach space E is called *k-boundedly complete*, if

a) in every $(k+1)$-dimensional linear subspace P_{k+1} of the Banach space of sequences of scalars

$$A_2 = \left\{ \{\alpha_n\} \subset K \;\middle|\; \sup_{1 \leqslant n < \infty} \left\| \sum_{i=1}^{n} \alpha_i x_i \right\| < \infty \right\}, \tag{6.1}$$

defined in Ch. I, § 5, remark 5.4, there exists a non-zero element $\{\alpha_j\} \in P_{k+1}$ such that the series $\sum_{i=1}^{\infty} \alpha_i x_i$ is convergent;

b) there exists a $(k+1)$-dimensional linear subspace P'_{k+1} of A_2 for which the above element $\{\alpha_j\}$ is unique up to a homothety.

The 0-boundedly complete bases, i.e. the bases $\{x_n\}$ such that the relation $\sup_{1 \leqslant n < \infty} \left\| \sum_{i=1}^{n} \alpha_i x_i \right\| < \infty$ implies the convergence of $\sum_{i=1}^{\infty} \alpha_i x_i$, are also called *boundedly complete bases.*

For instance, the natural basis of l^p $(p \geqslant 1)$ and the Haar basis of $L^p([0,1])$ $(p>1)$ are boundedly complete bases. We shall also give an example of a k-boundedly complete basis, for arbitrary $k>0$ (see example 6.1 below).

---

[1] See [217], theorem 6 and [26], corollary 3.

[2] See [26], theorem 1.

Some characterizations of k-boundedly complete bases are collected in

Theorem 6.1. *Let $\{x_n\}$ be a basis of a Banach space E, with the a.s.c.f. $\{f_n\}$, and let k be a non-negative integer. The following statements are equivalent:*

1°. *$\{x_n\}$ is k-boundedly complete.*

2°. *We have*[1]

$$\operatorname{codim}_{E^{**}}(\pi(E)\oplus[f_n]^{\perp})=k, \tag{6.2}$$

where π denotes the canonical mapping of E into E^{**}.

3°. *We have*

$$\operatorname{codim}_{[f_n]^*}\phi(E)=k, \tag{6.3}$$

where ϕ denotes the canonical mapping of E into $[f_n]^*$.

4°. *In every $(k+1)$-dimensional linear subspace Q_{k+1} of E^{**} there exists a non-zero element $\Phi\in Q_{k+1}$ such that the series $\sum_{i=1}^{\infty}\Phi(f_i)x_i$ is convergent, and there exists a $(k+1)$-dimensional linear subspace Q'_{k+1} of E^{**} for which this element Φ is unique up to a homothety.*

5°. *In every $(k+1)$-dimensional linear subspace R_{k+1} of $[f_n]^*$ there exists a non-zero element $\Psi\in R_{k+1}$ such that the series $\sum_{i=1}^{\infty}\Psi(f_i)x_i$ is convergent, and there exists a $(k+1)$-dimensional linear subspace R'_{k+1} of $[f_n]^*$ for which this element Ψ is unique up to a homothety.*

Proof. 1°⇒4°. Assume that we have 1°, and let Q_{k+1} be an arbitrary $(k+1)$-dimensional linear subspace of E^{**}. If $Q_{k+1}\cap[f_n]^{\perp}\neq\{0\}$, then for any $\Phi\in Q_{k+1}\cap[f_n]^{\perp}$ such that $\Phi\neq 0$, the series $\sum_{i=1}^{\infty}\Phi(f_i)x_i$ is convergent (to 0). If $Q_{k+1}\cap[f_n]^{\perp}=\{0\}$, then

$$\Phi\to\{\Phi(f_n)\}\qquad(\Phi\in Q_{k+1})$$

is a one-to-one linear mapping of Q_{k+1} into the space A_2 defined by (6.1) (we recall that by Ch. I, § 12, proposition 12.2c), $\{\Phi(f_n)\}\in A_2$ for every $\Phi\in E^{**}$). Hence $P_{k+1}=\{\{\Phi(f_n)\}\mid\Phi\in Q_{k+1}\}$ is a $(k+1)$-dimensional subspace of A_2 and thus by 1° there exists a non-zero element $\{\Phi(f_n)\}\in P_{k+1}$, i.e. a non-zero $\Phi\in Q_{k+1}$, such that the series $\sum_{i=1}^{\infty}\Phi(f_i)x_i$ is convergent.

[1] Since for any basis $\{x_n\}$ of E, $\pi(E)\oplus[f_n]^{\perp}$ and $\phi(E)$ are norm-closed linear subspaces of E^{**} and $[f_n]^*$ respectively (by Ch. I, § 12, theorem 12.5), the left hand sides of (6.2) and (6.3) have meaning for any basis $\{x_n\}$ of E.

On the other hand, by 1° there exists a $(k+1)$-dimensional linear subspace P'_{k+1} of A_2 such that the non-zero element $\{\alpha_j\}\in P'_{k+1}$ for which the series $\sum_{i=1}^{\infty}\alpha_i x_i$ is convergent, is unique up to a homothety. Let $\{\alpha_n^{(1)}\}, \{\alpha_n^{(2)}\}, \ldots, \{\alpha_n^{(k+1)}\}$ be a basis of P'_{k+1}. Then by $P'_{k+1}\subset A_2$ and Ch. I, § 12, proposition 12.2c), there exist $\Phi_1, \Phi_2, \ldots, \Phi_{k+1}\in E^{**}$ such that

$$\Phi_i(f_n)=\alpha_n^{(i)} \qquad (n=1,2,\ldots;\ i=1,2,\ldots,k+1),$$

whence

$$\Phi\to\{\Phi(f_n)\} \qquad (\Phi\in[\Phi_i]_{i=1}^{k+1})$$

is a linear mapping of $[\Phi_i]_{i=1}^{k+1}$ onto P'_{k+1}, and hence an isomorphism. Thus for $Q'_{k+1}=[\Phi_i]_{i=1}^{k+1}$ we have $\dim Q'_{k+1}=k+1$ and the non-zero element $\Phi\in Q'_{k+1}$, for which the series $\sum_{i=1}^{\infty}\Phi(f_i)x_i$ is convergent, is unique up to a homothety.

4°⇔2°. By Ch. I, § 12, theorem 12.5f), we have

$$\pi(E)\oplus[f_n]^{\perp}=\left\{\Phi\in E^{**}\;\middle|\;\sum_{i=1}^{\infty}\Phi(f_i)x_i \text{ converges}\right\}.$$

Hence, by § 4, lemma 4.1, we have 4°⇔2°.

2°⇒3°. If $E^{**}=\pi(E)\oplus[f_n]^{\perp}\oplus B$, where $\dim B=k$, then by restriction to $[f_n]$,

$$[f_n]^*=\phi(E)\oplus(B|_{[f_n]}),$$

where $\dim(B|_{[f_n]})=k$ (since $B\cap[f_n]^{\perp}=\{0\}$).

3°⇔5°. By Ch. I, § 12, theorem 12.5e), we have

$$\phi(E)=\left\{\Psi\in[f_n]^*\;\middle|\;\sum_{i=1}^{\infty}\Psi(f_i)x_i \text{ converges}\right\}.$$

Hence, by § 4, lemma 4.1, we have 3°⇔5°.

5°⇔1°. By Ch. I, § 12, theorem 12.5c), $[f_n]^*$ is isomorphic to the Banach space A_2, by the mapping

$$\Psi\to\{\Psi(f_n)\} \qquad (\Psi\in[f_n]^*).$$

Hence 5°⇔1°, which completes the proof of theorem 6.1.

From theorem 6.1 it follows that the k-boundedly complete bases have the following relation of duality with k-shrinking bases:

Corollary 6.1. *Let $\{x_n\}$ be a basis of a Banach space E, with the a.s.c.f. $[f_n]$, and let k be a non-negative integer. Then*

a) *$\{x_n\}$ is k-boundedly complete if and only if $\{f_n\}$ is a k-shrinking basis of $[f_n]$.*

b) *$\{x_n\}$ is k-shrinking if and only if $\{f_n\}$ is a k-boundedly complete basis of $[f_n]$.*

Proof. a) The a.s.c.f. $\{\Psi_n\} \subset [f_n]^*$ to $\{f_n\}$ is nothing else but $\Psi_n = \phi(x_n)$ $(n=1,2,\ldots)$, whence[1] $[\Psi_n] = [\phi(x_n)] = \phi(E)$. Consequently, by the equivalence $1^\circ \Leftrightarrow 3^\circ$ of theorem 6.1, $\{x_n\}$ is k-boundedly complete if and only if $\operatorname{codim}_{[f_n]^*}[\Psi_n] = k$. However, by § 4, definition 4.1, this condition means that $\{f_n\}$ is a k-shrinking basis of $[f_n]$.

b) Since ϕ is an isomorphism of $E = [x_n]$ onto $\phi(E) = [\Psi_n]$ (by Ch. I, § 12, theorem 12.2e)) and since any isomorphic image of a k-shrinking basis is obviously a k-shrinking basis, $\{x_n\}$ is a k-shrinking basis of E if and only if $\{\Psi_n\} = \{\phi(x_n)\}$ is a k-shrinking basis of $[\Psi_n] = \phi(E)$. However, by part a) (applied to the basis $\{f_n\}$ of the Banach space $[f_n]$), this happens if and only if $\{f_n\}$ is a k-boundedly complete basis of $[f_n]$. This completes the proof of corollary 6.1.

Combining theorems 4.1, 6.1 and corollary 6.1, one can obtain other characterizations of k-shrinking and k-boundedly complete bases.

Now we can also give the example announced at the beginning of this section.

Example 6.1. Let J be the Banach space of sequences of real numbers introduced in § 4, example 4.1. Then the basis $\{x_n\}$ of J defined by (4.22) is 1-boundedly complete. The corresponding basis $\{(x_n, 0)\} \cup \{(0, x_n)\}$ of $J \times J$ is 2-boundedly complete, etc. Furthermore, if $\{f_n\} \subset J^*$ is the a.s.c.f. to $\{x_n\}$, then $\{f_n\}$ is a boundedly complete basis of J, the sequence $\{g_n\} \subset J^*$ defined by (4.37) is a 1-boundedly complete basis of $[g_n]$, the corresponding basis $\{(g_n, 0)\} \cup \{(0, g_n)\}$ of $[g_n] \times [g_n]$ is 2-boundedly complete, etc.

Indeed, as shown in example 4.1, part f), $\{f_n\}$ is a 1-shrinking basis of $J^* = [f_n]$. Hence, by corollary 6.1 a), $\{x_n\}$ is 1-boundedly complete. Furthermore, as shown in example 4.1, part g), $\{(f_n, 0)\} \cup \{(0, f_n)\}$ is a 2-shrinking basis of $J^* \times J^* \equiv (J \times J)^*$. Since this is nothing else but the a.s.c.f. to the basis $\{(x_n, 0)\} \cup \{(0, x_n)\}$ of $J \times J$, from corollary 6.1 a) it follows that $\{(x_n, 0)\} \cup \{(0, x_n)\}$ is 2-boundedly complete, etc. On the other hand, as shown in example 4.1, part h), the sequence $\{y_n\} \subset J$ defined by (4.23) is a 1-shrinking basis of J, having the a.s.c.f. $\{g_n\}$ defined by (4.37). Hence, by corollary 6.1 b), $\{g_n\}$ is a 1-boundedly complete basis of $[g_n]$. Furthermore, as shown in example 4.1, part h), $\{(y_n, 0)\} \cup \{(0, y_n)\}$ is a 2-shrinking basis of J, with the a.s.c.f. $\{(g_n, 0)\} \cup \{(0, g_n)\}$. Hence, by corollary 6.1 b), $\{(g_n, 0)\} \cup \{(0, g_n)\}$ is a 2-boundedly complete basis of $[g_n] \times [g_n]$, etc. This completes the proof of the assertions stated in example 6.1.

---

[1] See the argument used in § 4, example 4.1, the footnote to part f).

The answer to the problem of existence of k-boundedly complete bases is negative, as shown by

Example 6.2. Let k be a non-negative integer and let E be a Banach space with a basis, which is not isomorphic to any subspace of codimension k of any conjugate Banach space (e.g. $E=c_0$, or $E=L^1([0,1])$, or $E=C([0,1])$). Then the space E has no k-boundedly complete basis.

Indeed, assume that $\{x_n\}$ is a k-boundedly complete basis of E, with the a.s.c.f. $\{f_n\}$. Then, by the implication 1°⇒3° of theorem 6.1, we have $\operatorname{codim}_{[f_n]^*}\phi(E)=k$, where ϕ is the canonical mapping of E into $[f_n]^*$. Since by Ch. I, § 12, theorem 12.2e), ϕ is an isomorphism, it follows that E is isomorphic to the subspace $\phi(E)$ of codimension k of the conjugate space $[f_n]^*$, which contradicts our assumption on E.

In the case when $k=0$, the answer to the problem of existence of non-k-boundedly complete bases is also negative, as shown by

Example 6.3. Let E be a reflexive Banach space with a basis. Then all bases of E are boundedly complete.

Indeed, assume that $\{x_n\}$ is a basis of E. Then, as we have seen in § 4, example 4.3, $\{x_n\}$ is shrinking, and thus the a.s.c.f. $\{f_n\}$ is a basis of $E^*=[f_n]$. Since E^* is also reflexive, again from § 4, example 4.3, it follows that $\{f_n\}$ is shrinking. Hence, by corollary 6.1a), $\{x_n\}$ is boundedly complete.

In Vol. II, Ch. IV we shall see that if k is a positive integer, then every infinite dimensional Banach space with a basis possesses a non-k-boundedly complete basis.

In view of the special importance of the particular case $k=0$, let us give separately

Theorem 6.2. *Let $\{x_n\}$ be a basis of a Banach space E, with the a.s.c.f. $\{f_n\}\subset E^*$. The following statements are equivalent:*

1°. *$\{x_n\}$ is boundedly complete (i.e. the relation $\sup_{1\leqslant n<\infty}\left\|\sum_{i=1}^{n}\alpha_i x_i\right\|<\infty$ implies that $\sum_{i=1}^{\infty}\alpha_i x_i$ converges).*

2°. *$\pi(E)\oplus[f_n]^{\perp}=E^{**}$, where π denotes the canonical mapping of E into E^{**}.*

3°. *$\phi(E)=[f_n]^*$, where ϕ denotes the canonical mapping of E into $[f_n]^*$ (hence, in this case E is canonically isomorphic—and, if $\{x_n\}$ is a monotone basis, E is canonically linearly isometric—to $[f_n]^*$).*

4°. *For every $\Phi\in E^{**}$ the series $\sum_{i=1}^{\infty}\Phi(f_i)x_i$ converges.*

5°. *For every $\Psi\in[f_n]^*$ the series $\sum_{i=1}^{\infty}\Psi(f_i)x_i$ converges.*

These statements are implied by the following statements, equivalent to each other:

6°. *For every number* $c>0$ *there exists a number* $r_c>0$ *(independent of* n *and* $\{\alpha_j\}\subset K$*), such that*

$$\left\|\sum_{i=1}^{n}\alpha_i x_i\right\|=1, \quad \left\|\sum_{i=n+1}^{\infty}\alpha_i x_i\right\|\geqslant c \quad \textit{imply} \quad \left\|\sum_{i=1}^{\infty}\alpha_i x_i\right\|\geqslant 1+r_c. \tag{6.4}$$

7°. *For every* $\varepsilon>0$ *there exists a* $\delta=\delta(\varepsilon)>0$ *(independent of* n *and* $\{\alpha_j\}\subset K$*), such that*

$$\left\|\sum_{i=1}^{n}\alpha_i x_i\right\|>1-\delta, \quad \left\|\sum_{i=1}^{\infty}\alpha_i x_i\right\|=1 \quad \textit{imply} \quad \left\|\sum_{i=n+1}^{\infty}\alpha_i x_i\right\|\leqslant\varepsilon. \tag{6.5}$$

Proof. The equivalences $1^\circ\Leftrightarrow\cdots\Leftrightarrow5^\circ$ are nothing else but the particular case $k=0$ of theorem 6.1.

To show the equivalence $6^\circ\Leftrightarrow7^\circ$, assume that 7° is not satisfied, i.e. there exists an $\varepsilon>0$ such that for every $\delta>0$ one can find $\{\alpha_j\}\subset K$ and n with

$$\left\|\sum_{i=1}^{n}\alpha_i x_i\right\|>1-\delta, \quad \left\|\sum_{i=1}^{\infty}\alpha_i x_i\right\|=1, \quad \left\|\sum_{i=n+1}^{\infty}\alpha_i x_i\right\|>\varepsilon. \tag{6.6}$$

Then for $c=\dfrac{\varepsilon}{v_{\{x_n\}}}$ (where $v_{\{x_n\}}$ is the norm of the basis $\{x_n\}$) there exists no $r_c>0$ so as to have (6.4). Indeed, let $r_c>0$ be arbitrary, let $\delta=\dfrac{r_c}{1+r_c}$, whence $r_c=\dfrac{\delta}{1-\delta}$, and let $\beta_i=\dfrac{\alpha_i}{\left\|\sum_{j=1}^{n}\alpha_j x_j\right\|}$ $(i=1,2,\ldots)$, where $\{\alpha_j\}\subset K$ and n are as in (6.6). Then

$$\left\|\sum_{i=1}^{n}\beta_i x_i\right\|=1, \quad \left\|\sum_{i=n+1}^{\infty}\beta_i x_i\right\|=\frac{\left\|\sum_{i=n+1}^{\infty}\alpha_i x_i\right\|}{\left\|\sum_{j=1}^{n}\alpha_j x_j\right\|}\geqslant\frac{\varepsilon}{v_{\{x_n\}}\left\|\sum_{j=1}^{\infty}\alpha_j x_j\right\|}$$

$$=\frac{\varepsilon}{v_{\{x_n\}}}=c,$$

$$\left\|\sum_{i=1}^{\infty}\beta_i x_i\right\|=\frac{1}{\left\|\sum_{j=1}^{n}\alpha_j x_j\right\|}<\frac{1}{1-\delta}=1+\frac{\delta}{1-\delta}=1+r_c,$$

which proves that non-7° $\Rightarrow$ non-6°.

Conversely, assume that 6° is not satisfied, i.e. there exists a $c>0$ such that for every $r_c>0$ one can find $\{\alpha_j\}\subset K$ and n with

$$\left\|\sum_{i=1}^{n}\alpha_i x_i\right\|=1,\quad \left\|\sum_{i=n+1}^{\infty}\alpha_i x_i\right\|\geqslant c,\quad \left\|\sum_{i=1}^{\infty}\alpha_i x_i\right\|<1+r_c. \tag{6.7}$$

Then for $\varepsilon=c(1-\eta)$, where $0<\eta<1$ is arbitrary, there exists no δ so as to have (6.5). Indeed, let $\delta>0$ be arbitrary, such that[1] $\delta\leqslant\eta$, let $r_c=\frac{\delta}{1-\delta}$, whence $\delta=\frac{r_c}{1+r_c}$, let $\{\alpha_j\}\subset K$ and n as in (6.7) and let

$$\beta_i=\frac{\alpha_i}{\left\|\sum_{j=1}^{\infty}\alpha_j x_j\right\|}\quad (i=1,2,\ldots). \text{ Then}$$

$$\left\|\sum_{i=1}^{n}\beta_i x_i\right\|=\frac{1}{\left\|\sum_{j=1}^{\infty}\alpha_j x_j\right\|}>\frac{1}{1+r_c}=1-\frac{r_c}{1+r_c}=1-\delta,\quad \left\|\sum_{i=1}^{\infty}\beta_i x_i\right\|=1,$$

$$\left\|\sum_{i=n+1}^{\infty}\beta_i x_i\right\|\geqslant\frac{c}{\left\|\sum_{j=1}^{\infty}\alpha_j x_j\right\|}>\frac{c}{1+r_c}=c(1-\delta)\geqslant c(1-\eta)=\varepsilon,$$

which shows that non-6° $\Rightarrow$ non-7°.

Finally, assume that we have 6° and let $\{\alpha_n\}\subset K$ be such that $\sup_{1\leqslant n<\infty}\left\|\sum_{i=1}^{n}\alpha_i x_i\right\|=A$. Choose a sequence of positive integers $\{n_k\}$ such that $\lim_{k\to\infty}\|y_{n_k}\|=\overline{\lim_{n\to\infty}}\|y_n\|=B$, where we have put $y_n=\sum_{i=1}^{n}\alpha_i x_i\,(n=1,2,\ldots)$. If $B=0$, then $\sum_{i=1}^{\infty}\alpha_i x_i$ converges (to 0). If $B\neq 0$, we shall show that $\{y_{n_k}\}$ is a Cauchy sequence. In fact, otherwise there would exist a $\delta>0$ and subsequences $\{y_{n_{k_j}}\}$, $\{y_{n_{l_j}}\}$ of $\{y_{n_k}\}$ with $n_{k_j}>n_{l_j}$ $(j=1,2,\ldots)$, such that

$$\|y_{n_{k_j}}-y_{n_{l_j}}\|\geqslant\delta\qquad (j=1,2,\ldots).$$

Then, since

$$\left\|\frac{y_{n_{k_j}}-y_{n_{l_j}}}{\|y_{n_{l_j}}\|}\right\|\geqslant\frac{\delta}{A}=c>0,$$

[1] This is no restriction of the generality, since if for a $\delta_0>0$ we have (6.5), then also for any $\delta\leqslant\delta_0$ we have (6.5).

we would have, by 6°,

$$\left\| \frac{y_{n_{l_j}}}{\|y_{n_{l_j}}\|} + \frac{y_{n_{k_j}} - y_{n_{l_j}}}{\|y_{n_{l_j}}\|} \right\| \geqslant 1 + r_c,$$

whence

$$\|y_{n_{k_j}}\| \geqslant \|y_{n_{l_j}}\| (1 + r_c),$$

and thus

$$B = \lim_{j \to \infty} \|y_{n_{k_j}}\| \geqslant \lim_{j \to \infty} \|y_{n_{l_j}}\| (1 + r_c) = B(1 + r_c),$$

which is impossible, since $B \neq 0$. Consequently, $\{y_{n_k}\}$ is a Cauchy sequence, whence $\lim_{k \to \infty} y_{n_k} = x \in E$. Then for the sequence of coefficient functionals $\{f_n\} \subset E^*$ associated to the basis $\{x_n\}$ we have

$$f_j(x) = \lim_{k \to \infty} f_j(y_{n_k}) = \lim_{k \to \infty} f_j\left(\sum_{i=1}^{n_k} \alpha_i x_i \right) = \alpha_j \qquad (j = 1, 2, \ldots),$$

whence $x = \sum_{i=1}^{\infty} f_i(x) x_i = \sum_{i=1}^{\infty} \alpha_i x_i$, which shows that $\sum_{i=1}^{\infty} \alpha_i x_i$ converges whenever $\sup_{1 \leqslant n < \infty} \left\| \sum_{i=1}^{n} \alpha_i x_i \right\| < \infty$. This completes the proof of theorem 6.2.

The converse implications $1° \Rightarrow 6°$, $1° \Rightarrow 7°$ in theorem 6.2 do not hold even if $\{x_n\}$ is monotone and normal, as shown by

Example 6.4. Let $E = l^1$ and let $\{x_n\}$ be the basis of E defined by

$$x_1 = \tfrac{1}{2} e_1 + \tfrac{1}{2} e_2, \qquad x_2 = -\tfrac{1}{2} e_1 + \tfrac{1}{2} e_2, \qquad x_n = e_n \quad (n = 3, 4, \ldots), \tag{6.8}$$

where $\{e_n\}$ is the natural basis of l^1. Then $\{x_n\}$ is a boundedly complete basis of E, which is monotone and normal, but $\{x_n\}$ does not satisfy conditions 6°, 7° of theorem 6.2.

Indeed, $\{x_n\}$ is a boundedly complete basis of E, since so is $\{e_n\}$. By § 2, example 2.1, $\{x_n\}$ is also normal. Since

$$\|\alpha_1 x_1\| = \left\| \frac{\alpha_1}{2} e_1 + \frac{\alpha_1}{2} e_2 \right\| = |\alpha_1|,$$

$$\|\alpha_1 x_1 + \alpha_2 x_2\| = \left\| \frac{\alpha_1 - \alpha_2}{2} e_1 + \frac{\alpha_1 + \alpha_2}{2} e_2 \right\| = \frac{|\alpha_1 - \alpha_2| + |\alpha_1 + \alpha_2|}{2}$$

$$= \max(|\alpha_1|, |\alpha_2|),$$

$$\left\| \sum_{i=1}^{n} \alpha_i x_i \right\| = \max(|\alpha_1|, |\alpha_2|) + \sum_{i=3}^{n} |\alpha_i| \qquad (n \geqslant 3),$$

$\{x_n\}$ is also monotone. Finally, for $\alpha_1=\alpha_2=1$, $\alpha_3=\alpha_4=\cdots=0$, we have

$$\|x_1\|=1, \quad \left\|x_2+\sum_{i=3}^{\infty} 0\cdot x_i\right\|=1, \quad \left\|x_1+x_2+\sum_{i=3}^{\infty} 0\cdot x_i\right\|=1,$$

which shows that $\{x_n\}$ does not satisfy conditions 6°, 7°.

Since this basis $\{x_n\}$ is equivalent to the unit vector basis $\{e_n\}$ of l^1, which obviously satisfies 6°, 7°, example 6.4 also shows that conditions 6°, 7° are not isomorphic properties (i.e. they are not invariant under an isomorphism of the space E onto another space E_1). The following problem arises naturally:

Problem 6.2. If $\{x_n\}$ is a boundedly complete basis of a Banach space E, is it possible to introduce on E an equivalent norm, in which $\{x_n\}$ satisfies conditions 6°, 7° of theorem 6.2?

§ 7. Bases of types wc_0, $(wc_0)^*$, swc_0 and $(swc_0)^*$

For $\{z_n\}$, z in a Banach space E, such that $\{z_n\}$ converges weakly to z, we shall use the notation $z_n \xrightarrow{w} z$.

Definition 7.1. A basis $\{x_n\}$ of a Banach space E is said to be

a) *of type* wc_0, if $\{x_n\}$ is a bounded basis[1], and $x_n \xrightarrow{w} 0$;

b) *of type* $(wc_0)^*$, if the a.s.c.f. $\{f_n\}$ is a basis of type wc_0 of $[f_n]$;

c) *of type* swc_0, if $\{x_n\}$ is a bounded basis and there exists a subsequence $\{x_{i_n}\}$ of $\{x_n\}$ such that $x_{i_n} \xrightarrow{w} 0$;

d) *of type* $(swc_0)^*$, if the a.s.c.f. $\{f_n\}$ is a basis of type swc_0 of $[f_n]$.

For instance, the natural bases of c_0 and l^p $(p>1)$ and the normalized Haar basis[2] of $L^p([0,1])$ $(p>1)$ are of type wc_0, while the basis of c_0 constructed in § 2, example 2.2, formula (2.24), and the natural basis of l^1 are of type non-wc_0. Dual examples of bases of types $(wc_0)^*$ and non-$(wc_0)^*$ follow from proposition 7.3 below (see also theorem 7.1).

Let us make the following remark concerning the hypothesis of boundedness of $\{x_n\}$ in definition 7.1 above. We have assumed that $\inf_{1\leqslant n<\infty}\|x_n\|>0$, since for every basis $\{x_n\}$ the basis $\left\{\frac{1}{n\|x_n\|}x_n\right\}$ converges (whence also weakly converges) to 0 and we have also assumed that $\sup_{1\leqslant n<\infty}\|x_n\|<\infty$, since for every basis $\{x_n\}$ the a.s.c.f. to the basis $\left\{\frac{n}{\|x_n\|}x_n\right\}$ converges (whence also weakly converges) to 0.

---

[1] I.e. satisfying $0<\inf_{1\leqslant n<\infty}\|x_n\|\leqslant\sup_{1\leqslant n<\infty}\|x_n\|<\infty$ (see Ch. I, § 3, definition 3.2).

[2] See § 2, formula (2.3).

Similar remarks are also valid for the definitions which we shall give in §§ 8, 9 and 10.

Proposition 7.1. *A subsequence* $\{x_{i_n}\}$ *of a bounded basis* $\{x_n\}$ *of a Banach space E satisfies* $x_{i_n} \xrightarrow{w} 0$ *if and only if* $\{x_{i_n}\}$ *is a basis of type* wc_0 *of* $[x_{i_n}]$. *Consequently*

a) *Every subsequence* $\{x_{i_n}\}$ *of a basis* $\{x_n\}$ *of type* wc_0 *is a basis of type* wc_0 *of its closed linear hull* $[x_{i_n}]$.

b) *A bounded basis* $\{x_n\}$ *is of type* swc_0 *if and only if it contains a subsequence* $\{x_{i_n}\}$ *which is a basis of type* wc_0 *of* $[x_{i_n}]$.

Proof. If $\{x_{i_n}\}$ is a subsequence of a basis $\{x_n\}$, then, by Ch. I, § 4, proposition 4.1, $\{x_{i_n}\}$ is a basis of $[x_{i_n}]$. On the other hand, by the Hahn-Banach theorem, $x_{i_n} \xrightarrow{w} 0$ if and only if $x_{i_n} \to 0$ for the w-topology of $[x_{i_n}]$ (i.e. for the weak topology $\sigma([x_{i_n}], [x_{i_n}]^*)$). Hence $x_{i_n} \xrightarrow{w} 0$ if and only if $\{x_{i_n}\}$ is a basis of type wc_0 of $[x_{i_n}]$. The statements a) and b) being obvious consequences of this result, the proof of proposition 7.1 is complete.

Proposition 7.1 b) makes possible to deduce certain properties of bases of type swc_0 from those of bases of type wc_0.

Theorem 7.1. a) *The Schauder basis of* $C([0,1])$ *is of types non-*wc_0, swc_0 *and non-*$(swc_0)^*$.

b) *The normalized Haar basis of* $L^1([0,1])$ *is of types non-*swc_0, *non-*$(wc_0)^*$ *and* $(swc_0)^*$.

Proof. a) Define $f \in C([0,1])^*$ by

$$f(x) = x(t_0) \qquad (x \in C([0,1])), \tag{7.1}$$

where

$$t_0 = \sum_{j=1}^{\infty} \frac{1}{2^{j^2}}, \tag{7.2}$$

and consider the subsequence $\{x_{2^{k_n}+l_n}\}$ of the Schauder basis[1] $\{x_n\}_{n=0}^{\infty}$ of $C([0,1])$, where

$$k_n = n^2 - 1 \qquad (n = 1, 2, \ldots), \tag{7.3}$$

$$l_n = \frac{1}{2} \sum_{j=1}^{n} 2^{n^2 - j^2} + \frac{1}{2} \qquad (n = 1, 2, \ldots). \tag{7.4}$$

Then

$$\frac{2l_n - 1}{2^{k_n+1}} = \frac{2l_n - 1}{2^{n^2}} = \sum_{j=1}^{n} \frac{1}{2^{j^2}} \qquad (n = 1, 2, \ldots), \tag{7.5}$$

[1] For the definition of the Schauder basis $\{x_n\}_{n=0}^{\infty}$ of $C([0,1])$ see Ch. I, § 2, example 2.2, formula (2.3).

whence, by (7.2),

$$\frac{2l_n-1}{2^{k_n+1}} < t_0 \qquad (n=1,2,\ldots). \tag{7.6}$$

On the other hand, since

$$\sum_{j=n+1}^{\infty} \frac{1}{2^{j^2}} < \sum_{j=(n+1)^2}^{\infty} \frac{1}{2^j} = \frac{1}{2^{(n+1)^2-1}} \leqslant \frac{1}{2^{n^2}} = \frac{1}{2^{k_n+1}} \qquad (n=1,2,\ldots), \tag{7.7}$$

we also have, taking into account (7.2) and (7.5),

$$t_0 = \frac{2l_n-1}{2^{k_n+1}} + \sum_{j=n+1}^{\infty} \frac{1}{2^{j^2}} < \frac{2l_n}{2^{k_n+1}} \qquad (n=1,2,\ldots). \tag{7.8}$$

Consequently, taking into account (7.1), (7.6), (7.8) and the definition of $x_{2^{k_n}+l_n}$ on the segment $\left[\frac{2l_n-1}{2^{k_n+1}}, \frac{2l_n}{2^{k_n+1}}\right]$, we obtain

$$\begin{aligned} |f(x_{2^{k_n}+l_n})-1| &= \left| x_{2^{k_n}+l_n}(t_0) - x_{2^{k_n}+l_n}\left(\frac{2l_n-1}{2^{k_n+1}}\right) \right| \\ &= \left| \frac{1}{\frac{1}{2^{k_n+1}}} \left(t_0 - \frac{2l_n-1}{2^{k_n+1}}\right) \right| = 2^{n^2} \sum_{j=n+1}^{\infty} \frac{1}{2^{j^2}} \\ &< 2^{n^2} \frac{1}{2^{(n+1)^2-1}} = \frac{1}{2^{2n}} \qquad (n=1,2,\ldots), \end{aligned}$$

whence

$$\lim_{n\to\infty} f(x_{2^{k_n}+l_n}) = 1, \tag{7.9}$$

which proves that $\{x_n\}_{n=0}^{\infty}$ is of type non-wc_0.

On the other hand, for the subsequence $\{x_{i_k}\}_{k=0}^{\infty} = \{x_{2^k+1}\}_{k=0}^{\infty}$ of $\{x_n\}_{n=0}^{\infty}$ we have $\|x_{i_k}\|=1$ $(k=0,1,2,\ldots)$ and $\lim_{k\to\infty} x_{i_k}(t)=0$ $(t\in[0,1])$, whence[1] $x_{i_k} \xrightarrow{w} 0$. Thus $\{x_n\}_{n=0}^{\infty}$ is of type swc_0.

Finally, by § 9, theorem 9.3 a) and the implication $1^\circ \Rightarrow 5^\circ$ of theorem 9.1, there exists a $\Phi \in C([0,1])^{**}$ such that

$$|\Phi(f_n)| = 1 \qquad (n=0,1,2,\ldots), \tag{7.10}$$

---

[1] See e.g. [10], p. 134.

where $\{f_n\}_{n=0}^{\infty} \subset C([0,1])^*$ is the a.s.c.f. to the Schauder basis $\{x_n\}_{n=0}^{\infty}$ of $C([0,1])$. Thus $\{x_n\}_{n=0}^{\infty}$ is of type non-$(swc_0)^*$.

b) Assume that the normalized Haar basis[1] $\{\tilde{z}_n\}$ of $L^1([0,1])$ has a subsequence $z_{i_n} \xrightarrow{w} 0$. Then, by a theorem of H. Lebesgue[2], for every $\varepsilon > 0$ there exists an $\eta > 0$ such that $\left|\int_e z_{i_n}(t)\,dt\right| < \varepsilon$ $(n=1,2,\ldots)$ whenever $e \subset [0,1]$ is of Lebesgue measure $< \eta$. However, this contradicts the definition of the normalized Haar basis of $L^1([0,1])$, according to which

$$\int_{\frac{2l-2}{2^{k+1}}}^{\frac{2l-1}{2^{k+1}}} z_{2^k+l}(t)\,dt = \int_{\frac{2l-2}{2^{k+1}}}^{\frac{2l-1}{2^{k+1}}} 2^k\,dt = \frac{1}{2} \qquad (l=1,\ldots,2^k;\ k=0,1,2,\ldots).$$

Consequently, $\{\tilde{z}_n\}$ is of type non-swc_0.

On the other hand, for the subsequence $\{\tilde{z}_{i_k}\}_{k=0}^{\infty} = \{\tilde{z}_{2^k+1}\}_{k=0}^{\infty}$ of $\{\tilde{z}_n\}$ we have

$$\left\|\sum_{k=0}^{n} \tilde{z}_{i_k}\right\| = 2 - \frac{1}{2^n} < 2 \qquad (n=0,1,2,\ldots). \tag{7.11}$$

Indeed, by the definition of $\{z_n\}$ we have

$$\sum_{k=0}^{n} z_{i_k}(t) = \begin{cases} \sum_{k=0}^{n} 2^k = 2^{n+1} - 1 & \text{for} \quad t \in \left[0, \frac{1}{2^{n+1}}\right), \\ \sum_{k=0}^{n-1} 2^k - 2^n = -1 & \text{for} \quad t \in \left[\frac{1}{2^{n+1}}, \frac{1}{2^n}\right), \\ \sum_{k=0}^{n-1} z_{i_k}(t) = -1 & \text{for} \quad t \in \left[\frac{1}{2^n}, 1\right] \end{cases}$$

$$(n=0,1,2,\ldots),$$

whence

$$\left\|\sum_{k=0}^{n} \tilde{z}_{i_k}\right\| = \int_0^1 \left|\sum_{k=0}^{n} z_{i_k}(t)\right| dt = \frac{1}{2^{n+1}}(2^{n+1}-1) + \frac{1}{2^{n+1}} + 1 - \frac{1}{2^n} = 2 - \frac{1}{2^n}$$

$$(n=0,1,2,\ldots).$$

---

[1] See § 2, formula (2.3) for $p=1$.

[2] See e.g. [10], pp. 7–8 and 136, or [175], Ch. VI, § 3, corollary 2.

Hence, by Ch. I, § 12, proposition 12.2c), there exists a $\Phi \in L^1([0,1])^{**}$ such that

$$\Phi(h_{2^k+1}) = 1 \qquad (k = 0, 1, 2, \ldots),$$

where $\{h_n\} \subset L^1([0,1])^*$ is the sequence of coefficient functionals associated to the basis $\{\tilde{z}_n\}$ of $L^1([0,1])$. This proves that $\{\tilde{z}_n\}$ is of type non-$(w c_0)^*$.

Finally, the subsequence $\{\tilde{z}_{j_k}\}_{k=1}^{\infty} = \{\tilde{z}_{2^k+2}\}_{k=1}^{\infty}$ of $\{\tilde{z}_n\}$ is equivalent to the unit vector basis of l^1, since for any finite sequence of scalars $\alpha_1, \ldots, \alpha_n$ we have, by the definition of $\{\tilde{z}_n\}$,

$$\left\| \sum_{k=1}^{n} \alpha_k \tilde{z}_{j_k} \right\| = \sum_{k=1}^{n} |\alpha_k|.$$

Hence $\{\tilde{z}_{j_k}\}$ is of type $(w c_0)^*$ and thus $\{\tilde{z}_n\}$ is of type $(s w c_0)^*$, which completes the proof of theorem 7.1.

Problem 7.1.[1] a) Does the space $C([0,1])$ possess a basis of type $w c_0$? What about a basis of type $(w c_0)^*$ or $(s w c_0)^*$?

b) Does the space $L^1([0,1])$ possess a basis of type $w c_0$ or $s w c_0$? What about a basis of type $(w c_0)^*$ or non-$(s w c_0)^*$?

We shall see in § 9, corollary 9.1 b), that the space $C([0,1])$ has a basis of type non-$s w c_0$.

The answer to the problem of existence of bases of types $s w c_0$ and $(s w c_0)^*$ (and hence also for those of types $w c_0$ and $(w c_0)^*$) is negative, since all bases of any finite-dimensional Banach space and all bounded bases of any subspace of l^1 are of type non-$s w c_0$ (because, in l^1, $y_n \xrightarrow{w} 0$ implies[2] $\|y_n\| \to 0$, whence all bounded bases of c_0 are of type non-$(s w c_0)^*$. The answer to the problem of existence of bases of types non-$w c_0$ and non-$(w c_0)^*$ (and hence also for those of types non-$s w c_0$ and non-$(s w c_0)^*$) is also negative, since in any infinite dimensional reflexive Banach space with a basis all bounded bases are of types $w c_0$ and $(w c_0)^*$ (this is a consequence of § 4, example 4.3 and theorem 4.2, but can be also deduced from weak sequential compactness and biorthogonality).

The next theorem gives some characterizations of bases of type $w c_0$.

Theorem 7.2. *Let $\{x_n\}$ be a bounded basis of an infinite dimensional Banach space E, with the a.s.c.f. $\{f_n\}$. The following statements are equivalent:*

---

[1] Recently, part of this problem has been solved in the affirmative (see the Notes and remarks).

[2] See e.g. [10], p. 137–139.

1°. $\{x_n\}$ *is of type* wc_0.

2°. *The set* $\{x_n\}$ *is weakly sequentially compact (i.e. every infinite subsequence* $\{x_{i_n}\}$ *of* $\{x_n\}$ *contains a subsequence* $\{x_{i_{j_n}}\}$ *such that* $x_{i_{j_n}} \xrightarrow{w} x \in E$*).*

3°. $\{x_n\}$ *is weakly countably compact (i.e. every infinite subsequence* $\{x_{i_n}\}$ *of* $\{x_n\}$ *has a weak limit point*[1] *in* E*).*

4°. $\{x_n\}$ *is weakly conditionally compact (i.e. the weak closure of* $\{x_n\}$ *is weakly compact)*[2].

5°. *The relations* $\{\beta_n\} \subset K$, $\sup_{1 \leq n < \infty} \left\| \sum_{i=1}^{n} \beta_i f_i \right\| < \infty$ *imply* $\lim_{n \to \infty} \beta_n = 0$.

Proof. The implication 1°⇒2° is obvious. The equivalences 2°⇔3°⇔4° are consequences of the Eberlein-Šmulian theorem on weak compactness[3] and can be seen also directly, while the equivalence 1°⇔5° is an immediate consequence of Ch. I, § 12, proposition 12.2 a). Finally, the implication 2°⇒1° is a consequence of the following more general result:

Proposition 7.2. *Let* $\{y_n\}$ *be a basic sequence in a Banach space* E. *If* $x \in E$ *is a weak limit point of* $\{y_n\}$ *(in particular, if* $y_n \xrightarrow{w} x$*), then* $x = 0$.

Proof. By the Hahn-Banach theorem there exists a sequence $\{g_n\} \subset E^*$ such that $g_i(y_j) = \delta_{ij}$ $(i, j = 1, 2, \ldots)$. Since x is a weak limit point of $\{y_n\}$, we have

$$g_i(x) = \lim_{n \to \infty} g_i(y_{k_n(i)}) = 0 \qquad (i = 1, 2, \ldots), \tag{7.12}$$

where $\{y_{k_n(i)}\}$ is some subsequence of $\{y_n\}$ depending on i. Since $[y_n]$ is weakly closed[4], we have $x \in [y_n]$, whence, by (7.12),

$$x = \sum_{i=1}^{\infty} g_i(x) y_i = 0.$$

This completes the proof of proposition 7.2. Now, if a basis $\{x_n\}$ does not satisfy 1° of theorem 7.2, then there exist an $f \in E^*$, an $\varepsilon > 0$, and a subsequence $\{x_{i_n}\}$ of $\{x_n\}$, such that

$$|f(x_{i_n})| > \varepsilon \qquad (n = 1, 2, \ldots). \tag{7.13}$$

---

[1] I.e. a point $x \in E$ such that every weak neighbourhood of x contains an element of the subsequence $\{x_{i_n}\}$, different from x. Let us mention that for this definition of a weak limit point there are weakly convergent sequences which have no weak limit point; indeed, such is e.g. any stationary sequence.

[2] Let us recall that we use the term "compact" in the sense: bicompact Hausdorff.

[3] See e.g. [50], p. 430, theorem 1.

[4] See e.g. [43], Ch. I, § 6, corollary 4, or [50], p. 422, theorem 13.

However, by 2° and proposition 7.2 applied to[1] $\{y_n\}=\{x_{i_n}\}$, there exists a subsequence $\{x_{i_{j_n}}\}$ of $\{x_{i_n}\}$ such that $x_{i_{j_n}} \xrightarrow{w} 0$, which contradicts (7.13) and completes the proof of theorem 7.2.

The bases of types wc_0 and $(wc_0)^*$ are in the following relation of duality:

Proposition 7.3. *Let $\{x_n\}$ be a bounded basis of a Banach space E, with the a.s.c.f. $\{f_n\}$. Then*

a) *$\{x_n\}$ is of type wc_0 if and only if $\{f_n\}$ is a basis of type $(wc_0)^*$ of $[f_n]$.*

b) *The following statements are equivalent:*

1°. *$\{x_n\}$ is of type $(wc_0)^*$.*

2°. *$\{f_n\}$ is a basis of type wc_0 of $[f_n]$.*

3°. *The relations $\{\alpha_n\}\subset K$, $\sup_{1\leqslant n<\infty}\left\|\sum_{i=1}^{n}\alpha_i x_i\right\|<\infty$ imply $\lim_{n\to\infty}\alpha_n=0$.*

Proof. a) The a.s.c.f. $\{\Psi_n\}\subset[f_n]^*$ to the basis $\{f_n\}$ of $[f_n]$ is nothing else but $\Psi_n=\phi(x_n)$ $(n=1,2,\ldots)$, where ϕ is the canonical mapping of E into $[f_n]^*$. Since by Ch. I, § 12, theorem 12.2e), ϕ is an isomorphism, we have $x_n \xrightarrow{w} 0$ if and only if $\phi(x_n) \xrightarrow{w} 0$. Consequently, $\{x_n\}$ is of type wc_0 if and only if $\{\phi(x_n)\}$ is a basis of type wc_0 of $[\phi(x_n)]$. However, by definition 7.1, this condition means that $\{f_n\}$ is a basis of type $(wc_0)^*$ of $[f_n]$.

b) The equivalence 1°⇔2° is nothing else but definition 7.1 b).

2°⇔3°. Since the canonical mapping $\phi: E\to[f_n]^*$ is an isomorphism, we have $\sup_{1\leqslant n<\infty}\left\|\sum_{i=1}^{n}\alpha_i x_i\right\|<\infty$ if and only if $\sup_{1\leqslant n<\infty}\left\|\sum_{i=1}^{n}\alpha_i \phi(x_i)\right\|<\infty$. Hence, by the equivalence 1°⇔5° of theorem 7.2, $\{f_n\}$ is of type wc_0 if and only if the relations $\{\alpha_n\}\subset K$, $\sup_{1\leqslant n<\infty}\left\|\sum_{i=1}^{n}\alpha_i x_i\right\|<\infty$ imply $\lim_{n\to\infty}\alpha_n=0$. This completes the proof of proposition 7.3.

Let us mention that one can also prove the equivalence b) 2°⇔b) 3° in a manner similar to the proof of the equivalence 1°⇔5° of theorem 7.2, i.e. directly with the aid of Ch. I, § 12, proposition 12.2b).

Combining theorem 7.2 and proposition 7.3, one can obtain other characterizations of bases of type $(wc_0)^*$.

Let us now turn our attention to w^*-convergence in conjugate Banach spaces E^*. For $\{h_n\}\subset E^*$, $h\in E^*$ such that $\{h_n\}$ converges to h for the w^*-topology $\sigma(E^*,E)$, we shall use the notation $h_n \xrightarrow{w^*} h$.

---

[1] See Ch. I, § 4, proposition 4.1a).

Proposition 7.4. *Let $\{x_n\}$ be a bounded basis of an infinite dimensional Banach space E, with the a.s.c.f. $\{f_n\}$. Then $f_n \xrightarrow{w^*} 0$.*

Proof. Since $\inf_{1 \leqslant n < \infty} \|x_n\| > 0$ and since $\alpha_n x_n \to 0$ for $\sum_{i=1}^{\infty} \alpha_i x_i \in E$, we have

$$f_n\left(\sum_{i=1}^{\infty} \alpha_i x_i\right) = \alpha_n \to 0 \qquad \left(\sum_{i=1}^{\infty} \alpha_i x_i \in E\right),$$

which completes the proof.

Proposition 7.4 suggests the question, whether every basis $\{f_n\}$ of E^* or every w^*-basis $\{f_n\}$ of E^* satisfies $f_n \xrightarrow{w^*} 0$. The answer is negative, as shown by

Example 7.1. Let $\{f_n\}$ be the basis of $E^* = l^1$ considered in Ch. I, § 14, example 14.1, formula (14.3). Then, as we have seen in that example, $\{f_n\}$ is a basis of E^* and a w^*-basis of E^* (but not a w^*-Schauder basis of E^*, or equivalently[1], not the a.s.c.f. to any basis $\{x_n\}$ of E). However, for $b_1 = \{1, 0, 0, \ldots\} \in E = c_0$ we have

$$f_1(b_1) = 1, \quad f_n(b_1) = (-1)^{n+1} \quad (n = 2, 3, \ldots), \tag{7.14}$$

which shows that $\{f_n\}$ does not converge for the w^*-topology $\sigma(E^*, E)$.

Proposition 7.4 and example 7.1 suggest the question, whether the condition of being a w^*-Schauder basis of E^* (or equivalently, the condition of being the a.s.c.f. to a suitable basis $\{x_n\}$ of E) or, at least a w^*-basis of E^*, is necessary for a basis $\{f_n\}$ of E^* to satisfy $f_n \xrightarrow{w^*} 0$. The answer is negative, as shown by

Example 7.2. Let $\{\chi_n\}$ be the basis $\Phi_0 \cup \{\pi(x_n)\}$ of $E^* = J^{**}$, where the notations J, $\{x_n\}$ and $\Phi_0 \in J^{**}$ are those of § 4, example 4.1. Then $\chi_n \xrightarrow{w^*} 0$, since $\{\chi_n\}_{n=2}^{\infty} = \{\pi(x_{n-1})\}_{n=2}^{\infty}$ is the a.s.c.f. to the natural basis $\{f_{n-1}\}_{n=2}^{\infty}$ of $E = J^*$ (see § 4, example 4.1). However, $\{\chi_n\}_{n=1}^{\infty}$ is not a w^*-basis (whence also not a w^*-Schauder basis) of $E^* = J^{**}$, since by $f = \sum_{i=1}^{\infty} f(x_i) f_i$ $(f \in E = J^*)$ and § 4, formula (4.31), we have

$$\chi_1(f) = \Phi_0(f) = \sum_{i=1}^{\infty} \Phi_0(f_i) f(x_i) = \sum_{i=1}^{\infty} f(x_i)$$

$$= \sum_{i=1}^{\infty} [\pi(x_i)](f) = \sum_{k=2}^{\infty} \chi_k(f) \quad (f \in E = J^*).$$

---

[1] See Ch. I, § 14, theorem 14.1.

§ 8. Some properties of the set of all elements of a basis. Weakly closed and (weakly closed)* bases

In § 7, theorem 7.2, we have given some characterizations of bases $\{x_n\}$ of type wc_0 in terms of w-topological properties of the set of all elements of the basis $\{x_n\}$, which we denote again by $\{x_n\}$. It is natural to ask about other properties of this set for an arbitrary basis $\{x_n\}$, both in the norm-topology and w-topology of E.

Proposition 8.1. *Let $\{x_n\}$ be a basis of a Banach space E. Then*

a) *Both in the norm-topology and w-topology of E all points of the set $\{x_n\}$ are isolated points.*

b) *The set $\{x_n\}$ is closed in the norm topology of E if and only if*

$$\inf_{1 \leqslant n < \infty} \|x_n\| > 0. \tag{8.1}$$

c) *The set $\{x_n\}$ is closed in the w-topology of E if and only if 0 is not a weak limit point of $\{x_n\}$.*

Proof. a) If $\{f_j\}$ is the a.s.c.f. to $\{x_j\}$, the cell $S\left(x_n, \frac{1}{2\|f_n\|}\right) = \left\{x \in E \,\middle|\, \|x - x_n\| \leqslant \frac{1}{2\|f_n\|}\right\}$ and the w-neighbourhood $V_{f_n; \frac{1}{2}}(x) = \left\{x \in E \,\middle|\, |f_n(x - x_n)| < \frac{1}{2}\right\}$ of x_n contain no point of the set $\{x_j\}$, different from x_n, since $\|f_n\| \, \|x_j - x_n\| \geqslant |f_n(x_j - x_n)| = 1$ for $j \neq n$.

b) If $\inf_{1 \leqslant n < \infty} \|x_n\| = 0$, then in every ε-cell $S(0, \varepsilon)$ there exists an element of $\{x_n\}$, i.e. $0 \in \overline{\{x_n\}}$ = the closure of $\{x_n\}$ in the norm-topology of E. However, since $\{x_n\}$ is a basis, $0 \notin \{x_n\}$, and thus $\{x_n\}$ is not closed.

Conversely, if $\{x_n\}$ is not closed in the norm-topology, there exists an $x \in \overline{\{x_n\}} \setminus \{x_n\}$. Then, by § 7, proposition 7.2, $x = 0$, whence $\inf_{1 \leqslant n < \infty} \|x_n\| = 0$.

c) If 0 is a w-limit point of $\{x_n\}$, then $0 \in \overline{\{x_n\}}$ = the closure of $\{x_n\}$ in the w-topology of E. However, since $\{x_n\}$ is a basis, $0 \notin \{x_n\}$, and thus $\{x_n\}$ is not w-closed.

Conversely, if $\{x_n\}$ is not w-closed, there exists an $x \in \overline{\{x_n\}} \setminus \{x_n\}$ in the w-topology. Then x is a w-limit point of $\{x_n\}$, whence, by § 7, proposition 7.2, $x = 0$. Thus 0 is a w-limit point of $\{x_n\}$, which completes the proof.

Definition 8.1. A basis $\{x_n\}$ of a Banach space E is said to be *weakly closed*, if $\{x_n\}$ is a bounded basis and the set $\{x_n\}$ is weakly closed; or, equivalently[1], if $\{x_n\}$ is a bounded basis and 0 is not a weak limit point

[1] See proposition 8.1 c) above.

of the set $\{x_n\}$. If the a.s.c.f. $\{f_n\}$ is a weakly closed basis of $[f_n]$, we shall say that $\{x_n\}$ is a (weakly closed)* basis.

For instance, the basis of c_0 constructed in § 2, example 2.2, formula (2.24), and the natural basis of l^1 are weakly closed. On the other hand, every basis of type swc_0 is non weakly closed, whence, in particular, the natural basis of $l^p (p>1)$, and the normalized Haar basis of $L^p([0,1])$ $(p>1)$ are non weakly closed. Dual examples of (weakly closed)* and non (weakly closed)* bases follow from the fact that for weakly closed bases we have a duality proposition similar to § 7, proposition 7.3 (the proof is similar, using again that the canonical mapping $\phi: E \to [f_n]^*$ is an isomorphism and that the a.s.c.f. to the basis $\{f_n\}$ of $[f_n]$ is $\{\phi(x_n)\}$).

Theorem 8.1. a) *The Schauder basis of* $C([0,1])$ *is non weakly closed but (weakly closed)*.*

b) *The normalized Haar basis of* $L^1([0,1])$ *is non weakly closed and non-(weakly closed)*.*

Proof. a) From § 7, theorem 7.1a) it follows that the Schauder basis $\{x_n\}_{n=0}^{\infty}$ of $C([0,1])$ is non weakly closed. On the other hand, from § 7, formula (7.10) it follows that the w-neighbourhood $V_{\Phi|[f_n]; \frac{1}{2}}(0)$ of 0 in $[f_n]$ corresponding to that Φ contains none of the elements f_n $(n=0,1,2,\ldots)$ of the a.s.c.f. $\{f_n\}_{n=0}^{\infty}$ to $\{x_n\}_{n=0}^{\infty}$. Hence 0 is not a w-limit point of $\{f_n\}_{n=0}^{\infty}$ and thus $\{x_n\}_{n=0}^{\infty}$ is (weakly closed)*.

b) Let $V_{g_1, g_2, \ldots, g_m; \varepsilon}(0)$ (where $g_1, g_2, \ldots, g_m \in L^1([0,1])^*$, $\varepsilon>0$) be an arbitrary w-neighbourhood of 0 in $L^1([0,1])$. Then there exist essentially bounded measurable functions $\gamma_1, \gamma_2, \ldots, \gamma_m$ on $[0,1]$ such that

$$g_i(\tilde{x}) = \int_0^1 x(t)\gamma_i(t)\,dt \qquad (\tilde{x} \in L^1([0,1]);\ i=1,\ldots,m).$$

Since γ_i is essentially bounded and measurable, it is also integrable, whence, by a theorem of Lebesgue[1], the derivative of

$$\beta_i(t) = \int_0^t \gamma_i(\tau)\,d\tau \qquad (t \in [0,1])$$

exists and coincides with $\gamma_i(t)$, almost everywhere on $[0,1]$ $(i=1,2,\ldots,m)$. Consequently, there is a point t_0 with $0<t_0<1$, such that all $\beta_i'(t_0)$ exist and satisfy

$$\beta_i'(t_0) = \gamma_i(t_0) \qquad (i=1,2,\ldots,m).\,^2 \tag{8.2}$$

---

[1] See e.g. [211], Ch. II, § 23, or [175], Ch. IX, § 4, theorem 2. For a simple proof, which makes use also of the essential boundedness of γ_i, see [257], Ch. 11, § 11.53.

[2] Indeed, since the sets $\{t \in [0,1] \mid \beta_i'(t) = \gamma_i(t)\}$ $(i=1,2,\ldots,m)$ are each of measure 1 and $\subset [0,1]$, their intersection must be of measure 1.

Let $\{z_{2^{k_n}+l_n}\}$ be any subsequence of the normalized Haar basis $\{\tilde{z}_n\}=\left\{\frac{\tilde{y}_n}{\|\tilde{y}_n\|}\right\}$ of $L^1([0,1])$, such that

$$\frac{2l_n-2}{2^{k_n+1}} \leqslant t_0 \leqslant \frac{2l_n}{2^{k_n+1}} \quad (n=1,2,\ldots). \tag{8.3}$$

Then, putting

$$a_n=\frac{2l_n-2}{2^{k_n+1}}, \quad b_n=\frac{2l_n-1}{2^{k_n+1}}, \quad c=\frac{2l_n}{2^{k_n+1}} \quad (n=1,2,\ldots),$$

$$\alpha_i(h)=\frac{\beta_i(t_0+h)-\beta_i(t_0)}{h}-\gamma_i(t_0) \quad (-t_0\leqslant h\leqslant 1-t_0;\ i=1,\ldots,m),$$

$$h_1^{(n)}=b_n-t_0, \quad h_2^{(n)}=a_n-t_0, \quad h_3^{(n)}=c_n-t_0 \quad (n=1,2,\ldots),$$

and taking into account that $2b_n-a_n-c_n=0$ $(n=1,2,\ldots)$, we obtain

$$\begin{aligned} g_i(\tilde{z}_{2^{k_n}+l_n}) &= \int_0^1 z_{2^{k_n}+l_n}(t)\gamma_i(t)\,dt \\ &= 2^{k_n}\left[\int_{a_n}^{b_n}\gamma_i(\tau)\,d\tau-\int_{b_n}^{c_n}\gamma_i(\tau)\,d\tau\right] \\ &= 2^{k_n}[\beta_i(b_n)-\beta_i(a_n)-\beta_i(c_n)+\beta_i(b_n)] \\ &= 2^{k_n}[2\alpha_i(h_1^{(n)})h_1^{(n)}-\alpha_i(h_2^{(n)})h_2^{(n)}-\alpha_i(h_3^{(n)})h_3^{(n)}]. \end{aligned}$$

Furthermore, by (8.2), there exists an $\eta=\eta(\varepsilon)>0$ such that $|\alpha_i(h)|<\frac{\varepsilon}{4}$ $(i=1,\ldots,m)$ whenever $|h|<\eta$. Since by (8.3) $|h_1^{(n)}|, |h_2^{(n)}|, |h_3^{(n)}| \leqslant \frac{2l_n}{2^{k_n+1}}-\frac{2l_n-2}{2^{k_n+1}}=\frac{1}{2^{k_n}}$ $(n=1,2,\ldots)$, it follows that for $N=N[\eta(\varepsilon)]$ such that $\frac{1}{2^{k_n}}<\eta$ $(n>N)$ we shall have

$$|g_i(\tilde{z}_{2^{k_n}+l_n})|<2^{k_n}\left(\frac{2\varepsilon}{4}\frac{1}{2^{k_n}}+\frac{\varepsilon}{4}\frac{1}{2^{k_n}}+\frac{\varepsilon}{4}\frac{1}{2^{k_n}}\right)=\varepsilon \quad (n>N;\ i=1,\ldots,m),$$

and thus

$$\tilde{z}_{2^{k_n}+l_n}\in V_{g_1,g_2,\ldots,g_m;\varepsilon}(0) \quad (n>N).$$

Since $V_{g_1,g_2,\ldots,g_m;\varepsilon}(0)$ has been arbitrary, it follows that 0 is a w-limit point of the set $\{\tilde{z}_n\}$, i.e. $\{\tilde{z}_n\}$ is a non weakly closed basis. On the other hand, from § 7, theorem 7.1 b) it follows that $\{\tilde{z}_n\}$ is also non-(weakly closed)*, which completes the proof.

Remark 8.1. Theorems 7.1 b) and 8.1 b) show that *there exist normalized bases* $\{x_n\}$ *such that* 0 *is a weak limit point of the set* $\{x_n\}$ *but no subsequence of* $\{x_n\}$ *converges weakly to* 0.

The answer to the problem of existence of weakly closed bases is negative, since in any infinite-dimensional reflexive Banach space with a basis all bounded bases are of type wc_0 (see § 7) and hence non weakly closed. By duality (see the remark after definition 8.1), a similar assertion is also valid for (weakly closed)* bases.

Problem 8.1. Does the space $E=L^1([0,1])$ possess a weakly closed basis? What about a (weakly closed)* basis?[1]

We shall see in § 9, corollary 9.1 a), that the space $C([0,1])$ has a weakly closed basis; on the other hand, we have seen in theorem 8.1 a) above that it also possesses a (weakly closed)* basis.

Obviously, the answer to the problem of existence of non weakly closed bases in finite-dimensional Banach spaces is also negative.

Problem 8.2. a) Does every infinite dimensional Banach space with a basis possess a non weakly closed basis? In particular, does the l^1 possess such a basis?

b) Does every infinite dimensional Banach space with a basis possess a non-(weakly closed)* basis? In particular, does one of the spaces c_0 or $C([0,1])$ possess such a basis?

In connection with problem 8.2 a) it may present some interest to observe that the space l^1 has a *subspace* with a non weakly closed basis. In order to give such an example, let us recall the following weakened version of a theorem A. Dvoretzky[2]: *For each positive integer* n, *every infinite dimensional Banach space* E *has an* n*-dimensional subspace* G_n *and an euclidean norm* $|||x|||$ *in it (i.e. a norm in which* G_n *will be an* n*-dimensional inner product space), such that*

$$\|x\| \leqslant |||x||| \leqslant 2\|x\| \qquad (x \in G_n). \tag{8.4}$$

Applying this theorem of Dvoretzky for every positive integer n and for $E=l^1$, it follows that the space[3]

---

[1] Recently this problem has been solved in the affirmative (see the Notes and remarks).

[2] See [52].

[3] We recall that if E is one of the spaces c_0, l^1 or m and E_n $(n=1,2,\dots)$ are Banach spaces, $(E_1 \times E_2 \times \cdots \times E_n \times \cdots)_E$ is the Banach space of all sequences $y=\{y_n\}$ with $y_n \in E_n$ $(n=1,2,\dots)$ and $\{\|y_n\|_{E_n}\} \in E$, endowed with the usual vector operations and with the norm $\|y\| = \|\{\|y_n\|_{E_n}\}\|_E$. For the definition of the space l_n^2 see § 1, the footnote to the example given after formula (1.6).

$$B=(l_1^2\times l_2^2\times\cdots\times l_n^2\times\cdots)_{l^1} \tag{8.5}$$

is isomorphic to a closed linear subspace of

$$B_1=(l^1\times l^1\times\cdots\times l^1\times\cdots)_{l^1}.$$

Indeed, if for each n we take in the n-th factor of B_1 a subspace G_n whose existence is assured by the above theorem of Dvoretzky, the closed linear subspace G of B_1 defined by

$$G=(G_1\times G_2\times\cdots\times G_n\times\cdots)_{l^1}$$

can be mapped isomorphically onto B by mapping suitably G_n onto l_n^2.

Since B_1 is isometric to l^1 (see § 18, lemma 18.5 b)), it follows that B is isomorphic to a closed linear subspace F of l^1.

Now we can give the example announced above:

Example 8.1. Let F be the closed linear subspace of l^1 constructed above. Then F has a non weakly closed basis.

Indeed, since F is isomorphic to the Banach space B defined by (8.5), it will be sufficient to prove that B has a non weakly closed basis. We claim that the sequence $\{z_j\}\subset B$ defined by

$$z_{\frac{n(n-1)}{2}+k}=\{\underbrace{0,\ldots,0}_{n-1},\ y_k^{(n)},0,0,\ldots\}\qquad(k=1,\ldots,n;\ n=1,2,\ldots), \tag{8.6}$$

where $\{y_k^{(n)}\}_{k=1}^n$ is the natural basis of l_n^2 $(n=1,2,\ldots)$, is a non weakly closed basis of B.

Indeed, let us first show that $\{z_n\}$ is a basis of B. Let $y=\{y_n\}\in B$ be arbitrary. Then, since $y_n\in l_n^2$ and since $\{y_k^{(n)}\}_{k=1}^n$ is the natural basis of l_n^2 $(n=1,2,\ldots)$, there exist scalars $\alpha_k^{(n)}$ $(k=1,\ldots,n;\ n=1,2,\ldots)$ such that $y_n=\sum_{k=1}^n\alpha_k^{(n)}y_k^{(n)}$ $(n=1,2,\ldots)$, whence, by (8.6),

$$\sum_{j=1}^n\sum_{k=1}^j\alpha_k^{(j)}z_{\frac{j(j-1)}{2}+k}+\sum_{k=1}^h\alpha_k^{(n+1)}z_{\frac{(n+1)n}{2}+k}$$
$$=\left\{y_1,y_2,\ldots,y_n,\sum_{k=1}^h\alpha_k^{(n+1)}y_k^{(n+1)},0,0,\ldots\right\}\qquad(h=1,\ldots,n+1;\ n=1,2,\ldots).$$

Consequently, we have

$$\left\|y-\sum_{j=1}^n\sum_{k=1}^j\alpha_k^{(j)}z_{\frac{j(j-1)}{2}+k}-\sum_{k=1}^h\alpha_k^{(n+1)}z_{\frac{(n+1)n}{2}+k}\right\|$$
$$=\left\|\left\{\underbrace{0,\ldots,0}_{n},\sum_{k=h+1}^{n+1}\alpha_k^{(n+1)}y_k^{(n+1)},y_{n+2},y_{n+3},\ldots\right\}\right\|$$
$$=\left(\sum_{k=h+1}^{n+1}|\alpha_k^{(n+1)}|^2\right)^{\frac{1}{2}}+\sum_{i=n+2}^{\infty}\|y_i\|\leqslant\sum_{i=n+1}^{\infty}\|y_i\|\to0\quad\text{as}\quad n\to\infty,$$

which proves that y has an expansion of the form $\sum_{i=1}^{\infty} \alpha_i z_i$. This expansion is unique, since $\sum_{n=1}^{\infty} \sum_{k=1}^{n} \alpha_k^{(n)} z_{\frac{n(n-1)}{2}+k} = 0$ implies $\left\{\sum_{k=1}^{n} \alpha_k^{(n)} y_k^{(n)}\right\}_{n=1}^{\infty} = 0$, whence $\sum_{k=1}^{n} \alpha_k^{(n)} y_k^{(n)} = 0$ $(n=1,2,\ldots)$, whence, since $\{y_k^{(n)}\}_{k=1}^{n}$ is a basis of l_n^2, we obtain $\alpha_k^{(n)}=0$ $(k=1,2,\ldots,n;\ n=1,2,\ldots)$. Thus $\{z_n\}$ is a basis of B. Alternatively, this fact can be also deduced from Ch. I, § 7, corollary 7.3.

Let us prove now that $\{z_n\}$ is non weakly closed. Let $V_{g_1,\ldots,g_j;\varepsilon}(0)$ (where $g_1, g_2, \ldots, g_j \in B^*$, $\varepsilon > 0$) be an arbitrary w-neighbourhood of 0 in B. Then, by the isometry $B^* \equiv (l_1^2 \times l_2^2 \times \cdots \times l_n^2 \times \cdots)_m$, there exist $\{g_n^{(1)}\}, \{g_n^{(2)}\}, \ldots, \{g_n^{(j)}\} \in (l_1^2 \times l_2^2 \times \cdots \times l_n^2 \times \cdots)_m$ such that

$$g_i(y) = \sum_{n=1}^{\infty} g_n^{(i)}(y_n) \qquad (y = \{y_n\} \in B;\ i=1,\ldots,j),$$

$$\|g_i\| = \sup_{1 \leqslant n < \infty} \|g_n^{(i)}\| \qquad (i=1,\ldots,j).$$

Hence, if $g_n^{(i)} = \{\beta_k^{(n,i)}\}_{k=1}^{n}$ $(n=1,2,\ldots;\ i=1,\ldots,j)$, we obtain

$$g_i\left(z_{\frac{n(n-1)}{2}+k}\right) = g_i(\{\underbrace{0,\ldots,0}_{n-1}, y_k^{(n)}, 0, 0, \ldots\}) = g_n^{(i)}(y_k^{(n)}) = \beta_k^{(n,i)} \tag{8.7}$$

$$(k=1,\ldots,n;\ n=1,2,\ldots;\ i=1,\ldots,j),$$

$$\|g_i\| = \sup_{1 \leqslant n < \infty} \left(\sum_{k=1}^{n} |\beta_k^{(n,i)}|^2\right)^{\frac{1}{2}} \qquad (i=1,2,\ldots,j). \tag{8.8}$$

Put $M = \sum_{i=1}^{j} \|g_i\| = \sum_{i=1}^{j} \sup_{1 \leqslant n < \infty} \left(\sum_{k=1}^{n} |\beta_k^{(n,i)}|^2\right)^{\frac{1}{2}}$. Then

$$\left(\sum_{i=1}^{j} \sum_{k=1}^{n} |\beta_k^{(n,i)}|^2\right)^{\frac{1}{2}} \leqslant \sum_{i=1}^{j} \left(\sum_{k=1}^{n} |\beta_k^{(n,i)}|^2\right)^{\frac{1}{2}} \leqslant M \qquad (n=1,2,\ldots),$$

whence for each $n=1,2,\ldots$ at least one of the numbers

$$\sum_{i=1}^{j} |\beta_1^{(n,i)}|^2,\ \sum_{i=1}^{j} |\beta_2^{(n,i)}|^2, \ldots,\ \sum_{i=1}^{j} |\beta_n^{(n,i)}|^2$$

must be $\leqslant \frac{M^2}{n}$, and thus

$$\min_{1 \leqslant k \leqslant n} \max_{1 \leqslant i \leqslant j} |\beta_k^{(n,i)}| \leqslant \min_{1 \leqslant k \leqslant n} \left(\sum_{i=1}^{j} |\beta_k^{(n,i)}|^2\right)^{\frac{1}{2}} \leqslant \frac{M}{\sqrt{n}}.$$

Consequently, taking into account (8.7), we have

$$\inf_{\substack{1\leqslant k\leqslant n\\ 1\leqslant n<\infty}} \max_{1\leqslant i\leqslant j} \left| g_i\left(z_{\frac{n(n-1)}{2}+k}\right)\right| = 0, \tag{8.9}$$

whence $z_{\frac{n(n-1)}{2}+k} \in V_{g_1,\ldots,g_j;\varepsilon}(0)$ for suitable n and k. Thus 0 is a weak limit point of $\{z_n\}$, i.e. $\{z_n\}$ is non weakly closed. This completes the proof of the assertions stated in example 8.1.

Remark 8.2. Actually, one can avoid the use of the above theorem of Dvoretzky, embedding $B=(l_1^2\times l_2^2\times\cdots)_{l^1}$ into l^1 with the method applied in § 18, the proof of theorem 18.3 (i.e. using Rademacher functions and the Khinchin inequality to embed l_n^2 into $l_{2^n}^1$, etc).

Dually, one can show that the space c_0 has a subspace with a non (weakly closed)* basis. Indeed, since $B_2=(l_1^2\times l_2^2\times\cdots\times l_n^2\times\cdots)_{c_0}$ is isomorphic to a closed linear subspace of c_0, it is sufficient to prove that B_2 has such a basis. However, the "natural" basis of B_2 has this property, since its a.s.c.f. is nothing else but the image of the basis $\{z_j\}$ of B defined by (8.6) under the canonical isometry $B\equiv B_2^*$.

Remark 8.3. Again, one can embed $B_2=(l_1^2\times l_2^2\times\cdots)_{c_0}$ into c_0 without using the above theorem of Dvoretzky, by embedding directly l_n^2 into $l_{2^{2^n}}^\infty$. For this purpose, observe first that one can map isometrically l_n^1 onto the subspace $[r_1,\ldots,r_n]$ of $L^\infty([0,1])$, where $r_1,\ldots,r_n$ are the first n Rademacher functions[1], by the mapping

$$e_i\to r_i \quad (i=1,\ldots,n), \tag{8.10}$$

where $\{e_i\}_{i=1}^n$ is the unit vector basis of l_n^1. Indeed, we have obviously

$$\left\|\sum_{i=1}^n \alpha_i r_i\right\|_{L^\infty} \leqslant \sum_{i=1}^n |\alpha_i| \tag{8.11}$$

for any scalars $\alpha_1,\ldots,\alpha_n$. On the other hand, for every pair $k_1\neq k_2$ with $1\leqslant k_1,\ k_2\leqslant 2^n$ there exists at least one i with $1\leqslant i\leqslant n$ such that $r_i\left(\frac{2k_1-1}{2^{n+1}}\right)\neq r_i\left(\frac{2k_2-1}{2^{n+1}}\right)$; for if $r_1\left(\frac{2k_1-1}{2^{n+1}}\right)=r_1\left(\frac{2k_2-1}{2^{n+1}}\right)$, then both $\frac{2k_1-1}{2^{n+1}}$ and $\frac{2k_2-1}{2^{n+1}}$ must belong to the same dyadic interval $\left(\frac{k-1}{2},\frac{k}{2}\right)$, where $1\leqslant k\leqslant 2$ and continuing in this way, by induction, we obtain finally that if $r_i\left(\frac{2k_1-1}{2^{n+1}}\right)=r_i\left(\frac{2k_2-1}{2^{n+1}}\right)$ $(i=1,\ldots,n)$, then both $\frac{2k_1-1}{2^{n+1}}$ and $\frac{2k_2-1}{2^{n+1}}$ must belong to the same dyadic interval

[1] See § 14, formula (14.1).

$\left(\frac{k-1}{2^n}, \frac{k}{2^n}\right)$, whence, since obviously $\frac{2k_j-1}{2^{n+1}} \in \left(\frac{k_j-1}{2^n}, \frac{k_j}{2^n}\right)$ $(j=1,2)$, we infer $k_1=k_2$. Consequently, for $k=1,2,\ldots,2^n$ we obtain 2^n distinct n-tuples $\left\{r_1\left(\frac{2k-1}{2^{n+1}}\right),\ldots,r_n\left(\frac{2k-1}{2^{n+1}}\right)\right\}$. Since there are exactly 2^n distinct n-tuples of signs $\varepsilon_i=\pm 1$, there exists a (unique) k with $1\leqslant k\leqslant 2^n$ such that $r_i\left(\frac{2k-1}{2^{n+1}}\right)=\varepsilon_i$ $(i=1,\ldots,n)$. Now, if $\{\alpha_i\}_{i=1}^n$ is arbitrary (we restrict ourselves to real scalars), then applying this remark for $\varepsilon_i=\operatorname{sign}\alpha_i$ if $\alpha_i\neq 0$ and 1 if $\alpha_i=0$, we find a k with $1\leqslant k\leqslant 2^n$ such that $r_i\left(\frac{2k-1}{2^{n+1}}\right)=\operatorname{sign}\alpha_i$ $(i=1,\ldots,n)$, whence

$$\left\|\sum_{i=1}^n \alpha_i r_i\right\|_{L^\infty} \geqslant \sum_{i=1}^n \alpha_i r_i\left(\frac{2k-1}{2^{n+1}}\right)=\sum_{i=1}^n |\alpha_i|, \tag{8.12}$$

which, together with (8.11), proves that (8.10) is an isometric linear mapping of l_n^1 onto the subspace $[r_1,\ldots,r_n]$ of $L^\infty([0,1])$. Since the subspace G_{2^n} of $L^\infty([0,1])$ spanned by the characteristic functions $\chi_{\left[\frac{k-1}{2^n},\frac{k}{2^n}\right)}$ $(k=1,\ldots,2^n)$ is obviously equivalent to $l_{2^n}^\infty$, the subspace $[r_1,\ldots,r_n]$ of G_{2^n} is equivalent to an n-dimensional subspace of $l_{2^n}^\infty$ and hence we have an isometric linear embedding of l_n^1 into $l_{2^n}^\infty$, whence, in particular, of $l_{2^n}^1$ into $l_{2^{2^n}}^\infty$. Composing this embedding with the embedding $l_n^2\to l_{2^n}^1$ of remark 8.2, we obtain the desired embedding $l_n^2\to l_{2^{2^n}}^\infty$.

Let us now turn our attention to w^*-topological properties in conjugate Banach spaces. Similarly to the proof of proposition 8.1 a) it follows that *if $\{f_n\}\subset E^*$ is the a.s.c.f. to a bounded basis $\{x_n\}$ of E, then both in the norm-topology and w^*-topology of E^*, all points of the set $\{f_n\}$ are isolated points.* Furthermore, from § 7, proposition 7.4 and Ch. I, § 12, theorem 12.1, it follows that in this case *the set $\{f_n\}$ is not w^*-closed.* However, if $\{f_n\}\subset E^*$ is a bounded basis of E^*, which is not the a.s.c.f. to any bounded basis of E, the situation is different. In fact, in this case the f_j are not necessarily w^*-isolated points of the set $\{f_n\}$, as shown by

Example 8.2. Let $\{f_n\}$ be the basis of $E^*=l^1$ considered in Ch. I, § 14, example 14.1, formula (14.3). Then

$$f_{2n-1}-f_1=e_{2n-1} \qquad (n=2,3,\ldots),$$

where $\{e_n\}$ is the unit vector basis of l^1. Since by § 7, proposition 7.4, $e_n \xrightarrow{w^*} 0$, it follows that

$$f_{2n-1}\xrightarrow{w^*} f_1, \tag{8.13}$$

and thus f_1 is not a w^*-isolated point of the set $\{f_n\}$.

Furthermore, if $\{f_n\}$ is a bounded basis of E^*, then[1] E is separable, whence[2] $\{f_n\}$ contains a subsequence $\{f_{i_n}\}$ which w^*-converges to an $f \in E^*$. However, it may also happen that $f \neq 0$, as shown by the preceding example, formula (8.13).

§ 9. Bases of types P, P^*, aP and aP^*

Definition 9.1. A basis $\{x_n\}$ of a Banach space E is said to be

a) *of type P*, if $\{x_n\}$ is a bounded basis and

$$\sup_{1 \leqslant n < \infty} \left\| \sum_{i=1}^{n} x_i \right\| < \infty; \tag{9.1}$$

b) *of type P^**, if the a.s.c.f. $\{f_n\}$ is a basis of type P of $[f_n]$;

c) *of type aP (of type aP^*)*, if there exists a sequence of scalars $\{\varepsilon_n\}$ with $|\varepsilon_n| = 1$ $(n = 1, 2, \ldots)$, such the basis $\{\varepsilon_n x_n\}$ of E is of type P (respectively, of type P^*).

For instance, the natural basis of c_0 is of type P, but not aP^*, the natural basis of l^1 is of type P^*, but not aP, while the natural basis of l^p $(p > 1)$ is of types non-aP and non-aP^*.

Let us observe that in definition 9.1 a) it would have been sufficient to assume that $\inf_{1 \leqslant n < \infty} \|x_n\| > 0$, since (9.1) implies $\sup_{1 \leqslant n < \infty} \|x_n\| < \infty$. Similarly, in definition 9.1 b) it would have been sufficient to assume that $\sup_{1 \leqslant n < \infty} \|x_n\| < \infty$.

Some characterizations of bases of type P are collected in

Theorem 9.1. *Let $\{x_n\}$ be a basis of a Banach space E, with $\inf_{1 \leqslant n < \infty} \|x_n\| > 0$, and let $\{f_n\} \subset E^*$ be the a.s.c.f. to $\{x_n\}$. The following statements are equivalent:*

1°. *$\{x_n\}$ is of type P.*

2°. *$\sup_{1 \leqslant n < \infty} \|x_n\| < \infty$ and the sequence $\{y_n\} \subset E$ defined by*

$$y_n = \sum_{i=1}^{n} x_i \quad (n = 1, 2, \ldots) \tag{9.2}$$

is a basis of E.

3°. *There exists a constant $M > 0$ such that for every monotonic sequence $\{\alpha_n\}$ tending to zero, the sum $\sum_{i=1}^{\infty} \alpha_i x_i$ exists and satisfies*

$$\left\| \sum_{i=1}^{\infty} \alpha_i x_i \right\| \leqslant M |\alpha_1|. \tag{9.3}$$

[1] See e.g. [10], p. 189, theorem 12.

[2] See e.g. [10], p. 123, theorem 3.

4°. *There exists a constant* $M>0$ *such that we have, for every finite sequence of scalars* $\alpha_1, \ldots, \alpha_n$

$$\left|\sum_{i=1}^{n} \alpha_i\right| \leqslant M \left\| \sum_{i=1}^{n} \alpha_i f_i \right\|. \tag{9.4}$$

5°. *There exists a* $\Phi \in E^{**}$ *such that*

$$\Phi(f_n)=1 \qquad (n=1,2,\ldots). \tag{9.5}$$

6°. *There exists a* $\Psi \in [f_n]^*$ *such that*

$$\Psi(f_n)=1 \qquad (n=1,2,\ldots), \tag{9.6}$$

7°. *The sequence* $\{f_1-f_2, f_2-f_3, \ldots\}$ *is not complete in* $[f_n]$.

8°. $\inf_{1 \leqslant n < \infty} \|f_n\| > 0$ *and* $\{f_1-f_2, f_2-f_3, \ldots\}$ *is a basic sequence.*

9°. *The sequence* $\{g_n\} \subset E^*$ *defined by*

$$g_1=f_1, \qquad g_n=f_{n-1}-f_n \qquad (n=2,3,\ldots) \tag{9.7}$$

is a basis of $[f_n]$.

Proof. The equivalence $1° \Leftrightarrow 2°$ is a consequence of Ch. I, § 4, proposition 4.3.

$1° \Rightarrow 3°$. Assume that we have 1°, and let $\alpha_1 \geqslant \alpha_2 \geqslant \cdots$ be a sequence tending to zero. Using the sequence (9.2), we have

$$\sum_{i=1}^{n} \alpha_i x_i = \alpha_1 y_1 + \sum_{i=2}^{n} \alpha_i (y_i - y_{i-1}) = \sum_{i=1}^{n-1} (\alpha_i - \alpha_{i+1}) y_i + \alpha_n y_n \qquad (n=1,2,\ldots). \tag{9.8}$$

Since $\{x_n\}$ is of type P, we have $\sup_{1 \leqslant n < \infty} \|y_n\| = M < \infty$, whence, since $\alpha_n \to 0$, we obtain $\alpha_n y_n \to 0$. On the other hand, from $\|y_n\| \leqslant M, \alpha_n \geqslant \alpha_{n+1}$ $(n=1,2,\ldots)$ and $\sum_{i=1}^{\infty} (\alpha_i - \alpha_{i+1}) = \alpha_1$ it follows that $\sum_{i=1}^{\infty} (\alpha_i - \alpha_{i+1}) y_i$ exists and satisfies $\left\| \sum_{i=1}^{\infty} (\alpha_i - \alpha_{i+1}) y_i \right\| \leqslant M |\alpha_1|$. Hence, by (9.8), we have 3°.

The implication $3° \Rightarrow 1°$ follows by applying (9.3) successively to the sequences $\{\alpha_j^{(n)}\}$ $(n=1,2,\ldots)$ defined by

$$\alpha_1^{(n)} = \cdots = \alpha_n^{(n)} = 1, \quad \alpha_{n+1}^{(n)} = \alpha_{n+2}^{(n)} = \cdots = 0 \qquad (n=1,2,\ldots).$$

The equivalence $1° \Leftrightarrow 5°$ follows from Ch. I, § 12, proposition 12.2c). Furthermore, the equivalence $4° \Leftrightarrow 5°$ is a particular case of a classical theorem of E. Helly[1], and the equivalence $5° \Leftrightarrow 6°$ follows from the

[1] See e.g. [10], p. 55, theorem 4.

Hahn-Banach theorem. The equivalence $6^\circ \Leftrightarrow 7^\circ$ is a consequence of a well known corollary of the Hahn-Banach theorem[1], since the relations (9.6) are equivalent to

$$\Psi(f_1)=1, \quad \Psi(f_{n-1}-f_n)=0 \quad (n=2,3,\ldots). \tag{9.9}$$

$2^\circ \Rightarrow 8^\circ$. Since $\sup_{1\leqslant n<\infty} \|x_n\|<\infty$, by Ch. I, § 3, corollary 3.1 b) we have $\inf_{1\leqslant n<\infty} \|f_n\|>0$. Furthermore, the sequence $\{f_1-f_2, f_2-f_3, \ldots\}$ is nothing else but the a.s.c.f. to the basis $\{y_n\}$ of E, since

$$(f_{i-1}-f_i)(y_j)=\delta_{i-1,j} \quad (i=2,3,\ldots;\ j=1,2,\ldots).$$

Hence, by Ch. I, § 12, theorem 12.1, $\{f_1-f_2, f_2-f_3, \ldots\}$ is a basic sequence.

$8^\circ \Rightarrow 9^\circ$. If we have 8°, then

$$f_1 \notin [f_1-f_2, f_2-f_3, \ldots]. \tag{9.10}$$

Indeed, otherwise there would exist a sequence $\{\alpha_n\}\subset K$ such that

$$f_1=\sum_{i=2}^{\infty} \alpha_{i-1}(f_{i-1}-f_i)=\alpha_1 f_1+\sum_{i=2}^{\infty}(\alpha_i-\alpha_{i-1}) f_i,$$

whence, since by Ch. I, § 12, theorem 12.1, $\{f_n\}$ is a basic sequence,

$$\alpha_1=1, \quad \alpha_2-\alpha_1=\alpha_3-\alpha_2=\cdots=0,$$

and thus

$$f_1=\sum_{i=2}^{\infty}(f_{i-1}-f_i). \tag{9.11}$$

Hence

$$f_n=f_1-\sum_{i=2}^{n}(f_{i-1}-f_i)\to 0 \quad \text{as} \quad n\to\infty,$$

which contradicts $\inf_{1\leqslant n<\infty}\|f_n\|>0$. This proves (9.10).

Thus $\{g_n\}_{n=2}^{\infty}$ is a basis of $[g_n]_{n=2}^{\infty}$ and $g_1\notin[g_n]_{n=2}^{\infty}$. Since obviously $[f_n]=[g_n]_{n=1}^{\infty}$, it follows that $\{g_n\}$ is a basis of $[f_n]$.

Finally, the implication $9^\circ \Rightarrow 7^\circ$ is obvious, since every basis is a minimal sequence (see Ch. I, § 6). This completes the proof of theorem 9.1.

The assumption $\sup_{1\leqslant n<\infty}\|x_n\|<\infty$ in 2° above is necessary for the validity of the implication $2^\circ \Rightarrow 1^\circ$, as shown by

Example 9.1. Let $E=c_0$ and $\{x_n\}=\{n e_n\}$, where $\{e_n\}$ is the natural basis of c_0. Then $\inf_{1\leqslant n<\infty}\|x_n\|>0$ and $\left\{\sum_{i=1}^{n} x_i\right\}$ is a basis of E, but $\{x_n\}$ is not of type P.

---

[1] See [10], p. 58, theorem 7.

Indeed, applying Ch. I, § 4, proposition 4.3 to $\{e_n\}$ and $\alpha_n=n$ $(n=1,2,\ldots)$, it follows that $\left\{\sum_{i=1}^{n} x_i\right\}$ is a basis of E (since $\dfrac{\left\|\sum_{j=1}^{n} j e_j\right\|}{n+1} = \dfrac{n}{n+1} < 1$). However, $\left\|\sum_{i=1}^{n} x_i\right\| = n$ $(n=1,2,\ldots)$.

Similarly, the assumption $\inf_{1\leqslant n<\infty} \|f_n\|>0$ in 8° above is essential.

By the duality methods used already in the preceding sections, i.e. taking into account that the canonical mapping $\phi: E\to[f_n]^*$ is an isomorphism and that the a.s.c.f. to the basis $\{f_n\}$ of $[f_n]$ is $\{\phi(x_n)\}$, we obtain

Theorem 9.2. *Let $\{x_n\}$ be a basis of a Banach space E, with $\sup_{1\leqslant n<\infty} \|x_n\| <\infty$, and let $\{f_n\}\subset E^*$ be the a.s.c.f. to $\{x_n\}$. Then*

a) *$\{x_n\}$ is of type P if and only if $\{f_n\}$ is a basis of type P^* of $[f_n]$.*

b) *The following statements are equivalent:*

1°. *$\{x_n\}$ is of type P^*.*

2°. *$\{f_n\}$ is a basis of type P of $[f_n]$.*

3°. *There exists a constant $L>0$ such that we have, for every finite sequence of scalars $\alpha_1,\ldots,\alpha_n$*

$$\left|\sum_{i=1}^{n} \alpha_i\right| \leqslant L\left\|\sum_{i=1}^{n} \alpha_i x_i\right\|. \tag{9.12}$$

4°. *There exists an $f\in E^*$ such that*

$$f(x_n)=1 \qquad (n=1,2,\ldots). \tag{9.13}$$

5°. *The sequence $\{x_1-x_2, x_2-x_3,\ldots\}$ is not complete in E.*

6°. *$\inf_{1\leqslant n<\infty} \|x_n\|>0$ and $\{x_1-x_2, x_2-x_3,\ldots\}$ is a basic sequence.*

7°. *The sequence $\{z_n\}\subset E$ defined by*

$$z_1=x_1, \qquad z_n=x_{n-1}-x_n \qquad (n=2,3,\ldots) \tag{9.14}$$

is a basis of E.

Naturally, one can also prove the equivalences b) 2°$\Leftrightarrow\cdots\Leftrightarrow$b) 7° directly, similarly to the proof of the corresponding equivalences of theorem 9.1.

Combining theorems 9.1 and 9.2 b), one can also obtain other characterizations of bases of type P^*. Obviously, all the preceding characterizations of bases of types P, P^* imply characterizations of bases of types aP and aP^* respectively.

Let us consider now the corresponding properties for the Schauder basis of $C([0,1])$ and the normalized Haar basis of $L^1([0,1])$.

Theorem 9.3. a) *The Schauder basis* $\{x_n\}_{n=0}^{\infty}$ *of* $C([0,1])$ *is of types non-a* P^* *and non-P, but* aP. *Namely, the sequence* $\{\varepsilon_n x_n\}_{n=0}^{\infty}$, *where*

$$\varepsilon_0=1, \quad \varepsilon_1=-1,$$

$$\varepsilon_{2^k+l}=-\operatorname{sign}\left[\sum_{i=0}^{2^k+l-1}\varepsilon_i x_i\left(\frac{2l-1}{2^{k+1}}\right)\right] \quad (l=1,\ldots,2^k;\ k=0,1,2,\ldots) \tag{9.15}$$

is a basis of type P of $C([0,1])$.

b) *The normalized Haar basis* $\{\tilde{z}_n\}$ *of* $L^1([0,1])$ *is of types non-a P and non-a* P^*.

Proof. a) Assume, a contrario, that $\{x_n\}_{n=0}^{\infty}$ is of type aP^*. Then, by the implication $1°\Rightarrow 4°$ of theorem 9.2 b), there exists an $f\in C([0,1])^*$ such that

$$|f(x_n)|=1 \quad (n=0,1,2,\ldots).$$

However, this contradicts § 7, theorem 7.1 a), according to which $\{x_n\}_{n=0}^{\infty}$ is of type swc_0. Consequently, $\{x_n\}_{n=0}^{\infty}$ is of type non-aP^*.

Furthermore, define $t_0\in[0,1]$ by (7.2) and k_n, l_n by (7.3), (7.4). Then, as we have seen in the proof of § 7, theorem 7.1 a), $\lim_{n\to\infty} x_{2^{k_n}+l_n}(t_0)=1$. Hence, since all $x_i(t_0)$ are $\geqslant 0$, we infer

$$\left\|\sum_{i=0}^{n} x_i\right\| \geqslant \sum_{i=0}^{n} x_i(t_0)\to\infty \quad \text{as} \quad n\to\infty,$$

which proves that $\{x_n\}_{n=0}^{\infty}$ is of type non-P.

Finally, we shall prove that for the sequence $\{y_n\}\subset C([0,1])$ defined by

$$y_n=\sum_{i=0}^{n}\varepsilon_i x_i \quad (n=0,1,2,\ldots), \tag{9.16}$$

where the ε_n $(n=0,1,2,\ldots)$ are defined by (9.15), we have

$$\|y_n\|=1 \quad (n=0,1,2,\ldots). \tag{9.17}$$

It is obvious that $\|y_0\|=\|y_1\|=\|y_2\|=1$. Since $y_n(0)=1$ $(n=3,4,\ldots)$, we also have

$$\|y_n\|\geqslant 1 \quad (n=3,4,\ldots). \tag{9.18}$$

Assume now that $\|y_{2^k+l-1}\|=1$. Then, since by (9.16), (9.15) and by the definition of $\{x_n\}_{n=0}^{\infty}$ we have

$$y_{2^k+l}(t)=y_{2^k+l-1}(t)-\left[\operatorname{sign} y_{2^k+l-1}\left(\frac{2l-1}{2^{k+1}}\right)\right]x_{2^k+l}(t)$$

$$=\begin{cases} y_{2^k+l-1}(t) & \text{for} \quad t\notin\left(\frac{2l-2}{2^{k+1}}, \frac{2l}{2^{k+1}}\right), \\ y_{2^k+l-1}\left(\frac{2l-1}{2^{k+1}}\right)-\operatorname{sign} y_{2^k+l-1}\left(\frac{2l-1}{2^{k+1}}\right) & \text{for} \quad t=\frac{2l-1}{2^{k+1}}, \\ \text{linear for the other } t, \end{cases}$$

it follows that $\|y_{2^k+l}\|\leqslant 1$, whence taking into account (9.18), $\|y_{2^k+l}\|=1$. Consequently, we have (9.17), which proves that $\{\varepsilon_n x_n\}_{n=0}^{\infty}$ is a basis of type P of $C([0,1])$.

b) Assume, a contrario, that $\{\tilde{z}_n\}$ is of type aP. Then, by the implication $1°\Rightarrow 5°$ of theorem 9.1, there must exist a $\Phi\in L^1([0,1])^{**}$ such that

$$|\Phi(h_n)|=1 \qquad (n=1,2,\ldots), \tag{9.19}$$

where $\{h_n\}\subset L^1([0,1])^*$ is the a.s.c.f. to $\{\tilde{z}_n\}$. However, this is impossible, since $\{\tilde{z}_n\}$ is of type $(swc_0)^*$ (by § 7, theorem 7.1 b)). Consequently, $\{\tilde{z}_n\}$ is of type non-aP.

Finally, from § 8, theorem 8.1 b) it follows that in every w-neighbourhood of 0 of the form $V_{f;\varepsilon}(0)$ $(f\in L^1([0,1])^*, \varepsilon>0)$ there exists at least one $\tilde{z}_n$, whence

$$\inf_{1\leqslant n<\infty}|f(\tilde{z}_n)|=0 \qquad (f\in L^1([0,1])^*),$$

and thus there exists no $f\in L^1([0,1])^*$ satisfying

$$|f(\tilde{z}_n)|=1 \qquad (n=1,2,\ldots).$$

Consequently, by the implication $1°\Rightarrow 4°$ of theorem 9.2 b), $\{\tilde{z}_n\}$ is of type non-aP. This completes the proof of theorem 9.3.

Remark 9.1. The following "Schauder basis" of $C([0,1])$ is sometimes also considered, because of its symmetry with respect to $t=\frac{1}{2}$:

$$x_0(t)=t, \qquad x_1(t)=1-t \qquad (t\in[0,1]), \tag{9.20}$$

and $x_{2^k+l}(t)=$ as before. For this basis, theorem 9.3 a) remains valid, with the same proof, provided that we replace in (9.15) $\varepsilon_1=-1$ by $\varepsilon_1=1$.

Let us also mention the following corollary of theorem 9.3:

Corollary 9.1. a) *If $\{x_n\}_{n=0}^{\infty}$ is the Schauder basis of $C([0,1])$, then the sequence (9.16), where the ε_n are defined by (9.15), is a weakly closed basis of type P^* of $C([0,1])$.*

b) *If* $\{\tilde{z}_n\}$ *is the normalized Haar basis of* $L^1([0,1])$, *then for any sequence of scalars* ε_n *with* $|\varepsilon_n|=1$ $(n=1,2,\ldots)$, *the sequence* $\left\{\sum_{i=1}^{n} \varepsilon_i \tilde{z}_i\right\}$ *is not a basis of* $L^1([0,1])$.

Proof. a) By theorem 9.3a), $\{\varepsilon_n x_n\}_{n=0}^{\infty}$ is a basis of type P of $C([0,1])$. Hence, by the implication $1^\circ \Rightarrow 2^\circ$ of theorem 9.1, the sequence $\{y_n\}_{n=0}^{\infty}$ defined by (9.16) is a basis of $C([0,1])$, and it is, obviously, of type P^*. Furthermore, since for the functional $f_0 \in C([0,1])^*$ defined by

$$f_0(x)=x(0) \quad (x\in C([0,1])) \tag{9.21}$$

we have $f_0(y_n)=1$ $(n=0,1,2,\ldots)$, it follows that 0 is not a weak limit point of the sequence $\{y_n\}_{n=0}^{\infty}$ and thus $\{y_n\}_{n=0}^{\infty}$ is weakly closed.

b) By theorem 9.3b) and the implication $2^\circ \Rightarrow 1^\circ$ of theorem 9.1, for any sequence of scalars $\{\varepsilon_n\}$ with $|\varepsilon_n|=1$ $(n=1,2,\ldots)$, the sequence $\left\{\sum_{i=1}^{n} \varepsilon_i \tilde{z}_i\right\}$ is not a basis of $L^1([0,1])$. This completes the proof of corollary 9.1.

Problem 9.1.[1] Does the space $L^1([0,1])$ possess a basis of type P? Or, equivalently: does it possess a basis of type P^*?

The answer to the problem of existence of bases of types aP, aP^* (and hence also for bases of types P, P^*) is negative, as shown by

Example 9.2. Let E be an infinite dimensional reflexive Banach space with a basis. Then E has no basis of type aP or aP^*.

Indeed, every basis $\{x_n\}$ of E is of type wc_0 (see § 7), whence, by the implication $1^\circ \Rightarrow 4^\circ$ of theorem 9.2b), $\{x_n\}$ is of type non-aP^*. Consequently, by the implication $1^\circ \Rightarrow 2^\circ$ of theorem 9.1, every basis of E is also of type non-aP.

The answer to the problem of existence of bases of types non-P, non-P^* (and hence also for bases of types non-aP, and non-aP^*) in finite-dimensional Banach spaces is obviously negative. On the other hand, the answer to the problem of existence of bases of types non-P, non-P^* in infinite dimensional Banach spaces with bases is affirmative. Indeed, this follows from the implication $1^\circ \Rightarrow 2^\circ$ of theorem 9.1 and the implication $1^\circ \Rightarrow 7^\circ$ of theorem 9.2b), since (9.2) is of type non-P and (9.14) of type non-P^* (since e.g. for (9.2) we have $\left\|\sum_{j=1}^{n} y_j\right\| = \left\|\sum_{i=1}^{n} (n-i+1)x_i\right\|$

$$\geqslant \frac{1}{\|f_n\|}\left|f_n\left(\sum_{i=1}^{n}(n-i+1)x_i\right)\right| = \frac{n}{\|f_n\|} \geqslant \frac{n}{\sup\limits_{1\leqslant j<\infty}\|f_j\|} \to \infty \quad \text{as} \quad n\to\infty).$$

---

[1] Recently, this problem has been solved in the affirmative (see the Notes and remarks).

Problem 9.2. a) Does every infinite dimensional Banach space with a basis possess a basis of type non-aP? In particular, does the space c_0 possess such a basis?

b) Does every infinite dimensional Banach space with a basis possess a basis of type non-aP^*? In particular, does the space l^1 possess such a basis?

Let us mention the following related problem. The unit vector basis $\{x_n\}$ of $E=c_0$ is of type P (whence aP), and the basis $\{y_n\}=\left\{\sum_{i=1}^{n} x_i\right\}$ of $E=c_0$ is again of type aP, since e.g. for $\varepsilon_n=(-1)^{n-1}$ the basis $\{\varepsilon_n y_n\}$ is of type P; indeed, we have $z_{2n-1}=\sum_{j=1}^{2n-1}(-1)^j y_j=\sum_{i=1}^{n} x_{2i-1}$, $z_{2n}=\sum_{j=1}^{2n}(-1)^j y_j=-\sum_{i=1}^{n} x_{2i}$. Again, it is easy to see that the basis $\{z_n\}$ of $E=c_0$ is of type aP, and that for the signs $\eta_n=\pm 1$ for which $\{\eta_n z_n\}$ is of type P, the basis $\left\{\sum_{i=1}^{n} \eta_i z_i\right\}$ is again of type aP. Can this process of itęration be continued indefinitely? If not, then in a finite number of steps one would arrive to a basis of type non-aP of $E=c_0$, obtaining thus a negative answer to problem 9.2 a).

In connection with problem 9.2 let us also mention that the space l^1 has a *subspace* with a basis of type non-aP^*. Indeed, such is e.g. the subspace F of l^1 defined in § 8, example 8.1, since by (8.9) and the implication $1^\circ \Rightarrow 4^\circ$ of theorem 9.2 b), the basis (8.6) of (8.5) is of type non-aP^*. Dually, the space c_0 has a subspace with a basis of type non-aP.

§ 10. Bases of types l_+, $(l_+)^*$, al_+ and $(al_+)^*$. The cone associated to a basis

Definition 10.1. A basis $\{x_n\}$ of a Banach space E is said to be

a) *of type l_+*, if $\{x_n\}$ is a bounded basis and there exists a constant $\eta>0$ such that we have, for all finite sequences $\alpha_1,\dots,\alpha_n\geqslant 0$,

$$\left\|\sum_{i=1}^{n}\alpha_i x_i\right\|\geqslant \eta\sum_{i=1}^{n}\alpha_i; \tag{10.1}$$

b) *of type $(l_+)^*$*, if the a.s.c.f. $\{f_n\}$ is a basis of type l_+ of $[f_n]$;

c) *of type al_+* (*of type $(al_+)^*$*), if there exists a sequence of scalars $\{\varepsilon_n\}$ with $|\varepsilon_n|=1$ $(n=1,2,\dots)$, such that the basis $\{\varepsilon_n x_n\}$ of E is of type l_+ (respectively, of type $(l_+)^*$).

For instance, the natural basis of l^1 is of type l_+, but not $(al_+)^*$, the natural basis of c_0 is of type $(l_+)^*$, but not al_+, while the natural basis of l^p $(p>1)$ is of types non-al_+ and non-$(al_+)^*$.

Let us observe that in definition 10.1 a) it would have been sufficient to assume that $\sup_{1\leqslant n<\infty}\|x_n\|<\infty$, since (10.1) implies $\inf_{1\leqslant n<\infty}\|x_n\|>0$. Similarly, in definition 10.1 b) it would have been sufficient to assume that $\inf_{1\leqslant n<\infty}\|x_n\|>0$.

Some characterizations of bases of type l_+ are collected in

Theorem 10.1. *Let $\{x_n\}$ be a basis of a Banach space E, with $\sup_{1\leqslant n<\infty}\|x_n\|<\infty$. The following statements are equivalent:*

1°. *$\{x_n\}$ is of type l_+.*

2°. *For $\alpha_n\geqslant 0$ $(n=1,2,\ldots)$*

$$\sum_{i=1}^{\infty}\alpha_i x_i \quad \textit{converges} \Leftrightarrow \sum_{i=1}^{\infty}\alpha_i<\infty. \tag{10.2}$$

3°. *There exists a constant $\eta>0$ such that we have, for every sequence $\alpha_n\geqslant 0$ $(n=1,2,\ldots)$ with $\sum_{i=1}^{\infty}\alpha_i x_i\in E$,*

$$\left\|\sum_{i=1}^{\infty}\alpha_i x_i\right\|\geqslant\eta\sum_{i=1}^{\infty}\alpha_i. \tag{10.3}$$

4°. *There exists a constant $\eta>0$ such that we have, for every[1] $x\in\operatorname{co}\{x_n\}$,*

$$\|x\|\geqslant\eta. \tag{10.4}$$

5°. *There exists an $f\in E^*$ such that*

$$\operatorname{Re} f(x_n)\geqslant 1 \qquad (n=1,2,\ldots). \tag{10.5}$$

6°. *$\{x_n\}$ admits a "contraction bounded from below" of type P^*, i.e. there exists a sequence of scalars $\{\beta_n\}$ satisfying*

$$0<\inf_{1\leqslant n<\infty}\beta_n\leqslant\sup_{1\leqslant n<\infty}\beta_n\leqslant 1, \tag{10.6}$$

such that $\{\beta_n x_n\}$ is a basis of type P^* of E.

7°. *There exists a sequence of scalars $\{\beta_n\}$ satisfying*

$$0<\beta_n\leqslant\sup_{1\leqslant j<\infty}\beta_j<\infty \qquad (n=1,2,\ldots), \tag{10.7}$$

such that $\{\beta_n x_n\}$ is a basis of type P^* of E.

8°–9°. *The same as 6°–7° respectively, with P^* replaced by l_+.*

10°. *For every sequence of scalars $\{\beta_n\}$ satisfying*

$$0<\inf_{1\leqslant n<\infty}\beta_n\leqslant\sup_{1\leqslant n<\infty}\beta_n<\infty, \tag{10.8}$$

$\{\beta_n x_n\}$ is a basis of type l_+ of E.

---

[1] We denote by co $\{x_n\}$ the convex hull of the set $\{x_n\}$.

Proof. $1^\circ \Rightarrow 3^\circ$. Assume that we have 1° and let $\alpha_n \geqslant 0$ $(n=1,2,\ldots)$, $\sum_{i=1}^{\infty} \alpha_i x_i \in E$. Then

$$0 \leqslant \sum_{i=n+1}^{n+p} \alpha_i \leqslant \frac{1}{\eta} \left\| \sum_{i=n+1}^{n+p} \alpha_i x_i \right\| < \varepsilon \qquad (n > N(\varepsilon),\ p=1,2,\ldots),$$

whence $\sum_{i=1}^{\infty} \alpha_i < \infty$. Furthermore, from (10.1) we infer

$$\left\| \sum_{i=1}^{\infty} \alpha_i x_i \right\| = \lim_{n\to\infty} \left\| \sum_{i=1}^{n} \alpha_i x_i \right\| \geqslant \lim_{n\to\infty} \eta \sum_{i=1}^{n} \alpha_i = \eta \sum_{i=1}^{\infty} \alpha_i,$$

i.e. (10.3).

The implication $3^\circ \Rightarrow 2^\circ$ is obvious, taking also into account the hypothesis $\sup_{1 \leqslant n < \infty} \|x_n\| < \infty$.

$2^\circ \Rightarrow 1^\circ$. Assume that $\{x_n\}$ is of type non-l_+. Then there exist an increasing sequence of positive integers $\{m_n\}$ and numbers $\alpha_i^{(n)} \geqslant 0$ $(m_{n-1}+1 \leqslant i \leqslant m_n;\ n=1,2,\ldots;\ m_0=0)$ such that

$$\left\| \sum_{i=m_{n-1}+1}^{m_n} \alpha_i^{(n)} x_i \right\| \leqslant \frac{1}{2^n}, \qquad \sum_{i=m_{n-1}+1}^{m_n} \alpha_i^{(n)} = 1 \qquad (n=1,2,\ldots). \tag{10.9}$$

Since $\{x_n\}$ is a basis, there exists, by Ch. I, § 7, theorem 7.1, a constant $C \geqslant 1$ such that

$$\left\| \sum_{i=m_{n-1}+1}^{k} \alpha_i^{(n)} x_i \right\| \leqslant C \left\| \sum_{i=m_{n-1}+1}^{m_n} \alpha_i^{(n)} x_i \right\| \qquad (m_{n-1}+1 \leqslant k \leqslant m_n;\ n=1,2,\ldots).$$

Hence, by (10.9), $\sum_{i=1}^{\infty} \alpha_i x_i$ converges but $\sum_{i=1}^{\infty} \alpha_i = \infty$, where we have put $\alpha_i = \alpha_i^{(n)}$ $(m_{n-1}+1 \leqslant i \leqslant m_n;\ n=1,2,\ldots)$. Thus, 2° is not satisfied, which proves that $2^\circ \Rightarrow 1^\circ$.

$1^\circ \Rightarrow 4^\circ$. Assume that we have 1° and let $x \in \mathrm{co}\{x_n\}$. Then there exist a positive integer n and scalars $\alpha_1, \ldots, \alpha_n \geqslant 0$ with $\sum_{i=1}^{n} \alpha_i = 1$, such that $x = \sum_{i=1}^{n} \alpha_i x_i$, whence, by 1°,

$$\|x\| = \left\| \sum_{i=1}^{n} \alpha_i x_i \right\| \geqslant \eta \sum_{i=1}^{n} \alpha_i = \eta.$$

$4^\circ \Rightarrow 1^\circ$. Assume that we have 4° and let $\alpha_1, \ldots, \alpha_n \geqslant 0$ be arbitrary. Then for $\gamma_i = \dfrac{\alpha_i}{\sum_{j=1}^{n} \alpha_j}$ $(i=1,\ldots,n)$ and $x = \sum_{i=1}^{n} \gamma_i x_i$ we have $x \in \mathrm{co}\{x_n\}$, whence, by 4°,

$$\left\| \sum_{i=1}^{n} \frac{\alpha_i}{\sum_{j=1}^{n} \alpha_j} x_i \right\| = \left\| \sum_{i=1}^{n} \gamma_i x_i \right\| = \|x\| \geqslant \eta,$$

which is nothing else than 1°.

The equivalence 1°⇔5° is an immediate consequence of a well known theorem of S. Mazur and W. Orlicz[1], if E is a real Banach space. In the case of complex scalars, the implication 5°⇒1° is again obvious, since for $\alpha_i \geqslant 0$ and $\operatorname{Re} f(x_i) \geqslant 1$ $(i=1, \ldots, n)$ we have

$$\left\| \sum_{i=1}^{n} \alpha_i x_i \right\| \geqslant \frac{1}{\|f\|} \left| \sum_{i=1}^{n} \alpha_i f(x_i) \right| \geqslant \frac{1}{\|f\|} \sum_{i=1}^{n} \alpha_i \operatorname{Re} f(x_i) \geqslant \frac{1}{\|f\|} \sum_{i=1}^{n} \alpha_i .$$

On the other hand, if we have 1° for a complex Banach space E, then, considering E as a real Banach space $E_{(r)}$ in the usual way, we can find, by the above, a functional $g \in E_{(r)}^*$ such that $g(x_n) \geqslant 1$ $(n=1,2,\ldots)$. Then the functional $f(x) = g(x) - i g(ix)$ $(x \in E)$, where $i = \sqrt{-1}$, satisfies (10.5).

5° ⇒ 6°. Assume that we have 5°. For $f \in E^*$ satisfying (10.5), define $g \in E_{(r)}^*$ by

$$g(x) = \operatorname{Re} f(x) \qquad (x \in E), \tag{10.10}$$

and put

$$\beta_n = \frac{1}{g(x_n)} \qquad (n=1,2,\ldots). \tag{10.11}$$

Then, by $\sup\limits_{1 \leqslant j < \infty} \|x_j\| < \infty$ and $g(x_n) \geqslant 1$ $(n=1,2,\ldots)$ we have

$$0 < \frac{1}{\|g\| \sup\limits_{1 \leqslant j < \infty} \|x_j\|} \leqslant \beta_n \leqslant 1 \qquad (n=1,2,\ldots),$$

whence also (10.6). Furthermore, by (10.11),

$$g(\beta_n x_n) = 1 \qquad (n=1,2,\ldots),$$

whence, by the implication 4° ⇒ 1° of § 9, theorem 9.2 b), $\{\beta_n x_n\}$ is a basis of P^* of E.

---

[1] See [158], p. 147, theorem 2.41; for other proofs see [228], [200], [57].

The implication $6^\circ \Rightarrow 7^\circ$ is obvious.

$7^\circ \Rightarrow 1^\circ$. If we have 7°, then $\sup_{1 \leqslant n < \infty} \left\| \sum_{i=1}^{n} \frac{1}{\beta_i} f_i \right\| = C < \infty$, where $\{f_n\} \subset E^*$ is the a.s.c.f. to $\{x_n\}$. Hence, for any finite sequence $\alpha_1, \ldots, \alpha_n \geqslant 0$ we have

$$\left\| \sum_{i=1}^{n} \alpha_i x_i \right\| \geqslant \frac{1}{C} \left| \left(\sum_{j=1}^{n} \frac{1}{\beta_j} f_j \right) \left(\sum_{i=1}^{n} \alpha_i x_i \right) \right| = \frac{1}{C} \sum_{i=1}^{n} \frac{\alpha_i}{\beta_i} \geqslant \frac{1}{C \sup_{1 \leqslant j < \infty} \beta_j} \sum_{i=1}^{n} \alpha_i .$$

The implications $1^\circ \Rightarrow 8^\circ \Rightarrow 9^\circ$ are obvious.

$9^\circ \Rightarrow 1^\circ$. If we have 9°, then for any $\alpha_1, \ldots, \alpha_n \geqslant 0$ we have

$$\left\| \sum_{i=1}^{n} \alpha_i x_i \right\| = \left\| \sum_{i=1}^{n} \frac{\alpha_i}{\beta_i} \beta_i x_i \right\| \geqslant \eta \sum_{i=1}^{n} \frac{\alpha_i}{\beta_i} \geqslant \frac{\eta}{\sup_{1 \leqslant j < \infty} \beta_j} \sum_{i=1}^{n} \alpha_i ,$$

with a suitable $\eta > 0$ corresponding to the basis $\{\beta_n x_n\}$.

$1^\circ \Rightarrow 10^\circ$. Assume that we have 1° and let $\{\beta_n\}$ be an arbitrary sequence of scalars satisfying (10.8). Then

$$\sup_{1 \leqslant n < \infty} \|\beta_n x_n\| \leqslant \sup_{1 \leqslant j < \infty} \beta_j \sup_{1 \leqslant n < \infty} \|x_n\| < \infty .$$

Furthermore, since $\left\{ \frac{1}{\beta_n} \beta_n x_n \right\} = \{x_n\}$ is of type l_+ and since

$$0 < \frac{1}{\sup_{1 \leqslant n < \infty} \beta_n} = \inf_{1 \leqslant n < \infty} \frac{1}{\beta_n} \leqslant \sup_{1 \leqslant n < \infty} \frac{1}{\beta_n} = \frac{1}{\inf_{1 \leqslant n < \infty} \beta_n} < \infty ,$$

applying the implication $9^\circ \Rightarrow 1^\circ$ above to the basis $\{\beta_n x_n\}$ and the sequence of scalars $\left\{ \frac{1}{\beta_n} \right\}$ we obtain that $\{\beta_n x_n\}$ is of type l_+.

Finally, the implication $10^\circ \Rightarrow 1^\circ$ is obvious. This completes the proof of theorem 10.1.

Remark 10.1. In contrast with $3^\circ - 5^\circ$, 7° and 9°, conditions 2°, 6°, 8° and 10° imply that $\sup_{1 \leqslant n < \infty} \|x_n\| < \infty$. Indeed, if we have 6° or 8°, then $\sup_{1 \leqslant n < \infty} \|x_n\| \leqslant \frac{1}{\inf_{1 \leqslant j < \infty} \beta_j} \sup_{1 \leqslant n < \infty} \|\beta_n x_n\| < \infty$, and it is obvious that 10° also implies $\sup_{1 \leqslant n < \infty} \|x_n\| < \infty$; for 2° see Ch. I, § 3, proof of lemma 3.1 b).

Combining § 9, theorem 9.2 b) and the equivalences $1^\circ \Leftrightarrow 6^\circ \Leftrightarrow 7^\circ$ of theorem 10.1, one obtains other characterizations of bases of type l_+, e.g. the following: *A basis $\{x_n\}$ of a Banach space E, with* $\sup_{1 \leqslant n < \infty} \|x_n\| < \infty$, *is of type* l_+ *if and only if there exists a sequence of scalars* $\{\beta_n\}$ *satisfying* (10.6) *such that the sequence*

$$z_1=\beta_1 x_1, \quad z_n=\beta_{n-1}x_{n-1}-\beta_n x_n \quad (n=2,3,\ldots) \tag{10.12}$$

is a basis of E.

Now we shall give some other characterizations of bases of type l_+ in terms of the "associated cone". Throughout the sequel by "cone" we shall understand "closed convex cone having the origin as extreme point", i.e. a closed set $\mathscr{K}$ such that $\mathscr{K}+\mathscr{K}\subset\mathscr{K}$, $\lambda\mathscr{K}\subset\mathscr{K}$ $(\lambda\geqslant 0)$ and $\mathscr{K}\cap(-\mathscr{K})=\{0\}$.

Definition 10.2. Let $\{x_n\}$ be a basis of a real[1] Banach space E. The set

$$\mathscr{K}_{\{x_n\}}=\left\{\sum_{i=1}^{\infty}\alpha_i x_i\in E \;\middle|\; \alpha_n\geqslant 0 \quad (n=1,2,\ldots)\right\} \tag{10.13}$$

is called *the cone associated to the basis* $\{x_n\}$.

It is easy to see that $\mathscr{K}_{\{x_n\}}$ coincides with the cone $\mathscr{C}_{\{x_n\}}$ generated by $\{x_n\}$ (i.e. the smallest cone containing the basis $\{x_n\}$). Indeed, the inclusion $\mathscr{K}_{\{x_n\}}\supset\mathscr{C}_{\{x_n\}}$ is obvious since $\mathscr{K}_{\{x_n\}}$ is a cone containing each x_j $(j=1,2,\ldots)$; on the other hand, for any cone $\mathscr{C}$ containing each x_j we have $\sum_{i=1}^{n}\alpha_i x_i\in\mathscr{C}$ $(\alpha_1,\ldots,\alpha_n\geqslant 0)$, whence also $\sum_{i=1}^{\infty}\alpha_i x_i=\lim_{n\to\infty}\sum_{i=1}^{n}\alpha_i x_i\in\mathscr{C}$ $\left(\sum_{i=1}^{\infty}\alpha_i x_i\in E,\ \alpha_n\geqslant 0,\ n=1,2,\ldots\right)$, and thus $\mathscr{K}_{\{x_n\}}\subset\mathscr{C}_{\{x_n\}}$, whence the equality.

We recall that a subset $\mathscr{B}$ of a cone $\mathscr{K}$ is said to be a *base* of the cone $\mathscr{K}$ if $\mathscr{B}$ is closed and convex and if every $x\in\mathscr{K}\setminus\{0\}$ has a unique representation of the form $x=\lambda y$, with $\lambda>0$, $y\in\mathscr{B}$. Furthermore, we

---

[1] We shall only consider cones in *real* Banach spaces. Whenever a complex Banach E with a basis $\{x_n\}$ is isomorphic to the complexification of the real Banach space $G=\left\{\sum_{j=1}^{\infty}\alpha_j x_j\in E_{(r)} \;\middle|\; \alpha_n=\text{real } (n=1,2,\ldots)\right\}$ (or, equivalently, the mapping $\sum_{j=1}^{\infty}\alpha_j x_j\to\sum_{j=1}^{\infty}(\operatorname{Re}\alpha_j)x_j$ is continuous—see Ch. I, § 1), one can define the associated cone by

$$\mathscr{K}_{\{x_n\}}=\left\{\sum_{j=1}^{\infty}\alpha_j x_j\in E \;\middle|\; \operatorname{Re}\alpha_j\geqslant 0,\ \operatorname{Im}\alpha_j\geqslant 0 \quad (i=1,2,\ldots)\right\},$$

and this may be regarded as the cartesian square of the cone

$$\mathscr{K}=\left\{\sum_{j=1}^{\infty}\alpha_j x_j\in G \;\middle|\; \alpha_n\geqslant 0 \quad (n=1,2,\ldots)\right\}$$

of the real Banach space G.

recall that a set $\mathscr{M}$ in a topological linear space $\mathscr{L}$ is called an *extremal subset* of a closed convex set $\mathscr{A}$, if it is a closed convex subset of $\mathscr{A}$ and if together with every interior point of a segment in $\mathscr{A}$ it contains the whole segment (i.e. the relations $x,y\in\mathscr{A}$, $\lambda x+(1-\lambda)y\in\mathscr{M}$ and $0<\lambda<1$ imply $x,y\in\mathscr{M}$); an extremal subset of $\mathscr{A}$ consisting of a single point is called an *extremal point* of $\mathscr{A}$. Finally, we recall that a cone $\mathscr{K}$ is called *solid*, if it contains an interior point.

Some properties of cones associated to general bases $\{x_n\}$ are collected in

Proposition 10.1. *Let $\{x_n\}$ be a basis of a real Banach space E and let $\mathscr{K}_{\{x_n\}}$ be the cone associated to the basis $\{x_n\}$. Then*

a) *$\mathscr{K}_{\{x_n\}}$ has an unbounded base, if* $\dim E=\infty$.

b) *$\mathscr{K}_{\{x_n\}}$ has no weakly compact (hence also no compact) base, if* $\dim E=\infty$.

c) *Each ray $r_j=\{\lambda x_j\,|\,0\leqslant\lambda<\infty\}$ is an "extremal ray" (i.e. an extremal subset) of $\mathscr{K}_{\{x_n\}}$ and these are the only extremal rays of $\mathscr{K}_{\{x_n\}}$. Consequently, the rays r_j intersect any base $\mathscr{B}$ of $\mathscr{K}_{\{x_n\}}$ in extremal points of $\mathscr{B}$.*

d) *$\mathscr{K}_{\{x_n\}}$ is not solid, if* $\dim E=\infty$.

Proof. a) Define $f\in E^*$ by

$$f(x)=\sum_{i=1}^{\infty}\frac{1}{2^i\|f_i\|}f_i(x)\qquad(x\in E),\tag{10.14}$$

where $\{f_n\}\subset E^*$ is the a.s.c.f. to $\{x_n\}$. We claim that the set

$$\mathscr{B}=\{y\in\mathscr{K}_{\{x_n\}}\,|\,f(y)=1\}\tag{10.15}$$

is an unbounded base of $\mathscr{K}_{\{x_n\}}$. Indeed, $\mathscr{B}$ is convex and closed, and for every $x\in\mathscr{K}_{\{x_n\}}\setminus\{0\}$ we have $x=\lambda y$, where $\lambda=f(x)>0$ and $y=\dfrac{1}{f(x)}x\in\mathscr{B}$. This representation is unique since the relations $x=\lambda_1 y_1=\lambda_2 y_2$, $\lambda_1,\lambda_2>0$, $y_1,y_2\in\mathscr{B}$ imply, by (10.15), $f(x)=\lambda_1=\lambda_2$, whence also $y_1=y_2$; therefore $\mathscr{B}$ is a base of $\mathscr{K}_{\{x_n\}}$. Furthermore, we have

$$f(2^n\|f_n\|x_n)=\sum_{i=1}^{\infty}\frac{1}{2^i\|f_i\|}f_i(2^n\|f_n\|x_n)=1\qquad(n=1,2,\ldots),$$

i.e. $2^n\|f_n\|x_n\in\mathscr{B}$ $(n=1,2,\ldots)$, and this sequence is unbounded, since

$$\|2^n\|f_n\|x_n\|=2^n\|f_n\|\;\|x_n\|\geqslant 2^n|f_n(x_n)|=2^n\qquad(n=1,2,\ldots).$$

b) Assume that $\mathscr{B}$ is a weakly compact base of $\mathscr{K}_{\{x_n\}}$. Since $\mathscr{B}$ is a base of $\mathscr{K}_{\{x_n\}}$, there exist $\lambda_n>0$ and $y_n\in\mathscr{B}$ such that $x_n=\lambda_n y_n$ $(n=1,2,\ldots)$.

Then, since $\mathscr{B}$ is a weakly compact set, whence also sequentially weakly compact[1], $\{y_n\}$ has a subsequence $\{y_{n_k}\}$ converging weakly to an element $y_0 \in \mathscr{B}$. For $n_k > j$ we have then $0 = f_j(y_{n_k}) \to f_j(y_0)$ as $k \to \infty$, whence $f_j(y_0) = 0$ $(j = 1, 2, \ldots)$, whence, since $\{f_n\}$ is total on E, $y_0 = 0$, and thus $0 \in \mathscr{B}$. However, this is impossible for a base $\mathscr{B}$ of a cone (since if $y \in \mathscr{B} \setminus \{0\}$, then, $\mathscr{B}$ being convex, the whole segment $\langle 0, y \rangle$ belongs to $\mathscr{B}$, whence the element $y \in \mathscr{K}_{\{x_n\}} \setminus \{0\}$ has two representations $y = 1 \cdot y$ and $y = 2\dfrac{y}{2}$), which proves that $\mathscr{K}_{\{x_n\}}$ has no weakly compact base $\mathscr{B}$.

c) Let us prove first that each ray $r_j = \{\lambda x_j \mid 0 \leqslant \lambda < \infty\}$ is an extremal ray of $\mathscr{K}_{\{x_n\}}$. Assume that $y, z \in \mathscr{K}_{\{x_n\}}$, $\alpha_0 y + (1 - \alpha_0) z = \lambda_0 x_j$, where $0 < \alpha_0 < 1$, $\lambda_0 \geqslant 0$. Then

$$\alpha_0 f_i(y) + (1 - \alpha_0) f_i(z) = \lambda_0 f_i(x_j) = \begin{cases} 0 & \text{for } i \neq j, \\ \lambda_0 & \text{for } i = j, \end{cases}$$

whence, since $f_i(y), f_i(z) \geqslant 0$ (by $y, z \in \mathscr{K}_{\{x_n\}}$) and $\alpha_0 > 0$, $1 - \alpha_0 > 0$, it follows that we must have $f_i(y) = f_i(z) = 0$ for $i \neq j$. Consequently,

$$f_i[f_j(y) x_j - y] = 0 \qquad (i = 1, 2, \ldots),$$

whence, since $\{f_n\}$ is total on E, $y = f_j(y) x_j$. Similarly, we have also $z = f_j(z) x_j$, which proves that r_j is extremal.

Now let $\mathscr{B}$ be an arbitrary base of $\mathscr{K}_{\{x_n\}}$. Then $x_j \in \mathscr{K}_{\{x_n\}} \setminus \{0\}$ admits a unique representation $x_j = \lambda y_j$, with $\lambda > 0$, $y_j \in \mathscr{B}$, whence $y_j = \dfrac{1}{\lambda} x_j$ is the unique point of intersection of the ray r_j with $\mathscr{B}$. To prove that this is an extremal point of $\mathscr{B}$, assume that there exist $y, z \in \mathscr{B}$, $y \neq z$ and α_0 with $0 < \alpha_0 < 1$, such that $\alpha_0 y + (1 - \alpha_0) z = \dfrac{1}{\lambda} x_j$. Then, by the above (with $\lambda_0 = \dfrac{1}{\lambda}$), we must have $y = f_j(y) x_j$, whence $x_j = \dfrac{1}{f_j(y)} y = \lambda y_j$, which, since $\lambda \neq \dfrac{1}{f_j(y)}$ (because otherwise $\alpha_0 y + (1 - \alpha_0) z = y$, whence $y = z$), contradicts the definition of a base $\mathscr{B}$ of $\mathscr{K}_{\{x_n\}}$. Thus, r_j intersects $\mathscr{B}$ in an extremal point of $\mathscr{B}$.

Now let $r = \{\lambda y \mid \lambda \geqslant 0\}$ be an arbitrary ray in $\mathscr{K}_{\{x_n\}}$, different from all rays r_j. Since $y \in \mathscr{K}_{\{x_n\}}$, we have $y = \sum_{i=1}^{\infty} \alpha_i x_i$, with $\alpha_n \geqslant 0$ $(n = 1, 2, \ldots)$, and with some $\alpha_j \neq 0$, whence $y = \frac{1}{2}(2 \alpha_j x_j) + \frac{1}{2}\left(2 \sum_{\substack{i=1 \\ i \neq j}}^{\infty} \alpha_i x_i\right)$, but $2 \alpha_j x_j \notin r$

[1] See e.g. [50], p. 430, theorem 1.

since by our hypothesis $r_j \neq r$. Hence r is not an extremal ray of $\mathscr{K}_{\{x_n\}}$, which proves that the r_j $(j=1,2,\ldots)$ are the only extremal rays of $\mathscr{K}_{\{x_n\}}$.

d) To prove that $\mathscr{K}_{\{x_n\}}$ is not solid if $\dim E = \infty$, let $x_0 = \sum_{i=1}^{\infty} \alpha_i^0 x_i \in \mathscr{K}_{\{x_n\}}$ and $c>0$ be arbitrary; it will be sufficient to prove that the cell $S(x_0,c) = \{x \in E \mid \|x-x_0\| \leqslant c\}$ is not contained in $\mathscr{K}_{\{x_n\}}$. We may assume, without loss of generality, that $\|x_n\|=1$ $(n=1,2,\ldots)$. Since $\alpha_n^0 \geqslant 0$ $(n=1,2,\ldots)$, $\alpha_n \to 0$ $(n\to\infty)$ and $\dim E=\infty$, there exists an index $N=N(c)$ such that $\alpha_N^0 < c$. Put

$$\alpha_n = \begin{cases} \alpha_n^0 & \text{for} \quad n \neq N, \\ \alpha_n^0 - c & \text{for} \quad n = N. \end{cases}$$

Then for $x = \sum_{i=1}^{\infty} \alpha_i x_i = \sum_{i \neq N} \alpha_i^0 x_i + (\alpha_N^0 - c) x_N \in E$ we have

$$\|x-x_0\| = \left\| \sum_{i=1}^{\infty} (\alpha_i - \alpha_i^0) x_i \right\| = |\alpha_N - \alpha_N^0| \, \|x_N\| = |\alpha_N - \alpha_N^0| = c,$$

whence $x \in S(x_0,c)$, but $\alpha_N = \alpha_N^0 - c < 0$, i.e. $x \notin \mathscr{K}_{\{x_n\}}$, which completes the proof of proposition 10.1.

We recall that two cones $\mathscr{K}'$ and $\mathscr{K}''$ in Banach spaces E and F respectively, are said to be *locally isomorphic*, if there exists a one-to-one linear mapping τ of $\mathscr{K}'$ onto $\mathscr{K}''$ such that for $\{z_n\} \subset \mathscr{K}'$, $z_0 \in \mathscr{K}'$,

$$\lim_{n\to\infty} z_n = z_0 \;\Leftrightarrow\; \lim_{n\to\infty} \tau(z_n) = \tau(z_0). \tag{10.16}$$

Now we shall give some characterizations of bases $\{x_n\}$ of type l_+ in terms of properties of the cone $\mathscr{K}_{\{x_n\}}$ associated to $\{x_n\}$.

Theorem 10.2. *Let $\{x_n\}$ be a bounded basis of a real Banach space E, with the a.s.c.f. $\{f_n\}$ and let $\mathscr{K}_{\{x_n\}}$ be the cone associated to the basis $\{x_n\}$. The following statements are equivalent:*

1°. *$\{x_n\}$ is of type l_+.*

2°. *We have*

$$\mathscr{K}_{\{x_n\}} = \left\{ \sum_{i=1}^{\infty} \alpha_i x_i \;\middle|\; \alpha_n \geqslant 0 \;(n=1,2,\ldots),\; \sum_{i=1}^{\infty} \alpha_i < \infty \right\}. \tag{10.17}$$

3°. *There exists a local isomorphism τ of the cone $\mathscr{K}_{\{e_n\}}$ associated to the natural basis $\{e_n\}$ of l^1, onto $\mathscr{K}_{\{x_n\}}$, such that*

$$\tau(e_n) = x_n \qquad (n=1,2,\ldots). \tag{10.18}$$

4°. *There exists a constant* $M>0$ *such that for every* $x\in\mathscr{K}_{\{x_n\}}$ *the series* $\sum_{i=1}^{\infty} f_i(x)x_i$ *is absolutely convergent (i.e.* $\sum_{i=1}^{\infty}\|f_i(x)x_i\|<\infty$*) and*

$$\sum_{i=1}^{\infty}\|f_i(x)x_i\|\leqslant M\|x\| \qquad (x\in\mathscr{K}_{\{x_n\}}). \tag{10.19}$$

5°. $\mathscr{K}_{\{x_n\}}$ *has a bounded base.*

Proof. The equivalence $1°\Leftrightarrow 2°$ is an immediate consequence of the equivalence $1°\Leftrightarrow 3°$ of theorem 10.1.

$2°\Rightarrow 3°$. Assume that we have 2° and define $\tau:\mathscr{K}_{\{e_n\}}\to\mathscr{K}_{\{x_n\}}$ (where $\{e_n\}$ is the natural basis of l^1) by

$$\tau\left(\sum_{i=1}^{\infty}\alpha_i e_i\right)=\sum_{i=1}^{\infty}\alpha_i x_i \qquad \left(\alpha_n\geqslant 0,\ n=1,2,\dots;\ \sum_{i=1}^{\infty}\alpha_i<\infty\right). \tag{10.20}$$

Then τ is linear, satisfies (10.18) and, by 2°, τ maps $\mathscr{K}_{\{e_n\}}$ *onto* $\mathscr{K}_{\{x_n\}}$. Since $\{x_n\}$ is a basis, τ is one-to-one. Let us prove (10.16).

If

$$z_n=\sum_{i=1}^{\infty}\alpha_i^{(n)}e_i\to z_0=\sum_{i=1}^{\infty}\alpha_i e_i \quad \text{as} \quad n\to\infty,$$

where $\alpha_i^{(n)},\alpha_i\geqslant 0$, $\sum_{i=1}^{\infty}\alpha_i^{(n)}$, $\sum_{i=1}^{\infty}\alpha_i<\infty$, then $\sum_{i=1}^{\infty}|\alpha_i^{(n)}-\alpha_i|\to 0$ as $n\to\infty$,

whence

$$\|\tau(z_n)-\tau(z_0)\|=\left\|\sum_{i=1}^{\infty}(\alpha_i^{(n)}-\alpha_i)x_i\right\|\leqslant \sup_{1\leqslant j<\infty}\|x_j\|\sum_{i=1}^{\infty}|\alpha_i^{(n)}-\alpha_i|\to 0$$

as $n\to\infty$.

Conversely, assume that

$$\tau(z_n)=\sum_{i=1}^{\infty}\alpha_i^{(n)}x_i\to\tau(z_0)=\sum_{i=1}^{\infty}\alpha_i x_i \quad \text{as} \quad n\to\infty,$$

where $\alpha_i^{(n)},\alpha_i\geqslant 0$, $\sum_{i=1}^{\infty}\alpha_i^{(n)}$, $\sum_{i=1}^{\infty}\alpha_i<\infty$ and let $\varepsilon>0$ be arbitrary. Then there exists a positive integer I_ε such that

$$\sum_{i=I_\varepsilon}^{\infty}\alpha_i<\frac{\varepsilon\eta}{\eta+1+\sup\limits_{1\leqslant j<\infty}\|x_j\|} \tag{10.21}$$

(where $\eta>0$ is as in (10.3)), whence

$$\left\| \sum_{i=I_\varepsilon}^{\infty} \alpha_i x_i \right\| < \frac{\varepsilon\eta \sup_{1 \leqslant j < \infty} \|x_j\|}{\eta+1+\sup_{1 \leqslant j < \infty} \|x_j\|}. \tag{10.22}$$

On the other hand,

$$\left\| \sum_{i=I_\varepsilon}^{\infty} \alpha_i^{(n)} x_i - \sum_{i=I_\varepsilon}^{\infty} \alpha_i x_i \right\| \leqslant \left\| \sum_{i=1}^{I_\varepsilon - 1} (\alpha_i^{(n)} - \alpha_i) x_i \right\| + \left\| \sum_{i=1}^{\infty} \alpha_i^{(n)} x_i - \sum_{i=1}^{\infty} \alpha_i x_i \right\|.$$

Here the right side $\to 0$ as $n \to \infty$, because of the continuity of the coefficient functionals, respectively by the hypothesis $\tau(z_n) \to \tau(z_0)$. Thus

$$\left\| \sum_{i=I_\varepsilon}^{\infty} \alpha_i^{(n)} x_i - \sum_{i=I_\varepsilon}^{\infty} \alpha_i x_i \right\| \leqslant \frac{\varepsilon\eta}{\eta+1+\sup_{1 \leqslant j < \infty} \|x_j\|} \qquad (n > N(\varepsilon)),$$

whence, taking into account (10.3) and (10.22),

$$\sum_{i=I_\varepsilon}^{\infty} \alpha_i^{(n)} \leqslant \frac{1}{\eta} \left\| \sum_{i=I_\varepsilon}^{\infty} \alpha_i^{(n)} x_i \right\| \leqslant \frac{1+\sup_{1 \leqslant j < \infty} \|x_j\|}{\eta+1+\sup_{1 \leqslant j < \infty} \|x_j\|} \varepsilon \qquad (n > N(\varepsilon)). \tag{10.23}$$

Now, from

$$\sum_{i=1}^{\infty} |\alpha_i^{(n)} - \alpha_i| \leqslant \sum_{i=1}^{I_\varepsilon - 1} |\alpha_i^{(n)} - \alpha_i| + \sum_{i=I_\varepsilon}^{\infty} \alpha_i^{(n)} + \sum_{i=I_\varepsilon}^{\infty} \alpha_i$$

for $n \to \infty$ we obtain, taking into account (10.21) and (10.23),

$$\overline{\lim_{n \to \infty}} \sum_{i=1}^{\infty} |\alpha_i^{(n)} - \alpha_i| \leqslant \varepsilon,$$

whence, since $\varepsilon > 0$ has been arbitrary,

$$\lim_{n \to \infty} \|z_n - z_0\| = \lim_{n \to \infty} \sum_{i=1}^{\infty} |\alpha_i^{(n)} - \alpha_i| = 0.$$

$3^\circ \Rightarrow 2^\circ$. Assume that we have 3° and let $\sum_{i=1}^{\infty} \alpha_i x_i \in \mathscr{K}_{\{x_n\}}$. Then, by (10.18),

$$\sum_{i=1}^{n} \alpha_i = \left\| \sum_{i=1}^{n} \alpha_i e_i \right\| = \left\| \tau^{-1}\left(\sum_{i=1}^{n} \alpha_i x_i \right) \right\|. \tag{10.24}$$

Since $\lim_{n \to \infty} \sum_{i=1}^{n} \alpha_i x_i = \sum_{i=1}^{\infty} \alpha_i x_i$ and since τ is a local isomorphism, $\lim_{n \to \infty} \tau^{-1}\left(\sum_{i=1}^{n} \alpha_i x_i \right)$ exists, whence, by (10.24), $\sum_{i=1}^{\infty} \alpha_i < \infty$.

Conversely, if $\alpha_n \geqslant 0\ (n=1,2,\ldots)$, $\sum_{i=1}^{\infty} \alpha_i < \infty$, then, by $\sup_{1 \leqslant n < \infty} \|x_n\| < \infty$, the series $\sum_{i=1}^{\infty} \alpha_i x_i$ converges, whence $\sum_{i=1}^{\infty} \alpha_i x_i \in \mathscr{K}_{\{x_n\}}$.

$1° \Rightarrow 4°$. If $\{x_n\}$ is of type l_+, then for every $x \in \mathscr{K}_{\{x_n\}}$ and $n=1,2,\ldots$ we have

$$\sum_{i=1}^{n} \|f_i(x) x_i\| \leqslant \sup_{1 \leqslant j < \infty} \|x_j\| \sum_{i=1}^{n} f_i(x) \leqslant \left(\frac{1}{\eta} \sup_{1 \leqslant j < \infty} \|x_j\|\right) \left\| \sum_{i=1}^{n} f_i(x) x_i \right\|,$$

whence, taking $n \to \infty$, we obtain (10.19) with $M = \frac{1}{\eta} \sup_{1 \leqslant j < \infty} \|x_j\|$.

$4° \Rightarrow 1°$. If $\{x_n\}$ is a bounded basis satisfying (10.19), then for any $\alpha_1, \ldots, \alpha_n \geqslant 0$ we have, setting $x = \sum_{i=1}^{n} \alpha_i x_i$ in (10.19),

$$\frac{\inf_{1 \leqslant j < \infty} \|x_j\|}{M} \sum_{i=1}^{n} \alpha_i \leqslant \frac{1}{M} \sum_{i=1}^{n} \|\alpha_i x_i\| \leqslant \left\| \sum_{i=1}^{n} \alpha_i x_i \right\|.$$

$1° \Rightarrow 5°$. Assume that $\{x_n\}$ is of type l_+ and put

$$\mathscr{B} = \{y \in \mathscr{K}_{\{x_n\}} \mid f(y) = 1\}, \tag{10.25}$$

where $f \in E^*$ is any functional satisfying (10.5). Then $\mathscr{B}$ is a base of $\mathscr{K}_{\{x_n\}}$ (see the argument in the proof of proposition 10.1 a), taking also into account that for any $x \in \mathscr{K}_{\{x_n\}} \backslash \{0\}$ we have $f(x) = f\left(\sum_{i=1}^{\infty} \alpha_i x_i\right) = \sum_{i=1}^{\infty} \alpha_i f(x_i) \geqslant \sum_{i=1}^{\infty} \alpha_i > 0$) and for every $y = \sum_{i=1}^{\infty} \alpha_i x_i \in \mathscr{B}$ we have $\alpha_n \geqslant 0$ $(n=1,2,\ldots)$, whence

$$\|y\| \leqslant \sup_{1 \leqslant n < \infty} \|x_n\| \sum_{i=1}^{\infty} \alpha_i \leqslant \sup_{1 \leqslant n < \infty} \|x_n\| \sum_{i=1}^{\infty} \alpha_i f(x_i)$$

$$= \sup_{1 \leqslant n < \infty} \|x_n\| f\left(\sum_{i=1}^{\infty} \alpha_i x_i\right) = \sup_{1 \leqslant n < \infty} \|x_n\|,$$

i.e. $\mathscr{B}$ is bounded.

$5° \Rightarrow 1°$. Assume that the cone $\mathscr{K}_{\{x_n\}}$ associated to the bounded basis $\{x_n\}$ has a bounded base $\mathscr{B}$. Then $\sup_{y \in \mathscr{B}} \|y\| = C < \infty$. Since $0 \notin \mathscr{B}$ (see the proof of proposition 10.1 b)) and $\mathscr{B}$ is closed, we have also $\inf_{y \in \mathscr{B}} \|y\| = \eta > 0$. Furthermore, each $x_j \in \mathscr{K}_{\{x_n\}}$ admits a unique representation $x_j = \lambda_j y_j$, with $\lambda_j > 0$, $y_j \in \mathscr{B}$, whence $\beta_n x_n = y_n \in \mathscr{B}$ $(n=1,2,\ldots)$, where we have put $\beta_n = \frac{1}{\lambda_n}$ $(n=1,2,\ldots)$. We have then $\|\beta_n x_n\| \leqslant C$ $(n=1,2,\ldots)$, whence $\beta_n \leqslant \frac{C}{\|x_n\|} \leqslant \frac{C}{\inf\limits_{1\leqslant j<\infty} \|x_j\|}$, whence $\sup\limits_{1\leqslant n<\infty} \beta_n < \infty$. Furthermore, if $x \in \operatorname{co}\{\beta_n x_n\}$, then, since $\mathscr{B}$ is convex, $x \in \mathscr{B}$, whence $\|x\| \geqslant \inf\limits_{y\in\mathscr{B}} \|y\| = \eta$. Consequently, by theorem 10.1 (implication $4^\circ \Rightarrow 1^\circ$), $\{\beta_n x_n\}$ is a basis of type l_+, whence, again by theorem 10.1 (implication $9^\circ \Rightarrow 1^\circ$), $\{x_n\}$ is a basis of type l_+, which completes the proof of theorem 10.2.

Remark 10.2. Let us also mention the following alternative proof of the implication $5^\circ \Rightarrow 1^\circ$: Assume that the cone $\mathscr{K}_{\{x_n\}}$ associated to the bounded basis $\{x_n\}$ has a bounded base $\mathscr{B}$. Then, since $\mathscr{B}$ is closed and $0 \notin \mathscr{B}$ (see the proof of proposition 10.1 b)), $\mathscr{B}$ can be[1] strictly separated from 0 by a hyperplane, i.e. there exists a functional $f \in E^*$ such that $\inf\limits_{y\in\mathscr{B}} f(y) = \delta > 0$, whence $f(\beta_n x_n) \geqslant \delta$, where $\beta_n > 0$ is the (unique) number for which $\beta_n x_n \in \mathscr{B}$ $(n=1,2,\ldots)$. Since $\|\beta_n x_n\| \leqslant \sup\limits_{y\in\mathscr{B}} \|y\| = C < \infty$ $(n=1,2,\ldots)$, we have $\beta_n \leqslant \frac{C}{\|x_n\|} \leqslant \frac{C}{\inf\limits_{1\leqslant j<\infty} \|x_j\|}$ $(n=1,2,\ldots)$, whence

$$f(x_n) = \frac{f(\beta_n x_n)}{\beta_n} \geqslant \frac{\delta}{C} \inf_{1\leqslant j<\infty} \|x_j\| > 0 \qquad (n=1,2,\ldots),$$

whence, by theorem 10.1 (implication $5^\circ \Rightarrow 1^\circ$), $\{x_n\}$ is of type l_+, which completes the proof.

Remark 10.3. Condition 3° implies that $\{x_n\}$ is already a bounded basis of E. Indeed, assume that we have 3°, but $\sup\limits_{1\leqslant n<\infty} \|x_n\| = \sup\limits_{1\leqslant n<\infty} \|\tau(e_n)\| = \infty$. Then there exists a subsequence $\{e_{i_n}\}$ of $\{e_n\}$ with $\tau(e_{i_n}) \neq 0$ $(n=1,2,\ldots)$, $\lim\limits_{n\to\infty} \|\tau(e_{i_n})\| = \infty$, whence

$$\lim_{n\to\infty} \frac{1}{\|\tau(e_{i_n})\|} e_{i_n} = 0, \qquad \left\|\tau\left(\frac{1}{\|\tau(e_{i_n})\|} e_{i_n}\right)\right\| = 1 \qquad (n=1,2,\ldots),$$

which contradicts (10.16). Similarly, if we have 3°, but $\inf\limits_{1\leqslant n<\infty} \|x_n\| = \inf\limits_{1\leqslant n<\infty} \|\tau(e_n)\| = 0$, then there exists a subsequence $\{e_{i_n}\}$ of $\{e_n\}$ with $\lim\limits_{n\to\infty} \tau(e_{i_n}) = 0 = \tau(0)$, which, since $\|e_{i_n}\| = 1$ $(n=1,2,\ldots)$, contradicts again (10.16).

Let us recall that a cone $\mathscr{K}$ is said to be *generating*, if $E = \mathscr{K} - \mathscr{K} = \{y - z \mid y, z \in \mathscr{K}\}$. From theorem 10.2 we obtain, in particular, the following characterization of bases equivalent to the unit vector basis of l^1:

---

[1] See e.g. [43], Ch. I, § 6, theorem 5, or [133], p. 245, theorem (1).

Corollary 10.1. *A bounded basis $\{x_n\}$ of a real Banach space E is equivalent to the unit vector basis of l^1 if and only if the associated cone $\mathscr{K}_{\{x_n\}}$ is generating and has a bounded base.*

Proof. The cone $\mathscr{K}_{\{e_n\}}$ associated to the unit vector basis $\{e_n\}$ of l^1 is generating and by theorem 10.2 (implication $1°\Rightarrow 5°$) it has a bounded base. Therefore, the cone $\mathscr{K}_{\{x_n\}}$ associated to any basis $\{x_n\}$ equivalent to $\{e_n\}$ has the same properties.

Conversely, assume that $\{x_n\}$ is a bounded basis such that $\mathscr{K}_{\{x_n\}}$ is generating and has a bounded base $\mathscr{B}$. Let $x \in E$ be arbitrary. Then, since $\mathscr{K}_{\{x_n\}}$ is generating, $x = y - z$, with $y, z \in \mathscr{K}_{\{x_n\}}$, whence, by theorem 10.2 (implication $5°\Rightarrow 4°$),

$$\sum_{i=1}^{\infty} |f_i(x)| \leqslant \sum_{i=1}^{\infty} |f_i(y)| + \sum_{i=1}^{\infty} |f_i(z)|$$

$$\leqslant \frac{1}{\inf\limits_{1 \leqslant n < \infty} \|x_n\|} \left(\sum_{i=1}^{\infty} \|f_i(y) x_i\| + \sum_{i=1}^{\infty} \|f_i(z) x_i\| \right) < \infty,$$

where $\{f_n\} \subset E^*$ is the a.s.c.f. to $\{x_n\}$. Conversely, if $\sum_{i=1}^{\infty} |\alpha_i| < \infty$, then, since $\sup\limits_{1 \leqslant n < \infty} \|x_n\| < \infty$ and since E is complete, $\sum_{i=1}^{\infty} \alpha_i x_i$ converges, which completes the proof.

Let us recall that a cone $\mathscr{K}$ induces a natural partial order relation on E, namely $x \geqslant y$ if and only if $x - y \in \mathscr{K}$ (in particular, $x \geqslant 0$ if and only if $x \in \mathscr{K}$). The cone $\mathscr{K}$ is said to be *normal*, if[1] the norm on E is "semi-monotone", i.e. if there exists a constant $L > 0$ such that

$$0 \leqslant x \leqslant y \Rightarrow \|x\| \leqslant L \|y\|. \tag{10.26}$$

Proposition 10.2. *If $\{x_n\}$ is a basis of type l_+ of a real Banach space E, then the associated cone $\mathscr{K}_{\{x_n\}}$ is normal.*

Proof. If $0 \leqslant \alpha_i \leqslant \beta_i$ $(i = 1, 2, \ldots)$, then, by theorem 10.1 (implication $1° \Rightarrow 3°$),

$$\left\| \sum_{i=1}^{\infty} \alpha_i x_i \right\| \leqslant \sup_{1 \leqslant n < \infty} \|x_n\| \sum_{i=1}^{\infty} \alpha_i \leqslant \sup_{1 \leqslant n < \infty} \|x_n\| \sum_{i=1}^{\infty} \beta_i$$

$$\leqslant \left(\frac{1}{\eta} \sup_{1 \leqslant n < \infty} \|x_n\| \right) \left\| \sum_{i=1}^{\infty} \beta_i x_i \right\|$$

---

[1] Initially, normal cones have been defined (see [139], [137]) by the following equivalent condition: there exists a constant $\delta > 0$ such that

$$\|x + y\| \geqslant \delta \qquad (x, y \in \mathscr{K}, \ \|x\| = \|y\| = 1).$$

(where $\eta>0$ is the constant occurring in 10.3)), which completes the proof.

Let us observe that proposition 10.2 can be also obtained as a corollary of theorem 10.2, since one can prove[1] that if a cone $\mathscr{K}$ in a Banach space E has a bounded base, then $\mathscr{K}$ is normal.

The converse of proposition 10.2 is not valid, since e.g. the cone $\mathscr{K}_{\{x_n\}}$ associated to the unit vector basis $\{x_n\}$ of c_0 is normal (and also generating), but $\{x_n\}$ is not of type l_+.

It is natural to raise the problem of characterizing other types of bases $\{x_n\}$ of a Banach space E by geometric properties of the associated cone $\mathscr{K}_{\{x_n\}}$. We shall give below some results of this kind, involving a special class of bases $\mathscr{B}$ of $\mathscr{K}_{\{x_n\}}$. Namely, a subset $\mathscr{B}$ of a cone $\mathscr{K}$ in a Banach space E will be called a *hyperbase* of $\mathscr{K}$, if there exists a strictly positive functional $f\in E^*$ (i.e. $f(x)>0$ for all $x\in\mathscr{K}\setminus\{0\}$) such that

$$\mathscr{B}=\{y\in\mathscr{K} \mid f(y)=1\}. \tag{10.27}$$

Lemma 10.1. *For a subset $\mathscr{B}$ of a cone $\mathscr{K}$ in a Banach space E the following statements are equivalent:*

1°. $\mathscr{B}$ is a hyperbase of $\mathscr{K}$.

2°. Every $x\in\mathscr{K}\setminus\{0\}$ has a representation $x=\lambda y$ with $\lambda>0$, $y\in\mathscr{B}$, and there exists a functional $f\in E^*$ such that

$$\mathscr{B}\subset\{y\in\mathscr{K} \mid f(y)=1\}. \tag{10.28}$$

3°. Every $x\in\mathscr{K}\setminus\{0\}$ has a representation $x=\lambda y$ with $\lambda>0$, $y\in\mathscr{B}$, and 0 does not belong to the closed linear manifold[2] spanned by $\mathscr{B}$.

Proof. The implication 1°⇒2° is obvious.

Conversely, if we have 2° and $x\in\mathscr{K}\setminus\{0\}$, then, by 2°, x has a representation $x=\lambda y$, with $\lambda>0$, $y\in\mathscr{B}$, whence, by (10.28), $f(x)=\lambda f(y)=\lambda>0$, and therefore f is strictly positive. If $y\in\mathscr{K}$ and $f(y)=1$, then there is a representation $y=\lambda z$, with $\lambda>0$, $z\in\mathscr{B}$, whence, by (10.28), $f(z)=1$, and thus $1=f(y)=\lambda f(z)=\lambda$ and $y=z\in\mathscr{B}$, which proves (10.27). Thus, 2°⇒1°.

Furthermore, if we have 2°, then $\mathscr{B}$, whence also the closed linear manifold spanned by $\mathscr{B}$, is contained in the hyperplane $\{y\in E \mid f(y)=1\}$, which does not contain 0. Thus, 2°⇒3°.

Finally, assume that we have 3° and let V be the closed linear manifold spanned by $\mathscr{B}$. Then for any fixed $y_0\in\mathscr{B}$ the set $V-y_0$ is a closed linear subspace of E, not containing $-y_0$, whence, by a well known

---

[1] See [163], proposition 2.

[2] We recall that a set $V\subset E$ is called a *linear manifold* if it is of the form $V=x_0+G=\{x_0+g \mid g\in G\}$, where G is a linear subspace of E.

corollary of the Hahn-Banach theorem, there exists a functional $f \in E^*$ such that $f(y-y_0)=0$ $(y \in V)$, $f(y_0)=1$, whence we infer (10.28). Thus, $3^\circ \Rightarrow 2^\circ$, which completes the proof of lemma 10.1.

From the proof of proposition 10.1 a) we see that *every hyperbase is a base*. However, the converse is not true, i.e. *there exist (even compact) bases which are not hyperbases*, as shown by the following example: Consider in the space $E=l^2$ the circled closed convex hull Q of the fundamental parallelotope of Hilbert, i.e. the compact set

$$Q=\left\{x=\{\xi_n\}\in l^2 \;\middle|\; |\xi_j| \leqslant \frac{1}{j} \quad (j=1,2,\ldots)\right\}. \tag{10.29}$$

Then the linear subspace $G=\bigcup_{n=1}^{\infty} nQ$ spanned by Q is dense in $E=l^2$ (since it contains all almost zero sequences), but does not coincide with E (since otherwise by the theorem of Baire some $n_0 Q$ would have an interior point, in contradiction with $\dim E=\infty$). Take an arbitrary $x \in E \setminus G = E \setminus \bigcup_{n=1}^{\infty} nQ$ and put

$$\mathscr{K}=\{\lambda(y-x) \mid y \in Q, \lambda \geqslant 0\}. \tag{10.30}$$

Then $\mathscr{K}$ is a cone and $\mathscr{B}=Q-x=\{y-x \mid y \in Q\}$ is a compact base of $\mathscr{K}$, but not a hyperbase of $\mathscr{K}$. Indeed, for any $y,z \in Q$, $\lambda,\mu \geqslant 0$ we have $\lambda(y-x)+\mu(z-x)=(\lambda+\mu)\left(\frac{\lambda}{\lambda+\mu}y+\frac{\mu}{\lambda+\mu}z-x\right)\in\mathscr{K}$ (since $\frac{\lambda}{\lambda+\mu}y + \frac{\mu}{\lambda+\mu}z \in Q$ by the convexity of Q), whence $\mathscr{K}+\mathscr{K}\subset\mathscr{K}$. For any $y \in Q$ and $\lambda,\mu \geqslant 0$ we have obviously $\mu\lambda(y-x)\in\mathscr{K}$, which shows that $\mu\mathscr{K}\subset\mathscr{K}$ for all $\mu \geqslant 0$. Furthermore, if for some $y,z \in Q$, $\lambda,\mu>0$ we have $\lambda(y-x)=-\mu(z-x)$, then $x=\frac{\lambda}{\lambda+\mu}y+\frac{\mu}{\lambda+\mu}z \in Q \subset G$ (by the convexity of Q), in contradiction with our assumption $x \in E\setminus G$, and thus $\mathscr{K}\cap(-\mathscr{K})=\{0\}$. To show that $\mathscr{K}$ is closed, let $y_n \in Q$, $\lambda_n \geqslant 0$ $(n=1,2,\ldots)$ be such that $\lambda_n(y_n-x)\to z \in E$ as $n\to\infty$. Then, since $x \in E\setminus G \subset E\setminus Q = E\setminus\bar{Q}$, we have $\inf_{1\leqslant n<\infty}\|y_n-x\|=\delta>0$, whence

$$\lambda_n \leqslant \frac{1}{\delta}\lambda_n\|y_n-x\| \leqslant \frac{1}{\delta}(\|\lambda_n(y_n-x)-z\|+\|z\|) \qquad (n=1,2,\ldots),$$

whence $\sup_{1\leqslant n<\infty}\lambda_n<\infty$. Since Q is compact, it follows that there exists a sequence of indices $\{n_k\}$ such that $\lambda_{n_k}\to\lambda\geqslant 0$, $y_{n_k}\to y \in Q$ as $k\to\infty$, whence $z=\lambda(y-x)\in\mathscr{K}$, which proves that $\mathscr{K}$ is closed. Thus, $\mathscr{K}$ is a cone.

Furthermore, by the definition of $\mathscr{K}$, every element of $\mathscr{K}\setminus\{0\}$ admits a representation $z=\lambda(y-x)$, with $\lambda>0$, $y\in Q$. To prove the uniqueness of these representations, assume that $\lambda(y-x)=\mu(z-x)$, where $\lambda,\mu>0$, $\lambda\neq\mu$, $y,z\in Q$. Then $x=\frac{\lambda}{\lambda-\mu}y-\frac{\mu}{\lambda-\mu}z\in G$, which contradicts the assumption $x\in E\setminus G$. Thus $\mathscr{B}=Q-x$ is a compact base of $\mathscr{K}$.

Finally, since G is dense in E, it follows that 0 belongs to the closed linear manifold $\overline{G-x}=E$ spanned by $\mathscr{B}=Q-x$, whence, by lemma 10.1 (implication $1°\Rightarrow 3°$), $\mathscr{B}$ is not a hyperbase of $\mathscr{K}$.

However, let us mention that the cone $\mathscr{K}$ in this example, and more generally, any cone $\mathscr{K}$ in a separable Banach space E, has a hyperbase, since it is known[1] that there exist strictly positive functionals $f\in E^*$ whenever E is separable.

Returning now to the cones $\mathscr{K}_{\{x_n\}}$ associated to bases $\{x_n\}$ of Banach spaces, we observe that by the above proof of proposition 10.1 a) $\mathscr{K}_{\{x_n\}}$ has an unbounded hyperbase. Some characterizations of other types of bases $\{x_n\}$ by geometric properties of the hyperbases $\mathscr{B}$ of the associated cone $\mathscr{K}_{\{x_n\}}$, or of the hyperbases $\mathscr{B}^{\{\varepsilon_n\}}$ of the cones $\mathscr{K}_{\{\varepsilon_n x_n\}}$ associated to the bases $\{\varepsilon_n x_n\}$ of E, where $\varepsilon_n=\pm 1$ $(n=1,2,\ldots)$, are given in

Proposition 10.3. *A bounded basis $\{x_n\}$ of a (real) Banach space E is*

a) *of type P^*, if and only if there exists a hyperbase $\mathscr{B}$ of the cone $\mathscr{K}_{\{x_n\}}$ associated to the basis $\{x_n\}$, containing all x_j $(j=1,2,\ldots)$.*

b) *of type wc_0, if and only if for every $\{\varepsilon_n\}$, $\varepsilon_j=\pm 1$ $(j=1,2,\ldots)$ and every hyperbase $\mathscr{B}^{\{\varepsilon_n\}}$ of the cone $\mathscr{K}_{\{\varepsilon_n x_n\}}$ associated to the basis $\{\varepsilon_n x_n\}$, the (unique) numbers $\beta_j>0$ for which $\mathscr{B}^{\{\varepsilon_n\}}\ni\beta_j\varepsilon_j x_j$ $(j=1,2,\ldots)$ satisfy $\lim_{n\to\infty}\beta_n=\infty$.*

c) *shrinking, only if for every $\{\varepsilon_n\}$, $\varepsilon_j=\pm 1$ $(j=1,2,\ldots)$ and every hyperbase $\mathscr{B}^{\{\varepsilon_n\}}$ of the cone $\mathscr{K}_{\{\varepsilon_n x_n\}}$ we have*

$$\operatorname{dist}(0,\mathscr{B}_n^{\{\varepsilon_j\}})\to\infty \quad \textit{as} \quad n\to\infty, \tag{10.31}$$

where

$$\mathscr{B}_n^{\{\varepsilon_j\}}=\mathscr{B}^{\{\varepsilon_j\}}\cap[x_n,x_{n+1},x_{n+2},\ldots] \quad (n=1,2,\ldots). \tag{10.32}$$

Proof. a) If $\{x_n\}$ is of type P^*, then, by § 9, theorem 9.2 b) (implication $1°\Rightarrow 4°$) there exists an $f\in E^*$ such that $f(x_j)=1$ $(j=1,2,\ldots)$. Then for any $x=\sum_{i=1}^{\infty}\alpha_i x_i\in\mathscr{K}_{\{x_n\}}\setminus\{0\}$ we have $f(x)=\sum_{i=1}^{\infty}\alpha_i>0$, whence $\mathscr{B}=\{y\in\mathscr{K}_{\{x_n\}}\mid f(y)=1\}$ is a hyperbase of $\mathscr{K}_{\{x_n\}}$ containing, obviously, all x_j $(j=1,2,\ldots)$. Conversely, if $\mathscr{B}$ is a hyperbase of $\mathscr{K}_{\{x_n\}}$ such that $x_j\in\mathscr{B}$ $(j=1,2,\ldots)$, then there exists a strictly positive $f\in E^*$ such that

[1] See e.g. [144], § 2, theorem 2.1.

$\mathscr{B}=\{y\in\mathscr{K}_{\{x_n\}} \mid f(y)=1\}$. Then $f(x_j)=1$ $(j=1,2,\ldots)$, and therefore, by § 9, theorem 9.2 b) (implication $4°\Rightarrow 1°$), $\{x_n\}$ is of type P^*.

b) If $\{x_n\}$ is of type wc_0, then for any $\{\varepsilon_n\}$, $\varepsilon_j=\pm 1$ $(j=1,2,\ldots)$, we have $\lim_{n\to\infty} f(\varepsilon_n x_n)=0$ $(f\in E^*)$. Let $\mathscr{B}^{\{\varepsilon_n\}}$ be an arbitrary hyperbase of the cone $\mathscr{K}_{\{\varepsilon_n x_n\}}$. Then there exists a strictly positive[1] $f\in E^*$ such that $\mathscr{B}^{\{\varepsilon_n\}}=\{y\in\mathscr{K}_{\{\varepsilon_n x_n\}} \mid f(y)=1\}$, whence $\dfrac{1}{f(\varepsilon_j x_j)}\varepsilon_j x_j\in\mathscr{B}^{\{\varepsilon_n\}}$ and thus $\beta_j=\dfrac{1}{f(\varepsilon_j x_j)}$ $(j=1,2,\ldots)$, whence $\lim_{n\to\infty}\beta_n=\infty$. Conversely, if a bounded basis $\{x_n\}$ is not of type wc_0, then there exists an $f\in E^*$ such that $\lim_{n\to\infty} f(x_n)\neq 0$. Let $\{n_k\}$ be the set of all indices for which $f(x_{n_k})\neq 0$ and let $\{m_k\}$ be the complementary set of indices. Put

$$\varepsilon_{n_k}=\operatorname{sign} f(x_{n_k}) \qquad (k=1,2,\ldots), \tag{10.33}$$

$$\varepsilon_{m_k}=1 \qquad (k=1,2,\ldots), \tag{10.34}$$

$$h(x)=f(x)+\sum_{i=1}^{\infty}\frac{1}{2^i\|f_i\|}\varepsilon_i f_i(x) \qquad (x\in E). \tag{10.35}$$

Then $\varepsilon_j=\pm 1$ $(j=1,2,\ldots)$, $h\in E^*$ and

$$h(\varepsilon_n x_n)=f(\varepsilon_n x_n)+\frac{1}{2^n\|f_n\|}>0 \qquad (n=1,2,\ldots), \tag{10.36}$$

whence h is strictly positive on $\mathscr{K}_{\{\varepsilon_n x_n\}}\setminus\{0\}$, whence $\mathscr{B}^{\{\varepsilon_n\}}=\{y\in\mathscr{K}_{\{\varepsilon_n x_n\}} \mid h(y)=1\}$ is a hyperbase of $\mathscr{K}_{\{\varepsilon_n x_n\}}$. Furthermore, $\dfrac{1}{h(\varepsilon_j x_j)}\varepsilon_j x_j\in\mathscr{B}^{\{\varepsilon_n\}}$ $(j=1,2,\ldots)$, but $\lim_{n\to\infty}\dfrac{1}{h(\varepsilon_n x_n)}\neq\infty$ (since by $\inf_{1\leqslant n<\infty}\|f_n\|>0$ and (10.36) we have $\lim_{n\to\infty} h(\varepsilon_n x_n)=\lim_{n\to\infty} f(\varepsilon_n x_n)=\lim_{n\to\infty}|f(x_n)|\neq 0$).

c) If $\{x_n\}$ is shrinking, let $\varepsilon_j=\pm 1$ $(j=1,2,\ldots)$ and let $\mathscr{B}^{\{\varepsilon_n\}}$ be an arbitrary hyperbase of the cone $\mathscr{K}_{\{\varepsilon_n x_n\}}$. Then there exists a strictly positive[2] $f\in E^*$ such that $\mathscr{B}^{\{\varepsilon_n\}}=\{y\in\mathscr{K}_{\{\varepsilon_n x_n\}} \mid f(y)=1\}$. Since $\{x_n\}$ is shrinking, for any $y\in\mathscr{B}_n^{\{\varepsilon_j\}}$, $y=\sum_{i=n}^{\infty}\alpha_i x_i$, $\alpha_i\geqslant 0$ $(i=n,n+1,\ldots)$ we have then $\dfrac{1}{\|y\|}=f\left(\dfrac{y}{\|y\|}\right)<\varepsilon$ for $n>N(\varepsilon)$. Therefore $\|y\|>\dfrac{1}{\varepsilon}$ for all $y\in\mathscr{B}_n^{\{\varepsilon_j\}}$ whenever $n>N(\varepsilon)$, which completes the proof of proposition 10.3.

---

[1] With respect to $\mathscr{K}_{\{\varepsilon_n x_n\}}$.

[2] With respect to $\mathscr{K}_{\{\varepsilon_n x_n\}}$.

Let us turn now our attention to the other types of bases introduced in definition 10.1. Some characterizations of bases of type al_+ are collected in

Theorem 10.3. *Let $\{x_n\}$ be a basis of a Banach space E, with $\sup_{1\leq n<\infty} \|x_n\| < \infty$. The following statements are equivalent:*

1°. $\{x_n\}$ is of type al_+.

2°. There exists an $f\in E^*$ such that

$$\inf_{1\leq n<\infty} |f(x_n)| > 0. \tag{10.37}$$

3°. There exists a basis $\{y_n\}$ of type P^* (of a Banach space F) such that $\{x_n\}$ is affinely equivalent[1] to $\{y_n\}$.

4°. There exists a basis $\{y_n\}$ of type P^* (of a Banach space F) such that $\{x_n\}$ affinely dominates[2] $\{y_n\}$.

5°. There exists a basis $\{y_n\}$ of type al_+ (of a Banach space F) such that $\{x_n\}$ affinely dominates $\{y_n\}$.

If E is a real Banach space, these statements are equivalent to the following:

6°. There exist a sequence $\{\varepsilon_n\}$, $\varepsilon_j = \pm 1$ $(j=1,2,\ldots)$ and a hyperbase $\mathscr{B}^{\{\varepsilon_n\}}$ of the cone $\mathscr{K}_{\{\varepsilon_n x_n\}}$ associated to the basis $\{\varepsilon_n x_n\}$ such that the unique numbers $\beta_j > 0$ for which $\mathscr{B}^{\{\varepsilon_n\}} \ni \beta_j \varepsilon_j x_j$ $(j=1,2,\ldots)$ satisfy $\sup_{1\leq n<\infty} \beta_n < \infty$.

Proof. 1°⇒2°. If $\{\varepsilon_n x_n\}$ is of type l_+, where $|\varepsilon_n|=1$ $(n=1,2,\ldots)$, then, by the implication 1°⇒5° of theorem 10.1, there exists an $f\in E^*$ such that

$$|f(x_n)| \geq \operatorname{Re} f(\varepsilon_n x_n) \geq 1 \qquad (n=1,2,\ldots).$$

2°⇒1°. If $f\in E^*$ satisfies (10.37), put

$$\varepsilon_n = \frac{\overline{f(x_n)}}{|f(x_n)|} \qquad (n=1,2,\ldots).$$

Then $|\varepsilon_n|=1$, $\|\varepsilon_n x_n\| = \|x_n\|$ $(n=1,2,\ldots)$ and for $h\in E^*$ defined by

$$h = \frac{1}{\inf_{1\leq j<\infty} |f(x_j)|} f$$

we have

$$h(\varepsilon_n x_n) = \frac{|f(x_n)|}{\inf_{1\leq j<\infty} |f(x_j)|} \geq 1 \qquad (n=1,2,\ldots),$$

whence, by the implication 5°⇒1° of theorem 10.1, $\{\varepsilon_n x_n\}$ is of type l_+.

[1] See Ch. I, § 8, definition 8.2.

[2] See Ch. I, § 8, definition 8.2.

$1^\circ \Rightarrow 3^\circ$. If $\{\varepsilon_n x_n\}$ is of type l_+, where $|\varepsilon_n|=1$ $(n=1,2,\ldots)$, then, by the implication $1^\circ \Rightarrow 6^\circ$ of theorem 10.1, there exists a sequence of scalars $\{\beta_n\}$ satisfying (10.6) and such that $\{\beta_n \varepsilon_n x_n\}$ is a basis of type P^* of E. Since $\{x_n\}$ is affinely equivalent to $\{\beta_n \varepsilon_n x_n\}$, it follows that we have 3° with $F=E$, $\{y_n\}=\{\beta_n \varepsilon_n x_n\}$.

The implication $3^\circ \Rightarrow 4^\circ$ is obvious. The implication $4^\circ \Rightarrow 5^\circ$ follows from the fact that every basis of type P^* is also of type l_+ (take $\beta_n=1$ in 7° of theorem 10.1 and apply the implication $7^\circ \Rightarrow 1^\circ$ of this theorem).

$5^\circ \Rightarrow 1^\circ$. Assume that $\{x_n\}$ affinely dominates a basis $\{y_n\}$ of type al_+ of a Banach space F. Then there exists a sequence of scalars $\{\lambda_n\}$ with $0<\inf_{1\leqslant n<\infty}|\lambda_n|\leqslant \sup_{1\leqslant n<\infty}|\lambda_n|<\infty$ such that $\{x_n\}$ dominates[1] the basis $\{\lambda_n y_n\}$ of F, whence, by the implication b) $6^\circ \Rightarrow$ b) 1° of Ch. I, § 8, theorem 8.1, there exists a continuous linear mapping $u\in L(E,F)$ such that

$$u(x_n)=\lambda_n y_n \qquad (n=1,2,\ldots).$$

Put $\varepsilon_n = \dfrac{\bar{\lambda}_n}{|\lambda_n|}$ $(n=1,2,\ldots)$. Then $|\varepsilon_n|=1$, $\|\varepsilon_n x_n\| = \|x_n\|$ $(n=1,2,\ldots)$ and for all finite sequences $\alpha_1,\ldots,\alpha_n\geqslant 0$ we have, taking into account that $\{\delta_n y_n\}$ is of type l_+ for suitable $\{\delta_n\}$ with $|\delta_n|=1$ $(n=1,2,\ldots)$,

$$\left\|\sum_{i=1}^n \alpha_i \varepsilon_i \delta_i x_i\right\| \geqslant \frac{1}{\|u\|}\left\|\sum_{i=1}^n \alpha_i \varepsilon_i \delta_i \lambda_i y_i\right\| = \frac{1}{\|u\|}\left\|\sum_{i=1}^n \alpha_i |\lambda_i| \delta_i y_i\right\|$$

$$\geqslant \frac{\eta}{\|u\|}\sum_{i=1}^n \alpha_i|\lambda_i| \geqslant \frac{\eta \inf_{1\leqslant j<\infty}|\lambda_j|}{\|u\|}\sum_{i=1}^n \alpha_i$$

(with a suitable $\eta>0$ corresponding to $\{\delta_n y_n\}$). Thus $\{\varepsilon_n \delta_n x_n\}$ is of type l_+, i.e. $\{x_n\}$ is of type al_+.

$1^\circ \Rightarrow 6^\circ$. If $\{x_n\}$ is a basis of type al_+ of a real Banach space E, then, by the implication $1^\circ \Rightarrow 2^\circ$ proved above, there exists an $f\in E^*$ such that $|f(x_n)|\geqslant 1$ $(n=1,2,\ldots)$. Put $\varepsilon_n=\operatorname{sign} f(x_n)$ $(n=1,2,\ldots)$. Then $f(\varepsilon_n x_n)\geqslant 1$ $(n=1,2,\ldots)$, whence f is strictly positive on $\mathscr{K}_{\{\varepsilon_n x_n\}}$, whence $\mathscr{B}^{\{\varepsilon_n\}}=\{y\in\mathscr{K}_{\{\varepsilon_n x_n\}} \mid f(y)=1\}$ is a hyperbase of $\mathscr{K}_{\{\varepsilon_n x_n\}}$. Furthermore, $\dfrac{1}{f(\varepsilon_j x_j)}\varepsilon_j x_j\in\mathscr{B}^{\{\varepsilon_n\}}$ $(j=1,2,\ldots)$, and $\beta_n = \dfrac{1}{f(\varepsilon_n x_n)}\leqslant 1$ $(n=1,2,\ldots)$.

$6^\circ \Rightarrow 1^\circ$. If $\{x_n\}$ is a basis of type non-al_+ of a real Banach space E, then, by the implication $2^\circ \Rightarrow 1^\circ$ proved above, we have $\inf_{1\leqslant n<\infty}|f(x_n)|=0$

---

[1] See Ch. I, § 8, definition 8.1; actually, we have replaced here the sequence $\{\lambda_n\}$ of Ch. I, § 8, definition 8.2 by $\left\{\dfrac{1}{\lambda_n}\right\}$.

$(f \in E^*)$. Let $\varepsilon_j = \pm 1$ $(j=1,2,\ldots)$ and let $\mathscr{B}^{\{\varepsilon_n\}}$ be an arbitrary hyperbase of the cone $\mathscr{K}_{\{\varepsilon_n x_n\}}$. Then there exists a strictly positive[1] $f \in E^*$ such that $\mathscr{B}^{\{\varepsilon_n\}} = \{y \in \mathscr{K}_{\{\varepsilon_n x_n\}} \mid f(y)=1\}$, whence $\frac{1}{f(\varepsilon_j x_j)} \varepsilon_j x_j \in \mathscr{B}^{\{\varepsilon_n\}}$, and thus $\beta_j = \frac{1}{f(\varepsilon_j x_j)}$ $(j=1,2,\ldots)$, whence $\sup_{1 \leqslant n < \infty} \beta_n = \infty$, which completes the proof of theorem 10.3.

Remark 10.4. The implication $5^\circ \Rightarrow 1^\circ$ of theorem 10.3 shows that the property of being a basis $\{y_n\}$ of type al_+ is conserved by all bases $\{x_n\}$ with $\sup_{1 \leqslant n < \infty} \|x_n\| < \infty$ which affinely dominate $\{y_n\}$. Other such properties are e.g. that of being a bounded basis of type non-wc_0 or a weakly closed bounded basis. Moreover, affine equivalence also conserves the opposite properties, i.e. those of being a bounded basis of type non-al_+ or wc_0 or non-weakly closed. On the other hand, the implication $3^\circ \Rightarrow 1^\circ$ of theorem 10.3 shows that the property of being a basis $\{y_n\}$ of type P^* is not conserved by all bases $\{x_n\}$ which are affinely equivalent to $\{y_n\}$ (since there are bases of type al_+ which are not of type P^*, as we shall see in § 12, example 12.4). Other properties of bases which are not conserved by affine equivalence are e.g. that of being of one of the types aP^*, l_+, P, aP or $(l_+)^*$.

By the duality methods used already in the preceding sections, it follows that $\{x_n\}$ *is of type* l_+ *if and only if the a.s.c.f.* $\{f_n\} \subset E^*$ *is of type* $(l_+)^*$. Furthermore, from definition 10.1 b) and theorem 10.1 one can obtain various characterizations of bases of type $(l_+)^*$. Combining two of them (namely, those obtained from the equivalences $1^\circ \Leftrightarrow 6^\circ \Leftrightarrow 7^\circ$ of theorem 10.1) with § 9, theorem 9.1, one can obtain other characterizations of bases of type $(l_+)^*$, e.g. the following: *A basis* $\{x_n\}$ *of a Banach space E, with* $\inf_{1 \leqslant n < \infty} \|x_n\| > 0$, *is of type* $(l_+)^*$ *if and only if there exists a sequence of scalars* $\{\beta_n\}$ *satisfying* (10.6) *and such that the sequence*

$$y_n = \sum_{i=1}^{n} \frac{1}{\beta_i} x_i \qquad (n=1,2,\ldots) \tag{10.38}$$

is a basis of E.

Obviously, all the preceding characterizations of bases of types $l_+, (l_+)^*$ imply characterizations of bases of types al_+ and $(al_+)^*$ respectively.

Let us consider now the corresponding properties for the Schauder basis of $C([0,1])$ and the normalized Haar basis of $L^1([0,1])$.

[1] With respect to $\mathscr{K}_{\{\varepsilon_n x_n\}}$.

Theorem 10.4. a) *The Schauder basis* $\{x_n\}_{n=0}^{\infty}$ *of* $C([0,1])$ *is of types non-*al_+, *non-*$(l_+)^*$, *but* $(al_+)^*$. *Namely, the sequence* $\{\varepsilon_n x_n\}$, *where the* ε_n *are defined by* (9.15), *is a basis of type* $(l_+)^*$ *of* $C([0,1])$.

b) *The normalized Haar basis* $\{\tilde{z}_n\}$ *of* $L^1([0,1])$ *is of types non-*al_+ *and non-*$(al_+)^*$.

Proof. a) The proof of the assertion that $\{x_n\}_{n=0}^{\infty}$ is of type non-al_+ is similar to the proof of the fact that it is of type non-aP^* (§ 9, theorem 9.3)), using the implication $1° \Rightarrow 2°$ of theorem 10.3.

Furthermore, the assertion that $\{x_n\}_{n=0}^{\infty}$ is of type non-$(l_+)^*$ is equivalent, by the above remarks on characterizations of bases of type $(l_+)^*$, to the assertion that for every sequence of scalars $\{\beta_n\}_{n=0}^{\infty}$ satisfying $0 < \inf_{0 \leqslant n < \infty} \beta_n \leqslant \sup_{0 \leqslant n < \infty} \beta_n \leqslant 1$ the basis $\left\{\frac{1}{\beta_n} x_n\right\}_{n=0}^{\infty}$ is of type non-P. Now, the proof of this latter assertion is similar to the proof of the fact that $\{x_n\}_{n=0}^{\infty}$ is of type non-P (§ 9, theorem 9.3 a)).

Finally, the assertion that $\{\varepsilon_n x_n\}_{n=0}^{\infty}$ is of type $(l_+)^*$ is a consequence of the fact that it is of type P (§ 9, theorem 9.3 a)).

b) The proof of the assertion that $\{\tilde{z}_n\}$ is of type non-al_+ is similar to the proof of the fact that it is of type non-aP^* (§ 9, theorem 9.3 b)), using the implication $1° \Rightarrow 2°$ of theorem 10.3.

Finally, the proof of the assertion that $\{\tilde{z}_n\}$ is of type non-$(al_+)^*$ is similar to the proof of the fact that it is of type non-aP (§ 9, theorem 9.3 b)), replacing (9.19) by

$$|\operatorname{Re} \Phi(h_n)| \geqslant 1 \quad (n = 1, 2, \ldots).$$

This completes the proof of theorem 10.4.

From the implication $7° \Rightarrow 1°$ of theorem 10.1 with $\beta_n = 1$ $(n = 1, 2, \ldots)$ and § 9, corollary 9.1 a) it follows that *the sequence* (9.16), *where the* ε_n *are defined by* (9.15), *is a basis of type* l_+ *of* $C([0,1])$.

Problem 10.1. Does the space $L^1([0,1])$ possess a basis of type l_+? Or, equivalently: does it possess a basis of type $(l_+)^*$?

By the equivalence $1° \Leftrightarrow 7°$ of theorem 10.1, this problem is equivalent to § 9, problem 9.1.[1]

The answer to the problem of existence of bases of types al_+, $(al_+)^*$ (and hence also for bases of types l_+, $(l_+)^*$) is negative, as shown by

Example 10.1. Let E be an infinite dimensional reflexive Banach space with a basis. Then E has no basis of type al_+ or $(al_+)^*$.

In fact, every basis $\{x_n\}$ of E is of type wc_0 (see § 5), whence, by the implication $1° \Rightarrow 5°$ of theorem 10.1, $\{x_n\}$ is of type non-al_+. Since E^*

[1] Hence, since problem 9.1 has been solved in the affirmative (see § 9, the footnote to problem 9.1), the answer to problem 10.1 is also affirmative.

is reflexive, too, applying this result to E^* it follows that the a.s.c.f. $\{f_n\}$ to $\{x_n\}$ is a basis of type non-al_+ of $[f_n]=E^*$, and thus $\{x_n\}$ is also of type non-$(al_+)^*$.

The answer to the problem of existence of bases of types non-l_+, non-$(l_+)^*$ (and hence also for bases of type non-al_+ and non-$(al_+)^*$) in finite-dimensional Banach spaces is obviously negative. On the other hand, the answer to the problem of existence of bases of types non-l_+, non-$(l_+)^*$ in infinite dimensional Banach spaces with bases is affirmative. Indeed, if $\{x_n\}$ is a basis of type l_+ of an infinite dimensional Banach space E, then, as we have observed in the above, there exists a sequence of scalars $\{\beta_n\}$ satisfying (10.6) and such that the sequence (10.12) is a basis of E. However, this latter basis is of type non-l_+, since otherwise there would exist an $\eta>0$ such that

$$\sup_{1\leqslant n<\infty} \eta n \leqslant \sup_{1\leqslant n<\infty} \left\|\beta_1 x_1 + \sum_{i=2}^{n} (\beta_{i-1} x_{i-1} - \beta_i x_i)\right\| = \sup_{1\leqslant n<\infty} \|2\beta_1 x_1 - \beta_n x_n\|,$$

which is impossible because of $\sup_{1\leqslant n<\infty} \|\beta_n x_n\| \leqslant \sup_{1\leqslant n<\infty} \|x_n\| < \infty$. Similarly, if $\{x_n\}$ is a basis of type $(l_+)^*$ of E, then there exists a sequence of scalars $\{\beta_n\}$ satisfying (10.6) and such that (10.38) is a basis of type non-$(l_+)^*$ of E.

Problem 10.2. a) Does every infinite dimensional Banach space with a basis possess a basis of type non-al_+? In particular, does the space l^1 possess such a basis?

b) Does every infinite dimensional Banach space with a basis possess a basis of type non-$(al_+)^*$? In particular, does the space c_0 possess such a basis?

By the equivalence $1^\circ \Leftrightarrow 7^\circ$ of theorem 10.1, this problem is equivalent to § 9, problem 9.2.

In connection with problem 10.2 let us mention that the space l^1 has a *subspace* with a basis of type non-al_+. Indeed, such is e.g. the subspace F of l^1 defined in § 8, example 8.1, since by (8.9) and the implication $1^\circ \Rightarrow 2^\circ$ of theorem 10.3, the basis (8.6) of (8.5) is of type non-al_+. Dually, the space c_0 has a subspace with a basis of type non-$(al_+)^*$.

§ 11. Besselian and Hilbertian bases. Stability theorems

Definition 11.1. A basis $\{x_n\}$ of a Banach space E is said to be

a) *Besselian*, if

$$\sum_{i=1}^{\infty} \alpha_i x_i \text{ is convergent } \Rightarrow \sum_{i=1}^{\infty} |\alpha_i|^2 < \infty, \tag{11.1}$$

i.e. if

$$\sum_{i=1}^{\infty} |f_i(x)|^2 < \infty \qquad (x \in E), \tag{11.2}$$

where $\{f_n\}$ is the a.s.c.f. to $\{x_n\}$;

b) *Hilbertian*, if

$$\sum_{i=1}^{\infty} |\alpha_i|^2 < \infty \Rightarrow \sum_{i=1}^{\infty} \alpha_i x_i \quad \text{is convergent}, \tag{11.3}$$

i.e. if for every $\{\alpha_n\} \subset K$ with $\sum_{i=1}^{\infty} |\alpha_i|^2 < \infty$ there exists an (obviously unique) $x \in E$ such that

$$f_n(x) = \alpha_n \qquad (n = 1, 2, \ldots). \tag{11.4}$$

For instance, the natural basis of l^1 is Besselian, the natural basis of c_0 is Hilbertian, and the natural basis of l^2 is simultaneously Besselian and Hilbertian.

Let us observe that by Ch. I, § 3, lemma 3.1, for every Besselian basis $\{x_n\}$ we have $\inf_{1 \leq n < \infty} \|x_n\| > 0$ and for every Hilbertian basis $\{x_n\}$ we have $\sup_{1 \leq n < \infty} \|x_n\| < \infty$.

The following theorem gives some characterizations of Besselian and Hilbertian bases and shows their relations of duality:

Theorem 11.1. *Let $\{x_n\}$ be a basis of a Banach space E and let $\{f_n\} \subset E^*$ be the a.s.c.f. to $\{x_n\}$. Then*

a) *The following statements are equivalent:*

1°. *$\{x_n\}$ is Besselian.*

2°. *$\{x_n\} \succ \{e_n\}$,[1] where $\{e_n\}$ is the unit vector basis of l^2.*

3°. *There exists a continuous linear mapping u of E into l^2 such that*

$$u(x_n) = e_n \qquad (n = 1, 2, \ldots). \tag{11.5}$$

4°. *There exists a constant $c > 0$ such that we have*

$$c \sqrt{\sum_{i=1}^{n} |\alpha_i|^2} \leq \left\| \sum_{i=1}^{n} \alpha_i x_i \right\| \tag{11.6}$$

for all finite sequences of scalars $\alpha_1, \ldots, \alpha_n$.

5°. *We have*

$$\sum_{i=1}^{\infty} |\Phi(f_i)|^2 < \infty \qquad (\Phi \in E^{**}). \tag{11.7}$$

---

[1] I.e. the convergence of $\sum_{i=1}^{\infty} \alpha_i x_i$ implies the convergence of $\sum_{i=1}^{\infty} \alpha_i e_i$ (see Ch. I, § 8, definition 8.1).

6°. $\{f_n\}$ is a Hilbertian basis of $[f_n]$.

b) *The following statements are equivalent:*

1°. $\{x_n\}$ is Hilbertian.

2°. $\{e_n\} \succ \{x_n\}$, where $\{e_n\}$ is the unit vector basis of l^2.

3°. There exists a continuous linear mapping v of l^2 into E, such that

$$v(e_n) = x_n \qquad (n = 1, 2, \ldots). \tag{11.8}$$

4°. There exists a constant $C > 0$ $(n = 1, 2, \ldots)$ such that we have

$$\left\| \sum_{i=1}^{n} \alpha_i x_i \right\| \leqslant C \sqrt{\sum_{i=1}^{n} |\alpha_i|^2} \tag{11.9}$$

for all finite sequences of scalars $\alpha_1, \ldots, \alpha_n$.

5°. We have

$$\sum_{i=1}^{\infty} |f(x_i)|^2 < \infty \qquad (f \in E^*). \tag{11.10}$$

6°. $\{f_n\}$ is a Besselian basis of $[f_n]$.

Proof. The equivalences a) 1°⇔a) 2° and b) 1°⇔b) 2° are obvious from the definitions. The equivalences a) 2°⇔a) 3°⇔a) 4° and b) 2°⇔b) 3°⇔b) 4° follow from the equivalences 6°⇔1°⇔2° of Ch. I, § 8, theorem 8.1 b).

a) 4°⇒a) 5°. By Ch. I, § 12, proposition 12.2 c), we have

$$\sup_{1 \leqslant n < \infty} \left\| \sum_{i=1}^{n} \Phi(f_i) x_i \right\| < \infty \qquad (\Phi \in E^{**}), \tag{11.11}$$

whence, by (11.6), we infer (11.7).

The implication a) 5°⇒a) 1° is obvious, by considering the canonical embedding π of E into E^{**}. Thus a) 1°⇔⋯⇔a) 5°.

a) 2°⇔a) 6°. By Ch. I, § 12, proposition 12.1, we have a) 2° if and only if[1] $\{e_n\} \succ \{f_n\}$, which, by the equivalence b) 2°⇔b) 1° observed above, happens if and only if we have a) 6°.

b) 4°⇒b) 5°. By (11.9) we have

$$\sum_{i=1}^{n} |f(x_i)|^2 = f\left(\sum_{i=1}^{n} \overline{f(x_i)} x_i \right) \leqslant \|f\| \left\| \sum_{i=1}^{n} \overline{f(x_i)} x_i \right\|$$

$$\leqslant C \|f\| \sqrt{\sum_{i=1}^{n} |f(x_i)|^2} \qquad (f \in E^*, n = 1, 2, \ldots)$$

[1] Here we identify canonically $(l^2)^*$ with l^2 and hence the a.s.c.f. to $\{e_n\}$ is identified with $\{e_n\}$.

and thus

$$\sqrt{\sum_{i=1}^{n} |f(x_i)|^2} \leqslant C\|f\| \qquad (f \in E^*, n=1,2,\ldots), \tag{11.12}$$

whence we infer (11.10).

b) 5°⇒b) 6°. The a.s.c.f. $\{\Psi_n\} \subset [f_n]^*$ to the basis $\{f_n\}$ of $[f_n]$ is nothing else but $\Psi_n = \phi(x_n)$ $(n=1,2,\ldots)$, where ϕ is the canonical mapping of E into $[f_n]^*$. Since by b) 5° we have

$$\sum_{i=1}^{\infty} |\Psi_i(f)|^2 < \infty \qquad (f \in [f_n]), \tag{11.13}$$

from definition 11.1 a) we see that $\{f_n\}$ is a Besselian basis of $[f_n]$.

b) 6°⇒b) 2°. If we have b) 6°, then, by the implication a) 1°⇒a) 2° observed above, $\{f_n\} \succ \{e_n\}$, whence, by Ch. I, § 12, proposition 12.1, we have b) 2°, which completes the proof of theorem 11.1.

Corollary 11.1. a) *The space c_0 has no bounded Besselian basis.*

b) *The space l^1 has no bounded Hilbertian basis.*

Proof. a) Assume that $\{x_n\}$ is a Besselian basis of c_0, with the a.s.c.f. $\{f_n\} \subset (c_0)^* \equiv l^1$. Then, by the implication a) 1°⇒a) 5° of theorem 11.1, $f_n \xrightarrow{w} 0$ in $(c_0)^* \equiv l^1$, whence also in the norm-topology[1] of l^1. Consequently (see Ch. I, § 3, corollary 3.1 b)), the basis $\{x_n\}$ is not bounded.

The proof of part b) is similar, making use of the implication b) 1°⇒b) 5° of theorem 11.1. This completes the proof.

By the same arguments one can also show that *no subspace of c_0 can have a bounded Besselian basis, and no subspace of l^1 can have a bounded Hilbertian basis.*

Remark 11.1. If we omit the requirement of boundedness of the basis $\{x_n\}$, then the situation is different. In fact, *every Banach space E with a basis has a Besselian basis and a Hilbertian basis, namely, if $\{x_n\}$ is any normalized basis of E, then $\{nx_n\}$ is a Besselian basis of E and $\left\{\frac{1}{n}x_n\right\}$ is a Hilbertian basis of E.*

Indeed, if $\sum_{i=1}^{\infty} \alpha_i i x_i$ converges, then $\lim_{n\to\infty} n\alpha_n x_n = 0$, whence

$$n|\alpha_n| < \varepsilon \qquad (n > N(\varepsilon)),$$

whence

$$|\alpha_n|^2 < \frac{\varepsilon^2}{n^2} \qquad (n > N(\varepsilon)),$$

[1] See e.g. [10], p. 137–139.

and thus $\sum_{i=N(\varepsilon)+1}^{\infty} |\alpha_i|^2$, whence also $\sum_{i=1}^{\infty} |\alpha_i|^2$, converges. The second assertion follows by duality.

Corollary 11.2. *For a basis $\{x_n\}$ of a Banach space E, with the a.s.c.f. $\{f_n\}$, the following statements are equivalent:*

1°. *$\{x_n\}$ is simultaneously Besselian and Hilbertian.*

2°. *$\{x_n\} \sim \{e_n\}$, where $\{e_n\}$ is the natural basis of l^2.*

3°. *There exists an isomorphism u of E onto l^2, such that we have* (11.5).

4°. *There exist two constants $c>0$ and $C \geqslant c$ such that we have*

$$c\sqrt{\sum_{i=1}^{n} |\alpha_i|^2} \leqslant \left\| \sum_{i=1}^{n} \alpha_i x_i \right\| \leqslant C\sqrt{\sum_{i=1}^{n} |\alpha_i|^2} \tag{11.14}$$

for all finite sequences of scalars $\alpha_1, \ldots, \alpha_n$.

5°. *$\{x_n\}$ is Besselian and $\{f_n\}$ is a Besselian basis of $[f_n]$.*

6°. *$\{x_n\}$ is Hilbertian and $\{f_n\}$ is a Hilbertian basis of $[f_n]$.*

Let us consider now the corresponding properties for the Schauder basis of $C([0,1])$ and the normalized Haar basis of $L^1([0,1])$.

Corollary 11.3. a) *The Schauder basis $\{x_n\}_{n=0}^{\infty}$ of $C([0,1])$ is both non-Besselian and non-Hilbertian.*

b) *The normalized Haar basis $\{\tilde{z}_n\}$ of $L^1([0,1])$ is both non-Besselian and non-Hilbertian.*

Proof. By § 7, theorem 7.1, both $\{x_n\}_{n=0}^{\infty}$ and $\{\tilde{z}_n\}$ are of type non-$w c_0$, whence, by the implication b) 1° ⇒ b) 5° of theorem 11.1, they are non-Hilbertian.

Furthermore, by § 9, theorem 9.3 a), there exists a sequence of scalars $\{\varepsilon_n\}_{n=0}^{\infty}$ with $\varepsilon_n = \pm 1$ $(n=0,1,2,\ldots)$, such that

$$\sup_{0 \leqslant n < \infty} \left\| \sum_{i=0}^{n} \varepsilon_i x_i \right\| < \infty,$$

whence, by the implication a) 1° ⇒ a) 4° of theorem 11.1, the Schauder basis $\{x_n\}_{n=0}^{\infty}$ of $C([0,1])$ is non-Besselian.

Finally, since by § 7, formula (7.11), we have

$$\left\| \sum_{k=0}^{n} \tilde{z}_{2^k+1} \right\| = 2 - \frac{1}{2^n} < 2 \quad (n=0,1,2,\ldots), \tag{11.15}$$

from the implication a) 1° ⇒ a) 4° of theorem 11.1 it follows that the normalized Haar basis $\{\tilde{z}_n\}$ of $L^1([0,1])$ is non-Besselian, which completes the proof.

Problem 11.1. a) Does the space $C([0,1])$ possess a bounded Besselian basis? What about a bounded Hilbertian basis?

b) Does the space $L^1([0,1])$ possess a bounded Besselian basis? What about a bounded Hilbertian basis?

By theorem 11.1 (implications a) $1^\circ \Rightarrow$ a) 5° and b) $1^\circ \Rightarrow$ b) 5°), an affirmative answer to one of these questions would imply an affirmative answer to the corresponding question of § 7, problem 7.1 (on the existence of bases of type wc_0 or $(wc_0)^*$ in $C([0,1])$ and $L^1([0,1])$).

If there existed in $C([0,1])$ a basis $\{x_n\}$ with $\sup_{1 \leq n < \infty} \|x_n\| < \infty$, which constitutes an orthonormal system, then the answer to the first question of problem 11.1 a) would be affirmative (indeed, in this case, from $\sum_{i=1}^{\infty} \alpha_i x_i \in C([0,1]) \subset L^2([0,1])$ it would follow that $\sum_{i=1}^{\infty} |\alpha_i|^2 < \infty$). However, in Volume II we shall see that $C([0,1])$ has no basis $\{x_n\}$ with these properties. Dually, if there existed in $L^1([0,1])$ a basis $\{x_n\}$ with $\inf_{1 \leq n < \infty} \|x_n\| > 0$, which constitutes an orthonormal system, then this would be a Hilbertian basis of $L^1([0,1])$ (since $\sum_{i=1}^{\infty} |\alpha_i|^2 < \infty$ would imply $\sum_{i=1}^{\infty} \alpha_i x_i \in L^2([0,1]) \subset L^1([0,1])$), and so the answer to the second question of problem 11.1 b) would be affirmative.

If we omit the requirement of boundedness of the basis $\{x_n\}$, then the situation is different. Namely, we shall see in Vol. II that there exists in $C([0,1])$ a basis $\{x_n\}$ which constitutes an orthonormal system and hence, by the above argument, $\{x_n\}$ has property (11.1) in $C([0,1])$. It is easy to verify (see Vol. II) that any such sequence $\{x_n\}$ is also a basis in $L^1([0,1])$ and hence, as observed above, $\{x_n\}$ has property (11.2) in $L^1([0,1])$. One can also show (using the interpolation theorem of M. Riesz) that the same $\{x_n\}$ is also a basis in each $L^p([0,1])$ $(1 \leq p < \infty)$, having property (11.1) if $p \geq 2$ and respectively property (11.2) if $1 \leq p \leq 2$. However, for the spaces $L^p([-\pi,\pi])$ with $1 < p < \infty$ it is known more, namely that they have a bounded Besselian basis if $p \geq 2$ and a bounded Hilbertian basis if $1 < p \leq 2$, as shown by

Example 11.1. In the space $E = L^p([-\pi,\pi])$ $(1 < p < \infty)$ the (orthogonal) sequence $\{\tilde{x}_n\}_{n=0}^{\infty}$, where

$$x_0(t) \equiv \tfrac{1}{2}, \quad x_{2n-1}(t) = \sin nt, \quad x_{2n}(t) = \cos nt \quad (t \in [-\pi,\pi],\ n = 1, 2, \ldots) \tag{11.16}$$

is a bounded Besselian basis if $p \geq 2$ and a bounded Hilbertian basis if $1 < p \leq 2$.

Indeed, let us first show that $\{\tilde{x}_n\}_{n=0}^{\infty}$ is a basis of E. Obviously, the sequence $\{f_n\} \subset E^*$ defined by

$$f_{2n}(\tilde{x}) = \frac{1}{\pi}\int_{-\pi}^{\pi} x(\tau)\cos n\tau\, d\tau, \quad f_{2n+1}(\tilde{x}) = \frac{1}{\pi}\int_{-\pi}^{\pi} x(\tau)\sin(n+1)\tau\, d\tau$$

$$(\tilde{x}\in E,\ n=0,1,2,\ldots), \tag{11.17}$$

satisfies $f_k(\tilde{x}_j)=\delta_{kj}$ $(k,j=1,2,\ldots)$. Consequently, taking into account that

$$\frac{\sin\left(n+\frac{1}{2}\right)\tau}{2\sin\frac{\tau}{2}} = \frac{\sin n\tau}{2\operatorname{tg}\frac{\tau}{2}} + \frac{1}{2}\cos n\tau, \tag{11.18}$$

we have[1], for the partial sum operators associated to the biorthogonal system $(\tilde{x}_n, f_n)$,

$$[s_{2n}(\tilde{x})](t) = \sum_{k=0}^{2n} f_k(\tilde{x})x_k(t) = \frac{f_0(\tilde{x})}{2} + \sum_{k=1}^{n} (f_{2k-1}(\tilde{x})\sin kt + f_{2k}(\tilde{x})\cos kt)$$

$$= \frac{1}{\pi}\int_{-\pi}^{\pi} x(t+\tau)\frac{\sin\left(n+\frac{1}{2}\right)\tau}{2\sin\frac{\tau}{2}}\,d\tau$$

$$= Q_n(t) + \frac{1}{2\pi}\int_{-\pi}^{\pi} x(t+\tau)\cos n\tau\, d\tau, \tag{11.19}$$

where we have put

$$Q_n(t) = \frac{1}{\pi}\int_{-\pi}^{\pi} x(t+\tau)\frac{\sin n\tau}{2\operatorname{tg}\frac{\tau}{2}}\,d\tau. \tag{11.20}$$

Now, taking into account that

$$\sin n\tau = \sin n(t+\tau)\cos nt - \cos n(t+\tau)\sin nt, \tag{11.21}$$

---

[1] We consider all functions extended beyond $[-\pi,\pi)$ by $x(t+2\pi)=x(t)$.

and putting

$$z_1(t)=x(t)\sin nt, \qquad z_2(t)=x(t)\cos nt \qquad (t\in[-\pi,\pi]), \tag{11.22}$$

we have, almost everywhere on $[-\pi,\pi]$,

$$Q_n(t)=\cos nt\,\bar{z}_1(t)-\sin nt\,\bar{z}_2(t),$$

where $\bar{z}_1(t)$, $\bar{z}_2(t)$ are the functions conjugate[1] to $z_1(t)$ and $z_2(t)$ respectively. Hence

$$|Q_n(t)|\leqslant|\bar{z}_1(t)|+|\bar{z}_2(t)|,$$

and thus

$$\left(\int_{-\pi}^{\pi}|\tilde{Q}_n(t)|^p\,dt\right)^{\frac{1}{p}}\leqslant\left(\int_{-\pi}^{\pi}(|\bar{z}_1(t)|+|\bar{z}_2(t)|)^p\,dt\right)^{\frac{1}{p}}$$

$$\leqslant\left(\int_{-\pi}^{\pi}|\bar{z}_1(t)|^p\,dt\right)^{\frac{1}{p}}+\left(\int_{-\pi}^{\pi}|\bar{z}_2(t)|^p\,dt\right)^{\frac{1}{p}}.$$

However, by a well known theorem of M. Riesz on conjugate functions[2], there exists a constant A_p depending only on p, such that

$$\left(\int_{-\pi}^{\pi}|\bar{z}(t)|^p\,dt\right)^{\frac{1}{p}}\leqslant A_p\left(\int_{-\pi}^{\pi}|z(t)|^p\,dt\right)^{\frac{1}{p}}.$$

Applying this to z_1 and z_2 defined by (11.22), we obtain

$$\left(\int_{-\pi}^{\pi}|Q_n(t)|^p\,dt\right)^{\frac{1}{p}}\leqslant A_p\left(\int_{-\pi}^{\pi}|z_1(t)|^p\,dt\right)^{\frac{1}{p}}+A_p\left(\int_{-\pi}^{\pi}|z_2(t)|^p\,dt\right)^{\frac{1}{p}}$$

$$\leqslant 2A_p\left(\int_{-\pi}^{\pi}|x(t)|^p\,dt\right)^{\frac{1}{p}}.$$

Consequently, taking into account (11.19), we get

$$\|s_{2n}(\tilde{x})\|\leqslant A'_p\|\tilde{x}\| \qquad (\tilde{x}\in E), \tag{11.23}$$

with a suitable A'_p depending only on p. Since

$$\|\tilde{x}_n\|\leqslant(2\pi)^{\frac{1}{p}}, \qquad \|f_n\|\leqslant\frac{1}{\pi}(2\pi)^{1-\frac{1}{p}} \qquad (n=1,2,\ldots), \tag{11.24}$$

---

[1] We recall that for $\tilde{z}\in L^2([-\pi,\pi])$ and $z\in\tilde{z}$ the function

$$\bar{z}(t)=-\frac{1}{2\pi}\int_{-\pi}^{\pi}z(t+\tau)\operatorname{ctg}\frac{\tau}{2}\,d\tau$$

is defined for almost all t and it is called the *conjugate function* to $z(t)$.

[2] See e.g. [274], Ch. VII, §2, theorem (2.4).

we have

$$\|f_n(\tilde{x})\tilde{x}_n\| \leqslant \|f_n\| \, \|\tilde{x}_n\| \, \|\tilde{x}\| \leqslant \frac{1}{\pi}(2\pi)^{\frac{1}{p}}(2\pi)^{1-\frac{1}{p}}\|\tilde{x}\| = 2\|\tilde{x}\|$$

$$(\tilde{x} \in E, \ n = 1, 2, \ldots),$$

whence, by (11.23) and $\|s_{2n+1}(\tilde{x})\| \leqslant \|s_{2n}(\tilde{x})\| + \|f_{2n+1}(\tilde{x})\tilde{x}_{2n+1}\|$,

$$\sup_{0 \leqslant n < \infty} \|s_n\| < \infty.$$

Since $\{\tilde{x}_n\}_{n=0}^{\infty}$ is complete in $E = L^p([-\pi, \pi])$ (extended Weierstrass theorem), it follows, by Ch. I, § 4, theorem 4.1, that $\{\tilde{x}_n\}_{n=0}^{\infty}$ is a basis of E.

Furthermore, assume that $p \geqslant 2$ and let $\tilde{x} \in E = L^p([-\pi, \pi])$. Then also $\tilde{x} \in L^2([-\pi, \pi])$, whence, since $\left\{\frac{1}{\sqrt{\pi}} x_n\right\}_{n=1}^{\infty}$ is an orthonormal system, $\sum_{i=0}^{\infty} |f_i(\tilde{x})|^2 < \infty$, and thus $\{\tilde{x}_n\}$ is a Besselian basis of E. By duality (theorem 11.1), it follows that for $1 < p \leqslant 2$ the sequence $\{\tilde{x}_n\}$ is a Hilbertian basis of $E = L^p([-\pi, \pi])$, which, together with (11.24), completes the proof of the assertions stated in example 11.1

We shall give now some stability properties for Besselian bases and then some characterizations of Besselian bases in terms of stability properties.

Theorem 11.2. *Let $\{x_n\}$ be a Besselian basis of a Banach space E, and let $\{y_n\}$ be a sequence in E, such that*

$$\sup_{\substack{f \in E^* \\ \|f\| \leqslant 1}} \left(\sum_{i=1}^{\infty} |f(x_i - y_i)|^2 \right)^{\frac{1}{2}} = M < c, \tag{11.25}$$

where c is any constant for which we have (11.6). *Then $\{y_n\}$ is a basis of E, equivalent to $\{x_n\}$ (whence Besselian).*

In particular, if $\{y_n\} \subset E$ is a sequence such that

$$\left(\sum_{i=1}^{\infty} \|x_i - y_i\|^2 \right)^{\frac{1}{2}} < c, \tag{11.26}$$

where c is any constant for which we have (11.6), *then the same conclusion holds.*

Proof. Let $\alpha_1, \ldots, \alpha_n$ be arbitrary scalars and take a $g_n \in E^*$ with $\|g_n\| = 1$ such that $g_n\left(\sum_{i=1}^{n} \alpha_i(x_i - y_i)\right) = \left\|\sum_{i=1}^{n} \alpha_i(x_i - y_i)\right\|$. Then

$$\left\| \sum_{i=1}^{n} \alpha_i (x_i - y_i) \right\| = \sum_{i=1}^{n} \alpha_i g_n(x_i - y_i) \leqslant \left(\sum_{i=1}^{n} |\alpha_i|^2 \right)^{\frac{1}{2}} \left(\sum_{i=1}^{n} |g_n(x_i - y_i)|^2 \right)^{\frac{1}{2}}$$
$$\leqslant \frac{M}{c} \left\| \sum_{i=1}^{n} \alpha_i x_i \right\|. \tag{11.27}$$

Since $\frac{M}{c} < 1$, this shows that $\{y_n\}$ is PW-near to $\{x_n\}$, whence, by the Paley-Wiener theorem (Ch. I, § 9, theorem 9.1), $\{y_n\}$ is a basis of E, equivalent to $\{x_n\}$, which completes the proof[1].

Definition 11.2. Let $\{x_n\}$, $\{y_n\}$ be sequences in a Banach space E. The sequence $\{y_n\}$ is said to be

a) *quadratically near* to $\{x_n\}$, if we have

$$\sum_{i=1}^{\infty} \|x_i - y_i\|^2 < \infty; \tag{11.28}$$

b) *weakly quadratically near* to $\{x_n\}$, if $\sum_{i=1}^{\infty} |f(x_i - y_i)|^2 < \infty$ uniformly with respect to $f \in E^*$, $\|f\| \leqslant 1$, i.e. if

$$\lim_{n\to\infty} \sup_{\substack{f\in E^* \\ \|f\|\leqslant 1}} \sum_{i=n+1}^{\infty} |f(x_i - y_i)|^2 = 0. \tag{11.29}$$

Obviously, quadratically near $\Rightarrow$ weakly quadratically near, but the converse implication is not valid. Indeed, e.g. if $p>2$ and $\{z_n\}$ is a normalized unconditional basis[2] of $E = L^p([0,1])$, then by § 14, corollary 14.3, $\{z_n\}$ is not Besselian, whence there exists a convergent series $\sum_{i=1}^{\infty} \alpha_i z_i \in E$ such that $\sum_{i=1}^{\infty} \|\alpha_i z_i\|^2 = \sum_{i=1}^{\infty} |\alpha_i|^2 = \infty$. On the other hand, since $\sum_{i=1}^{\infty} \alpha_i z_i$ is unconditionally convergent, by § 16, lemma 16.1 (implication $1^\circ \Rightarrow 6^\circ$), the series $\sum_{i=1}^{\infty} |f(\alpha_i z_i)|$, whence also the series $\sum_{i=1}^{\infty} |f(\alpha_i z_i)|^2$, converge uniformly with respect to $f \in E^*$, $\|f\| \leqslant 1$. Thus, any $\{x_n\}$, $\{y_n\}$ with $x_n - y_n = \alpha_n z_n$ $(n = 1, 2, \ldots)$ are weakly quadratically near but not quadratically near.

---

[1] Let us observe that in the particular case when $\{x_n\}$ is an orthonormal basis of a Hilbert space, we may take $c=1$; in this case, the second statement of theorem 11.2 is also a consequence of Ch. I, § 10, theorem 10.4, since

$$b = \sup_{|\varepsilon_i|=1} \sup_{1\leqslant n<\infty} \left\| \sum_{i=1}^{n} \varepsilon_i \gamma_i f_i \Big|_{[x_j]} \right\| = \left(\sum_{i=1}^{\infty} |\gamma_i|^2 \right)^{\frac{1}{2}}.$$

[2] See § 14, definition 14.1.

Now we can give the following corollary of theorem 11.2:

Corollary 11.4. *Let $\{x_n\}$ be a Besselian basis of a Banach space E and let $\{y_n\}$ be a sequence in E, weakly quadratically near (or, in particular, quadratically near) to $\{x_n\}$. Then there exists a positive integer n_0 such that the sequence $\{x_n\}_1^{n_0-1} \cup \{y_n\}_{n_0}^{\infty}$ is a basis of E, equivalent to $\{x_n\}$. Hence, in this case*

$$\operatorname{codim}_E [y_n]_{n_0}^{\infty} = n_0 - 1, \qquad \operatorname{codim}_E [y_n] = k \leqslant n_0 - 1. \tag{11.30}$$

Proof. If we have (11.29), then there exists a positive integer n_0 such that $\sup_{\substack{f \in E^* \\ \|f\| \leqslant 1}} \left(\sum_{i=n_0}^{\infty} |f(x_i - y_i)|^2 \right)^{\frac{1}{2}} = M < c$, where c is any constant for which we have (11.6). Hence, by theorem 11.2 applied to the sequences $\{x_n\}$, $\{x_n\}_1^{n_0-1} \cup \{y_n\}_{n_0}^{\infty}$, the desired conclusion follows.

Definition 11.3. A sequence $\{y_n\}$ in a Banach space E is said to be *l^2-linearly independent*, if

$$\{\alpha_n\} \in l^2, \qquad \sum_{i=1}^{\infty} \alpha_i y_i = 0 \quad \text{imply} \quad \alpha_i = 0 \qquad (i = 1, 2, \ldots). \tag{11.31}$$

Theorem 11.3. *Let $\{x_n\}$ be a Besselian basis of a Banach space E and let $\{y_n\}$ be a sequence in E, weakly quadratically near (or, in particular, quadratically near) to $\{x_n\}$. Then*

a) *The following statements are equivalent:*

1°. *$\{y_n\}$ is complete in E.*

2°. *$\{y_n\}$ is l^2-linearly independent.*

3°. *$\{y_n\}$ is ω-linearly independent.*

4°. *$\{y_n\}$ is minimal.*

5°. *$\{y_n\}$ is a basis of E.*

6°. *$\{x_n\} \approx \{y_n\}$.*

b) *The sequence $\{y_n\}$ can be modified to become a Besselian basis of E, by changing suitable k elements of it, where $k = \operatorname{codim}_E [y_n] < \infty$.*

Proof. a) $2° \Rightarrow 6°$. Assume that $\{x_n\}$ is Besselian and let $\{y_n\}$ be an l^2-linearly independent sequence in E, weakly quadratically near to $\{x_n\}$. Then by corollary 11.4 there exists a positive integer n_0 such that for every $x \in E$ the series $\sum_{i=n_0}^{\infty} f_i(x) y_i$, whence also the series $\sum_{i=1}^{\infty} f_i(x) y_i$, converges. Consequently, the mapping

$$u(x) = \sum_{i=1}^{\infty} f_i(x) y_i \tag{11.32}$$

is well defined on E and by the Banach-Steinhaus theorem it is continuous. Now, if $u(x)=0$, then, since $\sum_{i=1}^{\infty}|f_i(x)|^2<\infty$ (because $\{x_n\}$ is Besselian), it follows by the l^2-linearly independence of $\{y_n\}$ that $f_i(x)=0$ $(i=1,2,\ldots)$, whence, since $\{f_n\}$ is total on E, $x=0$. Thus, u is *one to one*.

We shall prove now that u maps E *onto* E. Let $x\in E$ and n be arbitrary and take a $g_{x,n}\in E^*$ with $\|g_{x,n}\|=1$ such that $g_{x,n}\left(\sum_{i=n+1}^{\infty} f_i(x)(x_i-y_i)\right) = \left\|\sum_{i=n+1}^{\infty} f_i(x)(x_i-y_i)\right\|$. Then, taking into account (11.6),

$$\begin{aligned}\left\|\sum_{i=n+1}^{\infty} f_i(x)(x_i-y_i)\right\| &= \sum_{i=n+1}^{\infty} f_i(x)g_{x,n}(x_i-y_i)\\ &\leqslant \left(\sum_{i=n+1}^{\infty}|f_i(x)|^2\right)^{\frac{1}{2}}\left(\sum_{i=n+1}^{\infty}|g_{x,n}(x_i-y_i)|^2\right)^{\frac{1}{2}}\\ &\leqslant \frac{1}{c}\left\|\sum_{i=n+1}^{\infty} f_i(x)x_i\right\|\left(\sum_{i=n+1}^{\infty}|g_{x,n}(x_i-y_i)|^2\right)^{\frac{1}{2}}\\ &\leqslant \frac{1}{c}(1+v_{\{x_j\}})\|x\|\left(\sum_{i=n+1}^{\infty}|g_{x,n}(x_i-y_i)|^2\right)^{\frac{1}{2}},\end{aligned}$$

where $v_{\{x_j\}}$ is the norm of the basis $\{x_j\}$. Hence for the continuous linear operators

$$s(x)=x-u(x)=\sum_{i=1}^{\infty} f_i(x)(x_i-y_i) \quad (x\in E), \tag{11.33}$$

$$v_n(x)=\sum_{i=1}^{n} f_i(x)(x_i-y_i) \quad (x\in E,\ n=1,2,\ldots), \tag{11.34}$$

we obtain, taking again into account that $\{y_n\}$ is weakly quadratically near to $\{x_n\}$,

$$\begin{aligned}\|s-v_n\| &= \sup_{\substack{x\in E\\ \|x\|\leqslant 1}}\|s(x)-v_n(x)\| = \sup_{\substack{x\in E\\ \|x\|\leqslant 1}}\left\|\sum_{i=n+1}^{\infty} f_i(x)(x_i-y_i)\right\|\\ &\leqslant \frac{1}{c}(1+v_{\{x_j\}})\left(\sum_{i=n+1}^{\infty}|g_{x,n}(x_i-y_i)|^2\right)^{\frac{1}{2}} \to 0 \quad \text{as} \quad n\to\infty,\end{aligned}$$

whence s is compact.

Consequently, $u = I_E - s$ maps E *onto*[1] E and hence, by the inversion theorem of Banach, u is an isomorphism of E onto E, which satisfies, by (11.32), $u(x_n) = y_n$ $(n = 1, 2, \ldots)$. Therefore $\{x_n\} \approx \{y_n\}$.

The implications $6^\circ \Rightarrow 5^\circ \Rightarrow 4^\circ \Rightarrow 3^\circ \Rightarrow 2^\circ$ and $6^\circ \Rightarrow 1^\circ$ are obvious.

Finally, the proof of the implication $1^\circ \Rightarrow 3^\circ$ and the proof of b) are similar to the proof of the corresponding statements in Ch. I, § 10, theorem 10.2, using corollary 11.4 above. This completes the proof of theorem 11.3.

Let us give now some characterizations of Besselian bases in terms of stability properties.

Definition 11.4. A complete minimal sequence $\{x_n\}$ in a Banach space E is called

a) *quadratically stable*, if every l^2-linearly independent sequence $\{y_n\} \subset E$, which is quadratically near to $\{x_n\}$, is complete in E;

b) *weakly quadratically stable*, if every l^2-linearly independent sequence $\{y_n\} \subset E$, which is weakly quadratically near to $\{x_n\}$, is complete in E.

Theorem 11.4. *Let $\{x_n\}$ be a complete minimal sequence in a Banach space E. The following statements are equivalent:*

1°. *$\{x_n\}$ is a Besselian basis of E.*

2°. *$\{x_n\}$ is quadratically stable.*

3°. *$\{x_n\}$ is weakly quadratically stable.*

Moreover, if one of these statements holds, then every l^2-linearly independent sequence $\{y_n\} \subset E$, which is weakly quadratically near (or, in particular, quadratically near) to $\{x_n\}$, is a Besselian basis of E, equivalent to $\{x_n\}$.

Proof. The implication $1^\circ \Rightarrow 3^\circ$ and also the last statement of the theorem, are contained in theorem 11.3a) (implication $2^\circ \Rightarrow 6^\circ$).

The implication $3^\circ \Rightarrow 2^\circ$ is obvious.

$2^\circ \Rightarrow 1^\circ$. Assume that $\{x_n\}$ is quadratically stable, but not a Besselian basis of E. Then there exists either an $x_0 \in E$ such that $\sum_{i=1}^{\infty} f_i(x_0) x_i$ does not converge, or an $x_0 \in E$ such that $\sum_{i=1}^{\infty} |f_i(x_0)|^2 = \infty$, where $\{f_n\} \subset E^*$, $f_i(x_j) = \delta_{ij}$ $(i, j = 1, 2, \ldots)$. In both cases there exists an index n such that $f_n(x_0) \neq 0$. Put

$$y_i = x_i - \frac{f_n(x_i)}{f_n(x_0)} x_0 \quad (i = 1, 2, \ldots). \tag{11.35}$$

[1] See e.g. S. Banach [10], p. 154, theorem 14.

Then $\sum_{i=1}^{\infty} \|x_i - y_i\|^2 = \frac{\|x_0\|^2}{|f_n(x_0)|^2} \sum_{i=1}^{\infty} |f_n(x_i)|^2 = \frac{\|x_0\|^2}{|f_n(x_0)|^2} < \infty$, i.e. $\{y_n\}$ is quadratically near to $\{x_n\}$. Furthermore, we show that $\{y_n\}$ is l^2-linearly independent. Indeed, let $\{\alpha_n\} \in l^2$, $\sum_{i=1}^{\infty} \alpha_i y_i = 0$. Then, applying f_i to the relations

$$\sum_{j=1}^{\infty} \alpha_j x_j = \sum_{j=1}^{\infty} \alpha_j (x_j - y_j) = \frac{x_0}{f_n(x_0)} \sum_{j=1}^{\infty} \alpha_j f_n(x_j) = \frac{x_0}{f_n(x_0)} \alpha_n,$$

we obtain

$$\alpha_i = \frac{f_i(x_0)}{f_n(x_0)} \alpha_n \quad (i = 1, 2, \ldots). \tag{11.36}$$

Since $\sum_{i=1}^{\infty} \alpha_i x_i$ converges and $\sum_{i=1}^{\infty} |\alpha_i|^2 < \infty$ and since either $\sum_{i=1}^{\infty} f_i(x_0) x_i$ does not converge or $\sum_{i=1}^{\infty} |f_i(x_0)|^2 = \infty$, the equalities (11.36) are possible only if $\alpha_n = 0$, whence $\alpha_i = 0$ $(i = 1, 2, \ldots)$, which proves that $\{y_n\}$ is l^2-linearly independent.

Since by our hypothesis $\{x_n\}$ is quadratically stable, it follows that $\{y_n\}$ must be complete in E. However, this contradicts the relations

$$f_n(y_i) = f_n(x_i) - \frac{f_n(x_i)}{f_n(x_0)} f_n(x_0) = 0 \quad (i = 1, 2, \ldots).$$

Thus, $2° \Rightarrow 1°$, which completes the proof of theorem 11.4.

The answer to the problem of existence of Besselian and Hilbertian bounded bases is negative, as shown by corollary 11.1 above. On the other hand, obviously, the answer to the problem of existence of non-Besselian and non-Hilbertian bounded bases in finite-dimensional Banach spaces is also negative.

Problem 11.2.[1] Does every infinite dimensional Banach space with a basis possess a non-Besselian bounded basis? What about a non-Hilbertian bounded basis?

Some information concerning this problem is given by the implication $1° \Rightarrow 2°$ of corollary 11.2 above: *every basis $\{x_n\}$ which is not equivalent to the natural basis $\{e_n\}$ of l^2 is either non-Besselian or non-Hilbertian.*

The answer to problem 11.2 in the usual concrete Banach spaces is affirmative. In fact, this is true even for Hilbert spaces (which constitute the most difficult case), as shown by

[1] Apparently, in the mean time this problem has been solved in the affirmative (see the Notes and remarks).

Example 11.2. Let $E=L^2([-\pi,\pi])$ and let $0<\beta<\frac{1}{2}$. Then $\{\tilde{x}_n\}_{n=0}^{\infty}$, where[1]

$$x_{2n}(t)=|t|^{-\beta}e^{-int}, \qquad x_{2n+1}(t)=|t|^{-\beta}e^{int} \qquad (t\in[-\pi,\pi];\ n=0,1,2,\ldots), \tag{11.37}$$

is a Besselian but non-Hilbertian bounded basis of E, and $\{\tilde{y}_n\}_{n=0}^{\infty}$, where

$$y_{2n}(t)=|t|^{\beta}e^{int}, \quad y_{2n+1}(t)=|t|^{\beta}e^{-int} \quad (t\in[-\pi,\pi];\ n=0,1,2,\ldots), \tag{11.38}$$

is a Hilbertian but non-Besselian bounded basis of E.

Indeed, the sequence $\{f_n\}\subset E^*$ defined by

$$f_{2n}(\tilde{x})=\frac{1}{2\pi}\int_{-\pi}^{\pi}x(\tau)|\tau|^{\beta}e^{in\tau}d\tau, \qquad f_{2n+1}(\tilde{x})=\frac{1}{2\pi}\int_{-\pi}^{\pi}x(\tau)|\tau|^{\beta}e^{-in\tau}d\tau$$
$$(\tilde{x}\in L^2([-\pi,\pi]),\ n=0,1,2,\ldots) \tag{11.39}$$

satisfies obviously $f_k(\tilde{x}_j)=\delta_{kj}$ $(k,j=0,1,2,\ldots)$. Consequently, taking into account (11.18), we have[2], for the partial sum operators associated to the biorthogonal system $(\tilde{x}_n, f_n)$,

$$\begin{aligned}
[s_{2n}(\tilde{x})](t) &= \sum_{k=0}^{2n} f_k(\tilde{x})x_k(t)=|t|^{-\beta}\sum_{k=0}^{n}[f_{2k}(\tilde{x})e^{-ikt}+f_{2k+1}(\tilde{x})e^{ikt}] \\
&= \frac{1}{\pi}|t|^{-\beta}\int_{-\pi}^{\pi}x(t+\tau)|t+\tau|^{\beta}\frac{\sin\left(n+\frac{1}{2}\right)\tau}{2\sin\frac{\tau}{2}}d\tau \\
&= |t|^{-\beta}Q_n(t)+\frac{1}{2\pi}|t|^{-\beta}\int_{-\pi}^{\pi}x(t+\tau)|t+\tau|^{\beta}\cos n\tau\, d\tau,
\end{aligned} \tag{11.40}$$

where we have put

$$Q_n(t)=\frac{1}{\pi}\int_{-\pi}^{\pi}x(t+\tau)|t+\tau|^{\beta}\frac{\sin n\tau}{2\operatorname{tg}\frac{\tau}{2}}d\tau. \tag{11.41}$$

Now, taking into account (11.21) and putting

$$z_1(t)=x(t)|t|^{\beta}\sin nt, \qquad z_2(t)=x(t)|t|^{\beta}\cos nt \qquad (t\in[-\pi,\pi]), \tag{11.42}$$

---

[1] Here we use i as $\sqrt{-1}$.

[2] We consider all functions extended beyond $[-\pi,\pi)$ by $x(t+2\pi)=x(t)$.

we have, almost everywhere on $[-\pi,\pi]$,

$$Q_n(t)=\cos nt\,\bar{z}_1(t)-\sin nt\,\bar{z}_2(t),$$

where $\bar{z}_1(t)$, $\bar{z}_2(t)$ are the functions conjugate[1] to $z_1(t)$ and $z_2(t)$ respectively. Hence

$$|Q_n(t)|\leqslant|\bar{z}_1(t)|+|\bar{z}_2(t)|,$$

and thus

$$\int_{-\pi}^{\pi}|t|^{-2\beta}|Q_n(t)|^2\,dt\leqslant 2\int_{-\pi}^{\pi}|t|^{-2\beta}|\bar{z}_1(t)|^2\,dt+2\int_{-\pi}^{\pi}|t|^{-2\beta}|\bar{z}_2(t)|^2\,dt.$$

However, let us recall the following theorem on conjugate functions[2]: If

$$\int_{-\pi}^{\pi}|t|^{\alpha}|z(t)|^2\,dt<\infty,\quad\text{where}\quad -1<\alpha<1,$$

then there exists a constant $A_\alpha>0$ depending only on α such that

$$\int_{-\pi}^{\pi}|t|^{\alpha}|\bar{z}(t)|^2\,dt\leqslant A_\alpha\int_{-\pi}^{\pi}|t|^{\alpha}|z(t)|^{\alpha}\,dt.$$

Applying this theorem to z_1 and z_2 defined by (11.42) and to $\alpha=-2\beta$, we obtain

$$\int_{-\pi}^{\pi}|t|^{-2\beta}|Q_n(t)|^2\,dt\leqslant 2A_{-2\beta}\int_{-\pi}^{\pi}|t|^{-2\beta}|z_1(t)|^2\,dt+2A_{-2\beta}\int_{-\pi}^{\pi}|t|^{-2\beta}|z_2(t)|^2\,dt$$

$$=4A_{-2\beta}\int_{-\pi}^{\pi}|x(t)|^2\,dt.$$

Consequently, taking into account (11.40), we get

$$\|s_{2n}(\tilde{x})\|\leqslant A'_\beta\|\tilde{x}\|\qquad(\tilde{x}\in E),\tag{11.43}$$

with a suitable A'_β depending only on β.

Since

$$\|\tilde{x}_n\|\leqslant\left(\int_{-\pi}^{\pi}|t|^{-2\beta}\,dt\right)^{\frac{1}{2}}=M_1<\infty,\quad \|f_n\|\leqslant\frac{1}{2\pi}\left(\int_{-\pi}^{\pi}|\tau|^{2\beta}\,d\tau\right)^{\frac{1}{2}}=M_2<\infty$$

$$(n=0,1,2,\ldots),$$

we have

$$\|f_n(\tilde{x})\tilde{x}_n\|\leqslant M_1M_2\|\tilde{x}\|\qquad(\tilde{x}\in E,\ n=0,1,2,\ldots),$$

---

[1] For the definition of conjugate functions see example 11.1.

[2] See [8].

whence, by (11.43) and $\|s_{2n+1}(\tilde{x})\| \leqslant \|s_{2n}(\tilde{x})\| + \|f_{2n+1}(\tilde{x})\tilde{x}_{2n+1}\|$,

$$\sup_{0 \leqslant n < \infty} \|s_n\| < \infty. \tag{11.44}$$

On the other hand, let us show that $\{\tilde{x}_n\}_{n=0}^{\infty}$ is complete in E. Let $\tilde{z} \in E = L^2([-\pi, \pi])$ be such that $(\tilde{z}, \tilde{x}_n) = 0$ $(n = 0, 1, 2, \ldots)$, i.e. such that

$$\int_{-\pi}^{\pi} z(t)|t|^{-\beta} e^{-int}\, dt = 0, \qquad \int_{-\pi}^{\pi} z(t)|t|^{-\beta} e^{int}\, dt = 0 \qquad (n = 0, 1, 2, \ldots).$$

Then, since $\widetilde{z(t)|t|^{-\beta}} \in L^1([-\pi, \pi])$ (because $\tilde{z} \in L^2([-\pi, \pi]), \widetilde{|t|^{-\beta}} \in L^2([-\pi, \pi])$) and since $\widetilde{\{e^{-int}\}}_{n=0}^{\infty} \cup \widetilde{\{e^{int}\}}_{n=1}^{\infty}$ is complete in $E = L^2([-\pi, \pi])$, it follows that

$$z(t)|t|^{-\beta} = 0 \quad \text{a.e. on} \quad [-\pi, \pi],$$

whence $\tilde{z} = 0$, which proves that $[\tilde{x}_n]_{n=0}^{\infty} = E$.

Thus, $(\tilde{x}_n, f_n)_{n=0}^{\infty}$ is an E-complete biorthogonal system satisfying (11.44). Hence, by Ch. I, § 4, theorem 4.1, $\{\tilde{x}_n\}_{n=0}^{\infty}$ is a basis of E. From (11.37) it is obvious that the basis $\{\tilde{x}_n\}_{n=0}^{\infty}$ is bounded.

Similarly, the sequence $\{\tilde{y}_n\}_{n=0}^{\infty} \subset E$ defined by (11.38) is a bounded basis of E. Let us prove now that: a) $\{\tilde{x}_n\}_{n=0}^{\infty}$ is Besselian but non-Hilbertian and b) $\{\tilde{y}_n\}_{n=0}^{\infty}$ is Hilbertian but non-Besselian. Since by (11.39) the image of $\{f_n\}_{n=0}^{\infty}$ under the canonical isometry $L^2([-\pi, \pi])^* \equiv L^2([-\pi, \pi])$ is nothing else but $\{\tilde{y}_n\}_{n=0}^{\infty}$, it will be sufficient, by virtue of theorem 11.1 above, to prove b).

Since the function $|t|^{-\beta}$ is not bounded above on $[-\pi, \pi]$, there exists a $\tilde{z} \in L^2([-\pi, \pi])$ such that $\widetilde{z(t)|t|^{-\beta}} \notin L^2([-\pi, \pi])$. Consequently, we have

$$\sum_{k=0}^{\infty} |g_k(\tilde{z})| = \infty,$$

where

$$g_{2n}(\tilde{x}) = \frac{1}{2\pi} \int_{-\pi}^{\pi} x(\tau)|\tau|^{-\beta} e^{-in\tau}\, d\tau, \qquad g_{2n+1}(\tilde{x}) = \frac{1}{2\pi} \int_{-\pi}^{\pi} x(\tau)|\tau|^{-\beta} e^{in\tau}\, d\tau$$

$$(\tilde{x} \in L^2([-\pi, \pi]),\ n = 0, 1, 2, \ldots) \tag{11.45}$$

is the a.s.c.f. to $\{\tilde{y}_n\}_{n=0}^{\infty}$. Thus $\{\tilde{y}_n\}_{n=0}^{\infty}$ is non-Besselian. On the other hand, let $\{\alpha_n\}_{n=0}^{\infty}$ be an arbitrary sequence of scalars such that $\sum_{k=0}^{\infty} |\alpha_k|^2 < \infty$. Then there exists an $\tilde{y} \in L^2([-\pi, \pi])$ such that

$$\alpha_{2n} = \frac{1}{2\pi} \int_{-\pi}^{\pi} y(\tau) e^{-in\tau}\, d\tau, \qquad \alpha_{2n+1} = \frac{1}{2\pi} \int_{-\pi}^{\pi} y(\tau) e^{in\tau}\, d\tau \qquad (n = 0, 1, 2, \ldots).$$

Hence for the $\tilde{x}\in L^2([-\pi,\pi])$ defined by

$$x(t)=|t|^{\beta}y(t) \qquad (t\in[-\pi,\pi])$$

we have, taking into account (11.45),

$$g_n(\tilde{x})=\alpha_n \qquad (n=0,1,2,\ldots),$$

and thus $\{\tilde{y}_n\}$ is Hilbertian. This completes the proof of the assertions stated in example 11.2.

Since the results on stability given in the foregoing, and also their proofs, are similar to those of Ch. I, § 10, it is natural to ask whether both series of results can be obtained as particular cases of a more general theory. We shall now show that this is indeed the case and that this general theory also has other applications.

Definition 11.5. Let F be a Banach space and let $\{e_n\}\subset F$ be a minimal sequence. A basis $\{x_n\}$ of a Banach space E is said to be

a) $(F,\{e_n\})$-*Besselian*, if $\{x_n\}\succ\{e_n\}$, i.e. if $\sum_{i=1}^{\infty} f_i(x)e_i$ converges for all $x\in E$, where $\{f_n\}$ is the a.s.c.f. to $\{x_n\}$;

b) $(F,\{e_n\})$-*Hilbertian*, if $\{e_n\}\succ\{x_n\}$, i.e. if for every convergent series $\sum_{i=1}^{\infty}\alpha_i e_i$ in F there exists an $x\in E$ such that $f_n(x)=\alpha_n$ $(n=1,2,\ldots)$.

In the particular case when $F=l^p$ $(1\leqslant p<\infty)$ or c_0 and $\{e_n\}=$ the unit vector basis of F, the terms *p-Besselian basis* and *p-Hilbertian basis*, respectively *∞-Besselian basis* and *∞-Hilbertian basis*, are also used.

The 2-Besselian (2-Hilbertian) bases are nothing else but the Besselian (Hilbertian) bases in the usual sense. By Ch. I, § 3, lemma 3.1, a basis $\{x_n\}$ is *∞-Besselian if and only if* $\inf_{1\leqslant n<\infty}\|x_n\|>0$ and $\{x_n\}$ is 1-*Hilbertian if and only if* $\sup_{1\leqslant n<\infty}\|x_n\|<\infty$. Furthermore, it is well known[1] that $\{x_n\}$ is *∞-Hilbertian if and only if* $\sum_{i=1}^{\infty} x_i$ *is weakly unconditionally Cauchy* (i.e. $\sum_{i=1}^{\infty}|f(x_i)|<\infty$ for all $f\in E^*$).

Let us observe that every basis $\{x_n\}$ of E is both $(E,\{x_n\})$-Besselian and $(E,\{x_n\})$-Hilbertian. Furthermore, if $\{e_n\}$ is a basis of F, a basis $\{x_n\}$ of E is $(F,\{e_n\})$-Besselian (respectively, $(F,\{e_n\})$-Hilbertian) if and only if $\{e_n\}$ is $(E,\{x_n\})$-Hilbertian (respectively, $(E,\{x_n\})$-Besselian).

If $\{e_n\}$ is a basis of F, theorem 11.1 (and also corollary 11.2) can be extended to $(F,\{e_n\})$-Besselian and $(F,\{e_n\})$-Hilbertian bases, replacing

[1] See Ch. I, § 17, corollary 17.5.

l^2 by F and $\sqrt{\sum_i |\alpha_i|^2}$ by $\left\|\sum_i \alpha_i e_i\right\|$, deleting[1] conditions a) 5°, b) 5° and modifying conditions a) 6°, b) 6° as follows: $\{f_n\}$ is a $(\{h_n\}, [h_n])$-Hilbertian (respectively, $(\{h_n\}, [h_n])$-Besselian) basis of $[f_n]$, where $\{h_n\} \subset F^*$ is the a.s.c.f. to $\{e_n\}$. Indeed, all proofs are the same, except that we give another proof of b) 3°⇒b) 6°, similar to the proof of a) 3°⇒a) 6°): if $v: F \to E$ is as in b) 3°, then $[v^*(f_i)](e_j) = f_i[v(e_j)] = f_i(x_j) = \delta_{ij}$ $(i,j = 1, 2, \ldots)$, whence $v^*(f_i) = h_i$ $(i = 1, 2, \ldots)$, whence, by the implication a) 3°⇒a) 1°, we obtain b) 6°.

Definition 11.6. Let F be a Banach space, let $\{e_n\} \subset F$ be a minimal sequence and let $\{x_n\}, \{y_n\}$ be two sequences in a Banach space E. The sequence $\{y_n\}$ is said to be

a) $(F, \{e_n\})$-*near* to $\{x_n\}$, if

$$\sum_{i=1}^{\infty} \|x_i - y_i\| e_i \quad \text{converges.} \tag{11.46}$$

b) *weakly* $(F, \{e_n\})$-*near* to $\{x_n\}$, if $\sum_{i=1}^{\infty} f(x_i - y_i) e_i$ converges uniformly with respect to $f \in E^*$, $\|f\| \leqslant 1$, i.e. if

$$\lim_{n \to \infty} \sup_{\substack{f \in E^* \\ \|f\| \leqslant 1}} \left\| \sum_{i=n+1}^{\infty} f(x_i - y_i) e_i \right\| = 0. \tag{11.47}$$

In the particular case when $F = l^p$ $(1 \leqslant p < \infty)$ or c_0 and $\{e_n\}$ = the unit vector basis of F, the terms *p-near* and *weakly p-near*, respectively ∞-*near* and *weakly* ∞-*near* are also used.

The 2-near (weakly 2-near) sequences are nothing else but the quadratically (weakly quadratically) near sequences and the 1-near (weakly 1-near) sequences are nothing else than the strictly (respectively, weakly) near sequences. It is also immediate that $\{x_n\}, \{y_n\}$ are weakly ∞-near if and only if they are ∞-near, i.e. $\lim_{n \to \infty} \|x_n - y_n\| = 0$.

In general there is no relation of implication between the notions $(F, \{e_n\})$-near and weakly $(F, \{e_n\})$-near. If $\{e_n\}$ is an unconditional basis[2] of F, then $(F, \{e_n\})$-near ⇒ weakly $(F, \{e_n\})$-near, but the converse implication is not true.

---

[1] They can be maintained (in the form a) 5° $\sum_{i=1}^{\infty} \Phi(f_i) e_i$ converges for all $\Phi \in E^{**}$, respectively b) 5° $\sum_{i=1}^{\infty} f(x_i) h_i$ converges for all $f \in E^*$) e.g. in the case when $\{e_n\}$ is a boundedly complete (respectively, shrinking) basis of E.

[2] See § 14, definition 14.1.

Definition 11.7. Let F be a Banach space and let $\{e_n\} \subset F$ be a minimal sequence. A sequence $\{y_n\}$ in a Banach space E is said to be $(F,\{e_n\})$-*linearly independent*, if

$$\sum_{i=1}^{\infty} \alpha_i e_i \in F, \quad \sum_{i=1}^{\infty} \alpha_i y_i = 0 \quad \text{imply} \quad \alpha_i = 0 \quad (i=1,2,\ldots). \tag{11.48}$$

In the particular case when $F = l^p$ $(1 \leqslant p < \infty)$ or c_0 and $\{e_n\}$ = the unit vector basis of F, the terms l^p-*linearly independent* and c_0-*linearly independent* are also used.

If $\inf_{1 \leqslant n < \infty} \|y_n\| > 0$, the sequence $\{y_n\}$ is c_0-linearly independent if and only if it is ω-linearly independent. Indeed, it is obvious from the definitions that ω-linearly independence implies $(F,\{e_n\})$-linearly independence for any F and $\{e_n\}$. Conversely, if $\inf_{1 \leqslant n < \infty} \|y_n\| > 0$ and $\{y_n\}$ is c_0-linearly independent, let $\{\alpha_n\}$ be an arbitrary sequence of scalars such that $\sum_{i=1}^{\infty} \alpha_i y_i = 0$. Then, since $\inf_{1 \leqslant n < \infty} \|y_n\| > 0$, we must have $\lim_{n \to \infty} \alpha_n = 0$, whence, since $\{y_n\}$ is c_0-linearly independent, $\alpha_i = 0$ $(i=1,2,\ldots)$, which proves our assertion.

It is also immediate that for any infinite dimensional F and any minimal sequence $\{e_n\}$ in F, with $\sup_{1 \leqslant n < \infty} \|e_n\| < \infty$, the $(F,\{e_n\})$-linearly independence implies l^1-linearly independence. Indeed, if $\{y_n\} \subset E$ is $(F,\{e_n\})$-linearly independent and $\sum_{i=1}^{\infty} |\alpha_i| < \infty$, $\sum_{i=1}^{\infty} \alpha_i y_i = 0$, then, since by hypothesis $\sup_{1 \leqslant n < \infty} \|e_n\| < \infty$, we have $\sum_{i=1}^{\infty} \alpha_i e_i \in F$, whence, since $\{y_n\}$ is $(F,\{e_n\})$-linearly independent, $\alpha_i = 0$ $(i=1,2,\ldots)$.

Definition 11.8. Let F be a Banach space, and let (e_n, h_n) $(\{e_n\} \subset F$, $\{h_n\} \subset F^*)$ be a biorthogonal system. A complete minimal sequence $\{x_n\}$ in a Banach space E is called

a) $(F,\{e_n\})$-*stable*, if every $(F,\{e_n\})$-linearly independent sequence $\{y_n\} \subset E$, which is $([h_n], \{h_n\})$-near to $\{x_n\}$, is complete in E;

b) *weakly* $(F,\{e_n\})$-*stable*, if every $(F,\{e_n\})$-linearly independent sequence $\{y_n\} \subset E$, which is weakly $([h_n], \{h_n\})$-near to $\{x_n\}$, is complete in E.

In the particular case when $F = l^p$ $(1 \leqslant p < \infty)$ or c_0 and $\{e_n\}$ = the unit vector basis of F, the terms *p-stable* and *weakly p-stable*, respectively ∞-*stable* and *weakly* ∞-*stable*, are also used.

The 2-stable (weakly 2-stable) complete minimal sequences are nothing else but the quadratically (weakly quadratically) stable ones. It is also immediate that a complete minimal sequence $\{x_n\} \subset E$ with $\inf_{1 \leqslant n < \infty} \|x_n\| > 0$ is ∞-stable (weakly ∞-stable) if and only if it is strictly stable (respectively, weakly stable) in the sense of Ch. I, § 10, definition 10.3. Indeed, the "only if" part is obvious even without the assumption $\inf_{1 \leqslant n < \infty} \|x_n\| > 0$, since ω-linearly independence implies c_0-linearly independence. Conversely, if a complete minimal sequence $\{x_n\} \subset E$ with $\inf_{1 \leqslant n < \infty} \|x_n\| > 0$ is strictly stable (respectively, weakly stable), let $\{y_n\}$ be a c_0-linearly independent sequence in E, 1-near (respectively, weakly 1-near) to $\{x_n\}$. Then, since $\sum_{i=1}^{\infty} (x_i - y_i)$ converges, we have $\lim_{n \to \infty} \|x_n - y_n\| = 0$, whence, since $\inf_{1 \leqslant n < \infty} \|x_n\| > 0$, we infer $\inf_{1 \leqslant n < \infty} \|y_n\| > 0$. Consequently, by the remark made after definition 11.7, $\{y_n\}$ is ω-linearly independent, whence, since $\{x_n\}$ is strictly (respectively, weakly) stable, $\{y_n\}$ is complete in E, which proves our assertion. By the remark made after definition 11.6 it is obvious that $\{x_n\}$ is 1-stable if and only if it is weakly 1-stable.

The assertions on weak quadratic nearness and weak quadratic stability, and in the particular case when $\{e_n\}$ is an unconditional basis of F, all assertions, of theorem 11.2, corollary 11.4 and theorems 11.3 and 11.4 can be extended to $(F, \{e_n\})$-Besselian bases, replacing conditions (11.25) and (11.26) by $\sup_{1 \leqslant n < \infty} \sup_{\substack{f \in E^* \\ \|f\| \leqslant 1}} \left\| \sum_{i=1}^{n} f(x_i - y_i) h_i \right\| = M < c$ and respectively, if $\{e_n\}$ is an unconditional basis of F, by $\sup_{1 \leqslant n < \infty} \left\| \sum_{i=1}^{n} \|x_i - y_i\| h_i \right\| = M' < \frac{c}{\nu_{\{e_n\}}^{(u)}}$, where $\{h_n\} \subset F^*$, $h_i(e_j) = \delta_{ij}$, $c > 0$ is any constant such that $c \left\| \sum_{i=1}^{n} \alpha_i e_i \right\| \leqslant \left\| \sum_{i=1}^{n} \alpha_i x_i \right\|$ for all finite sequences of scalars $\alpha_1, \ldots, \alpha_n$ and $\nu_{\{e_n\}}^{(u)} =$ the unconditional norm[1] of the basis $\{e_n\}$. Indeed, the proofs are essentially the same, using instead of (11.27) the inequalities

$$\begin{aligned} \left\| \sum_{i=1}^{n} \alpha_i (x_i - y_i) \right\| &= \sum_{i=1}^{n} \alpha_i g_n (x_i - y_i) = \left(\sum_{i=1}^{n} g_n (x_i - y_i) h_i \right) \left(\sum_{j=1}^{n} \alpha_j e_j \right) \\ &\leqslant \left\| \sum_{i=1}^{n} g_n (x_i - y_i) h_i \right\| \left\| \sum_{j=1}^{n} \alpha_j e_j \right\| \leqslant \frac{M}{c} \left\| \sum_{i=1}^{n} \alpha_i x_i \right\| \end{aligned} \tag{11.49}$$

and respectively, if $\{e_n\}$ is an unconditional basis of F, the inequalities

[1] For these notions see § 14 and § 17.

$$\left\|\sum_{i=1}^{n}\alpha_i(x_i-y_i)\right\| \leqslant \sum_{i=1}^{n}|\alpha_i|\,\|x_i-y_i\| = \left(\sum_{i=1}^{n}\|x_i-y_i\|\,h_i\right)\left(\sum_{j=1}^{n}|\alpha_j|\,e_j\right)$$
$$\leqslant \left\|\sum_{i=1}^{n}\|x_i-y_i\|\,h_i\right\| \left\|\sum_{j=1}^{n}|\alpha_j|\,e_j\right\| \leqslant \frac{M'\nu_{\{e_n\}}^{(u)}}{c}\left\|\sum_{i=1}^{n}\alpha_i x_i\right\|; \tag{11.50}$$

let us mention that the second result can be also obtained as a particular case of the first, since

$$\left\|\sum_{i=1}^{n} f(x_i-y_i)h_i\right\| \leqslant \nu_{\{e_n\}}^{(u)}\left\|\sum_{i=1}^{n}\|x_i-y_i\|\,h_i\right\| = \nu_{\{e_n\}}^{(u)}M' < c$$
$$(f\in E^*,\ \|f\|\leqslant 1).$$

In the particular case when $F=l^2$ and $\{e_n\}$ = the unit vector basis of F, from these extended results we obtain again the foregoing results on quadratic and weak quadratic stability of usual Besselian bases. In the particular case when $F=c_0$ and $\{e_n\}$ = the unit vector basis of F, from the extended results we obtain again the results of Ch. I, § 10 on strict and weak stability of usual bases $\{x_n\}$ with $\inf_{1\leqslant n<\infty}\|x_n\|>0$. In the particular case when $E=F=l^1$ and $\{x_n\}=\{e_n\}$ = the unit vector basis of l^1, from the extended theorem 11.2 we obtain the following result[1]:

Theorem 11.5. *Let $\{x_n\}$ be the unit vector basis of $E=l^1$, and let $\{y_n\}$ be a sequence in E such that*

$$\sup_{1\leqslant n<\infty}\|x_n-y_n\|=M<1. \tag{11.51}$$

Then $\{y_n\}$ is a basis of E, equivalent to $\{x_n\}$.

In the particular case when $F=l^1$ and $\{e_n\}$ = the unit vector basis of l^1, from the extended theorem 11.4 we obtain the following result:

Theorem 11.6. *A complete minimal sequence $\{x_n\}$ in a Banach space E, with $\sup_{1\leqslant n<\infty}\|x_n\|<\infty$, is a basis equivalent to the unit vector basis of l^1 if and only if every l^1-linearly independent sequence $\{y_n\}\subset E$ with $\lim_{n\to\infty}\|x_n-y_n\|=0$ is complete in E. Moreover, in this case every such sequence $\{y_n\}\subset E$ is a basis of E, equivalent to $\{x_n\}$.*

---

[1] Let us observe that theorem 11.5 is also an immediate consequence of the sufficiency part of Ch. I, § 10, theorem 10.4, since $b=\sup_{|\varepsilon_i|=1}\sup_{1\leqslant n<\infty}\left\|\sum_{i=1}^{n}\varepsilon_i\gamma_i f_i\right|_{[x_j]}\right\| = \sup_{1\leqslant i<\infty}|\gamma_i|$.

We leave to the reader to formulate the results obtained from the extended theorems in the particular case when $E=F$, $\{x_n\}=\{e_n\}$ and in the case when $F=l^p$ $(1<p\neq 2)$ and $\{e_n\}$ = the unit vector basis of F.

§ 12. Relations between various types of bases

Theorem 12.1. a) *For a bounded basis $\{x_n\}$ of an infinite dimensional Banach space E we have the following implications:*

Table 12.1

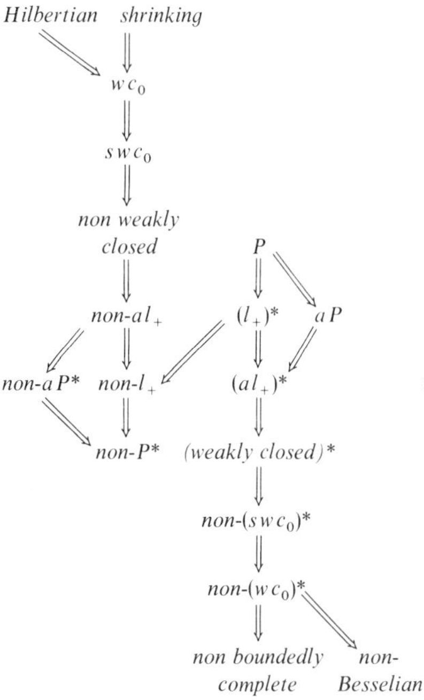

Table 12.2

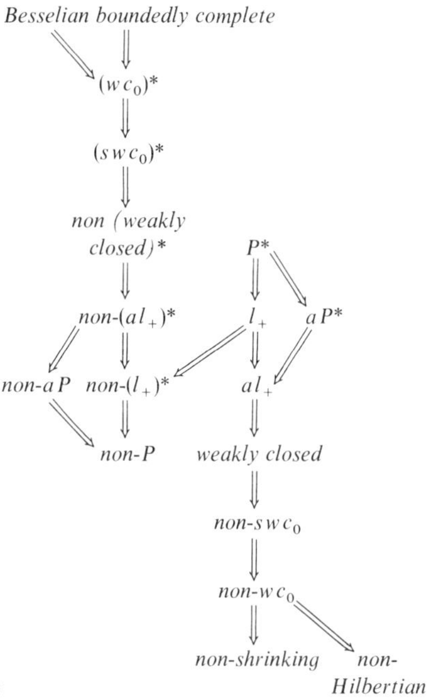

b) *The converse implications are not valid, except perhaps non-al_+ $\Rightarrow$ non weakly closed, (weakly closed)* $\Rightarrow (al_+)^*$, non-$(al_+)^*$ $\Rightarrow$ non-(weakly closed)* and weakly closed $\Rightarrow al_+$, which are unknown*[1].

c) *Between the 36 types of bounded bases considered in part* a), *the only implications are the 36 trivial ones (shrinking $\Rightarrow$ shrinking, etc.) and the 144 given by tables 12.1 and 12.2 above. All the other* $1296-36-144$

---

[1] After this monograph had been completed, we learned that this problem was recently solved, namely, these four implications do not hold either (see the Notes and remarks).

$=1116$ *relations between them, except perhaps the 4 considered in part* b), *are: non-implication.*

Proof. a) Shrinking $\Rightarrow wc_0$. Assume that $\{x_n\}$ is a shrinking bounded basis of an infinite dimensional Banach space E. Then, since $\sup_{1\leqslant n<\infty} \|x_n\|<\infty$, for the a.s.c.f. $\{f_n\}$ to $\{x_n\}$ we have, by, Ch. I, § 3, corollary 3.1, $\inf_{1\leqslant n<\infty} \|f_n\|>0$. Hence, since $\beta_n f_n \to 0$ for $\sum_{i=1}^{\infty} \beta_i f_i \in [f_n] = E^*$, it follows that

$$\left(\sum_{i=1}^{\infty} \beta_i f_i\right)(x_n) = \beta_n \to 0 \qquad \left(\sum_{i=1}^{\infty} \beta_i f_i \in [f_n] = E^*\right),$$

and thus $\{x_n\}$ is of type wc_0. Alternatively, the implication shrinking $\Rightarrow wc_0$ also follows from § 4, theorem 4.2, implication $1° \Rightarrow 2°$.

The implication Hilbertian $\Rightarrow wc_0$ follows from the implication b) $1° \Rightarrow$b) $5°$ of § 11, theorem 11.1.

The implications $wc_0 \Rightarrow swc_0 \Rightarrow$ non weakly closed are obvious.

Non weakly closed $\Rightarrow$ non-al_+. Assume that $\{x_n\}$ is a non weakly closed basis. Then, since 0 is a weak limit point of the set $\{x_n\}$, in every w-neighbourhood of 0 of the form $V_{f;\varepsilon}(0)$ (where $f \in E^*$, $\varepsilon>0$) there exists at least one x_n, whence we obtain

$$\inf_{1\leqslant n<\infty} |f(x_n)| = 0 \qquad (f \in E^*),$$

and thus, by the implication $1° \Rightarrow 2°$ of § 10, theorem 10.3, $\{x_n\}$ is of type non-al_+.

The implications non-$al_+ \Rightarrow$ non-l_+ and non-$aP^* \Rightarrow$ non-P^* are obvious.

The implications non-$l_+ \Rightarrow$non-P^* and non-$al_+ \Rightarrow$non-aP^* have been observed in § 10 (taking $\beta_n=1$ in $7°$ of theorem 10.1 and applying the implication $7° \Rightarrow 1°$ of this theorem).

$(l_+)^* \Rightarrow$ non-l_+. Assume that $\{x_n\}$ is a basis of type $(l_+)^*$. Then, applying the implication $1° \Rightarrow 6°$ of § 10, theorem 10.1, to the a.s.c.f. $\{f_n\}$ of $\{x_n\}$, it follows that there exists a sequence of scalars $\{\beta_n\}$ satisfying (10.6) and such that $\{\beta_n f_n\}$ is a basis of type P^* of $[f_n]$, whence, by § 9, theorem 9.2 a), $\left\{\frac{1}{\beta_n} x_n\right\}$ is a basis of type P of E. Now, $\left\{\frac{1}{\beta_n} x_n\right\}$ is of type non-l_+, since otherwise one would have, with a suitable $\eta>0$,

$$\sup_{1\leqslant n<\infty} \eta n \leqslant \sup_{1\leqslant n<\infty} \left\| \sum_{i=1}^{n} \frac{1}{\beta_i} x_i \right\| < \infty,$$

which is impossible. Consequently, by the implication $1° \Rightarrow 10°$ of § 10, theorem 10.1, $\{x_n\}$ is of type non-l_+.

The other implications of table 12.1 follow from the implications proved above, by duality (e.g. from the implication non-$l_+ \Rightarrow$ non-P^* we obtain $P \Rightarrow (l_+)^*$, etc.).

Finally, the implications of table 12.2 follow from those of table 12.1, either by duality (e.g. from the implication shrinking $\Rightarrow w c_0$ we obtain: boundedly complete $\Rightarrow (w c_0)^*$, etc.) or by a simple negation (e.g. from the implication shrinking $\Rightarrow w c_0$ we obtain non-$w c_0 \Rightarrow$ non-shrinking, i.e. the penultimate implication of table 12.2, etc.).

b) In order to prove that the converse implications are not valid, it is sufficient to consider only successive properties of table 12.1 and to give for them suitable examples.

$w c_0 \not\Rightarrow$ shrinking, as shown by

Example 12.1. Let d be the real linear space of all sequences of real numbers $x = \{\xi_n\}$ such that

$$\|x\| = \sup_{\sigma \in \Pi} \sum_{i=1}^{\infty} \frac{|\xi_{\sigma(i)}|}{i} < \infty, \tag{12.1}$$

where Π denotes the set of all permutations[1] of the set $\mathcal{N} = \{1, 2, 3, \ldots\}$. Then d is a Banach space, and the unit vectors $\{x_n\}$ are a non-shrinking basis of d, of type $w c_0$.

Indeed, let us first prove that d is complete. Let $z_j = \{\zeta_n^{(j)}\}_{n=1}^{\infty}$ $(j = 1, 2, \ldots)$ be a Cauchy sequence in d and let $\varepsilon > 0$ be arbitrary. Then there exists an $N = N(\varepsilon)$ such that

$$\|z_m - z_{m+p}\| = \sup_{\sigma \in \Pi} \sum_{i=1}^{\infty} \frac{|\zeta_{\sigma(i)}^{(m)} - \zeta_{\sigma(i)}^{(m+p)}|}{i} < \varepsilon \quad (m > N,\ p = 1, 2, \ldots). \tag{12.2}$$

Hence, taking $i = 1$ and $\sigma_n \in \Pi$ with $\sigma_n(1) = n$ $(n = 1, 2, \ldots)$, it follows that

$$|\zeta_n^{(m)} - \zeta_n^{(m+p)}| < \varepsilon \quad (m > N,\ p = 1, 2, \ldots;\ n = 1, 2, \ldots),$$

and thus each sequence $\{\zeta_n^{(m)}\}_{m=1}^{\infty}$ $(n = 1, 2, \ldots)$ is a Cauchy sequence, whence convergent, say to ζ_n $(n = 1, 2, \ldots)$. Now, by (12.2) we have

$$\sum_{i=1}^{k} \frac{|\zeta_{\sigma(i)}^{(m)} - \zeta_{\sigma(i)}^{(m+p)}|}{i} < \varepsilon \quad (m > N,\ p = 1, 2, \ldots;\ k = 1, 2, \ldots;\ \sigma \in \Pi),$$

whence for $p \to \infty$ we obtain

$$\sum_{i=1}^{k} \frac{|\zeta_{\sigma(i)}^{(m)} - \zeta_{\sigma(i)}|}{i} < \varepsilon \quad (m > N,\ k = 1, 2, \ldots;\ \sigma \in \Pi),$$

---

[1] I.e. the set of all one to one mappings of $\mathcal{N}$ *onto* $\mathcal{N}$.

whence, putting $z=\{\zeta_n\}$, we get $z_m-z\in d$, $\|z_m-z\|\leqslant\varepsilon$ $(m>N)$. Consequently, $z=(z_m-z)+z_m\in d$ and $z_m\to z$ as $m\to\infty$, which proves the completeness of d.

Observe that for every $x=\{\xi_n\}\in d$ we have $\lim_{n\to\infty}\xi_n=0$. Indeed, otherwise there would exist an infinite subsequence $\{\xi_{n_i}\}$ with $\inf_{1\leqslant i<\infty}|\xi_{n_i}|=\eta>0$, whence, taking for each k a permutation $\sigma_k\in\Pi$ with $\sigma_k(i)=n_i$ $(i=1,\dots,k)$, we would obtain $\|x\|\geqslant\sum_{i=1}^{k}\frac{|\xi_{\sigma_k(i)}|}{i}=\sum_{i=1}^{k}\frac{|\xi_{n_i}|}{i}\geqslant\eta\sum_{i=1}^{k}\frac{1}{i}$ $(k=1,2,\dots)$, whence $\|x\|=\infty$, contradicting the assumption $x\in d$. From this remark it follows that for every sequence $x=\{\xi_1,\xi_2,\dots\}\in d$ the subsequence $\{\xi_{j_1},\xi_{j_2},\dots\}$ consisting of all *non-zero* co-ordinates can be rearranged into a (finite or infinite) sequence $\{\xi_{j_1}^*,\xi_{j_2}^*,\dots\}$ such that $|\xi_{j_1}^*|\geqslant|\xi_{j_2}^*|\geqslant\cdots$ (note that if $\{\xi_n\}$ has an infinity of non-zero co-ordinates and also at least one zero co-ordinate, *then there is no permutation* $\sigma\in\Pi$ such that $\xi_{\sigma(n)}=\xi_{j_n}^*$ $(n=1,2,\dots)$). We claim that

$$\|x\|=\sum_{i=1}^{\infty}\frac{|\xi_{j_i}^*|}{i}\qquad(x=\{\xi_n\}\in d),\tag{12.3}$$

where $\xi_{j_n}^*$ are as defined above.

Indeed, let $x=\{\xi_n\}\in d$ and $\varepsilon>0$ be arbitrary. Then there exists a $\sigma=\sigma(x,\varepsilon)\in\Pi$ such that

$$\left|\,\|x\|-\sum_{i=1}^{\infty}\frac{|\xi_{\sigma(i)}|}{i}\right|<\frac{\varepsilon}{2},$$

and an index $N=N(x,\varepsilon,\sigma)$ such that

$$\sum_{i=N+1}^{\infty}\frac{|\xi_{\sigma(i)}|}{i}<\frac{\varepsilon}{2},$$

whence

$$\left|\,\|x\|-\sum_{i=1}^{N}\frac{|\xi_{\sigma(i)}|}{i}\right|<\varepsilon.\tag{12.4}$$

Now, if for an index $i<N$ we have $|\xi_{\sigma(i)}|<|\xi_{\sigma(i+1)}|$, then for the permutation $\sigma'\in\Pi$ defined by

$$\sigma'(i)=\sigma(i+1),\qquad\sigma'(i+1)=\sigma(i),\qquad\sigma'(k)=\sigma(k)\qquad(k\neq i,\ i+1)$$

we have, taking into account that $\frac{\alpha}{i}+\frac{\beta}{i+1}>\frac{\beta}{i}+\frac{\alpha}{i+1}$ for $\alpha>\beta\geqslant0$,

$$\sum_{i=1}^{N}\frac{|\xi_{\sigma'(i)}|}{i}>\sum_{i=1}^{N}\frac{|\xi_{\sigma(i)}|}{i}.$$

Thus, in a finite number of steps we obtain that for the permutation $\sigma_1 \in \Pi$ defined by

$$\sigma_1(k) = \sigma(k) \quad \text{for} \quad k = N+1, N+2, \ldots, \qquad |\xi_{\sigma_1(1)}| \geqslant |\xi_{\sigma_1(2)}| \geqslant \cdots \geqslant |\xi_{\sigma_1(N)}|, \tag{12.5}$$

we have

$$\sum_{i=1}^{N} \frac{|\xi_{\sigma_1(i)}|}{i} > \sum_{i=1}^{N} \frac{|\xi_{\sigma(i)}|}{i},$$

whence

$$\|x\| - \sum_{i=1}^{N} \frac{|\xi_{\sigma_1(i)}|}{i} < \|x\| - \sum_{i=1}^{N} \frac{|\xi_{\sigma(i)}|}{i} < \varepsilon. \tag{12.6}$$

Since there exists a permutation $\sigma_2 \in \Pi$ such that $\xi_{\sigma_2(i)} = \xi_{j_i}^*$ $(i = 1, \ldots, N)$ and since by (12.5) and the definition of $\xi_{j_i}^*$ we have $|\xi_{j_i}^*| \geqslant |\xi_{\sigma_1(i)}|$ $(i = 1, \ldots, N)$, it follows by (12.6) that

$$0 \leqslant \|x\| - \sum_{i=1}^{N} \frac{|\xi_{\sigma_2(i)}|}{i} = \|x\| - \sum_{i=1}^{N} \frac{|\xi_{j_i}^*|}{i} \leqslant \|x\| - \sum_{i=1}^{N} \frac{|\xi_{\sigma_1(i)}|}{i} < \varepsilon,$$

whence, since $x \in d$ and $\varepsilon > 0$ have been arbitrary, we infer (12.3).

Now we can prove that

$$\left\| x - \sum_{i=1}^{n} \xi_i x_i \right\| = \| \{ \underbrace{0, \ldots, 0}_{n}, \xi_{n+1}, \xi_{n+2}, \ldots \} \| \to 0 \quad \text{as} \quad n \to \infty \quad (x = \{\xi_n\} \in d). \tag{12.7}$$

Indeed, by the above remark we have, for any $x = \{\xi_n\} \in d$,

$$\| \{ \underbrace{0, \ldots, 0}_{n}, \xi_{n+1}, \xi_{n+2}, \ldots \} \| = \sum_{i=1}^{\infty} \frac{|\xi_{n+p_i}^{**}|}{i},$$

where $\{\xi_{n+p_1}^{**}, \xi_{n+p_2}^{**}, \ldots\}$ is the rearrangement with $|\xi_{n+p_1}^{**}| \geqslant |\xi_{n+p_2}^{**}| \geqslant \cdots$ of the subsequence $\{\xi_{n+p_1}, \xi_{n+p_2}, \ldots\}$ consisting of all non-zero coordinates of the sequence $\{\xi_{n+1}, \xi_{n+2}, \ldots\}$. Let

$$l_n = \max_{\{\xi_{j_1}^*, \xi_{j_2}^*, \ldots, \xi_{j_i}^*\} \subset \{\xi_1, \ldots, \xi_n\}} i.$$

Then $0 \leqslant l_n \leqslant n$, $\{\xi_{n+p_1}, \xi_{n+p_2}, \ldots\} \subset \{\xi_{j_{l_n+1}}^*, \xi_{j_{l_n+2}}^*, \ldots\}$ and $|\xi_{n+p_i}^{**}| \leqslant |\xi_{j_{l_n+i}}^*|$ $(i = 1, 2, \ldots)$, whence

$$\| \{ \underbrace{0, \ldots, 0}_{n}, \xi_{n+1}, \xi_{n+2}, \ldots \} \| = \sum_{i=1}^{\infty} \frac{|\xi_{n+p_i}^{**}|}{i} \leqslant \sum_{i=1}^{\infty} \frac{|\xi_{j_{l_n}+i}^*|}{i}. \tag{12.8}$$

Now let $\varepsilon > 0$ be arbitrary. Choose $N = N(x, \varepsilon)$ such that

$$\sum_{i=N+1}^{\infty} \frac{|\xi_{j_i}^*|}{i} < \frac{\varepsilon}{2}.$$

Since $\lim_{i\to\infty}\xi^*_{j_i}=0$, for this N we can choose a positive integer M such that $|\xi^*_{j_{m+1}}|<\frac{\varepsilon}{2N}$ for all $m>M$, whence

$$\max_{1\leqslant i\leqslant N}\frac{|\xi^*_{j_{m+i}}|}{i}=|\xi^*_{j_{m+1}}|<\frac{\varepsilon}{2N}\qquad(m>M).$$

Thus for $m>M$ we have, taking into account that $|\xi^*_{j_{m+i}}|\leqslant|\xi^*_{j_i}|$,

$$\sum_{i=1}^{\infty}\frac{|\xi^*_{j_{m+i}}|}{i}=\sum_{i=1}^{N}\frac{|\xi^*_{j_{m+i}}|}{i}+\sum_{i=N+1}^{\infty}\frac{|\xi^*_{j_{m+i}}|}{i}$$
$$\leqslant\sum_{i=1}^{N}\frac{|\xi^*_{j_{m+i}}|}{i}+\sum_{i=N+1}^{\infty}\frac{|\xi^*_{j_i}|}{i}<N\frac{\varepsilon}{2N}+\frac{\varepsilon}{2}=\varepsilon,$$

whence, taking into account (12.8) and that $l_n\to\infty$ as $n\to\infty$, we infer (12.7). Thus, every $x=\{\xi_n\}\in d$ has an expansion $x=\sum_{i=1}^{\infty}\xi_i x_i$. Since this expansion is obviously unique, $\{x_n\}$ is a basis of d.

Let us prove that the basis $\{x_n\}$ is non-shrinking. Put $\lambda_n=\sum_{i=1}^{n}\frac{1}{i}$ $(n=1,2,\ldots)$ and define $y_n\in d$ by

$$y_n=\{\underbrace{\frac{1}{\lambda_n},\ldots,\frac{1}{\lambda_n}}_{n},0,0,\ldots\}=\frac{1}{\lambda_n}\{\underbrace{1,\ldots,1}_{n},0,0,\ldots\}\qquad(n=1,2,\ldots).\tag{12.9}$$

Then clearly

$$\|y_n\|=\frac{1}{\lambda_n}\sum_{i=1}^{n}\frac{1}{i}=1\qquad(n=1,2,\ldots),\tag{12.10}$$

and, if $\{f_n\}\subset d^*$ is the a.s.c.f. to $\{x_n\}$, we have

$$f_i(y_n)=\frac{1}{\lambda_n}\qquad(n=i,\,i+1,\,i+2,\ldots;\,i=1,2,\ldots),$$

whence

$$\lim_{n\to\infty}f_i(y_n)=0\qquad(i=1,2,\ldots).\tag{12.11}$$

Define now a linear functional f on d by

$$f(x)=\sum_{i=1}^{\infty}\frac{\xi_i}{i}\qquad(x=\{\xi_n\}\in d).\tag{12.12}$$

Then by (12.1) we have $|f(x)|=\left|\sum_{i=1}^{\infty}\frac{\xi_i}{i}\right|\leqslant\|x\|$ $(x=\{\xi_n\}\in d)$, whence $f\in d^*$, but

$$f(y_n)=\frac{1}{\lambda_n}\sum_{i=1}^{n}\frac{1}{i}=1\qquad(n=1,2,\ldots).\tag{12.13}$$

Consequently, by § 4, theorem 4.2 (implication $1^\circ \Rightarrow 3^\circ$), the basis $\{x_n\}$ is non-shrinking.

Finally, let us prove that $\{x_n\}$ is of type wc_0. Since $\|x_n\|=1$ $(n=1,2,\ldots)$, it remains to prove that $x_n \xrightarrow{w} 0$. Assume the contrary, i.e. that there exist an $f \in d^*$, an infinite subsequence $\{x_{n_k}\}$ of $\{x_n\}$, and an $\varepsilon > 0$ such that

$$f(x_{n_k}) > \varepsilon \qquad (k=1,2,\ldots).$$

Then for $x = \sum_{k=1}^{\infty} \frac{1}{k} x_{n_k} = \{\underbrace{0,\ldots,0}_{n_1-1}, 1, \underbrace{0,\ldots,0}_{n_2-n_1-1}, \frac{1}{2}, 0, \ldots\} \in d$ we have

$$f(x) = \sum_{k=1}^{\infty} \frac{1}{k} f(x_{n_k}) > \varepsilon \sum_{k=1}^{\infty} \frac{1}{k} = \infty,$$

contradicting $f \in d^*$. Thus, $x_n \xrightarrow{w} 0$, which completes the proof of the assertions of example 12.1.

Since the existence of non-shrinking bases of type wc_0 is rather unexpected, let us also mention, without proof, another example of such bases.

Example 12.2. Let W be the real linear space of all sequences of real numbers $\{\xi_n\}$ such that

$$A_k(\{\xi_n\}) = \sup \left| \sum_{j=1}^{k+1} \xi_{i_j} \right| \to 0 \quad \text{as} \quad k \to \infty, \tag{12.14}$$

where the sup is taken over all sequences of $k+1$ positive integers such that

$$k = i_1 < i_2 < \cdots < i_{k+1}. \tag{12.15}$$

Then W endowed with the norm

$$\|\{\xi_n\}\| = \sup_{1 \leq k < \infty} A_k(\{\xi_n\}) \tag{12.16}$$

is a Banach space and the sequence $\{x_n\} \subset W$ defined by

$$x_n = \{\delta_{nj}\}_{j=1}^{\infty} \qquad (n=1,2,\ldots) \tag{12.17}$$

is a basis of W, such that the a.s.c.f. $\{f_n\} \subset W^*$ is a non-shrinking basis of W^*, of type wc_0.

$wc_0 \not\Rightarrow$ Hilbertian, as shown by § 11, example 11.2.

$swc_0 \not\Rightarrow wc_0$. Indeed, by § 7, theorem 7.1 a), the Schauder basis of $C([0,1])$ is of type swc_0, but non-wc_0.

Non weakly closed $\not\Rightarrow swc_0$. Indeed, by § 8, theorem 8.1 and § 7, theorem 7.1 b), the normalized Haar basis of $L^1([0,1])$ is non weakly closed and of type non-swc_0.

Non-$l_+ \not\Rightarrow$ non-al_+ and non-$P^* \not\Rightarrow$ non-aP^*, as shown by

Example 12.3. Let $E=l^1$ and let $\{h_n\}$ be the basis of l^1 considered in Ch. I, § 13, example 13.3, formula (13.14). Then $\{h_n\}$ is of type non-l_+ (and hence also non-P^*) but aP^* (whence also al_+).

Indeed, $\{h_n\}$ is of type P, whence non-l_+. On the other hand, for $\Phi=\{1,0,1,0,1,0,\ldots\}\in m\equiv(l^1)^*$ we have

$$\Phi(h_1)=1, \quad \Phi(h_{2n})=1, \quad \Phi(h_{2n+1})=-1 \quad (n=1,2,\ldots), \tag{12.18}$$

whence, by the implication $4°\Rightarrow 1°$ of § 9, theorem 9.2b), the sequence $\{g_n\}\subset l^1$ defined by

$$g_1=h_1, \quad g_{2n}=h_{2n}, \quad g_{2n+1}=-h_{2n+1} \quad (n=1,2,\ldots) \tag{12.19}$$

is a basis of type P^* of the space l^1, and thus $\{h_n\}$ is of type aP^*.

Non-$P^* \not\Rightarrow$ non-l_+ and non-$aP^* \not\Rightarrow$ non-al_+, as shown by

Example 12.4. Let $E=c_0$, and let $\{y_n\}\subset E$ be the sequence defined by

$$y_{2n-1}=\sum_{i=1}^{2n-1} x_i, \quad y_{2n}=2\sum_{i=1}^{2n} x_i \quad (n=1,2,\ldots), \tag{12.20}$$

where $\{x_n\}$ is the natural basis of c_0. Then $\{y_n\}$ is a basis of E, of type l_+ (and hence also al_+) but non-aP^* (whence also non-P^*).

Indeed, since $\{x_n\}$ is a basis of type P of E (because of $\|x_n\|=1$, $\left\|\sum_{i=1}^{n} x_i\right\|=1$, $n=1,2,\ldots$), the sequence $\left\{\sum_{i=1}^{n} x_i\right\}$ is, by the implication $1°\Rightarrow 2°$ of § 9, theorem 9.1, a basis of E, and thus the sequence $\{y_n\}$ defined by (12.20) is a basis of E. Furthermore, if $\{f_n\}\subset E^*$ is the a.s.c.f. to $\{x_n\}$, then

$$f_1(y_{2n-1})=1, \quad f_1(y_{2n})=2 \quad (n=1,2,\ldots), \tag{12.21}$$

whence, by the implication $5°\Rightarrow 1°$ of § 10, theorem 10.1, $\{y_n\}$ is of type l_+. Finally, let us show that $\{y_n\}$ is of type non-aP^*. Since the a.s.c.f. $\{g_n\}\subset E^*$ to $\{y_n\}$ is

$$g_{2n-1}=f_{2n-1}-f_{2n}, \quad g_{2n}=\tfrac{1}{2}(f_{2n}-f_{2n+1}) \quad (n=1,2,\ldots), \tag{12.22}$$

we have, for any sequence of scalars $\{\varepsilon_n\}$ with $|\varepsilon_n|=1$ $(n=1,2,\ldots)$,

$$\begin{aligned}\sum_{j=1}^{2n}\varepsilon_j g_j &= \sum_{j=1}^{n}\left[\varepsilon_{2j-1}(f_{2j-1}-f_{2j})+\tfrac{1}{2}\varepsilon_{2j}(f_{2j}-f_{2j+1})\right]\\ &=\varepsilon_1 f_1+\sum_{j=1}^{n}\left(\tfrac{1}{2}\varepsilon_{2j}-\varepsilon_{2j-1}\right)f_{2j}+\sum_{j=2}^{n}\left(\varepsilon_{2j-1}-\tfrac{1}{2}\varepsilon_{2j-2}\right)f_{2j-1}\\ &\quad -\tfrac{1}{2}\varepsilon_{2n}f_{2n+1},\end{aligned}$$

whence, taking into account that $|\varepsilon_n|=1$ $(n=1,2,\ldots)$,

$$\left\|\sum_{j=1}^{2n}\varepsilon_j g_j\right\| = |\varepsilon_1| + \sum_{j=1}^{n}\left|\tfrac{1}{2}\varepsilon_{2j}-\varepsilon_{2j-1}\right| + \sum_{j=2}^{n}\left|\varepsilon_{2j-1}-\tfrac{1}{2}\varepsilon_{2j-2}\right| + \tfrac{1}{2}|\varepsilon_{2n}|$$

$$\geqslant |\varepsilon_1| + \sum_{j=1}^{n}\left|\tfrac{1}{2}|\varepsilon_{2j}|-|\varepsilon_{2j-1}|\right| + \sum_{j=2}^{n}\left||\varepsilon_{2j-1}|-\tfrac{1}{2}|\varepsilon_{2j-2}|\right| + \tfrac{1}{2}|\varepsilon_{2n}|$$

$$= 1 + \frac{n}{2} + \left(\frac{n}{2}-1\right) + \tfrac{1}{2} = \tfrac{1}{2} + n \to \infty \quad \text{as} \quad n\to\infty,$$

and thus $\{y_n\}$ is of type non-aP^*.

Non-$l_+ \not\Rightarrow (l_+)^*$. Indeed, by § 10, example 10.1, every basis of an infinite dimensional reflexive Banach space is of types non-l_+ and non-$(l_+)^*$.

The non-validity of the converses of the other implications of table 12.1 follows by taking the a.s.c.f. to the bases of the above examples[1] and applying the duality results proved in the preceding sections (e.g. the basis $\{x_n\}$ of example 12.2 is non-boundedly complete, but of type $(wc_0)^*$, etc.).

Finally, as we have observed in the above proof of part a), table 12.2 is equivalent to table 12.1, both by duality and by negation. Hence the examples given above prove the non-validity of the converses of the implications of table 12.2, too.

c) The examples given in the above proof of part b), together with the natural bases of c_0 and l^1, are sufficient to show that all the other relations between the 36 types of bounded bases considered in tables 12.1, and 12.2, except perhaps the 4 considered in the statement of part b), are: non-implication.

Indeed, in part b) above we have proved $144-4=140$ non-implications. Furthermore, the natural basis of c_0 is Hilbertian, shrinking and of type P, whence, by part a) proved above, it has all the 18 properties of table 12.1 and none of the 18 properties of table 12.2. Thus, no property of table 12.2 is implied by any property of table 12.1, which gives 324 new non-implications. Similarly, the natural basis of l^1 is Besselian, boundedly complete and of type P^*, which gives other 324 non-implications. On the other hand, the natural basis of l^2 is Besselian, Hilbertian, shrinking and boundedly complete, whence, by part a), it has all the 18 properties of the left halfs of tables 12.1 and 12.2, and none

[1] Let us remark that for this purpose, instead of the a.s.c.f. $\{\Phi_n\}\subset(l^1)^*\equiv m$ to the basis $\{h_n\}$ of l^1 considered in example 12.3 it is more convenient to take the basis $\left\{\sum_{i=1}^{n} x_i\right\}$ of c_0, where $\{x_n\}$ is the natural basis of c_0 (or, equivalently, the basic sequence $\{\Phi_1+\Phi_{n+1}\}$ in m). This basis leads to the same results, since its a.s.c.f. is $\{h_{n+1}\}$.

of the 18 properties of the right halfs of these tables. Thus no property of the right half of table 12.1 is implied by any property of the left half of this table, and similarly for table 12.2. This gives 162 non-implications, but 4 of them (namely, non-$l_+ \not\Rightarrow (l_+)^*$, non-$l_+ \not\Rightarrow P$ and the dual ones) have been proved in part b) and hence counted above, and thus we obtain only 158 new non-implications. Furthermore, the basis $\{h_n\}$ of l^1 considered in example 12.3 is of types P and aP^*, and dually, the basis $\left\{\sum_{i=1}^{n} x_i\right\}$ of c_0 (where $\{x_n\}$ is the natural basis of c_0) is of types P^* and aP, whence, by part a), no property of the left halfs of tables 12.1 and 12.2 is implied by any property of the right halfs of these tables, except the 8 implications ($P \Rightarrow$ non-l_+, etc.) given in these tables. This gives $162-8=154$ new non-implications. Since $\{h_n\}$ above is also a basis of type non-l_+ of l^1, and dually, $\left\{\sum_{i=1}^{n} x_i\right\}$ is a basis of type non-$(l_+)^*$ of c_0 (where $\{x_n\}$ is the natural basis of c_0), none of the properties occurring in the couples (non-aP^*, non-l_+), ($(l_+)^*$, aP), (non-aP, non-$(l_+)^*$), (l_+, aP^*) is implied by the other property of the respective couple, which gives 8 new non-implications. Similarly, the bases of example 12.1 and § 11, example 11.2 and the natural bases of c_0 and l^1 show that none of the properties occurring in the couples (Hilbertian, shrinking), (non boundedly complete, non-Besselian), (Besselian, boundedly complete), (non-shrinking, non-Hilbertian), is implied by the other property of the respective couple, which gives again 8 new non-implications. Indeed, we have only to show that the basis $\{x_n\}$ of d is Hilbertian (whence its a.s.c.f. $\{f_n\}$ is Besselian). Let $\sum_{i=1}^{\infty} |\xi_i|^2 < \infty$. By (12.3) we may assume (omitting, if necessary, those ξ_i which are 0 and rearranging the remaining sequence) that $|\xi_1| \geqslant |\xi_2| \geqslant \cdots$ Put

$$\mathcal{M}_1 = \left\{ i \in \mathcal{N} \;\middle|\; |\xi_i| \leqslant \frac{1}{i} \right\}, \quad \mathcal{M}_2 = \left\{ i \in \mathcal{N} \;\middle|\; |\xi_i| > \frac{1}{i} \right\}.$$

Then $\dfrac{|\xi_i|}{i} \leqslant \dfrac{1}{i^2}$ for $i \in \mathcal{M}_1$ and $< |\xi_i|^2$ for $i \in \mathcal{M}_2$, whence

$$\sum_{i=1}^{\infty} \frac{|\xi_i|}{i} \leqslant \sum_{i \in \mathcal{M}_1} \frac{|\xi_i|}{i} + \sum_{i \in \mathcal{M}_2} \frac{|\xi_i|}{i} < \sum_{i=1}^{\infty} \frac{1}{i^2} + \sum_{i=1}^{\infty} |\xi_i|^2 < \infty,$$

whence, by (12.3), $\{\xi_n\} \in d$, which proves that $\{x_n\}$ is Hilbertian.

Thus, we have obtained finally

$$140+324+324+158+154+8+8=1116$$

non-implications between the 36 types of bounded bases considered in tables 12.1 and 12.2. This completes the proof of theorem 12.1.

Remark 12.1. a) The types of bases considered in § 1, § 2 and § 3, i.e. those depending on the metric structure of the space E, have not been included in theorem 12.1 above. It is immediate that all relations between them and the types of bases considered in theorem 12.1, are: non-implication.

b) Actually, several implications of tables 12.1 and 12.2 have been already used in the preceding sections, especially in the proofs of the theorems concerning the Schauder basis of $C([0,1])$ and the normalized Haar basis of $L^1([0,1])$.

c) The implications of tables 12.1 and 12.2 establish certain connections between some of the problems formulated in the preceding sections (e.g. an affirmative answer to § 9, problem 9.1, would imply an affirmative answer to § 8, problem 8.1, etc.).

By the duality results given in § 8 and § 10, the 4 implications occurring in the statement of part b) of theorem 12.1 above are mutually equivalent. Let us also formulate separately the problem of their validity:

Problem 12.1[1]. Is every weakly closed basis $\{x_n\}$ of type al_+?

By the equivalence $1° \Leftrightarrow 2°$ of § 10, theorem 10.3, this problem is equivalent to the following: If $\{x_n\}$ is a basis of a Banach space E, such that 0 is not a weak limit point of the set $\{x_n\}$, does there exist an $f \in E^*$ such that $\inf_{1 \leq n < \infty} |f(x_n)| > 0$?

Let us observe that there exist, in finite-dimensional Banach spaces E, sequences (but not bases) $\{x_n\}$ satisfying

$$\inf_{1 \leq n < \infty} |f(x_n)| = 0 \qquad (f \in E^*),$$

for which 0 is not a weak limit point; such is e.g. any dense subsequence of the unit sphere $\sigma_E = \{x \in E \mid \|x\| = 1\}$.

In spite of the negative results of theorem 12.1 and of the unsolved problem 12.1, one can also give some related positive results, e.g. the following:

Theorem 12.2. *Let $\{x_n\}$ be a basis of a Banach space E. Then*

a) *The following statements are equivalent:*

1°. *$\{x_n\}$ is shrinking.*

2°. *$\{x_n\}$ has no block basic sequence $\{y_n\}$ of type*[2] *l_+.*

---

[1] Recently, this problem has been solved in the negative (see the footnote to theorem 12.1 b)).

[2] I.e. such that $\{y_n\}$ is a basis of type l_+ of $[y_n]$.

3°. *$\{x_n\}$ has no block basic sequence $\{y_n\}$ of type P^*.*

b) *The following statements are equivalent:*

1°. *$\{x_n\}$ is boundedly complete.*

2°. *$\{x_n\}$ has no block basic sequence $\{y_n\}$ of type P.*

Proof. a) 1°⇒2°. If $\{x_n\}$ is shrinking, then, by § 4, theorem 4.2 (implication 1°⇒5°), every bounded block basic sequence $\{y_n\}$ (with respect to $\{x_n\}$) is of type wc_0, whence non-l_+.

The implication 2°⇒3° is obvious, since non-$l_+ \Rightarrow$ non-P^*.

3°⇒1°. Assume that $\{x_n\}$ is non-shrinking. Then, by the implication 2°⇒1° of § 4, theorem 4.2, there exists an $f \in E^*$ such that

$$\lim_{n\to\infty} \|f\|_n \neq 0,$$

where $\|f\|_n$ is defined by (4.3). Consequently, there exist a number $\varepsilon > 0$, an increasing sequence of positive integers $\{m_n\}$ ($m_0 = 0$), a sequence of scalars $\{\alpha_n\}$ and a sequence of elements $\{z_n\} \subset E$ such that

$$z_n = \sum_{i=m_{n-1}+1}^{m_n} \alpha_i x_i \qquad (n=1,2,\ldots), \tag{12.23}$$

$$\|z_n\| = 1 \qquad (n=1,2,\ldots), \tag{12.24}$$

$$f(z_n) > \varepsilon \qquad (n=1,2,\ldots). \tag{12.25}$$

By (12.23) and (12.24), $\{z_n\}$ is a block basic sequence with respect to $\{x_n\}$. Hence, by (12.25) and the implication 5°⇒6° of § 10, theorem 10.1, there exists a sequence of scalars $\{\beta_n\}$ satisfying (10.6) and such that $\{y_n\} = \{\beta_n z_n\}$ is a basis of type P^* of $[z_n] = [y_n]$. By (12.23), (12.24) and (10.6), $\{y_n\} = \{\beta_n z_n\}$ is also a block basic sequence with respect to $\{x_n\}$.

b) 1°⇒2°. Assume that $\{x_n\}$ admits a block basic sequence $y_n = \sum_{i=m_{n-1}+1}^{m_n} \alpha_i x_i$ ($n=1,2,\ldots$; $m_0 = 0$) of type P, i.e. such that $\inf_{1\leqslant n<\infty} \|y_n\| > 0$ and $\sup_{1\leqslant n<\infty} \left\|\sum_{l=1}^{n} y_l\right\| = M < \infty$. Then for any positive integer k and any n such that $m_n > k$ we have $\left\|\sum_{i=1}^{k} \alpha_i x_i\right\| \leqslant v_{\{x_j\}} \left\|\sum_{i=1}^{m_n} \alpha_i x_i\right\| = v_{\{x_j\}} \left\|\sum_{l=1}^{n} y_l\right\|$, whence

$$\sup_{1\leqslant k<\infty} \left\|\sum_{i=1}^{k} \alpha_i x_i\right\| \leqslant v_{\{x_j\}} M < \infty,$$

but $\sum_{i=1}^{\infty} \alpha_i x_i$ does not converge, since $\inf_{1\leqslant n<\infty} \left\|\sum_{i=m_{n-1}+1}^{m_n} \alpha_i x_i\right\| = \inf_{1\leqslant n<\infty} \|y_n\| > 0$, and thus $\{x_n\}$ is not boundedly complete.

$2^\circ \Rightarrow 1^\circ$. Assume that $\{x_n\}$ is not boundedly complete. Then there exists a sequence of scalars $\{\alpha_n\}$ such that

$$\sup_{1 \leqslant n < \infty} \left\| \sum_{i=1}^{n} \alpha_i x_i \right\| < \infty \tag{12.26}$$

and that $\sum_{i=1}^{\infty} \alpha_i x_i$ is not convergent. Since $\sum_{i=1}^{\infty} \alpha_i x_i$ is not convergent, there exists an increasing sequence of positive integers $\{m_n\}$ $(m_0 = 0)$, such that

$$\inf_{1 \leqslant n < \infty} \left\| \sum_{i=m_{n-1}+1}^{m_n} \alpha_i x_i \right\| > 0. \tag{12.27}$$

Put

$$y_n = \sum_{i=m_{n-1}+1}^{m_n} \alpha_i x_i \quad (n = 1, 2, \ldots). \tag{12.28}$$

Then, by (12.28) and (12.27), $\{y_n\}$ is a block basic sequence (with respect to $\{x_n\}$), which is, by (12.27) and (12.26), of type P. This completes the proof of theorem 12.2.

Combining theorem 12.2 and part a) of theorem 12.1, it follows

Corollary 12.1. *Every basis of a type T occurring in the right halfs of tables 12.1 and 12.2, except non-Besselian and non-Hilbertian bases, admits a block basic sequence $\{y_n\}$ of any of the types situated above T in the same half-tables.*

From corollary 12.1 we obtain, in particular, the following result related to problem 12.1: *every weakly closed basis $\{x_n\}$ of a Banach space E admits a block basic sequence $\{y_n\}$ of type al_+.* However, we also have the following slightly stronger result:

Proposition 12.1. *Every weakly closed basis $\{x_n\}$ of an infinite dimensional space E has an infinite subsequence $\{x_{i_n}\}$ of type al_+.*

Proof. Let $\{x_n\}$ be a weakly closed basis of a Banach space E. Then 0 is not a weak limit point of the set $\{x_n\}$, whence there exists a w-neighbourhood $V = V_{g_1, \ldots, g_m; \varepsilon}(0)$ (where $g_1, \ldots, g_m \in E^*$, $\varepsilon > 0$) such that $V \not\ni x_n$ $(n = 1, 2, \ldots)$, i.e. such that

$$\max_{1 \leqslant j \leqslant m} |g_j(x_n)| \geqslant \varepsilon \quad (n = 1, 2, \ldots).$$

Consequently, there exist a j_0 with $1 \leqslant j_0 \leqslant m$ and an infinite subsequence $\{x_{i_n}\}$ of $\{x_n\}$, such that

$$|g_{j_0}(x_{i_n})| \geqslant \varepsilon \quad (n = 1, 2, \ldots).$$

Hence, by the implication $2^\circ \Rightarrow 1^\circ$ of § 10, theorem 10.3, $\{x_{i_n}\}$ is a basis of type al_+ of $[x_{i_n}]$. This completes the proof.

Proposition 12.1 suggests the question, whether we can replace in corollary 12.1 above "block basic sequence" by "subsequence", at least for the first type situated above T. The answer is negative, as shown by example 12.1. Indeed, the basis $\{x_n\} \subset d$ of this example is non-shrinking, but every subsequence $\{x_{i_n}\}$ of $\{x_n\}$ is of type $w c_0$.

It is natural to ask whether corollary 12.1 remains valid if we replace in its formulation "the right halfs of tables 12.1 and 12.2" by "the left halfs of tables 12.1 and 12.2". However the answer is negative. Indeed, the basis $\{h_n\}$ of l^1 considered in example 12.3 is of type non-P^*, but admits no shrinking block basic sequence since the conjugate space of every subspace of l^1 is non-separable. Moreover, $\{h_n\}$ admits no block basic sequence of type $w c_0$ or $s w c_0$, since in every subspace of l^1 weak convergence of sequences implies their convergence in the norm-topology. Dually, the basis $\left\{\sum_{i=1}^{n} x_i\right\}$ of c_0 (where $\{x_n\}$ is the natural basis of c_0), is of type non-P, but admits no boundedly complete block basic sequence, since no subspace of c_0 is isomorphic to any conjugate Banach space, i.e. every subspace of c_0 satisfies the condition of § 6, example 6.2 (with $k=0$). Moreover, $\left\{\sum_{i=1}^{n} x_i\right\}$ admits no block basic sequence of type $(w c_0)^*$ or $(s w c_0)^*$, by virtue of the above remarks concerning subspaces of l^1.

In spite of these negative results, one can also give some positive results concerning the left halfs of tables 12.1 and 12.2, e.g. we have seen in § 7, proposition 7.1 b) that every basis $\{x_n\}$ of type $s w c_0$ contains a subsequence $\{x_{i_n}\}$ of type $w c_0$.

Finally, let us also mention the following natural question: if a Banach space E has a basis $\{x_n\}$ of a certain type T occurring in table 12.1 or table 12.2, what other types of bases does E possess? Theorem 12.1 a) can be also interpreted as a partial answer to this question. Another result in this sense is the following:

Proposition 12.2. *For a Banach space E the following statements are equivalent:*

1°. *E has a basis of type P.*
2°. *E has a basis of type P*.*
3°. *E has a basis of type aP.*
4°. *E has a basis of type aP^*.*
5°. *E has a basis of type l_+.*
6°. *E has a basis of type $(l_+)^*$.*
7°. *E has a basis of type al_+.*
8°. *E has a basis of type $(al_+)^*$.*

Proof. The equivalences $1°\Leftrightarrow 3°$, $2°\Leftrightarrow 4°$, $5°\Leftrightarrow 7°$ and $6°\Leftrightarrow 8°$ are immediate consequences of § 9, definition 9.1 and § 10, definition 10.1. The equivalence $2°\Leftrightarrow 5°$ follows from the equivalence $1°\Leftrightarrow 7°$ of § 10, theorem 10.1; the equivalence $1°\Leftrightarrow 6°$ follows from this by duality. Finally, the equivalence $1°\Leftrightarrow 2°$ follows from the implication $1°\Leftrightarrow 2°$ of § 9, theorem 9.1 and the implication $1°\Leftrightarrow 7°$ of § 9, theorem 9.2b), since (9.2) is of type P^* and (9.14) of type P. This completes the proof.

§ 13. Universal bases. Complementably universal bases. Block-universal bases

We have seen in Ch. I, § 4, proposition 4.1 a) that every subsequence $\{x_{i_n}\}$ of a basis $\{x_n\}$ is a basic sequence.

Definition 13.1. Let $\mathscr{B}$ be a family of bases. A basis $\{e_n\}$ of a Banach space F is said to be *universal* for $\mathscr{B}$ if every basis in $\mathscr{B}$ is equivalent to a subsequence of $\{e_n\}$.

For instance, every basis $\{x_n\}$ is universal for any family $\mathscr{B}$ of subsequences of $\{x_n\}$ (hence in particular, for the family of all subsequences of $\{x_n\}$).

Definition 13.2. A subsequence $\{x_{i_n}\}$ of a basis $\{x_n\}$ of a Banach space E is said to be *complemented* if for any sequence of scalars $\{\alpha_n\}$ the convergence of the series $\sum_{i=1}^{\infty} \alpha_i x_i$ implies the convergence of the series $\sum_{n=1}^{\infty} \alpha_{i_n} x_{i_n}$ (or, in other words, if $\sum_{n=1}^{\infty} f_{i_n}(x) x_{i_n}$ converges for every $x \in E$, where $\{f_n\} \subset E^*$ is the a.s.c.f. to $\{x_n\}$).

Remark 13.1. If $\{x_{i_n}\}$ is a complemented subsequence of a basis $\{x_n\}$ of a Banach space E, then $[x_{i_n}]$ is a complemented subspace of E, namely, $E=[x_{i_n}]\oplus[x_{l_n}]$, where $\{l_n\}$ is the set of indices complementary to $\{i_n\}$, since the mapping $u\colon E\to E$ defined by

$$u(x)=\sum_{n=1}^{\infty} f_{i_n}(x) x_{i_n} \qquad (x\in E) \tag{13.1}$$

is a well defined continuous linear projection of E onto $[x_{i_n}]$ (by the Banach-Steinhaus theorem). However, *the converse is not true*, since e.g. for a conditional basis[1] $\{x_n\}$ of the Hilbert space $E=l^2$ there exists,

[1] See § 14, examples 14.4 and 14.5.

by § 16, lemma 16.1, implication $1^\circ \Rightarrow 3^\circ$, a convergent series $\sum_{i=1}^{\infty} \alpha_i x_i \in E$ having a non-convergent subseries $\sum_{n=1}^{\infty} \alpha_{i_n} x_{i_n}$, but the subspace $[x_{i_n}]$ is complemented.

Definition 13.3. Let $\mathscr{B}$ be a family of bases. A basis $\{e_n\}$ of a Banach space E is said to be *complementably universal* for $\mathscr{B}$ if every basis in $\mathscr{B}$ is equivalent to a complemented subsequence of $\{e_n\}$.

For instance, by § 16, lemma 16.1, implication $1^\circ \Rightarrow 3^\circ$, every unconditional basis $\{x_n\}$ is complementably universal for any family $\mathscr{B}$ of subsequences of $\{x_n\}$.

Our next aim is to prove that the family $\mathscr{B}$ of all bounded bases contains a complementably universal element. For this purpose we need some preparation. For the sake of simplicity we shall consider only real scalars.

We shall use the following notations:

$R^\infty =$ the linear space of all real sequences $t = \{t(1), t(2), \ldots\}$,

$\pi_n =$ the mapping of R^∞ into R^∞ defined by

$$[\pi_n(t)](i) = \begin{cases} t(i) & \text{for} \quad i = 1, \ldots, n \\ 0 & \text{for} \quad i = n+1, n+2, \ldots \end{cases} \tag{13.2}$$

$R^n = \pi_n(R^\infty) =$ the linear space of all sequences $\{t(1), \ldots, t(n), 0, 0, \ldots\}$,

$R =$ the real line,

$$I^\infty = \{t \in R^\infty \,|\, |t(i)| \leqslant 1 \ (i = 1, 2, \ldots)\}, \tag{13.3}$$

$$I^n = I^\infty \cap R^n = \{t \in R^n \,|\, |t(i)| \leqslant 1 \ (i = 1, \ldots, n)\}, \tag{13.4}$$

$$\partial I^n = \{t \in I^n \,|\, \max_{1 \leqslant i \leqslant n} |t(i)| = 1\}, \tag{13.5}$$

$e_n =$ the element of R^∞ defined by

$$e_n(i) = \begin{cases} 0 & \text{for} \quad i = 1, \ldots, n-1, n+1, \ldots \\ 1 & \text{for} \quad i = n. \end{cases} \tag{13.6}$$

Furthermore, we shall denote by B_n (respectively B) the set of all non-negative functions $p(\cdot)$ on R^n (respectively, on $\bigcup_{n=1}^{\infty} R^n$) satisfying the following conditions:

(i) $p(t) = 0$ if and only if $t = 0$; $p(ct) = |c| p(t); p(t+s) \leqslant p(t) + p(s)$ for all $t, s \in R^n$ (respectively, $t, s \in \bigcup_{n=1}^{\infty} R^n$) and $c \in R$;

(ii) $p(e_i) = 1$ for $i = 1, \ldots, n$ (respectively, for $i = 1, 2, \ldots$);

(iii) $\max_{1 \leqslant i < \infty} |s(i)| \leqslant p(s)$ for all $s \in R^n$ (respectively, $s \in \bigcup_{n=1}^{\infty} R^n$);

(iv) $p(\pi_m(s)) \leqslant p(s)$ for $s \in R^n$ and $m=1,.,..,n$ (respectively, for $s \in \bigcup_{n=1}^{\infty} R^n$ and $m=1,2,\ldots$).

The elements of B_n and B will be called *norms*.

For $n=1,2,\ldots$ and $p,q \in B_n$ we shall put

$$d_n(p,q) = \log\left(\sup_{t \in \partial I^n} \frac{p(t)}{q(t)} \cdot \sup_{t \in \partial I^n} \frac{q(t)}{p(t)}\right) = \log \sup_{t \in \partial I^n} \frac{p(t)}{q(t)} - \log \inf_{t \in \partial I^n} \frac{p(t)}{q(t)}. \tag{13.7}$$

Lemma 13.1. *d_n is a metric on B_n. The metric space (B_n, d_n) is compact.*

Proof. Obviously, $d_n(p,p)=0$. Conversely, if $d_n(p,q)=0$, then $\frac{p(t)}{q(t)} = \lambda = \text{const.}$ ($t \in R^n$, $t \neq 0$) whence, by (ii), $\lambda = \frac{p(e_1)}{q(e_1)} = 1$, whence $p=q$. The relations $d_n(p,q)=d_n(q,p)$ and $d_n(p,q) \leqslant d_n(p,r)+d_n(r,q)$ are obvious.

To prove the second statement, put

$$\hat{p} = p|_{\partial I^n} \qquad (p \in B_n), \tag{13.8}$$

$$\hat{B}_n = \{\hat{p} \mid p \in B_n\}. \tag{13.9}$$

Then $\hat{B}_n$ consists of functions which are uniformly bounded on the compact space ∂I^n, since

$$\hat{p}(t) = p(t) = p\left(\sum_{i=1}^{n} t(i) e_i\right) \leqslant \sum_{i=1}^{n} p(t(i) e_i)$$

$$= \sum_{i=1}^{n} |t(i)| p(e_i) = \sum_{i=1}^{n} |t(i)| \leqslant n \qquad (t \in \partial I^n),$$

and equicontinuous on ∂I^n, since

$$|p(s)-p(t)| \leqslant p(s-t) \leqslant \sum_{i=1}^{n} |s(i)-t(i)| \leqslant n \max_{1 \leqslant i \leqslant n} |s(i)-t(i)| \qquad (s,t \in \partial I^n).$$

Furthermore, $\hat{B}_n$ is closed in $C(\partial I^n)$, since if a sequence $\{\hat{p}_k\} \subset \hat{B}_n$ converges uniformly to a function $f \in C(\partial I^n)$, then $f = \hat{p}$, where $p \in B_n$ is defined by

$$p(0)=0,\ p(s) = \max_{1 \leqslant i \leqslant n} |s(i)|\, f\left(\frac{s}{\max_{1 \leqslant i \leqslant n} |s(i)|}\right) \qquad (s \in R^n,\ s \neq 0). \tag{13.10}$$

Indeed, since $\max_{1 \leqslant i \leqslant n} |s(i)| = 1$ for $s \in \partial I^n$, we have $\hat{p}(s) = p(s) = f(s)$ for $s \in \partial I^n$, i.e. $\hat{p} = f$. Furthermore, let us show that $p \in B_n$. Since $p_k \in B_n$ and

$$\max_{s \in \partial I^n} |p_k(s) - p(s)| \to 0 \quad \text{as} \quad k \to \infty, \tag{13.11}$$

it follows that $p(s) \geqslant 0$ for all $s \in \partial I^n$, whence also for all $s \in R^n$. Assume now that $p(s)=0$. Then, by (iii) for p_k and (13.11),

$$\max_{1 \leqslant i < \infty} |s(i)| \leqslant p_k(s) < \varepsilon \qquad (k > N(\varepsilon)),$$

whence $s=0$. By (13.11) and (i)–(iv) for p_k, we have also (i)–(iv) for p, i. e. $p \in B_n$. Thus, $\hat{B}_n$ is closed in $C(\partial I^n)$.

Consequently, by the theorem of Arzelà, $\hat{B}_n$ is a compact set in $C(\partial I^n)$. We shall show that $\hat{B}_n$ is homeomorphic to B_n, by the mapping $\hat{p} \to p$, which will complete the proof. If $\hat{p}_1 = \hat{p}_2$, i. e. $p_1|_{\partial I^n} = p_2|_{\partial I^n}$, where $p_1, p_2 \in B^n$, then by the homogeneity of the norms p_1, p_2 we have $p_1 = p_2$, and thus the mapping $\hat{p} \to p$ is well defined. Furthermore, it is obviously one-to-one and maps $\hat{B}_n$ onto B_n. Finally, it is continuous, whence a homeomorphism, since $\max_{t \in \partial I^n} |p_k(t) - p(t)| \to 0$ implies

$$\left| \frac{p_k(t)}{p(t)} - 1 \right| < \frac{\varepsilon}{p(t)} \leqslant \frac{\varepsilon}{\inf_{s \in \partial I^n} p(s)} \qquad (t \in \partial I^n, k > N(\varepsilon)),$$

whence $d_n(p_k, p) \to 0$ as $k \to \infty$. This completes the proof of lemma 13.1.

For a fixed index n and for $p \in B_m$ with $m \geqslant n$ or $p \in B$, we shall put

$$J_n(p) = p|_{R^n}. \tag{13.12}$$

Lemma 13.2. *The restriction operator J_n has the following properties:*

a) *If $m \geqslant n$, then*

$$d_n(J_n(p), J_n(q)) \leqslant d_m(p, q) \qquad (p, q \in B_m). \tag{13.13}$$

b) *If $p \in B_{n+1}$ and $\tilde{q} \in B_n$, then there exists a $q \in B_{n+1}$ such that*

$$\tilde{q} = J_n(q), \tag{13.14}$$

$$d_n(J_n(p), J_n(q)) = d_{n+1}(p, q). \tag{13.15}$$

Hence, in particular,

$$J_n(B_{n+1}) = B_n. \tag{13.16}$$

Proof. a) If $m \geqslant n$, then $\partial I^m \supset \partial I^n$, whence, by the definition (13.7) of d_n, we infer (13.13).

b) Put

$$a = \inf_{t \in \partial I^n} \frac{p(t)}{\tilde{q}(t)}, \qquad b = \sup_{t \in \partial I^n} \frac{p(t)}{\tilde{q}(t)}. \tag{13.17}$$

Then, since by (ii) $p(e_1) = \tilde{q}(e_1) = 1$ and since all norms on R^n are equivalent, we have $b \geqslant 1 \geqslant a > 0$. Furthermore, obviously

$$d_n(J_n(p), \tilde{q}) = \log b - \log a, \tag{13.18}$$

$$\frac{p(t)}{b} \leqslant \tilde{q}(t) \leqslant \frac{p(t)}{a} \qquad (t \in R^n). \tag{13.19}$$

Put

$$\tilde{Q} = \{t \in R^n \,||\tilde{q}(t)| \leqslant 1\}, \tag{13.20}$$

$$P = \{s \in R^{n+1} \,||p(s)| \leqslant 1\}, \tag{13.21}$$

$$Q = \mathrm{co}(\tilde{Q}, aP, \pm e_{n+1}), \tag{13.22}$$

where co means "convex hull". Let q be the Minkovski functional of Q, i. e.

$$q(s) = \inf_{\substack{\lambda > 0 \\ \frac{1}{\lambda} s \in Q}} \lambda \qquad (s \in R^{n+1}). \tag{13.23}$$

We shall complete the proof by showing that q has the required properties, i. e. $q \in B_{n+1}$ and satisfies (13.14), (13.15).

Obviously, (13.19) is equivalent to the inclusions

$$bP \cap R^n \supset \tilde{Q} \supset aP \cap R^n. \tag{13.24}$$

Consequently, we have

$$Q \cap R^n = \tilde{Q}, \qquad bP \supset Q \supset aP, \tag{13.25}$$

whence, in particular, Q is a neighbourhood of 0 in R^{n+1}. Since Q is symmetric and convex, it follows that q satisfies condition (i) of the norm.

Furthermore, by (13.25), we have (13.14) and

$$\frac{p(s)}{b} \leqslant q(s) \leqslant \frac{p(s)}{a} \qquad (s \in R^{n+1}), \tag{13.26}$$

whence

$$d_{n+1}(p,q) = \log\left(\sup_{s \in \partial I^{n+1}} \frac{p(s)}{q(s)} \cdot \sup_{s \in \partial I^{n+1}} \frac{q(s)}{p(s)}\right) \leqslant \log\frac{b}{a} = \log b - \log a,$$

whence, by (13.18), (13.14) and (13.13), we infer (13.15).

Since $\tilde{q}$ and p satisfy (iii), we have $\tilde{Q} \subset I^n \subset I^{n+1}$, $aP \subset P \subset I^{n+1}$, whence, taking into account $\pm e_{n+1} \in I^{n+1}$, it follows that $Q \subset I^{n+1}$, and therefore q satisfies (iii).

Since $e_{n+1} \in Q$ and $ce_{n+1} \notin Q$ for $|c| > 1$ (because $Q \subset I^{n+1}$ by (iii)), we have $q(e_{n+1}) = 1$. Furthermore, since $\tilde{q}$ satisfies (ii), we have $q(e_i) = \tilde{q}(e_i) = 1$ for $i = 1, \ldots, n$. Thus, q satisfies (ii).

Finally, let us show that q satisfies (iv). Put

$$L_{m+1} = \{r \in R^{m+1} \,|\, \pi_m(r) = 0\} \qquad (m = 1, 2, \ldots). \tag{13.27}$$

Then, since p satisfies (iv), we have[1]

$$(P \cap R^n) + L_{n+1} \supset P;$$

[1] We use the notation $A + B = \{a + b \,|\, a \in A, b \in B\}$.

indeed, if $s \in P$, then $p(s) \leqslant 1$ and $s = \pi_n(s) + r$, where $\pi_n(s) \in R^n$, $r \in L_{n+1}$, and by (iv) for p we have $p(\pi_n(s)) \leqslant p(s) \leqslant 1$, whence $\pi_n(s) \in P \cap R^n$, whence $s \in (P \cap R^n) + L_{n+1}$. Hence, taking into account (13.24),

$$\tilde{Q} + L_{n+1} \supset (aP \cap R^n) + L_{n+1} \supset aP.$$

Since obviously $\tilde{Q} + L_{n+1} \supset \tilde{Q}$ and $\tilde{Q} + L_{n+1} \ni \pm e_{n+1}$, it follows that $\tilde{Q} \oplus L_{n+1} \supset Q$, whence the relations $s \in R^{n+1}$, $q(s) \leqslant 1$ imply $\tilde{q}(\pi_n(s)) \leqslant 1$. Consequently,

$$q(s) \geqslant \tilde{q}(\pi_n(s)) \qquad (s \in R^{n+1}). \tag{13.28}$$

Combining (13.28) with the assumption that $\tilde{q}$ satisfies (iv), we obtain, for any m with $1 \leqslant m \leqslant n$,

$$q(s) \geqslant \tilde{q}(\pi_n(s)) \geqslant \tilde{q}(\pi_m \pi_n(s)) = \tilde{q}(\pi_m(s)) = q(\pi_m(s)) \qquad (s \in R^{n+1}).$$

Thus, q satisfies (iv), which completes the proof of lemma 13.2.

Lemma 13.3. *Let* $\varepsilon > 0$. *Then there exists a sequence* $\{A_n\}$ *such that for each* $n = 1, 2, \ldots$

α) A_n *is a finite subset of* B_n;

β) A_n *is an* $\varepsilon \sum_{k=1}^{n} \frac{1}{2^k}$-*net (hence, in particular, an* ε-*net) for* B_n;

γ) $J_{n-1}(A_n) = A_{n-1}$.

Proof. Put $A_1 = B_1$ (observe that B_1 is a one-point set, namely, the norm $p(\alpha) = |\alpha|\,(\alpha \in R)$). Suppose that for some $m \geqslant 1$ we have already defined m sets $A_1, \ldots, A_m$ satisfying α)–γ). By virtue of lemma 13.1, let F be a finite $\frac{\varepsilon}{2^{m+1}}$-net for B_{m+1}. By lemma 13.2b), for each pair $(p, \tilde{q})$ with $p \in F$, $\tilde{q} \in A_m$ there exists a norm $q = q(p, \tilde{q}) \in B_{m+1}$ such that $J_m(q) = \tilde{q}$ and $d_{m+1}(p, q) = d_m(J_m(p), \tilde{q})$. We shall show that the set

$$A_{m+1} = \{q = q(p, \tilde{q}) \mid p \in F, \tilde{q} \in A_m\} \tag{13.29}$$

has the required properties α)–γ), which will complete the induction and the proof of lemma 13.3. Since F and A_m are finite, α) is obvious by the definition of A_{m+1}. Furthermore, let $p' \in B_{m+1}$ be arbitrary. Then, since F is an $\frac{\varepsilon}{2^{m+1}}$-net for B_{m+1}, there exists a $p \in F$ such that $d_{m+1}(p', p) < \frac{\varepsilon}{2^{m+1}}$. Since $J_m(p) \in B_m$ and since A_m is an $\varepsilon \sum_{k=1}^{m} \frac{1}{2^k}$-net for B_m, there exists a $\tilde{q} \in A_m$ such that $d_m(J_m(p), \tilde{q}) < \varepsilon \sum_{k=1}^{m} \frac{1}{2^k}$. By lemma 13.2b), there exists a $q = q(p, \tilde{q}) \in A_{m+1}$ such that $J_m(q) = \tilde{q}$ and

$d_{m+1}(p,q)=d_m(J_m(p),\tilde{q})$. Hence

$$d_{m+1}(p',q)\leqslant d_{m+1}(p',p)+d_{m+1}(p,q)<\varepsilon\sum_{k=1}^{m+1}\frac{1}{2^k},$$

and thus we have β) for A_{m+1}. Finally, since for any $\tilde{q}\in A_m$ we have $J_m(q(p,\tilde{q}))=\tilde{q}$ with $q(p,\tilde{q})\in A_{m+1}$ (where $p\in F$ is arbitrary), it follows that $J_m(A_{m+1})=A_m$, which completes the proof of lemma 13.3.

Lemma 13.4. *Let* $1\leqslant k\leqslant n$ *and let* $\tilde{p}\in B_n$, $q\in B_{k+1}$. *Suppose that there exists a sequence* $1\leqslant i_1<i_2<\cdots<i_k<i_{k+1}=n+1$ *such that*

$$q(t)=\tilde{p}\Big(\sum_{j=1}^{k}t(j)e_{i_j}\Big)\qquad(t\in R^k),\tag{13.30}$$

$$\tilde{p}(s)\geqslant\tilde{p}\Big(\sum_{j=1}^{k}s(i_j)e_{i_j}\Big)\qquad(s\in R^n).\tag{13.31}$$

Then there exists a norm $p\in B_{n+1}$ *such that* $J_n(p)=\tilde{p}$ *and*

$$q(t)=p\Big(\sum_{j=1}^{k+1}t(j)e_{i_j}\Big)\qquad(t\in R^{k+1}),\tag{13.32}$$

$$p(s)\geqslant p\Big(\sum_{j=1}^{k+1}s(i_j)e_{i_j}\Big)\qquad(s\in R^{n+1}).\tag{13.33}$$

Proof. Put

$$F(r,s,c)=\tilde{p}\Big(s-\sum_{j=1}^{k}r(j)e_{i_j}\Big)+q(r+ce_{k+1})\qquad(r\in R^k,s\in R^n,c\in R),\tag{13.34}$$

$$p(s+ce_{n+1})=\inf_{r\in R^k}F(r,s,c)\qquad(s\in R^n,c\in R).\tag{13.35}$$

We shall show that p has the required properties, which will complete the proof.

Observe that p satisfies (i), since it may be regarded as the natural norm in the quotient space X/Y, where $X=R^n\times R^{k+1}$ endowed with the norm $\|\{s,t\}\|=\tilde{p}(s)+q(t)$ and

$$Y=\Big\{\{s,r\}\in X\,\Big|\,s+\sum_{j=1}^{k}r(j)e_{i_j}=0,\ r(k+1)=0\Big\},$$

via the correspondence $s+ce_{n+1}\to\widehat{\{s,ce_{k+1}\}}$. Indeed,

$$\begin{aligned}\|\widehat{\{s,ce_{k+1}\}}\|&=\inf_{\{s_1,r_1\}\in Y}\|\{s+s_1,\ r_1+ce_{k+1}\}\|\\&=\inf_{\substack{s_1=-\sum_{j=1}^{k}r_1(j)e_{i_j}\\ r_1\in R^k}}\tilde{p}(s+s_1)+q(r_1+ce_{k+1})=\inf_{r_1\in R^k}F(r_1,s,c).\end{aligned}$$

Furthermore, we have $J_n(p)=\tilde{p}$, i.e.

$$p(s)=\tilde{p}(s) \qquad (s\in R^n). \tag{13.36}$$

Indeed, obviously $p(s)\leqslant F(0,s,0)=\tilde{p}(s)$. On the other hand, for any $r\in R^k$ we have, by (13.30),

$$F(r,s,0)=\tilde{p}\left(s-\sum_{j=1}^{k} r(j)e_{i_j}\right)+q(r)\geqslant\tilde{p}(s)-\tilde{p}\left(\sum_{j=1}^{k} r(j)e_{i_j}\right)+q(r)=\tilde{p}(s),$$

whence $p(s)=\inf_{r\in R^k} F(r,s,0)\geqslant\tilde{p}(s)$, and thus we obtain (13.36).

Let us show now (13.32). If $s=\sum_{j=1}^{k} t(j)e_{i_j}$ for some $t\in R^k$, then for any $c\in R$ we have, taking into account that $i_{k+1}=n+1$,

$$p\left(\sum_{j=1}^{k} t(j)e_{i_j}+ce_{i_{k+1}}\right)=p(s+ce_{n+1})\leqslant F(t,s,c)=q(t+ce_{k+1}).$$

On the other hand, for any $r\in R^k$ we have, by (13.30) applied to the element $t-r\in R^k$,

$$\begin{aligned} F\left(r,\sum_{j=1}^{k} t(j)e_{i_j},c\right)&=\tilde{p}\left(\sum_{j=1}^{k}(t-r)(j)e_{i_j}\right)+q(r+ce_{k+1})\\ &=q(t-r)+q(r+ce_{k+1})\geqslant q(t+ce_{k+1}), \end{aligned}$$

whence $p\left(\sum_{j=1}^{k} t(j)e_{i_j}+ce_{i_{k+1}}\right)\geqslant q(t+ce_{k+1})$, and thus we obtain (13.32) (with $t(k+1)=c$).

Taking now in (13.32) $t(1)=\cdots=t(k)=0$, $t(k+1)=1$, we obtain $p(e_{n+1})=p(e_{i_{k+1}})=q(e_{k+1})=1$, because q satisfies (ii). Since $\tilde{p}$ satisfies (ii), we also have, by (13.36), $p(e_m)=\tilde{p}(e_m)=1$ $(m=1,\ldots,n)$, and thus p satisfies (ii).

Furthermore, since q satisfies (iv), we have $q(r+ce_{k+1})\geqslant q(r)$ $(r\in R^k$, $c\in R)$, whence

$$p(s+ce_{n+1})=\inf_{r\in R^k} F(r,s,c)\geqslant\inf_{r\in R^k} F(r,s,0)=p(s)=\tilde{p}(s) \qquad (s\in R^n,\ c\in R). \tag{13.37}$$

Combining (13.37) with the assumption that $\tilde{p}$ satisfies (iv), we obtain, for any m with $1\leqslant m\leqslant n$,

$$\begin{aligned} p(s+ce_{n+1})&\geqslant\tilde{p}(s)=\tilde{p}(\pi_n(s+ce_{n+1}))\geqslant\tilde{p}(\pi_m\pi_n(s+ce_{n+1}))\\ &=\tilde{p}(\pi_m(s+ce_{n+1}))=p(\pi_m(s+ce_{n+1})), \end{aligned}$$

and thus p satisfies (iv).

Assume now that $p(s+ce_{n+1})\leqslant 1$ for some $s\in R^n$, $c\in R$. Then, by (13.37), $\tilde{p}(s)\leqslant 1$, whence, since $\tilde{p}$ satisfies (iii), $s\in I^n$. On the other hand, by the definition of p, for every $\varepsilon>0$ there exists an $r=r_\varepsilon\in R^k$ such that $q\left(\sum_{j=1}^{k} r(j)e_j+ce_{k+1}\right)\leqslant 1+\varepsilon$, whence, since q satisfies (iii), $|c|\leqslant 1+\varepsilon$, whence $|c|\leqslant 1$. Thus, $p(s+ce_{n+1})\leqslant 1$ implies $s\in I^n$, $|c|\leqslant 1$, whence it follows that p satisfies (iii). This proves that $p\in B_{n+1}$.

Finally, let $s\in R^{n+1}$ be arbitrary. Then for any $r\in R^k$ we have, by (13.31), (13.30), $i_{k+1}=n+1$ and (13.32),

$$\begin{aligned}
F\left(r,\sum_{i=1}^{n} s(i)e_i, s(n+1)\right)&=\tilde{p}\left(\sum_{i=1}^{n} s(i)e_i-\sum_{j=1}^{k} r(j)e_{i_j}\right)+q(r+s(n+1)e_{k+1})\\
&\geqslant\tilde{p}\left(\sum_{j=1}^{k}(s(i_j)-r(j))e_{i_j}\right)+q(r+s(n+1)e_{k+1})\\
&=q\left(\sum_{j=1}^{k} s(i_j)e_j-r\right)+q(r+s(n+1)e_{k+1})\\
&\geqslant q\left(\sum_{j=1}^{k+1} s(i_j)e_j\right)=p\left(\sum_{j=1}^{k+1} s(i_j)e_{i_j}\right),
\end{aligned}$$

whence we infer (13.33), which completes the proof of lemma 13.4.

The next lemma gives essentially the construction of the complementably universal basis.

Lemma 13.5. *Let* $\varepsilon>0$. *Then there exists a norm* $p\in B$ *with the property that for every* $q\in B$ *there is an increasing sequence of indices* $i_1<i_2<\cdots$ *such that*

$$\frac{1}{1+\varepsilon}q(t)\leqslant p\left(\sum_{j=1}^{\infty} t(j)e_{i_j}\right)\leqslant(1+\varepsilon)q(t)\qquad\left(t\in\bigcup_{k=1}^{\infty} R^k\right),\tag{13.38}$$

$$p(s)\geqslant p\left(\sum_{j=1}^{\infty} s(i_j)e_{i_j}\right)\qquad\left(s\in\bigcup_{k=1}^{\infty} R^k\right).\tag{13.39}$$

Proof. By lemma 13.3 (with $\log(1+\varepsilon)$ instead of ε) there exist an increasing sequence of indices $1=N_1<N_2<\cdots$ and a sequence $\{q_n\}$ such that for each $k=1,2,\ldots$

$$q_n\in B_k\quad\text{for}\quad n=N_k,N_k+1,\ldots,N_{k+1}-1,\tag{13.40}$$

$$A_k=\bigcup_{n=N_k}^{N_{k+1}-1}\{q_n\}\quad\text{forms a }\log(1+\varepsilon)\text{-net for }B_k,\tag{13.41}$$

$$J_k(A_{k+1})=A_k.\tag{13.42}$$

Now we shall define inductively a sequence of norms $\{p_n\}$ and a sequence of finite increasing sequences of indices $\{\alpha(n)\}$ with the following properties, for each $n=1,2,\ldots$:

a) $p_n \in B_n$, $J_n(p_{n+1}) = p_n$;

b) if $N_k \leqslant n \leqslant N_{k+1} - 1$ and $\alpha(n) = \{i_1(n), i_2(n), \ldots, i_{k_n}(n)\}$, then

$$k_n = k,\ i_{k_n}(n) = n; \tag{13.43}$$

c) if $1 \leqslant m \leqslant n$, $N_l \leqslant m \leqslant N_{l+1} - 1$ and $J_l(q_n) = q_m$, then

$$i_j(m) = i_j(n) \qquad (j = 1, \ldots, k_m = l); \tag{13.44}$$

d) if $1 \leqslant m \leqslant n$, then

$$p_n\left(\sum_{j=1}^{k_m} t(j) e_{i_j(m)}\right) = q_m(t) \qquad (t \in R^{k_m}), \tag{13.45}$$

$$p_n(s) \geqslant p_n\left(\sum_{j=1}^{k_m} s(i_j(m)) e_{i_j(m)}\right) \quad (s \in R^n). \tag{13.46}$$

Put $p_1 = q_1$ and $\alpha(1) = \{1\}$; then a)–d) are obviously satisfied. Assume that for some $n \geqslant 1$ we have already defined the norms p_m and the finite increasing sequences of indices $\alpha(m)$ so as to satisfy a)–d) for each $m \leqslant n$. Choose k such that $N_{k+1} \leqslant n+1 \leqslant N_{k+2} - 1$, whence, by (13.41), $q_{n+1} \in A_{k+1} \subset B_{k+1}$. Then, by (13.42) there exists an $m \leqslant n$ such that $q_m \in A_k$ (whence, by (13.41), $N_k \leqslant m \leqslant N_{k+1} - 1$) and $J_k(q_{n+1}) = q_m$. By the inductive hypothesis (conditions b) and d)), we have then $k_m = k \leqslant N_k \leqslant m \leqslant n$ and

$$p_n\left(\sum_{j=1}^{k_m} t(j) e_{i_j(m)}\right) = q_m(t) = q_{n+1}(t) \qquad (t \in R^k),$$

and also (13.46). Hence, by lemma 13.4, there exists a norm $p' \in B_{n+1}$ such that $J_n(p') = p_n$ and that

$$p'\left(\sum_{j=1}^{k} t(j) e_{i_j(m)} + t(k+1) e_{n+1}\right) = q_{n+1}(t) \qquad (t \in R^{k+1}),$$

$$p'(s) \geqslant p'\left(\sum_{j=1}^{k} s(i_j(m)) e_{i_j(m)} + s(n+1) e_{n+1}\right) \qquad (s \in R^{n+1}).$$

Put

$$p_{n+1} = p',\ \alpha(n+1) = \{i_1(m), i_2(m), \ldots, i_k(m), n+1\} = \alpha(m) \cup \{n+1\}.$$

Then obviously p_{n+1} and $\alpha(n+1)$ satisfy a)–d), which completes the induction.

Now put

$$p(t) = p_n(t) \qquad (t \in R^n). \tag{13.47}$$

Then by a) p is well defined and $p \in B$. We shall show that p has the required properties, which will complete the proof.

Let $q \in B$ be arbitrary. Then, by (13.41), there exists a sequence $\{q'_k\}$ such that

$$q'_k = q_{n_k} \in A_k, \quad \text{whence} \quad N_k \leqslant n_k \leqslant N_{k+1} - 1 \qquad (k = 1, 2, \ldots), \quad (13.48)$$

$$d_k(q'_k, J_k(q)) < \log(1+\varepsilon) \qquad (k = 1, 2, \ldots). \qquad (13.49)$$

We claim that without loss of generality we may also assume

$$J_k(q'_n) = q'_k \qquad (k = 1, \ldots, n;\ n = 1, 2, \ldots). \qquad (13.50)$$

Indeed, let $\{q'_n\}$ be an arbitrary sequence of norms such that $q'_n \in A_n$ $(n = 1, 2, \ldots)$. Then, by (13.42),

$$J_k(q'_n) = J_k J_{k+1} \ldots J_n(q'_n) \in A_k \qquad (n \geqslant k),$$

whence the set

$$F_k = \bigcup_{n \geqslant k} \{J_k(q'_n)\} \qquad (13.51)$$

is finite and therefore compact in the discrete topology. Consequently, the cartesian product

$$F = \mathop{\times}_{k=1}^{\infty} F_k = \{\{q''_k\} \mid q''_k \in F_k \ (k = 1, 2, \ldots)\} \qquad (13.52)$$

is compact. Put

$$Z_n = \{\{q''_k\} \in F \mid J_k(q''_n) = q''_k \ (k = 1, \ldots, n)\} \qquad (n = 1, 2, \ldots). \quad (13.53)$$

Then each Z_n is non-empty (since if we define $q''_k = J_k(q'_n)$ for $k = 1, \ldots, n$ and $q''_k =$ arbitrary in F_k for $k = n+1, n+2, \ldots$ then $\{q''_k\} \in Z_n$), closed (since the relations $\{q''^{(d)}_k\} \xrightarrow[d \in \Delta]{} \{q''_k\}$, $\{q''^{(d)}_k\} \in Z_n$ obviously imply $q''^{(d)}_k \xrightarrow[d \in \Delta]{} q''_k$, $J_k(q''^{(d)}_n) \xrightarrow[d \in \Delta]{} J_k(q''_n)$, whence $\{q''_k\} \in Z_n$) and $Z_1 \supset Z_2 \supset \cdots$ Hence, by the compactness of F, there exists a sequence $\{q''_k\} \in F$ such that $\{q''_k\} \in \bigcap_{n=1}^{\infty} Z_n$, whence $\{q''_k\}$ satisfies (13.50). Furthermore, by (13.52) and (13.51), for every index k there is an index $n = n(k) \geqslant k$ such that $J_k(q'_n) = q''_k$ (where $q'_n \in A_n$ by our hypothesis). Hence, if the sequence $\{q'_j\}$ satisfies (13.49), then, taking into account (13.13),

$$d_k(q''_k, J_k(q)) = d_k(J_k(q'_n), J_k(J_n(q))) \leqslant d_n(q'_n, J_n(q)) < \log(1+\varepsilon).$$

Thus the sequence $\{q''_k\}$ satisfies the conditions (13.48)–(13.50), which proves our claim on (13.50).

Now define the increasing sequence of indices $i_1 < i_2 < \cdots$ by

$$i_k = n_k \qquad (k = 1, 2, \ldots), \tag{13.54}$$

where n_k is defined by (13.48). Let us show that for each k

$$\alpha(n_k) = \{i_1, i_2, \ldots, i_k\}. \tag{13.55}$$

Since by (13.48) we have $N_k \leqslant n_k \leqslant N_{k+1} - 1$, by virtue of b) we have

$$\alpha(n_k) = \{i_1(n_k), i_2(n_k), \ldots, i_k(n_k) = n_k = i_k\}. \tag{13.56}$$

Furthermore, if $m \leqslant k$, then $n_m \leqslant n_k$, whence, by (13.50) and (13.48), $J_m(q_{n_k}) = J_m(q'_k) = q'_m = q_{n_m}$ and hence, by c), $i_j(n_k) = i_j(n_m)$ for $j = 1, \ldots, m$. Thus, in particular, for $j = m$ we obtain

$$i_m(n_k) = i_m(n_m) = n_m = i_m \qquad (m = 1, \ldots, k),$$

whence, by (13.56), there follows (13.55).

Consequently, by d), we get (13.39) and

$$p\left(\sum_{j=1}^{k} t(j) e_{i_j}\right) = p_{n_k}\left(\sum_{j=1}^{k} t(j) e_{i_j}\right) = q_{n_k}(t) = q'_k(t) \qquad (t \in R^k). \tag{13.57}$$

On the other hand, by (13.49) we have

$$\sup_{t \in \partial I^k} \frac{q'_k(t)}{q(t)} \cdot \sup_{t \in \partial I^k} \frac{q(t)}{q'_k(t)} \leqslant 1 + \varepsilon,$$

whence, since $q'_k(e_1) = q(e_1) = 1$ (by (ii)),

$$1 \leqslant \sup_{t \in \partial I^k} \frac{q'_k(t)}{q(t)} \leqslant (1+\varepsilon) \inf_{t \in \partial I^k} \frac{q'_k(t)}{q(t)} \leqslant 1 + \varepsilon,$$

whence

$$\frac{q(t)}{1+\varepsilon} \leqslant q'_k(t) \leqslant (1+\varepsilon) q(t) \qquad (t \in R^k). \tag{13.58}$$

From (13.57) and (13.58) it follows that we have (13.38), which completes the proof of lemma 13.5.

Now we can prove

Theorem 13.1. *The family $\mathscr{B}$ of all bounded bases contains a complementably universal element.*

Proof. Let $\varepsilon > 0$ be arbitrary. Let $p = p_\varepsilon \in B$ be as in lemma 13.5 and let $E = E_\varepsilon$ be the completion of the linear space $\left(\bigcup_{k=1}^{\infty} R^k, p\right)$; we shall indentify $\bigcup_{k=1}^{\infty} R^k$ with its canonical image in E. The unit vectors $\{e_n\}$ form a (monotone normalized) basis of E, since their linear combinations are dense in E and, by (iv),

$$p\left(\sum_{i=1}^{n} \alpha_i e_i\right) \leqslant p\left(\sum_{i=1}^{n+m} \alpha_i e_i\right)$$

for any finite sequence of real numbers $\alpha_1, \ldots, \alpha_{n+m}$. We shall show that this basis $\{e_n\} \in \mathscr{B}$ is complementably universal for the family $\mathscr{B}$ of all bounded bases, which will complete the proof.

Let $\{x_n\}$ be an arbitrary bounded basis (of a Banach space, say X). Put

$$a = \inf_{1 \leqslant n < \infty} \|x_n\|, \qquad b = \sup_{1 \leqslant n < \infty} \|x_n\|. \tag{13.59}$$

Then, since $\{x_n\}$ is a bounded basis, $0 < a \leqslant b < \infty$. Furthermore, put

$$q(t) = \max\left(\frac{1}{b} \max_{1 \leqslant n < \infty} \left\| \sum_{j=1}^{n} t(j) x_j \right\|, \max_{1 \leqslant n < \infty} |t(n)|\right) \quad \left(t \in \bigcup_{k=1}^{\infty} R^k\right). \tag{13.60}$$

Then obviously $q \in B$ and hence, by lemma 13.5, there exists an increasing sequence of indices $i_1 < i_2 < \cdots$ such that we have (13.38) and (13.39). By (13.39), $\{e_{i_n}\}$ is a complemented subsequence of $\{e_n\}$. Furthermore, since $\frac{1}{b} \leqslant \frac{1}{a}$, we have

$$\frac{1}{b}\left\| \sum_{j=1}^{m} t(j) x_j \right\| \leqslant \frac{v}{a}\left\| \sum_{j=1}^{n} t(j) x_j \right\| \qquad (t \in R^n,\ m = 1, \ldots, n;\ n = 1, 2, \ldots)$$

and by Ch. I, § 3, formula (3.8), we also have

$$|t(m)| \leqslant \frac{2v}{a}\left\| \sum_{j=1}^{n} t(j) x_j \right\| \qquad (t \in R^n,\ m = 1, \ldots, n;\ n = 1, 2, \ldots)$$

where $v = v_{\{x_n\}}$ is the norm of the basis $\{x_n\}$. Consequently, by (13.60) and (13.38),

$$\begin{aligned} \frac{1}{b(1+\varepsilon)}\left\| \sum_{j=1}^{n} t(j) x_j \right\| &\leqslant \frac{1}{1+\varepsilon} q(t) \leqslant p\left(\sum_{j=1}^{n} t(j) e_{i_j}\right) \\ &\leqslant (1+\varepsilon) q(t) \leqslant \frac{2v}{a}(1+\varepsilon)\left\| \sum_{j=1}^{n} t(j) x_j \right\| \qquad (t \in R^n, n = 1, 2, \ldots), \end{aligned}$$

whence by Ch. I, § 8, theorem 8.1 d), the basis $\{x_n\}$ is equivalent to the basis $\{e_{i_n}\}$ of $[e_{i_n}]$, which completes the proof of theorem 13.1.

Remark 13.2 Since the basis $\{e_n\}$ in the above proof is actually normalized, we obtain again that every bounded basis is equivalent to a normalized basis (see Ch. I, § 3, theorem 3.2) and that *the family* $\mathscr{B}$

of all normalized bases contains a complementably universal element. Conversely, these two results imply theorem 13.1, since from these results it follows that *every normalized basis which is complementably universal for the family of all normalized bases is also complementably universal for the family of all bounded bases.*

By remark 13.1 (made after definition 13.2) and by Ch. I, § 8, theorem 8.1 d), we have the following corollary of theorem 13.1:

Corollary 13.1. *There exists a Banach space E with a basis such that every Banach space with a basis is isomorphic to a complemented subspace of E.*

This result suggests naturally the following problem:

Problem 13.1. Does there exist a separable Banach space E_s such that every separable Banach space is isomorphic to a complemented subspace of E_s?

By corollary 13.1, *a negative answer to problem 13.1 would imply a negative answer to the basis problem.* This seems to be an "easier" approach towards a negative answer to the basis problem, than constructing a counter-example, since this approach is "purely existential". The answer to problem 13.1 is believed to be negative, namely, it is conjectured that there exists no separable Banach space E_s such that every subspace of $C([0,1])$ of the form $[e^{2\pi i n t}]_{n \in M}$, where M runs over all infinite subsets of the set $\mathcal{N}=\{1,2,3,\ldots\}$, be isomorphic to a complemented subspace of E_s.

An affirmative answer to problem 13.1 would also have interesting consequences. Indeed, if the space E_s had no basis, we would have a negative answer to the basis problem, while if the space E_s had a basis, then it would follow that every separable Banach space has the approximation property (since it is obvious that every complemented subspace of a Banach space with the approximation property has the approximation property) whence also every Banach space has the approximation property [89] and thus we would have an affirmative answer to the "problem of approximation" [89].

Let us consider now the problem, "how many" complementably universal bases and "how many" spaces with complementably universal bases exist. The answer is given by

Theorem 13.2. *Any two bounded bases which are complementably universal for the family $\mathcal{B}$ of all bounded bases are permutatively equivalent*[1]. *Hence they span isomorphic Banach spaces and therefore the space E constructed in the proof of theorem 13.1 is unique up to an isomorphism.*

[1] See Ch. I, § 8, definition 8.3.

Proof. Since every bounded basis is equivalent to a normalized basis (see remark 13.2), it will be sufficient to consider only the case of two normalized elements of $\mathscr{B}$. By "equivalence class" or "class" of a basis $\{x_n\}$ we shall mean the class of all bases which are permutatively equivalent to $\{x_n\}$.

Let $\mathscr{X}$ be the equivalence class of a normalized basis which is complementably universal for the family of all bounded bases. Then, by remark 13.1 and Ch. I, § 4, proposition 4.2 and lemma 4.1, for the class[1] $\mathscr{X}^\infty$ and for the class $\mathscr{Y}$ of an arbitrary normalized basis there exist classes $\mathscr{Z}$ and $\mathscr{Z}_1$ such that

$$\mathscr{X} = \mathscr{X}^\infty \times \mathscr{Z} = \mathscr{Y} \times \mathscr{Z}_1 .$$

Hence, by Ch. I, § 8, proposition 8.4, we obtain

$$\begin{aligned}\mathscr{X} &= \mathscr{X}^\infty \times \mathscr{Z} = (\mathscr{Y} \times \mathscr{Z}_1)^\infty \times \mathscr{Z} = \mathscr{Y}^\infty \times \mathscr{Z}_1^\infty \times \mathscr{Z} = \mathscr{Y} \times (\mathscr{Y}^\infty \times \mathscr{Z}_1 \times \mathscr{Z}) \\ &= \mathscr{Y} \times \mathscr{X}\end{aligned}$$

Now, assuming that $\mathscr{Y}$ above is also the class of a normalized basis which is complementably universal for the family of all bounded bases, we get, by symmetry, that $\mathscr{Y} = \mathscr{X} \times \mathscr{Y}$, and hence

$$\mathscr{X} = \mathscr{Y} \times \mathscr{X} = \mathscr{X} \times \mathscr{Y} = \mathscr{Y} .$$

Thus, since by Ch. I, § 8, theorem 8.1 d), implication $6^\circ \Rightarrow 1^\circ$, any two equivalent bases span isomorphic Banach spaces, the proof of theorem 13.2 is complete.

We shall denote by E_b the Banach space, unique up to an isomorphism, which has a bounded basis complementably universal for the family of all bounded bases. The following corollary shows that the property described in corollary 13.1 also characterizes E_b up to an isomorphism.

Corollary 13.2. *If E is a Banach space with a basis, such that every Banach space with a basis is isomorphic to a complemented subspace of E, then E is isomorphic to E_b.*

Proof. Again, by "equivalence class" or "class" of a basis $\{x_n\}$ we shall understand the class of all bases which are permutatively equivalent to $\{x_n\}$. Let $\mathscr{Y}$ denote the equivalence class of a normalized basis of E_b, which is complementably universal for the family of all bounded bases (again, we may assume that E_b has such a basis) and let $\mathscr{X}$ be the class of a normalized basis of E. Then, as observed in the proof of theorem 13.2, $\mathscr{Y} = \mathscr{Y} \times \mathscr{Y}$ and $\mathscr{Y} = \mathscr{Y} \times \mathscr{X}$, whence[2] $E_b \times E_b \cong E_b$ and $E_b \times E \cong E_b$. On the

[1] See Ch. I, § 8, definition 8.5.

[2] We write $E_1 \cong E_2$ if the spaces E_1 and E_2 are isomorphic.

other hand, by our assumption on E, there exists a Banach space F such that $E \cong E_b \times F$. Consequently,

$$E \cong E_b \times F \cong (E_b \times E_b) \times F \cong E_b \times (E_b \times F) \cong E_b \times E \cong E_b,$$

which completes the proof.

Remark 13.3. One can also prove, by an argument similar to that used in the proof of theorem 13.2 (considering $(E_s \times E_s \times \cdots)_{l^2}$ and applying § 18, lemma 18.5) that if the answer to problem 13.1 is affirmative, then the space E_s of that problem is also unique up to an isomorphism.

We shall consider the problem, "how many" universal bases and "how many" spaces with universal bases exist, together with the analogous problem for block-universal bases.

Definition 13.4. Let $\mathscr{B}$ be a family of bases. A basis $\{e_n\}$ of a Banach space E is said to be *block-universal* for $\mathscr{B}$ if every basis in $\mathscr{B}$ is equivalent to a block basic sequence with respect to $\{e_n\}$.

Obviously, every universal basis is block universal, since every subsequence of a basis is a block basic sequence with respect to that basis.

For the proof of the next theorem we shall need

Lemma 13.6. *Let E and F_1 be Banach spaces, such that E contains a subspace F isomorphic to F_1. Then there exists an equivalent norm $\|x\|_1$ on E such that the subspace $F \subset E$ endowed with this new norm is linearly isometric to F_1.*

Proof. Let u be an isomorphism of F onto F_1, such that

$$\|y\| \leqslant \|u(y)\| \leqslant C\|y\| \qquad (y \in F), \tag{13.61}$$

where $C \geqslant 1$ is a constant. Put

$$A = \operatorname{co}\left(\{y \in F \mid \|u(y)\| \leqslant 1\} \cup \left\{x \in E \,\middle|\, \|x\| \leqslant \frac{1}{C}\right\}\right), \tag{13.62}$$

$$\|x\|_1 = \inf_{\substack{\lambda > 0 \\ x \in \lambda A}} \lambda \qquad (x \in E). \tag{13.63}$$

Then $\|x\|_1$ is a norm on E (it is the Minkowski functional of A), equivalent to the original norm on E, since by (13.62) and (13.61) we have

$$\frac{1}{C} S_E \subset A \subset S_E = \{x \in E \mid \|x\| \leqslant 1\}.$$

Furthermore, by (13.63), (13.62) and (13.61) we readily have

$$\|u(y)\| = \|y\|_1 \qquad (y \in F),$$

i. e., u is a linear isometry of $(F, \|y\|_1)$ onto F_1, which completes the proof of lemma 13.6.

We recall that a Banach space E is said to be isomorphically universal for a family $(\mathscr{S})$ of Banach spaces if every space $X \in (\mathscr{S})$ is isomorphic to a subspace of E.

Theorem 13.3. *Let E be a Banach space, isomorphically universal for the family of all separable Banach spaces and assume that E has a basis. Then every bounded basis* $\{e_n\}$ *of E is block-universal for the family* $\mathscr{B}$ *of all bounded bases.*

Proof. By hypothesis, E contains a subspace isomorphic to $C([0,1])$. Hence, by lemma 13.6, there exists an equivalent norm on E such that E endowed with this new norm contains a subspace linearly isometric to $C([0,1])$; we shall assume that E is endowed with this new norm and we shall identify $C([0,1])$ with its image in E under this linear isometry.

Let $\{h_n\}$ be the a. s. c. f. to the basis $\{e_n\}$ of E and let

$$\mu_n = h_n|_{C([0,1])} \qquad (n=1,2,\ldots). \tag{13.64}$$

We claim that there exists a set $\Delta \subset [0,1]$ homeomorphic to the Cantor discontinuum and such that

$$|\mu_n|(\Delta)=0 \qquad (n=1,2,\ldots). \tag{13.65}$$

Indeed, let $Q_1 = D_1^{(1)} \cup D_2^{(1)}$, where $D_1, D_2 \subset [0,1]$ are two disjoint Cantor sets such that $|\mu_1|(Q_1) < \frac{1}{2}$, let $Q_2 = D_1^{(2)} \cup D_2^{(2)} \cup D_3^{(2)} \cup D_4^{(2)}$, where $D_1^{(2)}, D_2^{(2)} \subset D_1^{(1)}, D_3^{(2)}, D_4^{(2)} \subset D_2^{(1)}$ are four mutually disjoint Cantor sets such that $\sum_{j=1}^{2} |\mu_j|(Q_2) < \frac{1}{4}$, let $Q_3 = \bigcup_{j=1}^{8} D_j^{(3)}$, where $D_1^{(3)}$, $D_2^{(3)} \subset D_1^{(2)}, \ldots, D_7^{(3)}, D_8^{(3)} \subset D_4^{(2)}$ are eight mutually disjoint Cantor sets such that $\sum_{j=1}^{3} |\mu_j|(Q_3) < \frac{1}{8}$, and continue this process indefinitely; then the set $\Delta = \bigcap_{i=1}^{\infty} Q_i$ has the required properties (Δ is compact, zero-dimensional and perfect and by $\Delta \subset Q_j$ we have $|\mu_n|(\Delta) \leqslant |\mu_n|(Q_j) < \frac{1}{2^j}$ for $n, j = 1, 2, \ldots$ with $n \leqslant j$, whence (13.65)).

Now let $\{x_n\}$ be an arbitrary bounded basis of a Banach space X. By the remark made after proposition 1.3 of § 1, we may assume that $\{x_n\}$ is monotone (renorming, if necessary, the space X with a suitable equivalent norm); furthermore, there exists[1] a linear isometry w of X into $C(\Delta)$.

---

[1] See, e.g., [10], p. 93.

Let $\varepsilon>0$ be arbitrary. Then, since μ_1 is regular and $|\mu_1|(\Delta)=0$, there exists an open set $U_1\supset\Delta$ such that $|\mu_1|(U_1)<\frac{\varepsilon}{2}$. Put $F_1=[0,1]\setminus U_1$. Then F_1 is closed, $F_1\cap\Delta=\emptyset$ and $|\mu_1|([0,1]\setminus F_1)<\frac{\varepsilon}{2}$. Put

$$y_1(t)=\begin{cases}[w(x_1)](t) & \text{for } t\in\Delta,\\ 0 & \text{for } t\in F_1,\\ \text{linear} & \text{for the other } t.\end{cases} \tag{13.66}$$

Then we have, putting $M=\sup_{1\leqslant n<\infty}\|e_n\|$, $L=\sup_{1\leqslant n<\infty}\|x_n\|$,

$$\begin{aligned}\|\mu_1(y_1)e_1\|&\leqslant M|h_1(y_1)|\leqslant M\Big[\Big|\int_{F_1}y_1(t)d\mu_1(t)\Big|+\Big|\int_{[0,1]\setminus F_1}y_1(t)d\mu_1(t)\Big|\Big]\\&\leqslant M\|y_1\|\frac{\varepsilon}{2}=M\|x_1\|\frac{\varepsilon}{2}\leqslant\frac{ML\varepsilon}{2}.\end{aligned} \tag{13.67}$$

Furthermore, since $y_1=\sum_{j=1}^{\infty}h_j(y_1)e_j=\sum_{j=1}^{\infty}\mu_j(y_1)e_j$, there exists a positive integer n_1 such that

$$\Big\|\sum_{j=n_1+1}^{\infty}\mu_j(y_1)e_j\Big\|<\frac{ML\varepsilon}{2}. \tag{13.68}$$

Similarly, since $\mu_1|_{U_1},\dots,\mu_{n_1}|_{U_1}$ are regular and $\Delta\subset U_1$, $|\mu_1|(\Delta)=\cdots=|\mu_{n_1}|(\Delta)=0$, there exists an open set $U_2\subset U_1$ with $U_2\supset\Delta$ such that $\sum_{j=1}^{n_1}|\mu_j|(U_2)<\frac{\varepsilon}{4}$. Put $F_2=[0,1]\setminus U_2$. Then F_2 is closed, $F_2\supset F_1$, $F_2\cap\Delta=\emptyset$ and $\sum_{j=1}^{n_1}|\mu_j|([0,1]\setminus F_2)<\frac{\varepsilon}{4}$. Put

$$y_2(t)=\begin{cases}[w(x_2)](\mathrm{t}) & \text{for } t\in\Delta,\\ 0 & \text{for } t\in F_2,\\ \text{linear} & \text{for the other } t.\end{cases} \tag{13.69}$$

Then we have

$$\begin{aligned}\Big\|\sum_{j=1}^{n_1}\mu_j(y_2)e_j\Big\|&\leqslant M\sum_{j=1}^{n_1}|\mu_j(y_2)|\\&\leqslant M\sum_{j=1}^{n_1}\Big[\Big|\int_{F_2}y_2(t)d\mu_j(t)\Big|+\Big|\int_{[0,1]\setminus F_2}y_2(t)d\mu_j(t)\Big|\Big]\\&\leqslant M\|y_2\|\frac{\varepsilon}{4}=M\|x_2\|\frac{\varepsilon}{4}<\frac{ML\varepsilon}{4}.\end{aligned} \tag{13.70}$$

Furthermore, since $y_2 = \sum_{j=1}^{\infty} \mu_j(y_2) e_j$, there exists a positive integer $n_2 > n_1$ such that

$$\left\| \sum_{j=n_2+1}^{\infty} \mu_j(y_2) e_j \right\| < \frac{ML\varepsilon}{4}. \tag{13.71}$$

Continuing in this manner, we obtain an increasing sequence of positive integers $\{n_k\}$, a sequence of closed sets $\{F_k\} \subset [0,1]$ with $F_k \subset F_{k+1}$, $F_k \cap \Delta = \emptyset$, $\sum_{j=1}^{n_k} |\mu_j|([0,1] \setminus F_k) < \frac{\varepsilon}{2^k}$ $(k=1,2,\ldots)$, and a sequence $\{y_k\} \subset C([0,1])$ such that

$$y_k(t) = \begin{cases} [w(x_k)](t) & \text{for} \quad t \in \Delta \\ 0 & \text{for} \quad t \in F_k \\ \text{linear} & \text{for} \quad \text{the other } t \end{cases} \qquad (k=1,2,\ldots), \tag{13.72}$$

$$\left\| \sum_{j=1}^{n_{k-1}} \mu_j(y_k) e_j \right\| \leqslant \frac{ML\varepsilon}{2^k} \qquad (k=1,2,\ldots; n_0 = 1), \tag{13.73}$$

$$\left\| \sum_{j=n_k+1}^{\infty} \mu_j(y_k) e_j \right\| \leqslant \frac{ML\varepsilon}{2^k} \qquad (k=1,2,\ldots). \tag{13.74}$$

Put

$$u(x) = \sum_{k=1}^{\infty} \alpha_k y_k \qquad \left(x = \sum_{k=1}^{\infty} \alpha_k x_k \in X \right). \tag{13.75}$$

We shall prove that u is a linear isometry from X into $C([0,1])$. For this purpose it will be sufficient to prove that for any finite sequence of scalars $\alpha_1, \ldots, \alpha_k$ we have

$$\left\| \sum_{i=1}^{k} \alpha_i y_i \right\| = \left\| \sum_{i=1}^{k} \alpha_i x_i \right\|. \tag{13.76}$$

Formula (13.76) is obvious for $k=1$, since $\|\alpha_1 y_1\| = |\alpha_1| \|y_1\| = |\alpha_1| \|x_1\| = \|\alpha_1 x_1\|$ (by (13.72) and since w is an isometry). Assume now that (13.76) holds for some $m \geqslant 1$. Let $\alpha_1, \ldots, \alpha_{m+1}$ be arbitrary and put $x = \sum_{i=1}^{m} \alpha_i x_i$, $\alpha = \alpha_{m+1}$. Then, since by (13.72) $u(x) + a y_{m+1}$ is linear on each interval of the set $[0,1] \setminus (\Delta \cup F_{m+1})$, we have

$$\begin{aligned} \|u(x) + \alpha y_{m+1}\| &\leqslant \sup_{t \in \Delta \cup F_{m+1}} |[u(x) + \alpha y_{m+1}](t)| \\ &= \max\{\sup_{t \in \Delta} |[u(x) + \alpha y_{m+1}](t)|, \sup_{t \in F_{m+1}} |[u(x) + \alpha y_{m+1}(t)]|\}. \end{aligned} \tag{13.77}$$

However, by $y_{m+1}(t) = 0$ $(t \in F_{m+1})$ we have

$$\sup_{t \in F_{m+1}} |[u(x) + \alpha y_{m+1}(t)]| = \sup_{t \in F_{m+1}} |[u(x)](t)| \leqslant \|u(x)\|, \tag{13.78}$$

and by the induction hypothesis and the monotony of the basis $\{x_n\}$ we have

$$\|u(x)\| = \|x\| \leqslant \|x + \alpha x_{m+1}\|. \tag{13.79}$$

On the other hand, since by (13.75) and (13.72) $u(x)+\alpha y_{m+1}|_\Delta = w(x)+\alpha w(x_{m+1})$ and since w is a linear isometry, we have

$$\begin{aligned}\|u(x)+\alpha y_{m+1}\| &\geqslant \sup_{t\in\Delta} |[u(x)+\alpha y_{m+1}](t)| = \|w(x)+\alpha w(x_{m+1})\| \\ &= \|w(x+\alpha x_{m+1})\| = \|x+\alpha x_{m+1}\|. \end{aligned}\tag{13.80}$$

From (13.77)–(13.80) we infer $\|u(x)+\alpha y_{m+1}\| = \|x+\alpha x_{m+1}\|$, and this completes the induction. Thus, u is a linear isometry of X into $C([0,1])$, whence, in particular, $\{x_k\} \sim \{u(x_k)\}$.

Put $n_0 = 0$ and

$$z_k = \sum_{i=n_{k-1}+1}^{n_k} h_i[u(x_k)]e_i = \sum_{i=n_{k-1}+1}^{n_k} \mu_i(y_k)e_i \qquad (k=1,2,\ldots). \tag{13.81}$$

Then by (13.73), (13.74), (13.75) and (13.64) we have

$$\|u(x_k)-z_k\| = \left\| \sum_{j=1}^{n_{k-1}} \mu_j(y_k)e_j + \sum_{j=n_k+1}^{\infty} \mu_j(y_k)e_j \right\| \leqslant \frac{ML\varepsilon}{2^{k-1}} \qquad (k=1,2,\ldots), \tag{13.82}$$

whence, denoting by $\{g_k\}$ the a.s.c.f. to the bounded basis $\{u(x_k)\}$ of $u(X)$,

$$\sum_{k=1}^{\infty} \|g_k\| \, \|u(x_k)-z_k\| \leqslant \sup_{1\leqslant j<\infty} \|g_j\| \sum_{k=1}^{\infty} \frac{ML\varepsilon}{2^{k-1}} = 2ML\varepsilon \sup_{1\leqslant j<\infty} \|g_j\| < 1$$

(for ε sufficiently small), whence, by Ch. I, § 10, theorem 10.1, we infer that $\{z_k\} \sim \{u(x_k)\} \sim \{x_k\}$, which completes the proof of theorem 13.3.

Remark 13.4. One can also prove that under the assumptions of theorem 13.3 every bounded basis $\{x_n\}$ of any infinite dimensional Banach space X is equivalent to a *normalized* block basic sequence with respect to $\{e_n\}$. Indeed, by the remarks made after proposition 1.3 of § 1, we may assume that $\{x_n\}$ is normalized and monotone (renorming, if necessary, the space X with a suitable equivalent norm). Then, taking into account that u is an isometry, we get, by (13.82),

$$\begin{aligned}\left\| z_k - \frac{z_k}{\|z_k\|} \right\| &= \left| 1 - \frac{1}{\|z_k\|} \right| \|z_k\| = \big|\, \|z_k\| - 1 \big| = \big|\, \|z_k\| - \|u(x_k)\| \,\big| \\ &\leqslant \|z_k - u(x_k)\| < \frac{ML\varepsilon}{2^{k-1}} \qquad (k=1,2,\ldots),\end{aligned}$$

whence, as in the above, from Ch. I, § 10, theorem 10.1 we infer that $\left\{\frac{z_k}{\|z_k\|}\right\} \sim \{z_k\} \sim \{x_k\}$, which proves our assertion.

Remark 13.5. Since $C([0,1])$ is isomorphically universal[1], from theorem 13.3 it follows, in particular, that *every bounded basis* $\{x_n\}$ *of an infinite dimensional Banach space* X *is equivalent to a block basic sequence with respect to the usual Schauder basis of* $C([0,1])$. Consequently, *if a separable Banach space* X *is not isomorphic to any block subspace*[2] *with respect to the Schauder basis of* $C([0,1])$, *then* X *has no basis*, giving thus a negative answer to the basis problem. It seems to be promising to look for such subspaces X of $C([0,1])$, since e.g. in the spaces c_0 and $l^p(1 \leqslant p < \infty, p \neq 2)$ there exist subspaces which are not isomorphic to any block subspace with respect to the unit vector basis (such as, by § 18, proposition 18.1 a), any subspace which is not isomorphic to the whole space).

For this reason, we suggest

Problem 13.2. Describe, up to isomorphisms, all block subspaces with respect to the various usual bases in concrete Banach spaces.

We conjecture that except for spaces isomorphic to l^2, in every Banach space E with a basis $\{e_n\}$ there exists a subspace X not isomorphic to any block subspace with respect to $\{e_n\}$.

Returning to spaces with universal and block-universal bases, we shall now prove

Theorem 13.4. *Let* E *be a Banach space with a basis. The following statements are equivalent:*

1°. *E has a bounded basis which is universal for the family of all bounded bases.*

2°. *E has a bounded basis which is block-universal for the family of all bounded bases.*

3°. *E is isomorphically universal for the family of all separable Banach spaces.*

Proof. 1° ⇒ 3°. Assume that we have 1°. Then the Schauder basis of $C([0,1])$ is equivalent to some subsequence of the universal basis of E, whence, by Ch. I, § 8, theorem 8.1 d), $C([0,1])$ is isomorphic to the subspace G of E spanned by this subsequence. Since $C([0,1])$, whence also G, is isomorphically universal for the family of all separable Banach spaces, it follows that the space E is isomorphically universal for this family.

The implication 3° ⇒ 2° is an immediate consequence of theorem 13.3.

2° ⇒ 1°. Assume that we have 2° and let $\{e_n\}$ be a bounded basis of E which is block-universal for the family of all bounded bases. Then, in particular, the universal basis of theorem 13.1 is equivalent to a bounded

---

[1] See e. g. [10], p. 185, theorem 9.

[2] I.e., subspace spanned by a block basic sequence.

block basic sequence $\{z_n\}$ with respect to $\{e_n\}$. However, by Ch. I, § 7, theorem 7.2, $\{z_n\}$ can be extended to a bounded basis of E, which will obviously be universal for the family of all bounded bases. This completes the proof of theorem 13.4.

Remark 13.6. From theorem 13.4 it follows, in particular, that *the space $C([0,1])$ has a basis $\{e_n\}$ such that every bounded basis $\{x_n\}$ of any Banach space X is equivalent to some suitable subsequence $\{e_{i_n}\}$ of $\{e_n\}$*. Consequently, *if a separable Banach space X is not isomorphic to any subspace $[e_{i_n}]$ spanned by a subsequence $\{e_{i_n}\}$ of the above basis $\{e_n\}$ of $C([0,1])$, then X has no basis*, giving thus a negative answer to the basis problem; however, observe that the universal basis $\{e_n\}$ of $C([0,1])$ is not explicitly constructed, we have only proved its existence.

It is natural to raise

Problem 13.3. Describe, up to isomorphisms, the subspaces spanned by all infinite subsequences of the various usual bases in concrete Banach spaces.

Since every infinite subsequence of a basis is obviously a block basic sequence with respect to that basis, this problem is a part of problem 13.2. We also formulate the following weakening of the conjecture made after problem 13.2: Except for spaces isomorphic to l^2, in every Banach space E with a basis $\{e_n\}$ there exists a subspace X which is not isomorphic to any subspace spanned by a subsequence of $\{e_n\}$. Observe that this is certainly true for $E=C([0,1])$ and $\{e_n\}=$the usual Schauder basis of $C([0,1])$, since every subsequence of the Schauder basis of $C([0,1])$ contains either a subsequence which is a basic sequence equivalent to the unit vector basis of c_0, or a subsequence which is a basic sequence equivalent to the basis $\{1,1,1,\ldots\}$, $\{0,1,1,\ldots\}$, $\{0,0,1,1,\ldots\},\ldots$ of the space c; indeed, the proof of this latter statement is similar to that of § 18, proposition 18.4.

Corollary 13.3. *There exist a continuum of mutually non-equivalent normalized bases which are universal for the family of all bounded bases.*

Proof. By theorem 13.4 and by Ch. I, § 8, theorem 8.1 d), it is sufficient to prove that *there exist a continuum of mutually non-isomorphic Banach spaces which are isomorphically universal for all separable Banach spaces and which have bases*. For each p with $1<p<\infty$ let $E_p=C([0,1])\times l^p$. Then obviously each E_p is isomorphically universal for all separable Banach spaces (since it contains a subspace linearly isometric to $C([0,1])$. Furthermore, since $C([0,1])$ and each l^p have bases, each E_p has a basis as well (by Ch. I, § 4, proposition 4.2). Finally, let us show that if $p\neq r$, then E_p is not isomorphic to E_r. For this purpose it is sufficient to show that if $p\neq r$, then E_p has no complemented subspace isomorphic to l^r

(clearly the subspace $\{0\}\times l^r$ of E_r is complemented in l^r and it is isomorphic to l^r). Assume the contrary, i.e. that there exists a bounded linear projection u of $E_p = C([0,1])\times l^p$ onto some subspace G isomorphic to l^r and let v be an isomorphism of l^r onto G. Furthermore, let u_1 and u_2 denote the natural projections $E_p \to C([0,1])\times\{0\}$ and $E_p \to \{0\}\times l^p$ respectively, and let $l^{r,n} = \{x = \{\xi_j\}\in l^r \mid \xi_1 = \cdots = \xi_n = 0\}$. Then the mapping $u_2 v|_{l^{r,n}}: l^{r,n} \to \{0\}\times l^p$ has no bounded inverse[1], whence there exist elements $x_n \in l^{r,n}$ such that $\|u_2 v(x_n)\| < \frac{1}{n}$, $\|x_n\| = 1$ $(n=1,2,\ldots)$. Observe that by the well known[2] characterization of weak convergence in l^r, we have $x_n \xrightarrow{w} 0$. Furthermore, the mapping $u|_{C([0,1])\times\{0\}}: C([0,1])\times\{0\} \to G \cong l^r$ is weakly compact and hence[3] it takes weakly convergent sequences into convergent sequences. Since $x_n \xrightarrow{w} 0$, we have $u_1 v(x_n) \xrightarrow{w} 0$, where $\{u_1 v(x_n)\} \subset C([0,1])\times\{0\}$, and hence $\|u u_1 v(x_n)\| \to 0$. Since $\|u_2 v(x_n)\| < \frac{1}{n}$ for $n=1,2,\ldots$ we also have $\|u u_2 v(x_n)\| \to 0$. Consequently,

$$\|v(x_n)\| = \|u v(x_n)\| = \|u u_1 v(x_n) + u u_2 v(x_n)\| \to 0,$$

whence, since v is an isomorphism, we obtain $\|x_n\| \to 0$, which contradicts the condition $\|x_n\| = 1$ $(n=1,2,\ldots)$ above. This completes the proof of corollary 13.3.

Let us consider now the problem of existence of universal bases for other classes of bases. We have the following negative result:

Theorem 13.5. *There exists no shrinking basis which is universal for the family of all shrinking bases.*

Proof. Assume the contrary, i.e. that there exists a shrinking basis $\{e_n\}$ of a Banach space E, which is universal for the family of all shrinking bases. Let ω_1 denote the first uncountable ordinal number. We define inductively a family $\{X_a\}_{0\leqslant a<\omega_1}$ of Banach spaces as follows: $X_0 = l^2$; if $a = b+1$, then $X = (X_b \times l^2)_{l^2_1}$; if a is a limit ordinal number, then $X_a = \left(\prod_{0\leqslant b<a} X_b\right)_{l^2}$, i.e. the space of all families of elements $\{x_b\}_{0\leqslant b<a}$ with $x_b \in X_b (0\leqslant b<a)$, $\|\{x_b\}\| = \sqrt{\sum_{0\leqslant b<a} \|x_b\|^2} < \infty$. Then, since the unit vector basis in l^2 is a normalized monotone unconditional[4] basis, from

---

[1] See e.g. [10], p. 205, theorem 7.

[2] See e.g. [10], p. 137.

[3] See e.g. [50], p. 494, theorem 4.

[4] See § 14, definition 14.1.

Ch. I, § 8, proposition 8.3, it follows (by transfinite induction) that each X_a has a normalized monotone unconditional basis. Since all X_a are reflexive, the bases are shrinking (see § 4, example 4.3) and hence, by our hypothesis, equivalent to subsequences of $\{e_n\}$. Therefore, by Ch. I, § 8, theorem 8.1 d), each $X_a (0 \leqslant a < \omega_1)$ is isomorphic to a subspace of E, whence[1] E^* is non-separable, which contradicts the assumption that $\{e_n\}$ is a shrinking basis of E. This completes the proof of theorem 13.5.

Problem 13.4. Does a similar result to theorem 13.5 hold for the family $\mathscr{B}$ of all boundedly complete bases?

Problem 13.5. Does there exist a universal basis for the family $\mathscr{B}$ of all bases?

Problem 13.6. Does there exist a basis in Hilbert space l^2, which is universal for all bases of l^2?

II. Unconditional Bases and Some Classes of Unconditional Bases

§ 14. Unconditional bases. Conditional bases

Definition 14.1. A basis $\{x_n\}$ of a Banach space E is said to be *unconditional* if every convergent series of the form $\sum_{i=1}^{\infty} \alpha_i x_i$ is unconditionally convergent, i.e., if for every $x \in E$ the series $\sum_{i=1}^{\infty} f_i(x) x_i$ (where $\{f_n\}$ is the a.s.c.f. to $\{x_n\}$) converges unconditionally. The basis $\{x_n\}$ is said to be *conditional* if it is non-unconditional, i.e., if there exists a convergent series $\sum_{i=1}^{\infty} \alpha_i x_i$ which is not unconditionally convergent.

For instance, the natural bases of c_0 and $l^p (p \geqslant 1)$ are unconditional bases. We shall see in § 15 that the Schauder basis of $C([0,1])$ and the Haar basis of $L^1([0,1])$ are conditional bases. Now we shall prove that for $1 < p < \infty$ the Haar basis of $L^p([0,1])$ is an unconditional basis (theorem 14.1 below).

We recall that the Rademacher functions $r_k(t)$ on $[0,1]$ are defined by

$$r_k(t) = \operatorname{sign} \sin 2^k \pi t \qquad (k = 1, 2, \ldots). \tag{14.1}$$

[1] By [254], p. 57–60.

By the definitions of the Haar functions[1] $y_j(t)$ and of the Rademacher functions $r_k(t)$ we have, obviously, the following expression of the Rademacher functions with the aid of Haar functions:

$$r_k(t) = \frac{1}{\sqrt{2^{k-1}}} \sum_{j=1}^{2^{k-1}} y_{2^{k-1}+j}(t) \qquad (k=1,2,\ldots). \tag{14.2}$$

By (14.1) (or by (14.2) and the definition of Haar functions) we have, for each $k=1,2,\ldots$,

$$r_k(t)=\begin{cases} 1 \quad \text{for} \quad t\in\left(\dfrac{2l-2}{2^k}, \dfrac{2l-1}{2^k}\right) & (l=1,\ldots,2^k) \\ -1 \quad \text{for} \quad t\in\left(\dfrac{2l-1}{2^k}, \dfrac{2l}{2^k}\right) & (l=1,\ldots,2^k) \end{cases} \tag{14.3}$$

Since in the dyadic representation $t = \frac{t_1}{2} + \frac{t_2}{2^2} + \frac{t_3}{2^3} + \cdots$ of the points $t\in[0,1]$ the points of the open intervals $\left(\frac{2l-2}{2^k}, \frac{2l-1}{2^k}\right)$ $(l=1,\ldots,2^{k-1})$ are characterized by $t_k=0$, the points of $\left(\frac{2l-1}{2^k}, \frac{2l}{2^k}\right)$ $(l=1,\ldots,2^{k-1})$ by $t_k=1$, and the division points $\frac{j}{2^k}$ $(j=1,\ldots,2^k)$ by the fact that they admit two dyadic developments, one with $t_k=0$ (in this case $t_{k+1}=t_{k+2}=\cdots=1$) and one with $t_k=1$ (and $t_{k+1}=t_{k+2}=\cdots=0$), it follows that we have, for each $k=1,2,\ldots$

$$r_k(t)=r_k\left(\sum_{j=1}^{\infty}\frac{t_j}{2^j}\right)=\begin{cases} 1 & \text{if} \quad t_k=0 \\ -1 & \text{if} \quad t_k=1 \\ 0 & \text{if} \quad t_k=0 \text{ and } t_k=1 \text{ are both possible.} \end{cases} \tag{14.4}$$

Since $r_{k-1}(2t)=\operatorname{sign}\sin 2^{k-1}\pi 2t=\operatorname{sign}\sin 2^k\pi t$ for $0\leqslant t\leqslant\frac{1}{2}$ and $r_{k-1}(2t-1)=\operatorname{sign}\sin 2^{k-1}\pi(2t-1)=\operatorname{sign}\sin(2^k\pi t-2^{k-1}\pi)=\operatorname{sign}\sin 2^k\pi t$ for $\frac{1}{2}\leqslant t\leqslant 1$, we also have the following recursive formula for the Rademacher functions:

$$r_k(t) = \begin{cases} r_{k-1}(2t) & \text{for} \quad 0\leqslant t\leqslant\frac{1}{2} \\ r_{k-1}(2t-1) & \text{for} \quad \frac{1}{2}\leqslant t\leqslant 1 \end{cases} \qquad (k=2,3,\ldots). \tag{14.5}$$

---

[1] We now slightly change the definition of Haar functions, putting $y_{2^{k-1}+j}\left(\frac{l}{2^k}\right)=0$ for $j=1,\ldots,2^{k-1}$ and $l=0,1,\ldots,2^k$ $(k=1,2,\ldots)$.

We recall that the Walsh functions $w_j(t)$ on $[0,1]$ are defined by

$$w_1(t) \equiv 1,$$

$$w_{k+1}(t) = r_{n_1+1}(t) r_{n_2+1}(t) \dots r_{n_\nu+1}(t) \quad (k=1,2,\dots), \tag{14.6}$$

where

$$k = 2^{n_1} + 2^{n_2} + \dots + 2^{n_\nu}, \quad n_1 > n_2 > \dots > n_\nu \geqslant 0. \tag{14.7}$$

In particular, taking $k = 2^{n_1}$ $(n_1 = 0,1,2,\dots)$, we see that the Walsh system contains all the Rademacher functions, i.e., it is an extension of the Rademacher system.

Since the Walsh functions are defined by means of the Rademacher functions and since the Rademacher functions satisfy the recurrence formula (14.5), we also have the following recursive formula for the Walsh functions, for each $l = 2m+1$, $m = 1,2,\dots$:

$$w_l(t) = \begin{cases} w_{\frac{l+1}{2}}(2t) & \text{for } 0 \leqslant t \leqslant \frac{1}{2}, \\ w_{\frac{l+1}{2}}(2t-1) & \text{for } \frac{1}{2} \leqslant t \leqslant 1, \end{cases} \tag{14.8}$$

$$w_{l+1}(t) = \begin{cases} w_{\frac{l+1}{2}}(2t) & \text{for } 0 \leqslant t \leqslant \frac{1}{2}, \\ -w_{\frac{l+1}{2}}(2t-1) & \text{for } \frac{1}{2} \leqslant t \leqslant 1. \end{cases} \tag{14.9}$$

Indeed, since l is odd, we have $l = 2^{n_1} + 2^{n_2} + \dots + 2^{n_{\nu-1}} + 2^0$, whence, by (14.6), (14.5), and $\dfrac{l+1}{2} = \dfrac{l-1}{2} + 1 = 2^{n_1-1} + 2^{n_2-1} + \dots + 2^{n_{\nu-1}-1} + 1$, we obtain

$$w_l(t) = w_{(l-1)+1}(t) = r_{n_1+1}(t) r_{n_2+1}(t) \dots r_{n_{\nu-1}+1}(t)$$
$$= \begin{cases} r_{n_1}(2t) r_{n_2}(2t) \dots r_{n_{\nu-1}}(2t) = w_{\frac{l+1}{2}}(2t) & \text{for } 0 \leqslant t \leqslant \frac{1}{2}, \\ r_{n_1}(2t-1) r_{n_2}(2t-1) \dots r_{n_{\nu-1}}(2t-1) = w_{\frac{l+1}{2}}(2t-1) & \text{for } \frac{1}{2} \leqslant t \leqslant 1, \end{cases}$$

and

$$w_{l+1}(t) = r_{n_1+1}(t) r_{n_2+1}(t) \dots r_{n_{\nu-1}+1}(t) r_1(t)$$
$$= \begin{cases} r_{n_1}(2t) r_{n_2}(2t) \dots r_{n_{\nu-1}}(2t) = w_{\frac{l+1}{2}}(2t) & \text{for } 0 \leqslant t \leqslant \frac{1}{2}, \\ -r_{n_1}(2t-1) r_{n_2}(2t-1) \dots r_{n_{\nu-1}}(2t-1) = -w_{\frac{l+1}{2}}(2t-1) & \text{for } \frac{1}{2} \leqslant t \leqslant 1. \end{cases}$$

Let us also mention the following useful recurrence relation between the Walsh and Rademacher functions:

$$w_{2^k+l}(t) = r_{k+1}(t) w_l(t) \quad (l = 1,2,\dots,2^k;\ k = 0,1,2,\dots). \tag{14.10}$$

Indeed, writing $l-1=2^{n_2}+\cdots+2^{n_\nu}$, where $n_2>\cdots>n_\nu\geqslant 0$, we have, by (14.6),

$$\begin{aligned} w_{2^k+l}(t) &= w_{2^k+2^{n_2}+\cdots+2^{n_\nu}+1}(t) = r_{k+1}(t)\, r_{n_2+1}(t)\ldots r_{n_\nu+1}(t) \\ &= r_{k+1}(t)\, w_{2^{n_2}+\cdots+2^{n_\nu}+1}(t) = r_{k+1}\, w_l(t). \end{aligned}$$

Since the Walsh functions are defined by means of the Rademacher functions and the Rademacher functions can be expressed with the aid of Haar functions (see (14.2)), it is natural that the Walsh functions can be expressed directly with the aid of Haar functions. Indeed, by (14.10) and (14.2) we have

$$w_{2^k+l}(t)=\frac{1}{\sqrt{2^k}}\sum_{j=1}^{2^k} w_l(t)\, y_{2^k+j}(t) \qquad (l=1,2,\ldots,2^k;\ k=0,1,2,\ldots), \tag{14.11}$$

whence, since for $1\leqslant l\leqslant 2^k$ the function $w_l(t)$ is constant on each interval $\left(\frac{2j-2}{2^{k+1}},\frac{2j}{2^{k+1}}\right)=\left(\frac{j-1}{2^k},\frac{j}{2^k}\right)$ $(j=1,2,\ldots,2^k)$ (because writing $l-1=2^{n_1}+\cdots+2^{n_\nu}$, where $n_1>\cdots>n_\nu\geqslant 0$, we have, by (14.6), $w_l(t)=r_{n_1+1}(t)\ldots r_{n_\nu+1}(t)$, and here $n_1\leqslant k-1$ by $1\leqslant l\leqslant 2^k$, and thus $n_1+1\leqslant k$, whence w_l is constant on each $\left(\frac{j-1}{2^k},\frac{j}{2^k}\right)$), and since $\{t\in[0,1]\,|\,y_{2^k+j}(t)\neq 0\}\subset\left(\frac{j-1}{2^k},\frac{j}{2^k}\right)$, it follows that we can write

$$w_{2^k+l}(t)=\frac{1}{\sqrt{2^k}}\sum_{j=1}^{2^k} w_l\left(\left(\frac{j-1}{2^k},\frac{j}{2^k}\right)\right) y_{2^k+j}(t) \qquad (l=1,\ldots,2^k;\ k=0,1,2,\ldots), \tag{14.12}$$

or in other words,

$$w_{2^k+l}(t)=\frac{1}{\sqrt{2^k}}\left[y_{2^k+1}(t)\pm y_{2^k+2}(t)\pm\cdots\pm y_{2^k+l}(t)\right], \tag{14.13}$$

where the coefficient of y_{2^k+j} is $\frac{1}{\sqrt{2^k}}\, w_l\left(\left(\frac{j-1}{2^k},\frac{j}{2^k}\right)\right)$.

With the aid of (14.12), (14.8) and (14.9) one can also establish a recursive rule for the coefficients $\pm\frac{1}{\sqrt{2^k}}$ occurring in (14.13). Firstly, from (14.12) we have $w_3=\frac{1}{\sqrt{2}}(y_3+y_4)$, $w_4=\frac{1}{\sqrt{2}}(y_3-y_4)$, which we shall express by saying that $w_3, w_4=\frac{1}{\sqrt{2}}(y_3\pm y_4)$, with the matrix of coefficients

$$\begin{pmatrix} \frac{1}{\sqrt{2}} & \frac{1}{\sqrt{2}} \\ \frac{1}{\sqrt{2}} & -\frac{1}{\sqrt{2}} \end{pmatrix}. \tag{14.14}$$

Similarly, by (14.12) we have $w_5, w_6, w_7, w_8 = \frac{1}{2}(y_5 \pm y_6 \pm y_7 \pm y_8)$ with the matrix of coefficients

$$\begin{pmatrix} \frac{1}{2} & \frac{1}{2} & \frac{1}{2} & \frac{1}{2} \\ \frac{1}{2} & \frac{1}{2} & -\frac{1}{2} & -\frac{1}{2} \\ \frac{1}{2} & -\frac{1}{2} & \frac{1}{2} & -\frac{1}{2} \\ \frac{1}{2} & -\frac{1}{2} & -\frac{1}{2} & \frac{1}{2} \end{pmatrix}. \tag{14.15}$$

The matrix (14.14) can be obtained from (14.15) by the following rule: Denote by B_1 the first row and by B_2 the second row of (14.14), i.e., let the matrix (14.14) be $\begin{pmatrix} B_1 \\ B_2 \end{pmatrix}$. Then (14.15) will be the matrix

$$\begin{pmatrix} \frac{1}{\sqrt{2}} B_1 & \frac{1}{\sqrt{2}} B_1 \\ \frac{1}{\sqrt{2}} B_1 & -\frac{1}{\sqrt{2}} B_1 \\ \frac{1}{\sqrt{2}} B_2 & \frac{1}{\sqrt{2}} B_2 \\ \frac{1}{\sqrt{2}} B_2 & -\frac{1}{\sqrt{2}} B_2 \end{pmatrix}.$$

Actually, this shows also the desired general recursive rule for the coefficients in (14.13), namely, if $w_{2^{k-1}+1}, w_{2^{k-1}+2}, \ldots, w_{2^k} = \frac{1}{\sqrt{2^{k-1}}}(y_{2^{k-1}+1} \pm \cdots \pm y_{2^k})$ with the matrix of coefficients

$$\begin{pmatrix} B_1 \\ B_2 \\ \vdots \\ B_{2^{k-1}} \end{pmatrix}, \tag{14.16}$$

then $w_{2^k+1}, w_{2^k+2}, \ldots, w_{2^{k+1}} = \frac{1}{\sqrt{2^k}}(y_{2^k+1} \pm \cdots \pm y_{2^{k+1}})$, with the matrix of coefficients

$$\begin{pmatrix} \frac{1}{\sqrt{2}} B_1 & \frac{1}{\sqrt{2}} B_1 \\ \frac{1}{\sqrt{2}} B_1 & -\frac{1}{\sqrt{2}} B_1 \\ \frac{1}{\sqrt{2}} B_2 & \frac{1}{\sqrt{2}} B_2 \\ \frac{1}{\sqrt{2}} B_2 & -\frac{1}{\sqrt{2}} B_2 \\ \cdots & \cdots \\ \frac{1}{\sqrt{2}} B_{2^{k-1}} & \frac{1}{\sqrt{2}} B_{2^{k-1}} \\ \frac{1}{\sqrt{2}} B_{2^{k-1}} & -\frac{1}{\sqrt{2}} B_{2^{k-1}} \end{pmatrix}. \tag{14.17}$$

Indeed, from (14.12) for l and $l+1$ and from (14.8), (14.9) it follows that for odd l with $1 \leqslant l < 2^k$, $1 \leqslant k < \infty$, we have (taking also into account that $y_{2^k+j}=0$ on $[0,\frac{1}{2}]$ for $2^{k-1}+1 \leqslant j \leqslant 2^k$ and on $[\frac{1}{2},1]$ for $1 \leqslant j \leqslant 2^{k-1}$),

$$\begin{aligned} w_{2^k+l}(t) &= \frac{1}{\sqrt{2^k}} \sum_{j=1}^{2^k} w_l\left(\left(\frac{j-1}{2^k}, \frac{j}{2^k}\right)\right) y_{2^k+j}(t) \\ &= \frac{1}{\sqrt{2^k}} \sum_{j=1}^{2^{k-1}} w_{\frac{l+1}{2}}\left(\left(\frac{j-1}{2^{k-1}}, \frac{j}{2^{k-1}}\right)\right) y_{2^k+j}(t) \\ &\quad + \frac{1}{\sqrt{2^k}} \sum_{j=2^{k-1}+1}^{2^k} w_{\frac{l+1}{2}}\left(\left(\frac{j-1-2^{k-1}}{2^{k-1}}, \frac{j-2^{k-1}}{2^{k-1}}\right)\right) y_{2^k+j}(t), \end{aligned}$$

$$\begin{aligned} w_{2^k+l+1}(t) &= \frac{1}{\sqrt{2^k}} \sum_{j=1}^{2^k} w_{l+1}\left(\left(\frac{j-1}{2^k}, \frac{j}{2^k}\right)\right) y_{2^k+j}(t) \\ &= \frac{1}{\sqrt{2^k}} \sum_{j=1}^{2^{k-1}} w_{\frac{l+1}{2}}\left(\left(\frac{j-1}{2^{k-1}}, \frac{j}{2^{k-1}}\right)\right) y_{2^k+j}(t) \\ &\quad - \frac{1}{\sqrt{2^k}} \sum_{j=2^{k-1}+1}^{2^k} w_{\frac{l+1}{2}}\left(\left(\frac{j-1-2^{k-1}}{2^{k-1}}, \frac{j-2^{k-1}}{2^{k-1}}\right)\right) y_{2^k+j}(t), \end{aligned}$$

and, since $w_{\frac{l+1}{2}}$ is constant on $\left(\frac{j-1}{2^{k-1}}, \frac{j}{2^{k-1}}\right)$ (because from $1 \leqslant l \leqslant 2^k - 1$

it follows that $1 \leqslant \frac{l+1}{2} \leqslant 2^{k-1}$), we also have

$$w_{2^{k-1}+\frac{l+1}{2}}(t) = \frac{1}{\sqrt{2^{k-1}}} \sum_{j=1}^{2^{k-1}} w_{\frac{l+1}{2}}\left(\left(\frac{j-1}{2^{k-1}}, \frac{j}{2^{k-1}}\right)\right) y_{2^{k-1}+j},$$

and thus it remains only to observe that the sums $\sum_{j=2^{k-1}+1}^{2^k}$ above can be also written in the form $\sum_{j=1}^{2^{k-1}} w_{\frac{l+1}{2}}\left(\left(\frac{j-1}{2^{k-1}}, \frac{j}{2^{k-1}}\right)\right) y_{2^k+2^{k-1}+j}$. This completes the proof of our assertion concerning the recursive rule (14.17) for the coefficients occurring in (14.13), putting

$$B_m = \left(\frac{1}{\sqrt{2^{k-1}}} w_m\left(\left(\frac{j-1}{2^{k-1}}, \frac{j}{2^{k-1}}\right)\right)\right)_{j=1,\ldots,2^{k-1}} \qquad (m=1,2,\ldots,2^{k-1}).$$

We recall that a real matrix $A=(a_{ij})$ is called *symmetric*, if $A=A'$ (= the transposed matrix (a_{ji})), respectively *orthogonal* if $AA'=I$. We shall now prove that *all matrices of coefficients* $\begin{pmatrix} B_1 \\ B_2 \\ \vdots \\ B_{2^k} \end{pmatrix}$ *above are orthogonal and symmetric.*

Indeed, the matrix (14.14) is obviously orthogonal and symmetric, which shows that the assertions are valid for $k=1$. Let k be an arbitrary integer $\geqslant 2$ and assume that the first assertion is valid for $k-1$, i.e., that $\begin{pmatrix} B_1 \\ B_2 \\ \vdots \\ B_{2^{k-1}} \end{pmatrix}$ is orthogonal. Let us show that in this case the matrix (14.17) is also orthogonal. By our hypothesis we have

$$\begin{pmatrix} B_1 \\ B_2 \\ \vdots \\ B_{2^{k-1}} \end{pmatrix} (B_1' B_2' \ldots B_{2^{k-1}}') = \begin{pmatrix} B_1 B_1' & B_1 B_2' & \ldots & B_1 B_{2^{k-1}}' \\ B_2 B_1' & B_2 B_2' & \ldots & B_2 B_{2^{k-1}}' \\ \cdots & \cdots & \cdots & \cdots \\ B_{2^{k-1}} B_1' & B_{2^{k-1}} B_2' & \ldots & B_{2^{k-1}} B_{2^{k-1}}' \end{pmatrix}$$

$$= \begin{pmatrix} 1 & 0 & \ldots & 0 \\ 0 & 1 & \ldots & 0 \\ \cdots & \cdots & \cdots & \cdots \\ 0 & 0 & \ldots & 1 \end{pmatrix}$$

i.e.,

$$B_i B_j' = \delta_{ij} = \begin{cases} 1 & \text{for } i=j \\ 0 & \text{for } i \neq j \end{cases} \qquad (i,j=1,2,\ldots,2^{k-1}),$$

whence

$$\begin{pmatrix} \frac{1}{\sqrt{2}}B_1 & \frac{1}{\sqrt{2}}B_1 \\ \frac{1}{\sqrt{2}}B_1 & -\frac{1}{\sqrt{2}}B_1 \\ \cdots & \cdots \\ \frac{1}{\sqrt{2}}B_{2^{k-1}} & \frac{1}{\sqrt{2}}B_{2^{k-1}} \\ \frac{1}{\sqrt{2}}B_{2^{k-1}} & -\frac{1}{\sqrt{2}}B_{2^{k-1}} \end{pmatrix} \begin{pmatrix} \frac{1}{\sqrt{2}}B'_1 & \frac{1}{\sqrt{2}}B'_1 \ldots \frac{1}{\sqrt{2}}B'_{2^{k-1}} & \frac{1}{\sqrt{2}}B'_{2^{k-1}} \\ \frac{1}{\sqrt{2}}B'_1 & -\frac{1}{\sqrt{2}}B'_1 \ldots \frac{1}{\sqrt{2}}B'_{2^{k-1}} & -\frac{1}{\sqrt{2}}B'_{2^{k-1}} \end{pmatrix}$$

$$= \begin{pmatrix} \frac{1}{2}B_1B'_1+\frac{1}{2}B_1B'_1 & \frac{1}{2}B_1B'_1-\frac{1}{2}B_1B'_1 & \ldots \frac{1}{2}B_1B'_{2^{k-1}}-\frac{1}{2}B_1B'_{2^{k-1}} \\ \cdots & \cdots & \cdots \\ \frac{1}{2}B_{2^{k-1}}B'_1-\frac{1}{2}B_{2^{k-1}}B'_1 & \frac{1}{2}B_{2^{k-1}}B'_1+\frac{1}{2}B_{2^{k-1}}B'_1 & \ldots \frac{1}{2}B_{2^{k-1}}B'_{2^{k-1}}+\frac{1}{2}B_{2^{k-1}}B'_{2^{k-1}} \end{pmatrix}$$

$$= \begin{pmatrix} 1 & 0\ldots0 & 0 \\ 0 & 1\ldots0 & 0 \\ \cdots & \cdots & \cdots \\ 0 & 0\ldots0 & 1 \end{pmatrix},$$

which proves that (14.17) is orthogonal. Thus all matrices (14.16) are orthogonal ($k=2,3,\ldots$).

Let us prove now that all matrices (14.16) are also symmetric. For this purpose we recall that (14.16) is the matrix (b_{lj}), where

$$b_{lj}=w_l\left(\left(\frac{j-1}{2^{k-1}}, \frac{j}{2^{k-1}}\right)\right) \qquad (l,j=1,2,\ldots,2^{k-1}),$$

and thus we have to prove that $b_{lj}=b_{jl}$ $(l,j=1,2,\ldots,2^{k-1})$, i.e.,

$$w_l\left(\left(\frac{j-1}{2^{k-1}}, \frac{j}{2^{k-1}}\right)\right) = w_j\left(\left(\frac{l-1}{2^{k-1}}, \frac{l}{2^{k-1}}\right)\right) \qquad (l,j=1,2,\ldots,2^{k-1}). \tag{14.18}$$

Let us write

$$l-1=2^{n_1}+\cdots+2^{n_\nu}, \quad j-1=2^{m_1}+\cdots+2^{m_\mu},$$

where $n_1>\cdots>n_\nu\geqslant 0$, $m_1>\cdots>m_\mu\geqslant 0$, and put $J=\left(\frac{j-1}{2^{k-1}}, \frac{j}{2^{k-1}}\right)$, $L=\left(\frac{l-1}{2^{k-1}}, \frac{l}{2^{k-1}}\right)$. Then we have to prove that

$$r_{n_1+1}(J)\ldots r_{n_\nu+1}(J)=w_l(J)=w_j(L)=r_{m_1+1}(L)\ldots r_{m_\mu+1}(L).$$

The midpoint of J is $\frac{2j-1}{2^k}=\frac{2(j-1)}{2^k}+\frac{1}{2^k}=\frac{2^{m_1+1}+\cdots+2^{m_\mu+1}+1}{2^k}$ $=\left(\frac{1}{2}\right)^{k-1-m_1}+\cdots+\left(\frac{1}{2}\right)^{k-1-m_\mu}+\left(\frac{1}{2}\right)^k$. Thus by (14.4), we have $r_{n_\lambda+1}(J)=r_{n_\lambda+1}\left(\frac{2j-1}{2^k}\right)=-1$ if and only if $n_\lambda+1$ coincides with one of the numbers $k-1-m_1,\ldots,k-1-m_\mu,k$, and otherwise we have $r_{n_\lambda+1}(J)=1$ $(\lambda=1,\ldots,\nu)$. Let us observe that here $n_\lambda+1=k$ is impossible, since by $l\leqslant 2^{k-1}$ we have $2^{n_1}+\cdots+2^{n_\nu}=l-1\leqslant 2^{k-1}-1=2^{k-2}+2^{k-3}+\cdots+2$, whence $n_\lambda\leqslant k-2$ and thus $n_\lambda+1\leqslant k-1$ $(\lambda=1,\ldots,\nu)$. Similarly, $r_{m_\lambda+1}(L)=-1$ if and only if $m_\lambda+1$ coincides with one of the numbers $k-1-n_1,\ldots,k-1-n_\nu$ (the case $m_\lambda+1=k$ being again impossible), otherwise $r_{m_\lambda+1}(L)=1$. However, it is obvious that we have $n_\lambda+1=k-1-m_\alpha$ if and only if $m_\alpha+1=k-1-n_\lambda$. Hence $w_l(J)=w_j(L)$, which completes the proof of the symmetry of the matrix (14.16).

We have seen in Ch. I, § 2, that the Haar system is orthonormal on [0,1]. We shall prove now that the Rademacher and Walsh systems also have this property. This will follow from the next lemma, which we shall use also later (in the proof of lemma 14.4):

Lemma 14.1. *Let $m_1>m_2>\cdots>m_\mu\geqslant 1$ and let $\alpha_1,\ldots,\alpha_\mu$ be integers We have*

$$\int_0^1 r_{m_1}^{\alpha_1}(t)r_{m_2}^{\alpha_2}(t)\ldots r_{m_\mu}^{\alpha_\mu}(t)\,dt=\begin{cases}1 & \text{if all } \alpha_j \text{ are even,}\\ 0 & \text{otherwise.}\end{cases}\tag{14.19}$$

Proof. The first relation is obvious, since if all α_j are even, then each $r_{m_j}^{\alpha_j}(t)$, whence also the integrand, is $=1$ a.e. For the proof of the second relation we may assume that $\alpha_1=\cdots=\alpha_\mu=1$ (since for any even number β and any j we have $r_{m_j}^{\beta}(t)=1$ a.e.). The product $r_{m_2}(t)\ldots r_{m_\mu}(t)$ admits (by $m_2>\cdots>m_\mu$) 2^{m_2} dyadic intervals of constantness (each of length $\frac{1}{2^{m_2}}$), the union of which is [0,1]. Any such interval J is divided into $2^{m_1-m_2}$ parts (each of length $\frac{1}{2^{m_1}}$), in which $r_{m_1}(t)$ is alternatively $+1,-1$. Consequently, $\int_0^1 r_{m_1}(t)r_{m_2}(t)\ldots r_{m_\mu}(t)\,dt$ is decomposed into a sum of 2^{m_2} terms of the form $\text{const.}\int_J r_{m_1}(t)\,dt$, and each term is 0 since $\int_J r_{m_1}(t)\,dt=0$ (because of the above decomposition of J). This completes the proof of lemma 14.1.

Corollary 14.1. *The Walsh system $\{w_n\}$ (whence also, in particular, the Rademacher system $\{r_n\}$) constitutes an orthonormal system on $[0,1]$, i.e.,*

$$\int_0^1 w_m(t)w_n(t)dt=\begin{cases}1 & \text{for} \quad m=n\\ 0 & \text{for} \quad m\neq n\end{cases} \qquad (m,n=1,2,\ldots). \tag{14.20}$$

Proof. The first relation is obvious, since $w_n^2(t)=1$ a.e. Assume now that $m\neq n$. Then there exists at least one m_k such that r_{m_k} occurs as a factor in w_m, but not in w_n, or as a factor in w_n, but not in w_m. Consequently, $\int_0^1 w_m(t)w_n(t)dt$ is of the form (14.19) with $\alpha_k=1$, and thus by lemma 14.1 it is $=0$, which completes the proof.

As we have seen in Ch. I, § 2, the Haar system[1] $\{y_n\}$ is *complete* in the spaces $L^p([0,1])$ for $1\leqslant p<\infty$ (being even a basis of these spaces). We shall now show that the Walsh system $\{w_n\}$ is also *complete* in these spaces. Since both $\{y_n\}$ and $\{w_n\}$ are (finitely) linearly independent (being orthogonal systems), by (14.12) we have $[w_{2^k+1}, w_{2^k+2},\ldots,w_{2^{k+1}}]$ $=[y_{2^k+1},y_{2^k+2},\ldots,y_{2^{k+1}}]$ (because we have by (14.12) the inclusion $\subseteq$ and both subspaces are of the same dimension 2^k by the linear independence of $\{y_j\}$, $\{w_j\}$). Taking into account that $y_1=w_1$ and again the linear independence of $\{y_j\}$, $\{w_j\}$, it follows that $[w_n]_1^\infty=[y_n]_1^\infty=L^p([0,1])$ $(1\leqslant p<\infty)$, which completes the proof of our assertion.

Define now $\{g_n\}, \{h_n\}\subset L^p([0,1])^*\equiv L^q([0,1])$ $\left(1\leqslant p<\infty, \frac{1}{p}+\frac{1}{q}=1\right)$ by

$$g_n(x)=\int_0^1 x(t)y_n(t)dt \qquad (x\in L^p([0,1]), n=1,2,\ldots), \tag{14.21}$$

$$h_n(x)=\int_0^1 x(t)w_n(t)dt \qquad (x\in L^p([0,1]), n=1,2,\ldots). \tag{14.22}$$

Then, since $y_n, w_n\in L^q([0,1])$ $(1<q\leqslant\infty)$, we have $g_n,h_n\in L^p([0,1])^*$ $(1\leqslant p<\infty)$. Since $\{y_n\}$, $\{w_n\}$ are orthonormal systems, it follows that $(\{y_n\},\{g_n\})$ and $(\{w_n\},\{h_n\})$ are biorthogonal systems. Let us denote respectively by $\{s_n\}$ and $\{S_n\}$ the sequences of partial sum operators associated to these biorthogonal systems, i.e.,

$$s_n(x)=\sum_{i=1}^n g_i(x)y_i,\quad S_n(x)=\sum_{i=1}^n h_i(x)w_i \qquad (x\in L^p([0,1]), n=1,2,\ldots). \tag{14.23}$$

---

[1] From now on, we shall write y_n instead of $\tilde{y}_n$ and x instead of $\tilde{x}$. This will not lead to any confusion.

Lemma 14.2. *With the above notations, we have*

$$s_{2^k}(x)=S_{2^k}(x) \quad (x\in L^p([0,1]),\ k=0,1,2,\ldots). \tag{14.24}$$

Proof. Take an $x\in L^p([0,1])$ and a positive integer k and put

$$b_{lj}=\frac{1}{\sqrt{2^{k-1}}}\,w_l\left(\left(\frac{j-1}{2^{k-1}},\ \frac{j}{2^{k-1}}\right)\right) \quad (l,j=1,2,\ldots,2^{k-1}).$$

Then, by (14.12) (for $k-1$ instead of k), we have

$$\begin{aligned} S_{2^k}(x)-S_{2^{k-1}}(x)&=\sum_{l=1}^{2^{k-1}} h_{2^{k-1}+l}(x)\,w_{2^{k-1}+l}=\sum_{l=1}^{2^{k-1}} h_{2^{k-1}+l}(x)\sum_{j=1}^{2^{k-1}} b_{lj}y_{2^{k-1}+j}\\ &=\sum_{j=1}^{2^{k-1}} y_{2^{k-1}+j}\sum_{l=1}^{2^{k-1}} h_{2^{k-1}+l}(x)\,b_{lj}. \end{aligned}$$

Since the matrix $(b_{lj})_{l,j=1,\ldots,2^{k-1}}$ is orthogonal and symmetric (as shown in the above), we can express $y_{2^{k-1}+l}$ $(l=1,\ldots,2^{k-1})$ by $w_{2^{k-1}+j}$ $(j=1,\ldots,2^{k-1})$ with the same matrix that expressed $w_{2^{k-1}+l}$ by $y_{2^{k-1}+j}$ in (14.12) (for $k-1$ instead of k), whence

$$\begin{aligned} g_{2^{k-1}+j}(x)&=\int_0^1 x(t)\,y_{2^{k-1}+j}(t)\,dt=\int_0^1 x(t)\left[\sum_{l=1}^{2^{k-1}} b_{lj}w_{2^{k-1}+l}(t)\right]dt\\ &=\sum_{l=1}^{2^{k-1}} b_{lj}h_{2^{k-1}+l}(x), \end{aligned}$$

and thus

$$\begin{aligned} S_{2^k}(x)-S_{2^{k-1}}(x)&=\sum_{j=1}^{2^{k-1}} y_{2^{k-1}+j}\sum_{j=1}^{2^{k-1}} h_{2^{k-1}+l}(x)\,b_{lj}\\ &=\sum_{j=1}^{2^{k-1}} y_{2^{k-1}+j}\,g_{2^{k-1}+j}(x)=s_{2^k}(x)-s_{2^{k-1}}(x), \end{aligned} \tag{14.25}$$

for all $x\in L^p([0,1])$ and $k=1,2,\ldots$ On the other hand, we have

$$\begin{aligned} S_1(x)=h_1(x)\,w_1=w_1\int_0^1 x(t)\,w_1(t)\,dt=y_1\int_0^1 x(t)\,y_1(t)\,dt\\ =g_1(x)\,y_1=s_1(x) \quad (x\in L^p([0,1])). \end{aligned}$$

Hence, by (14.25) applied successively for $k=1,2,\ldots$ we obtain (14.24), which completes the proof of lemma 14.2.

Let us introduce the notation

$$\Delta_k(x)=S_{2^k}(x)-S_{2^{k-1}}(x)=\sum_{j=1}^{2^{k-1}} h_{2^{k-1}+j}(x)w_{2^{k-1}+j} \qquad (x\in L^p([0,1]),\ k=1,2,\dots). \tag{14.26}$$

Lemma 14.3. *For $p=2\nu$ $(\nu=1,2,\dots)$ there exist positive constants m_p and M_p depending only on p such that*

$$m_p\int_0^1\left\{\sum_{k=1}^{\infty}[\Delta_k(x)]^2(t)\right\}^{\frac{p}{2}}dt\leqslant\int_0^1|x(t)|^p\,dt\leqslant M_p\int_0^1\left\{\sum_{k=1}^{\infty}[\Delta_k(x)]^2(t)\right\}^{\frac{p}{2}}dt \qquad \left(x\in L^1([0,1]),\ \int_0^1 x(t)\,dt=0\right) \tag{14.27}$$

whenever one of the members exists.

Accepting, for the moment, lemma 14.3, we shall prove

Theorem 14.1. *For $1<p<\infty$ the Haar system $\{y_n\}$ is an unconditional basis of the space $L^p([0,1])$.*

Proof. Let $p=2\nu$ and $x\in L^p([0,1]$, $\int_0^1 x(t)\,dt=0$. By lemma 14.2 we have

$$\Delta_k^2(x)=[S_{2^k}(x)-S_{2^{k-1}}(x)]^2=[s_{2^k}(x)-s_{2^{k-1}}(x)]^2$$
$$=\left[\sum_{j=1}^{2^{k-1}} g_{2^{k-1}+j}(x)y_{2^{k-1}+j}\right]^2=\sum_{j=1}^{2^{k-1}} g^2_{2^{k-1}+j}(x)y^2_{2^{k-1}+j} \qquad (k=1,2,\dots)$$

(the other products are $=0$ because of the definition of Haar functions). Replacing this in (14.27), we obtain

$$m_p\int_0^1\left[\sum_{l=2}^{\infty} g_l^2(x)y_l^2(t)\right]^{\frac{p}{2}}dt\leqslant\int_0^1|x(t)|^p\,dt\leqslant M_p\int_0^1\left[\sum_{l=2}^{\infty} g_l^2(x)y_l^2(t)\right]^{\frac{p}{2}}dt. \tag{14.28}$$

Now let $x\in L^p([0,1])$ be arbitrary. Then the element $z=x-g_1(x)y_1$ $\in L^p([0,1])$ satisfies $\int_0^1 z(t)\,dt=0$, $g_l(z)=g_l(x)$ $(l=2,3,\dots)$, whence, by (14.28) applied to z,

$$m_p\int_0^1\left[\sum_{l=2}^{\infty} g_l^2(x)y_l^2(t)\right]^{\frac{p}{2}}dt\leqslant\int_0^1|x(t)-g_1(x)y_1(t)|^p\,dt$$
$$\leqslant M_p\int_0^1\left[\sum_{l=2}^{\infty} g_l^2(x)y_l^2(t)\right]^{\frac{p}{2}}dt \qquad (x\in L^p([0,1])).$$

Consequently, taking into account that $a^p + b^p \leqslant (a^2+b^2)^{\frac{p}{2}} \leqslant 2^{\frac{p}{2}-1}(a^p+b^p)$ and that $\int_0^1 |g_1(x)y_1(t)|^p dt = |g_1(x)|^p \leqslant \int_0^1 |x(t)|^p dt$ and denoting the constant $\frac{m_p}{2^{\frac{p}{2}-1}(m_p+2)}$ again by m_p, we get

$$m_p \int_0^1 \left[\sum_{l=1}^{\infty} g_l^2(x) y_l^2(t)\right]^{\frac{p}{2}} dt \leqslant \int_0^1 |x(t)|^p dt \leqslant M_p \int_0^1 \left[\sum_{l=1}^{\infty} g_l^2(x) y_l^2(t)\right]^{\frac{p}{2}} dt \tag{14.29}$$
$$(x \in L^p([0,1])).$$

Applying this for $s_n(x)$ instead of x and taking into account that $g_i(y_j) = \delta_{ij}$ (whence $g_l[s_n(x)] = g_l(x)$ for $1 \leqslant l \leqslant n$ and $=0$ for $l > n$), we obtain

$$\|s_n(x)\| \leqslant M_p \left\{\int_0^1 \left[\sum_{l=1}^{n} g_l^2(x) y_l^2(t)\right]^{\frac{p}{2}} dt\right\}^{\frac{1}{p}} \leqslant M_p \left\{\int_0^1 \left[\sum_{l=1}^{\infty} g_l^2(x) y_l^2(t)\right]^{\frac{p}{2}} dt\right\}^{\frac{1}{p}}$$

$$\leqslant \frac{M_p}{m_p}\left\{\int_0^1 |x(t)|^p dt\right\}^{\frac{1}{p}} = \frac{M_p}{m_p}\|x\| \qquad (x \in L^p([0,1]),\ n = 1, 2, \ldots),$$

and thus, by Ch. I, § 4, theorem 4.1, $\{y_n\}$ is a basis of $L^p([0,1])$. By virtue of (14.29) the same argument remains valid for any permutation $\{y_{\sigma(n)}\}$ of $\{y_n\}$ and therefore, obviously[1], $\{y_n\}$ is an unconditional basis of $L^p([0,1])$, for any $p = 2\nu$ $(\nu = 1, 2, \ldots)$.

Now, for any p with $2 \leqslant p < \infty$, consider s_n as a linear operator from $L^p([0,1])$ into itself and denote its norm by $\|s_n\|_p$. Then, by the above, for every even p there exists a constant $C_p \geqslant 1$ such that

$$\|s_n\|_p \leqslant C_p \qquad (n = 1, 2, \ldots). \tag{14.30}$$

Assume now that $2\nu \leqslant p < 2\nu + 2$. Then $\frac{1}{2\nu+2} < \frac{1}{p} \leqslant \frac{1}{2\nu}$, whence there exists a (unique) λ with $0 < \lambda \leqslant 1$ such that

$$\frac{1}{p} = \lambda \frac{1}{2\nu+2} + (1-\lambda)\frac{1}{2\nu},$$

[1] If every permutation of a basis $\{x_n\}$ is a basis, then $\{x_n\}$ is an unconditional basis (and conversely); indeed, see § 17, theorem 17.1.

whence, since by the Riesz convexity theorem[1] $\psi_n\left(\frac{1}{p}\right)=\log\|s_n\|_p$ is a convex function of $\frac{1}{p}$ for $0\leqslant\frac{1}{p}\leqslant 1$, it follows that

$$\log\|s_n\|_p=\psi_n\left(\frac{1}{p}\right)\leqslant\lambda\psi_n\left(\frac{1}{2\nu+2}\right)+(1-\lambda)\psi_n\left(\frac{1}{2\nu}\right)$$

$$=\lambda\log\|s_n\|_{2\nu+2}+(1-\lambda)\log\|s_n\|_{2\nu}.$$

and therefore

$$\|s_n\|_p\leqslant\|s_n\|_{2\nu+2}^{\lambda}\,\|s_n\|_{2\nu}^{1-\lambda}\leqslant C_p^{\lambda}C_p^{1-\lambda}=C_p.$$

Thus, we have proved that (14.30) holds for any p with $2\leqslant p<\infty$, whence, by Ch. I, § 4, theorem 4.1, $\{y_n\}$ is a basis of $L^p([0,1])$ for any p with $2\leqslant p<\infty$. Again, the same argument remains valid for any permutation $\{y_{\sigma(n)}\}$ of $\{y_n\}$ and therefore $\{y_n\}$ is an unconditional basis of $L^p([0,1])$ for any p with $2\leqslant p<\infty$.

Finally, assume that $1<p<2$. Then, by the above, $\{g_n\}$ is an unconditional basis of $L^q([0,1])$, where $\frac{1}{p}+\frac{1}{q}=1$, whence $\{y_n\}$ is an unconditional basis[2] of $L^p([0,1])$, which completes the proof of theorem 14.1, provided that we prove lemma 14.3.

In order to prove lemma 14.3, we shall need several lemmas.

Lemma 14.4. *Let* $x\in L^1([0,1])$ *and let* $m_1,m_2,\dots,m_\nu$ *be positive integers, with* $m_2,\dots,m_\nu$ *not necessarily distinct, such that* $m_1>\max(m_2,m_3,\dots,m_\nu)$. *Then for* $\Delta_k(x)$ *defined by* (14.26) *we have*

$$\int_0^1[\Delta_{m_1}(x)](t)[\Delta_{m_2}(x)](t)\dots[\Delta_{m_\nu}(x)](t)\,dt=0. \tag{14.31}$$

Proof. By the definitions of $\Delta_k(x)$ and w_n we have

$$[\Delta_{m_k}(x)](t)=\sum_{j=1}^{2^{m_k-1}}h_{2^{m_k-1}+j}(x)\,w_{2^{m_k-1}+(j-1)+1}(t)$$

$$=h_{2^{m_k-1}+1}(x)r_{m_k}(t)+h_{2^{m_k-1}+2}(x)r_{m_k}(t)r_1(t)+\cdots$$

$$+h_{2^{m_k}}(x)r_{m_k}(t)r_{m_k-1}(t)\dots r_1(t).$$

---

[1] See e.g. [50], p. 523, theorem 8.

[2] See § 17, theorem 17.7 and Ch. I, § 12, corollary 12.2 (or, alternatively, the final part of § 17, theorem 17.7 and Ch. I, § 2, example 2.3).

Hence, when we express $[\Delta_{m_2}(x)](t),\ldots,[\Delta_{m_\nu}(x)](t)$ as linear combinations of Walsh functions, none of the terms contains $r_{m_1}(t)$ as a factor, while in the expression of $[\Delta_{m_1}(x)](t)$ all terms contain r_{m_1} as a factor. Consequently, applying lemma 14.1, we get (14.31), which completes the proof.

Lemma 14.5. *For any* $t,\theta\in[0,1]$ *and* $k=1,2,\ldots$ *we have*

$$\sum_{m=1}^{2^k} w_m(t)w_m(\theta)=\prod_{m=1}^{k}\{1+r_m(t)r_m(\theta)\}. \tag{14.32}$$

Proof. Let us denote the left side by σ_{2^k} and the right side by π_k. Let $k=1$. Then (14.32) becomes

$$w_1(t)w_1(\theta)+w_2(t)w_2(\theta)=1+r_1(t)r_1(\theta),$$

which is true since by definition $w_1\equiv 1$, $w_2=r_1$.

Assume now that (14.32) is valid for a positive integer k. Since by (14.10) we have

$$w_{2^k+l}(t)w_{2^k+l}(\theta)=r_{k+1}(t)r_{k+1}(\theta)w_l(t)w_l(\theta),$$

it follows that

$$\sigma_{2^{k+1}}-\sigma_{2^k}=\sum_{l=1}^{2^k} w_{2^k+l}(t)w_{2^k+l}(\theta)$$

$$=r_{k+1}(t)r_{k+1}(\theta)\sum_{l=1}^{2^k} w_l(t)w_l(\theta)=r_{k+1}(t)r_{k+1}(\theta)\sigma_{2^k}.$$

Consequently, taking into account the induction hypothesis, we get

$$\sigma_{2^{k+1}}=\sigma_{2^k}[1+r_{k+1}(t)r_{k+1}(\theta)]=\pi_k[1+r_{k+1}(t)r_{k+1}(\theta)]=\pi_{k+1},$$

i.e., (14.32) for $k+1$. This completes the proof.

Corollary 14.2. *For* $x\in L^1([0,1])$, $k=1,2,\ldots$ *we have*

$$[\Delta_k(x)](t)=\int_0^1 x(\theta)r_k(t)r_k(\theta)\prod_{m=1}^{k-1}\{1+r_m(t)r_m(\theta)\}\,d\theta. \tag{14.33}$$

Proof. We have, by lemma 14.5,

$$[\Delta_k(x)](t) = \sum_{j=1}^{2^{k-1}} w_{2^{k-1}+j}(t) \int_0^1 x(\theta) w_{2^{k-1}+j}(\theta)\,d\theta$$

$$= \int_0^1 x(\theta) \left[\sum_{j=1}^{2^{k-1}} w_{2^{k-1}+j}(t) w_{2^{k-1}+j}(\theta) \right] d\theta$$

$$= \int_0^1 x(\theta) \left[\sum_{m=1}^{2^k} w_m(t) w_m(\theta) - \sum_{m=1}^{2^{k-1}} w_m(t) w_m(\theta) \right] d\theta$$

$$= \int_0^1 x(\theta) \left[\prod_{m=1}^{k} \{1 + r_m(t) r_m(\theta)\} - \prod_{m=1}^{k-1} \{1 + r_m(t) r_m(\theta)\} \right] d\theta$$

$$= \int_0^1 x(\theta) \prod_{m=1}^{k-1} \{1 + r_m(t) r_m(\theta)\} [1 + r_k(t) r_k(\theta) - 1]\,d\theta,$$

which completes the proof.

Lemma 14.6. *For* $2 \leqslant p \leqslant \infty$ *and* $x \in L^p([0,1])$ *we have*

$$\sum_{k=1}^{\infty} \int_0^1 |[\Delta_k(x)](t)|^p\,dt \leqslant \int_0^1 |x(t)|^p\,dt, \tag{14.34}$$

where for $p = \infty$ *we take* $\operatorname{ess\,sup}_{t \in [0,1]} |\cdot|$ *instead of* $\int_0^1 |\cdot|^p\,dt$ *and* $\sup_{1 \leqslant k < \infty}$ *instead of* $\sum_{k=1}^{\infty}$.

Proof. Let us first observe that for $t \neq \frac{j}{2^{k-1}}$, $\theta \neq \frac{j}{2^{k-1}}$ $(j = 1, \ldots, 2^{k-1})$ the function

$$\phi_t(\theta) = \prod_{m=1}^{k-1} \{1 + r_m(t) r_m(\theta)\} \qquad (0 \leqslant \theta \leqslant 1) \tag{14.35}$$

vanishes except on a dyadic interval J of length $\frac{1}{2^{k-1}}$, containing t, and in this interval it takes the value 2^{k-1}. Indeed, since $r_m = \pm 1$ (for the points $\neq \frac{j}{2^{k-1}}$), there are two cases: a) There exists an index l,

$1 \leqslant l \leqslant k-1$, such that $r_l(t) = -r_l(\theta)$. In this case, $\phi_t(\theta) = 0$. b) There exists no such l, i.e., $r_m(t) = r_m(\theta)$ $(m = 1, \ldots, k-1)$. Then, since $r_1(t) = r_1(\theta)$, t and θ are both in the same (open) dyadic interval of length $\frac{1}{2}$ (i.e., either in $(0, \frac{1}{2})$, or in $(\frac{1}{2}, 1)$). Furthermore, since we also have $r_2(t) = r_2(\theta)$, it follows that t and θ must be in the same dyadic interval of length $\frac{1}{4}$. Continuing in this manner, we see, finally (since also $r_{k-1}(t) = r_{k-1}(\theta)$), that t and θ are both in the same dyadic interval J of length $\frac{1}{2^{k-1}}$. Now in J all $1 + r_m(t) r_m(\theta) = 2$ $(m = 1, 2, \ldots, k-1)$, whence $\phi_t(\theta) = 2^{k-1}$ in J, which proves our assertion.

Consequently, by (14.33) we obtain

$$\begin{aligned} |[\Delta_k(x)](t)| &= \left| \int_0^1 x(\theta) r_k(t) r_k(\theta) \prod_{m=1}^{k-1} \{1 + r_m(t) r_m(\theta)\} \, d\theta \right| \\ &= 2^{k-1} \left| \int_J x(\theta) r_k(t) r_k(\theta) \, d\theta \right| \leqslant 2^{k-1} \int_J |x(\theta)| \, d\theta \\ &\leqslant \operatorname*{ess\,sup}_{\theta \in J} |x(\theta)| \leqslant \operatorname*{ess\,sup}_{t \in [0,1]} |x(t)| . \end{aligned}$$

whence

$$\sup_{1 \leqslant k < \infty} \operatorname*{ess\,sup}_{t \in [0,1]} |[\Delta_k(x)](t)| \leqslant \operatorname*{ess\,sup}_{t \in [0,1]} |x(t)| . \tag{14.36}$$

On the other hand, since by corollary 14.1 and the remark made after this corollary, $\{w_n\}$ is a complete orthonormal system on $[0, 1]$, we have

$$\int_0^1 |[\Delta_k(x)](t)|^2 \, dt = \sum_{n=1}^{\infty} |h_n[\Delta_k(x)]|^2 = \sum_{j=1}^{2^{k-1}} |h_{2^{k-1}+j}(x)|^2 ,$$

and hence

$$\begin{aligned} \left\{ \sum_{k=1}^{\infty} \int_0^1 |[\Delta_k(x)](t)|^2 \, dt \right\}^{\frac{1}{2}} &= \left\{ \sum_{k=1}^{\infty} \sum_{j=1}^{2^{k-1}} |h_{2^{k-1}+j}(x)|^2 \right\}^{\frac{1}{2}} = \left\{ \sum_{n=2}^{\infty} |h_n(x)|^2 \right\}^{\frac{1}{2}} \\ &\leqslant \left\{ \sum_{n=1}^{\infty} |h_n(x)|^2 \right\}^{\frac{1}{2}} = \left\{ \int_0^1 |x(t)|^2 \, dt \right\}^{\frac{1}{2}} . \end{aligned} \tag{14.37}$$

Now, let $L_0 = L_0([0, 1])$ be the linear subspace of $L^p([0, 1])$ consisting of all finite linear combinations of the Walsh functions w_n and let $\mathscr{X}_p = l^p_{L^p([0,1])}$ be the Banach space of all p-summable (respectively, bounded) sequences $\{x_k\}$, where $x_k \in L^p([0, 1])$ $(k = 1, 2, \ldots)$, $\|\{x_k\}\|_{\mathscr{X}_p} = \left[\sum_{k=1}^{\infty} \|x_k\|^p_{L^p([0,1])} \right]^{\frac{1}{p}}$ (respectively, $x_k \in L^\infty([0, 1])$ for $k = 1, 2, \ldots$ and

$\|\{x_k\}\|_{\mathscr{X}_\infty} = \sup\limits_{1 \leqslant k < \infty} \|x_k\|_{L^\infty([0,1])}$). Then, by biorthogonality, for every $x \in L_0$ the sum $\sum\limits_{k=1}^{\infty} \|\Delta_k(x)\|^p$ has only a finite number of non-zero terms, whence $\{\Delta_k(x)\} \in \mathscr{X}_p$. Consequently, we can define a linear mapping $u: L_0 \to \mathscr{X}_p$ by $x \to \{\Delta_k(x)\}$. Let $\|u\|_p$ be the norm of this mapping, i. e.,

$$\|u\|_p = \sup_{\substack{x \in L_0 \\ x \neq 0}} \frac{\|u(x)\|}{\|x\|} = \sup_{\substack{x \in L_0 \\ x \neq 0}} \frac{\left[\sum\limits_{k=1}^{\infty} \|\Delta_k(x)\|^p\right]^{\frac{1}{p}}}{\|x\|}$$

$$= \sup_{\substack{x \in L_0 \\ x \neq 0}} \frac{\left\{\sum\limits_{k=1}^{\infty} \int\limits_0^1 |[\Delta_k(x)](t)|^p \, dt\right\}^{\frac{1}{p}}}{\left\{\int\limits_0^1 |x(t)|^p \, dt\right\}^{\frac{1}{p}}}.$$

Then, by the convexity theorem of M. Riesz for vector-valued functions[1], $\psi\left(\frac{1}{p}\right) = \log \|u\|_p$ is a convex function of $\frac{1}{p}$ for $0 \leqslant \frac{1}{p} \leqslant 1$. Since by (14.36), (14.37) we have $\|u\|_\infty \leqslant 1$, $\|u\|_2 \leqslant 1$, it follows that $\|u\|_p \leqslant 1$ $(2 \leqslant p \leqslant \infty)$. Indeed, for any p with $0 \leqslant \frac{1}{p} \leqslant \frac{1}{2}$ there exists a λ with $0 \leqslant \lambda \leqslant 1$ such that $\frac{1}{p} = \lambda \cdot 0 + (1-\lambda)\frac{1}{2}$ (namely, $\lambda = \frac{p-2}{p}$). Consequently,

$$\log \|u\|_p = \psi\left(\frac{1}{p}\right) \leqslant \lambda \psi(0) + (1-\lambda)\psi\left(\frac{1}{2}\right) = \lambda \log \|u\|_\infty + (1-\lambda) \log \|u\|_2$$

whence $\|u\|_p \leqslant \|u\|_\infty^\lambda \|u\|_2^{1-\lambda} \leqslant 1$. Thus we can extend u by continuity to $\tilde{u}: L^p([0,1]) \to \mathscr{X}_p$ and we have (14.34), which completes the proof of lemma 14.6.

Lemma 14.7. *Let* $1 < p < \infty$ *and let* $n(t)$ *be an arbitrary non-negative integer-valued function of* $t \in [0,1]$. *Then there exists a constant* $C_p > 0$ *depending only on* p, *such that*

$$\int\limits_0^1 |[S_{2^{n(t)}}(x)](t)|^p \, dt \leqslant C_p \int\limits_0^1 |x(t)|^p \, dt \qquad (x \in L^p([0,1])). \tag{14.38}$$

Proof. Put

$$\beta_x(t) = \operatorname*{ess\,sup}_{s \in (0,1)} \frac{1}{2s} \int\limits_{t-s}^{t+s} |x(\theta)| \, d\theta, \tag{14.39}$$

with the convention that $x(\theta) = 0$ for $\theta \notin [0,1]$.

---

[1] See, e.g., [50], p. 536, exercise 39.

Then, by a "maximal theorem" of Hardy and Littlewood [104][1], there exists a constant $A_p>0$ such that

$$\int_0^1 [\beta_x(t)]^p\,dt \leqslant A_p \int_0^1 |x(t)|^p\,dt \qquad (x\in L^p([0,1])). \tag{14.40}$$

On the other hand, by lemma 14.5 we have, for any $k=1,2,\ldots$

$$[S_{2^k}(x)](t) = \sum_{m=1}^{2^k} w_m(t)\int_0^1 x(\theta)\,w_m(\theta)\,d\theta = \int_0^1 x(\theta)\left[\sum_{m=1}^{2^k} w_m(t)\,w_m(\theta)\right]d\theta$$

$$= \int_0^1 x(\theta)\prod_{m=1}^{k}\{1+r_m(t)\,r_m(\theta)\}\,d\theta,$$

and, as we have seen in the proof of lemma 14.6, the product $\prod_{m=1}^{k}$ as a function of θ vanishes except on a dyadic interval J of length $\frac{1}{2^k}$, containing t, and in this interval it takes the value 2^k. Consequently, by (14.39) and since $J\subset\left(t-\frac{1}{2^k},\, t+\frac{1}{2^k}\right)$,

$$\left|[S_{2^k}(x)](t)\right| \leqslant 2^k \int_J |x(\theta)|\,d\theta \leqslant 2\,\beta_x(t). \tag{14.41}$$

Since the right side of (14.41) does not depend on k, we can apply (14.41) for $k=n(t)$, whence, taking also into account (14.40), we obtain

$$\int_0^1 \left|[S_{2^{n(t)}}(x)](t)\right|^p dt \leqslant 2^p \int_0^1 [\beta_x(t)]^p\,dt \leqslant 2^p A_p \int_0^1 |x(t)|^p\,dt,$$

which, by putting $C_p=2^p A_p$, completes the proof of lemma 14.7.

Lemma 14.8. *For* $1\leqslant l<\infty$ *there exists a constant* $A_l>0$ *depending only on* l, *such that we have, for any* $\alpha_1,\ldots,\alpha_n\geqslant 0$,

$$\left(\sum_{k=1}^{n}\alpha_k\right)^l \leqslant A_l\left(\sum_{k_1<\cdots<k_l}\alpha_{k_1}\alpha_{k_2}\ldots\alpha_{k_l} + \sum_{k=1}^{n}\alpha_k^l\right). \tag{14.42}$$

Proof. Dividing both sides of (14.42) by $\left(\sum_{k=1}^{n}\alpha_k\right)^l$ and denoting the new variable $\frac{\alpha_j}{\sum_{k=1}^{n}\alpha_k}$ again by α_j ($j=1,\ldots,n$), we see that the existence

[1] For a shorter proof see, e.g., A. Zygmund [274], p. 30–32.

of A_l above is equivalent to the existence of a constant $C_l>0$ such that

$$\inf_{\substack{\alpha_1,\ldots,\alpha_n\geqslant 0\\ \sum_{k=1}^{n}\alpha_k=1}}\left[\sum_{k_1<\cdots<k_l}\alpha_{k_1}\alpha_{k_2}\ldots\alpha_{k_l}+\sum_{k=1}^{n}\alpha_k^l\right]\geqslant C_l. \tag{14.43}$$

Since this is obvious for $l=1$, we may assume that $l>1$.

Let $\varepsilon>0$ be such that $(l-1)\frac{\varepsilon}{2}<\frac{1}{l}$. Put

$$C_l=\min\left\{\varepsilon^l,\left(\frac{1}{l}-\frac{\varepsilon}{2}\right)^{l-1}\left[\frac{1}{l}-(l-1)\frac{\varepsilon}{2}\right]\right\}. \tag{14.44}$$

Let $\alpha_1,\ldots,\alpha_n\geqslant 0$, $\sum_{k=1}^{n}\alpha_k=1$. Then there are two cases.

a) $\max_{1\leqslant k\leqslant n}\alpha_k\geqslant\varepsilon$. In this case for a suitable k_0 with $1\leqslant k_0\leqslant n$ we have

$$\sum_{k_1<\cdots<k_l}\alpha_{k_1}\alpha_{k_2}\ldots\alpha_{k_l}+\sum_{k=1}^{n}\alpha_k^l\geqslant\alpha_{k_0}^l\geqslant\varepsilon^l\geqslant C_l.$$

b) $\max_{1\leqslant k\leqslant n}\alpha_k<\varepsilon$. In this case there exist positive integers n_j $(j=0,1,\ldots,l)$ with the following properties:

$$0=n_0<n_1<\cdots<n_{l-1}<n_l=n, \tag{14.45}$$

$$\frac{1}{l}-\frac{\varepsilon}{2}\leqslant\sum_{k=n_j+1}^{n_{j+1}}\alpha_k\leqslant\frac{1}{l}+\frac{\varepsilon}{2}\quad(j=0,1,\ldots,l-2), \tag{14.46}$$

$$\frac{1}{l}-(l-1)\frac{\varepsilon}{2}\leqslant\sum_{k=n_{l-1}+1}^{n}\alpha_k. \tag{14.47}$$

Indeed, there exists an n_1 with $0<n_1<n$ such that $\frac{1}{l}-\frac{\varepsilon}{2}\leqslant\sum_{k=1}^{n_1}\alpha_k$, since otherwise $\sum_{k=1}^{n-1}\alpha_k<\frac{1}{l}-\frac{\varepsilon}{2}$, whence $1=\sum_{k=1}^{n}\alpha_k<\frac{1}{l}-\frac{\varepsilon}{2}+\alpha_n$, and thus $\varepsilon>\alpha_n>1-\frac{1}{l}-\frac{\varepsilon}{2}=\frac{l-1}{l}-\frac{\varepsilon}{2}$, whence $\frac{\varepsilon}{2}>\frac{l-1}{l}$, and thus $\frac{1}{l}>(l-1)\frac{\varepsilon}{2}>\frac{(l-1)^2}{l}$, which implies $1>l-1$, i.e., $l-1=0$, $l=1$, in contradiction to our hypothesis. Now, take the smallest such n_1. Then $\sum_{k=1}^{n_1}\alpha_k\leqslant\frac{1}{l}+\frac{\varepsilon}{2}$, since otherwise $\sum_{k=1}^{n_1}\alpha_k>\frac{1}{l}+\frac{\varepsilon}{2}$, whence $\sum_{k=1}^{n_1-1}\alpha_k$ $=\sum_{k=1}^{n_1}\alpha_k-\alpha_{n_1}>\frac{1}{l}+\frac{\varepsilon}{2}-\varepsilon=\frac{1}{l}-\frac{\varepsilon}{2}$, i.e., n_1 would not be the smallest positive integer such that $\frac{1}{l}-\frac{\varepsilon}{2}\leqslant\sum_{k=1}^{n_1}\alpha_k$. Now, if $l=2$, then $1=\sum_{k=1}^{n}\alpha_k$ $=\sum_{k=1}^{n_1}\alpha_k+\sum_{k=n_1+1}^{n}\alpha_k\leqslant\frac{1}{2}+\frac{\varepsilon}{2}+\sum_{k=n_1+1}^{n}\alpha_k$, whence $\frac{1}{2}-\frac{\varepsilon}{2}\leqslant\sum_{k=n_1+1}^{n}\alpha_k$, i.e., we have (14.47) for $l=2$. On the other hand, if $l>2$, then there exists an n_2 with $n_1<n_2<n$ such that $\frac{1}{l}-\frac{\varepsilon}{2}\leqslant\sum_{k=n_1+1}^{n_2}\alpha_k$, since otherwise $\sum_{k=n_1+1}^{n-1}\alpha_k<\frac{1}{l}-\frac{\varepsilon}{2}$, whence $\sum_{k=1}^{n-1}\alpha_k<\frac{1}{l}+\frac{\varepsilon}{2}+\frac{1}{l}-\frac{\varepsilon}{2}=\frac{2}{l}$, whence $1=\sum_{k=1}^{n}\alpha_k<\frac{2}{l}+\alpha_n$, and thus $\varepsilon>\alpha_n>1-\frac{2}{l}$, whence $\frac{1}{l}>(l-1)\frac{\varepsilon}{2}$ $>\frac{l-1}{2}\left(1-\frac{2}{l}\right)=\frac{(l-1)(l-2)}{2l}$, and thus $1>\frac{(l-1)(l-2)}{2}$, which implies $(l-1)(l-2)=0$, i.e., $l=1$ or $l=2$, in contradiction to our hypothesis. Take the smallest such n_2, etc. Continuing in this manner, we obtain $n_3,\dots,n_{l-1}$ satisfying (14.45), (14.46). Finally, if (14.47) is not satisfied, i.e., $\sum_{k=n_{l-1}+1}^{n}\alpha_k<\frac{1}{l}-(l-1)\frac{\varepsilon}{2}$, then $1=\sum_{k=1}^{n}\alpha_k<(l-1)\left(\frac{1}{l}+\frac{\varepsilon}{2}\right)$ $+\frac{1}{l}-(l-1)\frac{\varepsilon}{2}=1-\frac{1}{l}+(l-1)\frac{\varepsilon}{2}+\frac{1}{l}-(l-1)\frac{\varepsilon}{2}=1$, an impossibility, which proves that we also have (14.47).

Consequently, we have

$$\sum_{k_1<\dots<k_l}\alpha_{k_1}\alpha_{k_2}\dots\alpha_{k_l}+\sum_{k=1}^{l}\alpha_k^l>\left(\sum_{k=n_0+1}^{n_1}\alpha_k\right)\left(\sum_{k=n_1+1}^{n_2}\alpha_k\right)\dots\left(\sum_{k=n_{l-1}+1}^{n}\alpha_k\right)$$

$$\geqslant\left(\frac{1}{l}-\frac{\varepsilon}{2}\right)^{l-1}\left[\frac{1}{l}-(l-1)\frac{\varepsilon}{2}\right]\geqslant C_l,$$

which completes the proof of lemma 14.8.

Now we can give the

Proof of lemma 14.3. Let $p=2v$. For simplicity, put

$$\begin{aligned}[F_n(x)](t)=[S_{2^n}(x)](t)&=\sum_{k=1}^{n}\sum_{j=1}^{2^{k-1}}h_{2^{k-1}+j}(x)w_{2^{k-1}+j}(t)\\&=\sum_{k=1}^{n}[\Delta_k(x)](t)\qquad(n=1,2,\dots)\end{aligned}\tag{14.48}$$

(we recall that $h_1(x)=\int_0^1 x(t)dt=0$ by the hypothesis of lemma 14.3).

Taking into account lemma 14.4, we have[1]

$$\int_0^1 (F_{n+1}^p - F_n^p)\,dt = \int_0^1 [(F_n + \varDelta_{n+1})^p - F_n^p]\,dt$$

$$= \int_0^1 \left\{ p\,\varDelta_{n+1} F_n^{p-1} + \frac{p(p-1)}{2} \varDelta_{n+1}^2 F_n^{p-2} + \cdots + \varDelta_{n+1}^p \right\} dt$$

$$= \int_0^1 \left\{ p\,\varDelta_{n+1} \left(\sum_{k=1}^n \varDelta_k \right)^{p-1} + \frac{p(p-1)}{2} \varDelta_{n+1}^2 F_n^{p-2} + \cdots + \varDelta_{n+1}^p \right\} dt$$

$$= \int_0^1 \left\{ \frac{p(p-1)}{2} \varDelta_{n+1}^2 F_n^{p-2} + \cdots + \varDelta_{n+1}^p \right\} dt.$$

Now, for $3 \leqslant \mu \leqslant p-1$ and $\alpha = \frac{p-2}{p-\mu}$ we have, by the Hölder inequality,

$$\left| \int_0^1 \varDelta_{n+1}^\mu F_n^{p-\mu}\,dt \right| = \left| \int_0^1 \left(\varDelta_{n+1}^{\frac{2(p-\mu)}{p-2}} F_n^{p-\mu} \right) \left(\varDelta_{n+1}^{\mu - \frac{2(p-\mu)}{p-2}} \right) dt \right|$$

$$\leqslant \left[\int_0^1 \left| \varDelta_{n+1}^{\frac{2(p-\mu)}{p-2}} F_n^{p-\mu} \right|^\alpha dt \right]^{\frac{1}{\alpha}} \left[\int_0^1 \left| \varDelta_{n+1}^{\mu - \frac{2(p-\mu)}{p-2}} \right|^{\frac{\alpha}{\alpha-1}} dt \right]^{\frac{\alpha-1}{\alpha}}$$

$$= \left[\int_0^1 \varDelta_{n+1}^2 F_n^{p-2}\,dt \right]^\theta \left[\int_0^1 \varDelta_{n+1}^p\,dt \right]^{1-\theta},$$

where we have put $\theta = \frac{1}{\alpha} = \frac{p-\mu}{p-2}$ (indeed, $\frac{\alpha}{\alpha-1} = \frac{\frac{p-2}{p-\mu}}{\frac{p-2}{p-\mu} - 1} = \frac{p-2}{\mu-2}$, whence $\left(\mu - \frac{2(p-\mu)}{p-2} \right) \frac{p-2}{\mu-2} = \frac{p\mu - 2\mu - 2p + 2\mu}{\mu - 2} = p$). Since $3 \leqslant \mu \leqslant p-1$,

[1] For the sake of simplicity, we don't write the arguments x and t; this will lead to no confusion.

we have $0<\theta=\dfrac{p-\mu}{p-2}<1$ (since $\mu\leqslant p-1<p$ and $p-\mu\leqslant p-3<p-2$). Hence, since $a^\theta b^{1-\theta}\leqslant a+b$ for $a,b\geqslant 0$, $0\leqslant\theta\leqslant 1$,[1] and since $p=2\nu$, it follows that

$$\left|\int_0^1 \Delta_{n+1}^{\mu} F_n^{p-\mu}\,dt\right| \leqslant \int_0^1 \Delta_{n+1}^2 F_n^{p-2}\,dt + \int_0^1 \Delta_{n+1}^p\,dt. \tag{14.49}$$

Consequently, we have

$$\left|\int_0^1 (F_{n+1}^p - F_n^p)\,dt\right| = \left|\int_0^1 \left\{\frac{p(p-1)}{2}\Delta_{n+1}^2 F_n^{p-2} + \cdots + \Delta_{n+1}^p\right\}dt\right|$$

$$\leqslant \frac{p(p-1)}{2}\left[\int_0^1 \Delta_{n+1}^2 F_n^{p-2}\,dt + \int_0^1 \Delta_{n+1}^p\,dt\right]$$

$$+\frac{p(p-1)(p-2)}{6}\left|\int_0^1 \Delta_{n+1}^3 F_n^{p-3}\,dt\right| + \cdots + p\left|\int_0^1 \Delta_{n+1}^{p-1} F_n\,dt\right|$$

$$\leqslant A_p\left[\int_0^1 \Delta_{n+1}^2 F_n^{p-2}\,dt + \int_0^1 \Delta_{n+1}^p\,dt\right],$$

where we have put $A_p=\dfrac{p(p-1)}{2}+\dfrac{p(p-1)(p-2)}{6}+\cdots+p$. Putting $F_0=0$, the above inequality is also valid, obviously, for $n=0$ (since $F_1=\Delta_1$).

Summing for $n=0,1,2,\ldots,N$ and applying the Hölder inequality, we obtain (taking also into account that p is even)

---

[1] Indeed, for $\frac{a}{b}\geqslant 1$ we have $\left(\frac{a}{b}\right)^\theta\leqslant\frac{a}{b}<\frac{a}{b}+1$, while for $\frac{a}{b}\leqslant 1$ we have $\left(\frac{a}{b}\right)^\theta\leqslant 1\leqslant\frac{a}{b}+1$, whence always $\left(\frac{a}{b}\right)^\theta\leqslant\frac{a}{b}+1$, and thus $a^\theta b^{1-\theta}\leqslant a+b$. In our case, we can put $\int_0^1 \Delta_{n+1}^2 F_n^{p-2}\,dt=a$, $\int_0^1 \Delta_{n+1}^p\,dt=b$ (because the integrands are $\geqslant 0$ by $p=2\nu$).

$$\int_0^1 F_{N+1}^p\,dt = \left|\int_0^1 F_{N+1}^p\,dt\right| = \left|\int_0^1 (F_1^p - F_0^p)\,dt + \cdots + \int_0^1 (F_{N+1}^p - F_N^p)\,dt\right|$$

$$\leqslant A_p \sum_{n=0}^{N}\left[\int_0^1 \Delta_{n+1}^2 F_n^{p-2}\,dt + \int_0^1 \Delta_{n+1}^p\,dt\right]$$

$$\leqslant A_p \int_0^1 \left\{\sum_{n=0}^{N} \Delta_{n+1}^2\right\} \max_{0\leqslant n\leqslant N}\{F_n^{p-2}\}\,dt + A_p \int_0^1 \sum_{n=0}^{N} (\Delta_{n+1}^2)^{\frac{p}{2}}\,dt$$

$$\leqslant A_p \left[\int_0^1 \left\{\sum_{n=0}^{N}\Delta_{n+1}^2\right\}^{\frac{p}{2}}\right]^{\frac{2}{p}} \left[\int_0^1 \max_{0\leqslant n\leqslant N}\{F_n^p\}\,dt\right]^{\frac{p-2}{p}}$$

$$+ A_p \left[\int_0^1 \left\{\sum_{n=0}^{N}\Delta_{n+1}^2\right\}^{\frac{p}{2}} dt\right].$$

Now, for each $t\in[0,1]$ let $n(t)$ be an integer, $0\leqslant n(t)\leqslant N$, such that $|[S_{2^{n(t)}}(x)](t)|^p = [F_{n(t)}(x)]^p(t) = \max_{0\leqslant n\leqslant N}\{[F_n(x)]^p(t)\}$. Then, by lemma 14.7 applied for $F_{N+1}(x)$ instead of x (taking into account that by biorthogonality $F_n(F_{N+1}(x)) = F_n(x)$ for $n=0,1,\ldots,N$) we have $\int_0^1 \max_{0\leqslant n\leqslant N}\{F_n^p\}\,dt \leqslant C_p \int_0^1 F_{N+1}^p\,dt$, whence

$$\int_0^1 F_{N+1}^p\,dt \leqslant A_p C_p \left[\int_0^1 \left\{\sum_{n=0}^{N}\Delta_{n+1}^2\right\}^{\frac{p}{2}}\right]^{\frac{2}{p}} \left[\int_0^1 F_{N+1}^p\,dt\right]^{\frac{p-2}{p}}$$

$$+ A_p \left[\int_0^1 \left\{\sum_{n=0}^{N}\Delta_{n+1}^2\right\}^{\frac{p}{2}} dt\right].$$

Denoting the left side by a and the factor of A_p in the second term of the right side by b, this inequality becomes

$$a \leqslant A_p C_p b^{\frac{2}{p}} a^{\frac{p-2}{p}} + A_p b,$$

whence

$$a\left(1 - A_p C_p b^{\frac{2}{p}} a^{-\frac{2}{p}}\right) \leqslant A_p b.$$

Put $\theta = \frac{b}{a}$. Then

$$1 - A_p C_p \theta^{\frac{2}{p}} \leqslant A_p \theta,$$

whence

$$1 \leqslant A_p C_p \theta^{\frac{2}{p}} + A_p \theta \leqslant \begin{cases} (A_p C_p + A_p)\theta & \text{for} \quad \theta \geqslant 1 \\ (A_p C_p + A_p)\theta^{\frac{2}{p}} & \text{for} \quad \theta \leqslant 1, \end{cases}$$

and thus in any case

$$\theta \geqslant \frac{1}{M_p} = \min\left(\frac{1}{A_p C_p + A_p}, \frac{1}{(A_p C_p + A_p)^{\frac{p}{2}}}\right) > 0,$$

whence

$$M_p b \geqslant a. \tag{14.50}$$

Consequently, if $\int_0^1 \left(\sum_{n=1}^{\infty} \Delta_n^2\right)^{\frac{p}{2}} dt$ exists, then we have

$$\int_0^1 F_{N+1}^p \, dt \leqslant M_p \int_0^1 \left\{\sum_{n=0}^{N} \Delta_{n+1}^2\right\}^{\frac{p}{2}} dt \leqslant M_p \int_0^1 \left\{\sum_{n=1}^{\infty} \Delta_n^2\right\}^{\frac{p}{2}} dt. \tag{14.51}$$

Applying (14.51) to $F_{N_2}(x) - F_{N_1}(x)$ instead of x, we get

$$\int_0^1 \{F_{N_2} - F_{N_1}\}^p \, dt \leqslant M_p \int_0^1 \left\{\sum_{n=N_1+1}^{N_2} \Delta_n^2\right\}^{\frac{p}{2}} dt$$

$(N_1 < N_2)$, whence $\|F_{N_2} - F_{N_1}\|_{L^p} \to \infty$, and thus there exists (since $L^p([0,1])$ is complete) an element $z \in L^p([0,1])$ such that $\|F_N - z\|_{L^p} \to 0$ as $N \to \infty$. Then, by (14.51), we also have

$$\int_0^1 z^p \, dt \leqslant M_p \int_0^1 \left\{\sum_{n=1}^{\infty} \Delta_n^2\right\}^{\frac{p}{2}} dt.$$

Furthermore, taking into account the continuity of the functionals h_n and corollary 14.1 (biorthogonality of $\{w_n\}, \{h_n\}$), we have

$$\begin{aligned} h_n(z) &= h_n\left\{\lim_{N\to\infty} [S_{2^N}(x)]\right\} = \lim_{N\to\infty} h_n\{S_{2^N}(x)\} \\ &= \lim_{N\to\infty} h_n\left\{\sum_{j=1}^{2^N} h_j(x) w_j\right\} = \lim_{N\to\infty} h_n(x) = h_n(x) \qquad (n = 1, 2, \ldots), \end{aligned}$$

whence, since $\{h_n\}$ is total on $L^p([0,1])$ (because $\{w_n\}$ is complete in $L^q([0,1])$, where $\frac{1}{p}+\frac{1}{q}=1$), we obtain $z=x$, and consequently[1]

$$\int_0^1 x^p\,dt \leqslant M_p \int_0^1 \left\{\sum_{n=1}^{\infty} \Delta_n^2\right\}^{\frac{p}{2}} dt,$$

which is nothing else but the second inequality of (14.27), in the case $p=2\nu$.

Let us prove now the first inequality of (14.27), in the case $p=2\nu$. Let $N-1 \geqslant n_1 > n_2 > \cdots > n_{\nu-1}$. Then we have

$$\int_0^1 \Delta_{n_1}^2 \Delta_{n_2}^2 \cdots \Delta_{n_{\nu-1}}^2 F_N^2\,dt = \int_0^1 \Delta_{n_1}^2 \Delta_{n_2}^2 \cdots \Delta_{n_{\nu-1}}^2 \left(\sum_{n=1}^{N} \Delta_n\right)^2 dt$$

$$= \int_0^1 \Delta_{n_1}^2 \Delta_{n_2}^2 \cdots \Delta_{n_{\nu-1}}^2 \left(F_{n_1} + \sum_{n=n_1+1}^{N} \Delta_n\right)^2 dt$$

$$= \int_0^1 \Delta_{n_1}^2 \Delta_{n_2}^2 \cdots \Delta_{n_{\nu-1}}^2 F_{n_1}^2\,dt + \sum_{n=n_1+1}^{N} \int_0^1 \Delta_{n_1}^2 \Delta_{n_2}^2 \cdots \Delta_{n_{\nu-1}}^2 \Delta_n^2\,dt$$

$$+ 2 \sum_{n=n_1+1}^{N} \int_0^1 \Delta_{n_1}^2 \cdots \Delta_{n_{\nu-1}}^2 \Delta_n F_{n_1}\,dt + 2 \sum_{\substack{n \neq m \\ n_1+1 \leqslant n,m \leqslant N}}\sum \int_0^1 \Delta_{n_1}^2 \cdots \Delta_{n_{\nu-1}}^2 \Delta_n \Delta_m\,dt.$$

By virtue of lemma 14.4, the last two sums vanish (the first, by putting $F_{n_1} = \sum_{m=1}^{n_1} \Delta_m$ and observing that n is greater than all other indices, and the second by observing that in each summand either m or n is greater than all other indices). Consequently, omitting also the first integral in the right side, we obtain

$$\sum_{n=n_1+1}^{N} \int_0^1 \Delta_{n_1}^2 \Delta_{n_2}^2 \cdots \Delta_{n_{\nu-1}}^2 \Delta_n^2\,dt \leqslant \int_0^1 \Delta_{n_1}^2 \Delta_{n_2}^2 \cdots \Delta_{n_{\nu-1}}^2 F_N^2\,dt.$$

[1] Alternatively, this follows also from (14.51), lemma 14.2 and Ch. I, § 2, example 2.3.

Summing over all possible combinations of positive integers $n_1, n_2, \ldots, n_{\nu-1}$ satisfying $N-1 \geqslant n_1 > n_2 > \cdots > n_{\nu-1}$, we get

$$\sum_{N \geqslant n_\nu > n_1 > n_2 > \cdots > n_{\nu-1}} \left\{ \int_0^1 \Delta_{n_1}^2 \Delta_{n_2}^2 \ldots \Delta_{n_\nu}^2 \, dt \right\}$$

$$\leqslant \sum_{N-1 \geqslant n_1 > \cdots > n_{\nu-1}} \left\{ \int_0^1 \Delta_{n_1}^2 \Delta_{n_2}^2 \ldots \Delta_{n_{\nu-1}}^2 F_N^2 \, dt \right\} \tag{14.52}$$

$$\leqslant \int_0^1 F_N^2 \left\{ \sum_{n=1}^N \Delta_n^2 \right\}^{\nu-1} dt = \int_0^1 F_N^2 \left\{ \sum_{n=1}^N \Delta_n^2 \right\}^{\frac{p-2}{2}} dt.$$

On the other hand, we have, by the biorthogonality of $\{y_n\}, \{h_n\}$,

$$h_l[F_N(x)] = h_l\left(\sum_{k=1}^N \sum_{j=1}^{2^{k-1}} h_{2^{k-1}+j}(x) w_{2^{k-1}+j} \right)$$

$$= \begin{cases} h_l(x) & \text{for } \quad l = 1, 2, \ldots, 2^N \\ 0 & \text{for } \quad l = 2^N+1, 2^N+2, \ldots \end{cases}$$

and thus

$$\Delta_k[F_N(x)] = \begin{cases} \sum_{j=1}^{2^{k-1}} h_{2^{k-1}+j}(x) w_{2^{k-1}+j} = \Delta_k(x) & \text{for } \quad k = 1, \ldots, N \\ 0 \quad \text{for } \quad k = N+1, N+2, \ldots \end{cases}$$

whence, applying lemma 14.6 for $F_N(x)$ instead of x, we obtain

$$\sum_{k=1}^N \left\{ \int_0^1 \Delta_k^p \, dt \right\} \leqslant \int_0^1 F_N^p \, dt. \tag{14.53}$$

Denoting by S the left side of (14.52) and taking into account lemma 14.8 (with $\alpha_k = [\Delta_k(x)]^2(t)$, $n = N$, $l = \frac{p}{2} = \nu$) and (14.52), (14.53), we get

$$\int_0^1 \left\{ \sum_{k=1}^N \Delta_k^2 \right\}^{\frac{p}{2}} dt \leqslant B_p \left[S + \sum_{k=1}^N \int_0^1 \Delta_k^p \, dt \right]$$

$$\leqslant B_p \left[\int_0^1 F_N^2 \left\{ \sum_{k=1}^N \Delta_k^2 \right\}^{\frac{p-2}{2}} dt + \int_0^1 F_N^p \, dt \right].$$

Applying the Hölder inequality with the exponents $\frac{p}{2}$ and $\frac{p}{p-2}$, we see that this is

$$\leqslant B_p\left[\left(\int_0^1 F_N^p dt\right)^{\frac{2}{p}}\left(\int_0^1\left\{\sum_{k=1}^N \Delta_k^2\right\}^{\frac{p}{2}} dt\right)^{\frac{p-2}{p}}+\int_0^1 F_N^p dt\right].$$

Hence, putting $a=\int_0^1\left\{\sum_{k=1}^N \Delta_k^2\right\}^{\frac{p}{2}} dt,\ b=\int_0^1 F_N^p dt$, we have

$$a \leqslant B_p\left[b^{\frac{2}{p}} a^{\frac{p-2}{p}}+b\right],$$

and thus, similarly to the proof of (14.50), we obtain

$$\gamma_p b \geqslant a,$$

where $\gamma_p>0$, whence, applying lemma 14.7 with $n(t)=n$,

$$\int_0^1\left\{\sum_{k=1}^N \Delta_k^2\right\}^{\frac{p}{2}} dt \leqslant \gamma_p \int_0^1 F_N^p dt \leqslant \gamma_p C_p \int_0^1 x^p dt \qquad (N=1,2,\ldots).$$

Consequently, the series $\sum_{k=1}^{\infty} \Delta_k^2$ converges almost everywhere and satisfies the first inequality of (14.27), with $m_p=\frac{1}{\gamma_p C_p}$, in the case $p=2v$. This completes the proof of lemma 14.3. Thus, the proof of theorem 14.1 is now complete.

Let us now turn our attention to conditional bases.

In finite dimensional Banach spaces all bases $\{x_n\}$ are unconditional. It is natural to ask whether the converse is also true, i.e., whether this property characterizes finite dimensional Banach spaces among Banach spaces with bases. We shall see in § 23 that the answer to this question is affirmative. In the present section we shall show only that in some concrete infinite dimensional Banach spaces which have unconditional bases, there also exist conditional bases.

Example 14.1. In $E=c_0$ the sequence

$$x_n=\{\underbrace{1,\ldots,1}_{n},0,0,\ldots\} \qquad (n=1,2,\ldots) \tag{14.54}$$

is a conditional basis of E.

Indeed, by Ch. I, § 4, proposition 4.3, $\{x_n\}$ is a basis of E. Furthermore, if $\{e_n\}$ denotes the unit vector basis of $E=c_0$, we have

$$\sum_{i=m}^{m+p} \alpha_i x_i = \sum_{i=m}^{m+p} \alpha_i \sum_{j=1}^{i} e_j$$
$$= \left(\sum_{i=m}^{m+p} \alpha_i\right)\left(\sum_{j=1}^{m} e_j\right) + \left(\sum_{i=m+1}^{m+p} \alpha_i\right) e_{m+1} + \cdots + \alpha_{m+p} e_{m+p},$$

whence

$$\left\|\sum_{i=m}^{m+p} \alpha_i x_i\right\| = \sup_{m\leqslant k\leqslant m+p} \left|\sum_{i=k}^{m+p} \alpha_i\right|.$$

Consequently, $\sum_{i=1}^{\infty} \alpha_i x_i$ converges if and only if $\sum_{i=1}^{\infty} \alpha_i$ converges and thus $\{x_n\}$ is a conditional basis of E (e.g., by § 16, lemma 16.1, equivalence $1°\Leftrightarrow 4°$).

This assertion also follows from the fact that $\{x_n\}$ is a basis of type P^*, taking into account § 17, corollary 17.1 b).

Example 14.2. In $E=l^1$ the sequence

$$h_1=\{1,0,0,\ldots\}, \qquad h_n=\{\underbrace{0,\ldots,0}_{n-2},1,-1,0,0,\ldots\} \qquad (n=2,3,\ldots) \tag{14.55}$$

is a conditional basis of E.

Indeed, we have seen in Ch. I, § 13, example 13.3, that $\{h_n\}$ is a basis of E. Furthermore, if $\{f_n\}$ denotes the unit vector basis of l^1, we have, for any $m\geqslant 2, p\geqslant 1$,

$$\sum_{i=m}^{m+p} \alpha_i h_i = \sum_{i=m}^{m+p} \alpha_i (f_{i-1}-f_i) = \alpha_m f_{m-1} + \sum_{i=m}^{m+p-1} (\alpha_{i+1}-\alpha_i) f_i - \alpha_{m+p} f_{m+p},$$

whence

$$\left\|\sum_{i=m}^{m+p} \alpha_i h_i\right\| = |\alpha_m| + \sum_{i=m}^{m+p-1} |\alpha_{i+1}-\alpha_i| + |\alpha_{m+p}|.$$

Consequently, $\sum_{i=1}^{\infty} \alpha_i h_i$ converges if and only if $\sum_{i=1}^{\infty} |\alpha_{i+1}-\alpha_i| < \infty$. In particular, for $\alpha_n = \frac{1}{n}$ $(n=1,2,\ldots)$ it follows that the series $\sum_{i=1}^{\infty} \alpha_i h_i$ converges but $\sum_{i=1}^{\infty} (-1)^i \alpha_i h_i$ diverges and thus $\{h_n\}$ is a conditional basis of E (by § 16, lemma 16.1, equivalence $1°\Leftrightarrow 4°$).

The assertion of example 14.2 also follows from the fact that $\{h_{n+1}\}$ is the a.s.c.f. to the basis $\{x_n\}$ of c_0 defined in example 14.1, taking into account § 17, theorem 17.7, or from the fact that $\{h_n\}$ is a basis of type P, taking into account § 17, corollary 17.1a).

Let us consider now the spaces $L^p([0,1])$ and $l^p (1 \leqslant p < \infty)$.

Let $1 \leqslant p < \infty$ and let $r_n(t)$ be the Rademacher functions (14.1) on $[0,1]$. We recall[1] that by the Khinchin inequality there exist two constants $A_p > 0$ and $B_p > 0$ such that we have, for any scalars $\alpha_1, \ldots, \alpha_n$

$$A_p \left(\sum_{k=1}^{n} |\alpha_k|^2 \right)^{\frac{1}{2}} \leqslant \left(\int_0^1 \left| \sum_{k=1}^{n} \alpha_k r_k(t) \right|^p dt \right)^{\frac{1}{p}} \leqslant B_p \left(\sum_{k=1}^{n} |\alpha_k|^2 \right)^{\frac{1}{2}}; \qquad (14.56)$$

in the particular case when $p=1$, one can take $A_1 = \frac{1}{8}$.

Lemma 14.9. a) *Let* $\sum_{k=1}^{\infty} x_k$ *be an unconditionally convergent series in the space* $L^p([0,1])$, *where* $1 \leqslant p < \infty$. *Then there exists a constant* $C = C_{p,\{x_n\}}$ *such that*

$$\int_0^1 \left(\sum_{k=1}^{n} |x_k(t)|^2 \right)^{\frac{p}{2}} dt \leqslant C \qquad (n=1,2,\ldots). \qquad (14.57)$$

b) *Let* $\sum_{k=1}^{\infty} x_k$ *be an unconditionally convergent series in the space* l^p, *where* $1 \leqslant p < \infty$, $x_k = \{\xi_n^{(k)}\}_{n=1}^{\infty}$ $(k=1,2,\ldots)$. *Then there exists a constant* $C > 0$ *such that*

$$\sum_{i=1}^{\infty} \left(\sum_{k=1}^{n} |\xi_i^{(k)}|^2 \right)^{\frac{p}{2}} \leqslant C \qquad (n=1,2,\ldots). \qquad (14.58)$$

Proof. b) Since $\sum_{k=1}^{\infty} x_k$ is unconditionally convergent, there exists, by § 15, corollary 15.1, a constant $M > 0$ such that

$$\left(\sum_{i=1}^{\infty} \left| \sum_{k=1}^{n} r_k(t) \xi_i^{(k)} \right|^p \right)^{\frac{1}{p}} = \left\| \sum_{k=1}^{n} r_k(t) x_k \right\| \leqslant M \qquad (t \in [0,1],\ n=1,2,\ldots), \qquad (14.59)$$

where r_k are the Rademacher functions. Hence, putting in the left side of the Khinchin inequality (14.56) $\alpha_k = \xi_i^{(k)}$ $(k=1,\ldots,n)$, summing over i, and applying the theorem of Lebesgue, we get

[1] See e.g. [117], p. 131—132.

$$\sum_{i=1}^{\infty}\left(\sum_{k=1}^{n}|\xi_i^{(k)}|^2\right)^{\frac{p}{2}} \leqslant \frac{1}{(A_p)^p}\sum_{i=1}^{\infty}\int_0^1\left|\sum_{k=1}^{n}\xi_i^{(k)}r_k(t)\right|^p dt$$

$$= \frac{1}{(A_p)^p}\int_0^1\sum_{i=1}^{\infty}\left|\sum_{k=1}^{n}\xi_i^{(k)}r_k(t)\right|^p dt$$

$$= \frac{1}{(A_p)^p}\int_0^1\left\|\sum_{k=1}^{n}r_k(t)x_k\right\|^p dt$$

$$\leqslant \frac{M^p}{(A_p)^p} = C \quad (n=1,2,\ldots),$$

which completes the proof of b). The proof of a) is similar, using the theorem of Fubini.

Lemma 14.10. *Let* $\sum_{k=1}^{\infty} x_k$ *be an unconditionally convergent series in the space* $L^p([0,1])$ *or* l^p, *where* $1\leqslant p\leqslant 2$. *Then*

$$\sum_{k=1}^{\infty}\|x_k\|^2<\infty. \tag{14.60}$$

Proof. It is sufficient to consider one of these spaces, e.g., l^p, since the proof for the other is similar. For any scalars $\beta_1,\ldots,\beta_n$ we have, by the Hölder inequality and by lemma 14.9,

$$\sum_{k=1}^{n}\|x_k\|^p\beta_k = \sum_{k=1}^{n}\sum_{i=1}^{\infty}|\xi_i^{(k)}|^p\beta_k = \sum_{i=1}^{\infty}\sum_{k=1}^{n}|\xi_i^{(k)}|^p\beta_k$$

$$\leqslant\left[\sum_{i=1}^{\infty}\left(\sum_{k=1}^{n}|\xi_i^{(k)}|^2\right)^{\frac{p}{2}}\right]\left[\sum_{k=1}^{n}|\beta_k|^{\frac{2}{2-p}}\right]^{\frac{2-p}{2}}$$

$$\leqslant C\left(\sum_{k=1}^{n}|\beta_k|^{\frac{2}{2-p}}\right)^{\frac{2-p}{2}}.$$

Consequently, for any sequence of scalars $\{\beta_n\}$ with $\sum_{k=1}^{\infty}|\beta_k|^{\frac{2}{2-p}}<\infty$ the series $\sum_{k=1}^{\infty}\|x_k\|^p\beta_k$ is convergent, whence, by the theorem of Landau[1], we have $\sum_{k=1}^{\infty}\|x_k\|^2=\sum_{k=1}^{\infty}(\|x_k\|^p)^{\frac{2}{p}}<\infty$, which completes the proof.

---

[1] See, e.g. [10], p. 86.

Proposition 14.1. *In the spaces $L^p([0,1])$ and l^p, where $1 \leqslant p \leqslant 2$, every bounded unconditional basic sequence is Besselian. Consequently, in the spaces $L^p([0,1])$ and l^p, where $2 \leqslant p < \infty$, as well as in the space c_0, every bounded unconditional basis is Hilbertian*[1].

Proof. Assume that $\{x_n\}$ is a bounded unconditional basic sequence in $L^p([0,1])$ or l^p, where $1 \leqslant p \leqslant 2$, and let $\{\alpha_n\}$ be an arbitrary sequence of scalars such that $\sum_{i=1}^{\infty} \alpha_i x_i$ converges. Then, since $\{x_n\}$ is an unconditional basis of $[x_n]$, $\sum_{i=1}^{\infty} \alpha_i x_i$ is unconditionally convergent, whence, by lemma 14.10 and since $\{x_n\}$ is a bounded basis of $[x_n]$, we obtain

$$\sum_{i=1}^{\infty} |\alpha_i|^2 \leqslant \frac{1}{\inf\limits_{1 \leqslant n \leqslant \infty} \|x_n\|_p^2} \sum_{i=1}^{\infty} \|\alpha_i x_i\|_p^2 < \infty,$$

and thus $\{x_n\}$ is Besselian.

Assume now that $\{x_n\}$ is a bounded unconditional basis of $E = L^p([0,1])$ or l^p, where $2 \leqslant p < \infty$, or of $E = c_0$. Then, by § 17, theorem 17.7, the a.s.c.f. $\{f_n\} \subset E^*$ is a bounded unconditional basic sequence in E^*, where $E^* \equiv L^q([0,1])$ or l^q, with $\frac{1}{p} + \frac{1}{q} = 1$, or $E^* \equiv l^1$, respectively. Hence, since $1 < q \leqslant 2$, by the above proved first assertion of proposition 14.1 it follows that $\{f_n\}$ is a Besselian basis of $[f_n] \subset E^*$. Consequently, by § 11, theorem 11.1 b) (implication $6^\circ \Rightarrow 1^\circ$), $\{x_n\}$ is a Hilbertian basis of E, which completes the proof of proposition 14.1.

Corollary 14.3. *In the spaces $L^p([0,1])$ and l^p, where $1 < p < 2$, there exists no bounded unconditional Hilbertian basis. In the spaces $L^p([0,1])$ and l^p, where $2 < p < \infty$, there exists no bounded unconditional Besselian basis*[2].

Proof. If there existed a bounded unconditional Hilbertian basis of $L^p([0,1])$ or l^p, where $1 < p < 2$, or a bounded unconditional Besselian basis of $L^p([0,1])$ or l^p, where $2 < p < \infty$, then by proposition 14.1 above and § 11, corollary 11.2, $\{x_n\}$ would be equivalent to the unit vector basis of l^2, whence $L^p([0,1])$ or l^p would be isomorphic to l^2, which is not true for $p \neq 2$. This completes the proof.

---

[1] We mention that the spaces $L^1([0,1])$ and $C([0,1])$ have no unconditional basis (see § 15).

[2] We recall that l^1 has no bounded Hilbertian basis and c_0 has no bounded Besselian basis (by § 11, corollary 11.1).

Now we can give

Example 14.3. In $E=L^p([-\pi,\pi])$ $(1<p\neq 2)$ the sequence $\{\tilde{x}_n\}_{n=0}^{\infty}$, where

$$x_0(t)\equiv\tfrac{1}{2}, x_{2n-1}(t)=\sin nt, x_{2n}(t)=\cos nt \quad (t\in[-\pi,\pi], n=1,2,\ldots), \tag{14.61}$$

is a conditional basis of E.

Indeed, this follows from corollary 14.3 above, taking into account § 11, example 11.1, according to which (14.61) is a bounded Besselian basis of E for $p>2$ and a bounded Hilbertian basis of E for $1<p<2$.

An example of a conditional basis of $E=l^p$ $(1<p\neq 2)$ can be obtained from § 23, proposition 23.2 and its proof, taking into account § 21, proposition 21.5 and § 18, theorem 18.3.

Although Hilbert spaces have "the best" geometric properties among all Banach spaces, the construction of conditional bases in separable Hilbert spaces appears to be more difficult than in the other concrete Banach spaces with unconditional bases. We shall give below two different ways of constructing such bases.

Example 14.4. Let $E=L^2([-\pi,\pi])$ and let $0<\beta<\frac{1}{2}$. Then $\{\tilde{x}_n\}_0^{\infty}$, where[1]

$$x_{2n}(t)=|t|^{-\beta}e^{-int}, \; x_{2n+1}(t)=|t|^{-\beta}e^{int} \quad (t\in[-\pi,\pi], n=0,1,2,\ldots), \tag{14.62}$$

and $\{\tilde{y}_n\}_0^{\infty}$, where

$$y_{2n}(t)=|t|^{\beta}e^{int}, \; y_{2n+1}(t)=|t|^{\beta}e^{-int} \quad (t\in[-\pi,\pi], n=0,1,2,\ldots), \tag{14.63}$$

are conditional bases of E.

Indeed, we have seen in § 11, example 11.2, that $\{\tilde{x}_n\}_0^{\infty}$ is a non-Hilbertian bounded basis of E and $\{\tilde{y}_n\}_0^{\infty}$ is a non-Besselian bounded basis of E, whereas by proposition 14.1 above every bounded unconditional basis of E must be both Hilbertian and Besselian.

Since the proof of the assertions of § 11, example 11.2, whence also the proof of the fact that the sequences (14.62) and (14.63) are conditional bases of $E=L^2([-\pi,\pi])$, leans heavily on analytic tools, it is natural to raise the problem of finding a simple geometric construction of conditional bases in $E=l^2$ of the form $x_n=\sum_{i=1}^{\infty}\gamma_i^{(n)}e_i$ $(n=1,2,\ldots)$, where $\{e_n\}$ is the unit vector basis of l^2.

Moreover, with the Gram-Schmidt process we can obtain from any conditional basis $\{x_n\}$ of l^2 or $L^2([0,1])$ an orthonormal basis $\{z_n\}$ of the form $z_n=\sum_{i=1}^{n}\alpha_i^{(n)}x_i$, $\alpha_n^{(n)}\neq 0$ $(n=1,2,\ldots)$, whence $x_n=\sum_{i=1}^{n}\beta_i^{(n)}z_i$, $\beta_n^{(n)}\neq 0$

[1] Here we use i as $\sqrt{-1}$.

$(n=1,2,\ldots)$. Since $\{z_n\}$ is orthonormal, there exists a linear isometry u of l^2 or $L^2([0,1])$, respectively, onto l^2, such that $u(z_n)=e_n$ $(n=1,2,\ldots)$, whence the sequence

$$y_n=u(x_n)=\sum_{i=1}^{n}\beta_i^{(n)}e_i,\quad \beta_n^{(n)}\neq 0\qquad (n=1,2,\ldots)\tag{14.64}$$

is a conditional basis of l^2. Thus, there arises the problem of finding explicitly a conditional basis of l^2 of the form (14.64). (Let us observe that one cannot obtain the desired basis by applying the above procedure to the basis $\{\tilde{x}_n\}_0^\infty$ or $\{\tilde{y}_n\}_0^\infty$ of example 14.4, since in this case the $\alpha_i^{(n)}, \beta_i^{(n)}$ above involve integrals which are not computable.)

A solution to this problem is given by

Example 14.5. The sequences $\{x_n\}\subset l^2$ and $\{y_n\}\subset l^2$ defined by

$$x_{2n-1}=e_{2n-1}+\sum_{i=n}^{\infty}\alpha_{i-n+1}e_{2i},\qquad x_{2n}=e_{2n}\qquad (n=1,2,\ldots)\tag{14.65}$$

$$y_{2n-1}=e_{2n-1},\quad y_{2n}=\sum_{i=1}^{n}(-\alpha_{n-i+1})e_{2i-1}+e_{2n}\quad (n=1,2,\ldots)\tag{14.66}$$

where $\{e_n\}$ denotes the unit vector basis of l^2 and $\alpha_n\geqslant 0$ $(n=1,2,\ldots)$, $\sum_{j=1}^{\infty}j\alpha_j^2<\infty$, $\sum_{j=1}^{\infty}\alpha_j=\infty$ (e.g., one can take $\alpha_n=\frac{1}{n\log n}$), are conditional bases of l^2.

Indeed, for any finite sequence of scalars $\beta_1,\ldots,\beta_{2n}$ we have

$$\begin{aligned}\sum_{j=1}^{2n}\beta_j x_j&=\sum_{j=1}^{n}\beta_{2j-1}e_{2j-1}+\sum_{j=1}^{n}\beta_{2j-1}\sum_{i=j}^{\infty}\alpha_{i-j+1}e_{2i}+\sum_{j=1}^{n}\beta_{2j}e_{2j}\\&=\sum_{j=1}^{n}\beta_{2j-1}e_{2j-1}+\sum_{j=1}^{n}\left(\sum_{k=1}^{j}\beta_{2k-1}\alpha_{j-k+1}+\beta_{2j}\right)e_{2j}\\&\quad+\sum_{j=n+1}^{\infty}\left(\sum_{k=1}^{n}\beta_{2k-1}\alpha_{j-k+1}\right)e_{2j},\end{aligned}$$

whence

$$\begin{aligned}\left\|\sum_{j=1}^{2n}\beta_j x_j\right\|^2&=\sum_{j=1}^{n}|\beta_{2j-1}|^2+\sum_{j=1}^{n}\left|\sum_{k=1}^{j}\beta_{2k-1}\alpha_{j-k+1}+\beta_{2j}\right|^2\\&\quad+\sum_{j=n+1}^{\infty}\left|\sum_{k=1}^{n}\beta_{2k-1}\alpha_{j-k+1}\right|^2.\end{aligned}\tag{14.67}$$

Since by the Hölder inequality we have

$$\sum_{j=n+1}^{\infty}\left|\sum_{k=1}^{n}\beta_{2k-1}\alpha_{j-k+1}\right|^2 \leqslant \sum_{j=n+1}^{\infty}\left(\sum_{k=j-n+1}^{j}|\alpha_k|^2\right)\sum_{i=1}^{n}|\beta_{2i-1}|^2$$

$$\leqslant \sum_{j=2}^{\infty}(j-1)|\alpha_j|^2\sum_{i=1}^{n}|\beta_{2i-1}|^2 \leqslant \sum_{j=1}^{\infty} j\alpha_j^2\sum_{i=1}^{n}|\beta_{2i-1}|^2,$$

it follows that for any finite sequence of scalars $\beta_1,\ldots,\beta_m$ with $m>2n$ we have

$$\left\|\sum_{j=1}^{2n}\beta_j x_j\right\|^2 \leqslant \left(1+\sum_{j=1}^{\infty} j\alpha_j^2\right)\sum_{j=1}^{n}|\beta_{2j-1}|^2+\sum_{j=1}^{n}\left|\sum_{k=1}^{j}\beta_{2k-1}\alpha_{j-k+1}+\beta_{2j}\right|^2$$

$$\leqslant\left(1+\sum_{j=1}^{\infty} j\alpha_j^2\right)\left\|\sum_{j=1}^{m}\beta_j x_j\right\|^2.$$

Similarly, for any finite sequences of scalars $\beta_1,\ldots,\beta_m$ with $m>2n-1$ we obtain

$$\left\|\sum_{j=1}^{2n-1}\beta_j x_j\right\|^2 \leqslant\left(1+\sum_{j=1}^{\infty} j\alpha_j^2\right)\left\|\sum_{j=1}^{m}\beta_j x_j\right\|^2.$$

Consequently, by Ch. I, § 7, theorem 7.1, $\{x_n\}$ is a basis of $E=l^2$ (with $v_{\{x_n\}}\leqslant\sqrt{1+\sum_{j=1}^{\infty} j\alpha_j^2}$). Let us also observe that $\{x_n\}$ is a bounded basis, since

$$\|x_{2n-1}\|=\sqrt{1+\sum_{j=1}^{\infty}\alpha_j^2},\qquad \|x_{2n}\|=1\qquad (n=1,2,\ldots). \tag{14.68}$$

On the other hand, by (14.67) we have

$$\left\|\sum_{j=1}^{n}\beta_{2j-1}x_{2j-1}\right\|=\sum_{j=1}^{n}|\beta_{2j-1}|^2+\sum_{j=1}^{n}\left|\sum_{k=1}^{j}\beta_{2k-1}\alpha_{j-k+1}\right|^2$$

$$+\sum_{j=n+1}^{\infty}\left|\sum_{k=1}^{n}\beta_{2k-1}\alpha_{j-k+1}\right|^2\geqslant\sum_{j=1}^{n}|\beta_{2j-1}|^2$$

$$+\sum_{j=1}^{n}\left|\sum_{k=1}^{j}\beta_{2k-1}\alpha_{j-k+1}\right|^2,$$

whence, in particular, for $\beta_{2j-1}=\frac{1}{\sqrt{n}}$ $(j=1,\dots,n)$ we obtain, taking into account that by our hypothesis $\left|\sum_{j=1}^{n}\alpha_j\right|^2\to\infty$ as $n\to\infty$,

$$\left\|\sum_{j=1}^{n}\frac{1}{\sqrt{n}}x_{2j-1}\right\|\geqslant 1+\frac{1}{n}\sum_{j=1}^{n}\left|\sum_{k=1}^{j}\alpha_{j-k+1}\right|^2\to\infty \quad \text{as} \quad n\to\infty.$$

Since $\left\|\sum_{j=1}^{n}\frac{1}{\sqrt{n}}e_{2j-1}\right\|=1$ $(n=1,2,\dots)$, from § 11, theorem 11.1b) (implication $1^\circ\Rightarrow 4^\circ$) it follows that $\{x_n\}$ is not Hilbertian, whence, by proposition 14.1 above, $\{x_n\}$ is a conditional basis of $E=l^2$.

Observe now that (x_n, y_n) is a biorthogonal system, since for all $m, n=1,2,\dots$ we obviously have

$$(x_{2n},y_{2m-1})=(e_{2n},e_{2m-1})=0,$$

$$(x_{2n-1},y_{2m-1})=\left(e_{2n-1}+\sum_{i=n}^{\infty}\alpha_{i-n+1}e_{2i},\, e_{2m-1}\right)=\delta_{nm},$$

$$(x_{2n},y_{2m})=\left(e_{2n},\sum_{i=1}^{m}(-\alpha_{m-i+1})e_{2i-1}+e_{2m}\right)=\delta_{nm},$$

$$(x_{2n-1},y_{2m})=\left(e_{2n-1}+\sum_{i=n}^{\infty}\alpha_{i-n+1}e_{2i},\sum_{i=1}^{m}(-\alpha_{m-i+1})e_{2i-1}+e_{2m}\right)$$

$$=\begin{cases}0 & \text{if} \quad m<n\\ -\alpha_{m-n+1}(e_{2n-1},e_{2n-1})+\alpha_{m-n+1}(e_{2m},e_{2m})=0 & \text{if} \quad m\geqslant n.\end{cases}$$

Hence, by Ch. I, § 12, corollaries 12.2 and 12.1, $\{y_n\}$ is a conditional[1] basis of $E^*\equiv l^2$, which completes the proof of the assertions of example 14.5.

Note that $\{y_n\}$ is obviously different from the basis of the form (14.64) obtained by starting with the basis $\{x_n\}$ defined by (14.65) (i.e., applying the Gram-Schmidt process to $\{x_n\}$, etc.).

The question answered by example 14.5 above also suggests the following more general problem:

---

[1] We again use the observation that a basis is unconditional if and only if every permutation of this basis is a basis. Alternatively, one could observe that $\{y_n\}$ is non-Besselian, whence, by proposition 14.1, conditional; or, alternatively, one could use § 17, theorem 17.7.

Problem 14.1. a) Let $\{e_n\}$ be an unconditional basis of a Banach space E. Does there exist a conditional basis $\{x_n\}$ of E of the form

$$x_n = \sum_{i=1}^{n} \beta_i^{(n)} e_i, \quad \beta_n^{(n)} \neq 0 \quad (n=1,2,\ldots)\,? \tag{14.69}$$

b) What about a conditional basis of the form

$$x_n = \sum_{i=n}^{\infty} \beta_i^{(n)} e_i, \quad \beta_n^{(n)} \neq 0 \quad (n=1,2,\ldots)\,? \tag{14.70}$$

Example 14.5 shows that the answer is affirmative if $E=l^2$. The answer to problem 14.1 a) is still affirmative if $E=c_0$ or l^1, by examples 14.1 and 14.2 above and by § 18, theorem 18.2.

§ 15. Some separable Banach spaces having no unconditional basis

In the present section we shall show that some concrete separable Banach spaces (among which are, in particular, the important spaces $C([0,1])$ and $L^1([0,1])$) have no unconditional basis. Some other separable Banach spaces having no unconditional basis will be given in § 17.

Let us first recall that a series $\sum_{i=1}^{\infty} x_i$ in a Banach space E is called $\sigma(E, E^*)$*-unconditionally Cauchy* or *weakly unconditionally Cauchy*, if $\sum_{i=1}^{\infty} |f(x_i)| < \infty$ for all $f \in E^*$. A series $\sum_{i=1}^{\infty} f_i$ in a conjugate Banach space E^* is called $\sigma(E^*, E)$*-unconditionally Cauchy* if $\sum_{i=1}^{\infty} |f_i(x)| < \infty$ for all $x \in E$.

Lemma 15.1. *For a series* $\sum_{i=1}^{\infty} f_i$ *in a conjugate Banach space* E^* *the following statements are equivalent:*

1°. $\sum_{i=1}^{\infty} f_i$ *is* $\sigma(E^*, E)$*-unconditionally Cauchy.*

2°. *There exists a constant* $C>0$ *such that*

$$\sum_{i=1}^{\infty} |f_i(x)| \leqslant C\,\|x\| \quad (x \in E). \tag{15.1}$$

3°. *There exists a constant* $C>0$ *such that*

$$\left\|\sum_{i=1}^{n} \alpha_i f_i\right\| \leqslant C \quad (\alpha_i = 0 \text{ or } 1, \quad i=1,\ldots,n; \quad n=1,2,\ldots). \tag{15.2}$$

4°. *There exists a constant* $C>0$ *such that*

$$\left\|\sum_{i=1}^{n} \varepsilon_i f_i\right\| \leqslant C \qquad (\varepsilon_i = \pm 1,\ i=1,\ldots,n;\ n=1,2,\ldots). \tag{15.3}$$

5°. *There exists a constant* $C>0$ *such that*

$$\left\|\sum_{i=1}^{n} \beta_i f_i\right\| \leqslant C \qquad (|\beta_i|\leqslant 1,\ i=1,\ldots,n;\ n=1,2,\ldots). \tag{15.4}$$

6°. $\sum_{i=1}^{\infty} f_i$ *is* $\sigma(E^*, E^{**})$*-unconditionally Cauchy.*

Proof. The implication $1^\circ \Rightarrow 2^\circ$ follows applying the uniform boundedness principle[1] to the sequence of continuous non-linear functionals $\{p_n\}$, where

$$p_n(x) = \sum_{i=1}^{n} |f_i(x)| \qquad (x\in E,\ n=1,2,\ldots).$$

Assume now that we have 2° and let β_i with $|\beta_i|\leqslant 1$ and $x\in E$ be arbitrary. Then

$$\left|\sum_{i=1}^{n} \beta_i f_i(x)\right| \leqslant \sum_{i=1}^{n} |f_i(x)| \leqslant C\,\|x\|,$$

whence we obtain (15.4). Thus $2^\circ \Rightarrow 5^\circ$.

The implication $5^\circ \Rightarrow 3^\circ$ is obvious.

The implication $3^\circ \Rightarrow 4^\circ$ follows from

$$\left\|\sum_{i=1}^{n} \varepsilon_i x_i\right\| \leqslant \left\|\sum_{i=1}^{n} \alpha_i x_i\right\| + \left\|\sum_{i=1}^{n} \alpha_i' x_i\right\|,$$

where $\alpha_i=1, \alpha_i'=0$ if $\varepsilon_i=1$ and $\alpha_i=0, \alpha_i'=1$ if $\varepsilon_i=-1$.

Assume now that we have 4° and let $\Phi\in E^{**}$ be arbitrary. Then, putting $\varepsilon_i = \operatorname{sign}\operatorname{Re}\Phi(f_i)$ if $\operatorname{Re}\Phi(f_i)\neq 0$ and $\varepsilon_i=1$ if $\operatorname{Re}\Phi(f_i)=0$, we have $\varepsilon_i=\pm 1$ $(i=1,2,\ldots)$ and, by 4°,

$$\begin{aligned}\sum_{i=1}^{n} |\operatorname{Re}\Phi(f_i)| &= \sum_{i=1}^{n} \varepsilon_i \operatorname{Re}\Phi(f_i) = \operatorname{Re}\Phi\left(\sum_{i=1}^{n} \varepsilon_i f_i\right)\\ &\leqslant \left|\Phi\left(\sum_{i=1}^{n} \varepsilon_i f_i\right)\right| \leqslant \|\Phi\| \left\|\sum_{i=1}^{n} \varepsilon_i f_i\right\| \leqslant C\|\Phi\| \qquad (n=1,2,\ldots),\end{aligned}$$

[1] See e.g. [50], p. 53, lemma 13.

whence $\sum_{i=1}^{\infty} |\mathrm{Re}\,\Phi(f_i)| < \infty$. Similarly, $\sum_{i=1}^{\infty} |\mathrm{Im}\,\Phi(f_i)| < \infty$, whence $\sum_{i=1}^{\infty} |\Phi(f_i)| \leqslant \sum_{i=1}^{\infty} |\mathrm{Re}\,\Phi(f_i)| + \sum_{i=1}^{\infty} |\mathrm{Im}\,\Phi(f_i)| < \infty$. Thus, $4^\circ \Rightarrow 6^\circ$.

Finally, the implication $6^\circ \Rightarrow 1^\circ$ is obvious, which completes the proof of lemma 15.1.

Corollary 15.1. *For a series $\sum_{i=1}^{\infty} x_i$ in a Banach space E the following statements are equivalent:*

1°. *$\sum_{i=1}^{\infty} x_i$ is weakly unconditionally Cauchy.*

2°. *There exists a constant $C>0$ such that*

$$\sum_{i=1}^{\infty} |f(x_i)| \leqslant C\|f\| \qquad (f \in E^*). \tag{15.5}$$

3°. *There exists a constant $C>0$ such that*

$$\left\|\sum_{i=1}^{n} \alpha_i x_i\right\| \leqslant C \qquad (\alpha_i = 0 \text{ or } 1,\ i=1,\dots,n;\ n=1,2,\dots). \tag{15.6}$$

4°. *There exists a constant $C>0$ such that*

$$\left\|\sum_{i=1}^{n} \varepsilon_i x_i\right\| \leqslant C \qquad (\varepsilon_i = \pm 1,\ i=1,\dots,n;\ n=1,2,\dots). \tag{15.7}$$

5°. *There exists a constant $C>0$ such that*

$$\left\|\sum_{i=1}^{n} \beta_i x_i\right\| \leqslant C \qquad (|\beta_i| \leqslant 1,\ i=1,\dots,n;\ n=1,2,\dots). \tag{15.8}$$

Proof. This follows by embedding E in E^{**} and applying lemma 15.1 in E^{**}.

Now we can give

Theorem 15.1. *Let E be a separable Banach space containing a subspace E_0 with E_0^* weakly complete*[1] *and non-separable (e.g., a subspace E_0 isomorphic to $C([0,1])$). Then E has no unconditional basis.*

Proof. Assume that E has an unconditional basis $\{x_n\}$ with the a.s.c.f. $\{f_n\} \subset E^*$ and put

$$\phi_n = f_n|_{E_0} \qquad (n=1,2,\dots). \tag{15.9}$$

[1] Actually, it is sufficient to assume that E_0^* contains no subspace isomorphic to c_0 (by lemma 15.8 below).

Let $\phi \in E_0^*$ be arbitrary and let $f \in E^*$ be an extension of ϕ to E. Then, since $\{x_n\}$ is an unconditional basis of E, the series $\sum_{i=1}^{\infty} f(x_i) f_i$ is $\sigma(E^*, E)$-unconditionally convergent to f (by Ch. I, § 14, formula (14.8)), whence, by lemma 15.1, there exists a constant $M > 0$ such that

$$\left\| \sum_{i=1}^{n} \varepsilon_i f(x_i) \phi_i \right\| \leqslant \left\| \sum_{i=1}^{n} \varepsilon_i f(x_i) f_i \right\| \leqslant M \qquad (\varepsilon_i = \pm 1, i = 1, \ldots, n; n = 1, 2, \ldots).$$

Consequently, by corollary 15.1, the series $\sum_{i=1}^{\infty} f(x_i) \phi_i$ is $\sigma(E_0^*, E_0^{**})$-unconditionally Cauchy, whence, since E_0^* is weakly complete, this series is $\sigma(E_0^*, E_0^{**})$-convergent to a $\psi \in E_0^*$, i.e.,

$$\sum_{i=1}^{\infty} f(x_i) \Phi(\phi_i) = \Phi(\psi) \qquad (\Phi \in E_0^{**}).$$

On the other hand, as observed above, we have

$$\sum_{i=1}^{\infty} f(x_i) \phi_i(x) = \sum_{i=1}^{\infty} f(x_i) f_i(x) = f(x) = \phi(x) \qquad (x \in E_0),$$

and hence, considering the functionals $\Phi_x(\gamma) = \gamma(x)$ $(\gamma \in E_0^*)$, we obtain $\psi = \phi$. Thus, the sequence $\left\{ \sum_{i=1}^{n} f(x_i) \phi_i \right\}$ converges to ϕ for $\sigma(E_0^*, E_0^{**})$. Since $\phi \in E_0^*$ was arbitrary, this proves that E_0^* is separable for $\sigma(E_0^*, E_0^{**})$, whence[1] also strongly separable, which contradicts the hypothesis. This completes the proof of theorem 15.1.

We shall say that a Banach space E_0 has *property* (p) if every subspace B of E_0^* such that for each $\phi_0 \in E_0^*$ there exists a sequence $\{\beta_n\} \subset B$ satisfying

$$\phi_0(x) = \lim_{n \to \infty} \beta_n(x) \qquad (x \in E_0), \tag{15.10}$$

$$\lim_{n \to \infty} \Phi(\beta_n) \quad \text{exists for each} \quad \Phi \in E_0^{**}, \tag{15.11}$$

(i.e., $\beta_n \to \phi_0$ for $\sigma(E_0^*, E_0)$ and $\{\beta_n\}$ is a Cauchy sequence for $\sigma(E_0^*, E_0^{**})$) is non-separable.

Theorem 15.2. *Let E be a separable Banach space containing a subspace E_0 which has property* (p) *(e.g., by lemma* 15.6 *below, a subspace E_0 isomorphic to $L^1([0,1])$). Then E has no unconditional basis.*

[1] See e.g. [10], p. 134, theorem 2.

Proof. Assume that E has an unconditional basis $\{x_n\}$ with the a.s.c.f. $\{f_n\} \subset E^*$ and put

$$\phi_n = f_n|_{E_0} \quad (n=1,2,\ldots), \tag{15.12}$$

$$B = [\phi_n]. \tag{15.13}$$

Then B is a separable subspace of E_0^* and we shall show that for each $\phi_0 \in E_0^*$ there exists a sequence $\{\beta_n\} \subset B$ satisfying (15.10) and (15.11), in contradiction with our assumption that E_0 has property (p), which will complete the proof of theorem 15.2.

Let $\phi_0 \in E_0^*$ be arbitrary and let $f_0 \in E^*$ be an extension of ϕ_0 to E. Then, since $\{x_n\}$ is an unconditional basis of E, the series $\sum_{i=1}^{\infty} f_0(x_i) f_i$ is $\sigma(E^*, E)$-unconditionally convergent to f_0 (by Ch. I, § 14, formula (14.8)), whence the sequence $\{\beta_n\} \subset B$ defined by

$$\beta_n = \sum_{i=1}^{n} f_0(x_i) \phi_i \quad (n=1,2,\ldots) \tag{15.14}$$

satisfies (15.10). Furthermore, by the above proof of theorem 15.1, the series $\sum_{i=1}^{\infty} f_0(x_i) \phi_i$ is $\sigma(E_0^*, E_0^{**})$-unconditionally Cauchy, whence $\{\beta_n\}$ also satisfies (15.11). This completes the proof of theorem 15.2.

Now we shall prove that a Banach space E_0 isomorphic to $L^1([0,1])$ has property (p) (lemma 15.6, below). We shall denote by v the Lebesgue measure on $[0,1]$.

Lemma 15.2. *Let B be a separable subspace of $L^\infty([0,1])$. Then there exists a perfect set $T \subset [0,1]$ with $v(T) > 0$, such that every equivalence class $\beta \in B$ contains a function α with $\alpha|_T$ continuous on T.*

Proof. Let $\{\beta_n\}$ be a countable dense set in B and let $0 < \varepsilon < 1$. By the theorem of Luzin there exists a decreasing sequence of closed sets

$$T_1 \supset T_2 \supset T_3 \supset \cdots \tag{15.15}$$

such that

$$v(T_1) > 1 - \frac{\varepsilon}{2}, \quad v(T_{n+1}) > v(T_n) - \frac{\varepsilon}{2^{n+1}} \quad (n=1,2,\ldots), \tag{15.16}$$

and that the equivalence class β_n contains a function α_n with $\alpha_n|_{T_n}$ continuous on T_n $(n=1,2,\ldots)$. Put

$$T_0 = \bigcap_{i=1}^{\infty} T_i. \tag{15.17}$$

Then T_0 is closed, and thus[1] we can write $T_0 = T \cup S$, where T is perfect and S is finite or countable. We shall prove that T also has the other required properties. By (15.16) we have

$$\nu(T_n) > 1 - \frac{\varepsilon}{2} - \frac{\varepsilon}{4} - \cdots - \frac{\varepsilon}{2^n} > 1 - \varepsilon \quad (n = 1, 2, \ldots),$$

whence, by (15.15) and (15.17),

$$\nu(T) = \nu(T_0) \geqslant 1 - \varepsilon > 0. \tag{15.18}$$

Now let $\beta \in B$ be arbitrary. Then, since $\{\beta_n\}$ is dense in B, it has a subsequence $\{\beta_{n_k}\}$ such that $\|\beta_{n_k} - \beta\| \to 0$ as $k \to \infty$. By the above, there exist functions $\alpha_{n_k} \in \beta_{n_k}$ with $\alpha_{n_k}|_{T_{n_k}}$ continuous; then, since $T \subset T_{n_k}$, the function $\alpha_{n_k}|_T$ is also continuous $(k = 1, 2, \ldots)$. Since[2] $\|\tilde{\alpha}_{n_k} - \beta\| \to 0$ as $k \to \infty$ and since T is perfect, $\{\alpha_{n_k}|_T\}$ is a Cauchy sequence in $C(T)$, whence it converges to an $\alpha_0 \in C(T)$. Putting $\alpha = \alpha_0$ on T and $\alpha =$ any function of the class β on $[0,1] \setminus T$, and taking into account that T is perfect, we obtain

$$\begin{aligned} \|\tilde{\alpha} - \beta\| &\leqslant \|\tilde{\alpha} - \beta_{n_k}\| + \|\beta_{n_k} - \beta\| \\ &\leqslant \max(\|\alpha_0 - \alpha_{n_k}\|_{C(T)}, \|\beta - \tilde{\alpha}_{n_k}\|_{L^\infty([0,1]\setminus T)}) + \|\beta_{n_k} - \beta\| \to 0 \text{ as } k \to \infty, \end{aligned}$$

whence $\tilde{\alpha} = \beta$, i.e., $\alpha \in \beta$, which completes the proof of lemma 15.2.

Lemma 15.3. *Let M be a ν-measurable subset of $[0,1]$, with $\nu(M) > 0$. If $\nu(I \cap M) > 0$ for some closed interval $I \subset [0,1]$, then there exists a compact set $Q \subset M$, nowhere dense in[3] M, such that $\nu(I \cap Q) > 0$.*

Proof. Let $\{t_n\}_{n=1}^\infty$ be a dense sequence in $I \cap M$. Then, by the regularity of the Lebesgue measure, there exists a compact set $Q \subset (I \cap M) \setminus \{t_n\}_{n=1}^\infty$ such that

$$\nu(Q) > \tfrac{1}{2}\nu((I \cap M) \setminus \{t_n\}_{n=1}^\infty) = \tfrac{1}{2}\nu(I \cap M) > 0.$$

Since Q is compact, it is closed. Furthermore, since Q is disjoint from the dense subset $\{t_n\}_{n=1}^\infty$ of $I \cap M$, it is nowhere dense in $I \cap M$, whence (since $Q \subset I \cap M$) also nowhere dense in M, which completes the proof.

---

[1] See e.g. [175], Ch. II, § 6, theorem 4.

[2] We denote by $\tilde{\alpha}$ the equivalence class of the function α.

[3] I.e., nowhere dense for the relative topology induced by $[0,1]$ in M.

Lemma 15.4. *Let T be a ν-measurable subset of $[0,1]$, with $\nu(T)>0$. Then there exists a compact set $T^* \subset T$ with $\nu(T^*)>0$, such that if $\nu(I \cap T^*)>0$ for a closed interval $I \subset [0,1]$, then $\nu((I \cap T^*) \setminus Q)>0$ for every compact subset $Q \subset T^*$ which is nowhere dense in T^*.*

Proof. Let $\nu|_T$ be the restriction of ν to T and let

$$T^* = S(\nu|_T) = \text{the carrier of } \nu|_T. \tag{15.19}$$

We shall prove that T^* has the required properties. Obviously, $\nu(T^*)>0$. Assume that $\nu(I \cap T^*)>0$ for a closed interval $I \subset [0, 1]$. Then $\nu(\mathring{I} \cap T^*) = \nu(I \cap T^*)>0$, where $\mathring{I}$ denotes the interior of I. Hence $\mathring{I} \cap T^* \neq \emptyset$ and consequently, $(\mathring{I} \cap T^*) \setminus Q \neq \emptyset$ for any compact subset $Q \subset T^*$ which is nowhere dense in T^* (because $(\mathring{I} \cap T^*) \setminus Q = \emptyset$ would imply $\mathring{I} \cap T^* \subset Q$, whence Q would not be nowhere dense in T^*). On the other hand, $(\mathring{I} \cap T^*) \setminus Q$ is open in $\mathring{I} \cap T^*$, whence also in T^*, and $(\mathring{I} \cap T^*) \setminus Q \subset T^* = S(\nu|_{T^*})$. Consequently, by the definition of the carrier, $\nu((\mathring{I} \cap T^*) \setminus Q)>0$, whence also $\nu((I \cap T^*) \setminus Q)>0$, which completes the proof.

Lemma 15.5. *Let T be a ν-measurable subset of $[0,1]$, with $\nu(T)>0$. Then there exists a bounded measurable real function α_0 such that $\alpha_0(t)=0$ for $t \in [0,1] \setminus T$ and that $\alpha_0|_T$ is not equivalent to any function belonging to the first Baire class on T.*

Proof. Let T^* be as in lemma 15.4. Obviously, it will be sufficient to construct an α_0 on T^* which is not equivalent to any function of the first Baire class on T^*.

Let $\{I_n\}$ be the sequence of all closed intervals $I_n \subset [0,1]$ with rational endpoints, such that $\nu(I_n \cap T^*)>0$.

By lemma 15.3 with $M = T^*, I = I_1$, we can choose a compact set $A_1 \subset T^*$, nowhere dense in T^*, such that $\nu(I_1 \cap A_1)>0$. Then, since by hypothesis $\nu(I_1 \cap T^*)>0$, by virtue of lemma 15.4 we have $\nu((I_1 \cap T^*) \setminus A_1)>0$. Since $I_1 \cap (T^* \setminus A_1) = (I_1 \cap T^*) \setminus A_1$, we can apply lemma 15.3 with $M = T^* \setminus A_1$, $I = I_1$, to obtain a compact set $B_1 \subset T^* \setminus A_1$, nowhere dense in $T^* \setminus A_1$, whence also in T^*, and such that $\nu(I_1 \cap B_1)>0$. We can continue this procedure to obtain two sequences of compact sets $\{A_n\}, \{B_n\}$ in T^*, with the following properties:

a) A_n and B_n are nowhere dense in T^* $(n=1,2,\ldots)$;

b) $A_n \cap B_n = \emptyset$, $(A_n \cup B_n) \cap \left(\bigcup_{j=1}^{n-1} (A_j \cup B_j) \right) = \emptyset$ $(n=1,2,\ldots)$;

c) $\nu(I_n \cap A_n)>0$, $\nu(I_n \cap B_n)>0$ $(n=1,2,\ldots)$.

Indeed, assume that $A_1, B_1, \ldots, A_n, B_n$ have been already constructed. Then, by a), $\bigcup_{j=1}^{n} (A_j \cup B_j)$ is nowhere dense in T^*. Hence, by lemma 15.4 and the hypothesis $\nu(I_{n+1} \cap T^*) > 0$, we have $\nu\left((I_{n+1} \cap T^*) \setminus \bigcup_{j=1}^{n} (A_j \cup B_j)\right) > 0$. Since $I_{n+1} \cap \left(T^* \setminus \bigcup_{j=1}^{n} (A_j \cup B_j)\right) = (I_{n+1} \cap T^*) \setminus \bigcup_{j=1}^{n} (A_j \cup B_j)$, we can apply lemma 15.3 with $M = T^* \setminus \bigcup_{j=1}^{n} (A_j \cup B_j)$, $I = I_{n+1}$, to obtain a compact set $A_{n+1} \subset T^* \setminus \bigcup_{j=1}^{n} (A_j \cup B_j)$, nowhere dense in $T^* \setminus \bigcup_{j=1}^{n} (A_j \cup B_j)$, whence also in T^*, and such that $\nu(I_{n+1} \cap A_{n+1}) > 0$. The construction of B_{n+1} is similar, starting with the nowhere dense at $A_{n+1} \cup \left(\bigcup_{j=1}^{n} (A_j \cup B_j)\right)$. This completes the induction.

Put

$$A = \bigcup_{j=1}^{\infty} A_j. \tag{15.20}$$

Then $A \subset T^*$ is ν-measurable and $\nu(A) > 0$. We claim that *for every closed interval $I \subset [0,1]$ with $I \cap T^* \neq \emptyset$ we have*

$$\nu(I \cap A) > 0, \qquad \nu(I \cap (T^* \setminus A)) > 0. \tag{15.21}$$

Indeed, since $\nu(I \cap T^*) > 0$, there exists an index n such that $I_n \subset I$. Hence $I_n \cap A_n \subset I \cap A_n \subset I \cap A$, and thus, by c), we have the first inequality of (15.21). Furthermore, by b) we have $B_n \cap A = B_n \cap \left(\bigcup_{j=1}^{\infty} A_j\right) = \emptyset$, whence $B_n \subset T^* \setminus A$, whence $I_n \cap B_n \subset I \cap B_n \subset I \cap (T^* \setminus A)$, and thus, by c), we also have the second inequality of (15.21).

Now we shall complete the proof by showing that the characteristic function α_0 of the set A has the required properties. This function is obviously bounded and measurable and $\alpha_0(t) = 0$ for $t \in [0,1] \setminus T$. Finally, since for every closed interval $I \subset [0,1]$ with $I \cap T^* \neq \emptyset$ the oscillation of α_0 on $I \cap T^*$ is equal to 1 and cannot be diminished by any change of the values of α_0 on a set of measure 0 (by (15.21)), $\alpha_0|_{T^*}$ is not equivalent to any function belonging to the first Baire class[1] on T^*, which completes the proof.

Lemma 15.6. *The space $L^1([0,1])$ (whence also every Banach space E_0 isomorphic to $L^1([0,1])$) has property (p).*

---

[1] Indeed, it is well known (see e.g. [175], Ch. XV, § 3, theorem 3) that if γ is of the first Baire class on T^*, then for any closed set $Q \subset T^*$ the function $\gamma|_Q$ must have at least one point of continuity.

Proof. Let B be a subspace of[1] $L^\infty([0,1])$, such that for every $\phi_0 \in L^\infty([0,1])$ there exists a sequence $\{\beta_n\} \subset B$ satisfying (15.10) and (15.11) (with $\Phi \in L^\infty([0,1])^*$) and assume that B is separable. Choose $T \subset [0,1]$ as in lemma 15.2 and for this T choose α_0 as in lemma 15.5. For $\phi_0 = \tilde{\alpha}_0 \in L^\infty([0,1])$ let $\{\beta_n\} \subset B$ be a sequence satisfying (15.10) and (15.11). Since by lemma 15.2 every $\beta \in B$ contains a function α with $\alpha|_T \in C(T)$, for each $t \in T$ we can define a functional $\phi_t \in B^*$ by

$$\phi_t(\beta) = \alpha(t) \qquad (\beta \in B), \tag{15.22}$$

where $\alpha \in \beta$, $\alpha|_T \in C(T)$; this functional is well defined, since the relations $\alpha_1, \alpha_2 \in \beta$, $\alpha_1|_T$, $\alpha_2|_T \in C(T)$ imply $\alpha_1|_T = \alpha_2|_T$, and it belongs to B^* since, T being perfect,

$$|\phi_t(\beta)| = |\alpha(t)| \leqslant \max_{t' \in T} |\alpha(t')| \leqslant \operatorname{ess\,sup}_{t' \in [0,1]} |\alpha(t')| = \|\beta\| \qquad (\beta \in B).$$

Consequently, by the Hahn-Banach theorem, for each $t \in T$ there exists a functional $\Phi_t \in L^\infty([0,1])^*$ such that

$$\Phi_t(\beta) = \alpha(t) \qquad (\beta \in B), \tag{15.23}$$

where $\alpha \in \beta$, $\alpha|_T \in C(T)$. Then, by (15.11), there exist the limits

$$\lim_{n\to\infty} \Phi_t(\beta_n) = \lim_{t\to\infty} \alpha_n(t) \qquad (t \in T), \tag{15.24}$$

where $\alpha_n \in \beta_n$, $\alpha_n|_T \in C(T)$ $(n = 1, 2, \ldots)$; furthermore, again by (15.11), we have

$$\sup_n \|\beta_n\| < \infty. \tag{15.25}$$

Now, let $x \in L^1([0,1])$ be an arbitrary element such that $\xi(t) = 0$ $(t \in [0,1] \setminus T,\ \xi \in x)$. Then, by (15.10), (15.24), (15.25) and the Lebesgue theorem on integration of sequences of functions, we have

$$\begin{aligned}\int_T \xi(t)\alpha_0(t)\,dt &= \int_0^1 \xi(t)\alpha_0(t)\,dt = \phi_0(x) = \lim_{n\to\infty} \beta_n(x) = \lim_{n\to\infty} \int_0^1 \xi(t)\alpha_n(t)\,dt \\ &= \lim_{n\to\infty} \int_T \xi(t)\alpha_n(t)\,dt = \int_T \xi(t) \lim_{n\to\infty} \alpha_n(t)\,dt,\end{aligned}$$

whence, since $\xi|_T$ is an arbitrary summable function on T, it follows that $\alpha_0|_T$ is equivalent to the function $\lim_{n\to\infty} \alpha_n(.)|_T$ belonging to the first Baire class on T, in contradiction with our choice of α_0. This completes the proof of lemma 15.6.

---

[1] We identify (canonically) the spaces $L^\infty([0,1])$ and $L^1([0,1])^*$.

Consequently, as we observed above, a separable Banach space E containing a subspace E_0 isomorphic to $L^1([0,1])$ has no unconditional basis[1]. In particular, for the space $E = L^1([0,1])$ itself one can give another simpler proof, as shown by

Theorem 15.3. *Let E be a weakly complete separable Banach space which is not isomorphic to any conjugate Banach space (e.g.[2], $E = L^1([0,1])$). Then E has no unconditional basis.*

Proof. Assume that E has an unconditional basis $\{x_n\}$ with the a.s.c.f. $\{f_n\} \subset E^*$. We claim that $\{x_n\}$ must be boundedly complete. Indeed, assume the contrary, i.e., that there exists a sequence of scalars $\{a_i\}$ such that $\sup_n \left\| \sum_{i=1}^{n} a_i x_i \right\| < \infty$ and that $\sum_{i=1}^{\infty} a_i x_i$ does not converge. Then, since $\{x_n\}$ is an unconditional basis, for any $\varepsilon_i = \pm 1$ we have, by § 17, theorem 17.1, $\sup_{1 \leqslant n < \infty} \left\| \sum_{i=1}^{n} \varepsilon_i a_i x_i \right\| \leqslant C \sup_{1 \leqslant n < \infty} \left\| \sum_{i=1}^{n} a_i x_i \right\| < \infty$, whence by corollary 15.1, the series $\sum_{i=1}^{\infty} a_i x_i$ is weakly unconditionally Cauchy. However, this series does not converge weakly to an element $y \in E$, since then one would have, by biorthogonality, $f_k(y) = \lim_{n \to \infty} f_k\left(\sum_{i=1}^{n} a_i x_i \right) = a_k$ $(k = 1, 2, \ldots)$, whence $y = \sum_{k=1}^{\infty} f_k(y) x_k = \sum_{k=1}^{\infty} a_k x_k$, in contradiction to the assumption that $\sum_{k=1}^{\infty} a_k x_k$ does not converge. Consequently, E is not weakly complete (the weak Cauchy sequence $\left\{ \sum_{i=1}^{n} a_i x_i \right\}$ does not converge weakly to any element $y \in E$), which contradicts our hypothesis. This proves that $\{x_n\}$ is boundedly complete, and hence, taking into account Ch. I, § 3, proposition 3.2, it follows that E_0 is isomorphic to the Banach space of sequences of scalars $\left\{ \{\alpha_n\} \subset K \;\middle|\; \sup_{1 \leqslant n < \infty} \left\| \sum_{i=1}^{n} a_i x_i \right\| < \infty \right\}$ endowed with the norm $\|\{\alpha_n\}\| = \sup_{1 \leqslant n < \infty} \left\| \sum_{i=1}^{n} \alpha_i x_i \right\|$. However, by Ch. I, § 12, theorem 12.5 c), this latter space, whence also E, is isomorphic to the conjugate space $[f_n]^*$, which contradicts our hypothesis. This completes the proof of theorem 15.3.

---

[1] Let us observe that the same assertion remains valid if we replace $L^1([0,1])$ by any space $L^1(Q, \nu)$, where Q is a compact metric space and ν a non-purely atomic measure defined on the Borel sets of Q.

[2] See [73], p. 265 and [46].

A Banach space E is said to have *property* (u) if for every weak Cauchy sequence $\{z_n\} \subset E$ there exists a sequence $\{y_n\} \subset E$ such that

a) the series $\sum_{i=1}^{\infty} y_i$ is weakly unconditionally Cauchy;

b) the sequence $\left\{z_n - \sum_{i=1}^{n} y_i\right\}$ converges weakly to 0.

Our next aim is to prove

Theorem 15.4. *Let E be a separable Banach space containing a subspace E_0 which does not have the property (u) (e.g., by corollary 15.4 below, a subspace E_0 isomorphic to the space J of § 4, example 4.1). Then E has no unconditional basis.*

For this purpose we shall need some auxiliary results. Let us prove first a result on selection of basic sequences.

Proposition 15.1. *Let $\{x_n\}$ be a basis of a Banach space E, with the a.s.c.f. $\{f_n\} \subset E^*$. If a sequence $\{y_n\} \subset E$ satisfies the conditions*

$$\inf_{1 \leqslant n < \infty} \|y_n\| = \varepsilon > 0, \tag{15.26}$$

$$\lim_{n \to \infty} f_i(y_n) = 0 \qquad (i = 1, 2, \ldots), \tag{15.27}$$

then $\{y_n\}$ has a subsequence $\{y_{p_{n+1}}\}$ which is a basic sequence, equivalent to a block basic sequence with respect to $\{x_n\}$.

Proof. Let $C = v_{\{x_n\}} = \sup_{1 \leqslant n < \infty} \|s_n\|$. By (15.27) we can construct two increasing sequences of positive integers $\{p_n\}$, $\{q_n\}$ such that

$$\frac{4C}{\varepsilon} \left\| \sum_{i=q_n+1}^{\infty} f_i(y_{p_n}) x_i \right\| \leqslant \frac{1}{2^{n+2}} \qquad (n = 1, 2, \ldots), \tag{15.28}$$

$$\frac{4C}{\varepsilon} \left\| \sum_{i=1}^{q_n} f_i(y_{p_{n+1}}) x_i \right\| \leqslant \frac{1}{2^{n+2}} \qquad (n = 1, 2, \ldots). \tag{15.29}$$

Indeed, take $p_1 = 1$. Since $\sum_{i=1}^{\infty} f_i(y_1) x_i = y_1$, there exists a q_1 such that $\left\| \sum_{i=q_1+1}^{\infty} f_i(y_1) x_i \right\| \leqslant \frac{\varepsilon}{4C \cdot 2^3}$. Assume that we have already constructed $p_1, \ldots, p_n, q_1, \ldots, q_n$. Then, by (15.27), there exists a p_{n+1} such that $|f_i(y_{p_{n+1}})| < \dfrac{\varepsilon}{4C \cdot 2^{n+2} \sum_{i=1}^{q_n} \|x_i\|}$ $(i = 1, \ldots, q_n)$, whence we get

$$\left\| \sum_{i=1}^{q_n} f_i(y_{p_{n+1}}) x_i \right\| \leqslant \sum_{i=1}^{q_n} |f_i(y_{p_{n+1}})| \, \|x_i\| < \frac{\varepsilon}{4C \cdot 2^{n+2}}$$

i.e., (15.29). Since $\sum_{i=1}^{\infty} f_i(y_{p_{n+1}}) x_i$ converges (to $y_{p_{n+1}}$), there also exists a q_{n+1} such that we have (15.28) for $n+1$ instead of n, which proves our assertion.

Put

$$z_n = \sum_{i=q_n+1}^{q_{n+1}} f_i(y_{p_{n+1}}) x_i \qquad (n=1,2,\ldots). \tag{15.30}$$

Then, by (15.26), (15.28), (15.29) and by $C \geqslant 1$,

$$\begin{aligned} \varepsilon \leqslant \|y_{p_{n+1}}\| &= \left\| \sum_{i=1}^{\infty} f_i(y_{p_{n+1}}) x_i \right\| \\ &< \frac{\varepsilon}{4C \cdot 2^{n+2}} + \left\| \sum_{i=q_n+1}^{q_{n+1}} f_i(y_{p_{n+1}}) x_i \right\| + \frac{\varepsilon}{4C \cdot 2^{n+3}} \\ &< \frac{\varepsilon}{2} + \left\| \sum_{i=q_n+1}^{q_{n+1}} f_i(y_{p_{n+1}}) x_i \right\| \qquad (n=1,2,\ldots), \end{aligned}$$

whence

$$\|z_n\| \geqslant \frac{\varepsilon}{2} \qquad (n=1,2,\ldots). \tag{15.31}$$

Consequently, for the sequence $\{h_n\} \subset [z_n]^*$ of coefficient functionals associated to the block basic sequence $\{z_n\}$ we have, by Ch. I, § 3, formula (3.8) and Ch. I, § 7, corollary 7.4,

$$\|h_n\| \leqslant \frac{4C}{\varepsilon} \qquad (n=1,2,\ldots). \tag{15.32}$$

On the other hand, again by (15.28) and (15.29), we have

$$\begin{aligned} \|z_n - y_{p_{n+1}}\| &= \left\| \sum_{i=1}^{q_n} f_i(y_{p_{n+1}}) x_i + \sum_{i=q_{n+1}+1}^{\infty} f_i(y_{p_{n+1}}) x_i \right\| \\ &\leqslant \frac{\varepsilon}{4C \cdot 2^{n+2}} + \frac{\varepsilon}{4C \cdot 2^{n+3}} < \frac{\varepsilon}{4C \cdot 2^{n+1}} \qquad (n=1,2,\ldots), \end{aligned}$$

whence, taking into account (15.32), we obtain

$$\sum_{i=1}^{\infty} \|h_i\| \, \|z_i - y_{p_{i+1}}\| < \frac{4C}{\varepsilon} \sum_{i=1}^{\infty} \frac{\varepsilon}{4C \cdot 2^{n+1}} = \frac{1}{2}.$$

Consequently, by Ch. I, § 10, theorem 10.1, $\{y_{p_{n+1}}\}$ is a basic sequence, equivalent to the block basic sequence $\{z_n\}$, which completes the proof of proposition 15.1.

Corollary 15.2. a) *Let $\{x_n\}$ be a basis of a Banach space E, with the a.s.c.f. $\{f_n\} \subset E^*$. If $\{z_n\} \subset E$ is a sequence with $\sup_{1 \leqslant n < \infty} \|z_n\| < \infty$, satisfying*

$$\lim_{n \to \infty} f_i(z_n) = 0 \quad (i = 1, 2, \ldots), \tag{15.33}$$

and if $z_n \not\to 0$ as $n \to \infty$, for the weak topology $\sigma(E, E^*)$, then $\{z_n\}$ has an infinite subsequence which is a basic sequence of type al_+, equivalent to a block basic sequence with respect to $\{x_n\}$.

b) *If $\{x_n\}$ is an unconditional basis of E, with the a.s.c.f. $\{f_n\} \subset E^*$, and if $\{z_n\} \subset E$ is as in* a), *then $\{z_n\}$ has a subsequence which is a basic sequence equivalent to the unit vector basis of l^1, and which is equivalent to a block basic sequence with respect to $\{x_n\}$.*

Proof. a) By our hypothesis, there exist an $f \in E^*$ and an infinite subsequence $\{z_{n_k}\}$ of $\{z_n\}$ such that[1]

$$\inf_{1 \leqslant k < \infty} |f(z_{n_k})| > 0. \tag{15.34}$$

Then $\inf_{1 \leqslant k < \infty} \|z_{n_k}\| \geqslant \frac{1}{\|f\|} \inf_{1 \leqslant k < \infty} |f(z_{n_k})| > 0$ and, by (15.33), $\lim_{k \to \infty} f_i(z_{n_k}) = 0$ $(i = 1, 2, \ldots)$, whence, by proposition 15.1, $\{z_{n_k}\}$ has a subsequence $\{z_{n_{k_m}}\}$ which is a basic sequence, equivalent to a block basic sequence with respect to $\{x_n\}$. By (15.34) and § 10, theorem 10.3 (implication $2^\circ \Rightarrow 1^\circ$), the basic sequence $\{z_{n_{k_m}}\}$ is of type al_+.

b) If $\{x_n\}$ is an unconditional basis of E, the basic sequence $\{z_{n_{k_m}}\}$ found in part a) above is unconditional, by § 17, corollary 17.2. Consequently, by § 17, corollary 17.1, $\{z_{n_{k_m}}\}$ is equivalent to the unit vector basis of l^1, which completes the proof.

Corollary 15.3. *Let $\{x_n\}$ be an unconditional basis of a Banach space E, with the a.s.c.f. $\{f_n\} \subset E^*$. If $\{z_n\} \subset E$ is a weak Cauchy sequence in E, satisfying* (15.33), *then $z_n \xrightarrow{w} 0$.*

Proof. If $z_n \not\to 0$ as $n \to \infty$, for the weak topology $\sigma(E, E^*)$, then, by corollary 15.2, $\{z_n\}$ has a subsequence, say $\{z_{n_p}\}$, which is a basic sequence equivalent to the unit vector basis of l^1. However, the unit vector basis in l^1 is not a weak Cauchy sequence (to see this, take e.g. the functional defined by $\{1, 0, 1, 0, \ldots\} \in m \equiv (l^1)^*$), whence $\{z_{n_p}\}$ is not a

---

[1] If the scalars are real, we can also find $\{z_{n_k}\} \subset \{z_n\}$ and $f \in E^*$ with $\inf_{1 \leqslant n < \infty} f(z_{n_k}) > 0$, whence every basic subsequence $\{z_{n_{k_m}}\}$ of $\{z_{n_k}\}$ will be of type l_+.

$\sigma([z_{n_p}], [z_{n_p}]^*)$-Cauchy sequence and therefore (by the Hahn-Banach theorem) not a $\sigma(E, E^*)$-Cauchy sequence, in contradiction with the hypothesis that the whole sequence $\{z_n\} \subset E$ is a weak Cauchy sequence. This completes the proof.

The hypothesis that $\{x_n\}$ is an unconditional basis in corollaries 15.2b) and 15.3 is essential, as shown by

Example 15.1. Let $E = c_0$ and let

$$x_n = z_n = \sum_{i=1}^{n} e_i \qquad (n = 1, 2, \ldots), \tag{15.35}$$

where $\{e_n\}$ is the unit vector basis of c_0. Then, by § 14, example 14.1, $\{x_n\}$ is a conditional basis of c_0. Furthermore, $\|z_n\| = 1$ $(n = 1, 2, \ldots)$, $\{z_n\}$ satisfies (15.33) (by biorthogonality) and $z_n \not\to 0$ as $n \to \infty$, for the weak topology $\sigma(c_0, c_0^*)$, since for the functional $f(x) = \xi_1$ $(x = \{\xi_n\} \in c_0)$ we have $f(z_n) = 1$ $(n = 1, 2, \ldots)$. However, $\{z_n\}$ has no basic subsequence equivalent to the unit vector basis of l^1 (since c_0 has no subspace isomorphic to l^1). Finally, $\{z_n\}$ is a weak Cauchy sequence, since for $f(x) = \sum_{i=1}^{\infty} \xi_i \eta_i$ $(x = \{\xi_n\} \in c_0)$, where $\sum_{i=1}^{\infty} |\eta_i| < \infty$, we have $\lim_{n \to \infty} f(z_n) = \sum_{i=1}^{\infty} \eta_i$.

Proposition 15.2. *Every Banach space E with an unconditional basis $\{x_n\}$ has property* (*u*).

Proof. We may assume, without loss of generality, that $\|x_n\| = 1$ $(n = 1, 2, \ldots)$. Let $\{f_n\} \subset E^*$ be the a.s.c.f. to $\{x_n\}$ and let $\{z_n\} \subset E$ be a weak Cauchy sequence. Then the limits

$$\alpha_i = \lim_{k \to \infty} f_i(z_k) \qquad (i = 1, 2, \ldots)$$

exist. Put

$$y_n = \alpha_n x_n = \left[\lim_{k \to \infty} f_n(z_k)\right] x_n \qquad (n = 1, 2, \ldots). \tag{15.36}$$

Since $\{x_n\}$ is an unconditional basis of E, for every $f \in E^*$ the series $\sum_{i=1}^{\infty} f(x_i) f_i$ is $\sigma(E^*, E)$-unconditionally convergent to f (by Ch. I, § 14, formula (14.8)). Consequently, by corollary 15.1, for every $f \in E^*$ there exists a constant $M_f > 0$ such that

$$\left\| \sum_{i=1}^{n} \beta_i f(x_i) f_i \right\| \leqslant M_f \qquad (n = 1, 2, \ldots; \ |\beta_i| \leqslant 1),$$

whence

$$\sum_{i=1}^{n} |f(y_i)| = \sum_{i=1}^{n} |f(x_i)\alpha_i|$$
$$= \lim_{k\to\infty} \sum_{i=1}^{n} [\operatorname{sign} f(x_i)\alpha_i]\, f(x_i)\, f_i(z_k)$$
$$\leqslant \left\| \sum_{i=1}^{n} [\operatorname{sign} f(x_i)\alpha_i]\, f(x_i)\, f_i \right\| \sup_{1\leqslant k<\infty} \|z_k\|$$
$$\leqslant M_f \sup_{1\leqslant k<\infty} \|z_k\| < \infty \qquad (f\in E^*,\ n=1,2,\ldots),$$

(since $\{z_n\}$ is a weak Cauchy sequence), and thus $\sum_{i=1}^{\infty} |f(y_i)| < \infty$ for all $f\in E^*$, i.e. $\sum_{i=1}^{\infty} y_i$ is weakly unconditionally Cauchy. Furthermore, we have

$$f_j\left(z_n - \sum_{i=1}^{n} y_i\right) = f_j(z_n) - \sum_{i=1}^{n} \left[\lim_{k\to\infty} f_i(z_k)\right] f_j(x_i)$$
$$= f_j(z_n) - \lim_{k\to\infty} f_j(z_k) \to 0 \quad \text{as} \quad n\to\infty \qquad (j=1,2,\ldots),$$

and $\left\{z_n - \sum_{i=1}^{n} y_i\right\}$ is a weak Cauchy sequence (indeed, $\{z_n\}$ is a weak Cauchy sequence by our hypothesis and $\left\{\sum_{i=1}^{n} y_i\right\}$ is a weak Cauchy sequence since we have shown above that $\sum_{i=1}^{\infty} y_i$ is even weakly unconditionally Cauchy). Consequently, by corollary 15.3, the sequence $\left\{z_n - \sum_{i=1}^{n} y_i\right\}$ converges weakly to 0, which completes the proof of proposition 15.2.

Lemma 15.7. *Every subspace E_0 of a Banach space E with property (u) has property (u).*

Proof. Let $\{z_n\} \subset E_0$ be a $\sigma(E_0, E_0^*)$-Cauchy sequence. Then, since $f|_{E_0} \in E^*$ for all $f\in E^*$, $\{z_n\}$ is also a $\sigma(E, E^*)$-Cauchy sequence, whence, since E has property (u), there exists a sequence $\{y_n\} \subset E$ such that the series $\sum_{i=1}^{\infty} y_i$ is $\sigma(E, E^*)$-unconditionally Cauchy and that the sequence $\{v_n\}$ is $\sigma(E, E^*)$-convergent to 0, where

$$v_n = z_n - \sum_{i=1}^{n} y_i \qquad (n=1,2,\ldots). \tag{15.37}$$

Then one can define inductively an increasing sequence of integers $0=p_0<p_1<p_2<\cdots$ and a sequence of scalars $\{\lambda_n\}$ with $\lambda_n\geqslant 0$, $\sum_{i=p_{n-1}+1}^{p_n}\lambda_i=1$ $(n=1,2,\ldots)$, such that

$$\|u_n\|\leqslant\frac{1}{2^n}\qquad(n=1,2,\ldots),\tag{15.38}$$

where

$$u_n=\sum_{i=p_{n-1}+1}^{p_n}\lambda_i v_i\qquad(n=1,2,\ldots).\tag{15.39}$$

Indeed, since $v_n\to 0$ for $\sigma(E,E^*)$, by a well known theorem of S. Mazur[1] there exists a sequence of finite convex combinations of the elements v_n which converges to 0 in the norm topology, whence there exist a positive integer p_1 and scalars $\lambda_1,\ldots,\lambda_{p_1}\geqslant 0$, $\sum_{i=1}^{p_1}\lambda_i=1$, such that $\left\|\sum_{i=1}^{p_1}\lambda_i v_i\right\|\leqslant\frac{1}{2}$; the elements $u_1,\ldots,u_n$ being constructed, $u_{n+1}=\sum_{i=p_n+1}^{p_{n+1}}\lambda_i v_i$ is obtained in a similar way from the fact that the sequence $\{v_{p_n+k}\}_{k=1}^{\infty}$ converges to 0 for $\sigma(E,E^*)$. Put

$$w_n=\sum_{i=p_{n-1}+1}^{p_n}\lambda_i z_i\qquad(n=1,2,\ldots),\tag{15.40}$$

$$y_0^0=w_1,\qquad y_n^0=w_{n+1}-w_n\qquad(n=1,2,\ldots).\tag{15.41}$$

Then, since $\{z_n\}\subset E_0$, we have $\{w_n\}\subset E_0$, whence $\{y_n^0\}\subset E_0$. We shall show that the series $\sum_{i=0}^{\infty}y_i^0$ is $\sigma(E_0,E_0^*)$-unconditionally Cauchy and that the sequence $\left\{z_n-\sum_{i=0}^{n}y_i^0\right\}$ is $\sigma(E_0,E_0^*)$-convergent to 0, which will complete the proof.

We have, by (15.41), (15.40), (15.37) and (15.39),

$$\begin{aligned}y_n^0=w_{n+1}-w_n&=\sum_{i=p_n+1}^{p_{n+1}}\lambda_i z_i-\sum_{i=p_{n-1}+1}^{p_n}\lambda_i z_i\\&=\sum_{i=p_n+1}^{p_{n+1}}\lambda_i\left(\sum_{j=1}^{i}y_j+v_i\right)-\sum_{i=p_{n-1}+1}^{p_n}\lambda_i\left(\sum_{j=1}^{i}y_j+v_i\right)\end{aligned}$$

[1] See e.g. [50], p. 422, corollary 14.

$$= \sum_{i=p_n+1}^{p_{n+1}} \lambda_i \sum_{j=1}^{i} y_j - \sum_{i=p_{n-1}+1}^{p_n} \lambda_i \sum_{j=1}^{i} y_j + u_{n+1} - u_n$$

$$= \sum_{j=1}^{p_{n+1}} \mu_j^n y_j + u_{n+1} - u_n,$$

where

$$\mu_j^n = \begin{cases} \sum_{i=p_n+1}^{p_{n+1}} \lambda_i - \sum_{i=p_{n-1}+1}^{p_n} \lambda_i = 1-1=0 & \text{for} \quad 1 \leqslant j \leqslant p_{n-1} \\ \sum_{i=p_n+1}^{p_{n+1}} \lambda_i - \sum_{i=j}^{p_n} \lambda_i = 1 - \sum_{i=j}^{p_n} \lambda_i & \text{for} \quad p_{n-1}+1 \leqslant j \leqslant p_n \\ \sum_{i=j}^{p_{n+1}} \lambda_i & \text{for} \quad p_n+1 \leqslant j \leqslant p_{n+1}, \end{cases}$$

and thus $\mu_j^n = 0$ $(1 \leqslant j \leqslant p_{n-1})$, $0 \leqslant \mu_j^n \leqslant 1$ $(p_{n-1}+1 \leqslant j \leqslant p_{n+1})$. Hence, taking also into account (15.38) and that $\sum_{i=1}^{\infty} y_i$ is $\sigma(E, E^*)$-unconditionally Cauchy, we obtain, for every $f \in E^*$,

$$\sum_{n=1}^{\infty} |f(y_n^0)| = \sum_{n=1}^{\infty} \left| f\left(\sum_{j=1}^{p_{n+1}} \mu_j^n y_j + u_{n+1} - u_n \right) \right|$$

$$\leqslant \sum_{n=1}^{\infty} \left[\sum_{j=p_{n-1}+1}^{p_n} |f(y_j)| + \sum_{j=p_n+1}^{p_{n+1}} |f(y_j)| \right] + \sum_{n=1}^{\infty} |f(u_{n+1})|$$

$$+ \sum_{n=1}^{\infty} |f(u_n)| \leqslant 2 \sum_{n=1}^{\infty} |f(y_n)| + 2 \|f\| \sum_{n=1}^{\infty} \frac{1}{2^n} < \infty .$$

Consequently, by the Hahn-Banach theorem, $\sum_{n=0}^{\infty} |\phi(y_n^0)| < \infty$ for all $\phi \in E_0^*$, i.e., $\sum_{n=0}^{\infty} y_n^0$ is $\sigma(E_0, E_0^*)$-unconditionally Cauchy.

Furthermore, since $\{z_n\}$ is a $\sigma(E_0, E_0^*)$-Cauchy sequence, for every $\phi \in E_0^*$ there exists the limit $\lim_{n \to \infty} \phi(z_n) = \Phi(\phi)$. Let $\varepsilon > 0$ be arbitrary, and choose $N = N(\varepsilon, \phi)$ so that $\Phi(\phi) - \varepsilon \leqslant \phi(z_n) \leqslant \Phi(\phi) + \varepsilon$ for $n \geqslant N$. Then

$$\Phi(\phi) - \varepsilon = [\Phi(\phi) - \varepsilon] \sum_{i=p_{n-1}+1}^{p_n} \lambda_i \leqslant \phi\left(\sum_{i=p_{n-1}+1}^{p_n} \lambda_i z_i \right) = \phi(w_n)$$

$$\leqslant [\Phi(\phi) + \varepsilon] \sum_{i=p_{n-1}+1}^{p_n} \lambda_i = \Phi(\phi) + \varepsilon \qquad (p_{n-1}+1 \geqslant N),$$

whence $\lim_{n\to\infty}\phi(w_n)=\Phi(\phi)\ (\phi\in E_0^*)$ and thus $z_n-w_{n+1}=z_n-\sum_{i=0}^{n}y_i^0$ converges to 0 for $\sigma(E_0,E_0^*)$, which completes the proof of lemma 15.8.

The main assertion of theorem 15.4 follows now from proposition 15.2 and lemma 15.7. In order to prove that the space $E_0=J$ of § 4, example 4.1, does not have property (u), we need

Lemma 15.8. *If a Banach space E contains no subspace isomorphic to c_0, then every weakly unconditionally Cauchy series in E is unconditionally convergent in the norm topology.*

Proof. Assume that there exists a series $\sum_{i=1}^{\infty}y_i$ in E which is weakly unconditionally Cauchy but not strongly unconditionally convergent. Then for some permutation σ of the set of all positive integers the series $\sum_{i=1}^{\infty}y_{\sigma(i)}$ is not convergent, whence there exists an increasing sequence $0=q_0<q_1<q_2<\cdots$ of integers such that

$$\inf_{1\leqq n<\infty}\left\|\sum_{i=q_{n-1}+1}^{q_n}y_{\sigma(i)}\right\|>0. \tag{15.42}$$

Put

$$z_n=\sum_{i=q_{n-1}+1}^{q_n}y_{\sigma(i)}\qquad(n=1,2,\ldots). \tag{15.43}$$

Then, since $\sum_{i=1}^{\infty}y_i$ is weakly unconditionally Cauchy, the series $\sum_{n=1}^{\infty}z_n$ is $\sigma(E_0,E_0^*)$-unconditionally Cauchy, where $E_0=[z_n]$, whence $\{z_n\}$ converges to 0 for $\sigma(E_0,E_0^*)$; furthermore, by (15.43) and (15.42), $\inf_{1\leqq n<\infty}\|z_n\|>0$. Since E_0 is separable, it can be embedded into a Banach space with a basis $\{x_n\}$ (e.g., into $C([0,1])$, by the Banach-Mazur theorem[1]). Consequently, by proposition 15.1, $\{z_n\}$ has a subsequence $\{z_{p_n}\}$ which is a basic sequence. We shall show that $\{z_{p_n}\}$ is equivalent to the unit vector basis of c_0, whence $[z_{p_n}]$ is a subspace of E, isomorphic to c_0, which will complete the proof.

If $\{\alpha_n\}$ is a sequence of scalars such that $\sum_{n=1}^{\infty}\alpha_n z_{p_n}$ is convergent, then $\lim_{n\to\infty}\alpha_n z_{p_n}=0$, whence, by $\inf_{1\leqq n<\infty}\|z_n\|>0$, we have $\lim_{n\to\infty}\alpha_n=0$. Conversely, assume that $\{\alpha_n\}$ is a sequence of scalars such that $\lim_{n\to\infty}\alpha_n=0$.

---

[1] See e.g. [10], p. 185, theorem 9.

Since $\sum_{n=1}^{\infty} z_{p_n}$ is $\sigma(E_0, E_0^*)$-unconditionally Cauchy, by corollary 15.1 we have

$$\sup_{\substack{\phi \in E_0^* \\ \|\phi\| \leqslant 1}} \sum_{n=1}^{\infty} |\phi(z_{p_n})| = C < \infty,$$

whence

$$\left\| \sum_{n=N}^{N+k} \alpha_n z_{p_n} \right\| = \sup_{\substack{\phi \in E_0^* \\ \|\phi\| \leqslant 1}} \left| \phi\left(\sum_{n=N}^{N+k} \alpha_n z_{p_n} \right) \right| \leqslant \sup_{\substack{\phi \in E_0^* \\ \|\phi\| \leqslant 1}} \sum_{n=N}^{N+k} |\phi(z_{p_n})| \sup_{N \leqslant n \leqslant N+k} |\alpha_n|$$

$$\leqslant C \sup_{N \leqslant n \leqslant N+k} |\alpha_n| \to 0 \quad \text{as} \quad N \to \infty \qquad (k = 1, 2, \ldots),$$

and thus, since E_0 is a closed subspace of the complete space E, $\sum_{n=1}^{\infty} \alpha_n z_{p_n}$ is convergent to an element of E_0, which completes the proof.

Corollary 15.4. *The space J does not have property* (u).

Proof. Since J^{**} is separable, J contains no subspace isomorphic to c_0, whence, by lemma 15.8, every weakly unconditionally Cauchy series in J is strongly unconditionally convergent. Assume now that J has property (u) and let $\{z_n\}$ be an arbitrary weak Cauchy sequence in J and $\{y_n\}$ a sequence in J such that $\sum_{i=1}^{\infty} y_i$ is weakly unconditionally Cauchy and $\left\{ z_n - \sum_{i=1}^{n} y_i \right\}$ converges weakly to 0. Then, by the above remark, $\sum_{i=1}^{\infty} y_i$ is strongly unconditionally convergent to an element $y \in J$. Hence, taking also into account that $z_n - \sum_{i=1}^{n} y_i \to 0$ for $\sigma(E, E^*)$, we have, for each $f \in E^*$,

$$|f(z_n) - f(y)| \leqslant \left| f(z_n) - f\left(\sum_{i=1}^{n} y_i \right) \right| + \|f\| \left\| \sum_{i=1}^{n} y_i - y \right\| \to 0 \quad \text{as} \quad n \to \infty,$$

i.e., $\{z_n\}$ converges weakly to y. Since $\{z_n\}$ has been an arbitrary weak Cauchy sequence in J, it follows that J is weakly complete, which is impossible (e.g., since[1] J^* is separable and J non-reflexive). This completes the proof of corollary 15.4.

Consequently, as we have observed above, a separable Banach space containing a subspace E_0 isomorphic to J has no unconditional basis. Again, for the space $E = J$ itself one can give other simpler proofs, e.g., from § 23, remark 23.1 and § 17, corollary 17.3 it follows that if a non-reflexive Banach space E has no subspace isomorphic to l^1 or c_0 (we

[1] See e.g. [43], Ch. III, § 4, remark (7).

have observed in the preceding that the space J satisfies these conditions), then E has no unconditional basis.

Remark 15.1. Each of theorems 15.1, 15.2 and 15.4 obviously implies that *a separable Banach space E containing a subspace E_0 which is isomorphically universal for all separable Banach spaces, has no unconditional basis.*

The examples of Banach spaces having no unconditional basis, given above, have the property that they cannot be embedded into any Banach space with an unconditional basis. It is natural to ask whether there exists a Banach space having no unconditional basis, which can be embedded into a space with an unconditional basis. We shall now show that the answer is affirmative, namely, that the space l^1 (which has an unconditional basis) contains a subspace having no unconditional basis. To this end, we need

Lemma 15.9. *Let E, F be Banach spaces such that E contains no subspace isomorphic to F and that there exists a bounded linear operator u of E onto F, and let $\{F_\alpha\}_{\alpha\in A}$ be a set of subspaces of F, directed by inclusion, such that $F=\overline{\bigcup_{\alpha\in A} F_\alpha}$. Assume that for every α there exists a bounded linear operator $v_\alpha\colon F_\alpha\to E$ such that*

$$u[v_\alpha(y)]=y \qquad (y\in F_\alpha), \tag{15.44}$$

$$\|v_\alpha\|\leqq\lambda, \tag{15.45}$$

for some $\lambda>0$ independent of α. Then the space $E_0=u^{-1}(0)$ (with the norm induced by E) is not isomorphic to any conjugate Banach space.

Proof. It is sufficient to prove that E_0 is not complemented in any conjugate Banach space, since if E_0 is isomorphic to a conjugate space, then there exists[1] a bounded linear projection $E_0^{**}\to E_0$, whence E_0 is complemented in the conjugate space $(E_0^*)^*$.

Assume now, a contrario, that E_0 is complemented in a conjugate Banach space B^*. We shall prove that in this case there exists an isomorphism v of F into E, in contradiction with our hypothesis, which will complete the proof. For this purpose it will be sufficient to prove that there exists a bounded linear mapping $v\colon F\to E$ such that

$$u[v(y)]=y \qquad (y\in F), \tag{15.46}$$

since then v is an isomorphism (indeed, v is one to one, because $v(y)=0$ implies $y=uv(y)=u(0)=0$ and v^{-1} is continuous on $v(F)$, since $v(y_n)\to v(y)$ implies $y_n=u[v(y_n)]\to u[v(y)]=y$).

---

[1] See e.g. [33], p. 122, exercise 16a).

Since u maps E *onto* F, there exists a mapping $\phi: F \to E$ (not necessarily linear or continuous) such that

$$u[\phi(y)] = y \qquad (y \in F), \tag{15.47}$$

$$\|\phi(y)\| \leqslant \eta \|y\| \qquad (y \in F), \tag{15.48}$$

for a suitable $\eta > 0$ independent of y (indeed, by the open mapping theorem there exists an $\eta > 0$ such that for every $y \in F$ one can find an $x \in E$ with $u(x) = y$, $\|x\| \leqslant \eta \|y\|$ and hence it is sufficient to take $\phi(y) =$ any such x).

Furthermore, since $u[\phi(y) - v_\alpha(y)] = y - y = 0$ for all $y \in F_\alpha$ (by (15.47) and (15.44)), we have

$$\phi(y) - v_\alpha(y) \in E_0 \qquad (y \in F_\alpha). \tag{15.49}$$

Let us consider the product space

$$\Pi = \prod_{y \in F} [(\lambda + \eta) \|y\|] S_{B^*}$$

(where $S_{B^*} =$ the unit cell of B^*, endowed with the weak* topology $\sigma(B^*, B)$). By the theorem of Tychonov, Π is compact.

Let us assign to every $\alpha \in A$ the point $\pi_\alpha \in \Pi$ defined by

$$\pi_\alpha(y) = \begin{cases} v_\alpha(y) - \phi(y) & \text{if} \quad y \in F_\alpha, \\ 0 & \text{if} \quad y \notin F_\alpha. \end{cases} \tag{15.50}$$

(This π_α belongs indeed to Π, since by (15.49) we have $\pi_\alpha(y) = v_\alpha(y) - \phi(y) \in E_0 \subset B^*$ for $y \in F_\alpha$ and since by (15.45) and (15.48) we have $\|\pi_\alpha(y)\| = \|v_\alpha(y) - \phi(y)\| \leqslant \|v_\alpha(y)\| + \|\phi(y)\| \leqslant (\lambda + \eta) \|y\|$ for $y \in F_\alpha$).

Let $y_1 \in F_{\alpha_1}$, $y_2 \in F_{\alpha_2}$. Then, since $\{F_\alpha\}_{\alpha \in A}$ is directed by inclusion, there exists an $\alpha \in A$ such that $F_{\alpha_1}, F_{\alpha_2} \subset F_\alpha$, whence $y_1, y_2, y_1 + y_2 \in F_\alpha$. Therefore, by (15.50), we have

$$\pi_\alpha(y_i) = v_\alpha(y_i) - \phi(y_i) \qquad (i = 1, 2),$$

$$\pi_\alpha(y_1 + y_2) = v_\alpha(y_1 + y_2) - \phi(y_1 + y_2),$$

whence, by the linearity of v_α, we obtain

$$\begin{aligned} \pi_\alpha(y_1) + \pi_\alpha(y_2) - \pi_\alpha(y_1 + y_2) &= v_\alpha(y_1) + v_\alpha(y_2) - v_\alpha(y_1 + y_2) \\ - \phi(y_1) - \phi(y_2) + \phi(y_1 + y_2) &= -\phi(y_1) - \phi(y_2) + \phi(y_1 + y_2). \end{aligned} \tag{15.51}$$

Now, since Π is compact, the net $\{\pi_\alpha\}_{\alpha \in A}$ has a cluster point π. Consider the neighborhood $V(\pi)$ of π defined by

$$\begin{aligned} V(\pi) = \{\pi' \in \Pi \mid &|[\pi'(y_i)](x) - [\pi(y_i)](x)| < \varepsilon \qquad (i = 1, 2), \\ &|[\pi'(y_1 + y_2)](x) - [\pi(y_1 + y_2)](x)| < \varepsilon\}, \end{aligned}$$

where $x \in E$ and $\varepsilon > 0$ are arbitrary. Then, since π is a cluster point of $\{\pi_\alpha\}_{\alpha \in A}$, for every $\alpha \in A$ there exists an $\alpha' \in A$ with $\alpha' \geqslant \alpha$, such that $\pi_{\alpha'} \in V(\pi)$, whence, by (15.51) (which also holds for every $\alpha' \geqslant \alpha$), we get

$$\begin{aligned}|[\pi(y_1)](x)+[\pi(y_2)](x)-[\pi(y_1+y_2)](x)+[\phi(y_1)](x)+[\phi(y_2)](x)\\ -[\phi(y_1+y_2)](x)| = |[\pi(y_1)](x)+[\pi(y_2)](x)-[\pi(y_1+y_2)](x)\\ -[\pi_{\alpha'}(y_1)]-[\pi_{\alpha'}(y_2)](x)+[\pi_{\alpha'}(y_1+y_2)](x)| < 3\varepsilon.\end{aligned}$$

Consequently, since $x \in E$ and $\varepsilon > 0$ were arbitrary,

$$\pi(y_1)+\pi(y_2)-\pi(y_1+y_2)=\phi(y_1+y_2)-\phi(y_1)-\phi(y_2). \tag{15.52}$$

Now, since E_0 is complemented in B^*, there exists a bounded linear projection p of B^* onto E_0. Hence, by (15.52), the linearity of v_α and (15.49), we get

$$\begin{aligned}p[\pi(y_1)]+p[\pi(y_2)]-p[\pi(y_1+y_2)] &= p[\phi(y_1+y_2)]-p[\phi(y_1)]-p[\phi(y_2)]\\ &= p[\phi(y_1+y_2)-v_\alpha(y_1+y_2)]-p[\phi(y_1)-v_\alpha(y_1)]-p[\phi(y_2)-v_\alpha(y_2)]\\ &= \phi(y_1+y_2)-v_\alpha(y_1+y_2)-[\phi(y_1)-v_\alpha(y_1)]-[\phi(y_2)-v_\alpha(y_2)]\\ &= \phi(y_1+y_2)-\phi(y_1)-\phi(y_2).\end{aligned}$$

Consequently, for the mapping $v_0 \colon \bigcup_{\alpha \in A} F_\alpha \to E$ defined by

$$v_0(y)=p[\pi(y)]+\phi(y) \qquad \Big(y \in \bigcup_{\alpha \in A} F_\alpha\Big) \tag{15.53}$$

we have

$$\begin{aligned}v_0(y_1+y_2) = p[\pi(y_1+y_2)]+\phi(y_1+y_2) = p[\pi(y_1)]+p[\pi(y_2)]\\ +\phi(y_1)+\phi(y_2) = v_0(y_1)+v_0(y_2) \qquad \Big(y_1, y_2 \in \bigcup_{\alpha \in A} F_\alpha\Big),\end{aligned}$$

i.e., v_0 is linear. Furthermore, by $p[\pi(y)] \in E_0 = u^{-1}(0)$ we have $u(p[\pi(y)])=0$ $\Big(y \in \bigcup_{\alpha \in A} F_\alpha\Big)$, whence, taking also into account (15.47),

$$u[v_0(y)] = u\{p[\pi(y)]+\phi(y)\} = y \qquad \Big(y \in \bigcup_{\alpha \in A} F_\alpha\Big).$$

Finally, by (15.48) and the inequality $\|\pi(y)\| \leqslant (\lambda+\eta)\|y\|$ $(y \in F)$ (because $\pi \in \Pi$), we have

$$\begin{aligned}\|v_0(y)\| \leqslant \|p[\pi(y)]\| + \|\phi(y)\| \leqslant \|p\|\,\|\pi(y)\| + \eta\,\|y\|\\ \leqslant [\|p\|(\lambda+\eta)+\eta]\,\|y\| \qquad \Big(y \in \bigcup_{\alpha \in A} F_\alpha\Big),\end{aligned}$$

and hence v_0 can be extended by continuity to a bounded linear mapping $v \colon F \to E$ satisfying (15.46), which completes the proof of lemma 15.9.

Now we can prove

Theorem 15.5. *The space l^1 has a subspace E_0 (e.g., $E_0=u^{-1}(0)$, where u is any continuous linear mapping of l^1 onto $L^1([0,1])$) which is not isomorphic to any conjugate Banach space. Hence this space E_0 has no unconditional basis.*

Proof. Let $F=L^1([0,1])$, let A be the set of all partitions of $[0,1]$ into a finite number of disjoint measurable sets, and for each $\alpha\in A$ let F_α be the subspace of $L^1([0,1])$ spanned by the equivalence classes of the characteristic functions of the sets in the decomposition. Then the set $\{F_\alpha\}_{\alpha\in A}$ is directed by inclusion and we have $F=\overline{\bigcup_{\alpha\in A} F_\alpha}$. Now let u be an arbitrary continuous linear mapping of $E=l^1$ onto[1] $F=L^1([0,1])$. Then by the open mapping theorem there exists an $\eta>0$ such that for every $y\in L^1([0,1])$ one can find an $x\in l^1$ with $u(x)=y$, $\|x\|\leqslant\eta\|y\|$. Consequently, for every $\alpha=\{e_1^{(\alpha)},\dots,e_{n_\alpha}^{(\alpha)}\}\in A$ we can define a continuous linear mapping $v_\alpha\colon F_\alpha\to E=l^1$ by

$$v_\alpha\left(\frac{1}{\nu(e_i^{(\alpha)})}\tilde{\chi}_{e_i^{(\alpha)}}\right)=x_i^{(\alpha)}\qquad(i=1,\dots,n_\alpha)\tag{15.54}$$

where ν denotes the Lebesgue measure, $\chi_{e_i^{(\alpha)}}$ denotes the characteristic function of the set $e_i^{(\alpha)}$, and where $x_i^{(\alpha)}\in E=l^1$ are such that

$$u(x_i^{(\alpha)})=\frac{1}{\nu(e_i^{(\alpha)})}\tilde{\chi}_{e_i^{(\alpha)}},\qquad\|x_i^{(\alpha)}\|\leqslant\eta\qquad(i=1,\dots,n_\alpha).\tag{15.55}$$

Then for every $\sum_{i=1}^{n_\alpha}\frac{\beta_i}{\nu(e_i^{(\alpha)})}\tilde{\chi}_{e_i^{(\alpha)}}\in F_\alpha$ we obviously have

$$u\left[v_\alpha\left(\sum_{i=1}^{n_\alpha}\frac{\beta_i}{\nu(e_i^{(\alpha)})}\tilde{\chi}_{e_i^{(\alpha)}}\right)\right]=u\left(\sum_{i=1}^{n_\alpha}\beta_i x_i^{(\alpha)}\right)=\sum_{i=1}^{n_\alpha}\frac{\beta_i}{\nu(e_i^{(\alpha)})}\tilde{\chi}_{e_i^{(\alpha)}},$$

$$\begin{aligned}\left\|v_\alpha\left(\sum_{i=1}^{n_\alpha}\frac{\beta_i}{\nu(e_i^{(\alpha)})}\tilde{\chi}_{e_i^{(\alpha)}}\right)\right\|&=\left\|\sum_{i=1}^{n_\alpha}\beta_i x_i^{(\alpha)}\right\|\leqslant\eta\sum_{i=1}^{n_\alpha}|\beta_i|\\&=\eta\left\|\sum_{i=1}^{n_\alpha}\frac{\beta_i}{\nu(e_i^{(\alpha)})}\tilde{\chi}_{e_i^{(\alpha)}}\right\|,\end{aligned}$$

i.e., (15.44) and (15.45) with $\lambda=\eta$. Finally, $E=l^1$ contains[2] no subspace isomorphic to $F=L^1([0,1])$, and thus all conditions of lemma 15.9

[1] Such mappings exist (see [12], theorem e) or [133], p. 283, theorem 1).
[2] See e.g. [10], Ch. XII, theorem 1.

are satisfied. Consequently, by this lemma, the space $E_0 = u^{-1}(0)$ is not isomorphic to any conjugate Banach space.

Since E_0 is weakly complete (being a subspace of l^1), by theorem 15.3 it follows that E_0 has no unconditional basis, which completes the proof of theorem 15.5.

One can also construct explicitly a subspace E_0 of l^1 as in theorem 15.5. Indeed, let $\{x_n\}$ be the unit vector basis of l^1 and put

$$u(x_{2^n+k-1}) = 2^n \tilde{\chi}_{\left[\frac{k}{2^n}, \frac{k+1}{2^n}\right]} \qquad (k=0,1,\ldots,2^n-1;\ n=1,2,\ldots), \tag{15.56}$$

where, as before, χ_e denotes the characteristic function of the set $e \subset [0,1]$. Then we can extend u by linearity and continuity to a mapping $u: l^1 \to L^1([0,1])$ with $\|u\|=1$, since for any finite sequence of scalars $\xi_1, \ldots, \xi_n$ we have

$$\left\| u\left(\sum_{i=1}^{n} \xi_i x_i\right)\right\| \leqslant \sum_{i=1}^{n} |\xi_i| \, \|u(x_i)\| = \sum_{i=1}^{n} |\xi_i| = \left\| \sum_{i=1}^{n} \xi_i x_i \right\|.$$

We claim that u maps l^1 *onto* $L^1([0,1])$. Indeed, let $y \in S_{L^1([0,1])}$ be arbitrary[1]. Then there exists a sequence $\{y_n\} \subset L^1([0,1])$ of the form $y_n = \sum_{k=0}^{2^n-1} \alpha_k^{(n)} \tilde{\chi}_{\left[\frac{k}{2^n}, \frac{k+1}{2^n}\right]}$ $(n=1,2,\ldots)$ such that $\lim_{n\to\infty} \|y - y_n\| = 0$ in $L^1([0,1])$; we may assume (omitting, if necessary, a finite number of the y_n) that $\|y_n\| \leqslant 2$. Then for $w_n = \sum_{k=0}^{2^n-1} \frac{1}{2^n} \alpha_k^{(n)} x_{2^n+k-1} \in l^1$ $(n=1,2,\ldots)$ we have $\|u(w_n)\| = \|y_n\| \leqslant 2$ $(n=1,2,\ldots)$ and $u(w_n) = y_n \to y$ as $n \to \infty$, whence $y \in 2\overline{u(S_{l^1})}$. Thus, $2\overline{u(S_{l^1})} \supset S_{L^1([0,1])}$, whence it follows[2] that u maps l^1 onto $L^1([0,1])$.

Now we shall prove that *the subspace* $E_0 = u^{-1}(0)$ *of* l^1 *has the following sequence* $\{z_n\}$ *as a monotone basis:*

$$z_n = x_n - \tfrac{1}{2} x_{2n+1} - \tfrac{1}{2} x_{2n+2} \qquad (n=1,2,\ldots). \tag{15.57}$$

Indeed, let us first observe that $z_n \in E_0$ $(n=1,2,\ldots)$ and thus $[z_n] \subset E_0$, since

$$\begin{aligned} u(z_{2^n+k-1}) &= u(x_{2^n+k-1}) - \tfrac{1}{2} u(x_{2^{n+1}+2k-1}) - \tfrac{1}{2} u(x_{2^{n+1}+(2k+1)-1}) \\ &= 2^n \tilde{\chi}_{\left[\frac{k}{2^n}, \frac{k+1}{2^n}\right]} - \tfrac{1}{2} 2^{n+1} \tilde{\chi}_{\left[\frac{2k}{2^{n+1}}, \frac{2k+1}{2^{n+1}}\right]} - \tfrac{1}{2} 2^{n+1} \tilde{\chi}_{\left[\frac{2k+1}{2^{n+1}}, \frac{2k+2}{2^{n+1}}\right]} = 0. \end{aligned} \tag{15.58}$$

---

[1] We recall that S_B denotes the unit cell $\{y \in B \mid \|y\| \leqslant 1\}$ of the Banach space B.

[2] See e.g. [263], p. 198, lemma 1 or [91], Ch. I, § 14, lemma.

Let us prove now the inclusion $E_0 \subset [z_n]$. Assume that for an $x = \sum_{i=1}^{\infty} \xi_i x_i \in l^1$ we have $u(x)=0$, and let $\varepsilon>0$ be arbitrary. Then there exists an index $N=N(\varepsilon,x)$ such that

$$\left\| x - \sum_{i=1}^{2^{N+1}-2} \xi_i x_i \right\| < \frac{\varepsilon}{2}, \tag{15.59}$$

whence, by $u(x)=0$ and $\|u\|=1$,

$$\left\| u\left(\sum_{i=1}^{2^{N+1}-2} \xi_i x_i \right) \right\| = \left\| u\left(x - \sum_{i=1}^{2^{N+1}-2} \xi_i x_i \right) \right\| < \frac{\varepsilon}{2} \|u\| = \frac{\varepsilon}{2}. \tag{15.60}$$

By the definition (15.57) of $\{z_n\}$ one can find, by induction, scalars $\alpha_1, \ldots, \alpha_{2^N-2}$, $\beta_{2^N-1}, \ldots, \beta_{2^{N+1}-2}$ such that

$$\sum_{i=1}^{2^{N+1}-2} \xi_i x_i - \sum_{i=1}^{2^N-2} \alpha_i z_i = \sum_{i=2^N-1}^{2^{N+1}-2} \beta_i x_i. \tag{15.61}$$

Indeed, taking $\alpha_1=\xi_1$, we have

$$\sum_{i=1}^{2^{N+1}-2} \xi_i x_i - \alpha_1 z_1 = \sum_{i=1}^{2^{N+1}-2} \xi_i x_i - \xi_1(x_1 - \tfrac{1}{2}x_3 - \tfrac{1}{2}x_4) = \sum_{i=2}^{2^{N+1}-1} \lambda_i x_i,$$

where $\lambda_3=\xi_3-\frac{1}{2}\xi_1$, $\lambda_4=\xi_4-\frac{1}{2}\xi_1$ and $\lambda_i=\xi_i$ for the other $i\geqslant 2$. Furthermore, if we have already found scalars $\alpha_1, \ldots, \alpha_k$, $\mu_{k+1}, \ldots, \mu_{2^{N+1}-2}$, where $k \leqslant 2^N-3$, such that

$$\sum_{i=1}^{2^{N+1}-2} \xi_i x_i - \sum_{i=1}^{k} \alpha_i z_i = \sum_{i=k+1}^{2^{N+1}-2} \mu_i x_i,$$

then, putting $\alpha_{k+1}=\mu_{k+1}$, and taking into account that $2k+4 \leqslant 2(2^N-3)+4=2^{N+1}-2$, we obtain

$$\sum_{i=1}^{2^{N+1}-2} \xi_i x_i - \sum_{i=1}^{k+1} \alpha_i z_i = \sum_{i=k+1}^{2^{N+1}-2} \mu_i x_i - \alpha_{k+1} z_{k+1}$$

$$= \sum_{i=k+1}^{2^{N+1}-2} \mu_i x_i - \mu_{k+1}(x_{k+1} - \tfrac{1}{2}x_{2k+3} - \tfrac{1}{2}x_{2k+4}) = \sum_{i=k+2}^{2^{N+1}-2} \nu_i x_i,$$

which proves (15.61). Hence, taking into account (15.58) and the definition (15.56) of u, we obtain

$$\left\| u\left(\sum_{i=1}^{2^{N+1}-2} \xi_i x_i\right)\right\| = \left\| u\left(\sum_{i=1}^{2^{N+1}-2} \xi_i x_i - \sum_{i=1}^{2^N-2} \alpha_i z_i\right)\right\| = \left\| u\left(\sum_{i=2^N-1}^{2^{N+1}-2} \beta_i x_i\right)\right\|$$

$$= 2^N \int_0^1 \left| \sum_{k=0}^{2^N-1} \beta_{2^N+k-1} \chi_{\left[\frac{k}{2^N}, \frac{k+1}{2^N}\right]}(t)\right| dt = 2^N \sum_{k=0}^{2^N-1} \frac{1}{2^N} |\beta_{2^N+k-1}|$$

$$= \sum_{k=0}^{2^N-1} |\beta_{2^N+k-1}| = \left\| \sum_{i=2^N-1}^{2^{N+1}-2} \beta_i x_i \right\| = \left\| \sum_{i=1}^{2^{N+1}-2} \xi_i x_i - \sum_{i=1}^{2^N-2} \alpha_i z_i \right\|,$$

whence, by (15.60) and (15.59),

$$\left\| x - \sum_{i=1}^{2^N-2} \alpha_i z_i \right\| \leqslant \left\| x - \sum_{i=1}^{2^{N+1}-2} \xi_i x_i \right\| + \left\| \sum_{i=1}^{2^{N+1}-2} \xi_i x_i - \sum_{i=1}^{2^N-2} \alpha_i z_i \right\| < \varepsilon,$$

which proves that we have $E_0 \subset [z_n]$ and thus $E_0 = [z_n]$.

Finally, let us prove that $\{z_n\}$ is a monotone basic sequence and hence a monotone basis of E_0. For any scalars $\alpha_1, \ldots, \alpha_n$ we have

$$\left\| \sum_{i=1}^{n} \alpha_i z_i \right\| = \left\| \sum_{i=1}^{n} \alpha_i (x_i - \tfrac{1}{2} x_{2i+1} - \tfrac{1}{2} x_{2i+2}) \right\| = \left\| \sum_{k=1}^{2n+2} \beta_k x_k \right\| = \sum_{k=1}^{2n+2} |\beta_k|,$$

with suitable scalars β_k. Hence, if α_{n+1} is an arbitrary scalar,

$$\left\| \sum_{i=1}^{n+1} \alpha_i z_i \right\| = \left\| \sum_{i=1}^{n} \alpha_i z_i + \alpha_{n+1} z_{n+1} \right\|$$

$$= \left\| \sum_{k=1}^{2n+2} \beta_k x_k + \alpha_{n+1}(x_{n+1} - \tfrac{1}{2} x_{2n+3} - \tfrac{1}{2} x_{2n+4}) \right\|$$

$$= \left\| \sum_{k=1}^{n} \beta_k x_k + (\beta_{n+1} + \alpha_{n+1}) x_{n+1} + \sum_{k=n+2}^{2n+2} \beta_k x_k \right.$$

$$\left. - \frac{\alpha_{n+1}}{2} x_{2n+3} - \frac{\alpha_{n+1}}{2} x_{2n+4} \right\|$$

$$= \sum_{k=1}^{2n+2} |\beta_k| - |\beta_{n+1}| + |\beta_{n+1} + \alpha_{n+1}| + \frac{|\alpha_{n+1}|}{2} + \frac{|\alpha_{n+1}|}{2}$$

$$\geqslant \sum_{k=1}^{2n+2} |\beta_k| = \left\| \sum_{i=1}^{n} \alpha_i z_i \right\|,$$

which completes the proof of our assertion.

§ 16. Some characterizations of unconditional bases among E-complete or total biorthogonal systems and among bases. Some characterizations by properties of the associated cone. Multipliers

Let us introduce the following notations, which we shall also use in the subsequent sections:

$$\mathcal{N} = \{1, 2, 3, \ldots\}, \tag{16.1}$$

$$\mathcal{O} = \{\{i_n\} \subset \mathcal{N} \mid i_1 < i_2 < \cdots\}, \tag{16.2}$$

$$\mathcal{D} = \{\{i_1, \ldots, i_n\} \subset \mathcal{N} \mid 1 \leqslant n < \infty\}, \tag{16.3}$$

$$\Pi = \text{the set of all permutations of the set } \mathcal{N}. \tag{16.4}$$

The set $\mathcal{D}$ defined by (16.3) is countable; we shall sometimes also consider it as a directed set (by inclusion). Let us observe that in (16.3) it is not assumed $i_1 < i_2 < \cdots < i_n$.

We recall some well known characterizations of unconditional convergence, which will be used in the sequel.

Lemma 16.1. *Let E be a Banach space and $\{x_n\}$ a sequence in E. The following statements are equivalent:*

1°. $\sum_{i=1}^{\infty} x_i$ *is unconditionally convergent (i.e., $\sum_{i=1}^{\infty} x_{\sigma(i)}$ is convergent for every $\sigma \in \Pi$).*

2°. *The limit $\lim_{d \in \mathcal{D}} \sum_{i \in d} x_i$ exists (i.e., for every $\varepsilon > 0$ there exists a $d = d(\varepsilon) \in \mathcal{D}$ such that $\left\| x - \sum_{i \in d'} x_i \right\| < \varepsilon$ for all $d' \in \mathcal{D}$ with $d' \supset d$).*

3°. *For every $\{i_n\} \subset \mathcal{O}$ the series $\sum_{j=1}^{\infty} x_{i_j}$ converges.*

4°. *For every $\{\varepsilon_n\} \subset K$ with $\varepsilon_n = \pm 1$ $(n = 1, 2, \ldots)$ the series $\sum_{i=1}^{\infty} \varepsilon_i x_i$ converges.*

5°. *For every $\{\beta_n\} \subset K$ with $|\beta_n| \leqslant 1$ $(n = 1, 2, \ldots)$ the series $\sum_{i=1}^{\infty} \beta_i x_i$ converges.*

6°. *The series $\sum_{i=1}^{\infty} |f(x_i)|$ converge uniformly with respect to $f \in E^*$, $\|f\| \leqslant 1$ (i.e., $\lim_{n \to \infty} \sup_{\substack{f \in E^* \\ \|f\| \leqslant 1}} \sum_{i=n+1}^{\infty} |f(x_i)| = 0$).*

Proof. Assume that we have 1°, but not 2°, and let $x = \sum_{i=1}^{\infty} x_i$. Then there exists an $\varepsilon_0 > 0$ such that for every $d \in \mathcal{D}$ one can find a $d' \in \mathcal{D}$

with $d' \supset d$ and $\left\| x - \sum_{i \in d'} x_i \right\| \geqslant \varepsilon_0$ (since otherwise we would have $\lim_{d \in \mathscr{D}} \sum_{i \in d} x_i = x$). Since $x = \sum_{i=1}^{\infty} x_i$, there exists a positive integer $N = N(\varepsilon_0)$ such that

$$\left\| x - \sum_{i=1}^{n} x_i \right\| < \frac{\varepsilon_0}{2} \qquad (n \geqslant N).$$

Not let $d_1 = \{1, \ldots, N\}$ and let $d_1' \in \mathscr{D}$ with $d_1' \supset d_1$ be such that $\left\| x - \sum_{i \in d_1'} x_i \right\| \geqslant \varepsilon_0$. Let $d_2 = \left\{1, \ldots, \max_{i \in d_1'} i\right\}$ and let $d_2' \in \mathscr{D}$ be such that $d_2' \supset d_2$, $\left\| x - \sum_{i \in d_2'} x_i \right\| \geqslant \varepsilon_0$. Construct in the same way $d_3, d_3', d_4, d_4', \ldots$ and define then a rearrangement $\{\sigma(n)\}$ of $\mathscr{N}$, enumerating the elements of the sets

$$d_1, d_1' \setminus d_1, d_2 \setminus d_1', d_2' \setminus d_2, \ldots$$

Then the series $\sum_{n=1}^{\infty} x_{\sigma(n)}$ is not convergent, since

$$\left\| \sum_{i \in d_n' \setminus d_n} x_i \right\| = \left\| \left(x - \sum_{i \in d_n} x_i\right) - \left(x - \sum_{i \in d_n'} x_i\right) \right\|$$

$$\geqslant \left| \left\| x - \sum_{i \in d_n} x_i \right\| - \left\| x - \sum_{i \in d_n'} x_i \right\| \right| \geqslant \varepsilon_0 - \frac{\varepsilon_0}{2} = \frac{\varepsilon_0}{2} \qquad (n = 1, 2, \ldots),$$

and hence $\sum_{i=1}^{\infty} x_i$ is not unconditionally convergent. Thus, $1° \Rightarrow 2°$.

Assume now that we have 2°, and let $x = \lim_{d \in \mathscr{D}} \sum_{i \in d} x_i$. Let $\varepsilon > 0$ be arbitrary and let $d = d(\varepsilon)$ be such that $\left\| x - \sum_{i \in d'} x_i \right\| < \frac{\varepsilon}{4}$ for all $d' \in \mathscr{D}$ with $d' \supset d$. Put $N = N(\varepsilon) = \max_{i \in d(\varepsilon)} i$ and let $n \geqslant N$, $l \geqslant 1$ and $f \in E^*$ with $\|f\| \leqslant 1$ be arbitrary. Put

$$d_1(f) = \{i \in \{n+1, \ldots, n+l\} \mid \operatorname{Re} f(x_i) \geqslant 0\},$$
$$d_2(f) = \{i \in \{n+1, \ldots, n+l\} \mid \operatorname{Re} f(x_i) < 0\}.$$

Then

$$\sum_{i=n+1}^{n+l} |\operatorname{Re} f(x_i)| = \sum_{j=1}^{2} \sum_{i \in d_j(f)} |\operatorname{Re} f(x_i)| = \sum_{j=1}^{2} \left| \operatorname{Re} f\left(\sum_{i \in d_j(f)} x_i \right) \right| \leqslant \sum_{j=1}^{2} \left\| \sum_{i \in d_j(f)} x_i \right\|$$

$$\leqslant \sum_{j=1}^{2} \left(\left\| x - \sum_{i \in d(\varepsilon) \cup d_j(f)} x_i \right\| + \left\| x - \sum_{i \in d(\varepsilon)} x_i \right\| \right) < \frac{\varepsilon}{2}.$$

By a similar argument we also have

$$\sum_{i=n+1}^{n+l} |\mathrm{Im} f(x_i)| < \frac{\varepsilon}{2}.$$

Consequently,

$$\sum_{i=n+1}^{n+l} |f(x_i)| \leqslant \sum_{i=n+1}^{n+l} |\mathrm{Re} f(x_i)| + \sum_{i=n+1}^{n+l} |\mathrm{Im} f(x_i)| < \varepsilon,$$

whence we infer $\lim_{n\to\infty} \sup_{\substack{f\in E^* \\ \|f\|\leqslant 1}} \sum_{i=n+1}^{\infty} |f(x_i)| = 0$. Thus, $2°\Rightarrow 6°$.

Assume now that we have 6° and let $\{\beta_i\}\subset K$, $|\beta_i|\leqslant 1$ $(i=1,2,\ldots)$, $n\geqslant 1$, $l\geqslant 1$. Then there exists an $f_{n,l}\in E^*$ with $\|f_{n,l}\|=1$ such that $f_{n,l}\left(\sum_{i=n+1}^{n+l} \beta_i x_i\right) = \left\|\sum_{i=n+1}^{n+l} \beta_i x_i\right\|$, whence, by $|\beta_i|\leqslant 1$ and 6°,

$$\left\|\sum_{i=n+1}^{n+l} \beta_i x_i\right\| = \left|f_{n,l}\left(\sum_{i=n+1}^{n+l} \beta_i x_i\right)\right| \leqslant \sum_{i=n+1}^{n+l} |f_{n,l}(x_i)| < \varepsilon \quad (n > N(\varepsilon),\ l=1,2,\ldots)$$

and therefore the series $\sum_{i=1}^{\infty} \beta_i x_i$ converges. Thus, $6°\Rightarrow 5°$.

The implication $5°\Rightarrow 4°$ is obvious.

Assume now that we have 4° and let $\{i_j\}\in\mathcal{O}$, $n\geqslant 1$, $l\geqslant 1$. Then $\sum_{j=n+1}^{n+l} x_{i_j} = \frac{1}{2}\left(\sum_{i=m}^{m+k} \varepsilon_i x_i + \sum_{i=m}^{m+k} \varepsilon_i' x_i\right)$, where $m=i_{n+1}$, $m+k=i_{n+l}$, $\varepsilon_{i_j}=\varepsilon_{i_j}'=1$ for $j=n+1,\ldots,n+l$ and $\varepsilon_i=1$, $\varepsilon_i'=-1$ for $i\in\{m,\ldots,m+k\}\setminus\{i_{n+1}, i_{n+2},\ldots,i_{n+l}\}$. Hence, by 4°,

$$\left\|\sum_{j=n+1}^{n+l} x_{i_j}\right\| \leqslant \frac{1}{2}\left(\left\|\sum_{i=m+1}^{m+k} \varepsilon_i x_i\right\| + \left\|\sum_{i=m+1}^{m+k} \varepsilon_i' x_i\right\|\right) < \varepsilon \quad (i_{n+1} > N(\varepsilon)),$$

and therefore the series $\sum_{j=1}^{\infty} x_{i_j}$ converges. Thus, $4°\Rightarrow 3°$.

Assume, finally, that we do not have 1°, i.e., that there exists a permutation $\sigma\in\Pi$ for which the series $\sum_{i=1}^{\infty} x_{\sigma(i)}$ is not convergent. Then there exist an $\varepsilon_0>0$ and a sequence $\{m_n\}\in\mathcal{O}$ such that

$$\left\|\sum_{i=m_n+1}^{m_{n+1}} x_{\sigma(i)}\right\| \geqslant \varepsilon_0 \quad (n=1,2,\ldots). \tag{16.5}$$

Choose an infinite subsequence $\{m_{n_j}\}\subset\{m_n\}$ such that

$$\min_{m_{n_{j+1}}+1\leqslant i\leqslant m_{n_{j+1}+1}} \sigma(i) > \max_{m_{n_j}+1\leqslant i\leqslant m_{n_j+1}} \sigma(i) \quad (j=1,2,\ldots),$$

and rearrange the integers $\sigma(i)$ $(m_{n_j}+1 \leqslant i \leqslant m_{n_{j+1}}, j=1,2,\ldots)$ into an increasing sequence, say $\{i_j\}$. Then, by (16.5), the series $\sum_{j=1}^{\infty} x_{i_j}$ is not convergent. Thus, $3^\circ \Rightarrow 1^\circ$, which completes the proof of lemma 16.1.

If E is a Banach space and $(x_n, f_n)(\{x_n\} \subset E, \{f_n\} \subset E^*)$ a biorthogonal system, let us introduce the operators

$$s_d(x) = s_{\{i_1, i_2, \ldots, i_n\}}(x) = \sum_{j=1}^{n} f_{i_j}(x) x_{i_j} \qquad (x \in E,\ d = \{i_1, i_2, \ldots, i_n\} \in \mathscr{D}), \tag{16.6}$$

$$s_{\{\beta_n\}, l}(x) = \sum_{i=1}^{l} \beta_i f_i(x) x_i \qquad (x \in E,\ \{\beta_n\} \subset K,\ l = 1, 2, \ldots), \tag{16.7}$$

$$s_{\sigma, n}(x) = \sum_{i=1}^{n} f_{\sigma(i)}(x) x_{\sigma(i)} \qquad (x \in E,\ \sigma \in \Pi,\ n = 1, 2, \ldots). \tag{16.8}$$

and the sets

$$\mathscr{U}_0 = \left\{ x \in E \;\middle|\; \lim_{d \in \mathscr{D}} s_d(x) = x \right\}, \tag{16.9}$$

$$\mathscr{U}_1 = \left\{ x \in E \;\middle|\; \lim_{d \in \mathscr{D}} s_d(x) \ \text{ exists} \right\}, \tag{16.10}$$

$$\mathscr{U}_2 = \left\{ x \in E \;\middle|\; \sup_{d \in \mathscr{D}} \|s_d(x)\| < \infty \right\}, \tag{16.11}$$

$$\begin{aligned} \mathscr{U}_3 = &\left\{ x \in E \;\middle|\; \lim_{d \in \mathscr{D}} \|s_d(x)\| \ \text{ exists and } \ < \infty \right\} \\ &\cup \left\{ x \in E \;\middle|\; \lim_{d \in \mathscr{D}} \|s_d(x)\| \ \text{ does not exist} \right\}. \end{aligned} \tag{16.12}$$

Theorem 16.1. *Let E be a Banach space and $(x_n, f_n)(\{x_n\} \subset E, \{f_n\} \subset E^*)$ an E-complete biorthogonal system. Then*

a) *The following statements are equivalent:*

1°. *$\{x_n\}$ is an unconditional basis of the space E.*

2°. *For every $x \in E$ the limit $\lim_{d \in \mathscr{D}} \sum_{i \in d} f_i(x) x_i$ exists.*

3°. *For every $\{i_n\} \in \mathcal{O}$ and $x \in E$ the series $\sum_{j=1}^{\infty} f_{i_j}(x) x_{i_j}$ converges.*

4°. *For every $\{\varepsilon_n\} \subset K$ with $\varepsilon_n = \pm 1$ $(n = 1, 2, \ldots)$ and $x \in E$ the series $\sum_{i=1}^{\infty} \varepsilon_i f_i(x) x_i$ converges.*

5°. *For every $\{\beta_n\} \subset K$ with $|\beta_n| \leqslant 1$ $(n = 1, 2, \ldots)$ and $x \in E$ the series $\sum_{i=1}^{\infty} \beta_i f_i(x) x_i$ converges.*

6°. *We have*

$$\sup_{1\leqslant n<\infty}\|s_{\sigma,n}(x)\|<\infty \qquad (x\in E,\ \sigma\in\Pi). \tag{16.13}$$

7°. *We have*

$$\sup_{1\leqslant n<\infty}\|s_{\{i_1,\ldots,i_n\}}(x)\|<\infty \quad (x\in E,\ \{i_n\}\in\mathcal{O}). \tag{16.14}$$

8°. *We have*

$$\sup_{1\leqslant l<\infty}\|s_{\{\varepsilon_n\},l}(x)\|<\infty \qquad (x\in E,\ \varepsilon_1,\varepsilon_2,\ldots=\pm 1). \tag{16.15}$$

9°. *We have*

$$\sup_{1\leqslant l<\infty}\|s_{\{\beta_n\},l}(x)\|<\infty \qquad (x\in E,\ |\beta_1|,|\beta_2|,\ldots\leqslant 1). \tag{16.16}$$

10° – 13°. *We have*

$$\sup_{\sigma\in\Pi}\ \sup_{1\leqslant n<\infty}\|s_{\sigma,n}(x)\|<\infty \qquad (x\in E), \tag{16.17}$$

and, respectively, similar conditions for $\{i_n\}$,[1] $\{\varepsilon_n\}$ *and* $\{\beta_n\}$.

14° – 17°. *We have*

$$\sup_{1\leqslant n<\infty}\|s_{\sigma,n}\|<\infty \qquad (\sigma\in\Pi), \tag{16.18}$$

and, respectively, similar conditions for $\{i_n\}$, $\{\varepsilon_n\}$ *and* $\{\beta_n\}$.

18° – 21°. *We have*

$$\sup_{\sigma\in\Pi}\ \sup_{1\leqslant n<\infty}\|s_{\sigma,n}\|<\infty \tag{16.19}$$

and, respectively, similar conditions for $\{i_n\}$,[2] $\{\varepsilon_n\}$ *and* $\{\beta_n\}$.

22°. *For every* $x\in E$ *and* $f\in E^*$ *the numerical series*

$$\sum_{i=1}^{\infty}|f_i(x)|\,|f(x_i)| \tag{16.20}$$

converges.

23°. *For every* $x\in E$ *the series* (16.20) *converge uniformly with respect to* $f\in E^*$, $\|f\|\leqslant 1$.

---

[1] The condition for $\{i_n\}$ can be also written in the following forms:

11′. $\sup_{d\in\mathcal{D}}\|s_d(x)\|<\infty \qquad (x\in E)$.

11″. $\mathcal{U}_2=E$.

[2] The condition for $\{i_n\}$ can be also written in the following form:

19′. $\sup_{d\in\mathcal{D}}\|s_d\|<\infty$.

24°. *For every* $x \in E$ *the series*

$$\sum_{i=1}^{\infty} |f_i(x)| x_i$$

converges, and there exists a constant $C \geqslant 1$ *such that*

$$\|x\| \leqslant C \left\| \sum_{i=1}^{\infty} |f_i(x)| x_i \right\| \qquad (x \in E). \tag{16.21}$$

25°. *We have*

$$\mathcal{U}_2 = \mathcal{U}_3. \tag{16.22}$$

b) *We have*[1]

$$\begin{aligned}\sup_{d \in \mathcal{D}} \|s_d(x)\| &= \sup_{\sigma \in \Pi} \sup_{1 \leqslant n < \infty} \|s_{\sigma,n}(x)\| \leqslant \sup_{\substack{f \in E^* \\ \|f\| \leqslant 1}} \sum_{i=1}^{\infty} |f_i(x)| |f(x_i)| \\ &= \sup_{\substack{\{\beta_n\} \subset K \\ |\beta_i| \leqslant 1}} \sup_{1 \leqslant l < \infty} \|s_{\{\beta_n\},l}(x)\| \leqslant 2 \sup_{\substack{\{\varepsilon_n\} \subset K \\ \varepsilon_i = \pm 1}} \sup_{1 \leqslant l < \infty} \|s_{\{\varepsilon_n\},l}(x)\| \\ &\leqslant 4 \sup_{d \in \mathcal{D}} \|s_d(x)\| \qquad (x \in E),\end{aligned} \tag{16.23}$$

and, if $\{x_n\}$ *is an unconditional basis of E, each of these numbers defines a norm on E, equivalent to the initial norm on E.*

Proof. a) The equivalences $1° \Leftrightarrow 2° \Leftrightarrow 3° \Leftrightarrow 4° \Leftrightarrow 5° \Leftrightarrow 23°$ are consequences of lemma 16.1.

The implications $1° \Rightarrow 6°$, $3° \Rightarrow 7°$, $4° \Rightarrow 8°$ and $5° \Rightarrow 9°$ are obvious.

$9° \Rightarrow 13°$.[2] For any fixed $x \in E$ put

$$u_l(\{\beta_n\}) = \sum_{i=1}^{l} \beta_i f_i(x) x_i = s_{\{\beta_n\},l}(x) \qquad (\{\beta_n\} \in m,\ l = 1, 2, \ldots).$$

Then each u_l is a linear mapping of m into E, which is also continuous, since

$$\|u_l(\{\beta_n\})\| \leqslant C_l \max_{1 \leqslant i \leqslant l} |\beta_i| \leqslant C_l \|\{\beta_n\}\| \qquad (\{\beta_n\} \in m,\ l = 1, 2, \ldots),$$

---

[1] The same proof shows that in the particular case when the scalars are real, we have

$$\sup_{|\beta_i| \leqslant 1} \sup_{1 \leqslant l < \infty} \|s_{\{\beta_n\},l}(x)\| = \sup_{\varepsilon_i = \pm 1} \sup_{1 \leqslant l < \infty} \|s_{\{\varepsilon_n\},l}(x)\| \qquad (x \in E).$$

[2] It is also easy to prove directly that non $13° \Rightarrow$ non $9°$. Indeed, if we have non 13°, then there exist sequences $\{l_k\} \in \mathcal{O}$ and $\{\beta_i^{(k)}\} \subset K$, $|\beta_i^{(k)}| \leqslant 1$, such that

$$\left\| \sum_{i=l_{k-1}+1}^{l_k} \beta_i^{(k)} f_i(x) x_i \right\| \geqslant 2^k + \left\| \sum_{n=1}^{k-1} \sum_{i=l_{n-1}+1}^{l_n} \beta_i^{(n)} f_i(x) x_i \right\| \qquad (k = 1, 2, \ldots;\ l_0 = 0),$$

whence, putting $\beta_i = \beta_i^{(k)}$ $(l_{k-1} + 1 \leqslant i \leqslant l_k;\ k = 1, 2, \ldots)$, we obtain

$$\left\| \sum_{i=1}^{l_k} \beta_i f_i(x) x_i \right\| \geqslant \left\| \left\| \sum_{i=l_{k-1}+1}^{l_k} \beta_i f_i(x) x_i \right\| - \left\| \sum_{i=1}^{l_{k-1}} \beta_i f_i(x) x_i \right\| \right\| \geqslant 2^k \qquad (k = 1, 2, \ldots),$$

i.e., non-9°.

where $C_l = \sum_{i=1}^{l} |f_i(x)| \, \|x_i\|$ $(l=1,2,\ldots)$. Furthermore, by 9° we have

$$\sup_{1 \leqslant l < \infty} \|u_l(\{\beta_n\})\| < \infty \qquad (\{\beta_n\} \in m).$$

Consequently, by the principle of uniform boundedness, we have

$$\sup_{1 \leqslant l < \infty} \|u_l\| = \sup_{\substack{\{\beta_n\} \in m \\ \|\{\beta_n\}\| \leqslant 1}} \sup_{1 \leqslant l < \infty} \|s_{\{\beta_n\},l}(x)\| < \infty .$$

The proof of the implications 7°⇒11° and 8°⇒12° is similar. The implication 6°⇒10° will be proved via 6°⇒7°⇒11°⇒10°.

6°⇒7°. Assume that we have 6° and non-7°. Then by non-7°, there exist an $x \in E$ and a sequence $\{i_n\} \in \mathcal{O}$ such that

$$\sup_{1 \leqslant n < \infty} \|s_{\{i_1, i_2, \ldots, i_n\}}(x)\| = \infty , \tag{16.24}$$

and by 6° we have $\sup_{1 \leqslant n < \infty} \|f_n(x) x_n\| = M < \infty$. Furthermore, by (16.24) there exists a sequence $\{m_n\} \in \mathcal{O}$ such that

$$\left\| \sum_{j=1}^{m_n - n} f_{i_j}(x) x_{i_j} \right\| \geqslant 2^n + nM \qquad (n=1,2,\ldots).$$

Let $\{l_j\} \in \mathcal{O}$ be the complementary set to $\{i_n\}$ in $\mathcal{N}$ and define a permutation $\sigma \in \Pi$ by

$$\sigma(j) = \begin{cases} i_{j-n+1} & \text{for } m_{n-1} + 1 \leqslant j \leqslant m_n - 1 \quad (n=1,2,\ldots;\ m_0 = 0), \\ l_n & \text{for } j = m_n \quad (n=1,2,\ldots). \end{cases}$$

Then

$$\begin{aligned} \left\| \sum_{j=1}^{m_n} f_{\sigma(j)}(x) x_{\sigma(j)} \right\| &= \left\| \sum_{j=1}^{m_n - n} f_{i_j}(x) x_{i_j} + \sum_{k=1}^{n} f_{l_k}(x) x_{l_k} \right\| \\ &\geqslant \left| \left\| \sum_{j=1}^{m_n - n} f_{i_j}(x) x_{i_j} \right\| - \left\| \sum_{k=1}^{n} f_{l_k}(x) x_{l_k} \right\| \right| \\ &\geqslant 2^n + nM - nM = 2^n \qquad (n=1,2,\ldots), \end{aligned}$$

which contradicts 6°.

11°⇒10°. Let $\sigma \in \Pi$ be arbitrary. For any fixed $n \in \mathcal{N}$ let $i_1, i_2, \ldots, i_n$ denote the numbers $\sigma(1), \sigma(2), \ldots, \sigma(n)$ arranged in increasing order and let $\{i_j\}_{j=n+1}^{\infty}$ be an arbitrary sequence in $\mathcal{N}$ such that $\{i_j\}_{j=1}^{\infty} \in \mathcal{O}$. Then $s_{\{i_1, i_2, \ldots, i_n\}}(x) = s_{\sigma,n}(x)$ $(x \in E)$. Consequently, we have

$$\sup_{\sigma \in \Pi} \sup_{1 \leqslant n < \infty} \|s_{\sigma,n}(x)\| \leqslant \sup_{\{i_n\} \in \mathcal{O}} \sup_{1 \leqslant n < \infty} \|s_{\{i_1, i_2, \ldots, i_n\}}(x)\| \qquad (x \in E), \tag{16.25}$$

whence we obtain the implication 11°⇒10°. Thus also 6°⇒10°.

The implications $10^\circ \Rightarrow 18^\circ$, $11^\circ \Rightarrow 19^\circ$, $12^\circ \Rightarrow 20^\circ$ and $13^\circ \Rightarrow 21^\circ$ are consequences of the principle of uniform boundedness[1], and the implications $18^\circ \Rightarrow 14^\circ \Rightarrow 1^\circ$, $19^\circ \Rightarrow 15^\circ \Rightarrow 3^\circ$, $20^\circ \Rightarrow 16^\circ \Rightarrow 4^\circ$ and $21^\circ \Rightarrow 17^\circ \Rightarrow 5^\circ$ are obvious. Furthermore, the implication $23^\circ \Rightarrow 22^\circ$ is obvious and the implication $22^\circ \Rightarrow 9^\circ$ is a consequence of § 15, corollary 15.1.

$5^\circ \cap 21^\circ \Rightarrow 24^\circ$. Putting[2] $\beta_n = \operatorname{sign} f_n(x)$ $(n=1,2,\ldots)$, from 5° it follows that for every $x \in E$ the series $\sum_{i=1}^{\infty} |f_i(x)| x_i = \sum_{i=1}^{\infty} \beta_i f_i(x) x_i$ converges. Furthermore, putting

$$\sum_{i=1}^{\infty} |f_i(x)| x_i = z$$

and taking into account that (by 5°) $\{x_n\}$ is a basis of E and 21°, we obtain

$$\|x\| = \left\| \sum_{i=1}^{\infty} f_i(x) x_i \right\| = \left\| \sum_{i=1}^{\infty} \bar{\beta}_i |f_i(x)| x_i \right\| = \left\| \sum_{i=1}^{\infty} \bar{\beta}_i f_i(z) x_i \right\|$$

$$\leqslant \sup_{1 \leqslant l < \infty} \left\| \sum_{i=1}^{l} \bar{\beta}_i f_i(z) x_i \right\| \leqslant C \|z\| = C \left\| \sum_{i=1}^{\infty} |f_i(x)| x_i \right\|,$$

where $C = \sup_{\substack{\{\beta_n\} \subset K \\ |\beta_n| \leqslant 1}} \sup_{1 \leqslant l < \infty} \|s_{\{\beta_n\},l}\|$.

$24^\circ \Rightarrow 23^\circ$. Let $\varepsilon > 0$, $x \in E$ and $f \in E^*$ with $\|f\| \leqslant 1$ be arbitrary. Since by 24° the series $\sum_{i=1}^{\infty} |f_i(x)| x_i$ converges, there exists a positive integer $N(\varepsilon, x)$ such that

$$\left\| \sum_{i \leqslant n+1}^{n+k} |f_i(x)| x_i \right\| < \frac{\varepsilon}{C} \qquad (n > N(\varepsilon, x);\ k = 1, 2, \ldots),$$

where C is as in (16.21). Consequently, putting

$$\beta_i = \operatorname{sign} f_i(x) f(x_i) \text{ if } f(x_i) \neq 0, \quad \beta_i = 1 \text{ if } f(x_i) = 0 \quad (i = n+1, \ldots, n+k),$$

$$y = \sum_{i=n+1}^{n+k} \beta_i f_i(x) x_i = \sum_{i=n+1}^{n+k} f_i(y) x_i,$$

[1] See e.g. [50], p. 66, corollary 21.

[2] For the definition of sign α (α complex) see Ch. I, § 12, footnote to example 12.2.

and taking into account (16.21), we obtain

$$\sum_{i=n+1}^{n+k} |f_i(x)|\,|f(x_i)| = \left| f\left(\sum_{i=n+1}^{n+k} \beta_i f_i(x) x_i\right)\right| = |f(y)| \leqslant \|y\|$$
$$\leqslant C \left\| \sum_{i=n+1}^{n+k} |f_i(y)| x_i \right\| = C \left\| \sum_{i=n+1}^{n+k} |f_i(x)| x_i \right\| < \varepsilon$$

for $n > N(\varepsilon, x)$ and $k = 1, 2, \ldots$ whence, since $\varepsilon > 0$, $x \in E$ and $f \in E^*$ with $\|f\| \leqslant 1$ were arbitrary, we infer 23°.

11° ⇒ 25°. If we have 11°, then, obviously, $\mathscr{U}_2 = \mathscr{U}_3 = E$.

25° ⇒ 11°. Assume that we have non-11°. Then for every positive integer n the set $\left\{x \in E \,\middle|\, \sup_{d \in \mathscr{D}} \|s_d(x)\| \leqslant n\right\}$ is nowhere dense (since $\mathscr{U}_2 \neq E$) and closed, whence the set

$$\mathscr{U}_2 = \bigcup_{n=1}^{\infty} \left\{x \in E \,\middle|\, \sup_{d \in \mathscr{D}} \|s_d(x)\| \leqslant n\right\}$$

is of the first category. On the other hand, for every $d \in \mathscr{D}$ the set

$$\mathscr{U}_d = \{x \in E \mid \|s_{d'}(x)\| \geqslant \|x\| + 1 \quad \text{for all} \quad d' \in \mathscr{D},\ d' \supset d\}$$

is closed and nowhere dense (since $\mathscr{U}_0$ is dense in E), whence, by

$$E \setminus \mathscr{U}_3 \subset \bigcup_{d \in \mathscr{D}} \mathscr{U}_d,$$

$\mathscr{U}_3$ is of the second category. Thus $\mathscr{U}_2 \neq \mathscr{U}_3$, which completes the proof of the equivalences 1° ⇔ … ⇔ 25°.

b) Let $\{i_n\} \in \mathscr{O}$ be arbitrary. For any fixed $n \in \mathscr{N}$, take a $\sigma \in \Pi$ such that

$$\sigma(j) = \begin{cases} i_j & \text{for} \quad 1 \leqslant j \leqslant n, \\ l_j \in \mathscr{N} \setminus \{i_1, i_2, \ldots, i_n\} & \text{for} \quad n+1 \leqslant j < \infty. \end{cases}$$

Then $s_{\sigma,n}(x) = s_{\{i_1, i_2, \ldots, i_n\}}(x)$ $(x \in E)$. Consequently, we have

$$\sup_{\{i_n\} \in \mathscr{O}} \sup_{1 \leqslant n < \infty} \|s_{\{i_1, i_2, \ldots, i_n\}}(x)\| \leqslant \sup_{\sigma \in \Pi} \sup_{1 \leqslant n < \infty} \|s_{\sigma,n}(x)\| \qquad (x \in E),$$

which, together with the inequality (16.25) proved above, yields the first equality of (16.23).

The inequality

$$\sup_{\{i_n\} \in \mathscr{O}} \sup_{1 \leqslant n < \infty} \|s_{\{i_1, i_2, \ldots, i_n\}}(x)\| \leqslant \sup_{\substack{\{\beta_n\} \subset K \\ |\beta_n| \leqslant 1}} \sup_{1 \leqslant l < \infty} \|s_{\{\beta_n\}, l}(x)\| \qquad (x \in E)$$

is obvious.

Now let $x \in E$, $\{\beta_n\} \subset K$ with $|\beta_i| \leqslant 1$ $(i=1,2,\ldots)$ and l be arbitrary. Take $g = g_{\{\beta_n\},l,x} \in E^*$ with $\|g\|=1$ such that $g\left(\sum_{i=1}^{l} \beta_i f_i(x) x_i\right) = \left\|\sum_{i=1}^{l} \beta_i f_i(x) x_i\right\|$. Then

$$\left\|\sum_{i=1}^{l} \beta_i f_i(x) x_i\right\| = g\left(\sum_{i=1}^{l} \beta_i f_i(x) x_i\right) \leqslant \sum_{i=1}^{l} |\beta_i|\,|f_i(x)|\,|g(x_i)|$$

$$\leqslant \sum_{i=1}^{\infty} |f_i(x)|\,|g(x_i)|$$

and hence

$$\sup_{\substack{\{\beta_n\} \subset K \\ |\beta_i| \leqslant 1}} \sup_{1 \leqslant l < \infty} \|s_{\{\beta_n\},l}(x)\| \leqslant \sup_{\substack{f \in E^* \\ \|f\| \leqslant 1}} \sum_{i=1}^{\infty} |f_i(x)|\,|f(x_i)| \qquad (x \in E). \tag{16.26}$$

Conversely, let $x \in E$ and $f \in E^*$, $\|f\|=1$, be arbitrary. Put

$$\beta_i = \operatorname{sign} f_i(x)\, f(x_i) \qquad (i=1,2,\ldots).$$

Then $|\beta_i| \leqslant 1$ $(i=1,2,\ldots)$ and

$$\sum_{i=1}^{l} |f_i(x)|\,|f(x_i)| = \left|f\left(\sum_{i=1}^{l} \beta_i f_i(x) x_i\right)\right| \leqslant \left\|\sum_{i=1}^{l} \beta_i f_i(x) x_i\right\|$$

$$= \|s_{\{\beta_n\},l}(x)\| \qquad (l=1,2,\ldots),$$

whence

$$\sup_{\substack{f \in E^* \\ \|f\| \leqslant 1}} \sum_{i=1}^{\infty} |f_i(x)|\,|f(x_i)| \leqslant \sup_{\substack{\{\beta_n\} \subset K \\ |\beta_i| \leqslant 1}} \sup_{1 \leqslant l < \infty} \|s_{\{\beta_n\},l}(x)\| \qquad (x \in E),$$

which, together with (16.26), proves the second equality in (16.23).

Furthermore, assuming that the scalars are complex, let $\{\beta_n\} \subset K$, $|\beta_n| \leqslant 1$ $(n=1,2,\ldots)$ be arbitrary. Take an $x \in E$ and a fixed $n \in \mathcal{N}$. Then, for[1] $\beta_k = \gamma_k + i\delta_k$, where $\gamma_k = \operatorname{Re}\beta_k$, $\delta_k = \operatorname{Im}\beta_k$ $(k=1,2,\ldots,n)$, we have $|\gamma_k| \leqslant 1$, $|\delta_k| \leqslant 1$ $(k=1,2,\ldots,n)$ and

$$\left\|\sum_{k=1}^{n} \beta_k f_k(x) x_k\right\| \leqslant \left\|\sum_{k=1}^{n} \gamma_k f_k(x) x_k\right\| + \left\|\sum_{k=1}^{n} \delta_k f_k(x) x_k\right\|.$$

Since $\{\gamma_k\}_1^n$ is in the unit cell S_{m_n} of the *real* Banach space[2] m_n, there exist, by the theorem of Carathéodory, $n+1$ extremal points of

[1] Here $i = \sqrt{-1}$.

[2] We recall that (the complex or real) m_n is the space of all (complex, respectively real) n-tuples $y = \{\eta_k\}_1^n$ endowed with the norm

$$\|y\| = \max_{1 \leqslant k \leqslant n} |\eta_k|.$$

S_{m_n}, i.e.[1] $n+1$ points $\{\varepsilon_k^{(j)}\}_{k=1}^n \in m_n$ $(j=1,2,\dots,n+1)$ with $\varepsilon_k^{(j)}=\pm 1$ $(k=1,2,\dots,n;\, j=1,2,\dots,n+1)$, and $n+1$ numbers $\lambda_j \geqslant 0$ $(j=1,2,\dots,n+1)$ with $\sum_{j=1}^{n+1} \lambda_j = 1$, such that

$$\gamma_k = \sum_{j=1}^{n+1} \lambda_j \varepsilon_k^{(j)} \qquad (k=1,2,\dots,n).$$

Hence

$$\left\| \sum_{k=1}^{n} \gamma_k f_k(x) x_k \right\| = \left\| \sum_{k=1}^{n} \sum_{j=1}^{n+1} \lambda_j \varepsilon_k^{(j)} f_k(x) x_k \right\| = \left\| \sum_{j=1}^{n+1} \lambda_j \sum_{k=1}^{n} \varepsilon_k^{(j)} f_k(x) x_k \right\|$$
$$\leqslant \max_{1 \leqslant j \leqslant n} \left\| \sum_{k=1}^{n} \varepsilon_k^{(j)} f_k(x) x_k \right\|.$$

Consequently, we have[2]

$$\left\| \sum_{k=1}^{n} \gamma_k f_k(x) x_k \right\| \leqslant \sup_{\substack{\{\varepsilon_n\} \subset K \\ \varepsilon_n = \pm 1}} \sup_{1 \leqslant l < \infty} \| s_{\{\varepsilon_n\},l}(x) \| . \tag{16.27}$$

Since we also have a similar relation for $\{\delta_k\}_1^n$, it follows that

$$\left\| \sum_{k=1}^{n} \beta_k f_k(x) x_k \right\| \leqslant 2 \sup_{\substack{\{\varepsilon_n\} \subset K \\ \varepsilon_n = \pm 1}} \sup_{1 \leqslant l < \infty} \| s_{\{\varepsilon_n\},l}(x) \| ,$$

whence we obtain the penultimate inequality of (16.23).

The last inequality of (16.23) follows from

$$\left\| \sum_{i=1}^{n} \varepsilon_i f_i(x) x_i \right\| \leqslant \left\| \sum_{i \in d_1} f_i(x) x_i \right\| + \left\| \sum_{i \in d_2} f_i(x) x_i \right\|,$$

where $d_1 = \{ i \in \mathcal{N} \mid 1 \leqslant i \leqslant n,\ \varepsilon_i = +1 \}$, $d_2 = \{ i \in \mathcal{N} \mid 1 \leqslant i \leqslant n,\ \varepsilon_i = -1 \}$.

---

[1] See e.g. [32], p. 82, example 2).

[2] One can also give the following proof of (16.27), which makes no use of the theorem of Carathéodory: Take a $g \in (E_{(r)})^*$ (see Ch. I, § 1) such that $\|g\| = 1$, $g\left(\sum_{k=1}^{n} \gamma_k f_k(x) x_k \right) = \left\| \sum_{k=1}^{n} \gamma_k f_k(x) x_k \right\|$ and put $\varepsilon_k = \operatorname{sign}[f_k(x) g(x_k)]$ if $f_k(x) g(x_k) \neq 0$ and $\varepsilon_k = 1$ if $f_k(x) g(x_k) = 0$. Then $\varepsilon_k = \pm 1$ $(k=1,\dots,n)$ and

$$\left\| \sum_{k=1}^{n} \gamma_k f_k(x) x_k \right\| = \sum_{k=1}^{n} \gamma_k f_k(x) g(x_k) \leqslant \sum_{k=1}^{n} |f_k(x) g(x_k)|$$
$$= g\left[\sum_{k=1}^{n} \varepsilon_k f_k(x) x_k \right] \leqslant \left\| \sum_{k=1}^{n} \varepsilon_k f_k(x) x_k \right\|,$$

whence we infer (16.27).

Finally, if $\{x_n\}$ is an unconditional basis of E, then, by the implication $1^\circ \Rightarrow 10^\circ$ of part a) proved above, the numbers (16.23) are finite for all $x \in E$. Since they are obviously norms on E and since

$$\|x\| = \left\| \sum_{i=1}^{\infty} f_i(x) x_i \right\| \leqslant \sup_{d \in \mathscr{D}} \|s_d(x)\| \qquad (x \in E),$$

from the inversion theorem of Banach[1] it follows that they are equivalent norms to the initial norm on E. This completes the proof of theorem 16.1.

Let us observe that the equivalences $1^\circ \Leftrightarrow \ldots \Leftrightarrow 5^\circ$ in theorem 16.1 a) also hold if the hypothesis of E-completeness is replaced by the assumption that $\{f_n\}$ *is total on* E. If E contains no subspace isomorphic to c_0, then, by § 15, lemma 15.8, a similar remark is also valid for the equivalences $1^\circ \Leftrightarrow \ldots \Leftrightarrow 24^\circ$. In this latter case the assumption that E contains no subspace isomorphic to c_0 is essential, as shown by the example $E = c$, $x_n =$ the n-th unit vector, and $f_n =$ the n-th coordinate functional (then $E \supset c_0$ isometrically and (x_n, f_n) is a total biorthogonal system satisfying $2^\circ, 3^\circ, \ldots, 24^\circ$ but not 1°, since $[x_n] = c_0 \neq E$).

The assumption (16.21) in 24° of theorem 16.1 a) is necessary for the validity of the implication $24^\circ \Rightarrow 1^\circ$, as shown by

Example 16.1. Let $E = l^1$ and let $\{h_n\}$ be the conditional basis of l^1 considered in § 14, example 14.2, i.e.,

$$h_1 = f_1, \qquad h_n = f_{n-1} - f_n \qquad (n = 2, 3, \ldots), \tag{16.28}$$

where $\{f_n\}$ is the unit vector basis of l^1. Then the convergence of $\sum_{i=1}^{\infty} \alpha_i h_i$ still implies the convergence of $\sum_{i=1}^{\infty} |\alpha_i| h_i$.

Indeed, as observed in § 14, example 14.2, $\sum_{i=1}^{\infty} \alpha_i h_i$ converges if and only if $\sum_{i=1}^{\infty} |\alpha_{i+1} - \alpha_i| < \infty$. Since $\big| |\alpha_{i+1}| - |\alpha_i| \big| \leqslant |\alpha_{i+1} - \alpha_i|$, it follows that the convergence of $\sum_{i=1}^{\infty} \alpha_i h_i$ implies the convergence of $\sum_{i=1}^{\infty} |\alpha_i| h_i$.

Comparing the equivalences $1^\circ \Leftrightarrow 11^\circ \Leftrightarrow 25^\circ$ of theorem 16.1 with Ch. I, § 5, theorem 5.1, it is natural to ask whether the condition $\mathscr{U}_1 = \mathscr{U}_2$ is also equivalent to 1°. The answer is negative. In fact, although 1° obviously implies $\mathscr{U}_1 = \mathscr{U}_2 (= E)$, the converse implication is not valid, as shown by

Example 16.2. Let $E = l^1$ and let $\{x_n\}$ be any conditional basis of l^1 (e.g., the basis considered in example 16.1 above). Then $\mathscr{U}_1 = \mathscr{U}_2$.

---

[1] See e.g. [10], p. 41, theorem 5.

Indeed, the inclusion $\mathscr{U}_1 \subset \mathscr{U}_2$ is obvious. Conversely, let $x \in \mathscr{U}_2$ be arbitrary. Then, by § 15, corollary 15.1 (implication $3^\circ \Rightarrow 1^\circ$) $\sum_{i=1}^{\infty} f_i(x) x_i$ is a weakly unconditionally Cauchy series. Hence, since l^1 is (sequentially) weakly complete and since in l^1 weak convergence of sequence implies norm convergence, it follows that $\sum_{i=1}^{\infty} f_i(x) x_i$ is unconditionally convergent[1] and therefore, by lemma 16.1 (implication $1^\circ \Rightarrow 2^\circ$), $x \in \mathscr{U}_1$. Thus, $\mathscr{U}_2 \subset \mathscr{U}_1$, whence, finally $\mathscr{U}_1 = \mathscr{U}_2$.

We shall give now a characterization of unconditional bases among E-complete biorthogonal systems in terms of a stability property.

Definition 16.1. A sequence $\{y_n\}$ in a Banach space E is said to be *unconditionally ω-linearly independent*, if for any unconditionally convergent series $\sum_{i=1}^{\infty} \alpha_i y_i$ with $\sum_{i=1}^{\infty} \alpha_i y_i = 0$ we have $\alpha_i = 0$ $(i = 1, 2, \ldots)$.

Obviously, every ω-linearly independent sequence $\{y_n\}$ is also unconditionally ω-linearly independent.

Definition 16.2. Let E be a Banach space. An E-complete biorthogonal system (x_n, f_n) $(\{x_n\} \subset E, \{f_n\} \subset E^*)$ is said to be *U-stable* if every unconditionally ω-linearly independent sequence $\{y_n\} \subset E$ for which the series $\sum_{i=1}^{\infty} \|x_i - y_i\| \, f_i$ converges, is complete in E.

Theorem 16.2. *Let E be a Banach space and $(x_n, f_n)(\{x_n\} \subset E, \{f_n\} \subset E^*)$ an E-complete biorthogonal system. $\{x_n\}$ is an unconditional basis of E if and only if (x_n, f_n) is U-stable. Moreover, in this case every unconditionally ω-linearly independent sequence $\{y_n\} \subset E$ for which the series $\sum_{i=1}^{\infty} \|x_i - y_i\| \, f_i$ converges, is an unconditional basis of E, equivalent to $\{x_n\}$.*

Proof. Assume that $\{x_n\}$ is an unconditional basis of E. Then, obviously, $\{f_n\}$ is total on E. Furthermore, let $\{y_n\}$ be an unconditionally ω-linearly independent sequence in E, such that the series $\sum_{i=1}^{\infty} \|x_i - y_i\| \, f_i$ converges. Then, since $\{x_n\}$ is an unconditional basis of E, $\{f_n\}$ is an unconditional basic sequence (see § 17, theorem 17.7), whence $\sum_{i=1}^{\infty} \|x_i - y_i\| \, f_i$ converges unconditionally. Consequently, by Ch. I, § 10, remark 10.4, there exists a positive integer n_0 such that $\{x_n\}_{n_0}^{\infty} \approx \{y_n\}_{n_0}^{\infty}$, whence

[1] Actually, by § 15, lemma 15.8, the same conclusion holds in any space E containing no subspace isomorphic to c_0.

$\{y_n\}_{n_0}^{\infty}$ is an unconditional basis of $[y_n]_{n_0}^{\infty}$. Since $\{y_n\}$ is unconditionally ω-linearly independent, it follows that $\{y_n\}$ is also ω-linearly independent and therefore, by Ch. I, § 10, remark 10.4, $\{x_n\} \approx \{y_n\}$, whence $\{y_n\}$ is an unconditional basis of E.

Let us observe that this part of theorem 16.2 also follows from an extended stability theorem mentioned at the end of § 11. Indeed, assume again that $\{x_n\}$ is an unconditional basis of E and $\{y_n\}$ an unconditionally ω-linearly independent sequence in E such that the series $\sum_{i=1}^{\infty} \|x_i - y_i\| f_i$ converges and let $\{\alpha_n\}$ be any sequence of scalars such that $\sum_{i=1}^{\infty} \alpha_i x_i = x_0 \in E$, $\sum_{i=1}^{\infty} \alpha_i y_i = 0$. Then, since $\{x_n\}$ is an unconditional basis, both $\sum_{i=1}^{\infty} \alpha_i x_i$ and $\sum_{i=1}^{\infty} \|x_i - y_i\| f_i$ are unconditionally convergent. Let β_i be any scalars with $|\beta_i| \leqslant 1$ $(i=1,2,\ldots)$. Then, by § 17, theorem 17.7, and § 17, theorem 17.1 applied to the unconditional basic sequence $\{f_n\}$, there exists a constant $C \geqslant 1$ such that

$$\left\| \sum_{i=n}^{n+p} \beta_i \alpha_i (x_i - y_i) \right\| = \left\| \sum_{i=n}^{n+p} \beta_i f_i(x_0)(x_i - y_i) \right\| = \sup_{\substack{f \in E^* \\ \|f\| \leqslant 1}} \left| \sum_{i=n}^{n+p} \beta_i f_i(x_0) f(x_i - y_i) \right|$$

$$\leqslant \sup_{\substack{f \in E^* \\ \|f\| \leqslant 1}} \left\| \sum_{i=n}^{n+p} \beta_i f(x_i - y_i) f_i \right\| \|x_0\| \leqslant C \left\| \sum_{i=n}^{n+p} \|x_i - y_i\| f_i \right\|.$$

Hence $\sum_{i=1}^{\infty} \beta_i \alpha_i (x_i - y_i)$ converges and thus, by lemma 16.1, $\sum_{i=1}^{\infty} \alpha_i (x_i - y_i)$, whence also $\sum_{i=1}^{\infty} \alpha_i y_i = \sum_{i=1}^{\infty} \alpha_i x_i + \sum_{i=1}^{\infty} \alpha_i (y_i - x_i)$, is unconditionally convergent. Therefore, by $\sum_{i=1}^{\infty} \alpha_i y_i = 0$ and by our hypothesis on $\{y_n\}$, it follows that $\alpha_i = 0$ $(i=1,2,\ldots)$. This proves that $\{y_n\}$ is $(E, \{x_n\})$-linearly independent. Consequently, since $\{y_n\}$ is $([f_n], \{f_n\})$-near to the $((E, \{x_n\})$-Besselian) basis $\{x_n\}$, from the extended theorem 11.4 it follows that $\{y_n\}$ is a basis of E, equivalent to $\{x_n\}$, whence an unconditional basis.

Conversely, assume now that (x_n, f_n) is U-stable, but $\{x_n\}$ is not an unconditional basis of E. Then there exists an $x_0 \in E$ such that either $\sum_{i=1}^{\infty} f_i(x_0) x_i$ is not convergent or $\sum_{i=1}^{\infty} f_i(x_0) x_i$ converges, but not unconditionally. In both cases there exists an index n such that $f_n(x_0) \neq 0$.

Put

$$y_i = x_i - \frac{f_n(x_i)}{f_n(x_0)} x_0 = \begin{cases} x_i & \text{for } i \neq n, \\ x_n - \dfrac{1}{f_n(x_0)} x_0 & \text{for } i = n. \end{cases} \tag{16.29}$$

Then, by an argument similar to that used in Ch. I, § 10, the proof of theorem 10.6, it follows that $\sum_{i=1}^{\infty} \|x_i - y_i\| f_i$ converges and $\{y_n\}$ is unconditionally ω-linearly independent (observing that if $\sum_{i=1}^{\infty} \alpha_i y_i$ converges unconditionally, then so does $\sum_{i=1}^{\infty} \alpha_i x_i$, by (16.29), but $\sum_{i=1}^{\infty} f_i(x_0) x_i$ does not converge unconditionally). Hence, by our hypothesis, $\{y_n\}$ must be complete in E, but this contradicts the relations

$$f_n(y_i) = f_n(x_i) - \frac{f_n(x_i)}{f_n(x_0)} f_n(x_0) = 0 \quad (i = 1, 2, \ldots),$$

completing the proof of theorem 16.2.

Now we shall give some characterizations of unconditional bases among E-complete total biorthogonal systems by properties of the "associated cone". We recall (see § 10) that by "cone" we understand "closed convex cone having the origin as extreme point", i.e., a closed set $\mathcal{K}$ such that $\mathcal{K} + \mathcal{K} \subset \mathcal{K}$, $\lambda \mathcal{K} \subset \mathcal{K}$ $(\lambda \geqslant 0)$ and $\mathcal{K} \cap (-\mathcal{K}) = \{0\}$.

Definition 16.3. Let E be a real[1] Banach space and (x_n, f_n) $(\{x_n\} \subset E$, $\{f_n\} \subset E^*)$ a total biorthogonal system (i.e., a biorthogonal system such that $\{f_n\}$ is total on E). The set

$$\mathcal{K}_{(x_n, f_n)} = \{x \in E \mid f_n(x) \geqslant 0 \ (n = 1, 2, \ldots)\} \tag{16.30}$$

is called *the cone associated to the biorthogonal system* (x_n, f_n).

It is easy to see that $\mathcal{K}_{(x_n, f_n)}$ is a cone. Indeed, $\mathcal{K}_{(x_n, f_n)}$ is obviously a closed set, satisfying $\mathcal{K}_{(x_n, f_n)} + \mathcal{K}_{(x_n, f_n)} \subset \mathcal{K}_{(x_n, f_n)}$ and $\lambda \mathcal{K}_{(x_n, f_n)} \subset \mathcal{K}_{(x_n, f_n)}$ $(\lambda \geqslant 0)$. Furthermore, if $x \in \mathcal{K}_{(x_n, f_n)} \cap (-\mathcal{K}_{(x_n, f_n)})$, then, by (16.30), $f_n(x) = 0$ $(n = 1, 2, \ldots)$, whence, since $\{f_n\}$ is total on E, $x = 0$ and thus $\mathcal{K}_{(x_n, f_n)} \cap (-\mathcal{K}_{(x_n, f_n)}) = \{0\}$.

In the particular case when $\{x_n\}$ is a basis of E and $\{f_n\}$ the a.s.c.f. to $\{x_n\}$, we obviously have[2] $\mathcal{K}_{(x_n, f_n)} = \mathcal{K}_{\{x_n\}}$. Let us observe that some properties proved in § 10 for $\mathcal{K}_{\{x_n\}}$ (e.g., proposition 10.1 a), b) and first half of c)) remain valid, with the same proof, for $\mathcal{K}_{(x_n, f_n)}$, where (x_n, f_n) is any total biorthogonal system.

[1] See § 10, footnote to definition 10.2.
[2] See § 10, definition 10.2.

A cone $\mathscr{K}$ is called *minihedral* if for every[1] $x,y\in\mathscr{K}$ there exists $z_0=\sup(x,y)$, i.e., the element $z_0\geqslant x,y$ with the property $z\geqslant x,y\Rightarrow z\geqslant z_0$ (we recall that $x\geqslant y$ if and only if $x-y\in\mathscr{K}$). This condition is equivalent to the following: for every $x,y\in\mathscr{K}$ there exists $z_0\in\mathscr{K}$ such that $(x+\mathscr{K})\cap(y+\mathscr{K})=z_0+\mathscr{K}$. For the definitions of normal cones and generating cones see § 10.

Theorem 16.3. *Let E be a (real) Banach space and (x_n,f_n) ($\{x_n\}\subset E$, $\{f_n\}\subset E^*$) an E-complete total biorthogonal system. The following statements are equivalent:*

1°. *$\{x_n\}$ is an unconditional basis of E.*

2°. *$\mathscr{K}_{(x_n,f_n)}$ is normal and generating.*

3°. *$\mathscr{K}_{(x_n,f_n)}$ is normal and for every $x\in E$ there exists an element $z\in E$ such that*

$$f_n(z)=|f_n(x)| \qquad (n=1,2,\ldots). \tag{16.31}$$

4°. *$\mathscr{K}_{(x_n,f_n)}$ is generating and for every $x\in\mathscr{K}_{(x_n,f_n)}$ the set*[2]

$$\mathscr{P}_x=\mathscr{K}_{(x_n,f_n)}\cap(x-\mathscr{K}_{(x_n,f_n)})=\{y\in E\mid 0\leqslant y\leqslant x\} \tag{16.32}$$

is bounded (in the norm).

5°. *$\mathscr{K}_{(x_n,f_n)}$ is generating and minihedral and for every $x\in\mathscr{K}_{(x_n,f_n)}$ the set $\mathscr{P}_x$ above is linearly homeomorphic either to a finite dimensional cube or to the fundamental parallelotope of Hilbert (hence compact).*

Proof. $1^\circ\Rightarrow3^\circ$. Assume that we have 1° and let $x,y\in E$ be arbitrary elements such that $0\leqslant y\leqslant x$. Then $0\leqslant f_i(y)\leqslant f_i(x)$ ($i=1,2,\ldots$), whence $f_i(y)=\beta_i f_i(x)$ with $0\leqslant\beta_i\leqslant1$ ($i=1,2,\ldots$). Consequently, by the implication $1^\circ\Rightarrow17^\circ$ of theorem 16.1,

$$\|y\|=\lim_{n\to\infty}\left\|\sum_{i=1}^{n}f_i(y)x_i\right\|=\lim_{n\to\infty}\left\|\sum_{i=1}^{n}\beta_i f_i(x)x_i\right\|\leqslant C\|x\|,$$

where $C=\sup\limits_{1\leqslant n<\infty}\|s_{\{\beta_j\},n}\|$. This proves that $\mathscr{K}_{(x_n,f_n)}$ is normal.

Furthermore, the existence of a $z\in E$ satisfying (16.31) is obvious from the implication $1^\circ\Rightarrow24^\circ$ of theorem 16.1.

$3^\circ\Rightarrow2^\circ$. Let $x\in E$ be arbitrary and let $z\in E$ satisfy (16.31). Put

$$y_1=\frac{z+x}{2},\qquad y_2=\frac{z-x}{2}. \tag{16.33}$$

[1] Some authors modify this definition replacing the condition $x,y\in\mathscr{K}$ by $x,y\in E$. It is readily seen that these latter cones $\mathscr{K}$ are nothing else than those which are minihedral and generating (in the terminology used here).

[2] For any cone $\mathscr{K}$ and any $x\in\mathscr{K}$ we have $\mathscr{K}\cap(x-\mathscr{K})=\{y\in E\mid 0\leqslant y\leqslant x\}$. Indeed, $y\in x-\mathscr{K}$ if and only if $x-y\in\mathscr{K}$ and thus $x-\mathscr{K}=\{y\in E\mid y\leqslant x\}$.

Then $x = y_1 - y_2$ and

$$f_n(y_1) = \frac{|f_n(x)| + f_n(x)}{2} \geqslant 0, \quad f_n(y_2) = \frac{|f_n(x)| - f_n(x)}{2} \geqslant 0 \quad (n = 1, 2, \ldots),$$

i.e., $y_1, y_2 \in \mathcal{K}_{(x_n, f_n)}$, which proves that $\mathcal{K}_{(x_n, f_n)}$ is generating.

The implication $2^\circ \Rightarrow 4^\circ$ is obvious.

$4^\circ \Rightarrow 1^\circ$. Assume that we have 4° and let $x \in \mathcal{K}_{(x_n, f_n)}$ be arbitrary. Since for $s_n(x) = \sum_{i=1}^{n} f_i(x) x_i$ we have, by biorthogonality,

$$f_i[s_n(x)] = \begin{cases} f_i(x) & \text{for} \quad i = 1, \ldots, n \\ 0 & \text{for} \quad i = n+1, n+2, \ldots \end{cases} \quad (n = 1, 2, \ldots), \quad (16.34)$$

it follows that

$$0 \leqslant s_n(x) \leqslant x \quad (n = 1, 2, \ldots) \quad (16.35)$$

whence, by the boundedness of the set $\mathcal{P}_x$ defined by (16.32),

$$\sup_{1 \leqslant n < \infty} \|s_n(x)\| = M_x < \infty. \quad (16.36)$$

Now let $x \in E$ be arbitrary. Then, since $\mathcal{K}_{(x_n, f_n)}$ is generating, we have $x = y - z$, where $y, z \in \mathcal{K}_{(x_n, f_n)}$, whence, by (16.36) applied to y and z,

$$\sup_{1 \leqslant n < \infty} \|s_n(x)\| \leqslant \sup_{1 \leqslant n < \infty} \|s_n(y)\| + \sup_{1 \leqslant n < \infty} \|s_n(z)\| \leqslant M_y + M_z < \infty.$$

Consequently, by $[x_n] = E$ and Ch. I, § 4, theorem 4.1, $\{x_n\}$ is a basis of E.

Now let $(x_{\sigma(n)}, f_{\sigma(n)})$ be an arbitrary permutation of (x_n, f_n). Then $[x_{\sigma(n)}] = E$, $\{f_{\sigma(n)}\}$ is total on E and $\mathcal{K}_{(x_{\sigma(n)}, f_{\sigma(n)})} = \mathcal{K}_{(x_n, f_n)}$ is normal and generating, whence, by the above, $\{x_{\sigma(n)}\}$ is a basis of E. Consequently, $\{x_n\}$ is an unconditional basis of E.

$1^\circ \Rightarrow 5^\circ$. Assume that we have 1° and let $x, y \in \mathcal{K}_{(x_n, f_n)}$ be arbitrary. Then the series $\sum_{i=1}^{\infty} [f_i(x) + f_i(y)] x_i$ is unconditionally convergent, whence, by lemma 16.1, so is the series $\sum_{i=1}^{\infty} \max[f_i(x), f_i(y)] x_i$ and the sum of this latter is obviously $\sup(x, y)$. This proves that $\mathcal{K}_{(x_n, f_n)}$ is minihedral. By the implication $1^\circ \Rightarrow 4^\circ$ proved above, $\mathcal{K}_{(x_n, f_n)}$ is also generating.

Now let $x \in \mathcal{K}_{(x_n, f_n)}$ be arbitrary, such that the set

$$\{n_j\} = \{n \in \mathcal{N} \mid f_n(x) > 0\} \quad (16.37)$$

is infinite; in the case when the set $\{n_j\}$ is finite, the argument is simpler, with obvious modifications.

Define a mapping Φ of the set $\mathscr{P}_x$ (see (16.32)) into the fundamental parallelotope of Hilbert

$$Q_0 = \left\{ x = \{\xi_n\} \in l^2 \;\middle|\; 0 \leqslant \xi_j \leqslant \frac{1}{j} \quad (j = 1, 2, \ldots) \right\}, \tag{16.38}$$

by

$$\Phi(y) = \left\{ \frac{f_{n_j}(y)}{j f_{n_j}(x)} \right\} \qquad (y \in \mathscr{P}_x). \tag{16.39}$$

Then the relations $y_1, y_2 \in \mathscr{P}_x$, $\Phi(y_1) = \Phi(y_2)$ imply $y_1 = y_2$ (since for $n \neq n_j$ we have $f_n(x) = 0$, whence, by $0 \leqslant y_k \leqslant x$, $f_n(y_k) = 0$), i.e., Φ is one to one. Furthermore, y_1, y_2, $\alpha_1 y_1 + \alpha_2 y_2 \in \mathscr{P}_x$ imply $\Phi(\alpha_1 y_1 + \alpha_2 y_2) = \alpha_1 \Phi(y_1) + \alpha_2 \Phi(y_2)$, i.e., Φ is linear.

Let $\{\xi_j\} \in Q_0$ be arbitrary. Then, since $\sum_{i=1}^{\infty} f_{n_i}(x) x_{n_i}$ is unconditionally convergent and since $0 \leqslant j\xi_j \leqslant 1$, the series $\sum_{i=1}^{\infty} i\xi_i f_{n_i}(x) x_{n_i}$ is unconditionally convergent (by lemma 16.1) to an element $y \in E$. Since $0 \leqslant f_{n_j}(y) = j\xi_j f_{n_j}(x) \leqslant f_{n_j}(x)$ and $0 = f_n(y) = f_n(x)$ for $n \neq n_j$, we have $y \in \mathscr{P}_x$ and

$$\Phi(y) = \left\{ \frac{j\xi_j f_{n_j}(x)}{j f_{n_j}(x)} \right\} = \{\xi_j\},$$

which proves that Φ maps $\mathscr{P}_x$ *onto* Q_0.

Now, since Q_0 is compact, in order to prove that Φ is a homeomorphism it will be sufficient to prove that Φ^{-1} is continuous. Let $y_0 \in \mathscr{P}_x$ and let $U = U(y_0) = \{y \in \mathscr{P}_x \mid \|y - y_0\| < \varepsilon\}$ be an arbitrary neighborhood of y_0 in $\mathscr{P}_x$. We shall find a neighborhood $V = V(z_0)$ of $z_0 = \{\zeta_j^{(0)}\} = \Phi(y_0) = \left\{ \frac{f_{n_j}(y_0)}{j f_{n_j}(x)} \right\}$ in Q_0, such that $\Phi^{-1}(V) \subset U$. By 1° and theorem 16.1 there exists an $N = N(\varepsilon, x)$ such that for all $\{\beta_n\} \subset K$ with $|\beta_n| \leqslant 1$ we have

$$\left\| \sum_{i=N+1}^{\infty} \beta_{n_i} f_{n_i}(x) x_{n_i} \right\| = \lim_{l \to \infty} \left\| s_{\{\beta_n\}, l} \left(\sum_{i=N+1}^{\infty} f_{n_i}(x) x_{n_i} \right) \right\|$$

$$\leqslant C \left\| \sum_{i=N+1}^{\infty} f_{n_i}(x) x_{n_i} \right\| < \frac{\varepsilon}{3},$$

where $C = \sup_{|\beta_1|, |\beta_2|, \ldots, \leqslant 1} \sup_{1 \leqslant l < \infty} \|s_{\{\beta_n\}, l}\|$. Consequently, for any $z = \{\zeta_j\}$

$$= \Phi(y) = \left\{ \frac{f_{n_j}(y)}{j f_{n_j}(x)} \right\} \in Q_0 \text{ (where } y \in \mathscr{P}_x\text{) with } |\zeta_j - \zeta_j^0| < \delta = \frac{\varepsilon}{3N^2 \max_{1 \leqslant i \leqslant N} \|f_{n_i}(x) x_{n_i}\|}$$

$(j = 1, \ldots, N)$ we have, by $0 \leqslant j\zeta_j, j\zeta_j^{(0)} \leqslant 1$,

$$\|y - y_0\| = \left\| \sum_{i=1}^{\infty} f_{n_i}(y) x_{n_i} - \sum_{i=1}^{\infty} f_{n_i}(y_0) x_{n_i} \right\| = \left\| \sum_{i=1}^{\infty} (i\zeta_i - i\zeta_i^{(0)}) f_{n_i}(x) x_{n_i} \right\|$$

$$\leqslant \left\| \sum_{i=1}^{N} (i\zeta_i - i\zeta_i^{(0)}) f_{n_i}(x) x_{n_i} \right\| + \left\| \sum_{i=N+1}^{\infty} i\zeta_i f_{n_i}(x) x_{n_i} \right\|$$

$$+ \left\| \sum_{i=N+1}^{\infty} i\zeta_i^{(0)} f_{n_i}(x) x_{n_i} \right\| < \frac{\varepsilon}{3} + \frac{\varepsilon}{3} + \frac{\varepsilon}{3} = \varepsilon,$$

i.e., $y \in U(y_0)$, and thus we may take $V(z_0) = \{z \in Q_0 \mid \|z - z_0\| < \delta\}$.

Finally, the implication $5^\circ \Rightarrow 4^\circ$ is obvious, which completes the proof of theorem 16.3.

Remark 16.1. One can also prove the implication $4^\circ \Rightarrow 1^\circ$ using directly $s_d(x)$ $(d \in \mathscr{D})$ instead of $s_n(x)$ (obviously, (16.35) also holds for $s_d(x)$) and applying theorem 16.1. However, we have given the above proof in order to show what is the difficulty in the problem of characterizing (not necessarily unconditional) bases among E-complete total biorthogonal systems (x_n, f_n) by properties of the associated cone $\mathscr{K}_{(x_n, f_n)}$: namely, these should be such properties of $\mathscr{K}_{(x_n, f_n)}$ which depend on the order of $\{x_n\}$.

In connection with this problem let us mention that the conditions occurring in theorem 16.3 are not necessary in order that $\{x_n\}$ be a *basis* of E, as shown by the following two examples:

Example 16.3. Let $\{x_n\}$ be the conditional basis

$$x_n = \{\underbrace{1, \ldots, 1}_{n}, 0, 0, \ldots\} \qquad (n = 1, 2, \ldots) \tag{16.40}$$

of $E = c_0$, considered in § 14, example 14.1. Then the a. s. c. f. is

$$f_n(x) = \xi_n - \xi_{n+1} \qquad (x = \{\xi_n\} \in c_0),$$

whence

$$\mathscr{K}_{(x_n, f_n)} = \{x = \{\xi_n\} \in c_0 \mid \xi_1 \geqslant \xi_2 \geqslant \cdots\}.$$

Since the relations $0 \leqslant x = \{\xi_n\} \leqslant y = \{\eta_n\}$ imply $\|x\| = \sup_{1 \leqslant n < \infty} |\xi_n| \leqslant \sup_{1 \leqslant n < \infty} |\eta_n| = \|y\|$, the cone $\mathscr{K}_{(x_n, f_n)}$ is normal, whence, by theorem 16.3, $\mathscr{K}_{(x_n, f_n)}$ is not generating and does not satisfy (16.31).

Example 16.4. Let $E=c_0$ and let

$$x_n=\{(-1)^{n+1},(-1)^{n+2},\dots,(-1)^{2n},0,0,\dots\}$$
$$=\sum_{i=1}^{n}(-1)^{n+i}e_i \quad (n=1,2,\dots), \tag{16.41}$$

where $\{e_n\}$ denotes the unit vector basis of $E=c_0$. Then $(-1)^n x_n = \sum_{i=1}^{n}(-1)^i e_i\,(n=1,2,\dots)$, whence, by Ch. I, § 4, proposition 4.3, $\{(-1)^n x_n\}$ is a basis of E and therefore $\{x_n\}$ is a basis of E. Moreover, as in § 14, example 14.1, we see that $\{(-1)^n x_n\}$, whence also $\{x_n\}$, is a conditional basis. The a. s. c. f. to $\{x_n\}$ is

$$f_n(x)=\xi_n+\xi_{n+1} \quad (x=\{\xi_n\}\in c_0),$$

whence

$$\mathscr{K}_{(x_n,f_n)}=\{x=\{\xi_n\}\in c_0\,|\,\xi_1+\xi_2\geqslant 0,\ \xi_2+\xi_3\geqslant 0,\dots\}. \tag{16.42}$$

The cone $\mathscr{K}_{(x_n,f_n)}$ is generating, since every $x=\{\xi_n\}\in c_0$ can be written as $x=y-z=\{\eta_n\}-\{\zeta_n\}$, with $\eta_n\geqslant 0$, $\zeta_n\geqslant 0$ $(n=1,2,\dots)$ and then, by (16.42), $y,z\in\mathscr{K}_{(x_n,f_n)}$. Hence, by theorem 16.3, $\mathscr{K}_{(x_n,f_n)}$ is not normal and there exists an $x\in\mathscr{K}_{(x_n,f_n)}$ with $\mathscr{P}_x$ unbounded, whence non-compact and therefore not linearly homeomorphic to a finite dimensional cube or to the fundamental parallelotope of Hilbert.

Furthermore, $\mathscr{K}_{(x_n,f_n)}$ does not satisfy the second condition in 3°. Indeed, if for every $x=\{\xi_n\}\in c_0$ there existed an element $z=\{\zeta_n\}\in c_0$ satisfying (16.31), then we would have

$$\sum_{k=1}^{\infty}(-1)^k|\xi_k+\xi_{k+1}|=\sum_{k=1}^{\infty}(-1)^k|f_k(x)|=\sum_{k=1}^{\infty}(-1)^k f_k(z)$$
$$=\sum_{k=1}^{\infty}(-1)^k(\zeta_k+\zeta_{k+1})=-\zeta_1,$$

whence $\sum_{k=1}^{\infty}(-1)^k|\xi_k+\xi_{k+1}|$ would converge for every $x=\{\xi_n\}\in c_0$. However, this is not satisfied e. g. for $x=\{\xi_n\}\in c_0$ defined by

$$\xi_1=0,\ \xi_{2n}=-\xi_{2n+1}=\sum_{k=n+1}^{\infty}\frac{(-1)^k}{k} \quad (n=1,2,\dots),$$

since for each $n=1,2,\dots$ we have

$$|\xi_{2n}+\xi_{2n+1}|=0,$$
$$|\xi_{2n+1}+\xi_{2n+2}|=\left|-\sum_{k=n+1}^{\infty}\frac{(-1)^k}{k}+\sum_{k=n+2}^{\infty}\frac{(-1)^k}{k}\right|=\frac{1}{n+1},$$

whence $\sum_{k=1}^{\infty}(-1)^k|\xi_k+\xi_{k+1}|=-\infty$.

Finally, $\mathscr{K}_{(x_n,f_n)}$ is not minihedral. Indeed, assume that $\mathscr{K}_{(x_n,f_n)}$ is minihedral and let $x \in E$ be arbitrary. Then, since $\mathscr{K}_{(x_n,f_n)}$ is generating, we have $x = y - z$, with $y, z \in \mathscr{K}_{(x_n,f_n)}$. Since $\mathscr{K}_{(x_n,f_n)}$ is minihedral, $\sup(y,z)$ exists and hence there also exist the elements

$$x_+ = \sup(x,0) = \sup(y-z,0) = \sup(y,z) - z, \; x_- = \sup(-x,0).$$

Since[1] $f_n(x_+) = \max(f_n(x),0), f_n(x_-) = \max(-f_n(x),0)$, it follows that for the element $z = x_+ + x_-$ we have (16.31), in contradiction with the above. This proves our assertion.

The hypothesis $[x_n] = E$ in theorem 16.3 is essential, as shown by

Example 16.5. Let $E = c$ and let

$$x_n = \{\underbrace{0,\ldots,0}_{n-1},1,0,\ldots\} \qquad (n=1,2,\ldots), \tag{16.43}$$

$$f_n(x) = \xi_n \qquad (x = \{\xi_j\} \in c;\, n=1,2,\ldots). \tag{16.44}$$

Then $[x_n] = c_0 \neq E$ (and thus $\{x_n\}$ is not a basis of E), but $\{f_n\}$ is total on E and $\mathscr{K}_{(x_n,f_n)}$ is both normal and generating.

Indeed, obviously,

$$\mathscr{K}_{(x_n,f_n)} = \{x = \{\xi_j\} \in c \mid \xi_n \geqslant 0\,(n=1,2,\ldots)\}.$$

Now, if $0 \leqslant x = \{\xi_n\} \leqslant y = \{\eta_n\}$, then $\|x\| \leqslant \|y\|$, and thus $\mathscr{K}_{(x_n,f_n)}$ is normal. Furthermore, let $x = \{\xi_n\} \in E$ be arbitrary. Then $x = x_+ - x_- = \{\xi_n^+\} - \{\xi_n^-\}$, where $\xi_n^+ = \max(\xi_n, 0)$, $\xi_n^- = \max(-\xi_n, 0)$; here $x_+, x_- \in c$, since if $\lim_{n\to\infty} \xi_n = \xi \neq 0$, then $\operatorname{sign}\xi_n = \operatorname{sign}\xi$ for all sufficiently large n. Thus, $\mathscr{K}_{(x_n,f_n)}$ is generating.

However, such an example is no longer possible if E is reflexive, i.e., in this case one can omit the hypothesis $[x_n] = E$ in the equivalence $1^\circ \Leftrightarrow 2^\circ$ of theorem 16.3. Indeed, we have

Proposition 16.1. *Let E be a reflexive Banach space and let (x_n, f_n) be a total biorthogonal system such that $\mathscr{K}_{(x_n,f_n)}$ is normal and generating. Then $\{x_n\}$ is an unconditional basis of E.*

Proof. Let $x \in \mathscr{K}_{(x_n,f_n)}$ be arbitrary. Then by (16.34) we have

$$f_i[s_n(x)] \leqslant f_i[s_{n+1}(x)] \leqslant f_i(x) \qquad (i, n = 1,2,\ldots),$$

---

[1] Indeed, $x_+ \geqslant x, 0$ implies $f_n(x_+) \geqslant f_n(x), 0$ and the relations $\alpha_n \geqslant f_n(x), 0$ imply $z = \alpha_n x_n + \sum_{i \neq n} f_i(x_+) x_i \geqslant x, 0$, whence $z \geqslant x_+$, whence $\alpha_n = f_n(z) \geqslant f_n(x_+)$, which proves that $f_n(x_+) = \max(f_n(x), 0)$.

whence the limits $\lim_{n\to\infty} f_i[s_n(x)]$ exist. Since E is reflexive and $\{f_n\}$ is total on E, $\{f_n\}$ is also complete in E^*, and thus $\lim_{n\to\infty} f[s_n(x)]$ exists for all f in a dense subset of E^*. Since $\mathscr{K}_{(x_n,f_n)}$ is normal, by (16.35) we also have $\|s_n(x)\| \leqslant L\|x\|$ $(n=1,2,\ldots)$. Consequently, $\{s_n(x)\}$ is a weak Cauchy sequence. Since E is reflexive, whence weakly complete, it follows that $s_n(x)$ converges weakly to an element $y\in E$. By biorthogonality and since $\{f_n\}$ is total on E, we have $y=x$, i.e. $\{s_n(x)\}$ converges weakly to x.

Now, let $x\in E$ be arbitrary. Then, since $\mathscr{K}_{(x_n,f_n)}$ is generating, we have $x=y-z$, where $y,z\in\mathscr{K}_{(x_n,f_n)}$, whence, by the above, $s_n(x) = s_n(y)-s_n(z)$ converges weakly to $y-z=x$. Consequently[1], $x\in[x_n]$, and thus $[x_n]=E$ and the desired result follows by theorem 16.3. This completes the proof.

Let us recall that a cone $\mathscr{K}$ is called *regular* if the relations $y_1\leqslant y_2\leqslant\cdots\leqslant z$ imply that $\{y_n\}$ is norm-convergent. One may assume here that $0\leqslant y_1\leqslant y_2\leqslant\cdots\leqslant z$ (by considering, if necessary, the sequence $\{y_n-y_1\}$). Replacing the normality of $\mathscr{K}_{(x_n,f_n)}$ by the stronger[2] condition of regularity of $\mathscr{K}_{(x_n,f_n)}$, one can again omit the hypothesis $[x_n]=E$ in the equivalence $1^\circ\Leftrightarrow 2^\circ$ of theorem 16.3. Indeed, we have

Proposition 16.2. *Let E be a Banach space and (x_n,f_n) ($\{x_n\}\subset E$, $\{f_n\}\subset E^*$) a total biorthogonal system. The sequence $\{x_n\}$ is an unconditional basis of E if and only if $\mathscr{K}_{(x_n,f_n)}$ is regular and generating.*

Proof. Assume that $\{x_n\}$ is an unconditional basis of E. Then, by the implication $1^\circ\Rightarrow 2^\circ$ of theorem 16.3, $\mathscr{K}_{(x_n,f_n)}$ is normal and generating. We shall now prove that whenever $\{x_n\}$ is a basis and $\mathscr{K}_{(x_n,f_n)}$ is normal, then $\mathscr{K}_{(x_n,f_n)}$ is also regular.

Let $0\leqslant y_1\leqslant y_2\leqslant\cdots\leqslant z$. Then $0\leqslant f_i(y_n)\leqslant f_i(y_{n+1})\leqslant f_i(z)$ $(i,n=1,2,\ldots)$, whence the limits $\lim_{n\to\infty} f_i(y_n)=\alpha_i$ $(i=1,2,\ldots)$ exist and we have $0\leqslant\alpha_i\leqslant f_i(z)$ $(i=1,2,\ldots)$. Hence, since $\mathscr{K}_{(x_n,f_n)}$ is normal,

$$\left\|\sum_{i=n+1}^{n+m}\alpha_i x_i\right\| \leqslant L\left\|\sum_{i=n+1}^{n+m} f_i(z)x_i\right\| \qquad (n,m=1,2,\ldots)$$

and therefore $\sum_{i=1}^{\infty}\alpha_i x_i = y\in E$. We shall show that $\lim_{n\to\infty}\|y-y_n\|=0$, which will prove that $\mathscr{K}_{(x_n,f_n)}$ is regular.

---

[1] See e.g. [10], p. 134, theorem 2.

[2] One can show (see e.g. M. A. Krasnoselskiĭ [137], Ch. I, section 1.2.2), that every regular cone is normal, but not conversely. However, we shall not use this result here.

Let $\varepsilon>0$ be arbitrary and take $N=N(\varepsilon,z)$ such that $\left\|\sum_{i=N+1}^{\infty} f_i(z)x_i\right\| < \frac{\varepsilon}{3L}$. Then, by the normality of $\mathscr{K}_{(x_n,f_n)}$,

$$\left\|\sum_{i=N+1}^{\infty} f_i(y)x_i\right\|, \left\|\sum_{i=N+1}^{\infty} f_i(y_n)x_i\right\| \leqslant L\left\|\sum_{i=N+1}^{\infty} f_i(z)x_i\right\| < \frac{\varepsilon}{3} \quad (n=1,2,\ldots).$$

On the other hand, since $\lim_{n\to\infty} f_i(y_n)=\alpha_i=f_i(y)$ $(i=1,2,\ldots)$, there exists an $M=M(\varepsilon)$ such that

$$\left\|\sum_{i=1}^{N} [f_i(y)-f_i(y_n)]x_i\right\| < \frac{\varepsilon}{3} \qquad (n>M(\varepsilon)).$$

Consequently,

$$\|y-y_n\| = \left\|\sum_{i=1}^{\infty} [f_i(y)-f_i(y_n)]x_i\right\| \leqslant \left\|\sum_{i=1}^{N} [f_i(y)-f_i(y_n)]x_i\right\|$$

$$+\left\|\sum_{i=N+1}^{\infty} f_i(y)x_i\right\| + \left\|\sum_{i=N+1}^{\infty} f_i(y_n)x_i\right\| < \varepsilon \qquad (n>M(\varepsilon)),$$

which proves that $\lim_{n\to\infty}\|y-y_n\|=0$. Thus, $\mathscr{K}_{(x_n,f_n)}$ is regular.

Conversely, assume that $\mathscr{K}_{(x_n,f_n)}$ is regular and generating and let $x\in\mathscr{K}_{(x_n,f_n)}$ be arbitrary. Then, by (16.34) and since $\mathscr{K}_{(x_n,f_n)}$ is regular, $\lim_{n\to\infty} s_n(x)=y\in E$. Hence, by biorthogonality and since $\{f_n\}$ is total on E, it follows that $y=x$, and thus $\lim_{n\to\infty} s_n(x)=x$.

Now let $x\in E$ be arbitrary. Then, since $\mathscr{K}_{(x_n,f_n)}$ is generating, we have $x=y-z$, where $y,z\in\mathscr{K}_{(x_n,f_n)}$. Hence, by the above,

$$\lim_{n\to\infty} s_n(x)=\lim_{n\to\infty} s_n(y)-\lim_{n\to\infty} s_n(z)=y-z=x,$$

and thus $\{x_n\}$ is a basis of E. Considering an arbitrary permutation $(x_{\sigma(n)},f_{\sigma(n)})$ of (x_n,f_n) it follows, as in the proof of the implication $4^\circ\Rightarrow1^\circ$ of theorem 16.3, that $\{x_n\}$ is an unconditional basis of E, which completes the proof.

In the particular case when $\dim E=n<\infty$ and $\{x_j\}_1^n$ is a basis of E, by theorem 16.1 the associated cone $\mathscr{K}_{\{x_j\}_1^n}$ is minihedral. Furthermore, $\mathscr{K}_{\{x_j\}_1^n}$ is obviously solid, since any $x=\sum_{i=1}^{n}\alpha_i x_i$ with $\alpha_i>0$ $(i=1,\ldots,n)$ is an interior point of $\mathscr{K}_{\{x_j\}_1^n}$. The following converse is also true:

Theorem 16.4. *Let E be a (real) Banach space with* $\dim E = n < \infty$ *and let* $\mathcal{K} \subset E$ *be a solid minihedral cone. Then there exists a basis* $\{x_j\}_1^n$ *of E such that* $\mathcal{K}_{\{x_j\}_1^n} = \mathcal{K}$.

For the proof we shall make use of the following "decomposition lemma":

Lemma 16.2. *If* $\mathcal{K}$ *is a minihedral cone in a linear space E, then*[1] *for every pair of finite sequences* $\{y_i\}_{i=1}^r, \{z_k\}_{k=1}^s \subset \mathcal{K}$, *with* $r \geqslant 2$, $s \geqslant 2$, *satisfying*

$$\sum_{i=1}^{r} y_i = \sum_{k=1}^{s} z_k, \tag{16.45}$$

there exist $r \cdot s$ *elements* $w_{ik} \in \mathcal{K}$ $(i=1,\dots,r;\ k=1,\dots,s)$ *such that*

$$y_i = \sum_{k=1}^{s} w_{ik} \quad (i=1,\dots,r), \qquad z_k = \sum_{i=1}^{r} w_{ik} \quad (k=1,\dots,s). \tag{16.46}$$

Proof. Assume first that $r=s=2$. Let $y_1, y_2, z_1, z_2 \in \mathcal{K}$ be such that

$$y_1 + y_2 = z_1 + z_2.$$

Put

$$w_{11} = \inf(y_1, z_1),\ w_{12} = y_1 - w_{11},\ w_{21} = z_1 - w_{11}. \tag{16.47}$$

Obviously, $w_{11}, w_{12}, w_{21} \geqslant 0$. Furthermore,

$$\inf(w_{12}, w_{21}) = \inf(y_1 - w_{11}, z_1 - w_{11}) = \inf(y_1, z_1) - w_{11} = 0.$$

Since

$$w_{12} + y_2 = (y_1 + y_2) - w_{11} = (z_1 + z_2) - w_{11} = w_{21} + z_2 \geqslant w_{21},$$

we also have

$$w_{21} = \inf(w_{21}, w_{12} + y_2) \leqslant \inf(w_{21}, w_{12}) + \inf(w_{21}, y_2) = \inf(w_{21}, y_2),$$

whence $w_{21} \leqslant y_2$ and thus for the element w_{22} defined by

$$w_{22} = y_2 - w_{21} = z_2 - w_{12} \tag{16.48}$$

we have $w_{22} \geqslant 0$. Since by (16.47) and (16.48) we have (16.46) with $r=s=2$, the lemma is proved for $r=s=2$.

Assume now that the lemma is true for all pairs of finite sequences $\{y_i\}_{i=1}^p, \{z_k\}_{k=1}^q \subset \mathcal{K}$ with $2 \leqslant p \leqslant r, 2 \leqslant q \leqslant s$ and let $y_1, \dots, y_r, z_1, \dots, z_{s+1} \in \mathcal{K}$ be arbitrary, satisfying $\sum_{i=1}^{r} y_i = \sum_{k=1}^{s+1} z_k$. Then by the induction hypothesis applied to the pair of finite sequences $y_1, \dots, y_r$, $z = z_1 + z_2$,

[1] One can also prove the converse statement and thus this decomposition lemma actually characterizes the minihedral cones.

$z_3,\ldots,z_{s+1}\in\mathscr{K}$, there exist $r\cdot s$ elements $w_{ik}\in\mathscr{K}$ $(i=1,\ldots,r;\ k=1,\ldots,s)$ such that

$$y_i=\sum_{k=1}^{s} w_{ik} \qquad (i=1,\ldots,r),$$

$$z=z_1+z_2=\sum_{i=1}^{r} w_{i1},\ z_{k+1}=\sum_{i=1}^{r} w_{ik} \qquad (k=2,3,\ldots,s).$$

Furthermore, by the induction hypothesis applied to the pair of sequences $w_{11},\ldots,w_{r1},\ z_1,z_2\in\mathscr{K}$, there exist $2r$ elements $u_{ik}\in\mathscr{K}$ $(i=1,\ldots,r;\ k=1,2)$ such that

$$w_{i1}=\sum_{j=1}^{2} u_{ij} \quad (i=1,\ldots,r), \qquad z_j=\sum_{i=1}^{r} u_{ij} \quad (j=1,2).$$

Consequently, we have

$$y_i=\sum_{j=1}^{2} u_{ij}+\sum_{k=2}^{s} w_{ik} \quad (i=1,\ldots,r)$$

$$z_j=\sum_{i=1}^{r} u_{ij} \quad (j=1,2), \qquad z_{k+1}=\sum_{i=1}^{r} w_{ik} \quad (k=2,3,\ldots,s),$$

i.e., a decomposition of type (16.46) for the pair $y_1,\ldots,y_r,\ z_1,\ldots,z_{s+1}\in\mathscr{K}$, which, since the roles of y_i and z_k are symmetric, completes the proof of lemma 16.2.

Now we can give the

Proof of theorem 16.4. Since $\dim E=n<\infty$, by the classical theorem of Minkowski-Weyl $\mathscr{K}$ coincides with the convex hull of its extremal rays[1], whence, since $\mathscr{K}$ is solid and $\dim E=n$, it follows that $\mathscr{K}$ has at least n distinct extremal rays. We shall show that $\mathscr{K}$ has exactly n distinct extremal rays $[0,x_1[,\ldots,[0,x_n[$ which will complete the proof (since then $\{x_j\}_1^n$ will be a basis of E such that $\mathscr{K}_{\{x_j\}_1^n}=\mathscr{K}$).

Assume, a contrario, that $\mathscr{K}$ has at least $n+1$ extremal rays, say $[0,x_1[,\ldots,[0,x_{n+1}[$. Then, since $\dim E=n$, $x_1,\ldots,x_{n+1}$ are linearly dependent, i.e., there exist real numbers $\alpha_1,\ldots,\alpha_{n+1}$ with $\sum_{i=1}^{n+1}|\alpha_i|\neq 0$, such that $\sum_{i=1}^{n+1}\alpha_i x_i=0$. Put

$$R=\{i\,|\,\alpha_i\geqslant 0\},\quad S=\{k\,|\,\alpha_k<0\} \tag{16.49}$$

[1] For the definition of extremal rays of $\mathscr{K}$ see § 10.

and let r, s denote the number of elements of R and S, respectively. Then $r \geqslant 0$, $s \geqslant 0$ and $r+s=n+1 \geqslant 2$. Put

$$y_i = \alpha_i x_i \ (i \in R), \quad z_k = (-\alpha_k) x_k \ (k \in S). \tag{16.50}$$

Then $\{y_i\}_{i \in R}, \{z_k\}_{k \in S} \subset \mathscr{K}$ and $\sum_{i \in R} y_i = \sum_{k \in S} z_k$. Let us prove that $r \geqslant 2$, $s \geqslant 2$. If we had $r=0$, then $0 = \sum_{j \in S} z_j \geqslant z_k \geqslant 0 \ (k \in S)$, whence $\alpha_k = 0 \ (k \in S)$, which, together with $R = \emptyset$, would contradict the assumption $\sum_{i=1}^{n+1} |\alpha_i| \neq 0$. Assume now that $r=1$, say $R=\{i_1\}$. Then, if $\alpha_{i_1}=0$, we arrive, as in the case $r=0$, to a contradiction with the assumption $\sum_{i=1}^{n+1} |\alpha_i| \neq 0$. If $\alpha_{i_1} \neq 0$, then $0 \neq y_{i_1} = \sum_{k \in S} z_k = \frac{1}{2}(2 z_{k_1}) + \frac{1}{2}\Big(2 \sum_{k \in S \setminus \{k_1\}} z_k\Big)$, where $2 z_{k_1} \notin [0, y_{i_1}[$ (since the rays $[0, z_{k_1}[= [0, x_{k_1}[$ and $[0, y_{i_1}[= [0, x_{i_1}[$ are distinct by our hypothesis), which contradicts the assumption that $[0, y_{i_1}[= [0, x_{i_1}[$ is an extremal ray of $\mathscr{K}$. Thus, $r \geqslant 2$ and, similarly, $s \geqslant 2$.

Consequently, by lemma 16.2, there exist $r \cdot s$ elements $w_{ik} \in \mathscr{K}$ $(i \in R, k \in S)$ such that

$$y_i = \sum_{k \in S} w_{ik} \ (i \in R), \qquad z_k = \sum_{i \in R} w_{ik} \ (k \in S). \tag{16.51}$$

Then for any $w_{i_0 k_0} \neq 0$ the relations (16.51) for y_{i_0}, z_{k_0} imply, taking into account that $[0, y_{i_0}[= [0, x_{i_0}[$ and $[0, z_{k_0}[= [0, x_{k_0}[$ are extremal rays,

$$w_{i_0 k_0} = \lambda_0 y_{i_0} = \mu_0 z_{k_0},$$

where $\lambda_0, \mu_0 > 0$. However, this contradicts the assumption that the extremal rays $[0, y_{i_0}[= [0, x_{i_0}[$ and $[0, z_{k_0}[= [0, x_{k_0}[$ are distinct and thus the proof of theorem 16.4 is complete.

Theorem 16.4 cannot be extended to infinite dimensional Banach spaces, as shown by

Example 16.6. Let $\mathscr{K}$ be the natural positive cone in the space $E = C([0,1])$ (i.e., $x \in \mathscr{K}$ if and only if $x(t) \geqslant 0$ for all $t \in [0,1]$). Then $\mathscr{K}$ is solid and minihedral, but there exists no E-complete total biorthogonal system (x_n, f_n) such that $\mathscr{K}_{(x_n, f_n)} = \mathscr{K}$.

Indeed, otherwise, since $\mathscr{K}$ is obviously normal and generating, from theorem 16.3 it would follow that $\{x_n\}$ is an unconditional basis of $E = C([0,1])$, in contradiction with § 15, theorem 15.1.

We shall now characterize unconditional bases among total biorthogonal systems in terms of properties of the set of multipliers[1].

[1] For the definition of the sets of multipliers $M(x, (x_n, f_n))$ and $M(E, (x_n, f_n))$ and for characterizations of bases by properties of the set $M(E, (x_n, f_n))$, see Ch. I, § 5.

Theorem 16.5. *Let E be a Banach space, (x_n, f_n) $(\{x_n\}\subset E, \{f_n\}\subset E^*)$ a total biorthogonal system and $x\in E$. Then*

a) *If E is separable, the following statements are equivalent:*

1°. $\sum_{i=1}^{\infty} f_i(x)x_i$ *is unconditionally convergent.*

2°. $M(x,(x_n, f_n))\supset m(=l^\infty)$.

If E contains no subspace isomorphic to c_0, statement 1° is equivalent to the following:

3°. $M(x,(x_n, f_n))\supset c_0$.

b) *If E is separable, the following statements are equivalent:*

1°. $\{x_n\}$ *is an unconditional basis of E.*

2°. $M(E,(x_n, f_n))\supset m$. *Moreover, in this case* $M(E,(x_n, f_n))=m$ *and the identity mapping is an isomorphism of* $M(E,(x_n, f_n))$ *onto* m.

If E contains no subspace isomorphic to c_0, statement 1° is equivalent to the following:

3°. $M(E,(x_n, f_n))\supset c_0$.

c) *If* $[x_n]=E$, *the statements* b) 1°, b) 2°, *and* b) 3° *are equivalent.*

Proof. a) The implication $1^\circ\Rightarrow 2^\circ$ is a consequence of lemma 16.1 (implication $1^\circ\Rightarrow 5^\circ$).

Conversely, assume that E is separable and that $M(x,(x_n, f_n))\supset m$. Then an argument similar to that used in Ch. I, § 5, the proof of theorem 5.2 (implication $2^\circ\Rightarrow 1^\circ$) shows that the mapping u_x: $\{\gamma_n\}\to x_{\{\gamma_n\}}$ of m into E is continuous. Now, for any $\{\gamma_n\}\in m$ and any n the sequence $\left\{\sum_{i=1}^{n}\gamma_i e_i\right\}_{n=1}^{\infty}=\{\{\gamma_1,\ldots,\gamma_n,0,0,\ldots\}\}_{n=1}^{\infty}$ is a Cauchy sequence for the weak topology $\sigma(c_0,c_0^*)$, whence also for $\sigma(m,m^*)$. Since every continuous linear mapping of m into a separable Banach space transforms weak Cauchy sequences into norm convergent sequences[1], it follows that the sequence $\left\{\sum_{i=1}^{n}\gamma_i f_i(x)x_i\right\}=\left\{u_x\left(\sum_{i=1}^{n}\gamma_i e_i\right)\right\}$ is convergent, whence, by lemma 16.1 (implication $5^\circ\Rightarrow 1^\circ$), the series $\sum_{i=1}^{\infty} f_i(x)x_i$ is unconditionally convergent. Thus, $2^\circ\Rightarrow 1^\circ$.

The implication $2^\circ\Rightarrow 3^\circ$ is obvious.

Conversely, assume now that E contains no subspace isomorphic to c_0 and that $M(x,(x_n, f_n))\supset c_0$. Then, by the above, the mapping u_x: $\{\gamma_n\}\to x_{\{\gamma_n\}}$ of c_0 into E is continuous. Hence, since the series $\sum_{i=1}^{\infty} e_i$ in c_0 is weakly unconditionally Cauchy, so is the series $\sum_{i=1}^{\infty} f_i(x)x_i=\sum_{i=1}^{\infty} u_x(e_i)$

---

[1] See e.g. [88], p. 168 or [91], Ch. V, §4, exercise 12 and [50], p. 494, theorem 4.

in E. Consequently, by § 15, lemma 15.8, $\sum_{i=1}^{\infty} f_i(x)x_i$ is unconditionally convergent in the norm topology.

b) The implication $1^\circ \Rightarrow 2^\circ$ is a consequence of part a) above (implication $1^\circ \Rightarrow 2^\circ$) and of Ch. I, § 5, proposition 5.4c) and the inversion theorem of Banach. Alternatively, instead of the inversion theorem of Banach one can also observe directly that in this case for any $\{\gamma_n\} \in m$ we have, by theorem 16.1,

$$\|\{\gamma_n\}\|_{M(E,(x_n,f_n))} = \sup_{\substack{x\in E\\ \|x\|\leqslant 1}} \left\| \sum_{i=1}^{\infty} \gamma_i f_i(x) x_i \right\| \leqslant \sup_{1\leqslant l<\infty} \sup_{\substack{x\in E\\ \|x\|\leqslant 1}} \|s_{\{\gamma_n\},l}(x)\|$$

$$\leqslant C \sup_{1\leqslant n<\infty} |\gamma_n|, \qquad (16.52)$$

where $C = \sup_{|\beta_1|,|\beta_2|,\ldots,\leqslant 1} \sup_{1\leqslant l<\infty} \|s_{\{\beta_n\},l}\|$. The other implications are immediate consequences of part a), taking into account that $\{f_n\}$ is total on E.

c) Assume now that $[x_n] = E$ and that we have 3°. Let σ be an arbitrary permutation of $\mathcal{N} = \{1, 2, 3, \ldots\}$. Then, since

$$M(E,(x_{\sigma(n)}, f_{\sigma(n)})) = \{\{\gamma_{\sigma(n)}\} \mid \{\gamma_n\} \in M(E,(x_n, f_n))\},$$

from 3° it follows that $M(E,(x_{\sigma(n)}, f_{\sigma(n)}))$ contains every non-increasing sequence tending to zero, whence, by Ch. I, § 5, theorem 5.2 (implication $3^\circ \Rightarrow 1^\circ$) $\{x_{\sigma(n)}\}$ is a basis of E. Consequently, $\{x_n\}$ is an unconditional basis of E, which completes the proof of theorem 16.5.

The assumptions on E made in theorem 16.5 are essential, as shown by the following two examples:

Example 16.7. Let $E = m$, $x_n =$ the n-th unit vector, and $f_n =$ the n-th coordinate functional $(n = 1, 2, \ldots)$. Then (x_n, f_n) is a total biorthogonal system, E is non-separable, $[x_n] = c_0 \neq E$ and for any $x \in m \setminus c_0$ we have a) 2°, but not a) 1°. Hence we have b) 2°, but not b) 1°.

Example 16.8. Let $E = c$, $x_n =$ the n-th unit vector, and $f_n =$ the n-th coordinate functional $(n = 1, 2, \ldots)$. Then (x_n, f_n) is a total biorthogonal system, $E \supset c_0$ isometrically, $[x_n] = c_0 \neq E$, and for any $x \in c \setminus c_0$ we have a) 3°, but not a) 1°. Hence we have b) 3°, but not b) 1°.

Problem 16.1[1]. Let E be separable, $(x_n, f_n)(\{x_n\} \subset E, \{f_n\} \subset E^*)$ a total biorthogonal system and $x \in E$. a) If $M(x,(x_n, f_n))$ contains all sequences $\{\varepsilon_n\}$ with $\varepsilon_n = \pm 1$ $(n = 1, 2, \ldots)$, is the series $\sum_{i=1}^{\infty} f_i(x) x_i$ uncon-

[1] Recently problem 16.1 b) has been solved in the affirmative (see the Notes and remarks).

ditionally convergent? b) If $M(E,(x_n,f_n))$ contains all sequences $\{\varepsilon_n\}$ with $\varepsilon_n=\pm 1$ $(n=1,2,\ldots)$, is $\{x_n\}$ an unconditional basis of E? What happens if $[x_n]=E$?

We shall now show that in the particular case when the linear subspace V_0 of E^* spanned by $\{f_n\}$ is of characteristic[1] $r(V_0)>0$, the answer to problem 16.1a), whence also the answer to problem 16.1b), is affirmative (theorem 16.6 below).

Lemma 16.3. *Let E be a separable normed linear space and V a linear subspace of E^* with $r(V)>0$. Then there exists a norm $|||x|||$ on E, equivalent to the initial norm $\|x\|$ on E and having the following two properties:*

(K$_1$) *If*[2] $x_n \xrightarrow{(V)} x_0$, *then* $\underline{\lim}_{n\to\infty} |||x_n||| \geqslant |||x_0|||$.

(K$_2$) *If* $x_n \xrightarrow{(V)} x_0$ *and* $\lim_{n\to\infty} |||x_n||| = |||x_0|||$, *then* $\lim_{n\to\infty} |||x_n - x_0||| = 0$.

Proof. Since E is separable, the unit cell $S_{E^*}=\{f\in E^* \mid \|f\|\leqslant 1\}$ is compact and metrizable[3] for the topology $\sigma(E^*,E)$. Let ρ be a metric on S_{E^*}, inducing a topology equivalent to $\sigma(E^*,E)$. Put

$$\omega_0(x) = \|x\|_V = \sup_{f\in V\cap S_{E^*}} |f(x)| \quad (x\in E), \tag{16.53}$$

$$\omega_k(x) = \sup_{\substack{f,g\in V\cap S_{E^*}\\ \rho(f,g)\leqslant \frac{1}{k}}} |f(x)-g(x)| \quad (x\in E,\ k=1,2,\ldots), \tag{16.54}$$

$$|||x||| = \sum_{k=0}^{\infty} \frac{1}{2^k}\,\omega_k(x) \quad (x\in E). \tag{16.55}$$

We shall prove that $|||x|||$ has the required properties. Indeed, $|||x|||$ is a norm on E, equivalent to the initial norm $\|x\|$ on E, since by Ch. I, § 12, formula (12.13) and by (16.53), (16.54), (16.55) we have

$$r(V)\|x\| \leqslant \|x\|_V \leqslant |||x||| \leqslant 3\|x\|_V \leqslant 3\|x\| \quad (x\in E). \tag{16.56}$$

Furthermore, let $x_n \xrightarrow{(V)} x_0$ and let $\varepsilon>0$ and $k\geqslant 0$ (integer) be arbitrary. Then there exist functionals $f', g'\in V\cap S_{E^*}$ with $\rho(f',g')\leqslant \dfrac{1}{k}$

---

[1] See Ch. I, § 12.

[2] We recall (see Ch. I, § 13, the proof of proposition 13.1) that we write $x_n \xrightarrow{(V)} x$ or $(V)-\lim x_n = x$ if $f(x_n)\to f(x)$ for all $f\in V$.

[3] See e.g. [50], p. 424, theorem 2 and p. 426, theorem 1.

and $|f'(x_0)-g'(x_0)| \geqslant \omega_k(x_0) - \frac{\varepsilon}{3}$ (respectively, $|f'(x_0)| \geqslant \omega_0(x_0) - \frac{\varepsilon}{3}$ if $k=0$) and a positive integer $N = N(\varepsilon, k)$ such that $|f'(x_0) - f'(x_n)| < \frac{\varepsilon}{3}$, $|g'(x_0) - g'(x_n)| < \frac{\varepsilon}{3}$ for all $n > N$. Hence

$$\omega_k(x_0) \leqslant |f'(x_0) - g'(x_0)| + \frac{\varepsilon}{3} \leqslant |f'(x_0) - f'(x_n)| + |f'(x_n) - g'(x_n)| + |g'(x_n) - g'(x_0)| < |f'(x_n) - g'(x_n)| + \varepsilon \qquad (n > N)$$

(respectively, $\omega_0(x_0) \leqslant |f'(x_0)| + \frac{\varepsilon}{3} \leqslant |f'(x_0) - f'(x_n)| + |f'(x_n)| + \frac{\varepsilon}{3} < |f'(x_n)| + \varepsilon$ for $n > N$), and therefore

$$\omega_k(x_n) > \omega_k(x_0) - \varepsilon \qquad (n > N(\varepsilon, k);\ k = 0, 1, 2, \ldots).$$

Consequently, we have

$$\varliminf_{n \to \infty} \omega_k(x_n) \geqslant \omega_k(x_0) \qquad (k = 0, 1, 2, \ldots) \tag{16.57}$$

and hence it follows that $|||x|||$ satisfies (K_1). Indeed, by the remark made below on the functions (16.59), for every $\varepsilon > 0$ there exists a positive integer $p = p(\varepsilon)$ such that $\omega_k(x_0) < \frac{\varepsilon}{2}$ for all $k > p$. By (16.57), let $N = N(\varepsilon, p(\varepsilon))$ be such that

$$\omega_k(x_n) > \omega_k(x_0) - \frac{\varepsilon}{2} \qquad (n > N;\ k = 0, 1, \ldots, p).$$

Then

$$|||x_n||| = \sum_{k=0}^{\infty} \frac{1}{2^k} \omega_k(x_n) \geqslant \sum_{k=0}^{p} \frac{1}{2^k} \omega_k(x_n) > \sum_{k=0}^{p} \frac{1}{2^k} \left[\omega_k(x_0) - \frac{\varepsilon}{2} \right]$$

$$> \sum_{k=0}^{\infty} \frac{1}{2^k} \omega_k(x_0) - \sum_{k=p+1}^{\infty} \frac{1}{2^k} \frac{\varepsilon}{2} - \sum_{k=0}^{p} \frac{\varepsilon}{2^{k+1}} = |||x_0||| - \varepsilon \qquad (n > N),$$

and consequently we have (K_1).

Finally, let $x_n \xrightarrow{(V)} x_0$ and $\lim_{n \to \infty} |||x_n||| = |||x_0|||$. Then

$$\lim_{n \to \infty} \sum_{k=0}^{\infty} \frac{1}{2^k} [\omega_k(x_n) - \omega_k(x_0)] = \lim_{n \to \infty} (|||x_n||| - |||x_0|||) = 0,$$

whence, by (16.57), it follows that

$$\lim_{n \to \infty} \omega_k(x_n) = \omega_k(x_0) \qquad (k = 0, 1, 2, \ldots) \tag{16.58}$$

(indeed, if (16.58) is not satisfied, then by (16.57) there exist an index k_0, an infinite subsequence $\{x_{n_p}\}$ of $\{x_n\}$ and a $\delta > 0$ such that

$\omega_{k_0}(x_{n_p}) - \omega_{k_0}(x_0) \geqslant \delta$ $(p=1,2,\ldots)$, whence $\sum_{k=0}^{\infty} \frac{1}{2^k} \omega_k(x_{n_p}) \geqslant \sum_{k=0}^{\infty} \frac{1}{2^k} \omega_k(x_0)$ $+2\delta$ $(p=1,2,\ldots)$, contradicting $\lim_{n\to\infty} \sum_{k=0}^{\infty} \frac{1}{2^k} [\omega_k(x_n) - \omega_k(x_0)] = 0$).

Since for each $x \in E$ the function $f \to f(x)$ $(f \in S_{E^*})$ is continuous on S_{E^*} for the topology $\sigma(E^*, E)$, whence also uniformly continuous for this topology, the function

$$\phi_x(f) = f(x) \quad (f \in V \cap S_{E^*}) \tag{16.59}$$

is uniformly continuous on $V \cap S_{E^*}$ for the topology induced by $\sigma(E^*, E)$. Hence, in particular. for anv $\varepsilon > 0$ there exists an integer $k_0 = k_0(\varepsilon) > 0$ such that $\omega_{k_0}(x_0) < \frac{\varepsilon}{2}$. By (16.58) there exists an integer $N = N(\varepsilon, k_0) > 0$ such that $|\omega_{k_0}(x_n) - \omega_{k_0}(x_0)| < \frac{\varepsilon}{2}$ for all $n > N$, whence

$$\omega_{k_0}(x_n) < \omega_{k_0}(x_0) + \frac{\varepsilon}{2} < \varepsilon \quad (n > N)$$

and thus, by (16.54), the sequence of functions $\{\phi_{x_n}\}$ is equicontinuous on $V \cap S_{E^*}$. Furthermore, by (16.56) and $\lim_{n\to\infty} |||x_n||| = |||x_0|||$ we have

$$\sup_{1 \leqslant n < \infty} \sup_{f \in V \cap S_{E^*}} |\phi_{x_n}(f)| = \sup_{1 \leqslant n < \infty} \|x_n\|_V \leqslant \sup_{1 \leqslant n < \infty} |||x_n||| < \infty,$$

and thus the sequence $\{\phi_{x_n}\}$ is uniformly bounded on $V \cap S_{E^*}$. Since by $x_n \xrightarrow{(V)} x_0$ we have $\phi_{x_n}(f) \to \phi_{x_0}(f)$ $(f \in V \cap S_{E^*})$, from the theorem of Arzelà it follows that $\|x_n - x_0\|_V = \sup_{f \in V \cap S_{E^*}} |\phi_{x_n}(f) - \phi_{x_0}(f)| \to 0$ as $n \to \infty$, whence, by (16.56), $\lim_{n\to\infty} |||x_n - x_0||| = 0$. Thus, $|||x|||$ satisfies (K_2), which completes the proof of lemma 16.3.

Lemma 16.4. *Let E be a normed linear space and let $\{f_n\}$ be a sequence in E^* such that the linear subspace V of E^* spanned by $\{f_n\}$ satisfies condition* (K_1) *(with the initial norm on E). Then for every finite dimensional linear subspace G of E and every $\varepsilon > 0$ there exists a positive integer $N = N(G, \varepsilon)$ such that*

$$\|y + x\| \geqslant (1 - \varepsilon) \|y\| \quad (y \in G,\, x \in [f_1, \ldots, f_N]_\perp). \tag{16.60}$$

Proof. Let G be an arbitrary finite dimensional linear subspace of E. We claim that for every $y \in G$ and $\varepsilon > 0$ there exists a positive integer $N = N\left(y, \frac{\varepsilon}{2}\right)$ such that

$$\|y + x\| \geqslant \left(1 - \frac{\varepsilon}{2}\right) \|y\| \quad (x \in [f_1, \ldots, f_N]_\perp). \tag{16.61}$$

Indeed, assume the contrary, i.e., that there exist an element $y_0 \in G$ and an $\varepsilon_0 > 0$ for which there is no such N. Then for each $N = 1, 2, \ldots$ there exists an element $x_N \in E$ such that

$$f_n(x_N) = 0 \qquad (n = 1, \ldots, N), \tag{16.62}$$

$$\|y_0 + x_N\| < (1 - \tfrac{1}{2}\varepsilon_0)\, \|y_0\|. \tag{16.63}$$

However, by (16.62) and since V is the linear subspace spanned by $\{f_n\}$, we have $x_N \xrightarrow{(V)} 0$, whence $y_0 + x_N \xrightarrow{(V)} y_0$ and hence, by (K_1), $\varliminf_{N \to \infty} \|y_0 + x_N\| \geqslant \|y_0\|$, in contradiction with (16.63). This proves the claim (16.61).

Since $\dim G < \infty$, the unit sphere $\sigma_G = \{y \in G \mid \|y\| = 1\}$ is compact and hence it has a finite ε-net, say $y_1, \ldots, y_m$. Put

$$N(G, \varepsilon) = \max_{1 \leqslant i \leqslant m} N\left(y_i, \frac{\varepsilon}{2}\right) \tag{16.64}$$

Let $y \in \sigma_G$ be arbitrary and let y_i be such that $\|y - y_i\| < \dfrac{\varepsilon}{2}$. Then for any $x \in E$ with $f_n(x) = 0$ $(n = 1, \ldots, N(G, \varepsilon))$ we have, by (16.61) applied to y_i,

$$\|y + x\| \geqslant \|y_i + x\| - \|y_i - y\| > \left(1 - \frac{\varepsilon}{2}\right) \|y_i\| - \frac{\varepsilon}{2} = 1 - \varepsilon.$$

Consequently, for any $y \in G$ with $y \neq 0$ and $x \in E$ with $f_n(x) = 0$ $(n = 1, \ldots, N(G, \varepsilon))$ we have

$$\left\| \frac{y}{\|y\|} + \frac{x}{\|y\|} \right\| \geqslant 1 - \varepsilon,$$

whence we get (16.60) for $y \neq 0$. Since (16.60) is obviously valid for $y = 0$, the proof of lemma 16.4 is complete.

Proposition 16,3. *Let E be a separable Banach space and (x_n, f_n) ($\{x_n\} \subset E$, $\{f_n\} \subset E^*$) a biorthogonal system such that for the linear subspace V_0 of E^* spanned by $\{f_n\}$ we have $r(V_0) > 0$. Then there exist two sequences of integers $\{m_n\}$, $\{k_n\}$ with*

$$0 = m_0 < k_1 < m_1 < k_2 < m_2 < \cdots \tag{16.65}$$

such that for every $x \in E$ satisfying $f_i(x) = 0$ for $i = k_n + 1, \ldots, m_n$, $n = 1, 2, \ldots$ (or, respectively, $f_i(x) = 0$ for $i = m_{n-1} + 1, \ldots, k_n$; $n = 1, 2, \ldots$), we have

$$x = \sum_{n=1}^{\infty} \left(\sum_{i=m_{n-1}+1}^{k_n} f_i(x) x_i \right) \tag{16.66}$$

$$\left(\textit{respectively}, \; x = \sum_{n=1}^{\infty} \left(\sum_{i=k_n+1}^{m_n} f_i(x) x_i \right)\right).$$

Proof. By lemma 16.3 there exists an equivalent norm $|||x|||$ on E, with properties (K_1), (K_2). Let $1 > \varepsilon_0 > \varepsilon_1 > \cdots$, $\lim_{n\to\infty} \varepsilon_n = 0$ and let $P_{(n)} = [x_1, \ldots, x_n]$ $(n = 1, 2, \ldots)$. Put

$$k_n = N(P_{(m_{n-1}+1)}, \varepsilon_{n-1}), \quad m_n = N(P_{(k_n+1)}, \varepsilon_n) \quad (n = 1, 2, \ldots), \tag{16.67}$$

where the integers $N(G, \varepsilon)$ are chosen according to lemma 16.4. Since for $\varepsilon < 1$ and each $n = 1, 2, \ldots$ we have $N(P_{(n)}, \varepsilon) > n - 1$ (because for $y = x = x_n$ we have $f_1(x) = \cdots = f_{n-1}(x) = 0$, but $0 = \|y - x\| < (1 - \varepsilon)\|y\|$), it follows that $\{m_n\}$, $\{k_n\}$ satisfy (16.65).

Now let $x \in E$ be an arbitrary element satisfying

$$f_i(x) = 0 \quad (i = k_n + 1, \ldots, m_n; \; n = 1, 2, \ldots). \tag{16.68}$$

Put

$$y_r = \sum_{n=1}^{r} \left(\sum_{i=m_{n-1}+1}^{k_n} f_i(x) x_i \right) \quad (r = 1, 2, \ldots). \tag{16.69}$$

Then, since by (16.69) and (16.68) $f_j(y_r) = f_j(x)$ $(j = 1, \ldots, k_r; \; r = 1, 2, \ldots)$, we have $y_r \xrightarrow{(V_0)} x$, whence, by (K_1),

$$\underline{\lim}_{r\to\infty} |||y_r||| \geqslant |||x|||. \tag{16.70}$$

On the other hand, since by (16.69) and (16.68) $y_r \in P_{(k_r)} \subset P_{(k_r+1)}$ and $f_j(x - y_r) = 0$ $(j = 1, \ldots, m_r)$ and since $m_r = N(P_{(k_r+1)}, \varepsilon_r)$, from lemma 16.4 it follows that

$$|||x||| = |||y_r + (x - y_r)||| \geqslant (1 - \varepsilon_r)|||y_r|||,$$

whence

$$|||x||| \geqslant \overline{\lim}_{r\to\infty} (1 - \varepsilon_r)|||y_r||| = \overline{\lim}_{r\to\infty} |||y_r|||,$$

which, together with (16.70), gives $\lim_{r\to\infty} |||y_r||| = |||x|||$. Since $y_r \xrightarrow{(V_0)} x$, from (K_2) we obtain

$$\lim_{r\to\infty} y_r = x$$

in the norm $|||x|||$, whence also in the initial norm $\|x\|$ on E.

The argument for the case when $f_i(x) = 0$ $(i = m_{n-1} + 1, \ldots, k_n; n = 1, 2, \ldots)$ is similar, which completes the proof of proposition 16.3.

Now we are ready to prove

Theorem 16.6. *Let E be a separable Banach space and (x_n, f_n) $(\{x_n\}\subset E$, $\{f_n\}\subset E^*)$ a biorthogonal system such that the linear subspace V_0 of E^* spanned by $\{f_n\}$ is of characteristic $r(V_0)>0$. Then*

a) *For an element $x\in E$ the set $M(x,(x_n, f_n))$ contains all sequences $\{\varepsilon_n\}$ with $\varepsilon_n=\pm 1$ $(n=1,2,\ldots)$ if and only if the series $\sum_{i=1}^{\infty} f_i(x)x_i$ is unconditionally convergent.*

b) *$\{x_n\}$ is an unconditional basis of E if and only if $M(E,(x_n, f_n))$ contains all sequences $\{\varepsilon_n\}$ with $\varepsilon_n=\pm 1$ $(n=1,2,\ldots)$.*

Proof. a) If $\sum_{i=1}^{\infty} f_i(x)x_i$ is unconditionally convergent, then, by theorem 16.5, $M(x,(x_n, f_n))\supset m\supset\{\{\varepsilon_n\}\mid \varepsilon_n=\pm 1\ (n=1,2,\ldots)\}$.

Conversely, assume that for an element $x\in E$ the set $M(x,(x_n, f_n))$ contains all sequences $\{\varepsilon_n\}$ with $\varepsilon_n=\pm 1$ $(n=1,2,\ldots)$. Let $\{m_n\}$, $\{k_n\}$ be as in proposition 16.3 and let $\{\varepsilon_n\}$ be an arbitrary sequence with $\varepsilon_n=\pm 1$ $(n=1,2,\ldots)$. Put

$$\varepsilon_i^{(1)}=\begin{cases}\varepsilon_i & \text{for } i=m_{n-1}+1,\ldots,k_n;\ n=1,2,\ldots\\ 1 & \text{for } i=k_n+1,\ldots,m_n;\ n=1,2,\ldots\end{cases}\tag{16.71}$$

$$\varepsilon_i^{(2)}=\begin{cases}\varepsilon_i & \text{for } i=m_{n-1}+1,\ldots,k_n;\ n=1,2,\ldots\\ -1 & \text{for } i=k_n+1,\ldots,m_n;\ n=1,2,\ldots.\end{cases}\tag{16.72}$$

Then, since by our assumption $\{\varepsilon_n\}$, $\{\varepsilon_n^{(1)}\}$, $\{\varepsilon_n^{(2)}\}\in M(E,(x_n, f_n))$, there exist elements $x_{\{\varepsilon_n\}}, x_{\{\varepsilon_n^{(1)}\}}, x_{\{\varepsilon_n^{(2)}\}}\in E$ such that

$$f_i(x_{\{\varepsilon_n\}})=\varepsilon_i f_i(x),\quad f_i(x_{\{\varepsilon_n^{(j)}\}})=\varepsilon_i^{(j)} f_i(x)\qquad (i=1,2,\ldots;j=1,2).\tag{16.73}$$

Put

$$x'=\tfrac{1}{2}(x_{\{\varepsilon_n^{(1)}\}}+x_{\{\varepsilon_n^{(2)}\}}).\tag{16.74}$$

Then, by (16.73), (16.71) and (16.72),

$$f_i(x')=\begin{cases}\varepsilon_i f_i(x) & \text{for } i=m_{n-1}+1,\ldots,k_n;\ n=1,2,\ldots\\ 0 & \text{for } i=k_n+1,\ldots,m_n;\ n=1,2,\ldots\end{cases}$$

whence for $x''=x_{\{\varepsilon_n\}}-x'$ we have

$$f_i(x'')=\begin{cases}0 & \text{for } i=m_{n-1}+1,\ldots,k_n;\ n=1,2,\ldots\\ \varepsilon_i f_i(x) & \text{for } i=k_n+1,\ldots,m_n;\ n=1,2,\ldots\end{cases}$$

Consequently, by proposition 16.3,

$$x'=\sum_{n=1}^{\infty}\Big(\sum_{i=m_{n-1}+1}^{k_n}\varepsilon_i f_i(x)x_i\Big),\quad x''=\sum_{n=1}^{\infty}\Big(\sum_{i=k_n+1}^{m_n}\varepsilon_i f_i(x)x_i\Big),$$

and hence, putting $l_{2n-1}=k_n$, $l_{2n}=m_n$ $(n=1,2,\ldots)$, we obtain

$$x_{\{\varepsilon_n\}}=x'+x''=\lim_{n\to\infty}\sum_{i=1}^{l_n}\varepsilon_i f_i(x)x_i \quad (\varepsilon_j=\pm 1; j=1,2,\ldots). \quad (16.75)$$

Assume now that there exists a sequence $\{\varepsilon_n\}$ with $\varepsilon_n=\pm 1$ $(n=1,2,\ldots)$ such that the series $\sum_{i=1}^{\infty}\varepsilon_i f_i(x)x_i$ does not converge. Then there exist a $\delta>0$ and positive integers $p_1<p_2<\cdots$ such that

$$\left\|\sum_{i=p_n+1}^{p_{n+1}}\varepsilon_i f_i(x)x_i\right\|\geqslant\delta \quad (n=1,2,\ldots). \quad (16.76)$$

Let $\{l'_n\}$ be a subsequence of $\{l_n\}$ with the property that for each n there exists a $p'_n\in\{p_n\}$ such that $l'_n+1\leqslant p'_n+1\leqslant p'_{n+1}\leqslant l'_{n+1}$ and let $\varepsilon'_n=\pm 1$ $(n=1,2,\ldots)$ be such that $\left\|\sum_{i=l'_n+1}^{l'_{n+1}}\varepsilon'_i f_i(x)x_i\right\|=\max_{\varepsilon_i=\pm 1}\left\|\sum_{i=l'_n+1}^{l'_{n+1}}\varepsilon_i f_i(x)x_i\right\|$ $(n=1,2,\ldots)$. Then, by (16.76),

$$\left\|\sum_{i=l'_n+1}^{l'_{n+1}}\varepsilon'_i f_i(x)x_i\right\|\geqslant\frac{1}{2}\left(\left\|\sum_{i=l'_n+1}^{l'_{n+1}}\varepsilon_i f_i(x)x_i\right\|\right.$$

$$\left.+\left\|\sum_{i=p'_n+1}^{p'_{n+1}}\varepsilon_i f_i(x)x_i-\sum_{i=l'_n+1}^{p'_n}\varepsilon_i f_i(x)x_i-\sum_{i=p'_{n+1}+1}^{l'_{n+1}}\varepsilon_i f_i(x)x_i\right\|\right)$$

$$\geqslant\left\|\sum_{i=p'_n+1}^{p'_{n+1}}\varepsilon_i f_i(x)x_i\right\|\geqslant\delta \quad (n=1,2,\ldots),$$

which contradicts (16.75). This proves that all series $\sum_{i=1}^{\infty}\varepsilon_i f_i(x)x_i$ with $\varepsilon_n=\pm 1$ $(n=1,2,\ldots)$ converge, whence, by lemma 16.1 (implication $4^\circ\Rightarrow 1^\circ$), $\sum_{i=1}^{\infty} f_i(x)x_i$ converges unconditionally.

b) is an immediate consequence of a), taking into account that $\{f_n\}$ is total on E (by $r(V_0)>0$), which completes the proof of theorem 16.6.

The following theorem on unconditional bases corresponds to Ch. I, § 5, theorem 5.3 (on general bases):

Theorem 16.7. *Let E be a Banach space and (x_n, f_n) $(\{x_n\}\subset E, \{f_n\}\subset E^*)$ an E-complete biorthogonal system. The following statements are equivalent:*

1°. *$\{x_n\}$ is an unconditional basis of E.*

2°. *There exists a continuous linear mapping* $\{\gamma_n\}\to v_{\{\gamma_n\}}$ *of* m *into* $L(E,E)$, *such that*

$$v_{e_n}(x)=f_n(x)x_n \qquad (x\in E, n=1,2,\ldots), \tag{16.77}$$

where e_n *is the* n*-th unit vector* $\{\delta_{ni}\}_{i=1}^{\infty}$ *in* m.

3°. *There exists a continuous linear mapping* $\{\gamma_n\}\to v_{\{\gamma_n\}}$ *of* c_0 *into* $L(E,E)$, *satisfying* (16.77).

Moreover, if $\{x_n\}$ *is an unconditional basis of* E, *then the mapping* $\{\gamma_n\}\to v_{\{\gamma_n\}}$ *is* a) *an isomorphism of* m *onto a complemented subspace of* $L(E,E)$ *and* b) *an isomorphism of* c_0 *onto the subspace* $[v_{e_n}]$ *of*[1] $\mathscr{C}(E,E)$ *and the subspace* $[v_{e_n}]$ *is complemented in* $\mathscr{C}(E,E)$.

Proof. Assume that $\{x_n\}$ is an unconditional basis of E. Then, by theorem 16.5 (implication b) 1°⇒b) 2°), we have $M(E,(x_n,f_n))=m$ and the identity mapping $m\to M(E,(x_n,f_n))$ is an isomorphism. Since by Ch. I, § 5, proposition 5.4b), the mapping $\{\gamma_n\}\to v_{\{\gamma_n\}}$, where $v_{\{\gamma_n\}}(x)=x_{\{\gamma_n\}}$ $(x\in E)$, is a linear isometry of $M(E,(x_n,f_n))$ into $L(E,E)$, it follows that this mapping is an isomorphism of m into $L(E,E)$ (and hence[2] onto a complemented subspace of $L(E,E)$), satisfying (16.77).

The implication 2°⇒3° is obvious.

Assume now that we have 3°. Then there exists a constant $C\geqslant 1$ such that we have, for any finite sequence of scalars $\gamma_1,\ldots,\gamma_n$

$$\sup_{\substack{x\in E\\ \|x\|\leqslant 1}}\left\|\sum_{i=1}^{n}\gamma_i f_i(x)x_i\right\|=\left\|v_{\sum\limits_{i=1}^{n}\gamma_i e_i}\right\|\leqslant C\sup_{1\leqslant i\leqslant n}|\gamma_i|,$$

whence[3], by theorem 16.1, $\{x_n\}$ is an unconditional basis of E.

Furthermore, assume again that $\{x_n\}$ is an unconditional basis of E. Then, by the above, the mapping $\{\gamma_n\}\to v_{\{\gamma_n\}}$ is an isomorphism of c_0 onto the subspace $[v_{e_n}]$ of $L(E,E)$. Let C be the norm of this isomorphism and let $\{\gamma_n\}\in c_0$ be arbitrary. Then

$$\left\|v_{\{\gamma_n\}}-\sum_{i=1}^{l}\gamma_i v_{e_i}\right\|=\left\|v_{\{\gamma_n\}}-v_{\sum\limits_{i=1}^{l}\gamma_i e_i}\right\|=\left\|v_{\{\gamma_n\}-\sum\limits_{i=1}^{l}\gamma_i e_i}\right\|$$

$$\leqslant C\sup_{l+1\leqslant i<\infty}|\gamma_i|\to 0 \quad \text{as} \quad l\to\infty,$$

---

[1] We recall (see Ch. I, § 18) that we denote by $\mathscr{C}(E,E)$ the subspace of $L(E,E)$ consisting of all compact linear mappings of E onto E.

[2] See e.g. [43], Ch. V, § 4, remark (9)(a) or theorem 3.

[3] If E contains no subspace isomorphic to c_0 and (x_n,f_n) is assumed to be total (but not necessarily E-complete), the same conclusion holds (by § 15, lemma 15.8).

whence, since $\sum_{i=1}^{l} \gamma_i v_{e_i}$ is of finite rank, we infer that $v_{\{\gamma_n\}}$ is compact. Thus, $[v_{e_n}] \subset \mathscr{C}(E,E)$ and it remains to prove that $[v_{e_n}]$ is complemented in $\mathscr{C}(E,E)$.

We may assume, without loss of generality, that $\inf_{1 \leqslant n < \infty} \|x_n\| > 0$ (whence $\sup_{1 \leqslant n < \infty} \|f_n\| < \infty$). We claim that

$$\{f_n[u(x_n)]\} \in c_0 \qquad (u \in \mathscr{C}(E,E)). \tag{16.78}$$

Indeed, let $\{i_n\}$ be an arbitrary sequence of positive integers. Then, since u is compact, $\{u(x_{i_n})\}$ has a subsequence, say $\{u(x_{i_{n_k}})\}$, converging to an element $y \in E$. Consequently,

$$|f_{i_{n_k}}[u(x_{i_{n_k}})]| \leqslant \sup_{1 \leqslant j < \infty} \|f_j\| \, \|u(x_{i_{n_k}}) - y\| + |f_{i_{n_k}}(y)| \to 0 \quad \text{as} \quad k \to \infty,$$

whence, since $\{f_{i_n}[u(x_{i_n})]\}$ was an arbitrary subsequence of $\{f_n[u(x_n)]\}$, we infer (16.78).

Thus, since by the above $\{\gamma_n\} \to v_{\{\gamma_n\}}$ is an isomorphism of c_0 onto the subspace $[v_{e_n}]$ of $\mathscr{C}(E,E)$, the mapping

$$\pi(u) = \sum_{i=1}^{\infty} f_i[u(x_i)] \, v_{e_i} \qquad (u \in \mathscr{C}(E,E)) \tag{16.79}$$

of $\mathscr{C}(E,E)$ into $[v_{e_n}]$ is well defined and continuous (actually, we have $\|\pi(u)\| \leqslant C \sup_{1 \leqslant i < \infty} \|x_i\| \, \|f_i\| \, \|u\|$ for all $u \in \mathscr{C}(E,E)$). Since by (16.77)

$$\pi(v_{e_j}) = \sum_{i=1}^{\infty} f_i[v_{e_j}(x_i)] \, v_{e_i} = \sum_{i=1}^{\infty} f_i(x_j) \, v_{e_i} = v_{e_j} \qquad (j = 1, 2, \ldots),$$

it follows that π is a continuous linear projection of $\mathscr{C}(E,E)$ onto $[v_{e_n}]$, which completes the proof of theorem 16.7.

Finally, let us give some characterizations of unconditional bases among bases.

Theorem 16.8. *Let $\{x_n\}$ be a basis of a Banach space E. The following statements are equivalent:*

1°. *$\{x_n\}$ is an unconditional basis of the space E.*

2°. *For every increasing sequence of indices $\{i_n\}$ the subspaces $[x_{i_n}]$ and $[x_j]_{j \in \mathscr{N} \setminus \{i_n\}}$ are complementary to each other (i. e., $E = [x_{i_n}] \oplus [x_j]_{j \in \mathscr{N} \setminus \{i_n\}}$).*

3°. *For every sequence of scalars $\{\varepsilon_n\}$ with $\varepsilon_n = \pm 1$ $(n = 1, 2, \ldots)$, $\{x_n\}$ is equivalent to the basis $\{\varepsilon_n x_n\}$ of E.*

4°. *For every sequence of scalars $\{\beta_n\}$ with $0 < \alpha \leqslant |\beta_n| \leqslant \beta < \infty$ $(n = 1, 2, \ldots)$, $\{x_n\}$ is equivalent to the basis $\{\beta_n x_n\}$ of E.*

Proof. 1°⇒2°. Assume that $\{x_n\}$ is an unconditional basis of E and let $\{f_n\} \subset E^*$ be the a. s. c. f. to $\{x_n\}$ and $\{i_n\}$ an arbitrary increasing sequence of indices. Then every $x=\sum_{i=1}^{\infty} f_i(x)x_i$ can be written in the form

$$x=\sum_{j=1}^{\infty} f_{i_j}(x)x_{i_j}+\sum_{j\in\mathcal{N}\setminus\{i_n\}} f_j(x)x_j, \tag{16.80}$$

where both series in the right hand side are convergent (by lemma 16.1, implication $1^\circ\Rightarrow 3^\circ$), the first to an element $y\in[x_{i_n}]$ and the second to an element $z\in[x_j]_{j\in\mathcal{N}\setminus\{i_n\}}$ (because these subspaces are closed). Since obviously $[x_{i_n}]\cap[x_j]_{j\in\mathcal{N}\setminus\{i_n\}}=\{0\}$, it follows that[1] $E=[x_{i_n}]\oplus[x_j]_{j\in\mathcal{N}\setminus\{i_n\}}$.

$2^\circ\Rightarrow 1^\circ$. Assume that $\{x_n\}$ is a basis of E, satisfying 2° and let $\{f_n\}\subset E^*$ be the a. s. c. f. to $\{x_n\}$ and $\{i_n\}$ an arbitrary increasing sequence of indices. Then, by our hypothesis, every $x\in E$ can be written, in a unique way, in the form $x=y+z$, where $y\in[x_{i_n}]$ and $z\in[x_j]_{j\in\mathcal{N}\setminus\{i_n\}}$. Hence

$$f_{i_j}(x)=f_{i_j}(y+z)=f_{i_j}(y) \quad (x\in E, j=1,2,\ldots),$$

and therefore, by Ch. I, § 4, proposition 4.1, the series $\sum_{j=1}^{\infty} f_{i_j}(x)x_{i_j}=\sum_{j=1}^{\infty} f_{i_j}(y)x_{i_j}=y$ converges. Consequently, by lemma 16.1 (implication $3^\circ\Rightarrow 1^\circ$), $\sum_{i=1}^{\infty} f_i(x)x_i$ is unconditionally convergent for every $x\in E$, i. e., the basis $\{x_n\}$ is an unconditional basis of E.

$1^\circ\Rightarrow 4^\circ$. Assume that $\{x_n\}$ is an unconditional basis of E and let $0<\alpha\leqslant|\beta_n|\leqslant\beta<\infty$. Then $\{\beta_n x_n\}$ is a basis of E. Furthermore, if $\sum_{i=1}^{\infty}\alpha_i x_i$ converges, by lemma 16.1 (implication $1^\circ\Rightarrow 5^\circ$) the series $\sum_{i=1}^{\infty}\frac{1}{\beta}\beta_i\alpha_i x_i$, whence also the series $\sum_{i=1}^{\infty}\beta_i\alpha_i x_i$, converges, Conversely, if $\sum_{i=1}^{\infty}\beta_i\alpha_i x_i$ converges, then again by lemma 16.1 (implication $1^\circ\Rightarrow 5^\circ$) the series $\sum_{i=1}^{\infty}\alpha\alpha_i x_i=\sum_{i=1}^{\infty}\frac{\alpha}{\beta_i}\beta_i\alpha_i x_i$, whence also the series $\sum_{i=1}^{\infty}\alpha_i x_i$, converges.

The implication $4^\circ\Rightarrow 3^\circ$ is obvious.

$3^\circ\Rightarrow 1^\circ$. Assume that we have 3°. Then $\sum_{i=1}^{\infty}\alpha_i x_i$ is convergent if and only if $\sum_{i=1}^{\infty}\varepsilon_i\alpha_i x_i$ is convergent for all $\{\varepsilon_n\}$ with $\varepsilon_n=\pm 1$ $(n=1,2,\ldots)$,

---

[1] Moreover, from theorem 16.1 (implication $1^\circ\Rightarrow 19^\circ$) it follows that the projections $x\to\sum_{j=1}^{\infty} f_{i_j}(x)x_{i_j}$ $(x\in E)$, where $\{i_n\}\in\mathcal{O}$, are uniformly bounded.

which happens, by lemma 16.1 (equivalence 1°⇔4°), if and only if $\sum_{i=1}^{\infty} \alpha_i x_i$ is unconditionally convergent. Thus $\{x_n\}$ is an unconditional basis of E, which completes the proof of theorem 16.8.

Remark 16.3. The condition 2° of theorem 16.8 cannot be replaced by the weaker condition that for every increasing sequence of indices $\{i_n\}$ the subspace $[x_{i_n}]$ be complemented in E, since e. g. in the Hilbert space l^2 every subspace is complemented (admitting even a projection of norm 1), but there exist conditional bases of l^2 (see § 14, examples 14.4 and 14.5). Next, it is natural to ask whether every basis $\{x_n\}$ of any Banach space E satisfies this weakened condition (i. e., that for every $\{i_n\} \in \mathcal{O}$ the subspace $[x_{i_n}]$ be complemented in E). We shall see that the answer is negative, in Vol. II, Ch. IV, where we shall also study other relations between bases and complemented subspaces.

It is also natural to ask the "unconditional analogue" of Ch. I, § 4, problem 4.1, i. e., the question, whether in a Banach space E with an unconditional basis every unconditional basic sequence can be extended to an unconditional basis of the whole space E. The answer is negative, since e. g. in $E = l^1$ the natural basis $\{z_n\}$ of the subspace $B_2 = (l_1^2 \times l_2^2 \times \cdots)_{l^1}$ (see § 8, example 8.1) is a bounded unconditional basic sequence which is not equivalent to the unit vector basis of $E = l^1$ and hence $\{z_n\}$ cannot be extended to an unconditional basis of E (since by § 18, theorem 18.2, every bounded unconditional basis of $E = l^1$ is equivalent to the unit vector basis of l^1, which is obviously equivalent to each of its subsequences).

However, the "unconditional analogue" of Ch. I, § 4, proposition 4.2, is true, namely, we have

Proposition 16.4. *Let G, F be two Banach spaces with unconditional bases $\{y_n\}$ and $\{z_n\}$, respectively. Then the sequence $\{x_n\} \subset G \times F$ defined by*

$$x_{2n-1} = \{y_n, 0\}, \quad x_{2n} = \{0, z_n\} \qquad (n = 1, 2, \ldots), \tag{16.81}$$

whence also every permutation $\{x_{\sigma(n)}\}$ of this sequence, is an unconditional basis of $G \times F$.

Proof. Let $\varepsilon_n = \pm 1$ be arbitrary. Then, since $\{y_n\}$, $\{z_n\}$ are unconditional bases of G and F respectively, by virtue of theorem 16.8 (implication 1°⇒3°) $\{\varepsilon_{2n-1} y_n\}$ is a basis of G, equivalent to $\{y_n\}$, and $\{\varepsilon_{2n} z_n\}$ is a basis of F, equivalent to $\{z_n\}$.

Now, if $\sum_{i=1}^{\infty} \alpha_i x_i$ converges, then, as we have observed in Ch. I, § 4, the proof of proposition 4.2, $\sum_{i=1}^{\infty} \alpha_{2i-1} y_i$ and $\sum_{i=1}^{\infty} \alpha_{2i} z_i$ converge, whence, by the preceding remark, $\sum_{i=1}^{\infty} \varepsilon_{2i-1}\alpha_{2i-1}y_i$ and $\sum_{i=1}^{\infty} \varepsilon_{2i}\alpha_{2i}z_i$ converge and hence the series $\sum_{i=1}^{\infty} \varepsilon_i\alpha_i x_i$ converges. Consequently, by lemma 16.1 (implication $4^\circ \Rightarrow 1^\circ$), $\sum_{i=1}^{\infty} \alpha_i x_i$ is unconditionally convergent, and thus $\{x_n\}$ is an unconditional basis of $G \times F$. Hence, by § 17, theorem 17.1 (equivalence $1^\circ \Leftrightarrow 2^\circ$), every permutation $\{x_{\sigma(n)}\}$ of $\{x_n\}$ is also an unconditional basis of $G \times F$, which completes the proof of proposition 16.4.

In other words, proposition 16.4 says that *the cartesian product of two unconditional bases is an unconditional basis*. For tensor products of bases the situation is different, as shown by the following example (for more complete results on tensor products, see § 17):

Example 16.9. The tensor square[1] $\{x_i \otimes x_j)$ of the unit vector basis of l^2 is not an unconditional basis of $l^2 \otimes_\gamma l^2$, nor of $l^2 \otimes_\lambda l^2$.

Indeed, to prove the first assertion we shall exhibit a functional $h \in (l^2 \otimes_\gamma l^2)^*$ and an element $\sum_{i,j=1}^{\infty} \alpha_{ij} x_i \otimes x_j \in l^2 \otimes_\gamma l^2$ such that $\sum_{i,j=1}^{\infty} |h(\alpha_{ij} x_i \otimes x_j)| = \infty$. Let $\{A_n\}$ be the sequence of $2^n \times 2^n$ matrices defined by

$$A_1 = \begin{pmatrix} \frac{1}{\sqrt{2}} & \frac{1}{\sqrt{2}} \\ \frac{1}{\sqrt{2}} & -\frac{1}{\sqrt{2}} \end{pmatrix},$$

$$A_2 = \begin{pmatrix} \frac{1}{2} & \frac{1}{2} & \frac{1}{2} & \frac{1}{2} \\ \frac{1}{2} & -\frac{1}{2} & \frac{1}{2} & -\frac{1}{2} \\ \frac{1}{2} & \frac{1}{2} & -\frac{1}{2} & -\frac{1}{2} \\ \frac{1}{2} & -\frac{1}{2} & -\frac{1}{2} & \frac{1}{2} \end{pmatrix}, \ldots, A_n = \frac{1}{\sqrt{2}} \begin{pmatrix} A_{n-1} & A_{n-1} \\ A_{n-1} & -A_{n-1} \end{pmatrix}, \ldots \qquad (16.82)$$

let $u: l^2 \to l^2$ be the continuous linear mapping defined by the cartesian product of these matrices (on $l^2 \equiv (l_2^2 \times l_{2^2}^2 \times \cdots \times l_{2^n}^2 \times \cdots)_{l^2}$), i. e.,

[1] See Ch. I, § 18, the remark made after the proof of theorem 18.1.

$$u(x)=\left\{\frac{\xi_1+\xi_2}{\sqrt{2}},\frac{\xi_1-\xi_2}{\sqrt{2}},\frac{\xi_3+\xi_4+\xi_5+\xi_6}{2},\frac{\xi_3-\xi_4+\xi_5-\xi_6}{2},\right.$$

$$\left.\frac{\xi_3+\xi_4-\xi_5-\xi_6}{2},\frac{\xi_3-\xi_4-\xi_5+\xi_6}{2},\dots\right\} \quad (x=\{\xi_n\}\in l^2), \tag{16.83}$$

and let $h\in(l^2\otimes_\gamma l^2)^*$ be the image of u by the canonical linear isometry $L(l^2,l^2)\equiv(l^2\otimes_\gamma l^2)^*$, i.e.,

$$h(x\otimes y)=(x,u(y)) \quad (x\in l^2, y\in l^2). \tag{16.84}$$

Furthermore, let

$$y_n=\{\underbrace{0,\dots,0}_{2^n-2},\underbrace{\frac{1}{\sqrt{2^n}},\frac{1}{\sqrt{2^n}},\dots,\frac{1}{\sqrt{2^n}}}_{2^n},0,0,\dots\} \quad (n=1,2,\dots), \tag{16.85}$$

and let

$$z=\sum_{n=1}^{\infty}\frac{1}{n}y_n=\{\zeta_n\}=\left\{\frac{1}{\sqrt{2}},\frac{1}{\sqrt{2}},\frac{1}{4},\frac{1}{4},\frac{1}{4},\frac{1}{4},\frac{1}{3\sqrt{8}},\dots\right\}. \tag{16.86}$$

Then, since the supports of y_i, y_j are disjoint for $i\neq j$, we have $(y_i,y_j)=0$ $(i\neq j)$, whence $\|z\|^2=\sum_{n=1}^{\infty}\frac{1}{n^2}\|y_n\|^2=\sum_{n=1}^{\infty}\frac{1}{n^2}<\infty$, i.e., $z\in l^2$, whence $z\otimes z=\sum_{i=1}^{\infty}\zeta_i x_i\otimes\sum_{j=1}^{\infty}\zeta_j x_j=\sum_{i,j=1}^{\infty}\zeta_i\zeta_j x_i\otimes x_j\in l^2\otimes_\gamma l^2$. On the other hand, since $\zeta_i=\zeta_j=\frac{1}{n\sqrt{2^n}}$, $|(x_i,u(x_j))|=\frac{1}{\sqrt{2^n}}$ $(i,j=2^n-1,2^n,\dots 2^{n+1}-2;\ n=1,2,\dots)$, we obtain

$$\sum_{i,j=1}^{\infty}|h(\zeta_i\zeta_j x_i\otimes x_j)|=\sum_{i,j=1}^{\infty}\zeta_i\zeta_j|(x_i,u(x_j))|$$

$$\geqslant\sum_{n=1}^{\infty}\sum_{i,j=2^n-1}^{2^{n+1}-2}\zeta_i\zeta_j|(x_i,u(x_j))|=\sum_{n=1}^{\infty}(2^n)^2\left(\frac{1}{n\sqrt{2^n}}\right)^2\frac{1}{\sqrt{2^n}}=\sum_{n=1}^{\infty}\frac{\sqrt{2^n}}{n^2}=\infty,$$

which completes the proof of the first assertion of example 16.9. Now, if $\{x_i\otimes x_j\}$ were an unconditional basis of $l^2\otimes_\lambda l^2$, then its associated sequence of coefficient functionals, which is again $\{x_i\otimes x_j\}$, would be an unconditional basic sequence in $(l^2\otimes_\lambda l^2)^*\equiv l^2\otimes_\gamma l^2$ (by § 17, theorem 17.7), which is impossible by the above.

§ 17. Intrinsic characterizations of unconditional bases. Some more separable Banach spaces having no unconditional basis. Properties of strong duality. Unconditional bases and sequence spaces

Let E be a Banach space, $\{x_n\}$ a sequence in E and $d=\{i_1,\ldots,i_n\}\in\mathscr{D}$ (see § 16, formula (16.3)). We shall use the notations

$$P_{(d)}=P_{(i_1,\ldots,i_n)}=[x_{i_1},x_{i_2},\ldots,x_{i_n}], \tag{17.1}$$

$$P^{(d)}=P^{(i_1,\ldots,i_n)}=[x_j]_{j\in\mathscr{N}\setminus d}, \tag{17.2}$$

$$\sigma_{(d)}=\sigma_{(i_1,\ldots,i_n)}=\{x\in P_{(d)} \mid \|x\|=1\}=\sigma_{P_{(d)}}, \tag{17.3}$$

$$\sigma^{(d)}=\sigma^{(i_1,\ldots,i_n)}=\{x\in P^{(d)} \mid \|x\|=1\}=\sigma_{P^{(d)}}, \tag{17.4}$$

$$S_d\left(\sum_{i=1}^{k}\alpha_i x_i\right)=S_{\{i_1,\ldots,i_n\}}\left(\sum_{i=1}^{k}\alpha_i x_i\right)=\begin{cases}0 & (i_1>k)\\ \sum_{j=1}^{m}\alpha_{i_j}x_{i_j} & (m<n, i_m\leqslant k, i_{m+1}>k)\\ \sum_{j=1}^{n}\alpha_{i_j}x_{i_j} & (i_n\leqslant k),\end{cases} \tag{17.5}$$

and the notation P of Ch. I, § 6, formula (6.18).

Theorem 17.1. *Let E be a Banach space and $\{x_n\}$ a complete sequence in E such that $x_n\neq 0$ $(n=1,2,\ldots)$. The following statements are equivalent:*

1°. *$\{x_n\}$ is an unconditional basis of the space E.*

2°. *Every permutation $\{x_{\sigma(n)}\}$ $(\sigma\in\Pi)$ of the sequence $\{x_n\}$ is a basis of the space E.*

3°. *There exists a countable directed set of endomorphisms $\{u_d\}=\{u_{\{i_1,\ldots,i_n\}}\}\subset L(E,E)$, where $d=\{i_1,\ldots,i_n\}$ runs over the countable directed set $\mathscr{D}$ (§ 16, formula (16.3)), satisfying*

$$u_d(x)=x \qquad (x\in P_{(d)},\ d\in\mathscr{D}), \tag{17.6}$$

$$u_d(x)=0 \qquad (x\in P^{(d)},\ d\in\mathscr{D}), \tag{17.7}$$

$$1\leqslant M_1=\sup_{d\in\mathscr{D}}\|u_d\|<\infty. \tag{17.8}$$

In this case, the set $\{u_d\}_{d\in\mathscr{D}}$ is uniquely determined and coincides with the set of partial sums $\{s_d\}_{d\in\mathscr{D}}$ defined in § 16, formula (16.6).

4°. *There exists a constant* M_2 *with* $1 \leqslant M_2 < \infty$, *such that*[1]

$$\left\| \sum_{j=1}^{n} \alpha_{i_j} x_{i_j} \right\| \leqslant M_2 \left\| \sum_{j=1}^{n} \alpha_{i_j} x_{i_j} + \sum_{j=1}^{m} \alpha_{l_j} x_{l_j} \right\| \tag{17.9}$$

for any $n, m \in \mathcal{N}$ *and any scalars* $\alpha_{i_1}, \ldots, \alpha_{i_n}, \alpha_{l_1}, \ldots, \alpha_{l_m} \in K$ *whose indices satisfy* $\{i_1, \ldots, i_n\} \cap \{l_1, \ldots, l_m\} = \emptyset$.

5°. *There exists a constant* M_3 *with* $1 \leqslant M_3 < \infty$, *such that*

$$\left\| \sum_{i=1}^{n} \varepsilon_i \alpha_i x_i \right\| \leqslant M_3 \left\| \sum_{i=1}^{n} \alpha_i x_i \right\| \tag{17.10}$$

for any $n \in \mathcal{N}$ *and any* $\varepsilon_1, \ldots, \varepsilon_n$, $\alpha_1, \ldots, \alpha_n \in K$ *with* $\varepsilon_i = \pm 1$ $(i = 1, \ldots, n)$.

6°. *There exist two constants* m_4, M_4 *with* $0 < m_4 \leqslant 1 \leqslant M_4 < \infty$, *such that*

$$m_4 \left\| \sum_{i=1}^{n} |\alpha_i| x_i \right\| \leqslant \left\| \sum_{i=1}^{n} \alpha_i x_i \right\| \leqslant M_4 \left\| \sum_{i=1}^{n} |\alpha_i| x_i \right\| \tag{17.11}$$

for any $n \in \mathcal{N}$ *and any scalars* $\alpha_1, \ldots, \alpha_n \in K$.

7°. *There exists a constant* M_5 *with* $1 \leqslant M_5 < \infty$, *such that*

$$\left\| \sum_{i=1}^{n} \beta_i \alpha_i x_i \right\| \leqslant M_5 \left\| \sum_{i=1}^{n} \alpha_i x_i \right\| \tag{17.12}$$

for any $n \in \mathcal{N}$ *and* $\beta_1, \ldots, \beta_n$, $\alpha_1, \ldots, \alpha_n \in K$ *with* $|\beta_i| \leqslant 1$ $(i = 1, \ldots, n)$.

8°. *We have*

$$M_6 = \inf_{d \in \mathcal{D}} \operatorname{dist}(\sigma_{(d)}, P^{(d)}) > 0. \tag{17.13}$$

9°. *We have*

$$\inf_{d \in \mathcal{D}} \operatorname{dist}(\sigma_{(d)}, \sigma^{(d)}) > 0. \tag{17.14}$$

10°. *There exists a constant* M_7, $1 \leqslant M_7 < \infty$, *with the following property: for every* $d \in \mathcal{D}$ *and* $p_0 \in P_{(d)}$ *there exists an* $f \in E^*$ *such that*

$$f(p_0) = \|p_0\|, \tag{17.15}$$

$$f(y) = 0 \qquad (y \in P^{(d)}), \tag{17.16}$$

$$1 \leqslant \|f\| \leqslant M_7. \tag{17.17}$$

11°. *We have*

$$M_8 = \sup_{d \in \mathcal{D}} \sup_{\substack{p \in P \\ \|p\| \leqslant 1}} \|S_d(p)\| < \infty. \tag{17.18}$$

[1] One can also write condition 4° in the following form:

$$\left\| \sum_{i \in d_1} \alpha_i x_i \right\| \leqslant M_2 \left\| \sum_{i \in d_2} \alpha_i x_i \right\|$$

for any $d_1, d_2 \in \mathcal{D}$ with $d_1 \subset d_2$ and any $\{\alpha_i\}_{i \in d_2} \subset K$.

Furthermore, for the above constants we have[1]

$$1\leqslant M_1=\inf M_2=\frac{1}{M_6}=\inf M_7=M_8\leqslant\inf M_3\leqslant\inf M_5\leqslant 2\inf M_3$$
$$\leqslant 4\inf M_2\leqslant\infty, \tag{17.19}$$

where $<\infty$ *holds if and only if* $\{x_n\}$ *is an unconditional basis of* E.

Proof. $1°\Rightarrow 2°$. If $\{x_n\}$ is an unconditional basis of E with the a.s.c.f. $\{f_n\}$ and if $\sigma\in\Pi$, then $\sum_{i=1}^{\infty} f_{\sigma(i)}(x)x_{\sigma(i)}$ is convergent, obviously to x (since $f_j\left(\sum_{i=1}^{\infty} f_{\sigma(i)}(x)x_{\sigma(i)}\right)=f_j(x)$, $j=1,2,\ldots$) and thus, by Ch. I, § 4, theorem 4.1 (implication $2°\Rightarrow 1°$), $\{x_{\sigma(n)}\}$ is a basis of E.

$2°\Rightarrow 1°$. If every permutation $\{x_{\sigma(n)}\}$ $(\sigma\in\Pi)$ of the sequence $\{x_n\}$ is a basis of E, then for every $x=\sum_{i=1}^{\infty} f_i(x)x_i\in E$ (where $\{f_n\}\subset E^*$ is the a.s.c.f. to $\{x_n\}$) we have $x=\sum_{i=1}^{\infty} f_{\sigma(i)}(x)x_{\sigma(i)}$ (since the a.s.c.f. to $\{x_{\sigma(n)}\}$ is obviously $\{f_{\sigma(n)}\}$), and therefore $\sum_{i=1}^{\infty} f_i(x)x_i$ is unconditionally convergent. Thus, $\{x_n\}$ is an unconditional basis of E.

$1°\Rightarrow 3°$. If $\{x_n\}$ is an unconditional basis of E, then, by the implication $1°\Rightarrow 19°$ of § 16, theorem 16.1, the countable directed set of partial sum operators $\{s_d\}_{d\in\mathscr{D}}$ (defined by formula (16.6)) satisfies (17.6), (17.7) and (17.8), and obviously it is the only set having these properties (since u_d is the projection of E onto $P_{(d)}$ along $P^{(d)}$).

$3°\Rightarrow 4°$. By (17.6), (17.7) and (17.8) we have

$$\left\|\sum_{j=1}^{n}\alpha_{i_j}x_{i_j}\right\|=\left\|u_{\{i_1,\ldots,i_n\}}\left(\sum_{j=1}^{n}\alpha_{i_j}x_{i_j}+\sum_{j=1}^{m}\alpha_{l_j}x_{l_j}\right)\right\|$$
$$\leqslant M_1\left\|\sum_{j=1}^{n}\alpha_{i_j}x_{i_j}+\sum_{j=1}^{m}\alpha_{l_j}x_{l_j}\right\|$$

whenever $\{i_1,\ldots,i_n\}\cap\{l_1,\ldots,l_m\}=\emptyset$.

$4°\Rightarrow 5°$. If we have $4°$, then, for any $\varepsilon_i=\pm 1$ $(i=1,\ldots,n)$,

$$\left\|\sum_{i=1}^{n}\varepsilon_i\alpha_i x_i\right\|\leqslant\left\|\sum_{i\in d_1}\alpha_i x_i\right\|+\left\|\sum_{i\in d_2}\alpha_i x_i\right\|\leqslant 2M_2\left\|\sum_{i=1}^{n}\alpha_i x_i\right\|,$$

where $d_1=\{i\in\mathscr{N}\,|\,1\leqslant i\leqslant n,\ \varepsilon_i=+1\}$, $d_2=\{i\in\mathscr{N}\,|\,1\leqslant i\leqslant n,\ \varepsilon_i=-1\}$.

---

[1] The same proof shows that in the particular case when the scalars are real, we have $\inf M_5=\inf M_3$.

$5^\circ \Rightarrow 7^\circ$. If we have 5°, then for any scalars $\alpha_1, \ldots, \alpha_n$ and $\beta_1, \ldots, \beta_n$ with $|\beta_k| \leqslant 1$ $(k=1, \ldots, n)$ we have, by the argument used in the proof of formula (16.27) of § 16,

$$\left\| \sum_{k=1}^{n} \beta_i \alpha_i x_i \right\| \leqslant 2 \sup_{\varepsilon_i = \pm 1} \left\| \sum_{i=1}^{n} \varepsilon_i \alpha_i x_i \right\| \leqslant 2 M_3 \left\| \sum_{i=1}^{n} \alpha_i x_i \right\|$$

(respectively, $\left\| \sum_{i=1}^{n} \beta_i \alpha_i x_i \right\| \leqslant M_3 \left\| \sum_{i=1}^{n} \alpha_i x_i \right\|$ if the scalars are real).

$7^\circ \Rightarrow 6^\circ$. If we have 7°, then

$$\frac{1}{M_5} \left\| \sum_{i=1}^{n} |\alpha_i| x_i \right\| = \frac{1}{M_5} \left\| \sum_{i=1}^{n} (\operatorname{sign} \alpha_i) \alpha_i x_i \right\| \leqslant \left\| \sum_{i=1}^{n} \alpha_i x_i \right\|$$

$$= \left\| \sum_{i=1}^{n} (\operatorname{sign} \bar{\alpha}_i) |\alpha_i| x_i \right\| \leqslant M_5 \left\| \sum_{i=1}^{n} |\alpha_i| x_i \right\|.$$

$6^\circ \Rightarrow 5^\circ$. If we have 6°, then for any $\varepsilon_i = \pm 1$

$$\left\| \sum_{i=1}^{n} \varepsilon_i \alpha_i x_i \right\| \leqslant M_4 \left\| \sum_{i=1}^{n} |\varepsilon_i \alpha_i| x_i \right\| = M_4 \left\| \sum_{i=1}^{n} |\alpha_i| x_i \right\| \leqslant \frac{M_4}{m_4} \left\| \sum_{i=1}^{n} \alpha_i x_i \right\|.$$

$7^\circ \Rightarrow 2^\circ$. Assume that we have 7° and let $\sigma \in \Pi$, $n, m \in \mathcal{N}$ and $\alpha_{\sigma(1)}, \ldots, \alpha_{\sigma(n+m)} \in K$ be arbitrary. Then, putting $l = \max_{1 \leqslant i \leqslant n+m} \sigma(i)$, $\alpha_j = 0$ $(j \in \{1, \ldots, l\} \setminus \{\sigma(1), \ldots, \sigma(n+m)\})$, $\beta_{\sigma(i)} = 1$ $(i = 1, \ldots, n)$ and $\beta_j = 0$ $(j \in \{1, \ldots, l\} \setminus \{\sigma(1), \ldots, \sigma(n)\})$, we obtain

$$\left\| \sum_{i=1}^{n} \alpha_{\sigma(i)} x_{\sigma(i)} \right\| = \left\| \sum_{j=1}^{l} \beta_j \alpha_j x_j \right\| \leqslant M_5 \left\| \sum_{j=1}^{l} \alpha_j x_j \right\| = M_5 \left\| \sum_{i=1}^{n+m} \alpha_{\sigma(i)} x_{\sigma(i)} \right\|,$$

and hence, by Ch. I, § 7, theorem 7.1 (implication $4^\circ \Rightarrow 1^\circ$), $\{x_{\sigma(n)}\}$ is a basis of E.

$4^\circ \Rightarrow 8^\circ$. If we have 4°, then for any $\sum_{j=1}^{n} \alpha_{i_j} x_{i_j} \in \sigma_{(i_1, \ldots, i_n)}$ and any $\sum_{j=1}^{m} \alpha_{l_j} x_{l_j} \in P^{(i_1, \ldots, i_n)}$ we have

$$\left\| \sum_{j=1}^{n} \alpha_{i_j} x_{i_j} - \sum_{j=1}^{m} \alpha_{l_j} x_{l_j} \right\| \geqslant \frac{1}{M_2} \left\| \sum_{j=1}^{n} \alpha_{i_j} x_{i_j} \right\| = \frac{1}{M_2}.$$

The equivalence $8^\circ \Leftrightarrow 9^\circ$ is an immediate consequence of Ch. I, § 6, lemma 6.1 a) applied to $F = P_{(d)}$, $G = P^{(d)}$.

$8^\circ \Rightarrow 10^\circ$. Assume that we have 8° and let $p_0 \in P_{(d)}$. If $p_0 = 0$, there exists an $f \in E^*$ satisfying (17.15), (17.16) and $\|f\| =$ arbitrary. If $p_0 \neq 0$, we have, by 8°, $\operatorname{dist}\left(\frac{p_0}{\|p_0\|}, P^{(d)}\right) \geqslant M_6 > 0$, whence there exists[1] an $f \in E^*$ satisfying (17.15), (17.16) and

$$\|f\| = \frac{1}{\operatorname{dist}\left(\frac{p_0}{\|p_0\|}, P^{(d)}\right)} \leqslant \frac{1}{M_6}.$$

$10^\circ \Rightarrow 11^\circ$. By (17.15), (17.16) and (17.17), for every $p = \sum_{i=1}^{k} \alpha_i x_i \in P$ and $\{i_1, \ldots, i_n\} \in \mathscr{D}$ we have, putting $p_0 = S_{\{i_1, \ldots, i_n\}}(p)$,

$$\|S_{\{i_1, \ldots, i_n\}}(p)\| = \|p_0\| = |f(p_0)| = |f(p)| \leqslant M_7 \|p\|.$$

$11^\circ \Rightarrow 2^\circ$. If we have 11°, and $\sigma \in \Pi$, then

$$\sup_{1 \leqslant n < \infty} \sup_{\substack{p \in P \\ \|p\| \leqslant 1}} \|S_{\{\sigma(1), \ldots, \sigma(n)\}}(p)\| \leqslant \sup_{d \in \mathscr{D}} \sup_{\substack{p \in P \\ \|p\| \leqslant 1}} \|S_d(p)\| < \infty,$$

whence, by Ch. I, § 7, theorem 7.1 (implication $10^\circ \Rightarrow 1^\circ$), $\{x_{\sigma(n)}\}$ is a basis of E. Thus, $1^\circ \Leftrightarrow \cdots \Leftrightarrow 11^\circ$.

Furthermore, from the above proofs of the implications $3^\circ \Rightarrow 4^\circ \Rightarrow 8^\circ \Rightarrow 10^\circ \Rightarrow 11^\circ$ it follows that

$$M_1 \geqslant \inf M_2 \geqslant \frac{1}{M_6} \geqslant \inf M_7 \geqslant M_8 \geqslant 1.$$

Let us prove that $M_8 = M_1$. If $\{x_n\}$ is not an unconditional basis of E, then, by the above, we have $M_8 = M_1 = \infty$. On the other hand, if $\{x_n\}$ is an unconditional basis of E, then, by the above proof of the implication $1^\circ \Rightarrow 3^\circ$, we have $M_1 = \sup_{d \in \mathscr{D}} \|s_d\| < \infty$, where $\{s_d\}$ are the partial sum operators defined by (16.6). Hence, by (17.18) and $s_d(p) = S_d(p)$ $(p \in P, d \in \mathscr{D})$, $M_8 = M_1 < \infty$.

Furthermore, the inequality $\inf M_3 \geqslant \inf M_2$ follows from

$$\left\|\sum_{j=1}^{n} \alpha_{i_j} x_{i_j}\right\| \leqslant \frac{1}{2}\left(\left\|\sum_{k=1}^{l} \varepsilon_k \alpha_k x_k\right\| + \left\|\sum_{k=1}^{l} \varepsilon'_k \alpha_k x_k\right\|\right) \leqslant M_3 \left\|\sum_{k=1}^{n} \alpha_i x_i\right\|,$$

where $l = \max_{1 \leqslant j \leqslant n} i_j$, $\varepsilon_{i_j} = \varepsilon'_{i_j} = 1$ for $j = 1, \ldots, n$ and $-\varepsilon_k = \varepsilon'_k = -1$ for $k \in \{1, \ldots, l\} \setminus \{i_1, \ldots, i_n\}$.

---

[1] See e.g. [10], p. 57, lemma.

Finally, from the above proofs of the implications $4^\circ \Rightarrow 5^\circ \Rightarrow 7^\circ$ it follows that we have the inequalities $4 \inf M_2 \geqslant 2 \inf M_3 \geqslant \inf M_5$, which completes the proof of theorem 17.1[1].

Definition 17.1. Let $\{x_n\}$ be a complete sequence in a Banach space E, such that $x_n \neq 0$ $(n=1,2,\ldots)$. Then the number $\nu^{(u)}_{\{x_n\}} = \sup\limits_{d \in \mathscr{D}} \|s_d\|$ is called *the unconditional norm of the sequence* $\{x_n\}$. The number $\gamma^{(u)}_{\{x_n\}} = \dfrac{1}{\nu^{(u)}_{\{x_n\}}}$ is called *the unconditional index of the sequence* $\{x_n\}$.

We have $1 \leqslant \nu_{\{x_n\}} \leqslant \nu^{(u)}_{\{x_n\}} \leqslant \infty$; the sequence $\{x_n\}$ is an unconditional basis of E if and only if $\nu^{(u)}_{\{x_n\}} < \infty$. Moreover, by theorem 17.1 (equivalence $1^\circ \Leftrightarrow 2^\circ$), for an unconditional basis $\{x_n\}$ we have

$$1 \leqslant \sup_{\sigma \in \Pi} \nu_{\{x_{\sigma(n)}\}} \leqslant \sup_{\sigma \in \Pi} \nu^{(u)}_{\{x_{\sigma(n)}\}} = \nu^{(u)}_{\{x_n\}}. \tag{17.20}$$

Let us give now some corollaries of theorem 17.1.

Corollary 17.1. a) *Every unconditional basis of type aP is equivalent to the unit vector basis of c_0.*

b) *Every unconditional basis of type al_+ (whence also every unconditional basis of type aP^*) is equivalent to the unit vector basis of l^1.*

Proof. By § 16, theorem 16.8 (implication $1^\circ \Rightarrow 4^\circ$), it is sufficient to prove the statements a) and b) for unconditional bases of types P and l_+, respectively.

a) Let $\{x_n\}$ be an unconditional basis of type P and let $\alpha_1, \ldots, \alpha_n$ be arbitrary scalars with $\sup\limits_{1 \leqslant j \leqslant n} |\alpha_j| \neq 0$. Put

$$\beta_i = \frac{\alpha_i}{\sup\limits_{1 \leqslant j \leqslant n} |\alpha_j|} \qquad (i=1,\ldots,n).$$

Then $|\beta_j| \leqslant 1$ $(j=1,\ldots,n)$, whence, by theorem 17.1 (implication $1^\circ \Rightarrow 7^\circ$),

---

[1] Alternatively, instead of the last part of the above proof one can also prove directly that

$$\inf M_3 = \sup_{\substack{\{\varepsilon_n\} \subset K \\ \varepsilon_i = \pm 1}} \sup_{1 \leqslant l < \infty} \|s_{\{\varepsilon_n\},l}\|, \qquad \inf M_5 = \sup_{\substack{\{\beta_n\} \subset K \\ |\beta_i| \leqslant 1}} \sup_{1 \leqslant l < \infty} \|s_{\{\beta_n\},l}\|$$

and then apply § 16, formula (16.23). Let us also mention that by the above proofs of the implications $7^\circ \Rightarrow 6^\circ \Rightarrow 5^\circ$ we have $\inf M_5 \geqslant \max\left(\inf M_4, \inf \dfrac{1}{m_4}\right)$ and $\inf \dfrac{M_4}{m_4} \geqslant \inf M_3$.

$$\left\|\sum_{i=1}^{n}\alpha_i x_i\right\| = \left\|\sum_{i=1}^{n}\beta_i \sup_{1\leqslant j\leqslant n}|\alpha_j| x_i\right\| \leqslant M_5\left\|\sum_{i=1}^{n}\sup_{1\leqslant j\leqslant n}|\alpha_j| x_i\right\|$$
$$\leqslant\left(M_5\sup_{1\leqslant l<\infty}\left\|\sum_{i=1}^{l}x_i\right\|\right)\sup_{1\leqslant j\leqslant n}|\alpha_j|. \tag{17.21}$$

On the other hand, by theorem 17.1 (implication $1^\circ\Rightarrow 4^\circ$) we have

$$\left\|\sum_{i=1}^{n}\alpha_i x_i\right\| \geqslant \frac{1}{M_2}\|\alpha_j x_j\| \geqslant \frac{1}{M_2}\inf_{1\leqslant l<\infty}\|x_l\|\,|\alpha_j| \qquad (j=1,\dots,n),$$

whence

$$\left\|\sum_{i=1}^{n}\alpha_i x_i\right\| \geqslant \left(\frac{1}{M_2}\inf_{1\leqslant l<\infty}\|x_l\|\right)\sup_{1\leqslant j\leqslant n}|\alpha_j|.$$

From this inequality and (17.21) it follows, by Ch. I, § 8, theorem 8.1d) (implication $2^\circ\Rightarrow 6^\circ$) that $\{x_n\}$ is equivalent to the unit vector basis of c_0.

b) Let $\{x_n\}$ be an unconditional basis of type l_+ and let $\alpha_1,\dots,\alpha_n$ be arbitrary scalars. Then, by theorem 17.1 (implication $1^\circ\Rightarrow 6^\circ$), we have

$$\sup_{1\leqslant l<\infty}\|x_l\|\sum_{i=1}^{n}|\alpha_i| \geqslant \left\|\sum_{i=1}^{n}\alpha_i x_i\right\| \geqslant m_4\left\|\sum_{i=1}^{n}|\alpha_i| x_i\right\| \geqslant m_4\eta\sum_{i=1}^{n}|\alpha_i|,$$

where $\eta>0$ is the constant occurring in the definition of bases of type l_+ (§ 10, definition 10.1). Consequently, by Ch. I, § 8, theorem 8.1d) (implication $2^\circ\Rightarrow 6^\circ$), $\{x_n\}$ is equivalent to the unit vector basis of l^1.

The same assertion for unconditional bases of type P^* follows from the fact that every basis of type P^* is also of type l_+ (§ 10, theorem 10.1), which completes the proof of corollary 17.1.

Corollary 17.2. *Let $\{x_n\}$ be an unconditional basis of a Banach space E, let $\{m_n\}$ be an increasing sequence of positive integers, $m_0=0$, and let*

$$y_n=\sum_{i=m_{n-1}+1}^{m_n}\alpha_i x_i,\quad y_n\neq 0 \qquad (n=1,2,\dots). \tag{17.22}$$

Then $\{y_n\}$ is an unconditional (block) basic sequence[1], *with $v_{\{y_n\}}^{(u)}\leqslant v_{\{x_n\}}^{(u)}$.*

Proof. By the implication $1^\circ\Rightarrow 4^\circ$ of theorem 17.1, we have

$$\left\|\sum_{i\in d_1}\beta_i x_i\right\| \leqslant v_{\{x_n\}}^{(u)}\left\|\sum_{i\in d_2}\beta_i x_i\right\|$$

for any $d_1,d_2\in\mathscr{D}$ with $d_1\subset d_2$ and any $\{\beta_i\}_{i\in d_2}\subset K$. Hence, by (17.22),

[1] I.e., $\{y_n\}$ is an unconditional basis of $[y_n]$.

$$\left\|\sum_{j\in d_1'}\gamma_j y_j\right\| = \left\|\sum_{j\in d_1'}\gamma_j \sum_{i=m_{j-1}+1}^{m_j}\alpha_i x_i\right\| = \left\|\sum_{j\in d_1'}\sum_{i=m_{j-1}+1}^{m_j}\gamma_j\alpha_i x_i\right\|$$

$$\leqslant v_{\{x_n\}}^{(u)}\left\|\sum_{j\in d_2'}\sum_{i=m_{j-1}+1}^{m_j}\gamma_j\alpha_i x_i\right\| = v_{\{x_n\}}^{(u)}\left\|\sum_{j\in d_2'}\gamma_j y_j\right\|$$

for any $d_1', d_2' \in \mathscr{D}$ with $d_1' \subset d_2'$ and any $\{\gamma_j\}_{j\in d_2'} \subset K$. Consequently, by the implication $4° \Rightarrow 1°$ of theorem 17.1, $\{y_n\}$ is an unconditional basic sequence, with $v_{\{y_n\}}^{(u)} \leqslant v_{\{x_n\}}^{(u)}$, which completes the proof.

The next natural question is whether we have a result on the extension of every block basic sequence with respect to an unconditional basis $\{x_n\}$ to an unconditional basis of the whole space E, similar to Ch. I, § 7, theorem 7.2, or at least whether in the case of an unconditional basis $\{x_n\}$ every block subspace with respect to $\{x_n\}$ (i. e., every subspace spanned by a block basic sequence) is complemented in E. The answer to this latter question, whence also to the first question, is negative, as shown by

Example 17.1. Let $1<p<2$, let $\{x_n\}$ be an unconditional basis of $E=L^p([0,1])$, e. g., the Haar system (§ 14, theorem 14.1), and let r be an arbitrary number such that $p<r<2$. Then there exists[1] a sequence $\{y_n\} \subset E$ equivalent to the unit vector basis of l^r. Since obviously $\{y_n\}$ converges weakly to 0 and $\inf_{1\leqslant n<\infty} \|y_n\|>0$, there exists, by § 15, proposition 15.1, a block basic sequence (with respect to $\{x_n\}$)

$$z_n = \sum_{i=m_{n-1}+1}^{m_n} \alpha_i x_i \neq 0 \quad (n=1,2,\ldots) \tag{17.23}$$

which is equivalent to some subsequence $\{y_{i_n}\}$ of $\{y_n\}$. Since by § 18, proposition 18.1a) $\{y_{i_n}\}$ is equivalent to $\{y_n\}$, it follows that $\{z_n\} \sim \{y_n\}$, whence $[z_n]$ is isomorphic to l^r. Consequently, since for $1<p<r<2$ the space $E=L^p([0,1])$ has[2] no complemented subspace isomorphic to l^r, it follows that $[z_n]$ is not complemented in E.

Another useful result on block basic sequences with respect to unconditional bases is

Corollary 17.3. a) *An unconditional basis $\{x_n\}$ is shrinking if and only if it has no block basic sequence equivalent to the unit vector basis of l^1.*

b) *An unconditional basis $\{x_n\}$ is boundedly complete if and only if it has no block basic sequence equivalent to the unit vector basis of c_0.*

---

[1] See [119] or [150], corollary 1 to theorem 7.2 and [10], p. 206, theorem 9.

[2] See [119], p. 98.

Proof. a) By § 12, theorem 12.2a), $\{x_n\}$ is shrinking if and only if it has no block basic sequence of type l_+, which, by corollaries 17.2 and 17.1b), happens if and only if $\{x_n\}$ has no block basic sequence equivalent to the unit vector basis of l^1.

The proof of part b) is similar, using § 12, theorem 12.2b) and corollaries 17.2 and 17.1a). This completes the proof of corollary 17.3.

We have the following "unconditional analogue" of Ch. I, § 8, proposition 8.3:

Corollary 17.4. *Let $\{E_n\}$ be a sequence of Banach spaces and for each n let E_n have a normalized unconditional basis $\{x_k^{(n)}\}$, such that $\sup_{1\leqslant n<\infty} v_{\{x_k^{(n)}\}}^{(u)} < \infty$. Then the sequence $\{e_k\}$ defined by*

$$e_{N^2+j} = \begin{cases} \{\underbrace{0,\ldots,0}_{N}, x_j^{(N+1)}, 0, 0, \ldots\} & \text{for } j=1,\ldots,N+1;\ N=0,1,2,\ldots \\ \{\underbrace{0,\ldots,0}_{j-N-2}, x_{N+1}^{(j-N-1)}, 0, 0, \ldots\} & \text{for } j=N+2,\ldots,2N+1;\ N=0,1,2,\ldots \end{cases} \tag{17.24}$$

is a normalized unconditional basis of the space $E=(E_1\times E_2\times\cdots)_{l^2}$, with $v_{\{e_k\}}^{(u)} \leqslant \sup_{1\leqslant n<\infty} v_{\{x_k^{(n)}\}}^{(u)}$.

Proof. By Ch. I, § 8, proposition 8.3, $\{e_k\}$ is a basis of E. The proof of the fact that $\{e_k\}$ satisfies (17.9) with $M_2 = \sup_{1\leqslant n<\infty} v_{\{x_k^{(n)}\}}^{(u)}$ is analogous to the second part of the proof of Ch. I, § 8, proposition 8.3.

As an application of theorem 17.1 we shall now give some more separable Banach spaces having no unconditional basis (among which are some tensor products of Banach spaces with unconditional bases, spaces of compact linear operators, and a reflexive Banach space). For this purpose it will be convenient to work with spaces of matrices.

Definition 17.2. Let M denote the linear space of all scalar valued infinite matrices $a=(a_{ij})_{i,j=1}^{\infty}$ such that $a_{ij}=0$ for "almost all" i,j (i. e., such that the set $\{\{i,j\} | a_{ij}\neq 0\}$ is finite). A non-negative function α on M is called a *matrix norm* if it satisfies the following conditions:

(i) $\alpha(a)=0 \Leftrightarrow a=0$; $\alpha(ta)=|t|\alpha(a)$ $(a\in M, t\in K)$; $\alpha(a+b)\leqslant\alpha(a)+\alpha(b)$ $(a,b\in M)$.

(ii) $\alpha(u_{nm})=1$ $(n,m=1,2,\ldots)$, where u_{nm} denotes the matrix in M defined by

$$(u_{nm})_{ij} = \begin{cases} 1 & \text{for } \ i=n, j=m \\ 0 & \text{for } \ \text{the other } i,j=1,2,\ldots \end{cases} \tag{17.25}$$

(iii) $\alpha(P_{n,m}(a)) \leqslant \alpha(a)$ $(a \in M, n, m = 1, 2, \ldots)$, where $P_{n,m}$ denotes the linear mapping of M into M defined by

$$P_{nm}(a) = \sum_{i=1}^{n} \sum_{j=1}^{m} a_{ij} u_{ij} = \begin{pmatrix} a_{11} \ldots a_{1m} & 0 & 0 \ldots \\ \ldots\ldots\ldots\ldots\ldots\ldots & & \\ a_{n1} \ldots a_{nm} & 0 & 0 \ldots \\ 0 \; \ldots \; 0 & 0 & 0 \ldots \\ \ldots\ldots\ldots\ldots\ldots\ldots & & \end{pmatrix}. \tag{17.26}$$

A matrix norm α is called *symmetric*, if

(iv) $\alpha(s_i t_j a_{\sigma_1(i), \sigma_2(j)}) = \alpha(a)$ $(a \in M, \sigma_1, \sigma_2 \in \Pi, |s_i| = |t_j| = 1$ for $i, j = 1, 2, \ldots)$.

For any matrix norm α we shall denote by M_α the completion of the linear space M with respect to the norm α. The norm in the Banach space M_α will be also denoted by α.

Proposition 17.1. *If α is a matrix norm, then the double sequence $\{u_{nm}\}$, arranged into a single sequence $\{z_k\} = \{u_{i(k), j(k)}\}$ by the numbering*

$i(1) = j(1) = 1$,

$$i(k) = \begin{cases} m & \text{for} \quad k = n^2 + m, \; m = 1, \ldots, n+1 \; (n = 1, 2, \ldots) \\ n+1 & \text{for} \quad k = n^2 + n + 1 + m, \; m = 1, \ldots, n \; (n = 1, 2, \ldots), \end{cases} \tag{17.27}$$

$$j(k) = \begin{cases} n+1 & \text{for} \quad k = n^2 + m, \; m = 1, \ldots, n+1 \; (n = 1, 2, \ldots) \\ n+1-m & \text{for} \quad k = n^2 + n + 1 + m, \; m = 1, \ldots, n \; (n = 1, 2, \ldots), \end{cases} \tag{17.28}$$

is a basis of M_α.

Proof. Put

$$S_n(a) = \sum_{k=1}^{n} a_{i(k), j(k)} u_{i(k), j(k)} \qquad (a \in M, n = 1, 2, \ldots). \tag{17.29}$$

Then, by (17.26), for each $n = 1, 2, \ldots$ we have

$$S_{n^2} = P_{n,n},$$
$$S_{n^2+l} = P_{n,n} + P_{l,n+1} - P_{l,n} \qquad (l = 1, \ldots, n+1),$$
$$S_{n^2+n+1+l} = P_{n+1,n+1} - P_{n+1,n-l} + P_{n,n-l} \qquad (l = 1, \ldots, n; \; P_{n,0} = P_{n+1,0} = 0),$$

whence, by condition (iii) of definition 17.2,

$$\sup_{1 \leqslant n < \infty} \sup_{\substack{a \in M \\ \alpha(a) \leqslant 1}} \alpha(S_n(a)) \leqslant 3. \tag{17.30}$$

Consequently, since M is dense in M_α, from Ch. I, § 7, theorem 7.1 (implication $10^\circ \Rightarrow 1^\circ$) it follows that $\{u_{i(k), j(k)}\}$ is a basis of M_α, which completes the proof of proposition 17.1.

We shall show below that for some important concrete matrix norms α the spaces M_α, and even their superspaces, have no unconditional basis.

Let us observe that for every matrix norm α the space M_α can be identified in a natural way with a space of matrices (by the correspondence $a \to (h_{ij}(a))_{i,j=1}^{\infty}$, where $\{h_{i(k),j(k)}\}$ is the a.s.c.f. to the basis $\{u_{i(k),j(k)}\}$ of M_α).

We return now, for a moment, to the linear space M.

Definition 17.3. For each $n=1,2,\ldots$ the linear mapping $T_n\colon M \to M$ defined by

$$T_n(a) = \sum_{i+j=2}^{n+1} a_{ij} u_{ij} = \begin{pmatrix} a_{11} & a_{12} & \ldots & a_{1,n-1} & a_{1n} & 0 & 0 & \ldots \\ a_{21} & a_{22} & \ldots & a_{2,n-1} & 0 & 0 & 0 & \ldots \\ \ldots & \ldots & \ldots & \ldots & \ldots & \ldots & \ldots & \ldots \\ a_{n1} & 0 & \ldots & 0 & 0 & 0 & 0 & \ldots \\ 0 & 0 & \ldots & 0 & 0 & 0 & 0 & \ldots \\ \ldots & \ldots & \ldots & \ldots & \ldots & \ldots & \ldots & \ldots \end{pmatrix} \quad (a \in M) \tag{17.31}$$

is called the n-th *main triangle projection*. If α is a matrix norm, we shall put

$$t_n(\alpha) = \|T_n\|_\alpha = \sup_{\substack{a \in M \\ \alpha(a) \leq 1}} \alpha(T_n(a)) \quad (n=1,2,\ldots). \tag{17.32}$$

For each $n=1,2,\ldots$ we also define a projection $D_n\colon M \to M$ by

$$D_{2m-1}(a) = \sum_{k=1}^{m} \sum_{\max(i,j)=2k-1} a_{ij} u_{ij}$$

$$= \begin{pmatrix} a_{11} & 0 & a_{13} & \ldots & 0 & a_{1,2m-1} & 0 & 0 & \ldots \\ 0 & 0 & a_{23} & \ldots & 0 & a_{2,2m-1} & 0 & 0 & \ldots \\ a_{31} & a_{32} & a_{33} & \ldots & 0 & a_{3,2m-1} & 0 & 0 & \ldots \\ \ldots & \ldots & \ldots & \ldots & \ldots & \ldots & \ldots & \ldots & \ldots \\ 0 & 0 & 0 & \ldots & 0 & a_{2m-2,2m-1} & 0 & 0 & \ldots \\ a_{2m-1,1} & a_{2m-1,2} & a_{2m-1,3} & \ldots & a_{2m-1,2m-2} & a_{2m-1,2m-1} & 0 & 0 & \ldots \\ 0 & 0 & 0 & \ldots & 0 & 0 & 0 & 0 & \ldots \\ \ldots & \ldots & \ldots & \ldots & \ldots & \ldots & \ldots & \ldots & \ldots \end{pmatrix}$$

$$(a \in M), \tag{17.33}$$

$$D_{2m}(a) = \sum_{k=1}^{m} \sum_{\max(i,j)=2k} a_{ij} u_{ij} \quad (a \in M). \tag{17.34}$$

If α is a matrix norm, we shall put

$$d_n(\alpha) = \|D_n\|_\alpha = \sup_{\substack{a \in M \\ \alpha(a) \leq 1}} \alpha(D_n(a)) \quad (n=1,2,\ldots). \tag{17.35}$$

Lemma 17.1. *If α is a symmetric matrix norm, we have*

$$t_n(\alpha)=d_n(\alpha) \qquad (n=1,2,\ldots). \tag{17.36}$$

Proof. We shall consider only the case when $n=2m-1$, since for $n=2m$ the proof is similar. Define $\sigma_n \in \Pi$ by

$$\sigma_n(i) = \begin{cases} \dfrac{n-i+2}{2} & \text{for} \quad i=1,3,5,\ldots,n-2,n \\ \dfrac{n+1+i}{2} & \text{for} \quad i=2,4,6,\ldots,n-3,n-1 \\ i & \text{for} \quad i=n+1,n+2,\ldots \end{cases} \tag{17.37}$$

and a linear mapping $U_{\sigma_n}\colon M\to M$ by

$$U_{\sigma_n}(a)=\sum_i \sum_j a_{ij} u_{\sigma_n(i),\sigma_n(j)} \qquad (a\in M). \tag{17.38}$$

Then we have

$$T_n[U_{\sigma_n}(a)]=U_{\sigma_n}[D_n(a)] \qquad (a\in M), \tag{17.39}$$

since by (17.38), (17.31) and (17.33) both sides are equal to

$$\begin{pmatrix} a_{nn} & a_{n,n-2} & a_{n,n-4} & \ldots a_{n1} & a_{n2} & a_{n4} & \ldots a_{n,n-3} & a_{n,n-1} & 0 \; 0\ldots \\ a_{n-2,n} & a_{n-2,n-2} & a_{n-2,n-4} & \ldots a_{n-2,1} & a_{n-2,2} & a_{n-2,4} & \ldots a_{n-2,n-3} & 0 & 0 \; 0\ldots \\ \cdots & \cdots & \cdots & \cdots & \cdots & \cdots & \cdots & \cdots & \cdots \\ a_{1n} & a_{1,n-2} & a_{1,n-4} & \ldots a_{11} & 0 & 0 & \ldots \; 0 & 0 & 0 \; 0\ldots \\ a_{2n} & a_{2,n-2} & a_{2,n-4} & \ldots \; 0 & 0 & 0 & \ldots \; 0 & 0 & 0 \; 0\ldots \\ \cdots & \cdots & \cdots & \cdots & \cdots & \cdots & \cdots & \cdots & \cdots \\ a_{n-3,n} & a_{n-3,n-2} & 0 & \ldots \; 0 & 0 & 0 & \ldots \; 0 & 0 & 0 \; 0\ldots \\ a_{n-1,n} & 0 & 0 & \ldots \; 0 & 0 & 0 & \ldots \; 0 & 0 & 0 \; 0\ldots \\ 0 & 0 & 0 & \ldots \; 0 & 0 & 0 & \ldots \; 0 & 0 & 0 \; 0\ldots \end{pmatrix}.$$

Hence, since U_{σ_n} is an isometry (because α is a symmetric matrix norm),

$$\begin{aligned} t_n(\alpha) &= \sup_{\substack{a\in M \\ \alpha(U_{\sigma_n}(a))\leqslant 1}} \alpha(T_n[U_{\sigma_n}(a)]) = \sup_{\substack{a\in M \\ \alpha(a)\leqslant 1}} \alpha(U_{\sigma_n}[D_n(a)]) \\ &= \sup_{\substack{a\in M \\ \alpha(a)\leqslant 1}} \alpha(D_n(a)) = d_n(\alpha), \end{aligned}$$

which completes the proof of lemma 17.1.

We return now to the spaces M_α.

Lemma 17.2. *Let E be a Banach space, let $\{f_n\}\subset E^*$ and let α be a symmetric matrix norm. If there exists an isomorphic embedding U_0: $M_\alpha\to E$, then there exists another isomorphic embedding $U: M_\alpha\to E$ satisfying*

$$\lim_{i\to\infty} f_n[U(u_{ij})]=0 \quad (j,n=1,2,\ldots), \tag{17.40}$$

$$\lim_{j\to\infty} f_n[U(u_{ij})]=0 \quad (i,n=1,2,\ldots). \tag{17.41}$$

Proof. Since for each fixed pair of indices (n,r) the sequence $\{f_n[U_0(u_{rs})]\}_{s=1}^\infty$ is bounded (because $\alpha(u_{rs})=1$), one can find, by the standard diagonal procedure, an increasing sequence of indices $\{s_j\}_{j=1}^\infty$ such that the limits $\lim\limits_{j\to\infty} f_n[U_0(u_{r,s_j})]$ $(n,r=1,2,\ldots)$ exist. Then, with a similar argument, one can find an increasing sequence of indices $\{r_i\}_{i=1}^\infty$ such that the limits $\lim\limits_{i\to\infty} f_n[U_0(u_{r_i,s_j})]$ $(n,j=1,2,\ldots)$ exist. Put

$$V(a)=\sum_i\sum_j a_{ij}(u_{r_{2i},s_{2j}}+u_{r_{2i-1},s_{2j-1}}-u_{r_{2i},s_{2j-1}}-u_{r_{2i-1},s_{2j}}) \quad (a\in M). \tag{17.42}$$

Observe now that, since α is a symmetric matrix norm, for any $b=\sum\limits_{i=1}^{n_0}\sum\limits_{j=1}^{m_0} b_{ij}u_{ij}\in M$ and any pair of increasing sequences of indices $\{p_i\}_{i=1}^k$, $\{q_j\}_{j=1}^l$ with $p_k\leqslant n_0$, $q_l\leqslant m_0$, we have

$$\begin{aligned}\alpha\left(\sum_{i=1}^k\sum_{j=1}^l b_{p_i,q_j}u_{ij}\right)&=\alpha\left(\sum_{i=1}^k\sum_{j=1}^l b_{\sigma_1(i),\sigma_2(j)}u_{ij}\right)\\&=\alpha\left(\sum_{i=1}^k\sum_{j=1}^l b_{ij}u_{ij}\right)=\alpha(P_{kl}(b))\leqslant\alpha(b),\end{aligned} \tag{17.43}$$

where $\sigma_1,\sigma_2\in\Pi$ are permutations such that $\sigma_1(i)=p_i$ $(i=1,\ldots,k)$, $\sigma_2(j)=q_j$ $(j=1,\ldots,l)$ and for any $\{p_i\}_{i=1}^{n_0}$, $\{q_j\}_{j=1}^{m_0}$ we have

$$\alpha\left(\sum_{i=1}^{n_0}\sum_{j=1}^{m_0} b_{ij}u_{p_i,q_j}\right)=\alpha\left(\sum_{i=1}^{n_0}\sum_{j=1}^{m_0} b_{ij}u_{ij}\right)=\alpha(b). \tag{17.44}$$

Applying (17.43) to the matrix $b=V(a)=\sum\limits_{i=1}^{n_0}\sum\limits_{j=1}^{m_0} b_{ij}u_{ij}$ (where $b_{r_{2i},s_{2j}}=b_{r_{2i-1},s_{2j-1}}=-b_{r_{2i},s_{2j-1}}=-b_{r_{2i-1},s_{2j}}=a_{ij}$ and $b_{ij}=0$ for the other i,j) and to $p_i=r_{2i}$, $q_j=s_{2j}$, we obtain

$$\alpha(a)=\alpha\Big(\sum_i\sum_j a_{ij}u_{ij}\Big)=\alpha\Big(\sum_i\sum_j b_{r_{2i},s_{2j}}u_{ij}\Big)$$
$$\leqslant\alpha\Big(\sum_{i=1}^{n_0}\sum_{j=1}^{m_0} b_{ij}u_{ij}\Big)=\alpha(V(a))\qquad(a\in M). \tag{17.45}$$

On the other hand, applying (17.44) to $b=a$, we get

$$\alpha(V(a))\leqslant\alpha\Big(\sum_i\sum_j a_{ij}u_{r_{2i},s_{2j}}\Big)+\alpha\Big(\sum_i\sum_j a_{ij}u_{r_{2i-1},s_{2j-1}}\Big)$$
$$+\alpha\Big(\sum_i\sum_j a_{ij}u_{r_{2i},s_{2j-1}}\Big)+\alpha\Big(\sum_i\sum_j a_{ij}u_{r_{2i-1},s_{2j}}\Big)=4\alpha(a)\qquad(a\in M). \tag{17.46}$$

Thus, by (17.45) and (17.46), V is an isomorphism of M into M and hence it can be extended, by continuity, to an isomorphism V_1 of M_α into M_α. Put

$$U=U_0V_1. \tag{17.47}$$

Then U is an isomorphic embedding of M_α into E, satisfying

$$f_n[U(u_{ij})]=f_n[U_0V_1(u_{ij})]=f_n[U_0(u_{r_{2i},s_{2j}})-U_0(u_{r_{2i},s_{2j-1}})]$$
$$+f_n[U_0(u_{r_{2i-1},s_{2j-1}})-U_0(u_{r_{2i-1},s_{2j}})]\to 0\quad\text{as}\quad j\to\infty$$
$$(n,i=1,2,\dots)$$

and, similarly, $f_n[U(u_{ij})]\to 0$ as $i\to\infty$ $(n,j=1,2,\dots)$, which completes the proof of lemma 17.2.

Now we can prove

Theorem 17.2. *Let E be a Banach space with a basis and let α be a symmetric matrix norm. If there exists an isomorphic embedding $U: M_\alpha\to E$, then for every basis $\{x_n\}$ of E there exists an equivalent norm on E such that in this norm*

$$\nu_{\{x_n\}}^{(u)}\geqslant\sup_{1\leqslant n<\infty}t_n(\alpha). \tag{17.48}$$

Consequently, if E contains a subspace isomorphic to M_α, where α is a symmetric matrix norm such that $\sup_{1\leqslant n<\infty}t_n(\alpha)=\infty$, then E has no unconditional basis.

Proof. Let $\varepsilon>0$ and let s be a positive integer. Then, by lemma 17.1, there exists a matrix $a\in M$ with $\alpha(a)\leqslant 1$, such that

$$\alpha(D_s(a))\geqslant t_s(\alpha)-\varepsilon. \tag{17.49}$$

We shall prove that for any basis $\{x_n\}$ of E we have, in a suitable equivalent norm on E,

$$\nu_{\{x_n\}}^{(u)}\geqslant\alpha(D_s(a))-\varepsilon, \tag{17.50}$$

whence, by (17.49), $v_{\{x_n\}}^{(u)} \geqq t_s(\alpha) - 2\varepsilon$, which, since $\varepsilon > 0$ and $s \in \mathcal{N}$ are arbitrary, will complete the proof.

Let $\{x_n\}$ be an arbitrary basis of E. By lemma 17.2 we may assume, without loss of generality, that U satisfies (17.40) and (17.41), where $\{f_n\} \subset E^*$ is the a.s.c.f. to the basis $\{x_n\}$. Furthermore, by § 13, lemma 13.6, we may assume, replacing, if necessary, the initial norm of E with a suitable equivalent norm, that U is a linearly isometric embedding.

We shall consider only the case when $s = 2n-1$, since for $s = 2n$ the proof is similar.

By (17.40) and (17.41) one can construct, inductively, three increasing finite sequences of indices $\{m_k\}_{k=1}^{s+1}$, $\{p_k\}_{k=1}^{s}$, and $\{q_k\}_{k=1}^{s}$ such that for the matrices

$$b_k = \sum_{\max(i,j)=k} a_{ij} u_{p_i q_j} \qquad (k = 1, \ldots, s) \tag{17.51}$$

the following inequalities hold:

$$\sum_{r=1}^{m_k} |f_r[U(b_k)]| \, \|x_r\| < \frac{\varepsilon}{2s^2} \qquad (k = 1, \ldots, s) \tag{17.52}$$

$$\left\| \sum_{r=m_{k+1}+1}^{m'} f_r[U(b_k)] x_r \right\| < \frac{\varepsilon}{2s^2} \qquad (m' = m_{k+1}+1, m_{k+1}+2, \ldots; k = 1, \ldots, s). \tag{17.53}$$

Indeed, put $m_1 = p_1 = 1$ and, by (17.41), choose q_1 such that we have

$$\begin{aligned} |f_1[U(b_1)]| \, \|x_1\| &= |f_1[U(a_{11} u_{p_1,q_1})]| \, \|x_1\| \\ &= |a_{11}| \, \|x_1\| \, |f_1[U(u_{p_1,q_1})]| < \frac{\varepsilon}{2s^2}, \end{aligned}$$

i.e., (17.52) for $k=1$. Then, since $b_1 = \sum_{r=1}^{\infty} f_r(b_1) x_r$ (because $\{x_n\}$ is a basis of E), there exists an $m_2 > m_1$ such that we have

$$\left\| \sum_{r=m_2+1}^{m'} f_r[U(b_1)] x_r \right\| < \frac{\varepsilon}{2s^2} \qquad (m' = m_2+1, m_2+2, \ldots),$$

i.e., (17.53) for $k=1$. Next, by (17.40) we can choose p_2 such that

$$|a_{21}| \sum_{r=1}^{m_2} |f_r[U(u_{p_2,q_1})]| \, \|x_r\| < \frac{\varepsilon}{4s^2}$$

and then, by (17.41), q_2 such that

$$|a_{12}| \sum_{r=1}^{m_2} |f_r[U(u_{p_1,q_2})]| \, \|x_r\| + |a_{22}| \sum_{r=1}^{m_2} |f_r[U(u_{p_2,q_2})]| \, \|x_r\| < \frac{\varepsilon}{4s^2},$$

whence we obtain

$$\sum_{r=1}^{m_2} |f_r[U(b_2)]| \, \|x_r\| = \sum_{r=1}^{m_2} |f_r[U(a_{12}u_{p_1,q_2}+a_{21}u_{p_2,q_1}+a_{22}u_{p_2,q_2})]| \, \|x_r\| < \frac{\varepsilon}{2s^2},$$

i.e., (17.52) for $k=2$. Since $b_2=\sum_{r=1}^{\infty} f_r(b_2)x_r$, we can choose $m_3>m_2$ so as to have (17.53) for $k=2$. Continuing in this way, in a finite number of steps we achieve the construction of $\{m_k\}_{k=1}^{s+1}$, $\{p_k\}_{k=1}^{s}$ and $\{q_k\}_{k=1}^{s}$.

Conditions (17.52) and (17.53) imply

$$\left\| \sum_{r=m_k+1}^{m_{k+1}} f_r[U(b_k)]x_r - U(b_k) \right\|$$

$$= \left\| \sum_{r=1}^{m_k} f_r[U(b_k)]x_r + \sum_{r=m_{k+1}+1}^{\infty} f_r[U(b_k)]x_r \right\| < \frac{\varepsilon}{s^2} \quad (k=1,\dots,s) \tag{17.54}$$

and

$$\left\| \sum_{r=m_k+1}^{m_{k+1}} f_r[U(b_l)]x_r \right\| \leqslant \sum_{r=1}^{m_{k+1}} |f_r[U(b_l)]| \, \|x_r\| + \sum_{r=1}^{m_k} |f_r[U(b_l)]| \, \|x_r\|$$

$$\leqslant 2\sum_{r=1}^{m_l} |f_r[U(b_l)]| \, \|x_r\| < \frac{\varepsilon}{s^2} \quad (l\geqslant k+1;\, k,l=1,\dots,s),$$

$$\left\| \sum_{r=m_k+1}^{m_{k+1}} f_r[U(b_l)]x_r \right\| \leqslant \left\| \sum_{r=m_{l+1}+1}^{m_{k+1}} f_r[U(b_l)]x_r \right\| + \left\| \sum_{r=m_{l+1}+1}^{m_k} f_r[U(b_l)]x_r \right\|$$

$$\leqslant 2 \sup_{m_{l+1}\leqslant m'<\infty} \left\| \sum_{r=m_{l+1}+1}^{m'} f_r[U(b_l)]x_r \right\| < \frac{\varepsilon}{s^2}$$

$$(k\geqslant l+1^1;\, k,l=1,\dots,s),$$

whence

$$\left\| \sum_{r=m_k+1}^{m_{k+1}} f_r[U(b_l)]x_r \right\| < \frac{\varepsilon}{s^2} \quad (k\neq l;\, k,l=1,\dots,s). \tag{17.55}$$

Now, put

$$x=\sum_{k=1}^{s} U(b_k). \tag{17.56}$$

Then, since U is a linear isometry and since the symmetric matrix norm α satisfies (17.44), we have

---

[1] If $k=l+1$, the term $\left\| \sum_{i=m_{l+1}+1}^{m_k} f_r[U(b_l)]x_r \right\|$ will not occur.

$$\|x\| = \left\| \sum_{k=1}^{s} U(b_k) \right\| = \left\| U\left(\sum_{k=1}^{s} b_k \right) \right\| = \alpha\left(\sum_{k=1}^{s} b_k \right) = \alpha\left(\sum_{i=1}^{s} \sum_{j=1}^{s} a_{ij} u_{p_i, q_j} \right)$$

$$= \alpha\left(\sum_{i=1}^{s} \sum_{j=1}^{s} a_{ij} u_{ij} \right) = \alpha[P_{s,s}(a)] \leqslant \alpha(a) \leqslant 1 \tag{17.57}$$

and, similarly, recalling that $s = 2n-1$,

$$\left\| \sum_{l=1}^{n} U(b_{2l-1}) \right\| = \left\| U\left(\sum_{l=1}^{n} b_{2l-1} \right) \right\| = \alpha\left(\sum_{l=1}^{n} b_{2l-1} \right)$$

$$= \alpha\left(\sum_{l=1}^{n} \sum_{\max(i,j)=2l-1} \alpha_{ij} u_{p_i, q_j} \right)$$

$$= \alpha\left(\sum_{l=1}^{n} \sum_{\max(i,j)=2l-1} a_{ij} u_{ij} \right) = \alpha(D_s(a)). \tag{17.58}$$

Consequently, by (17.57), (17.58), (17.56), (17.54) and (17.55) we obtain

$$\nu^{(u)}_{\{x_n\}} \geqslant \nu^{(u)}_{\{x_n\}} \|x\| \geqslant \left\| \sum_{l=1}^{n} \sum_{r=m_{2l-1}+1}^{m_{2l}} f_r(x) x_r \right\| \geqslant \left\| \sum_{l=1}^{n} U(b_{2l-1}) \right\|$$

$$- \sum_{l=1}^{n} \left\| U(b_{2l-1}) - \sum_{r=m_{2l-1}+1}^{m_{2l}} f_r(x) x_r \right\| \geqslant \alpha(D_s(a))$$

$$- \sum_{l=1}^{n} \left(\left\| U(b_{2l-1}) - \sum_{r=m_{2l-1}+1}^{m_{2l}} f_r[U(b_{2l-1})] x_r \right\| \right.$$

$$\left. + \sum_{\substack{k=1 \\ k \neq 2l-1}}^{s} \left\| \sum_{r=m_{2l-1}+1}^{m_{2l}} f_r[U(b_k)] x_r \right\| \right)$$

$$\geqslant \alpha(D_s(a)) - n\left(\frac{\varepsilon}{s^2} + (s-1)\frac{\varepsilon}{s^2} \right) = \alpha(D_s(a)) - \frac{n}{s}\varepsilon \geqslant \alpha(D_s(a)) - \varepsilon,$$

i.e., (17.50), which completes the proof of theorem 17.2.

Now we shall give some concrete symmetric matrix norms α having the property $\sup_{1 \leqslant n < \infty} t_n(\alpha) = \infty$.

Let $1 \leqslant p_1, p_2 \leqslant \infty$. Put

$$\lambda_{p_1, p_2}(a) = \sup_{\substack{\sum_i |\xi_i|^{q_1} \leqslant 1 \\ \sum_j |\eta_j|^{q_2} \leqslant 1}} \left| \sum_i \sum_j a_{ij} \xi_i \eta_j \right| \quad (a \in M), \tag{17.59}$$

where $\frac{1}{p_k}+\frac{1}{q_k}=1$ $(k=1,2;\frac{1}{\infty}=0)$ and where in the case $q_k=\infty$ the sum $\sum_i |\xi_i|^{q_k}$ (or $\sum_j |\eta_j|^{q_k}$) is replaced by $\sup_i |\xi_i|$ (respectively, $\sup_j |\eta_j|$). Then, obviously, λ_{p_1,p_2} is a symmetric matrix norm.

Lemma 17.3. *If* $1 \leqslant p_1 \leqslant p_3 \leqslant \infty$ *and* $1 \leqslant p_2 \leqslant p_4 \leqslant \infty$, *then*

$$\lambda_{p_1,p_2}(P_{n,m}(a)) \leqslant \lambda_{p_3,p_4}(a) n^{\frac{1}{q_3}-\frac{1}{q_1}} m^{\frac{1}{q_4}-\frac{1}{q_2}} \qquad (a \in M, n, m = 1,2,\dots), \tag{17.60}$$

where $\frac{1}{p_k}+\frac{1}{q_k}=1$ $(k=1,2,3,4)$.

Proof. Since $\left(\frac{1}{n}\sum_{i=1}^{n}|\xi_i|^q\right)^{\frac{1}{q}}$ is[1] an increasing function of q and $q_1 \geqslant q_3$, $q_2 \geqslant q_4$, we have

$$\begin{aligned}
n^{\frac{1}{q_1}} m^{\frac{1}{q_2}} \lambda_{p_1,p_2}(P_{n,m}(a)) &= n^{\frac{1}{q_1}} m^{\frac{1}{q_2}} \sup_{\substack{\sum_{i=1}^{n}|\xi_i|^{q_1}\leqslant 1 \\ \sum_{j=1}^{m}|\eta_j|^{q_2}\leqslant 1}} \left|\sum_{i=1}^{n}\sum_{j=1}^{m} a_{ij}\xi_i\eta_j\right| \\
&= \sup_{\substack{\left(\frac{1}{n}\sum_{i=1}^{n}|\xi_i|^{q_1}\right)^{\frac{1}{q_1}}\leqslant 1 \\ \left(\frac{1}{m}\sum_{j=1}^{m}|\eta_j|^{q_2}\right)^{\frac{1}{q_2}}\leqslant 1}} \left|\sum_{i=1}^{n}\sum_{j=1}^{m} a_{ij}\xi_i\eta_j\right| \\
&\leqslant \sup_{\substack{\left(\frac{1}{n}\sum_{i=1}^{n}|\xi_i|^{q_3}\right)^{\frac{1}{q_3}}\leqslant 1 \\ \left(\frac{1}{m}\sum_{j=1}^{m}|\eta_j|^{q_4}\right)^{\frac{1}{q_4}}\leqslant 1}} \left|\sum_{i=1}^{n}\sum_{j=1}^{m} a_{ij}\xi_i\eta_j\right| \\
&= n^{\frac{1}{q_3}} m^{\frac{1}{q_4}} \lambda_{p_3,p_4}(P_{n,m}(a)) \leqslant n^{\frac{1}{q_3}} m^{\frac{1}{q_4}} \lambda_{p_3,p_4}(a),
\end{aligned}$$

whence (17.60), which completes the proof of lemma 17.3.

Let us recall that the *Hilbert matrices* h_n $(n=1,2,\dots)$ are defined by

$$(h_n)_{ij} = \begin{cases} \dfrac{1}{n+1-i-j} & \text{for} \quad i+j \neq n+1,\ i,j=1,\dots,n \\ 0 & \text{for} \quad \text{the other } i,j \end{cases} \tag{17.61}$$

[1] See e.g. [263], p. 8, problem 2.

and that for each p with $1<p<\infty$ there exists[1] a constant $B_p>0$ such that

$$\lambda_{p,q}(h_n)\leqslant B_p \quad (n=1,2,\ldots), \tag{17.62}$$

where $\frac{1}{p}+\frac{1}{q}=1$.

Lemma 17.4. *Let* $1\leqslant p_1,p_2<\infty$, $\frac{1}{p_1}+\frac{1}{p_2}\geqslant 1$. *Then there exists a constant* C_{p_1,p_2} *such that*

$$t_n(\lambda_{p_1,p_2})\geqslant C_{p_1,p_2}\ln n \quad (n=1,2,\ldots). \tag{17.63}$$

Consequently, $\lim_{n\to\infty} t_n(\lambda_{p_1,p_2})=\infty$.

Proof. By (17.61), (17.31) and (17.59) we have

$$\begin{aligned}
\lambda_{p_1,p_2}[T_n(h_n)] &= \lambda_{p_1,p_2}\left(\sum_{i+j=2}^{n}\frac{1}{n+1-i-j}u_{ij}\right)\\
&= \sup_{\substack{\sum_{i=1}^{n}|\xi_i|^{q_1}\leqslant 1\\ \sum_{j=1}^{n}|\eta_j|^{q_2}\leqslant 1}}\left(\sum_{i+j=2}^{n}\frac{1}{n+1-i-j}\xi_i\eta_j\right)\\
&\geqslant \frac{1}{n^{\frac{1}{q_1}}}\frac{1}{n^{\frac{1}{q_2}}}\sum_{i+j=2}^{n}\frac{1}{n+1-i-j}\\
&= n^{-\frac{1}{q_1}-\frac{1}{q_2}}\sum_{j=1}^{n-1}(n-j)\frac{1}{j}=n^{-\frac{1}{q_1}-\frac{1}{q_2}}\left[n\sum_{j=1}^{n-1}\frac{1}{j}-(n-1)\right]\\
&\geqslant n^{-\frac{1}{q_1}-\frac{1}{q_2}}[n\ln n-(n-1)]\geqslant Cn^{1-\frac{1}{q_1}-\frac{1}{q_2}}\ln n = Cn^{\frac{1}{p_1}-\frac{1}{q_2}}\ln n,
\end{aligned} \tag{17.64}$$

with a suitable constant $C>0$ (independent on n,p_1,p_2). We shall consider three cases:

a) $1<p_1<\infty$. Then, since $P_{n,n}(h_n)=h_n$ and since the hypothesis $\frac{1}{p_2}\geqslant 1-\frac{1}{p_1}=\frac{1}{q_1}$ implies $p_2\leqslant q_1$, from (17.60) with $p_3=p_1$, $p_4=q_1$ and (17.62) we obtain

$$\lambda_{p_1,p_2}(h_n)=\lambda_{p_1,p_2}(P_{n,n}(h_n))\leqslant\lambda_{p_1,q_1}(h_n)n^{\frac{1}{p_1}-\frac{1}{q_2}}\leqslant B_{p_1}n^{\frac{1}{p_1}-\frac{1}{q_2}},$$

[1] See e.g. [51], Ch. XI, § 7 or [79], Ch. III, § 10.

whence, by (17.64),

$$t_n(\lambda_{p_1,p_2}) \geqslant \frac{\lambda_{p_1,p_2}(T_n(h_n))}{\lambda_{p_1,p_2}(h_n)} \geqslant \frac{C n^{\frac{1}{p_1}-\frac{1}{q_2}} \ln n}{B_{p_1} n^{\frac{1}{p_1}-\frac{1}{q_2}}} = \frac{C}{B_{p_1}} \ln n.$$

b) $p_1=1$ and $1<p_2<\infty$. Then, by (17.60) with $p_1=1$, $p_3=q_2$, $p_4=p_2$ and (17.62) we obtain

$$\lambda_{1,p_2}(h_n) = \lambda_{1,p_2}(P_{n,n}(h_n)) \leqslant \lambda_{q_2,p_2}(h_n) n^{\frac{1}{p_2}} \leqslant B_{q_2} n^{\frac{1}{p_2}},$$

whence, by (17.64) with $p_1=1$,

$$t_n(\lambda_{1,p_2}) \geqslant \frac{\lambda_{1,p_2}(T_n(h_n))}{\lambda_{1,p_2}(h_n)} \geqslant \frac{C n^{1-\frac{1}{q_2}} \ln n}{B_{q_2} n^{\frac{1}{p_2}}} = \frac{C}{B_{q_2}} \ln n.$$

c) $p_1=p_2=1$. Then, from (17.60) with $p_3=p_4=2$ and (17.62) we obtain

$$\lambda_{1,1}(h_n) = \lambda_{1,1}(P_{n,n}(h_n)) \leqslant \lambda_{2,2}(h_n) n \leqslant B_2 n,$$

whence, by (17.64) with $p_1=1$, $q_2=\infty$,

$$t_n(\lambda_{1,1}) \geqslant \frac{\lambda_{1,1}(T_n(h_n))}{\lambda_{1,1}(h_n)} \geqslant \frac{C n \ln n}{B_2 n} = \frac{C}{B_2} \ln n,$$

which completes the proof of lemma 17.4.

Remark 17.1. The assumptions $p_1, p_2 < \infty$ in lemma 17.4 are essential. Indeed, we have

$$\lambda_{\infty,p_2}(a) = \sup_{\substack{\sum_i |\xi_i| \leqslant 1 \\ \sum_j |\eta_j|^{q_2} \leqslant 1}} \left| \sum_i \Big(\sum_j a_{ij} \eta_j \Big) \xi_i \right| = \sup_i \sup_{\sum_j |\eta_j|^{q_2} \leqslant 1} \left| \sum_j a_{ij} \eta_j \right|$$

$$= \sup_i \Big(\sum_j |a_{ij}|^{p_2} \Big)^{\frac{1}{p_2}} \qquad (a \in M,\ 1 \leqslant p_2 \leqslant \infty),$$

whence

$$t_n(\lambda_{\infty,p_2}) = \sup_{\substack{a \in M \\ \lambda_{\infty,p_2}(a) \leqslant 1}} \lambda_{\infty,p_2}(T_n(a)) = 1 \qquad (1 \leqslant p_2 \leqslant \infty;\ n=1,2,\ldots) \tag{17.65}$$

and, similarly, $\lambda_{p_1,\infty}(a) = \sup_j \Big(\sum_i |a_{ij}|^{p_1} \Big)^{\frac{1}{p_1}}$ $(a \in M,\ 1 \leqslant p_1 \leqslant \infty)$, whence

$$t_n(\lambda_{p_1,\infty}) = 1 \qquad (1 \leqslant p_1 \leqslant \infty;\ n=1,2,\ldots). \tag{17.66}$$

If α is a matrix norm, the *conjugate norm* α^* is defined by

$$\alpha^*(a) = \sup_{\substack{b \in M \\ \alpha(b) \leqslant 1}} \left| \sum_i \sum_j a_{ij} b_{ji} \right| \qquad (a \in M). \tag{17.67}$$

Since for $a \in M$ fixed, $g_a(b) = \sum_i \sum_j a_{ij} b_{ji}$ may be regarded as a linear functional on a suitable finite dimensional space containing a, and since $\alpha^*(P_{n,m}(b)) \leqslant \alpha^*(b)$, we obtain

$$\alpha^{**}(a) = \sup_{\substack{b \in M \\ \alpha^*(b) \leqslant 1}} \left| \sum_i \sum_j a_{ij} b_{ji} \right| = \alpha(a) \qquad (a \in M), \tag{17.68}$$

whence

$$\begin{aligned} t_n(\alpha^*) &= \sup_{\substack{b \in M \\ \alpha^*(b) \leqslant 1}} \alpha^*(T_n(b)) = \sup_{\substack{b \in M \\ \alpha^*(b) \leqslant 1}} \sup_{\substack{a \in M \\ \alpha(a) \leqslant 1}} \left| \sum_{i+j=2}^{n+1} a_{ij} b_{ji} \right| \\ &= \sup_{\substack{a \in M \\ \alpha(a) \leqslant 1}} \alpha^{**}(T_n(a)) = \sup_{\substack{a \in M \\ \alpha(a) \leqslant 1}} \alpha(T_n(a)) = t_n(\alpha) \qquad (n = 1, 2, \ldots). \end{aligned} \tag{17.69}$$

Consequently, from lemma 17.4 it follows

Corollary 17.5. *Let* $1 \leqslant p_1, p_2 < \infty$, $\frac{1}{p_1} + \frac{1}{p_2} \geqslant 1$. *Then there exists a constant* C_{p_1, p_2} *such that*

$$t_n(\lambda^*_{p_1, p_2}) \geqslant C_{p_1, p_2} \ln n \qquad (n = 1, 2, \ldots). \tag{17.70}$$

From theorem 17.2, lemma 17.4, and corollary 17.5 we obtain immediately

Theorem 17.3. *Let* $1 \leqslant p_1, p_2 < \infty$, $\frac{1}{p_1} + \frac{1}{p_2} \geqslant 1$ *and let E be a Banach space with a basis, containing a subspace isomorphic to* $M_{\lambda_{p_1, p_2}}$ *or* $M_{\lambda^*_{p_1, p_2}}$. *Then E has no unconditional basis.*

Remark 17.2. In particular, like the space J (see § 15, theorem 15.4), the space $M_{\lambda_{1,1}}$ is an example of a space with a basis which is isomorphic to a conjugate Banach space and which cannot be embedded into any space with an unconditional basis. Indeed, since for any infinite matrix $a = (a_{ij})$ the relations $\sup_{1 \leqslant n, m < \infty} \lambda_{1,1}\left(\sum_{i=1}^{n} \sum_{j=1}^{m} a_{ij} u_{ij} \right) = \sup_{1 \leqslant n, m < \infty} \sup_{\substack{|\xi_1|, |\xi_2|, \ldots \leqslant 1 \\ |\eta_1|, |\eta_2|, \ldots \leqslant 1}} \left| \sum_{i=1}^{n} \sum_{j=1}^{m} a_{ij} \xi_i \eta_j \right| < \infty$ imply $a \in M_{\lambda_{1,1}}$, the basis $\{u_{i(k), j(k)}\}$ of $M_{\lambda_{1,1}}$ (see proposition 17.1) is boundedly complete, whence, from § 6, theorem 6.2 (implication $1^\circ \Rightarrow 3^\circ$) it follows that $M_{\lambda_{1,1}}$ is isomorphic to the conjugate space $[g_{ij}]^*$, where $\{g_{i(k), j(k)}\} \subset M^*_{\lambda_{1,1}}$ is the a. s. c. f. to the basis $\{u_{i(k), j(k)}\}$. Moreover, one can also show[1] that in $M_{\lambda_{1,1}}$, weak and strong convergence of sequences coincide.

[1] See [145], example 2.1.

Let us observe now that if E and F are Banach spaces with normalized monotone bases $\{x_n\}$ and $\{y_n\}$, respectively, and if α is a uniform crossnorm on their algebraic tensor product $E \otimes F$, then one can define in a natural way a matrix norm $\tilde{\alpha}=\tilde{\alpha}(E,F,\{x_n\},\{y_n\},\alpha)$ by

$$\tilde{\alpha}(a)=\tilde{\alpha}\Big(\sum_i \sum_j a_{ij}u_{ij}\Big)=\alpha\Big(\sum_i \sum_j a_{ij}x_i \otimes y_j\Big) \qquad (a \in M). \qquad (17.71)$$

Indeed, $\tilde{\alpha}(u_{ij})=\alpha(x_i \otimes y_j)=\|x_i\|\,\|y_j\|=1$ $(i,j=1,2,\ldots)$ and, if s_n^1, s_m^2 denote the partial sum operators associated to the bases $\{x_n\}$ and $\{y_n\}$, respectively,

$$\begin{aligned}\tilde{\alpha}(P_{n,m}(a)) &= \alpha\left(\sum_{i=1}^{n}\sum_{j=1}^{m} a_{ij}x_i \otimes y_j\right)=\alpha\left(\sum_{i=1}^{n} x_i \otimes \sum_{j=1}^{m} a_{ij}y_j\right)\\ &= \alpha\Big[s_n^1\Big(\sum_i x_i\Big)\otimes s_m^2\Big(\sum_j a_{ij}y_j\Big)\Big]=\alpha\Big[(s_n^1 \otimes s_m^2)\Big(\sum_i x_i \otimes \sum_j a_{ij}y_j\Big)\Big]\\ &\leqslant \|s_n^1\|\,\|s_m^2\|\,\alpha\Big(\sum_i x_i \otimes \sum_j a_{ij}y_j\Big)=\alpha\Big(\sum_i\sum_j a_{ij}x_i \otimes y_j\Big)=\tilde{\alpha}(a)\\ &\qquad\qquad (a \in M).\end{aligned}$$

Consequently, by (17.71), proposition 17.1 and Ch. I, § 18, theorem 18.1, $E \otimes_\alpha F$ is linearly isometric to $M_{\tilde{\alpha}}$.

In particular, to $E=l^{p_1}$, $F=l^{p_2}$ $(1 \leqslant p_1, p_2 \leqslant \infty$ and l^∞ stands for $c_0)$, $\{x_n\}, \{y_n\}=$ the unit vector bases of E and F, respectively, and[1] $\alpha=\lambda$ there corresponds the matrix norm

$$\begin{aligned}\tilde{\lambda}(a) &= \lambda\Big(\sum_i\sum_j a_{ij}x_i \otimes y_j\Big)=\sup_{\substack{f \in E^*,\ \|f\| \leqslant 1\\ g \in F^*,\ \|g\| \leqslant 1}} \Big|\sum_i f(x_i) \otimes \sum_j g(a_{ij}y_j)\Big|\\ &= \sup_{\substack{\sum_{i=1}^\infty |\xi_i|^{q_1} \leqslant 1\\ \sum_{j=1}^\infty |\eta_j|^{q_2} \leqslant 1}} \Big|\sum_i\sum_j a_{ij}\xi_i\eta_j\Big| = \lambda_{p_1,p_2}(a) \qquad (a \in M),\end{aligned}$$

where $\dfrac{1}{p_k}+\dfrac{1}{q_k}=1$ $(k=1,2)$ and where in the case $q_k=\infty$ the sum $\sum_i |\xi_i|^{q_k}$ (or $\sum_j |\eta_j|^{q_k}$) is replaced by $\sup_i |\xi_i|$ $\Big(\sup_j |\eta_j|\Big)$. Hence $l^{p_1} \otimes_\lambda l^{p_2}$ *is linearly isometric to* $M_{\lambda_{p_1,p_2}}$ and consequently $l^{p_1} \otimes_\gamma l^{p_2}$ *is linearly isometric to* $M_{\lambda^*_{q_1,q_2}}$, where $\dfrac{1}{p_k}+\dfrac{1}{q_k}=1$ $(k=1,2)$ and where by l^∞ we mean c_0. Therefore, theorem 17.3 can be also reformulated as follows:

[1] As in Ch. I, § 18, we denote by λ the least crossnorm whose associate is a crossnorm and by γ the greatest crossnorm.

Theorem 17.4. *Let* $1 \leqslant p_1, p_2 < \infty$, $\frac{1}{p_1} + \frac{1}{p_2} \geqslant 1$ *and let E be a Banach space with a basis, containing a subspace isomorphic to one of the spaces*[1] $l^{p_1} \otimes_\lambda l^{p_2} \equiv \mathscr{C}(l^{q_1}, l^{p_2})$ *or* $l^{q_1} \otimes_\gamma l^{q_2}$, *where* $\frac{1}{p_k} + \frac{1}{q_k} = 1$ $(k=1,2)$ *and where* l^∞ *stands for* c_0. *Then E has no unconditional basis.*

In order to give a similar result for tensor products of the spaces $L^p([0,1])$ and $C([0,1])$, we need

Proposition 17.2. *Let* $1 < p < \infty$, *and let* r_n $(n=1,2,\ldots)$ *be the Rademacher functions on* $[0,1]$. *Then*

a) $\{r_n\}$ *is a basic sequence in* $L^p([0,1])$, *equivalent to the unit vector basis of* l^2 *and hence the subspace* $[r_n]$ *of* $L^p([0,1])$ *is isomorphic to* l^2.

b) *The mapping*

$$u(x) = \sum_{i=1}^{\infty} \left(\int_0^1 x(t) r_i(t) dt \right) r_i \quad (x \in L^p([0,1])) \tag{17.72}$$

is a continuous linear projection of $L^p([0,1])$ *onto* $[r_n]$ *and hence the subspace* $[r_n]$ *is complemented in* $L^p([0,1])$.

Proof. a) By § 14, formula (14.2), $\{r_n\}$ is a block basic sequence with respect to the Haar basis $\{\tilde{y}_n\}$. By the Khinchin inequality (14.56), $\{r_n\}$ is equivalent to the unit vector basis of l^2.

b) Let $p \geqslant 2$ and let $x \in L^p([0,1])$. Then $x \in L^2([0,1])$ and $r_n \in L^2([0,1])$, whence, by the Bessel inequality, $\left[\sum_{i=1}^{\infty} |(x, r_i)|^2 \right]^{\frac{1}{2}} \leqslant \left(\int_0^1 |x(t)|^2 dt \right)^{\frac{1}{2}} < \infty$ and thus, by the Khinchin inequality (14.56),

$$\left(\int_0^1 \left| \sum_{i=n+1}^{n+k} (x, r_i) r_i(t) \right|^p dt \right)^{\frac{1}{p}} \leqslant B_p \left[\sum_{i=n+1}^{n+k} |(x, r_i)|^2 \right]^{\frac{1}{2}} < \varepsilon \quad (n > N(\varepsilon), k=1,2,\ldots).$$

Consequently, for every $x \in L^p([0,1])$ the series $\sum_{i=1}^{\infty} (x, r_i) r_i$ converges in $L^p([0,1])$ and hence (17.72) is a continuous linear mapping of $L^p([0,1])$ into $[r_n]$. Finally, by the orthogonality of the Rademacher functions (§ 14, corollary 14.1), we have

$$u(r_k) = \sum_{i=1}^{\infty} \left(\int_0^1 r_k(t) r_i(t) dt \right) r_i = r_k \quad (k=1,2,\ldots)$$

and hence u is a continuous linear projection of $L^p([0,1])$ onto $[r_n]$.

---

[1] See Ch. I, § 18, footnote to corollary 18.5.

Finally, assume that $1<p\leqslant 2$ and let $\frac{1}{p}+\frac{1}{q}=1$. Then $q\geqslant 2$, whence, by the above, for any $f\in L^p([0,1])^*\equiv L^q([0,1])$ the series $\sum_{i=1}^{\infty}(r_i,f)r_i$ (where the r_i are considered as elements of $L^q([0,1])$) is convergent in $L^q([0,1])$ and by a) this convergence is unconditional. Hence

$$\sum_{i=1}^{\infty}|f((x,r_i)r_i)|=\sum_{i=1}^{\infty}\varepsilon_i f((x,r_i)r_i)=\sum_{i=1}^{\infty}[\varepsilon_i(r_i,f)r_i](x)<\infty$$

for any $x\in L^p([0,1])$, $f\in L^p([0,1])^*=L^q([0,1])$, where we have put $\varepsilon_i=\operatorname{sign} f((x,r_i)r_i)$. Thus the series $\sum_{i=1}^{\infty}(x,r_i)r_i$ in $L^p([0,1])$ is weakly unconditionally Cauchy, whence, by § 15, lemma 15.8, it is convergent in the norm topology of $L^p([0,1])$. Consequently, again the mapping (17.72) is a continuous linear projection of $L^p([0,1])$ onto $[r_n]$, which completes the proof of proposition 17.2.

Theorem 17.5. *Let $1\leqslant p_1,p_2\leqslant\infty$ and let E be a Banach space with a basis, containing a subspace isomorphic to one of the spaces $L^{p_1}([0,1])\otimes_\lambda L^{p_2}([0,1])$ or $L^{p_1}([0,1])\otimes_\gamma L^{p_2}([0,1])$, where $L^\infty([0,1])$ stands for $C([0,1])$. Then E has no unconditional basis.*

Proof. If $1<p_1,p_2<\infty$, then, by proposition 17.2, the space $l^2\otimes_\lambda l^2$, respectively $l^2\otimes_\gamma l^2$, can be embedded[1] isomorphically into $L^{p_1}([0,1])\otimes_\lambda L^{p_2}([0,1])$, respectively into $L^{p_1}([0,1])\otimes_\gamma L^{p_2}([0,1])$, whence the desired conclusion follows by theorem 17.4 applied to $p_1=p_2=2$. The remaining tensor products (with $L^1([0,1])$ or $C([0,1])$) contain a subspace isomorphic to $L^1([0,1])$, respectively $C([0,1])$, whence the desired conclusion follows by § 15, theorems 15.2 and 15.1. This completes the proof of theorem 17.5.

Let us consider now the symmetric matrix norms $\mathfrak{S}_p$ defined by[2]

$$\mathfrak{S}_p(a)=\begin{cases}\left[\operatorname{tr}(aa^*)^{\frac{p}{2}}\right]^{\frac{1}{p}} & \text{for}\quad 1\leqslant p<\infty\\ \lambda_{2,2}(a) & \text{for}\quad p=\infty\end{cases}\qquad (a\in M),\tag{17.73}$$

where $a_{ij}^*=\bar{a}_{ji}$ $(i,j=1,2,\ldots)$ and where $\operatorname{tr} b=\sum_i b_{ii}$.

---

[1] See [89], p. 93 and p. 40, corollary 1, or [223], Ch. II, lemma 2.12 and Ch. III. corollary 3.4.

[2] For a detailed study of the spaces $M_{\mathfrak{S}_p}$ see e.g. [164] or [51], Ch. XI, § 9 or [79], Ch. III.

Lemma 17.5. *We have*

$$\lim_{p\to 1} t_n(\mathfrak{S}_p) = \lim_{p\to\infty} t_n(\mathfrak{S}_p) \geqslant C' \ln n \qquad (n=1,2,\ldots), \tag{17.74}$$

with a suitable constant $C'>0$.

Proof. Since $\lim_{p\to\infty} \mathfrak{S}_p(a) = \mathfrak{S}_\infty(a)\ (a\in M)$, we have, by (17.62) and (17.64),

$$\lim_{p\to\infty} \mathfrak{S}_p(h_n) = \mathfrak{S}_\infty(h_n) = \lambda_{2,2}(h_n) \leqslant B_2 \qquad (n=1,2,\ldots),$$

$$\lim_{p\to\infty} \mathfrak{S}_p(T_n(h_n)) = \lambda_{2,2}\,(T_n(h_n)) \geqslant C \ln n \quad (n=1,2,\ldots),$$

whence

$$\lim_{p\to\infty} t_n(\mathfrak{S}_p) \geqslant \lim_{p\to\infty} \frac{\mathfrak{S}_p(T_n(h_n))}{\mathfrak{S}_p(h_n)} \geqslant \frac{C}{B_2} \ln n \qquad (n=1,2,\ldots).$$

Since it is well known[1] that $\mathfrak{S}_p^* = \mathfrak{S}_q$, where $\dfrac{1}{p} + \dfrac{1}{q} = 1$, we also have, by (17.69) and the above,

$$\lim_{p\to 1} t_n(\mathfrak{S}_p) = \lim_{p\to 1} t_n(\mathfrak{S}_p^*) = \lim_{q\to\infty} t_n(\mathfrak{S}_q) \geqslant \frac{C}{B_2} \ln n \qquad (n=1,2,\ldots),$$

which completes the proof of lemma 17.5.

Let us mention that by (17.69) and (17.63) we also have

$$t_n(\mathfrak{S}_1) = t_n(\mathfrak{S}_1^*) = t_n(\mathfrak{S}_\infty) = t_n(\lambda_{2,2}) \geqslant C \ln n \qquad (n=1,2,\ldots). \tag{17.75}$$

Theorem 17.6. *Let* $\{p_k\}$ *be a sequence of positive numbers such that* $1<p_k<\infty$ $(k=1,2,\ldots)$ *and that either* $\lim_{k\to\infty} p_k = \infty$ *or* $\lim_{k\to\infty} p_k = 1$ *and let E be a Banach space containing for each* $k=1,2,\ldots$ *a subspace isomorphic to the space* $M_{\mathfrak{S}_{p_k}}$, *such that these isomorphisms and their inverses are uniformly bounded. Then E has no unconditional basis.*

Consequently, in particular the space $E_0 = (M_{\mathfrak{S}_{p_1}} \times M_{\mathfrak{S}_{p_2}} \times \cdots)_{l^2}$ *is a reflexive Banach space with a basis, which cannot be embedded into any space with an unconditional basis.*

Proof. For every basis $\{x_n\}$ of E we have, by our hypothesis and by the proof of theorem 17.2,

$$\nu_{\{x_n\}}^{(u)} \geqslant A \sup_{1\leqslant n<\infty} t_n(\mathfrak{S}_{p_k}) \qquad (k=1,2,\ldots),$$

where $A>0$ depends only on $\{x_n\}$, whence, by lemma 17.5,

$$\nu_{\{x_n\}}^{(u)} \geqslant A \sup_{1\leqslant k<\infty} \sup_{1\leqslant n<\infty} t_n(\mathfrak{S}_{p_k}) = \infty,$$

and thus E has no unconditional basis.

---

[1] See e.g. [79], Ch. III, § 1 or [164], p. 265, § 4.

Furthermore, by proposition 17.1 each $M_{\mathfrak{S}_{p_k}}$ has a basis, whence, by Ch. I, § 8, proposition 8.3, the product space E_0 has a basis and thus, by the above, E_0 has no unconditional basis. Since each $M_{\mathfrak{S}_{p_k}}$ is reflexive[1], their l^2-product space E_0 is reflexive, which completes the proof of theorem 17.6.

Problem 17.1. Let $1<p<\infty, p\neq 2$. Does the space $M_{\mathfrak{S}_p}$ have an unconditional basis?

Let us give now some properties of strong duality for unconditional bases.

Theorem 17.7. *Let $\{x_n\}$ be an unconditional basis of a Banach space E and let $\{f_n\}\subset E^*$ be the a.s.c.f. Then $\{f_n\}$ is an unconditional basic sequence and we have*

$$1\leqslant r([f_n])\, v^{(u)}_{\{x_n\}}\leqslant v^{(u)}_{\{f_n\}}\leqslant v^{(u)}_{\{x_n\}}. \tag{17.76}$$

Hence, in particular, if $r([f_n])=1$,[2] then

$$v^{(u)}_{\{f_n\}}=v^{(u)}_{\{x_n\}}. \tag{17.77}$$

Conversely, if $\{x_n\}$ is a basis of E, such that the a.s.c.f. $\{f_n\}\subset E^*$ is an unconditional basic sequence, then $\{x_n\}$ is an unconditional basis of E.

Proof. If $\{x_n\}$ is an unconditional basis of E, then every permutation $\{x_{\sigma(n)}\}$ is a basis of E, whence, by Ch. I, § 12, theorem 12.1, every permutation $\{f_{\sigma(n)}\}$ is a basic sequence in E^*, and thus $\{f_n\}$ is an unconditional basic sequence in E^*.

The first inequality in (17.76) follows from $v_{\{x_n\}}\leqslant v^{(u)}_{\{x_n\}}$ and Ch. I, § 12, theorem 12.2a). The proof of the other inequalities in (17.76) is similar to that of the corresponding inequalities for $v_{\{x_n\}}, v_{\{f_n\}}$ (Ch. I, § 12, theorem 12.3) considering instead of s_n the operators s_d.

Finally, assume that $\{x_n\}$ is a basis of E, such that the a.s.c.f. $\{f_n\}\subset E^*$ is an unconditional basic sequence. Then, by the first statement proved above, the a.s.c.f. $\{\phi(x_n)\}\subset [f_n]^*$ to the basis $\{f_n\}$ of $[f_n]$ is an unconditional basic sequence in $[f_n]^*$ and by Ch. I, § 12, theorem 12.2, the canonical mapping ϕ of E into $[f_n]^*$ is an isomorphism, whence $\{x_n\}$ is an unconditional basis of E. This completes the proof of theorem 17.7.

The inequalities (17.76) are, in a certain sense, the best possible, since for orthogonal bases (i.e., for which $v^{(u)}_{\{x_n\}}=1$) they all become equalities. However, they are, in general, strict inequalities, as shown by example 12.3 of Ch. I, §12, since in that example the basis $\{x_n\}$ is

---

[1] See e.g. [79], Ch. III, §1 or [164], p. 265, § 4.

[2] This happens e.g. when the basis $\{x_n\}$ is monotone, in particular when $v^{(u)}_{\{x_n\}}=1$ (such bases are called orthogonal, see § 20), or when $\{x_n\}$ is shrinking.

unconditional, $r([f_n])=\lambda$, $v^{(u)}_{\{x_n\}}\geqslant v_{\{x_n\}}=2+\frac{1}{\lambda}$ and $v^{(u)}_{\{f_n\}}$ does not depend on λ (by Ch. I, § 12, formula (12.33)). Furthermore, as shown by Ch. I, § 12, example 12.4, there exists a biorthogonal system (x_n, f_n) with $[x_n]=E$ and $\{f_n\}$ total on E, such that $\{f_n\}$ is an unconditional basic sequence, but $\{x_n\}$ is not even a basis of E.

Let us also mention that if $\{x_n\}$ is an unconditional basis of E and ϕ the canonical mapping of E into $[f_n]^*$, then, similarly to Ch. I, § 12, theorem 12.4 and corollary 12.3, we obtain

$$v^{(u)}_{\{\phi(x_n)\}}=v^{(u)}_{\{f_n\}}, \tag{17.78}$$

$$1\leqslant r([f_n])\, v^{(u)}_{\{x_n\}}\leqslant v^{(u)}_{\{\phi(x_n)\}}\leqslant v^{(u)}_{\{x_n\}}. \tag{17.79}$$

We also observe that the "unconditional analogue" of Ch. I, § 12, problem 12.1, i.e., the problem "if E^* has an unconditional basis, does E have an unconditional basis?" has a negative answer, as shown by the example $E=C(\omega^\omega)$ considered in § 5, example 5.3. Indeed, as observed there, $E^*\equiv l^1$ and so E^* has an unconditional basis, but we shall see in Vol. II that $E=C(\omega^\omega)$ has no unconditional basis.

The following property of strong duality will be used in § 21:

Proposition 17.3. *Let $\{x_n\}$ be an unconditional basis of a Banach space E, with the a.s.c.f. $\{f_n\}\subset E^*$, let $\{x_{n_k}\}$ be a subsequence of $\{x_n\}$ and let $E_0=[x_{n_j}]$, $\phi_{n_j}=f_{n_j}|_{E_0}\in E_0^*$ $(j=1,2,\ldots)$. Then $\{\phi_{n_j}\}$ is a basic sequence, equivalent to the basic sequence $\{f_{n_j}\}$.*

Proof. By Ch. I, § 12, theorem 12.1 and Ch. I, § 4, proposition 4.1 a), $\{f_{n_j}\}$ and $\{\phi_{n_j}\}$ are basic sequences. Since $\{x_n\}$ is an unconditional basis, by § 16, theorem 16.8 the mapping

$$u(x)=\sum_{j=1}^{\infty} f_{n_j}(x)x_{n_j} \qquad (x\in E)$$

is a well defined bounded linear projection of E onto E_0 and thus for any scalars $\alpha_1,\ldots,\alpha_k$ we have

$$\begin{aligned}\Big\|\sum_{i=1}^{k}\alpha_i\phi_{n_i}\Big\| &= \Big\|\Big(\sum_{i=1}^{k}\alpha_i f_{n_i}\Big)\Big|_{E_0}\Big\| \leqslant \Big\|\sum_{i=1}^{k}\alpha_i f_{n_i}\Big\| = \sup_{\substack{x\in E\\ \|x\|\leqslant 1}}\Big|\sum_{i=1}^{k}\alpha_i f_{n_i}(x)\Big| \\ &= \sup_{\substack{x\in E\\ \|x\|\leqslant 1}}\Big|\sum_{i=1}^{k}\alpha_i f_{n_i}\Big[\sum_{j=1}^{\infty} f_{n_j}(x)x_{n_j}\Big]\Big| = \sup_{\substack{x\in E\\ \|x\|\leqslant 1}}\Big|\sum_{i=1}^{k}\alpha_i\phi_{n_i}[u(x)]\Big| \\ &\leqslant \|u\|\,\Big\|\sum_{i=1}^{k}\alpha_i\phi_{n_i}\Big\|,\end{aligned}$$

whence $\{\phi_{n_j}\}\sim\{f_{n_j}\}$, which completes the proof.

In the above proposition the assumption that $\{x_n\}$ is unconditional is essential, as shown by

Example 17.2. Let $\{e_n\}$ be the natural basis of $E=c_0$ and let

$$x_n=\sum_{j=1}^{n} e_j \quad (n=1,2,\dots) \tag{17.80}$$

be the conditional basis of E considered in § 14, example 14.1. Then every subsequence $\{x_{n_j}\}$ of $\{x_n\}$ is equivalent to the basis $\{x_n\}$, but $\{f_{2j}\}$ is equivalent to the unit vector basis of l^1, whence unconditional, whence not equivalent to $\{\phi_{2j}\}$.

Indeed, we have seen in § 14, example 14.1, that for any increasing sequence of indices $\{n_j\}$ and any scalars $\alpha_l,\dots,\alpha_{l+k}$ we have, putting $\beta_{n_i}=\alpha_i$ $(i=l,\dots,l+k)$, $\beta_j=0$ for $j\neq n_l,\dots,n_{l+k}$

$$\left\|\sum_{i=l}^{l+k}\alpha_i x_{n_i}\right\|=\left\|\sum_{j=n_l}^{n_{l+k}}\beta_j x_j\right\|=\sup_{n_l\leq p\leq n_{l+k}}\left|\sum_{j=p}^{n_{l+k}}\beta_j\right|=\sup_{l\leq p\leq l+k}\left|\sum_{i=p}^{l+k}\alpha_i\right|.$$

Consequently, $\sum_{i=1}^{\infty}\alpha_i x_{n_i}$ converges if and only if $\sum_{i=1}^{\infty}\alpha_i$ converges, and thus $\{x_{n_j}\}$ is equivalent to $\{x_n\}$.

On the other hand, we have

$$f_j(x)=\xi_j-\xi_{j+1} \quad (x=\{\xi_n\}\in c_0,\ j=1,2,\dots)$$

whence

$$\left\|\sum_{i=m}^{m+p}a_i f_{2i}\right\|=\sup_{\substack{\{\xi_n\}\in c_0\\ |\xi_1|,|\xi_2|,\dots\leq 1}}\left|\sum_{i=m}^{m+p}a_i(\xi_{2i}-\xi_{2i+1})\right|=2\sum_{i=m}^{m+p}|a_i|.$$

Consequently, $\sum_{i=1}^{\infty}a_i f_{2i}$ converges if and only if $\sum_{i=1}^{\infty}|a_i|$ converges, and thus $\{f_{2j}\}$ is equivalent to the unit vector basis of l^1, which completes the proof of the assertions of example 17.2.

Finally, we shall give a relation between unconditional bases and sequence spaces, similar to Ch. I, § 12, theorem 12.6 (on general bases and sequences spaces).

We recall that the *α-dual* of a sequence space S is the sequence space S^α defined by

$$S^\alpha=\left\{\{\beta_n\}\subset K \;\middle|\; \sum_{i=1}^{\infty}|\beta_i\alpha_i|<\infty \quad \text{for all} \quad \{\alpha_n\}\in S\right\}. \tag{17.81}$$

For every sequence space S we have, obviously, $S\subset S^{\alpha\alpha}$. A sequence space S is said to be *α-perfect* if $S^{\alpha\alpha}=S$.

It is also obvious that for any sequence space S we have $S^\alpha \subset S^\gamma$, where S^γ is the γ-dual of S defined in Ch. I, § 12, formula (12.58). We shall consider now a class of sequence spaces for which we have the equality $S^\gamma = S^\alpha$.

A sequence space S is called *normal* if for every $\{\alpha_n\} \in S$ and every $\{\gamma_n\} \in m (= l^\infty)$ we have $\{\gamma_n \alpha_n\} \in S$.

Lemma 17.6. *If S is a normal sequence space, then $S^\gamma = S^\alpha$ and hence*

$$S^{\gamma\gamma} = S^{\alpha\alpha}. \tag{17.82}$$

Proof. If $\{\beta_n\} \in S^\gamma$ and $\{\alpha_n\} \in S$, then, since S is normal, $\{[\operatorname{sign}(\beta_n \alpha_n)]\alpha_n\} \in S$, whence

$$\sum_{i=1}^{\infty} |\beta_i \alpha_i| = \sup_{1 \leqslant n < \infty} \left| \sum_{i=1}^{n} [\operatorname{sign}(\beta_i \alpha_i)] \beta_i \alpha_i \right| < \infty,$$

i.e., $\{\beta_n\} \in S^\alpha$. Thus, $S^\gamma \subset S^\alpha$ and hence, by the opposite inclusion $S^\alpha \subset S^\gamma$ observed above, $S^\gamma = S^\alpha$.

Since for any sequence space S the α-dual S^α is obviously normal, it follows that for any sequence space S we have $S^{\alpha\gamma} = S^{\alpha\alpha}$. Consequently, if S is normal,

$$S^{\gamma\gamma} = S^{\alpha\gamma} = S^{\alpha\alpha},$$

which completes the proof of lemma 17.6.

Theorem 17.8. *A sequence space S is associated to an unconditional basis of a Banach space if and only if S contains all unit vectors e_n and there exists an α-perfect BK-space T such that $[e_n]_T = S$. In this case, $T = S^{\alpha\alpha}$.*

Proof. Assume that S is associated to an unconditional basis $\{x_n\}$ of a Banach space E, i.e., $S = \left\{ \{\alpha_n\} \subset K \,\middle|\, \sum_{i=1}^{\infty} \alpha_i x_i \text{ converges} \right\}$. Then, by Ch. I, § 12, theorem 12.6, S contains all unit vectors e_n and there exists a γ-perfect BK-space T such that $[e_n]_T = S$ and $T = S^{\gamma\gamma}$.

Now, since $\{x_n\}$ is an unconditional basis of E, from § 16, theorem 16.1 (implication $1^\circ \Rightarrow 5^\circ$) it follows that S is normal. Hence, by lemma 17.6, $S^{\alpha\alpha} = S^{\gamma\gamma} = T$. Since $T = (S^\alpha)^\alpha$ is normal and since T is γ-perfect, by lemma 17.6 we also have $T^{\alpha\alpha} = T^{\gamma\gamma} = T$, i.e., T is α-perfect.

Conversely, assume now that S is a sequence space containing all unit vectors e_n and that there exists an α-perfect BK-space T such that $[e_n]_T = S$. We shall prove that $\{e_n\}$ is an unconditional basic sequence in T, whence S is associated to the unconditional basis $\{x_n\} = \{e_n\}$ of the Banach space $E = S$.

Since $T = (T^\alpha)^\alpha$ is normal, by lemma 17.6 we have $T^{\gamma\gamma} = T^{\alpha\alpha} = T$, i.e., T is also γ-perfect. Hence, by Ch. I, § 12, theorem 12.6, $\{e_n\}$ is a basis of $S = [e_n]$.

Now let $\{\gamma_j\}\in m$ be arbitrary. Define a linear mapping $v_{\{\gamma_j\}}: T\to T$ by

$$v_{\{\gamma_j\}}(\{\alpha_j\})=\{\gamma_j\alpha_j\} \qquad (\{\alpha_j\}\in T); \tag{17.83}$$

since $T=(T^\alpha)^\alpha$ is normal, we have indeed $\{\gamma_j\alpha_j\}\in T$. Let $\{\alpha_j^{(n)}\}$, $\{\alpha_j\}\in T$ be such that

$$\lim_{n\to\infty}\{\alpha_j^{(n)}\}=\{\alpha_j\},\ \lim_{n\to\infty} v_{\{\gamma_j\}}(\{\alpha_j^{(n)}\})=\lim_{n\to\infty}\{\gamma_j\alpha_j^{(n)}\}=\{\delta_j\}.$$

Then, since T is a BK-space, we have

$$\lim_{n\to\infty}\alpha_j^{(n)}=\alpha_j,\ \lim_{n\to\infty}\gamma_j\alpha_j^{(n)}=\delta_j \qquad (j=1,2,\ldots),$$

whence

$$\delta_j=\lim_{n\to\infty}\gamma_j\alpha_j^{(n)}=\gamma_j\lim_{n\to\infty}\alpha_j^{(n)}=\gamma_j\alpha_j \qquad (j=1,2,\ldots),$$

i.e., $\{\delta_j\}=v_{\{\gamma_j\}}(\{\alpha_j\})$. Thus, $v_{\{\gamma_j\}}$ is closed, whence continuous and therefore (since $S=[e_n]$ and $v_{\{\gamma_j\}}(e_n)=\gamma_n e_n$ for all $n=1,2,\ldots$)

$$v_{\{\gamma_j\}}(S)\subset S \qquad (\{\gamma_j\}\in m),$$

i.e., S is normal. Consequently, by § 16, theorem 16.1 (implication $5°\Rightarrow 1°$), $\{e_n\}$ is an unconditional basis of S, which completes the proof of theorem 17.8.

One can also prove the sufficiency part in theorem 17.8 with an argument similar to the proof of Ch. I, § 12, theorem 12.6, by introducing on T^α the norm

$$\|\{\beta_j\}\|=\sup_{\substack{\{\alpha_j\}\in T\\ \|\{\alpha_j\}\|\leq 1}}\ \sup_{1\leq n<\infty}\sum_{i=1}^{n}|\beta_i\alpha_i|=\sup_{\substack{\{\alpha_j\}\in T\\ \|\{\alpha_j\}\|\leq 1}}\sum_{i=1}^{\infty}|\beta_i\alpha_i| \qquad (\{\beta_j\}\in T^\alpha) \tag{17.84}$$

(we have $\|\{\beta_j\}\|<\infty$ by the principle of uniform boundedness applied to the sequence of continuous non-linear functionals $F_n(\{\alpha_j\})=\sum_{i=1}^{n}|\beta_i\alpha_i|$, where $\{\alpha_j\}\in T$, $n=1,2,\ldots$) and on $T=T^{\alpha\alpha}$ the equivalent norm

$$|||\{\alpha_j\}|||=\sup_{\substack{\{\beta_j\}\in T^\alpha\\ \|\{\beta_j\}\|\leq 1}}\sum_{i=1}^{\infty}|\beta_i\alpha_i| \qquad (\{\alpha_j\}\in T) \tag{17.85}$$

and observing that for any scalars $\alpha_1,\ldots,\alpha_n,\gamma_1,\ldots,\gamma_n$ with $|\gamma_i|\leq 1$ $(i=1,\ldots,n)$ we have

$$\left|\left|\left|\sum_{i=1}^{n}\gamma_i\alpha_i e_i\right|\right|\right|=\sup_{\substack{\{\beta_j\}\in T^\alpha\\ \|\{\beta_j\}\|\leq 1}}\sum_{i=1}^{n}|\beta_i\gamma_i\alpha_i|\leq\sup_{\substack{\{\beta_j\}\in T^\alpha\\ \|\{\beta_j\}\|\leq 1}}\sum_{i=1}^{n}|\beta_i\alpha_i|=\left|\left|\left|\sum_{i=1}^{n}\alpha_i e_i\right|\right|\right|.$$

§ 18. Equivalence and permutative equivalence of unconditional bases. Universal unconditional bases

In finite dimensional Banach spaces all bases are unconditional and equivalent, hence also permutatively equivalent. It is natural to ask whether in an infinite dimensional Banach space with an unconditional basis there exist two non-permutatively equivalent bounded[1] unconditional bases, or at least two non-equivalent bounded unconditional bases. We shall give now the answer in some concrete Banach spaces with unconditional bases. In § 24 we shall determine the Banach spaces which have, up to equivalence, a unique bounded unconditional basis.

Theorem 18.1. *In the space l^2 all bounded unconditional bases are equivalent.*

Proof. By § 14, proposition 14.1, every bounded unconditional basis of l^2 is both Besselian and Hilbertian and hence, by § 11, corollary 11.3, equivalent to the unit vector basis of l^2, which completes the proof.

Now we shall show that a result similar to theorem 18.1 holds also in the spaces l^1 and c_0. We shall first prove some lemmas, culminating in lemma 18.3.

Let $S^{n-1}=\{x\in E^n \mid \|x\|=1\}$ denote the $(n-1)$-dimensional sphere in the n-dimensional *real* euclidean space E^n. Let μ be the rotation invariant Borel measure on S^{n-1} normalized so that $\mu(S^{n-1})=1$. Let $(x,y)=\sum_{i=1}^{n}\xi_i\eta_i$ ($x=\{\xi_i\}_{i=1}^n$, $y=\{\eta_i\}_{i=1}^n\in E^n$) denote the usual inner product.

Lemma 18.1. *Let $x,y\in S^{n-1}$. Then*

$$\int_{S^{n-1}} \operatorname{sign}(x,z)\operatorname{sign}(y,z)\,d\mu(z)=1-\frac{2}{\pi}\theta(x,y), \tag{18.1}$$

where $\theta=\theta(x,y)$ is the unique number satisfying $\cos\theta=(x,y)$ and $0<\theta<\pi$ (i.e., $\theta=\arccos(x,y)$).

Proof. Applying, if necessary, a rotation, we can choose the basis in E^n in such a way that $x=\{1,0,0,\ldots,0\}$ and $y=\{\cos\theta,\sin\theta,0,\ldots,0\}$ (i.e., that x is the first basis element, y is in the two-dimensional plane spanned by the first two basis elements and $(x,y)=1\cdot\cos\theta+0\cdot\sin\theta+0=\cos\theta$). This being done, we shall use polar coordinates $\phi=\{\phi_1,\ldots,\phi_{n-1}\}$ to write the points $z=\{\zeta_1,\ldots,\zeta_n\}\in S^{n-1}$ as $z(\phi)=\{\zeta_1(\phi),\ldots,\zeta_n(\phi)\}$, where

[1] We have already observed in a footnote in Ch. I, § 8, that a boundedness condition is necessary in problems of this type.

$$\zeta_1(\phi)=\prod_{i=1}^{n-1}\sin\phi_i, \tag{18.2}$$

$$\zeta_k(\phi)=\cos\phi_{k-1}\prod_{i=k}^{n-1}\sin\phi_i \qquad (k=2,3,\dots,n-1), \tag{18.3}$$

$$\zeta_n(\phi)=\cos\phi_{n-1}. \tag{18.4}$$

For any bounded measurable function g on S^{n-1} we have[1] then

$$\int\limits_{S^{n-1}} g(z)\,d\mu(z)=\frac{1}{|S^{n-1}|}\int\limits_{I^{n-1}} g(z(\phi))J(\phi)\,d(\phi), \tag{18.5}$$

where

$$I^{n-1}=\{\phi\,|\,0\leqslant\phi_1<2\pi,\ 0\leqslant\phi_i\leqslant\pi\ (i=2,3,\dots,n-1)\}, \tag{18.6}$$

$$J(\phi)=\prod_{i=2}^{n-1}(\sin\phi_i)^{i-1}, \tag{18.7}$$

$$|S^{n-1}|=\int\limits_{I^{n-1}} J(\phi)\,d(\phi)=2\pi\prod_{i=2}^{n-1}\int\limits_0^{\pi}(\sin\phi_i)^{i-1}\,d\phi_i. \tag{18.8}$$

Let $h(z)=(x,z)\,(y,z)$. Then, by our choice of the basis in E^n we have $h(z)=\zeta_1(\zeta_1\cos\theta+\zeta_2\sin\theta)$, whence, by (18.2) and (18.3),

$$h(z(\phi))=\left(\prod_{i=2}^{n-1}\sin\phi_i\right)^2\sin\phi_1(\sin\phi_1\cos\theta+\cos\phi_1\sin\theta).$$

Consequently, for the function

$$g(z)=\operatorname{sign}(x,z)\operatorname{sign}(y,z)=\operatorname{sign}h(z) \qquad (z\in S^{n-1}) \tag{18.9}$$

we obtain

$$g(z(\phi))=\operatorname{sign}[\sin\phi_1\sin(\phi_1+\theta)]=f(\phi_1,\theta). \tag{18.10}$$

Since $0<\theta<\pi$, we have

$$f(\phi_1,\theta)=\operatorname{sign}[\sin\phi_1\sin(\phi_1+\theta)]=\begin{cases}\ \ 1 & \text{if}\quad \phi_1\in(0,\pi-\theta)\cup(\pi,2\pi-\theta),\\ -1 & \text{if}\quad \phi_1\in(\pi-\theta,\pi)\cup(2\pi-\theta,2\pi),\end{cases}$$

(indeed, e.g. for $\phi_1\in(0,\pi-\theta)$ we have $\phi_1+\theta\in(\theta,\pi)\subset(0,\pi)$, whence $\sin\phi_1>0$, $\sin(\phi_1+\theta)>0$, etc.). Consequently, by (18.5)–(18.8) we get

[1] See e.g. [59], p. 401–402.

$$\int_{S^{n-1}} g(z)\,d\mu(z) = \frac{1}{|S^{n-1}|}\int_{I^{n-1}} f(\phi_1,\theta)J(\phi)d(\phi)$$

$$= \frac{1}{|S^{n-1}|}\int_0^{2\pi} f(\phi_1,\theta)d\phi_1 \prod_{i=2}^{n-1}\int_0^{2\pi}(\sin\phi_i)^{i-1}d\phi_i$$

$$= \frac{1}{2\pi}\int_0^{2\pi} f(\phi_1,\theta)d\phi_1 = \frac{1}{2\pi}[(\pi-\theta)-\theta+(\pi-\theta)-\theta] = 1-\frac{2\theta}{\pi},$$

which completes the proof of lemma 18.1.

Lemma 18.2. *Let* $(a_{ij})_{i,j=1}^N$ *be a real-valued* $N\times N$ *matrix and let* $M>0$ *be such that*

$$\left|\sum_{i=1}^N \sum_{j=1}^N a_{ij}t_i s_j\right| \leqslant M \qquad (|t_i|\leqslant 1,\ |s_j|\leqslant 1;\ i,j=1,\dots,N). \qquad (18.11)$$

Then for arbitrary elements x_i, y_j $(i,j=1,\dots,N)$ *in a real inner product space* $\mathscr{H}$ *we have*

$$\left|\sum_{i=1}^N \sum_{j=1}^N a_{ij}(x_i,y_j)\right| \leqslant C M \sup_i \|x_i\| \sup_j \|y_j\|, \qquad (18.12)$$

where C is a constant independent of N and x_i, y_j, *satisfying*

$$C \leqslant \sinh\frac{\pi}{2} = \frac{e^{\frac{\pi}{2}}-e^{-\frac{\pi}{2}}}{2}. \qquad (18.13)$$

Proof. Let us first observe that if a matrix (a_{ij}) satisfies (18.11) then for arbitrary real numbers c_i', c_j'' $(i,j=1,\dots,N)$, the matrix (a_{ij}'), where

$$a_{ij}' = c_i' a_{ij} c_j'' \qquad (i,j=1,\dots,N), \qquad (18.14)$$

also satisfies (18.11), with the constant $M' = M\sup_i|c_i'|\sup_j|c_j''|$. Indeed, this is obvious if $\sup_i|c_i'|\sup_j|c_j''|=0$; if $\sup_i|c_i'|\sup_j|c_j''|\neq 0$, then by (18.11) we have

$$\left|\sum_{i=1}^N \sum_{j=1}^N a_{ij}\frac{c_i't_i}{\sup_k|c_k'|}\,\frac{c_j''s_j}{\sup_l|c_l''|}\right| \leqslant M \qquad (|t_i|,|s_j|\leqslant 1;\ i,j=1,\dots,N),$$

whence the assertion follows.

Let us proceed now to the proof of lemma 18.2. Since any $2N$ elements in $\mathscr{H}$ belong to some $2N$-dimensional subspace of $\mathscr{H}$ which is isometric to E^{2N}, we may assume without loss of generality that $x_i, y_j \in E^{2N}$

$(i,j=1,\ldots,N)$. Furthermore, we may also assume that $\|x_i\|=\|y_j\|=1$ $(i,j=1,\ldots,N)$. Indeed, if lemma 18.2 is true in this case, then for arbitrary elements $x_i, y_i \in \mathscr{H}$ we have, by the above remark applied with $c_i'=\|x_i\|$, $c_j''=\|y_j\|$ $(i,j=1,\ldots,N)$,

$$\left|\sum_{i=1}^{N}\sum_{j=1}^{N} a_{ij}\|x_i\|\,\|y_j\|\,t_i s_j\right| \leqslant M \sup_i \|x_i\| \sup_j \|y_j\|$$

$$(|t_i|,\ |s_j|\leqslant 1;\ i,j=1,\ldots,N),$$

whence, by our assumption applied to the matrix $(a_{ij}\|x_i\|\,\|y_j\|)_{i,j=1}^N$ and to the elements $x_i'=\dfrac{x_i}{\|x_i\|}$, $y_j'=\dfrac{y_j}{\|y_j\|}$ if $x_i, y_j \neq 0$ and x_i', y_j' $=$arbitrary elements of norm 1 of E^{2N} for the other i,j, we obtain

$$\left|\sum_{i=1}^{N}\sum_{j=1}^{N} a_{ij}(x_i,y_j)\right| = \left|\sum_{i=1}^{N}\sum_{j=1}^{N} a_{ij}\|x_i\|\,\|y_j\|(x_i',y_j')\right|$$
$$\leqslant C M \sup_i \|x_i\| \sup_j \|y_j\|,$$

which proves the assertion.

Thus, let $x_i, y_j \in E^{2N}$, $\|x_i\|=\|y_j\|=1$ $(i,j=1,\ldots,N)$. Define

$$t_i(z)=\operatorname{sign}(x_i,z),\ s_j(z)=\operatorname{sign}(y_j,z) \qquad (z\in S^{2N-1};\ i,j=1,\ldots,N). \tag{18.15}$$

Then, by (18.11),

$$-M \leqslant \sum_{i=1}^{N}\sum_{j=1}^{N} a_{ij}t_i(z)s_j(z) \leqslant M \qquad (z\in S^{2N-1}), \tag{18.16}$$

whence, integrating over S^{2N-1} with respect to the normalized rotation invariant measure μ we get, by lemma 18.1,

$$-\frac{\pi}{2}M \leqslant \sum_{i=1}^{N}\sum_{j=1}^{N} a_{ij}\left[\frac{\pi}{2}-\theta(x_i,y_j)\right] \leqslant \frac{\pi}{2}M, \tag{18.17}$$

for any matrix (a_{ij}) satisfying (18.11).

We claim that the matrix $(a_{ij}^{(1)})$, where

$$a_{ij}^{(1)}=a_{ij}\left[\frac{\pi}{2}-\theta(x_i,y_j)\right] \qquad (i,j=1,\ldots,N), \tag{18.18}$$

also satisfies (18.11) with M replaced by $\dfrac{\pi M}{2}$. Indeed, if $|t_i|\leqslant 1$, $|s_j|\leqslant 1$ $(i,j=1,\ldots,N)$, then by the remark made at the beginning of this proof, the matrix $(a_{ij}^{(t_i,s_j)})$, where $a_{ij}^{(t_i,s_j)}=a_{ij}t_i s_j$ $(i,j=1,\ldots,N)$, also satisfies (18.11), whence, applying (18.17) for this matrix instead of (a_{ij}),

$$-\frac{\pi}{2}M\leqslant\sum_{i=1}^{N}\sum_{j=1}^{N}a_{ij}t_is_j\left[\frac{\pi}{2}-\theta(x_i,y_j)\right]=\sum_{i=1}^{N}\sum_{j=1}^{N}a_{ij}^{(1)}t_is_j\leqslant\frac{\pi}{2}M,$$

which proves our assertion. Therefore, repeating the argument of integration over S^{2N-1}, we obtain

$$\begin{aligned}-\left(\frac{\pi}{2}\right)^2 M&\leqslant\sum_{i=1}^{N}\sum_{j=1}^{N}a_{ij}^{(1)}\left[\frac{\pi}{2}-\theta(x_i,y_j)\right]\\&=\sum_{i=1}^{N}\sum_{j=1}^{N}a_{ij}\left[\frac{\pi}{2}-\theta(x_i,y_j)\right]^2\leqslant\left(\frac{\pi}{2}\right)^2 M.\end{aligned}$$

Putting $a_{ij}^{(2)}=a_{ij}^{(1)}\left[\frac{\pi}{2}-\theta(x_i,y_j)\right]$ and continuing in the same manner, we obtain inductively

$$-\left(\frac{\pi}{2}\right)^n M\leqslant\sum_{i=1}^{N}\sum_{j=1}^{N}a_{ij}\left[\frac{\pi}{2}-\theta(x_i,y_j)\right]^n\leqslant\left(\frac{\pi}{2}\right)^n M\quad(n=1,2,\ldots).\quad(18.19)$$

Applying this for $2n+1$ instead of n and multiplying by $\frac{(-1)^n}{(2n+1)!}$, we get

$$\begin{aligned}-\left(\frac{\pi}{2}\right)^{2n+1}\frac{M}{(2n+1)!}&\leqslant\sum_{i=1}^{N}\sum_{j=1}^{N}a_{ij}\frac{(-1)^n\left[\frac{\pi}{2}-\theta(x_i,y_j)\right]^{2n+1}}{(2n+1)!}\\&\leqslant\left(\frac{\pi}{2}\right)^{2n+1}\frac{M}{(2n+1)!}\qquad(n=1,2,\ldots),\end{aligned}$$

and hence

$$\begin{aligned}-M\sinh\frac{\pi}{2}&=-\sum_{n=0}^{\infty}\left(\frac{\pi}{2}\right)^{2n+1}\frac{M}{(2n+1)!}\\&\leqslant\sum_{n=0}^{\infty}\sum_{i=1}^{N}\sum_{j=1}^{N}a_{ij}\frac{(-1)^n\left[\frac{\pi}{2}-\theta(x_i,y_j)\right]^{2n+1}}{(2n+1)!}\\&=\sum_{i=1}^{N}\sum_{j=1}^{N}a_{ij}\sum_{n=0}^{\infty}\frac{(-1)^n\left[\frac{\pi}{2}-\theta(x_i,y_j)\right]^{2n+1}}{(2n+1)!}\\&=\sum_{i=1}^{N}\sum_{j=1}^{N}a_{ij}\sin\left[\frac{\pi}{2}-\theta(x_i,y_j)\right]=\sum_{i=1}^{N}\sum_{j=1}^{N}a_{ij}\cos\theta(x_i,y_j)\\&=\sum_{i=1}^{N}\sum_{j=1}^{N}a_{ij}(x_i,y_j)\leqslant M\sinh\frac{\pi}{2},\end{aligned}$$

which completes the proof of lemma 18.2.

Corollary 18.1. *Let (a_{ij}) be a real valued $N \times N$ matrix satisfying* (18.11). *Then for arbitrary elements y_j $(j=1,\dots,N)$ in a real inner product space $\mathscr{H}$ we have*

$$\sum_{i=1}^{N} \left\| \sum_{j=1}^{N} a_{ij} y_j \right\| \leqslant C M \sup_j \|y_j\| . \tag{18.20}$$

Proof. For each $i=1,\dots,N$ choose an element $x_i \in \mathscr{H}$ such that

$$\left(\sum_{j=1}^{N} a_{ij} y_j, x_i \right) = \left\| \sum_{j=1}^{N} a_{ij} y_j \right\|, \quad \|x_i\| = 1.$$

Then, by using these elements x_i, y_j in (18.12), we obtain (18.20), which completes the proof.

Lemma 18.3. *Every continuous linear mapping $u: l^1 \to l^2$ is "absolutely summing", i.e., carries each unconditionally convergent series $\sum_{i=1}^{\infty} z_i$ in l^1 into an absolutely convergent series $\sum_{i=1}^{\infty} u(z_i)$ in l^2.*

Proof. It is clearly sufficient to prove the statement for real spaces.

Let $\sum_{i=1}^{\infty} z_i$ be an unconditionally convergent series in l^1. Then, by § 15, corollary 15.1, there exists a constant $M>0$ such that

$$\sum_{i=1}^{\infty} |f(z_i)| \leqslant M \|f\| \qquad (f \in (l^1)^* \equiv l^\infty). \tag{18.21}$$

Let $\{x_n\}$ be the unit vector basis of l^1 and $\{f_n\} \subset (l^1)^* \equiv l^\infty$ the a. s. c. f. Then

$$z_i = \sum_{j=1}^{\infty} a_{ij} x_j \qquad (i=1,2,\dots), \tag{18.22}$$

where $a_{ij} = f_j(z_i)$ $(i,j=1,2,\dots)$.

Let t_i, s_j be arbitrary scalars with $|t_i|, |s_j| \leqslant 1$ $(i,j=1,2,\dots,N)$. Then for $f = \sum_{j=1}^{N} s_j f_j \in (l^1)^* \equiv l^\infty$ we have $f(z_i) = \sum_{j=1}^{N} a_{ij} s_j$ $(i=1,\dots,N)$, $\|f\| = \max_{1 \leqslant j \leqslant N} |s_j| \leqslant 1$, whence

$$\left| \sum_{i,j=1}^{N} a_{ij} t_i s_j \right| \leqslant \sum_{i=1}^{N} |t_i| \left| \sum_{j=1}^{N} a_{ij} s_j \right| \leqslant \sum_{i=1}^{N} \left| \sum_{j=1}^{N} a_{ij} s_j \right|$$

$$= \sum_{i=1}^{N} |f(z_i)| \leqslant M \|f\| \leqslant M .$$

Hence, by corollary 18.1 applied for $y_j = u(x_j) \in l^2$ $(j=1,\dots,N)$, and by $\|x_j\| = 1$ $(j=1,2,\dots)$ we infer

$$\sum_{i=1}^{n} \left\| \sum_{j=1}^{N} a_{ij} u(x_j) \right\| \leqslant \sum_{i=1}^{N} \left\| \sum_{j=1}^{N} a_{ij} u(x_j) \right\| \leqslant CM \sup_{1 \leqslant j \leqslant N} \|u(x_j)\| \leqslant CM \|u\|$$
$$(n \leqslant N = 1,2,\dots),$$

whence, for $N \to \infty$, $\sum_{i=1}^{n} \|u(z_i)\| \leqslant CM \|u\|$ $(n = 1,2,\dots)$. Consequently, $\sum_{i=1}^{\infty} \|u(z_i)\| < \infty$, which completes the proof of lemma 18.3.

Now we can prove

Theorem 18.2. *In each of the spaces l^1 and c_0 all bounded unconditional bases are equivalent.*

Proof. Let $\{x_n\}$ be an arbitrary bounded unconditional basis of $E = l^1$, with the a. s. c. f. $\{f_n\}$. Then, by § 14, proposition 14.1, $\{x_n\}$ is a Besselian basis, and hence, by § 11, theorem 11.1a) (implication $1^\circ \Rightarrow 3^\circ$), there exists a continuous linear mapping $u: l^1 \to l^2$ such that

$$u(x_n) = e_n \quad (n = 1,2,\dots), \tag{18.23}$$

where $\{e_n\}$ is the unit vector basis of l^2. By lemma 18.3, u is absolutely summing, whence, since $\|u(x_i)\| = \|e_i\|_{l^2} = 1$ $(i = 1,2,\dots)$, we obtain

$$\sum_{i=1}^{\infty} |f_i(x)| = \sum_{i=1}^{\infty} |f_i(x)| \, \|u(x_i)\| = \sum_{i=1}^{\infty} \|u[f_i(x) x_i]\| < \infty \quad (x \in l^1). \tag{18.24}$$

Conversely, since $\{x_n\}$ is a bounded basis, by Ch. I, § 3, lemma 3.1b) we have the implication

$$\sum_{i=1}^{\infty} |\alpha_i| < \infty \Rightarrow \sum_{i=1}^{\infty} \alpha_i x_i \quad \text{converges}, \tag{18.25}$$

which, together with (18.24), proves that $\{x_n\}$ is equivalent to the unit vector basis of l^1. This proves the assertion of theorem 18.2 for the space $E = l^1$.

Now let $\{x_n\}$ be an arbitrary bounded unconditional basis of $E = c_0$, with the a. s. c. f. $\{f_n\}$. Then, by § 17, corollary 17.3a), $\{x_n\}$ is a shrinking basis and hence, by § 17, theorem 17.7 and Ch. I, § 3, corollary 3.1, $\{f_n\}$ is a bounded unconditional basis of $E^* \equiv l^1$. Therefore, by the above, $\{f_n\}$ is equivalent to the unit vector basis of l^1. Consequently, by Ch. I, § 12, proposition 12.1, $\{x_n\}$ is equivalent to the unit vector basis of c_0, which completes the proof of theorem 18.2.

Our next aim is to show that the fact that in l^2 and $L^2([0,1])$ all bounded unconditional bases are equivalent (theorem 18.1) characterizes the space l^2 among the spaces l^p (the space $L^2([0,1])$ among the spaces $L^p([0,1])$) for $1<p<\infty$; moreover, in each of the spaces l^p and $L^p([0,1])$, where $1<p<\infty$, $p\neq 2$, there exist two non-permutatively equivalent bounded unconditional bases. For this purpose we need some propositions and lemmas. By "projection" we shall understand "continuous linear projection."

Proposition 18.1. *In $E=l^p$ $(1\leqslant p<\infty)$ let $\{x_n\}$ be the unit vector basis and let $\{z_n\}$ be a block basic sequence with respect to $\{x_n\}$, i.e.,*

$$z_n=\sum_{i=m_{n-1}+1}^{m_n}\alpha_i x_i\neq 0 \qquad (n=1,2,\ldots), \tag{18.26}$$

where $0=m_0<m_1<\cdots$ Then

a) *There exists a linear isometry u of l^p onto $[z_n]$, such that*

$$u(x_n)=\frac{z_n}{\|z_n\|} \qquad (n=1,2,\ldots). \tag{18.27}$$

b) *There exists a projection v of norm 1 of l^p onto $[z_n]$.*

Proof. a) For any scalars $\xi_1,\ldots,\xi_n$ we have

$$\left\|\sum_{j=1}^{n}\xi_j\frac{z_j}{\|z_j\|}\right\|=\left\|\sum_{j=1}^{n}\frac{\xi_j}{\|z_j\|}\sum_{i=m_{j-1}+1}^{m_j}\alpha_i x_i\right\|=\left(\sum_{j=1}^{n}\sum_{i=m_{j-1}+1}^{m_j}\left|\frac{\xi_j}{\|z_j\|}\alpha_i\right|^p\right)^{\frac{1}{p}}$$

$$=\left(\sum_{j=1}^{n}\frac{|\xi_j|^p}{\|z_j\|^p}\sum_{i=m_{j-1}+1}^{m_j}|\alpha_i|^p\right)^{\frac{1}{p}}=\left(\sum_{j=1}^{n}|\xi_j|^p\right)^{\frac{1}{p}}=\left\|\sum_{j=1}^{n}\xi_j x_j\right\| \tag{18.28}$$

whence the assertion follows.

b) Put $E_n=[x_i]_{i=m_{n-1}+1}^{m_n}$ $(n=1,2,\ldots)$. Since $z_n\in E_n$, there exists a functional $\phi_n\in E_n^*$ such that $\phi_n(z_n)=1$, $\|\phi_n\|=\dfrac{1}{\|z_n\|}$ $(n=1,2,\ldots)$. Put

$$v(x)=\sum_{j=1}^{\infty}\phi_j\left(\sum_{i=m_{j-1}+1}^{m_j}\xi_i x_i\right)z_j \qquad \left(x=\sum_{i=1}^{\infty}\xi_i x_i\in l^p\right). \tag{18.29}$$

This series is convergent and $\|v\|=1$, since by (18.28) and $\|\phi_n\|=\dfrac{1}{\|z_n\|}$ we have

$$\left\|\sum_{j=1}^{\infty}\phi_j\left(\sum_{i=m_{j-1}+1}^{m_j}\xi_i x_i\right)z_j\right\|=\left(\sum_{j=1}^{\infty}\left|\phi_j\left(\sum_{i=m_{j-1}+1}^{m_j}\xi_i x_i\right)\right|^p\|z_j\|^p\right)^{\frac{1}{p}}$$

$$\leqslant\left(\sum_{j=1}^{\infty}\frac{1}{\|z_j\|^p}\left\|\sum_{i=m_{j-1}+1}^{m_j}\xi_i x_i\right\|^p\|z_j\|^p\right)^{\frac{1}{p}}$$

$$=\left(\sum_{j=1}^{\infty}\sum_{i=m_{j-1}+1}^{m_j}|\xi_i|^p\right)^{\frac{1}{p}}=\|x\|.$$

Finally, since $\phi_n(z_n)=1$,

$$v(z_n)=\phi_n\left(\sum_{i=m_{n-1}+1}^{m_n}\alpha_i x_i\right)z_n=\phi_n(z_n)z_n=z_n \quad (n=1,2,\ldots)$$

whence $v(x)=x$ for all $x\in[z_n]$, which completes the proof.

Proposition 18.2. *Let $\{z_n\}$ be a basic sequence in a Banach space E with the a.s.c.f. $\{h_n\}\subset[z_n]^*$ and let $\{y_n\}$ be a sequence in E. If there exists a projection v of E onto $[z_n]$ such that*

$$\|v\|\sum_{n=1}^{\infty}\|h_n\|\,\|y_n-z_n\|<1, \tag{18.30}$$

then $\{y_n\}$ is a basic sequence equivalent to $\{z_n\}$ and there exists a projection w of E onto $[y_n]$.

Proof. By (18.30) we have

$$\sum_{n=1}^{\infty}\|h_n\|\,\|y_n-z_n\|<\frac{1}{\|v\|}\leqslant 1,$$

whence, by Ch. I, § 10, theorem 10.1, $\{y_n\}$ is a basic sequence equivalent to $\{z_n\}$. Put

$$u(x)=x-v(x)+\sum_{i=1}^{\infty}h_i[v(x)]y_i \quad (x\in E) \tag{18.31}$$

(the series in (18.31) is convergent since by $v(x)\in[z_n]$ the series $\sum_{i=1}^{\infty}h_i[v(x)]z_i$ is convergent and since $\{y_n\}$ is equivalent to $\{z_n\}$). Then, by (18.30),

$$\|I_E-u\|=\sup_{\|x\|\leqslant 1}\|x-u(x)\|=\sup_{\|x\|\leqslant 1}\left\|v(x)-\sum_{i=1}^{\infty}h_i[v(x)]y_i\right\|$$

$$=\sup_{\|x\|\leqslant 1}\left\|\sum_{i=1}^{\infty}h_i[v(x)](z_i-y_i)\right\|\leqslant\|v\|\sum_{i=1}^{\infty}\|h_i\|\,\|z_i-y_i\|<1,$$

whence u is an isomorphism of E onto E. Furthermore, by $v(z_n)=z_n$ and $h_i(z_j)=\delta_{ij}$ we have

$$u(z_n)=z_n-v(z_n)+\sum_{i=1}^{\infty}h_i[v(z_n)]y_i=\sum_{i=1}^{\infty}h_i(z_n)y_i=y_n \quad (n=1,2,\ldots),$$

whence $u([z_n])=[y_n]$. Put

$$w=uvu^{-1}. \tag{18.32}$$

Then

$$w(E)=(uvu^{-1})(E)=(uv)(E)=u([z_n])=[y_n]$$

and

$$w(y_n)=(uvu^{-1})(y_n)=(uv)(z_n)=u(z_n)=y_n \quad (n=1,2,\ldots),$$

whence w is a projection of E onto $[y_n]$, which completes the proof.

Lemma 18.4. *Let E be an infinite dimensional subspace of l^p $(1\leqslant p<\infty)$. Then E contains a subspace F which is complemented in l^p and isomorphic to l^p.*

Proof. Let $\{x_n\}$ be the unit vector basis of l^p. Since $\dim E=\infty$, we can construct a sequence $\{y_n\}\subset E$ and a sequence of indices $0=m_0<m_1<\cdots$ such that

$$y_n=\sum_{i=m_{n-1}+1}^{\infty}\alpha_i^{(n)}x_i \quad (n=1,2,\ldots), \tag{18.33}$$

$$\|y_n\|=1 \quad (n=1,2,\ldots), \tag{18.34}$$

$$\left\|\sum_{i=m_n+1}^{\infty}\alpha_i^{(n)}x_i\right\|\leqslant\frac{1}{2^{n+1}} \quad (n=1,2,\ldots). \tag{18.35}$$

Indeed, take an arbitrary $y_1=\sum_{i=1}^{\infty}\alpha_i^{(1)}x_i$ with $\|y_1\|=1$ and take m_1 such that $\left\|\sum_{i=m_1+1}^{\infty}\alpha_i^{(1)}x_i\right\|\leqslant\frac{1}{4}$. Assume that we have constructed $y_1,\ldots,y_n$ and $m_1,\ldots,m_n$. Then, since $\dim E=\infty$ and $\operatorname{codim}[x_{m_n+k}]_{k=1}^{\infty}=m_n<\infty$, there exists an element $y_{n+1}=\sum_{i=m_n+1}^{\infty}\alpha_i^{(n+1)}x_i\in[x_{m_n+k}]_{k=1}^{\infty}\cap E$ with $\|y_{n+1}\|=1$, whence also an index m_{n+1} with $\left\|\sum_{i=m_{n+1}+1}^{\infty}\alpha_i^{(n+1)}x_i\right\|\leqslant\frac{1}{2^{n+2}}$, which proves our assertion.

Put

$$z_n=\sum_{i=m_{n-1}+1}^{m_n}\alpha_i^{(n)}x_i \quad (n=1,2,\ldots). \tag{18.36}$$

Then, by (18.33), (18.36) and (18.35) we have

$$\|y_n-z_n\|=\left\|\sum_{i=m_{n-1}+1}^{\infty}\alpha_i^{(n)}x_i-\sum_{i=m_{n-1}+1}^{m_n}\alpha_i^{(n)}x_i\right\|=\left\|\sum_{i=m_n+1}^{\infty}\alpha_i^{(n)}x_i\right\|\leqslant\frac{1}{2^{n+1}}$$
$$(n=1,2,\ldots), \tag{18.37}$$

whence, taking into account (18.34),

$$\|z_n\|\geqslant\|y_n\|-\|y_n-z_n\|\geqslant 1-\frac{1}{2^{n+1}}>0 \quad (n=1,2,\ldots), \tag{18.38}$$

and thus $\{z_n\}$ is a block basic sequence with respect to x_n. Let $\{h_n\}\subset[z_n]^*$ be its a.s.c.f. Then for the sequence of coefficient functionals $\{h_n\|z_n\|\}$ associated to the basis $\left\{\frac{z_n}{\|z_n\|}\right\}$ of $[z_n]$ we have, by proposition 18.1a), $\|h_n\|z_n\|\|=1$ $(n=1,2,\ldots)$, whence, by (18.38),

$$\|h_n\|=\frac{1}{\|z_n\|}\leqslant\frac{1}{1-\frac{1}{2^{n+1}}}\qquad(n=1,2,\ldots). \tag{18.39}$$

Furthermore, by proposition 18.1b) there exists a projection v of norm 1 of l^p onto $[z_n]$ and we have, by (18.39) and (18.37),

$$\|v\|\sum_{n=1}^{\infty}\|h_n\|\,\|y_n-z_n\|\leqslant\sum_{n=1}^{\infty}\frac{1}{1-\frac{1}{2^{n+1}}}\cdot\frac{1}{2^{n+1}}<1.$$

Consequently, by proposition 18.2, the subspace $F=[y_n]$ is complemented in l^p and isomorphic to $[z_n]$, whence also to l^p (by proposition 18.1a)), which completes the proof of lemma 18.4.

Let $1\leqslant p<\infty$ and let $\{E_n\}$ be a sequence of Banach spaces. We denote by $(E_1\times E_2\times\cdots)_{l^p}$ the space of all sequences $\{x_n\}$ with $x_n\in E_n$ $(n=1,2,\ldots)$, for which $\{\|x_n\|\}\in l^p$, endowed with the norm

$$\|\{x_n\}\|=\left(\sum_{i=1}^{\infty}\|x_i\|^p\right)^{\frac{1}{p}}. \tag{18.40}$$

It is easy to verify that $(E_1\times E_2\times\cdots)_{l^p}$ is a Banach space. The finite product $(E_1\times\cdots\times E_m)_{l^p}$ is defined similarly. By $E_1\times\cdots\times E_m$ (where $m<\infty$) we shall denote the product space endowed with an arbitrary norm equivalent to the norm of $(E_1\times\cdots\times E_m)_{l^p}$.

We shall write $E\cong F$, respectively, $E\equiv F$, if the spaces E and F are isomorphic, respectively equivalent (i. e., linearly isometric).

Lemma 18.5. *Let* $1\leqslant p<\infty$. *Then*

a) *If* E,F *are Banach spaces such that* $E\cong F$, *then*

$$(E\times E\times\cdots)_{l^p}\cong(F\times F\times\cdots)_{l^p}. \tag{18.41}$$

b) *We have*

$$(l^p\times l^p\times\cdots)_{l^p}\equiv l^p. \tag{18.42}$$

c) *For any two Banach spaces* E,F *we have*

$$((E\times F)_{l^p}\times(E\times F)_{l^p}\times\cdots)_{l^p}\equiv((E\times E\times\cdots)_{l^p}\times(F\times F\times\cdots)_{l^p})_{l^p} \tag{18.43}$$

and hence

$$((E\times F)_{l^p}\times(E\times F)_{l^p}\times\cdots)_{l^p}\cong(E\times E\times\cdots)_{l^p}\times(F\times F\times\cdots)_{l^p}. \tag{18.44}$$

d) *For any Banach space E,*

$$(E \times (E \times E \times \cdots)_{l^p})_{l^p} \equiv (E \times E \times \cdots)_{l^p} \tag{18.45}$$

and hence

$$E \times (E \times E \times \cdots)_{l^p} \cong (E \times E \times \cdots)_{l^p}. \tag{18.46}$$

Proof. a) If u' is an isomorphism of E onto F, the required isomorphism may be given by the formula

$$u(\{x_n\}) = \{u'(x_n)\} \qquad (x_n \in E, n = 1, 2, \ldots). \tag{18.47}$$

Indeed, u is one to one, linear and onto, and since $C_1 \|x\| \leqslant \|u'(x)\| \leqslant C_2 \|x\|$ $(x \in E)$ for suitable constants $C_1, C_2 > 0$, we have

$$C_1 \|\{x_n\}\| = C_1 \left(\sum_{i=1}^{\infty} \|x_i\|^p \right)^{\frac{1}{p}} \leqslant \left(\sum_{i=1}^{\infty} \|u'(x_i)\|^p \right)^{\frac{1}{p}} = \|u(\{x_n\})\|$$

$$\leqslant C_2 \left(\sum_{i=1}^{\infty} \|x_i\|^p \right)^{\frac{1}{p}} = C_2 \|\{x_n\}\| \qquad (\{x_n\} \in (E \times E \times \cdots)_{l^p}).$$

b) Let $\{e_n\}$ be the unit vector basis of l^p and let u be an arbitrary one to one mapping of the set of elements

$$z_{ij} = \{\underbrace{0, \ldots, 0}_{j-1}, e_i, 0, 0, \ldots\} \in (l^p \times l^p \times \cdots)_{l^p} \qquad (i, j = 1, 2, \ldots) \tag{18.48}$$

onto the set $\{e_n\}$, extended by linearity to the set of all finite linear combinations. Then we have, for any scalars α_{ij} $(i = 1, \ldots, n; j = 1, \ldots, m)$,

$$\left\| u\left(\sum_{i=1}^{n} \sum_{j=1}^{m} \alpha_{ij} z_{ij} \right) \right\| = \left\| \sum_{i=1}^{n} \sum_{j=1}^{m} \alpha_{ij} u(z_{ij}) \right\| = \left(\sum_{i=1}^{n} \sum_{j=1}^{m} |\alpha_{ij}|^p \right)^{\frac{1}{p}}$$

$$= \left(\sum_{j=1}^{m} \left\| \sum_{i=1}^{n} \alpha_{ij} e_i \right\|^p \right)^{\frac{1}{p}}$$

$$= \left\| \left\{ \sum_{i=1}^{n} \alpha_{i1} e_i, \ldots, \sum_{i=1}^{n} \alpha_{im} e_i, 0, 0, \ldots \right\} \right\|$$

$$= \left\| \sum_{j=1}^{m} \left\{ \underbrace{0, \ldots, 0}_{j-1}, \sum_{i=1}^{n} \alpha_{ij} e_i, 0, 0, \ldots \right\} \right\| = \left\| \sum_{i=1}^{n} \sum_{j=1}^{m} \alpha_{ij} z_{ij} \right\|,$$

whence the assertion follows, since $\{z_{ij}\}$ is linearly independent and $[z_{ij}] = (l^p \times l^p \times \cdots)_{l^p}$, $[e_n] = l^p$.

c) The required isometry may be given by the formula

$$u(\{\{x_n, y_n\}\}) = \{\{x_n\}, \{y_n\}\} \qquad (x_n \in E,\ y_n \in F,\ n = 1, 2, \ldots). \tag{18.49}$$

Indeed, u is one to one, linear and onto, and we have

$$\|u(\{\{x_n, y_n\}\})\| = \|\{\{x_n\}, \{y_n\}\}\| = (\|\{x_n\}\|^p + \|\{y_n\}\|^p)^{\frac{1}{p}}$$
$$= \left(\sum_{i=1}^{\infty} \|x_i\|^p + \sum_{i=1}^{\infty} \|y_i\|^p\right)^{\frac{1}{p}} = \left(\sum_{i=1}^{\infty} (\|x_i\|^p + \|y_i\|^p)\right)^{\frac{1}{p}}$$
$$= \left(\sum_{i=1}^{\infty} \|\{x_i, y_i\}\|^p\right)^{\frac{1}{p}} = \|\{\{x_n, y_n\}\}\|.$$

Hence, we also have the isomorphism (18.44).

d) The required isometry may be given by the formula

$$u(\{x, \{x_n\}\}) = \{x, x_1, x_2, \ldots\} \qquad (x, x_n \in E,\ n = 1, 2, \ldots). \tag{18.50}$$

Indeed, u is one to one, linear and onto, and we have

$$\|u(\{x, \{x_n\}\})\| = \|\{x, x_1, x_2, \ldots\}\| = \left(\|x\|^p + \sum_{i=1}^{\infty} \|x_i\|^p\right)^{\frac{1}{p}}$$
$$= (\|x\|^p + \|\{x_n\}\|^p)^{\frac{1}{p}} = \|\{x, \{x_n\}\}\|.$$

Hence, we also have the isomorphism (18.46), which completes the proof of lemma 18.5.

Lemma 18.6. *If E is an infinite dimensional subspace of l^p, complemented in l^p, then E is isomorphic to l^p* $(1 \leqslant p < \infty)$.

Proof. By lemma 18.4, E contains a subspace F which is complemented in l^p, whence also in E, and which is isomorphic to l^p. Thus there exist[1] subspaces E_1 of l^p and F, F_1 of E such that

$$l^p \cong E \times E_1, \tag{18.51}$$
$$E \cong F \times F_1, \tag{18.52}$$
$$F \cong l^p. \tag{18.53}$$

Hence, taking into account that the operation of product of Banach spaces is associative and commutative and applying lemma 18.5, we obtain

$$l^p \cong E \times E_1 \cong (F \times F_1) \times E_1 \cong (l^p \times F_1) \times E_1 \cong l^p \times (F_1 \times E_1)$$
$$\cong (l^p \times l^p \times \cdots)_{l^p} \times (F_1 \times E_1) \cong ((E \times E_1) \times (E \times E_1) \times \cdots)_{l^p} \times (F_1 \times E_1)$$
$$\cong (E \times E \times \cdots)_{l^p} \times (E_1 \times E_1 \times \cdots)_{l^p} \times (E_1 \times F_1)$$

---

[1] We use the well known fact (see Ch. I, § 4, lemma 4.1) that if F has a complement F_1 in E, then $E \cong F \times F_1$.

$$\cong (E \times E \times \cdots)_{l^p} \times (E_1 \times E_1 \times \cdots)_{l^p} \times F_1$$

$$\cong ((E \times E_1) \times (E \times E_1) \times \cdots)_{l^p} \times F_1 \cong (l^p \times l^p \times \cdots)_{l^p} \times F_1$$

$$\cong l^p \times F_1 \cong F \times F_1 \cong E,$$

which completes the proof of lemma 18.6.

We shall write $E \overset{C}{\cong} F$, if there exists an isomorphism u of E onto F satisfying

$$\|x\| \leqslant \|u(x)\| \leqslant C\|x\| \qquad (x \in E). \tag{18.54}$$

Lemma 18.7. *Let* $1 \leqslant p < \infty$. *Then*

a) *If* $\{E_n\}$ *and* $\{F_n\}$ *are sequences of Banach spaces such that there exists a constant* $C \geqslant 1$ *for which* $E_n \overset{C}{\cong} F_n$ $(n=1,2,\ldots)$, *then*

$$(E_1 \times E_2 \times \cdots)_{l^p} \overset{C}{\cong} (F_1 \times F_2 \times \cdots)_{l^p}. \tag{18.55}$$

b) *If* $\{E_n\}$ *is a sequence of Banach spaces and* F_n *is a subspace of* E_n *such that there exists a projection* v_n *of* E_n *onto* F_n $(n=1,2,\ldots)$ *and* $\sup_{1 \leqslant n < \infty} \|v_n\| < \infty$, *then there exists a projection* v *of* $(E_1 \times E_2 \times \cdots)_{l^p}$ *onto* $(F_1 \times F_2 \times \cdots)_{l^p}$.

Proof. a) For each $n=1,2,\ldots$ let u_n be an isomorphism of E_n onto F_n satisfying

$$\|x\| \leqslant \|u_n(x)\| \leqslant C\|x\| \qquad (x \in E_n). \tag{18.56}$$

Then the required isomorphism can be given by the formula

$$u(\{x_n\}) = \{u_n(x_n)\} \qquad (\{x_n\} \in (E_1 \times E_2 \times \cdots)_{l^p}). \tag{18.57}$$

Indeed, u is one to one, linear and onto (since so is each u_n) and for any $\{x_n\} \in (E_1 \times E_2 \times \cdots)_{l^p}$ we have, by (18.56),

$$\|\{x_n\}\| = \left(\sum_{i=1}^{\infty} \|x_i\|^p\right)^{\frac{1}{p}} \leqslant \left(\sum_{i=1}^{\infty} \|u_i(x_i)\|^p\right)^{\frac{1}{p}} = \|\{u_n(x_n)\}\| = \|u(\{x_n\})\|$$

$$\leqslant C\left(\sum_{i=1}^{\infty} \|x_i\|^p\right)^{\frac{1}{p}} = C\|\{x_n\}\|.$$

b) The required projection can be given by the formula

$$v(\{x_n\}) = \{v_n(x_n)\} \qquad (\{x_n\} \in (E_1 \times E_2 \times \cdots)_{l^p}). \tag{18.58}$$

Indeed, v is linear and continuous, since

$$\|v(\{x_n\})\| = \|\{v_n(x_n)\}\| = \left(\sum_{i=1}^{\infty} \|v_i(x_i)\|^p\right)^{\frac{1}{p}} \leqslant \left(\sum_{i=1}^{\infty} \|v_i\|^p \|x_i\|^p\right)^{\frac{1}{p}}$$

$$\leqslant \sup_{1 \leqslant n < \infty} \|v_n\| \left(\sum_{i=1}^{\infty} \|x_i\|^p\right)^{\frac{1}{p}} = \left(\sup_{1 \leqslant n < \infty} \|v_n\|\right) \|\{x_n\}\|$$

$$(\{x_n\} \in (E_1 \times E_2 \times \cdots)_{l^p}),$$

and v is a projection onto $(F_1 \times F_2 \times \cdots)_{l^p}$ since for $\{x_n\} \in (F_1 \times F_2 \times \cdots)_{l^p}$ we have $v_n(x_n) = x_n$ $(n = 1, 2, \ldots)$, whence

$$v(\{x_n\}) = \{v_n(x_n)\} = \{x_n\}.$$

This completes the proof of lemma 18.7.

Lemma 18.8. *Let* $1 < p < \infty$ *and let* r_n $(n = 1, 2, \ldots)$ *be the Rademacher functions on* $[0, 1]$. *Then for each n the mapping*

$$v_n(x) = \sum_{i=1}^{n} \left[\int_0^1 x(t) r_i(t) dt\right] r_i \qquad (x \in L^p([0,1])) \tag{18.59}$$

is a continuous linear projection of $L^p([0,1])$ *onto* $[r_1, \ldots, r_n]$ *and there exists a constant* B_p *depending only on p, such that*

$$\|v_n\| \leqslant B_p \qquad (n = 1, 2, \ldots). \tag{18.60}$$

Proof. Since the functionals $f_i(x) = \int_0^1 x(t) r_i(t) dt$ are continuous on $L^p([0,1])$, each v_n is continuous. Furthermore, each v_n is a projection of $L^p([0,1])$ onto $[r_1, \ldots, r_n]$, since for $x = \sum_{k=1}^{n} \alpha_k r_k \in [r_1, \ldots, r_n]$ we have, by the orthogonality of the Rademacher functions (§ 14, corollary 14.1),

$$v_n(x) = \sum_{i=1}^{n} \left\{ \int_0^1 \left[\sum_{k=1}^{n} \alpha_k r_k(t)\right] r_i(t) dt \right\} r_i = \sum_{i=1}^{n} \alpha_i r_i = x.$$

Finally, by § 17, proposition 17.2b), for every $x \in L^p([0,1])$ the limit $u(x) = \lim_{n \to \infty} v_n(x)$ exists, whence, by the principle of uniform boundedness, we obtain (18.60), which completes the proof.

Theorem 18.3. *Let* $1 < p < \infty, p \neq 2$. *Then the space* l^p *has two bounded unconditional bases which are not permutatively equivalent.*

Proof. We shall prove that l^p has a bounded unconditional basis which is not equivalent to the unit vector basis of l^p; since every permutation of this latter is equivalent to the unit vector basis of l^p, it follows that our basis is not permutatively equivalent to the unit vector basis of l^p.

Let F_p be the space of all sequences of scalars $\{\xi_n\}$ such that

$$|||\{\xi_n\}||| = \left[\sum_{k=1}^{\infty}\left(\sum_{i=\frac{(k-1)k}{2}+1}^{\frac{k(k+1)}{2}} |\xi_i|^2\right)^{\frac{p}{2}}\right]^{\frac{1}{p}} < \infty. \tag{18.61}$$

It is easy to verify that F_p endowed with the norm (18.61) is a Banach space[1]. Furthermore, the unit vectors $y_n = \{\underbrace{0,\dots,0}_{n-1},1,0,0,\dots\}$ $(n=1,2,\dots)$ form a normalized unconditional basis of F_p (because $[y_n]=F_p$ and by § 17, theorem 17.1, $\{y_n\}$ is an unconditional basic sequence in F_p, with $v^{(u)}_{\{x_n\}}=1$).

Since for each p with $1<p<2$ there exists a sequence $\{\xi_n\}\in F_p$ such that $\sum_{i=1}^{\infty}|\xi_i|^p=\infty$ (e.g. take $\xi_j=\sqrt[p]{\frac{1}{2^n}}$ for $j=\frac{(2^n-1)2^n}{2}+1, \dots, \frac{2^n(2^n+1)}{2}$, $n=1,2,\dots$ and $\xi_j=0$ for the other j) and for each $p>2$ there exists a sequence $\{\xi_j\}$ such that $|||\{\xi_n\}|||=\infty$ but $\sum_{i=1}^{\infty}|\xi_i|^p<\infty$, it follows that the basis $\{y_n\}$ of F_p is not equivalent (and hence, as observed above, not permutatively equivalent) to the unit vector basis $\{x_n\}$ of l^p. Therefore, in order to complete the proof it is sufficient to prove that the spaces F_p and l^p are isomorphic. For this purpose it will be sufficient, by lemma 18.6, to prove that there exists a subspace E_p of l^p isomorphic to F_p and complemented in l^p.

Let $[r_1,\dots,r_n]$ be the subspace of $L^p([0,1])$ spanned by the equivalence classes of the Rademacher functions $r_1,\dots,r_n$ and let G_{2^n} be the subspace of $L^p([0,1])$ spanned by the equivalence classes of the characteristic functions $\chi_{\left[\frac{k-1}{2^n},\frac{k}{2^n}\right]}$ $(k=1,\dots,2^n;\, n=1,2,\dots)$. Since G_{2^n} is equivalent to $l^p_{2^n}$, the subspace $[r_1,\dots,r_n]$ of G_{2^n} is equivalent to an n-dimensional subspace R_n of $l^p_{2^n}$ $(n=1,2,\dots)$.

Define E_p to be the image of the subspace $(R_1\times R_2\times\cdots)_{l^p}$ of $(l^p_2\times l^p_{2^2}\times\cdots)_{l^p}$ in l^p by the canonical linear isometry[2] $(l^p_2\times l^p_{2^2}\times\cdots)_{l^p}\equiv l^p$. We shall show that E_p has the required properties.

---

[1] Actually, $F_p\equiv(l_1^2\times l_2^2\times\cdots)_{l^p}$, by the mapping $\{\xi_n\}\to\{\xi_1,\{\xi_2,\xi_3\},\{\xi_4,\xi_5,\xi_6\},\dots\}$.

[2] This isometry is given by the formula $\{\{\xi_1,\xi_2\},\{\xi_3,\xi_4,\xi_5,\xi_6\},\dots\}\to\{\xi_n\}$.

By the Khinchin inequality (§ 14, formula (14.56)) there exists a constant $C_p>0$ such that $R_n \overset{C_p}{\cong} l_n^2$ $(n=1,2,\ldots)$. Consequently, by lemma 18.7a), the spaces E_p and $(l_1^2\times l_2^2\times\cdots)_{l^p}$ are isomorphic. On the other hand, it is obvious that $F_p\equiv(l_1^2\times l_2^2\times\cdots)_{l^p}$. Thus $E_p\cong F_p$.

Since $G_{2^n}\supset[r_1,\ldots,r_n]$, from lemma 18.8 it follows that there exists a projection[1] u_n' of G_{2^n} onto $[r_1,\ldots,r_n]$, of norm $\|u_n'\|\leqslant B_p$ $(n=1,2,\ldots)$, whence, by the definition of R_n, there exists a projection u_n of $l_{2^n}^p$ onto R_n of norm $\|u_n\|\leqslant B_p$ $(n=1,2,\ldots)$. Consequently, by lemma 18.7b), $(R_1\times R_2\times\cdots)_{l^p}$ is complemented in $(l_2^p\times l_{2^2}^p\times\cdots)_{l^p}$ and thus E_p is complemented in l^p, which completes the proof of theorem 18.3.

In order to prove that a result similar to theorem 18.3 also holds in the spaces $L^p([0,1])$, where $1<p<\infty$, $p\neq 2$, we shall need some propositions and lemmas.

Proposition 18.3. *If $\{\tilde{z}_n\}$ is a sequence in $L^p([0,1])$ $(1\leqslant p<\infty)$, such that $z_n\neq 0$ $(n=1,2,\ldots)$ and that the sets $A_n=\{t\in[0,1]\,|\,z_n(t)\neq 0\}$ $(n=1,2,\ldots)$ are mutually disjoint, then $\{\tilde{z}_n\}$ is a basic sequence and the subspace $[\tilde{z}_n]$ is isometrically isomorphic to l^p, by the mapping $\dfrac{\tilde{z}_n}{\|\tilde{z}_n\|}\to x_n$ $(n=1,2,\ldots)$, where $\{x_n\}$ is the unit vector basis of l^p.*

Proof. For any scalars $\xi_1,\ldots,\xi_n$ we have, by our hypothesis on the sets A_n,

$$\left\|\sum_{j=1}^n \xi_j\frac{\tilde{z}_j}{\|\tilde{z}_j\|}\right\|^p=\int_0^1\left|\sum_{j=1}^n \xi_j\frac{z_j(t)}{\|\tilde{z}_j\|}\right|^p dt=\int_0^1\sum_{j=1}^n|\xi_j|^p\frac{|z_j(t)|^p}{\|\tilde{z}_j\|^p}dt$$

$$=\sum_{j=1}^n|\xi_j|^p\int_0^1\frac{|z_j(t)|^p}{\|\tilde{z}_j\|^p}dt=\sum_{j=1}^n|\xi_j|^p=\left\|\sum_{j=1}^n\xi_jx_j\right\|^p,$$

and hence the mapping $u\colon x_n\to\dfrac{\tilde{z}_n}{\|\tilde{z}_n\|}$ $(n=1,2,\ldots)$ can be extended, by linearity and continuity, to an isometric isomorphism of l^p onto $[\tilde{z}_n]$.

Lemma 18.9. *Let $\{\Delta_k\}$ be an infinite sequence of distinct sets $\Delta_k\subset[0,1]$ $(k=1,2,\ldots)$, such that if $k_1<k_2$ and $\Delta_{k_1}\cap\Delta_{k_2}\neq\emptyset$, then $\Delta_{k_1}\supset\Delta_{k_2}$. Then there exists a subsequence $\{\Delta_{k_\nu}\}$ of $\{\Delta_k\}$ such that either*

$$\Delta_{k_\nu}\cap\Delta_{k_\mu}=\emptyset\quad\textit{whenever}\quad \nu\neq\mu, \tag{18.62}$$

or

$$\Delta_{k_1}\supset\Delta_{k_2}\supset\Delta_{k_3}\supset\cdots \tag{18.63}$$

[1] Namely, $u_n'=v_n|_{G_{2^n}}$ $(n=1,2,\ldots)$, where v_n is defined by (18.59).

Proof. Assume that no subsequence of $\{\Delta_k\}$ satisfies (18.63). Then for every index k there exists an index $\phi(k)$ such that $\Delta_{\phi(k)} \subset \Delta_k$ and that $\Delta_{\phi(k)}$ contains no other $\Delta_j \neq \Delta_{\phi(k)}$. Since $\{\Delta_k\}$ is infinite, the set $Z = \{\phi(1), \phi(2), \ldots\}$ is infinite. From the hypothesis on $\{\Delta_k\}$ and from the definition of Z it follows that if $k_\mu, k_\nu \in Z$ are such that $k_\mu \neq k_\nu$, then $\Delta_{k_\mu} \cap \Delta_{k_\nu} = \emptyset$. Hence the subsequence $\{\Delta_{k_\nu}\}_{k_\nu \in Z}$ of $\{\Delta_k\}$ satisfies (18.62), which completes the proof of lemma 18.9.

Proposition 18.4. *Let $\{\tilde{y}_n\}$ be the Haar basis of $L^p([0,1])$ $(1 \leqq p < \infty)$. Then for every sequence of indices $n_1 < n_2 < \cdots$ the space $[\tilde{y}_{n_k}]$ contains a subspace isomorphic to l^p and hence $[\tilde{y}_{n_k}]$ is not isomorphic to l^2 if $p \neq 2$.*

Proof. Put

$$\Delta_k = \{t \in [0,1] \mid y_{n_k}(t) \neq 0\} \quad (k = 1, 2, \ldots), \tag{18.64}$$

where we assume, for convenience, that each Haar function y_{n_k} is 0 in its dyadic dividion points (i.e., in the endpoints and midpoint of $\bar{\Delta}_k$). Then the sequence $\{\Delta_k\}$ satisfies the assumption of lemma 18.9 and hence it must have a subsequence $\{\Delta_{k_\nu}\}$ satisfying either (18.62) or (18.63). Now, if $\{\Delta_{k_\nu}\}$ satisfies (18.62), then the sequence $\{\tilde{y}_{n_{k_\nu}}\}$ satisfies the assumption of proposition 18.3, whence the subspace $[\tilde{y}_{n_{k_\nu}}]$ is isometrically isomorphic to l^p. On the other hand, if $\{\Delta_{k_\nu}\}$ satisfies (18.63), then the sequence $\left\{\tilde{y}_{n_{k_{2\nu-1}}} - \dfrac{y_{n_{k_{2\nu-1}}}(t_\nu)}{|y_{n_{k_{2\nu}}}(t_\nu)|} \tilde{y}_{n_{k_{2\nu}}}\right\}$, where $t_\nu \in \Delta_{k_{2\nu}}$ $(\nu = 1, 2, \ldots)$, satisfies the assumption of proposition 18.3, whence it spans a subspace isometrically isomorphic to l^p, which completes the proof of proposition 18.4.

Lemma 18.10. *Let $1 < p < \infty$. Then the product space $L^p([0,1]) \times l^2$ is isomorphic to the space $L^p([0,1])$.*

Proof. By § 17, proposition 17.2 b), there exists a Banach space E such that $L^p([0,1]) \cong E \times l^2$. Since $l^2 \times l^2 \cong l^2$, we obtain

$$L^p([0,1]) \times l^2 \cong (E \times l^2) \times l^2 \cong E \times (l^2 \times l^2) \cong E \times l^2 \cong L^p([0,1]),$$

which completes the proof.

Now we can prove

Theorem 18.4. *Let $1 < p < \infty$, $p \neq 2$. Then the space $L^p([0,1])$ has two bounded unconditional bases which are not permutatively equivalent.*

Proof. Let $\{\tilde{y}_n\}$ be the Haar basis of $L^p([0,1])$ and let $\{x_n\}$ be the unit vector basis of l^2. Then, by § 14, theorem 14.1, and § 16, proposition 16.4, the sequence $\{z_n\} \subset L^p([0,1]) \times l^2$ defined by

$$z_{2n-1} = \left\{\frac{\tilde{y}_n}{\|\tilde{y}_n\|}, 0\right\}, \quad z_{2n} = \{0, x_n\} \quad (n = 1, 2, \ldots) \tag{18.65}$$

is a bounded unconditional basis of $L^p([0,1]) \times l^2$. Since by lemma 18.10 the space $L^p([0,1]) \times l^2$ is isomorphic to $L^p([0,1])$, it follows that the image $\{z'_n\}$ of $\{z_n\}$ in $L^p([0,1])$ under this isomorphism is a bounded unconditional basis of $L^p([0,1])$ containing a subsequence equivalent to the unit vector basis of l^2. On the other hand, for the bounded unconditional basis $\left\{\frac{\tilde{y}_n}{\|\tilde{y}_n\|}\right\}$ of $L^p([0,1])$ no permutation $\left\{\frac{\tilde{y}_{\sigma(n)}}{\|\tilde{y}_{\sigma(n)}\|}\right\}$ has such a subsequence (by proposition 18.4), and therefore $\{z'_n\}$ and $\left\{\frac{\tilde{y}_n}{\|\tilde{y}_n\|}\right\}$ are not permutatively equivalent, which completes the proof of theorem 18.4.

The spaces c_0, l^1 and l^2 are not the only ones (up to an isomorphism) in which all normalized unconditional bases are permutatively equivalent. Indeed, one can prove[1] that e. g. the product space $l^1 \times l^2$ also has this property. However, if we replace "permutative equivalence" by the more restrictive "equivalence," then the situation is already different. Namely, we shall see in § 24 that the only spaces (up to an isomorphism) in which all normalized unconditional bases are equivalent, are c_0, l^1 and l^2.

Finally, we shall use permutative equivalence to prove the uniqueness, up to an isomorphism, of Banach spaces having an unconditional basis which is universal (in the sense of § 13) for the family $\mathscr{B}_u$ of all bounded unconditional bases. To prove first the existence of such bases, we shall need the following "unconditional analogue" of Ch. I, § 5, proposition 5.3 a), b):

Proposition 18.5. *Let E be a Banach space, (x_n, f_n) ($\{x_n\} \subset E$, $\{f_n\} \subset E^*$) a biorthogonal system and let*

$$U_1 = \left\{ x \in E \,\middle|\, \sum_{i=1}^{\infty} f_i(x) x_i \text{ is unconditionally convergent} \right\}, \tag{18.66}$$

endowed with the usual vector operations and with the norm

$$|||x||| = \sup_{|\beta_1|, |\beta_2|, \ldots \leqq 1} \; \sup_{1 \leqq l < \infty} \|s_{\{\beta_n\}, l}(x)\|. \tag{18.67}$$

Then

a) *U_1 is a Banach space.*

b) *$\{x_n\}$ is an unconditional basis of U_1.*

Proof. a) Consider the linear space of sequences of scalars

$$A_1^{(u)} = \left\{ \{\alpha_n\} \subset K \,\middle|\, \sum_{i=1}^{\infty} \alpha_i x_i \text{ is unconditionally convergent} \right\}, \tag{18.68}$$

[1] See [53].

endowed with the norm

$$\|\{\alpha_n\}\| = \sup_{|\beta_1|, |\beta_2|, \ldots \leqslant 1} \sup_{1 \leqslant l < \infty} \left\| \sum_{i=1}^{l} \beta_i \alpha_i x_i \right\|. \tag{18.69}$$

Then, by an argument similar to that used in the proof of Ch. I, § 3, proposition 3.1, $A_1^{(u)}$ is a Banach space. Since the mapping

$$\{\alpha_n\} \to \sum_{i=1}^{\infty} \alpha_i x_i \tag{18.70}$$

is a linear isometry of $A_1^{(u)}$ onto U_1, it follows that U_1 is a Banach space.

b) Let

$$\phi_n(x) = f_n(x) \qquad (x \in U_1, n = 1, 2, \ldots). \tag{18.71}$$

Then, by (18.67),

$$|\phi_n(x)| = |f_n(x)| = \frac{1}{\|x_n\|} \|f_n(x) x_n\| \leqslant \frac{1}{\|x_n\|} |||x||| \qquad (x \in U_1, n = 1, 2, \ldots)$$

and hence $\phi_n \in U_1^*$ $(n = 1, 2, \ldots)$. Since (x_n, f_n) is a biorthogonal system, it follows that (x_n, ϕ_n) is a biorthogonal system. Furthermore, by (18.67), (18.71), (16.26), (18.66) and § 16, lemma 16.1 (implication $1° \Rightarrow 6°$) we have

$$\begin{aligned} \left|\left|\left| x - \sum_{i=1}^{n} \phi_i(x) x_i \right|\right|\right| &= \sup_{|\beta_j| \leqslant 1} \sup_{1 \leqslant l < \infty} \left\| \sum_{j=1}^{l} \beta_j f_j \left[x - \sum_{i=1}^{n} f_i(x) x_i \right] x_j \right\| \\ &= \sup_{|\beta_j| \leqslant 1} \sup_{n+1 \leqslant l < \infty} \left\| \sum_{j=n+1}^{l} \beta_j f_j(x) x_j \right\| \\ &\leqslant \sup_{\substack{f \in E^* \\ \|f\| \leqslant 1}} \sum_{i=n+1}^{\infty} |f_i(x)| \, |f(x_i)| \to 0 \qquad (x \in U_1), \end{aligned}$$

whence, by Ch. I, § 4, theorem 4.1, $\{x_n\}$ is a basis of U_1. Since by the same argument[1] every permutation $\{x_{\sigma(n)}\}$ of $\{x_n\}$ is a basis of U_1, it follows that $\{x_n\}$ is an unconditional basis of U_1, which completes the proof of proposition 18.5.

From the above proof it follows that *the unit vectors* $e_n = \{\delta_{nj}\}_{j=1}^{\infty}$ $(n = 1, 2, \ldots)$ *constitute an unconditional basis of the Banach space* $A_1^{(u)}$ defined by (18.68), (18.69). However, in contrast to Ch. I, § 8, proposition 8.1b), $\{x_n\}$, considered as a sequence in E, need not be equivalent to $\{e_n\}$, as shown e. g. by any conditional basis $\{x_n\}$ of E.

Let us also observe that by (18.67) we obviously have

$$|||x_n||| = \|x_n\| \qquad (n = 1, 2, \ldots). \tag{18.72}$$

---

[1] Alternatively, one can observe that $\left|\left|\left| \sum_{i=1}^{n} \delta_i \alpha_i x_i \right|\right|\right| \leqslant \left|\left|\left| \sum_{i=1}^{n} \alpha_i x_i \right|\right|\right|$ for every finite sequence of scalars $\alpha_1, \ldots, \alpha_n$, $\delta_1, \ldots, \delta_n$ with $|\delta_i| \leqslant 1$ $(i = 1, \ldots, n)$ and apply § 17, theorem 17.1.

Theorem 18.5. *The family $\mathscr{B}_u$ of all bounded unconditional bases contains a universal element.*

Proof. Let $\{x_n\}$ be a bounded basis of a Banach space E, which is universal for the family of all bounded bases (by § 13, corollary 13.3, there exist a continuum of mutually non-equivalent such bases) and let $\{f_n\} \subset E^*$ be the a. s. c. f. to $\{x_n\}$. For this biorthogonal system (x_n, f_n) define the Banach space U_1 as in proposition 18.5 above. We shall prove that the bounded unconditional basis $\{x_n\}$ of U_1 is universal for the family $\mathscr{B}_u$ of all bounded unconditional bases.

Let $\{y_n\}$ be an arbitrary bounded unconditional basis (of a Banach space, say F). Then, since $\{x_n\}$ is universal for the family of all bounded bases, there exists a subsequence $\{x_{i_n}\}$ of $\{x_n\}$ such that $\{y_n\} \sim \{x_{i_n}\}$. Hence, since $\{y_n\}$ is an unconditional basis, from Ch. I, § 8, theorem 8.1d) (implication $6^\circ \Rightarrow 1^\circ$) it follows that $\{x_{i_n}\}$ is an unconditional basis of the subspace $[x_{i_n}]$ of E. Therefore, $[x_{i_n}] \subset U_1$ and by § 16, theorem 16.1b), the norm on $[x_{i_n}]$ induced by the norm (18.67) of U_1 is equivalent to the norm on $[x_{i_n}]$ induced by the initial norm of E. Consequently, $\{y_n\}$ is equivalent to the subsequence $\{x_{i_n}\}$ of the bounded unconditional basis $\{x_n\}$ of U_1, which completes the proof of theorem 18.5.

Let us observe that by § 16, theorem 16.8 (implication $1^\circ \Rightarrow 2^\circ$) every universal unconditional basis is complementably universal and hence theorem 18.5 also shows the existence of a complementably universal element in the family $\mathscr{B}_u$ of all bounded unconditional bases. Moreover, the "unconditional analogue" of the uniqueness theorem for complementably universal bases (§ 13, theorem 13.2) is also valid, yielding the uniqueness of universal unconditional bases (in contrast to § 13, corollary 13.3 on the non-uniqueness of universal bases):

Theorem 18.6. *Any two bounded unconditional bases, which are universal for the family $\mathscr{B}_u$ of all bounded unconditional bases, are permutatively equivalent. Hence they span isomorphic Banach spaces and therefore the Banach space U_1 in the above proof of theorem* 18.5 *is unique up to an isomorphism.*

The proof is analogous to that of § 13, theorem 13.2, taking into account § 16, proposition 16.4, § 17, corollary 17.4 and the above remark that every universal unconditional basis is complementably universal.

Corollary 18.2. *There exists a Banach space E with an unconditional basis such that every Banach space with an unconditional basis is isomorphic to a complemented subspace of E.*

We shall denote by E_u the Banach space, unique up to an isomorphism, which has a bounded unconditional basis universal for the family of all bounded unconditional bases. The following corollary shows that the property described in corollary 18.2 also characterizes E_u up to an isomorphism:

Corollary 18.3. *If E is a Banach space with an unconditional basis, such that every Banach space with an unconditional basis is isomorphic to a complemented subspace of E, then E is isomorphic to E_u.*

The proof is analogous to that of § 13, corollary 13.2.

Problem 18.1. Are all normalized unconditional bases of the space E_u permutatively equivalent?

Problem 18.2. Let a Banach space E with an unconditional basis contain a subspace isomorphic to E_u. Is E isomorphic to E_u?

Concerning the problem of existence of universal bases for other classes of unconditional bases we mention that by § 13, the proof of theorem 13.5, we actually have the following sharpening of that theorem: *There exists no shrinking basis which is universal for the family of all normalized unconditional shrinking bases.* An affirmative answer to § 17, problem 17.1 would again imply, by § 17, theorem 17.6, that the family of all normalized unconditional shrinking bases has no universal element.

§ 19. Best approximation in Banach spaces with unconditional bases

In the present section we shall consider for unconditional bases some notions and problems analogous to those studied in Ch. I, § 19 for arbitrary bases. We shall see in theorem 19.1 and example 19.1 below that the relations between the "unconditional analogues" of T-norms and K-norms are different from those between T-norms and K-norms.

Definition 19.1. The norm in a Banach space E with an unconditional basis $\{x_n\}$ is called an *NT-norm* (with respect to the basis $\{x_n\}$), if

a) for every $x \in E$ and $d = \{i_1, \ldots, i_n\} \in \mathscr{D}$ there exists a unique polynomial $y_0 = \pi_{P_{(d)}}(x) \in P_{(d)} = [x_{i_1}, \ldots, x_{i_n}]$ of best approximation of x;

b) this polynomial coincides with the d-th partial sum of the expansion of the element x with respect to the bais $\{x_n\}$, i. e.

$$\pi_{P_{(d)}}(x) = s_d(x) = \sum_{j=1}^{n} f_{i_j}(x) x_{i_j} \qquad (x \in E, d = \{i_1, \ldots, i_n\} \in \mathscr{D}). \tag{19.1}$$

We shall denote the NT-norms $((\ ((_u$.

Definition 19.2. The norm in a Banach space E with an unconditional basis $\{x_n\}$ is called an *NK-norm* (with respect to the basis $\{x_n\}$), if

a) for every $x \in E$ and $d=\{i_1,\dots,i_n\}\in\mathscr{D}$ there exists a unique polynomial complement $y_0=\pi_{P^{(d)}}(x)\in P^{(d)}=[x_j]_{j\in\mathscr{N}\setminus d}$ of best approximation of x;

b) this polynomial complement coincides with the d-th rest of the expansion of the element x with respect to the basis $\{x_n\}$, i.e.

$$\pi_{P^{(d)}}(x)=r_d(x)=x-s_d(x) \quad (x\in E,\, d=\{i_1,\dots,i_n\}\in\mathscr{D}). \tag{19.2}$$

We shall denote the NK-norms by $)) \;))_u$.

Definition 19.3. The norm in a Banach space E with an unconditional basis $\{x_n\}$ is called an *NTK-norm* (with respect to the basis $\{x_n\}$) if it is simultaneously an NT-norm and an NK-norm with respect to this basis.

We shall denote the NTK-norms by $((\;))_u$.

For instance, the natural norm in a separable Hilbert space H is an NTK-norm with respect to any orthogonal basis $\{x_n\}$ of the space H.

Let us first give a useful characterization for each of these norms.

Proposition 19.1. *Let E be a Banach space with an unconditional basis $\{x_n\}$. Then*

a) *The norm in E is an NT-norm if and only if*

$$\Big\|\sum_{i\in\mathscr{N}\setminus d_2}\alpha_i x_i\Big\|<\Big\|\sum_{i\in\mathscr{N}\setminus d_1}\alpha_i x_i\Big\| \tag{19.3}$$

for every pair $d_1, d_2\in\mathscr{D}$ with $d_1\subset d_2$ and every sequence of scalars $\{\alpha_i\}_{i\in\mathscr{N}\setminus d_1}$ with $\sum_{i\in d_2\setminus d_1}|\alpha_i|\neq 0$, for which the series in (19.3) are convergent.

b) *The norm in E is an NK-norm if and only if*

$$\Big\|\sum_{i\in d_1}\alpha_i x_i\Big\|<\Big\|\sum_{i\in d_2}\alpha_i x_i\Big\| \tag{19.4}$$

for every pair $d_1, d_2\in\mathscr{D}$ with $d_1\subset d_2$ and every finite sequence of scalars $\{\alpha_i\}_{i\in d_2}$ with $\sum_{i\in d_2\setminus d_1}|\alpha_i|\neq 0$.[1]

Proof. a) Assume that the norm in E is an NT-norm. Let $\sum_{i\in\mathscr{N}\setminus d_2}\alpha_i x_i$ be convergent, where $d_2\in\mathscr{D}$, and let $d_1\subset d_2$ be arbitrary. Then $\sum_{i\in\mathscr{N}\setminus d_1}\alpha_i x_i$ has a unique element of best approximation in $P_{(d_2)}=[x_i]_{i\in d_2}$, namely

$$\pi_{P_{(d_2)}}\Big(\sum_{i\in\mathscr{N}\setminus d_1}\alpha_i x_i\Big)=s_{d_2}\Big(\sum_{i\in\mathscr{N}\setminus d_1}\alpha_i x_i\Big)=\sum_{i\in d_2\setminus d_1}\alpha_i x_i,$$

[1] The bases $\{x_n\}$ satisfying (19.4) (i.e. such that the norm in E is an NK-norm with respect to $\{x_n\}$) are called strictly orthogonal (see § 20).

whence, since $0 \in P_{(d_2)}$,

$$\Big\|\sum_{i\in\mathcal{N}\setminus d_2}\alpha_i x_i\Big\| = \Big\|\sum_{i\in\mathcal{N}\setminus d_1}\alpha_i x_i - \pi_{P_{(d_2)}}\Big(\sum_{i\in\mathcal{N}\setminus d_1}\alpha_i x_i\Big)\Big\| < \Big\|\sum_{i\in\mathcal{N}\setminus d_1}\alpha_i x_i\Big\|,$$

i. e. we have (19.3).

Conversely, assume now that we have (19.3). Then for every $x=\sum_{i=1}^{\infty}\alpha_i x_i \in E$, $d\in\mathscr{D}$ and $p=\sum_{i\in d}\beta_i x_i \in P_{(d)}$ with $p\neq s_d(x)$, we have (by (19.3) with $d_2=d$, $d_1=\emptyset$)

$$\|x-s_d(x)\| = \Big\|\sum_{i\in\mathcal{N}\setminus d}\alpha_i x_i\Big\| < \Big\|\sum_{i\in\mathcal{N}\setminus d}\alpha_i x_i - \sum_{i\in d}(\beta_i-\alpha_i)x_i\Big\| = \|x-p\|,$$

and thus the norm in E is an NT-norm.

b) Assume that the norm in E is an NK-norm. Let $\{\alpha_i\}_{i\in d_2}$ be a finite sequence of scalars and let $d_1\subset d_2$ be such that $\sum_{i\in d_2\setminus d_1}|\alpha_i|\neq 0$. Then $\sum_{i\in d_2}\alpha_i x_i$ has a unique element of best approximation in $P^{(d_1)}=[x_i]_{i\in\mathcal{N}\setminus d_1}$, namely

$$\pi_{P^{(d_1)}}\Big(\sum_{i\in d_2}\alpha_i x_i\Big) = \sum_{i\in d_2}\alpha_i x_i - s_{d_1}\Big(\sum_{i\in d_2}\alpha_i x_i\Big) = \sum_{i\in d_2\setminus d_1}\alpha_i x_i,$$

whence, since $0\in P^{(d_1)}$,

$$\Big\|\sum_{i\in d_1}\alpha_i x_i\Big\| = \Big\|\sum_{i\in d_2}\alpha_i x_i - \pi_{P^{(d_1)}}\Big(\sum_{i\in d_2}\alpha_i x_i\Big)\Big\| < \Big\|\sum_{i\in d_2}\alpha_i x_i\Big\|,$$

i. e. we have (19.4).

Conversely, assume now that we have (19.4) and let $x=\sum_{i=1}^{\infty}\alpha_i x_i$, $d\in\mathscr{D}$ and $p=\sum_{i\in\mathcal{N}\setminus d}\beta_i x_i\in P^{(d)}$ with $p\neq r_d(x)=\sum_{i\in\mathcal{N}\setminus d}\alpha_i x_i$ be arbitrary. Then there exists in $\mathcal{N}\setminus d$ a smallest index, say i_0, such that $\beta_{i_0}\neq\alpha_{i_0}$. Hence, applying (19.4) successively, we obtain

$$\begin{aligned}\|x-r_d(x)\| &= \Big\|\sum_{i\in d}\alpha_i x_i\Big\| = \Big\|\sum_{i\in d}\alpha_i x_i + \sum_{\substack{i\in\mathcal{N}\setminus d\\ i<i_0}}(\beta_i-\alpha_i)x_i\Big\| \\ &< \Big\|\sum_{i\in d}\alpha_i x_i + \sum_{\substack{i\in\mathcal{N}\setminus d\\ i\leqslant i_0}}(\beta_i-\alpha_i)x_i\Big\| \leqslant \Big\|\sum_{i\in d}\alpha_i x_i + \sum_{\substack{i\in\mathcal{N}\setminus d\\ i\leqslant i_0+1}}(\beta_i-\alpha_i)x_i\Big\| \\ &\ldots \leqslant \Big\|\sum_{i\in d}\alpha_i x_i - \sum_{i\in\mathcal{N}\setminus d}(\beta_i-\alpha_i)x_i\Big\| = \|x-p\|,\end{aligned}$$

and thus the norm in E is an NK-norm, which completes the proof of proposition 19.1.

Let us consider now the relations between $N\,T$-norms and $N\,K$-norms.

Theorem 19.1. *Let E be a Banach space with an unconditional basis $\{x_n\}$. Then every $N\,T$-norm with respect to $\{x_n\}$ is an $N\,K$-norm (whence also an $N\,TK$-norm) with respect to $\{x_n\}$.*

Proof. We shall prove that the relations (19.3) imply the relations (19.4). The idea is to complete the α_i's in (19.4) with zeros until we obtain suitable sums of the form $\sum_{i\in\mathcal{N}\setminus d}\alpha_i x_i$, to which we can apply (19.3).

Let $d_1, d_2\in\mathcal{D}$, $d_1\subset d_2$ and $\{\alpha_i\}_{i\in d_2}\subset K$ be such that $\sum_{i\in d_2\setminus d_1}|\alpha_i|\neq 0$. Put

$$\beta_i=\begin{cases}\alpha_i & \text{for } i\in d_2\\ 0 & \text{for } i\in\mathcal{N}\setminus d_2,\end{cases} \tag{19.5}$$

$$d_1'=\emptyset,\ d_2'=d_2\setminus d_1. \tag{19.6}$$

Then $d_1', d_2'\in\mathcal{D}$, $d_1'\subset d_2'$, $\sum_{i\in d_2'\setminus d_1'}|\beta_i|=\sum_{i\in d_2\setminus d_1}|\alpha_i|\neq 0$, and the series $\sum_{i\in\mathcal{N}\setminus d_1'}\beta_i x_i=\sum_{i\in d_2}\alpha_i x_i$ is convergent. Hence, since the norm is an $N\,T$-norm with respect to $\{x_n\}$, by (19.3) we have

$$\Big\|\sum_{i\in d_1}\alpha_i x_i\Big\|=\Big\|\sum_{i\in d_2\setminus d_2'}\alpha_i x_i\Big\|=\Big\|\sum_{i\in\mathcal{N}\setminus d_2'}\beta_i x_i\Big\|<\Big\|\sum_{i\in\mathcal{N}\setminus d_1'}\beta_i x_i\Big\|=\Big\|\sum_{i\in d_2}\alpha_i x_i\Big\|,$$

i.e. (19.4), which completes the proof of theorem 19.1

The converse implication is not true, i.e. there exist $N\,K$-norms which are not $N\,T$-norms, as shown by

Example 19.1. Let E be the space c_0 endowed with the equivalent norm $))x))$ given in Ch. I, § 19, example 19.2, formula (19.10). Then $))x))_u=))x))$ is an $N\,K$-norm, but not an $N\,T$-norm, with respect to the unit vector basis $\{x_n\}$ of c_0.

Indeed, proposition 19.1 and Ch. I, § 19, formulas (19.13), (19.14) show that $))x))_u$ is an $N\,K$-norm, but not an $N\,T$-norm, with respect to $\{x_n\}$.

Theorem 19.2. *Let E be a Banach space with an unconditional basis $\{x_n\}$ and let $\{f_n\}\subset E^*$ be the a.s.c.f. Then one can introduce on E an $N\,TK$-norm equivalent to the initial norm on E, by the formula*

$$((x))_u=\sum_{i=1}^{\infty}\frac{1}{2^i}\|f_i(x)x_i\|+\sup_{\{i_1,\dots,i_n\}\in\mathcal{D}}\Big\|\sum_{j=1}^{n}f_{i_j}(x)x_{i_j}\Big\|. \tag{19.7}$$

Proof. It is obvious that $((x))_u$ is a norm on E. This norm is equivalent to the initial norm on E, since for every $x\in E$ we have

$$\|x\| \leqslant ((x))_u \leqslant \max_{1 \leqslant i < \infty} \|f_i(x)x_i\| + \sup_{\{i_1,\ldots,i_n\} \in \mathscr{D}} \left\| \sum_{j=1}^{n} f_{i_j}(x)x_{i_j} \right\|$$

$$\leqslant \max_{1 \leqslant i < \infty} \left(\left\| \sum_{j=1}^{i} f_j(x)x_j \right\| + \left\| \sum_{j=1}^{i-1} f_j(x)x_j \right\| \right) + \sup_{\{i_1,\ldots,i_n\} \in \mathscr{D}} \left\| \sum_{j=1}^{n} f_{i_j}(x)x_{i_j} \right\|$$

$$\leqslant 3\, v_{\{x_n\}}^{(u)} \|x\|.$$

Finally, in order to prove that $((x))_u$ is an NTK-norm it will be sufficient, by theorem 19.1, to prove that it is an NT-norm. Let $d_1, d_2 \in \mathscr{D}$ with $d_1 \subset d_2$ and $\{\alpha_i\}_{i \in \mathscr{N} \setminus d_2}$ with $\sum_{i \in d_2 \setminus d_1} |\alpha_i| \neq 0$ be such that $\sum_{i \in \mathscr{N} \setminus d_2} \alpha_i x_i$ converges. We have then $\mathscr{N} \setminus d_1 = (\mathscr{N} \setminus d_2) \cup (d_2 \setminus d_1)$, whence

$$\left(\left(\sum_{i \in \mathscr{N} \setminus d_2} \alpha_i x_i \right)\right)_u = \sum_{i \in \mathscr{N} \setminus d_2} \frac{1}{2^i} \|\alpha_i x_i\| + \sup_{\{i_1,\ldots,i_n\} \in \mathscr{D} \cap (\mathscr{N} \setminus d_2)} \left\| \sum_{j=1}^{n} \alpha_{i_j} x_{i_j} \right\|$$

$$< \sum_{i \in \mathscr{N} \setminus d_1} \frac{1}{2^i} \|\alpha_i x_i\| + \sup_{\{i_1,\ldots,i_n\} \in \mathscr{D} \cap (\mathscr{N} \setminus d_1)} \left\| \sum_{j=1}^{n} \alpha_{i_j} x_{i_j} \right\|$$

$$= \left(\left(\sum_{i \in \mathscr{N} \setminus d_1} \alpha_i x_i \right)\right)_u,$$

and thus, by proposition 19.1 a), $((x))_u$ is an NT-norm, which completes the proof.

Let us mention that one can also prove, with a similar argument, that in the conditions of theorem 19.2 it is possible to introduce another equivalent NTK-norm on E, by the formula

$$\begin{aligned} ((x))_u &= \sum_{i=1}^{\infty} \frac{1}{2^i} \|f_i(x)x_i\| + \sup_{\substack{1 \leqslant n < \infty \\ \varepsilon_i = \pm 1}} \left\| \sum_{i=1}^{n} \varepsilon_i f_i(x)x_i \right\| \\ &= \sum_{i=1}^{\infty} \frac{1}{2^i} \|f_i(x)x_i\| + \sup_{\substack{f \in E^* \\ \|f\| \leqslant 1}} \sum_{i=1}^{\infty} |f_i(x) f(x_i)|. \end{aligned} \tag{19.8}$$

Remark 19.1. Theorem 19.2 admits, in a certain sense, a converse. Namely, one can define *weak NT-norms* ($=$*weak NK-norms*$=$*weak NTK-norms*) with respect to an arbitrary sequence $\{x_n\} \subset E$ with $x_n \neq 0$ $(n = 1, 2, \ldots)$ and $[x_n] = E$, similarly to remark 19.1 of Ch. I, § 19. If one can introduce on E such a norm, equivalent to the initial norm on E, then by § 17, theorem 17.1, $\{x_n\}$ is an unconditional basis of E endowed with this new norm, and therefore also of E with its initial norm.

§ 20. Orthogonal bases. Strictly orthogonal bases. Hyperorthogonal and strictly hyperorthogonal bases

In this section and the subsequent ones we shall study some special classes of unconditional bases. In the present section we shall consider some "unconditional analogues" of monotone bases (defined by the condition that one of the constants of § 17, theorem 17.1, be equal to 1) and of strictly monotone bases.

Definition 20.1. A basis $\{x_n\}$ of a Banach space E is said to be *orthogonal*, if we have

$$\left\| \sum_{j=1}^{n} \alpha_{i_j} x_{i_j} \right\| \leqslant \left\| \sum_{j=1}^{n} \alpha_{i_j} x_{i_j} + \sum_{j=1}^{m} \alpha_{l_j} x_{l_j} \right\| \tag{20.1}$$

for any $n, m \in \mathcal{N}$ and any scalars $\alpha_{i_1}, \ldots, \alpha_{i_n}, \alpha_{l_1}, \ldots, \alpha_{l_m} \in K$ whose indices satisfy $\{i_1, \ldots, i_n\} \cap \{l_1, \ldots, l_m\} = \emptyset$. The basis $\{x_n\}$ is said to be *strictly orthogonal*, if we have

$$\left\| \sum_{j=1}^{n} \alpha_{i_j} x_{i_j} \right\| < \left\| \sum_{j=1}^{n} \alpha_{i_j} x_{i_j} + \sum_{j=1}^{m} \alpha_{l_j} x_{l_j} \right\| \tag{20.2}$$

for any $n, m \in \mathcal{N}$ and any scalars $\alpha_{i_1}, \ldots, \alpha_{i_n}, \alpha_{l_1}, \ldots, \alpha_{l_m} \in K$ with $\{i_1, \ldots, i_m\} \cap \{l_1, \ldots, l_m\} = \emptyset$, such that $\sum_{j=1}^{m} |\alpha_{l_j}| \neq 0$.[1]

For instance, the unit vector basis of l^p $(p \geqslant 1)$ is strictly orthogonal, while the unit vector basis of c_0 is orthogonal, but not strictly orthogonal.

Similarly to the geometric interpretations of monotone and strictly monotone bases, given in § 1, we see that *a basis $\{x_n\}$ of E is orthogonal (respectively, strictly orthogonal) if and only if we have $[x_{i_1}, \ldots, x_{i_n}] \perp [x_{l_1}, \ldots, x_{l_m}]$ (respectively, $[x_{i_1}, \ldots, x_{i_n}] \perp\!\!\perp [x_{l_1}, \ldots, x_{l_m}]$) for any indices such that $\{i_1, \ldots, i_n\} \cap \{l_1, \ldots, l_m\} = \emptyset$.* From this remark and from § 1, lemma 1.1, it results that *for a basis $\{x_n\}$ of a Hilbert space E the following statements are equivalent*: 1°. *$\{x_n\}$ is orthogonal.* 2°. *$\{x_n\}$ is strictly orthogonal.* 3°. *$\{x_n\}$ is orthogonal in the usual Hilbert space sense (i.e., $(x_i, x_j) = 0$ for all $i \neq j$).*

---

[1] One may also express these conditions in the following form:

$$\left\| \sum_{i \in d_1} \alpha_i x_i \right\| \leqslant \left\| \sum_{i \in d_2} \alpha_i x_i \right\|$$

for any $d_1, d_2 \in \mathcal{D}$ with $d_1 \subset d_2$, respectively

$$\left\| \sum_{i \in d_1} \alpha_i x_i \right\| < \left\| \sum_{i \in d_2} \alpha_i x_i \right\|$$

for any $d_1, d_2 \in \mathcal{D}$ with $d_1 \subset d_2$, such that $\sum_{i \in d_2 \setminus d_1} |\alpha_i| \neq 0$.

Let us also observe that *for every orthogonal basis* $\{x_n\}$ *with the a.s.c.f.* $\{f_n\}$ *we have*

$$\|x_n\| \, \|f_n\| = 1 \qquad (n=1,2,\ldots), \tag{20.3}$$

and hence, in particular, *every normalized orthogonal basis* $\{x_n\}$ *is normal.* Indeed, by (20.1),

$$\left| f_n\left(\sum_{i=1}^{\infty} \alpha_i x_i\right)\right| = |\alpha_n| = \frac{1}{\|x_n\|} \|\alpha_n x_n\| \leqslant \frac{1}{\|x_n\|} \left\| \sum_{i=1}^{\infty} \alpha_i x_i \right\|$$
$$\left(\sum_{i=1}^{\infty} \alpha_i x_i \in E,\ n=1,2,\ldots\right),$$

whence (20.3) follows.

By § 17, theorem 17.1, *a basis* $\{x_n\}$ *of a Banach space E is orthogonal if and only if it is of unconditional norm* $v_{\{x_n\}}^{(u)} = \sup_{d \in \mathscr{D}} \|s_d\| = 1$ *(whence, in particular, unconditional).* From this remark it follows that the problem of existence of orthogonal and strictly orthogonal bases has a negative answer in the class of all infinite dimensional Banach spaces with bases. Furthermore, from this remark and § 17, theorem 17.1, formula (17.19), there follow at once several intrinsic characterizations of orthogonal bases. Some other intrinsic characterizations of such bases are given in

Theorem 20.1. *Let E be a Banach space and* $\{x_n\}$ *a complete sequence in E such that* $x_n \neq 0$ $(n=1,2,\ldots)$. *The following statements are equivalent:*

1°. $\{x_n\}$ *is an orthogonal basis of E.*

2°. *Every permutation* $\{x_{\sigma(n)}\}$ $(\sigma \in \Pi)$ *of* $\{x_n\}$ *is a monotone basis of E.*

3°. *Every subsequence* $\{x_{i_n}\}$ *of* $\{x_n\}$ *is a monotone basic sequence.*

Moreover, in this case the relations $\sum_{i=1}^{\infty} \alpha_i x_i \in E$ *and* $0 \leqslant \gamma_n \leqslant \alpha_n$ $(n=1,2,\ldots)$ *imply* $\sum_{i=1}^{\infty} \gamma_i x_i \in E$ *and*

$$\left\| \sum_{i=1}^{\infty} \gamma_i x_i \right\| \leqslant \left\| \sum_{i=1}^{\infty} \alpha_i x_i \right\|. \tag{20.4}$$

Proof. 1°⇒2°. If $\{x_n\}$ is an orthogonal basis and $\sigma \in \Pi$, then by (20.1) we have, for any $n, m \in \mathscr{N}$ and any scalars $\alpha_{\sigma(1)}, \alpha_{\sigma(2)}, \ldots, \alpha_{\sigma(n+m)}$,

$$\left\| \sum_{i=1}^{n} \alpha_{\sigma(i)} x_{\sigma(i)} \right\| \leqslant \left\| \sum_{i=1}^{n+m} \alpha_{\sigma(i)} x_{\sigma(i)} \right\|,$$

and thus $\{x_{\sigma(n)}\}$ is a monotone basis.

$2^\circ \Rightarrow 1^\circ$. Assume that we have 2° and let $\{i_1,\dots,i_n\} \cap \{l_1,\dots,l_m\} = \emptyset$, $\alpha_{i_1},\dots,\alpha_{i_n}$, $\alpha_{l_1},\dots,\alpha_{l_m} \in K$ be arbitrary. Take any $\sigma \in \Pi$ such that

$$\sigma(j) = i_j \quad (j=1,\dots,n),$$
$$\sigma(n+j) = l_j \quad (j=1,\dots,m).$$

Then, since $\{x_{\sigma(n)}\}$ is a monotone basis,

$$\left\| \sum_{j=1}^{n} \alpha_{i_j} x_{i_j} \right\| = \left\| \sum_{j=1}^{n} \alpha_{\sigma(j)} x_{\sigma(j)} \right\| \leqslant \left\| \sum_{j=1}^{n+m} \alpha_{\sigma(j)} x_{\sigma(j)} \right\|$$
$$= \left\| \sum_{j=1}^{n} \alpha_{i_j} x_{i_j} + \sum_{j=1}^{m} \alpha_{l_j} x_{l_j} \right\|$$

and thus $\{x_n\}$ is an orthogonal basis.

The proof of the equivalence $1^\circ \Leftrightarrow 3^\circ$ is similar.

Finally, for any $k \in \mathcal{N}$ and $\gamma_1,\dots,\gamma_{k-1}, \gamma_{k+1},\dots,\gamma_n \in K$ the continuous real function

$$\phi(\lambda) = \left\| \sum_{i=1}^{k-1} \gamma_i x_i + \lambda x_k + \sum_{i=k+1}^{n} \gamma_i x_i \right\| \qquad (-\infty < \lambda < \infty) \tag{20.5}$$

is convex (obviously, $\phi(\alpha\lambda_1 + (1-\alpha)\lambda_2) \leqslant \alpha\phi(\lambda_1) + (1-\alpha)\phi(\lambda_2)$ for any $0 \leqslant \alpha \leqslant 1$ and any $-\infty < \lambda_1, \lambda_2 < \infty$) and, by 1°, $\phi(\lambda)$ has a minimum at $\lambda = 0$, whence for any α_k, γ_k with $0 \leqslant \gamma_k \leqslant \alpha_k$ $(k=1,\dots,n)$

$$\left\| \sum_{i=1}^{n} \gamma_i x_i \right\| = \phi(\gamma_k) \leqslant \phi(\alpha_k) = \left\| \sum_{i=1}^{k-1} \gamma_i x_i + \alpha_k x_k + \sum_{i=k+1}^{n} \gamma_i x_i \right\|.$$

Applying this successively for $k=1,\dots,n$ it follows that for any $n \in \mathcal{N}$ and $0 \leqslant \gamma_k \leqslant \alpha_k$ $(k=1,\dots,n)$

$$\left\| \sum_{i=1}^{n} \gamma_i x_i \right\| \leqslant \left\| \sum_{i=1}^{n} \alpha_i x_i \right\|, \tag{20.6}$$

whence, taking $n \to \infty$, we infer (20.4), which completes the proof of theorem 20.1.

Remark 20.1. With a similar argument to the above proof of (20.4), it follows that *for a strictly orthogonal basis $\{x_n\}$ of E the relations $\{\alpha_n\}$, $\{\gamma_n\} \in K$, $\sum_{i=1}^{\infty} \alpha_i x_i \in E$, $0 \leqslant \gamma_n \leqslant \alpha_n$ $(n=1,2,\dots)$ and $\gamma_{n_0} < \alpha_{n_0}$ for some index n_0 imply $\sum_{i=1}^{\infty} \gamma_i x_i \in E$ and, for any $n \in \mathcal{N}$, $n \geqslant n_0$,*

$$\left\| \sum_{i=1}^{n} \gamma_i x_i \right\| < \left\| \sum_{i=1}^{n} \alpha_i x_i \right\|; \tag{20.7}$$

conversely, if $\{x_n\}$ has this property, then it is obviously strictly orthogonal. However, (20.7) need not hold for $n=\infty$, as shown by § 19, example 19.1.

Definition 20.2. A basis $\{x_n\}$ of a Banach space E is said to be *hyperorthogonal* if we have

$$\left\| \sum_{i=1}^{n} \beta_i \alpha_i x_i \right\| \leqslant \left\| \sum_{i=1}^{n} \alpha_i x_i \right\| \tag{20.8}$$

for every finite sequence of scalars $\alpha_1, \ldots, \alpha_n$, $\beta_1, \ldots, \beta_n$ with $|\beta_i| \leqslant 1$ $(i=1, \ldots, n)$. The basis $\{x_n\}$ is said to be *strictly hyperorthogonal* if we have

$$\left\| \sum_{i=1}^{n} \beta_i \alpha_i x_i \right\| < \left\| \sum_{i=1}^{n} \alpha_i x_i \right\| \tag{20.9}$$

for every finite sequence of scalars $\alpha_1, \ldots, \alpha_n$, $\beta_1, \ldots, \beta_n$ with $|\beta_i| \leqslant 1$ $(i=1, \ldots, n)$ such that $|\beta_{i_0}| < 1$, $\alpha_{i_0} \neq 0$ for some index i_0.

For instance, the unit vector basis of l^p $(p \geqslant 1)$ is strictly hyperorthogonal, while the unit vector basis of c_0 is hyperorthogonal but not strictly hyperorthogonal.

Some intrinsic characterizations of hyperorthogonal bases are collected in

Theorem 20.2. *Let E be a Banach space and $\{x_n\}$ a complete sequence in E such that $x_n \neq 0$ $(n=1,2,\ldots)$. The following statements are equivalent:*

1°. *$\{x_n\}$ is a hyperorthogonal basis of E.*

2°. *The relations $\{\alpha_n\}, \{\gamma_n\} \subset K$, $|\gamma_n| \leqslant |\alpha_n|$ $(n=1,2,\ldots)$, $\sum_{i=1}^{\infty} \alpha_i x_i \in E$ imply $\sum_{i=1}^{\infty} \gamma_i x_i \in E$ and*

$$\left\| \sum_{i=1}^{\infty} \gamma_i x_i \right\| \leqslant \left\| \sum_{i=1}^{\infty} \alpha_i x_i \right\|. \tag{20.10}$$

3°. *We have*

$$\left\| \sum_{i=1}^{\infty} \alpha_i x_i \right\| = \left\| \sum_{i=1}^{\infty} |\alpha_i| x_i \right\| \quad \left(\sum_{i=1}^{\infty} \alpha_i x_i \in E \right). \tag{20.11}$$

4°. *For every finite sequence of scalars $\alpha_1, \ldots, \alpha_n$, $\varepsilon_1, \ldots, \varepsilon_n$ with $|\varepsilon_i| = 1$ $(i=1, \ldots, n)$ we have*[1]

$$\left\| \sum_{i=1}^{n} \varepsilon_i \alpha_i x_i \right\| \leqslant \left\| \sum_{i=1}^{n} \alpha_i x_i \right\|. \tag{20.12}$$

---

[1] Naturally, a corresponding condition could be also added in § 16, theorem 16.1 and § 17, theorem 17.1, but there it would be an obvious consequence of the equivalences 4° ⇔ 5° and 5° ⇔ 7° respectively.

If E is a real Banach space, these statements are equivalent to the following:

5°. *$\{x_n\}$ admits a total sequence $\{f_n\} \subset E^*$ such that $f_i(x_j) = \delta_{ij}$ and the space E, endowed with the natural order induced by $\mathscr{K}_{(x_n, f_n)}$ (i.e., $x \geqslant 0$ if and only if $f_n(x) \geqslant 0$ for $n = 1, 2, \ldots$) is a KB-lineal (that is*[1]*, a normed vector lattice in which the relations $x, y \in E$, $|x| \leqslant |y|$ imply $\|x\| \leqslant \|y\|$).*

Furthermore, in this case all constants occurring in § 17, *theorem* 17.1, *are equal to* 1.

Proof. 1° ⇒ 2°. If $|\gamma_n| \leqslant |\alpha_n|$ $(n = 1, 2, \ldots)$, then $\gamma_n = \beta_n \alpha_n$ for suitable β_n with $|\beta_n| \leqslant 1$ $(n = 1, 2, \ldots)$ and hence, by definition 20.1,

$$\left\| \sum_{i=1}^{n} \gamma_i x_i \right\| = \left\| \sum_{i=1}^{n} \beta_i \alpha_i x_i \right\| \leqslant \left\| \sum_{i=1}^{n} \alpha_i x_i \right\| \qquad (n = 1, 2, \ldots),$$

whence, taking $n \to \infty$, we obtain 2°.

The implication 2° ⇒ 3° is obvious, since $|\alpha_i| \leqslant ||\alpha_i|| \leqslant |\alpha_i|$.

3° ⇒ 4°. If we have 3°, then for any $|\varepsilon_i| = 1$ $(i = 1, \ldots, n)$

$$\left\| \sum_{i=1}^{n} \varepsilon_i \alpha_i x_i \right\| = \left\| \sum_{i=1}^{n} |\varepsilon_i \alpha_i| x_i \right\| = \left\| \sum_{i=1}^{n} |\alpha_i| x_i \right\| = \left\| \sum_{i=1}^{n} \alpha_i x_i \right\|.$$

4° ⇒ 1°. If we have 4° and the scalars are complex, then, since every complex number β_i with $|\beta_i| \leqslant 1$ can be written in the form $\beta_i = \dfrac{\varepsilon_i^{(1)} + \varepsilon_i^{(2)}}{2}$, where $|\varepsilon_i^{(1)}| = |\varepsilon_i^{(2)}| = 1$, we have

$$\left\| \sum_{i=1}^{n} \beta_i \alpha_i x_i \right\| \leqslant \frac{1}{2} \left(\left\| \sum_{i=1}^{n} \varepsilon_i^{(1)} \alpha_i x_i \right\| + \left\| \sum_{i=1}^{n} \varepsilon_i^{(2)} \alpha_i x_i \right\| \right) \leqslant \left\| \sum_{i=1}^{n} \alpha_i x_i \right\|.$$

If we have 4° and the scalars are real, then, by the proof of § 16, inequality (16.27), we again obtain (20.8). Thus, 1° ⇔ ⋯ ⇔ 4°.

Assume now that E is a real Banach space.

1° ∩ 2° ⇒ 5°. If we have 1°, then, by § 16, theorem 16.3, E is a normed vector lattice. By virtue of 1° and 2°, the relations $x, y \in E$, $|x| \leqslant |y|$ imply $\|x\| \leqslant \|y\|$, and thus E is a KB-lineal.

The implication 5° ⇒ 1° is obvious, by § 17, theorem 17.1.

Finally, the assertion concerning the constants of § 17, formula (17.19) is an obvious consequence of that formula. This completes the proof of theorem 20.2.

If $\{x_n\}$ is a hyperorthogonal basis, then, obviously, every sequence $\{\varepsilon_n x_n\}$, where $|\varepsilon_n| = 1$ $(n = 1, 2, \ldots)$, is an orthogonal basis (in particular, *every hyperorthogonal basis is orthogonal*). The converse is not true, as shown by

---

[1] See e.g. [129], p. 211.

Example 20.1. Let E be the two-dimensional Banach space of all pairs of scalars $x=\{\xi_1,\xi_2\}$ endowed with the norm

$$\|x\|=\max(|\xi_1|,|\xi_2|,|\xi_1+\xi_2|) \quad (x=\{\xi_1,\xi_2\}\in E) \tag{20.13}$$

and let $x_1=\{1,0\}$, $x_2=\{0,1\}$. Then for any scalars $\alpha_1,\alpha_2,\varepsilon_1,\varepsilon_2$ with $|\varepsilon_1|=|\varepsilon_2|=1$ we have

$$\|\alpha_1\varepsilon_1 x_1\|=|\alpha_1|, \|\alpha_2\varepsilon_2 x_2\|=|\alpha_2|\leqslant\max(|\alpha_1|,|\alpha_2|,|\alpha_1\varepsilon_1+\alpha_2\varepsilon_2|)$$
$$=\|\alpha_1\varepsilon_1 x_1+\alpha_2\varepsilon_2 x_2\|$$

whence every $\{\varepsilon_1 x_1,\varepsilon_2 x_2\}$ (where $|\varepsilon_1|=|\varepsilon_2|=1$) is an orthogonal basis of E, but

$$\|x_1+x_2\|=\max(1,1,2)=2, \qquad \|x_1-x_2\|=\max(1,1,0)=1,$$

and hence $\{x_1,x_2\}$ is not hyperorthogonal (by theorem 20.2, implication $1^\circ\Rightarrow 3^\circ$).

Proposition 20.1. *A basis $\{x_n\}$ of a Banach space E is strictly hyperorthogonal if and only if $\{x_n\}$ is both strictly orthogonal and hyperorthogonal.*

Proof. Clearly, every strictly hyperorthogonal basis is both strictly orthogonal and hyperorthogonal.

Conversely, assume that $\{x_n\}$ is both strictly orthogonal and hyperorthogonal. Let $\alpha_1,\dots,\alpha_n$, $\beta_1,\dots,\beta_n$ be arbitrary scalars with $|\beta_i|\leqslant 1$ $(i=1,\dots,n)$ and $|\beta_{i_0}|<1$, $\alpha_{i_0}\neq 0$. Then $|\beta_{i_0}\alpha_{i_0}|<|\alpha_{i_0}|$ and hence, by remark 20.1 and theorem 20.1, implication $1^\circ\Rightarrow 3^\circ$, we have

$$\left\|\sum_{i=1}^n \beta_i\alpha_i x_i\right\|=\left\|\sum_{i=1}^n |\beta_i\alpha_i| x_i\right\|<\left\|\sum_{i=1}^n |\alpha_i| x_i\right\|=\left\|\sum_{i=1}^n \alpha_i x_i\right\|,$$

which completes the proof.

We have the following theorem of "orthogonalization":

Theorem 20.3. *Let E be a Banach space with a basis $\{x_n\}$. The following statements are equivalent:*

1°. *There exists on E a norm $\|x\|_1$, equivalent to the initial norm on E, such that in the norm $\|x\|_1$ the basis $\{x_n\}$ is strictly hyperorthogonal (hence also hyperorthogonal, strictly orthogonal, orthogonal).*

2°. *There exists a Banach space F with a strictly hyperorthogonal basis $\{y_n\}$, such that $\{x_n\}\sim\{y_n\}$.*

3°. *$\{x_n\}$ is unconditional.*

Proof. 1°⇒3°. If we have 1°, then, as observed in the above, $\{x_n\}$ is unconditional in the norm $\|x\|_1$, whence also in the initial norm of E.

3°⇒2°. Assume now that we have 3° and let[1]

$$\left\|\sum_{i=1}^{\infty}\alpha_i x_i\right\|_1=\sum_{i=1}^{\infty}\frac{1}{2^i}\|\alpha_i x_i\|+\sup_{\substack{\{\beta_n\}\subset K\\ |\beta_i|\leqslant 1}}\sup_{1\leqslant n<\infty}\left\|\sum_{i=1}^{n}\beta_i\alpha_i x_i\right\|\quad\left(\sum_{i=1}^{\infty}\alpha_i x_i\in E\right). \tag{20.14}$$

Then, by § 17, theorem 17.1, $\|x\|_1<\infty$ $(x\in E)$ and

$$\|x\|\leqslant\|x\|_1\leqslant\max_{1\leqslant i<\infty}\|\alpha_i x_i\|+M_5\|x\|\leqslant 2M_5\|x\|\qquad\left(x=\sum_{i=1}^{\infty}\alpha_i x_i\in E\right)$$

and hence the norm $\|x\|_1$ on E is equivalent to the initial norm of E. Furthermore, in the norm $\|x\|_1$ the basis $\{x_n\}$ is obviously strictly hyperorthogonal. Hence, for $F=E$ endowed with the norm $\|x\|_1$ and $\{y_n\}=\{x_n\}$ (in F), we have 2°.

2°⇒1°. If we have 2°, then, by Ch. I, § 8, theorem 8.1 d) (implication 6°⇒1°), there exists an isomorphism u of E onto F, such that $u(x_n)=y_n$ $(n=1,2,\ldots)$. Then, defining

$$\|x\|_1=\|u(x)\|\qquad(x\in E),$$

we obtain a norm on E, equivalent to the initial norm and such that $\{x_n\}$ is strictly hyperorthogonal in the norm $\|x\|_1$ (since $\{u(x_n)\}=\{y_n\}$ is a strictly hyperorthogonal basis of F), which completes the proof.

Finally, let us give some relations between positive bases, the disjoint support condition (§ 3, remark 3.2) and various types of orthogonal bases.

Proposition 20.2. *Let $\{x_n\}$ be a positive strictly hyperorthogonal basis of a Banach space E, satisfying (20.7) for $n=\infty$. Then $(E,\{x_n\})$ satisfies the disjoint support condition, i.e., for any linear isometry $T\colon E\to E$ with $T(x_j)=\sum_{i=1}^{\infty}a_{ij}x_i$ $(j=1,2,\ldots)$ we have*

$$a_{ij}\cdot a_{im}=0\qquad(i,j,m=1,2,\ldots;\ j\neq m). \tag{20.15}$$

Proof. Assume the contrary, i.e., that for some indices i_0,j_0,m_0 we have $a_{i_0j_0}a_{i_0m_0}\neq 0$. Then $|a_{i_0j_0}|+|a_{i_0m_0}|>|a_{i_0j_0}|-|a_{i_0m_0}|$, whence

[1] Let us also mention that by § 19, theorem 19.2, one can introduce on E an NTK-norm equivalent to the initial norm and, by § 19, already in every NK-norm (with respect to $\{x_n\}$) the basis $\{x_n\}$ is strictly orthogonal.

for the linear isometry T_+ with $T_+(x_j)=\sum_{i=1}^{\infty}|a_{ij}|x_i$ $(j=1,2,\ldots)$ we have, by (20.7) for $n=\infty$,

$$\|T_+(x_j+x_m)\| = \left\|\sum_{i=1}^{\infty}(|a_{ij}|+|a_{im}|)x_i\right\| > \left\|\sum_{i=1}^{\infty}(|a_{ij}|-|a_{im}|)x_i\right\|$$
$$= \|T_+(x_j-x_m)\|,$$

which is impossible, since by the hyperorthogonality of $\{x_n\}$ we have $\|x_j+x_m\| = \|x_j-x_m\|$. This completes the proof.

Let us observe that in proposition 20.2 one cannot replace "strictly hyperorthogonal" by "hyperorthogonal", since e.g. the natural basis $\{x_n\}$ of c_0 is positive and hyperorthogonal, but $(c_0,\{x_n\})$ does not satisfy the disjoint support condition (by § 3, lemma 3.1).

Proposition 20.3. *Let $\{x_n\}$ be a hyperorthogonal basis of a Banach space E, such that $(E,\{x_n\})$ satisfies the disjoint support condition. Then $\{x_n\}$ is a positive basis.*

Proof. Let $T: E\to E$ be a linear isometry, with $T(x_j)=\sum_{i=1}^{\infty}a_{ij}x_i$ $(j=1,2,\ldots)$, and let $\alpha_1,\ldots,\alpha_n$ be arbitrary scalars. Then, by the disjoint support condition, for each i there is, among the n numbers $a_{i1},\ldots,a_{in}$, at most one $\neq 0$, whence $\left|\sum_{j=1}^{n}|a_{ij}|\alpha_j\right|=\left|\sum_{j=1}^{n}a_{ij}\alpha_j\right|$, whence, taking into account the hyperorthogonality of $\{x_n\}$ (the implication $1^\circ\Rightarrow 4^\circ$ of theorem 20.2), we get

$$\left\|\sum_{i=1}^{\infty}\sum_{j=1}^{n}|a_{ij}|\alpha_j x_i\right\| = \left\|\sum_{i=1}^{\infty}\sum_{j=1}^{n}a_{ij}\alpha_j x_i\right\|$$

and hence, by § 3, proposition 3.4, $\{x_n\}$ is positive, which completes the proof.

From propositions 20.2 and 20.3 we infer

Proposition 20.4. *A strictly hyperorthogonal basis $\{x_n\}$ of a Banach space E, satisfying* (20.7) *for $n=\infty$, is positive if and only if $(E,\{x_n\})$ satisfies the disjoint support condition.*

Note that if $\{x_n\}$ is a hyperorthogonal basis of a Banach space E, such that $(E,\{x_n\})$ satisfies the disjoint support condition, $\{x_n\}$ need not be strictly hyperorthogonal, as shown by the unit vector basis of l_n^∞.

Let us observe, finally, that a positive basis need not be orthogonal, since we have seen in § 3 that there exist finite dimensional Banach spaces in which all bases are positive.

§ 21. Subsymmetric bases

Definition 21.1. A basis $\{x_n\}$ of a Banach space E is said to be *subsymmetric*, if it is an unconditional basis and for every increasing sequence of positive integers $\{n_i\}$ the basis $\{x_{n_i}\}$ of the space $[x_{n_i}]$ is equivalent to the basis $\{x_n\}$.

For instance, the natural bases of c_0 and l^p $(p \geqslant 1)$ are subsymmetric. The normalized Haar basis $\{\tilde{z}_n\}$ in $L^p([0,1])$ $(1<p\neq 2)$ is not subsymmetric, since the support of the function $z_{2^{k+1}-1}$ is contained[1] in $\left(1-\frac{1}{2^{k-1}}, 1-\frac{1}{2^k}\right)$ $(k=1,2,\dots)$, whence by § 18, proposition 18.3, $[\tilde{z}_{2^{k+1}-1}]_{k=1}^{\infty}$ is isometrically isomorphic to l^p, which is not isomorphic to $L^p([0,1])$ for $p\neq 2$.

Moreover, we shall now show that the space $L^p([0,1])$ $(1<p\neq 2)$ has no subsymmetric basis at all.

For $1\leqslant p<\infty$ and $\varepsilon>0$ put[2]

$$M_\varepsilon^p = \{x\in L^p([0,1]) \,|\, v(\{t \,|\, |x(t)| \geqslant \varepsilon \|x\|_p\}) \geqslant \varepsilon\}, \tag{21.1}$$

where v is the Lebesgue measure and where $\|x\|_p = \left(\int_0^1 |x(t)|^p dt\right)^{\frac{1}{p}}$.

Lemma 21.1. *The classes M_ε^p have the following properties:*

a) *If* $2\leqslant p<\infty, \varepsilon>0$, *then*

$$\|x\|_p \geqslant \|x\|_2 \geqslant \varepsilon^{\frac{3}{2}} \|x\|_p \qquad (x\in M_\varepsilon^p). \tag{21.2}$$

b) *If* $2\leqslant p<\infty, \varepsilon>0$, *and* $\{x_n\}$ *is a sequence in* M_ε^p *such that the series* $\sum_{n=1}^{\infty} x_n$ *is unconditionally convergent in* $L^p([0,1])$, *then* $\sum_{n=1}^{\infty} \|x_n\|_p^2<\infty$.

c) *If* $x\in L^p([0,1])\setminus\{0\}$ $(1\leqslant p<\infty)$ *does not belong to* M_ε^p, *then there exists a set* $A\subset[0,1]$ *such that*

$$v(A)<\varepsilon, \qquad \int_A \left|\frac{x(t)}{\|x\|}\right|^p dt \geqslant 1-\varepsilon^p. \tag{21.3}$$

Proof. a) If $x\in L^p([0,1])$, where $2\leqslant p<\infty$, then by the Hölder inequality we have

$$\int_0^1 |x(t)|^2 dt \leqslant \left(\int_0^1 (|x(t)|^p dt\right)^{\frac{2}{p}} \left(\int_0^1 1^{\frac{p}{p-2}} dt\right)^{\frac{p-2}{p}} = \left(\int_0^1 |x(t)|^p dt\right)^{\frac{2}{p}}.$$

---

[1] We assume now that $z_{2^{k+1}-1}(t)=0$ for $t=1-\frac{1}{2^{k+1}}$, $t=1-\frac{1}{2^k}$; this does not change the class $\tilde{z}_{2^{k+1}-1}$.

[2] For simplicity, we shall denote both x and $\tilde{x}$ by x; this will lead to no confusion.

Consequently, whenever $2 \leqslant p < \infty$,

$$\|x\|_2 = \left[\int_0^1 |x(t)|^2 dt\right]^{\frac{1}{2}} \leqslant \left[\int_0^1 |x(t)|^p dt\right]^{\frac{1}{p}} = \|x\|_p \quad (x \in L^p([0,1])), \tag{21.4}$$

and thus, in particular, we have the first inequality in (21.2).

Put

$$S_p^\varepsilon(x) = \{t \in [0,1] \mid |x(t)| \geqslant \varepsilon \|x\|_p\}. \tag{21.5}$$

Then, if $x \in M_\varepsilon^p$, we have $v[S_p^\varepsilon(x)] \geqslant \varepsilon$, whence

$$\|x\|_2 = \left(\int_0^1 |x(t)|^2 dt\right)^{\frac{1}{2}} \geqslant \left(\int_{S_\varepsilon^p(x)} |x(t)|^2 dt\right)^{\frac{1}{2}} \geqslant (\varepsilon^2 \|x\|_p^2 v[S_p^\varepsilon(x)])^{\frac{1}{2}} \geqslant \varepsilon^{\frac{3}{2}} \|x\|_p,$$

which proves the second inequality in (21.2).

b) Assume that $2 \leqslant p < \infty, \varepsilon > 0$ and let $\{x_n\}$ be a sequence in M_ε^p such that the series $\sum_{n=1}^\infty x_n$ is unconditionally convergent in $L^p([0,1])$. Then, by (21.4), $\sum_{n=1}^\infty x_n$ is also unconditionally convergent in $L^2([0,1])$, whence, by § 14, lemma 14.10, $\sum_{n=1}^\infty \|x_n\|_2^2 < \infty$. Hence, since $x_n \in M_\varepsilon^p$ $(n = 1, 2, \ldots)$, by part a) proved above we get

$$\sum_{n=1}^\infty \|x_n\|_p^2 \leqslant \varepsilon^{\frac{3}{2}} \sum_{n=1}^\infty \|x_n\|_2^2 < \infty.$$

c) Assume that $x \in L^p([0,1]) \setminus \{0\}$, $x \notin M_\varepsilon^p$ and let A be the set $S_p^\varepsilon(x)$ defined by (21.5). Then, since $x \notin M_\varepsilon^p$, we have $v(A) < \varepsilon$. Furthermore, since $|x(t)|^p \leqslant \varepsilon^p \|x\|^p$ for $t \in [0,1] \setminus A$, we have

$$\|x\|^p = \int_A |x(t)|^p dt + \int_{[0,1] \setminus A} |x(t)|^p dt \leqslant \int_A |x(t)|^p dt + \varepsilon^p \|x\|^p,$$

whence the second inequality in (21.3), which completes the proof of lemma 21.1.

Proposition 21.1. *Let $2 \leqslant p < \infty$ and let $\{x_n\}$ be an unconditional basis of $L^p([0,1])$ for which there exists an $\varepsilon > 0$ such that*

$$x_n \in M_\varepsilon^p \quad (n = 1, 2, \ldots). \tag{21.6}$$

Then $\left\{\frac{x_n}{\|x_n\|}\right\}$ is equivalent to the unit vector basis of l^2.

Proof. Let $\{\alpha_n\}$ be an arbitrary sequence of scalars such that $\sum_{i=1}^{\infty} \alpha_i \frac{x_i}{\|x_i\|}$ converges. Then, since $\left\{\frac{x_n}{\|x_n\|}\right\}$ is an unconditional basis, $\sum_{i=1}^{\infty} \alpha_i \frac{x_i}{\|x_i\|}$ is unconditionally convergent. Furthermore, by (21.6) and the definition (21.1) of M_ε^p we have $\alpha_n \frac{x_n}{\|x_n\|} \in M_\varepsilon^p$ $(n=1,2,\ldots)$. Consequently, by lemma 21.1 b),

$$\sum_{i=1}^{\infty} |\alpha_i|^2 = \sum_{i=1}^{\infty} \left\| \alpha_i \frac{x_i}{\|x_i\|_p} \right\|_p^2 < \infty .$$

Conversely, let $\{\alpha_n\}$ be an arbitrary sequence of scalars such that $\sum_{i=1}^{\infty} |\alpha_i|^2 < \infty$. Then, since by § 14, proposition 14.1, $\left\{\frac{x_n}{\|x_n\|}\right\}$ is a Hilbertian basis, $\sum_{i=1}^{\infty} \alpha_i \frac{x_i}{\|x_i\|}$ converges, which completes the proof of proposition 21.1.

Proposition 21.2. *Let* $1 \leqslant p < \infty$ *and let* $\{x_n\}$ *be a sequence in* $L^p([0,1])$ *with the property that for every* $\varepsilon > 0$ *there exists an index* n_ε *such that* $x_{n_\varepsilon} \notin M_\varepsilon^p$. *Then there exists a subsequence* $\{x'_n\} = \{x_{k_n}\}$ *of* $\{x_n\}$, *such that* $\left\{\frac{x'_n}{\|x'_n\|}\right\}$ *is a basic sequence equivalent to the unit vector basis of* l^p.

Proof. We first show that one can construct successively a subsequence $\{x'_n\}$ of $\{x_n\}$ and a sequence of sets $\{A_n\}$ in $[0,1]$ such that

$$\int_{A_n} \left| \frac{x'_n(t)}{\|x'_n\|} \right|^p dt \geqslant 1 - \frac{1}{4^{(n+1)p}} \qquad (n=1,2,\ldots). \tag{21.7}$$

$$\int_{A_{n+1}} \sum_{i=1}^{n} \left| \frac{x'_i(t)}{\|x'_i\|} \right|^p dt < \frac{1}{4^{(n+1)p}} \quad (n=1,2,\ldots). \tag{21.8}$$

Indeed, take $\varepsilon = \frac{1}{4^2}$. Then by our assumption there exists an index $k_1 = k_1(\varepsilon)$ such that $x'_1 = x_{k_1} \notin M_{4^2}^p$. Hence, by lemma 21.1 c) there exists a set $A_1 \subset [0,1]$ such that

$$v(A_1) < \frac{1}{4^2}, \qquad \int_{A_1} \left| \frac{x'_1(t)}{\|x'_1\|} \right|^p dt \geqslant 1 - \frac{1}{4^{2p}} .$$

Assume that we have already constructed $x_1', \ldots, x_n'$ and $A_1, \ldots, A_n$. Then, since the set function $\phi(A) = \int_A \sum_{i=1}^{n} \left| \frac{x_i'(t)}{\|x_i'\|} \right|^p dt$ is absolutely continuous (because each summand is absolutely continuous), there exists an $\varepsilon > 0$ such that $\phi(A) < \frac{1}{4^{(n+1)p}}$ whenever $v(A) < \varepsilon$. We may assume, without loss of generality, that $\varepsilon \leqslant \frac{1}{4^{n+2}}$. Then, by our assumption, there exists an index $k_{n+1} = k_{n+1}(\varepsilon)$ such that $x_{n+1}' = x_{k_{n+1}} \notin M_\varepsilon^p$. Hence, by lemma 21.1 c), there exists a set $A_{n+1} \subset [0,1]$ such that

$$v(A_{n+1}) < \varepsilon, \qquad \int_{A_{n+1}} \left| \frac{x_{n+1}'(t)}{\|x_{n+1}'\|} \right|^p dt \geqslant 1 - \varepsilon^p \geqslant 1 - \frac{1}{4^{(n+2)p}},$$

and by the first inequality we also have $\int_{A_{n+1}} \sum_{i=1}^{n} \left| \frac{x_i'(t)}{\|x_i'\|} \right|^p dt = \phi(A_{n+1}) < \frac{1}{4^{(n+1)p}}$, which proves our assertion.

Now we shall prove that the sequence $\left\{ \frac{x_n'}{\|x_n'\|} \right\}$ has the required properties. Put

$$A_n' = A_n \setminus \bigcup_{i=n+1}^{\infty} A_i \qquad (n = 1, 2, \ldots), \tag{21.9}$$

$$z_n(t) = \begin{cases} \dfrac{x_n'(t)}{\|x_n'\|} & \text{for } t \in A_n' \\ 0 & \text{for } t \notin A_n' \end{cases} \qquad (n = 1, 2, \ldots), \tag{21.10}$$

$$y_n = \frac{z_n}{\|z_n\|} \qquad (n = 1, 2, \ldots). \tag{21.11}$$

Obviously, $A_n' \cap A_m' = \emptyset$ for $n \neq m$ and $\{t \in [0,1] \mid y_n(t) \neq 0\} \subset A_n'$ $(n = 1, 2, \ldots)$, whence, by § 18, proposition 18.3, $\{y_n\}$ is a basic sequence and the subspace $[y_n]$ is isometrically isomorphic to l^p, by the mapping $y_n \to e_n$, where $\{e_n\}$ is the unit vector basis of l^p; hence for the a.s.c.f. $\{g_n\} \subset [y_n]^*$ to $\{y_n\}$ we have $\|g_n\| = 1$ $(n = 1, 2, \ldots)$. Furthermore, by (21.7)–(21.10) we have, for every $n = 1, 2, \ldots$

$$\left\| \frac{x_n'}{\|x_n'\|} - z_n \right\|^p = \int_{[0,1]\setminus A_n'} \left| \frac{x_n'(t)}{\|x_n'\|} \right|^p dt = \int_{[0,1]\setminus A_n} \left| \frac{x_n'(t)}{\|x_n'\|} \right|^p dt + \int_{A_n \setminus A_n'} \left| \frac{x_n'(t)}{\|x_n'\|} \right|^p dt$$

$$\leqslant 1 - \int_{A_n} \left| \frac{x_n'(t)}{\|x_n'\|} \right|^p dt + \sum_{i=n+1}^{\infty} \int_{A_i} \left| \frac{x_n'(t)}{\|x_n'\|} \right|^p dt \tag{21.12}$$

$$\leqslant \frac{1}{4^{(n+1)p}} + \sum_{i=n+1}^{\infty} \frac{1}{4^{ip}} = \frac{1}{4^{(n+1)p}} + \frac{4}{3\cdot 4^{(n+1)p}} < \frac{1}{4^{np}},$$

$$1 \geqslant \|z_n\|^p = \int_{A_n} \left| \frac{x_n'(t)}{\|x_n'\|} \right|^p dt \geqslant \int_{A_n} \left| \frac{x_n'(t)}{\|x_n'\|} \right|^p dt - \sum_{i=n+1}^{\infty} \int_{A_i} \left| \frac{x_n'(t)}{\|x_n'\|} \right|^p dt$$

$$\geqslant 1 - \frac{1}{4^{(n+1)p}} - \sum_{i=n+1}^{\infty} \frac{1}{4^{ip}} > 1 - \frac{1}{4^{np}}. \tag{21.13}$$

From (21.11)–(21.13) we infer

$$\left\| \frac{x_n'}{\|x_n'\|} - y_n \right\| \leqslant \left\| \frac{x_n'}{\|x_n'\|} - z_n \right\| + \|z_n - y_n\| < \frac{1}{4^n} + \left\| z_n - \frac{z_n}{\|z_n\|} \right\|$$

$$= \frac{1}{4^n} + \left\| \frac{z_n}{\|z_n\|} (\|z_n\| - 1) \right\| = \frac{1}{4^n} + (1 - \|z_n\|)$$

$$< \frac{1}{4^n} + \left(1 - \sqrt[p]{1 - \frac{1}{4^{np}}}\right) < \frac{2}{4^n} \qquad (n = 1, 2, \ldots),$$

which, together with the relation $\|g_n\| = 1$ $(n = 1, 2, \ldots)$ above, implies

$$\sum_{n=1}^{\infty} \|g_n\| \left\| \frac{x_n'}{\|x_n'\|} - y_n \right\| \leqslant \sum_{n=1}^{\infty} \frac{2}{4^n} < 1.$$

Consequently, by Ch. I, § 10, theorem 10.1, $\left\{ \frac{x_n'}{\|x_n'\|} \right\}$ is a basic sequence equivalent to $\{y_n\}$, whence also to the unit vector basis of l^p, which completes the proof of proposition 21.2.

Remark 21.1. Assume, in particular, that $\{x_n\}$ is the Haar system in $L^p([0,1])$ $(1 \leqslant p < \infty)$. Then for any $\varepsilon > 0$ there is only a finite number of indices n such that $x_n \in M_\varepsilon^p$ (because even the number of all indices n such that $\nu(\{t \in [0,1] \mid x_n(t) \neq 0\}) \geqslant \varepsilon$ is finite for any $\varepsilon > 0$), whence, by proposition 21.2, for any $n_1 < n_2 < \cdots$ there exists a subsequence $\{x_n'\}$ of $\{x_{n_k}\}$ such that $\left\{ \frac{x_n'}{\|x_n'\|} \right\}$ is a basic sequence equivalent to the unit vector basis of l^p. This result implies again proposition 18.4 of § 18.

We shall also need the following result on duality of subsymmetric bases:

Proposition 21.3. *If $\{x_n\}$ is a subsymmetric basis of a Banach space E, then the a.s.c.f. $\{f_n\} \subset E^*$ is a subsymmetric basis of $[f_n]$.*

Proof. By § 17, theorem 17.7, $\{f_n\}$ is an unconditional basis of $[f_n]$. Let $\{n_k\}$ be an arbitrary increasing sequence of positive integers. Then, since $\{x_n\}$ is subsymmetric, $\{x_n\}$ is equivalent to the basis $\{x_{n_k}\}$ of $[x_{n_k}]$, whence, by Ch. I, § 12, proposition 12.1, $\{f_n\} \sim \{\phi_{n_k}\}$, where $\{\phi_{n_k}\} \subset [x_{n_k}]^*$ is the a.s.c.f. to the basis $\{x_{n_k}\}$ of $[x_{n_k}]$. Since $\{x_n\}$ is unconditional, $\{\phi_{n_k}\} = \{f_{n_k} | _{[x_{n_j}]}\} \sim \{f_{n_k}\}$, by virtue of § 17, proposition 17.3. Thus $\{f_n\} \sim \{\phi_{n_k}\} \sim \{f_{n_k}\}$, which completes the proof.

This being said, we can prove now

Theorem 21.1. *The space $L^p([0,1])$ $(1<p\neq 2)$ has no subsymmetric basis.*

Proof. Assume first that $2<p<\infty$ and that $\{x_n\}$ is a subsymmetric basis of $L^p([0,1])$. Then there are two cases: a) If there exists an $\varepsilon_0>0$ such that $x_n \in M^p_{\varepsilon_0}$ $(n=1,2,\ldots)$, then, by proposition 21.1, $\left\{\frac{x_n}{\|x_n\|}\right\}$ is equivalent to the unit vector basis of l^2, whence we obtain that $L^p([0,1])$ is isomorphic to l^2, contradicting the assumption $p\neq 2$. b) If there exists no such $\varepsilon_0>0$, i.e., if for every $\varepsilon>0$ there exists an index n_ε such that $x_{n_\varepsilon} \notin M^p_\varepsilon$, then, by proposition 21.2, there exists a subsequence $\{x'_n\}$ of $\{x_n\}$ such that $\left\{\frac{x'_n}{\|x'_n\|}\right\}$ is equivalent to the unit vector basis of l^p, whence we obtain that $L^p([0,1])$ is isomorphic to l^p, contradicting[1] again the assumption $p\neq 2$. Thus theorem 21.1 is true for the spaces $L^p([0,1])$ with $2<p<\infty$.

Assume now that $1<p<2$ and that $\{x_n\}$ is a subsymmetric basis of $E=L^p([0,1])$. Then, since every basis of a reflexive space is shrinking (see § 4, example 4.3), by proposition 21.3 the a.s.c.f. $\{f_n\} \subset E^*$ is a subsymmetric basis of $[f_n]=E^* \equiv L^q([0,1])$, where $\frac{1}{p}+\frac{1}{q}=1$, which, since $2<q<\infty$, contradicts the fact proved above, that theorem 21.1 is true for the spaces $L^p([0,1])$ with $2<p<\infty$. This completes the proof of theorem 21.1.

Let us observe that in definition 21.1 of a subsymmetric basis it is essential to assume separately that $\{x_n\}$ is unconditional, as shown by § 17, example 17.2.

---

[1] See e.g. S. Banach [10], p. 206, theorem 9.

Proposition 21.4. *Every subsymmetric basis $\{x_n\}$ is bounded.*

Proof. Assume that for a basis $\{x_n\}$ there exists an infinite sequence of positive integers $\{m_k\}$ such that

$$\|x_{m_k}\| > k \qquad (k=1,2,\ldots). \tag{21.14}$$

Then, taking an $x = \sum_{i=1}^{\infty} f_i(x) x_i \in E$ such that

$$f_n(x) \neq 0 \qquad (n=1,2,\ldots) \tag{21.15}$$

and an infinite subsequence $\{m_{k_n}\}$ of $\{m_k\}$ such that

$$\|f_n(x) x_{m_{k_n}}\| \geqslant 1 \qquad (n=1,2,\ldots), \tag{21.16}$$

the series $\sum_{i=1}^{\infty} f_i(x) x_{m_{k_i}}$ will be divergent, whence $\{x_n\}$ is not subsymmetric. We have thus proved that for every subsymmetric basis $\{x_n\}$ we have $\sup_{1 \leqslant n < \infty} \|x_n\| < \infty$.

Assume now that there exists an infinite sequence of indices $\{m_k\}$ such that

$$\|x_{m_k}\| < \frac{1}{k} \qquad (k=1,2,\ldots). \tag{21.17}$$

Then, taking an infinite subsequence $\{m_{k_n}\}$ of $\{m_k\}$ such that

$$\|x_{m_{k_n}}\| \leqslant \frac{\|x_n\|}{2^n} \qquad (n=1,2,\ldots), \tag{21.18}$$

the series $\sum_{n=1}^{\infty} \frac{1}{\|x_n\|} x_{m_{k_n}}$ will be convergent, while the series $\sum_{n=1}^{\infty} \frac{1}{\|x_n\|} x_n$ is, of course, divergent, whence $\{x_n\}$ is not subsymmetric. Thus for every subsymmetric basis $\{x_n\}$ we have $\inf_{1 \leqslant n < \infty} \|x_n\| > 0$, which completes the proof of proposition 21.4.

As before, we shall use the following notation:

$$\mathcal{O} = \text{the set of all increasing sequences of positive integers.} \tag{21.19}$$

If $\{x_n\}$ is a subsymmetric basis of E, then for every $x \in E$ and $(\{m_j\}, \{n_j\}) \in \mathcal{O} \times \mathcal{O}$ the series $\sum_{i=1}^{\infty} f_{m_i}(x) x_{n_i}$ converges. Indeed, since $\{x_n\}$ is unconditional, $\sum_{i=1}^{\infty} f_{m_i}(x) x_{m_i}$ converges, whence, since $\{x_{m_j}\} \sim \{x_n\} \sim \{x_{n_j}\}$, $\sum_{i=1}^{\infty} f_{m_i}(x) x_{n_i}$ converges. Therefore the linear operators

$$A_{\{m_j\},\{n_j\}}(x) = \sum_{i=1}^{\infty} f_{m_i}(x) x_{n_i} \qquad (x \in E) \tag{21.20}$$

are well defined on E, and below we shall see that they are uniformly bounded.

Some characterizations of subsymmetric bases are given in

Theorem 21.2. *Let $\{x_n\}$ be a basis of a Banach space E, with the a.s.c.f. $\{f_n\}$. The following statements are equivalent:*

1°. *$\{x_n\}$ is a subsymmetric basis of E.*

2°. *We have*

$$\sup_{(\{m_j\},\{n_j\})\in\mathcal{O}\times\mathcal{O}} \|A_{\{m_j\},\{n_j\}}\| < \infty. \tag{21.21}$$

3°. *We have*

$$\sup_{(\{m_j\},\{n_j\})\in\mathcal{O}\times\mathcal{O}} \; \sup_{1\leqslant k<\infty} \left\| \sum_{i=1}^{k} f_{m_i}(x) x_{n_i} \right\| < \infty \qquad (x\in E). \tag{21.22}$$

4°. *We have*

$$\sup_{1\leqslant k<\infty} \left\| \sum_{i=1}^{k} f_{m_i}(x) x_{n_i} \right\| < \infty \quad (x\in E,\ (\{m_j\},\{n_j\})\in\mathcal{O}\times\mathcal{O}). \tag{21.23}$$

5°. *For every $x\in E$ and $(\{m_j\},\{n_j\})\in\mathcal{O}\times\mathcal{O}$ the series $\sum_{i=1}^{\infty} f_{m_i}(x)x_{n_i}$ is convergent.*

6°. *We have*

$$\sup_{\{m_i\}\in\mathcal{O}} \; \sup_{1\leqslant k<\infty} \left\| \sum_{i=1}^{k} f_{m_i}(x) x_i \right\| < \infty \qquad (x\in E) \tag{21.24}$$

and

$$\sup_{\{n_i\}\in\mathcal{O}} \; \sup_{1\leqslant k<\infty} \left\| \sum_{i=1}^{k} f_i(x) x_{n_i} \right\| < \infty \qquad (x\in E). \tag{21.25}$$

7°. *We have*

$$\sup_{1\leqslant k<\infty} \left\| \sum_{i=1}^{k} f_{m_i}(x) x_i \right\| < \infty \qquad (x\in E,\ \{m_i\}\in\mathcal{O}) \tag{21.26}$$

and

$$\sup_{1\leqslant k<\infty} \left\| \sum_{i=1}^{k} f_i(x) x_{n_i} \right\| < \infty \qquad (x\in E,\ \{n_i\}\in\mathcal{O}). \tag{21.27}$$

8°. *For every $x\in E$ and $(\{m_j\},\{n_j\})\in\mathcal{O}\times\mathcal{O}$ the series $\sum_{i=1}^{\infty} f_{m_i}(x)x_i$ and $\sum_{i=1}^{\infty} f_i(x)x_{n_i}$ are convergent.*

Proof. Assume that 2° is not valid. Then, by the principle of uniform boundedness, there exists an $x\in E$ such that

$$\sup_{(\{m_j\},\{n_j\})\in\mathcal{O}\times\mathcal{O}} \|A_{\{m_j\},\{n_j\}}(x)\| = \sup_{(\{m_j\},\{n_j\})\in\mathcal{O}\times\mathcal{O}} \left\| \sum_{i=1}^{\infty} f_{m_i}(x) x_{n_i} \right\| = \infty. \tag{21.28}$$

Then there exist sequences $\{m_{i1}\}, \{n_{i1}\} \in \mathcal{O}$ and a positive integer l_1 such that

$$\left\| \sum_{i=1}^{l_1} f_{m_{i1}}(x) x_{n_{i1}} \right\| \geqslant 1. \tag{21.29}$$

Let us denote

$$\mathcal{N}_p = \{p, p+1, p+2, \ldots\} \qquad (p=1,2,\ldots), \tag{21.30}$$

$$\mathcal{O}(\mathcal{A}) = \text{the set of all increasing sequences of elements of } \mathcal{A}, \tag{21.31}$$

where $\mathcal{A} \subset \mathcal{N}$. Observe that

$$\sup_{(\{m_j\},\{n_j\}) \in \mathcal{O}(\mathcal{N}_p) \times \mathcal{O}(\mathcal{N}_p)} \left\| \sum_{i=1}^{\infty} f_{m_i}(x) x_{n_i} \right\| = \infty \qquad (p=1,2,\ldots), \tag{21.32}$$

since otherwise for some p one would have, by proposition 21.4,

$$\begin{aligned} \sup_{(\{m_j\},\{n_j\}) \in \mathcal{O} \times \mathcal{O}} \left\| \sum_{i=1}^{\infty} f_{m_i}(x) x_{n_i} \right\| &\leqslant (p-1) \sup_{1 \leqslant n < \infty} \|x_n\| \sup_{1 \leqslant n < \infty} \|f_n\| \, \|x\| \\ &+ \sup_{(\{m_j\},\{n_j\}) \in \mathcal{O}(\mathcal{N}_p) \times \mathcal{O}(\mathcal{N}_p)} \left\| \sum_{i=1}^{\infty} f_{m_i}(x) x_{n_i} \right\| < \infty, \end{aligned}$$

in contradiction with (21.28).

Putting $k_1 = \max\limits_{1 \leqslant i \leqslant l_1} (m_{i1}, n_{i1})$ and applying (21.32) for $p = k_1 + 1$, it follows that there exist sequences $\{m_{i2}\}, \{n_{i2}\} \in \mathcal{O}(\mathcal{N}_{k_1+1})$ and a positive integer l_2 such that

$$\left\| \sum_{i=1}^{l_2} f_{m_{i2}}(x) x_{n_{i2}} \right\| \geqslant 1. \tag{21.33}$$

Putting $k_2 = \max\limits_{1 \leqslant i \leqslant l_2} (m_{i2}, n_{i2})$ and continuing in this manner, we obtain sequences $\{l_j\} \in \mathcal{O}$, $\{k_j\} \in \mathcal{O}$ and sequences $\{m_{ij}\}_{i=1}^{l_j} \in \mathcal{O}(\mathcal{N}_{k_{j-1}+1} \setminus \mathcal{N}_{k_j+1})$, $\{n_{ij}\}_{i=1}^{l_j} \in \mathcal{O}(\mathcal{N}_{k_{j-1}+1} \setminus \mathcal{N}_{k_j+1})$ $(j=1,2,\ldots; k_0=0)$, such that

$$\left\| \sum_{i=1}^{l_j} f_{m_{ij}}(x) x_{n_{ij}} \right\| \geqslant 1 \qquad (j=1,2,\ldots). \tag{21.34}$$

Define now the following sequences $\{m_j\}, \{n_j\} \in \mathcal{O}$:

$$\begin{gathered} m_{l_1 + \cdots + l_{j-1} + i} = m_{ij}; \qquad n_{l_1 + \cdots + l_{j-1} + i} = n_{ij} \\ (i=1,\ldots,l_j;\ j=1,2,\ldots;\ l_0=0). \end{gathered} \tag{21.35}$$

For these sequences, by (21.34) the series $\sum_{i=1}^{\infty} f_{m_i}(x) x_{n_i}$ is divergent, whence $\{x_n\}$ is not subsymmetric. Thus $1^\circ \Rightarrow 2^\circ$.

Assume now that we have 2°. Since $\{x_n\}$ is a basis, there exists a constant $C \geqslant 1$ such that

$$\sup_{1 \leqslant k < \infty} \left\| \sum_{i=1}^{k} f_{m_i}(x) x_{n_i} \right\| \leqslant C \left\| \sum_{i=1}^{\infty} f_{m_i}(x) x_{n_i} \right\| = C \| A_{\{m_j\},\{n_j\}}(x) \|$$

$$\leqslant C \| A_{\{m_j\},\{n_j\}} \| \, \|x\| \qquad (x \in E, (\{m_j\}, \{n_j\}) \in \mathcal{O} \times \mathcal{O}),$$

whence, by 2°, we obtain (21.22). Thus, $2^\circ \Rightarrow 3^\circ$.

The implication $3^\circ \Rightarrow 4^\circ$ is obvious.

Assume now that we have 4°. Then, by the principle of uniform boundedness applied to the operators $u_{\{m_j\},\{n_j\},k}(x) = \sum_{i=1}^{k} f_{m_i}(x) x_{n_i}$ $(x \in E, (\{m_j\}, \{n_j\}) \in \mathcal{O} \times \mathcal{O}, k = 1, 2, \ldots)$, we have

$$\sup_{1 \leqslant k < \infty} \| u_{\{m_j\},\{n_j\},k} \| < \infty \qquad ((\{m_j\}, \{n_j\}) \in \mathcal{O} \times \mathcal{O}), \tag{21.36}$$

whence, since for every finite linear combination $p = \sum_{j=1}^{l} \alpha_j x_j$ and every $(\{m_j\}, \{n_j\}) \in \mathcal{O} \times \mathcal{O}$ we have $\lim_{k \to \infty} u_{\{m_j\},\{n_j\},k}(p) = \sum_{i=1}^{l} \alpha_{m_i} x_{n_i}$, where $\alpha_{m_i} = 0$ for $m_i \geqslant l+1$ and since $[x_n] = E$, it follows that we have 5°. Thus, $4^\circ \Rightarrow 5^\circ$.

Assume now that we have 5°. Then, in particular, for every $x \in E$ and $\{n_i\} \in \mathcal{O}$ the series $\sum_{i=1}^{\infty} f_{n_i}(x) x_{n_i}$ converges and hence, by § 16, theorem 16.1, $\{x_n\}$ is an unconditional basis. Furthermore, if $\sum_{i=1}^{\infty} \alpha_i x_i$ converges, then, by 5°, the series $\sum_{i=1}^{\infty} \alpha_i x_{n_i}$ converges for $\{n_i\} \in \mathcal{O}$. Conversely, if $\sum_{i=1}^{\infty} \alpha_i x_{n_i}$ converges, say to $x \in E$, then $\alpha_i = f_{n_i}(x)$, whence, by 5°, $\sum_{i=1}^{\infty} \alpha_i x_i = \sum_{i=1}^{\infty} f_{n_i}(x) x_i$ also converges, which proves that $\{x_n\}$ is equivalent to $\{x_{n_i}\}$. Thus, $5^\circ \Rightarrow 1^\circ$.

The implications $3^\circ \Rightarrow 6^\circ \Rightarrow 7^\circ$ are obvious, and the proof of the implication $7^\circ \Rightarrow 8^\circ$ is similar to the above proof of $4^\circ \Rightarrow 5^\circ$.

Assume, finally, that we have 8° and let $x \in E$, $\{n_i\} \in \mathcal{O}$ be arbitrary. Then by 8° $\sum_{i=1}^{\infty} f_{n_i}(x) x_i$ converges, say to $y \in E$, whence $f_i(y) = f_{n_i}(x)$

$(i=1,2,\ldots)$, whence, again by 8°, $\sum_{i=1}^{\infty} f_{n_i}(x)x_{n_i} = \sum_{i=1}^{\infty} f_i(y)x_{n_i}$ converges. Consequently, by § 16, theorem 16.1, $\{x_n\}$ is an unconditional basis. Now, as the above proof of the implication 5°⇒1°, we see that $\{x_n\}$ is equivalent to every subsequence $\{x_{n_i}\}$, and therefore $\{x_n\}$ is subsymmetric. Thus, 8°⇒1°, which completes the proof of theorem 21.2.

Let us observe that the conditional basis $\{x_n\}$ of c_0 given in § 17, example 17.2, has the property that $\sum_{i=1}^{\infty} f_i(x)x_{n_i}$ converges for every $x \in E$ and $\{n_i\} \in \mathcal{O}$ (and conversely, the convergence of $\sum_{i=1}^{\infty} \alpha_i x_{n_i}$ implies that of $\sum_{i=1}^{\infty} \alpha_i x_i$), since this amounts to the convergence of $\sum_{i=1}^{\infty} f_i(x)$, but $\sum_{i=1}^{\infty} f_{m_i}(x)x_i$ is not necessarily convergent for every $x \in E$ and $\{m_i\} \in \mathcal{O}$ (since $\sum_{i=1}^{\infty} f_{m_i}(x)$ need not be convergent, or, alternatively, by theorem 21.2, implication 8°⇒1° and the conditionality of $\{x_n\}$). Dually, the conditional basis $\{h_n\}$ of l^1 given in § 14, example 14.2, has the property that $\sum_{i=1}^{\infty} \alpha_{m_i} h_i$ converges for every $\sum_{i=1}^{\infty} \alpha_i h_i \in l^1$ and $\{m_i\} \in \mathcal{O}$ (since $\sum_{i=1}^{\infty} |\alpha_i - \alpha_{i+1}| < \infty$ implies $\sum_{i=1}^{\infty} |\alpha_{m_i} - \alpha_{m_{i+1}}| < \infty$), but $\sum_{i=1}^{\infty} \alpha_i h_{n_i}$ is not necessarily convergent for every $\sum_{i=1}^{\infty} \alpha_i h_i \in l^1$ and $\{n_i\} \in \mathcal{O}$. However, we don't know any example of an *unconditional* basis $\{x_n\}$ such that $\sum_{i=1}^{\infty} f_i(x)x_{n_i}$ converges for every $x \in E$ and $\{n_i\} \in \mathcal{O}$ (or such that $\sum_{i=1}^{\infty} f_{m_i}(x)x_i$ converges for every $x \in E$ and $\{m_i\} \in \mathcal{O}$) but $\{x_n\}$ is not subsymmetric.

In the usual concrete Banach spaces with a subsymmetric basis all subsymmetric bases are equivalent. In fact, by the results of § 18, in c_0, l^1 and l^2 all bounded unconditional bases are equivalent, while for the other spaces l^p we have

Proposition 21.5. *In the space* $E = l^p$ $(1 < p < \infty, p \neq 2)$ *all subsymmetric bases are equivalent.*

Proof. Let $\{y_n\}$ be an arbitrary subsymmetric basis of E. Then, by proposition 21.4, $\{y_n\}$ is bounded, and since E is reflexive, by § 7, the remark made before theorem 7.2, $y_n \to 0$ weakly. Consequently, by § 15, proposition 15.1, one can extract a subsequence $\{y_{n_k}\}$ which is a basic sequence, equivalent to a block basic sequence $\{z_n\}$ with respect to the unit vector basis $\{x_n\}$ of $E = l^p$. Since $\{y_n\}$ is subsymmetric, $\{y_n\} \sim \{y_{n_k}\}$

and, on the other hand, by § 16, theorem 16.8 and § 18, proposition 18.1, $\{z_n\} \sim \left\{\frac{z_n}{\|z_n\|}\right\} \sim \{x_n\}$. Thus $\{y_n\} \sim \{y_{n_k}\} \sim \{z_n\} \sim \{x_n\}$, which completes the proof of proposition 21.5.

Problem 21.1. In a Banach space E with a subsymmetric basis are all subsymmetric bases of E equivalent?

§ 22. Symmetric bases. Symmetric spaces

Definition 22.1. A basis $\{x_n\}$ of a Banach space E is called *symmetric* if

$$\sup_{\sigma\in\Pi} \sup_{\substack{|\beta_i|\leqslant 1\\ 1\leqslant k<\infty}} \left\| \sum_{i=1}^{k} \beta_i f_i(x) x_{\sigma(i)} \right\| < \infty \qquad (x\in E), \tag{22.1}$$

where $\{f_n\} \subset E^*$ is the a.s.c.f. to $\{x_n\}$, and where, as before, Π denotes the set of all permutations of the set $\mathcal{N}=\{1,2,3,\ldots\}$.

For instance, the natural bases of c_0 and l^p $(p\geqslant 1)$ are symmetric. The normalized Haar basis $\{\tilde{z}_n\}$ in $L^p([0,1])$ $(1<p\neq 2)$ is not symmetric, and moreover, the space $L^p([0,1])$ $(1<p\neq 2)$ has no symmetric basis at all, since it has no subsymmetric basis (by § 21, theorem 21.1) and since we shall see below that every symmetric basis is subsymmetric.

Taking in (22.1) the identical permutation $\sigma=\iota$ it follows (by § 16, theorem 16.1) that *every symmetric basis* $\{x_n\}$ *is unconditional*. Furthermore, if $\{x_n\}$ is a symmetric basis of E, then for every $x\in E$ and $(\rho,\sigma) \in \Pi\times\Pi$ the series $\sum_{i=1}^{\infty} f_{\rho(i)}(x) x_{\sigma(i)}$ converges. Indeed, by (22.1) we have

$$\|x\|_\sigma \overset{df}{=} \sup_{1\leqslant k<\infty} \left\| \sum_{i=1}^{k} f_i(x) x_{\sigma(i)} \right\| < \infty \qquad (x\in E,\ \sigma\in\Pi), \tag{22.2}$$

whence, by the principle uniform boundedness applied to the operators $u_{\sigma,n}(x)=\sum_{i=1}^{n} f_i(x) x_{\sigma(i)}$ $(x\in E,\ \sigma\in\Pi,\ n=1,2,\ldots)$, we have

$$\|x\|_\sigma \leqslant C_\sigma \|x\| \qquad (x\in E,\ \sigma\in\Pi), \tag{22.3}$$

where the constant C_σ depends only on σ, and thus

$$\left\| \sum_{i=n}^{n+p} f_i(x) x_{\sigma(i)} \right\| \leqslant \sup_{1\leqslant k<\infty} \left\| \sum_{i=1}^{k} f_i\left[\sum_{j=n}^{n+p} f_j(x) x_j\right] x_{\sigma(i)} \right\|$$
$$= \left\| \sum_{i=n}^{n+p} f_i(x) x_i \right\|_\sigma \leqslant C_\sigma \left\| \sum_{i=n}^{n+p} f_i(x) x_i \right\|,$$

which shows that $\sum_{i=1}^{\infty} f_i(x) x_{\sigma(i)}$ converges for every $x \in E, \sigma \in \Pi$. Applying this to the permutation $\rho^{-1}\sigma \in \Pi$, we infer that $\sum_{i=1}^{\infty} f_i(x) x_{\rho^{-1}\sigma(i)}$ converges for every $x \in E, (\rho, \sigma) \in \Pi \times \Pi$, whence, since $\{x_n\}$ is unconditional, it follows that $\sum_{i=1}^{\infty} f_{\rho(i)}(x) x_{\sigma(i)} = \sum_{i=1}^{\infty} f_i(x) x_{\rho^{-1}\sigma(i)}$ converges for every $x \in E, (\rho, \sigma) \in \Pi \times \Pi$. Therefore the linear operators

$$A_{\rho,\sigma}(x) = \sum_{i=1}^{\infty} f_{\rho(i)}(x) x_{\sigma(i)} \qquad (x \in E) \tag{22.4}$$

are well defined on E and we shall see below that they are uniformly bounded isomorphisms of E onto E.

Some characterizations of symmetric bases are given in

Theorem 22.1. *Let $\{x_n\}$ be a basis of a Banach space E, with the a.s.c.f. $\{f_n\}$. The following statements are equivalent:*

1°. *$\{x_n\}$ is a symmetric basis of E.*

2°. *We have*

$$\sup_{(\rho,\sigma) \in \Pi \times \Pi} \|A_{\rho,\sigma}\| < \infty. \tag{22.5}$$

3°. *We have*

$$\sup_{(\rho,\sigma) \in \Pi \times \Pi} \sup_{1 \leqslant k < \infty} \left\| \sum_{i=1}^{k} f_{\rho(i)}(x) x_{\sigma(i)} \right\| < \infty \qquad (x \in E). \tag{22.6}$$

4°. *We have*

$$\sup_{1 \leqslant k < \infty} \left\| \sum_{i=1}^{k} f_{\rho(i)}(x) x_{\sigma(i)} \right\| < \infty \qquad (x \in E, \ (\rho,\sigma) \in \Pi \times \Pi). \tag{22.7}$$

5°. *For every $x \in E$ and $(\rho, \sigma) \in \Pi \times \Pi$ the series $\sum_{i=1}^{\infty} f_{\rho(i)}(x) x_{\sigma(i)}$ converges.*

6°. *We have*

$$\sup_{\sigma \in \Pi} \sup_{1 \leqslant k < \infty} \left\| \sum_{i=1}^{k} f_i(x) x_{\sigma(i)} \right\| < \infty \qquad (x \in E). \tag{22.8}$$

7°. *We have*

$$\sup_{1 \leqslant k < \infty} \left\| \sum_{i=1}^{k} f_i(x) x_{\sigma(i)} \right\| < \infty \qquad (x \in E, \ \sigma \in \Pi). \tag{22.9}$$

8°. *For every $x \in E$ and $\sigma \in \Pi$ the series $\sum_{i=1}^{\infty} f_i(x) x_{\sigma(i)}$ converges.*

9°. *We have*

$$\sup_{\rho\in\Pi}\ \sup_{1\leqslant k<\infty}\left\|\sum_{i=1}^{k} f_{\rho(i)}(x)x_i\right\|<\infty \qquad (x\in E). \qquad (22.10)$$

10°. *We have*

$$\sup_{1\leqslant k<\infty}\left\|\sum_{i=1}^{k} f_{\rho(i)}(x)x_i\right\|<\infty \qquad (x\in E,\ \rho\in\Pi). \quad (22.11)$$

11°. *For every* $x\in E$ *and* $\rho\in\Pi$ *the series* $\sum_{i=1}^{\infty} f_{\rho(i)}(x)x_i$ *converges.*

12°. *Every permutation* $\{x_{\sigma(n)}\}$ *of the basis* $\{x_n\}$ *is a basis of the space E, equivalent to the basis* $\{x_n\}$.

Proof. Assume that we have 1°. Then, since $\{x_n\}$ is unconditional, we have

$$\begin{aligned}\sup_{(\rho,\sigma)\in\Pi\times\Pi}\|A_{\rho,\sigma}(x)\| &= \sup_{(\rho,\sigma)\in\Pi\times\Pi}\left\|\sum_{i=1}^{\infty} f_{\rho(i)}(x)x_{\sigma(i)}\right\|\\ &= \sup_{(\rho,\sigma)\in\Pi\times\Pi}\left\|\sum_{i=1}^{\infty} f_i(x)x_{\rho^{-1}\sigma(i)}\right\|\\ &\leqslant \sup_{\sigma\in\Pi}\ \sup_{\substack{|\beta_i|\leqslant 1\\ 1\leqslant k<\infty}}\left\|\sum_{i=1}^{k}\beta_i f_i(x)x_{\sigma(i)}\right\|<\infty \qquad (x\in E),\end{aligned}$$

whence, by the principle of uniform boundedness, we infer (22.5). Thus, $1°\Rightarrow 2°$.

Assume now that we have 2°. Since $\{x_n\}$ is a basis, there exists a constant $C\geqslant 1$ such that

$$\begin{aligned}\sup_{1\leqslant k<\infty}\left\|\sum_{i=1}^{k} f_{\rho(i)}(x)x_{\sigma(i)}\right\| &\leqslant C\left\|\sum_{i=1}^{\infty} f_{\rho(i)}(x)x_{\sigma(i)}\right\|\\ &= C\,\|A_{\rho,\sigma}(x)\|\leqslant C\|A_{\rho,\sigma}\|\,\|x\| \quad (x\in E),\end{aligned}$$

whence, by 2°,

$$\sup_{(\rho,\sigma)\in\Pi\times\Pi}\ \sup_{1\leqslant k<\infty}\left\|\sum_{i=1}^{k} f_{\rho(i)}(x)x_{\sigma(i)}\right\|\leqslant C\sup_{(\rho,\sigma)\in\Pi\times\Pi}\|A_{\rho,\sigma}\|\,\|x\|<\infty \quad (x\in E),$$

and thus $2°\Rightarrow 3°$.

The implications $3°\Rightarrow 4°\Rightarrow 7°$ and $3°\Rightarrow 6°\Rightarrow 7°$ are obvious. The implication $7°\Rightarrow 8°$ has been already proved in the above (before the statement of theorem 22.1).

Assume now that we have 8°. Then the series $\sum_{i=1}^{\infty} f_{\sigma(i)}(x)x_{\sigma(i)} = \sum_{i=1}^{\infty} f_i\left[\sum_{j=1}^{\infty} f_j(x)x_{\sigma^{-1}(j)}\right]x_{\sigma(i)}$ also converges for every $x\in E$, $\sigma\in\Pi$, i.e., $\{x_n\}$ is an unconditional basis, and therefore $\sum_{i=1}^{\infty} f_{\rho(i)}(x)x_{\sigma(i)} = \sum_{i=1}^{\infty} f_i(x)x_{\rho^{-1}\sigma(i)}$ converges for every $x\in E$, $(\rho,\sigma)\in\Pi\times\Pi$. Thus, $8°\Rightarrow 5°$.

The implications $3°\Rightarrow 9°\Rightarrow 10°$ are obvious.

Assume now that we have 10°. Then, by the principle of uniform boundedness applied to the operators $u_{\rho,k}(x) = \sum_{i=1}^{k} f_{\rho(i)}(x)x_i$ ($x\in E$, $\rho\in\Pi$, $k=1,2,\ldots$), we have

$$\sup_{1\leqslant k<\infty} \|u_{\rho,k}\| < \infty \qquad (\rho\in\Pi), \tag{22.12}$$

whence, since for every finite linear combination $p = \sum_{j=1}^{m} \alpha_j x_j$ and every $\rho\in\Pi$ we have $\lim_{k\to\infty} u_{\rho,k}(p) = \sum_{i=1}^{m} \alpha_{\rho(i)}x_i$, where $\alpha_{\rho(i)}=0$ for $\rho(i)\geqslant m+1$ and since $[x_n]=E$, it follows that we have 11°. Thus, $10°\Rightarrow 11°$.

Assume now that we have 11°. Then, as in the proof of Ch. I, § 12, theorem 12.1, it follows that for every $f\in[f_n]$ and $\rho\in\Pi$ the series $\sum_{i=1}^{\infty} f(x_i)f_{\rho(i)}$ converges, whence, by the implication $8°\Rightarrow 5°$ proved above, $\{f_n\}$ is an unconditional basis of $[f_n]$ and hence, by the last statement of § 17, theorem 17.2, $\{x_n\}$ is an unconditional basis. Hence, again by 11°, it follows that the series $\sum_{i=1}^{\infty} f_{\rho(i)}(x)x_{\sigma(i)} = \sum_{i=1}^{\infty} f_{\sigma^{-1}\rho(i)}(x)x_i$ is convergent for every $x\in E$ and $(\rho,\sigma)\in\Pi\times\Pi$, i.e., we have 5°. Thus, $11°\Rightarrow 5°$.

Assume now that we have 5°. Then, applying 5° for all $\rho=\sigma\in\Pi$ we obtain that $\{x_n\}$ is an unconditional basis, whence every permutation $\{x_{\sigma(n)}\}$ is a basis of E. If $\sum_{i=1}^{\infty} \alpha_i x_i$ converges, then, by 5°, $\sum_{i=1}^{\infty} \alpha_i x_{\sigma(i)}$ also converges. Conversely, if $\sum_{i=1}^{\infty} \alpha_i x_{\sigma(i)}$ converges, then, since $\{x_n\}$ is unconditional, we can write $\sum_{i=1}^{\infty} \alpha_i x_{\sigma(i)} = \sum_{i=1}^{\infty} \alpha_{\sigma^{-1}(i)}x_i$, whence, by 5°, the series $\sum_{i=1}^{\infty} \alpha_{\sigma\sigma^{-1}(i)}x_i = \sum_{i=1}^{\infty} \alpha_i x_i$ also converges. Thus, $5°\Rightarrow 12°$.

Assume, finally, that we have 12°. Let us first show that in this case the basis $\{x_n\}$ is bounded. Assume that there exists an infinite sequence $\{m_k\}\in\mathcal{O}$ such that

$$\|x_{m_k}\|>k \qquad (k=1,2,\ldots); \tag{22.13}$$

we may also assume, without loss of generality, that the set $\mathcal{N}\setminus\{m_k\}$ is infinite. Take an $x=\sum_{i=1}^{\infty} f_i(x)x_i\in E$ such that

$$f_n(x)\neq 0 \qquad (n=1,2,\ldots) \tag{22.14}$$

and an infinite subsequence $\{m_{k_n}\}$ of $\{m_k\}$ such that

$$\|f_n(x)x_{m_{k_n}}\|\geqslant 1 \qquad (n=1,2,\ldots). \tag{22.15}$$

Then, taking a $\sigma\in\Pi$ such that $\sigma(m_j)=m_{k_j}$ for $j=1,2,\ldots$, $\sigma(n)\in\mathcal{N}\setminus\{m_{k_j}\}$ for $n\in\mathcal{N}\setminus\{m_k\}$, the series $\sum_{i=1}^{\infty} f_i(x)x_{\sigma(i)}$ will be divergent, contradicting the assumption 12°. Thus, $12°\Rightarrow \sup_{1\leqslant n<\infty}\|x_n\|<\infty$.

On the other hand, assume that there exists an infinite sequence $\{m_k\}\in\mathcal{O}$ such that

$$\|x_{m_k}\|<\frac{1}{k} \qquad (k=1,2,\ldots) \tag{22.16}$$

and that the set $\mathcal{N}\setminus\{m_k\}$ is infinite. Then, taking an infinite subsequence $\{m_{k_n}\}$ of $\{m_k\}$ such that

$$\|x_{m_{k_n}}\|\leqslant\frac{\|x_{m_n}\|}{2^n} \qquad (n=1,2,\ldots), \tag{22.17}$$

and a permutation $\sigma\in\Pi$ such that $\sigma(m_j)=m_{k_j}$ for $j=1,2,\ldots$, $\sigma(n)\in\mathcal{N}\setminus\{m_{k_j}\}$ for $n\in\mathcal{N}\setminus\{m_k\}$, and putting

$$\alpha_n=\begin{cases}\dfrac{1}{\|x_n\|} & \text{for} \quad n=m_j, \quad j=1,2,\ldots\\[2ex] 0 & \text{for the other } n,\end{cases} \tag{22.18}$$

the series $\sum_{n=1}^{\infty}\alpha_n x_{\sigma(n)}=\sum_{j=1}^{\infty}\frac{1}{\|x_{m_j}\|}x_{m_{k_j}}$ will be convergent, while the series $\sum_{n=1}^{\infty}\alpha_n x_n=\sum_{j=1}^{\infty}\frac{1}{\|x_{m_j}\|}x_{m_j}$ is divergent, contradicting the assumption 12°. Thus $12°\Rightarrow \inf_{1\leqslant n<\infty}\|x_n\|>0$.

Now we can show that $\{x_n\}$ satisfies 1°. Indeed, assume the contrary, i.e., that there exists an $x \in E$ such that

$$\sup_{\sigma \in \Pi} \sup_{\substack{|\beta_i| \leqslant 1 \\ 1 \leqslant k < \infty}} \left\| \sum_{i=1}^{k} \beta_i f_i(x) x_{\sigma(i)} \right\| = \infty . \tag{22.19}$$

We claim that in this case

$$\sup_{\tau \in \Pi(\mathcal{N}_{m+1})} \sup_{\substack{|\beta_i| \leqslant 1 \\ m+1 \leqslant k < \infty}} \left\| \sum_{i=m+1}^{k} \beta_i f_i(x) x_{\tau(i)} \right\| = \infty \quad (m = 1, 2, \ldots), \tag{22.20}$$

where $\Pi(\mathcal{N}_{m+1})$ denotes the set of all permutations of the set[1] $\mathcal{N}_{m+1}$.

Indeed, by the boundedness of the basis $\{x_n\}$, proved in the above, we have, for any fixed $m \in \mathcal{N}$,

$$\sup_{\sigma \in \Pi} \sup_{\substack{|\beta_i| \leqslant 1 \\ 1 \leqslant k \leqslant m}} \left\| \sum_{i=1}^{k} \beta_i f_i(x) x_{\sigma(i)} \right\| \leqslant \sum_{i=1}^{m} |f_i(x)| \sup_{1 \leqslant n < \infty} \|x_n\| < \infty , \tag{22.21}$$

$$\begin{aligned} &\sup_{\sigma \in \Pi} \sup_{\substack{|\beta_i| \leqslant 1 \\ m+1 \leqslant k < \infty}} \left\| \sideset{}{'}\sum_{\substack{i=m+1 \\ \sigma(i) \in \mathcal{N} \setminus \mathcal{N}_{m+1}}}^{k} \beta_i f_i(x) x_{\sigma(i)} \right\| \\ &\qquad \leqslant m \|x\| \sup_{1 \leqslant i < \infty} \|f_i\| \sup_{1 \leqslant j < \infty} \|x_j\| < \infty . \end{aligned} \tag{22.22}$$

Now, let $\sigma \in \Pi$ be arbitrary. Then the cardinal numbers of the finite sets $\mathcal{M}_1 = \{i \in \mathcal{N}_{m+1} \mid \sigma(i) \in \mathcal{N} \setminus \mathcal{N}_{m+1}\}$, $\mathcal{M}_2 = \{j \in \mathcal{N} \setminus \mathcal{N}_{m+1} \mid \sigma(j) \in \mathcal{N}_{m+1}\}$ are equal. Let ϕ be an arbitrary one to one mapping of $\mathcal{M}_1$ onto $\mathcal{M}_2$ and let $\tau_\sigma(i) = \sigma(i)$ for $i \in \mathcal{N}_{m+1} \setminus \mathcal{M}_1$, $\tau_\sigma(i) = \sigma[\phi(i)]$ for $i \in \mathcal{M}_1$. Then $\tau_\sigma \in \Pi(\mathcal{N}_{m+1})$ and

$$\sideset{}{'}\sum_{\substack{i=m+1 \\ \sigma(i) \in \mathcal{N}_{m+1}}}^{k} \beta_i f_i(x) x_{\sigma(i)} = \sideset{}{'}\sum_{\substack{i=m+1 \\ \sigma(i) \in \mathcal{N}_{m+1}}}^{k} \beta_i f_i(x) x_{\tau_\sigma(i)} \quad (|\beta_i| \leqslant 1, \ m+1 \leqslant k < \infty).$$

Therefore if (22.20) were not true, one would have, taking into account that by 12° the basis $\{x_n\}$ is unconditional,

$$\begin{aligned} &\sup_{\sigma \in \Pi} \sup_{\substack{|\beta_i| \leqslant 1 \\ m+1 \leqslant k < \infty}} \left\| \sideset{}{'}\sum_{\substack{i=m+1 \\ \sigma(i) \in \mathcal{N}_{m+1}}}^{k} \beta_i f_i(x) x_{\sigma(i)} \right\| \\ &\qquad \leqslant C \sup_{\tau \in \Pi(\mathcal{N}_{m+1})} \sup_{\substack{|\beta_i| \leqslant 1 \\ m+1 \leqslant k < \infty}} \left\| \sum_{i=m+1}^{k} \beta_i f_i(x) x_{\tau(i)} \right\| < \infty , \end{aligned}$$

[1] For the definition of the sets $\mathcal{N}_p$ see § 21, formula (21.30).

where C depends only on the basis $\{x_n\}$, and this, together with (22.21) and (22.22), would contradict (22.19). Thus, (22.19)$\Rightarrow$(22.20).

Using this remark one can inductively construct two infinite sequences $\{n_j\}, \{m_j\} \in \mathcal{O}$, a sequence of permutations $\tau_j \in \Pi(\mathcal{N}_{m_{j-1}+1})$ $(j=1,2,\ldots;\ m_0=0)$ and a sequence of scalars $\{\beta_i\}$ of modulus $|\beta_i| \leqslant 1$ $(i=1,2,\ldots)$, with the following properties:

$$n_j \geqslant m_{j-1}+1 \qquad (j=1,2,\ldots), \tag{22.23}$$

$$m_j = \max_{m_{j-1}+1 \leqslant i \leqslant n_j} \tau_j(i) \qquad (j=1,2,\ldots), \tag{22.24}$$

$$\left\| \sum_{i=m_{j-1}+1}^{n_j} \beta_i f_i(x) x_{\tau_j(i)} \right\| \geqslant 1 \qquad (j=1,2,\ldots). \tag{22.25}$$

Since by $\tau_j \in \Pi(\mathcal{N}_{m_{j-1}+1})$, (22.23) and (22.24) the cardinal numbers of the finite sets $\{i \in \mathcal{N} \mid n_j+1 \leqslant i \leqslant m_j\}$ and $\{i \in \mathcal{N} \mid m_{j-1}+1 \leqslant i \leqslant m_j\} \setminus \{\tau_j(m_{j-1}+1), \ldots, \tau_j(n_j)\}$ are equal, let ϕ_j be a one to one mapping of the first of these sets onto the second $(j=1,2,\ldots)$. Let

$$\sigma(i) = \begin{cases} \tau_j(i) & \text{for } m_{j-1}+1 \leqslant i \leqslant n_j \quad (j=1,2,\ldots), \\ \phi_j(i) & \text{for } n_j+1 \leqslant i \leqslant m_j \quad (j=1,2,\ldots). \end{cases} \tag{22.26}$$

Then $\sigma \in \Pi$, and by (22.25) the series $\sum_{i=1}^{\infty} \beta_i f_i(x) x_{\sigma(i)}$ is divergent. On the other hand, since by 12° the basis $\{x_n\}$ is unconditional and $\sum_{i=1}^{\infty} f_i(x) x_{\sigma(i)}$ converges, the series $\sum_{i=1}^{\infty} \beta_i f_i(x) x_{\sigma(i)}$ must be convergent, a contradiction. Thus, 12°$\Rightarrow$1°, which completes the proof of theorem 22.1.

The following proposition is useful for applications:

Proposition 22.1. *Let $\{x_n\}$ be a symmetric basis of a Banach space E. Then*

a) *There exists a constant $C \geqslant 1$ such that*

$$\left\| \sum_{i=1}^{n} \beta_i \alpha_i x_{\sigma(i)} \right\| \leqslant C \left\| \sum_{i=1}^{n} \alpha_i x_i \right\| \tag{22.27}$$

for any $\sigma \in \Pi$ and any finite sequence of scalars $\alpha_1, \ldots, \alpha_n, \beta_1, \ldots, \beta_n$ with $|\beta_1|, \ldots, |\beta_n| \leqslant 1$.

b) *The numbers*

$$|||x||| = \sup_{\sigma \in \Pi} \sup_{\substack{|\beta_i| \leqslant 1 \\ 1 \leqslant k < \infty}} \left\| \sum_{i=1}^{k} \beta_i f_i(x) x_{\sigma(i)} \right\| \qquad (x \in E) \tag{22.28}$$

give a new norm on E, equivalent to the original norm of E. In this new norm we have (22.27) *with* $C=1$ *and*

$$\left|\left|\left| \sum_{i=1}^{\infty} \varepsilon_i f_{\rho(i)}(x) x_{\sigma(i)} \right|\right|\right| = |||x||| \qquad (x\in E, (\rho,\sigma)\in \Pi\times\Pi, |\varepsilon_i|=1), \tag{22.29}$$

$$|||x_1||| = |||x_2||| = \cdots = \sup_{1\leqslant n<\infty} \|x_n\|. \tag{22.30}$$

Proof. a) Since $\{x_n\}$ is a symmetric basis, for the operators

$$u_{\sigma,\{\beta_i\},k}(x) = \sum_{i=1}^{k} \beta_i f_i(x) x_{\sigma(i)}$$
$$(x\in E, \sigma\in\Pi, |\beta_1|, \ldots, |\beta_k|\leqslant 1, k=1,2,\ldots) \tag{22.31}$$

we have, by the principle of uniform boundedness,

$$\sup_{\sigma\in\Pi} \sup_{\substack{|\beta_i|\leqslant 1\\ 1\leqslant k<\infty}} \|u_{\sigma,\{\beta_i\},k}\| = C < \infty \tag{22.32}$$

and hence

$$\left\| \sum_{i=1}^{n} \beta_i \alpha_i x_{\sigma(i)} \right\| = \left\| u_{\sigma,\{\beta_i\},n}\left(\sum_{i=1}^{n} \alpha_i x_i \right) \right\| \leqslant C \left\| \sum_{i=1}^{n} \alpha_i x_i \right\|.$$

b) Obviously, $|||x|||$ is a norm on E. Since $\{x_n\}$ is a symmetric basis, by part a) proved above we have

$$\|x\| \leqslant |||x||| \leqslant C\|x\| \qquad (x\in E), \tag{22.33}$$

and thus the norm $|||x|||$ is equivalent to the original norm of E.

Furthermore, for any $\sigma\in\Pi$ and any finite sequence of scalars $\alpha_1, \ldots, \alpha_n, \beta_1, \ldots, \beta_n$ with $|\beta_i|\leqslant 1$ $(i=1,\ldots,n)$ we have

$$\left|\left|\left| \sum_{i=1}^{n} \beta_i \alpha_i x_{\sigma(i)} \right|\right|\right| = \sup_{\tau\in\Pi} \sup_{\substack{|\delta_i|\leqslant 1\\ 1\leqslant k\leqslant n}} \left\| \sum_{i=1}^{k} \delta_i \beta_i \alpha_i x_{\tau\sigma(i)} \right\| \leqslant \left|\left|\left| \sum_{i=1}^{n} \alpha_i x_i \right|\right|\right|, \tag{22.34}$$

i.e., (22.27) with $C=1$. Hence it follows that for any $\sigma\in\Pi$ and any $\alpha_1, \ldots, \alpha_n, \varepsilon_1, \ldots, \varepsilon_n$ with $|\varepsilon_i|=1$ $(i=1,\ldots,n)$ we also have

$$\left|\left|\left| \sum_{i=1}^{n} \alpha_i x_i \right|\right|\right| \leqslant \left|\left|\left| \sum_{i=1}^{n} \varepsilon_i \alpha_i x_{\sigma(i)} \right|\right|\right|. \tag{22.35}$$

Indeed, since the cardinal numbers of the finite sets $\mathcal{M}_1 = \left\{1,2,\ldots, \max_{1\leqslant i\leqslant n} \sigma(i)\right\} \setminus \{\sigma(1),\ldots,\sigma(n)\}$ and $\mathcal{M}_2 = \left\{1,2,\ldots, \max_{1\leqslant i\leqslant n} \sigma(i)\right\} \setminus \{1,\ldots,n\}$ are equal, let ϕ be an arbitrary one to one mapping of $\mathcal{M}_1$ onto $\mathcal{M}_2$ and let

$$\tau(j) = \begin{cases} \sigma^{-1}(j) & \text{for } j\in\{\sigma(1),\ldots,\sigma(n)\}, \\ \phi(j) & \text{for } j\in\mathcal{M}_1, \\ j & \text{for } j\in\mathcal{N}_{l+1}, \end{cases}$$

where $l = \max_{1 \leqslant i \leqslant n} \sigma(i)$. Then, putting

$$\gamma_j = \begin{cases} \varepsilon_{\sigma^{-1}(j)} \alpha_{\sigma^{-1}(j)} & \text{for} \quad j \in \{\sigma(1), \ldots, \sigma(n)\} \\ 0 & \text{for} \quad j \in \mathcal{M}_1, \end{cases}$$

from (22.34) we obtain

$$\left|\left|\left| \sum_{i=1}^{n} \alpha_i x_i \right|\right|\right| = \left|\left|\left| \sum_{j=1}^{l} \bar{\varepsilon}_{\sigma^{-1}(j)} \gamma_j x_{\tau(j)} \right|\right|\right| \leqslant \left|\left|\left| \sum_{j=1}^{l} \gamma_j x_j \right|\right|\right| = \left|\left|\left| \sum_{i=1}^{n} \varepsilon_i \alpha_i x_{\sigma(i)} \right|\right|\right|,$$

i.e., (22.35).

Now, by theorem 22.1, $\sum_{i=1}^{\infty} f_{\rho(i)}(x) x_{\sigma(i)}$ converges for every $x \in E$, $(\rho, \sigma) \in \Pi \times \Pi$. Since $\{x_n\}$ is unconditional, $\sum_{i=1}^{\infty} \varepsilon_i f_{\rho(i)}(x) x_{\sigma(i)}$ also converges for $|\varepsilon_i| = 1$ $(i = 1, 2, \ldots)$ and we have, taking into account (22.34) and (22.35),

$$\left|\left|\left| \sum_{i=1}^{\infty} \varepsilon_i f_{\rho(i)}(x) x_{\sigma(i)} \right|\right|\right| = \left|\left|\left| \sum_{i=1}^{\infty} \varepsilon_{\rho^{-1}(i)} f_i(x) x_{\rho^{-1}\sigma(i)} \right|\right|\right| = \left|\left|\left| \sum_{i=1}^{\infty} f_i(x) x_i \right|\right|\right| = |||x|||.$$

Finally, applying (22.28) to $x_j = \sum_{i=1}^{\infty} f_i(x_j) x_i = \sum_{i=1}^{\infty} \delta_{ij} x_i$ $(j = 1, 2, \ldots)$, we get

$$|||x_j||| = \sup_{\sigma \in \Pi} \sup_{|\beta_j| \leqslant 1} \|\beta_j x_{\sigma(j)}\| = \sup_{1 \leqslant n < \infty} \|x_n\| \qquad (j = 1, 2, \ldots),$$

which completes the proof of proposition 22.1.

Let us mention that the converse statements are also true, i.e., each of (22.27) and (22.29) above characterize symmetric bases, as shown by the implications $8^\circ \Rightarrow 1^\circ$ and $5^\circ \Rightarrow 1^\circ$ of theorem 22.1.

Definition 22.2. Let $\{x_n\}$ be a basis of a Banach space E, with the a.s.c.f. $\{f_n\}$. The least constant $C \geqslant 1$ for which (22.27) holds i.e., the number $\nu_{\{x_n\}}^{(s)} = \sup_{\sigma \in \Pi} \sup_{\substack{|\beta_i| \leqslant 1 \\ 1 \leqslant k < \infty}} \|u_{\sigma, \{\beta_i\}, k}\| = \sup_{\sigma \in \Pi} \sup_{\substack{|\beta_i| \leqslant 1 \\ 1 \leqslant k < \infty}} \sup_{\substack{x \in E \\ \|x\| \leqslant 1}} \left\| \sum_{i=1}^{k} \beta_i f_i(x) x_{\sigma(i)} \right\|$ is called the *symmetric constant* (or *symmetric norm*) of the basis $\{x_n\}$. The number $\nu_s(E) = \inf_{\{x_n\} \in \mathcal{B}_s} \nu_{\{x_n\}}^{(s)}$, where $\mathcal{B}_s$ denotes the set of all symmetric bases of the space E, is called the *symmetric constant* of the space E. Any norm $|||x|||$ on E for which we have (22.29) or, equivalently, (22.27) with $C = 1$, is called *symmetric* with respect to the basis $\{x_n\}$. We shall call *symmetric space* any couple $(E, \{x_n\})$, where E is a Banach space with a symmetric basis and $\{x_n\}$ a symmetric basis of E, such that the original norm of E is symmetric with respect to $\{x_n\}$ (or, equivalently, $\nu_s(E) = \nu_{\{x_n\}}^{(s)} = 1$).

Proposition 22.2. *Every symmetric basis $\{x_n\}$ is subsymmetric.*

Proof. Let $\{n_i\}\in\mathcal{O}$ be arbitrary. Then by (22.33) and (22.29) we have

$$\left\|\sum_{i=m}^{m+p}\alpha_i x_i\right\| \leqslant \left|\left\|\sum_{i=m}^{m+p}\alpha_i x_i\right\|\right| = \left|\left\|\sum_{i=m}^{m+p}\alpha_i x_{n_i}\right\|\right| \leqslant C\left\|\sum_{i=m}^{m+p}\alpha_i x_{n_i}\right\|,$$

$$\left\|\sum_{i=m}^{m+p}\alpha_i x_{n_i}\right\| \leqslant \left|\left\|\sum_{i=m}^{m+p}\alpha_i x_{n_i}\right\|\right| = \left|\left\|\sum_{i=m}^{m+p}\alpha_i x_i\right\|\right| \leqslant C\left\|\sum_{i=m}^{m+p}\alpha_i x_i\right\|,$$

whence $\{x_n\}$ is equivalent to the basis $\{x_{n_i}\}$ of $[x_{n_i}]$, which completes the proof.

The converse of proposition 22.2 is not valid, as shown by

Example 22.1. Let G be the space of all sequences of real numbers $x=\{\xi_n\}$ such that

$$\|x\| \stackrel{df}{=} \sup_{\{n_j\}\in\mathcal{O}}\sum_{i=1}^{\infty}\frac{|\xi_{n_i}|}{\sqrt{i}}<\infty. \tag{22.36}$$

Then G is a Banach space, and the unit vectors $\{x_n\}$ are a subsymmetric but non-symmetric basis of G.

Indeed, let us first prove that G is complete. Let $z_n=\{\zeta_n^{(m)}\}_{n=1}^{\infty}$ $(m=1,2,\ldots)$ be a Cauchy sequence in G and let $\varepsilon>0$ be arbitrary. Then there exists an $N=N(\varepsilon)$ such that

$$\|z_m-z_{m+p}\| = \sup_{\{n_j\}\in\mathcal{O}}\sum_{i=1}^{\infty}\frac{|\zeta_{n_i}^{(m)}-\zeta_{n_i}^{(m+p)}|}{\sqrt{i}}<\varepsilon \quad (m>N, p=1,2,\ldots). \tag{22.37}$$

Hence, taking $i=1$, $n_1=1,2,\ldots$ it follows that

$$|\zeta_n^{(m)}-\zeta_n^{(m+p)}|<\varepsilon \quad (m>N, p=1,2,\ldots; n=1,2,\ldots),$$

and therefore the limits $\lim_{p\to\infty}\zeta_n^{(p)}=\zeta_n$ $(n=1,2,\ldots)$ exist. Now, by (22.37) we have

$$\sum_{i=1}^{k}\frac{|\zeta_{n_i}^{(m)}-\zeta_{n_i}^{(m+p)}|}{\sqrt{i}}<\varepsilon \quad (m>N, p=1,2,\ldots; k=1,2,\ldots; \{n_j\}\in\mathcal{O}),$$

whence, for $p\to\infty$ we obtain

$$\sum_{i=1}^{k}\frac{|\zeta_{n_i}^{(m)}-\zeta_{n_i}|}{\sqrt{i}}<\varepsilon \quad (m>N, k=1,2,\ldots; \{n_j\}\in\mathcal{O}),$$

and hence, putting $z=\{\zeta_n\}$, we get $z_m-z\in G$, $\|z_m-z\|\leqslant\varepsilon$ $(m>N)$. Consequently, $z=(z_m-z)+z_m\in G$ and $z_m\to z$ as $m\to\infty$, which proves the completeness of G.

Let us prove now that

$$\left\|x-\sum_{i=1}^{n}\xi_i x_i\right\|=\|\{\underbrace{0,\ldots,0}_{n},\xi_{n+1},\xi_{n+2},\ldots\}\|\to 0 \quad \text{as} \quad n\to\infty \quad (x=\{\xi_n\}\in G). \tag{22.38}$$

Let $x\in G$ and $\varepsilon>0$ be arbitrary. Then by (22.36) there exists a sequence $\{n_i\}\in\mathcal{O}$ such that

$$\|x\|\geqslant\sum_{i=1}^{\infty}\frac{|\xi_{n_i}|}{\sqrt{i}}>\|x\|-\frac{\varepsilon}{3},$$

whence also a $k\in\mathcal{N}$ such that

$$\|x\|\geqslant\sum_{i=1}^{k}\frac{|\xi_{n_i}|}{\sqrt{i}}>\|x\|-\frac{\varepsilon}{3}. \tag{22.39}$$

Observe now that we must have $\xi_n\to 0$ as $n\to\infty$, since otherwise there would exist an infinite subsequence $\{\xi_{m_j}\}$ with $\inf\limits_{1\leqslant j<\infty}|\xi_{m_j}|=\delta>0$, whence $\|x\|\geqslant\sum\limits_{i=1}^{\infty}\frac{|\xi_{m_i}|}{\sqrt{i}}\geqslant\sum\limits_{i=1}^{\infty}\frac{\delta}{\sqrt{i}}=\infty$, which is impossible. Therefore there exists an $N=N(\varepsilon)$ such that

$$|\xi_i|\leqslant\frac{\varepsilon}{3\|x\|}\min_{1\leqslant j\leqslant k}|\xi_{n_j}| \qquad (i>N). \tag{22.40}$$

We claim that

$$\rho_n\overset{df}{=}\left\|x-\sum_{i=1}^{n}\xi_i x_i\right\|=\|\{\underbrace{0,\ldots,0}_{n},\xi_{n+1},\xi_{n+2},\ldots\}\|<\varepsilon \qquad (n>N). \tag{22.41}$$

Indeed, take a fixed $n>N$. Then, by (22.36), there exists a sequence $\{l_j\}\in\mathcal{O}(\mathcal{N}_{n+1})$ such that

$$\rho_n\geqslant\sum_{i=1}^{\infty}\frac{|\xi_{l_i}|}{\sqrt{i}}>\rho_n-\frac{\varepsilon}{3}. \tag{22.42}$$

By virtue of (22.40) we have then

$$\sum_{i=1}^{k}\frac{|\xi_{l_i}|}{\sqrt{i}}<\frac{\varepsilon}{3\|x\|}\sum_{i=1}^{k}\frac{|\xi_{n_i}|}{\sqrt{i}}\leqslant\frac{\varepsilon}{3\|x\|}\|x\|=\frac{\varepsilon}{3}$$

and we also have

$$\sum_{i=k+1}^{\infty}\frac{|\xi_{l_i}|}{\sqrt{i}}\leqslant\frac{\varepsilon}{3},$$

since otherwise for the sequence $\{m_j\}\in\mathcal{O}$ defined by $m_j=n_j$ $(j=1,\ldots,k)$, $m_j=l_j$ $(j=k+1,k+2,\ldots)$ one would get, by (22.36) and (22.39),

$$\|x\|\geqslant\sum_{i=1}^{\infty}\frac{|\xi_{m_i}|}{\sqrt{i}}=\sum_{i=1}^{k}\frac{|\xi_{m_i}|}{\sqrt{i}}+\sum_{i=k+1}^{\infty}\frac{|\xi_{m_i}|}{\sqrt{i}}>\|x\|-\frac{\varepsilon}{3}+\frac{\varepsilon}{3}=\|x\|,$$

which is impossible. Consequently,

$$\sum_{i=1}^{\infty}\frac{|\xi_{l_i}|}{\sqrt{i}}=\sum_{i=1}^{k}\frac{|\xi_{l_i}|}{\sqrt{i}}+\sum_{i=k+1}^{\infty}\frac{|\xi_{l_i}|}{\sqrt{i}}<\frac{2\varepsilon}{3},$$

whence, by (22.42), we infer

$$\rho_n<\sum_{i=1}^{\infty}\frac{|\xi_{l_i}|}{\sqrt{i}}+\frac{\varepsilon}{3}<\frac{2\varepsilon}{3}+\frac{\varepsilon}{3}=\varepsilon,$$

which proves (22.41) and thus also (22.38).

By (22.36) it is clear that for any finite sequence of scalars $\xi_1,\ldots,\xi_k$ and any sequence of indices $1\leqslant m_1<\cdots<m_p\leqslant k$ we have

$$\left\|\sum_{i=1}^{p}\xi_{m_i}x_{m_i}\right\|\leqslant\left\|\sum_{i=1}^{k}\xi_i x_i\right\| \tag{22.43}$$

and thus, by § 17, theorem 17.1, $\{x_n\}$ is an unconditional basic sequence. This, together with (22.38), implies that $\{x_n\}$ is an unconditional basis of G. If $\{f_n\}\subset G^*$ is the a.s.c.f. to $\{x_n\}$, then by (22.36) it is also clear that

$$\sup_{1\leqslant k<\infty}\left\|\sum_{i=1}^{k}f_{m_i}(x)x_{n_i}\right\|<\infty\qquad (x\in G,(\{m_j\},\{n_j\})\in\mathcal{O}\times\mathcal{O}),$$

whence, by § 21, the implication $3^\circ\Rightarrow1^\circ$ of theorem 21.1, $\{x_n\}$ is a subsymmetric basis of G.

Finally, let us prove that $\{x_n\}$ is not symmetric, by showing that condition 2° of theorem 22.1 is violated. Consider the sequence $\{y_k\}\subset G$ defined by

$$y_k=\left\{\frac{1}{\sqrt{k}},\frac{1}{\sqrt{k-1}},\ldots,1,0,0,\ldots\right\}\qquad (k=1,2,\ldots). \tag{22.44}$$

We claim that this sequence is bounded in G. Indeed, for a fixed k, there exists a finite set of positive integers $1\leqslant n_1<\cdots<n_s\leqslant k$ such that

$$\|y_k\|=\sum_{i=1}^{s}\frac{1}{\sqrt{i(k+1-n_i)}}. \tag{22.45}$$

Let

$$m_i=\begin{cases}n_i & \text{for } i=1,\ldots,s-1,\\ k & \text{for } i=s.\end{cases}$$

Then by (22.36) and (22.45),

$$0 \leqslant \|y_k\| - \sum_{i=1}^{s} \frac{1}{\sqrt{i(k+1-m_i)}} = \frac{1}{\sqrt{s(k+1-n_s)}} - \frac{1}{\sqrt{s}} = \frac{1}{\sqrt{s}}\left(\frac{1}{\sqrt{k+1-n_s}} - 1\right),$$

which, together with $n_s \leqslant k$, implies that $n_s = k$. Continuing with a similar argument, we obtain $n_{s-1} = k-1$, $n_{s-2} = k-2, \ldots, n_1 = k+1-s$, i.e., $n_i = k+i-s$ $(i=1,\ldots,s)$, whence, by (22.45),

$$\|y_k\| = \sum_{i=1}^{s} \frac{1}{\sqrt{i(k+1-k-i+s)}} = \sum_{i=1}^{s} \frac{1}{\sqrt{i(s+1-i)}}. \tag{22.46}$$

However, we have

$$\sum_{i=1}^{s} \frac{1}{\sqrt{i(s+1-i)}} = \begin{cases} \dfrac{2}{\sqrt{s}} + 2 \displaystyle\sum_{i=2}^{\frac{s-1}{2}} \frac{1}{\sqrt{i(s+1-i)}} + \frac{1}{s+1} & \text{for } s=3,5,7,\ldots \\ \dfrac{2}{\sqrt{s}} + 2 \displaystyle\sum_{i=2}^{\frac{s}{2}} \frac{1}{\sqrt{i(s+1-i)}} & \text{for } s=2,4,6,\ldots \end{cases}$$

and hence, since the function $\phi_s(t) = \dfrac{1}{\sqrt{t(s+1-t)}}$ is non-increasing for $t \in \left[1, \left[\frac{s}{2}\right]\right]$,

$$\|y_k\| = \sum_{i=1}^{s} \frac{1}{\sqrt{i(s+1-i)}} \leqslant \frac{2}{\sqrt{s}} + 2 \sum_{i=1}^{\left[\frac{s}{2}\right]-1} \int_i^{i+1} \frac{dt}{\sqrt{t(s+1-t)}} + \frac{2}{s+1}$$

$$\leqslant \frac{2}{\sqrt{s}} + 2 \int_1^{\frac{s+1}{2}} \frac{dt}{\sqrt{t(s+1-t)}} + \frac{2}{s+1} = \frac{2}{\sqrt{s}} - 2 \int_{s-1}^{0} \frac{dv}{\sqrt{(s+1)^2 - v^2}}$$

$$+ \frac{2}{s+1} = \frac{2}{\sqrt{s}} + 2 \arcsin \frac{s-1}{s+1} + \frac{2}{s+1} \leqslant 3 + \pi,$$

which proves the assertion that $\{y_k\}$ is bounded.

Now for each $k \in \mathcal{N}$ define a permutation $\rho_k \in \Pi$ by

$$\rho_k(n) = \begin{cases} k+1-n & \text{for} \quad n=1,2,\ldots,k \\ n & \text{for} \quad n=k+1,\ k+2,\ldots \end{cases} \tag{22.47}$$

and let $\iota \in \Pi$ be the identical permutation $\iota(n)=n$ $(n=1,2,\ldots)$. Then

$$A_{\rho_k,\iota}(y_k) = \sum_{i=1}^{\infty} f_{\rho_k(i)}(y_k)x_i = \left\{1, \frac{1}{\sqrt{2}}, \ldots, \frac{1}{\sqrt{k}}, 0, 0, \ldots\right\} \qquad (k=1,2,\ldots),$$

whence

$$\|A_{\rho_k,\iota}(y_k)\| = \sum_{j=1}^{k} \frac{1}{j} \to \infty \quad \text{as} \quad k \to \infty,$$

which, together with the relation $\sup_{1 \leqslant k < \infty} \|y_k\| < \infty$ proved above, shows that the condition 2° of theorem 22.1 is violated, and this completes the proof of the assertions of example 22.1.

In the usual concrete Banach spaces with a symmetric basis all symmetric bases are equivalent. Indeed, we have made a similar remark for subsymmetric bases in § 21, and by proposition 22.2 above every symmetric basis is subsymmetric. Therefore it is natural to ask

Problem 22.1. In a Banach space E with a symmetric basis are all symmetric bases of E equivalent?

We have seen in § 17, example 17.1, that a block subspace with respect to an unconditional basis $\{x_n\}$ of a Banach space E need not be complemented in E, whence a block basic sequence with respect to an unconditional basis $\{x_n\}$ of E cannot be extended, in general, to an unconditional basis of E. It is natural to raise

Problem 22.2. Can every block basic sequence, with respect to a symmetric basis $\{x_n\}$ of a Banach space E, be extended to a symmetric basis of the whole space E? Or, at least, to an unconditional basis of the whole space E? Or, at least, is every block subspace with respect to a symmetric basis $\{x_n\}$ of a Banach space E complemented in E?

We shall give now some partial results concerning these questions, which will be useful in determining the Banach spaces which have, up to equivalence, a unique normalized unconditional basis (§ 24). First we shall show that in a symmetric space $(E, \{x_n\})$ the subspace spanned by a (finite or infinite) sequence of non-overlapping finite sums of the basis elements x_n, is complemented in E, admitting a projection of norm 1.

Proposition 22.3. *Let $(E,\{x_n\})$ be a symmetric space, $\{A_n\}$ a (finite or infinite) sequence of disjoint finite sets of positive integers and $l_n \neq 0$ the cardinality of A_n $(n=1,2,\ldots)$. Then the block subspace*[1] $\left[\sum_{j\in A_k} x_j\right]_{k=1}^{\infty}$ *is complemented in E, namely, the operator $u_{\{A_n\}}$ defined by*

$$u_{\{A_n\}}\left(\sum_{i=1}^{\infty}\alpha_i x_i\right)=\sum_{k=1}^{\infty}\frac{1}{l_k}\left(\sum_{j\in A_k}\alpha_j\right)\left(\sum_{j\in A_k}x_j\right)\qquad\left(\sum_{i=1}^{\infty}\alpha_i x_i\in E\right)\tag{22.48}$$

is a projection of norm 1 *of E onto* $\left[\sum_{j\in A_k} x_j\right]_{k=1}^{\infty}$.

Proof. Let us first show that for every positive integer N and every sequence of scalars $\{\alpha_i\}$ for which $\sum_{i=1}^{\infty}\alpha_i x_i$ converges, we have

$$\left\|\sum_{k=1}^{N}\frac{1}{l_k}\left(\sum_{j\in A_k}\alpha_j\right)\left(\sum_{j\in A_k}x_j\right)\right\|\leqslant\left\|\sum_{i=1}^{\infty}\alpha_i x_i\right\|.\tag{22.49}$$

Let Σ denote the set of all permutations σ of $\bigcup_{k=1}^{N} A_k$ such that $\sigma(A_k)=A_k$ $(k=1,\ldots,N)$. The cardinality of Σ is $M=\prod_{k=1}^{N}(l_k!)$ (since each $\sigma\in\Sigma$ is an N-tuple $\{\sigma_1,\ldots,\sigma_N\}$, where σ_k is a permutation of A_k and since the cardinality of the set $\Pi(A_k)$ of all permutations σ_k of A_k is $l_k!$).

Furthermore, we have

$$\frac{1}{M}\sum_{\sigma\in\Sigma}\sum_{k=1}^{N}\sum_{j\in A_k}\alpha_j x_{\sigma(j)}=\sum_{k=1}^{N}\frac{1}{l_k}\left(\sum_{j\in A_k}\alpha_j\right)\left(\sum_{j\in A_k}x_j\right),\tag{22.50}$$

since the coefficient of α_i in the left hand side is $\frac{1}{M}\sum_{\sigma\in\Sigma}x_{\sigma(i)}$ and since for any pair $i,j\in\bigcup_{k=1}^{N}A_k$ with $j\in A_{k_0}$ there are exactly $(l_{k_0}-1)!\prod_{\substack{k=1\\k\neq k_0}}^{N}(l_k!)=\frac{M}{l_{k_0}}$ permutations $\sigma\in\Sigma$ such that $\sigma(i)=j$ if $i\in A_{k_0}$ and no such permutation if $i\notin A_{k_0}$.

---

[1] For the simplicity of notation, we shall unify the finite and infinite cases, by writing always $[\quad]_{k=1}^{\infty}$ and $\sum_{k=1}^{\infty}$.

Now, since $(E,\{x_n\})$ is a symmetric space, for every $\sigma\in\Sigma$ we have

$$\left\|\sum_{k=1}^{N}\sum_{j\in A_k}\alpha_j x_{\sigma(j)}\right\|\leqslant\left\|\sum_{i=1}^{\infty}\alpha_i x_i\right\|,$$

whence, by (22.50), we obtain

$$\left\|\sum_{k=1}^{N}\frac{1}{l_k}\Big(\sum_{j\in A_k}\alpha_j\Big)\Big(\sum_{j\in A_k}x_j\Big)\right\|=\left\|\frac{1}{M}\sum_{\sigma\in\Sigma}\sum_{k=1}^{N}\sum_{j\in A_k}\alpha_j x_{\sigma(j)}\right\|\leqslant\left\|\sum_{i=1}^{\infty}\alpha_i x_i\right\|,$$

i. e., (22.49).

Applying (22.49) to $\sum\limits_{k=n+1}^{n+m}\sum\limits_{j\in A_k}\alpha_j x_j$, we get

$$\left\|\sum_{k=n+1}^{n+m}\frac{1}{l_k}\Big(\sum_{j\in A_k}\alpha_j\Big)\Big(\sum_{j\in A_k}x_j\Big)\right\|\leqslant\left\|\sum_{k=n+1}^{n+m}\sum_{j\in A_k}\alpha_j x_j\right\|,$$

whence $\sum\limits_{k=1}^{\infty}\frac{1}{l_k}\Big(\sum\limits_{j\in A_k}\alpha_j\Big)\Big(\sum\limits_{j\in A_k}x_j\Big)$ converges, i. e., $u_{\{A_k\}}$ is well defined. Furthermore, $u_{\{A_k\}}$ is obviously linear and, by (22.49) for $N\to\infty$, we have $\|u_{\{A_n\}}\|\leqslant 1$.

On the other hand, obviously $u_{\{A_n\}}$ maps E into $\Big[\sum\limits_{j\in A_k}x_j\Big]_{k=1}^{\infty}$ and

$u_{\{A_n\}}\Big(\sum\limits_{j\in A_k}x_j\Big)=\frac{1}{l_k}\Big(\sum\limits_{j\in A_k}1\Big)\Big(\sum\limits_{j\in A_k}x_j\Big)=\sum\limits_{j\in A_k}x_j$ $(k=1,2,\ldots)$, whence $u_{\{A_n\}}$ is a projection of E onto $\Big[\sum\limits_{j\in A_k}x_j\Big]_{k=1}^{\infty}$. Since any projection has norm $\geqslant 1$, by the above it follows that $\|u_{\{A_n\}}\|=1$, which completes the proof of proposition 22.3.

Definition 22.3. Let $(E,\{x_n\})$ be a symmetric space and let $\{A_n\}$, $l_n\neq 0$ be as in proposition 22.3. The projection $u_{\{A_n\}}$ above is called the *averaging projection* with respect to $\{A_n\}$.

Now we shall show that in a symmetric space $(E,\{x_n\})$ certain unconditional basic sequences, consisting of weighted finite sums of the basis elements x_n, span complemented subspaces with complements having unconditional bases and hence they can be extended to unconditional bases of the whole space E.

Proposition 22.4. *Let $(E,\{x_n\})$ be a symmetric space, let $\{n_{k,j}\}_{k,j=1}^{\infty}$ be an enumeration of the set $\mathcal{N}=\{1,2,3,\ldots\}$ as a double sequence and let $\{p_k\}_{k=1}^{\infty}\subset\mathcal{N}\setminus\{1\}$. Furthermore, let*

$$z_{k,m}=\sum_{j=1}^{p_k}x_{n_{k,(m-1)p_k+j}}\qquad(k,m=1,2,3,\ldots)\tag{22.51}$$

and for $k=1,2,\ldots$ *let*

$$y_{k,i}=\begin{cases}\dfrac{1}{p_k}z_{k,1}=\dfrac{1}{p_k}\sum_{j=1}^{p_k}x_{n_{k,j}} & \text{for}\quad i=1\\[2mm] x_{n_{k,i}}-\dfrac{1}{p_k-1}(z_{k,1}-x_{n_{k,1}})-\dfrac{1}{p_k}z_{k,2} & \\ \quad =x_{n_{k,i}}-\dfrac{1}{p_k-1}\sum_{j=2}^{p_k}x_{n_{k,j}}-\dfrac{1}{p_k}\sum_{j=1}^{p_k}x_{n_{k,p_k+j}} & \text{for}\quad i=2,\ldots,p_k\\[2mm] x_{n_{k,i}}-\dfrac{1}{p_k}(z_{k,m}+z_{k,m+1}) & \\ \quad =x_{n_{k,i}}-\dfrac{1}{p_k}\left(\sum_{j=1}^{p_k}x_{n_{k,(m-1)p_k+j}}+\sum_{j=1}^{p_k}x_{n_{k,mp_k+j}}\right) & \\ \quad \text{for}\quad i=(m-1)p_k+1,\ldots,mp_k;\ m=2,3,\ldots & \end{cases} \tag{22.52}$$

Finally, let $E_1=[x_{n_{k,1}}]_{k=1}^{\infty}$, $E_2=[y_{k,1}]_{k=1}^{\infty}$ *and* $E_3=[y_{k,i}]_{k=1,i=2}^{\infty}$. *Then* a) $E=E_2\oplus E_1\oplus E_3$; b) $\{y_{k,i}\}_{k=1,i=2}^{\infty}$ *is an unconditional basis of* E_3; *hence* c) $\{y_{k,1}\}_{k=1}^{\infty}\cup\{x_{n_{k,1}}\}_{k=1}^{\infty}\cup\{y_{k,i}\}_{k=1,i=2}^{\infty}$ *is an unconditional basis of* E.

Proof. a) Let us first observe that the algebraic sum $E_1+E_2+E_3$ contains all basis elements $x_n (n=1,2,\ldots)$ and hence is dense in E. Indeed, for every $k=1,2,\ldots$ we have $x_{n_{k,1}}\in E_1$, $z_{k,1}\in E_2$ and

$$\begin{aligned}-\frac{1}{p_k}z_{k,2}&=\frac{1}{p_k-1}\sum_{j=2}^{p_k}x_{n_{k,j}}-\frac{1}{p_k-1}(z_{k,1}-x_{n_{k,1}})-\frac{1}{p_k}z_{k,2}\\ &=\frac{1}{p_k-1}\sum_{j=2}^{p_k}\left[x_{n_{k,j}}-\frac{1}{p_k-1}(z_{k,1}-x_{n_{k,1}})-\frac{1}{p_k}z_{k,2}\right]\\ &=\frac{1}{p_k-1}\sum_{j=2}^{p_k}y_{k,j}\in E_3,\end{aligned}$$

whence, for $i=2,\ldots,p_k$ and $k=1,2,\ldots$

$$x_{n_{k,i}}=y_{k,i}+\frac{1}{p_k-1}(z_{k,1}-x_{n_{k,1}})+\frac{1}{p_k}z_{k,2}\in E_1+E_2+E_3. \tag{22.53}$$

Similarly, if we assume that $\{x_{n_{k,i}}\}_{i=1}^{(m-1)p_k}\subset E_1+E_2+E_3$ and $\{z_{k,j}\}_{j=1}^{m}\subset E_3$ for a pair $k\geqslant 1, m\geqslant 2$, then, taking into account the relations

$$-z_{k,m+1}=\sum_{j=1}^{p_k} x_{n_{k,(m-1)p_k+j}}-(z_{k,m}+z_{k,m+1})$$
$$=\sum_{j=1}^{p_k}\left[x_{n_{k,(m-1)p_k+j}}-\frac{1}{p_k}(z_{k,m}+z_{k,m+1})\right]=\sum_{j=1}^{p_k} y_{k,(m-1)p_k+j}\in E_3,$$

we obtain, for $i=(m-1)p_k+1,\dots,mp_k$

$$x_{n_{k,i}}=y_{k,i}+\frac{1}{p_k}(z_{k,m}+z_{k,m+1})\in E_3\subset E_1+E_2+E_3, \tag{22.54}$$

which proves the assertion that $x_n\in E_1+E_2+E_3$ $(n=1,2,\dots)$; actually, the above argument shows that we have even $x_{n_{k,i}}\in E_3$ for $i=p_k+1$, $p_k+2,\dots$ and $k=1,2,\dots$.

We show next that there exists a bounded linear projection of E onto E_1 which carries E_2 and E_3 into 0, whence $E=E_1\oplus(E_2\oplus E_3)$. Indeed, let $\{f_n\}\subset E^*$ be the a.s.c.f. to $\{x_n\}$ and put

$$g_k=f_{n_{k,1}}-\frac{1}{p_k-1}\sum_{i=2}^{p_k} f_{n_{k,i}} \quad (k=1,2,\dots). \tag{22.55}$$

Then, by the biorthogonality relations $f_i(x_j)=\delta_{ij}$, we obtain

$$g_k(x_{n_{h,1}})=\delta_{kh},\ g_k(y_{h,i})=0 \quad (i,h,k=1,2,\dots). \tag{22.56}$$

Since $g_k=\dfrac{p_k}{p_k-1}\left(f_{n_{k,1}}-\dfrac{1}{p_k}\sum_{i=1}^{p_k} f_{n_{k,i}}\right)$ and $\dfrac{p_k}{p_k-1}\leqslant 2$ $(k=1,2,\dots)$ and since $(E,\{x_n\})$ is a symmetric space, we have, for every $x\in E$ and every positive integer N,

$$\left\|\sum_{k=1}^{N} g_k(x)x_{n_{k,1}}\right\|\leqslant 2\left\|\sum_{k=1}^{N} f_{n_{k,1}}(x)x_{n_{k,1}}\right\|+2\left\|\sum_{k=1}^{N}\sum_{i=1}^{p_k}\frac{1}{p_k} f_{n_{k,i}}(x)x_{n_{k,1}}\right\|$$
$$\leqslant 2\|x\|+2\left\|\sum_{k=1}^{N}\frac{1}{p_k}\left(\sum_{i=1}^{p_k} f_{n_{k,i}}(x)\right)\left(\sum_{i=1}^{p_k} x_{n_{k,i}}\right)\right\|$$
$$=2\|x\|+2\|u_{\{A_k^{(0)}\}}(x)\|,$$

where $u_{\{A_k^{(0)}\}}$ denotes the averaging projection with respect to the sequence $A_k^{(0)}=\{n_{k,i}\}_{i=1}^{p_k}$ $(k=1,2,\dots)$ of disjoint finite sets of positive integers. Hence, by proposition 22.3,

$$\left\|\sum_{k=1}^{N} g_k(x)x_{n_{k,1}}\right\|\leqslant 4\|x\| \quad (x\in E,\ N=1,2,\dots), \tag{22.57}$$

i.e., $\sup_{1\leqslant N<\infty}\|u_N\|<\infty$, where $u_N(x)=\sum_{k=1}^{N} g_k(x)x_{n_{k,1}}$ $(x\in E,\ N=1,2,\dots)$.

Consequently, since for every finite linear combination $p=\sum_{j=1}^{m}\alpha_j x_j$ the limit $\lim_{N\to\infty} u_N(p)$ exists (by (22.55) and $f_i(x_j)=\delta_{ij}$) and since $[x_n]=E$, it follows that $\sum_{k=1}^{\infty} g_k(x)x_{n_{k,1}}$ converges and thus

$$u(x)=\sum_{k=1}^{\infty} g_k(x)x_{n_{k,1}} \qquad (x\in E) \tag{22.58}$$

is a well defined bounded linear mapping of E into E_1. By (22.56), we get that u is a projection of E onto E_1 such that $u(E_2+E_3)=0$.

Finally, to complete the proof of a), we shall show that $\overline{E_2+E_3}$ $=E_2\oplus E_3$. Let $u_{\{A_k^{(0)}\}}$ be the averaging projection which was considered above, i.e.,

$$u_{\{A_k^{(0)}\}}(x)=\sum_{k=1}^{\infty}\frac{1}{p_k}\left(\sum_{j=1}^{p_k} f_{n_{k,j}}(x)\right)\left(\sum_{j=1}^{p_k} x_{n_{k,j}}\right) \qquad (x\in E). \tag{22.59}$$

Then for every $h=1,2,\ldots$ we obtain

$$u_{\{A_k^{(0)}\}}(y_{h,1})=y_{h,1}, \quad u_{\{A_k^{(0)}\}}(y_{h,i})=0 \qquad (i=2,3,\ldots), \tag{22.60}$$

whence the restriction of $u_{\{A_k^{(0)}\}}$ to E_2 is the identity and $u_{\{A_k^{(0)}\}}(E_3)=0$, which proves a).

b) Let u_1 be the averaging projection with respect to $\{A_{k,m}\}_{k,m=1}^{\infty}$ where $A_{k,m}=\{n_{k,(m-1)p_k+j}\}_{j=1}^{p_k}$, i.e.,

$$u_1(x)=\sum_{k,m=1}^{\infty}\frac{1}{p_k}\left(\sum_{j=1}^{p_k} f_{n_{k,(m-1)p_k+j}}(x)\right)\left(\sum_{j=1}^{p_k} x_{n_{k,(m-1)p_k+j}}\right) \qquad (x\in E). \tag{22.61}$$

Then for every $h=1,2,\ldots$ and $i\geqslant 2$ we obtain, by biorthogonality,

$$u_1(y_{h,i})=-\frac{1}{p_h}\sum_{j=1}^{p_h} x_{n_{h,mp_h+j}}=-\frac{1}{p_h}z_{h,m+1} \qquad (i=(m-1)p_h+1,\ldots,mp_h). \tag{22.62}$$

Indeed, for $2\leqslant i\leqslant p_h$ we have

$$u_1(y_{h,i})=\sum_{k,m=1}^{\infty}\frac{1}{p_k}\left(\sum_{j=1}^{p_k} f_{n_{k,(m-1)p_k+j}}\left(x_{n_{h,i}}-\frac{1}{p_h-1}\sum_{l=2}^{p_h} x_{n_{h,l}}\right.\right.$$
$$\left.\left.-\frac{1}{p_h}\sum_{l=1}^{p_h} x_{n_{h,p_h+l}}\right)\right)\left(\sum_{j=1}^{p_k} x_{n_{k,(m-1)p_k+j}}\right)$$

$$= \sum_{m=1}^{\infty} \frac{1}{p_h} \left(\sum_{j=1}^{p_h} f_{n_h,(m-1)p_h+j} \left(x_{n_h,i} - \frac{1}{p_h - 1} \sum_{l=2}^{p_h} x_{n_h,l} \right.\right.$$

$$\left.\left. - \frac{1}{p_h} \sum_{l=1}^{p_h} x_{n_h,p_h+l} \right)\right) \left(\sum_{j=1}^{p_h} x_{n_h,(m-1)p_h+j} \right)$$

$$= \frac{1}{p_h} \left[\left(\sum_{j=1}^{p_h} f_{n_h,j} \left(x_{n_h,i} - \frac{1}{p_h - 1} \sum_{l=2}^{p_h} x_{n_h,l} \right)\right) \sum_{j=1}^{p_h} x_{n_h,j} \right.$$

$$\left. + \sum_{j=1}^{p_h} f_{n_h,p_h+j} \left(- \frac{1}{p_h} \sum_{l=1}^{p_h} x_{n_h,p_h+l} \right) \sum_{j=1}^{p_h} x_{n_h,p_h+j} \right]$$

$$= \frac{1}{p_h} \left[\left(1 - \frac{1}{p_h - 1} (p_h - 1) \right) \sum_{j=1}^{p_h} x_{n_h,j} + \left(- \frac{1}{p_h} p_h \right) \sum_{j=1}^{p_h} x_{n_h,p_h+j} \right]$$

$$= - \frac{1}{p_h} \sum_{j=1}^{p_h} x_{n_h,p_h+j} = - \frac{1}{p_h} z_{h,2}$$

and for the other i the computation is similar.

Hence, taking into account proposition 22.3 and that $(E, \{x_n\})$ is a symmetric space, we have, for any scalars $\alpha_{k,i}$,

$$\left\| \sum_{k=1}^{N} \sum_{i=2}^{Np_k} \alpha_{k,i} y_{k,i} \right\| \geqslant \left\| u_1 \left(\sum_{k=1}^{N} \sum_{i=2}^{Np_k} \alpha_{k,i} y_{k,i} \right) \right\| \tag{22.63}$$

$$= \left\| \sum_{k=1}^{N} \frac{1}{p_k} \left[\sum_{i=2}^{p_k} \alpha_{k,i} z_{k,2} + \sum_{m=1}^{N-1} \left(\sum_{j=1}^{p_k} \alpha_{k,mp_k+j} \right) z_{k,m+2} \right] \right\|$$

$$= \left\| \sum_{k=1}^{N} \frac{1}{p_k} \left[\sum_{i=2}^{p_k} \alpha_{k,i} z_{k,1} + \sum_{m=1}^{N-1} \left(\sum_{j=1}^{p_k} \alpha_{k,mp_k+j} \right) z_{k,m+1} \right] \right\|$$

$$\geqslant \left\| \sum_{k=1}^{N} \frac{1}{p_k} \left[\sum_{i=2}^{p_k} \alpha_{k,i} (z_{k,1} - x_{n_k,1}) \right.\right.$$

$$\left.\left. + \sum_{m=1}^{N-1} \left(\sum_{j=1}^{p_k} \alpha_{k,mp_k+j} \right) z_{k,m+1} \right] \right\|$$

$$\geqslant \frac{1}{2} \left\| \sum_{k=1}^{N} \frac{1}{p_k - 1} \sum_{i=2}^{p_k} \alpha_{k,i} (z_{k,1} - x_{n_k,1}) \right.$$

$$\left. + \sum_{k=1}^{N} \sum_{m=1}^{N-1} \frac{1}{p_k} \left(\sum_{j=1}^{p_k} \alpha_{k,mp_k+j} \right) z_{k,m+1} \right\| .$$

Consequently,

$$\begin{aligned}\left\|\sum_{k=1}^{N}\sum_{i=2}^{Np_k}\alpha_{k,i}(y_{k,i}-x_{n_{k,i}})\right\| &= \left\|\sum_{k=1}^{N}\left[\sum_{i=2}^{p_k}\alpha_{k,i}\left(-\frac{1}{p_k-1}(z_{k,1}-x_{n_{k,1}})-\frac{1}{p_k}z_{k,2}\right)\right.\right.\\
&\quad\left.\left.+\sum_{m=2}^{N}\sum_{i=(m-1)p_k+1}^{mp_k}\alpha_{k,i}\left(-\frac{1}{p_k}\right)(z_{k,m}+z_{k,m+1})\right]\right\|\\
&\leqslant \left\|\sum_{k=1}^{N}\frac{1}{p_k}\left[\sum_{i=2}^{p_k}\alpha_{k,i}z_{k,2}\right.\right.\\
&\quad\left.\left.+\sum_{m=1}^{N-1}\left(\sum_{j=1}^{p_k}\alpha_{k,mp_k+j}\right)z_{k,m+2}\right]\right\|\\
&\quad+\left\|\sum_{k=1}^{N}\frac{1}{p_k-1}\sum_{i=2}^{p_k}\alpha_{k,i}(z_{k,1}-x_{n_{k,1}})\right.\\
&\quad\left.+\sum_{k=1}^{N}\sum_{m=1}^{N-1}\frac{1}{p_k}\left(\sum_{j=1}^{p_k}\alpha_{k,mp_k+j}\right)z_{k,m+1}\right\|\\
&\leqslant 3\left\|\sum_{k=1}^{N}\sum_{i=2}^{Np_k}\alpha_{k,i}y_{k,i}\right\|,\end{aligned}\tag{22.64}$$

and thus

$$\left\|\sum_{k=1}^{N}\sum_{i=2}^{Np_k}\alpha_{k,i}x_{n_{k,i}}\right\|\leqslant 4\left\|\sum_{k=1}^{N}\sum_{i=2}^{Np_k}\alpha_{k,i}y_{k,i}\right\|.\tag{22.65}$$

On the other hand, again by proposition 22.3, we have

$$\begin{aligned}\left\|\sum_{k=1}^{N}\sum_{i=2}^{Np_k}\alpha_{k,i}x_{n_{k,i}}\right\| &\geqslant \left\|u_1\left(\sum_{k=1}^{N}\sum_{i=2}^{Np_k}\alpha_{k,i}x_{n_{k,i}}\right)\right\|\\
&= \left\|\sum_{k=1}^{N}\frac{1}{p_k}\left[\sum_{i=2}^{p_k}\alpha_{k,i}z_{k,1}+\sum_{m=1}^{N-1}\left(\sum_{j=1}^{p_k}\alpha_{k,mp_k+j}\right)z_{k,m+1}\right]\right\|\end{aligned}$$

and thus, by using (22.63) we obtain, as in (22.64), that

$$\left\|\sum_{k=1}^{N}\sum_{i=2}^{Np_k}\alpha_{k,i}y_{k,i}\right\|\leqslant 4\left\|\sum_{k=1}^{N}\sum_{i=2}^{Np_k}\alpha_{k,i}x_{n_{k,i}}\right\|.\tag{22.66}$$

From the inequalities (22.65) and (22.66) it follows, by Ch. I, § 8, theorem 8.1d) (implication $2^\circ\Rightarrow 1^\circ$) that $\{y_{k,i}\}_{k=1,i=2}^{\infty}$ is strictly equivalent to the unconditional basic sequence $\{x_{n_{k,i}}\}_{k=1,i=2}^{\infty}$, whence $\{y_{k,i}\}_{k=1,i=2}^{\infty}$ is an unconditional basis of E_3.

c) Since $\{x_n\}$ is unconditional, the subsequence $\{x_{n_{k,1}}\}_{k=1}^{\infty}$ is an unconditional basis of E_1 and, by applying § 17, theorem 17.1, it follows that the sequence $\{y_{k,1}\}_{k=1}^{\infty}$ is an unconditional basis of E_2 (same proof as for § 17, corollary 17.2). Hence, by a), b) and § 16, proposition 16.4 on cartesian products of unconditional bases,

$\{y_{k,1}\}_{k=1}^{\infty} \cup \{x_{n_{k,1}}\}_{k=1}^{\infty} \cup \{y_{k,i}\}_{k=1,i=2}^{\infty}$ is an unconditional basis of E, which completes the proof of proposition 22.4.

Proposition 22.5. *If $\{x_n\}$ is a symmetric basis of a Banach space E, then the a. s. c. f. $\{f_n\} \subset E^*$ is a symmetric basis of $[f_n]$ and*

$$v_{\{f_n\}}^{(s)} \leqslant v_{\{x_n\}}^{(s)}. \tag{22.67}$$

Hence, in particular, if $(E, \{x_n\})$ is a symmetric space, then so is $([f_n], \{f_n\})$.

Proof. Consider the operators

$$\begin{gathered} v_{\sigma,\{\beta_i\},k}(g) = \sum_{i=1}^{k} \beta_i [\phi(x_i)](g) f_{\sigma(i)} = \sum_{i=1}^{k} \beta_i g(x_i) f_{\sigma(i)} \\ (g \in E^*, \sigma \in \Pi, |\beta_1|, \ldots, |\beta_k| \leqslant 1, k = 1, 2, \ldots), \end{gathered} \tag{22.68}$$

where ϕ denotes the canonical mapping of E into $[f_n]^*$. We have

$$\begin{aligned} [v_{\sigma,\{\beta_i\},k}(g)](x) &= \sum_{i=1}^{k} \beta_i g(x_i) f_{\sigma(i)}(x) = g\left[\sum_{i=1}^{k} \beta_i f_{\sigma(i)}(x) x_i\right] \\ &= g\left[\sum_{\sigma^{-1}(j)=1}^{k} \beta_{\sigma^{-1}(j)} f_j(x) x_{\sigma^{-1}(j)}\right] = g\left[\sum_{j=1}^{l} \beta'_j f_j(x) x_{\sigma^{-1}(j)}\right] \\ &= g\left[u_{\sigma^{-1},\{\beta'_j\},l}(x)\right] = \left[u^*_{\sigma^{-1},\{\beta'_j\},l}(g)\right](x) \end{aligned}$$

$(x \in E, g \in E^*, \sigma \in \Pi, |\beta_1|, \ldots, |\beta_k| \leqslant 1, k = 1, 2, \ldots)$, where we have put $l = \max_{1 \leqslant i \leqslant k} \sigma(i)$ and

$$\beta'_j = \begin{cases} \beta_{\sigma^{-1}(j)} & \text{for } j \in \{\sigma(1), \ldots, \sigma(k)\}, \\ 0 & \text{for } j \in \{1, \ldots, l\} \setminus \{\sigma(1), \ldots, \sigma(k)\}. \end{cases}$$

Consequently, taking into account that $\|v|_{[f_n]}\| \leqslant \|v\|$ and $\|u^*\| = \|u\|$,

$$v_{\{f_n\}}^{(s)} = \sup_{\sigma \in \Pi} \sup_{\substack{|\beta_i| \leqslant 1 \\ 1 \leqslant k < \infty}} \|v_{\sigma,\{\beta_i\},k}|_{[f_n]}\| \leqslant \sup_{\sigma \in \Pi} \sup_{\substack{|\beta_i| \leqslant 1 \\ 1 \leqslant k < \infty}} \|u_{\sigma,\{\beta_i\},k}\| = v_{\{x_n\}}^{(s)},$$

which completes the proof.

Remark 22.1. In (22.67) the sign $<$ may also occur. Indeed, the bounded basis $\{x_n\}$ of l^1 given in Ch. I, § 12, example 12.3 is unconditional, whence, by § 18, theorem 18.2, equivalent to the unit vector basis of l^1, and thus symmetric. By the expression of $\left\|\sum_{i=1}^{n} \alpha_i f_i\right\|$ (Ch. I, § 12, formula (12.33)), $v_{\{f_n\}}^{(s)}$ does not depend on $r([f_n]) = \lambda$, but $v_{\{x_n\}}^{(s)} \geqslant v_{\{x_n\}} = 2 + \frac{1}{\lambda} \to \infty$ as $\lambda \to 0$, which proves our assertion.

Let us give now a relation between symmetric bases and sequence spaces, similar to § 17, theorem 17.8 (on unconditional bases) and to Ch. I, § 12, theorem 12.6 (on general bases).

We recall that the *σ-dual* of a sequence space S is the sequence space S^σ defined by

$$S^\sigma = \left\{ \{\beta_n\} \subset K \,\middle|\, \sum_{i=1}^{\infty} |\beta_i \alpha_{\rho(i)}| < \infty \quad \text{for all} \quad \{\alpha_n\} \in S, \rho \in \Pi \right\}. \tag{22.68}$$

For every sequence space S we have, obviously, $S \subset S^{\sigma\sigma}$. A sequence space S is said to be *σ-perfect* if $S^{\sigma\sigma} = S$.

It is also obvious that for any sequence space S we have $S^\sigma \subset S^\alpha$, where S^α is the α-dual of S defined in § 17, formula (17.81).

A sequence space S is called *symmetric* if for every $\{\alpha_n\} \in S$ and every $\rho \in \Pi$ we have $\{\alpha_{\rho(n)}\} \in S$.

Lemma 22.1. *If S is a symmetric sequence space, then $S^\sigma = S^\alpha$ and hence*

$$S^{\sigma\sigma} = S^{\alpha\alpha}. \tag{22.69}$$

Proof. If $\{\beta_n\} \in S^\alpha$, $\{\alpha_n\} \in S$ and $\rho \in \Pi$, then, since S is symmetric, $\{\alpha_{\rho(n)}\} \in S$, whence

$$\sum_{i=1}^{\infty} |\beta_i \alpha_{\rho(i)}| < \infty,$$

i.e., $\{\beta_n\} \in S^\sigma$. Thus, $S^\alpha \subset S^\sigma$ and hence, by the opposite inclusion $S^\sigma \subset S^\alpha$ observed above, $S^\sigma = S^\alpha$.

Since for any sequence space S the σ-dual S^σ is symmetric (because $\{\beta_n\} \in S^\sigma$ implies $\sum_{i=1}^{\infty} |\beta_{\tau(i)} \alpha_{\rho(i)}| = \sum_{i=1}^{\infty} |\beta_i \alpha_{\tau^{-1}\rho(i)}| < \infty$ for all $\{\alpha_n\} \in S$ and $\rho, \tau \in \Pi$), it follows that for any sequence space S we have $S^{\sigma\sigma} = S^{\sigma\alpha}$. Consequently, if S is symmetric,

$$S^{\sigma\sigma} = S^{\sigma\alpha} = S^{\alpha\alpha},$$

which completes the proof of lemma 22.1.

Theorem 22.2. *A sequence space S is associated to a symmetric basis of a Banach space if and only if S contains all unit vectors e_n and there exists a σ-perfect BK-space T such that $[e_n]_T = S$. In this case, $T = S^{\sigma\sigma}$.*

Proof. Assume that S is associated to a symmetric basis $\{x_n\}$ of a Banach space E, i. e., $S = \left\{ \{\alpha_n\} \subset K \,\middle|\, \sum_{i=1}^{\infty} \alpha_i x_i \text{ converges} \right\}$. Then, since $\{x_n\}$ is unconditional, by § 17, theorem 17.8, S contains all unit vectors e_n and there exists an α-perfect BK-space T such that $[e_n]_T = S$ and $T = S^{\alpha\alpha}$. Since $\{x_n\}$ is symmetric, $\{\alpha_{\rho(n)}\} \in S$ for all $\{\alpha_n\} \in S$, $\rho \in \Pi$ (by theorem 22.1, implication $1^\circ \Rightarrow 11^\circ$) and thus S is symmetric. Hence, by lemma 22.1, $S^{\sigma\sigma}=S^{\alpha\alpha}=T$. Since $T=(S^\sigma)^\sigma$ is symmetric and since T is α-perfect, by lemma 22.1 we also have $T^{\sigma\sigma}=T^{\alpha\alpha}=T$, i. e., T is σ-perfect.

Conversely, assume now that S is a sequence space containing all unit vectors e_n and that there exists a σ-perfect BK-space T such that $[e_n]_T=S$. We shall prove that $\{e_n\}$ is a symmetric basic sequence in T, whence S is associated to the symmetric basis $\{x_n\}=\{e_n\}$ of the Banach space $E=S$.

Since $T=(T^\sigma)^\sigma$ is symmetric, by lemma 22.1 we have $T^{\alpha\alpha}=T^{\sigma\sigma}=T$, i. e., T is also α-perfect. Hence, by § 17, theorem 17.8, $\{e_n\}$ is an unconditional basis of $S=[e_n]$.

Now let $\rho\in\Pi$ be arbitrary. Define a linear mapping $u_\rho\colon T\to T$ by

$$u_\rho(\{\alpha_n\})=\{\alpha_{\rho(n)}\} \qquad (\{\alpha_n\}\in T); \tag{22.70}$$

since $T=(T^\sigma)^\sigma$ is symmetric, we have indeed $\{\alpha_{\rho(n)}\}\in T$. Then, since T is a BK-space, the mapping u_ρ is closed, whence continuous and therefore (since $S=[e_n]$ and $u_\rho(e_n)=e_{\rho(n)}$ for all $n=1,2,\ldots$),

$$u_\rho(S)\subset S \qquad (\rho\in\Pi),$$

i. e., S is symmetric. Consequently, by theorem 22.1 (implication $11^\circ\Rightarrow 1^\circ$), $\{e_n\}$ is a symmetric basis of S, which completes the proof of theorem 22.2.

One can also prove the sufficiency part in theorem 22.2 with an argument similar to the proof of Ch. I, § 12, theorem 12.6, by introducing on T^σ the norm

$$\|\{\beta_n\}\| = \sup_{\substack{\{\alpha_n\}\in T\\ \|\{\alpha_n\}\|\leqslant 1}} \sup_{\rho\in\Pi} \sum_{i=1}^{\infty} |\beta_i\alpha_{\rho(i)}| \qquad (\{\beta_n\}\in T^\sigma) \tag{22.71}$$

and on $T=T^{\sigma\sigma}$ the equivalent norm

$$|||\{\alpha_n\}||| = \sup_{\substack{\{\beta_n\}\in T^\sigma\\ \|\{\beta_n\}\|\leqslant 1}} \sup_{\tau\in\Pi} \sum_{i=1}^{\infty} |\beta_{\tau(i)}\alpha_i| = \sup_{\substack{\{\beta_n\}\in T^\sigma\\ \|\{\beta_n\}\|\leqslant 1}} \sup_{\rho\in\Pi} \sum_{i=1}^{\infty} |\beta_i\alpha_{\rho(i)}| \qquad (\{\alpha_n\}\in T) \tag{22.72}$$

and observing that for any $|\gamma_i|\leqslant 1$ $(i=1,2,\ldots)$, $\sigma\in\Pi$ and $n=1,2,\ldots$ we have

$$\left|\left|\left|\sum_{i=1}^{n}\gamma_i\alpha_i e_{\sigma(i)}\right|\right|\right| \leqslant \left|\left|\left|\sum_{i=1}^{\infty}\gamma_i\alpha_i e_{\sigma(i)}\right|\right|\right| = \left|\left|\left|\sum_{i=1}^{\infty}\gamma_{\sigma^{-1}(i)}\alpha_{\sigma^{-1}(i)} e_i\right|\right|\right|$$
$$\leqslant |||\{\alpha_n\}||| \qquad (\{\alpha_n\}\in S). \tag{22.73}$$

Moreover, this remark makes also possible the following proof of the implication $12^\circ\Rightarrow 1^\circ$ of theorem 22.1: If every permutation $\{x_{\sigma(n)}\}$

of a basis $\{x_n\}$ is a basis, equivalent to $\{x_n\}$ and if $S=\left\{\{\alpha_n\}\subset K \,\middle|\, \sum_{i=1}^{\infty}\alpha_i x_i\right.$ converges$\Big\}$, then $\{x_n\}$ is an unconditional basis and hence, by § 17, theorem 17.8, there exists an α-perfect BK-space T such that $[e_n]_T=S$ and $T=S^{\alpha\alpha}$. Furthermore, by our assumption, for every $\{\alpha_n\}\in S$ and $\rho\in\Pi$ the series $\sum_{i=1}^{\infty}\alpha_i x_{\rho^{-1}(i)}$ converges, whence, since $\{x_n\}$ is unconditional, $\sum_{i=1}^{\infty}\alpha_{\rho(i)}x_i=\sum_{i=1}^{\infty}\alpha_i x_{\rho^{-1}(i)}$ converges, i.e., $\{\alpha_{\rho(n)}\}\in S$ and thus S is symmetric Therefore, by lemma 22.1, $T=S^{\alpha\alpha}=S^{\sigma\sigma}$ and thus T is symmetric. Consequently, by the sufficiency part of theorem 22.2, $\{e_n\}$ is a symmetric basis of S and hence $\{x_n\}$ is a symmetric basis of E (since $\{e_n\}\sim\{x_n\}$), which completes the proof.

Finally, we shall consider finite-dimensional symmetric spaces. In such spaces one can introduce analogues of the function systems of Haar and Rademacher and prove a certain abstract analogue of the Khinchin inequality (see § 14, formula (14.56)), which is useful for applications.

Definition 22.5. Let $(E_{2^n},\{x_j\})$ be a 2^n-dimensional symmetric space, where $n<\infty$. We shall call *Haar system* (with respect to $\{x_n\}$) the sequence $\{y_j\}_{j=1}^{2^n}$ defined by

$$y_1=\sum_{i=1}^{2^n}x_i,\quad y_{2^k+l}=\sum_{i=1}^{2^n}\beta_i^{(k,l)}x_i\quad (l=1,\dots,2^k;\,k=0,1,\dots,n-1),\tag{22.74}$$

where

$$\beta_i^{(k,l)}=\begin{cases}1 & \text{for}\quad (2l-2)2^{n-k-1}+1\leqslant i\leqslant(2l-1)2^{n-k-1},\\ -1 & \text{for}\quad (2l-1)2^{n-k-1}+1\leqslant i\leqslant 2l\cdot 2^{n-k-1},\\ 0 & \text{for}\quad 1\leqslant i\leqslant(2l-2)2^{n-k-1}\ \text{and}\ 2l\cdot 2^{n-k-1}+1\leqslant i\leqslant 2^n.\end{cases}\tag{22.75}$$

We shall call *Rademacher system* the sequence $\{r_k\}_{k=1}^{n}$ defined by

$$r_k=\sum_{l=1}^{2^{k-1}}y_{2^{k-1}+l}\quad (k=1,\dots,n),\tag{22.76}$$

where $\{y_j\}$ is the Haar system in $(E_{2^n},\{x_j\})$.

Proposition 22.6. *Let $(E_{2^n},\{x_j\})$ be a 2^n-dimensional symmetric space. Then the Haar system $\{y_j\}_{j=1}^{2^n}$ is a monotone basis of E_{2^n}. Consequently, the Rademacher system $\{r_k\}_{k=1}^{n}$ is a monotone block basic sequence with respect to the Haar system $\{y_j\}_{j=1}^{2^n}$.*

Proof. Let m be an arbitrary integer such that $1 \leqslant m \leqslant 2^n - 1$, and let $\alpha_1, \ldots, \alpha_{m+1}$ be arbitrary scalars. Then, since $\{x_j\}$ is a basis of E_{2^n}, there exists a sequence of scalars $\{\gamma_j\}_{j=1}^{2^n}$ such that

$$\sum_{j=1}^{m} \alpha_j y_j = \sum_{i=1}^{2^n} \gamma_i x_i. \tag{22.77}$$

Let (k, l) be the couple of non-negative integers determined by the following properties: $1 \leqslant l \leqslant 2^k$, $2^k + l = m + 1$. Then, by (22.77), (22.74), and (22.75),

$$\begin{aligned} \left\| \sum_{j=1}^{m+1} \alpha_j y_j \right\| &= \left\| \sum_{i=1}^{2^n} \gamma_i x_i + \alpha_{m+1} \sum_{i=1}^{2^n} \beta_i^{(k,l)} x_i \right\| \\ &= \left\| \sum_{i=1}^{(2l-2)2^{n-k-1}} \gamma_i x_i + \sum_{i=(2l-2)2^{n-k-1}+1}^{(2l-1)2^{n-k-1}} (\gamma_i + \alpha_{m+1}) x_i \right. \\ &\quad \left. + \sum_{i=(2l-1)2^{n-k-1}+1}^{2l \cdot 2^{n-k-1}} (\gamma_i - \alpha_{m+1}) x_i + \sum_{i=2l \cdot 2^{n-k-1}+1}^{2^n} \gamma_i x_i \right\|. \end{aligned}$$

Since $(E_{2^n}, \{x_j\})$ is a symmetric space, this number is equal to

$$\begin{aligned} &\left\| \sum_{i=1}^{(2l-2)2^{n-k-1}} \gamma_i x_i + \sum_{i=(2l-2)2^{n-k-1}+1}^{(2l-1)2^{n-k-1}} (\gamma_i - \alpha_{m+1}) x_i \right. \\ &\quad \left. + \sum_{i=(2l-1)2^{n-k-1}+1}^{2l \cdot 2^{n-k-1}} (\gamma_i + \alpha_{m+1}) x_i + \sum_{i=2l \cdot 2^{n-k-1}+1}^{2^n} \gamma_i x_i \right\|. \end{aligned}$$

Adding these equalities and multiplying by $\frac{1}{2}$, we get

$$\left\| \sum_{j=1}^{m+1} \alpha_j y_j \right\| \geqslant \left\| \sum_{i=1}^{2^n} \gamma_i x_i \right\| = \left\| \sum_{j=1}^{m} \alpha_j y_j \right\|,$$

which proves the first assertion of proposition 22.6. By (22.76) and Ch. I, § 7, corollary 7.4, the second assertion is a consequence of the first one, which completes the proof of proposition 22.6.

Proposition 22.7. *Let* $(E_{2^n}, \{x_j\})$ *be a* 2^n*-dimensional symmetric space and let* $\lambda_n = \left\| \sum_{i=1}^{2^n} x_i \right\|$, $\mu_n = \left\| \sum_{i=1}^{2^n} f_i \right\|$, *where* $\{f_n\} \subset E^*$ *is the a.s.c.f. to* $\{x_n\}$. *Then we have*

$$\lambda_n \mu_n = 2^n. \tag{22.78}$$

Proof. We have, obviously,

$$\lambda_n \mu_n \geqslant \left(\sum_{i=1}^{2^n} f_i \right) \left(\sum_{j=1}^{2^n} x_j \right) = 2^n. \tag{22.79}$$

On the other hand, let $f=\sum_{i=1}^{2^n} f_i$ and let $x=\sum_{j=1}^{2^n} \alpha_j x_j \in E$ be such that $f(x)=\mu_n$, $\|x\|=1$. Then

$$\mu_n=f(x)=\left(\sum_{i=1}^{2^n} f_i\right)\left(\sum_{j=1}^{2^n} \alpha_j x_j\right)=\sum_{j=1}^{2^n} \alpha_j. \tag{22.80}$$

Let u_A be the averaging projection with respect to the sequence consisting of the single set $A=\{1,\ldots,2^n\}$ (see definition 22.3). Then, by (22.80),

$$u_A(x)=\frac{1}{2^n}\left(\sum_{j=1}^{2^n} \alpha_j\right)\left(\sum_{j=1}^{2^n} x_j\right)=\frac{1}{2^n}\mu_n \sum_{j=1}^{2^n} x_j, \tag{22.81}$$

whence, since $\|u_A\|=1$ (by proposition 22.3),

$$\frac{1}{2^n}\lambda_n \mu_n=\|u_A(x)\| \leqslant \|x\|=1 \tag{22.82}$$

which, together with (22.79), gives (22.78). This completes the proof of proposition 22.7.

Theorem 22.3. *Let* $(E_{2^n},\{x_j\}_{j=1}^{2^n})$ *be a* 2^n*-dimensional symmetric space and let* $\{r_k\}_{k=1}^n$ *be the Rademacher system in this space. Then for any scalars* $\alpha_1,\ldots,\alpha_n$ *we have*

$$\left\|\sum_{k=1}^n \alpha_k \frac{r_k}{\|r_k\|}\right\| \geqslant \frac{1}{8}\sqrt{\sum_{k=1}^n |\alpha_k|^2}. \tag{22.83}$$

Proof. Since $\{x_j\}$ is a basis of E_{2^n}, there exists for each integer k with $1\leqslant k\leqslant n$ a unique sequence of scalars $\{r_{kj}\}_{j=1}^{2^n}$ such that

$$r_k=\sum_{i=1}^{2^n} r_{ki} x_i \qquad (k=1,\ldots,n). \tag{22.84}$$

Precisely, by (22.74), (22.75) and (22.76) we have

$$r_{ki}=\begin{cases} 1 & \text{for} \quad (2l-2)2^{n-k}+1\leqslant i\leqslant(2l-1)2^{n-k} \\ -1 & \text{for} \quad (2l-1)2^{n-k}+1\leqslant i\leqslant 2l\cdot 2^{n-k} \end{cases} \tag{22.85}$$

$(l=1,2,\ldots,2^{k-1};\ k=1,\ldots,n)$.

Let $r_k(\cdot)$ be the usual Rademacher functions on $[0,1]$ (see § 14, formula (14.2)). We claim that for any scalars $\alpha_1,\ldots,\alpha_n$ we have

$$\int_0^1 \left|\sum_{k=1}^n \alpha_k r_k(t)\right| dt=\frac{1}{2^n}\sum_{i=1}^{2^n}\left|\sum_{k=1}^n \alpha_k r_{ki}\right|. \tag{22.86}$$

Indeed, let us denote by $(l^1_{2^n}, \{e_j\})$ the 2^n-dimensional symmetric space in which the norm is defined by $\left\| \sum_{i=1}^{2^n} \xi_i e_i \right\| = \sum_{i=1}^{2^n} |\xi_i|$ and by $\chi_i(\cdot)$ the characteristic function of the interval $\left(\frac{i-1}{2^n}, \frac{i}{2^n}\right)$ $(i=1, 2, \dots, 2^n)$. Then the mapping v defined by $v\left(\sum_{i=1}^{2^n} \xi_i \chi_i\right) = \frac{1}{2^n} \sum_{i=1}^{2^n} \xi_i e_i$ is obviously a linear isometry of the 2^n-dimensional subspace $[\chi_1, \dots, \chi_{2^n}] \subset L^1([0,1])$ onto $l^1_{2^n}$. Since $r_k(t) = \sum_{i=1}^{2^n} r_{ki} \chi_i(t)$ $(t \in [0,1], k=1, \dots, n)$, it follows that

$$v\left[\sum_{k=1}^{n} \alpha_k r_k(\cdot)\right] = \sum_{k=1}^{n} \alpha_k v[r_k(\cdot)] = \sum_{k=1}^{n} \frac{\alpha_k}{2^n} \sum_{i=1}^{2^n} r_{ki} e_i,$$

whence since v is an isometry, we infer (22.86).

By (22.86) and the usual Khinchin inequality in $L^1([0,1])$ (see § 14, the remark after formula (14.56)) we have, for any scalars $\alpha_1, \dots, \alpha_n$,

$$\frac{1}{2^n} \sum_{i=1}^{2^n} \left| \sum_{k=1}^{n} \alpha_k r_{ki} \right| \geqslant \frac{1}{8} \sqrt{\sum_{k=1}^{n} |\alpha_k|^2}. \tag{22.87}$$

On the other hand, since $(E_{2^n}, \{x_j\})$ is a symmetric space, we have, by (22.84) and (22.85),

$$\|r_1\| = \|r_2\| = \cdots = \|r_n\| = \left\| \sum_{k=1}^{2^n} x_k \right\| = \lambda_n. \tag{22.88}$$

Now let $\alpha_1, \dots, \alpha_n$ be arbitrary scalars and let $\varepsilon_i = \operatorname{sign} \sum_{k=1}^{n} \alpha_k r_{ki}$ $(i=1, \dots, 2^n)$. Then, by proposition 22.5, $(E^*_{2^n}, \{f_j\})$ is a symmetric space (where $\{f_j\} \subset E^*_{2^n}$ is the a.s.c.f. to $\{x_j\}$), whence $\mu_n = \left\| \sum_{i=1}^{2^n} f_i \right\| = \left\| \sum_{i=1}^{2^n} \varepsilon_i f_i \right\|$. Consequently, taking into account (22.88), proposition 22.7, (22.84) and (22.87), we get

$$\left\| \sum_{k=1}^{n} \alpha_k \frac{r_k}{\|r_k\|} \right\| = \frac{1}{\lambda_n} \left\| \sum_{k=1}^{n} \alpha_k r_k \right\| = \frac{\mu_n}{2^n} \left\| \sum_{k=1}^{n} \alpha_k r_k \right\|$$

$$= \frac{1}{2^n} \left\| \sum_{i=1}^{2^n} \varepsilon_i f_i \right\| \left\| \sum_{k=1}^{n} \alpha_k r_k \right\| \geqslant \frac{1}{2^n} \left| \left(\sum_{i=1}^{2^n} \varepsilon_i f_i\right) \left(\sum_{k=1}^{n} \alpha_k r_k\right) \right|$$

$$= \frac{1}{2^n} \left| \sum_{i=1}^{2^n} \varepsilon_i \sum_{k=1}^{n} \alpha_k r_{ki} \right|$$

$$= \frac{1}{2^n} \sum_{i=1}^{2^n} \left| \sum_{k=1}^{n} \alpha_k r_{ki} \right| \geqslant \frac{1}{8} \sqrt{\sum_{k=1}^{n} |\alpha_k|^2},$$

which completes the proof of theorem 22.3.

§ 23. Applications: Existence of non-equivalent normalized bases and conditional bases in infinite dimensional Banach spaces with bases

As a first application of symmetric bases we shall prove that the equivalence of all normalized[1] bases characterizes finite dimensional spaces among Banach spaces with bases. The idea of the proof is to reduce the problem first to symmetric spaces and then to one of the spaces c_0, l^1 or l^2, where we know the existence of both unconditional and conditional bases (§ 14), whence also of non-equivalent normalized bases.

Proposition 23.1. *Let E be a Banach space with a basis, in which all normalized bases are equivalent, and let $\{x_n\}$ be a normalized basis of E. Then*

a) *$\{x_n\}$ is a symmetric basis.*

b) *If $\{y_n\}$ is a bounded block basic sequence with respect to $\{x_n\}$, then*[2] *$\{x_n\} \sim \{y_n\}$.*

c) *$\{x_n\}$ is a Besselian basis.*

d) *E is reflexive.*

Proof. a) By our hypothesis, for any $\varepsilon_n = \pm 1$ $(n = 1, 2, \ldots)$, $\{x_n\}$ is equivalent to the normalized basis $\{\varepsilon_n x_n\}$ of E, whence, by § 16, theorem 16.8 (implication $3^\circ \Rightarrow 1^\circ$), $\{x_n\}$ is a normalized unconditional basis of E. Hence, by § 17, theorem 17.1 (implication $1^\circ \Rightarrow 2^\circ$), every permutation $\{x_{\sigma(n)}\}$ of $\{x_n\}$ is a normalized basis of E. Since by our hypothesis the bases $\{x_n\}$ and $\{x_{\sigma(n)}\}$ must be equivalent, it follows by § 22, theorem 22.1 (implication $12^\circ \Rightarrow 1^\circ$) that $\{x_n\}$ is a symmetric basis of E.

---

[1] As we have observed in a footnote in Ch. I, § 8, some boundedness condition is necessary, since otherwise it is easy to find non-equivalent bases (e.g. $\left\{\frac{x_n}{\|x_n\|}\right\}$ and $\left\{n \frac{x_n}{\|x_n\|}\right\}$) in any infinite dimensional Banach space with a basis $\{x_n\}$.

[2] Such bases $\{x_n\}$ are called *perfectly homogeneous* (see § 24, definition 24.1).

b) Let

$$y_n = \sum_{m_{n-1}+1}^{m_n} \alpha_i x_i \neq 0 \qquad (n=1,2\ldots;\, m_0=0) \tag{23.1}$$

be an arbitrary bounded block basic sequence with respect to $\{x_n\}$. Then, by Ch. I, § 7, theorem 7.2, we can extend $\left\{\frac{y_n}{\|y_n\|}\right\}$ to a normalized basis $\{z_n\}$ of E. By our hypothesis we have $\{x_n\} \sim \{z_n\}$ and by part a) $\{z_n\} \sim \left\{\frac{y_n}{\|y_n\|}\right\}$. Furthermore, since $0 < \inf_{1 \leqslant n < \infty} \|y_n\| \leqslant \sup_{1 \leqslant n < \infty} \|y_n\| < \infty$, by § 16, theorem 16.8 we have $\left\{\frac{y_n}{\|y_n\|}\right\} \sim \{y_n\}$. Thus, $\{x_n\} \sim \{z_n\} \sim \left\{\frac{y_n}{\|y_n\|}\right\} \sim \{y_n\}$, which proves b).

c) By the assertion a) proved above, $\{x_n\}$ is a symmetric basis. We may assume, without loss of generality, that $(E, \{x_n\})$ is a symmetric space (introducing, if necessary, the equivalent norm $|||x|||$ defined in § 22, proposition 22.1b); the basis $\{x_n\}$ will remain normalized in this new norm).

Assume now that $\{x_n\}$ is not Besselian, i. e., that there exists a sequence of scalars $\{\gamma_n\}$ for which $\sum_{i=1}^{\infty} \gamma_i x_i$ is convergent, but $\sum_{i=1}^{\infty} |\gamma_i|^2 = \infty$. Then there exists an increasing sequence of positive integers $\{m_n\}$ such that

$$\sum_{i=m_{n-1}+1}^{m_n} |\gamma_i|^2 \geqslant 1 \qquad (n=1,2,\ldots;\, m_0=0). \tag{23.2}$$

Let

$$p_n = m_n - m_{n-1} \qquad (n=1,2,\ldots), \tag{23.3}$$

$$q_0 = 0,\; q_n = \sum_{j=1}^{n} 2^{p_j} \qquad (n=1,2,\ldots), \tag{23.4}$$

and let $E_{2^{p_n}}$ denote the 2^{p_n}-dimensional subspace $[x_j]_{j=q_{n-1}+1}^{q_n}$ of E $(n=1,2,\ldots)$. Furthermore, let $\{y_j\}_{j=q_{n-1}+1}^{q_n}$ denote the Haar system and $\{r_i\}_{i=m_{n-1}+1}^{m_n}$ the Rademacher system in the symmetric space $(E_{2^{p_n}}, \{x_j\}_{j=q_{n-1}+1}^{q_n})$. Then, by § 22, proposition 22.6, $\{y_j\}_{j=q_{n-1}+1}^{q_n}$ is a monotone basis of $E_{2^{p_n}}$, whence, by Ch. I, § 7, corollary 7.3, the sequence

$$\{y_j\} = \bigcup_{n=1}^{\infty} \{y_j\}_{j=q_{n-1}+1}^{q_n} \tag{23.5}$$

is a basis of the space E. Since by our hypothesis the normalized bases $\{x_n\}$ and $\left\{\frac{y_n}{\|y_n\|}\right\}$ of E are equivalent and since $\sum_{i=1}^{\infty} \gamma_i x_i$ is convergent, the series $\sum_{i=1}^{\infty} \gamma_i \frac{y_i}{\|y_i\|}$ is convergent.

Put $z_i = \frac{r_i}{\|r_i\|}$ $(i=1,2,\ldots)$. Since by § 22, proposition 22.6, the sequence $\{z_i\}$ is a normalized block basic sequence with respect to the normalized basis $\left\{\frac{y_n}{\|y_n\|}\right\}$ of E and since $\sum_{i=1}^{\infty} \gamma_i \frac{y_i}{\|y_i\|}$ converges, from the assertion b) proved above it follows that the series $\sum_{i=1}^{\infty} \gamma_i z_i$ is convergent, whence

$$\lim_{n\to\infty} \sum_{i=m_{n-1}+1}^{m_n} \gamma_i z_i = 0. \tag{23.6}$$

On the other hand, by § 22, theorem 22.3 and by (23.2) we have

$$\left\| \sum_{i=m_{n-1}+1}^{m_n} \gamma_i z_i \right\| \geqslant \frac{1}{8} \sqrt{\sum_{i=m_{n-1}+1}^{m_n} |\gamma_i|^2} \geqslant \frac{1}{8},$$

which contradicts (23.6). This proves c).

d) Let us first show that $\{x_n\}$ is both shrinking and boundedly complete. Indeed, assume that $\{x_n\}$ is non-shrinking. Since by the assertion a) proved above, $\{x_n\}$ is unconditional, it follows from § 17, corollary 17.3a) that $\{x_n\}$ admits a block basic sequence $\{y_n\}$ equivalent to the unit vector basis of l^1. Now there are two cases:

1) If the subspace $[y_n]$ is complemented in E, then, denoting by F an arbitrary complementary subspace of $[y_n]$, we have the isomorphisms $E \times l^1 \cong (F \times l^1) \times l^1 \cong F \times (l^1 \times l^1) \cong F \times l^1 \cong E$. Hence, taking a conditional basis of l^1 (e. g. see § 14, example 14.2) we obtain, by Ch. I, § 4, proposition 4.2, a conditional basis of E, whence also a normalized conditional basis of E, which, since $\{x_n\}$ is unconditional, contradicts the assumption that all normalized bases in E are equivalent.

2) If the subspace $[y_n]$ is not complemented[1] in E, then extend $\{y_n\}$, by Ch. I, § 7, theorem 7.2, to a basis $\{z_n\}$ of E. Since $[y_n]$ is non-complemented in E, by virtue of § 16, theorem 16.8 (implication $1° \Rightarrow 2°$), $\{z_n\}$ is a conditional basis of E, which again contradicts the assumption that all normalized bases in E are equivalent. Thus, $\{x_n\}$ must be shrinking. The proof of the assertion that $\{x_n\}$ is boundedly complete, is similar, using § 17, corollary 17.3b), which leads to a block basic sequence $\{y_n\}$ equivalent to the unit vector basis of c_0.

Now the reflexivity of E follows easily. Indeed, since $\{x_n\}$ is boundedly complete, by § 6, theorem 6.2 (implication $1° \Rightarrow 2°$) we have

[1] One can show (as we shall see in Vol. II, Ch. IV) that this second case cannot occur, i. e., every subspace spanned by a block basic sequence of type l_+, with respect to an unconditional basis, is complemented. However, we shall not use this remark here.

$E^{**}=\pi(E)\oplus[f_n]^\perp$, where π denotes the canonical mapping of E into E^{**} and $\{f_n\}$ the a. s. c. f. to $\{x_n\}$. Since $\{x_n\}$ is shrinking, we have $[f_n]=E^*$, whence $[f_n]^\perp=\{0\}$ and hence $E^{**}=\pi(E)$, i. e., E is reflexive, which completes the proof of proposition 23.1.

Remark 23.1. The final part of the above proof of proposition 23.1 d), together with § 4, example 4.3 and § 6, example 6.3, shows that *a Banach space E with a basis $\{x_n\}$ is reflexive if and only if $\{x_n\}$ is both shrinking and boundedly complete.* We shall study the reflexivity of Banach spaces with bases in more detail in Vol. II, Ch. IV.

Now we can prove

Theorem 23.1. *In every infinite dimensional Banach space with a basis there exist two non-equivalent normalized bases.*

Proof. Let E be an infinite dimensional Banach space with a basis, such that all normalized bases of E are equivalent. Then, by proposition 23.1 d), E is reflexive, whence, by Ch. I, § 12, corollary 12.2, E^* has a basis. Let $\{g_n\}$, $\{h_n\}$ be two normalized bases of E^*, with the a. s. c. f. $\{\Phi_n\}\subset E^{**}=\pi(E)$ and $\{\Psi_n\}\subset E^{**}=\pi(E)$, respectively, where $[\pi(x)](f)=f(x)$ $(x\in E, f\in E^*)$. Then, by Ch. I, § 12, corollary 12.1, the sequences $\{y_n\}=\{\pi^{-1}(\Phi_n)\}$, $\{z_n\}=\{\pi^{-1}(\Psi_n)\}$ are bases of E, such that $g_i(y_j)=\delta_{ij}$, $h_i(z_j)=\delta_{ij}$. Since by Ch. I, § 3, theorem 3.1, we have $1\leqslant\|y_j\|$, $\|z_j\|\leqslant M<\infty$ $(j=1,2,\ldots)$ and since by proposition 23.1 a) $\{y_n\}$ and $\{z_n\}$ are unconditional bases of E, by virtue of § 16, theorem 16.8 (implication $1^\circ\Rightarrow 4^\circ$) $\{y_n\}$, $\{z_n\}$ are equivalent to $\left\{\frac{y_n}{\|y_n\|}\right\}$ and $\left\{\frac{z_n}{\|z_n\|}\right\}$, respectively. Since by our hypothesis $\left\{\frac{y_n}{\|y_n\|}\right\}$ and $\left\{\frac{z_n}{\|z_n\|}\right\}$ are equivalent, it follows that $\{y_n\}$ and $\{z_n\}$ are equivalent, whence, by Ch. I, § 12, proposition 12.1, $\{g_n\}$ and $\{h_n\}$ are also equivalent. Thus all normalized bases of E^* are equivalent.

Now, let $\{x_n\}$ be a normalized basis of E and let $\{f_n\}\subset E^*$ be the a. s. c. f. to $\{x_n\}$. Then, by the above arguments, $\{f_n\}$ is a basis of E^*, equivalent to the normalized basis $\left\{\frac{f_n}{\|f_n\|}\right\}$. Since all normalized bases of E^* are equivalent, it follows, by proposition 23.1 c) applied to E^* and $\left\{\frac{f_n}{\|f_n\|}\right\}$, that $\left\{\frac{f_n}{\|f_n\|}\right\}$, whence also $\{f_n\}$, is Besselian. Since by proposition 23.1 c) $\{x_n\}$ is Besselian as well, from § 11, corollary 11.2 (implication $5^\circ\Rightarrow 2^\circ$) we infer that E is isomorphic to l^2. Consequently, E has a normalized conditional basis (§ 14, examples 14.4 and 14.5), in contradiction with the hypothesis that in E all normalized bases are equivalent. This completes the proof of theorem 23.1.

We shall also give two other proofs of theorem 23.1 in § 24, remarks 24.1 and 24.2.

As a second application of symmetric bases we shall prove now that the unconditionality of all bases characterizes finite-dimensional spaces among Banach spaces with bases.

Proposition 23.2. *Let $\{x_n\}$ be a normalized non-symmetric unconditional basis of a Banach space E. Then there exists a block perturbation of a suitable permutation of $\{x_n\}$, which is a conditional basis of E.*

Proof. We claim that there exists a permutation of the basic sequence $\{x_{2j}\}$ which is not equivalent to the basic sequence $\{x_{2j-1}\}$. Indeed, assume that all permutations of $\{x_{2j}\}$ are equivalent to $\{x_{2j-1}\}$. Then, by § 22, theorem 22.1 (implication $12^\circ \Rightarrow 1^\circ$), $\{x_{2j}\}$ is a symmetric, whence also subsymmetric, basic sequence and thus $\{x_{2j}\}$ is equivalent to its subsequences $\{x_{4j-2}\}$ and $\{x_{4j}\}$. We shall show that the mapping $x_{2j-1} \to x_{4j-2}$, $x_{2j} \to x_{4j}$ defines an equivalence of the basis $\{x_n\}$ with its subsequence $\{x_{2j}\}$, which is a contradiction since $\{x_n\}$ is non-symmetric. Indeed, since $\{x_n\}$ is unconditional, $\sum_{i=1}^{\infty} \alpha_i x_i$ converges if and only if both $\sum_{i=1}^{\infty} \alpha_{2i-1} x_{2i-1}$ and $\sum_{i=1}^{\infty} \alpha_{2i} x_{2i}$ converge. Since $\{x_{2j-1}\}$, $\{x_{2j}\}$ are equivalent to $\{x_{4j-2}\}$ and $\{x_{4j}\}$, respectively, this happens if and only if $\sum_{i=1}^{\infty} \alpha_{2i-1} x_{4i-2}$ and $\sum_{i=1}^{\infty} \alpha_{2i} x_{4i}$ are convergent, i. e., (since $\{x_n\}$ is unconditional) if and only if $\sum_{i=1}^{\infty} \alpha_i x_{2i}$ is convergent.

Thus, let $\{x_{\tau(2j)}\}$ be a permutation of $\{x_{2j}\}$ such that $\{x_{2j-1}\}$ and $\{x_{\tau(2j)}\}$ are not equivalent. Let $\{x_{\sigma(n)}\}$ be the permutation of $\{x_n\}$ defined by

$$x_{\sigma(n)} = \begin{cases} x_n & \text{for } n = 2j-1 \\ x_{\tau(n)} & \text{for } n = 2j \end{cases} \qquad (j = 1, 2, \ldots) \tag{23.7}$$

and let $\{z'_k\}$, $\{z''_k\}$ be the following two block perturbations[1] of the basis $\{x_{\sigma(n)}\}$:

$$z'_k = \begin{cases} x_{\sigma(k)} & \text{for } k = 2n-1 \\ x_{\sigma(k-1)} + x_{\sigma(k)} & \text{for } k = 2n \end{cases} \qquad (n = 1, 2, \ldots), \tag{23.8}$$

$$z''_k = \begin{cases} x_{\sigma(k)} + x_{\sigma(k+1)} & \text{for } k = 2n-1 \\ x_{\sigma(k)} & \text{for } k = 2n \end{cases} \qquad (n = 1, 2, \ldots). \tag{23.9}$$

[1] See Ch. I, § 4, definition 4.7.

By Ch. I, § 4, proposition 4.4, $\{z'_k\}$ and $\{z''_k\}$ are bases of the space E. We shall complete the proof by showing that at least one of these bases must be conditional.

Assume that both $\{z'_k\}$ and $\{z''_k\}$ are unconditional bases of E. Then, since $\{z'_k\}$ is unconditional, by § 17, theorem 17.1 there exists a constant $C' \geqslant 1$ such that we have, for any scalars $\alpha_1, \ldots, \alpha_n \in K$,

$$\left\| \sum_{i=1}^{n} \alpha_i x_{\sigma(2i)} \right\| = \left\| \sum_{i=1}^{n} \alpha_i x_{\sigma(2i-1)} - \sum_{i=1}^{n} \alpha_i (x_{\sigma(2i-1)} + x_{\sigma(2i)}) \right\|$$
$$= \left\| \sum_{i=1}^{n} \alpha_i z'_{2i-1} - \sum_{i=1}^{n} \alpha_i z'_{2i} \right\| \geqslant C' \left\| \sum_{i=1}^{n} \alpha_i z'_{2i-1} \right\|$$
$$= C' \left\| \sum_{i=1}^{n} \alpha_i x_{\sigma(2i-1)} \right\| .$$

Similarly, since $\{z''_k\}$ is unconditional, there exists a constant $C'' \geqslant 1$ such that we have, for any scalars $\alpha_1, \ldots, \alpha_n \in K$,

$$\left\| \sum_{i=1}^{n} \alpha_i x_{\sigma(2i-1)} \right\| \geqslant C'' \left\| \sum_{i=1}^{n} \alpha_i x_{\sigma(2i)} \right\| .$$

Hence, by Ch. I, § 8, theorem 8.1d) (implication $2^\circ \Rightarrow 6^\circ$), the basic sequences $\{x_{\sigma(2j-1)}\}$ and $\{x_{\sigma(2j)}\}$ are equivalent, which contradicts the construction (23.7) of the permutation $\{x_{\sigma(n)}\}$ and completes the proof of proposition 23.2.

Theorem 23.2. *In every infinite dimensional Banach space E with a basis there exists a conditional basis.*

Proof. Assume that all bases of E are unconditional and let $\{x_n\}$ be a normalized basis of E. Then, by virtue of theorem 23.1, there exists a normalized basis $\{y_{2j}\}$ of the subspace $[x_{2j}]$ which is not equivalent to $\{x_{2j}\}$. Since by § 16, theorem 16.8 (implication $1^\circ \Rightarrow 2^\circ$), $[x_{2j-1}] \oplus [x_{2j}] = E$, the sequence $\{z_n\} \subset E$ defined by

$$z_{2j-1} = x_{2j-1}, \; z_{2j} = y_{2j} \quad (j = 1, 2, \ldots) \tag{23.10}$$

is a normalized basis of E (by Ch. I, § 4, proposition 4.2 and lemma 4.1) and hence, by our hypothesis, $\{z_n\}$ is an unconditional basis of E. Consequently (since either $\{x_{2j-1}\} \nsim \{x_{2j}\}$ or $\{x_{2j-1}\} \sim \{x_{2j}\} \nsim \{y_{2j}\}$), either $\{x_n\}$ or $\{z_n\}$ is a normalized unconditional non-subsymmetric, whence non-symmetric, basis of E and hence, by proposition 23.2, the space E has a conditional basis, in contradiction with our hypothesis that in E all bases are unconditional. This completes the proof of theorem 23.2.

We shall give another proof of theorem 23.2 in § 24, remark 24.3.

Now we can also prove the existence of "many" mutually non-equivalent normalized conditional bases.

Theorem 23.3. *In every infinite dimensional Banach space E with a basis there exist a continuum of mutually non-equivalent normalized conditional bases.*

Proof. By theorem 23.2, there exists in E a normalized conditional basis $\{x_n\}$. Then, by § 16, theorem 16.1 (implication $8^\circ \Rightarrow 1^\circ$), there exist sequences of scalars $\{\alpha_n\}$, $\{\varepsilon_n\}$ with $\varepsilon_n = \pm 1$ $(n=1,2,\ldots)$ such that $\sum_{i=1}^{\infty} \alpha_i x_i$ is convergent but

$$\sup_{1 \leqslant n < \infty} \left\| \sum_{i=1}^{n} \varepsilon_i \alpha_i x_i \right\| = \infty. \tag{23.11}$$

Hence there exists an increasing sequence of positive integers $\{m_n\}$ with the following properties:

$$\left\| \sum_{i=m_{n-1}+1}^{l} \alpha_i x_i \right\| \leqslant \frac{1}{2^n} \quad (m_{n-1}+1 \leqslant l \leqslant m_n;\ n=2,3,\ldots) \tag{23.12}$$

$$\left\| \sum_{i=m_{n-1}+1}^{m_n} \varepsilon_i \alpha_i x_i \right\| \geqslant 1 \quad (n=1,2,\ldots;\ m_0=0). \tag{23.13}$$

Now, for each increasing infinite sequence of positive integers $\{p_i\}$ let us define a normalized conditional basis $\{y_j^{\{p_i\}}\}$ of E by

$$y_j^{\{p_i\}} = \begin{cases} \varepsilon_j x_j & \text{for} \quad m_{p_n-1}+1 \leqslant j \leqslant m_{p_n} \quad (n=1,2,\ldots), \\ x_j & \text{for the other } j. \end{cases} \tag{23.14}$$

We claim that for any increasing infinite $\{p_i'\}$ and $\{p_i''\}$ such that the set $(\{p_i'\} \setminus \{p_i''\}) \cup (\{p_i''\} \setminus \{p_i'\})$ is infinite[1], the bases $\{y_j^{\{p_i'\}}\}$ and $\{y_j^{\{p_i''\}}\}$ are not equivalent. Indeed, assume that $\{p_i'\} \setminus \{p_i''\} = \{p_{i_k}'\}$ is infinite (the treatment of the case when $\{p_i''\} \setminus \{p_i'\}$ is infinite is similar) and let

$$\beta_j = \begin{cases} \varepsilon_j \alpha_j & \text{for} \quad m_{p_{i_k}'-1}+1 \leqslant j \leqslant m_{p_{i_k}'}, \quad (k=1,2,\ldots), \\ 0 & \text{for the other } j. \end{cases} \tag{23.15}$$

Then, by (23.15) and (23.14) we have

$$\sum_{j=1}^{\infty} \beta_j y_j^{\{p_i'\}} = \sum_{j=1}^{\infty} \gamma_j x_j, \tag{23.16}$$

where

$$\gamma_j = \begin{cases} \alpha_j & \text{for} \quad m_{p_{i_k}'-1}+1 \leqslant j \leqslant m_{p_{i_k}'} \quad (k=1,2,\ldots), \\ 0 & \text{for the other } j \end{cases} \tag{23.17}$$

[1] Here the symbol $\{p_i\}$ denotes the set of all elements of the sequence $\{p_i\}$.

and, by (23.12), the series (23.16) is convergent. On the other hand, by (23.15), (23.14) and the definition of $\{p'_{i_k}\}$, we have

$$\sum_{j=1}^{\infty} \beta_j y_j^{\{p''_i\}} = \sum_{j=1}^{\infty} \delta_j x_j, \tag{23.18}$$

where

$$\delta_j = \begin{cases} \varepsilon_j \alpha_j & \text{for } \quad m_{p'_{i_k}-1}+1 \leqq j \leqq m_{p'_{i_k}} \qquad (k=1,2,\ldots), \\ 0 & \text{for the other } j, \end{cases} \tag{23.19}$$

whence, by (23.13) and since $\{p'_{i_k}\}$ is infinite, the series (23.18) is divergent. Thus the bases $\{y_j^{\{p'_i\}}\}$ and $\{y_j^{\{p''_i\}}\}$ are not equivalent. Since there exists a continuum of increasing sequences of positive integers $\{p_i\}$ such that for $\{p'_i\} \neq \{p''_i\}$ even both $\{p'_i\} \setminus \{p''_i\}$ and $\{p''_i\} \setminus \{p'_i\}$ are infinite[1], the corresponding collection of bases $\{y_j^{\{p_i\}}\}$ of E has the properties required in theorem 23.3, which completes the proof.

§ 24. Perfectly homogeneous bases. Application: Banach spaces with an unique normalized unconditional basis

Definition 24.1. A basis $\{x_n\}$ of a Banach space E is called *perfectly homogeneous* if it is a bounded basis and every bounded block basic sequence $\{y_n\}$ (with respect to $\{x_n\}$) is equivalent to $\{x_n\}$.

For instance, by § 18, proposition 18.1 and § 16, theorem 16.8, the unit vector basis in l^p $(1 \leqq p < \infty)$ is perfectly homogeneous and, by a similar argument, so is the unit vector basis in c_0. Now we shall prove that these are, up to an equivalence, the only perfectly homogeneous bases.

Proposition 24.1. *Every perfectly homogeneous basis $\{x_n\}$ of a Banach space E is an unconditional basis.*

Proof. Since for any sequence $\{\varepsilon_n\}$ with $\varepsilon_n = \pm 1$ $(n=1,2,\ldots)$ the sequence $\{\varepsilon_n x_n\}$ is a bounded block basis with respect to $\{x_n\}$, we have $\{\varepsilon_n x_n\} \sim \{x_n\}$ and hence, by § 16, theorem 16.8 (implication $3^\circ \Rightarrow 1^\circ$), $\{x_n\}$ is an unconditional basis of E.

---

[1] Indeed, let ϕ be a one-to-one mapping of $\mathcal{N} = \{1,2,3,\ldots\}$ onto the set of all rational numbers. Take, for each real number α a sequence of rational numbers $\{q_n^{(\alpha)}\}$ such that $\lim_{n\to\infty} q_n^{(\alpha)} = \alpha$ and let $p_n^{(\alpha)} = \phi^{-1}(q_n^{(\alpha)})$ $(n=1,2,\ldots)$. Then the collection of all sequences $\{p_i^{(\alpha)}\}$ (rearranged, if necessary, to form increasing sequences) has the required properties.

Proposition 24.2. *Let $\{x_n\}$ be a perfectly homogeneous orthogonal normalized basis of a Banach space E, let $\{\{y_n^d\}\}_{d\in\Delta}$ be the set of all normalized block basic sequences with respect to $\{x_n\}$, where Δ is the suitable index set, and for each $d\in\Delta$ let u_d be*[1] *the isomorphism of E onto $[y_n^d]$ satisfying*

$$u_d(x_n)=y_n^d \qquad (n=1,2,\ldots). \tag{24.1}$$

Then there exists a constant $M\geqslant 1$ such that

$$\|u_d\|\leqslant M, \ \|u_d^{-1}\|\leqslant M \qquad (d\in\Delta). \tag{24.2}$$

Proof. Assume first that $\sup_{d\in\Delta}\|u_d\|=\infty$. Then, by the principle of uniform boundedness, there exists an element $x=\sum_{i=1}^{\infty}\alpha_i x_i\in E$ with $\|x\|=1$, such that $\sup_{d\in\Delta}\|u_d(x)\|=\infty$ and hence there exists a sequence $\{d_n\}\subset\Delta$ such that

$$\left\|\sum_{i=1}^{\infty}\alpha_i y_i^{d_n}\right\|=\|u_{d_n}(x)\|\geqslant n+1 \qquad (n=1,2,\ldots). \tag{24.3}$$

We shall construct inductively two sequences of positive integers $\{p_k\}$, $\{q_k\}$, as follows: Take $p_1=1$ and take $q_1\geqslant p_1$ so that $\left\|\sum_{i=p_1}^{q_1}\alpha_i y_i^{d_{p_1}}\right\|\geqslant 1$. Assume that we have already constructed $p_1,\ldots,p_k$, $q_1,\ldots,q_k$ such that

$$\left\|\sum_{i=p_j}^{q_j}\alpha_i y_i^{d_{p_j}}\right\|\geqslant 1 \qquad (j=1,\ldots,k), \tag{24.4}$$

$$q_{j-1}<p_j\leqslant q_j \qquad (j=2,\ldots,k), \tag{24.5}$$

and that if M_j (respectively, N_j) is the least (respectively, the largest) index of the x_i's which appear in the representation of $y_{p_j}^{d_{p_j}}, y_{p_j+1}^{d_{p_j}},\ldots,y_{q_j}^{d_{p_j}}$, then

$$N_j<M_{j+1} \qquad (j=1,\ldots,k-1). \tag{24.6}$$

Put

$$p_{k+1}=\max(N_k,q_k)+1. \tag{24.7}$$

Since $\{x_n\}$ is a normalized orthogonal basis, we have $|\alpha_j|=\|\alpha_j x_j\|\leqslant\left\|\sum_{i=1}^{\infty}\alpha_i x_i\right\|=\|x\|=1\ (j=1,2,\ldots)$, whence, since $\|y_j^{d_{p_{k+1}}}\|=1\ (j=1,2,\ldots)$,

$$\left\|\sum_{j=1}^{p_{k+1}-1}\alpha_j y_j^{d_{p_{k+1}}}\right\|\leqslant p_{k+1}-1$$

---

[1] By Ch. I, § 8, theorem 8.1d) (implication 6°⇒1°).

and thus, taking into account (24.3),

$$\left\| \sum_{j=p_{k+1}}^{\infty} \alpha_j y_j^{d_{p_{k+1}}} \right\| \geqslant \left\| \sum_{j=1}^{\infty} \alpha_j y_j^{d_{p_{k+1}}} \right\| - \left\| \sum_{j=1}^{p_{k+1}-1} \alpha_j y_j^{d_{p_{k+1}}} \right\|$$
$$\geqslant p_{k+1} + 1 - p_{k+1} + 1 = 2.$$

Hence we can choose $q_{k+1} \geqslant p_{k+1}$ such that

$$\left\| \sum_{j=p_{k+1}}^{q_{k+1}} \alpha_j y_j^{d_{p_{k+1}}} \right\| \geqslant 1,$$

i.e., that we have (24.4) for $j=k+1$. Since by (24.7) $p_{k+1} > q_k$, we also have (24.5) for $j=k+1$. Finally, since the representation of each $y_j^{d_{p_{k+1}}}$ $(j=1,\dots,p_{k+1}-1)$ contains at least one x_i, we have $p_{k+1} \leqslant M_{k+1}$ and by (24.7) we have $N_k < p_{k+1}$, whence $N_k < M_{k+1}$, i.e., we also have (24.6) for $j=k$.

Now, by (24.6), the sequence $\bigcup_{k=1}^{\infty} \{y_i^{d_{p_k}}\}_{i=p_k}^{q_k}$ constitutes a normalized block basic sequence. By (24.4) the series $\sum_{k=1}^{\infty} \sum_{i=p_k}^{q_k} \alpha_i y_i^{d_{p_k}}$ does not converge, while the series $\sum_{k=1}^{\infty} \sum_{i=p_k}^{q_k} \alpha_i x_i$ converges, since $\{x_n\}$ is an unconditional basis. Thus the normalized block basic sequence $\bigcup_{k=1}^{\infty} \{x_i\}_{i=p_k}^{q_k}$ is not equivalent to the normalized block basic sequence $\bigcup_{k=1}^{\infty} \{y_i^{d_{p_k}}\}_{i=p_k}^{q_k}$, which contradicts the hypothesis that $\{x_n\}$ is a perfectly homogeneous basis.

Assume now that $\sup_{d \in \Delta} \|u_d^{-1}\| = \infty$. Then there exists a sequence $\{d_n\} \subset \Delta$ such that $\|u_{d_n}^{-1}\| \geqslant (n+1)2^n + 1$ $(n=1,2,\dots)$, whence also a sequence $z_n = \sum_{j=1}^{\infty} \alpha_j^{(n)} x_j \in E$ $(n=1,2,\dots)$ such that

$$\left\| \sum_{j=1}^{\infty} \alpha_j^{(n)} y_j^{d_n} \right\| = \|u_{d_n}(z_n)\| \leqslant \frac{1}{2^n}, \quad \left\| \sum_{j=1}^{\infty} \alpha_j^{(n)} x_j \right\| = \|z_n\| \geqslant n+1 \qquad (n=1,2,\dots). \tag{24.8}$$

We shall construct again sequences $\{p_j\}$, $\{q_j\}$ such that we have (24.5), (24.6) (for $j=1,2,\dots$ and $j=2,3,\dots$, respectively) and

$$\left\| \sum_{j=p_k}^{q_k} \alpha_j^{(p_k)} x_j \right\| \geqslant 1 \qquad (k=1,2,\dots). \tag{24.9}$$

Put $p_1 = 1$ and take $q_1 \geqslant p_1$ such that $\left\| \sum_{j=p_1}^{q_1} \alpha_j^{(p_1)} x_j \right\| \geqslant 1$. Assume that we have already constructed $p_1,\dots,p_k$, $q_1,\dots,q_k$. Define p_{k+1} by

(24.7). Since $\{x_n\}$ is an orthogonal basis, $\{y_j^{d_n}\}$ is an orthogonal basis of $[y_j^{d_n}]$ (by § 17, corollary 17.2), whence, since it is also normalized, we have $|\alpha_i^{(n)}| = \|\alpha_i^{(n)} y_i^{d_n}\| \leqslant \left\| \sum_{j=1}^{\infty} \alpha_j^{(n)} y_j^{d_n} \right\| \leqslant \frac{1}{2^n} < 1$ $(i, n = 1, 2, \ldots)$, and thus, since $\{x_n\}$ is normalized, $\left\| \sum_{j=1}^{p_{k+1}-1} \alpha_j^{(p_{k+1})} x_j \right\| \leqslant p_{k+1} - 1$. Therefore we have, taking again into account (24.8),

$$\left\| \sum_{j=p_{k+1}}^{\infty} \alpha_j^{(p_{k+1})} x_j \right\| \geqslant \left\| \sum_{j=1}^{\infty} \alpha_j^{(p_{k+1})} x_j \right\| - \left\| \sum_{j=1}^{p_{k+1}-1} \alpha_j^{(p_{k+1})} x_j \right\|$$

$$\geqslant p_{k+1} + 1 - p_{k+1} + 1 = 2$$

and hence we can choose $q_{k+1} \geqslant p_{k+1}$ such that

$$\left\| \sum_{j=p_{k+1}}^{q_{k+1}} \alpha_j^{(p_{k+1})} x_j \right\| \geqslant 1.$$

Now, since $\{y_j^{d_{p_k}}\}$ is an orthogonal basis of $[y_j^{d_{p_k}}]$, for any l_k with $p_k \leqslant l_k \leqslant q_k$ we have

$$\left\| \sum_{j=p_k}^{l_k} \alpha_j^{(p_k)} y_j^{d_{p_k}} \right\| \leqslant \left\| \sum_{j=1}^{\infty} \alpha_j^{(p_k)} y_j^{d_{p_k}} \right\| \leqslant \frac{1}{2^{p_k}} \qquad (k = 1, 2, \ldots), \tag{24.10}$$

and thus the series $\sum_{k=1}^{\infty} \sum_{j=p_k}^{q_k} \alpha_j^{(p_k)} y_j^{d_{p_k}}$ converges. On the other hand, by (24.9), the series $\sum_{k=1}^{\infty} \sum_{j=p_k}^{q_k} \alpha_j^{(p_k)} x_j$ does not converge, and thus the normalized block basic sequence $\bigcup_{k=1}^{\infty} \{y_j^{d_{p_k}}\}_{j=p_k}^{q_k}$ is not equivalent to the normalized block basic sequence $\bigcup_{k=1}^{\infty} \{x_j\}_{j=p_k}^{q_k}$, in contradiction with the hypothesis that $\{x_n\}$ is a perfectly homogeneous basis. This completes the proof of proposition 24.2.

Theorem 24.1. *A basis $\{x_n\}$ of a Banach space E is perfectly homogeneous if and only if it is equivalent to the unit vector basis of c_0 or l^p for some p with $1 \leqslant p < \infty$.*

Proof. The sufficiency part is obvious, since we have observed above that the unit vector bases in c_0 and l^p $(1 \leqslant p < \infty)$ are perfectly homogeneous.

Conversely, assume that $\{x_n\}$ is a perfectly homogeneous basis of a Banach space E. Then, since $\{x_n\}$ is a bounded unconditional basis, we may assume that it is hyperorthogonal and normalized (by considering, if necessary, a suitable equivalent norm on E; indeed, by Ch. I, § 3, theorem

3.2, one can take an equivalent norm $\|x\|'$ in which $\{x_n\}$ is normalized and then the equivalent norm

$$\left|\left|\left|\sum_{i=1}^{\infty} \alpha_i x_i\right|\right|\right| = \sup_{\substack{\{\beta_n\}\subset K \\ |\beta_n|\leqslant 1}} \sup_{1\leqslant n<\infty} \left\|\sum_{i=1}^{n} \beta_i \alpha_i x_i\right\|' \quad \left(\sum_{i=1}^{\infty} \alpha_i x_i \in E\right), \tag{24.11}$$

in which $|||x_n||| = \|x_n\|' = 1$ and $\{x_n\}$ is hyperorthogonal). Hence, by proposition 24.2, there exists an $M \geqslant 1$ such that for every normalized block basic sequence $\{z_n\}$ and all scalars $\alpha_1, \ldots, \alpha_n$

$$M\left\|\sum_{i=1}^{n} \alpha_i z_i\right\| \geqslant \left\|\sum_{i=1}^{n} \alpha_i x_i\right\| \geqslant \frac{1}{M}\left\|\sum_{i=1}^{n} \alpha_i z_i\right\|. \tag{24.12}$$

Put

$$\lambda_k = \left\|\sum_{i=1}^{k} x_i\right\| \quad (k=1,2,\ldots). \tag{24.13}$$

Then, since $\{x_n\}$ is a normalized monotone basis,

$$1=\lambda_1 \leqslant \lambda_2 \leqslant \lambda_3 \leqslant \cdots \tag{24.14}$$

Taking in (24.12) the normalized block basic sequence $z_j = x_{p_j}$ and the scalars $\alpha_j = 1$, it follows that for every increasing sequence $\{p_j\}_{j=1}^n$ of positive integers we have

$$\frac{1}{M} \leqslant \frac{\left\|\sum_{j=1}^{n} x_j\right\|}{\left\|\sum_{j=1}^{n} x_{p_j}\right\|} \leqslant M. \tag{24.15}$$

Taking in the right side of (24.12) the normalized block basic sequence $z_i = \dfrac{\sum_{j=1}^{n^{k-1}} x_{(i-1)n^{k-1}+j}}{\left\|\sum_{j=1}^{n^{k-1}} x_{(i-1)n^{k-1}+j}\right\|}$ and using the left side of (24.15) we obtain

$$M^2\left\|\sum_{i=1}^{n} x_i\right\| \geqslant \frac{M^2}{M}\left\|\sum_{i=1}^{n} \frac{\sum_{j=1}^{n^{k-1}} x_{(i-1)n^{k-1}+j}}{\left\|\sum_{j=1}^{n^{k-1}} x_{(i-1)n^{k-1}+j}\right\|}\right\|$$

$$\geqslant \frac{M^2}{M^2}\left\|\frac{\sum_{i=1}^{n}\sum_{j=1}^{n^{k-1}} x_{(i-1)n^{k-1}+j}}{\left\|\sum_{j=1}^{n^{k-1}} x_j\right\|}\right\| = \frac{\left\|\sum_{j=1}^{n^k} x_j\right\|}{\left\|\sum_{j=1}^{n^{k-1}} x_j\right\|}. \tag{24.16}$$

Hence, by induction,

$$M^{2k}\left\|\sum_{i=1}^{n} x_i\right\|^k \geqslant \left\|\sum_{j=1}^{n^k} x_j\right\| \qquad (n,k=1,2,\ldots); \tag{24.17}$$

indeed, for $k=1$ this is trivial, and if $M^{2k-2}\left\|\sum_{i=1}^{n} x_i\right\|^{k-1} \geqslant \left\|\sum_{j=1}^{n^{k-1}} x_j\right\|$, then, multiplying by the respective sides of (24.16), we obtain (24.17).

On the other hand, by the left side of (24.12) and the right side of (24.15), we have

$$\left\|\sum_{i=1}^{n} x_i\right\| \leqslant M \left\|\sum_{i=1}^{n} \frac{\sum_{j=1}^{n^{k-1}} x_{(i-1)n^{k-1}+j}}{\left\|\sum_{j=1}^{n^{k-1}} x_{(i-1)n^{k-1}+j}\right\|}\right\| \leqslant M^2 \frac{\left\|\sum_{j=1}^{n^k} x_i\right\|}{\left\|\sum_{j=1}^{n^{k-1}} x_j\right\|},$$

whence, again by induction,

$$\left\|\sum_{i=1}^{n} x_i\right\|^k \leqslant M^{2k}\left\|\sum_{i=1}^{n^k} x_i\right\| \qquad (n,k=1,2,\ldots). \tag{24.18}$$

By (24.13), we can write (24.17) and (24.18) in the form

$$\frac{1}{M^{2k}}\lambda_{n^k} \leqslant \lambda_n^k \leqslant M^{2k}\lambda_{n^k} \qquad (n,k=1,2,\ldots). \tag{24.19}$$

For any positive integers N,n and k let $h=h(N,n,k)$ be the non-negative integer for which

$$N^h \leqslant n^k < N^{h+1}. \tag{24.20}$$

Then, by (24.19), (24.20), (24.14) and again (24.19), we have

$$\begin{aligned} h\log\lambda_N &\leqslant \log(M^{2h}\lambda_{N^h}) = 2h\log M + \log\lambda_{N^h} \\ &\leqslant 2h\log M + \log\lambda_{n^k} \leqslant 2h\log M + \log(M^{2k}\lambda_n^k) \\ &= 2h\log M + 2k\log M + k\log\lambda_n. \end{aligned}$$

Since by (24.20) $h \leqslant k\dfrac{\log n}{\log N} < h+1$, if follows that we have

$$\left[k\frac{\log n}{\log N} - 1\right]\log\lambda_N < h\log\lambda_N \leqslant 2k\frac{\log n}{\log N}\log M + 2k\log M + k\log\lambda_n,$$

whence, dividing by $k\log n$,

$$\left[\frac{1}{\log N} - \frac{1}{k\log n}\right]\log\lambda_N < \frac{2\log M}{\log N} + \frac{2\log M}{\log n} + \frac{\log\lambda_n}{\log n}$$

and hence, for $k \to \infty$,

$$\frac{\log \lambda_N}{\log N} \leqslant 2 \log M \left(\frac{1}{\log N} + \frac{1}{\log n} \right) + \frac{\log \lambda_n}{\log n}. \tag{24.21}$$

Interchanging the roles of n and N, we obtain

$$\frac{\log \lambda_n}{\log n} \leqslant 2 \log M \left(\frac{1}{\log N} + \frac{1}{\log n} \right) + \frac{\log \lambda_N}{\log N}, \tag{24.22}$$

which, together with (24.21), yields

$$\left| \frac{\log \lambda_n}{\log n} - \frac{\log \lambda_N}{\log N} \right| \leqslant 2 \log M \left(\frac{1}{\log N} + \frac{1}{\log n} \right).$$

Consequently, the sequence $\left\{ \frac{\log \lambda_n}{\log n} \right\}$ converges to a limit c. Since $1 \leqslant \lambda_n \leqslant n$, we have $0 \leqslant \log \lambda_n \leqslant \log n$, and thus $0 \leqslant c \leqslant 1$. Taking $N \to \infty$ in (24.21), we obtain

$$c \log n \leqslant 2 \log M + \log \lambda_n = \log(M^2 \lambda_n)$$

and similarly, from (24.22) for $N \to \infty$ we obtain

$$\log \frac{\lambda_n}{M^2} \leqslant c \log n,$$

whence

$$\frac{n^c}{M^2} \leqslant \lambda_n \leqslant M^2 n^c \qquad (n = 1, 2, \ldots). \tag{24.23}$$

Consequently, if $c = 0$, we have $\lambda_n \leqslant M^2$ $(n = 1, 2, \ldots)$, i.e., $\{x_n\}$ is an unconditional basis of type P and hence, by § 17, corollary 17.1, $\{x_n\}$ is equivalent to the unit vector basis of c_0.

Assume now that $0 < c \leqslant 1$ and put $p = \frac{1}{c}$. Then we can write (24.23) in the form

$$\frac{n^{\frac{1}{p}}}{M^2} \leqslant \left\| \sum_{i=1}^{n} x_i \right\| \leqslant M^2 n^{\frac{1}{p}} \qquad (n = 1, 2, \ldots). \tag{24.24}$$

Let $r_1, \ldots, r_n$ be arbitrary positive rational numbers, say $r_i = \frac{k_i}{m}$ $(i = 1, \ldots, n)$, where k_i, m are positive integers and let $k = \sum_{i=1}^{n} k_i$. Then, using that $\{x_n\}$ is a hyperorthogonal basis, the right side of (24.24) for

$n=k_i$, (24.12) for the normalized block basic sequence $z_i = \dfrac{\sum_{j=1}^{k_i} x_{\sum_{m=1}^{i-1} k_m+j}}{\left\| \sum_{j=1}^{k_i} x_{\sum_{m=1}^{i-1} k_m+j} \right\|}$ and for the scalars $\alpha_i = \left\| \sum_{j=1}^{k_i} x_j \right\|$, the left side of (24.15) and the left side of (24.24) for n replaced by $k=\sum_{i=1}^{n} k_i$, we get

$$\left\| \sum_{i=1}^{n} r_i^{\frac{1}{p}} x_i \right\| = \frac{1}{m^{\frac{1}{p}}} \left\| \sum_{i=1}^{n} k_i^{\frac{1}{p}} x_i \right\| \geqslant \frac{1}{M^2 m^{\frac{1}{p}}} \left\| \sum_{i=1}^{n} \left\| \sum_{j=1}^{k_i} x_j \right\| x_i \right\|$$

$$\geqslant \frac{1}{M^3 m^{\frac{1}{p}}} \left\| \sum_{i=1}^{n} \left\| \sum_{j=1}^{k_i} x_j \right\| \frac{\sum_{j=1}^{k_i} x_{\sum_{m=1}^{i-1} k_m+j}}{\left\| \sum_{j=1}^{k_i} x_{\sum_{m=1}^{i-1} k_m+j} \right\|} \right\|$$

$$\geqslant \frac{1}{M^4 m^{\frac{1}{p}}} \left\| \sum_{i=1}^{n} \sum_{j=1}^{k_i} x_{\sum_{m=1}^{i-1} k_m+j} \right\| = \frac{1}{M^4 m^{\frac{1}{p}}} \left\| \sum_{j=1}^{k} x_j \right\|$$

$$\geqslant \frac{1}{M^6 m^{\frac{1}{p}}} k^{\frac{1}{p}} = \frac{1}{M^6 m^{\frac{1}{p}}} \left(\sum_{i=1}^{n} k_i \right)^{\frac{1}{p}} = \frac{1}{M^6} \left(\sum_{i=1}^{n} r_i \right)^{\frac{1}{p}}.$$

Hence, taking into account the hyperorthogonality of the basis $\{x_n\}$, it follows that for any scalars $\alpha_1, \ldots, \alpha_n$ we have

$$\left\| \sum_{i=1}^{n} \alpha_i x_i \right\| = \left\| \sum_{i=1}^{n} |\alpha_i| x_i \right\| \geqslant \frac{1}{M^6} \left(\sum_{i=1}^{n} |\alpha_i|^p \right)^{\frac{1}{p}}. \tag{24.25}$$

Similar arguments yield

$$\left\| \sum_{i=1}^{n} r_i^{\frac{1}{p}} x_i \right\| = \frac{1}{m^{\frac{1}{p}}} \left\| \sum_{i=1}^{n} k_i^{\frac{1}{p}} x_i \right\| \leqslant \frac{M^2}{m^{\frac{1}{p}}} \left\| \sum_{i=1}^{n} \left\| \sum_{j=1}^{k_i} x_j \right\| x_i \right\|$$

$$\leqslant \frac{M^3}{m^{\frac{1}{p}}} \left\| \sum_{i=1}^{n} \left\| \sum_{j=1}^{k_i} x_j \right\| \frac{\sum_{j=1}^{k_i} x_{\sum_{m=1}^{i-1} k_m+j}}{\left\| \sum_{j=1}^{k_i} x_{\sum_{m=1}^{i-1} k_m+j} \right\|} \right\|$$

$$\leqslant \frac{M^4}{m^{\frac{1}{p}}} \left\| \sum_{j=1}^{k} x_j \right\| \leqslant \frac{M^6}{m^{\frac{1}{p}}} \left(\sum_{i=1}^{n} k_i \right)^{\frac{1}{p}} = M^6 \left(\sum_{i=1}^{n} r_i \right)^{\frac{1}{p}},$$

whence, by the hyperorthogonality of $\{x_n\}$, for any scalars $\alpha_1,\dots,\alpha_n$ we obtain

$$\left\|\sum_{i=1}^{n} \alpha_i x_i\right\| \leqslant M^6 \left(\sum_{i=1}^{n} |\alpha_i|^p\right)^{\frac{1}{p}}. \tag{24.26}$$

Thus, by (24.25) and (24.26), $\{x_n\}$ is equivalent to the unit vector basis of l^p, which completes the proof of theorem 24.1.

Remark 24.1. With the aid of theorem 24.1, one can give the following proof of § 23, theorem 23.1: Let E be an infinite dimensional Banach space with a basis, such that all normalized bases of E are equivalent. Then, by § 23, proposition 23.1b), every such basis is perfectly homogeneous. Consequently, by theorem 24.1 above, E is isomorphic to one of the spaces c_0 or l^p $(p \geqslant 1)$, in contradiction with the hypothesis that in E all normalized bases are equivalent. This completes the proof.

Now we shall give a sharpening of theorem 24.1, which will be useful in determining the Banach spaces which have, up to equivalence, a unique normalized unconditional basis. For this purpose, let us first observe the following sharpening of proposition 24.2:

Proposition 24.3. *Let $\{x_n\}$ be a subsymmetric orthogonal normalized basis of a Banach space E, with the property that for every increasing sequence of positive integers $\{m_n\}$ the basis $\{x_n\}$ is equivalent to the normalized block basic sequence*

$$y_n = \frac{\sum_{i=m_{n-1}+1}^{m_n} x_i}{\left\|\sum_{i=m_{n-1}+1}^{m_n} x_i\right\|} \qquad (n=1,2,\dots;\ m_0=0), \tag{24.27}$$

let $\{\{y_n^d\}\}_{d\in\Delta}$ be the set of all normalized block basic sequences of the form (24.27), *where Δ is the suitable index set, and for each $d\in\Delta$ let u_d be the isomorphism of E onto $[y_n^d]$ satisfying* (24.1). *Then there exists a constant $M \geqslant 1$ such that we have* (24.2).

Proof. Since $\{x_n\}\sim\{y_n\}$ and $\{x_n\}$ is subsymmetric, it follows that $\{y_n\}$ is subsymmetric and thus for any two increasing sequences of positive integers $\{n_i\}$, $\{m_i\}$ we have $\{x_{n_i}\}\sim\{x_n\}\sim\{y_n\}\sim\{y_{m_i}\}$, whence $\{x_{n_i}\}\sim\{y_{m_i}\}$.

Now, assuming that $\sup_{d\in\Delta}\|u_d\|=\infty$, it follows, as in the above proof of proposition 24.2, that there exists a normalized block basic sequence of the form $\bigcup_{k=1}^{\infty}\{y_i^{d_{p_k}}\}_{i=p_k}^{q_k}$ (where each $\{y_n^d\}$ is of the form (24.27)), which is not equivalent to $\bigcup_{k=1}^{\infty}\{x_i\}_{i=p_k}^{q_k}$, in contradiction with the above (since

$\bigcup_{k=1}^{\infty} \{y_i^{d_{p_k}}\}_{i=p_k}^{q_k}$ can be extended to a sequence $\{y_n\}$ of the form (24.27)). Similarly, assuming that $\sup_{d \in \Delta} \|u_d^{-1}\| = \infty$, we obtain again a contradiction with the above, which completes the proof.

Now we can give

Theorem 24.2. *Let $\{x_n\}$ be a subsymmetric basis of a Banach space E, with the property that for every increasing sequence of integers $\{m_n\}_{n=0}^{\infty}$ with $m_0 = 0$ the basis $\{x_n\}$ is equivalent to the normalized block basic sequence* (24.27). *Then $\{x_n\}$ is equivalent to the unit vector basis of c_0 or l^p for some p with $1 \leq p < \infty$.*

Proof. Observe that for any equivalent norm on E, the basis $\{x_n\}$ is still equivalent to (24.27), by § 16, theorem 16.8 (implication $1^\circ \Rightarrow 4^\circ$) and that in the above proof of theorem 24.1 only subsequences of $\{x_n\}$ and normalized block basic sequences of the form (24.27) have been used, for a suitable equivalent norm. Thus, applying, in that argument, proposition 24.3 and § 21, theorem 21.2 (implication $1^\circ \Rightarrow 2^\circ$) instead of proposition 24.2, the desired conclusion follows.

Proposition 24.3 and theorem 24.2 are sharpenings of proposition 24.2 and theorem 24.1, respectively, since every perfectly homogeneous basis $\{x_n\}$ is subsymmetric (it is unconditional by proposition 24.1 and equivalent to any subsequence because any subsequence is a bounded block basic sequence) and equivalent to any normalized block basic sequence of the form (24.27) and since à priori we don't know whether the converse is also true; however, it turns out that this is indeed the case, namely, every such basis is already perfectly homogeneous (by theorem 24.2).

As an application of the above we shall prove now that up to an isomorphism, the only Banach spaces with an unconditional basis in which all normalized unconditional bases are equivalent are those of § 18, theorems 18.2 and 18.1, i.e., c_0, l^1 and l^2. If the answer to the first or second question of § 22, problem 22.2, were affirmative, then the desired result would follow easily, with the same argument as that used in remark 24.1 above, taking into account also § 18, theorem 18.3. We shall obtain the same result by using theorem 24.2 above and § 22, proposition 22.4.

Theorem 24.3. *The only infinite dimensional Banach spaces with an unconditional basis, in which all normalized unconditional bases are equivalent, are (up to an isomorphism) c_0, l^1 and l^2.*

Proof. Let E be an infinite dimensional Banach space with an unconditional basis, in which all normalized unconditional bases are equivalent, and let $\{x_n\}$ be an arbitrary normalized unconditional basis of E. Then, by § 17, theorem 17.1 (implication $1^\circ \Rightarrow 2^\circ$), every permutation $\{x_{\sigma(n)}\}$ of $\{x_n\}$ is a normalized unconditional basis of E. Since by our hypothesis the bases $\{x_n\}$ and $\{x_{\sigma(n)}\}$ must be equivalent, it follows by § 22, theorem 22.1 (implication $12^\circ \Rightarrow 1^\circ$) that $\{x_n\}$ is a symmetric basis of E. Consequently, by § 22, proposition 22.1 b), we may assume that $(E, \{x_n\})$ is a symmetric space. Let $\{m_n\}_{n=0}^{\infty}$ be an arbitrary increasing sequence of integers with $m_0=0$, and let

$$y_n = \frac{\sum_{i=m_{n-1}+1}^{m_n} x_i}{\left\| \sum_{i=m_{n-1}+1}^{m_n} x_i \right\|} \qquad (n=1,2,\ldots;\ m_0=0). \tag{24.28}$$

We shall show that $\{x_n\}$ is equivalent to the normalized block basic sequence $\{y_n\}$, which, by theorem 24.2 above and § 18, theorem 18.3, will complete the proof.

Let

$$p_n = m_n - m_{n-1} \qquad (n=1,2,\ldots). \tag{24.29}$$

Observe that if $p_n=1$ for an index n, then $y_n = x_{m_n}$. Hence, if $p_n>1$ only for a finite number of indices n, say $n_1,\ldots,n_k$, then $\{y_j\}_{j\in\mathcal{N}\setminus\{n_1,\ldots,n_k\}}$ is a subsequence of $\{x_n\}$, and thus, since $\{x_n\}$ is symmetric, $\{y_j\}_{j\in\mathcal{N}\setminus\{n_1,\ldots,n_k\}}$ is equivalent to $\{x_j\}_{j=k+1}^{\infty}$, whence the basic sequence $\{y_n\}$ is equivalent to $\{x_n\}$.

Assume now that there are infinitely many indices n with $p_n>1$, say $n_1,n_2,\ldots$ We claim that in this case we may assume

$$p_n>1 \qquad (n=1,2,\ldots). \tag{24.30}$$

Indeed, putting

$$q_0=0, \qquad q_k = \sum_{j=1}^{k} p_{n_j} \qquad (k=1,2,\ldots), \tag{24.31}$$

$$z_k = \frac{\sum_{i=q_{k-1}+1}^{q_k} x_i}{\left\| \sum_{i=q_{k-1}+1}^{q_k} x_i \right\|} \qquad (k=1,2,\ldots), \tag{24.32}$$

we obtain a normalized block basic sequence $\{z_k\}$ with $q_k - q_{k-1} = p_{n_k} > 1$ $(k=1,2,\ldots)$ and, since $(E,\{x_n\})$ is a symmetric space, $\{z_k\}$ is equivalent to the normalized block basic sequence $\{y_{n_k}\}$. Now, if $\{x_n\}$ is equivalent to $\{z_k\}$ and if the set $\mathcal{N}\setminus\{n_1,n_2,\ldots\}$ has a finite number of elements, say l, then, since $\{x_n\}$ is symmetric, $\{x_j\}_{j=l+1}^{\infty} \sim \{x_n\} \sim \{z_k\} \sim \{y_{n_k}\}$, whence the basic sequence $\{y_n\}$ is equivalent to $\{x_n\}$. On the other hand, if $\{x_n\}$ is equivalent to $\{z_k\}$ and the set $\mathcal{N}\setminus\{n_1,n_2,\ldots\}$ is infinite, then $\{x_n\}\sim\{z_k\}\sim\{y_{n_k}\}$ and, since $\{x_n\}$ is symmetric, it is also equivalent to its infinite subsequence $\{y_j\}_{j\in\mathcal{N}\setminus\{n_1,n_2,\ldots\}}$, whence, again since $\{x_n\}$ is symmetric, it follows[1] that $\{x_n\}$ is equivalent to $\{y_n\}$. This proves the claim that we may assume (24.30).

Now let $\{n_{k,j}\}_{k,j=1}^{\infty}$ be an enumeration of the set $\mathcal{N}=\{1,2,3,\ldots\}$ as a double sequence. For this double sequence and for $\{p_n\}$ defined by (24.29) and satisfying (24.30), construct $y_{k,i}$ as in § 22, proposition 22.4. Then, by that proposition, $\{y_{k,1}\}_{k=1}^{\infty}\cup\{x_{n_{k,1}}\}_{k=1}^{\infty}\cup\{y_{k,i}\}_{k=1,i=2}^{\infty}$ is an unconditional basis of E. Since by our hypothesis all normalized unconditional bases of E are equivalent, after normalization this basis becomes equivalent to $\{x_n\}$, whence, by the symmetry of $\{x_n\}$ we get that $\{x_n\}$ is equivalent to $\left\{\frac{y_{k,1}}{\|y_{k,1}\|}\right\}_{k=1}^{\infty}$. Since by the symmetry of $\{x_n\}$ we also have the equivalence $\left\{\frac{y_{k,1}}{\|y_{k,1}\|}\right\}_{k=1}^{\infty}\sim\{y_n\}$ (observe that both $y_{k,1}$ and y_k have $p_k=m_k-m_{k-1}$ summands), it follows that $\{x_n\}\sim\{y_n\}$, which completes the proof of theorem 24.3.

Remark 24.2. From theorem 24.3 there follows again § 23, theorem 23.1 on the existence of non-equivalent normalized bases in infinite dimensional Banach spaces with bases. Indeed, if in E with $\dim E=\infty$ all normalized bases are equivalent, then, as we have seen in § 23, proposition 23.1 a), they are all unconditional, whence, by theorem 24.3 above, E must be isomorphic to c_0, l^1 or l^2, but in each of these spaces we know examples of conditional bases. This contradiction completes the proof.

Remark 24.3. The above proof of theorem 24.3 shows also that *if a Banach space E has an unconditional basis and every normalized unconditional basis of E is symmetric, then E is isomorphic to c_0, l^1 or l^2* (since $\{x_n\}\sim\{x_{n_{k,1}}\}_{k=1}^{\infty}\sim\left\{\frac{y_{k,1}}{\|y_{k,1}\|}\right\}_{k=1}^{\infty}$ by the symmetry of $\{x_n\}$ and

---

[1] Indeed, let $\mathcal{N}_1=\{l_i\}\subset\mathcal{N}$, $\mathcal{N}_2=\{r_i\}\subset\mathcal{N}$ be any pair of infinite sequences of indices such that $\mathcal{N}_1\cup\mathcal{N}_2=\mathcal{N}$, $\mathcal{N}_1\cap\mathcal{N}_2=\emptyset$. Then, $\{x_n\}$ being unconditional, $\sum_{i=1}^{\infty}\alpha_i x_i$ converges if and only if both $\sum_{i=1}^{\infty}\alpha_{l_i}x_{l_i}$ and $\sum_{i=1}^{\infty}\alpha_{r_i}x_{r_i}$ converge. Since $\{x_n\}$ is symmetric, this happens if and only if $\sum_{i=1}^{\infty}\alpha_{l_i}x_i$ and $\sum_{i=1}^{\infty}\alpha_{r_i}x_i$ converge. Since $\{x_n\}\sim\{y_{l_i}\}$, $\{x_n\}\sim\{y_{r_i}\}$, this happens if and only if both $\sum_{i=1}^{\infty}\alpha_{l_i}y_{l_i}$ and $\sum_{i=1}^{\infty}\alpha_{r_i}y_{r_i}$ converge, which (since $\{y_n\}$ is an unconditional basic sequence) happens if and only if $\sum_{i=1}^{\infty}\alpha_i y_i$ converges.

$\left\{\frac{y_{k,1}}{\|y_{k,1}\|}\right\}_{k=1}^{\infty} \cup \{x_{n_{k,1}}\}_{k=1}^{\infty} \cup \left\{\frac{y_{k,i}}{\|y_{k,i}\|}\right\}_{k=1,i=2}^{\infty}$ respectively). In other words, the only spaces with an unconditional basis, which do not satisfy the hypothesis of § 23, proposition 23.2, are (up to an isomorphism) c_0, l^1 and l^2. From this remark we obtain again § 23, theorem 23.2.

§ 25. Absolutely convergent bases. Uniform bases

One can obtain some new classes of bases by requiring the convergence of all expansions $\sum_{i=1}^{\infty} f_i(x)x_i$ in a stronger sense, e.g. this idea yields the very useful notion of unconditional bases. However, a further strengthening of the mode of convergence of all expansions $\sum_{i=1}^{\infty} f_i(x)x_i$ leads, in general, to very restrictive classes of bases. In the present section we shall consider two such instances.

Definition 25.1. A basis $\{x_n\}$ of a Banach space E is said to be an *absolutely convergent basis* if every convergent expansion $\sum_{i=1}^{\infty} \alpha_i x_i$ is absolutely convergent, i.e., $\sum_{i=1}^{\infty} \|\alpha_i x_i\| < \infty$.

Obviously, this amounts to

$$\sum_{i=1}^{\infty} \| f_i(x)x_i \| < \infty \qquad (x \in E), \tag{25.1}$$

where $\{f_n\} \subset E^*$ is the a.s.c.f. to the basis $\{x_n\}$.

For instance, every basis in a finite dimensional space E and the unit vector basis of $E = l^1$, are absolutely convergent bases. However, the converse statement is also true, i.e., these are, up to a normalization and an equivalence, the only absolutely convergent bases, as shown by

Theorem 25.1. *Let $\{x_n\}$ be an absolutely convergent basis of an infinite dimensional Banach space E. Then $\left\{\frac{x_n}{\|x_n\|}\right\}$ is equivalent to the unit vector basis of l^1.*

Proof. If $\sum_{i=1}^{\infty} \alpha_i \frac{x_i}{\|x_i\|} = \sum_{i=1}^{\infty} \frac{\alpha_i}{\|x_i\|} x_i$ is convergent, then, since $\{x_n\}$ is an absolutely convergent basis, we have $\sum_{i=1}^{\infty} |\alpha_i| = \sum_{i=1}^{\infty} \left\| \frac{\alpha_i}{\|x_i\|} x_i \right\| < \infty$.

Conversely, if $\sum_{i=1}^{\infty} |\alpha_i| < \infty$, then from

$$\left\| \sum_{i=n}^{n+p} \alpha_i \frac{x_i}{\|x_i\|} \right\| \leqslant \sum_{i=n}^{n+p} |\alpha_i|$$

and from the completeness of E it follows that $\sum_{i=1}^{\infty} \alpha_i \frac{x_i}{\|x_i\|}$ converges, which completes the proof.

Definition 25.2. A basis $\{x_n\}$ of a Banach space E, with the a.s.c.f. $\{f_n\}$, is said to be a *uniform basis*, if the series $\sum_{i=1}^{\infty} f_i(x) x_i$ converge uniformly in the unit cell $S_E = \{x \in E \mid \|x\| \leqslant 1\}$, i.e.,

$$\sup_{\substack{x \in E \\ \|x\| \leqslant 1}} \left\| \sum_{i=n+1}^{\infty} f_i(x) x_i \right\| = \sup_{\substack{x \in E \\ \|x\| \leqslant 1}} \left\| x - \sum_{i=1}^{n} f_i(x) x_i \right\| \to 0 \quad \text{as} \quad n \to \infty. \tag{25.2}$$

For instance, every basis in a finite dimensional space E is a uniform basis. However, the converse statement is also true, i.e., these are the only uniform bases, as shown by

Theorem 25.2. *Let E be a Banach space with a uniform basis $\{x_n\}$. Then* $\dim E < \infty$.

Proof. By (25.2) we have

$$\|I_E - s_n\| \to 0 \quad \text{as} \quad n \to \infty, \tag{25.3}$$

where I_E is the identical mapping of E onto itself and s_n is the n-th partial sum operator associated to the basis $\{x_n\}$. Since by $\dim s_n(E) = \dim[x_1, \ldots, x_n] = n < \infty$ each s_n is compact, from (25.3) it follows that I_E is compact, whence the unit cell $S_E = I_E(S_E)$ of E is compact. Consequently, by a classical theorem of F. Riesz[1], $\dim E < \infty$, which completes the proof.

Notes and remarks

Part I. § 1. Bases satisfying (1.1), respectively bases with $v_{\{x_n\}} = 1$, have been considered, independently, by R. C. James [114] and V. Ya. Kozlov [135] and called by them "orthogonal bases"; this term is still used by some authors, e.g. V. I. Gurarii. The term "monotone basis" was suggested by M. M. Day [43]. Strictly monotone bases were introduced in [141], definition 2.1 and proposition 1.1 was given in the same paper ([141], proposition 2.1).

[1] See e.g. [10], p. 84, theorem 8.

Theorem 1.1 and example 1.1 are due to M. Z. Solomiak [253]. As we already observed in § 1, H. F. Bohnenblust [29] proved that for every integer $n \geqslant 3$ there exists an n-dimensional Banach space E_n satisfying (1.6), namely, a suitable subspace of a space l_k^p with $k > 2(2n-3)$ and $p \neq$ integer. The construction culminating in theorem 1.2 is due, essentially, to V. I. Gurariĭ [92], [96] to whom we are indebted for sending us a part of [92]; we slightly modified here the arguments of [92], following L. S. Pontryagin [199], Ch. I. For the definition and properties of the opening $\theta(G_1, G_2)$ (formula (1.15)) see M. G. Krein, M. A. Krasnoselskiĭ and D. P. Milman [141]. The modified "opening" $\tilde{\theta}(G_1, G_2)$ (formula (1.17)) was introduced by I. C. Gohberg and A. S. Markus [80]. For lemma 1.10 see I. C. Gohberg and M. G. Krein [77], theorem 6.2.

Lemma 1.12 and theorem 1.3 were given by V. I. Gurariĭ [95].

Lemmas 1.13 – 1.15 and theorem 1.4 are due to V. I. Gurariĭ [94], [97] and so are lemma 1.16, proposition 1.2 (verbal communication) and corollary 1.1 (see [97a] for the first part and [94], [97] for the second part of corollary 1.1).

Problem 1.1 was raised by M. M. Grinblium [82].

§ 2. Normal bases were already considered by S. Banach ([10], p. 238), but the term "normal basis" has been introduced by S. Karlin [131].

Lemma 2.1 is due to G. Ascoli [6].

In the particular case of finite dimensional Banach spaces, the equivalence $1° \Leftrightarrow 2°$ of theorem 2.1 was observed by M. M. Day [42], the equivalence $1° \Leftrightarrow 3°$ by A. E. Taylor [255] and the implication $1° \Rightarrow 4°$ by A. Yu. Levin and Yu. I. Petunin [146], who have also observed the equivalence $1° \Leftrightarrow 2°$ in the infinite dimensional case.

Proposition 2.1 has been given, with a different proof, by S. Karlin ([131], theorem 7).

Theorem 2.2 for real Banach spaces was stated, without proof, in the monograph of S. Banach ([10], p. 238), with the mention that it is due to H. Auerbach; again for real Banach spaces, it has been rediscovered independently, with different proofs, by M. M. Day [42] and A. E. Taylor ([255], theorem 2). The proof given here can be found in A. F. Timan ([256], p. 407) with the mention that it is due to M. I. Kadec; later, the same proof was also given by A. F. Ruston [218].

Theorem 2.3 is due to C. Bessaga (verbal communication). Corresponding to § 1, where it was observed (after proposition 1.3) that every basis of a Banach space E can be "monotonized", one could express theorem 2.3 by saying that every bounded basis of a Banach space E can be "normalized"; however, this might be misleading, since the term "normalized basis" is used in a different sense (Ch. I, § 3, definition 3.2).

§ 3. The notion of positive basis was introduced by W. J. Davis [37]. Lemma 3.1, proposition 3.1, lemma 3.2, lemma 3.3, remark 3.2, proposition 3.2, lemma 3.4 and proposition 3.3 are given in the same paper [37]. The example of an n-dimensional Banach space with only two linear isometries, given before problem 3.1, was communicated to us by C. Kottman. Problem 3.2 was raised by W. J. Davis [37].

§ 4. Bases $\{x_n\}$ such that $[f_n]=E^*$ (where $\{f_n\}$ is the a.s.c.f. to $\{x_n\}$) have been considered by M. M. Grinblium and L. A. Gurevich [87] and S. Karlin [131] who have proved the result given in Ch. I, § 12, corollary 12.2. Bases with the property that $\lim_{n\to\infty}\|f\|_n=0$ for all $f\in E^*$, where $\|f\|_n$ is defined by (4.3), were considered by R. C. James [113]. The term "shrinking basis" was suggested by M. M. Day [43]. For $k>0$, the k-shrinking bases were introduced in [244].

Lemma 4.1 was given in [244], lemma 1. Proposition 4.1 was proved in [237]; however, later V. F. Gaposhkin [66] observed that a more general result had been given by S. M. Nikolskiĭ ([178], theorem 1). Proposition 4.2, except its last equality, and theorem 4.1, except its equivalence $1°\Leftrightarrow 5°$, were given in [244], lemma 2 and theorem 2. The equivalences $1°\Leftrightarrow 2°\Leftrightarrow 5°\Leftrightarrow 6°$ of theorem 4.2 were proved by R. C. James ([113], theorem 3) and the equivalences $1°\Leftrightarrow 3°\Leftrightarrow 4°$ by A. Wilansky ([262], theorems 1 and 2). For monotone bases the implication $1°\Rightarrow 7°$ of theorem 4.2 has been given by R. C. James [113] and in the general case in [239], corollary 3. The implication $1°\Rightarrow$ the first part of $8°$ is due to B. R. Gelbaum ([69], theorem 6). The equivalences $1°\Leftrightarrow 7°\Leftrightarrow 8°$ were observed by J. R. Retherford [205].

Example 4.1, except its parts g) and h), was given by R. C. James [113]; part g) was observed in [241], example 2.3. Example 4.2 was observed in [241], example 2.4. As we already mentioned, example 4.3 was found by M. M. Grinblium and L. A. Gurevich [87] (see also S. Karlin [131], lemma 2).

§ 5. Retro-bases in conjugate Banach spaces were introduced by B. R. Gelbaum [69].

The equivalences $1°\Leftrightarrow 2°\Leftrightarrow 4°$ of proposition 5.1 were given by B. R. Gelbaum ([69], theorems 17 and 5). Problem 5.1 was raised in [241], problems 2.4, 2.3′ and 2.4′.

The problem of the existence of normal non-retro-bases in conjugate Banach spaces was raised by S. Karlin ([131], problem 1). Example 5.3 was given by A. Pelczynski (verbal communication) and, later, proposition 5.2 and example 5.2 by J. R. Retherford [207]. Lemma 5.1 is due to J. P. Williams ([264], lemma).

§ 6. Boundedly complete bases have been considered by N. Dunford and A. P. Morse [49], L. Alaoglu [1 a] and, later, by S. Karlin [131] and R. C. James [113]. The term "boundedly complete basis" was suggested by M. M. Day [43]. For $k>0$, the k-boundedly complete bases were introduced in [244].

Theorem 6.1 and corollary 6.1 were given in [244], theorems 1 and 4. Examples 6.1 and 6.2 were observed in [241], Ch. II, § 5. Example 6.3 is due to R. C. James ([113], theorem 1).

The implications $1^\circ \Rightarrow 4^\circ \Rightarrow 2^\circ \Rightarrow 3^\circ$ of theorem 6.2 were given, in a slightly weaker form and with different proofs, by S. Karlin ([131], theorems 9 and 10) and the implication $3^\circ \Rightarrow 1^\circ$ in [235], theorem 5; see also M. M. Day [43], Ch. IV, § 3, lemma 2 for the equivalences $1^\circ \Leftrightarrow 2^\circ \Leftrightarrow 4^\circ$ in the case of monotone bases and [239], theorem 5). The implication $1^\circ \Rightarrow 6^\circ$ was stated, without proof, by A. R. Lovaglia ([154], p. 234); the proof given here and, essentially, example 6.4, were communicated to us by A. Pelczynski. The implication $1^\circ \Rightarrow 7^\circ$ of theorem 6.2 as well as problem 6.2 were communicated to us by C. Bessaga, with the mention that they are due to J. Lindenstrauss.

§ 7. Bases of type $wc_0, swc_0, (wc_0)^*$ and $(swc_0)^*$ were introduced in [62], p. 932. For bases of type wc_0, A. Pelczynski and W. Szlenk [197] have suggested the term "semi-shrinking basis"; however, W. Ruckle [215 a] has used the term "semi-shrinking basis" for bases of type $(wc_0)^*$.

Theorem 7.1 was given in [62], theorem 1 and corollary 2. Problem 7.1, except the last question of part b), was raised in [62], p. 940, problem 2. Recently part of this problem has been solved in the affirmative, namely, J. R. Holub [108] has proved that $C([0,1])$ has a basis of type $(swc_0)^*$ and $L^1([0,1])$ has a basis of type swc_0. Actually, the following more general result holds ([108], proposition (4.2)): *Let E be a Banach space with a basis $\{x_n\}$, in which weak and norm convergence of sequences are not equivalent. Then E has a basis $\{z_n\}$ of type swc_0, such that $[h_n]=[f_n]$, where $\{f_n\}$, $\{h_n\}$ E^* are the a.s.c.f. to the bases $\{x_n\}$ and $\{z_n\}$ respectively. Consequently, $\{h_n\}$ is a basis of type $(swc_0)^*$ of $[f_n]$.*

Indeed, by hypothesis there exists a sequence $\{y_n\} \subset E$ such that $y_n \xrightarrow{w} 0$, $\inf_{1 \leqq n < \infty} \|y_n\| > 0$. Then, by § 15, proposition 15.1, $\{y_n\}$ has a subsequence $\{y_{p_{n+1}}\}$ which is a basic sequence, equivalent to a block basic sequence $\{y'_{p_{n+1}}\}$ with respect to $\{x_n\}$. Extend $\{y'_{p_{n+1}}\}$, by Ch. I, § 7, theorem 7.2, to a basis $\{z_n\}$ of E, such that $[h_n]=[f_n]$. Then, since

$y'_{p_{n+1}} \xrightarrow{w} 0$ (because $y_{p_{n+1}} \xrightarrow{w} 0$), the basis $\{z_n\}$ is of type swc_0. Hence, by the "swc_0-analogue" of Ch. II, § 7, proposition 7.3, $\{h_n\}$ is a basis of type $(swc_0)^*$ of $[f_n]$, which completes the proof.

Since e.g. the sequence $\{\widetilde{\sin 2n\pi t}\} \subset L^1([0,1])$ is weakly convergent to zero but not norm convergent, it follows that $L^1([0,1])$ has a basis $\{z_n\}$ of type swc_0. Since for the a.s.c.f. $\{f_n\}$ to the Haar basis of $L^1([0,1])$ we have $[f_n] \cong C([0,1])$, (see Vol. II, Ch. V), it follows that $C([0,1])$ has a basis of type $(swc_0)^*$.

Let us also mention that if weak and norm convergence of sequences in E are equivalent, then E clearly has no basis of type swc_0 (this was observed in § 7 for $E = l^1$).

Proposition 7.2 is due to A. Pelczynski ([192], p. 544, lemma).

Recently W. Ruckle ([215a], lemma 1) has observed that a basis $\{x_n\}$ of a Banach space is of type $(wc_0)^*$ if and only if $l^1 \subset S^{\gamma\gamma} \subset c_0$, where $S = A_1(\{x_n\})$ is the sequence space associated to the basis $\{x_n\}$; this amounts, essentially, to the equivalence $1° \Leftrightarrow 3°$ of proposition 7.3 (see Ch. I, § 12, proof of theorem 12.6).

§ 8. Weakly closed and (weakly closed)* bases were introduced in [62], p. 932.

Proposition 8.1c) was given in [62], lemma 1a). Theorem 8.1 was given in [62], theorem 1 and corollaries 1 and 2. Remark 8.1 was made in [62], p. 937. Problem 8.1 was raised in [62], p. 940, problem 2b). Since every basis of type P^* (respectively P) is weakly closed (respectively (weakly closed)*) (see § 12, theorem 12.1) and since $L^1([0,1])$ has bases of types P^* and P (see the Notes and remarks to § 9, problem 9.1) it follows that the answer to problem 8.1 is affirmative (J. R. Holub [108], remark 1).

Remarks 8.2 and 8.3 are due to A. Pelczynski (verbal communication).

§ 9. Bases of types P and P^* were introduced in [237]. For bounded biorthogonal systems conditions of this type occur in V. Pták [201], theorems 1 and 2. In aP and aP^* the letter a stands for "almost". For every family $\mathscr{B}$ of bases one can define in a natural way the families $a\mathscr{B}$ and $\mathscr{B}^*$, as follows: $\{x_n\} \in a\mathscr{B}$ if there exists a sequence $\{\varepsilon_n\} \subset K$ with $|\varepsilon_n| = 1$ $(n = 1, 2, \ldots)$ such that $\{\varepsilon_n x_n\} \in \mathscr{B}$, respectively $\{x_n\} \in \mathscr{B}^*$ if $\{f_n\} \in \mathscr{B}$, where $\{f_n\} \subset E^*$ is the a.s.c.f. to $\{x_n\}$. Then, obviously, $a(P^*) = (aP)^*$ and therefore we use for this family the notation aP^*. Actually, the families of bases of type $(wc_0)^*$, $(swc_0)^*$ and (weakly closed)* considered in § 7 and § 8, as well as the bases of types $(l_+)^*$ and $(al_+)^*$ in § 10, are families $\mathscr{B}^*$ for the corresponding families $\mathscr{B}$. One can also introduce the family $s\mathscr{B}$ as follows: $\{x_n\} \in s\mathscr{B}$ if there exists a subsequence $\{x_{i_n}\}$ of $\{x_n\}$ such that $\{x_{i_n}\} \in \mathscr{B}$.

Theorems 9.1 and 9.2 were proved in [237]. Let us also mention the following characterization of bases of type P: *A basis* $\{x_n\}$ *of a Banach space E, with* $\inf_{1 \leqslant n < \infty} \|x_n\| > 0$, *is of type P if and only if* $\{e_n\} \succ \{x_n\}$, *where* $\{e_n\}$ *is the unit vector basis of the space* $bv_0 = \{x \in \{\xi_n\} \in bv \mid \lim_{n \to \infty} \xi_n = 0\}$. Indeed, if $\{x_n\}$ is of type P, then $\{e_n\} \succ \{x_n\}$ by the implication $1^\circ \Rightarrow 3^\circ$ of theorem 9.1 and by an argument similar to that used in Ch. I, § 5, proof of theorem 5.2, implication $3^\circ \Rightarrow 2^\circ$. Conversely, if $\{e_n\} \succ \{x_n\}$, then, by Ch. I, § 8, theorem 8.1 d), implication $6^\circ \Rightarrow 1^\circ$, there exists a continuous linear mapping $u: bv_0 \to E$ such that $u(e_n) = x_n$ $(n = 1, 2, \dots,)$, whence, since $\sup_{1 \leqslant n < \infty} \left\| \sum_{i=1}^{n} e_i \right\|_{bv_0} \leqslant 2$, it follows that $\sup_{1 \leqslant n < \infty} \left\| \sum_{i=1}^{n} x_i \right\| < \infty$, which completes the proof. Hence it follows, using Ch. I, § 12, proposition 12.1, that *a basis* $\{x_n\}$, *with* $\sup_{1 \leqslant n < \infty} \|x_n\| < \infty$, *is of type* P^* *if and only if* $\{x_n\} \succ \{e_n\}$, *where* $\{e_n\}$ *is the unit vector basis of the space* $cs = \left\{ x = \{\xi_n\} \,\middle|\, \sum_{i=1}^{\infty} \xi_i \text{ converges} \right\}$, *with* $\|x\| = \sup_{1 \leqslant n < \infty} \left| \sum_{i=1}^{n} \xi_i \right|$. Similar results, in terms of sequence spaces, have been given, with a different proof, by W. Ruckle ([215], proposition 5.6).

If $\{x_n\}$, $\{y_n\}$ are bases of Banach spaces E and F respectively, then, obviously, $\{x_n\} \times \{y_n\}$ is of type P (or P^*) if and only if both $\{x_n\}$ and $\{y_n\}$ are; a related result for sequence spaces has been given by W. Ruckle ([215], theorem 5.7). H. Joiner has observed [116] that a similar result also holds for the basis $\{x_i \otimes y_j\}$ of $E \otimes_\alpha F$, where α is any uniform crossnorm on $E \otimes F$, such that $\lambda \leqslant \alpha \leqslant \gamma$; the proof of the sufficiency part is analogous to that of Ch. I, § 18, theorem 18.1, expressing the finite sums $\sum_{i,j} x_i \otimes y_j$ with the aid of $\sum_i x_i$ and $\sum_j y_j$.

Theorem 9.3 was given in [62], theorem 2 and corollary 1. Remark 9.1 was made in [62], p. 940, remark 2. Corollary 9.1 is a part of [62], theorem 2. Problem 9.1 was raised in [62], problem 2. Recently J. R. Holub [108] has given the following affirmative solution of this problem: Let $\{z_n^{(1)}\} = \left\{ \frac{\tilde{y}_n}{\|\tilde{y}_n\|} \right\}$ be the normalized Haar basis of $L^1([0,1])$ (Ch. II, § 2, formula (2.3)) and let $\{h_n\} \subset E^* \equiv L^\infty([0,1])$ be the a.s.c.f. to $\{z_n^{(1)}\}$ (Ch. II, § 2, formula (2.4)). Put

$$f_i = \begin{cases} h_i & \text{for } \ i \neq 2^m - 1,\ m = 3, 4, \dots \\ h_{2^m - 1} - \sum_{j = 2^{m-1}}^{2^m - 2} h_j & \text{for } \ i = 2^m - 1,\ m = 3, 4, \dots \end{cases}$$

Then, since $\|h_i\|=1$ $(i=1,2,\dots)$ and $\left\|\sum_{j=2^{m-1}}^{2^m-2} h_j\right\|=1$ $(m=3,4,\dots)$, $\{f_i\}$ is basis of $[h_n]$ (by virtue of Ch. I, § 4, proposition 4.4 on block perturbations of bases). Furthermore, $\|f_i\|\geqslant 1$ $(i=1,2,\dots)$ and for any $m=3,4,\dots$ we have

$$\left\|\sum_{i=1}^{2^m+l} f_i\right\| \leqslant \begin{cases} \left\|\sum_{i=1}^{3} f_i\right\| + \left\|\sum_{i=4}^{2^m-1} f_i\right\| + \left\|\sum_{i=2^m}^{2^m+l} f_i\right\| \\ = \left\|\sum_{i=1}^{3} f_i\right\| + \left\|\sum_{j=3}^{m} h_{2^j-1}\right\| + \left\|\sum_{i=2^m}^{2^m+l} h_i\right\| & \text{for } l=1,\dots,2^m-2, \\ \left\|\sum_{i=1}^{3} f_i\right\| + \left\|\sum_{j=3}^{m+1} h_{2^j-1}\right\| & \text{for } l=2^m-1, \\ \left\|\sum_{i=3}^{3} f_i\right\| + \left\|\sum_{j=3}^{m+1} h_{2^j-1}\right\| + \|h_{2^{m+1}}\| & \text{for } l=2^m, \end{cases}$$

whence, since $\left\|\sum_{j=3}^{p} h_{2^j-1}\right\|=1$ $(p=3,4,\dots)$ and $\left\|\sum_{i=2^m}^{2^m+l} h_i\right\|=1$ $(l=1,\dots,2^m-2)$, it follows that $\{f_i\}$ is a basis of type P of $[h_n]$. Now let $\{\Phi_n\}, \{\Psi_n\}\subset[f_n]^* =[h_n]^*$ be the a.s.c.f. to $\{f_n\}$ and $\{h_n\}$ respectively. Then $\{\Phi_n\}$ is a basic sequence of type P^* and by the proof of Ch. I, § 4, proposition 4.4, we have $[\Phi_n]=[\Psi_n]$. However, since the canonical mapping $\phi: L^1([0,1]) \to[h_n]^*$ is an isomorphism (by Ch. I, § 12, theorem 12.2e)), we have $[\Psi_n]=[\phi(\tilde{z}_n^{(1)})]=\phi(L^1([0,1]))\cong L^1([0,1])$, whence $[\Phi_n]\cong L^1([0,1])$, and consequently $L^1([0,1])$ has a basis of type P^*, whence also a basis of type P, which completes the proof.

Example 9.2 was observed in [237].

§ 10. Bases of type l_+ were introduced in [237] and bases of type al_+ in [62], lemma 1 and problem 1, but the term "basis of type al_+" was used only in [163]. Bases satisfying conditions 2°, 3° of theorem 10.2 were also considered later by D. P. Milman and V. D. Milman [168] (who were apparently unaware of the paper [237]) and bases satisfying condition 4° of theorem 10.1 were studied later by R. C. James ([115], p. 116, condition 35).

The equivalences 1°⇔3°⇔5°⇔6° of theorem 10.1 were given, for real Banach spaces, in [237].

The cartesian product $\{x_n\}\times\{y_n\}$ or the tensor product $\{x_i\otimes y_j\}$ of two bases $\{x_n\}, \{y_n\}$ is of type l_+ if and only if both $\{x_n\}$ and $\{y_n\}$ are; see the corresponding results for bases of types P, P^* in the Notes and remarks to § 9.

The cone associated to a basis of an infinite dimensional Banach space has been studied by H. H. Schaefer [220], R. E. Fullerton [63] and others. Proposition 10.1 a) was given in [163], proposition 1 and the rest of proposition 10.1 is due, essentially, to R. E. Fullerton [63].

The equivalences $1^\circ \Leftrightarrow 5^\circ \Leftrightarrow 4^\circ$ of theorem 10.2 were proved in [163], theorem 3 and proposition 3. The proof of the implication $2^\circ \Rightarrow 3^\circ$ was communicated to us by C. Foiaş.

Corollary 10.1 was given in [163], theorem 4. Proposition 10.2 was observed in [163], corollary 1. The example of a compact base of a cone, which is not a hyperbase, given after lemma 10.1, was exhibited in [163]. Proposition 10.3 was given in [163], proposition 4a), c), d). The equivalence $1^\circ \Leftrightarrow 6^\circ$ of theorem 10.3 was proved in [163], proposition 4b).

Theorem 10.4 is partially contained in [62], corollary 1. The fact that the sequence (9.16), where the ε_n are defined by (9.15), is a basis of type l_+ of $C([0,1])$, was pointed out in [62], theorem 2.

Example 10.1 was observed, essentially, in [237].

§ 11. Besselian and Hilbertian bases have been studied first in Hilbert spaces (see e.g. N. K. Bari [15] and the references mentioned therein). In general Banach spaces they have been considered, independently, by several authors (see e.g. [196], p. 23, definition 6 and Z. A. Čanturija [35], B. E. Veic [260]).

In the particular case when E is a Hilbert space, the equivalences a), b) $1^\circ \Leftrightarrow 3^\circ \Leftrightarrow 6^\circ$ and the implications a), b) $1^\circ \Rightarrow 4^\circ$ of theorem 11.1 were given by N. K. Bari [15]. As we already mentioned, the equivalences a), b) $2^\circ \Leftrightarrow 3^\circ \Leftrightarrow 4^\circ$ in general Banach spaces are consequences of the equivalences b) $6^\circ \Leftrightarrow 1^\circ \Leftrightarrow 2^\circ$ of Ch. I, § 8, theorem 8.1 (therefore, see the Notes and remarks to Ch. I, § 8; see also B. E. Veic [260], for the implications a), b) $1^\circ \Rightarrow 3^\circ$ and $1^\circ \Rightarrow 6^\circ$). Corollary 11.1 has been proved in [251], corollary 1.

Problem 11.1 was raised by A. Pelczynski ([193], problem 13); see also B. E. Veic [260].

Example 11.1 was given by S. Karlin ([131], theorem 1.2).

Theorem 11.2 is due, essentially, to B. E. Veic [260]. Some of the implications of theorem 11.3 have been given by N. K. Bari [15], I. C. Gohberg and A. S. Markus [81] and B. E. Veic [260]. In the case when E is a Hilbert space, the equivalence $1^\circ \Leftrightarrow 2^\circ$ of theorem 11.4 was given by I. C. Gohberg and A. S. Markus [81]; for arbitrary Banach spaces theorem 11.4 was proved, with a different argument, by B. E. Veic [260].

Problem 11.2 was raised in [196], problem 4. Recently, J. R. Retherford has communicated to us in a letter that he was able to answer both questions in the affirmative, by combining the method indicated in [196], remark 3, a perturbation lemma and Ch. I, § 7, theorem 7.2.

Example 11.2 is due to K. I. Babenko [8] (see also B. R. Gelbaum [70], M. Š. Altman [2a] and V. F. Gaposhkin [64]; the proof given here is based on V. F. Gaposhkin [64] and N. K. Bari [16]. There also exist bounded bases of $L^2([-\pi, \pi])$ which are simultaneously non-Besselian and non-Hilbertian, e.g. the sequence $\{\tilde{x}_n\}$, where

$$x_{2n}(t) = \frac{1}{\sqrt{\pi}} |t|^\alpha \cos nt, \qquad x_{2n+1}(t) = \frac{1}{\sqrt{\pi}} |t|^\alpha \sin nt \qquad (n=0,1,2,\ldots)$$

and where $0<\alpha<\frac{1}{2}$ (M. Š. Altman [2a]).

If $\{e_n\}$ is a symmetric basis, with symmetric norm $v^{(s)}_{\{e_n\}}=1$, of its closed linear span $[e_n]$ (in the sense of § 22, definitions 22.1 and 22.2), then $\Phi(\{\alpha_n\}) = \left\| \sum_{i=1}^{\infty} \alpha_i e_i \right\|$ is a "symmetric gauge function" [223] and conversely; the class of Besselian bases with respect to such a function was studied by I. C. Gohberg and A. S. Markus [81]. In the case when F is an arbitrary Banach space with a basis $\{e_n\}$, the $(F, \{e_n\})$-Besselian and $(F, \{e_n\})$-Hilbertian bases were introduced by Z. A. Čanturija [35]. For the extension of theorem 11.1 to such bases, see the Notes and remarks to Ch. I, § 8, theorem 8.1 (see also B. E. Veic [260] for the case when $F=l^p$, $1\leqslant p<\infty$, $\{e_n\}$ = the unit vector basis of F and Z. A. Čanturija [35]; for the general case, see [251], theorem 4). The extensions of the results of § 11 to $(F, \{e_n\})$-Besselian bases, mentioned at the end of § 11, were given in [251] (see also the references of [251], for some particular cases). Theorem 11.5 can be found in the paper of V. D. Milman [167], with the mention that it is due to V. I. Gurariĭ.

§ 12. The part of theorem 12.1 concerning shrinking bases, boundedly complete bases, bases of types P, P^*, l_+ and the five complementary classes of bases was given in [237]. The implication $al_+ \Rightarrow$ weakly closed was observed in [62], lemma 1b) and the implication (weakly closed)* $\Rightarrow$ non-$(swc_0)^*$ in [62], corollary 2. We raised the problem of the existence of a non-shrinking basis of type wc_0 in a letter to A. Pelczynski. The affirmative answer to this problem, given in example 12.2, is due to A. Pelczynski and W. Szlenk [197]; however, one can give a direct proof, simpler than that of [197], of the fact that $\{f_n\}$ is a non-shrinking basis of W^*, of type wc_0. The space d of example 12.1 is a particular case of a space introduced by G. G. Lorentz [153] (see also [39], [78]). The proof of formula (12.7) is based, essentially, on the argument of [78]. The fact that $\{x_n\}$ is a non-shrinking basis of type wc_0 of the space d has been proved by J. R. Retherford [208]. Recently, W. Ruckle has communicated to us the following more general example (unpublished): Let $\{x_n\}$ be a symmetric basis of a Banach space E (see § 22, definition 22.1), which is not equivalent to the unit vector basis of c_0 or l^1 and let $\{f_n\} \subset E^*$ be the a.s.c.f. to $\{x_n\}$. Then both $\{x_n\}$ and $\{f_n\}$ are bases of type wc_0 of E and $[f_n]$ respectively, but, whenever E is non-reflexive, either $\{x_n\}$ or $\{f_n\}$ is non-shrinking (by § 23, remark 23.1 and § 6, corollary 6.1); this is a corrected version of [215a], theorem 3.

Example 12.4 was given in [237].

Recently J. Lindenstrauss and J. R. Retherford have informed us that problem 12.1 (concerning part b) of theorem 12.1) has been solved in the negative by J. R. Holub [108]. The original construction of J. R. Holub involved perturbed members of the Schauder basis of $C([0,1])$, but in the referee's report to that paper a simpler construction has been given, which is equivalent, essentially, to the following: Let $\{a_n\}$ be any dense sequence in $[-1,1]$ and let $\{e_n\}$ be the unit vector basis of c_0. Then, by Ch. II, § 14, example 14.1 and Ch. I, § 4, proposition 4.2 (on cartesian products of bases) and lemma 4.1, the sequence $\{y_n\} \subset E$ defined by

$$y_{4n-3} = \sum_{i=1}^{2n-1} e_{2i-1}, y_{4n-2} = \sum_{i=1}^{2n-1} e_{2i}, y_{4n-1} = \sum_{i=1}^{2n} e_{2i}, y_{4n} = \sum_{i=1}^{2n} e_{2i-1}$$

$$(n=1,2,\ldots)$$

is a basis of c_0. Hence, by Ch. I, § 4, proposition 4.4 (on block perturbations of bases), the sequence $\{x_n\} \subset c_0$ defined by

$$x_{4n-3} = \sum_{i=1}^{2n-1} e_{2i-1}, \quad x_{4n-2} = a_n \sum_{i=1}^{2n-1} e_{2i-1} + \sum_{i=1}^{2n-1} e_{2i}, \quad x_{4n-1} = \sum_{i=1}^{2n} e_{2i},$$

$$x_{4n} = \sum_{i=1}^{2n} e_{2i-1} + a_n \sum_{i=1}^{2n} e_{2i} \quad (n=1,2,\ldots)$$

is a basis of c_0. We claim that this is a *weakly closed basis of type non-al_+*. Indeed, $\|x_j\| = 1$ $(j=1,2,\ldots)$ and for the functionals $f_0, g_0 \in (c_0)^*$ defined by

$$f_0(x) = \xi_1, \quad g_0(x) = \xi_2 \quad (x = \{\xi_n\} \in c_0)$$

we have $\max(|f_0(x_n)|, |g_0(x_n)|) = 1$ $(n=1,2,\ldots)$, whence $x_n \notin V_{f_0, g_0; \frac{1}{2}}(0)$ $(n=1,2,\ldots)$ and thus $\{x_n\}$ is weakly closed. Furthermore, if

$$f(x) = \sum_{i=1}^{\infty} \eta_i \xi_i \quad (x = \{\xi_n\} \in c_0)$$

is an arbitrary continuous linear functional on c_0 (hence $\|f\| = \sum_{i=1}^{\infty} |\eta_i| < \infty$), then

$$\lim_{n\to\infty} |f(x_{4n-2})| = \lim_{n\to\infty} \left| a_n \sum_{i=1}^{2n-1} \eta_{2i-1} + \sum_{i=1}^{2n-1} \eta_{2i} \right|,$$

$$\lim_{n\to\infty} |f(x_{4n})| = \lim_{n\to\infty} \left| \sum_{i=1}^{2n} \eta_{2i-1} + a_n \sum_{i=1}^{2n} \eta_{2i} \right|,$$

and since the sequence $\{a_n\}$ is dense in $[-1, 1]$, one of these two numbers is $=0$ (because either

$$\sum_{i=1}^{\infty} \eta_{2i-1} = \sum_{i=1}^{\infty} \eta_{2i} = 0 \quad \text{or} \quad \min\left(\left|\frac{\sum_{i=1}^{\infty} \eta_{2i-1}}{\sum_{i=1}^{\infty} \eta_{2i}}\right|, \left|\frac{\sum_{i=1}^{\infty} \eta_{2i}}{\sum_{i=1}^{\infty} \eta_{2i-1}}\right|\right) \leqslant 1).$$

Thus, $\{x_n\}$ is of type non-al_+ (by § 10, theorem 10.3), which completes the proof.

Let us observe that the construction of this example is similar to that of Ch. II, § 12, example 12.4 (of a bounded basis $\{z_n\}$ of c_0 such that there exists an $f \in (c_0)^*$ with $f(z_n) \geqslant 1$ $(n=1,2,\ldots)$ but there exists no $f \in (c_0)^*$ with $|f(z_n)|=1$ $(n=1,2,\ldots)$).

The implications a) $2^\circ \Leftrightarrow 3^\circ \Rightarrow 1^\circ$ and b) $2^\circ \Rightarrow 1^\circ$ of theorem 12.2 were given in [237].

§ 13. Universal and complementably universal bases were introduced by A. Pelczynski [195]; block-universal bases also occur in the same paper [195], without bearing any denomination. The construction culminating in theorem 13.1 as well as remark 13.2, corollary 13.1, problem 13.1 (and the comments to it), theorem 13.2, corollary 13.2 and remark 13.3 are due to A. Pelczynski [195]. Lemma 13.6 was observed by A. Pelczynski ([190], proposition 1). The particular case of theorem 13.3 when $E = C([0,1])$ and $\{e_n\}$ = the Schauder basis of $C([0,1])$, occurs in [195]; in the general case, theorem 13.3 was communicated to us by A. Pelczynski.

Remark 13.5 and problem 13.2 were given in [250].

The equivalence $1^\circ \Leftrightarrow 3^\circ$ of theorem 13.4 was given, with a different proof, by A. Pelczynski [195], theorem 2; actually, that proof also contains, implicitly, the equivalence $2^\circ \Leftrightarrow 3^\circ$. The proof presented here was communicated to us by A. Pelczynski.

Remark 13.6 and problem 13.3 were given in [250].

The remark concerning subsequences of the Schauder basis of $C([0,1])$, made before corollary 13.3, is due to A. Pelczynski ([195], proposition 4). Corollary 13.3, theorem 13.5 and problems 13.4, 13.5 were given by A. Pelczynski [195]. Problem 13.6 is also due to A. Pelczynski (verbal communication).

Part II. § 14. Some authors, e.g. S. Karlin [131], B. R. Gelbaum [69], R. E. Fullerton [63] have used the term "absolute basis" instead of unconditional basis and "non-absolute basis" instead of conditional basis, but to-day the terms "unconditional basis" and "conditional basis" are adopted by the majority of specialists in the theory of bases.

The fact that for $1<p<\infty$ the Haar system is an unconditional basis of $L^p([0,1])$ (theorem 14.1) was proved by J. Marcinkiewicz [155]. The crucial lemma in this proof, namely, lemma 14.3, is due to R. E. A. C. Paley [187]; for another proof of this lemma see S. Yano [269]. The proof of lemma 14.3, given here, is essentially the one given by R. E. A. C. Paley [187], with some simplifications. The proof of lemma 14.8 has been communicated to us by G. Albinus. It would be desirable to have a simpler proof of theorem 14.1. In Vol. II, Ch. V we shall consider the problem of unconditionality of the Haar basis in Orlicz spaces.

Examples 14.1 and 14.2 were given in [237]; the first example of a conditional basis of c_0 was produced by B. R. Gelbaum [69].

Lemmas 14.9 and 14.10 are due to W. Orlicz [186]. For the space l^2, proposition 14.1 has been obtained, independently, by B. R. Gelbaum ([69], p. 193, theorem 14) and N. K. Bari [15]; see also B. E. Veic [260].

Example 14.3 was given by S. Karlin ([131], theorem 12).

Example 14.4 is due to K. I. Babenko [8] (see also B. R. Gelbaum [70], M. Š. Altman [2a]). V. F. Gaposhkin has proved the following more general result ([64], theorem 2 and p. 370): *Let $z(\cdot)$ be a measurable function on $[-\pi,\pi]$ for which the function $y(\cdot)=z^2(\cdot)$ has the properties*

a) $y(-\pi)=y(\pi)$,

b) $y(r,t)\geqslant c\eta(r,t)$ $(0\leqslant r<1,\ -\pi\leqslant t\leqslant\pi)$, *where* $y(\zeta)\equiv y(r,t)$ *denotes the harmonic function defined by*

$$y(\zeta)\equiv y(r,t)=\frac{1}{2\pi}\int_{-\pi}^{\pi} y(s)\frac{1-r^2}{1+r^2-2r\cos(s-t)}\,ds \qquad (0\leqslant r<1,\ \zeta=re^{it}),$$

and $\eta(\zeta)\equiv\eta(r,t)$ *its conjugate function and where c is a constant such that*

$$c>0 \quad \textit{for} \quad p\geqslant 2,$$

$$c>\left|\operatorname{tg}\frac{p\pi}{2}\right| \quad \textit{for} \quad 1<p\leqslant 2.$$

In this case the equivalence classes $\{\tilde{x}_n\}$ *and* $\{\tilde{x}'_n\}$ *of the sequences* $\{x_n(\cdot)$ *and* $\{x'_n(\cdot)\}$ *defined by*

$$x_n(t)=\frac{1}{\sqrt{2\pi}}\,z(t)\,e^{int} \qquad (n=0,\pm1,\pm2,\ldots)$$

$$x'_n(t)=\frac{1}{\sqrt{2\pi}}\,\frac{1}{z(t)}\,e^{int} \qquad (n=0,\pm1,\pm2,\ldots)$$

form bounded bases of the space $L^2([-\pi,\pi])$. *If there exist constants* $m>0$ *and* $M\geqslant m$ *such that* $m\leqslant|z(t)|\leqslant M$ *a.e. (on* $[-\pi,\pi]$*), then* $\{\tilde{x}_n\}$ *is simultaneously Besselian and Hilbertian; if there exists* $m>0$ *such that* $m\leqslant|z(t)|$ *a.e., but there does not exist* M *such that* $|z(t)|\leqslant M$ *a.e., then the basis* $\{\tilde{x}_n\}$ *is Besselian and non-Hilbertian; if there exists* M *such that* $|z(t)|\leqslant M$ *a.e., but there exists no* $m>0$ *such that* $m\leqslant|z(t)|$ *a.e., then the basis* $\{\tilde{x}_n\}$ *is Hilbertian and non-Besselian; if there exists no* $m>0$ *such that* $m\leqslant|z(t)|$ *a.e. and no* M *such that* $|z(t)|\leqslant M$ *a.e., then the basis* $\{\tilde{x}_n\}$ *is non-Besselian and non-Hilbertian.*

Taking in this theorem the function $z(t)=|t|^{\alpha}$, where $0<\alpha<\frac{1}{2}$ (in [64] it is shown that this function satisfies the conditions of the theorem), one obtains again the non-Besselian bounded Hilbertian basis $\{\tilde{x}_n\}$ of $L^2([-\pi,\pi])$ constructed by Babenko.

For conditional bases in $L^2([-\pi,\pi])$ see also H. Helson and G. Szegö [105].

J. Lindenstrauss and A. Pelczynski have called to our attention that from a recent paper of C. A. McCarthy and J. Schwartz [165] follows the explicit construction of a conditional basis of l^2 of the form $x_n=\sum_{i=1}^{\infty}\gamma_i^{(n)}e_i$ $(n=1,2,\ldots)$, where $\{e_n\}$ is the unit vector basis of l^2. Namely, from [165] one can obtain, for each n, a basis $x_1^{(n)},\ldots,x_n^{(n)}$ of the n-dimensional Hilbert space l_n^2 such that $v_{\{x_j^{(n)}\}_{j=1}^n}\leqslant M<\infty$ $(n=1,2,\ldots)$ but $v^{(u)}_{\{x_j^{(n)}\}_{j=1}^n}\to\infty$ as $n\to\infty$, whence, by Ch. I, § 7, corollary 7.3, the sequence

$$\{x_1^{(1)},0,0,\ldots\}\ \{0,x_1^{(2)},0,\ldots\}\ \{0,x_2^{(2)},0,\ldots\},\ldots$$
$$\ldots\{\underbrace{0,\ldots,0}_{n-1},\ x_1^{(n)},0,\ldots\},\ldots,\{\underbrace{0,\ldots,0}_{n-1},\ x_n^{(n)},0,\ldots\},\ldots$$

is a conditional basis of $(l_1^2\times l_2^2\times\cdots\times l_n^2\times\cdots)_{l^2}$. Hence, by the natural canonical equivalence of $(l_1^2\times l_2^2\times\cdots\times l_n^2\times\cdots)_{l^2}$ with l^2, we obtain a conditional basis of l^2 of the form $x_n=\sum_{i=1}^{m_n}\gamma_i^{(n)}e_i$. Example 14.5 gives somewhat more, namely, conditional bases of l^2 of the "triangular forms" (14.69) and (14.70). These bases of l^2 were obtained in [147a].

Problem 14.1 a) was raised by A. Pelczynski (verbal communication).

Recently J. Lindenstrauss has communicated to us that *the space* c_0 *has a monotone conditional basis*, namely, the following: Let $\{t_n\}$ be a sequence of numbers such that

$$0<t_n<1\quad(n=1,2,\ldots),\qquad\lim_{n\to\infty}t_n=1,\qquad\prod_{n=1}^{\infty}t_n=0$$

(e.g., one can take $t_n = 1 - \dfrac{1}{n+1}$, $n = 1, 2, \ldots$) and let

$$x_n = \{\underbrace{0, \ldots, 0}_{n-1}, t_n, t_n t_{n+1}, t_n t_{n+1} t_{n+2}, \ldots\} = \sum_{i=n}^{\infty} \left(\prod_{j=n}^{i} t_j\right) e_i \qquad (n = 1, 2, \ldots),$$

where $\{e_n\}$ is the unit vector basis of c_0. Then, obviously, $x_n \in c_0$ $(n = 1, 2, \ldots)$. Since

$$x_n - t_n x_{n+1} = \{\underbrace{0, \ldots, 0}_{n-1}, t_n, 0, \ldots\} = t_n e_n \qquad (n = 1, 2, \ldots),$$

we have $[x_n] = c_0$. Furthermore, since $0 < t_n < 1$ $(n = 1, 2, \ldots)$, for any scalars $\alpha_1, \ldots, \alpha_n$ we have

$$\left\| \sum_{i=1}^{n} \alpha_i x_i \right\|$$

$$= \max\left(|\alpha_1 t_1|, |\alpha_1 t_1 t_2 + \alpha_2 t_2|, \ldots, \left| \sum_{k=1}^{n-1} \alpha_k \prod_{j=k}^{n-1} t_j \right|, \max_{n \leqslant l < \infty} \left| \sum_{k=1}^{n} \alpha_k \prod_{j=k}^{l} t_j \right| \right)$$

$$= \max\left(|\alpha_1| t_1, |\alpha_1 t_1 + \alpha_2| t_2, \ldots, \left| \sum_{k=1}^{n-2} \alpha_k \prod_{j=k}^{n-2} t_j + \alpha_{n-1} \right| t_{n-1}, \right.$$

$$\left. \left| \sum_{k=1}^{n-1} \alpha_k \prod_{j=k}^{n-1} t_j + \alpha_n \right| \max_{n \leqslant l < \infty} \prod_{j=n}^{l} t_j \right)$$

$$= \max\left(|\alpha_1| t_1, |\alpha_1 t_1 + \alpha_2| t_2, \ldots, \left| \sum_{k=1}^{n-1} \alpha_k \prod_{j=k}^{n-1} t_j + \alpha_n \right| t_n \right),$$

whence $\left\| \sum_{i=1}^{n} \alpha_i x_i \right\| \leqslant \left\| \sum_{i=1}^{n+1} \alpha_i x_i \right\|$ for all $\alpha_1, \ldots, \alpha_{n+1} \in K$ and thus $\{x_n\}$ is a monotone basis of c_0. Finally, for every $n > m$ we have

$$\left\| \sum_{i=1}^{n} x_i \right\| = \max\left(t_1, (t_1 + 1) t_2, \ldots, \left(\sum_{k=1}^{n-1} \prod_{j=k}^{n-1} t_j + 1 \right) t_n \right)$$

$$= \max\left(t_1, t_1 t_2 + t_2, \ldots, \sum_{k=1}^{n} \prod_{j=k}^{n} t_j \right) \geqslant \sum_{k=n-m}^{n} \prod_{j=k}^{n} t_j .$$

Keeping m fixed and letting $n \to \infty$, the right hand side tends to $m+1$ (since $\lim_{n\to\infty} t_n = 1$) and hence $\sup_{1 \leqslant n < \infty} \left\| \sum_{i=1}^{n} x_i \right\| = \infty$, i.e. $\{x_n\}$ is of type non-P. Consequently, by § 18, theorem 18.2, $\{x_n\}$ is a conditional basis of c_0, which completes the proof.

From the computation made in example 14.2 it follows that the conditional basis $\{h_n\}$ of l^1 given in that example is monotone (for the subsequence $\{h_n\}_2^\infty$ this was also observed in § 1, after proposition 1.4). However, we don't know whether the space l^p, where $1<p<\infty$, has a monotone conditional basis.

Recently J. R. Holub and J. R. Retherford have observed ([109], example 1) that *the space c_0 has a conditional shrinking basis of type P*, namely, the following:

$$x_1=e_1, \quad x_n=(-1)^{n-1}\left(\frac{1}{n-1}e_1+\sum_{j=2}^{n}(-1)^{j-1}\frac{j-1}{n-1}e_j\right) \quad (n=2,3,\dots),$$

where $\{e_n\}$ is the unit vector basis of c_0.

Indeed, if $\{h_n\}\subset(c_0)^*$ is the a.s.c.f. to $\{e_n\}$ (i.e. the sequence of coordinate functionals on c_0), then for the sequence $\{f_n\}\subset c_0^*$ defined by

$$f_1=h_1+h_2, \quad f_n=h_n+\frac{n-1}{n}h_{n+1} \quad (n=2,3,\dots)$$

we have $f_i(x_j)=\delta_{ij}$ $(i,j=1,2,\dots)$ and

$$\left\|x-\sum_{i=1}^{n}f_i(x)x_i\right\|=\left\|\frac{\xi_{n+1}}{n}(-1)^{n-1}\left[e_1+\sum_{j=2}^{n}(-1)^{j-1}(j-1)e_j\right]+\sum_{i=n+1}^{\infty}\xi_i e_i\right\|\to 0$$

for each $x=\{\xi_n\}\in c_0$, whence $\{x_n\}$ is a basis of c_0. Furthermore,

$$\left\|h_1-f_1-\sum_{j=2}^{n}\frac{(-1)^{j-1}}{j-1}f_j\right\|=\frac{1}{n},$$

whence $h_1\in[f_j]$. Consequently, $h_n\in[f_j]$ (n $=2,3,\dots$), whence $(c_0)^*=[h_n]\subset[f_j]\subset(c_0)^*$ and thus $\{x_n\}$ is shrinking. Since by the above the series $\sum_{j=2}^{\infty}\frac{(-1)^{j-1}}{j-1}f_j$ converges in $(c_0)^*$, but the series $\sum_{j=2}^{\infty}\frac{1}{j-1}f_j$ $=\sum_{j=2}^{\infty}\left(\frac{1}{j-1}h_j+\frac{1}{j}h_{j+1}\right)$ clearly does not converge in $(c_0)^*$, it follows that $\{f_n\}$ is a conditional basic sequence and hence, by § 17, theorem 17.7, $\{x_n\}$ is a conditional basis of c_0. Finally, a simple computation shows that $\left\|\sum_{i=1}^{n}x_i\right\|=1$ $(n=1,2,\dots)$ and thus $\{x_n\}$ is of type P, which completes the proof.

From the above it follows immediately, by duality, that $\{f_n\}$ is a conditional boundedly complete basis of type P^* of the space l^1 [109].

Let us also mention the following special class of (conditional) bases, introduced by V. G. Vinokurov [261]: A conditional basis $\{x_n\}$ of a Banach space E is called a *basis with a discontinuity of the uncondition-*

ality, if $\{x_n\}$ is the union of two disjoint subsequences $\{x_{i_n}\}$ and $\{x_{l_n}\}$ which are unconditional bases of $[x_{i_n}]$ and $[x_{l_n}]$ respectively. The discontinuity of the unconditionality of the basis $\{x_n\}$ is called *simple*, if every permutation of at least one of the sequences $\{x_{i_n}\}$, $\{x_{l_n}\}$ is a part[1] of a permutation $\{x_{\sigma(n)}\}$ of the basis $\{x_n\}$, with the property that $\{x_{\sigma(n)}\}$ is also a basis of E. For instance, the unit vector basis of the space J (§ 4, example 4.1) is a basis with a discontinuity of the unconditionality. The conditional basis of l^p $(1<p\neq 2)$, mentioned after example 14.3, is a basis with a simple discontinuity of the unconditionality. Such bases were introduced for the study of the problem of existence of conditional bases in Banach spaces with bases; since in the meantime this problem has been solved (see § 23, theorem 23.2 and § 24, remark 24.3), we have omitted them.

§ 15. Lemma 15.1 and corollary 15.1 have been given, with different proofs, by W. Orlicz [184] and I. M. Gelfand [73].

In the particular case when $E=C([0,1])$, theorem 15.1 was proved by S. Karlin [131]; the fact that a separable Banach space containing a subspace isomorphic to $C([0,1])$ has no unconditional basis, was proved by C. Bessaga and A. Pelczynski [23] (see also A. Pelczynski [189]).

Theorem 15.2 is due to A. Pelczynski [191] and so are the lemmas 15.2, 15.5 and 15.6. Since the proof of lemma 15.5, given in [191], was not accurate (it claimed that a certain set is closed, but this was not the case), the proof given here is partially based on some arguments communicated to us by M. I. Kadec and A. Pelczynski. For some results in certain quotient spaces of L^1-spaces similar to those of [191] and with analogous proofs, see also B. M. Byčkov and V. M. Grober [34].

Theorem 15.3 has been proved, independently, in [230], [231] and in the paper of A. Pelczynski [190]. In Vol. II, Ch. V, we shall also consider the problem of existence of unconditional bases in Orlicz spaces.

The notion of a Banach space having property (u) was introduced by A. Pelczynski [189]. Theorem 15.4, proposition 15.2, lemma 15.7 and corollary 15.4 were stated, without proof, by A. Pelczynski [189] (see also the paper of C. Bessaga and A. Pelczynski [23], which contains a proof of the fact that a separable Banach space containing a subspace E_0 isomorphic to J has no unconditional basis); the proof of lemma 15.7 was communicated to us by A. Pelczynski. Proposition 15.1 on selection of basic sequences is due to C. Bessaga and A. Pelczynski [22] and so are corollary 15.2b) and corollary 15.3 [23]. In the particular

---

[1] A permutation $\{x_{\tau(i_n)}\}$ of the elements of the subsequence $\{x_{i_n}\}$ is said to be a part of a permutation $\{x_{\sigma(n)}\}$ of the elements of the sequence $\{x_n\}$, if $\tau(i_n)=\sigma(i_n)$ $(n=1,2,\ldots)$.

case when E is weakly complete, lemma 15.8 has been proved by W. Orlicz ([184], theorem 2) and in the general case by C. Bessaga and A. Pelczynski [22].

The problem of the existence of a Banach space E_0 having no unconditional basis, which is a subspace of a Banach space with an unconditional basis, was raised by C. Bessaga and A. Pelczynski ([23], problem 5.3), who conjectured that the answer is affirmative [23]. Theorem 15.5, which substantiates this conjecture and lemma 15.9, were proved by J. Lindenstrauss [149]. Recently, J. R. Holub and J. R. Retherford ([109], example 2) have proved that the monotone basic sequence $\{z_n\} \subset l^1 \equiv (c_0)^*$ defined by (15.57) is the sequence of coefficient functionals for a suitable basis of c_0, whence $\{z_n\}$ is a w^*-Schauder basis of l^1.

§ 16. The equivalence $1^\circ \Leftrightarrow 3^\circ$ of lemma 16.1 is due to W. Orlicz [186]. The equivalences $1^\circ \Leftrightarrow 2^\circ \Leftrightarrow 6^\circ$ of this lemma have been proved by T. H. Hildebrandt [106] and the implication $6^\circ \Rightarrow 3^\circ$ also by C. W. McArthur [160]. The equivalences $1^\circ \Leftrightarrow 4^\circ \Leftrightarrow 5^\circ$ are also well known. There exist many other characterizations of unconditional convergence, but we did not need them here.

The equivalence $1^\circ \Leftrightarrow 3^\circ$ of theorem 16.1 was given by M. M. Grinblium ([82], theorems B) and Б)). The implication $1^\circ \Rightarrow 7^\circ$ can be found in the book of M. M. Day ([43], Ch. IV, § 4, theorem 1). The implication $12^\circ \Rightarrow 20^\circ$ was proved by I. M. Gelfand [74]. The equivalences $1^\circ \Leftrightarrow 22^\circ \Leftrightarrow 23^\circ \Leftrightarrow 24^\circ$ and example 16.1 were given by B. E. Veic [258] (see also [260a]). The equivalence $1^\circ \Leftrightarrow 25^\circ$ and example 16.2 are due to J. R. Retherford [202].

Definition 16.1 was given by B. E. Veic [258]. Theorem 16.2, with a different proof, is due, essentially, to B. E. Veic [258]; the observation that the necessity part of this theorem also follows from a extended stability theorem mentioned at the end of § 11, was made in [251].

The equivalences $1^\circ \Leftrightarrow 4^\circ \Leftrightarrow 5^\circ$ of theorem 16.3 have been given, in a slightly different form, by R. E. Fullerton [63]. The equivalence $1^\circ \Leftrightarrow 2^\circ$ of this theorem is due to H. H. Schaefer ([220], propositions (5.3), (5.4); see also [221], p. 251, exercise 8) and has been rediscovered later by L. A. Gurevich [102], who also gave the equivalence $1^\circ \Leftrightarrow 3^\circ$ and, essentially, examples 16.3 and 16.4. Propositions 16.1 and 16.2 were proved by L. A. Gurevich [102]. Some other characterizations of unconditional bases among E-complete total biorthogonal systems by properties of the associated cone and partial order have been given recently by Ya. M. Ceĭtlin [35a] and Nguen Van Khue [177]. The problem of characterization of general (i.e. not necessarily unconditional) bases by cone and order properties was raised by R. E. Fullerton [63].

Theorem 16.4 was proved by A. Yudin [271]. Lemma 16.2 and the proof of theorem 16.4, given here, are due to B. Sz.-Nagy [172]. Another proof of this theorem has been given by M. G. Krein and M. A. Rutman [144].

The implications a) $2^\circ \Rightarrow 1^\circ$ and a) $3^\circ \Rightarrow 1^\circ$ of theorem 16.5 were proved by M. I. Kadec and A. Pelczynski ([128], theorems 5 and 6); under the more restrictive assumption that $\{x_n\}$ is a basis of E, with the a.s.c.f. $\{f_n\}$, theorem 16.5c) was observed by S. Yamazaki [268]. This latter result, together with the results of Ch. 1, § 5, suggests naturally the question, whether there exist bases $\{x_n\}$ for which $M(E,(x_n,f_n))$ is distinct from bv and m. By Ch. I, § 12, theorem 12.8, this problem is equivalent to that of the existence of γ-perfect BK-algebras containing all unit vectors e_n and the identity $e=\{1,1,\ldots\}$, distinct from bv and m. An affirmative answer has been given by R. J. McGivney and W. Ruckle ([166], theorem 6.2 and example). Recently, L. Sternbach has communicated to us the following simple way of obtaining bases $\{x_n\}$ for which $M(E,(x_n,f_n))$ is distinct from bv and m: Let $\{x_n\}$ be a conditional basis of a Banach space E, such that there exists a complemented infinite subsequence $\{x_{i_n}\}$ of $\{x_n\}$ (in the sence of Ch. II, § 13, definition 13.2); such a basis is e.g. any cartesian product $\{x_n\}=\{y_n\}\times\{z_n\}$ of two bases $\{y_n\}$, $\{z_n\}$ (see Ch. I, § 4, definition 4.6), one of which is conditional. Then $M(E,(x_n,f_n))$ (where $\{f_n\}$ is the a.s.c.f. to $\{x_n\}$) is distinct from bv and m, since it contains the sequence $\{\gamma_j\}\notin bv$ defined by $\gamma_j=1$ for $j=i_1,i_2,\ldots$ and $\gamma_j=0$ for $j\neq i_1,i_2,\ldots$ and since $\{x_n\}$ is a conditional basis of E. There remains open the problem of classifying bases $\{x_n\}$ in terms of their multiplier algebras $M(E,(x_n,f_n))$.

Problem 16.1 was raised by B. S. Mityagin [169] (see also A. Pelczynski [194], problem 1). M. I. Kadec and A. Pelczynski [128] expressed the opinion that "it is very probable" that the answer to this problem is affirmative. Indeed, recently problem 16.1 b) has been solved in the affirmative [41]; we shall give the proof of this result in Vol. II, Ch. III. Lemma 16.1 is due to M. I. Kadec and A. Pelczynski [128] (see also M. I. Kadec [120], [121], [122] and [123] and V. Klee [132]). The sufficiency parts of theorem 16.6a), b) were given by M. I. Kadec [124]; for this theorem, lemma 16.4 and proposition 16.3 see also M. I. Kadec and A. Pelczynski [128] (for a version of lemma 16.4 in the particular case when $r(V)>0$, see also A. Pelczynski [191 a], p. 371, lemma).

The final part of theorem 16.7 has been proved, independently, by S. Yamazaki [266] and by A. Pelczynski (verbal communication, 1964, unpublished).

The equivalence $1^\circ \Leftrightarrow 2^\circ$ of theorem 16.8 was observed by B. E. Veic [258] (although he stated it in a slightly different form, this is what he actually proved; see also B. E. Veic [260a]); the implication $1^\circ \Rightarrow 2^\circ$ was observed also by M. I. Kadec and A. Pelczynski [127]. The implication $3^\circ \Rightarrow 1^\circ$ occurs, implicitly, in [196], proof of proposition 4(a).

Proposition 16.4 was observed by A. Pelczynski ([195], lemma 11). Example 16.9 is due to B. R. Gelbaum and J. Gil de Lamadrid [72].

Various generalizations of unconditional bases and of spaces with unconditional bases, as well as corresponding extensions of the characterizations of unconditional bases and of the properties of spaces having unconditional bases, will be given in Vol. II.

§ 17. The equivalences $1^\circ \Leftrightarrow 8^\circ \Leftrightarrow 10^\circ$ of theorem 17.1 were given, in a weaker form, by M. M. Grinblium [82] and the equivalences $1^\circ \Leftrightarrow 5^\circ \Leftrightarrow 7^\circ$, in a weaker form, by L. A. Gurevich [98] (namely, both in [82] and [98] it is assumed that $\{x_n\}$ is a basis of E). The equivalence $1^\circ \Leftrightarrow 4^\circ$ has been given by R. C. James [113] and the equivalence $1^\circ \Leftrightarrow 7^\circ$ in the general case by C. Bessaga and A. Pelczynski [22]. The equivalence $8^\circ \Leftrightarrow 9^\circ$ is the "unconditional analogue" of a result of S. Yamazaki [265] (see the Notes and remarks to Ch. I, § 7, theorem 7.1); this equivalence was also given by J. R. Holub and J. R. Retherford in [110].

Corollary 17.1 was observed in [237]. Corollary 17.2 is due, essentially, to C. Bessaga and A. Pelczynski [22].

Example 17.1 was given by A. Pelczynski [195].

The sufficiency parts of corollary 17.3 a), b) are contained, implicitly, in a paper of R. C. James ([113], the proofs of lemmas 1 and 2). Corollary 17.4 is due to A. Pelczynski ([195], lemma 12).

The material of § 17 on Banach spaces of matrices and other Banach spaces having no unconditional basis (definitions 17.2, 17.3, proposition 17.1, lemmas 17.1 – 17.5, theorems 17.2 – 17.6, corollary 17.5 and problem 17.1) are due to S. Kwapien and A. Pelczynski [145]. Proposition 17.2 was given by A. Pelczynski ([190], proposition 5).

The relations (17.76) and (17.77) of theorem 17.7 were given in [249].

Proposition 17.3 is due to M. I. Kadec and A. Pelczynski [127]. An example that in this proposition the unconditionality of $\{x_n\}$ is essential, different from example 17.2, has been given by M. I. Kadec and A. Pelczynski [127]. The first part of example 17.2 was observed in [238]. Recently, J. R. Holub and J. R. Retherford ([110], theorem (3.4)) have proved that the property occurring in proposition 17.3 actually characterizes unconditional bases, namely, if this property holds for every subsequence $\{x_{n_k}\}$ of $\{x_n\}$, then $\{x_n\}$ is an unconditional basis.

Lemma 17.6 and theorem 17.8 are due to W. Ruckle [214].

§ 18. Theorem 18.1 has been proved by N. K. Bari [15] and I. M. Gelfand [74]. Other proofs of this theorem have been given by M. I. Kadec and A. Pelczynski ([127], p. 168, remark 3) and J. Lindenstrauss ([148], p. 246, remark); the short proof presented here was given in [248].

The problem, whether in the spaces c_0 and l^1 there exist two non-equivalent bounded unconditional bases, was raised by A. Pelczynski ([190], p. 224]; the same problem, for Banach spaces non-isomorphic to l^2 and having an unconditional basis, was raised in [196], p. 22, problem 3. Lemmas 18.1 and 18.2 are due to A. Grothendieck ([90], p. 59 – 64) and so are, essentially, corollary 18.1 and lemma 18.3; we have reproduced them from the paper of J. Lindenstrauss and A. Pelczynski [150]. Theorem 18.2, which solves the above problem, was given by J. Lindenstrauss and A. Pelczynski ([150], corollary 1 to theorem 6.1).

Proposition 18.1 is due to A. Pelczynski ([190], lemma 1) and proposition 18.2 to C. Bessaga and A. Pelczynski ([22], theorem 2). Lemmas 18.4 – 18.8 and theorem 18.3 were proved by A. Pelczynski ([190], lemma 2, propositions 3, 4 and 5 and theorem 7). Propositions 18.3, 18.4, lemmas 18.9, 18.10 and theorem 18.4 were given in the same paper of A. Pelczynski ([190], lemma 4, proposition 8 and theorem 7); let us mention that, independently, V. F. Gaposhkin [65] has proved that for $1<p<2, p\neq 2$, the Haar system $\{\tilde{y}_n\}$ has a permutation $\{\tilde{y}_{\sigma(n)}\}$ which is not equivalent to $\{\tilde{y}_n\}$.

Theorem 18.5 was proved by A. Pelczynski ([195], theorem 1); the proof given here (via proposition 18.5), communicated to us by A. Pelczynski, is due to M. Zippin.

Theorem 18.6 and corollaries 18.2, 18.3 have been given by A. Pelczynski ([195], theorem 3 and corollaries 3, 4). Problems 18.1 and 18.2 were raised by A. Pelczynski ([195], problems 4 and 5).

§ 19. NT-norms, NK-norms and NTK-norms were introduced in [236], but this terminology has been used in [243]; here N stands for the Romanian word for unconditional.

The necessity part of proposition 19.1 b) is contained, implicitly, in [236], proof of theorem 3. The other statements of theorem 19.1 were proved, with a slightly different argument, by J. R. Retherford and R. C. James ([210], theorem (2.4) (b) and (d)); for bases satisfying (19.3) they have used the term "strictly co-orthogonal bases").

Theorem 19.1 was given by J. R. Retherford and R. C. James ([210], remark (2.2) (c)). Example 19.1 is a slightly changed version of an example due to J. R. Retherford and R. C. James ([210], example (2.5)); see the Notes and remarks to Ch. I, § 19, example 19.2.

Theorem 19.2 was proved, partially, in [236], theorem 2, namely, there it was proved that (19.7) is an NK-norm on E, equivalent to the initial norm and the problem was raised, whether there exists on E an equivalent NT-norm. This latter question has been answered in the affirmative by J. R. Retherford [206], who has used, instead of (19.7), the norm defined by the right hand side of (19.8); however, as shown by theorem 19.2, the norm (19.7) given in [236], theorem 2, also yields an affirmative answer to the problem.

§20. Bases with property (20.1) have been considered by S. Karlin [131] and M. M. Day [43]; the term "orthogonal basis" for such bases was introduced in [236]. Strictly orthogonal bases were introduced in [236]. In Banach sequence spaces which are "symmetric" in the sense of A. Sobczyk [252], the unit vectors form a hyperorthogonal basic sequence.

A part of theorem 20.2 was given, without proof, in [241], theorem 2.13. A slightly weaker version of theorem 20.3 was given in [236], theorems 1 and 3; let us mention that M. M. Day [43] has observed that in every Banach space with an unconditional basis one can introduce an equivalent norm in which the basis $\{x_n\}$ is orthogonal.

Propositions 20.2 – 20.4 are due to W. J. Davis [37].

§21. Bases with the property described in definition 21.1 were considered in [238], condition (SB_2), but the term "subsymmetric basis" was not used there (see the Notes and remarks to §22).

The sets M_ε^p (formula (21.1)) were introduced by M. I. Kadec [118]. Lemma 21.1, propositions 21.1, 21.2 and remark 21.1 are due to M. I. Kadec and A. Pelczynski ([127], theorems 1, 2 and lemma 1). Theorem 21.1 was proved by M. I. Kadec and A. Pelczynski ([127], corollary 8).

The remark that the condition of unconditionality of $\{x_n\}$ in definition 21.1 is essential, was made in [238], remark 1.

Proposition 21.4 was proved in [238], lemma 1. Theorem 21.2 was given, essentially, in [238], lemma 2 and corollary 1 (equivalences $(SB_2)\Leftrightarrow(SB_6)\Leftrightarrow(SB_7)$).

Proposition 21.5 and problem 21.1, with "symmetric basis" instead of "subsymmetric basis", are due to I. Edelstein (verbal communication).

§22. Symmetric bases in Banach spaces were introduced in [232], [233]. Independently, M. I. Kadec and A. Pelczynski [127] have considered bases satisfying condition 12° of theorem 22.1 and called them "permutatively homogeneous bases". In Banach sequence spaces which are both "symmetric" and "permutable" in the sense of A. Sobczyk [252], the unit vectors form a symmetric basic sequence, with symmetric constant 1.

The implications $1^\circ \Rightarrow 5^\circ$ and $1^\circ \Rightarrow 12^\circ$ of theorem 22.1 were proved in [233], theorem 1 and p. 161, remark 2°. The equivalences $1^\circ \Leftrightarrow 6^\circ \Leftrightarrow 7^\circ \Leftrightarrow 12^\circ$ of theorem 22.1 was proved in [238], theorem (equivalence

$(SB_1) \Leftrightarrow (SB_3)$) and corollary 1 (equivalences $(SB_1) \Leftrightarrow (SB_4) \Leftrightarrow (SB_5)$). The equivalences $1^\circ \Leftrightarrow 9^\circ \Leftrightarrow 10^\circ$ are due, essentially, to D. J. H. Garling [68].

Recently W. Ruckle has observed ([215a], theorem 1) that if S is the sequence space associated to a symmetric basis of a locally convex F-space, then either $S=s$ (the space of all sequences of scalars) or $l^1 \subset S \subset c_0$. However, for Banach spaces this merely says that every symmetric basis is bounded (by Ch. I, § 3, lemma 3.1), a fact which was observed in the proof of theorem 22.1 (implication $12^\circ \Rightarrow 1^\circ$). For similar remarks on the sequence space associated to a symmetric basis see also D. J. H. Garling [68].

Proposition 22.1 b) was given in [233], theorem 1. The "subsymmetric analogue" of proposition 22.1 a) appears in [196], p. 22, formula (37). Definition 22.2, with a slightly different notion of symmetric constant, was given in [196], p. 23 and p. 6, definition 1.

Proposition 22.2 was proved in [238], theorem (implication $(SB_1) \Rightarrow (SB_2)$); independently, M. I. Kadec and A. Pelczynski [127] proved the equivalent result that every "permutatively homogeneous basis" (see above) is subsymmetric. In [238] it was claimed that the converse implication also holds, i.e., every subsymmetric basis is symmetric. This claim has been disproved by D. J. H. Garling [68], who has given as a counterexample the closed linear span of the unit vectors in the Banach space G of example 22.1. The proof of the fact that this subspace of G actually coincides with G, has been communicated to us by V. I. Gurarii; later, a similar proof was given to us, independently, by K. I. Oskolkov, who also observed that the same fact remains valid if the sequence $\frac{1}{\sqrt{i}}$ in (22.36) is replaced by any sequence $a_i \geqq 0$ with $\sum_{i=1}^{\infty} a_i = \infty$ satisfying the following additional condition: there exists a $p<1$ such that for every positive integer n one can find an index $i_0 = i_0(n)$ with the property that $a_i(1-p) - a_{i+n} \leqq 0$ $(i \geqq i_0)$.

As we already mentioned (in the Notes and remarks to § 21), problem 22.1 is due to I. Edelstein.

Problem 22.2 was raised in [250].

In the particular case when the sequence $\{A_n\}$ consists of one single set A and the space E is of dimension 2^n, proposition 22.3 is contained, implicitly, in [196], proof of lemma 1. In the general case, proposition 22.3, definition 22.3 and proposition 22.4 were given by J. Lindenstrauss and M. Zippin ([151], lemmas 4, 5 and 6).

Lemma 22.1 and theorem 22.2 were proved by W. Ruckle [214]. The proof of the implication $12^\circ \Rightarrow 1^\circ$ of theorem 22.1, given after theorem 22.2, is due, essentially, to W. Ruckle [214], [213].

Definition 22.5 of Haar system and Rademacher system with respect to a symmetric basis of a finite dimensional symmetric space, propositions 22.6, 22.7 and theorem 22.3 (which may be considered as an "abstract analogue" of the classical Khinchin inequality) were given in [196], proposition 1, corollary, lemma 1 and proposition 2.

§ 23. Proposition 23.1 a), c), d) was proved in [196], proposition 4 and proposition 23.1 b) in [250]. Theorem 23.1 was established in [196], theorem. Proposition 23.2 and theorem 23.2 were given in [196], proposition 3 and theorem. Finally, theorem 23.3 was proved in [196], remark 1.

§ 24. Perfectly homogeneous bases were introduced by C. Bessaga and A. Pelczynski ([122], definition 2).

Propositions 24.1, 24.2 and theorem 24.1 are due to M. Zippin [272] (see also H. F. Bohnenblust [28 a]); for some previous results on perfectly homogeneous bases, which are covered now by theorem 24.1, see [238], corollary 2 and remark 2 and W. J. Davis and D. W. Dean [39].

The proof of § 23, theorem 23.1, presented in remark 24.1, was given in [250].

Proposition 24.2, theorems 24.2, 24.3 and remarks 24.2, 24.3 are due to J. Lindenstrauss and M. Zippin ([151], lemma 2, theorem 1 and p. 124, note (1)).

§ 25. Absolutely convergent bases and uniform bases were introduced by S. Karlin [131].

Theorems 25.1 and 25.2 were proved by S. Karlin [131].

Finally, we mention a generalization of uniform bases, due to C. W. McArthur and J. R. Retherford [161]. Let E be a Banach space which is also a topological linear space for a topology T, where T is not necessarily the norm-topology. A T-Schauder basis $\{x_n\}$ of E, with the a.s.c.f. $\{f_n\}$, is called [161] *T-uniform* if $x=\lim_{n\to\infty}\sum_{i=1}^{n} f_i(x)x_i$ for the topology T, uniformly in the unit cell $S_E=\{x\in E \mid \|x\|\leqslant 1\}$, i.e., if for every T-neighbourhood V of 0 in E there exists a positive integer $N=N(V)$ such that

$$x-\sum_{i=1}^{n} f_i(x)x_i\in V \qquad (x\in S_E,\ n>N).$$

Theorem 25.2 says that an infinite dimensional Banach space admits no norm-uniform norm-Schauder basis. In contrast to this, C. W. McArthur and J. R. Retherford have observed ([161], theorem 2) that *every w^*-Schauder basis $\{f_n\}$ of a conjugate Banach space E^* is w^*-uniform.* Indeed, by Ch. I, § 14, theorem 14.1, $\{f_n\}$ is the a.s.c.f. to a basis $\{x_n\}$ of E, whence, given a w^*-neighbourhood $V = V_{y_1,\dots,y_m;\varepsilon}(0)$ of 0 in E^*, there exists a positive integer $N = N(V)$ such that

$$\left\| \sum_{i=n+1}^{\infty} f_i(y_j) x_i \right\| < \varepsilon \qquad (n > N, j = 1, \dots, m)$$

and hence for any $f \in S_{E^*}, n > N$ and $j = 1, \dots, m$ we have

$$\left| \left[f - \sum_{i=1}^{n} f(x_i) f_i \right](y_j) \right| = \left| f\left(\sum_{i=n+1}^{\infty} f_i(y_j) x_i \right) \right| \leqslant \|f\| \left\| \sum_{i=n+1}^{\infty} f_i(y_j) x_i \right\| < \varepsilon,$$

which completes the proof. From this result it follows that *every norm-basis $\{x_n\}$ of E is $\sigma(E,[f_n])$-uniform, where $\{f_n\} \subset E^*$ is the a.s.c.f. to $\{x_n\}$* (clearly, $\{x_n\}$ is also a $\sigma(E,[f_n])$-Schauder basis); indeed, $\{\phi(x_n)\}$ is a w^*-Schauder basis of $[f_n]^*$, where ϕ is the canonical mapping of E into $[f_n]^*$ and ϕ is an isomorphism (by Ch. I, § 12, theorem 12.2e)).

On the other hand, C. W. McArthur and J. R. Retherford proved ([161], theorem 3) that *a w-Schauder basis $\{x_n\}$ of a Banach space E is w-uniform if and only if $\{x_n\}$ is shrinking.* Indeed, by Ch. I, § 13, theorem 13.1, $\{x_n\}$ is a norm-basis of E. Let $f \in E^*$ and $\varepsilon > 0$ be arbitrary and consider the w-neighbourhood $V = V_{f;\varepsilon}(0)$ of 0 in E. Then, if $\{x_n\}$ is w-uniform, there exists a positive integer $N = N(V)$ such that

$$\left| f(x) - f\left(\sum_{i=1}^{n} f_i(x) x_i \right) \right| < \varepsilon \ (x \in S_E, n > N), \text{ whence } \left\| f - \sum_{i=1}^{n} f(x_i) f_i \right\| < \varepsilon$$

$(n > N)$ and thus $[f_n] = E^*$. Conversely, if $\{x_n\}$ is shrinking, then, by the above remark, $\{x_n\}$ is a $\sigma(E,E^*) = \sigma(E,[f_n])$-uniform basis of E, which completes the proof.

Bibliography

1. Akutowicz, E.: Construction of a Schauder basis in some spaces of holomorphic functions in the unit disk. Colloq. Math. **15,** 287—296 (1966).

1a. Alaoglu, L.: Weak topologies in normed linear spaces. Ann. of Math. (2) **41,** 252—267 (1940).

2. Altman, M. Š.: On biorthogonal systems. Doklady Akad. Nauk SSSR **67,** 413—416 (1949) [Russian].

2a. — On bases in Hilbert space. Doklady Akad. Nauk SSSR **69,** 483—485 (1949) [Russian].

3. Arsove, M. G.: Similar bases and isomorphisms in Fréchet spaces. Math. Annalen **135,** 283—293 (1958).

4. — The Paley-Wiener theorem in metric linear spaces. Pacific J. Math. **10,** 365—379 (1960).

5. — Edwards, R. E.: Generalized bases in topological linear spaces. Studia Math. **19,** 95—113 (1960).

6. Ascoli, G.: Sugli spazi lineari metrici e le loro varietà liniari. Ann. Math. Pura Appl. (4) **10,** 33—81, 203—232 (1932).

7. Babenko, K. I.: On bases in Hilbert space. Doklady Akad. Nauk SSSR **57,** 427—430 (1947) [Russian].

8. — On conjugate functions. Doklady Akad. Nauk SSSR **62,** 157—160 (1948) [Russian].

9. Banach, S.: Sur une propriété caractéristique des fonctions orthogonales. Comptes Rendus Acad. Sci. (Paris) **180,** 1637—1640 (1925).

10. — Théorie des opérations linéaires. Warszawa: Monografje Matematyczne 1932.

11. — Sur la divergence des séries orthogonales. Studia Math. **9,** 139—155 (1940).

12. — Mazur, S.: Zur Theorie der linearen Dimension. Studia Math. **4,** 100—112 (1933).

13. Bari, N. K.: Sur la stabilité de certaines propriétés des systèmes orthogonaux. Matem. Sbornik **12,** 3—27 (1943).

14. — Sur les systèmes complets de fonctions orthogonales. Matem. Sbornik **14,** 51—108 (1944).

15. — Biorthogonal systems and bases in Hilbert space. Uč. Zap. Moskov. Gos. Univ. 148, Matematika **4,** 69—107 (1951) [Russian].

16. — Trigonometrical series. Moscow 1961 [Russian]. English translation, New York 1964.

17. Baric, L. W.: Some notes on sequences which are similar or related to a Schauder basis. Duke Math. J. **35,** 1—7 (1968).

18. — Ruckle, W.: Matrix transformations of Schauder bases. Studia Math. **28,** 275—278 (1967).

19. Berkson, E.: Some metrics on the subspaces of a Banach space. Pacific J. Math. **13,** 7—22 (1963).

20. Bessaga, C.: Bases in certain spaces of continuous functions. Bull. Acad. Pol. Sci. **5,** 11—14 (1957).

21. — Pelczynski, A.: An extension of the Krein-Milman-Rutman theorem concerning bases to the case of B_0-spaces. Bull. Acad. Polon. Sci. **4,** 379—383 (1959).

22. — — On bases and unconditional convergence of series in Banach spaces. Studia Math. **17,** 151—164 (1958).

23. — — A generalization of results of R. C. James concerning absolute bases in Banach spaces. Studia Math. **17,** 165—174 (1958).

24. — — Properties of bases in spaces of type B_0. Prace Mat. **3,** 123—142 (1959) [Polish].

25. — — Banach spaces non isomorphic to their cartesian squares, I. Bull. Acad. Polon. Sci. **8,** 757—761 (1960).

26. — — Spaces of continuous functions, IV (On isomorphical characterizations of spaces $C(S)$). Studia Math. **19,** 53—62 (1960).

27. Boas, R. P.: General expansion theorems. Proc. Nat. Acad. Sci. U.S.A. **26,** 139—143 (1940).

28. Bohnenblust, H. F.: Convex regions and projections in Minkowski spaces. Ann. of Math. (2) **39,** 301—308 (1938).

28a. — An axiomatic characterization of L_p-spaces. Duke Math. J. **6,** 627—640 (1940).

29. — Subspaces of $l_{p,n}$ spaces. Amer. J. Math. **63,** 64—72 (1941).

30. Bourbaki, N.: Algèbre, Ch. II. Paris: Hermann et Cie 1955.

31. — Algèbre, Ch. III. Paris: Hermann et Cie 1958.

32. — Espaces vectoriels topologiques, Ch. I—II. Paris: Hermann et Cie 1953.

33. — Espaces vectoriels topologiques, Ch. III—V. Paris: Hermann et Cie 1955.

34. Byčkov, B. M., Grober, V. M.: The absence of an unconditional basis in the factor spaces L^1/H_0^1 and L'/H_0'. Doklady Akad. Nauk SSSR **179,** 511—514 (1968) [Russian].

35. Čanturija, Z. A.: On some properties of biorthogonal systems in a Banach space and their applications in spectral theory. Soobsč. Akad. Nauk Gruz. SSR **34,** 2, 271—276 (1964) [Russian].

35a. Ceĭtlin, Ya. M.: Unconditionality of a basis and partial order. Izv. Vysš. Uč. Zaved. Matematika **2**(51), 98—104 (1966) [Russian].

36. Davies, R. O.: A norm satisfying the Bernstein condition. Studia Math. **29,** 219—220 (1968).

36a. Davis, P. J., Fan Ky: Complete sequences and approximations in normed linear spaces. Duke Math. J. **24,** 183—192 (1957).

37. Davis, W. J.: Positive bases in Banach spaces. Rev. Roumaine math. pures et appl. (to appear).

38. — Basis preserving maps. Proc. Amer. Math. Soc. (to appear).

39. — Dean, D. W.: The direct sum of Banach spaces with respect to a basis. Studia Math. **28,** 209—219 (1967).

40. Davis, W. J., Dean, D. W., Singer, I.: Complemented subspaces and Λ systems in Banach spaces. Israel J. Math. **6,** 303—309 (1968).
41. — — — Multipliers and unconditional convergence of biorthogonal expansions. (to appear).
42. Day, M. M.: Polygons circumscribed about closed convex curves. Trans. Amer. Math. Soc. **62,** 315—319 (1947).
43. — Normed linear spaces. Berlin-Göttingen-Heidelberg: Springer 1958.
44. Dean, D. W., Bor-Luh Lin, Singer, I.: On k-shrinking and k-boundedly complete bases in Banach spaces. Pacific J. Math. (to appear).
45. Dieudonné, J.: Complex structures on real Banach spaces. Proc. Amer. Math. Soc. **3,** 162—164 (1952).
46. — Sur les espaces L^1. Arch. Math. **10,** 151—152 (1959).
47. Dixmier, J.: Sur un théorème de Banach. Duke Math. J. **15,** 1057—1071 (1948).
48. Duffin, R. J., Eachus, J. J.: Some notes on an expansion theorem of Paley and Wiener. Bull. Amer. Math. Soc. **48,** 850—855 (1942).
49. Dunford, N., Morse, A. P.: Remarks on the preceding paper of James A. Clarkson. Trans. Amer. Math. Soc. **40,** 415—420 (1936).
50. — Schwartz, J. T.: Linear operators. Part I: General theory. New York: Intersci. Publ. 1958.
51. — — Linear operators. Part II: Spectral theory. New York: Intersci. Publ. 1963.
52. Dvoretzky, A.: Some results on convex bodies and Banach spaces. Proc. Internat. Sympos. on Linear Spaces (July 5—12, 1960), 123—161, Jerusalem: Academic Press 1961.
53. Edelsteĭn, I.: Candidate thesis. Harkov State University 1968 [Russian].
54. Edwards, R. E.: Functional analysis. New York: Holt, Rinehart and Winston Inc. 1965.
55. Ellis, H. W., Halperin, I.: Haar functions and the basis problem for Banach spaces. J. London Math. Soc. **31,** 28—39 (1956).
56. Ezrohi, I. A.: General forms of linear operations in spaces with a countable basis. Doklady Akad. Nauk SSSR **59,** 1537—1540 (1948) [Russian].
57. Fan, Ky: On systems of linear inequalities. Linear inequalities and related problems, 99—156. Ann. of Math. Studies **38.** Princeton: Princeton Univ. Press 1956.
58. Fihtengolc, G. M.: Course of differential and integral calculus, I. Moscow-Leningrad 1949 [Russian].
59. — Course of differential and integral calculus, III. Moscow 1966 [Russian].
60. Foguel, S. R.: Biorthogonal systems in Banach spaces. Pacific J. Math. **7,** 1065—1072 (1957).
61. Foiaş, C., Singer, I.: Some remarks on strongly linearly independent sequences and bases in Banach spaces. Rev. math. pures et appl. **6,** 589—594 (1961).
62. — — On bases in $C([0,1])$ and $L^1([0,1])$. Rev. Roumaine math. pures et appl. **10,** 931—960 (1965).
63. Fullerton, R. E.: Geometric structure of absolute basis systems in a linear topological space. Pacific J. Math. **12,** 137—147 (1962).

64. Gaposhkin, V. F.: A generalization of a theorem of M. Riesz on conjugate functions. Matem. Sbornik **46** (88), 359—372 (1958) [Russian].

65. — On a property of unconditional bases in the space L^p. Uspehi Matem. Nauk 14, **4** (88), 143—148 (1959) [Russian].

66. — Review **3** Б 379. Refer. Ž. Matematika **3**Б, 84 (1963) [Russian].

67. — Kadec, M. I.: Operational bases in Banach space. Matem. Sbornik **61** (103), 3—12 (1963) [Russian].

68. Garling, D. J. H.: Symmetric bases of locally convex spaces. Studia Math. **30,** 163—181 (1968).

69. Gelbaum, B. R.: Expansions in Banach spaces. Duke Math. J. **17,** 187—196 (1950).

70. — A nonabsolute basis for Hilbert space. Proc. Amer. Math. Soc. **2,** 720—721 (1951).

71. — Notes on Banach spaces and bases. An. Acad. Brasil. Ci. **30,** 29—36 (1958).

72. — Gil de Lamadrid, J.: Bases of tensor products of Banach spaces. Pacific J. Math. **11,** 1281—1286 (1961).

73. Gelfand, I. M.: Abstrakte Funktionen und lineare Operatoren. Matem. Sbornik **4** (46), 235—286 (1938).

74. — Remark on the work of N. K. Bari "Biorthogonal systems and bases in Hilbert space". Uč. Zap. Moskov. Gos. Univ. 148, Matematika **4,** 224—225 (1951) [Russian].

75. Gil de Lamadrid, J.: Uniform crossnorms and tensor products of Banach algebras. Duke Math. J. **32,** 359—368 (1965).

76. Goffman, C., Waterman, D.: Basic sequences in the space of measurable functions. Proc. Amer. Math. Soc. **11,** 211—213 (1960).

77. Gohberg, I. C., Krein, M. G.: Fundamental aspects of defect numbers, root numbers and indexes of linear operators. Uspehi Matem. Nauk **12,** 2 (74), 43—118 (1957) [Russian].

78. — — Introduction to the theory of linear non-selfadjoint operators in Hilbert space. Moscow 1965 [Russian].

79. — — Theory of Volterra operators in Hilbert space and its applications. Moscow 1967 [Russian].

80. — Markus, A. S.: Two theorems on the opening of subspaces of a Banach space. Uspehi Matem. Nauk **14,** 5 (89), 135—140 (1959) [Russian].

81. — — On stability of bases of Banach and Hilbert spaces. Izv. Akad. Nauk Mold. SSR **5,** 17—35 (1962) [Russian].

82. Grinblium, M. M.: Certains théorèmes sur la base dans un espace du type (*B*). Doklady Akad. Nauk SSSR **31,** 428—432 (1941).

83. — Biorthogonal systems in Banach space. Doklady Akad. Nauk SSSR **45,** 75—78 (1945).

84. — Concerning my note "Biorthogonal systems in Banach Space". Doklady Akad. Nauk SSSR **52,** 387 (1946).

85. — Sur la théorie des systèmes biorthogonaux. Doklady Akad. Nauk SSSR **55,** 287—290 (1947).

86. — On a criterion of a basis. Doklady Akad. Nauk SSSR **59,** 9—11 (1948) [Russian].

87. — Gurevich, L. A.: Sur une propriété de la base dans l'espace de Hilbert. Doklady Akad. Nauk SSSR **30,** 289—291 (1941).

88. Grothendieck, A.: Sur les applications linéaires faiblement compactes d'espaces du type $C(K)$. Canad. J. Math. **5,** 129—173 (1953).

89. — Produits tensoriels topologiques et espaces nucléaires. Memoirs Amer. Math. Soc. **16,** 1—191 (1955).

90. — Résumé de le théorie métrique des produits tensoriels topologiques. Bol. Soc. Mat. São Paulo **8,** 1—79 (1956).

91. — Espaces vectoriels topologiques. 2nd ed., São Paulo 1958.

92. Gurariĭ, V. I.: Candidate thesis. Harkov State University 1963 [Russian].

93. — On inclinations of subspaces and conditional bases in Banach space. Doklady Akad. Nauk SSSR **145,** 504—506 (1962) [Russian].

94. — On bases in spaces of continuous functions. Doklady Akad. Nauk SSSR **148,** 493—495 (1963) [Russian].

95. — Some geometric characteristics of subspaces and bases in Banach spaces. Colloq. Math. **13,** 59—63 (1964) [Russian].

96. — On inclinations of subspaces and the existence of an orthogonal basis in Banach space. Zap. Meh.-Mat. Fak. Harkovsk. Gos. Univ. i Harkovsk. Mat. Obšč. (4) **30,** 34—37 (1964) [Russian].

97. — On indexes of sequences in $\tilde{C}$ and on the existence of infinite dimensional separable Banach spaces having no orthogonal basis. Rev. Roumaine math. pures appl. **10,** 967—971 (1965) [Russian].

97a. — Macaev, V. I.: Lacunary power sequences in spaces C and L_p. Izv. Akad. Nauk SSSR, Ser. Mat. **30,** 3—14 (1966) [Russian].

98. Gurevich, L. A.: On the basis of unconditional convergence. Uspehi Matem. Nauk 8, **5** (57), 153—156 (1953) [Russian].

99. — On equivalent systems. Trudy Semin. Funkc. Anal. Voronežsk, Gos. Univ. **2,** 47—54 (1956) [Russian].

100. — On the basis in a conjugate space. Trudy Semin. Funkc. Anal. **6,** 42—43 (1958) [Russian].

101. — A basis in the space of abstract functions. Doklady Akad. Nauk SSSR **136,** 12—15 (1961) [Russian].

102. — Conical criteria for bases of absolute convergence. Problems of Math. Phys. and Theory of Functions. II, 12—21. Kiev: Naukova Dumka 1964 [Russian].

103. Halperin, I.: Function spaces. Proc. Internat. Sympos. on Linear Spaces (July 5—12, 1960), 242—250, Jerusalem: Academic Press 1961.

104. Hardy, G. H., Littlewood, J.: A maximal theorem with function theoretic applications. Acta Math. **54,** 81—116 (1930).

105. Helson, H., Szegö, G.: A problem in prediction theory. Ann. Mat. Pura Appl. (4) **51,** 107—138 (1960).

106. Hildebrandt, T. H.: On unconditional convergence in normed vector spaces. Bull. Amer. Math. Soc. **46,** 959—962 (1940).

107. Hilding, S. H.: Note on completeness theorems of Paley-Wiener type. Ann. of Math. (2) **49,** 953—955 (1948).

107a. — Linear methods in the theory of complete sets in Hilbert space. Ark. för Mat., Astr. och Fys. **35** A, 1—44 (1948).

108. Holub, J. R.: Some problems concerning bases in Banach spaces. Proc. Amer. Math. Soc. (to appear).

109. — Retherford, J. R.: Some curious bases for c_0 and $C([0,1])$. Studia Math. (to appear).

110. — — Unconditional bases in Banach spaces. Dittoed notes. Louisiana State University 1969.

111. Ishii, J., Shimogaki, T.: On Haar functions in the space $L_{M(\xi,t)}$. J. Fac. Sci. Hokkaido Univ. (Ser. I) **17,** 55—63 (1963).

112. Istrăţescu, V.: Über die Banachräume mit zählbarer Basis, I. Rev. math. pures et appl. **7,** 481—482 (1962).

113. James, R. C.: Bases and reflexivity of Banach spaces. Ann. of Math. (2) **52,** 518—527 (1950).

114. — A non-reflexive Banach space isometric with its second conjugate space. Proc. Nat. Acad. Sci. U.S.A. **37,** 174—177 (1951).

115. — Weak compactness and reflexivity. Israel J. Math. **2,** 101—119 (1964).

116. Joiner, H.: Thesis, Florida State University, 1968.

117. Kaczmarz, S., Steinhaus, H.: Theorie der Orthogonalreihen. Warszawa-Lwów: Monografje Matematyczne 1935.

118. Kadec, M. I.: On the linear dimension of the space L_p ($p>2$). Naučn. Doklady Vysš. Školy **6,** 104—108 (1958) [Russian].

119. — On the linear dimension of the spaces L_p and l_q. Uspehi Matem. Nauk **13,** 6 (84), 95—98 (1958) [Russian].

120. — On strong and weak convergence. Doklady Akad. Nauk SSSR **122,** 13—16 (1958) [Russian].

121. — On the connection between weak and strong convergence. Dopovidi Akad. Nauk Ukr. RSR **9,** 949—952 (1959) [Ukrainian].

122. — Spaces isomorphic to a locally uniformly convex space. Izv. Vysš. Uč. Zaved. Matematika **6** (13), 51—57 (1959) [Russian].

123. — Letter to the editor. Izv. Vysš. Uč. Zaved. Matematika **6** (25), 186—187 (1961) [Russian].

124. — On biorthogonal systems and summation bases. Funkc. Anal. i ego Primen., 106—108, Akad. Nauk Azerb. SSR, Baku 1961 [Russian].

125. — On Lozinsky-Haršiladze systems. Uspehi Matem. Nauk **18,** 5 (113), 167—169 (1963) [Russian].

126. — Bases and their spaces of coefficients. Dopovidi Akad. Nauk Ukr. RSR **9,** 1139—1141 (1964) [Ukrainian].

127. — Pelczynski, A.: Bases lacunary sequences and complemented subspaces in the spaces L_p. Studia Math. **21,** 161—176 (1962).

128. — — Basic sequences, biorthogonal systems and norming sets in Banach and Fréchet spaces. Studia Math. **25,** 297—323 (1965) [Russian].

129. Kantorovich, L. V., Vulih, B. Z., Pinsker, A. G.: Functional analysis in semi-ordered spaces. Moscow-Leningrad 1950 [Russian].

130. Kaplansky, I.: Functional Analysis. Surveys in Applied Math. **IV,** 3—34 (1957).

131. Karlin, S.: Bases in Banach spaces. Duke Math. J. **15,** 971—985 (1948).

132. Klee, V.: Mappings into normed linear spaces. Fundam. Math. **49,** 25—34 (1960).

133. Köthe, G.: Topologische lineare Räume, I. Berlin-Heidelberg-New York: Springer 1960.

134. — Probleme der linearen Algebra in topologischen Vektorräumen. Proc. Internat. Sympos. on Linear Spaces (July 5—12, 1960), 290—298, Jerusalem: Academic Press 1961.

135. Kozlov, V. Ya.: On bases in the space $L_2[0,1]$. Matem. Sbornik **26** (68), 85—102 (1950) [Russian].

136. — On a generalization of the notion of basis. Doklady Akad. Nauk SSSR **73,** 643—646 (1950) [Russian].

137. Krasnoselskiĭ, M. A.: Positive solutions of operator equations. Moscow 1962 [Russian]. English translation, Groningen 1964.

138. — Rutickiĭ, Ya. B.: Convex functions and Orlicz spaces. Moscow 1958 [Russian]. English translation, Groningen 1961.

139. Krein, M. G.: Propriétés fondamentales des ensembles coniques normaux dans l'espace de Banach. Doklady Akad. Nauk SSSR **28,** 13—17 (1940).

140. — On Bari bases of Hilbert space. Uspehi Matem. Nauk **12,** 3 (75), 333—341 (1967) [Russian].

141. — Krasnoselskiĭ, M. A., Milman, D. P.: On the defficiency indices of linear operators in Banach spaces and some geometrical questions. Sbornik Trudov Inst. Akad. Nauk Ukr. SSR **11,** 97—112 (1948) [Russian].

142. — Liusternik, L. A.: Functional Analysis. Matem. v SSSR za 30 let (1917—1947), 608—697, Moscow-Leningrad 1947 [Russian].

143. — Milman, D. P., Rutman, M. A.: On a property of the basis in Banach space. Zapiski Mat. T-va (Harkov) **16,** 106—108 (1940) [Russian].

144. — Rutman, M. A.: Linear operators leaving invariant a cone in a Banach space. Uspehi Matem. Nauk **3,** 1 (23), 3—95 (1948) [Russian]. Amer. Math. Soc. Translation no. 26 (1950).

145. Kwapien, S., Pelczynski, A.: The main triangle projection in matrix spaces and its applications. Studia Math. (to appear).

146. Levin, A. Yu., Petunin, Yu. I.: Some problems related to the concept of orthogonality in a Banach space. Uspehi Matem. Nauk **18,** 3 (111), 167—170 (1963) [Russian].

147. Lichnerowicz, A.: Théorie globale des connexions et des groupes d'holonomie. Roma: Ediz. Cremonese 1957.

147a. Lin, Bor-Luh, Singer, I.: On conditional bases of l^2. Prace Mat. (to appear).

148. Lindenstrauss, J.: On the modulus of smoothness and divergent series in Banach spaces. Mich. Math. J. **10,** 241—252 (1963).

149. — On a certain subspace of l^1. Bull. Acad. Polon. Sci. **12,** 539—542 (1964).

150. — Pelczynski, A.: Absolutely summing operators in $\mathscr{L}_p$ spaces and their applications. Studia Math. **29,** 275—326 (1968).

151. — Zippin, M.: Banach spaces with a unique unconditional basis. J. of Functional Analysis **3,** 115—125 (1969).

152. Liusternik, L. A., Sobolev, V. I.: Elements of functional analysis. Moscow-Leningrad 1951 [Russian]. English translation, New York 1961.

153. Lorentz, G. G.: On the theory of spaces Λ. Pacific J. Math. **1,** 411—429 (1951).

154. Lovaglia, A. R.: Locally uniformly convex spaces. Trans. Amer. Math. Soc. **78,** 225—238 (1955).

155. Marcinkiewicz, J.: Quelques théorèmes sur les séries orthogonales. Ann. Soc. Polon. Math. **16,** 84—96 (1937).

156. Markushevich, A.: Sur les bases (au sens large) dans les espaces linéaires. Doklady Akad. Nauk SSSR **41,** 227—229 (1943).

157. Mazur, S., Orlicz, W.: Sur les espaces métriques linéaires, I. Studia Math. **10,** 184—208 (1948).

158. — — Sur les espaces métriques linéaires, II. Studia Math. **13,** 137—179 (1953).

159. Mazurkiewicz, S.: Sur la dérivée faible d'un ensemble de fonctionnelles linéaires. Studia Math. **2,** 68—71 (1930).

160. McArthur, C. W.: A note on subseries convergence. Proc. Amer. Math. Soc. **12,** 540—545 (1961).

161. — Retherford, J. R.: Uniform and equicontinuous Schauder bases of subspaces. Canad. J. Math. **17,** 207—212 (1965).

162. — — Some applications of an inequality in locally convex spaces. Trans. Amer. Math. Soc. **137,** 115—123 (1969).

163. — Singer, I., Lewin, M.: On the cones associated to biorthogonal systems and bases in Banach spaces. Canad. J. Math. **21,** 1206—1217 (1969).

164. McCarthy, C. A.: c_p. Israel J. Math. **5,** 249—271 (1967).

165. — Schwartz, J.: On the norm of a finite Boolean algebra of projections, and applications to theorems of Kreiss and Morton. Comm. Pure and Appl. Math. **18,** 191—201 (1965).

166. McGivney, R. J., Ruckle, W.: Multiplier algebras of biorthogonal systems. Pacific J. Math. **29,** 375—387 (1969).

167. Milman, V. D.: Perturbations of sequences of elements of a Banach space. Sibirsk. Mat. Ž. **6,** 398—412 (1965) [Russian].

168. Milman, D. P., Milman, V. D.: Some properties of non-reflexive Banach spaces. Matem. Sbornik **65** (107), 486—497 (1964) [Russian].

169. Mityagin, B. S.: Approximative dimension and bases in nuclear spaces. Uspehi Matem. Nauk. 16, **4** (100), 63—132 (1961) [Russian].

170. Murray, F. J.: On complementary manifolds and projections in spaces L_p and l_p. Trans. Amer. Math. Soc. **41,** 138—152 (1937).

171. Naimark, M. A.: Normed rings. Moscow 1956 [Russian]. English translation, revised edition, Groningen 1964.

172. Nagy, B. Sz.-: Sur les lattis linéaires de dimension finie. Comment. Math. Helv. **17,** 209—213 (1945).

173. — Expansion theorems of PaleyWiener type. Duke Math. J. **14,** 975—978 (1947).

174. Natanson, I. P.: Constructive theory of functions. Moscow-Leningrad 1949 [Russian]. English translation, New York 1964—1965.

175. — Theory of functions of a real variable. 2nd ed. Moscow 1957 [Russian]. English translation, New York 1961.

176. Newns, W. F.: On the representation of analytic functions by infinite series. Phil. Trans. Royal Soc. London (A) **245,** 429—468 (1953).

177. Nguen, Van Khue: Criteria of bases of unconditional convergence. Izv. Vysš. Učebn. Zaved. Matematika **2** (69), 68—74 (1968) [Russian].

178. Nikolskiĭ, S. M.: Approximation of functions by trigometrical polynomials. Izvestiya Akad. Nauk SSSR, ser. mat. **10,** 207—256 (1946) [Russian].

179. Nikolskiĭ, V. N.: Best approximation and basis in Fréchet space. Doklady Akad. Nauk SSSR **59,** 639—642 (1948) [Russian].

180. — Some problems of best approximation in a function space. Uč. Zap. Kalininsk. Gos. Ped. In.-ta **16,** 119—160 (1954) [Russian].

181. — Operational properties of polynomials of best approximation. Uspehi Matem. Nauk **12,** 3 (75), 353—358 (1957) [Russian].

182. — On properties of the operators of best approximation. Uč. Zap. Kalininsk. Gos. Ped. In.-ta **26,** 143—146 (1958) [Russian].

183. Orlicz, W.: Beiträge zur Theorie der Orthogonalentwicklungen, I, Studia Math. **1,** 1—39 (1929).

184. — Beiträge zur Theorie der Orthogonalentwicklungen, II. Studia Math. **1,** 241—255 (1929).

185. — Über eine gewisse Klasse von Räumen von Typus *B*. Bull. Int. Acad. Polon. Sci. Ser. A. 207—220 (1932).

186. — Über unbedingte Konvergenz in Funktionenräumen, I. Studia Math. **4,** 33—37 (1933).

187. Paley, R. E. A. C.: A remarkable series of orthogonal functions, I. Proc. London Math. Soc. (2) **34,** 241—264 (1932).

188. — Wiener, N.: Fourier transforms in the complex domain. Amer. Math. Soc. Coll. Publ. **19,** New York 1934.

189. Pelczynski, A.: A connection between weakly unconditional convergence and weakly completeness of Banach spaces. Bull. Acad. Polon. Sci. **6,** 251—253 (1958).

190. — Projections in certain Banach spaces. Studia Math. **19,** 209—228 (1960).

191. — On the impossibility of embedding of the space *L* in certain Banach spaces. Colloq. Math. **8,** 199—203 (1961).

191a. — A note on the paper of I. Singer "Basic sequences and reflexivity of Banach spaces". Studia Math. **21,** 371—374 (1962).

192. — A proof of Eberlein-Šmulian theorem by an application of basic sequences. Bull. Acad. Polon. Sci. **12,** 543—548 (1964).

193. — Some problems on bases in Banach and Fréchet spaces. Israel J. Math. **2,** 132—138 (1964).

194. — Some open questions in functional analysis (A lecture given to Louisiana State University). Dittoed notes (1966).

195. — Universal bases. Studia Math. **32,** 247—268 (1969).

196. — Singer, I.: On non-equivalent bases and conditional bases in Banach spaces. Studia Math. **25,** 5—25 (1964).

197. — Szlenk, W.: An example of a non-shrinking basis. Rev. Roumaine math. pures et appl. **10,** 961—966 (1965).

198. Pollard, H.: Completeness theorems of Paley-Wiener type. Ann. of Math. (2) **45,** 738—739 (1944).

199. Pontryagin, L. S.: Foundations of combinatorial topology. Moscow-Leningrad 1947 [Russian]. English translation, Rochester 1952.

200. Pták, V.: On a theorem of Mazur and Orlicz. Studia Math. **15,** 365—366 (1956).

201. — Biorthogonal systems and reflexivity of Banach spaces. Czechoslovak Math. J. **9,** 319—326 (1959).

202. Retherford, J. R.: A note on unconditional convergence. Proc. Amer. Math. Soc. **15,** 899—901 (1964).

203. — Basic sequences and the Paley-Wiener criterion. Pacific J. Math. **14,** 1019—1027 (1964).

204. — w^*-bases and bw^*-bases in Banach spaces. Studia Math. **25,** 65—71 (1964).

205. — Shrinking bases in Banach spaces. Amer. Math. Monthly **73,** 841—846 (1966).

206. — On Čebyšev subspaces and unconditional bases in Banach space. Bull. Amer. Math. Soc. **73,** 238—241 (1967).

207. — Two counterexamples to a conjecture of S. Karlin. Bull. Acad. Polon. Sci. **16,** 293—295 (1968).

208. — A semishrinking basis which is not shrinking. Proc. Amer. Math. Soc. **19,** 766 (1968).

209. — The Paley-Wiener criterion. Scripta Math. (to appear).

210. — James, R. C.: Unconditional bases and best approximation in Banach spaces. Bull. Amer. Math. Soc. **75,** 108—112 (1969).

211. Riesz, F., Nagy, B. Sz.-: Leçons d'analyse fonctionnelle. Budapest: Akadémiai Kiadó 1952.

212. Ruckle, W.: Infinite matrices which preserve Schauder bases. Duke Math. J. **33,** 547—550 (1966).

213. — Symmetric coordinate spaces and symmetric bases. Canad. J. Math. **19,** 828—838 (1967).

214. — On the characterization of sequence spaces associated with Schauder bases. Studia Math. **28,** 279—288 (1967).

215. — Lattices of sequence spaces. Duke Math. J. **35,** 491—504 (1968).

215a. — Sequence spaces associated with symmetric bases. Dittoed notes, Lehigh University 1968.

216. — Decompositions of operator spaces. Rev. Roumaine math. pures et appl. (to appear).

217. Rudin, W.: Continuous functions on compact spaces without perfect subsets. Proc. Amer. Math. Soc. **8,** 39—42 (1957).

218. Ruston, A. F.: Auerbach's theorem and tensor products of Banach spaces. Proc. Cambr. Philos. Soc. **58,** 476—480 (1962).

219. Sanders, B. L.: A bibliography of papers related to (Schauder) bases. Texas Christian University 1964.

220. Schaefer, H. H.: Halbgeordnete lokalkonvexe Vektorräume. Math. Ann. **135,** 115—141 (1958).

221. — Topological vector spaces. New York: MacMillan 1966.

222. Schäfke, F. W.: Das Kriterium von Paley und Wiener im Banachschen Raum. Math. Nachr. **3,** 59—61 (1949).

223. Schatten, R.: A theory of cross-spaces. Ann. of Math. Studies 26. Princeton: Princeton Univ. Press 1950.

224. Schauder, J.: Zur Theorie stetiger Abbildungen in Funktionalräumen. Math. Zeitschr. **26,** 47—65 (1927).

225. — Eine Eigenschaft des Haarschen Orthogonalsystems. Math. Zeitschr. **28,** 317—320 (1928).

226. Schonefeld, S.: Schauder bases in spaces of differentiable functions. Bull. Amer. Math. Soc. **75,** 586—590 (1969).

227. Semadeni, Z.: Product Schauder bases and approximation with nodes in spaces of continuous functions. Bull. Acad. Polon. Sci. **11,** 387—391 (1963).

228. Sikorski, R.: On a theorem of Mazur and Orlicz. Studia Math. **13,** 180—182 (1953).

229. Singer, I.: Elementary proof of a theorem of S. R. Foguel on biorthogonal systems in Banach spaces. Rev. math. pures et appl. **3,** 305—307 (1958).

230. — Sur les espaces de Banach à base absolue, canoniquement equivalents à un dual d'espace de Banach. Comptes Rendus Acad. Sci. (Paris) **251,** 620—621 (1960).

231. — Sur les espaces de Banach à base absolue, canoniquement equivalents à un dual d'espace de Banach, II. Comptes Rendus Acad. Sci. (Paris) **251,** 2456—2458 (1960).

232. — Quelques propriétés des espaces de Banach à base absolue. 2-ème Congrès Math. Hongrois (Budapest, 24—31 august 1960) 2, sec. III b, 84—86. Budapest: Akadémiai Kiadó 1961.

233. — On Banach spaces with a symmetric basis. Rev. math. pures appl. **6,** 159—166 (1961) [Russian].

234. — Weak* bases in conjugate Banach spaces. Studia Math. **21,** 75—81 (1961).

235. — On Banach spaces reflexive with respect to a linear subspace of their conjugate space, II. Math. Annalen **145,** 64—76 (1962).

236. — On a theorem of I. M. Gelfand. Uspehi Matem. Nauk **17,** 1 (103), 169—176 (1962) [Russian].

237. — Basic sequences and reflexivity of Banach spaces. Studia Math. **21,** 351—369 (1962).

238. — Some characterizations of symmetric bases in Banach spaces. Bull. Acad. Polon. Sci. **10,** 185—192 (1962).

239. — On Banach spaces reflexive with respect to a linear subspace of their conjugate space, III. Rev. math. pures et appl. **8,** 139—150 (1963).

240. — Weak* bases in conjugate Banach spaces, II. Rev. math. pures et appl. **8,** 575—584 (1963).

241. — Bases in Banach spaces, I. Studii și cercet. mat. **14,** 539—585 (1963) [Romanian].

242. — Bases in Banach spaces, II. Studii și cercet. mat. **15,** 157—208 (1964) [Romanian].

243. — Bases in Banach spaces, III. Studii și cercet. mat. **15,** 675—725 (1964) [Romanian].

244. — Bases and quasi-reflexivity of Banach spaces. Math Annalen **153,** 199—209 (1964).

245. — On the basis problem in topological linear spaces. Rev. math. pures et appl. **10,** 453—457 (1965).

246. — Best approximation in normed linear spaces by elements of linear subspaces. Edit. Acad. R. S. Romania, București 1967 [Romanian]. English translation to appear in Edit. Acad. R. S. Romania and Springer-Verlag 1970.

247. — On metric projections onto linear subspaces of normed linear spaces. Proc. Confer. on "Projections and related topics" held in Clemson, Aug. 1967. Preliminary Edition, January 1968.

248. — On a theorem of N. K. Bari and I. M. Gelfand. Arch. Math. **19,** 508—510 (1968).

249. — On the constants of basic sequences in Banach spaces. Studia Math. **31,** 125—134 (1968).

250. — Some remarks and problems on bases in Banach spaces. Proc. Conf. on "Abstract Spaces and Approximation" held in Oberwolfach, July 1968, pp. 130—139. Stuttgart: Birkhäuser Verlag 1969.

251. — Some remarks on domination of sequences. Math. Annalen **184,** 113—132 (1970).

252. Sobczyk, A.: Projections in Minkowski and Banach spaces. Duke Math. J. **8,** 78—106 (1941).

253. Solomyak, M. Z.: On the orthogonal basis in Banach space. Vestnik Leningradsk. Gos. Un.-ta **1,** 27—36 (1957) [Russian].

254. Szlenk, W.: The non-existence of a separable reflexive Banach space universal for all separable reflexive Banach spaces. Studia Math. **30,** 53—61 (1968).

255. Taylor, A. E.: A geometric theorem and its applications to biorthogonal systems. Bull. Amer. Math. Soc. **53,** 614—616 (1947).

256. Timan, A. F.: Theory of approximation of functions of a real variable. Moscow 1960 [Russian]. English translation, New York 1963.

257. Titchmarsh, E. C.: The theory of functions. Oxford: Clarendon Press 1932.

258. Veic, B. E.: On some characteristic properties of unconditional bases. Doklady Akad. Nauk SSSR **155,** 509—512 (1964) [Russian].

259. — On some stability properties of bases. Doklady Akad. Nauk SSSR **158,** 13—16 (1964) [Russian].

260. — Besselian and Hilbertian systems in Banach spaces and problems of stability. Izv. Vysš. Uč. Zaved. Matematika **2** (45), 7—23 (1965) [Russian].

260a. — On some characteristic properties of unconditional bases and stability theorems. Izv. Vysš. Učebn. Zaved. Matematika **4** (47), 24—36 (1965) [Russian].

261. Vinokurov, V. G.: On biorthogonal systems passing through given subspaces. Doklady Akad. Nauk SSSR **85,** 685—687 (1952) [Russian].

262. Wilansky, A.: The basis in Banach space. Duke Math. J. **18,** 785—791 (1951).

263. — Functional analysis. New York-Toronto-London: Blaisdell Publ. Co. 1964.

264. Williams, J. P.: A "metric" characterization of reflexivity. Proc. Amer. Math. Soc. **18,** 163—165 (1967).

265. Yamazaki, S.: On bases in Banach spaces. Sci. Papers Coll. Gen. Ed. Univ. Tokyo **10,** 163—169 (1960).

266. — Normed ring and unconditional bases in Banach space. Sci. Papers Coll. Gen. Ed. Univ. Tokyo **14,** 1—10 (1964).

267. — Normed rings and bases in Banach spaces. Sci. Papers Coll. Gen. Ed. Univ. Tokyo **15,** 1—13 (1965).

268. — Remark to "Normed rings and bases in Banach spaces". Sci. Papers Coll. Gen. Ed. Univ. Tokyo **16,** 25—26 (1966).

269. Yano, S.: On a lemma of Marcinkiewicz and its applications to Fourier series. Tôhoku Math. J. (2) **11,** 191—215 (1959).

270. Yosida, K.: Functional analysis. Berlin-Heidelberg-New York: Springer 1966.

271. Yudin, A. N.: Solution des deux problèmes de la théorie des espaces semi-ordonnés. Doklady Akad. Nauk SSSR **27,** 418—422 (1939).

272. Zippin, M.: On perfectly homogeneous bases in Banach spaces. Israel J. Math. **4,** 265—272 (1966).

273. — A remark on bases and reflexivity in Banach spaces. Israel J. Math. **6,** 74—79 (1968).

274. Zygmund, A.: Trigonometric series. Vol. I. Cambridge: Cambridge Univ. Press 1959.

275. Ciesielski, Z.: A construction of basis in $C^{(1)}(I^2)$. Studia Math. **33,** 243—247 (1969).

276. Marti, J. T.: Introduction to the theory of bases. Berlin-Heidelberg-New York: Springer 1969.

Notation Index

Author Index

Subject Index

Satz und Druck: Zechnersche Buchdruckerei, Speyer

Die Grundlehren der mathematischen Wissenschaften in Einzeldarstellungen mit besonderer Berücksichtigung der Anwendungsgebiete

---

74. Boerner: Darstellung von Gruppen. DM 58,—; US $ 16.00
75. Rado/Reichelderfer: Continuous Transformations in Analysis, with an Introduction to Algebraic Topology. DM 59,60; US $ 16.40
76. Tricomi: Vorlesungen über Orthogonalreihen. DM 68,—; US $ 18.70
77. Behnke/Sommer: Theorie der analytischen Funktionen einer komplexen Veränderlichen. DM 79,—; US $ 21.80
78. Lorenzen: Einführung in die operative Logik und Mathematik. DM 54,—; US $ 14.90
80. Pickert: Projektive Ebenen. DM 48,60; US $ 13.40
81. Schneider: Einführung in die transzendenten Zahlen. DM 24,80; US $ 6.90
82. Specht: Gruppentheorie. DM 69,60; US $ 19.20
84. Conforto: Abelsche Funktionen und algebraische Geometrie. DM 41,80; US $ 11.50
86. Richter: Wahrscheinlichkeitstheorie. DM 68,—; US $ 18.70
88. Müller: Grundprobleme der mathematischen Theorie elektromagnetischer Schwingungen. DM 52,80; US $ 14.60
89. Pfluger: Theorie der Riemannschen Flächen. DM 39,20; US $ 10.80
90. Oberhettinger: Tabellen zur Fourier-Transformation. DM 39,50; US $ 10.90
91. Prachar: Primzahlenverteilung. DM 58.—; US $ 16.00
93. Hadwiger: Vorlesungen über Inhalt, Oberfläche und Isoperimetrie. DM 49,80; US $ 13.70
94. Funk: Variationsrechnung und ihre Anwendung in Physik und Technik. DM 120,—; US $ 33.00
95. Maeda: Kontinuierliche Geometrien. DM 39,—; US $ 10.80
97. Greub: Linear Algebra. DM 39,20; US $ 9.80
98. Saxer: Versicherungsmathematik. 2. Teil. DM 48,60; US $ 13.40
99. Cassels: An Introduction to the Geometry of Numbers. DM 69,—; US $ 19.00
100. Koppenfels/Stallmann: Praxis der konformen Abbildung. DM 69,—; US $ 19.00
101. Rund: The Differential Geometry of Finsler Spaces. DM 59,60; US $ 16.40
103. Schütte: Beweistheorie. DM 48,—; US $ 13.20
104. Chung: Markov Chains with Stationary Transition Probabilities. DM 56,—; US $ 14.00
105. Rinow: Die innere Geometrie der metrischen Räume. DM 83,—; US $ 22.90
106. Scholz/Hasenjaeger: Grundzüge der mathematischen Logik. DM 98,—; US $ 27.00
107. Köthe: Topologische lineare Räume I. DM 78,—; US $ 21.50
108. Dynkin: Die Grundlagen der Theorie der Markoffschen Prozesse. DM 33,80; US $ 9.30
110. Dinghas: Vorlesungen über Funktionentheorie. DM 69,—; US $ 19.00
111. Lions: Equations différentielles opérationnelles et problèmes aux limites. DM 64,—; US $ 17.60
112. Morgenstern/Szabó: Vorlesungen über theoretische Mechanik. DM 69,—; US $ 19.00
113. Meschkowski: Hilbertsche Räume mit Kernfunktion. DM 58,—; US $ 16.00
114. MacLane: Homology. DM 62,—; US $ 15.50
115. Hewitt/Ross: Abstract Harmonic Analysis. Vol. 1: Structure of Topological Groups, Integration Theory, Group Representations. DM 76,—; US $ 20.90
116. Hörmander: Linear Partial Differential Operators. DM 42,—; US $ 10.50
117. O'Meara: Introduction to Quadratic Forms. DM 48,—; US $ 13.20
118. Schäfke: Einführung in die Theorie der speziellen Funktionen der mathematischen Physik. DM 49,40; US $ 13.60
119. Harris: The Theory of Branching Processes. DM 36,—; US $ 9.90

120. Collatz: Funktionalanalysis und numerische Mathematik. DM 58,—; US $ 16.00
121.
122. Dynkin: Markov Processes. DM 96,—; US $ 26.40
123. Yosida: Functional Analysis. DM 66,—; US $ 16.50
124. Morgenstern: Einführung in die Wahrscheinlichkeitsrechnung und mathematische Statistik. DM 38,—; US $ 10.50
125. Itô/McKean: Diffusion Processes and Their Sample Paths. DM 58,—; US $ 16.00
126. Lehto/Virtanen: Quasikonforme Abbildungen. DM 38,—; US $ 10.50
127. Hermes: Enumerability, Decidability, Computability. DM 39,—; US $ 10.80
128. Braun/Koecher: Jordan-Algebren. DM 48,—; US $ 13.20
129. Nikodým: The Mathematical Apparatus for Quantum-Theories. DM 144,—; US $ 36.00
130. Morrey: Multiple Integrals in the Calculus of Variations. DM 78,—; US $ 19.50
131. Hirzebruch: Topological Methods in Algebraic Geometry. DM 38,—; US $ 9.50
132. Kato: Perturbation Theory for Linear Operators. DM 79,20; US $ 19.80
133. Haupt/Künneth: Geometrische Ordnungen. DM 68,—; US $ 18.70
134. Huppert: Endliche Gruppen I. DM 156,—; US $ 42.90
135. Handbook for Automatic Computation. Vol. 1/Part a: Rutishauser: Description of ALGOL 60. DM 58,—; US $ 14.50
136. Greub: Multilinear Algebra. DM 32,—; US $ 8.00
137. Handbook for Automatic Computation. Vol. 1/Part b: Grau/Hill/Langmaack: Translation of ALGOL 60. DM 64,—; US $ 16.00
138. Hahn: Stability of Motion. DM 72,—; US $ 19.80
139. Mathematische Hilfsmittel des Ingenieurs. Herausgeber: Sauer/Szabó. 1. Teil. DM 88,—; US $ 24.20
140. Mathematische Hilfsmittel des Ingenieurs. Herausgeber: Sauer/Szabó. 2. Teil. DM 136,—; US $ 37.40
141. Mathematische Hilfsmittel des Ingenieurs. Herausgeber: Sauer/Szabó. 3. Teil. DM 98,—; US $ 27.00
142. Mathematische Hilfsmittel des Ingenieurs. Herausgeber: Sauer/Szabó. 4. Teil. DM 124,—; US $ 34.10
143. Schur/Grunsky: Vorlesungen über Invariantentheorie. DM 32,—; US $ 8.80
144. Weil: Basic Number Theory. DM 48,—; US $ 12.00
145. Butzer/Berens: Semi-Groups of Operators and Approximation. DM 56,—; US $ 14.00
146. Treves: Locally Convex Spaces and Linear Partial Differential Equations. DM 36,—; US $ 9.90
147. Lamotke: Semisimpliziale algebraische Topologie. DM 48,—; US $ 13.20
148. Chandrasekharan: Introduction to Analytic Number Theory. DM 28,—; US $ 7.00
149. Sario/Oikawa: Capacity Functions. DM 96,—; US $ 24.00
150. Iosifescu/Theodorescu: Random Processes and Learning. DM 68,—; US $ 18.70
151. Mandl: Analytical Treatment of One-dimensional Markov Processes. DM 36,—; US $ 9.80
152. Hewitt/Ross: Abstract Harmonic Analysis. Vol. II. DM 140,—; US $ 38.50
153. Federer: Geometric Measure Theory. DM 118,—; US $ 29.50
154. Singer: Bases in Banach Spaces I. DM 112,—; US $ 30.80
155. Müller: Foundations of the Mathematical Theory of Electromagnetic Waves. DM 58,—; US $ 16.00
156. van der Waerden: Mathematical Statistics. DM 68,—; US $ 18.70
157. Prohorov/Rozanov: Probability Theory. DM 68,—; US $ 18.70

159. Köthe: Topological Vector Spaces I. DM 78,—; US $ 21.50
160. Agrest/Maksimov: Theory of Incomplete Cylindrical Functions and their Applications. In preparation
161. Bhatia/Szegö: Stability Theory of Dynamical Systems. In preparation
162. Nevanlinna: Analytic Functions. DM 76,—; US $ 20.90
163. Stoer/Witzgall: Convexity and Optimization in Finite Dimensions I. DM 54,—; US $ 14.90
164. Sario/Nakai: Classification Theory of Riemann Surfaces. DM 98,—; US $ 27.00
165. Mitrinovič: Analytic Inequalities. DM 88,—; US $ 24.20
166. Grothendieck/Dieudonné: Eléments de Géometrie Algébrique. En préparation
167. Chandrasekharan: Arithmetical Functions. DM 58,—; US $ 16.00
168. Palamadov: Linear Differential Operators with Constant Coefficients. DM 98,—; US $ 27.00
169. Rademacher: Topics in Analytic Number Theory. In preparation
170. Lions: Optimal Control Systems Governed by Partial Differential Equations. In preparation
171. Singer: Best Approximation in Normed Linear Spaces by Elements of Linear Subspaces. DM 60,—; US $ 16.50